EDEXCEL
A LEVEL MATHS
YEAR 1 + YEAR 2

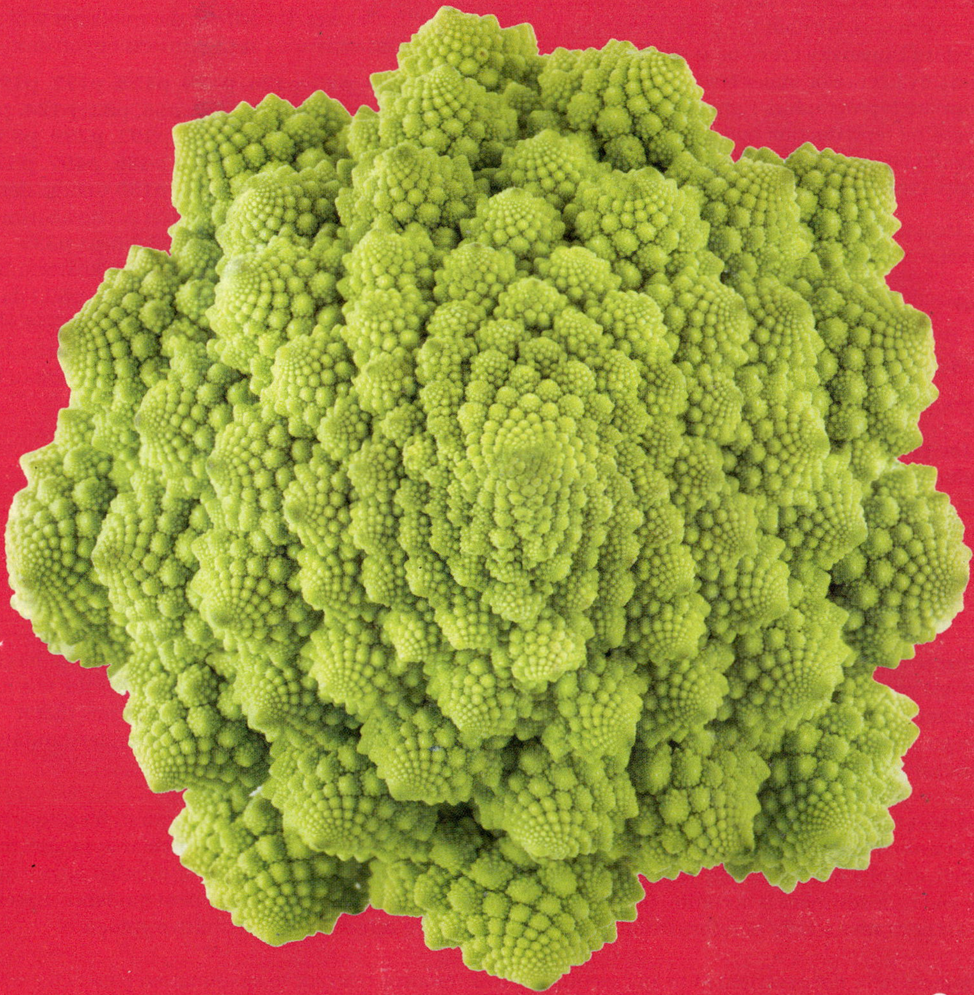

Series Editor
David Baker

Authors
Brian Jefferson, David Bowles, Eddie Mullan, Garry Wiseman, John Rayneau, Katie Wood, Mark Rowland, Mike Heylings, Paul Williams, Rob Wagner

OXFORD
UNIVERSITY PRESS

OXFORD

UNIVERSITY PRESS

Great Clarendon Street, Oxford, OX2 6DP, United Kingdom

Oxford University Press is a department of the University of Oxford.

It furthers the University's objective of excellence in research, scholarship, and education by publishing worldwide. Oxford is a registered trade mark of Oxford University Press in the UK and in certain other countries.

British Library Cataloguing in Publication Data
Data available

978-0-19-841315-8

10 9 8 7 6 5 4 3 2 1

Paper used in the production of this book is a natural, recyclable product made from wood grown in sustainable forests.
The manufacturing process conforms to the environmental regulations of the country of origin.

Printed in Italy by L.E.G.O SpA

Acknowledgements

Series Editor: David Baker

Authors
Brian Jefferson, David Bowles, Eddie Mullan, Garry Wiseman, John Rayneau, Katie Wood, Mark Rowland, Mike Heylings, Paul Williams, Rob Wagner

Editorial team
Dom Holdsworth, Ian Knowles, Matteo Orsini Jones, Felicity Ounsted, Anna Clarke, Anna Gupta, Rosie Day

With thanks also to Anna Cox, Karuna Boyum, Katherine Bird, Keith Gallick, Linnet Bruce and Pete Sherran for their contribution.

Index compiled by Marian Preston, Preston Indexing

Although we have made every effort to trace and contact all copyright holders before publication, this has not been possible in all cases. If notified, the publisher will rectify any errors or omissions at the earliest opportunity.

p3, **p43**, **p69**, **p85**, **p124**, **p129**, **p153**, **p177**, **p199**, **p229**, **p263**, **p279** iStock; **p40**, **p66**, **p82**, **p124**, **p150**, **p166**, **p196**, **p220**, **p258**, **p274**, **p292**, **p299**, **p333**, **p363**, **p391**, **p428**, **p433**, **p463**, **p486**, **p497**, **p523**, **p544**, **p555**, **p579** Shutterstock; **p150** George Bernard/Science Photo Library; **p166** Hulton Archive/Stringer/Getty images; **p292** Science Photo Library; **p333** Woodkern/iStockphoto; **p433** avelSmilyk/iStockphoto; **p463** memoriesarecaptured/iStockphoto; **p523** da-kuk/iStockphoto; **p555** kemalbas/iStockphoto; **p544** vkilikov/Shutterstock.com; **p576** Deborah Feingold/Getty Images; **p590** National Library Of Medicine/Science Photo Library; **p590** Nicku/Shutterstock

Contents

About this book

This book has been specifically written for those studying the Edexcel 2017 Mathematics AS and A Levels. It's been written by a team of experienced authors and teachers, using a carefully selected range of features and exercises to build understanding and to help you get the most out of your course.

Every section starts by covering the basic **Fluency and skills** (AO1), then builds on these techniques by looking at **Reasoning and problem-solving** (AO2 and AO3).

Strategy boxes help build problem-solving techniques.

Worked examples provide a model answer and commentary to realistic practice questions. The circled numbers show how each step is linked to the strategy box.

Challenge questions in each section stretch you with questions at the highest level of demand. **Answers to all questions** are in the back of this book, and **full solutions are available free** online.

Links to **MyMaths** provide a quick route to **extra support and practice**. Just log in and key the code into the search bar.

At the end of each chapter, a **What Next** box provides links to further support based on how well you've understood the content.

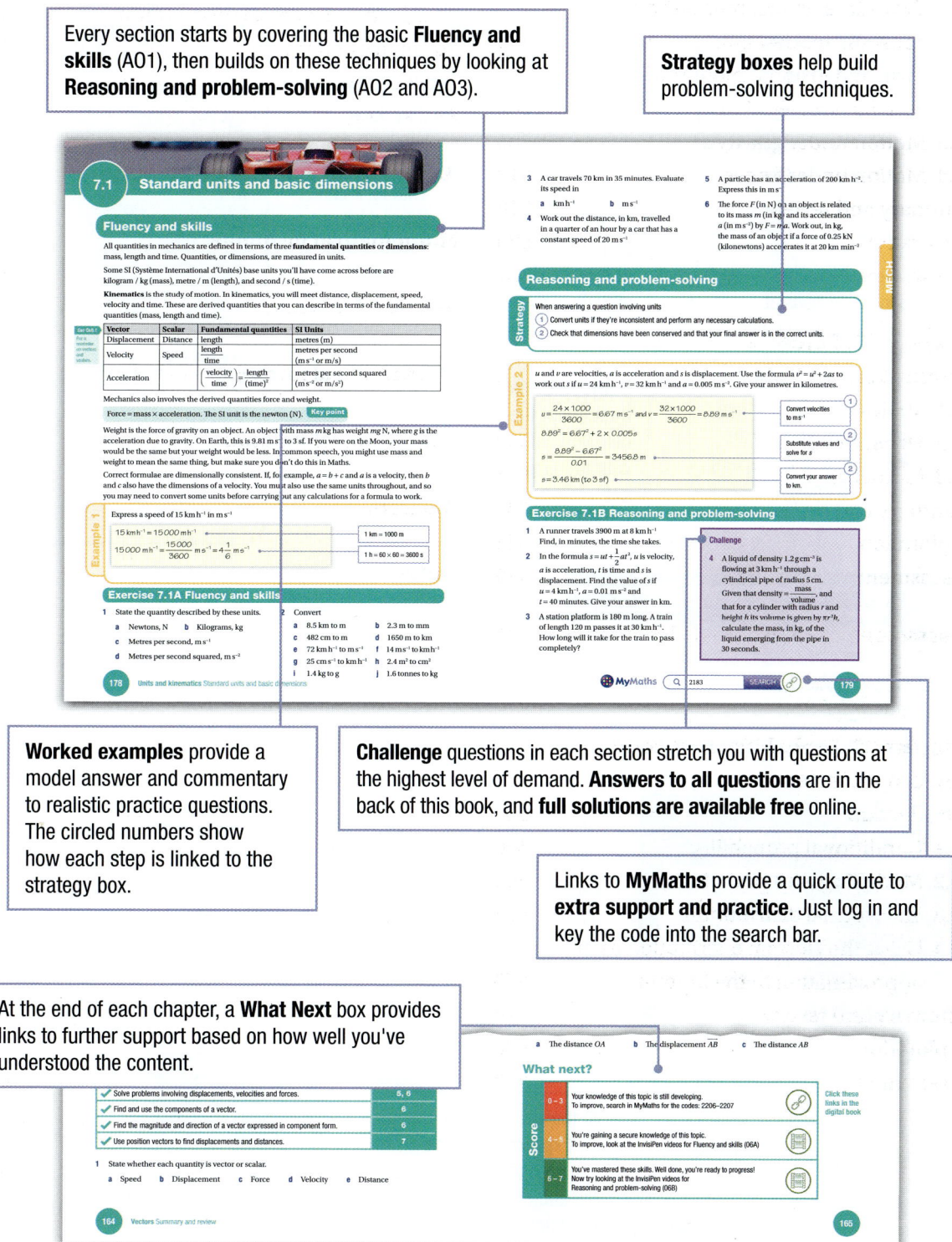

Links to **ICT resources** on Kerboodle show how technology can be used to help understand the maths involved.

ICT Resource online

To investigate gradients of chords for a graph, click this link in the digital book.

Support for when and how to use **calculators** is available throughout this book, with links to further demonstrations in the **digital book**. Unless otherwise stated, this book assumes that your calculator can do the required minimum according to specification guidelines. That is, it can perform an iterative function and it can compute summary statistics and access probabilities from standard statistical distributions.

Try it on your calculator

You can use a calculator to evaluate the gradient of the tangent to a curve at a given point.

$$d/dx(5X^2 - 2X, 3)$$

28

Activity

Find out how to calculate the gradient of the tangent to the curve $y = 5x^2 - 2x$ where $x = 3$ on *your* calculator.

Assessment sections at the end of each chapter test everything covered within that chapter. Further **synoptic assessments** for each of Pure, Mechanics and Statistics can be found at the end of chapters 6, 8, 11, 17, 19 and 21

Dedicated questions throughout the book help you get familiar with the **large data set.**

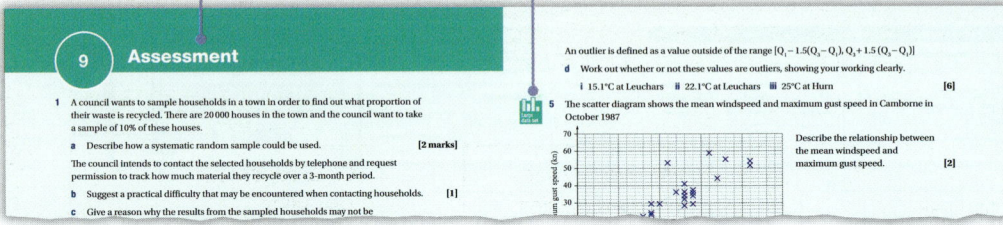

On the chapter **Introduction page**, the **Orientation box** explains what you should already know, what you will learn in this chapter, and what this leads to.

At the end of every chapter, an **Exploration page** gives you an opportunity to explore the subject beyond the specification.

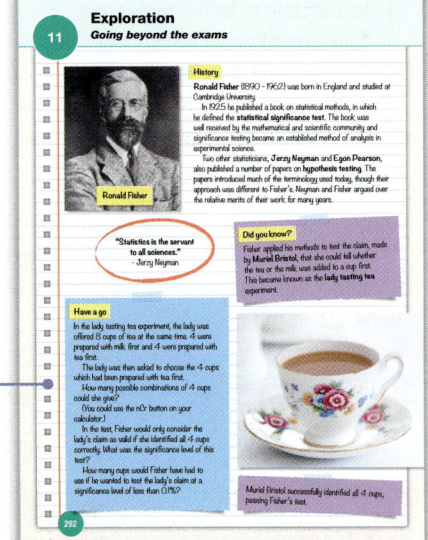

1 Algebra 1

Supply and demand is a well-known example of how maths helps us model real situations that occur in the world. Economists and business analysts use simultaneous equations to model how changes in price will affect both the supply of, and demand for, a particular product. This allows them to forecast the optimum price–the price at which both supply and demand are optimised.

Algebra is a branch of maths that includes simultaneous equations, along with many other topics such as inequalities, surds and polynomial functions. Algebra is used to model real world occurrences in fields such as economics, engineering and the sciences, and so an understanding of algebra is important in a wide range of different situations.

Orientation

What you need to know

KS4

- Understand and use algebraic notation and vocabulary.
- Simplify and manipulate algebraic expressions.
- Rearrange formulae to change the subject.
- Solve linear equations.

What you will learn

- To use direct proof, proof by exhaustion and counter examples.
- To use and manipulate index laws.
- To manipulate surds and rationalise a denominator.
- To solve quadratic equations and sketch quadratic curves.
- To understand and use coordinate geometry.
- To understand and solve simultaneous equations.
- To understand and solve inequalities.

What this leads to

Ch12 Algebra 2
Functions.
Parametric equations.
Algebraic and partial fractions.

Ch17 Numerical methods
Simple and iterative root finding.
Newton-Raphson root finding.

 **MyMaths** Practise before you start

Q 1170, 1171, 1928, 1929

Fluency and skills

A **proof** is a logical argument for a mathematical statement. It shows that something *must* be either true or false.

The most simple method of proving something is called **direct proof**. It's sometimes also called deductive proof. In direct proof, you rely on statements that are already established, or statements that can be assumed to be true, to show by deduction that another statement is true (or untrue).

> Statements that can be assumed to be true are sometimes known as **axioms**.

Examples of statements that can be assumed to be true include 'you can draw a straight line segment joining any two points', and 'you can write all even numbers in the form $2n$ and all odd numbers in the form $2n - 1$'.

> **Key point**
>
> To use direct proof you
> - Assume that a statement, P, is true.
> - Use P to show that another statement, Q, must be true.

Example 1

Use direct proof to prove that the square of any integer is one more than the product of the two integers either side of it.

> Let the integer be n
> The two numbers on each side of n are $n - 1$ and $n + 1$ •———— Assume that this statement is true.
> The product of these two numbers is $(n - 1)(n + 1)$
>
> $(n - 1)(n + 1) = n^2 - 1$ •———— Expand the brackets to get the square number n^2
> So $n^2 = (n - 1)(n + 1) + 1$ •———— Rearrange to show the required result.
>
> So the square of any integer is one more than the product of the two integers either side of it.

Another method of proof is called **proof by exhaustion**. In this method, you list all the possible cases and test each one to see if the result you want to prove is true. All cases must be true for proof by exhaustion to work, since a single counter example would disprove the result.

> **Key point**
>
> To use proof by exhaustion you
> - List a set of cases that exhausts all possibilities.
> - Show the statement is true in each and every case.

Example 2

Prove, by exhaustion, that $p^2 + 1$ is not divisible by 3, where p is an integer and $6 \le p \le 10$

p	$p^2 + 1$	Divisible by 3?
6	37	NO
7	50	NO
8	65	NO
9	82	NO
10	101	NO

In all cases, $p^2 + 1$ is not divisible by 3

> Write down every possible value of p and calculate $p^2 + 1$

> Show that the statement is true in each case.

You can also *disprove* a statement with **disproof by counter example**, in which you need to find just one example that does not fit the statement.

Example 3

Prove, by counter example, that the statement '$n^2 + n + 1$ is prime for all integers n' is false.

Let $n = 4$
$4^2 + 4 + 1 = 21 = 3 \times 7$
21 has factors 1, 3, 7 and 21, so is not prime.
This *disproves* the statement '$n^2 + n + 1$ is prime for all integers n'

> Show that the statement is false for one value of n

Exercise 1.1A Fluency and skills

Use direct proof to answer these questions.

1   Prove that the number 1 is *not* a prime number.

2   Prove that the sum of two odd numbers is always even.

3   Prove that the product of two consecutive odd numbers is one less than a multiple of four.

4   Prove that the mean of three consecutive integers is equal to the middle number.

5   **a**   Prove that the sum of the squares of two consecutive integers is odd.

   b   Prove that the sum of the squares of two consecutive even numbers is always a multiple of four.

6   Show that the sum of four consecutive positive integers has both even factors and odd factors greater than one.

7   Prove that the square of the sum of any two positive numbers is greater than the sum of the squares of the numbers.

8   Prove that the perimeter of an isosceles right-angled triangle is always greater than three times the length of one of the equal sides.

9   a and b are two numbers such that $a = b - 2$ and the sum and product of a and b are equal. Prove that neither a nor b is an integer.

10   If $(5y)^2$ is even for an integer y, prove that y must be even.

Use proof by exhaustion in the following questions.

11   Prove that there is exactly one square number and exactly one cube number between 20 and 30

12   Prove that, if a month has more than five letters in its name, a four letter word can be made using those letters.

 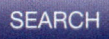

13 Prove that, for an integer x, $(x+1)^3 \geq 3^x$ for $0 \leq x \leq 4$

14 Prove that no square numbers can have a last digit 2, 3, 7 or 8

Give counter examples to disprove these statements.

15 The product of two prime numbers is always odd.

16 When you throw two six-sided dice, the total score shown is always greater than six.

17 When you subtract one number from another, the answer is always less than the first number.

18 Five times any number is always greater than that number.

19 If $a > b$, then $a^b > b^a$

20 The product of three consecutive integers is always divisible by four.

Reasoning and problem-solving

Strategy

To prove or disprove a statement

(**1**) Decide which method of proof to use.

(**2**) Follow the steps of your chosen method.

(**3**) Write a clear conclusion that proves/disproves the statement.

Example 4

Prove that the sum of the interior angles in any convex quadrilateral is 360°

The sum of the interior angles of any quadrilateral can be found by breaking the quadrilateral into two triangles.

The sum of the interior angles of any triangle equals 180°, and each of the two triangles will contribute 180° to the total sum of all angles in the quadrilateral.

So the interior angle sum of a convex quadrilateral is the same as the sum of the interior angles of two triangles, which is 360°

180°

180°

1 Since you know that the interior angles of a triangle sum to 180°, you use this to try to prove the result by direct proof.

2 Apply the result about angles in a triangle to angles in a quadrilateral.

3 Write your conclusion clearly.

Example 5

Jane says that there are exactly three prime numbers between the numbers 15 and 21 (inclusive).

Is she correct? Use a suitable method of proof to justify your answer.

NUMBER	PRIME?
15	NO → $15 = 1 \times 15, 3 \times 5$
16	NO → $16 = 1 \times 16, 2 \times 8, 4 \times 4$
17	YES → $17 = 1 \times 17$
18	NO → $18 = 1 \times 18, 2 \times 9, 3 \times 6$
19	YES → $19 = 1 \times 19$
20	NO → $20 = 1 \times 20, 2 \times 10, 4 \times 5$
21	NO → $21 = 1 \times 21, 3 \times 7$

There are exactly two prime numbers between 15 and 21, so Jane is wrong (she said there were three).

① ② Use proof by exhaustion to check all the numbers within the range of values.

③ Write a clear conclusion that proves or disproves the statement.

Exercise 1.1B Reasoning and problem-solving

1 P is a prime number and Q is an odd number.

   a Sue says PQ is even, Liz says that PQ is odd and Graham says PQ could be either. Who is right? Use a suitable method of proof to justify your answer.

   b Sue now says that $P(Q + 1)$ is always even. Is she correct? Use a suitable method of proof to justify your answer.

2 Use a suitable method of proof to show whether the statement '*Any odd number between 90 and 100 is either a prime number or the product of only two prime numbers*' is true or false.

3 Use a suitable method of proof to prove that the value of $9^n - 1$ is divisible by 8 for $1 \leq n \leq 6$

4 Is it true that 'all triangles are obtuse'? Use a suitable method of proof to justify your answer.

5 Prove that the sum of the interior angles of a convex hexagon is 720°

6 A student says that 'All quadrilaterals with equal sides are squares'. Use a suitable method of proof to show if this statement is true or false.

7 Prove that the sum of the interior angles of a convex n-sided polygon is $180(n - 2)°$

8 Use a suitable method of proof to prove or disprove the statement 'If $m^2 = n^2$ then $m = n$'.

9 The hypotenuse of a right-angled triangle is $(2s + a)$ cm and one other side is $(2s - a)$ cm. Use a suitable method of proof to show that the square of the remaining side is a multiple of eight.

10 Use a suitable method of proof to show that, for $1 \leq n \leq 5$,

$$\frac{1}{1 \times 2} + \frac{1}{2 \times 3} + \frac{1}{3 \times 4} + \ldots + \frac{1}{n \times (n+1)} = \frac{n}{n+1}$$

Challenge

11 A teacher tells her class that any number is divisible by three if the sum of its digits is divisible by three. Use a suitable method to prove this result for two-digit numbers.

Fluency and skills

The algebraic term $3x^5$ is written in **index form**. The 3 is called the **coefficient**. The x part of the term is called the **base**. The 5 is called the **power**, or **index**, or **exponent**. $3x^5$ means $3 \times x \times x \times x \times x \times x$

Indices follow some general rules.

Key point

Rule 1: Any number raised to the power zero is 1 $x^0 = 1$

Rule 2: Negative powers may be written as reciprocals. $x^{-n} = \dfrac{1}{x^n}$

Rule 3: Any base raised to the power of a unit fraction is a root. $x^{\frac{1}{n}} = \sqrt[n]{x}$

> Rule 1 has an exception when $x = 0$, as 0^0 is undefined.

> $x^{\frac{1}{2}} = \sqrt{x}$ and $x^{\frac{1}{3}} = \sqrt[3]{x}$
>
> You don't normally write the '2' in a square root.

You can combine terms in index form by following this simple set of rules called the **index laws**.

To use the index laws, the bases of all the terms must be the same.

Key point

Law 1: To multiply terms you add the indices. $x^a \times x^b = x^{a+b}$

Law 2: To divide terms you subtract the indices. $x^a \div x^b = x^{a-b}$

Law 3: To raise one term to another power you multiply the indices. $(x^a)^b = x^{a \times b}$

By combining the third general rule and the third index law you can see that $\sqrt[b]{(x^a)} = x^{\frac{a}{b}} = (\sqrt[b]{x})^a$

So $\sqrt[3]{(125^4)} = (\sqrt[3]{125})^4 = 5^4 = 625$

Example 1

Simplify these expressions, leaving your answers in index form.

a $2m^4n^2 \times 3m^3n^9$ **b** $4d^{\frac{5}{3}} \div 2d^{\frac{1}{3}}$

a $2 \times 3 \times m^{4+3} \times n^{2+9} = 6m^7n^{11}$ Use $x^a \times x^b = x^{a+b}$

b $4d^{\frac{5}{3} - \frac{1}{3}} \div 2 = 2d^{\frac{4}{3}}$ Use $x^a \div x^b = x^{a-b}$

Exercise 1.2A Fluency and skills

Simplify the expressions in questions **1** to **44**. Show your working.

1 4^3

2 $(-3)^5$

3 $7^8 \div 7^4$

4 $c^7 \times c^4$

5 $(-p^3)^4$

6 $-(p^3)^4$

7 $(2c^{-3})^6$

8 $d^7 \times d^3 \times d^4$

9 $2e^3 \times 5e^4 \times 7e^2$

10 $4f^2 \times -3f^4 \times 9f^6$

11 $24g^{12} \div 6g^6$

12 $-44k^{44} \div 11k^{-11}$

13 $12f^2 \times 4f^4 \div 6f^3$

14 $12e^{13} \div 6e^4 \div 3e^7$

15 $3a \times 5b$

16 $5w \times 4x \times (-6x)$

17 $2d \times 3e \times 4f^2$

18 $3h^6 \times (-3h^8)$

19 $5r^5s^6 \times r^3s^4$

20 $5r^5s^6 \div r^3s^4$

21 $(g^2h^3) \times (-g^7h^5) \times (ghi^4)$

22 $(g^2h^3) \times (-g^7h^5) \div (ghi^4)$

23 $(-20z^9y^6) \div (-4z^4y)$

24 $\sqrt{36u^{36}}$

25 $(36u^{36})^{\frac{1}{2}}$

26 $\sqrt[3]{125t^{27}}$

27 $\sqrt[3]{-125t^{27}c^{12}}$

28 $(5)^{-1}$

29 $\left(\dfrac{1}{5}\right)^{-1}$

30 $6u^0$

31 $-50f^0$

32 $7y^0 - 4z^0$

33 4^{-2}

34 2^{-10}

35 $(3w)^{-2}$

36 $(3w^{-2})^{-2}$

37 $64^{\frac{3}{2}}$

38 $1024^{\frac{1}{5}}$

39 $1024^{\frac{4}{5}}$

40 $16^{\frac{3}{4}}$

41 $16^{\frac{-3}{4}}$

42 $\left(\dfrac{36}{49}\right)^{\frac{1}{2}}$

43 $\left(\dfrac{36}{49}\right)^{-\frac{1}{2}}$

44 $\left(\dfrac{36}{49}\right)^{\frac{3}{2}}$

45 If $5^n = 625$, find the value of n

46 If $3^m = 243$, find the value of m

47 If $6^{2t+1} = 216$, find the value of t

48 If $(2^{2b})(2^{-6b}) = 256$, find the value of b

Reasoning and problem-solving

To solve problems involving indices

(1) Use the information in the question to write an expression or equation involving indices.

(2) Apply the laws of indices correctly.

(3) Simplify expressions as much as possible.

(4) Give your answer in an appropriate format that is relevant to the question.

Example 2

A swimming pool has a volume of $16s^2$ cubic metres.

a How long does it take to fill, from empty, if water is pumped in at a rate of $4s^{-3}$ cubic metres per minute?

b If it takes 128 minutes to fill the swimming pool, calculate the value of s

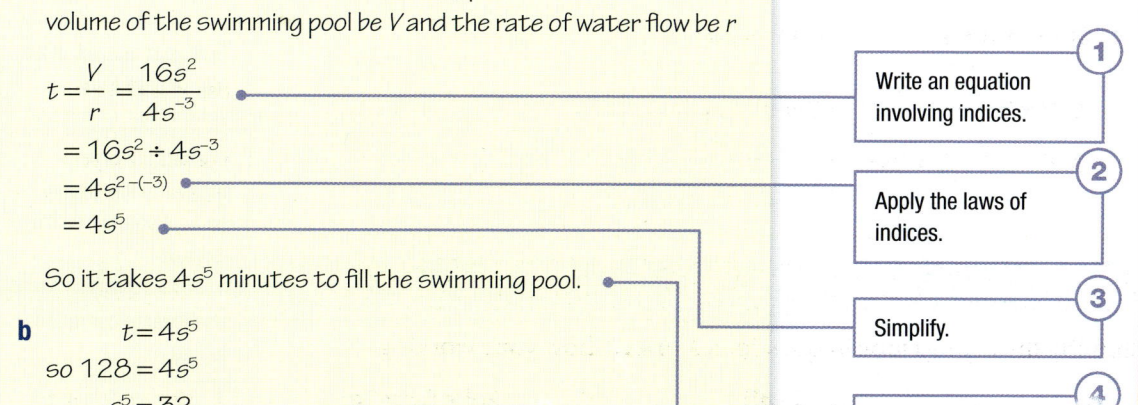

a Let the time taken to fill the swimming pool be t minutes, the volume of the swimming pool be V and the rate of water flow be r

$$t = \frac{V}{r} = \frac{16s^2}{4s^{-3}}$$ ① Write an equation involving indices.

$$= 16s^2 \div 4s^{-3}$$

$$= 4s^{2-(-3)}$$ ② Apply the laws of indices.

$$= 4s^5$$

So it takes $4s^5$ minutes to fill the swimming pool. ③ Simplify.

b $\qquad t = 4s^5$

$\qquad$ so $128 = 4s^5$

$\qquad\qquad s^5 = 32$ ④ Give your answer in an appropriate format.

$\qquad\qquad s = 2$

Example 3

A rectangular flower bed has sides of length x and $8x$

Around it are 6 further flower beds, each with an area equal to the cube root of the larger flower bed.

Calculate the total area covered by the 6 smaller flower beds, giving your answer in index form.

Area of large flower bed $= x \times 8x$

Total area of smaller flower beds

$$= 6 \times \sqrt[3]{x \times 8x}$$

$$= 6 \times \sqrt[3]{8x^2}$$

$$= 6 \times 8^{\frac{1}{3}} \times x^{\frac{2}{3}}$$

$$= 12x^{\frac{2}{3}}$$

1 Write an equation for the total area.

2 Apply the laws of indices.

3 **4** Simplify and give the final answer in index form.

Example 4

The brain mass (kg) of a species of animal is approximately one hundredth of the cube root of the square of its total body mass (kg).

a Write a formula relating brain mass, B, to total body mass, m, using index form.
b Use your formula to calculate
   i The approximate brain mass of an animal of mass 3.375 kg,
   ii The approximate total body mass of an animal with brain mass 202.5 g.

a $B = \dfrac{\sqrt[3]{m^2}}{100}$

$= \dfrac{m^{\frac{2}{3}}}{100}$

b i $B = \dfrac{3.375^{\frac{2}{3}}}{100}$

$= 0.0225\,\text{kg or } 22.5\,g$

ii $0.2025 = \dfrac{m^{\frac{2}{3}}}{100}$

$m^{\frac{2}{3}} = 20.25$

$m = 20.25^{\frac{3}{2}}$

$= 91.125\,\text{kg}$

1 Write a formula for the brain mass.

2 Apply the index laws correctly.

4 Give your answer in an appropriate form.

Convert the brain mass into kilograms and substitute into the formula.

Use inverse operations to find m

1 **a** Work out the area of a square of side $2s^2$ inches.

   b Work out the side length of a square of area $25p^4q^6\,\text{cm}^2$.

2 **a** Work out the circumference and area of a circle of radius $3w^5$ ft.

   b The volume of a sphere is $\dfrac{4}{3}\pi \times \text{radius}^3$ and the surface area is $4\pi \times \text{radius}^2$. Work out the surface area and volume of a sphere of radius $3w^4$ ft.

3 What term multiplies with $4c^2d^3$, $5de^2$ and $3c^2e^3$ to give 360?

4 Work out the volume of a cuboid with dimensions $4p^2q^3$, $3pq^2$ and $\sqrt{9p^4q^0}$ Give your answer in index form.

5 **a** The area of a rectangle is $8y^5z^7$ and its length is $4y^2z^3$. Work out its width.

   b Explain why the area of a rectangle of sides $\sqrt[3]{8m^{-3}n^6}$ and $\sqrt{16m^2n^{-4}}$ is independent of m and n. What is the area?

6 A cyclist travels $4b^2c^{\frac{1}{2}}$ miles in $3b^2c$ hours. What is her average speed.

7 **a** The volume of a cylinder is $8\pi c^2 d\,\text{cm}^3$. The radius of the cylinder is $2cd^{-1}$ cm. What is its height?

   b Explain why the volume of a cylinder of radius $3s^2t^{-1}$ and height $(5st)^2$ is independent of t. What is the volume?

8 A disc of radius $3v^2z^{-2}$ cm is removed from a disc of radius $4v^2z^{-2}$ cm. What is the remaining area?

9 **a** Work out the hypotenuse of a right-angled triangle with perpendicular sides of length $5n^{\frac{1}{2}}$ and $12n^{\frac{1}{2}}$

   b Work out the area of the right-angled triangle described in part **a**.

10 **a** To work out the voltage, V volts, in a circuit with current i amps and resistance r ohms, you multiply the current and resistance together.

Work out the voltage in a circuit of resistance $3\,m^4n^{-4}$ ohms carrying a current of $6\,m^{-2}n^{-3}$ amps.

   b The power, W watts, in a circuit with current i amps and resistance r ohms, is found by multiplying the resistance by the square of the current. Work out the power when the current and resistance are the same as in the circuit in part **a**.

11 The kinetic energy of a body is given by $\dfrac{1}{2} \times \text{mass} \times \text{velocity}^2$
Work out the kinetic energy when mass $= m$ and velocity $= 9x^{\frac{3}{4}}c^{\frac{3}{4}}$

12 You can find your Body Mass Index (BMI) by dividing your mass (kg) by the square of your height (m). If your mass is $3gt$ kg and your height is $4gt^{-1}$ m, what is your BMI?

13 In an electrical circuit, the total resistance of two resistors, t_1 and t_2, connected in parallel, is found by dividing the product of their resistances by the sum of their resistances. Work out the total resistance when $t_1 = 3rs^2$ ohms and $t_2 = 5rs^2$ ohms.

Challenge

14 In a triangle with sides of length a, b and c the semi-perimeter, s, is half the sum of the three sides. The area of the triangle can be found by subtracting each of the sides from the semi-perimeter in turn (to give three values), multiplying these expressions and the semi-perimeter altogether and then square rooting the answer.

   a Write a formula for the area of the triangle involving s, a, b and c

   b Use your formula to work out the area of a triangle with sides, $12xy$, $5xy$ and $13xy$

   c You could have found the area of this triangle in a much easier way. Explain why.

Fluency and skills

> **Key point**
>
> A **rational number** is one that you can write exactly in the form
> $$\frac{p}{q}$$
> where p and q are integers, $q \neq 0$

Numbers that you cannot write exactly in this form are **irrational numbers**. If you express them as decimals, they have an infinite number of non-repeating decimal places.

Some roots of numbers are irrational, for example, $\sqrt{3} = 1.732\ldots$ and $\sqrt[3]{10} = 2.15443\ldots$ are irrational numbers.

> **Key point**
>
> Irrational numbers involving roots, $\sqrt[n]{\ }$ or $\sqrt{\ }$, are called **surds**.
>
> You can use the following laws to simplify surds
> $$\sqrt{a} \times \sqrt{b} = \sqrt{ab}$$
> $$\frac{\sqrt{a}}{\sqrt{b}} = \sqrt{\frac{a}{b}}$$

You usually write surds in their simplest form, with the smallest possible number written inside the root sign.

You can simplify surds by looking at their factors.

You should look for factors that are square numbers.

For example $\sqrt{80} = \sqrt{16} \times \sqrt{5} = 4\sqrt{5}$

> If $\sqrt{a}$ and $\sqrt{b}$ cannot be simplified, then you cannot simplify $\sqrt{a} + \sqrt{b}$ or $\sqrt{a} - \sqrt{b}$ for $a \neq b$

Example 1

Simplify these expressions. Show your working.

a $\sqrt{7} \times \sqrt{7}$ b $\sqrt{5} \times \sqrt{20}$ c $\sqrt{\frac{1}{9}} \times \sqrt{9}$

d $\sqrt{80} - \sqrt{20}$ e $\sqrt{63} + \sqrt{112}$ f $\sqrt{\frac{4}{3}} + \sqrt{\frac{25}{3}}$

a $\sqrt{7} \times \sqrt{7} = 7$ b $\sqrt{5} \times \sqrt{20} = \sqrt{100} = 10$ c $\sqrt{\frac{1}{9}} \times \sqrt{9} = \frac{1}{3} \times 3 = 1$

d $4\sqrt{5} - 2\sqrt{5} = 2\sqrt{5}$ e $3\sqrt{7} + 4\sqrt{7} = 7\sqrt{7}$ f $2\sqrt{\frac{1}{3}} + 5\sqrt{\frac{1}{3}} = 7\sqrt{\frac{1}{3}}$

Calculations are often more difficult if surds appear in the denominator. You can simplify such expressions by removing any surds from the denominator. To do this, you multiply the numerator and denominator by the same value to find an **equivalent fraction** with surds in the numerator only. This is easier to simplify.

This process is called **rationalising the denominator**.

Key point

If the fraction is in the form

$\dfrac{k}{\sqrt{a}}$, multiply numerator and denominator by $\sqrt{a}$

$\dfrac{k}{a\pm\sqrt{b}}$, multiply numerator and denominator by $a\mp\sqrt{b}$

$\dfrac{k}{\sqrt{a}\pm\sqrt{b}}$, multiply numerator and denominator by $\sqrt{a}\mp\sqrt{b}$

Example 2

Rationalise the denominators of these expressions. Show your working.

a $\dfrac{4}{\sqrt{5}}$ **b** $\dfrac{6+\sqrt{7}}{9-\sqrt{7}}$

a $\dfrac{4}{\sqrt{5}} = \dfrac{4}{\sqrt{5}} \times \dfrac{\sqrt{5}}{\sqrt{5}} = \dfrac{4\times\sqrt{5}}{\sqrt{5}\times\sqrt{5}} = \dfrac{4\sqrt{5}}{5}$

Multiply top and bottom by $\sqrt{5}$

b $\dfrac{6+\sqrt{7}}{9-\sqrt{7}} = \dfrac{6+\sqrt{7}}{9-\sqrt{7}} \times \dfrac{9+\sqrt{7}}{9+\sqrt{7}}$

Multiply top and bottom by $9+\sqrt{7}$

$= \dfrac{54+9\sqrt{7}+6\sqrt{7}+7}{\left(9-\sqrt{7}\right)(9+\sqrt{7})}$

$= \dfrac{61+15\sqrt{7}}{81-9\sqrt{7}+9\sqrt{7}-\sqrt{7}\,\sqrt{7}}$

$= \dfrac{61+15\sqrt{7}}{81-9\sqrt{7}+9\sqrt{7}-7} = \dfrac{61+15\sqrt{7}}{74}$

Exercise 1.3A Fluency and skills

Complete this exercise without a calculator.

1 Classify these numbers as rational or irrational.

a $1+\sqrt{25}$ **b** π^2 **c** $4-\sqrt{3}$

d $\sqrt{21}$ **e** $\sqrt{169}$ **f** $\left(\sqrt{8}\right)^2$

g $\left(\sqrt{17}\right)^3$

2 For each of these expressions, show that they can be written in the form $a\sqrt{b}$ where a and b are integers.

a $\sqrt{4}\times\sqrt{21}$ **b** $\sqrt{8}\times\sqrt{7}$

c $\sqrt{75}$ **d** $\sqrt{27}$

e $\dfrac{\sqrt{800}}{10}$ **f** $\left(\sqrt{8}\right)^3$

g $\left(\sqrt{17}\right)^3$ **h** $2\sqrt{3}\times3\sqrt{2}$

i $5\sqrt{6}\times7\sqrt{18}$ **j** $4\sqrt{24}\times6\sqrt{30}$

3 Show that these expressions can be expressed as positive integers.

a $\dfrac{\sqrt{128}}{\sqrt{2}}$ **b** $\dfrac{\sqrt{125}}{\sqrt{5}}$

4 Show that these expressions can be written in the form $\dfrac{a}{b}$, where a and b are positive integers.

a $\dfrac{\sqrt{32}}{\sqrt{200}}$ **b** $\dfrac{\sqrt{50}}{\sqrt{72}}$

5 Show that these expressions can be written in the form $a\sqrt{b}$, where a and b are integers.

a $\sqrt{54}$ **b** $\sqrt{432}$

c $\sqrt{1280}$ **d** $\sqrt{3388}$

e $\sqrt{2} \times \sqrt{20}$ **f** $\sqrt{2} \times \sqrt{126}$

g $\sqrt{20} + \sqrt{5}$ **h** $\sqrt{18} - \sqrt{2}$

i $\sqrt{150} - \sqrt{24}$ **j** $\sqrt{75} + \sqrt{12}$

k $\sqrt{27} - \sqrt{3}$ **l** $\sqrt{5} + \sqrt{45}$

m $\sqrt{363} - \sqrt{48}$ **n** $\sqrt{72} - \sqrt{288} + \sqrt{200}$

6 Show these expressions can be written in the form $a + b\sqrt{c}$, where a, b and c are integers.

a $\left(3\sqrt{6} + \sqrt{5}\right)^2$

b $\left(\sqrt{2} + 3\right)\left(4 + \sqrt{2}\right)$

c $\left(\sqrt{2} - 3\right)\left(4 - \sqrt{2}\right)$

d $\left(3\sqrt{5} + 4\right)\left(2\sqrt{5} - 6\right)$

e $\left(5\sqrt{3} + 3\sqrt{2}\right)\left(4\sqrt{27} - 5\sqrt{8}\right)$

7 Rationalise the denominators in these expressions and leave your answers in their simplest form. Show your working.

a $\dfrac{1}{\sqrt{13}}$ **b** $\dfrac{8}{\sqrt{6}}$

c $\dfrac{\sqrt{11}}{2\sqrt{5}}$ **d** $\dfrac{3}{\sqrt{2} - 1}$

e $\dfrac{3\sqrt{7} \times 5\sqrt{4}}{6\sqrt{7}}$ **f** $\dfrac{13\sqrt{15} - 2\sqrt{10}}{4\sqrt{75}}$

g $\dfrac{5}{8 - \sqrt{5}}$ **h** $\dfrac{2\sqrt{2}}{4 + \sqrt{2}}$

i $\dfrac{\sqrt{6} - \sqrt{5}}{\sqrt{6} + \sqrt{5}}$ **j** $\dfrac{3\sqrt{11} - 4\sqrt{7}}{\sqrt{11} - \sqrt{7}}$

8 Rationalise the denominators and simplify these expressions. a, b and c are integers.

a $\dfrac{a + \sqrt{b}}{\sqrt{b}}$ **b** $\dfrac{a + \sqrt{b}}{a - \sqrt{b}}$

c $\dfrac{\sqrt{a} + b\sqrt{c}}{b\sqrt{c}}$ **d** $\dfrac{\sqrt{a} - b\sqrt{c}}{\sqrt{a} + \sqrt{b}}$

Reasoning and problem-solving

To solve problems involving surds

1 Use the information given to form an expression involving surds.

2 If possible, simplify surds. They should have the lowest possible number under the root sign.

3 Rationalise any denominator containing surds.

Example 3

The sides of a parallelogram are $\sqrt{27}$ m and $2\sqrt{12}$ m, and it has a perpendicular height of $\dfrac{10}{\sqrt{3}}$ m.

a Work out the perimeter of the parallelogram.

b Work out the area of the parallelogram.
Give your answers in their simplest form.

$2\sqrt{12}$ m

$\sqrt{27}$ m

a Perimeter

$2(\sqrt{27} + 2\sqrt{12}) = 2(3\sqrt{3} + 4\sqrt{3}) = 14\sqrt{3}$ m

b Area = base × perpendicular height

$\sqrt{27} \times \dfrac{10}{\sqrt{3}} = \dfrac{3\sqrt{3} \times 10}{\sqrt{3}} = 30 \text{ m}^2$

1 Form an expression involving surds.

2 Simplify.

2 **3** Simplify and rationalise the denominator.

Exercise 1.3B Reasoning and problem-solving

1 A rectangle has sides of length $2\sqrt{3}$ cm and $3\sqrt{2}$ cm. What is its area? Show your working.

2 **a** A circle has radius $9\sqrt{3}$ cm. Show that its area is 243π cm².

   b A circle has area 245π m². What is its diameter? Show your working.

3 **a** A car travels $18\sqrt{35}$ m in $6\sqrt{7}$ s. Work out its speed, showing your working.

   b A runner travels for 5 s at $\dfrac{8}{\sqrt{5}}$ m s⁻¹. Work out how far she ran in simplified form. Show your working.

4 A cube has sides of length $(2 + \sqrt{7})$ m. Work out its volume in simplified surd form.

5 Rectangle A has sides of length $3\sqrt{3}$ m and $\sqrt{5}$ m. Rectangle B has sides of length $\sqrt{5}$ m and $\sqrt{7}$ m. How many times larger is rectangle A than rectangle B? Give your answer in its simplest surd form, showing your working.

6 A right-angled triangle has perpendicular sides of $2\sqrt{3}$ cm and $3\sqrt{7}$ cm. Calculate the length of the hypotenuse. Show your working and give your answer in simplified form.

7 Base camp is $5\sqrt{5}$ miles due east and $5\sqrt{7}$ miles due north of a walker. What is the exact distance from the walker to the camp? Show your working.

8 The arc of a bridge forms part of a larger circle with radius $\dfrac{12}{\sqrt{3}}$ m. If the arc of the bridge subtends an angle of 45°, show that the length of the bridge is $\sqrt{3}\,\pi$ m.

9 The equation of a parabola is $y^2 = 4ax$ Find y when $a = 6 - \sqrt{6}$ and $x = \dfrac{6 + \sqrt{6}}{10}$ Show your working and give your answer in simplified form.

10 Show that the ratio of the volumes of two cubes of sides $6\sqrt{8}$ cm and $4\sqrt{2}$ cm is 27

11 The top speeds, in m s⁻¹, of two scooters are given as $\dfrac{12}{\sqrt{a}}$ and $\dfrac{17\sqrt{a}}{3a}$, where a is the volume of petrol in the tank. Find the difference in top speed between the two scooters if they both contain the same volume of petrol. Give your answer in surd form, showing your working.

12 The area of an ellipse with semi-diameters a and b is given by the formula πab Work out the area of an ellipse where $a = \dfrac{5}{4 + \sqrt{3}}$ m and $b = \dfrac{8}{4 - \sqrt{3}}$ m Show your working.

13 The force required to accelerate a particle can be calculated using $F = ma$, where F is the force, m is the mass of the particle and a is the acceleration. Showing your working, work out F when $m = 8\sqrt{6}$ and $a = \dfrac{5}{2 + \sqrt{6}}$

14 An equilateral triangle with side length $5\sqrt{6}$ inches has one vertex at the origin and one side along the positive x-axis.

The centre is on the vertical line of symmetry, $\dfrac{1}{3}$ of the way from the x-axis to the vertex. Work out the distance from the origin to the centre of mass of the triangle. Show your working.

Challenge

15 If the sides of a cyclic quadrilateral are a, b, c and d, and $s = \dfrac{a+b+c+d}{2}$, the area is $A = \sqrt{(s-a)(s-b)(s-c)(s-d)}$

A quadrilateral with side lengths $5 + 5\sqrt{2}$, $3 + 3\sqrt{2}$, $6 + 4\sqrt{2}$ and $2 + 4\sqrt{2}$ cm is inscribed in a circle.

Prove that $A = 2(1 + \sqrt{2})\sqrt{15(11 + 8\sqrt{2})}$

Fluency and skills

A **quadratic function** can be written in the form $ax^2 + bx + c$, where a, b and c are constants and $a \neq 0$

A **quadratic equation** can be written in the general form $ax^2 + bx + c = 0$

Curves of quadratic functions, $y = ax^2 + bx + c$, have the same general shape. The curve crosses the y-axis when $x = 0$, and the curve crosses the x-axis at any **roots** (or solutions) of the equation $ax^2 + bx + c = 0$

Quadratic curves are symmetrical about their **vertex** (the turning point). For $a > 0$, this vertex is always a **minimum** point, and for $a < 0$ this vertex is always a **maximum** point.

When $a > 0$, a quadratic graph looks like this.

When $a < 0$, a quadratic graph looks like this.

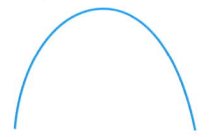

Example 1

The quadratic equation $3x^2 - 20x - 7 = 0$ has solutions $x = -\dfrac{1}{3}$ and $x = 7$
Sketch the curve of $f(x) = 3x^2 - 20x - 7$, showing where it crosses the axes.

When $x = 0$, $y = 3(0)^2 - 20(0) - 7 = -7$

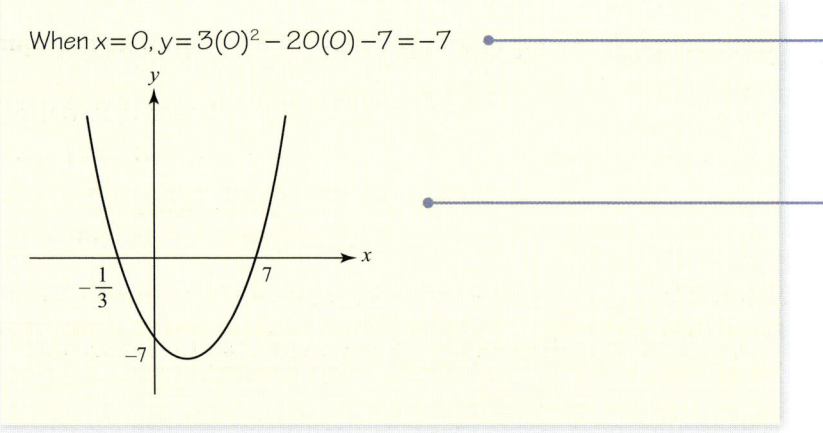

Find the y-intercept.

The curve $y = f(x)$ crosses the x-axis at the solutions to $f(x) = 0$

Try it on your calculator

You can sketch a curve on a graphics calculator.

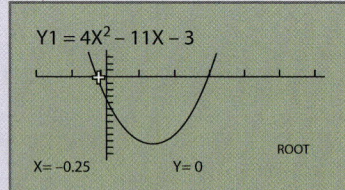

Activity

Find out how to sketch the curve $y = 4x^2 - 11x - 3$ on *your* graphics calculator.

You can solve some quadratics in the form $ax^2 + bx + c = 0$ by **factorisation**. To factorise a quadratic, try to write it in the form $(mx + p)(nx + q) = 0$

A quadratic equation that can be written in the form $(mx + p)(nx + q) = 0$ has solutions $x = -\dfrac{p}{m}$ or $x = -\dfrac{q}{n}$

Example 2

Find the solutions of the quadratic equation
$6x^2 + 17x + 7 = 0$ by factorisation.

$6x^2 + 17x + 7 = 0$
$6x^2 + 3x + 14x + 7 = 0$
$3x(2x + 1) + 7(2x + 1) = 0$
$(2x + 1)(3x + 7) = 0$
$x = -\dfrac{7}{3}$ or $x = -\dfrac{1}{2}$

Split the x term so that the two coefficients multiply to give ac.
$6 \times 7 = 42$ and $3 \times 14 = 42$

Factorise the first pair of terms, then the second pair of terms. Take out a factor which is common to both pairs.

Factorise the full expression.

Sometimes a quadratic will not factorise easily. In these cases you may need to **complete the square**.

Any quadratic expression can be written in the following way. This is called completing the square.
$$ax^2 + bx + c \equiv a\left(x + \frac{b}{2a}\right)^2 + q$$

You'll need to find the value of q yourself. It will be equal to $c - \dfrac{b^2}{4a}$

When $a = 1$ and $q = 0$, the expression is known as a **perfect square**. For example, $x^2 + 6x + 9 = (x + 3)^2$

Perfect squares have only one root, so a graph of the quadratic function touches the x-axis only once, at its vertex.

Example 3

By completing the square, find all the solutions of $4 - 3x^2 - 6x = 0$

$3x^2 + 6x - 4 = 0$
$3[x^2 + 2x] - 4 = 0$
$3[(x + 1)^2 - 1] - 4 = 0$
$3(x + 1)^2 - 7 = 0$
$(x + 1)^2 = \dfrac{7}{3} \Rightarrow x = -1 + \sqrt{\dfrac{7}{3}}$ or $-1 - \sqrt{\dfrac{7}{3}}$

Multiply both sides by -1

Manipulate the expression to obtain a bracket containing x^2 and the x term.

Complete the square and expand. Substitute this into the previous equation.

Completing the square is a useful tool for determining the maximum or minimum point of a quadratic function.

PURE

Example 4

Determine the minimum point of the graph of $f(x) = 2x^2 + 12x + 16$

$$f(x) = 2x^2 + 12x + 16 = 2(x^2 + 6x) + 16 = 2[(x+3)^2 - 9] + 16$$
$$= 2(x+3)^2 - 2$$

At minimum point $x = -3$

$f(-3) = 2(0)^2 - 2 = -2 \Rightarrow$ minimum point $(-3, -2)$

$(x+3)^2 \geq 0$, so the minimum point is when $(x+3)^2 = 0$

By writing the equation, $ax^2 + bx + c = 0$, $a \neq 0$, in completed square form you can derive the **quadratic formula** for solving equations.

$$ax^2 + bx + c = 0$$

$$a\left[x^2 + \frac{b}{a}x\right] + c = 0$$

$$a\left[\left(x + \frac{b}{2a}\right)^2 - \frac{b^2}{4a^2}\right] + c = 0$$

$$\left(x + \frac{b}{2a}\right)^2 - \frac{b^2}{4a^2} = \frac{-c}{a}$$

$$\left(x + \frac{b}{2a}\right)^2 = \frac{b^2}{4a^2} - \frac{c}{a}$$

$$\left(x + \frac{b}{2a}\right)^2 = \frac{b^2 - 4ac}{4a^2}$$

$$x + \frac{b}{2a} = \frac{\pm\sqrt{b^2 - 4ac}}{2a}$$

$$x = \frac{-b \pm \sqrt{b^2 - 4ac}}{2a}$$

Key point

For constants a, b and c, $a > 0$, the solutions to the equation $ax^2 + bx + c = 0$ are
$$x = \frac{-b \pm \sqrt{b^2 - 4ac}}{2a}$$

The expression inside the square root is called the **discriminant**, Δ

Key point

$$\Delta = b^2 - 4ac$$

If the discriminant, Δ, is positive, it has two square roots. If Δ is 0, it has one square root. If it is negative, it has no real square roots. The value of Δ tells you whether a quadratic equation $ax^2 + bx + c = 0$ has two, one or no real solutions. This result is useful for curve sketching.

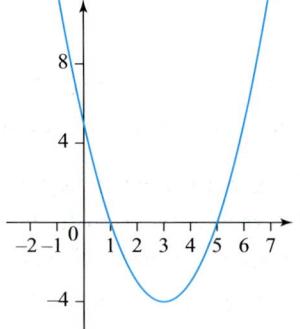

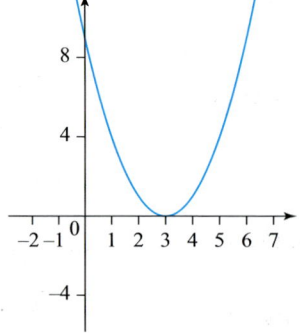

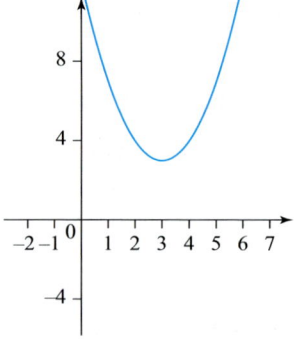

If $\Delta > 0$, the quadratic $y = ax^2 + bx + c$ has two distinct roots and the curve crosses the x-axis at two distinct points.

If $\Delta = 0$, the quadratic $y = ax^2 + bx + c$ has one repeated root and the x-axis is a tangent to the curve at this point.

If $\Delta < 0$, the quadratic $y = ax^2 + bx + c$ has no (real) roots and the curve does not cross the x-axis at any point.

Example 5

Use the discriminant $\Delta = b^2 - 4ac$ to determine how many roots each of these quadratic equations have.

a $x^2 + 2x + 1 = 0$ **b** $x^2 + 2x - 8 = 0$ **c** $x^2 + 6x + 10 = 0$

a $\Delta = b^2 - 4ac = 2^2 - 4 \times 1 \times 1 = 0$

So $x^2 + 2x + 1 = 0$ has one repeated root.

b $\Delta = b^2 - 4ac = 2^2 - 4 \times 1 \times -8 = 36 > 0$

So $x^2 + 2x - 8 = 0$ has two distinct roots.

c $\Delta = b^2 - 4ac = 6^2 - 4 \times 1 \times 10 = -4 < 0$

So $x^2 + 6x + 10 = 0$ has no real roots.

Calculator

Try it on your calculator

You can solve a quadratic equation on a calculator.

Activity

Find out how to solve $2x^2 - 3x - 20 = 0$ on *your* calculator.

Exercise 1.4A Fluency and skills

1 Solve these quadratic equations by factorisation.

   a $x^2 - 18 = 0$ **b** $2x^2 - 6 = 0$

   c $4x^2 + 5x = 0$ **d** $x^2 + 2\sqrt{3}x + 3 = 0$

   e $2x^2 + 5x - 3 = 0$ **f** $3x^2 - 23x + 14 = 0$

   g $16x^2 - 24x + 9 = 0$ **h** $18 + x - 4x^2 = 0$

2 For each quadratic function

   i Factorise the equation,

   ii Use your answer to part **i** to sketch a graph of the function.

   a $f(x) = x^2 + 3x + 2$ **b** $f(x) = x^2 + 6x - 7$

   c $f(x) = -x^2 - x + 2$ **d** $f(x) = -x^2 - 7x - 12$

   e $f(x) = 2x^2 - x - 1$ **f** $f(x) = -3x^2 + 11x + 20$

3 Solve these quadratic equations using the formula. Write your answers both exactly (in surd form) and also, where appropriate, correct to 2 decimal places.

   a $3x^2 + 9x + 5 = 0$ **b** $4x^2 + 5x - 1 = 0$

   c $x^2 + 12x + 5 = 0$ **d** $28 - 2x - x^2 = 0$

   e $x^2 + 15x - 35 = 0$ **f** $34 + 3x - x^2 = 0$

   g $4x^2 - 36x + 81 = 0$ **h** $3x^2 - 23x + 21 = 0$

   i $5x^2 + 16x + 9 = 0$ **j** $10x^2 - x - 1 = 0$

4 For each quadratic function, complete the square and thus determine the coordinates of the minimum or maximum point of the curve.

   a $f(x) = x^2 - 14x + 49$ **b** $f(x) = x^2 + 2x - 5$

   c $f(x) = -x^2 - 6x - 5$ **d** $f(x) = -x^2 + 4x + 3$

   e $f(x) = 9x^2 - 6x - 5$ **f** $f(x) = -2x^2 - 28x - 35$

5 Complete the square to work out the exact solutions to these quadratic equations.

   a $x^2 - 2x = 0$ **b** $3 - 4x - x^2 = 0$

c $x^2 - 14x + 33 = 0$ **d** $x^2 + 8x + 10 = 0$

e $x^2 - 6x + 9 = 0$ **f** $x^2 + 10x + 24 = 0$

g $x^2 + 22x + 118 = 0$ **h** $x^2 - 16x + 54 = 0$

i $4x^2 - 12x + 2 = 0$ **j** $9x^2 + 12x - 2 = 0$

k $x^2 + 11x + 3 = 0$ **l** $9x^2 - 30x - 32 = 0$

6 Solve these quadratic equations using your calculator.

a $2x^2 - 6x = 0$ **b** $x^2 + 2x - 15 = 0$

c $x^2 - 5x - 6 = 0$ **d** $8 + 2x - x^2 = 0$

e $2x^2 - x - 15 = 0$ **f** $6 + 8x - 8x^2 = 0$

7 By evaluating the discriminant, identify the number of real roots of these equations.

a $x^2 + 2x - 5 = 0$ **b** $13 + 3x - x^2 = 0$

c $x^2 + 5x + 5 = 0$ **d** $-3 + 2x - x^2 = 0$

e $4x^2 + 12x + 9 = 0$ **f** $-35 + 2x - x^2 = 0$

g $9x^2 - 66x + 121 = 0$

h $-100 - 100x - 100x^2 = 0$

Reasoning and problem-solving

To solve a problem involving a quadratic curve

1 Factorise the equation or complete the square and solve as necessary.

2 Sketch the curve using appropriate axes and scale.

3 Mark any relevant points in the context of the question.

The motion of a body, which has an initial velocity u and acceleration a, is given by the formula $s = ut + \frac{1}{2}at^2$, where s is the displacement after a time t

a By completing the square and showing all intermediate steps, sketch the graph of s against t when $u = 8$ and $a = -4$

b What is happening to the body at the turning point of the graph?

a $s = 8t - 2t^2$
$\quad = -2(t^2 - 4t)$
$\quad = -2[(t-2)^2 - 4] = -2(t-2)^2 + 8$

So there is a turning point at $(2, 8)$
Where the curve crosses the x-axis,

$2t(4 - t) = 0$
$t = 0$ or 4

Therefore the curve cuts the x-axis at $(0, 0)$ and $(4, 0)$

b The body is reversing direction. At this point it has zero velocity.

1 Factorise and complete the square.

2 **3** Sketch the curve and mark on it any relevant points.

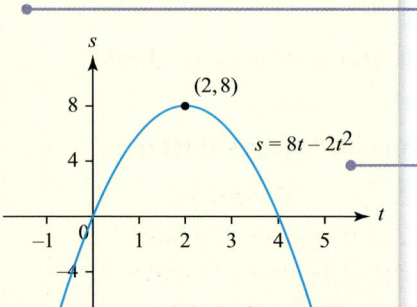

Exercise 1.4B Reasoning and problem-solving

1 A designer is lining the base and sides of a rectangular drawer, dimensions $2x$ cm by $3x$ cm by 5 cm, with paper.

The total area of paper is 4070 cm².

   a Write and solve an equation to find x

   b Hence work out the volume of the drawer in litres.

2 Sam and his mother Jane were both born on January 1st.

In 2002, Sam was x years old and Jane was $2x^2 + 11x$ years old. In 2007, Jane was five times as old as Sam.

   a Form and solve a quadratic equation in x

   b Hence work out Jane's age when Sam was born.

3 A photo is to be pasted onto a square of white card with side length x cm. The photo is $\frac{3}{4}x$ cm long and its width is 20 cm less than the width of the card. The area of the remaining card surrounding the photo is 990 cm². Work out the dimensions of the card and photo.

4 A piece of wire is bent into a rectangular shape with area 85 in².

The total perimeter is 60 inches and the rectangle is x^2 in long.

   a Form an expression for the area of the rectangle.

   b By substituting z for x^2, form a quadratic equation in z

   c Hence work out all possible values of x

5 A man stands on the edge of a cliff and throws a stone out over the sea. The height, h m, above the sea that the stone reaches after t seconds is given by the formula $h = 50 + 25t - 5t^2$

   a Complete the square.

   b Sketch the graph of $h = 50 + 25t - 5t^2$

   c Use your graph to estimate

      i The maximum height of the stone above the sea and the time at which it reaches this height,

      ii The time when the stone passes the top of the cliff on the way down,

      iii The time when the stone hits the sea.

6 A firm making glasses makes a profit of y thousand pounds from x thousand glasses according to the equation $y = -x^2 + 5x - 2$

   a Sketch the curve.

   b Use your graph to estimate

      i The value of x to give maximum profit,

      ii The value of x not to make a loss,

      iii The range of values of x which gives a profit of more than £3250

7 The mean braking distance, d yards, for a car is given by the formula $d = \dfrac{v^2}{50} + \dfrac{v}{3}$, where v is the speed of the car in miles per hour.

   a Sketch this graph for $0 \leq v \leq 80$

   b Use your sketch to estimate the safe braking distance for a car driving at

      i 15 mph   **ii** 45 mph   **iii** 75 mph

   c A driver just stops in time in a distance of 50 yards.
How fast was the car travelling when the brakes were applied?

Challenge

8 Use a suitable substitution to solve $2(k^6 - 11k^3) = 160$ for k. Give your answers in exact form.

Fluency and skills

The equation of a straight line can be written in the form $y = mx + c$

where m is the gradient and c is the y-intercept.

> **Key point**
>
> A straight line can also be written in the form
> $$y - y_1 = m(x - x_1)$$
> where (x_1, y_1) is a point on the line and m is the gradient.

You can rearrange the general equation of a straight line to get a formula for the gradient.

> **Key point**
>
> The gradient of a straight line through two points
> (x_1, y_1) and (x_2, y_2) is $m = \dfrac{y_2 - y_1}{x_2 - x_1}$

You can use Pythagoras' theorem to find the distance between two points.

> **Key point**
>
> The distance between two points (x_1, y_1) and (x_2, y_2) is
> given by the formula $\sqrt{(x_1 - x_2)^2 + (y_1 - y_2)^2}$

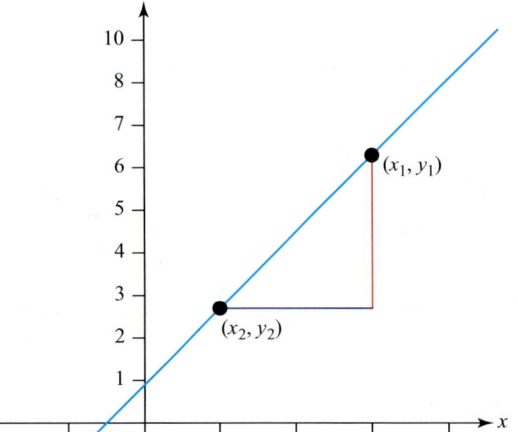

> **Key point**
>
> The coordinates of the midpoint of the line joining
> (x_1, y_1) and (x_2, y_2) are given by the formula
> $$\left(\dfrac{x_1 + x_2}{2}, \dfrac{y_1 + y_2}{2} \right)$$

You can use the gradients of two lines to decide if they are **parallel** or **perpendicular**.

Key point

Two lines are described by the equations
$y_1 = m_1 x + c_1$ and $y_2 = m_2 x + c_2$

If $m_1 = m_2$, the two lines are parallel.

If $m_1 \times m_2 = -1$, the two lines are perpendicular.

Example 1

A straight line segment joins the points $(-2, -3)$ and $(4, 9)$

a Work out the midpoint of the line segment.

b Work out the equation of the perpendicular bisector of the line segment.
   Give your answer in the form $ay + bx + c = 0$ where a, b and c are integers.

a $\left(\dfrac{-2+4}{2}, \dfrac{-3+9}{2} \right) = (1, 3)$

b Gradient of line segment $= \dfrac{9 - -3}{4 - -2} = 2$
  — Use gradient $= \dfrac{y_2 - y_1}{x_2 - x_1}$

Gradient of perpendicular bisector $= -\dfrac{1}{2}$
  — Use $m_1 \times m_2 = -1$

$y - 3 = -\dfrac{1}{2}(x - 1)$
  — Use $y - y_1 = m(x - x_1)$

$2y - 6 = -x + 1$

$2y + x - 7 = 0$
  — Multiply through by 2 and rearrange to the required form.

On a graph, the equation of any circle with centre (a, b) and radius r has the same general form.

The diagram shows a circle, centre $(1, 4)$ and radius 5, with a general point (x, y) shown on the circumference.

The vertical distance of the point (x, y) from the centre is $y - 4$ and the horizontal distance of the point (x, y) from the centre is $x - 1$

Using Pythagoras' theorem for the right-angled triangle shown, you get $(x - 1)^2 + (y - 4)^2 = 5^2$

Notice that this equation is in the form $(x - a)^2 + (y - b)^2 = r^2$
This is the equation for any circle.

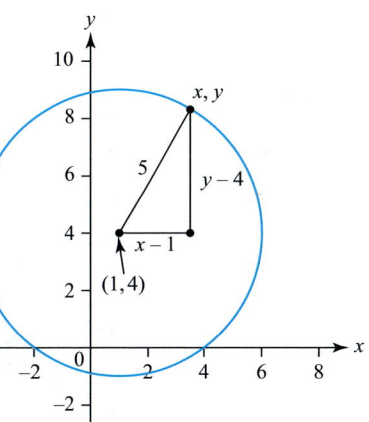

Key point

The equation of a circle, centre (a, b) and radius r, is
$$(x - a)^2 + (y - b)^2 = r^2$$

This can also be written in the form
$$x^2 + y^2 - 2ax - 2by + c = 0$$
where $c = a^2 + b^2 - r^2$

Key point

For a circle centred at the origin, $a = 0$ and $b = 0$, so the equation of the circle is simply
$$x^2 + y^2 = r^2$$

 MyMaths   🔍 2001–2004, 2020, 2021   SEARCH

Example 2

Work out the equation of the circle with centre $(-4, 9)$, radius $\sqrt{8}$

Write your answer without brackets.

$(x+4)^2 + (y-9)^2 = 8$ ———————— Use $(x-a)^2 + (y-b)^2 = r^2$

$x^2 + 8x + 16 + y^2 - 18y + 81 - 8 = 0$

$x^2 + y^2 + 8x - 18y + 89 = 0$

Example 3

Work out the centre and radius of the circle $4x^2 - 4x + 4y^2 + 3y - 6 = 0$

$4x^2 - 4x + 4y^2 + 3y - 6 = 0$

$x^2 - x + y^2 + \dfrac{3}{4}y - \dfrac{3}{2} = 0$ ———————— Divide by 4

$\left(x - \dfrac{1}{2}\right)^2 - \dfrac{1}{4} + \left(y + \dfrac{3}{8}\right)^2 - \dfrac{9}{64} - \dfrac{3}{2} = 0$ ———————— Complete the square terms using both x and y terms.

$\left(x - \dfrac{1}{2}\right)^2 + \left(y + \dfrac{3}{8}\right)^2 = \dfrac{121}{64}$ ———————— Use $(x-a)^2 + (y-b)^2 = r^2$

Hence the circle has centre $\left(\dfrac{1}{2}, -\dfrac{3}{8}\right)$ and radius $\dfrac{11}{8}$ ———————— You can check by drawing the circle on a graphics calculator.

When you're working with equations of circles, it's useful to remember some facts about the lines and angles in a circle. You should have come across these before in your studies.

Key point

- If a triangle passes through the centre of the circle, and all three corners touch the circumference of the circle, then the triangle is right-angled.
- The perpendicular line from the centre of the circle to a chord bisects the chord (Point A in the diagram).
- Any tangent to a circle is perpendicular to the radius at the point of contact (B).

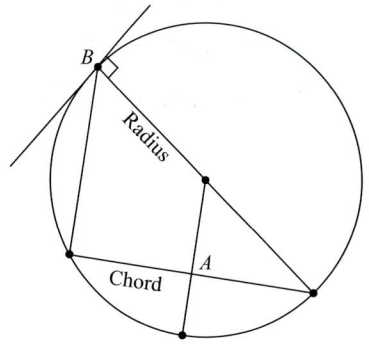

Exercise 1.5A Fluency and skills

1 **a** Write down the equation of the straight line with gradient $-\dfrac{2}{3}$ that passes through the point $(-4, 7)$. Give your answer in the form $ax + by + c = 0$, where a, b and c are integers.

   b Does the point $(13, 3)$ lie on the line described in part **a**?

2 Find the gradient and y-intercept of the line $4x - 3y = 8$

3 **a** Show that the lines $2x - 3y = 4$ and $6x + 4y = 7$ are perpendicular.

   b Show that the lines $2x - 3y = 4$ and $8x - 12y = 7$ are parallel.

4 Write down the gradient and y-intercept of the line $\dfrac{2}{3}x + \dfrac{3}{4}y + \dfrac{7}{8} = 0$

5 Calculate the gradient of the straight-line segment joining the points $(-5, -6)$ and $(4, -1)$

   Hence write down the equation of the line.

6 Write, in both the form $y = mx + c$ and the form $ax + by + c = 0$, the equation of the line with gradient -3 passing through $(-8, -1)$

7 Work out the midpoint and length of the line segment joining each of these pairs of points.

   a $(2, 2)$ and $(6, 10)$

   b $(-3, -4)$ and $(2, -3)$

   c $(0, 0)$ and $(\sqrt{5}, 2\sqrt{3})$

8 Which of these lines are parallel or perpendicular to each other?

   $2x + 3y = 4$ $4x - 5y = 6$ $y = 4x + 8$
   $10x - 8y = 5$ $10x + 8y = 5$ $3y - 12x = 7$
   $6x + 9y = 12$

9 **a** Write down the equation of the straight line through the point $(5, -4)$ which is parallel to the line $2x + 3y - 6 = 0$

   b Write down the equation of the straight line through the point $(-2, -3)$ which is perpendicular to the line $3x + 6y + 5 = 0$

10 Write down the equations of each of these circles.

   Expand your answers into the form $ax^2 + bx + cy^2 + dy + e = 0$

   a Centre $(1, 8)$; radius 5

   b Centre $(6, -7)$; radius 3

   c Centre $\left(\sqrt{5}, \sqrt{2}\right)$; radius $\sqrt{11}$

11 Work out the centre and radius of each of these circles.

   a $x^2 + 18x + y^2 - 14y + 30 = 0$

   b $x^2 + 12x + y^2 + 10y - 25 = 0$

   c $x^2 - 2\sqrt{3}x + y^2 + 2\sqrt{7}y - 1 = 0$

12 Prove that the points $A(-10, -12)$, $B(6, 18)$ and $C(-2, -14)$ lie on a semicircle.

13 Write the equation of the tangent to the circle with centre $(4, -3)$ at the point $P(-2, -1)$

14 $(-3, 9)$ is the midpoint of a chord within a circle with centre $(7, -1)$ and radius 18

   a Calculate the equation of the circle.

   b Calculate the length of the chord.

   c Complete the square to find the exact coordinates of the ends of the chord.

15 Write down the equations of each of the circles with diameters from

   a $(0, 0)$ to $(0, 20)$

   b $(2, 6)$ to $(6, 2)$

   c $(4, -2)$ to $(-3, 16)$

   d $(-4, -5)$ to $\left(-\sqrt{2}, \sqrt{5}\right)$

Strategy

To solve a problem involving a straight line or a circle

1. Choose the appropriate formulae.

2. Apply any relevant rules and theorems. Draw a sketch if it helps.

3. Show your working and give your answer in the correct form.

Example 4

a A diagonal of a rhombus has equation $2x - 3y + 8 = 0$ and midpoint $(-3, 7)$

Work out the equation of the other diagonal.

b One vertex of the rhombus on the original diagonal is $(14, 12)$

Work out the coordinates of the opposite vertex.

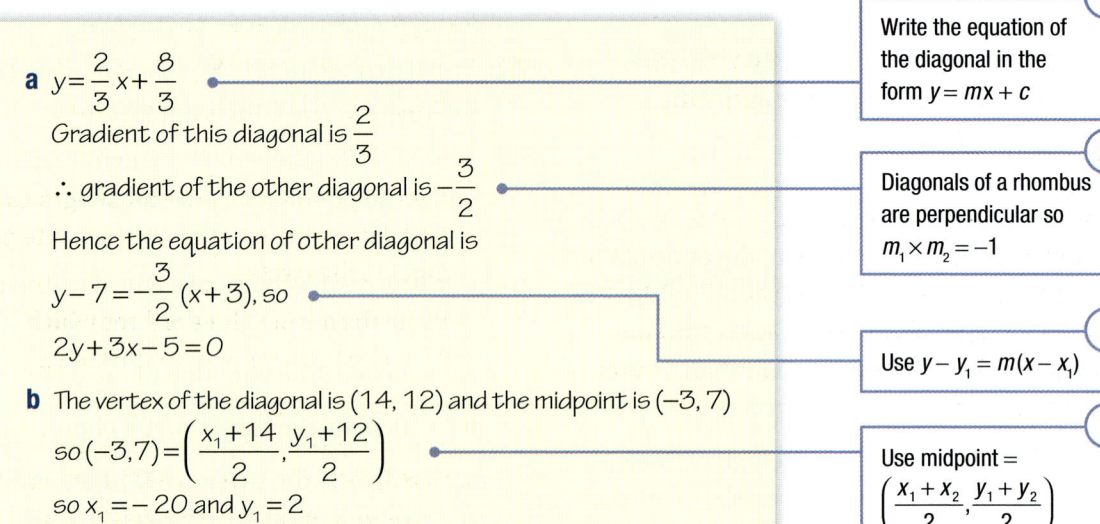

a $y = \dfrac{2}{3}x + \dfrac{8}{3}$

Gradient of this diagonal is $\dfrac{2}{3}$

∴ gradient of the other diagonal is $-\dfrac{3}{2}$

Hence the equation of other diagonal is

$y - 7 = -\dfrac{3}{2}(x + 3)$, so

$2y + 3x - 5 = 0$

b The vertex of the diagonal is $(14, 12)$ and the midpoint is $(-3, 7)$

so $(-3, 7) = \left(\dfrac{x_1 + 14}{2}, \dfrac{y_1 + 12}{2} \right)$

so $x_1 = -20$ and $y_1 = 2$

∴ the other vertex is $(-20, 2)$

1. Write the equation of the diagonal in the form $y = mx + c$

2. Diagonals of a rhombus are perpendicular so $m_1 \times m_2 = -1$

1. Use $y - y_1 = m(x - x_1)$

1. Use midpoint = $\left(\dfrac{x_1 + x_2}{2}, \dfrac{y_1 + y_2}{2} \right)$

Example 5

$A(-7, 1)$, $B(11, 13)$ and $C(19, 1)$ are three points on a circle. Prove that AC is a diameter.

Gradient of $AB = \left(\dfrac{13 - 1}{11 + 7} \right) = \dfrac{2}{3}$

Gradient of $BC = \left(\dfrac{1 - 13}{19 - 11} \right) = -\dfrac{3}{2}$

$\dfrac{2}{3} \times -\dfrac{3}{2} = -1$ so AB is perpendicular to BC

Therefore ABC is $90°$ and, since the angle in a semicircle is a right angle, ABC is a semicircle and thus AC is a diameter.

1. Use gradient = $\dfrac{y_2 - y_1}{x_2 - x_1}$

2. Use $m_1 \times m_2 = -1$ to prove that the lines are perpendicular.

2. 3. Use angle in a semicircle theorem and answer the question.

Example 6

The straight line, L, has equation $3x - 4y = -5$ and intersects the circle, with centre $C(4, -2)$ and radius 5, at the point $A(1, 2)$. Without finding the equation of the circle prove that L is a tangent to C

Gradient of $CA = \left(\dfrac{2+2}{1-4}\right) = -\dfrac{4}{3}$ Gradient of $L = \dfrac{3}{4}$

1. Use gradient $= \left(\dfrac{y_2 - y_1}{x_2 - x_1}\right)$

$-\dfrac{4}{3} \times \dfrac{3}{4} = -1$ so CA is perpendicular to L

2. Use $m_1 \times m_2 = -1$ to prove that the lines are perpendicular.

Since any tangent is perpendicular to the radius at the point of contact, L must be a tangent to the circle at the point A

Exercise 1.5B Reasoning and problem-solving

1 A quadrilateral has vertices $P(-15, -1)$, $Q(-3, 4)$, $R(12, 12)$ and $S(0, 7)$. Write the equation of each side and identify the nature of the quadrilateral.

2 A giant kite is constructed using bamboo for the edges and diagonals and card for the sail. When mapped on a diagram, the ends of the long diagonal are at $P(2, -2)$ and $R(-14, -14)$ and the diagonals intersect at M. The short diagonal QS divides RP in the ratio $3:1$ and $MP = MQ = MS$

Calculate

a The coordinates of M,

b The coordinates of Q and S,

c The equations of the diagonals,

d The equations of the sides of the kite,

e The area of card needed to make the kite,

f The total length of bamboo required for the structure.

3 On a map, three villages are situated at points $A(2, -5)$, $B(10, 1)$ and $C(9, -6)$, and all lie on the circumference of a circle.

a Find the equations of the perpendicular bisectors of AB and AC

b Hence work out the centre and equation of the circle and show that the triangle formed by the villages is right-angled.

4 The equation of a circle, centre C, is $x^2 + y^2 - 4x - 12y + 15 = 0$

a Prove the circle does not intersect the x-axis.

b P is the point $(8, 1)$. Find the length CP and determine whether P lies inside or outside the circle.

c Write the set of values of k for which $3y - 4x = k$ is a tangent to the circle.

5 A circle, centre C, has equation $x^2 + y^2 - 20x + 10y + 25 = 0$ and meets the y-axis at Q. The tangent at $P(16, 3)$ meets the y-axis at R. Work out the area of the triangle PQR

Challenge

6 a Show that if (a, c) and (b, d) are the ends of a diameter of a circle, the equation of the circle is

$$(x - a)(x - b) + (y - c)(y - d) = 0$$

b The line segment with endpoints $(-3, 12)$ and $(13, 0)$ is the diameter of a circle. Work out the equation of the circle. Give your answer without brackets.

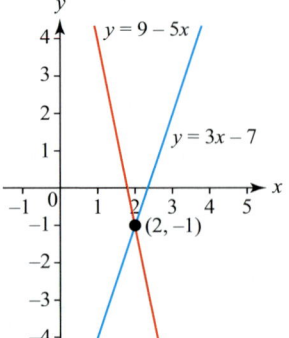

Fluency and skills

The graphs of $3x - y = 7$ and $5x + y = 9$ are shown. Only one pair of values for (x, y) satisfies both equations. This corresponds to the point of intersection of the two graphs. In this example it is $x = 2$ and $y = -1$

Equations solved together like this are called **simultaneous equations.**

> **Key point**
>
> You can solve a pair of linear simultaneous equations
> 1. Graphically.   2. By eliminating one of the **variables**.
> 3. By substituting an expression for one of the variables from one equation into the other.

Example 1

Solve the simultaneous equations $2a - 5b = -34$ and $3a + 4b = -5$ by elimination.

$$2a - 5b = -34 \quad (1) \quad \text{and} \quad 3a + 4b = -5 \quad (2)$$
$$8a - 20b = -136 \quad \text{and} \quad 15a + 20b = -25$$
$$23a = -161$$
$$a = -7$$
$$-14 - 5b = -34$$
$$-14 + 34 = 5b \implies b = 4$$
$$a = -7 \text{ and } b = 4$$

Label the equations 1 and 2

Multiply equation (1) by four and equation (2) by five.

Add equations to eliminate the b terms.

To find b, substitute $a = -7$ into equation (1)

You can check your answers by solving the simultaneous equations on your calculator.

A straight line can intersect a quadratic curve at either two, one or zero points.

You should equate expressions for the curve and the line to find the points of intersection. You will then obtain a quadratic equation in the form $ax^2 + bx + c = 0$

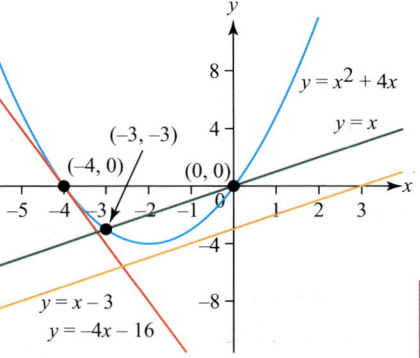

You can use the discriminant, $\Delta = b^2 - 4ac$, to show if this equation has two, one or no solutions.

The diagram shows the graphs of $y = x^2 + 4x$, $y = -4x - 16$, $y = x$ and $y = x - 3$

$y = -4x - 16$ touches $y = x^2 + 4x$ at the point $(-4, 0)$

$y = x$ intersects $y = x^2 + 4x$ at the points $(0, 0)$ and $(-3, -3)$

$y = x - 3$ does not intersect $y = x^2 + 4x$ at any point.

ICT Resource online

To investigate simultaneous equations, click this link in the digital book.

Example 2

Solve the simultaneous equations $y = x^2 + 4x$ and $y + 4x + 16 = 0$. Interpret your answers graphically.

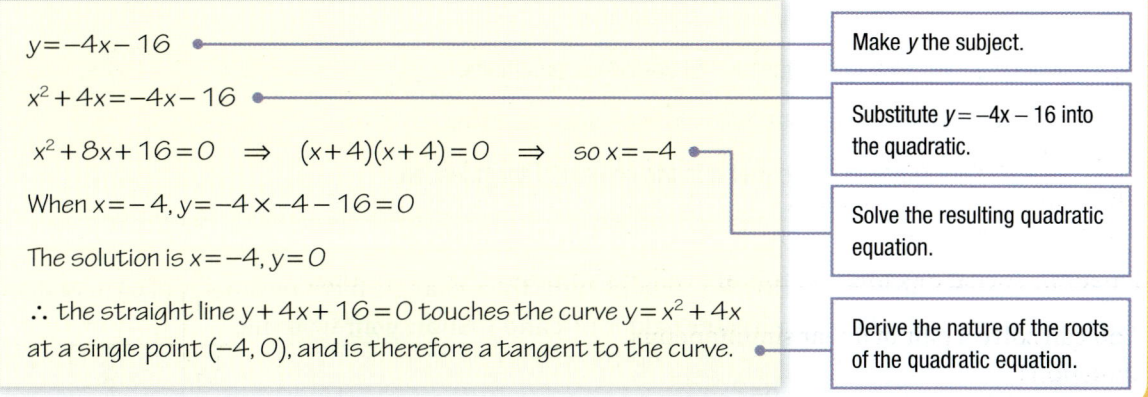

$y = -4x - 16$ — Make y the subject.

$x^2 + 4x = -4x - 16$ — Substitute $y = -4x - 16$ into the quadratic.

$x^2 + 8x + 16 = 0 \implies (x+4)(x+4) = 0 \implies$ so $x = -4$ — Solve the resulting quadratic equation.

When $x = -4$, $y = -4 \times -4 - 16 = 0$

The solution is $x = -4$, $y = 0$

$\therefore$ the straight line $y + 4x + 16 = 0$ touches the curve $y = x^2 + 4x$ at a single point $(-4, 0)$, and is therefore a tangent to the curve. — Derive the nature of the roots of the quadratic equation.

Exercise 1.6A Fluency and skills

Solve the simultaneous equations from 1 to 12
You must show your working.

1. $x + y = 7$
   $2x + y = 11$

2. $2a + 3b = 8$
   $a + 2b = 5$

3. $4c + 2d = 2$
   $5c + 2d = 7$

4. $2e - f = 13$
   $e + f = 5$

5. $7x + 4y = 12$
   $-5x - 4y = 12$

6. $-3m - 4n = -15$
   $-3m - n = 0$

7. $5c + 2d = 9$
   $3c = d - 10$

8. $2e = 13 - 3f$
   $6f = e - 4$

9. $x + y = 3$
   $x^2 + y = 3$

10. $g + h = 1$
    $g^2 - h = 5$

11. $3g + 2h = 13$
    $h + 2g^2 = 20$

12. $10m = 7n + 17$
    $m = n^2$

13. Find the point(s) of intersection of the graphs $y^2 = 5x$ and $y = x$ and show your working.

14. Find the point(s) of intersection of the graphs $y^2 = 6x + 7$ and $y = x + 2$ and show your working.

15. Solve the simultaneous equations $x + y^2 = 2$ and $2 = 3x + y$, showing your working. Find down the points of intersection.

16. Solve the simultaneous equations $y^2 = -1 - 5x$ and $y = 2x + 1$. Find the points of intersection, showing your working.

17. A curve has equation $xy = 20$
    A straight line has equation $y = 8 + x$

Solve the two equations simultaneously and show that the points of intersection are $(2, 10)$ and $(-10, -2)$

18. The circle with equation $x^2 + y^2 = 25$ crosses the line $y = 7 - x$ at two points. Solve these simultaneous equations and find the points of intersection. Show your working.

19. a. Write down the equation of the straight line with gradient $-\dfrac{1}{2}$ that passes through the point $(1, 1)$

    b. Write down the equation of the circle with radius 3 and centre $(2, 2)$

    c. The line in part **a** crosses the circle in part **b** at two points. Solve these simultaneous equations and find the coordinates of these two points. Show your working.

20. a. Write down the equation of the straight line that passes through the points $(3, 5)$ and $(-1, -3)$

    b. Write down the equation of the circle with centre $(1, 0)$ and radius $\dfrac{17}{2}$

    c. The line in part **a** crosses the circle in part **b** at two points. Solve these simultaneous equations and find the coordinates of these two points. Show your working.

Reasoning and problem-solving

Strategy

To solve a problem involving simultaneous equations

(1) Use the information in the question to create the equations.

(2) Use either elimination or substitution to solve your equations.

(3) Check your solution and interpret it in the context of the question.

Example 3

A rectangle has sides of length $(x + y)$ m and $2y$ m. The rectangle has a perimeter of 64 m and an area of 240 m². Calculate the possible values of x and y. Show your working.

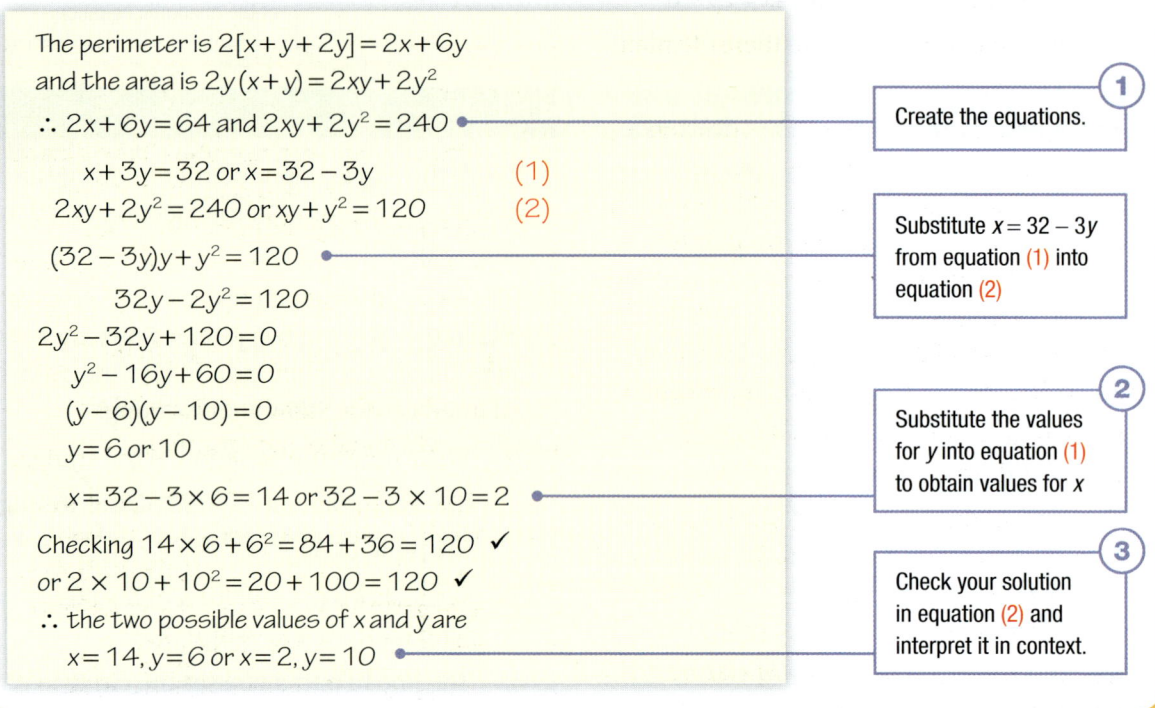

The perimeter is $2[x + y + 2y] = 2x + 6y$
and the area is $2y(x + y) = 2xy + 2y^2$

∴ $2x + 6y = 64$ and $2xy + 2y^2 = 240$ ●——— Create the equations. ①

$\quad x + 3y = 32$ or $x = 32 - 3y$ ········ (1)
$\quad 2xy + 2y^2 = 240$ or $xy + y^2 = 120$ ········ (2)

$(32 - 3y)y + y^2 = 120$ ●——— Substitute $x = 32 - 3y$ from equation (1) into equation (2)

$\quad 32y - 2y^2 = 120$
$2y^2 - 32y + 120 = 0$
$\quad y^2 - 16y + 60 = 0$
$\quad (y - 6)(y - 10) = 0$
$\quad y = 6$ or 10

$\quad x = 32 - 3 \times 6 = 14$ or $32 - 3 \times 10 = 2$ ●——— Substitute the values for y into equation (1) to obtain values for x ②

Checking $14 \times 6 + 6^2 = 84 + 36 = 120$ ✓
or $2 \times 10 + 10^2 = 20 + 100 = 120$ ✓
∴ the two possible values of x and y are
$\quad x = 14, y = 6$ or $x = 2, y = 10$ ●——— Check your solution in equation (2) and interpret it in context. ③

Exercise 1.6B Reasoning and problem-solving

1 In a recent local election, the winning candidate had an overall majority of 257 votes over her only opponent. There were 1619 votes cast altogether.

Form a pair of simultaneous linear equations.

How many votes did each candidate poll?

2 A fisherman is buying bait. He can either buy 6 maggots and 4 worms for £1.14 or 4 maggots and 7 worms for £1.28

How much do maggots and worms cost individually? Show your working.

3 The straight line $y = mx + c$ passes through the points $(3, -10)$ and $(-2, 5)$

Find the values of m and c

4 This triangle is equilateral.
Find the values of m
and n

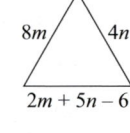

5 This triangle is isosceles.
It has a perimeter of
150 cm.
Find the values of p and q

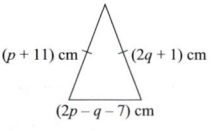

6 The ages of Florence and Zebedee are in the ratio $2:3$

In 4 years' time, their ages will be in the ratio $3:4$

Use simultaneous equations to calculate how old they are now. Show your working.

7 a Try to solve the simultaneous linear equations $y - 2x = 3$ and $4x = 2y - 6$

How many solutions are there? Explain your answer.

b Try to solve the simultaneous linear equations $y - 2x = 3$ and $4x = 2y - 8$

How many solutions are there? Explain your answer.

8 The equations of three straight lines and a parabola are $y + 2x + 4 = 0$, $y + 11x - 27 = 0$, $x - y + 3 = 0$ and $y = 2x^2 - 19x + 35$. One of the lines intersects the curve at two points, one 'misses' the curve and one is a tangent to the curve. Investigate the nature of the relationship between each of these lines and the curve, and calculate any real points of intersection.

9 Prove that the line $y = 2x - 9$ does not intersect the parabola $y = x^2 - x - 6$

10 The sums of the first n terms of two sequences of numbers are given by formulae $S_1 = 2n + 14$ and $2S_2 = n(n + 1)$. For which values of n does $S_1 = S_2$? Explain your results carefully.

11 A farmer has 600 m of fencing. He wants to use it to make a rectangular pen of area 16 875 m².

Calculate the possible dimensions of this pen.

12 The equation $x^2 + y^2 = 25$ represents a circle of radius 5 units. Prove that the line $3x + 4y = 25$ is a tangent to this circle and find the coordinates of the point where the tangent touches the circle.

13 An ellipse has the equation $4y^2 + 9x^2 = 36$

Show that the line $y = 2x + 1$ intersects this ellipse at the points $\left(\dfrac{-8 \pm 12\sqrt{6}}{25}, \dfrac{9 \pm 24\sqrt{6}}{25} \right)$

14 A park contains a circular lawn with a radius of 50 m. If the park is mapped on a set of axes with the y-axis due north, this is centred on the point $(-40, -80)$

A straight, underground water pipe runs through the park and its position is represented by the line $y + 2x = -60$

The town council wants to install a drinking fountain in the park. The fountain must be directly above the underground pipe, it must lie on the outer edge of the lawn, and it must be as close to the east side of the park as possible.

Determine the coordinates of the only possible location for the new drinking fountain.

Challenge

15 One diagonal of a rhombus has equation $2y - x = 20$. The two corners that form the other diagonal in the rhombus touch the edges of a circle with equation $x^2 + y^2 - 8x - 24y + 144 = 0$

a Find the radius of the circle.

b Find the equation of the other diagonal of the rhombus.

16 Two particles, A and B, move along a straight line. At a time, t, the position of A from a fixed point, O, on the line is given by the formula $x = 2 + 8t - t^2$ and that of B by $x = 65 - 8t$ points

a How far from O is each particle initially?

b Explain how you know that B is initially moving towards O

c Explain why A moves away from O and then moves back towards O

d What is the maximum distance of A from O?

e Calculate the first time when both particles are at the same distance from O

f In which directions are A and B moving at the time you calculated in part **e**?

Fluency and skills

See p.614

For a list of mathematical notation.

You can express **inequalities** using the symbols < (less than), > (greater than), ≤ (less than or equal to) and ≥ (greater than or equal to).

> **Key point**
>
> You can represent inequalities on a number line.

For example

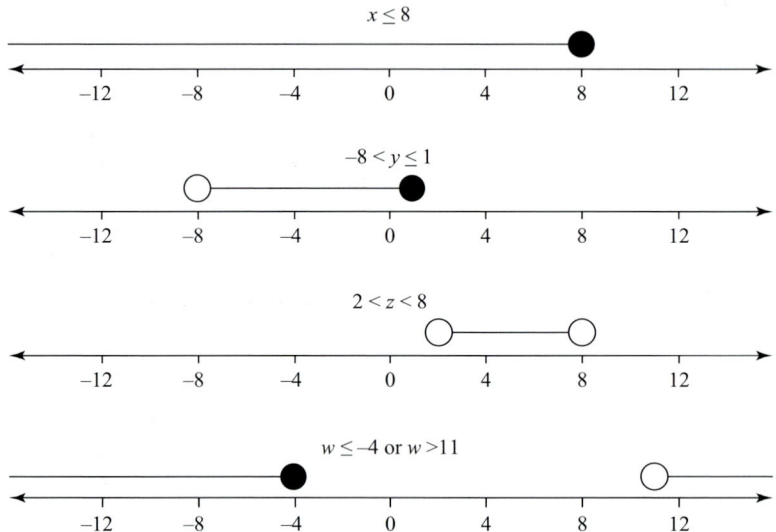

On a number line, you use a dot, ●, when representing ≤ or ≥, and you use an empty circle, ○, when representing < or >

You can also use set notation to represent inequalities.

For example, the last inequality could be represented in any of the following ways.

- $w \in \{w: w \leq -4 \text{ or } w > 11\}$
  w is an element of the set of values that are less than or equal to -4 or greater than 11

- $w \in \{w: w \leq -4\} \cup \{w: w > 11\}$
  w is an element of the union of two sets. This means w is in one set or the other.

- $w \in (-\infty, -4] \cup (11, \infty)$
  w is in the union of two intervals. Square brackets indicate the end value is included in the interval, round brackets indicate that the end value is not included in the interval.

To solve **linear inequalities** you follow the same rules for solving linear equations, but with one exception.

> **Key point**
>
> When you multiply or divide an inequality by a negative number, you reverse the inequality sign.

Example 1

Solve the inequality $4(3z + 12) \leq 5(4z - 8)$

$$4(3z + 12) \leq 5(4z - 8)$$
$$12z + 48 \leq 20z - 40$$
$$12z - 20z \leq -40 - 48$$
$$-8z \leq -88$$
$$z \geq 11$$

Expand the brackets.

Subtract $20z$ and 48 from each side.

Divide by -8
Remember to reverse the inequality sign.

Example 2

Shade each of these regions on a graph.

a $y - 2x < 6$

b $y + 3x \leq 8$; $y - 2x < 4$; $y > 1$

a $y - 2x < 6$

Sketch the line $y - 2x = 6$

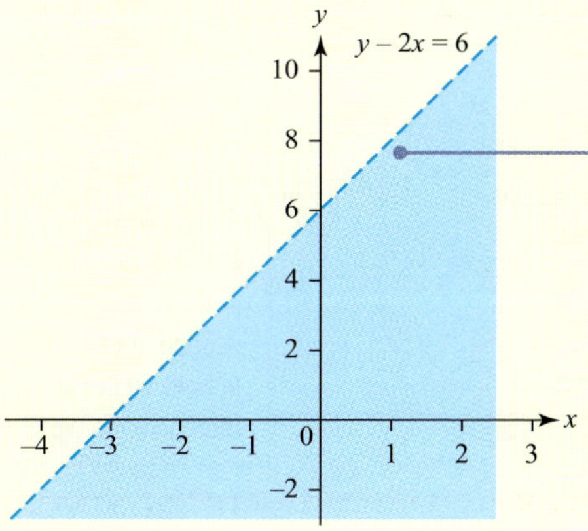

Use a dashed line to represent $<$ or $>$

Test a point on one side of the line $y - 2x = 6$ and shade the region that is needed.

b $y + 3x \leq 8$
$y - 2x < 4$
$y > 1$

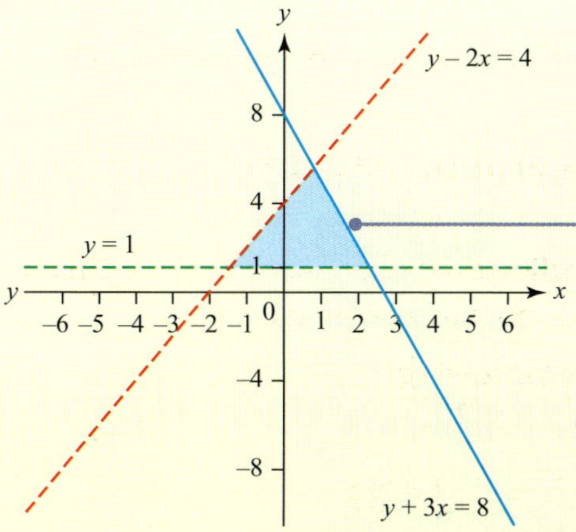

Test points and shade the correct region.

Use a solid line to represent $\leq$ or $\geq$

You can check your sketches using a graphics calculator. Use the graphing function and select the appropriate inequality symbol.

A **quadratic inequality** looks similar to a quadratic equation except it has an inequality sign instead of the '=' sign.

You can solve quadratic inequalities by starting the same way you would to solve quadratic equations. The answer, however, will be a range of values rather than up to two specific values.

Example 3

Solve the equation $x^2 + 4x - 5 = 0$ and sketch the graph $y = x^2 + 4x - 5$

Use your sketch to solve the inequality $x^2 + 4x - 5 \geq 0$

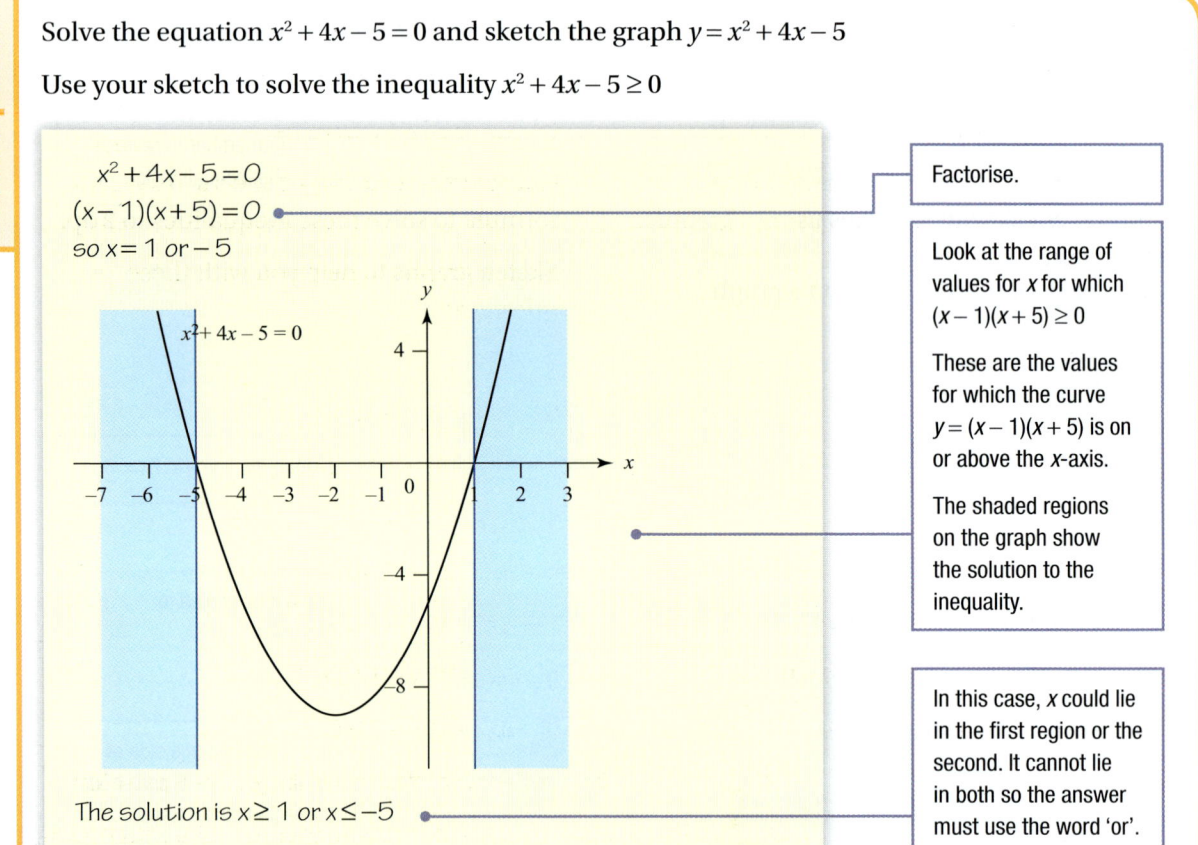

$x^2 + 4x - 5 = 0$
$(x - 1)(x + 5) = 0$
so $x = 1$ or -5

Factorise.

Look at the range of values for x for which $(x - 1)(x + 5) \geq 0$

These are the values for which the curve $y = (x - 1)(x + 5)$ is on or above the x-axis.

The shaded regions on the graph show the solution to the inequality.

In this case, x could lie in the first region or the second. It cannot lie in both so the answer must use the word 'or'.

The solution is $x \geq 1$ or $x \leq -5$

You could also have solved the question in Example 3 using the factorised form, by considering signs.

The product of the two brackets is positive if they are both positive or both negative.

$x - 1 \geq 0$ *and* $x + 5 \geq 0$ only if $x \geq 1$

$x - 1 \leq 0$ *and* $x + 5 \leq 0$ only if $x \leq -5$

The solution is $x \geq 1$ or $x \leq -5$

You can represent combinations of inequalities using set notation.

Key point

Use the union symbol ($\cup$) to represent 'or'. For example,

$0 < x < 3$ or $-1 < x < 2$ can be written $\{x : 0 < x < 3\} \cup \{x : -1 < x < 2\}$
$= \{x : -1 < x < 3\}$

Use the intersection symbol ($\cap$) to represent 'and'. For example,

$0 < x < 3$ and $-1 < x < 2$ can be written $\{x : 0 < x < 3\} \cap \{x : -1 < x < 2\}$
$= \{x : 0 < x < 2\}$

1 Show the following inequalities on a number line.

   a $r > 7$ and $r \leq 12$

   b $s \geq 14$

   c $3 < u \leq 9$

   d $v < 5$ and $v > 14$

2 Draw graphs to show these inequalities. You can check your sketches using a graphics calculator.

   a $x > -4$

   b $y \geq 5$

   c $y + x < 6$

   d $2y - 3x < 5$

   e $3y + 4x \leq 8$

   f $2y > 10 - 4x$

   g $y < x + 4;\ y + x + 1 > 0;\ x \leq 5$

   h $y \geq 2;\ x + y < 7;\ y - 2x - 4 \leq 0$

3 Find the values of x for which

   a $2x - 9 > -6$

   b $15 - 2x \geq 8x + 34$

   c $2(4x - 1) + 6 < 15 - 3x$

   d $3(x - 3) + 6(5 - 4x) \leq 54$

   e $4(2x + 1) - 7(3x + 2) > 5(4 - 2x) - 6(3 - x)$

   f $4\left(3x - \dfrac{1}{2}\right) + 2(8 - 3x) < 6\left(x + \dfrac{3}{2}\right) - 2\left(x - \dfrac{5}{2}\right)$

4 For each part **a** to **d**, sketch a suitable quadratic graph and use your sketch to solve the given inequality.

   a $x^2 + x - 6 > 0$

   b $x^2 + 11x + 28 < 0$

   c $x^2 - 11x + 24 \leq 0$

   d $x^2 - 2x - 24 \geq 0$

5 Sketch graphs to solve each of these inequalities.

   a $2x^2 - 3x - 2 > x + 4$

   b $3x^2 + 19x - 14 < 2x - 8$

   c $-3 + 13x - 4x^2 \leq 5x - 15$

   d $6x^2 + 16x + 8 \geq 8 - 2x$

6 Complete the square or use the quadratic formula to solve these inequalities to 2 dp.

Sketch graphs to help you with these questions.

   a $x^2 + 2x - 7 > 0$

   b $x^2 + 7x + 8 < 0$

   c $x^2 - 12x + 18 \leq 0$

   d $x^2 - 3x - 21 \geq 0$

   e $3x^2 - 5x - 7 > 0$

   f $4x^2 + 17x - 4 < 0$

   g $5x^2 - 17x + 12 \leq 0$

   h $6x^2 - 16x - 7 \geq 0$

7 For each of the pairs of inequalities below

   i Solve the inequalities simultaneously,

   ii Record the points of intersection,

   iii Shade the appropriate areas graphically.

   a $y < 2x + 3;\ y > x^2$

   b $x + y \leq 4;\ y > x^2 - 5x + 4$

   c $y - 4x \leq 17;\ y \leq 4x^2 - 4x - 15;\ x \leq 4$

   d $y - 2x - 20 < 0;\ y + 4x - 6 < 0;$ $y > x^2 - 5x - 24$

Reasoning and problem-solving

To solve a problem involving inequalities

1. Use the information in the question to write the inequalities.

2. Solve the inequalites and, if requested, show them on a suitable diagram.

3. Write a clear conclusion that answers the question.

Example 4

A man travels a journey of 200 miles in his car. He is travelling in an area with a speed limit of 70 mph.

Write down and solve an inequality in t (hours) to represent the time his journey takes.

$$70 \geq \frac{200}{t}$$
    Use speed limit $\geq \dfrac{\text{distance}}{\text{time}}$ ①

$$t \geq \frac{200}{70}$$

$$t \geq 2.857\ldots$$
    Solve the inequality. ②

$$0.857\ldots \times 60 = 51.42\ldots$$

The journey will take at least 2 hours 51 minutes.
    Write a clear conclusion. ③

Example 5

An illustration in a book is a rectangle $(x-7)$ cm wide and $(x+1)$ cm long.

It must have an area less than 65 cm²

Work out the range of possible values of x. Justify your answer.

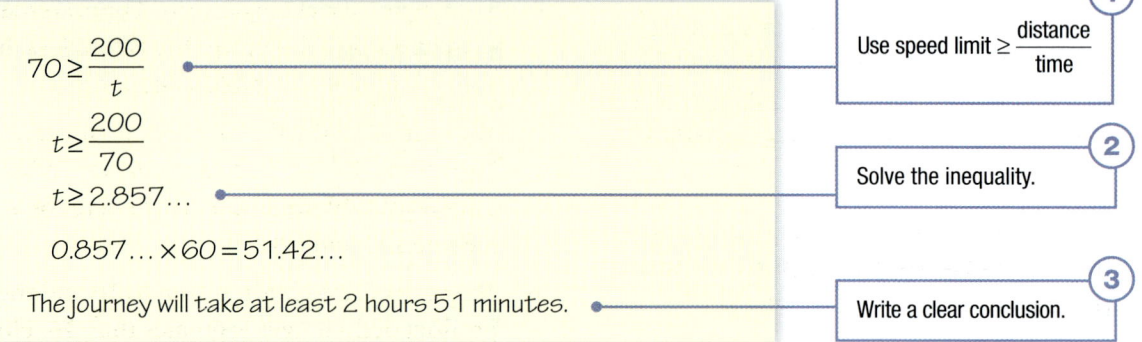

$$(x-7)(x+1) < 65$$
$$x^2 - 6x - 72 < 0$$
$$(x-12)(x+6) < 0$$
    Write the inequality. ①

$$-6 < x < 12$$
    Solve the inequality. ②

However, since any side of a rectangle must be positive, it follows that $x - 7 > 0$, so $x > 7$
    x is both greater than 7 *and* smaller than 12 so you combine the two inequalities. ③

So the solution is $7 < x < 12$

1 **a** Children in a nursery range from six months old to 4 years and six months old inclusive.

  Represent this information on a number line.

  b The range of temperatures outside the Met Office over a 24 hour period ranged from −4°C to 16°C.

  Represent this information on a number line.

2 On a youth athletics club trip there must be at least one trainer for every six athletes and the trip is not viable unless at least eight athletes travel. Due to illness there are fewer than six trainers available to travel. Represent this information as a shaded area on a graph.

3 In an exam, students take a written paper (marked out of 100) and a practical paper (marked out of 25).

  The total mark, T, awarded is gained by adding together twice the written mark and three times the practical mark. To pass the exam, T must be at least 200. A student scores w marks in the written paper and p in the practical.

  a Write an inequality in w and p

  b Solve this inequality for

   i $w = 74$

   ii $p = 9$

  c Can a student pass if she misses the practical exam?

4 The length of a rectangle, $(5m + 7)$ cm, is greater than its width, $(2m + 16)$ cm

  What values can m take?

5 For a student's 18th birthday party, 110 family and friends have been invited and at most 10% will not be able to come. Food has been prepared for 105 people.

Write down inequalities for the number of people, n, who come to the party and have enough to eat.

Solve them and find all possible solutions.

6 A bag contains green and red discs. There are r red discs and three more green than red. The total number of discs is not more than twenty. Write appropriate inequalities and find all solutions.

7 A girl is five years older than her brother. The product of their ages is greater than 50. What ages could the sister be?

8 The length of a rectangle, $(5b - 1)$ cm, is greater than its width, $(2b + 9)$ cm. The area is less than 456 cm². Find the possible values of b

9 The sum, S, of the first n positive integers is given by the formula $2S = n(n + 1)$. What are the possible values of n for values of S between 21 and 820?

10 The ages of two children sum to 10 and the product of their ages is greater than 16. Find all possible values of the children's ages.

Challenge

11 A firm makes crystal decanters.

  The profit, £P, earned on x thousand decanters is given by the formula $P = -20x^2 + 1200x - 2500$

  a Solve the equation $-20x^2 + 1200x - 2500 = 0$ giving your answer to two decimal places.

  b Sketch the graph of $y = -20x^2 + 1200x - 2500$

  c Use your graph to estimate

   i The values of x where the firm makes a loss,

   ii The range of values of x for which the profit is at least £10 000. Check this algebraically.

Chapter summary

- To use direct proof, assume P is true and then use P to show that Q must be true.
- To use proof by exhaustion, show that the cases are exhaustive and then prove each case.
- To use proof by counter example, give an example that disproves the statement.
- $x^a \times x^b = x^{a+b}, \quad x^a \div x^b = x^{a-b}, \quad (x^a)^b = x^{ab}$
- $x^0 = 1, \quad x^{-n} = \dfrac{1}{x^n}, \quad x^{\frac{1}{n}} = \sqrt[n]{x}, \quad x^{\frac{p}{r}} = \sqrt[r]{(x^p)} \text{ or } (\sqrt[r]{x})^p$
- You can write any rational number exactly in the form $\dfrac{p}{q}$, where p and q are integers.
- $\sqrt{A} \times \sqrt{B} = \sqrt{AB}; \quad \dfrac{\sqrt{A}}{\sqrt{B}} = \sqrt{\dfrac{A}{B}}$
- You rationalise a fraction in the form $\dfrac{k}{\sqrt{a}}$ by multiplying top and bottom by $\sqrt{a}$
- You rationalise a fraction in the form $\dfrac{k}{a \pm \sqrt{b}}$ by multiplying top and bottom by $a \mp \sqrt{b}$
- You rationalise a fraction in the form $\dfrac{k}{\sqrt{a} \pm \sqrt{b}}$ by multiplying top and bottom by $\sqrt{a} \mp \sqrt{b}$
- Any function of x in the form $ax^2 + bx + c$ where $a \neq 0$ is called a quadratic function and $ax^2 + bx + c = 0$ is called a quadratic equation.
- You can solve a quadratic equation $ax^2 + bx + c = 0$ using a calculator, by factorisation, by completing the square, by using the quadratic formula $x = \dfrac{-b \pm \sqrt{b^2 - 4ac}}{2a}$, and graphically.
- If the discriminant $\Delta = b^2 - 4ac > 0$, the quadratic has two different roots. If $\Delta = b^2 - 4ac = 0$, the quadratic has one repeated root. If $\Delta = b^2 - 4ac < 0$, the quadratic has no real roots.
- You can use gradients of two straight lines to decide if they are parallel, perpendicular, or neither.
- The equation of a circle, centre (a, b) and radius r, is $(x - a)^2 + (y - b)^2 = r^2$
- If you multiply or divide an inequality by a negative number you reverse the inequality sign.

Check and review

You should now be able to...	Try Questions
✓ Use direct proof, proof by exhaustion and counter examples to prove results.	1, 2, 3
✓ Use and manipulate the index laws for all powers.	4
✓ Manipulate surds and rationalise a denominator.	5, 6
✓ Solve quadratic equations using a variety of methods.	7, 8
✓ Understand and use the coordinate geometry of the straight line and of the circle.	9, 10
✓ Understand and solve simultaneous equations involving only linear or a mix of linear and non-linear equations.	11, 12
✓ Solve linear and quadratic inequalities algebraically and graphically.	13, 14

1   Prove that the product of two odd numbers must be odd.

2   Prove that there is at least one prime number between the numbers 40 and 48

3   Is it true that for every number n, $\frac{1}{n} < n$? Give a reason for your answer.

4   Simplify

   a $(-s^4)^3$ **b** $\sqrt{64c^{64}}$ **c** 3^{-4} **d** $(k^2)^{\frac{-3}{4}}$

5   **a** Express $\sqrt{275}$ in its simplest form.

   b Rationalise the denominator of $\dfrac{3-\sqrt{a}}{\sqrt{a}+1}$

6   What is the length of the hypotenuse of a right-angled triangle with sides containing the right angle of length $3\sqrt{3}$ and $3\sqrt{5}$ cm?

7   **a** Solve the equation $2c^2 + 9c - 5 = 0$ by factorisation.

   b Solve, to 2 dp, the equation $5x^2 + 9x - 28 = 0$

8   Sketch the quadratic curve $y = x^2 - 4x - 1$

9   **a** Write down the equations of these lines.

     i Gradient –6 passing through $(6, -7)$

     ii Gradient $\dfrac{2}{3}$ passing through $(-3, 4)$

   b A square joins the points $(-2, 1)$, $(2, 4)$, $(5, 0)$ and $(1, -3)$. Write the equations of its diagonals. Hence prove that they are perpendicular.

10 **a** Write the equations of the circles. The centre and radius are given for each.

     i $(3, 6)$; 8   **ii** $(-3, 9)$; 4   **iii** $(-2, -7)$; 11

   b Write the equation of the tangent, at point $P(-9, 19)$, to the circle with centre $(-4, 7)$ and radius 13

   c A circle with midpoint $C(5, 6)$ and radius 10 has $M(8, 5)$ as the centre of a chord. Work out the coordinates of the ends of the chord.

11 Solve these equations simultaneously.

   a $2x - 5y = 11$; $4x + 3y = 9$

   b $2x - 3y = 5$; $x^2 - y^2 + 5 = 0$

12 The line $y = 3x + 4$ intersects the curve $xy = 84$ at two points. Work out their coordinates.

13 **a** Writing your answers in set notation, solve these inequalities.

     i $12 - 3x \geq 7x + 2$

     ii $2(x - 7) + 5(6 - 3x) \leq 10$

   b Solve these inequalities, giving your answers to 2dp.

     i $x^2 - 14x + 16 \leq 0$

     ii $5x^2 - 13x - 11 \geq 0$

14 Shade the regions represented by these inequalities.

   a $2y - 2x - 7 \leq 0$; $x + y - 7 < 0$; $y > 0$

   b $x + y \leq 8$; $y > (x - 2)^2 - 4$

What next?

Score			
	0 – 7	Your knowledge of this topic is still developing. To improve, search in MyMaths for the codes: 2001–2005, 2008, 2009, 2014–2018, 2020, 2021, 2025, 2026, 2033–2037, 2252–2253, 2255–2257	 **Click these links in the digital book**
	8 – 10	You're gaining a secure knowledge of this topic. To improve, look at the InvisiPen videos for Fluency and skills (01A)	
	11 – 14	You've mastered these skills. Well done, you're ready to progress! Now try looking at the InvisiPen videos for Reasoning and problem-solving (01B)	

History

Pierre de Fermat was a lawyer in 17th century France who studied mathematics as a hobby. He often wrote comments in the margins of the maths books that he read and, on one occasion, wrote about a problem set over a thousand years ago by Greek mathematician **Diophantus**.

The problem was to find solutions to the equation $x^n + y^n = z^n$ where x, y, z and n are all positive integers. Fermat wrote that he had discovered 'the most remarkable proof' that the equation has no solutions if $n \geq 3$, but that the margin was too small to contain it.

Despite many attempts, a copy of Fermat's proof was never found. No one else was able to prove or disprove it for over 350 years.

Have a go

Find the flaw in the following 'proof'.

$$x = 1$$
$$x^2 = 1 \qquad \text{Square both sides}$$
$$x^2 - 1 = 0 \qquad \text{Subtract 1 from both sides}$$
$$(x + 1)(x - 1) = 0 \qquad \text{Factorise}$$
$$x + 1 = 0 \qquad \text{Divide both sides by } (x - 1)$$
$$2 = 0 \qquad \text{Substitute } x = 1$$

Research

Who eventually proved the result known as **Fermat's last Theorem**?

How long did it take them to complete the proof?

Investigation

How does Fermat's last theorem relate to **Pythagoras' Theorem**?

How can you use these diagrams to prove Pythagoras' Theorem?

What are Pythagorean triples? How many are there?

"We cannot solve our problems with the same level of thinking that created them."
- Einstein

In questions that tell you to show your working, you shouldn't depend solely on a calculator. For these questions, solutions based entirely on graphical or numerical methods are not acceptable.

1 a Simplify these expressions.

     i $2^m \times 2^n$ **ii** $\dfrac{5^{m+1}}{5^{2n}}$ **iii** $\left(3^m\right)^2 \times \sqrt{\left(3^m\right)}$ **[5 marks]**

  b Given $\dfrac{16^p \times 8^q}{4^{p+q}} = 2^n$, write down an expression for n in terms of p and q **[3]**

2 a Simplify these surds. You must show your working.

     i $\dfrac{12}{\sqrt{3}}$ **ii** $\left(3+2\sqrt{2}\right)\left(5-4\sqrt{2}\right)$ **iii** $\dfrac{3-\sqrt{7}}{1+3\sqrt{7}}$ **[8]**

  b A rectangle $ABCD$ has an area of $8\,\text{cm}^2$ and length $\left(3-\sqrt{5}\right)$cm.

    Work out its width, giving your answer as a surd in simplified form. Show your working.   **[3]**

3 What is the equation of the straight line that is perpendicular to $3x+2y=5$ and that passes through the point $(4, 5)$?   **[5]**

4 a Express $x^2 + 6x + 13$ in the form $(x+a)^2 + b$ **[2]**

  b Hence sketch the curve $y = x^2 + 6x + 13$ and label the vertex, and the point where the curve cuts the y-axis.   **[3]**

5 Solve these simultaneous equations. You must show your working.

  $2x + y = 3$ $3x^2 + 2xy + 7 = 0$ **[8]**

6 Prove that the equation $x = 1 + \dfrac{2x-5}{x+4}$ has no real solutions.   **[4]**

7 a Solve these inequalities. You must show your working.

     i $3x - 5 < 11 - x$ **ii** $x^2 - 6x + 5 \leq 0$ **[5]**

  b Show on a graph the set of values of x that satisfy both $3x - 5 < 11 - x$ and $x^2 - 6x + 5 \leq 0$ **[2]**

8 PQR is a right-angled triangle. Write an exact expression for x, show your working.   **[6]**

9 The equation of a circle is $x^2 + y^2 - 10x + 2y - 23 = 0$

  a Showing your working clearly, work out

     i Its centre,   **ii** Its radius.   **[5]**

  b The line $y = x + 2$ meets the circle at the points P and Q

    Work out, in exact form, the coordinates of P and Q. Show your working.   **[5]**

10 The quadratic equation $(k+1)x^2 - 4kx + 9 = 0$ has distinct real roots. What range of values can k take?   **[6]**

11 Prove that $\dfrac{a+b}{2} \geq \sqrt{ab}$ for all positive numbers a and b [4]

12 a Factorise the expression $2u^2 - 17u + 8$ [2]

    b Hence solve the equation $2^{2x+1} - 17 \times 2^x + 8 = 0$. Show your working. [3]

13 The straight line $y = mx + 2$ meets the circle $x^2 + y^2 + 4x - 6y + 10 = 0$

    a Prove that the x-values of the points of intersection satisfy the equation
$(m^2 + 1)x^2 + 2(2 - m)x + 2 = 0$ [4]

    b The straight line $y = mx + 2$ is a tangent to the circle $x^2 + y^2 + 4x - 6y + 10 = 0$

       What are the possible values of m? Give your answers in exact form. [5]

14 a Given $9^2 = 3^n$, write down the value of n [1]

    b Solve these simultaneous equations. Show your working.

       $3^{x+y} = 9^2$ $4^{x-2y} = 8^4$ [4]

15 Decide which of these statements are true and which are false.
For those that are true, prove that they are true.
For those that are false, give a counter-example to show that they are false.

    a If $a > b$ then $a^2 > b^2$ [2]

    b $n^2 + n$ is an even number for all positive integers n [3]

    c If a and b are real numbers then $b^2 \geq 4a(b - a)$ [3]

    d $2^n - 1$ is prime for all positive integers n [2]

16 The diagram shows the parabola $y = 2x^2 - 8x + 9$ and the straight line $y = 4x - 5$

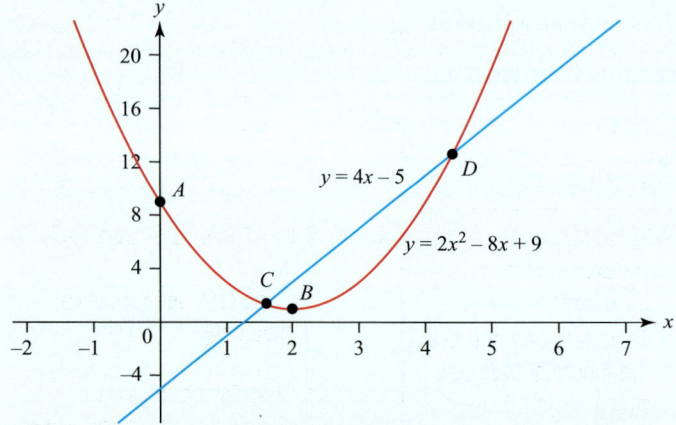

Showing your working clearly, work out the coordinates of the following points.

    a A, the y-intercept of the parabola. [1]

    b B, the vertex of the parabola. [2]

    c C and D, the points of intersection of the line and parabola. [4]

17 Prove that the circle $x^2 + y^2 + 6x - 4y - 2 = 0$
lies completely inside the circle $x^2 + y^2 - 2x - 10y - 55 = 0$ [9]

2 Polynomials and the binomial theorem

Natural disasters, like earthquakes, can strike at any time and cause major crises if they're not prepared for. We can't say exactly when they'll hit, but their behaviour can be predicted with mathematical models that use the binomial theorem. These models allow us to process huge amounts of data, which would otherwise be meaningless or impossible to use, in a limited amount of time.

The theorem can be applied to probability calculations because it simplifies the maths involved in expansions with many terms, and many possible outcomes. Because of this it's also used in a number of other fields that rely heavily on chance and probability: things like weather forecasting, modelling and predicting the behaviour of the economy, and planning large-scale projects with multiple possible outcomes at each stage.

Orientation

What you need to know	What you will learn	What this leads to
KS4 • To simplify and manipulate algebraic expressions, including collecting like terms and use of brackets.	• To manipulate, simplify and factorise polynomials. • To understand and use the binomial theorem. • To divide polynomials by algebraic expressions. • To understand and use the factor theorem. • To analyse a function and sketch its graph.	**Ch4 Differentiation and integration** Using calculus for curve sketching. **Ch10 Probability and discrete random variables** The binomial probability distribution. **Ch13 Sequences** Binomial expansions. Position-to-term and term-to-term rules.

  **Practise before you start**   Q 1150, 1155, 1156

Fluency and skills

A **polynomial** is an algebraic expression that can have constants, variables, **coefficients** and **powers** (also known as **exponents**), all combined using addition, subtraction, multiplication and division.

Key point

The highest power in a polynomial is called its **degree**.

All quadratics are of degree two.

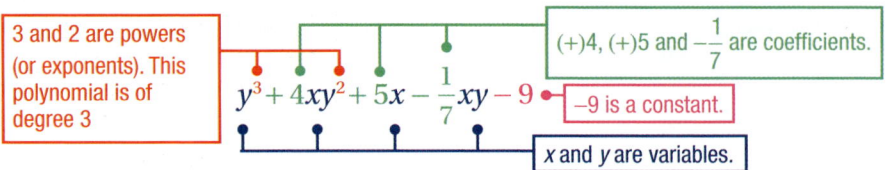

3 and 2 are powers (or exponents). This polynomial is of degree 3

$(+)4$, $(+)5$ and $-\dfrac{1}{7}$ are coefficients.

$$y^3 + 4xy^2 + 5x - \frac{1}{7}xy - 9$$

-9 is a constant.

x and y are variables.

You can simplify polynomials by collecting (adding or subtracting) **like terms**. You must *never* attempt to simplify a polynomial by dividing by a variable and the exponent of a variable can only be 0, 1, 2, 3, ... etc.

You can manipulate polynomials by **expanding**, **simplifying** and **factorising** them.

Example 1

Expand and simplify $(3x + 2y)^2 - (2x - 3y)^2$

$(3x + 2y)^2 = (3x + 2y)(3x + 2y)$
$\qquad\qquad = 9x^2 + 12xy + 4y^2$

$(2x - 3y)^2 = 4x^2 - 12xy + 9y^2$ so $-(2x - 3y)^2 = -4x^2 + 12xy - 9y^2$

$\qquad 9x^2 + 12xy + 4y^2$
$\quad -4x^2 + 12xy - 9y^2$

$\qquad = 5x^2 + 24xy - 5y^2$

Expand the brackets before adding or subtracting polynomials.

Multiply each term in the first bracket by each term in the second bracket.

You may find it useful to write like terms vertically under each other.

Collect like terms to simplify the polynomial.

Key point

A statement that is true for all values of the variable(s) is called an **identity**. You write an identity using the symbol $\equiv$

For example, $15x^3 + 8x^2 - 26x + 8 \equiv (3x^2 + 4x - 2)(5x - 4)$ is true for all values of x

It follows that $(3x^2 + 4x - 2)$ and $(5x - 4)$ are **factors** of $15x^3 + 8x^2 - 26x + 8$

Factorising is the opposite process to expanding brackets.

You can factorise polynomials by comparing coefficients.

Example 2

$(4x - 5)$ is a factor of the polynomial $12x^3 + 21x^2 - 61x + 20$

Factorise the polynomial completely.

$12x^3 + 21x^2 - 61x + 20 \equiv (4x - 5)(Ax^2 + Bx + C)$

Use the fact that $4x - 5$ is a factor to write an identity.

$(4x - 5)(Ax^2 + Bx + C)$

$\equiv 4Ax^3 + 4Bx^2 + 4Cx$

$\qquad -5Ax^2 - 5Bx - 5C$

To expand, multiply each term in the first bracket by each term in the second bracket.

$\equiv 4Ax^3 + (4B - 5A)x^2 + (4C - 5B)x - 5C$

To collect like terms write them under each other.

This is identical to $12x^3 + 21x^2 - 61x + 20$

so the coefficients must all be the same.

so $\quad 4A = 12 \qquad$ ①

$4B - 5A = 21 \qquad$ ②

$4C - 5B = -61 \qquad$ ③

$\qquad -5C = 20 \qquad$ ④

Equate and compare coefficients for x^3, x^2, x and compare the constants.

$A = 3$ and $C = -4$

Rearrange ① and ④

$4B - 5 \times 3 = 21$

Substitute $A = 3$ into ②

$4B = 36$ so $B = 9$

$4(-4) - 5(9) = -61$ ✓

Check by substituting the values into ③

So $12x^3 + 21x^2 - 61x + 20 \equiv (4x - 5)(3x^2 + 9x - 4)$

State your answer clearly and check it by expanding the brackets. $(3x^2 + 9x - 4)$ cannot be factorised so this is the final answer.

Exercise 2.1A Fluency and skills

1 Write the degree of each of these expressions.

  a $3 - 2x + x^2$ **b** $1 - 3x + 5x^4$ **c** $2x^2 - x + 1 - 4x^3$

2 Expand and simplify each of these expressions.

  a $2x(3x + 8)$ **b** $2x(3x^2 + 8x - 9)$ **c** $(3y + 2)(4y - 7)$

  d $3y(4y^2 + 8y - 7)$ **e** $(t - 5)^2$ **f** $(t + 3)(t - 5)^2$

3 Expand and simplify each of these expressions.

  a $(x + 4)^2 + (x - 4)^2$ **b** $(5p + q)^2 - (5p - q)^2$

4 Factorise each of these expressions.

  a $4m^3 + 6m^2$ **b** $16n^4 - 12n$ **c** $5p^4 - 2p^2 + 6p$ **d** $9y^2 - 15xy$

  e $6x^2 - 3xy + 9x$ **f** $7yz - 21z^3$ **g** $4e(e - 2f) - 12ef$ **h** $p^2 - 100$

  i $6q(3 - 2q) + 9q$ **j** $\dfrac{y}{5} - \dfrac{y^2}{15} + \dfrac{3y}{25}$ **k** $(d + 1)(d + 3) + (d + 1)(d - 5)$

  l $w(2w + 3)(3w + 9) + w(2w - 11)(2w + 3)$

5 Fully factorise these expressions.

    a $4m^3 + 4m^2 - 15m$ **b** $7n^3 - 15n^2 + 2n$

6 Factorise this expression $3x(x+2)^2 + (x+2)(5x^2 + 2x - 6)$

7 Expand and simplify these expressions.

    a $(5p + 4q)^2 - (5p - 4q)^2$ **b** $(x + y + z)^2 - (x - y - z)^2$

    c $(x\sqrt{3} + 4)^2 + (x\sqrt{3} - 4)^2$ **d** $(x\sqrt{5} + 4)^2 + (x\sqrt{3} - 4)^2$

8   **a** $(x^2 + 3x + 9)$ is a factor of $x^3 + 2x^2 + 6x - 9$. Work out the other factor.

    b $(x^2 - 2x + 3)$ is a factor of $2x^3 - 11x^2 + 20x - 21$. Work out the other factor.

    c $(y^2 + 2y - 15)$ is a factor of $2y^3 + 3y^2 - 32y + 15$. Work out the other factor.

    d $(z - 2)$ is a factor of $z^3 + z^2 - 2z - 8$. Work out the other factor.

    e $(2a + 5)$ is a factor of $6a^3 + 7a^2 - 2a + 45$. Work out the other factor.

    f $(x^2 - 4x + 7)$ is a factor of $2x^3 - 5x^2 + 2x + 21$. Factorise the polynomial fully.

    g $(k^2 - 3k + 1)$ is a factor of $k^4 + 3k^3 - 24k^2 + 27k - 7$. Work out the other factors.

Reasoning and problem-solving

Strategy

To factorise polynomials

(**1**) Look for obvious common factors and factorise them out.

(**2**) Write an identity and expand to compare coefficients.

(**3**) Write your solution clearly and use suitable units where appropriate.

Example 3

The volume of a cylinder is $y^2 - 25y + 24$ ft³.

The base area is $(y - 1)$ ft².

Write an expression for its height.

Let the height be $(Ay + B)$ ft

So $y^2 - 25y + 24 \equiv (Ay + B)(y - 1)$

$y^2 - 25y + 24 \equiv Ay^2 + By - Ay - B$

$y^2 - 25y + 24 \equiv Ay^2 + (B - A)y - B$

So $A = 1$ ①

     $(B - A) = -25$ ②

       $-B = 24 \therefore B = -24$ ③

       $-24 - 1 = -25$ ✓

So the height is $(y - 24)$ ft

The height must be linear because it multiplies with $(y - 1)$ to give a quadratic.

(**2**) Write an identity and expand to compare coefficients.

Check by substituting into ②

(**3**) Write your solution clearly and use suitable units.

1

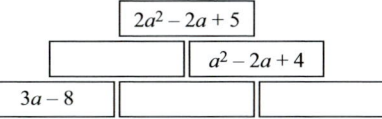

In this pyramid, each block is the sum of the two blocks vertically beneath.

Copy and complete the pyramid.

2 A square has side length $(4b - 7a)$ cm.

Write an expression for its area in expanded form.

3 A cuboid has sides of length $(c + 2)$, $(2c - 1)$ and $(3c - 7)$ cm. Write an expression for its volume in expanded form.

4 A square hole of side length $(a + 2)$ cm is cut from a larger square of side length $(2a + 5)$ cm. Without expanding any brackets, write the remaining part of the large square as a pair of factors.

5 A rectangle, sides $2a$ cm by a cm, has a square of side x cm cut from each corner. The sides are then folded up to make an open box. Work out the volume of this box.

6 A ball is thrown from ground level and its height, h ft, at time t s is given by the polynomial $h = 25t - 5t^2$

 a When does the ball next return to ground level?

 b What is the maximum height reached by the ball?

7 A body moves along a straight line from a point O where its position, x metres at time, t seconds is given by the equation $x = 3t^3 - 28t^2 + 32t$. Its velocity $v \, \text{m s}^{-1}$ and acceleration $a \, \text{m s}^{-2}$ at time t are given by the equations $v = 9t^2 - 56t + 32$ and $a = 18t - 56$

 a Find the values of t when the body is at O, and find its velocity and acceleration at these times.

 b Find the distance of the body from O and its velocity when its acceleration is zero.

 c Find the value(s) of t when its velocity is zero, and find its acceleration at these times.

8 A rectangle has the dimensions shown. All lengths are given in centimetres.

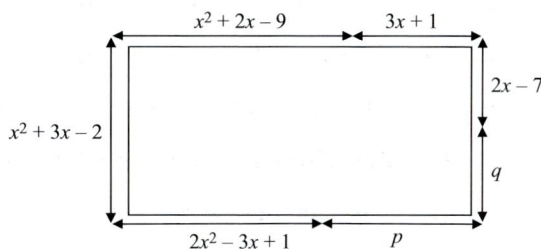

 a Find p and q in terms of x

 b Write expanded expressions for the rectangle's perimeter and area in terms of x

9 Show how these calculations can be completed without a calculator.

 a $66.89^2 - 33.11^2$

 b $(\sqrt{8})^3 - (\sqrt{2})^3$

10 A cuboid has volume $(2h^3 + 3h^2 - 23h - 12) \, \text{cm}^3$. Its length is $(h + 4)$ cm and its width is $(h - 3)$ cm. Work out the height of the cuboid.

11 The area of a trapezium is given by the polynomial $(2s^3 - 17s^2 + 41s - 30) \, \text{cm}^2$. The perpendicular height is $(4s - 6)$ cm. Write an expression for the sum of the parallel sides.

12 The area of an ellipse is given by the formula πab, where a and b are half the lengths of the axes of symmetry. The area is $\pi(6t^3 - 5t^2 + 15t + 14)$ and $a = (3t + 2)$ Write an expression for b

Challenge

13 $V = \pi I[r^2 - (r - a)^2]$ is the volume of a circular pipe.

Find an expression for V in terms of p when $I = 4p + 5$, $r = 3p - 4$ and $a = p + 1$

The binomial theorem

Fluency and skills

You can expand $(1 + x)^n$ where $n = 0, 1, 2, 3, \ldots$

EXPANSION	COEFFICIENTS
$(1 + x)^0 \equiv 1$	1
$(1 + x)^1 \equiv 1 + 1x$	1 1
$(1 + x)^2 \equiv 1 + 2x + 1x^2$	1 2 1
$(1 + x)^3 \equiv 1 + 3x + 3x^2 + 1x^3$	1 3 3 1
$(1 + x)^4 \equiv 1 + 4x + 6x^2 + 4x^3 + 1x^4$	1 4 6 4 1
$(1 + x)^5 \equiv 1 + 5x + 10x^2 + 10x^3 + 5x^4 + 1x^5$	1 5 10 10 5 1

The coefficients form a pattern known as **Pascal's triangle**.

Each coefficient in the triangle is the sum of the two coefficients above it.

> Pascal's Triangle was published in 1654, but was known to the Chinese and the Persians in the 11th century.

Example 1

Use Pascal's triangle to write the expansion of $(1 + 2y)^6$ in ascending powers of y.

The coefficients are $1, 6, 15, 20, 15, 6, 1$ ●──── Write down the 6th row of Pascal's triangle.

$(1 + (2y))^6$ ●────

$\equiv 1 + 6(2y) + 15(2y)^2 + 20(2y)^3 + 15(2y)^4 + 6(2y)^5 + (2y)^6$ ──── Use the expansion of $(1 + x)^n$, substituting $2y$ for x

$\equiv 1 + 12y + 60y^2 + 160y^3 + 240y^4 + 192y^5 + 64y^6$

Replacing 1 with a and x with b gives the **binomial expansion** $(a + b)^n$ where $n = 0, 1, 2, 3, \ldots$

As n increases you can see that again the coefficients form Pascal's triangle.

> A binomial expression has two terms.

See Ch10.2 For more examples of the binomial expansion.

$(a + b)^0 \equiv 1$

$(a + b)^1 \equiv 1a + 1b$

$(a + b)^2 \equiv 1a^2 + 2ab + 1b^2$

$(a + b)^3 \equiv 1a^3 + 3a^2b + 3ab^2 + 1b^3$

$(a + b)^4 \equiv 1a^4 + 4a^3b + 6a^2b^2 + 4ab^3 + 1b^4$

$(a + b)^5 \equiv 1a^5 + 5a^4b + 10a^3b^2 + 10a^2b^3 + 5ab^4 + 1b^5$

In each expansion, the power of a starts at n and decreases by 1 each term, so the powers are n, $n - 1$, $n - 2$, ..., 0

The power of b starts at 0 and increases by 1 each term, so the powers are 0, 1, 2, ..., n

The sum of the powers of any individual term is always n

Example 2

Expand $(2 + 3t)^4$

$(2 + 3t)^4$

$\equiv 2^4 + 4 \times 2^3 \times (3t) + 6 \times 2^2 \times (3t)^2 + 4 \times 2 \times (3t)^3 + (3t)^4$

$\equiv 16 + 96t + 216t^2 + 216t^3 + 81t^4$

> Use Pascal's triangle and the expansion of $(a + b)^4$ substituting 2 for a and $3t$ for b

It would be impractical to use Pascal's triangle every time you need to work out a coefficient—say, for example, you want to find the coefficient of x^6 in $(x + a)^{10}$

There is a general rule for finding this coefficient without needing to write out Pascal's triangle up to the tenth row.

> Note that the first coefficient in each row is the 0th coefficient.

Key point

The rth coefficient in the nth row is $^nC_r \equiv \dfrac{n!}{(n-r)!r!}$

> ^nC_r is sometimes written as $\binom{n}{r}$ or $_nC_r$

Key point

$n!$ stands for the product of all integers from 1 to n. You read it as n **factorial**.

For example, $6! = 6 \times 5 \times 4 \times 3 \times 2 \times 1 = 720$

> Look for the factorial button on your calculator. It may be denoted $x!$

^nC_r is the **choose function** and you read it as 'n choose r'. It gives the number of possible ways of choosing r elements from a set of n elements when the order of choosing does not matter. For example, the number of combinations in which you can choose 2 balls from a bag of 5 balls is $^5C_2$

You use the choose function because there are several ways of getting certain powers from an expansion. For example, there are 3 ways of getting ab^2 from the expansion of $(a + b)^3$: a from either the first, second or third bracket and b from the other two brackets in each case. The term in ab^2 for the expansion of $(a + b)^3$ is therefore $^3C_1\, ab^2 = 3ab^2$

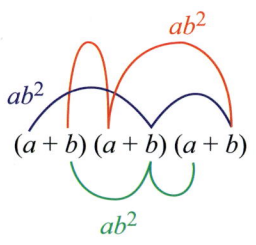

Example 3

A term in the expansion of $(y + 2x)^9$ is given by ky^3x^6

Find the value of k

$$^9C_6 \times y^3 \times (2x)^6 \equiv 84 \times y^3 \times 64x^6$$

Use your calculator to find $^9C_6$ and work out 2^6

$$\equiv 5376y^3x^6$$

Simplify to find the value of k

$$k = 5376$$

The formula for the binomial expansion of $(a + b)^n$ is sometimes called the **binomial theorem**.

Key point

$$(a + b)^n \equiv a^n + {}^nC_1a^{n-1}b + {}^nC_2a^{n-2}b^2 + \ldots + {}^nC_ra^{n-r}b^r + \ldots + b^n$$

For the expansion of $(1 + x)^n$ this gives

Key point

$$(1+x)^n \equiv 1 + nx + \frac{n(n-1)}{2!}x^2 + \frac{n(n-1)(n-2)}{3!}x^3 + \ldots + x^n$$

Example 4

Write the term in z^4 in the expression $(2z - 1)^{15}$. Simplify your answer.

Take $a = 2z$ and $b = -1$

$$^{15}C_{11}(2z)^{15-11}(-1)^{11}$$

The powers add to 15 so the second power must be 11. Use the coefficient $^{15}C_{11}$

$$\equiv 1365 \times 16z^4 \times (-1)$$

$(2z)^4 \equiv 2^4z^4$

$$\equiv -21840z^4$$

Exercise 2.2A Fluency and skills

1 Calculate the values of

   a $5!$ **b** $7!$ **c** $11!$

2 Calculate the values of

   a $^5C_2$ **b** $^9C_3$ **c** $^{11}C_7$ **d** $^{13}C_8$

3 Work out the values of

   a $\binom{5}{3}$ **b** $\binom{10}{1}$ **c** $\binom{13}{5}$ **d** $\binom{20}{6}$

4 Use Pascal's triangle to find the expansions of each of these expressions.

   a $(1 + 3x)^3$ **b** $\left(1 - \frac{z}{2}\right)^5$

   c $\left(1 - \frac{m}{3}\right)^4$ **d** $\left(1 + \frac{3x}{2}\right)^5$

5 Find the first four terms of these binomial expansions in ascending powers of x

   a $(1 + x)^8$ **b** $(1 - 3x)^7$

   c $(1 + 2x)^9$ **d** $(2 - 3x)^6$

   e $(x - 2)^8$ **f** $(2x - 1)^{10}$

6 Use Pascal's triangle to expand each of these expressions.

   a $(2 - 4y)^3$ **b** $(3b + 5)^4$ **c** $\left(4z - \frac{y}{3}\right)^5$

7 Find the first three terms of these binomial expansions in descending powers of x

   a $(2 + x)^6$ **b** $(1 - 2x)^8$

   c $(3 - x)^9$ **d** $(x + 4)^7$

   e $(2x + 3)^{10}$ **f** $\left(\frac{x}{2} + 4\right)^{11}$

8 Use the binomial theorem to expand each of these expressions.

   a $(2+3t)^4$ **b** $(3-2p)^4$

   c $(4p+3q)^5$ **d** $(3p-4q)^5$

   e $(3z-2)^4$ **f** $\left(2z-\dfrac{1}{2}\right)^6$

   g $\left(2+\dfrac{2x}{3}\right)^3$ **h** $\left(\dfrac{r}{3}+\dfrac{s}{4}\right)^8$

   i $\left(\dfrac{x}{2}+\dfrac{y}{3}\right)^3$

9 Find the terms indicated in each of these expansions and simplify your answers.

   a $(p+5)^5$ term in p^2

   b $(4+y)^9$ term in y^5

   c $(3+q)^{12}$ term in q^7

   d $(4-3m)^5$ term in m^3

   e $(2z-1)^{15}$ term in z^4

   f $\left(z+\dfrac{3}{2}\right)^8$ term in z^6

   g $(3x+4y)^5$ term in y

   h $(2a-3b)^{10}$ terms in **i** a^5 and **ii** b^4

   i $\left(4p+\dfrac{1}{4}\right)^3$ term in p^2

   j $\left(4a-\dfrac{3b}{4}\right)^{11}$ terms in **i** a^5 and **ii** b^5

   k $\left(\dfrac{a}{2}-\dfrac{2b}{3}\right)^{11}$ terms in **i** a^7 and **ii** b^5

10 Use the binomial theorem to expand each of these expressions.

   a $(c^2+d^2)^4$ **b** $(v^2-w^2)^5$

   c $(2s^2+5t^2)^3$ **d** $(2s^2-5t^2)^3$

   e $\left(d+\dfrac{1}{d}\right)^3$ **f** $\left(2w+\dfrac{3}{w}\right)^4$

11 Use the binomial theorem to expand each of these brackets.

   a $\left(x+\dfrac{2}{x}\right)^3$ **b** $(x^2-2)^4$

   c $\left(x^2-\dfrac{1}{x}\right)^5$ **d** $\left(\dfrac{1}{x^2}+3x\right)^6$

12 Expand and simplify each of these expressions.

   a $3x(2x-5)^5$ **b** $(2+x)^4(1+x)$

13 Expand and simplify each of these expressions.

   a $(5-2x)^3+(3+2x)^4$

   b $(1+3x)^5-(1-4x)^3$

14 Expand and fully simplify each of these expressions. Show your working.

   a $\left(2+\sqrt{3}\right)^4+\left(1-\sqrt{3}\right)^4$

   b $\left(1-\sqrt{5}\right)^5-\left(2\sqrt{5}+3\right)^3$

15 Write down the first four terms of the expansion of each of these in ascending powers of x

   a $(1+2x)^n$ **b** $(1-3x)^n$

   where $n \in \mathbb{N}, n>3$

16 a Expand $(1+4x)^6$ in ascending powers of x up to and including the term in x^2

   b Use your answer to part **a** to estimate the value of $(1.04)^6$

17 a Expand $(1-2x)^7$ in ascending powers of x up to and including the term in x^3

   b Use your answer to part **a** to estimate the value of $(0.99)^7$

18 Use the binomial expansion to simplify each of these expressions. Give your final solutions in the form $a+b\sqrt{2}$

   a $\left(1+\sqrt{2}\right)^3$ **b** $\left(1-\sqrt{2}\right)^5$

   c $\left(3+2\sqrt{2}\right)^4$ **d** $\left(\sqrt{2}-2\right)^6$

   e $\left(1-\dfrac{1}{\sqrt{2}}\right)^3$ **f** $\left(\dfrac{\sqrt{2}}{3}+3\right)^4$

19 Use the binomial expansion to fully simplify each of these expressions.

   Give your final answers in surd form.

   a $\left(1+\sqrt{3}\right)^4$ **b** $\left(1-\sqrt{5}\right)^6$

   c $\left(5-\sqrt{7}\right)^5$ **d** $\left(2\sqrt{6}+5\right)^3$

   e $\left(\sqrt{2}+\sqrt{6}\right)^4$ **f** $\left(\sqrt{3}-\sqrt{2}\right)^6$

Reasoning and problem-solving

Strategy

To construct a binomial expansion

1. Create an expression in the form $(1 + x)^n$ or $(a + b)^n$

2. Use Pascal's triangle or the binomial theorem to find the required terms of the binomial expansion.

3. Use your expansion to answer the question in context.

Example 5

A football squad consists of 13 players. Use the formula ${}^nC_r \equiv \dfrac{n!}{(n-r)!r!}$ to show that there are 78 possible combinations of choosing a team of 11 players from this squad.

$$
\begin{aligned}
{}^{13}C_{11} &= \frac{13!}{(13-11)!11!} \\
&= \frac{13 \times 12 \times 11 \times 10 \times \ldots \times 2 \times 1}{2! \times 11 \times 10 \times \ldots \times 2 \times 1} \\
&= \frac{13 \times 12}{2!} \\
&= \frac{156}{2} = 78
\end{aligned}
$$

2 — Cancel the common factor 11!

Example 6

a Using the first *three* terms of the binomial expansion, estimate the value of 1.003^8

b By calculating the fourth term in the expansion show that the estimate from part **a** is accurate to 3 decimal places.

a $1.003^8 = (1 + 0.003)^8$

1 — Rewrite in the form $(1 + x)^n$

First 3 terms

$$
\begin{aligned}
&= 1 + nx + \frac{n(n-1)}{2!}x^2 \\
&= 1 + 8(0.003) + 28(0.003)^2 \\
&= 1 + 0.024 + 0.000252 \\
&= 1.024252 \ (= 1.024 \text{ to 3 sf})
\end{aligned}
$$

2 — Use the first 3 terms of the general expansion.

3 — Substitute values and simplify.

b $\dfrac{n(n-1)(n-2)}{3!}x^3 = 56(0.003)^3$

$= 0.000001512$

Adding this term will not affect the first three decimal places.

Exercise 2.2B Reasoning and problem-solving

1 How many possible ways are there to pick a 7's rugby team from a squad of 10 players?

2 How many possible ways are there to choose half of the people in a group of 20?

3 A cube has side length $(2s - 3w)$. Use the binomial expansion to find its volume.

4 Use Pascal's triangle to find the value of

   a 1.05^6 correct to six decimal places

   b 1.96^3 correct to four decimal places.

5 Use the binomial theorem to work out the value of

   a 1.015^5 correct to 4 decimal places,

   b $\left(\dfrac{199}{100}\right)^{10}$ correct to five significant figures.

6 Use the binomial theorem to work out the value of $\left(\dfrac{13}{4}\right)^5$ correct to five decimal places.

7 Work out the exact value of the middle term in the expansion of $\left(\sqrt{3} + \sqrt{5}\right)^{10}$

8 a Find the coefficient of x^4 in the expansion of $(1 + x)(2x - 3)^5$

   b Find the coefficient of x^3 in the expansion of $(x - 2)(3x + 5)^4$

9 Find, in the expansion of $\left(x^2 - \dfrac{1}{2x}\right)^6$, the coefficient of

   a x^3 b x^6

10 Find, in the expansion of $\left(\dfrac{1}{t^2} + t^3\right)^{10}$, the coefficient of

   a t^{10} b t^{-5}

11 The first three terms in the expansion of $(1 + ax)^n$ are $1 + 35x + 490x^2$. Given that n is a positive integer find the value of

   a n b a

12 Given that $(1 + bx)^n \equiv 1 - 24x + 252x^2 + \ldots$ for a positive integer n find the value of

   a n b b

13 In the expansion of $(1 + 2x)^n$, n a positive integer, the coefficient of x^2 is eight times the coefficient of x. Find the value of n

14 In the expansion of $\left(1 + \dfrac{x}{2}\right)^n$, n a positive integer, the coefficients of x^4 and x^5 are equal. Calculate the value of n

15 Find an expression for

   a $\dbinom{n}{n-1}$ b $\dbinom{n}{3}$

   c $\dbinom{n}{n-2} - \dbinom{n+1}{n-1}$

Write your answers as polynomials in n with simplified coefficients.

16 Fully simplify these expressions.

   a $\dfrac{n!}{(n+1)!}$ b $\dfrac{(n+3)!}{n(n+1)!}$

17 Find the constant term in the expansion of $(2 + 3x)^3 \left(\dfrac{1}{x} - 4\right)^4$

18 Find the coefficient of y^3 in the expansion of $(y + 5)^3 (2 - y)^5$

Challenge

19 A test involves 6 questions.

   For each question there is a 25% chance that a student will answer it correctly.

   a How many ways are there of getting exactly two of the questions correct?

   b What is the probability of getting the first two questions correct then the next four questions incorrect?

   c What is the probability of getting exactly two questions correct?

   d What is the probability of getting exactly half of the questions correct?

2.3 Algebraic division

Fluency and skills

In Section **2.1** you learned how to factorise a polynomial by writing the identity and comparing and evaluating constants.

You can also use the method of dividing the polynomial by a known factor. You can divide algebraically using the same method as 'long division' in arithmetic.

It is an easier method than comparing coefficients when the polynomials are of degree 3 or higher.

ICT Resource online

To investigate algebraic division, click this link in the digital book.

Example 1

Use long division to divide $2x^4 + 7x^3 - 14x^2 - 3x + 15$ by $(x + 5)$

Give your answer in the form of a quotient and remainder.

$$
\begin{array}{r}
2x^3 - 3x^2 + x - 8 \\
(x+5)\overline{)\,2x^4 + 7x^3 - 14x^2 - 3x + 15} \\
\underline{2x^4 + 10x^3} \\
-3x^3 - 14x^2 \\
\underline{-3x^3 - 15x^2} \\
x^2 - 3x \\
\underline{x^2 + 5x} \\
-8x + 15 \\
\underline{-8x - 40} \\
55
\end{array}
$$

So $(2x^4 + 7x^3 - 14x^2 - 3x + 15) \div (x + 5)$
$= (2x^3 - 3x^2 + x - 8)$ remainder 55

Divide the first term $2x^4$ by x. Write the answer, $2x^3$, on the top.

Write $(x + 5) \times 2x^3 \equiv 2x^4 + 10x^3$ on this line and subtract from the line above to give $-3x^3$

Write the $-3x^3$ and bring down the next term, $-14x^2$, to make $-3x^3 - 14x^2$ here.

Repeat this process until you get a quotient (and a remainder if there is one).

$(2x^3 - 3x^2 + x - 8)$ is the quotient. 55 is the remainder.

Example 2

Use long division to show that $(x - 2)$ is a factor of $f(x) = x^3 + 10x^2 + 11x - 70$

$$
\begin{array}{r}
x^2 + 12x + 35 \\
(x-2)\overline{)\,x^3 + 10x^2 + 11x - 70} \\
\underline{x^3 - 2x^2} \\
12x^2 + 11x \\
\underline{12x^2 - 24x} \\
35x - 70 \\
\underline{35x - 70} \\
0
\end{array}
$$

There is no remainder when f(x) is divided by $x - 2$ so $x - 2$ is a factor of f(x)

Divide the first term x^3 by x. Write the answer, x^2, on top.

Multiply x^2 by $(x - 2)$ Write the answer, $x^3 - 2x^2$, underneath and subtract from the line above.

Write the answer, $12x^2$, and bring the next term down.

Repeat the process.

Example 1 shows that dividing f(x) by ($x - a$) leaves you with a remainder, R

In general, for a polynomial f(x) of degree $n \geq 1$ and any constant a

$$f(x) \equiv (x - a)\,g(x) + R$$

Where g(x) is a polynomial of order $n - 1$ and R is a constant.

For the particular case when $x = a$, this gives

$$f(a) = (a - a)\,g(a) + R$$

$$f(a) = R$$

You can see from this that f(a) = 0 implies there is no remainder when f(x) is divided by ($x - a$)

The **factor theorem** states that if f(a) = 0, ($x - a$) is a factor of f(x)   Key point

In Example 2 you saw that there is no remainder when $x^3 + 10x^2 + 11x - 70$ is divided by ($x - 2$), which is equivalent to saying that ($x - 2$) is a factor of $x^3 + 10x^2 + 11x - 70$

If you substitute $x = 2$ into the expression, the factor ($x - 2$) is zero so the value of f(x) is zero.

You can check this by substitution, which gives $f(2) = 2^3 + 10(2)^2 + 11(2) - 70$
$$= 8 + 40 + 22 - 70 = 0$$

Example 3

Show that ($x + 3$) is a factor of $2x^4 + 2x^3 - 9x^2 - 4x - 39$

$f(-3) = 2(-3)^4 + 2(-3)^3 - 9(-3)^2 - 4(-3) - 39$
$\quad = 0$
$(x + 3)$ is a factor since $f(-3) = 0$

> ($x - a$) is a factor if f(a) = 0, so if ($x + 3$) is a factor you need to show that f(-3) = 0

Example 4

Fully factorise the polynomial $2x^3 + 17x^2 - 13x - 168$

$f(x) = 2x^3 + 17x^2 - 13x - 168$

$f(1) = 2(1)^3 + 17(1)^2 - 13(1) - 168 = -162$
$f(1) \neq 0$ so ($x - 1$) is not a factor

$f(2) = 2(2)^3 + 17(2)^2 - 13(2) - 168 = -110$
$f(2) \neq 0$ so ($x - 2$) is not a factor

$f(3) = 2(3)^3 + 17(3)^2 - 13(3) - 168 = 0$
$f(3) = 0$ so ($x - 3$) is a factor

$$
\require{enclose}
\begin{array}{r}
2x^2 + 23x + 56 \\
(x - 3) \enclose{longdiv}{2x^3 + 17x^2 - 13x - 168} \\
\underline{2x^3 - 6x^2} \\
23x^2 - 13x \\
\underline{23x^2 - 69x} \\
56x - 168 \\
\underline{56x - 168} \\
0
\end{array}
$$

So $2x^3 + 17x^2 - 13x - 168 \equiv (x - 3)(2x^2 + 23x + 56)$
$\equiv (x - 3)(2x + 7)(x + 8)$

> Use trial and error with different values of a to find a case where f(a) = 0

> Use long division to divide the polynomial by the factor to get a quadratic expression in x

> Use the result from the long division to express the polynomial in a partially factorised form.

> Factorise the quadratic to fully factorise the polynomial.

1 Divide

    a $x^2 - x - 90$ by $(x + 9)$

    b $3x^2 - 19x - 14$ by $(x - 7)$

    c $8x^2 + 14x - 15$ by $(2x + 5)$

2 Divide each polynomial by the given factor by comparing coefficients.

    a $x^3 + 3x^2 - 11x + 7$ by $(x - 1)$

    b $x^3 + 2x^2 - 4x - 3$ by $(x + 3)$

    c $2x^3 + 9x^2 - 17x - 45$ by $(2x - 5)$

    d $3x^3 - 14x^2 + 16x + 7$ by $(3x + 1)$

    e $2x^4 - 17x^3 + 22x^2 + 65x - 9$ by $(2x - 9)$

3 Use long division to divide
$3x^4 + 4x^3 + 4x^2 - 8x + 5$ by $(x - 4)$

4 Use long division to show that
$2x^4 - 5x^3 + 5x^2 - 5x + 3$ is divisible by $(2x - 3)$

5 Divide using long division

    a $x^3 - 2x + 1$ by $(x - 1)$

    b $x^3 - 10x^2 - 10x - 11$ by $(x - 11)$

    c $6x^3 - 13x^2 - 19x + 12$ by $(3x + 4)$

    d $6x^4 - 19x^3 + 23x^2 - 26x + 21$ by $(2x - 3)$

    e $10x^4 + 33x^3 - 57x^2 + 5x + 1$ by $(5x - 1)$

6 Work out the values of **i** f(0) **ii** f(1) **iii** f(−1) **iv** f(2) **v** f(−2) when

    a $f(x) = x^3 - 2x^2 + 10x$

    b $f(x) = x^3 - 2x^2 - 2x - 2$

    c $f(x) = x^3 - 3x^2 + x + 2$

    d $f(x) = 2x^3 + x^2 - 5x + 2$

    e $f(x) = x^3 - x^2 - 4x + 4$

7 **a** Show that $(x + 6)$ is a factor of
$x^3 + 4x^2 - 9x + 18$

    b Show that $(x - 8)$ is a factor of
$2x^3 - 13x^2 - 20x - 32$

    c Show that $(3x - 1)$ is a factor of
$3x^3 + 11x^2 - 25x + 7$. HINT: Find a value
of x that makes the factor equal 0

    d Show that $(5x + 2)$ is a factor of
$10x^3 + 19x^2 - 39x - 18$

8 Fully factorise the polynomial $4x^3 + 27x^2 - 7x$

9 Fully factorise the polynomial $2x^3 + 9x^2 - 2x - 9$

10 a Factorise fully $x^3 + 3x^2 - 16x + 12$

    b Factorise fully $x^3 - 6x^2 - 55x + 252$

    c Factorise fully $6x^3 + 19x^2 + x - 6$

    d Factorise fully $x^4 - 13x^2 - 48$

Reasoning and problem-solving

Strategy

To factorise a polynomial

(1) Apply the factor theorem as necessary to find your first factor.

(2) Divide the polynomial by the factor to get a quadratic quotient.

(3) Factorise the quadratic quotient to fully factorise the polynomial.

Example 5

$(x + 1)$ is a factor of the polynomial $3x^3 + 8x^2 + ax - 28$. Fully factorise the polynomial.

$f(-1) = 0$

Use the factor theorem, $f(a) = 0$, to form an expression in a ①

$\Rightarrow 3(-1)^3 + 8(-1)^2 + a(-1) - 28 = 0$

$-3 + 8 - a - 28 = 0 \Rightarrow a = -23$

Simplify to find the value of a

$\qquad\qquad 3x^2 + 5x - 28$
$(x + 1) \overline{)\, 3x^3 + 8x^2 - 23x - 28}$

Use long division to get a quadratic quotient (the full calculation isn't shown here). ②

$(x + 1)(3x^2 + 5x - 28) \equiv (x + 1)(3x - 7)(x + 4)$

Factorise the quadratic. ③

Example 1

a For a constant $a > 1$ sketch these curves on one set of axes.

   i $f(x) = (a - x)(x + 1)(x + 2a)$ **ii** $g(x) = \dfrac{2}{x - a}$

b Show that there are no positive solutions to the equation $-(a - x)^2 (x + 1)(x + 2a) - 2 = 0$

a **i** x-intercepts: $a - x = 0 \Rightarrow x = a$

                  $x + 1 = 0 \Rightarrow x = -1$

                  $x + 2a = 0 \Rightarrow x = -2a$

   y-intercept: $x = 0 \Rightarrow y = a \times 1 \times 2a = 2a^2$

   The coefficient of x^3 is

   $-1 \times 1 \times 1 = -1 < 0$

a **ii** Undefined when $x - a = 0 \Rightarrow x = a$

b $-(a - x)^2 (x + 1)(x + 2a) - 2 = 0$

   $(a - x)(x + 1)(x + 2a) = \dfrac{2}{-(a - x)}$

   $(a - x)(x + 1)(x + 2a) = \dfrac{2}{(x - a)} \Rightarrow f(x) = g(x)$

The equation is satisfied at the points of intersection of $f(x)$ and $g(x)$.
From the graph, the curves have two points of intersection and both have
negative x-coordinates, so there are no positive solutions.

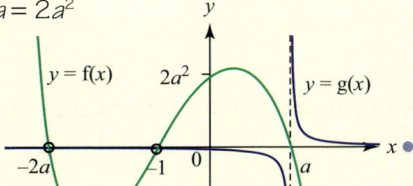

As the magnitude of x gets bigger and bigger the value of y gets closer to 0

Negative cubic shape, $a > 1$ and $-2a < -1$

You cannot divide by 0 so as x gets closer to a, y gets closer to ∞ or $-\infty$

Transformations can help you to see how different functions relate to one another.

You will work with four common transformations in this chapter.

$y = a\mathrm{f}(x)$ is a vertical stretch of $y = \mathrm{f}(x)$ with scale factor a	$y = \mathrm{f}(ax)$ is a horizontal stretch of $y = \mathrm{f}(x)$ with scale factor $\dfrac{1}{a}$

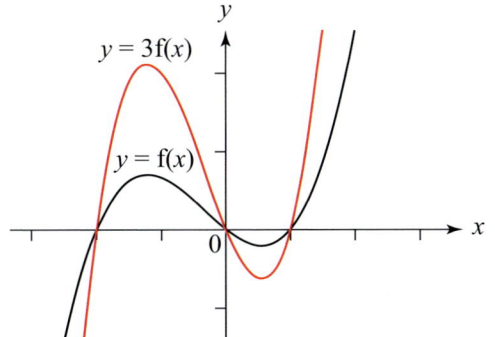 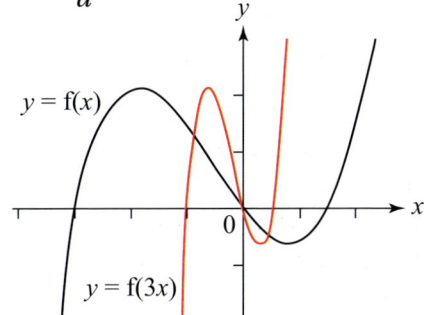

In the transformation $y = a\mathrm{f}(x)$, the x-values remain unchanged and each y-value is multiplied by a

Every point $(x, \mathrm{f}(x))$ becomes $(x, a\mathrm{f}(x))$

Key point

If $a < 0$ the transformation $y = a\mathrm{f}(x)$ reflects the curve in the x-axis.

In the transformation $y = \mathrm{f}(ax)$, each x value is multiplied by a before the corresponding y-value is calculated.

Every point $(x, \mathrm{f}(x))$ becomes $(x, \mathrm{f}(ax))$

Key point

If $a < 0$ the transformation $y = \mathrm{f}(ax)$ reflects the curve in the y-axis. If $-1 < a < 1$ the curve gets wider.

$y = \mathbf{f(x)} + \mathbf{a}$ is a translation of $y = \text{f}(x)$ by the vector $\begin{pmatrix} 0 \\ a \end{pmatrix}$

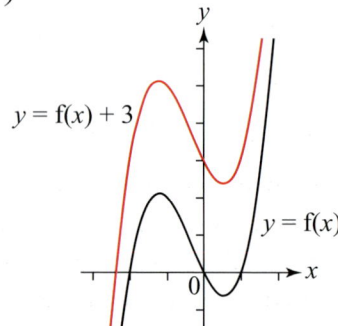

$y = \text{f}(x) + 3$

$y = \text{f}(x)$

In the transformation $y = \text{f}(x) + a$, the x-values remain unchanged and each y-value is increased by a

Every point $(x, \text{f}(x))$ becomes $(x, \text{f}(x) + a)$

Key point

If $a < 0$ the transformation $y = \text{f}(x) + a$ translates the curve downwards.

$y = \mathbf{f(x + a)}$ is a translation of $y = \text{f}(x)$ by the vector $\begin{pmatrix} -a \\ 0 \end{pmatrix}$

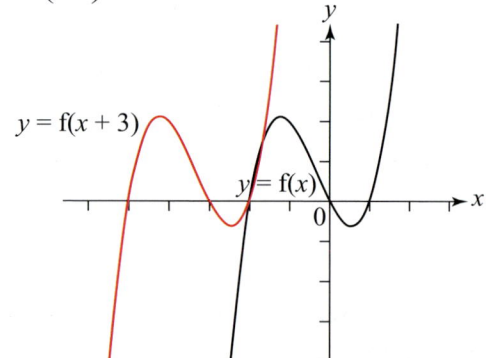

$y = \text{f}(x + 3)$

$y = \text{f}(x)$

In the transformation $y = \text{f}(x + a)$, a is added to each x-value before the corresponding y-value is calculated.

Every point $(x, \text{f}(x))$ becomes $(x, \text{f}(x + a))$

Key point

If $a < 0$ the transformation $y = \text{f}(x + a)$ translates the curve to the right.

Example 2

The graph shows a sketch of the curve $y = \text{f}(x)$

Sketch the curves

a $y = 2\text{f}(x)$ **b** $y = \text{f}(x) - 1$ **c** $y = \text{f}(x - 1)$ **d** $y = \text{f}(-x)$

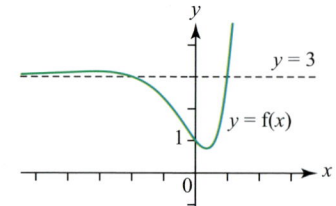

$y = 3$

$y = \text{f}(x)$

The translated curve has asymptote $y = 2 \times 3 = 6$

a

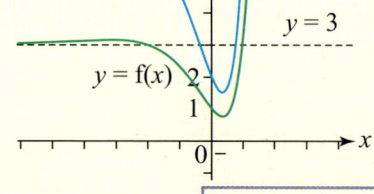

$y = 6$

$y = 2\text{f}(x)$

$y = 3$

$y = \text{f}(x)$

b

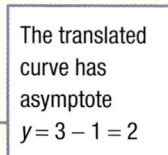

$y = \text{f}(x)$

$y = 3$

$y = \text{f}(x) - 1$

$y = 2$

The translated curve has asymptote $y = 3 - 1 = 2$

You don't have enough information to mark the y-intercept.

c

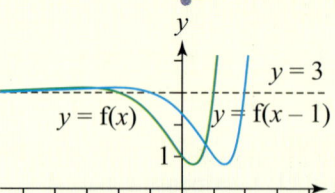

$y = 3$

$y = \text{f}(x)$ $y = \text{f}(x - 1)$

d

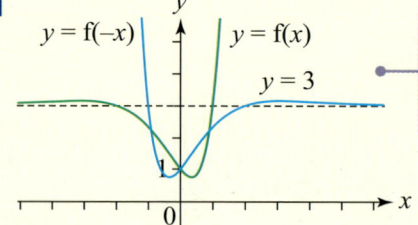

$y = \text{f}(-x)$ $y = \text{f}(x)$

$y = 3$

The curve is reflected in the y-axis.

1 Evaluate all the intercepts on the axes for these graphs. Show your working.

   a $y = x^2 - x - 6$ **b** $y = 2x^2 - 9x - 35$

   c $y = x^3 + 8$ **d** $y = 2x^3 - 54$

   e $y = (x - 3)^3$ **f** $y = (2x + 5)^3 - 7$

2 Identify all the vertical and horizontal asymptotes for $y = \dfrac{3}{x-1}$. Show your working.

3 Evaluate all axes of symmetry in these graphs. Show your working.

   a $y = x^2 - 8x - 9$ **b** $y = (x + 2)^4$

   c $(y - 3)^2 = x + 4$ **d** $y = (x - 4)^2(x + 3)^2$

   Hence sketch the graph of each function.

4 Sketch the graphs of these functions.

   a $y = x^3 + 3$ **b** $y = (x - 3)^3$

   c $y = -2x^3 + 3$ **d** $y = 2(x + 3)^3 - 1$

   e $y = (2x + 1)^3$ **f** $y = 5 + (3x - 4)^3$

   g $y = x^3 - 5x^2 - 14x$

   h $y = (x + 5)(x - 6)(2x + 1)$

   i $y = \dfrac{-2}{x}$ **j** $y = \dfrac{4}{x+2}$

   k $y = \dfrac{-1}{x+5}$

5 Sketch the graphs of these functions.

   a $y = 5x^3 - 2x^4$ **b** $y = 5x^2 + 2x^3$

   c $y = x^3 - 3x^2$ **d** $y = (1 - x)(x + 3)^2$

   e $y = x^2(x - 3)^2$ **f** $y = x^2(x + 3)^2$

   g $y = x^4 - 7x^3$ **h** $y = (x^2 - 4)(x^2 - 9)$

6 The graph of $y = f(x)$ is shown.

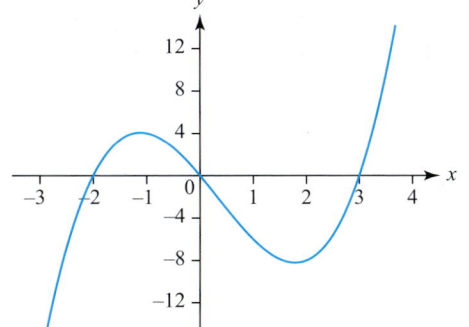

Sketch the graphs of

   a $y = f(2x)$ **b** $y = f(x - 2)$

   c $y = f\left(\dfrac{x}{3}\right)$ **d** $y = f(x + 3)$

   e $y = f(-x)$ **f** $y = -f(x)$

7 The graph of $y = g(x)$ has a maximum point at $(-2, 5)$ and a minimum point at $(8, -4)$ State the coordinates of the maximum and minimum points of these transformed graphs.

   a $y = g(4x)$ **b** $y = 3g(x)$

   c $y = g(x + 7)$ **d** $y = g(x) + 4$

   e $y = \dfrac{1}{2}g(x)$ **f** $y = -g(x)$

   g $y = g(-x)$ **h** $y = g\left(\dfrac{x}{2}\right)$

8 Describe each of the transformations in question **7**

9 $f(x) = x^3$. Write down the equation when the graph of $y = f(x)$ is

   a   Translated 3 units left,

   b   Translated 2 units up,

   c   Stretched vertically by scale factor 2,

   d   Stretched horizontally by scale factor 3

10 The graph of $y = f(x)$ is shown.

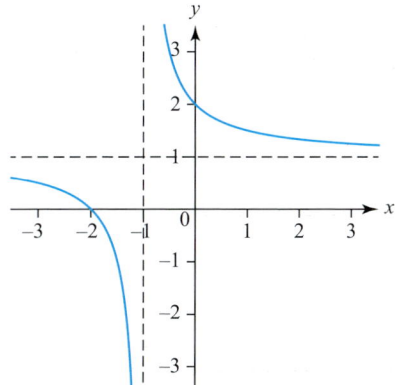

Sketch the graphs of

   a $y = f(x + 3)$ **b** $y = 3f(x)$

   c $y = f\left(\dfrac{x}{2}\right)$ **d** $y = f(x) + 1$

Strategy

When sketching a graph

(1) Define the function using any variables supplied in the question.

(2) Identify the standard shape of the curve and identify any symmetry.

(3) Identify any x- and y-intercepts and any asymptotes.

(4) Apply any suitable transformations.

(5) Show all relevant information on your sketch.

You can use graphs to show proportional relationships.

If y is proportional to x, you write $y \propto x$. This can be converted to an equation using a constant of proportionality, giving $y = kx$. The graph of y against x is a straight line through the origin with gradient k

If y is inversely proportional to x you write $y \propto \dfrac{1}{x}$ or $y = \dfrac{k}{x}$. The graph of $y = \dfrac{k}{x}$ is a vertical stretch, scale factor k, of the graph $y = \dfrac{1}{x}$

Example 3

A rectangle has a fixed area of 36 m². Its length, y m is inversely proportional to its width, x m.

a Write a formula for y in terms of x

b Without plotting exact points, sketch the graph of your function.

c Explain any asymptotes that the graph has.

a $y \propto \dfrac{1}{x}$ so $y = \dfrac{k}{x}$

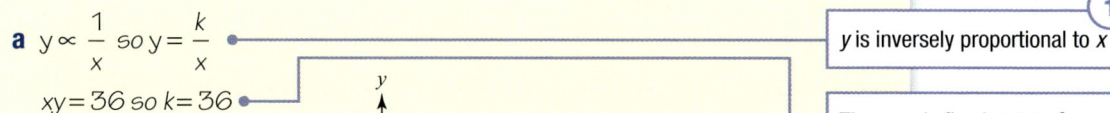

(1) y is inversely proportional to x

$xy = 36$ so $k = 36$

The area is fixed at 36 m².

$y = \dfrac{36}{x}$

b $y = \dfrac{36}{x}$

When $x = 0$, y is not defined.

The line $x = 0$ is an asymptote.

$x = \dfrac{36}{y}$

When $y = 0$, x is not defined.

The line $y = 0$ is an asymptote.

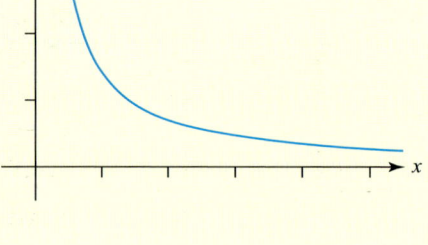

(2)(3)(5) Apply what you know about graphs of the form $y = \dfrac{k}{x}$

c y and x are actual lengths, so they must be positive and the curve approaches the asymptotes as shown.

1 The radius, r, of a container is inversely proportional to its height, h

A container of radius 4 cm will have a height of 14 cm.

a Write an equation linking h and r

b Sketch a graph to illustrate this relationship.

2 The volume, v cm³ of water in a tank is proportional to the square-root of the time, t seconds. After 15 minutes the tank has 1800 cm³ of water in it.

a Write an equation linking v and t

b Sketch a graph to illustrate this relationship.

3 a Sketch the graphs of $y = \dfrac{1}{x+2}$ and $y = x^2(x-3)$ on the same axes.

b Use your answer to part **a** to explain how many solutions there are to the equation $x^2(x-3) = \dfrac{1}{x+2}$

4 i Sketch both functions on the same set of axes.

ii Use your sketch to find approximate solutions to the equation.

a $y = 3x - 1;\ y = 2x^3 + x^2 - 4x + 1;\ 3x - 1 = 2x^3 + x^2 - 4x + 1$

b $y = \dfrac{1}{x};\ y = x^4 + 2;\ \dfrac{1}{x} = x^4 + 2$

c $y = \dfrac{1}{x-2};\ y = x^3 + 5;\ \dfrac{1}{x} - 2 = x^3 + 5$

5 The graph of $y = f(x)$ is shown.

Give the equations for each of these transformations in terms of $f(x)$

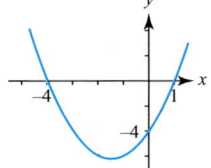

a

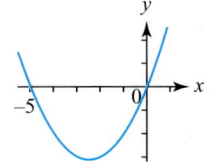

b

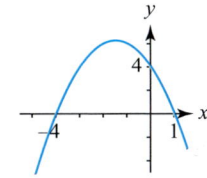

6 The graph of $y = x^3 + Ax^2 + Bx + C$ is shown.

Find the values of the constants A, B and C

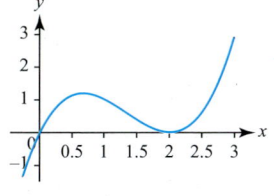

7 This is the graph of $y = f(x)$ where $f(x) = x^4 + Ax^3 + Bx^2 + Cx - 10$

a Find the values of the constants A, B and C

b Describe the transformation that maps $f(x)$ to

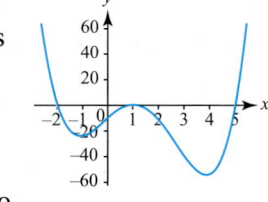

i $g(x) = -x^4 - Ax^3 - Bx^2 - Cx + 10$

ii $h(x) = x^4 + Ax^3 + Bx^2 + Cx + 10$

8 The graph shown has the equation $y = A + \dfrac{B}{x+C}$

Find the values of A, B and C

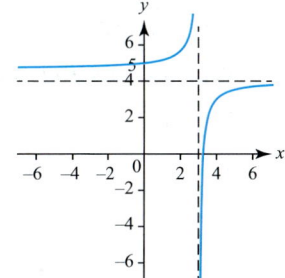

Challenge

9 For the graph of $y = ax^2 + bx + c$ where a, b and c are constants

a Explain the conditions for the graph to have a minimum point and the conditions for the graph to have a maximum point,

b Write down the coordinates of the maximum or minimum point,

c Write down the coordinates where the curve intersects the axes,

d Write down the equation of the line of symmetry of the curve.

Chapter summary

- The highest power in a polynomial expression is called its degree.
- When adding or subtracting polynomials, expand brackets before collecting like terms.
- Identities use the $\equiv$ sign. Identities are true for all values of the variable(s).
- For $n = 0, 1, 2, 3, ...$, the binomial expansions are

$$(1+x)^n \equiv 1 + nx + \frac{n(n-1)}{2!}x^2 + \frac{n(n-1)(n-2)}{3!}x^3 + ... + x^n$$

and

$$(a+b)^n \equiv a^n + {}^n C_1 a^{n-1}b + {}^n C_2 a^{n-2}b^2 + ... + {}^n C_r a^{n-r}b^r + ... + b^n$$

- The coefficients of these expansions can be found from Pascal's triangle or from ${}^n C_r \equiv \dfrac{n!}{(n-r)!r!}$
- You can divide algebraically using the same technique, as for long division in arithmetic.
- The factor theorem states that if $f(a) = 0$, then $(x - a)$ is a factor of $f(x)$
- When $f(x)$ is divided by $(x - a)$, the remainder is $f(a)$
- To sketch a graph you need to consider the symmetry, x- and y-intercepts, asymptotes, behaviour as x and/or y approaches $\pm\infty$, and any other obvious critical points. You can also apply your knowledge of transformations.

Check and review

You should now be able to...	Try Questions
✔ Manipulate, simplify and factorise polynomials.	1–4, 17
✔ Understand and use the binomial theorem.	7–11
✔ Divide polynomials by algebraic expressions.	6, 12, 14, 15
✔ Understand and use the factor theorem.	5, 13, 16
✔ Use a variety of techniques to analyse a function and sketch its graph.	18–23

1 Add together $2x^3 + 9x^5 + 11x^2 - 3x - 5x^4 - 12$ and $4x^2 - x^4 - 7x^5 + 3 + 12x - 5x^3$

2 Fully factorise $4n^3 + 4n^2 - 15n$

3 Expand and simplify these expressions.

    a $(y-1)(y+3)(2y+5)$ **b** $(2z+1)(z-2)^2$

4 Factorise these expressions.

    a $m(m+4) - (m+4)^2$

    b $(d+1)^2 - 4(d+1)(d-1)$

5 The equation $2x^3 + ax^2 + bx + c = 0$ has roots $-4, 3$ and $\dfrac{7}{2}$. Find the values of a, b and c

6 Find the function that, when divided by $(x+3)$, gives a quotient of $(2x-3)$ and a remainder of -4

7 Use Pascal's triangle to write the expansion of $\left(1+\dfrac{m}{10}\right)^4$
Use your answer to evaluate the value of 1.1^4 to 4 decimal places.

8 Use the binomial theorem to expand $(2s^2 - 4t)^4$

9 Use Pascal's triangle to expand and simplify these expressions.

    a $(1+\sqrt{3})^4$ **b** $(3-\sqrt{5})^5 - (3+\sqrt{5})^5$

10 Find the constant term in the binomial expansion of $\left(w - \dfrac{3}{2w}\right)^{14}$

11 a $(2 - ax)^9 \equiv 512 + 2304x + bx^2 + cx^3 + \ldots$ Find the values of a, b, and c

b Use your values of a, b, and c to find the first four terms in the expansion of $(1 - x)(2 - ax)^9$

12 Divide $2x^3 - 3x^2 - 26x + 3$ by $x + 3$

13 By successively evaluating f(1), f(−1), f(2), f(−2) and so on, find all the factors of $x^3 - 4x^2 + x + 6$

14 Divide $8x^3 + 14x^2 - x + 35$ by $(2x + 5)$

15 Divide $x^3 - 2x^2 + 3x + 4$ by $(x - 2)$

16 Show that $(2x - 3)$ is a factor of $4x^3 - 8x^2 + x + 3$

17 Factorise fully $2x^3 + x^2 - 18x - 9$

18 A student attempted to transform the graph of $y = \text{f}(x)$ into $y = \text{f}(x - 1)$

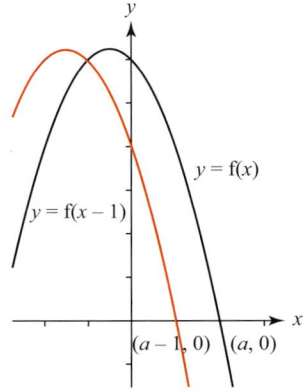

$y = \text{f}(x)$
$y = \text{f}(x - 1)$
$(a - 1, 0)$ $(a, 0)$

a Explain what mistake she has made.

b Sketch the graph of $y = \text{f}(x - 1)$

19 Sketch these curves on the same set of axes.

a $y = \dfrac{1}{x}$ **b** $y = \dfrac{4}{x}$ **c** $y = 2 + \dfrac{1}{x}$

20 Sketch the graph of $y = (x - 6)^3$

21 A particle moves along a straight line from O, so that, at time t s, it is s m from O, given by the equation $s = t(2t - 7)^2$

Sketch the graph and describe its motion fully.

22 A rectangular metal sheet, 16 in by 10 in, has squares of side x in removed from its corners.

The edges are turned up to form an open box.

a Show that the volume of this box is $V = 160x - 52x^2 + 4x^3 \, \text{in}^3$

b Sketch a graph to evaluate the value of x that gives the highest volume.

23 A particle moves along a straight line from O, so that, at time t seconds, it is s metres from O, given by the equation $s = t(t - 4)^2$. Sketch the graph and describe its motion.

What next?

Score		
0 – 11	Your knowledge of this topic is still developing. To improve, search in MyMaths for the codes 2006, 2022–2024, 2027, 2041–2043, 2258	🔗
12 – 17	You're gaining a secure knowledge of this topic. To improve, look at the InvisiPen videos for Fluency and skills (02A)	🎞
18 – 23	You've mastered these skills. Well done, you're ready to progress! Now try looking at the InvisiPen videos for Reasoning and problem-solving (02B)	🎞

Click these links in the digital book

History

The binomial theorem is a formula for finding any power of a two-term bracket without having to multiply them all out. It has existed in various forms for centuries and special cases, for low powers, were known in Ancient Greece, India and Persia.

The triangular arrangement of the binomial coefficients is known as **Pascal's triangle**. It took its name from the 17th century mathematician **Blaise Pascal**, who studied its properties in great depth. Although the triangle is named after Pascal, it had been known about much earlier. A proof linking it to the binomial theorem was given by an Iranian mathematician Al-Karaji in the 11th century.

Around 1665, **Isaac Newton** developed the binomial theorem further by applying it to powers other than positive whole numbers. He showed that a general formula worked with any rational value, positive or negative.

Newton showed how the binomial theorem could be used to simplify the calculation of roots and also used it in a calculation of π, which he found to 16 decimal places.

Pascal's triangle

```
            1
          1   1
        1   2   1
      1   3   3   1
    1   4   6   4   1
  1   5   10   10   5   1
```

Have a go

For small values of x, $(1 + x)^n \approx 1 + nx$

Use this result to estimate the value of
a) $(1.02)^4$
b) $(0.99)^5$
c) $(2.01)^5$

Find these values on a calculator and compare your results.

$(2.01)^5$ can be written in the form $2^5(1 + ...)^5$

> "If I have seen further than others, it is by standing on the shoulders of giants."
> – Isaac Newton

1 a   Simplify these expressions.

     i $(2x-3)(6x+1)$ **ii** $(2a-3b)^2$ **iii** $(5x+2y)(x^2-3xy-y^2)$ **[6 marks]**

   b   Given $\dfrac{ax^2+bx+c}{(3x+4)} \equiv (3x-4)$, evaluate the values of the constants a and c,

      and show that $b=0$ **[3]**

2   Write down the binomial expansion of $\left(1+\dfrac{1}{2}x\right)^8$ in ascending powers of x, up to

   and including the term in x^3. Simplify the terms as much as possible.   **[6]**

3 a   Factorise p^3-10p^2+25p **[2]**

   b   Deduce that $(2x+5)^3-10(2x+5)^2+25(2x+5) \equiv ax^2(2x+5)$, where a is a

      constant that should be stated.   **[2]**

4   Show that $(x-3)$ is *not* a factor of $2x^3-5x^2+6x-7$ **[3]**

5   Show how the binomial expansion can be used to work out each of these without a calculator.

   a 268^2-232^2 **[2]**

   b $469 \times 548 + 469^2 - 469 \times 17$ **[2]**

   c $\dfrac{65.1 \times 29.2 + 65.1 \times 35.9 - 91.7 \times 26.4 + 65.3 \times 26.4}{18.3^2 - 18.3 \times 5.4}$ **[5]**

6   Given that $(1+cx)^7 = 1+21x+Ax^2+Bx^3+........$

   a   Work out   **i** The value of c **ii** The value of A **iii** The value of B **[4]**

   b   Using your values of c, A and B, evaluate the coefficient of x^3 in the

      expansion of $(2+x)(1+cx)^7$ **[2]**

7   Express x^3-3x^2+5x+1 in the form $(x-2)(x^2+ax+b)+c$ **[3]**

8 a   Write down the expansions of   **i** $(x+y)^4$ **ii** $(x-y)^4$ **[4]**

   b   Show that $(\sqrt{5}+\sqrt{2})^4 + (\sqrt{5}-\sqrt{2})^4 = n$, where n is an integer to be found.   **[4]**

9   Write down the term which is independent of x in the expansion of $\left(x^2+\dfrac{2}{x}\right)^9$ **[3]**

10 a   Expand each of these in ascending powers of x up to and including the term in x^2

     i $(1+2x)^6$ **ii** $(2-x)^6$ **[5]**

   b   Hence write down the first three terms in the binomial expansion of $(2+3x-2x^2)^6$ **[4]**

11 a   Show that $(x-2)$ is a factor of $2x^3+x^2-7x-6$ **[2]**

   b   Show that the equation $2x^3+x^2-7x-6=0$ has the solutions 2, $-\dfrac{3}{2}$, and -1 **[5]**

12   Given that both $(x-1)$ and $(x+3)$ are factors of $ax^3+bx^2-16x+15$

   a   Evaluate the values of a and b **[6]**

   b   Fully factorise $ax^3+bx^2-16x+15$ **[3]**

c Sketch the graph of $y = ax^3 + bx^2 - 16x + 15$ [3]

d Solve the inequality $ax^3 + bx^2 - 16x + 15 \geq 0$ [2]

13 a Expand $\left(x + \dfrac{1}{x}\right)^6$, simplifying the terms. [7]

b Hence write down the expansion of $\left(x - \dfrac{1}{x}\right)^6$ [1]

c Prove that the equation $\left(x + \dfrac{1}{x}\right)^6 - \left(x - \dfrac{1}{x}\right)^6 = 64$ has precisely two real solutions. [5]

14 Prove these results

a $^{n+1}C_r \equiv {^n}C_r + {^n}C_{r-1}$ [6]

b $^{n+2}C_3 - {^n}C_3 \equiv n^2$ [8]

15 Here are five equations, labelled **i** – **v**, and five graphs, labelled **A** – **E**

i $y = \dfrac{1}{(x-2)^2}$ **ii** $y = 1 + \dfrac{1}{(x-1)}$ **iii** $y = -\dfrac{1}{x+1}$ **iv** $y = -\dfrac{1}{(x+1)^2}$ **v** $y = 1 + \dfrac{1}{x^2}$

A
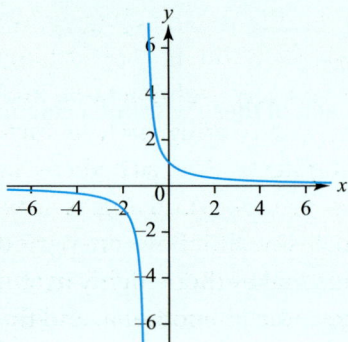

B
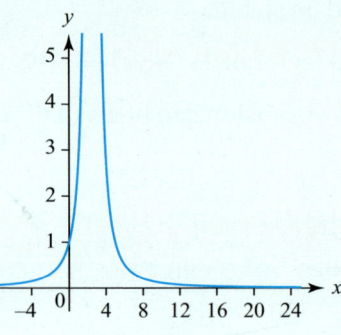

C
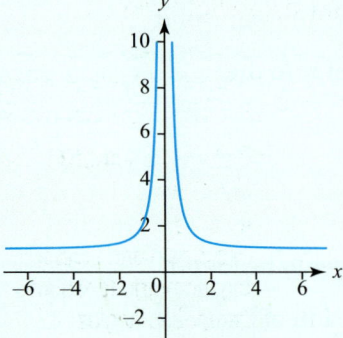

D
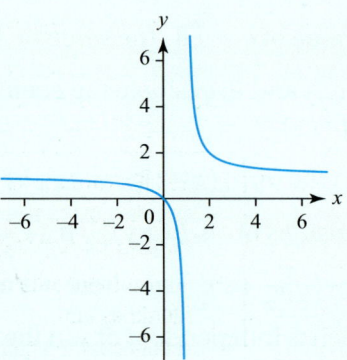

E
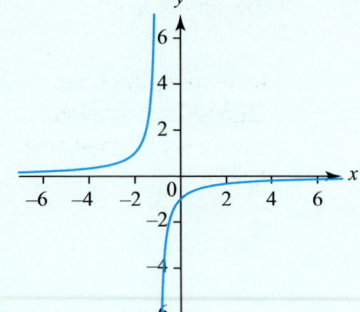

Four of the equations correspond to four of the graphs.

a Match the four equations to their graphs. [4]

b For the graph that has no equation, write down a possible equation. [1]

c For the equation that has no graph, sketch its graph. [3]

3 Trigonometry

GPS uses a technique called trilateration to calculate positions. The receiver, a phone for example, receives direct signals from four different satellites simultaneously. The imaginary lines between the satellites and the receiver form the sides of triangles, which are then used by the phone to calculate its position. Trilateration is a hi-tech version of triangulation, a technique that requires the use of trigonometry.

Trigonometry is the study of the relationships between angles and the sides of a triangle. Thus, it is immensely useful in fields such as astronomy, engineering, architecture, geography and navigation, as it allows easy calculation of distances and angles or bearings. The sine and cosine functions are periodic in nature. This makes them highly useful in modelling periodic phenomena, and they can be used to describe different types of wave, including sound and light waves.

Orientation

What you need to know	What you will learn	What this leads to
KS4 • Apply and derive Pythagoras' theorem. • Recognise graphs of trigonometric functions. • Apply some properties of angles and sides of a triangle.	• To calculate the values of sine, cosine and tangent for any angle. • To use trigonometric identities and recognise the equation of a circle. • To sketch and describe trigonometric functions. • To solve trigonometric equations. • To use the sine and cosine rule and the area formula for a triangle.	**Ch14 Trigonometric identities** Radians. Reciprocal and inverse trigonometric functions. Compound angles. Equivalent forms for $a\cos\theta + b\sin\theta$
Ch1 Algebra 1 • Working with surds.		**Ch16 Integration and differential equations** Integrating trigonometric functions.

 Practise before you start   🔍 1112, 1131, 1133, 2024, 2036

69

3.1 Sine, cosine and tangent

Fluency and skills

You can use trigonometry to find lengths and angles in right-angled triangles. This branch of mathematics is used in engineering, technology and many sciences.

Pythagoras' theorem for right-angled triangles is $a^2 + b^2 = c^2$

Dividing by c^2 gives $\dfrac{a^2}{c^2} + \dfrac{b^2}{c^2} = \dfrac{c^2}{c^2}$ or $\left(\dfrac{a}{c}\right)^2 + \left(\dfrac{b}{c}\right)^2 = 1$

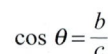

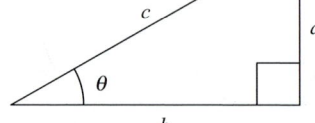

$$\sin\theta = \dfrac{a}{c}$$

$$\cos\theta = \dfrac{b}{c}$$

$$\tan\theta = \dfrac{a}{b}$$

See p.614
For a list of mathematical notation.

Key point
$$\sin^2\theta + \cos^2\theta \equiv 1$$

$$\tan\theta = \dfrac{a}{b}$$

Dividing numerator and denominator of $\tan\theta$ by c gives a definition for $\tan\theta$ in terms of $\sin\theta$ and $\cos\theta$

Key point
$$\tan\theta \equiv \dfrac{\frac{a}{c}}{\frac{b}{c}} \equiv \dfrac{\sin\theta}{\cos\theta}$$

These two identities are true for all values of θ

Example 1

Calculate **a** $\sin\theta$ **b** $\tan\theta$ as surds, given that θ is acute and $\cos\theta = \dfrac{1}{\sqrt{3}}$

See Ch1.3
For a reminder of simplifying surds.

a $\sin^2\theta = 1 - \cos^2\theta = 1 - \left(\dfrac{1}{\sqrt{3}}\right)^2 = \dfrac{2}{3}$

$\sin\theta = \sqrt{\dfrac{2}{3}}$

b $\tan\theta = \dfrac{\sin\theta}{\cos\theta}$

$= \sqrt{\dfrac{2}{3}} \times \dfrac{\sqrt{3}}{1} = \sqrt{\dfrac{2\times 3}{3\times 1}}$

$\tan\theta = \sqrt{2}$

— θ is acute, so ignore $-\sqrt{\dfrac{2}{3}}$

— Use the identities.

— Substitute values and simplify.

Example 2

Prove that $\tan\theta + \dfrac{1}{\tan\theta} \equiv \dfrac{1}{\sin\theta\cos\theta}$

$\tan\theta + \dfrac{1}{\tan\theta} \equiv \dfrac{\sin\theta}{\cos\theta} + \dfrac{\cos\theta}{\sin\theta}$

$\equiv \dfrac{\sin^2\theta + \cos^2\theta}{\sin\theta\cos\theta}$

$\equiv \dfrac{1}{\sin\theta\cos\theta}$

— Use $\tan\theta = \dfrac{\sin\theta}{\cos\theta}$

— Add the two fractions.

— Use $\sin^2\theta + \cos^2\theta = 1$

You can use the unit circle to draw graphs of the trigonometric ratios. The point P moving around the circle, centre O, has coordinates $x = \cos\theta$ and $y = \sin\theta$

As $P(x, y)$ moves around the circle you can plot graphs of the values of y, x and $\dfrac{y}{x}$ for each value of θ

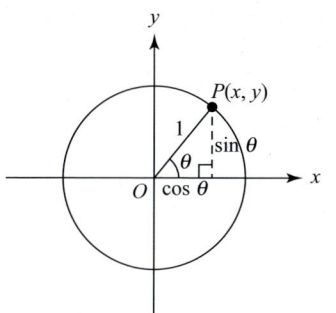

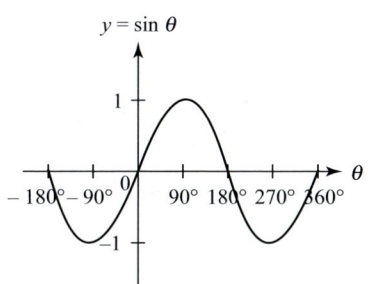

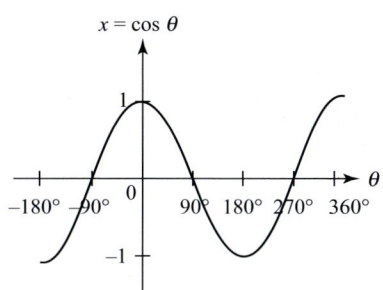

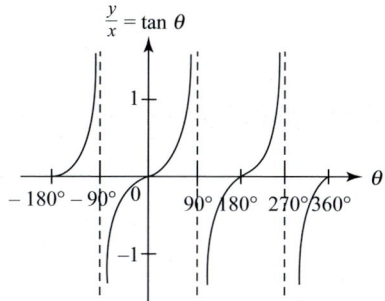

Extending the graphs for higher and lower values of θ shows they are all **periodic functions**, with a period of 360° for sine and cosine and 180° for tangent.

You can also see the symmetries from the graphs. For example, $y = \sin\theta$ has lines of symmetry at $\theta = -90°$, $\theta = 90°$, $\theta = 270°$, ... and it has rotational symmetry (order 2) about every point where the graph intersects the θ-axis.

Example 3

Express   **a** $\sin 127°$ **b** $\cos 132°$ in terms of acute angles.

a

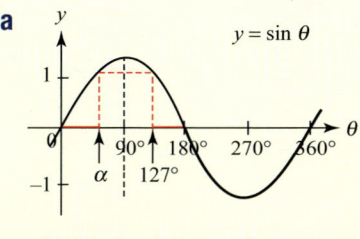

$\sin 127°$ is positive.

$127° = 180° - 53°$

Use the line of symmetry $\theta = 90°$

$\alpha = 53°$

$\sin 127° = \sin 53°$

b

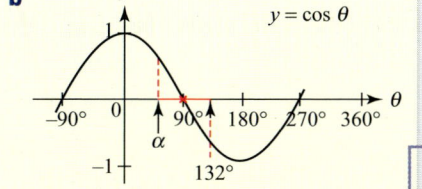

$\cos 132°$ is negative.

$132° - 90° = 42°$

Use rotational symmetry about $(90°, 0)$

$\alpha = 90° - 42° = 48°$

$\cos 132° = -\cos 48°$

> Find the given values.

> Use the symmetry of the graph to find the acute angle with the same numeric value.

> Write the sign and acute angle of the trigonometric ratio.

The trigonometric graphs and equations can be transformed in the same way as quadratic and polynomial graphs.

See Ch2.4 For a reminder on transforming graphs.

Another method for finding equivalent acute angles is to use a **quadrant diagram**.

Imagine a radius *OP* rotating about *O* through an angle *θ* from the positive *x*-axis. Whichever quadrant *OP* lies in, you can form an acute triangle with the *x*-axis as its base. Depending on the quadrant, the *x* and *y* coordinates of point *P* are positive or negative and so the sine, cosine and tangent of *θ* are also positive or negative.

You can use the word '**CAST**' as a mnemonic to help you remember where each ratio is positive.

CAST starts from the 4th quadrant and moves anticlockwise, as all the ratios are positive in the 1st quadrant.

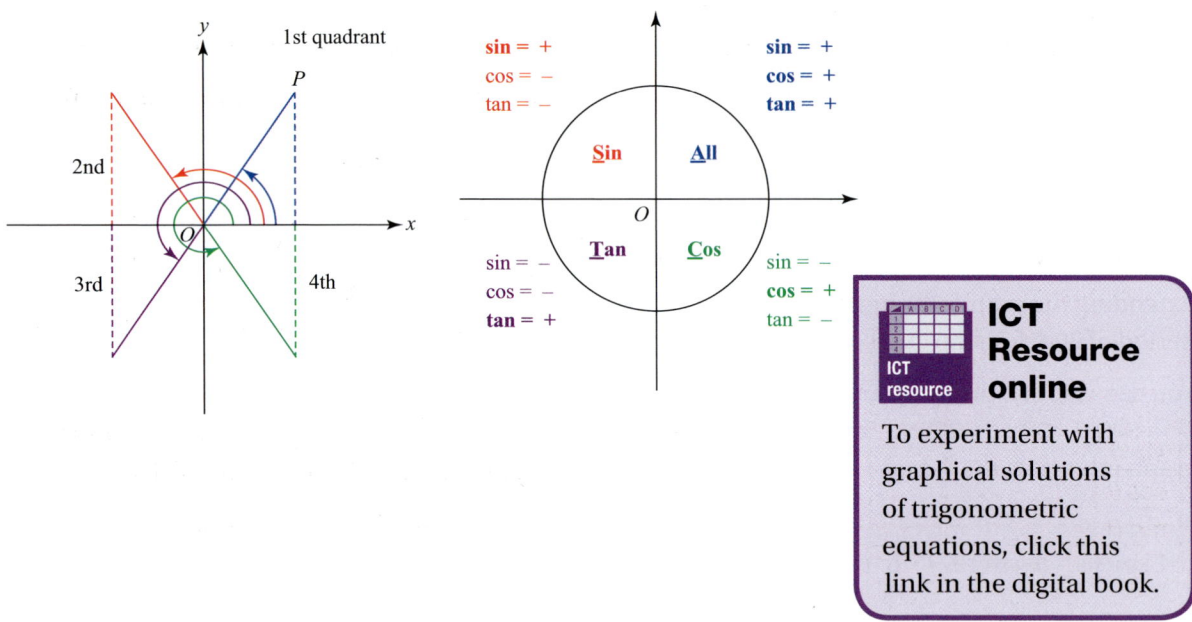

ICT Resource online

To experiment with graphical solutions of trigonometric equations, click this link in the digital book.

Example 4

Express **a** cos 132° **b** tan 683° in terms of acute angles.

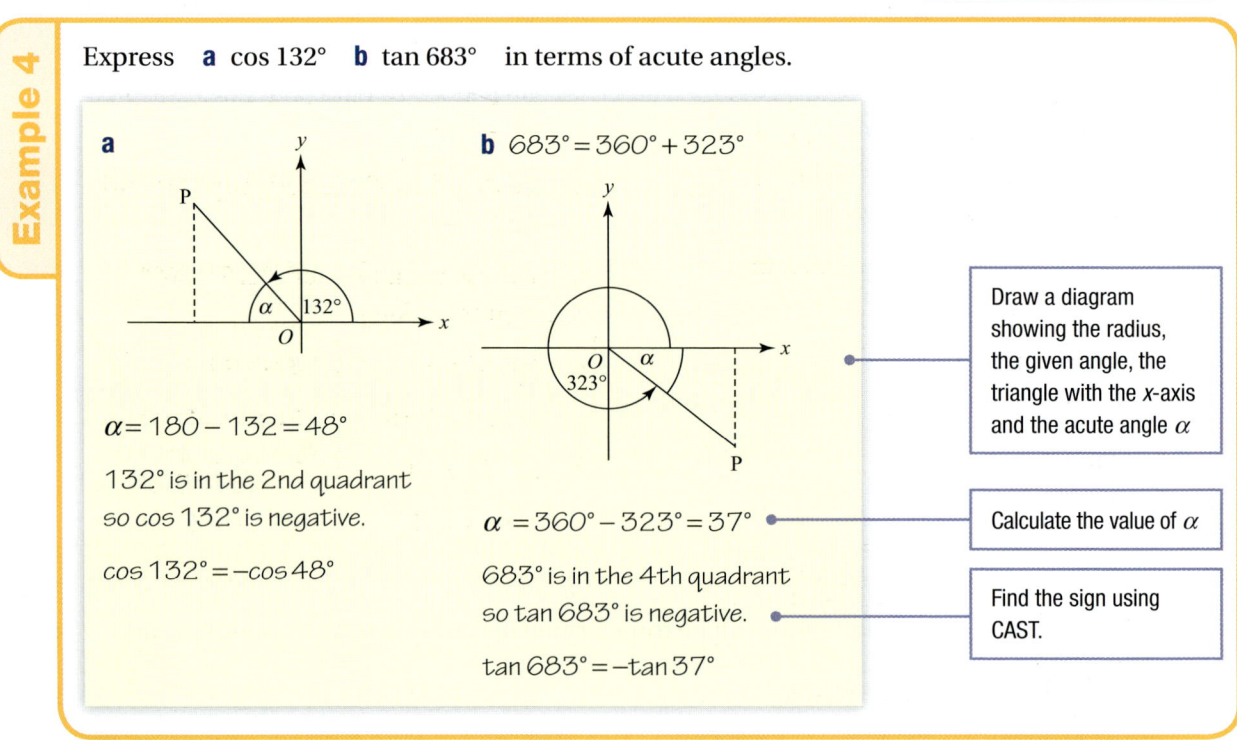

a

$\alpha = 180 - 132 = 48°$

132° is in the 2nd quadrant so cos 132° is negative.

$\cos 132° = -\cos 48°$

b $683° = 360° + 323°$

$\alpha = 360° - 323° = 37°$

683° is in the 4th quadrant so tan 683° is negative.

$\tan 683° = -\tan 37°$

Draw a diagram showing the radius, the given angle, the triangle with the *x*-axis and the acute angle α

Calculate the value of α

Find the sign using CAST.

1 Use $\sin^2\theta + \cos^2\theta = 1$ and $\tan\theta = \dfrac{\sin\theta}{\cos\theta}$ to calculate the value of $\sin\theta$ and $\tan\theta$, given that θ is acute and

   a $\cos\theta = \dfrac{3}{5}$ **b** $\cos\theta = 0.8$

   c $\cos\theta = \dfrac{12}{13}$

2 Use the quadrant or graphical method to find these values in terms of acute angles.

   a $\cos 190°$ **b** $\tan 160°$

   c $\sin 340°$ **d** $\cos 158°$

   e $\tan 215°$ **f** $\sin 285°$

   Check your answers using the method that you didn't use the first time.

3 Copy and complete this table.

θ	$-90°$	$0°$	$90°$	$180°$	$270°$	$360°$
$\sin\theta$						
$\cos\theta$						
$\tan\theta$						

4   **i** Sketch the graph $y = \mathrm{f}(x)$ for $-360° \le x \le 360°$. Describe the line and rotational symmetries of the graph.

    ii On the same set of axes, sketch the graph $y = \mathrm{g}(x)$ for $-360° \le x \le 360°$ You can check your sketches on a calculator.

    iii Describe the transformation that maps $\mathrm{f}(x)$ to $\mathrm{g}(x)$

   a $\mathrm{f}(x) = \sin x;\ \mathrm{g}(x) = \sin 3x$

   b $\mathrm{f}(x) = \cos x;\ \mathrm{g}(x) = \cos x - 1$

   c $y = \tan\dfrac{1}{2}x$

5 Simplify

   a $\dfrac{\sin\theta}{\sqrt{1-\sin^2\theta}}$ **b** $\sqrt{\dfrac{1-\cos^2\theta}{1-\sin^2\theta}}$

   c $\dfrac{\sqrt{1-\sin^2\theta}}{\cos\theta}$ **d** $\tan\theta\cos\theta$

   e $\dfrac{\sin\theta}{\tan\theta}$ **f** $\sin\theta\cos\theta\tan\theta$

6 Express, in terms of acute angles,

   a $\sin 380°$ **b** $\tan 390°$

   c $\cos 700°$ **d** $\tan(-42°)$

   e $\cos(-158°)$ **f** $\sin(-203°)$

7 Solve these equations for $-180° \le \theta \le 180°$

   a $\sin\theta = \cos\theta$ **b** $\sin\theta + \cos\theta = 0$

8 Use the triangle to write these trigonometric ratios in surd form.

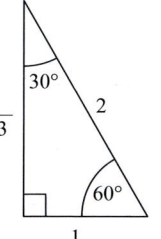

   a $\sin 150°$ **b** $\cos 300°$

   c $\tan 120°$ **d** $\sin 240°$

   e $\cos(-60°)$ **f** $\tan(-150°)$

9 Use a calculator and give all the values of θ in the range $-360°$ to $360°$ for which

   a $\sin\theta = 0.4$ **b** $\tan\theta = 1.5$

   c $\cos\theta = -0.5$

10 Use a calculator to find the smallest positive angle for which

   a $\sin\theta$ and $\cos\theta$ are both positive and $\sin\theta = 0.8$

   b $\sin\theta$ and $\tan\theta$ are both negative and $\sin\theta = -0.6$

11 Solve these equations for $0° \le \theta \le 360°$ Show your working.

   a $4\sin\theta = 3$ **b** $3\tan\theta = 4$

   c $2\sin\theta + 1 = 0$ **d** $3\cos\theta + 2 = 0$

   e $\tan\theta + 3 = 0$ **f** $7 + 10\sin\theta = 0$

   g $4\cos\theta = -3$ **h** $4 + 9\tan\theta = 0$

Reasoning and problem-solving

To solve problems involving trigonometric ratios

(1) Use trigonometric identities to simplify expressions.

(2) Draw either a quadrant diagram or a trigonometric graph to show the information.

(3) Use your knowledge of graphs, quadrant diagrams, symmetry and transformations to help you answer the question.

Example 5

Solve $5\cos 2\theta + 3 = 0$ for $0° \leq \theta \leq 180°$. Show your working.

$5\cos 2\theta = -3$

$\cos 2\theta = -\dfrac{3}{5} = -0.6$ ●————— Rearrange and simplify. **①**

For $0° \leq 2\theta \leq 360°$

either ●———————— or ●———— Draw either a quadrant diagram or a trigonometric graph. **②**

$\cos 2\theta$ is negative in 2nd and 3rd quadrants

$\cos \alpha = 0.6$

$\alpha = 53.1°$

$2\theta = 180° - \alpha$ or $180° + \alpha$

$= 126.9°$ or $233.1°$

$\cos 2\theta = -0.6$ at a and b

$2\theta = \cos^{-1}(-0.6)$

$= 126.9° (= a)$ ●————— Use a calculator to give the principal value.

Or $2\theta = 360° - 126.9°$

$= 233.1° (= b)$

$\theta = \dfrac{126.9°}{2}$

Or $\theta = \dfrac{233.1°}{2}$ ●————— Use the quadrant diagram or symmetry of graph to work out the values of θ **③**

$\theta = 63.5°$ or $116.6°$

You can check your solution by solving the equation on a calculator.

Exercise 3.1B Reasoning and problem-solving

1 Solve these equations for $0° \leq \theta \leq 360°$

Show your working.

a $\sin(\theta + 30°) = \dfrac{1}{2}$ **b** $\cos(\theta - 30°) = \dfrac{1}{2}$

c $\tan(\theta + 20°) = 1$ **d** $2\sin(\theta + 30°) = -1$

2 Solve these equations for $0° \leq \theta \leq 180°$

a $2\sin 2\theta = 1$ **b** $3\tan 2\theta = 2$

c $5\cos 3\theta = 2$ **d** $5\sin 3\theta + 3 = 0$

e $\dfrac{1}{2}\tan 2\theta + 3 = 0$ **f** $4\sin\left(\dfrac{1}{2}\theta\right) = 3$

3 Give full descriptions of any two transformations, which map the graph of

a $y = \sin \theta$ onto $y = \cos \theta$

b $y = \tan \theta$ onto itself.

4

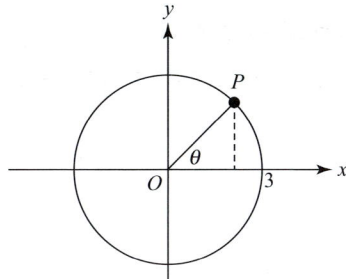

a Given a circle, centre O and radius 3, write the coordinates of point P

b Show these coordinates satisfy the equation of the circle $x^2 + y^2 = r^2$ and write the value of r

c Another circle has the equation $4x^2 + 4y^2 = 25$. What is its radius?

5 A circle, with centre at the origin, passes through the point $(6, 8)$. What is its equation in its simplest form?

6 Solve these equations for $-180° \leq \theta \leq 180°$ Show your working.

a $2 \sin \theta = \cos \theta$

b $4 \cos \theta = 5 \sin \theta$

c $3 \sin 2\theta - \cos 2\theta = 0$

d $3 \sin 2\theta = 2 \tan 2\theta$

e $3 \sin \theta + \tan \theta = 0$

f $\sin \theta \cos \theta - \cos \theta = 0$

7 Solve these equations for $0° \leq \theta \leq 360°$

a $\sin^2 \theta - 2 \sin \theta + 1 = 0$

b $\tan^2 \theta - \tan \theta - 2 = 0$

c $2 \sin \theta + 2 = 3 \cos^2 \theta$

d $2 \cos^2 \theta + \sin^2 \theta = 2$

e $2 \cos \theta + 2 = 4 \sin^2 \theta$

f $5 \sin \theta - 4 \cos^2 \theta = 2$

8 The graph of $y = a \sin b\theta$ has a maximum value of 5 and a period of 45°. Find the values of a and b. Show your working.

9 The depth of water, h metres, at point P on the seabed changes with the tide and is given by $h = 3 + 2 \sin (30° \times t)$, where t is the time in hours after midnight.

a What is the greatest and least depth of water at P?

b What is the period of the oscillation of the tide?

c At what times do the high tides occur on this day?

10 Solve these equations for $0 \leq \theta \leq 360°$

a $7 \cos \theta + 6 \sin^2 \theta - 8 = 0$

b $4 \cos^2 \theta + 5 \sin \theta = 3$

11 a Draw an accurate graph of the function $y = 2 \cos \theta + 3 \sin \theta$ for $-180° \leq \theta \leq 180°$

b Solve the equation $2 \cos \theta + 3 \sin \theta = 0$ for this range of θ by using

i your graph,

ii an algebraic method.

12 Prove these identities.

a $\cos^4 x - \sin^4 x \equiv \cos^2 x - \sin^2 x$

b $\tan x + \dfrac{1}{\tan x} \equiv \dfrac{1}{\sin x \cos x}$

Challenge

13 Prove that $\dfrac{1 - \tan^2 x}{1 + \tan^2 x} \equiv 1 - 2 \sin^2 x$ and

hence solve the equation $\dfrac{1 - \tan^2 x}{1 + \tan^2 x} = \dfrac{1}{2}$

for $0 \leq x \leq 360°$

The sine and cosine rules

Fluency and skills

You can use the sine and cosine rules to calculate lengths and angles in any triangle — not just right-angled triangles.

When you know a pair of opposite sides and angles, you can calculate other sides and angles using the **sine rule**.

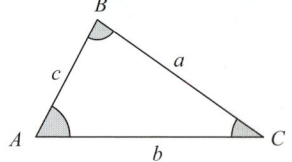

> **Key point**
>
> The sine rule states that, for triangle ABC,
>
> $$\frac{a}{\sin A} = \frac{b}{\sin B} = \frac{c}{\sin C} \qquad \text{or} \qquad \frac{\sin A}{a} = \frac{\sin B}{b} = \frac{\sin C}{c}$$

> Use the left-hand version for sides and the right-hand one for angles.

> Triangle ABC can be written $\triangle ABC$

Example 1

In $\triangle ABC$, angle $A = 50°$, side $a = 8$ cm and side $c = 10$ cm. Calculate angles B and C, given that the triangle is acute.

$$\frac{\sin C}{c} = \frac{\sin A}{a}$$

As c is known, use the sine rule to calculate angle C first.

$$\frac{\sin C}{10} = \frac{\sin 50°}{8}$$

Substitute in the correct values.

$$\sin C = \frac{\sin 50°}{8} \times 10 = 0.9575$$

Sine is positive between 0° and 180°, so there are two possible values of C

Rearrange to solve for $\sin C$

Do not round answers during a calculation.

The given triangle is acute, so $C = 73.2°$

Angle $B = 180° - 50° - 73.2° = 56.8°$

Use the angle sum of a triangle.

The data in Example 1 ($A = 50°$, $a = 8$ cm, $c = 10$ cm) can also describe an obtuse triangle where $C' = 180° - 73.2° = 106.8°$ and $B = 180° - 50° - 106.8° = 23.2°$. This is an example of the **ambiguous case**, where the initial data gives two possible triangles.

When you know two sides and the angle between them, you can use the **cosine rule** to calculate the third side. You can also use this rule to calculate angles when you know all three sides but no angles.

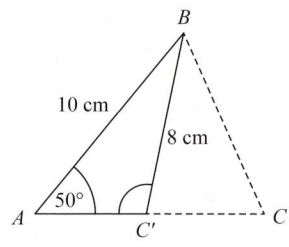

> **Key point**
>
> The cosine rule states that, for triangle ABC, $a^2 = b^2 + c^2 - 2bc \cos A$

Alternatively, $b^2 = a^2 + c^2 - 2ac \cos B$ or $c^2 = a^2 + b^2 - 2ab \cos C$

Example 2

In $\triangle ABC$, $a = 4\,\text{cm}$, $b = 9\,\text{cm}$ and $c = 6\,\text{cm}$. Calculate angle B

$b^2 = c^2 + a^2 - 2ca\cos B$ You need to find angle B, so use the formula which has b^2 as the subject.

$9^2 = 6^2 + 4^2 - 2 \times 6 \times 4 \times \cos B$

$48\cos B = 36 + 16 - 81 = -29$

$\cos B = -\dfrac{29}{48}$ Rearrange to make $\cos B$ the subject.

Angle $B = 127.2°$ $\cos B$ is negative, so angle B is obtuse.

You can calculate the area of any triangle when you know any two sides and the angle between them.

Key point

Area of triangle $ABC = \dfrac{1}{2}ab\sin C$

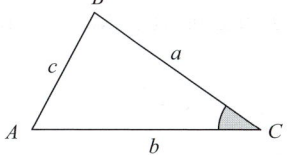

Exercise 3.2A Fluency and skills

1 Calculate the lengths BC and PR in these triangles.

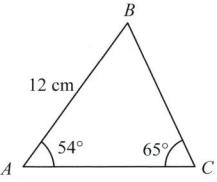 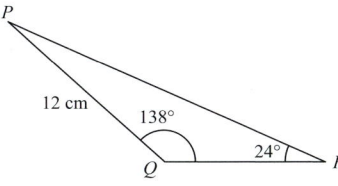

2 a Triangle ABC is acute with $AB = 9\,\text{cm}$, $BC = 8\,\text{cm}$ and angle $A = 52°$. Calculate angle C

   b Triangle EFG is obtuse with $EG = 11\,\text{cm}$, $EF = 7\,\text{cm}$ and angle $G = 35°$. Calculate obtuse angle F

   c Triangle HIJ is obtuse with $HI = 10\,\text{cm}$, $IJ = 5\,\text{cm}$ and angle $H = 28°$. Calculate obtuse angle J

3 a Calculate the lengths BC and PR in these triangles.

   b Calculate the areas of the triangles.

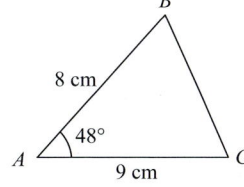 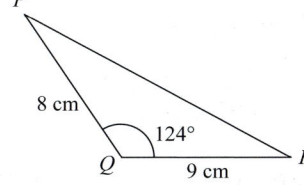

4 Calculate all the unknown sides and angles in these triangles. Give both solutions if the data is ambiguous.

a 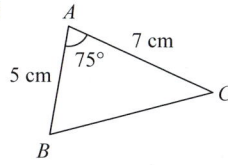   **b**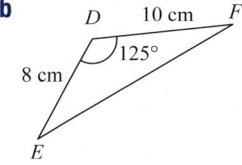

c 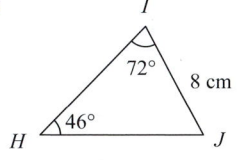   **d**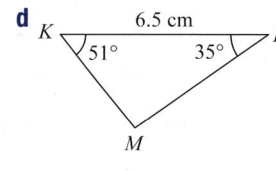

e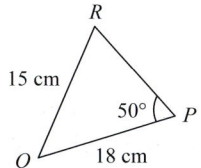

5 Calculate the unknown sides and angles in

   a $\triangle ABC$ where $a = 11.1\,\text{cm}$, $b = 17.3\,\text{cm}$ and $c = 21.2\,\text{cm}$

   b $\triangle DEF$ where $d = 75.3\,\text{cm}$, $e = 56.2\,\text{cm}$ and angle $F = 51°$

   c $\triangle HIJ$ where with $h = 44.2\,\text{cm}$, $i = 69.7\,\text{cm}$ and angle $J = 33°$

6 Calculate the length x and the area of the triangle, giving your answers as surds.

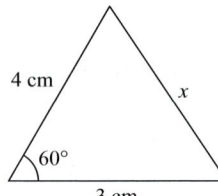

4 cm

x

60°

3 cm

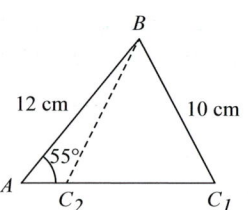

B

12 cm

10 cm

55°

A

C_2

C_1

7 a In ΔABC, use the sine rule to show that there are two possible positions for vertex C (C_1 and C_2). Calculate the two possible sizes of angle C

b In ΔXYZ, calculate the size of angle Z and explain why it is the only possible value.

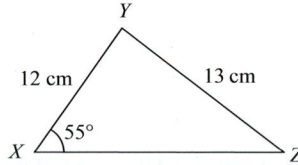

Y

12 cm

13 cm

55°

X

Z

Reasoning and problem-solving

Strategy

To solve problems involving sine and cosine rules or the area formula

① Draw a large diagram to show the information you have and what you need to work out.

② Decide which rule or combination of rules you need to use.

③ Calculate missing values and add them to your diagram as you solve the problem.

Example 3

In ΔABC, angle $A = 49°$, angle $B = 76°$ and $c = 12$ cm. Calculate the unknown sides and angles, and calculate the area of the triangle.

Angle $C = 180° - 76° - 49° = 55°$

The sine rule gives $\dfrac{a}{\sin 49°} = \dfrac{12}{\sin 55°}$

$a = \dfrac{12\sin 49°}{\sin 55°} = 11.055\ldots$

$\quad = 11.1$ cm (to 3 sf)

The sine rule gives

$\dfrac{b}{\sin 76°} = \dfrac{12}{\sin 55°}$

$b = \dfrac{12 \times \sin 76°}{\sin 55°}$

$\quad = 14.2$ cm (to 3 sf)

The cosine rule gives

$b^2 = 12^2 + 11.055^2 - 2 \times 12$
$\qquad \times 11.055 \times \cos 76°$

$\quad = 202.03$

$b = 14.213\ldots$

$\quad = 14.2$ cm (to 3 sf)

The two unknown sides are 11.1 cm and 14.2 cm.

The area of triangle $= \dfrac{1}{2} bc \sin A$

$\qquad = \dfrac{1}{2} \times 14.2 \times 12 \times \sin 49° = 64.3$ cm^2

C

b

a

49°

76°

B

A

12 cm

① Draw a diagram to show the information.

② Choose the sine rule because side c and angle C are now both known.

③ Rearrange and calculate a

② ③ You can use either the sine rule or the cosine rule to calculate b
You can decide which rule is easier to use.

You could also use $\dfrac{1}{2} ac \sin B$ or $\dfrac{1}{2} ab \sin C$

PURE

1   Calculate the area of ΔEFG, given that $e = 5$ cm, $f = 6$ cm and $g = 10$ cm.

2   **a** ΔABC has $AB = 5$ cm, $BC = 6$ cm and $AC = 7$ cm. Calculate the size of the smallest angle in the triangle.

    b ΔEGF has $EF = 5$ cm, $FG = 7$ cm and $EG = 10$ cm. Calculate the size of the largest angle in the triangle.

3   ΔABC has $b = 4\sqrt{3}$ cm, $c = 12$ cm and angle $A = 30°$. Prove the triangle is isosceles.

4   A parallelogram has diagonals 10 cm and 16 cm long and an angle of 42° between them. Calculate the lengths of its sides.

5   **a** ΔDEF has sides $d = 3x$, $e = x + 2$ and $f = 2x + 1$. If angle $D = 60°$, show that the triangle is equilateral. Calculate its area as a surd.

    b ΔPQR has an area of $\dfrac{3}{4}$ m². If $p = x + 1$, $q = 2x + 1$ and angle $R = 30°$, what is the value of x?

6   **a**   Find two different expressions for the height h using ΔACP and ΔBCP

      Hence, prove the sine rule. Also prove that the area of $\Delta ABC = \dfrac{1}{2}bc \sin A = \dfrac{1}{2}ca \sin B$

    b   Find expressions for CP, AP and BP in terms of the sides a, b, c and angle A. Hence, use Pythagoras' theorem to prove the cosine rule.

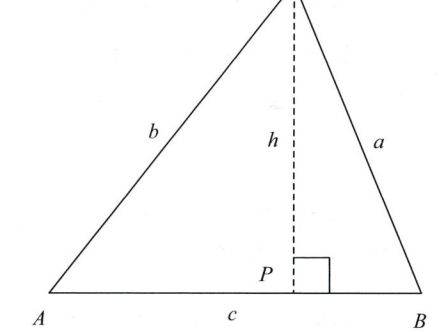

7   In ΔXYZ, $y = 2$ cm, $z = 2\sqrt{3}$ cm and angle $Y = 30°$. Prove that there are two triangles that satisfy this data and prove that one is isosceles and the other is right-angled.

8   The side opposite the smallest angle in a triangle is 8 cm long. If the angles are in the ratio $5 : 10 : 21$, find the length of the other two sides.

9   Two circles, radii 7 cm and 9 cm, intersect with centres 11 cm apart. What is the length of their common chord?

10 The circumcircle of ΔABC has centre O and radius r, as shown in the diagram. Point P is the foot of the perpendicular from O to BC. Consider ΔBOP and prove that $2r = \dfrac{a}{\sin A}$. Hence, prove the sine rule.

11 A triangle has base angles of 22.5° and 112.5°. Prove that the height of the triangle is half the length of the base.

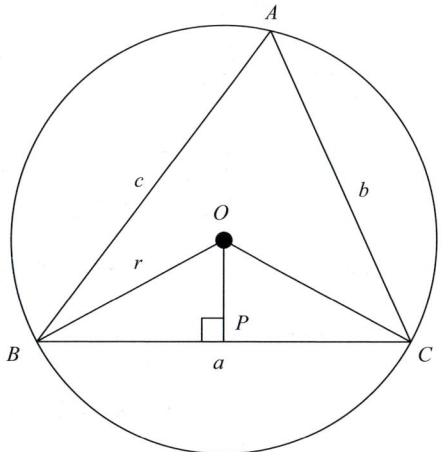

Challenge

12 In ΔXYZ, $x = n^2 - 1$, $y = n^2 - n + 1$ and $z = n^2 - 2n$
    Prove that angle $Y = 60°$

Chapter summary

- Sine, cosine and tangent are periodic functions. Their graphs have line and rotational symmetry.
- The sine, cosine and tangent of any angle can be expressed in terms of an acute angle.
- The sign and size of the sine, cosine or tangent of any angle can also be found using a sketch graph of the function.
- The two identities $\sin^2\theta + \cos^2\theta \equiv 1$ and $\tan\theta \equiv \dfrac{\sin\theta}{\cos\theta}$ help you manipulate trigonometric expressions.
- The quadrant diagram can be used to find the sign and size of the sine, cosine or tangent of any angle. The mnemonic **CAST** helps you to remember in which quadrants the trigonometric ratios are positive.
- To solve a trigonometric equation, use identities to simplify it and then use a quadrant diagram or graph to find all possible angles.
- The sine and cosine rules are used to calculate unknown sides and angles in any triangle.
- The sine rule, $\dfrac{a}{\sin A} = \dfrac{b}{\sin B} = \dfrac{c}{\sin C}$, is used when you know a pair of opposite sides and angles.
- The cosine rule, $a^2 = b^2 + c^2 - 2bc\cos A$, is used when you know either two sides and the angle between them or all three sides.
- The area formula for a triangle, Area $= \dfrac{1}{2}ab\sin C$, uses two sides and the included angle.

Check and review

You should now be able to...	Try Questions
✔ Calculate the values of sine, cosine and tangent for angles of any size.	1, 3, 7
✔ Use the two identities $\sin^2\theta + \cos^2\theta \equiv 1$ and $\tan\theta \equiv \dfrac{\sin\theta}{\cos\theta}$, and recognise $x^2 + y^2 = r^2$ as the equation of a circle.	2, 6
✔ Sketch graphs of trigonometric functions and describe their main features.	4–5
✔ Solve various types of trigonometric equations.	8–12
✔ Use the sine and cosine rules and the area formula for a triangle.	13

1 Given that θ is acute, calculate the value of

    a $\sin\theta$ and $\tan\theta$ when $\cos\theta = 0.8$ **b** $\cos\theta$ and $\tan\theta$ when $\sin\theta = \dfrac{5}{13}$

2 Simplify   **a** $\dfrac{\cos\theta\sqrt{1-\cos^2\theta}}{1-\sin^2\theta}$ **b** $\dfrac{\tan\theta(1-\sin^2\theta)}{\cos\theta}$

3 Express, in terms of acute angles,

    a $\sin 190°$ **b** $\tan 260°$ **c** $\cos 140°$ **d** $\tan 318°$ **e** $\sin 371°$

    f $\cos 480°$ **g** $\tan(-150°)$ **h** $\cos(-200°)$ **i** $\sin(-280°)$

4 Find the maximum value of y and the period for these functions, showing your working.

    a $\quad y = 4\sin x$ **b** $\quad y = 5\sin 2x$ **c** $\quad y = 6\cos 5x$

5 Sketch the graphs of $y = \sin x$ and $y = \cos x$ for $0° \le x \le 180°$ on the same axes.

    a Use your graph to solve the equation $\sin x = \cos x$

    b Solve the same equation algebraically to check your solutions.

6   **a** Show that the point $P(2\cos\theta, 2\sin\theta)$ lies on a circle and find its radius.

    b Show that the point $Q(1, \sqrt{3})$ lies on the circle and write the value of θ at Q

7 Use the triangle to write these trig ratios in surd form.

    a $\quad \sin 135°$ **b** $\quad \cos 225°$ **c** $\quad \tan 315°$

    d $\quad \cos 405°$ **e** $\quad \sin(-135°)$ **f** $\quad \tan(-225°)$

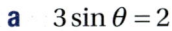

8 Give all the values of θ in the range $-360°$ to $360°$ for which

    a $\quad \cos\theta = 0.7$ **b** $\quad \tan\theta = 2.5$ **c** $\quad \sin\theta = -0.5$

9 Solve these equations for $0° \le \theta \le 360°$, showing your working.

    a $\quad 3\sin\theta = 2$ **b** $\quad 2\tan\theta = 7$ **c** $\quad 2\cos\theta + 1 = 0$ **d** $\quad \cos\theta\tan\theta = -0.5$ **e** $\quad \sin\theta\tan\theta = \dfrac{1}{4}$

10 Solve these equations for $-180° \le \theta \le 180°$. Show your working.

    a $\quad 4\sin\theta = 3\cos\theta$ **b** $\quad 4\sin\theta = 3\tan\theta$ **c** $\quad 3\sin^2\theta = \tan\theta\cos\theta$

    d $\quad \sin(\theta - 20°) = \dfrac{\sqrt{3}}{2}$ **e** $\quad \cos(\theta + 30°) = \dfrac{1}{2}$ **f** $\quad \tan(\theta - 10°) = -1$

11 Solve these equations for $0° \le \theta \le 180°$, showing your working.

    a $\quad 3\sin 2\theta = 1$ **b** $\quad 5\tan 2\theta - 2 = 0$ **c** $\quad 5\sin 3\theta - 1 = 0$

    d $\quad 3\cos 3\theta - 2 = 0$ **e** $\quad 3\sin 2\theta - \cos 2\theta = 0$ **f** $\quad 2\sin\left(\dfrac{1}{2}\theta\right) - \cos\left(\dfrac{1}{2}\theta\right) = 0$

12 Solve these equations for $0° \le \theta \le 360°$

    a $\quad 2\cos^2\theta + \sin\theta = 1$ **b** $\quad \cos^2\theta + \cos\theta = \sin^2\theta$ **c** $\quad 6\sin^2\theta + 5\cos\theta = 5$

    d $\quad \tan^2\theta = 2 + \dfrac{1}{\cos\theta}$ **e** $\quad 1 + \sin\theta\cos^2\theta = \sin\theta$ **f** $\quad 4\sin^2\theta = 2 + \cos\theta$

13 Calculate the side BC, the angle E, and the area of each triangle.

    a $\triangle ABC$ where $AC = 8\,\text{cm}$, Angle $A = 42°$ and Angle $B = 56°$

    b $\triangle DEF$ where $DF = 6\,\text{cm}$, $EF = 11\,\text{cm}$ and Angle $D = 124°$

What next?

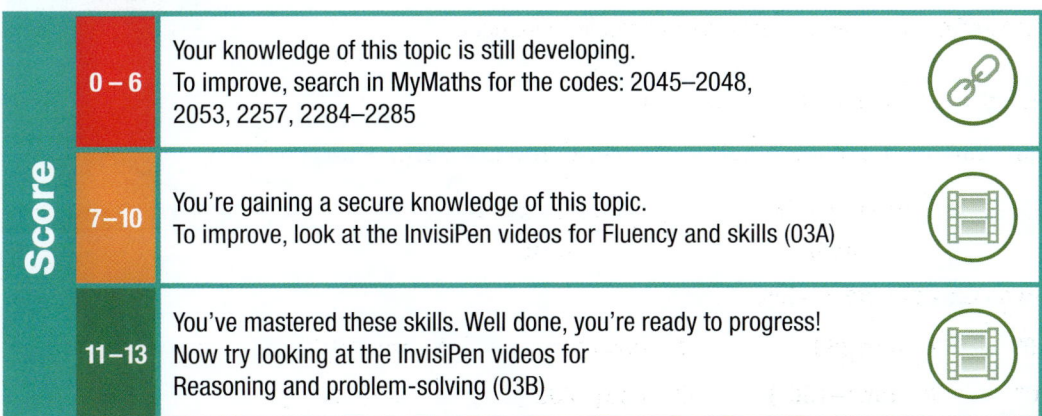

Score			
	0 – 6	Your knowledge of this topic is still developing. To improve, search in MyMaths for the codes: 2045–2048, 2053, 2257, 2284–2285	Click these links in the digital book
	7 – 10	You're gaining a secure knowledge of this topic. To improve, look at the InvisiPen videos for Fluency and skills (03A)	
	11 – 13	You've mastered these skills. Well done, you're ready to progress! Now try looking at the InvisiPen videos for Reasoning and problem-solving (03B)	

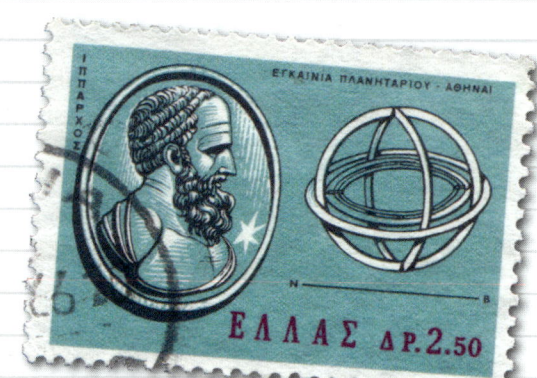

History

Trigonometry, as we know it today, was largely developed between the 16th and 18th centuries. However, the foundations of trigonometry were laid as long ago as the 3rd century BC.

Hipparchus (190 BC – 120 BC), a Greek mathematician and astronomer regarded by many as the founder of trigonometry, constructed the first known trigonometric tables based on the lengths of chords in circles. Contributions to the early development of the theory were made by scholars from a number of countries, including Greece, Turkey, India, Egypt and China.

ICT

Use a spreadsheet to calculate and compare the values of $\sin(x+y)$ and $\sin x \cos y + \cos x \sin y$ for different values of x and y

x	y	$\sin(x+y)$	$\sin x \cos y + \cos x \sin y$

Use a 3D graph plotter to draw the graphs of $\sin(x+y)$ and $\sin x \cos y + \cos x \sin y$
What do you notice about the two graphs?

Have a go

The result $\sin(A+B) = \sin A \cos B + \cos A \sin B$ has been known in various forms since ancient times.

Prove that this result is the case, where A and B are acute angles, by considering the areas of the three triangles shown in the diagram below.

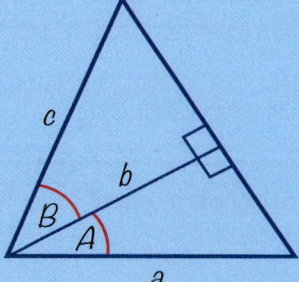

Have a go

Adapt your proof from the other 'Have a go' for the case where one of the angles is acute and the other is obtuse.

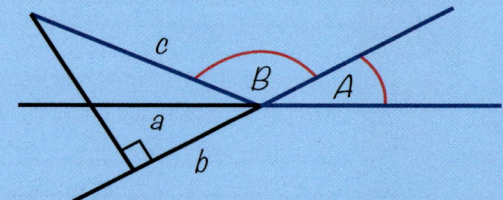

In questions that tell you to show your working, you shouldn't depend solely on a calculator. For these questions, solutions based entirely on graphical or numerical methods are not acceptable.

1 Sketch these pairs of graphs for $-180° \le x \le 180°$

    a $y = \sin x$; $y = \sin 2x$ **[3 marks]**   **b** $y = \tan x$; $y = \tan(x-20°)$ **[4]**

On each diagram, show the coordinates where the curve crosses the x-axis and give the equations of any asymptotes.

2 Solve these equations for $0 \le x \le 360°$. Show your working.

    a $\cos x = 0.2$ **[3]**   **b** $2\tan(x-80°) = 3$ **[4]**

3 $f(x) = \sin(x+45°)$ for $0 \le x \le 360°$

    a Sketch the graph of $y = f(x)$ and label the coordinates of intersection with the axes.   **[3]**

    b Write down the coordinates of the minimum and maximum points in this interval.   **[3]**

    c Solve the equation $\sin(x+45°) = 0.3$ for x in the interval $0° \le x \le 360°$. Show your working.   **[4]**

4 For the curve with equation $y = \cos\dfrac{x}{2}$, give the coordinates of a

    a Maximum point,   **b** Minimum point.   **[3]**

5 The triangle DEF has $DE = 8$ m, $EF = 6$ m and $DF = 7$ m.

    a Calculate the size of $\angle DEF$ **[3]**

    b Calculate the area of the triangle.   **[3]**

6 A triangle has side lengths 19 cm, 13 cm and x and angle 55° as shown.

Calculate the size of x **[3]**

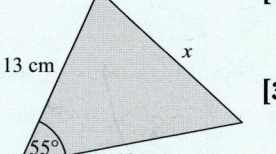

7 An equilateral triangle has area $\sqrt{3}$ square units.

Calculate the side lengths of the triangle.   **[3]**

8 Solve these equations for $-180° \le x \le 180°$. Show your working.

    a $2\sin(x-10) = -0.4$ **[4]**   **b** $\tan 3x = 0.7$ **[6]**

9 The curve C has equation $y = \tan(x-\alpha)$ with $-180° \le x \le 180°$ and $0° < \alpha < 45°$

    a Sketch C and label the points of intersection with the x-axis.   **[3]**

    b Write down the equations of the asymptotes.   **[2]**

    c Solve the equation $\tan(x-\alpha) = \sqrt{3}$ for $-180° \le x \le 180°$. Show your working.   **[4]**

10 $f(x) = k\cos x$ where k is a positive constant.

    a Sketch the graph of $y = f(x)$ for $0 \le x \le 360°$
       Label the points of intersection with the coordinate axes.   **[3]**

    b Given that $k = 3$, solve the equation $k\cos x = \sin x$ for $0° \le x \le 360°$. Show your working.   **[4]**

11 $f(x) = \cos 2x$, $g(x) = 1 - \dfrac{x}{45}$

   a   Sketch $y = f(x)$ and $y = g(x)$ on the same axes for $0° \leq x \leq 360°$ **[5]**

   b   How many solutions are there to $f(x) = g(x)$? Justify your answer.   **[1]**

12 A triangle ABC has $AB = 10\,\text{cm}$, $BC = 16\,\text{cm}$ and $\angle BCA = 30°$

   a   Calculate the possible lengths of AC **[5]**

   b   What is the minimum possible area of the triangle?   **[3]**

13 In the triangle CDE, $CD = 9\,\text{cm}$, $CE = 14\,\text{cm}$ and $\angle CDE = 54°$

   a   Calculate the size of $\angle DEC$ **[3]**

   b   Explain why there is only one possible value of $\angle DEC$ **[2]**

14 In the triangle ABC, $BC = 9\,\text{cm}$, $CA = 14\,\text{cm}$ and $\angle BCA = 46°$

   a   Calculate the length of side AB **[3]**

   b   Calculate the size of the largest angle in the triangle.   **[3]**

15 A triangle has side lengths $12\,\text{cm}$, $8\,\text{cm}$ and $6\,\text{cm}$.

Calculate the size of the largest angle in the triangle.   **[4]**

16 Solve the inequality $\sin x > \dfrac{\sqrt{2}}{2}$ for $0 \leq x \leq 360°$. Show your working.   **[4]**

17 **a**   Show that $\cos\theta + \tan\theta\sin\theta \equiv \dfrac{1}{\cos\theta}$ **[3]**

   b   Hence solve $\cos\theta + \tan\theta\sin\theta = 2.5$ for $-180° \leq \theta \leq 180°$ **[3]**

18 Solve these equations for $0° \leq x \leq 360°$. Show your working.

   a $\sin^2 x = 0.65$ **[5]**

   b $\tan^2 x - 2\tan x - 3 = 0$ **[4]**

19 Solve, for $0 \leq x \leq 180°$, the equation $3\cos^2 2x = 2\sin 2x + 3$. Show your working.   **[8]**

20 $f(\theta) = 2\tan 3\theta + 5\sin 3\theta$

Find all the solutions to $f(\theta) = 0$ in the range $0° \leq \theta \leq 180°$. Show your working.   **[9]**

21 Given that $\sin x = \dfrac{1}{5}$ and x is acute, find the exact value of

   a $\cos x$ **[2]**   **b** $\tan x$ **[3]**

Give your answers in the form $a\sqrt{b}$ where a is rational and b is the smallest possible integer.

22 $2\cos^2\theta - k\sin\theta = 2 - k$

   a   Find the range of values of k for which the equation has no solutions.   **[6]**

   b   Find the solutions in the range $0° \leq \theta \leq 360°$ when $k = 1$. Show your working.   **[4]**

23 Find an expression for $\tan\theta$ in terms of α, given that θ is acute and $\sin\theta = \alpha$ **[3]**

24 Solve the equation $\sin^4 x - 5\cos^2 x + 1 = 0$ for $0° \leq x \leq 360°$. Show your working.   **[8]**

25 The area of an isosceles triangle is $400\,\text{cm}^2$

Calculate the perimeter of the triangle given that one of the angles is $150°$ **[7]**

4 Differentiation and integration

Differentiation enables us to calculate rates of change. This is very useful for finding expressions for displacement, velocity and acceleration. For example, an expression for the displacement of an airplane in the sky informs us of the distance and direction of the airplane from its original position, at a given time. The first derivative of this expression gives the rate of change of displacement, that is, the velocity. The second derivative gives an expression for the rate of change of velocity, that is, the acceleration.

Differentiation and integration (which can be considered as reverse differentiation) belong to a branch of mathematics called calculus. Calculus is the study of change, and it's a powerful tool in modelling real-world situations. Calculus has many applications in a variety of fields including quantum mechanics, thermodynamics, engineering and economics, in modelling growth and movement.

Orientation

What you need to know	What you will learn	What this leads to
Ch1.5 Lines and circles • The equation of a straight line where m is the gradient. • When two lines are perpendicular to each other: $m_1 \times m_2 = -1$	• To differentiate from first principles. • To differentiate terms of the form ax^n • To calculate rates of change. • To work out and interpret equations, tangents, normals, turning points and second derivatives. • To work out the integral of a function, calculate definite integrals and use these to calculate the area under a curve.	**Ch7 Units and kinematics** Velocity and acceleration as rates of change. Acceleration as a second derivative. Area under a velocity-time graph.
Ch2 Polynomials and the binomial theorem • Binomial expansion formula. • Curve sketching.		**Ch15 Differentiation 2** Points of inflection. The product, quotient and chain rules.
		Ch16 Integration and differential equations Integration by parts, by substitution and using partial fractions.

p.22
p.43

 **Practise before you start** 🔍 2002, 2004, 2022, 2041

Differentiation from first principles

Fluency and skills

When looking at the graph of a function, the gradient of its curve at any given point tells you the rate of change. Differentiation from first principles is a method of calculating the gradient and, therefore, the rate of change.

For example, you can work out the gradient of the function $y = x^2$ at the point $P(2, 4)$ using differentiation from first principles.

Define a point Q that lies on the curve, a tiny horizontal distance h from P, so that Q has coordinates $(2 + h, (2 + h)^2)$

PQ is the chord that connects the points.

The gradient of the chord PQ is given by

$$m_{PQ} = \frac{y_Q - y_P}{x_Q - x_P} = \frac{(2+h)^2 - 4}{(2+h) - 2}$$

$$= \frac{4 + 4h + h^2 - 4}{(2+h) - 2} = \frac{4h + h^2}{h}$$

$$= 4 + h$$

As the distance between P and Q becomes very small, h very small and m_{PQ} approaches 4

Gradient at $P = 4$

This method can be generalised for any function.

Consider the graph $y = f(x)$
Let the point P lie on the curve and have x-coordinate x
Its y-coordinate is then $f(x)$
Let the point Q also lie on the curve, h units to the right of P
Its coordinates are therefore $(x + h, f(x + h))$

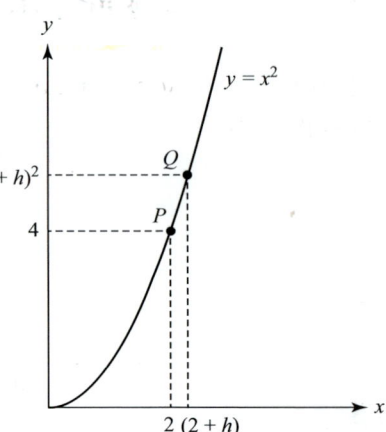

ICT Resource online

To investigate gradients of chords for a graph, click this link in the digital book.

Key point

The **gradient** of PQ is given by

$$m_{PQ} = \frac{y_Q - y_P}{x_Q - x_P} = \frac{f(x+h) - f(x)}{(x+h) - x} = \frac{f(x+h) - f(x)}{h}$$

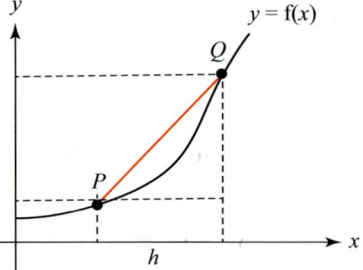

As h approaches 0, the point Q approaches P and the gradient of the chord PQ gets closer to the gradient of the curve at P

The gradient of the curve at P is defined as the **limiting value** of the gradient of PQ as h approaches 0. This limit is denoted by $f'(x)$ and is called the **derived function** or **derivative** of $f(x)$. See p.309 for a list of mathematical notation.

> **Key point**
> $$f'(x) = \lim_{h \to 0} \frac{f(x+h) - f(x)}{h}$$

A limiting value, or **limit**, is a specific value that a function approaches or tends towards. "$\lim_{h \to 0}$" followed by a function means the limit of the function as h tends to zero.

PURE

Differentiation with this method is referred to as finding the derivative from **first principles**.

Example 1

Use differentiation from first principles to work out the derivative of $y = x^2$ and the gradient at the point $(3, 9)$

$f(x) = x^2$

So, $f(x + h) = (x + h)^2$

$f'(x) = \lim_{h \to 0} \dfrac{f(x+h) - f(x)}{h}$

$\quad = \lim_{h \to 0} \dfrac{(x+h)^2 - x^2}{h} = \lim_{h \to 0} \dfrac{x^2 + 2xh + h^2 - x^2}{h}$

Substitute the function $f(x) = x^2$

$\quad = \lim_{h \to 0} (2x + h)$

Expand and simplify the expression.

$\quad = 2x$

The gradient at the point $(3, 9)$ is $f'(3) = 2 \times 3 = 6$

Let h tend towards zero.

Derivatives give the rate of change and **constants** don't change. So a function multiplied by a constant will differentiate to give the derived function multiplied by the *same constant*.

If a function is itself a constant, then the derivative will be zero.

> **Key point**
> For a function $af(x)$, where a is a constant, the derived function is given by $af'(x)$
>
> For a function $f(x) = a$, where a is a constant, the derived function $f'(x)$ is zero

Example 2

Differentiate   **a** $3x^2$ **b** 7

a $f(x) = x^2$, $a = 3$

$3x^2$ is in the form $af(x)$, where a is a constant.

So differentiating $3x^2$:

$3 \times 2x = 6x$

From Example 1, $f'(x) = 2x$

b $f(x) = 7$

So $f'(x) = 0$

$f(x) = a$, where a is a constant.

MyMaths

Q 2028   SEARCH

1 Use the method shown in Example 1 to work out the gradient of these functions at the points given.

   a $y = x^2$ at $x = 3$ **b** $y = x^2$ at $x = 4$

   c $y = 2x^2$ at $x = 2$ **d** $y = 5x^2$ at $x = 1$

   e $y = \frac{1}{2}x^2$ at $x = 4$ **f** $y = \frac{3}{4}x^2$ at $x = 10$

   g $y = x^3$ at $x = 1$ **h** $y = 3x^3$ at $x = 2$

2 Use the method of differentiation from first principles to work out the derivative and hence the gradient of the curve.

   a $y = x^2$ at the point $(1, 1)$

   b $y = 3x^2$ at the point $(2, 12)$

   c $y = x^3$ when $x = 3$

   d $y = 4x - 1$ at the point $(5, 19)$

   e $y = \frac{1}{2}x^2$ at the point $(6, 18)$

   f $y = x^2 + 1$ at the point $(2, 5)$

   g $y = 2x^3$ at the point $(1, 2)$

   h $y = x^3 + 2$ at the point $(2, 10)$

3 Use differentiation from first principles to work out the gradient of the tangent to

   a $y = 2x^3$ at the point where $x = 2$

   b $y = 3x^2 + 2$ at $(3, 29)$

   c $y = \frac{x^2}{2}$ at the point $(4, 8)$

   d $y = \frac{1}{2}x^2 - 4$ at the point $(2, -2)$

   e $y = 2x^2 - 1$ at the point $(-2, 7)$

   f $y = x^2 + x$ at the point where $x = 1$

   g $y = x^3 + x$ at the point $(2, 10)$

4 Work out, from first principles, the derived function when

   a $f(x) = 2x^2$ **b** $f(x) = 4x^2$

   c $f(x) = 6x^2$ **d** $f(x) = \frac{1}{2}x^2$

   e $f(x) = x^2 + 1$ **f** $f(x) = 2x - 3$

   g $f(x) = \frac{1}{3}x^3$ **h** $f(x) = 2x^3 + 1$

5 **a** Work out, from first principles, the derived function where

   i $f(x) = x + x^2$

   ii $f(x) = x^2 + x + 1$

   iii $f(x) = x^2 + x - 5$

   iv $f(x) = x^2 + 2x + 3$

   v $f(x) = x^2 - 3x - 1$

   vi $f(x) = 2x^2 + 5x - 3$

   b Work out, from first principles, the derived function of

   i $f(x) = 6$ **ii** $f(x) = 0$

   iii $f(x) = -2$ **iv** $f(x) = \pi$

   c Work out, from first principles, the derived function of

   i $f(x) = x$ **ii** $f(x) = -x$

   iii $f(x) = 2x + 1$ **iv** $f(x) = 4 - 3x$

6 Differentiate

   a $y = 2x^3 - 3$ **b** $y = 3x^4 - 2x^2$

   c $y = 2x^5$ **d** $y = 1 - x^3$

   e $y = x^2 - 2x^3$ **f** $y = x - x^2 + x^3$

Reasoning and problem-solving

Strategy

To solve problems involving differentiation from first principles

(1) Substitute your function into the formula $f'(x) = \lim_{h \to 0} \dfrac{f(x + h) - f(x)}{h}$

(2) Expand and simplify the expression.

(3) Let h tend towards 0 and write down the limit of the expression, $f'(x)$

(4) Find the value of the gradient at a point (a, b) on the curve by evaluating $f'(a)$

Example 3

At which point on the curve $y = 2x^3$ is the gradient equal to 24?

$f(x) = 2x^3$ so, $f'(x) = \lim\limits_{h \to 0} \dfrac{2(x+h)^3 - 2x^3}{h}$

$= \lim\limits_{h \to 0} \dfrac{2x^3 + 6x^2h + 6xh^2 + 2h^3 - 2x^3}{h} = \lim\limits_{h \to 0} (6x^2 + 6xh + 2h^2)$

$f'(x) = 6x^2$

$f'(x) = 24$ So, $6x^2 = 24$ $x = \pm 2$

If $x = \pm 2$ then $y = f(\pm 2) = 2 \times (\pm 2)^3 = \pm 16$

The gradient is 24 at the points $(2, 16)$ and $(-2, -16)$

1. Apply the algebra.

2. Find the derivative.

3. Use the derivative and the initial conditions to form an equation.

4. Solve the equation you formed.

Exercise 4.1B Reasoning and problem-solving

1 At which point on the curve $y = 5x^2$ does the gradient take the value given? Show your working.

  a 20

  b 100

  c 0.5

  d Half of what it is at $(2, 20)$

  e A third of what it is at $(3, 45)$

  f Four times what it is at $(1, 5)$

2 At which **two** points on the curve $y = x^3$ does the gradient equal the value given? Show your working.

  a 3   **b** 12   **c** 27

  d 0.03   **e** $\dfrac{1}{3}$ **f** 1.47

3 For what value of x will these pairs of curves have the same gradient? Show your working.

  a $y = x^2$ and $y = 2x$

  b $y = 2x^2$ and $y = 12x$

  c $y = 3x^2$ and $y = 15x$

  d $y = ax^2$ and $y = bx$ where a and b are constants.

4 For what value(s) of x will these pairs of curves have the same gradient? Show your working.

  a $y = x^3$ and $y = x^2 + 5x$

  b $y = x^3$ and $y = 3x^2 + 9x$

  c $y = x^3$ and $y = 2x^2 - x$

  d $y = x^3$ and $y = 4x^2 - 4x$

  e $y = x^2 + 3x + 1$ and $y = 7x + 1$

  f $y = 2x^2 + x - 4$ and $y = 11x + 2$

  g $y = x^2 - x$ and $y = 5x - 2$

  h $y = 2x^3 + 3x^2$ and $y = 12x$

5 **a** Consider the derivatives of 1, x, x^2, x^3, x^4, and hence suggest a general rule about the derivative of x^n

  b Consider the derivatives of 2, $4x$, $3x^2$, $5x^3$, $-2x^4$ and hence suggest a general rule about the derivative of ax^n

  c Test your rule by considering the derivative of $2x^3$ and $3x^2$

  d Can you say you have proved your rule?

Challenge

6 Can you show from first principles that

  a The curve $y = \dfrac{1}{x}$ has a gradient of -4 when $x = \dfrac{1}{2}$

  b The derivative of $\dfrac{1}{x}$ is $-\dfrac{1}{x^2}$?

Fluency and skills

You can use a simple rule to differentiate functions in the form ax^n

Key point

If $f(x) = ax^n$ then $f'(x) = nax^{n-1}$

If a function is a sum of two other functions, you can differentiate each function one at a time and then add the results.

Key point

If $h(x) = f(x) + g(x)$ then $h'(x) = f'(x) + g'(x)$

Isaac Newton is credited with having come up with the idea of Calculus first, but **Gottfried Leibniz**, a German mathematician and contemporary of Newton, also developed the concept and devised an alternative notation which is commonly used.

See p.614

For a list of mathematical notation.

For $y_Q - y_P$ he used the symbol δy and for $x_Q - x_P$ he used the symbol δx

So $\dfrac{y_Q - y_P}{x_Q - x_P} = \dfrac{\delta y}{\delta x}$ and he wrote that $\lim\limits_{\delta x \to 0} \dfrac{\delta y}{\delta x} = \dfrac{dy}{dx}$

Key point

If $y = f(x)$ then $\dfrac{dy}{dx} = f'(x)$

Differentiating a function, or finding $\dfrac{dy}{dx}$, gives a formula for the gradient of the graph of the function at a point. This is also the gradient of the tangent to the curve at this point.

Try it on your calculator

You can use a calculator to evaluate the gradient of the tangent to a curve at a given point.

$$\frac{d}{dx}(5X^2 - 2X)\Big|_{x=3}$$

28

Activity

Find out how to calculate the gradient of the tangent to the curve $y = 5x^2 - 2x$ where $x = 3$ on *your* calculator.

Example 1

Differentiate $y = 3x^5 + 4x^2 + 2x + 3$

$y = 3x^5 + 4x^2 + 2x + 3 = 3x^5 + 4x^2 + 2x^1 + 3x^0$ Write each term in the form ax^n

$\dfrac{dy}{dx} = 5 \times 3x^4 + 2 \times 4x^1 + 1 \times 2x^0 + 0 \times 3x^{-1}$ Use $f'(x) = nax^{n-1}$ on each term.

$\dfrac{dy}{dx} = 15x^4 + 8x + 2$

Example 2

Work out the derived function when $f(x) = 4 + \dfrac{3}{x} + \sqrt{x} + 2x^7$

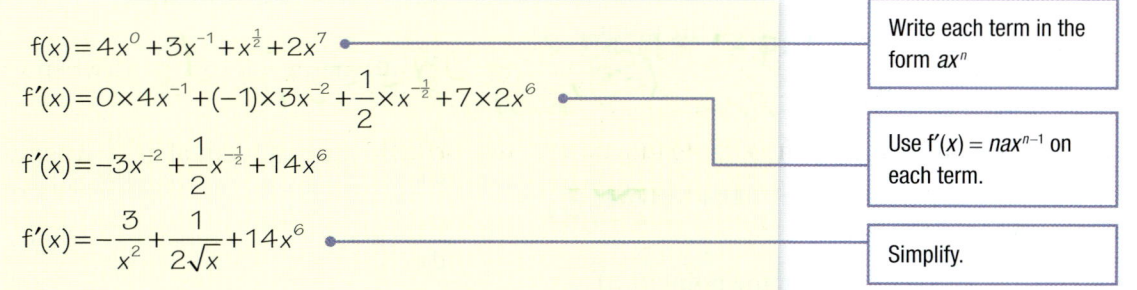

$f(x) = 4x^0 + 3x^{-1} + x^{\frac{1}{2}} + 2x^7$ — Write each term in the form ax^n

$f'(x) = 0 \times 4x^{-1} + (-1) \times 3x^{-2} + \dfrac{1}{2} \times x^{-\frac{1}{2}} + 7 \times 2x^6$ — Use $f'(x) = nax^{n-1}$ on each term.

$f'(x) = -3x^{-2} + \dfrac{1}{2}x^{-\frac{1}{2}} + 14x^6$

$f'(x) = -\dfrac{3}{x^2} + \dfrac{1}{2\sqrt{x}} + 14x^6$ — Simplify.

Exercise 4.2A Fluency and skills

1 Differentiate

  a $3x^7$ **b** $6x^4$ **c** $5x$

  d 8 **e** -4 **f** $\dfrac{1}{3}$

  g $-2x^5$ **h** $-x^7$ **i** $2x^{-1}$

  j $-x^{-3}$ **k** $x^{\frac{1}{2}}$ **l** $x^{\frac{2}{3}}$

  m $5x^{\frac{5}{3}}$ **n** $-x^{-\frac{1}{4}}$ **o** 0

2 Work out the derivative of

  a $\sqrt{x}$ **b** $\sqrt[3]{x}$ **c** $\sqrt[5]{x}$

  d $\sqrt{x^3}$ **e** $\sqrt[3]{x^2}$ **f** $\sqrt[3]{3x}$

  g $\dfrac{1}{x}$ **h** $\dfrac{3}{x^2}$ **i** $-\dfrac{6}{x^4}$

  j $\dfrac{1}{\sqrt{x}}$ **k** $\dfrac{3}{\sqrt{x^3}}$ **l** $\dfrac{1}{\sqrt{5}}$

  m π

3 Calculate $\dfrac{dy}{dx}$ when $y =$

  a $x^2 + 2x - 3$ **b** $1 - 2x - 5x^2$

  c $x^3 + 2x^2 - 3x + 1$ **d** $x^2 - x^4 + \pi$

  e $x - \dfrac{1}{x}$ **f** $3 + x + \dfrac{2}{x^2}$

  g $x^3 - \dfrac{1}{x^3} + \dfrac{3}{x} + 5$ **h** $\dfrac{10}{x} - 1 - \dfrac{x}{10}$

  i $\sqrt{x} + \dfrac{1}{\sqrt{x}}$ **j** $x^5 - \dfrac{5}{\sqrt[3]{x}}$

  k $3 + \dfrac{2}{\sqrt{x}} - \dfrac{5}{\sqrt[3]{x^2}}$ **l** $\dfrac{3}{\pi} - \dfrac{2}{x^2} - x^{\frac{3}{2}}$

4 For each question part **a** to **d**

  i Find an expression in x for the gradient function,

  ii Find the value of the gradient at the given point.

  a Given that $f(x) = 3x^2 + 4x - 6$ work out the value of $f'(-2)$

  b Given that $y = 2x^3 - 5x^2 - 1$ work out the value of $\dfrac{dy}{dx}$ when $x = 1$

  c A curve has equation $y = 10x + \dfrac{8}{x}$

  Calculate the gradient of the curve at the point where $x = 2$

  d Calculate the gradient of the curve $y = \dfrac{8}{x} + \dfrac{4}{x^2}$ at the point $(2, 5)$

  e i Expand $y = x(x - 1)$

   ii Hence evaluate $\dfrac{dy}{dx}$ when $x = 4$ and $y = x(x-1)$

   iii Hence state the gradient of the tangent to the curve at $(4, 12)$

5 Write an expression in x for $\dfrac{dy}{dx}$ and thus calculate the gradient of the tangent to each curve at the point given.

  a $y = 2x^2 - 5x + 1$ at $(1, -2)$

  b $y = 1 - 5x + \dfrac{10}{x}$ at $(2, -4)$

  c $y = x(2x + 1)$ at $(-3, 15)$

  d $y = \sqrt{x} + \dfrac{2}{\sqrt{x}}$ at $(4, 3)$

6 For each pair of functions, find which has the greater gradient at the given point.

a $y = x^2$ and $y = 20 - x$ at the point $(4, 16)$

b $y = x^2 + 3x - 8$ and $y = 6 - 2x$ at the point $(2, 2)$

c $y = 2x^2 + 13x - 18$ and $y = 2x + 3$ at the point $(-7, -11)$

d $y = 3x^2 - 5x - 2$ and $y = x^2 - 2x + 3$ at the point $(-1, 6)$

e $y = \sqrt{x}$ and $y = 2x - 15$ at the point $(9, 3)$

7 A curve is given by $y = x^3 + 5x^2 - 8x + 1$
Which of the following statements are true?

a $\dfrac{dy}{dx} > 4$ when $x = 1$ **b** $\dfrac{dy}{dx} < 0$ when $x = 0$

c $\dfrac{dy}{dx} = 0$ when $x = 4$ **d** $\dfrac{dy}{dx} > 0$ when $x = -2$

e $\dfrac{dy}{dx}$ when $x = -4$ is equal to $\dfrac{dy}{dx}$ when $x = \dfrac{2}{3}$

f $\dfrac{dy}{dx}$ when $x = -1$ is equal to $\dfrac{dy}{dx}$ when $x = 1$

Reasoning and problem-solving

Strategy

To solve differentiation problems involving polynomials with rational powers

1. Use the laws of algebra to make your expression the sum of terms of the form ax^n

2. Apply $f(x) = ax^n \Rightarrow f'(x) = nax^{n-1}$ to each term to find the derivative.

3. Substitute any numbers required and answer the question.

Example 3

Given that $f(x) = \dfrac{x+1}{\sqrt{x}}$, find the expression $f'(x)$ and hence find $f'(4)$

$f(x) = \dfrac{x+1}{\sqrt{x}}$

$= \dfrac{x}{x^{\frac{1}{2}}} + \dfrac{1}{x^{\frac{1}{2}}}$

$f(x) = x^{\frac{1}{2}} + x^{-\frac{1}{2}}$

$f'(x) = \dfrac{1}{2}x^{-\frac{1}{2}} - \dfrac{1}{2}x^{-\frac{3}{2}}$

$f'(x) = \dfrac{1}{2x^{\frac{1}{2}}} - \dfrac{1}{2x^{\frac{3}{2}}}$

$f'(x) = \dfrac{1}{2\sqrt{x}} - \dfrac{1}{2(\sqrt{x})^3}$

$f'(4) = \dfrac{1}{2\sqrt{4}} - \dfrac{1}{2(\sqrt{4})^3}$

$f'(4) = \dfrac{1}{4} - \dfrac{1}{16}$

$= \dfrac{3}{16}$

Write in index form. **1**

Divide by $x^{\frac{1}{2}}$ to make your expression a sum of terms of the form ax^n

Apply $f'(x) = nax^{n-1}$ to each term. **2**

Substitute $x = 4$ **3**

Exercise 4.2B Reasoning and problem-solving

1 Work out the derivative with respect to x of

a $f(x) = x(x+1)$

b $g(x) = (x-1)(x+1)$

c $h(x) = x^2(1-3x)$

d $k(x) = \sqrt{x}(x+3)$

e $m(x) = 3x(2x^2 + x - 3)$

f $n(x) = x(\sqrt{x}+1)$

g $p(x) = x^{-1}(4+2x)$

h $q(x) = (x^{-1} - 2)(x+1)$

i $r(x) = \dfrac{1}{4x^{\frac{1}{2}}}(x^2 + 3x + 1)$

2 Given y, find $\dfrac{dy}{dx}$

a $y = x$

b $y = x + \sqrt{x}$

c $y = \dfrac{x + \sqrt{x}}{x}$

d $y = \dfrac{1 - x - 2\sqrt{x}}{x}$

e $y = \dfrac{x - x^2}{\sqrt{x}}$

f $y = \dfrac{1 + 2\sqrt{x}}{x}$

g $y = \dfrac{x^2 + 3x - 1}{3x^{\frac{1}{2}}}$

h $y = \dfrac{1}{x} + \dfrac{1}{x^2}$

i $y = \dfrac{1}{\sqrt[3]{x}}$

3 Work out the derivative with respect to x of each of these functions.

a $f(x) = (x - 3)(x + 1)$

b $g(x) = (x - 4)(x - 5)$

c $h(x) = (1 - x)(3 + x)$

d $j(x) = (x + 1)(2x + 1)$

e $k(x) = x(x + 1)(x - 1)$

f $m(x) = (x + 1)(\sqrt{x} + 1)$

4 The sketch shows part of the curve $y = (x - 1)(x - 7)$ near the origin.

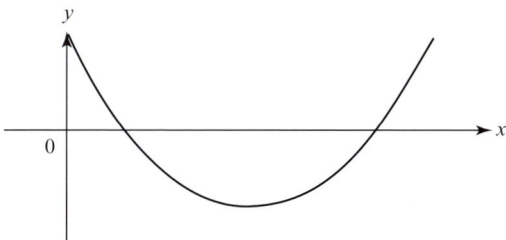

a Work out $\dfrac{dy}{dx}$

b Identify where the curve crosses the x-axis.

c Work out the point where the gradient of the curve is zero.

d **i** Where is $\dfrac{dy}{dx} < 0$?

   ii Where is $\dfrac{dy}{dx} > 0$?

e Choose the correct word to make each statement true.

   i When $\dfrac{dy}{dx} < 0$, the curve is (rising/ falling) from left to right.

   ii When $\dfrac{dy}{dx} > 0$, the curve is (rising/ falling) from left to right.

5 The sketch shows the curve $y = \dfrac{10x^2 + 40}{x}$, $0.5 \le x \le 5.5$

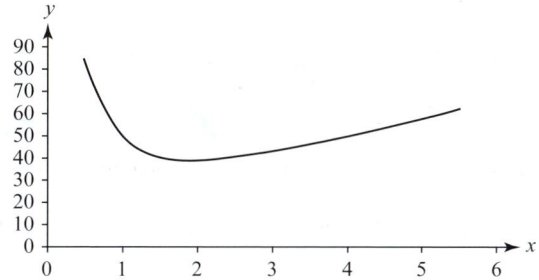

a Work out $\dfrac{dy}{dx}$

b **i** For which value of x is the gradient of the curve zero? Show your working.

   ii What is the value of y at this point?

c For which values of x is the curve

   i Decreasing,   **ii** Increasing?

See Ch2.2
For a reminder on expanding brackets and the binomial theorem.

Challenge

6 Prove that the gradient of the function $f(x) = (x + 1)^4$ at the point $(1, 16)$ is 32

7 **a** Differentiate the function $f(x) = (x - 1)(x + 2)$

b Work out the gradient of the curve $y = f(x)$ at $x = 5$

c At which point is the gradient equal to zero?

d A line with equation $y = 2x - k$, where k is a constant, is a tangent to the curve.

   i At what point does the tangent touch the curve?

   ii What is the value of k?

PURE

Fluency and skills

When y is a function of x, the gradient of a graph of y against x tells you how the y-measurement is changing per unit x-measurement.

> **Key point**
>
> The rate of change of y with respect to x can be written $\dfrac{dy}{dx}$

See Ch7.2 For more on distance–time graphs.

The gradient of a distance-time graph is a measure of the rate of change of distance (r) with respect to time (t), this is called **velocity** (v).

> **Key point**
>
> $v = \dfrac{dr}{dt}$

If v metres per second represents velocity and t seconds represents time, then the gradient, $\dfrac{dv}{dt}$, is a measure of the rate at which velocity is changing with time, in metres per second per second.

> **Key point**
>
> The rate of change of velocity is called **acceleration**, $a = \dfrac{dv}{dt}$

See Ch7.4 For more on acceleration as a derivative.

Acceleration is the derivative of a derivative, which is called the **second derivative**.
A similar notation is used for the second derivative, $f''(x)$, as for the first derivative, $f'(x)$

> **Key point**
>
> $y = f(x) \Rightarrow \dfrac{dy}{dx} = f'(x) \Rightarrow \dfrac{d^2y}{dx^2} = f''(x)$

Example 1

A particle is moving on the y-axis such that its distance, r cm, from the origin is given by $r = t^3 + 2t^2 + t$, where t is the time measured in seconds.

a Use the fact that the velocity $v = \dfrac{dr}{dt}$ to find an expression for the particle's velocity.

b Use the fact that the acceleration $a = \dfrac{dv}{dt} = \dfrac{d^2r}{dt^2}$ to find an expression for the particle's acceleration.

a $r = t^3 + 2t^2 + t$ $\dfrac{dr}{dt} = 3t^2 + 4t + 1$ ●————— Use $f'(t) = nat^{n-1}$

b $\dfrac{dr}{dt} = 3t^2 + 4t + 1$ $\dfrac{d^2r}{dt^2} = 6t + 4$

By differentiating a function and sketching a graph of the derivative, you can find out whether

the function is increasing $\left(\dfrac{dy}{dx} > 0\right)$,

the function is decreasing $\left(\dfrac{dy}{dx} < 0\right)$,

or the function is neither increasing nor decreasing $\left(\dfrac{dy}{dx} = 0\right)$, in which case you call it **stationary**.

> A positive gradient means an increasing function. A negative gradient means a decreasing function.

Example 2

Consider the curve $y = 2x^3 - 3x^2 - 36x + 2$

a Work out $\dfrac{dy}{dx}$

b Use your equation from part **a** to sketch a graph of $\dfrac{dy}{dx}$. You must show your working.

c Using your sketch, determine where the value of y is increasing and decreasing.

d Work out where the value of y is stationary.

> **See Ch2.4**
> For a reminder on curve sketching.

a $y = 2x^3 - 3x^2 - 36x + 2$

$\dfrac{dy}{dx} = 6x^2 - 6x - 36$

> Differentiate to find the gradient function.

b $\dfrac{dy}{dx} = 6(x^2 - x - 6)$

$= 6(x-3)(x+2)$

> Factorise.

When $x = 0$, $\dfrac{dy}{dx} = 6(-3)(2) = -36$

When $\dfrac{dy}{dx} = 0$, $x = 3$ or -2

> Determine where the gradient function crosses the axes.

> Use this information to sketch a graph. You can also use your calculator to sketch the function.

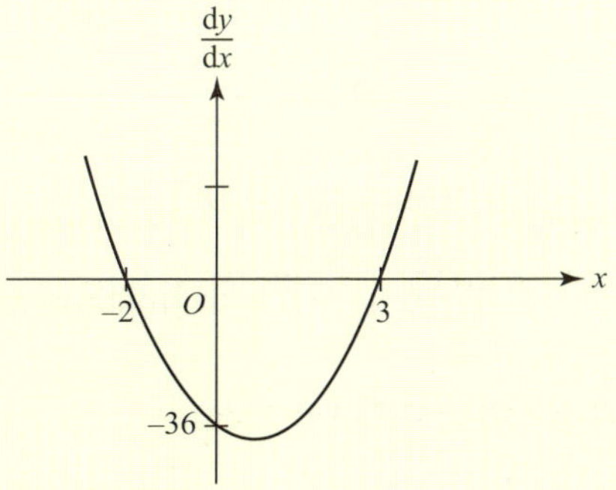

c From the sketch, $\dfrac{dy}{dx} > 0$ (it's increasing) when $x < -2$ or when $x > 3$

From the sketch, $\dfrac{dy}{dx} < 0$ (it's decreasing) when $-2 < x < 3$

> Interpret the graph.

d The value of y will be neither increasing nor decreasing (it is stationary) at both $x = -2$ and $x = 3$

Calculator

Try it on your calculator

You can use a graphics calculator to sketch a function and its derivative.

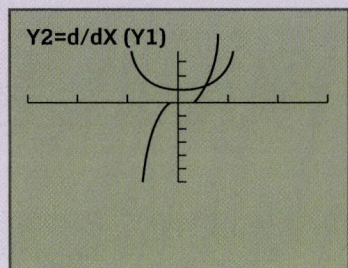

Y2=d/dX (Y1)

Activity

Find out how to sketch the curve $f(x) = 3x^4 - x^3 + 1$ and its derivative $f'(x)$ on *your* graphics calculator.

Exercise 4.3A Fluency and skills

1   Calculate the rate of change of the following functions at the given points. You must show all your working.

a $f(x) = x^2 + 10x$ at $x = 4$

b $g(x) = x^3 + 2x^2 + x + 1$ at $x = -1$

c $h(x) = 5x + 6$ at $x = 1$

d $k(x) = x + \dfrac{1}{x}$ at $x = 3$

e $m(x) = 9x^2 + 6x + 1$ at $x = 1$

f $n(x) = x^{-1} + x^{-2}$ at $x = -2$

g $p(x) = \dfrac{1}{\sqrt{x}}$ at $x = \dfrac{1}{9}$

h $q(x) = x^{\frac{3}{2}} + 1$ at $x = 36$

i $r(x) = x^4 - 8x^2$ at $x = -2$

j $s(x) = \sqrt{x} - x$ at $x = 4$

k $t(x) = 2\pi$ at $x = 7$

l $u(x) = 4 - 3x$ at $x = -10$

m $v(x) = \dfrac{1}{2x^4}$ at $x = -1$

n $w(x) = 1 - 3x - 2x^2$ at $x = 0$

o $y(x) = \dfrac{162}{x^2} + 2x^2$ at $x = 3$

p $z(x) = 20\sqrt{x} + \dfrac{1000}{\sqrt{x}}$ at $x = 25$

2   Work out the rate of change of the rate of change, $\left(\dfrac{d^2 y}{dx^2}\right)$, of the following functions at the given points. You must show all your working.

a $y = x^3 + x$ at $x = 3$

b $y = \dfrac{10}{x}$ at $x = 2$

c $y = \dfrac{1}{\sqrt{x}}$ at $x = 1$

d $y = x^4 - x^2$ at $x = -2$

e $y = x^2 - \dfrac{4}{x}$ at $x = 2$

f $y = x^3 + 4x^2 + 3x$ at $x = -1$

g $y = \sqrt{x}$ at $x = 0.25$

h $y = 3x - 4\sqrt{x}$ at $x = 0.16$

i $y = 4 - ax$, where a is a constant, at $x = -3$

j $y = x + 2 + \dfrac{1}{x}$ at $x = \dfrac{1}{2}$

k $y = \dfrac{1}{x^4}$ at $x = 0.5$

l $y = 1 + x + \dfrac{x^2}{2} + \dfrac{x^3}{6} + \dfrac{x^4}{24} + \dfrac{x^5}{120}$ at $x = -2$

3   Using the gradient function of each curve, determine where the curve is

i   Stationary,

ii   Increasing,

iii   Decreasing.

a $y = x^2 + 4x - 12$

b $y = 3 - 5x - 2x^2$

c $y = x^2 - 1$

d $y = \dfrac{x^3}{3} - 4x + 1$

e $y = 1 + 21x - 2x^2 - \dfrac{1}{3}x^3$

f $y = \dfrac{x^3}{3} - \dfrac{3x^2}{2} + 2x - 1$

g $y = \dfrac{2}{3}x^{\frac{3}{2}}, x > 0$

h $y = 1 - \dfrac{1}{x}, x > 0$

i $y = \dfrac{x^3}{3}$

j $y = \dfrac{x^4}{4}$

4 By finding expressions for $\dfrac{dy}{dx}$, determine which of each pair of functions has the greater rate of change with respect to x at the given x-value.

a $y = x^2$ and $y = -x^2$ at $x = -1$

b $y = 5 - 3x$ and $y = \dfrac{1}{x}$ at $x = 0.5$

c $y = x^2 - 2$ and $y = x^2 - x - 2$ at $x = 3$

d $y = 1 + 10x - x^3$ and $y = x - x^2$ at $x = 2$

Reasoning and problem-solving

Strategy

To answer a problem involving rate of change

1. Read and understand the context, identifying any function or relationship.
2. Differentiate to get a formula for the relevant rate of change.
3. Evaluate under the given conditions.
4. Apply this to answer the initial question, being mindful of the context and units.

Example 3

A conical vat is filled with water. The volume in the vat at time t seconds is V ml.

V is a function of t such that the volume at time t is found by multiplying the cube of t by 50π

a Work out $V'(t)$, the rate at which the vat is filled, in millilitres per second.

b Calculate the rate at which the vat is filling with water after 3 seconds.

a $V(t) = 50\pi t^3$

$V'(t) = 150\pi t^2$

b $V'(3) = 150\pi \times 3^2$

$\quad\quad = 1350\pi$

$V'(3) \approx 4241$

After 3 seconds, the vat is filling at the rate of 4241 ml s^{-1} (4 sf)

① Identify the function.

② Differentiate to get a formula for the relevant rate of change.

③ Evaluate under the required conditions.

④ Answer the question in context.

MyMaths 🔍 2269, 2270 SEARCH

Example 4

An object is thrown into the air. Its height after t seconds is given by $h = 1 + 30t - 5t^2$ where h is its height in metres.

a Write down an expression for the rate at which the object is climbing, in metres per second.

b Work out

   i When the object is rising,

   ii When the object is falling.

c After how many seconds does the object reach its maximum height?

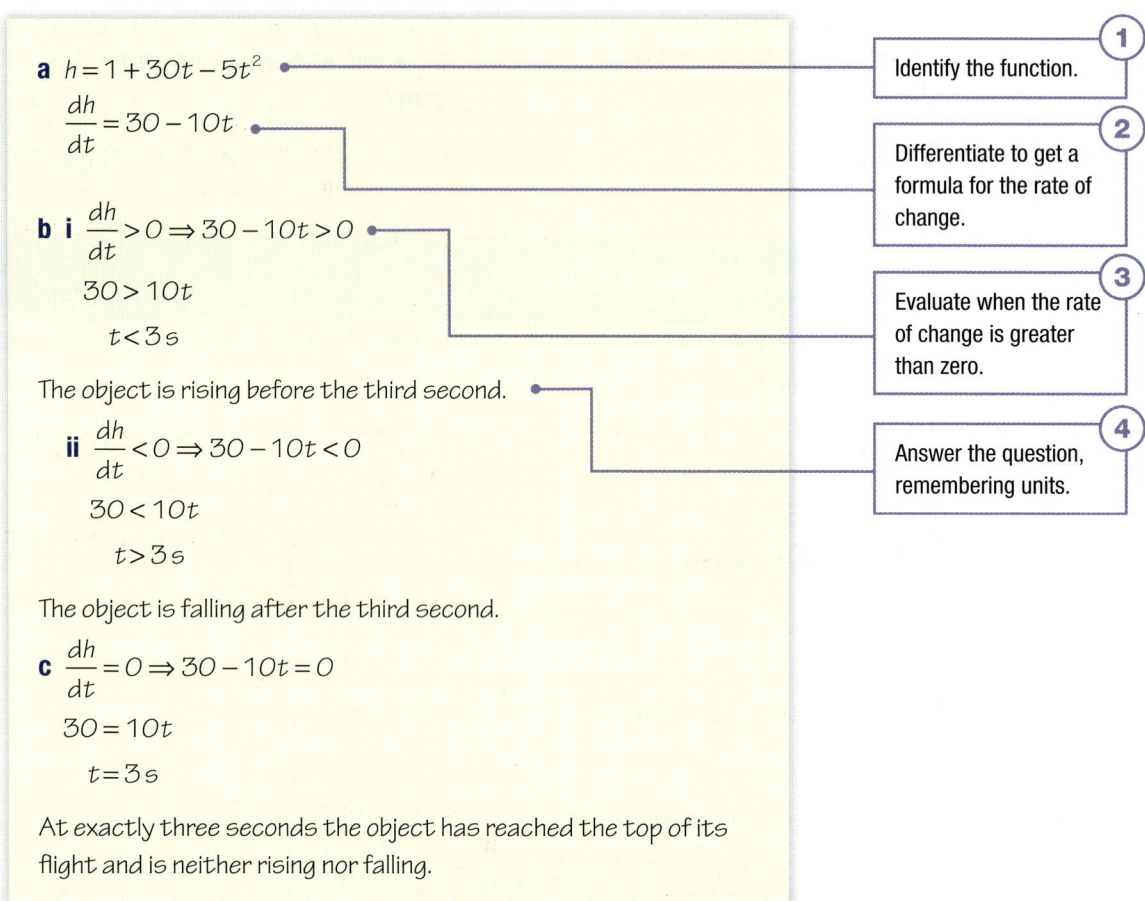

a $h = 1 + 30t - 5t^2$ ① Identify the function.

$\dfrac{dh}{dt} = 30 - 10t$ ② Differentiate to get a formula for the rate of change.

b i $\dfrac{dh}{dt} > 0 \Rightarrow 30 - 10t > 0$ ③ Evaluate when the rate of change is greater than zero.

$30 > 10t$

$t < 3\,s$

The object is rising before the third second.

ii $\dfrac{dh}{dt} < 0 \Rightarrow 30 - 10t < 0$ ④ Answer the question, remembering units.

$30 < 10t$

$t > 3\,s$

The object is falling after the third second.

c $\dfrac{dh}{dt} = 0 \Rightarrow 30 - 10t = 0$

$30 = 10t$

$t = 3\,s$

At exactly three seconds the object has reached the top of its flight and is neither rising nor falling.

Exercise 4.3B Reasoning and problem-solving

1 A taxi driver charges passengers according to the formula $C(x) = 2.5x + 6$, where C is the cost in £ and x is the distance travelled in km.

Work out the rate in £ per km that the taxi driver charges.

2 A block of ice melts so that its dry mass M, at a time t minutes after it has been removed from the freezer, is given by $M(t) = 500 - 7.5t$

a Work out $\dfrac{dM}{dt}$, the rate at which the mass is changing.

b Your answer should have a negative sign in it. Explain why this is.

3 The volume of a sphere is $V = \frac{4}{3}\pi r^3$

The surface area of a sphere is $S = 4\pi r^2$

A spherical bubble is expanding.

a Find an expression for the rate of change of the volume as r increases.

b Calculate the rate of change of the volume of the bubble when its radius is $3\,\text{cm}$.

c Calculate the rate of change of the surface area with respect to the radius when the radius is $3\,\text{cm}$.

4 A parabola has equation $y = 35 - 2x - x^2$

a Work out the rate of change of y with respect to x when x is equal to

   i -2 **ii** 2 **iii** 5

b When will this rate of change be zero?

5 A skydiver jumps from an ascending plane. His height, $h\,\text{m}$ above the ground, is given by $h = 4000 + 3t - 4.9t^2$, where t seconds is the time since leaving the plane.

a Work out $\frac{dh}{dt}$, the rate at which the skydiver is falling.

b How fast is he falling after 5 seconds?

c How fast is he falling after 10 seconds?

d Calculate his acceleration at this time.

6 A golf ball was struck on the moon in 1974. Its height, $h\,\text{m}$, is modelled by $h = 10t - 1.62t^2$, where t seconds is the time since striking the ball.

a Calculate $\frac{dh}{dt}$, the rate at which the height of the ball is changing.

b After how much time will the ball be falling?

c Calculate the acceleration of the ball in the gravitational field of the moon.

7 A cistern fills from empty. A valve opens and the volume of water, $V\,\text{ml}$, in the cistern t seconds after the valve opens is given by $V = 360t - 6t^2$

a Write down an expression for the rate at which the cistern is filling after t seconds.

b Calculate the rate at which the cistern is filling after

   i 10 seconds,   **ii** 20 seconds.

c When the rate is zero, a ballcock shuts off the valve. At what time does this occur?

d What volume of water is in the tank when the valve closes?

8 For each function

i Work out an expression for the rate of change of y with respect to x

ii Evaluate this rate of change when $x = 0$

iii By expressing the rate of change in the form $\frac{dy}{dx} = (x+a)^2 + b$, establish that each function is increasing for all values of x

a $y = \frac{x^3}{3} + 5x^2 + 30x + 1$

b $y = \frac{x^3}{3} - 3x^2 + 12x - 4$

c $y = \frac{x^3}{3} - \frac{5}{2}x^2 + 8x + \frac{1}{2}$

9 The derivative of a function is $\frac{dy}{dx} = 8x - x^2 - 17$

Show that the function is always decreasing.

Challenge

10 A curve has equation $y = x^4 + \frac{1}{x^2}, x \geq 1$

a Work out **i** $\frac{dy}{dx}$ **ii** $\frac{d^2y}{dx^2}$

b Show that the gradient of the curve is an increasing function.

Tangents and normals

Fluency and skills

See Ch1.5
For a reminder on the equation of a straight line.

When lines with gradient m_1 and m_2 are **perpendicular** to each other, $m_1 \times m_2 = -1$

Key point

$m_1 = -\dfrac{1}{m_2}$ for perpendicular lines.

The **tangent** to the curve $y = f(x)$, which touches the curve at the point $(x, f(x))$, has the same gradient as the curve at that point, giving $m_T = f'(x)$

The **normal** to the curve $y = f(x)$, which passes through the point $(x, f(x))$, is perpendicular to the tangent at that point.

giving $m_N = -\dfrac{1}{m_T} = -\dfrac{1}{f'(x)}$

A line with gradient m passing through the point (a, b) has equation
$(y - b) = m(x - a)$

Example 1

A curve has equation $y = 2x^2 - 3x - 10$

a Work out the equation of the tangent to the curve at the point $(4, 10)$

b Work out the equation of the normal to the curve when $x = -2$

a $y = 2x^2 - 3x - 10$ so $\dfrac{dy}{dx} = 4x - 3$ ——— Differentiate.

At the point $(4, 10)$ the tangent has gradient

$\dfrac{dy}{dx} = 4 \times 4 - 3 = 13$ ——— Substitute.

The equation of the tangent is

$(y - 10) = 13(x - 4)$ ——— Use $(y - b) = m(x - a)$

$y = 13x - 42$

b When $x = -2$

$y = 2 \times (-2)^2 - 3 \times (-2) - 10$ ——— Substitute $x = -2$ into the original equation.

$y = 4$

So $(-2, 4)$ is a point on the normal.

At $(-2, 4)$ the tangent has gradient

$\dfrac{dy}{dx} = 4 \times (-2) - 3 = -11$

So the normal has a gradient of $\dfrac{1}{11}$ ——— Use $m_1 = -\dfrac{1}{m_2}$

The equation of the normal is

$(y - 4) = \dfrac{1}{11}(x + 2)$ ——— Use $(y - b) = m(x - a)$

$11y - x = 46$

1 Work out the equation of the tangent to each of these curves at the given points. Show your working.

   a $y = 2x^2 + 3x - 1$ at $(1, 4)$

   b $y = 3 - x - x^2$ at $(-2, 1)$

   c $y = 2x^3 + 3x^2$ at $(1, 5)$

   d $y = 5x^2 - x^4$ at $(2, 4)$

   e $y = \dfrac{3}{x}$ at $(3, 1)$

   f $y = \dfrac{16}{x^2}$ at $(4, 1)$

   g $y = \sqrt{x}$ at $(9, 3)$

   h $y = \dfrac{25}{\sqrt{x}}$ at $(25, 5)$

   i $y = \sqrt{x} + \dfrac{2}{\sqrt{x}}$ at $(4, 3)$

   j $y = \dfrac{1}{\sqrt{x}} + \dfrac{1}{x}$ at $(1, 2)$

2 Work out the equation of the normal to each curve at the given points. Show your working.

   a $y = x^2 + 2x - 7$ at $(2, 1)$

   b $y = 4 - 5x - x^2$ at $(-3, 10)$

   c $y = 3 - x^3$ at $(2, -5)$

   d $y = x^4 + 2x^3 + x^2$ at $(1, 4)$

   e $y = \dfrac{6}{x}$ at $(2, 3)$

   f $y = \sqrt{x} + x$ at $(4, 6)$

   g $y = \dfrac{1}{x} + \dfrac{1}{\sqrt{x}}$ at $(4, \dfrac{3}{4})$

   h $y = \dfrac{1}{x^2} + \dfrac{\sqrt{x}}{x^2}$ at $(1, 2)$

3 A curve has equation $y = 3x^2$

   a Work out the point on the curve with x-coordinate 3

   b Work out the gradient of the curve at this point. Show your working.

   c Work out the equation of the tangent to the curve at this point.

4 A line is a tangent to the curve $y = x^2 + 3x - 1$ at the point $(1, k)$

   a What is the value of k?

   b What is the gradient of the tangent at this point? Show your working.

   c Work out the equation of the line.

5 A curve has equation $y = x^2 + 7x - 9$

   a Calculate the point on this curve with x coordinate 1

   b Calculate the gradient of the tangent to the curve at this point. Show your working.

   c Hence state the gradient of the normal to the curve at this point.

   d Work out the equation of the normal to the curve at this point.

6 A curve has equation $y = 2x + \dfrac{2}{x}$

   a Calculate the point on the curve with x-coordinate 2

   b Calculate the gradient of the tangent to the curve at this point. Show your working.

   c Hence state the gradient of the normal to the curve at this point.

   d Work out the equation of the normal to the curve at this point.

7 A parabola has equation $y = x^2 + 6x + 5$

   a **i** Work out the gradient of the tangent at the point where $x = -3$. Show your working.

      ii Give the equation of the tangent.

   b Give the equation of the normal to the curve at this point.

   c In a similar way, work out the equation of the tangent and normal to each curve at the given point. Show your working.

      i $y = x^2 + 2x - 24$ at $x = -1$

      ii $y = x^2 + 10x$ at $x = -5$

      iii $y = 21 + 4x - x^2$ at $x = 2$

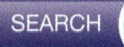

Reasoning and problem-solving

To work out where a tangent or normal meets a curve

(1) Differentiate the function for the curve.

(2) Equate this to the gradient of the tangent or normal (remember $m_T = -\dfrac{1}{m_N}$).

(3) Rearrange and solve for x

(4) Substitute x in the function and solve for y

The line $y = 3x + b$ is a tangent to the curve $y = 2x + 4\sqrt{x}$, $x > 0$

a Work out the point where the tangent meets the curve, thus find the value of the constant b

b Work out the equation of the normal to the curve at this point.

a $\quad y = 2x + 4\sqrt{x} = 2x + 4x^{\frac{1}{2}}$

$\dfrac{dy}{dx} = 2 + \dfrac{1}{2} \times 4x^{-\frac{1}{2}}$

> Differentiate the function of the curve. (1)

$\quad = 2 + \dfrac{2}{\sqrt{x}}$

$m_{tan} = 3$

so $2 + \dfrac{2}{\sqrt{x}} = 3$

> Set $\dfrac{dy}{dx}$ equal to the gradient of the tangent. (2)

$\sqrt{x} = 2$

$x = 4$

> Rearrange and solve for x (3)

x cannot be negative.

When $x = 4$,

$y = 2 \times 4 + 4 \times \sqrt{4} = 16$

> Solve for y (4)

So the tangent touches the curve at $(4, 16)$

$3(4) + b = 16$

$b = 4$

b $\quad m_N = -\dfrac{1}{m_T} = -\dfrac{1}{3}$

> Find the gradient of the normal.

$y - 16 = -\dfrac{1}{3}(x - 4)$

> Use $y - b = m(x - a)$

$y = \dfrac{52}{3} - \dfrac{x}{3}$

To work out the area bound between a tangent, a normal and the x-axis/y-axis

1. Work out the equation of the tangent and, from it, the equation of the normal.
2. Work out where each line crosses the required axis. Lines cut the x-axis when $y = 0$, and the y-axis when $x = 0$
3. Sketch the situation if required.
4. Use $A = \dfrac{1}{2} \times$ base $\times$ height, where the base is the length between the intercepts on the x-axis or y-axis and height is the y-coordinate or x-coordinate respectively.

Example 3

The point $T(1, 2)$ lies on the curve $y = x^3 + x$

Work out the triangular area trapped between the tangent and the normal to the curve at this point and the x-axis.

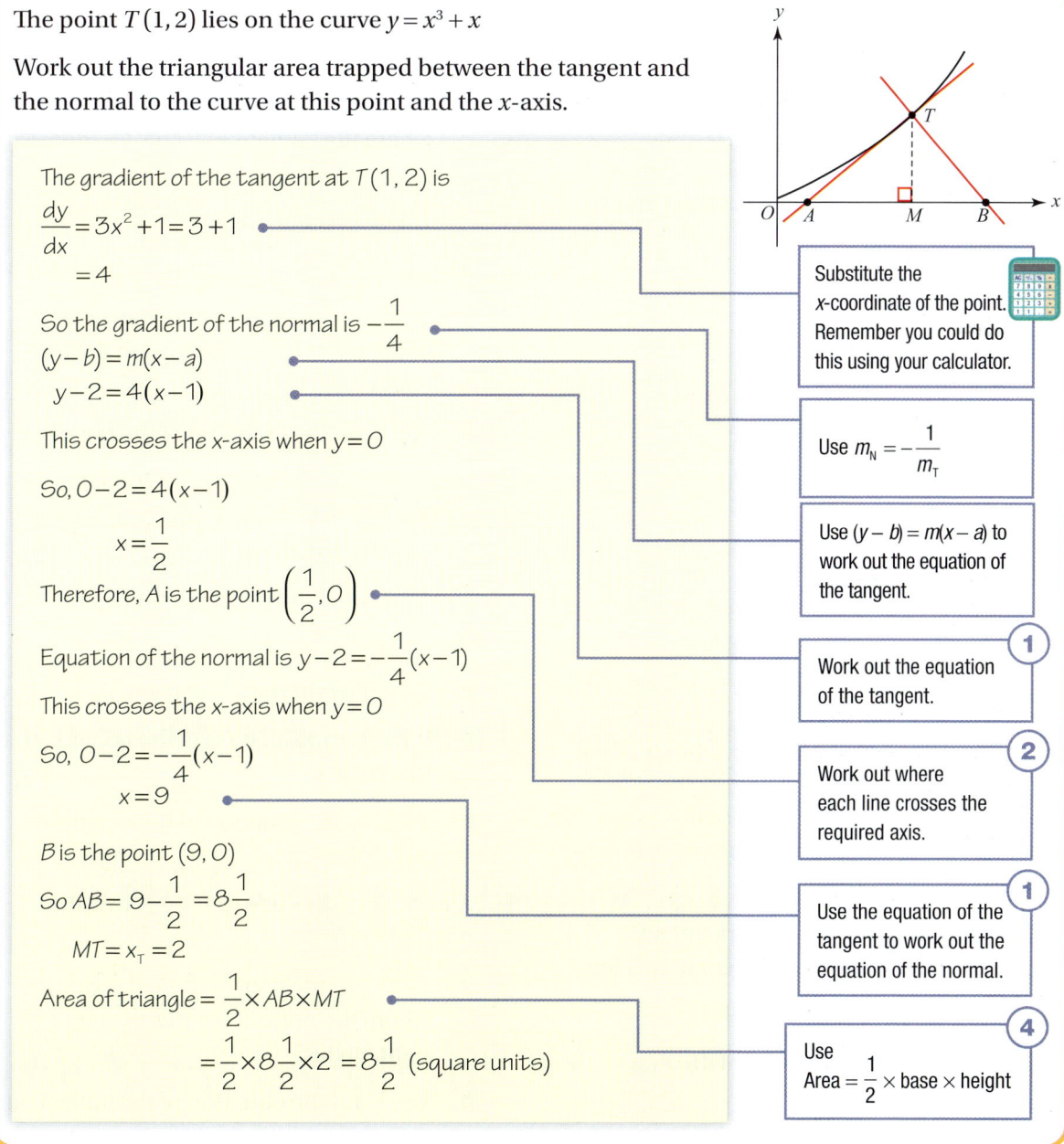

The gradient of the tangent at $T(1, 2)$ is

$$\frac{dy}{dx} = 3x^2 + 1 = 3 + 1$$

$$= 4$$

Substitute the x-coordinate of the point. Remember you could do this using your calculator.

So the gradient of the normal is $-\dfrac{1}{4}$

$(y - b) = m(x - a)$

$y - 2 = 4(x - 1)$

Use $m_N = -\dfrac{1}{m_T}$

This crosses the x-axis when $y = 0$

So, $0 - 2 = 4(x - 1)$

$$x = \frac{1}{2}$$

Therefore, A is the point $\left(\dfrac{1}{2}, 0\right)$

Use $(y - b) = m(x - a)$ to work out the equation of the tangent.

Equation of the normal is $y - 2 = -\dfrac{1}{4}(x - 1)$

This crosses the x-axis when $y = 0$

So, $0 - 2 = -\dfrac{1}{4}(x - 1)$

$$x = 9$$

(1) Work out the equation of the tangent.

(2) Work out where each line crosses the required axis.

B is the point $(9, 0)$

So $AB = 9 - \dfrac{1}{2} = 8\dfrac{1}{2}$

$MT = x_T = 2$

(1) Use the equation of the tangent to work out the equation of the normal.

Area of triangle $= \dfrac{1}{2} \times AB \times MT$

$$= \frac{1}{2} \times 8\frac{1}{2} \times 2 = 8\frac{1}{2} \text{ (square units)}$$

(4) Use Area $= \dfrac{1}{2} \times$ base $\times$ height

1 The line with equation $y = 1 - 3x$ is a tangent to the curve $y = x^2 - 7x + k$ where k is a constant.

    a Calculate the value of x at the point where the tangent meets the curve.

    b Hence calculate the value of k

2 The curve $y = x + \dfrac{3}{x}, x > 0$ has a normal that runs parallel to the line

$$y = 3x + 4$$

Work out the point where the normal crosses the curve at right angles.

3 **a** Work out the equation of the normal to the curve $y = x^2 - 3x + 1$ at the point A, where $x = 1$

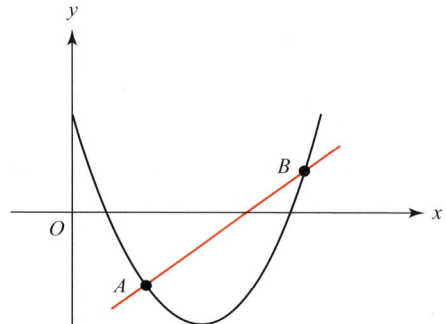

    b Work out the point B where the normal at A crosses the curve again.

    c Prove that the line AB is *not* normal to the curve at B

4 The curves $y = \dfrac{2x^2 + 1}{2}$ and $y = \dfrac{1 + 4x}{x}, x \neq 0$ intersect at the point (a, b). At this point, the line that is the tangent to one curve is the normal to the other line.

    a Use two methods to work out an expression for the gradient of this line.

    b Work out the point (a, b)

    c Work out the equation of the line.

5 **a** Work out the equation of the tangent to the curve $y = 2 - \dfrac{9}{x}$ when $x = 3$

    b Work out the equation of the normal to the curve at the same point.

    c Calculate the area of the triangle bound by the tangent, the normal and the y-axis.

6 A cubic curve $y = x^3 + x^2 + 2x + 1$ has a tangent at $x = 0$

    a Work out the equation of this tangent.

    b Work out the coordinates of the point B where the tangent crosses the curve again.

    c Work out the coordinates of the point C where the tangent at B crosses the x-axis.

    d Calculate the area of the triangle BCO where O is the origin.

7 A parabola has the equation $y = 2x^2 - 3x + 1$

    a Work out the equation of the tangent to the curve that is parallel to the line $y = 5x$

    b Work out the equation of the normal at this point.

8 A point in the first quadrant (p, q) lies on the curve $y = x^3 + 1$

The tangent at this point is perpendicular to the line $y = -\dfrac{x}{12}$

    a Calculate the values of p and q

    b What is the equation of the tangent at this point?

    c What is the equation of the normal at this point?

9 A normal to the curve $y = x + \dfrac{18}{x}, x > 0$, is parallel to $y = x$

    a Work out the coordinates of the point where the normal crosses the curve at right angles.

    b Work out the equation of the tangent at this point.

10 For each parabola

i Express the equation in the form $b \pm (x+a)^2$

ii Hence deduce the coordinates of the turning point on the parabola.

iii Work out the equation of the tangent and the normal at this point.

iv Comment on your answers.

a $y = x^2 + 6x - 1$

b $y = x^2 - 10x + 5$

c $y = x^2 - 3x - 7$

d $y = 4 - 8x - x^2$

e $y = x^2 + 4$

f $y = 1 - 2x - x^2$

11 A parabolic mirror has a cross-section as shown.

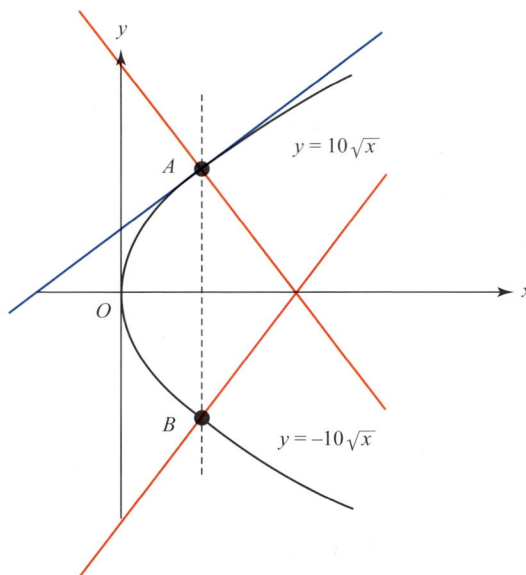

The branch above the x-axis has equation $y = 10\sqrt{x}$

The branch below it has equation $y = -10\sqrt{x}$

The line AB is parallel to the y-axis.

Let $x_A = x_B = p$

a Work out the equation of the normal to the curve

i At A **ii** At B

b Show that these two normals intersect on the x-axis.

12 The parabola $y = 4 + 2x - 2x^2$ crosses the y-axis at the point (p, q)

a State the values of p and q

b Work out the gradient of the tangent of the curve at this point.

c Work out the equation of the normal to this point.

d The curve crosses the x-axis at two points.

i Work out the coordinates of these points.

ii Work out the equation of the normal at both points.

e A related curve, $y = 4 - 2x^2$ crosses the y-axis at $(0, 4)$

Work out the equation of the normal to this curve at this point.

Challenge

13 a Work out the equation of the tangent to the curve $y = \dfrac{360}{x}$ $(x \geq 1)$ at the point where $x = 30$

b Work out the equation of the normal at the point where $x = 60$

c **i** At what point is the gradient of the normal $\dfrac{1}{10}$?

ii Give the equation of this normal.

d The line $y = -40x + k$ is a tangent to the curve.

i At what point does this line touch the curve?

ii What is the value of k?

Turning points

Fluency and skills

When a curve changes from an increasing function to a decreasing function or vice versa, it passes through a point where it is stationary. This is called a **turning point** or **stationary point**.

> **Key point**
>
> At a turning point, the gradient of the tangent is zero. Therefore, you can work out the coordinates of the turning point by equating the derivative to zero.

A turning point is a stationary point, but a stationary point is not necessarily a turning point. You will learn about other types of stationary point in **Section 15.1**

Here are examples of a **maximum** turning point and a **minimum** turning point.

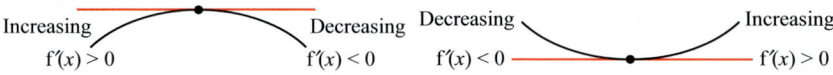

Increasing f'(x) > 0   Decreasing f'(x) < 0   Decreasing f'(x) < 0   Increasing f'(x) > 0

At a maximum turning point, as x increases, the gradient changes from positive through zero to negative.

At a minimum turning point, as x increases, the gradient changes from negative through zero to positive.

Example 1

Work out the coordinates of the turning point on the curve $y = x^2 + 4x - 12$ and determine its nature by inspection of the derivative either side of the point. Show your working.

$y = x^2 + 4x - 12$ so $\dfrac{dy}{dx} = 2x + 4$

Differentiate.

At a turning point,

$\dfrac{dy}{dx} = 0 \Rightarrow 2x + 4 = 0 \Rightarrow x = -2$

Find the value of x when the derivative is equal to zero.

$y = (-2)^2 + 4 \times (-2) - 12 \Rightarrow y = -16$

Substitute into the original equation.

The turning point is at $(-2, -16)$

At $x = -2.1$, $\dfrac{dy}{dx} = 2(-2.1) + 4 = -0.2$

At $x = -1.9$, $\dfrac{dy}{dx} = 2(-1.9) + 4 = 0.2$

Consider the value of the derivative either side of the turning point.

The gradient is increasing from negative to positive, so the point $(-2, -16)$ is a minimum turning point.

Calculator

Try it on your calculator

You can find turning points on a graphics calculator.

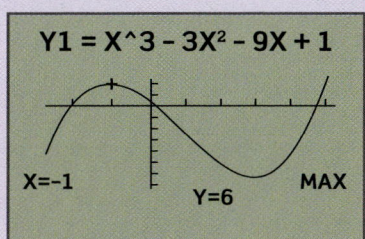

Y1 = X^3 – 3X² – 9X + 1

X=–1   Y=6   MAX

Activity

Find out how to find the turning points of $y = x^3 - 3x^2 - 9x + 1$ on *your* graphics calculator.

As well as by inspection, the nature of a turning point can be determined by finding the second derivative with respect to x, $\dfrac{d^2y}{dx^2}$

If the gradient, $\dfrac{dy}{dx}$, is *decreasing* and the second derivative is negative, the turning point is a *maximum*.

Key point

At a *maximum* turning point, $\dfrac{d^2y}{dx^2} < 0$

If the gradient, $\dfrac{dy}{dx}$, is *increasing* and the second derivative is positive, the turning point is a *minimum*.

Key point

At a *minimum* turning point, $\dfrac{d^2y}{dx^2} > 0$

Remember that you can also represent the second derivative using the notation $f''(x)$

Example 2

Use calculus to work out the coordinates of the turning point on the curve $f(x) = x + \dfrac{1}{x}$, $x > 0$ and determine its nature.

$f(x) = x + \dfrac{1}{x}$ so $f'(x) = 1 - \dfrac{1}{x^2}$

At a turning point, $f'(x) = 0$ so $1 - \dfrac{1}{x^2} = 0$

Solving for x gives $x = \pm 1$
but $x > 0$, so $x = 1$

When $x = 1$,
$f(x) = 1 + \dfrac{1}{1} = 2$

So the turning point is at $(1, 2)$

$f'(x) = 1 - \dfrac{1}{x^2} = 1 - x^{-2}$

$f''(x) = \dfrac{2}{x^3}$

When $x = 1$,

$f''(x) = \dfrac{2}{1^3} = 2$

The second derivative is positive, so the turning point is a minimum.

> Substitute into the original equation.

> Find the second derivative to determine the nature of the turning point.

MyMaths   🔍 2270   SEARCH

1 For each curve, work out the coordinates of the stationary point(s) and determine their nature by inspection. Show your working.

a $y = x^2 + 4x - 5$ **b** $y = x^2 + 4x - 32$

c $y = x^2 - 6x - 7$ **d** $y = 1 - x^2$

e $y = 2x^2 + 7x + 6$ **f** $y = 20 - x - x^2$

g $y = 6x^2 - x - 1$ **h** $y = 2 - 13x - 7x^2$

i $y = 6 - x - 2x^2$ **j** $y = x + \dfrac{4}{x}, x > 0$

k $y = 2x + \dfrac{18}{x}, x \neq 0$

l $y = 10 - x - \dfrac{1}{x}, x \neq 0$

m $y = x - 10\sqrt{x}, x \geq 0$

n $y = x^2 - 32\sqrt{x}, x \geq 0$

2 The curve $y = x^3 - 6x^2$ has two turning points.

a Work out the coordinates of both turning points. Show your working.

b Use the second derivative to determine the nature of each.

c Sketch the curve $y = x^3 - 6x^2$ Show the coordinates of the turning points and any intersections with axes on your sketch.

3 **a** Work out the coordinates of the turning points of $y = 2x^3 + 30x^2 + 1$ and determine their nature. Show your working.

b Sketch the curve $y = 2x^3 + 30x^2 + 1$ Show the coordinates of the turning points and any intersections with axes on your sketch.

4 $f(x) = 2x^3 - 9x^2 + 12x + 7$

a Differentiate $f(x)$

b The curve $y = f(x)$ has two turning points. Work out the coordinates of them both and determine their natures. Show your working.

c Repeat this process with the following functions.

i $f(x) = x^3 - 3x^2 - 24x + 1$

ii $f(x) = x^3 + 3x^2 - 45x - 45$

iii $f(x) = 1 - 36x - 21x^2 - 2x^3$

iv $f(x) = 2x^3 - 11x^2 - 8x + 2$

v $f(x) = 3 - 4x + 5x^2 - 2x^3$

vi $f(x) = 5 + x - 2x^2 - 4x^3$

5 The function $f(x) = 3x^4 + 8x^3 - 6x^2 - 24x - 1$ has three turning points.

a Show that stationary points can be found at $x = 1$, $x = -1$ and $x = -2$

b Work out the coordinates of the three stationary points.

c Use the second derivative to help you establish the nature of each.

6 The function $f(x) = 3x^4 - 4x^3 - 36x^2$ has three turning points. A sketch of the graph of the function is shown.

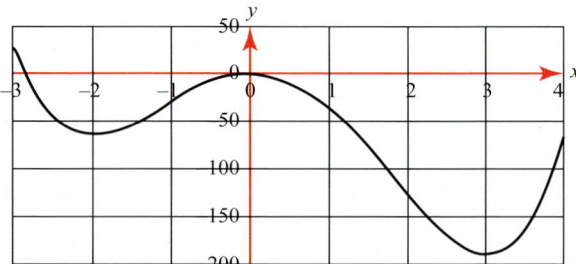

a The sketch suggests that these turning points can be found at $x = -2$, $x = 0$ and $x = 3$

Show that this is the case.

b Use the second derivative to verify the nature of each turning point.

7 A function is defined by $f(x) = 8x + \dfrac{72}{x}, x > 0$

a Show that the *only* stationary point on the curve $y = f(x)$ is at $x = 3$

b State the coordinates of the stationary point and determine its nature. Show your working.

c A related function is defined by $f(x) = 8x + \dfrac{72}{x}, x \neq 0$

It has a stationary point at $x = -3$ Determine the nature of this stationary point. Show your working.

Reasoning and problem-solving

Strategy

To identify the main features of a curve

1. Work out where it crosses the axes ($x = 0$ and $y = 0$)
2. Consider the behaviour of the curve as x tends to infinity and identify any asymptotes.
3. Work out the coordinates of the turning points $\left(\dfrac{dy}{dx} = 0\right)$ and determine their nature.
4. Use the information you have found to sketch the function.

See Ch 2.4
For a reminder about sketching curves.

Example 3

$y = (x - 10)(x + 5)(x + 14)$

a Show that the curve crosses the y-axis at $(0, -700)$

b Show that there is a maximum turning point at $(-10, 400)$ and a minimum turning point at $(4, -972)$

c Sketch the curve.

a The curve crosses the y-axis when $x = 0$

So $y = -10 \times 5 \times 14$

$= -700$, giving $(0, -700)$ as a point on the curve.

Find where the graph crosses the axes.

Expand.

b $y = (x - 10)(x + 5)(x + 14)$

$y = x^3 - 9x^2 - 120x - 700$

$\dfrac{dy}{dx} = 3x^2 - 18x - 120 = 3(x - 4)(x + 10)$

Differentiate and factorise.

Stationary points occur when $\dfrac{dy}{dx} = 0 \Rightarrow x = 4$ and $x = -10$

When $x = 4$

$y = (4 - 10)(4 + 5)(4 + 14) = -972$, giving $(4, -972)$

When $x = -10$

$y = (-10 - 10)(-10 + 5)(-10 + 14) = 400$, giving $(-10, 400)$

Work out the coordinates of the turning points.

$\dfrac{dy}{dx} = 3x^2 - 18x - 120$

So $\dfrac{d^2y}{dx^2} = 6x - 18$

Find the second derivative.

When $x = 4$, the second derivative is positive so $(4, -972)$ is a minimum turning point.

When $x = -10$, it is negative so $(-10, 400)$ is a maximum turning point.

Use the second derivative to determine the nature of the turning points.

c

[graph showing curve with labelled points $(-10, 400)$, crossings at -14, -5, 10, -700 on the y-axis, and minimum at $(4, -972)$]

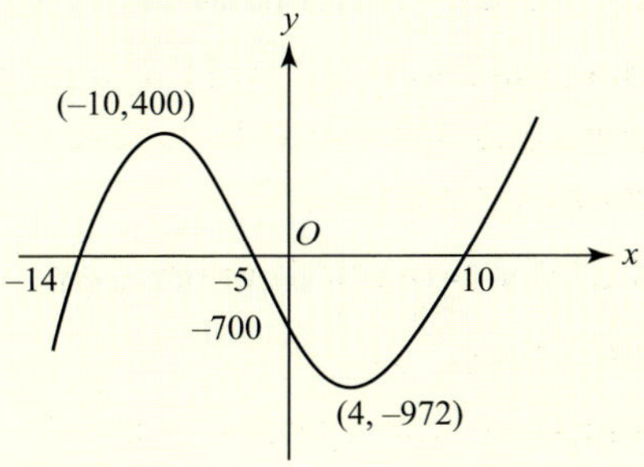

Use this information to sketch the function. You could use your graphics calculator to check your sketch.

To optimise a given situation

1. Express the dependent variable (say y) as a function of the independent variable (say x).
2. Differentiate y with respect to x
3. Let the derivative be zero and find the value of x that optimises the value of y
4. Examine the nature of the turning points to decide if it is a maximum or minimum.
5. Put your turning point in the context of the question.

Example 4

A piece of rope 120 m long is to be used to draw out a rectangle on the ground.

What is the biggest area that can be enclosed in the rectangle?

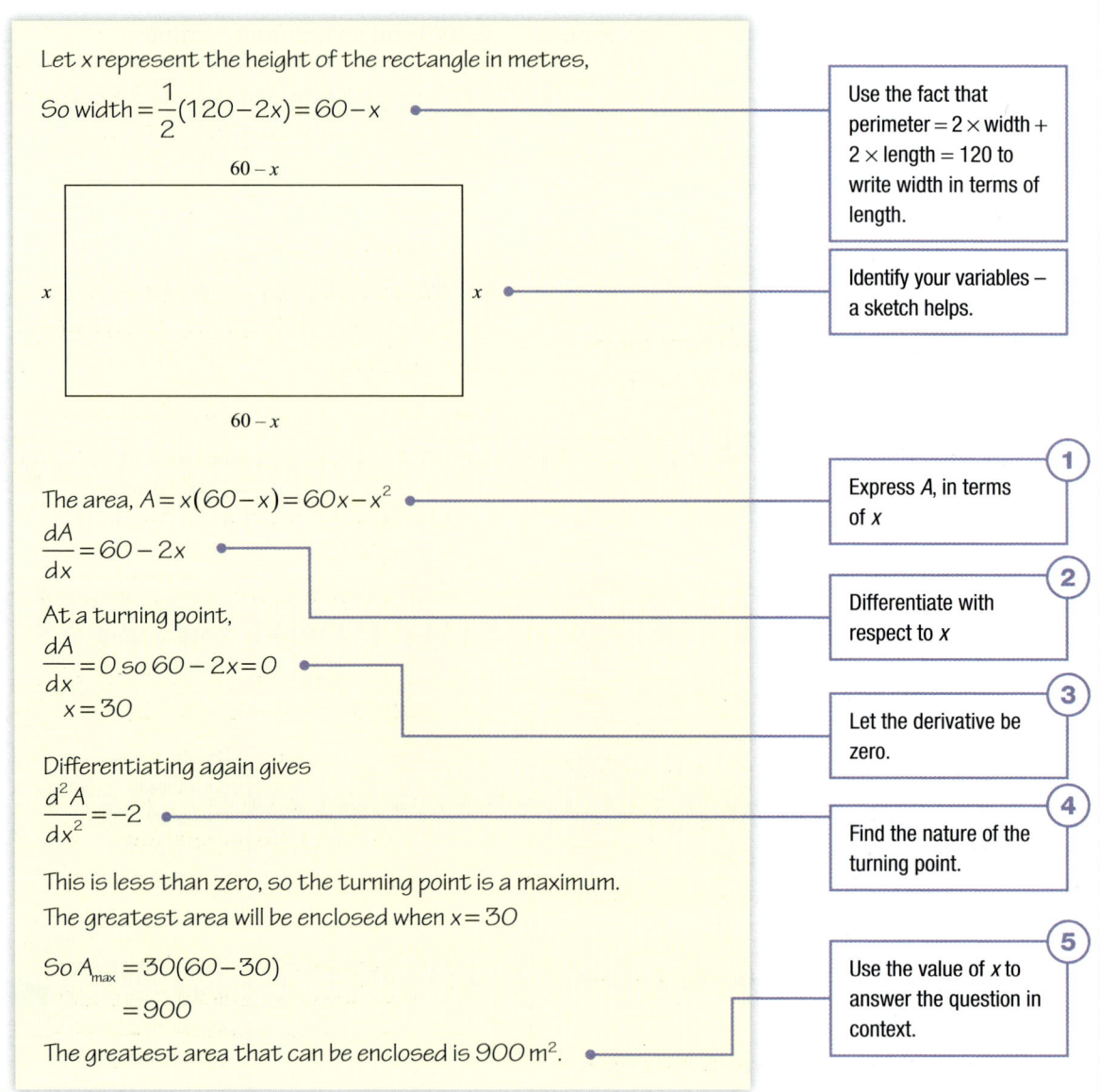

Let x represent the height of the rectangle in metres,

So width $= \dfrac{1}{2}(120-2x) = 60-x$

Use the fact that perimeter $= 2 \times$ width $+ 2 \times$ length $= 120$ to write width in terms of length.

Identify your variables – a sketch helps.

The area, $A = x(60-x) = 60x - x^2$

1 Express A, in terms of x

$\dfrac{dA}{dx} = 60 - 2x$

2 Differentiate with respect to x

At a turning point,

$\dfrac{dA}{dx} = 0$ so $60 - 2x = 0$

$x = 30$

3 Let the derivative be zero.

Differentiating again gives

$\dfrac{d^2A}{dx^2} = -2$

4 Find the nature of the turning point.

This is less than zero, so the turning point is a maximum.

The greatest area will be enclosed when $x = 30$

So $A_{max} = 30(60-30)$

$= 900$

5 Use the value of x to answer the question in context.

The greatest area that can be enclosed is 900 m².

1 Two numbers, x and $1000 - x$, add to make 1000. What would they need to be to maximise their product?

2 The product of two positive whole numbers, x and $\dfrac{3600}{x}$, is 3600. Their sum is the smallest it can be. What are the two numbers?

3 Two numbers, x and y, have a sum of 12. What values of x and y will make x^2y a maximum?

4 A wire model of a cuboid is made. The total length of wire used is 600 cm.

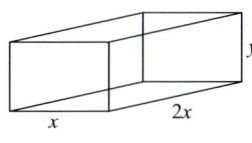

  a Express y in terms of x

  b Express the volume of the cuboid in terms of x

  c Find the values of x and y which maximise the volume.

5 A rope of length 16 m is used to form three sides of a rectangular pen against a wall.

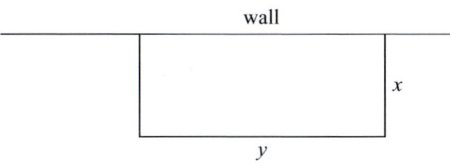

  a How should the rope be arranged to maximise the area enclosed?

  b Another length of rope is used to enclose an area of 50 m² in a similar way. What is the shortest length of rope that is needed?

6 The point $A(x, y)$ lies in the first quadrant on the line with equation $y = 6 - 5x$

A rectangle with two sides on the coordinate axes has A as one vertex.

  a Work out an expression in x for the area of the rectangle.

  b Work out the point for which this area is a maximum.

7 A small tray is made from a 12 cm square of metal.

Small squares are cut from each corner and the edges left are turned up.

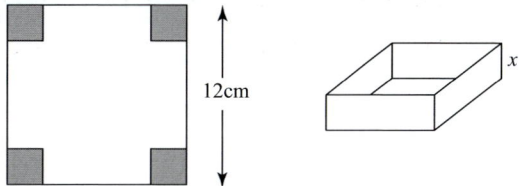

What should the side of the small square be so as to maximise the volume of the tray?

8 A fruit drink container is a cuboid with a square base. It has to hold 1000 ml of juice. Let one side of the square base be x cm and the height of the container be h cm.

  a Express h in terms of x

  b Find an expression for the surface area of the cuboid in terms of x

  c Find the value of x that will minimise the surface area (and hence the cost of the container).

Challenge

9 A cylindrical container has to hold 440 ml of juice.

  a Express its height h cm in terms of the radius x cm.

  b The surface area of a cylinder with a lid and a base is given by
$$2\pi x^2 + 2\pi xh$$
Calculate the value of x that minimises this surface area.

Show your working and write your answer correct to 2 decimal places.

  c If the container doesn't need a lid, how does this affect the answer?

Fluency and skills

The reverse process of differentiation is known is **integration**.

To differentiate sums of terms of the form ax^n you multiply by the power then reduce the power by 1 to get nax^{n-1}

To **integrate** you do the exact opposite: you add 1 to the power then divide by the new power.

When you differentiate a constant, the result is zero. So when you perform an integration, you should add a constant, c, to allow for this. This is referred to as the **constant of integration**. The value of this constant can only be determined if further information is given.

> When you perform an integration you can check your result by differentiating it – you should get back to what you started with.

Key point

Integrating x^n with respect to x is written as

$$\int x^n \, dx = \frac{x^{n+1}}{n+1} + c, \; n \neq -1$$

When a function is multiplied by a constant, the constant can be moved outside the integral.

Key point

$$\int a f(x) \, dx = a \int f(x) \, dx \text{ where } a \text{ is a constant.}$$

When integrating the sum of two functions, each function can be integrated separately.

Given that displacement is a function of time r(t), then velocity v(t) = r$'(t)$. Reversing this, we get

Key point

$$\int v(t) \, dt = r(t) + c$$

> You may see this rule referred to as the *sum rule*.
> $$\int (f(x) + g(x)) \, dx =$$
> $$\int f(x) \, dx + \int g(x) \, dx$$

Given that velocity is a function of time v(t), then acceleration a(t) = v$'(t)$. Reversing this we get

Key point

$$\int a(t) \, dt = v(t) + c$$

The Fundamental Theorem of Calculus shows how integrals and derivatives are linked to one another. The theorem states that, if f(x) is a continuous function on the interval $a \leq x \leq b$, then

Key point

$$\int_a^b f(x) \, dx = F(b) - F(a) \text{ where } \frac{d}{dx}(F(x)) = f(x)$$

Example 1

$f'(x) = 6x^3 + 3x^2 + \dfrac{1}{\sqrt{x}} + 4$

a Integrate to find $f(x)$

b Given that the point $(1, 4)$ lies on the curve $y = f(x)$, find the constant of integration.

a $f(x) = \int \left(6x^3 + 3x^2 + \dfrac{1}{\sqrt{x}} + 4 \right) dx$

$= \int 6x^3 dx + \int 3x^2 dx + \int x^{-\frac{1}{2}} dx + \int 4x^0 dx$

$= \dfrac{6x^4}{4} + \dfrac{3x^3}{3} + \dfrac{x^{\frac{1}{2}}}{\frac{1}{2}} + \dfrac{4x^1}{1} + c$

$= \dfrac{3}{2}x^4 + x^3 + 2x^{\frac{1}{2}} + 4x + c$

| Express all the terms in index form and use the sum rule to isolate functions. |

| Integrate using $\int ax^n \, dx = \dfrac{ax^{n+1}}{n+1} + c$ |

| Simplify. |

b $y = \dfrac{3}{2}x^4 + x^3 + 2x^{\frac{1}{2}} + 4x + c$

$4 = \dfrac{3}{2} + 1 + 2 + 4 + c$

$c = -\dfrac{9}{2}$

| Substitute in coordinate values for x and y |

| Rearrange and evaluate. |

Exercise 4.6A Fluency and skills

1 Work out the integral of each function with respect to x, remembering the constant of integration.

a 10

b $3x$

c $6x^2$

d $12x^3$

e $25x^4$

f x^6

g $3x+1$

h $5-4x$

i $3x^2 + 6x + 2$

j $12x^2 + 6$

k $3 - 4x - 6x^2$

l $x^3 + x^2 + x + 1$

m $3 - x - 24x^3$

n $2x^3 + 4x + 1$

o $x^3 + 3x^{-2}$

p $2x^{-3} - x^{-2}$

q $x^{\frac{1}{2}} - x^3 + 1$

r $x^{\frac{1}{2}} - x^{-\frac{1}{2}} + x + 6$

s $3x^{\frac{1}{3}} - x^{\frac{2}{3}}$

t $x^{\frac{1}{4}} - x^{\frac{3}{4}} + x^{-\frac{1}{2}}$

u $1 - x^{-2} - x^{-\frac{2}{3}}$

v $\dfrac{1}{3}x^{-\frac{2}{3}} - x^4 + 4$

w $\dfrac{4}{3}\pi x^3 - \dfrac{1}{3}\pi x^{\frac{1}{3}} + 2\pi x$

x $\dfrac{(x+1)^2}{x^{\frac{1}{3}}}$

2 Find

a $\int 1 \, dx$

b $\int x \, dx$

c $\int (6x + 7) \, dx$

d $\int (3 - 2x) \, dx$

e $\int (x^2 + x + 1) \, dx$

f $\int (1 - 4x - 3x^2) \, dx$

g $\int (4x^3 + 2x - 7) \, dx$

h $\int (2 + 9x^2 - 12x^3) \, dx$

i $\int \sqrt{x} \, dx$

j $\int \sqrt[3]{x} \, dx$

k $\int \sqrt{\dfrac{1}{x}} \, dx$

l $\int x^{\frac{2}{3}} \, dx$

3 Work out

a $\int \pi \, dx$

b $\int (3\pi + x) \, dx$

c $\int (x^2 \sin 30) \, dx$

d $\int (x^2 + 6x) \, dx$

e $\int 4x^2 + 4x - 28 \, dx$

f $\int \left(x^2 + \dfrac{1}{x^2} \right) dx$

g $\int \left(\sqrt{x} - \dfrac{2}{\sqrt{x}} \right) dx$

h $\int \left(8x - \dfrac{3}{x^2} \right) dx$

i $\int \left(\dfrac{1}{x^3} - \dfrac{1}{x^4} \right) dx$

j $\int \left(\dfrac{1}{x^2} - x - \dfrac{1}{x^3} \right) dx$

k $\int x + \dfrac{1}{\sqrt{x}} \, dx$

l $\int \left(x^2 - \dfrac{3}{\sqrt{x}} + 1 \right) dx$

m $\int \left(\dfrac{1}{x^2} + \dfrac{3}{x^3} \right) dx$

n $\int x \sin \dfrac{\pi}{3} + x^2 \sin \dfrac{\pi}{6} \, dx$

o $\int 3x^2 - \dfrac{1}{\sqrt{x^3}} \, dx$

p $\int \left(\dfrac{x^2 - x}{x} \right) dx$

q $\int \dfrac{(x+2)^2}{x^{\frac{1}{2}}} \, dx$

r $\int \dfrac{x+1}{\sqrt{x}} \, dx$

4 The derivative and a point on the curve $y = f(x)$ are given.

Work out the function $f(x)$

a $f'(x) = 6x^2$ at $(1, 5)$

b $f'(x) = 12x^3$ at $(2, 18)$

c $f'(x) = 5\sqrt{x}$ at $(9, 100)$

5 Work out the function, $f(x)$ for the given $f'(x)$ and the point $(x, f(x))$

a $f'(x) = 4x + 3; (2, 4)$

b $f'(x) = 10; (1, 12)$

c $f'(x) = 3x^2 + 2x + 1; (-2, 1)$

d $f'(x) = 4x + 3\sqrt{x}; (1, 5)$

Reasoning and problem-solving

Strategy

To solve problems that require integration

(1) Identify the variables and express the problem as a mathematical equation.

(2) Integrate.

(3) Use initial conditions to work out the constant of integration.

(4) Substitute c into the equation and answer the question.

Example 2

A moving particle has acceleration -10 cms^{-1}

The particle starts from rest 50 cm to the right of the origin.

> 'The particle starts from rest' means when $t = 0$, $v = 0$

a Express the velocity as a function of time.

b Express the displacement from the origin as a function of time.

c State the acceleration, velocity and displacement after 3 seconds.

a $v(t) = \int a(t) dt$

$= \int -10 \, dt$

$= -10 \int 1 \, dt$

$= -10t + c$

$v(0) = -10 \times 0 + c = 0$

$c = 0$

So $v(t) = -10t$

> **1** Identify the variables and express the rate of change as a mathematical equation.

> **2** Integrate to work out the general form of the function.

> **3** Use the initial conditions to work out c

b $s(t) = \int v(t) dt$

$= \int (-10t) dt$

$= -5t^2 + c$

$s(0) = -5 \times 0^2 + c = 50$

$c = 50$

So $s(t) = -5t^2 + 50$

c $a(3) = -10 \text{ cm s}^{-2}$

$v(3) = -10 \times 3$

$= -30 \text{ cm s}^{-1}$

$s(3) = -5 \times 3^2 + 50$

$= 5 \text{ cm}$

> **4** Use the value of c to answer the question.

1 Given that the rate of change of P with respect to t is 7, and that when $t = 4$, $P = 2$

   a Express P in terms of t

   b Work out P when $t = 5$

   c Work out t when $P = 16$

2 It is known that $f'(x) = 1 - 6x$ and that $f(3) = 6$

   a Calculate $f(-2)$

   b For what values of x does $f(x) = 0$?

3 Kepler was an astronomer who studied the relationship between the orbital period, in y years, of a celestial body and the mean distance from the Sun, R (measured in astronomical units AU). The rate at which the period increases as the distance from the Sun increases is given by $\dfrac{dy}{dr} = \dfrac{3}{2}r^{\frac{1}{2}}$

   a Express y in terms of r and c, the constant of integration.

   b The Earth is 1 astronomical unit from the Sun and takes one year to orbit it. Find the relation between y and r

   c Mars is 1.5 AU from the Sun. How long does it take to orbit the Sun?

   d Saturn takes 29.4 years to orbit the Sun. What is its mean distance form the Sun?

4 The rate at which the depth, h metres, in a reservoir drops as time passes is given by $\dfrac{dh}{dt} = \dfrac{8}{3}t^3 - 4t, t > 0$ where t is the time in days. When $t = 0$, $h = 4$

   a Express h as a function of t

   b What is the depth after half a day?

   c When is the depth 16 m?

5 The second derivative of a function is given by $\dfrac{d^2y}{dx^2} = 12x + 1$ When $x = 1$, $\dfrac{dy}{dx} = 4$

   a Work out an expression in x for $\dfrac{dy}{dx}$

   b When $x = 1$, $y = \dfrac{1}{2}$. What is the value of y when $x = 2$?

6 The second derivative of a function is given by $\dfrac{d^2y}{dx^2} = 6$ When $x = 2$, $y = 1$ and $\dfrac{dy}{dx} = 3$ What is the value of y when $x = 4$?

7 **a** A particle moves on the x-axis so that its acceleration is a function of time, $a(t) = 2t$. Initially (at $t = 0$) the particle was 2 cm to the right of the origin travelling with a velocity, v, of 2 cm s^{-1}.

     i Express the velocity and the displacement, s, as a function of t

     ii Calculate both velocity and displacement when $t = 1$

   b Repeat part **a** for particles with the following initial conditions.

     i $a(t) = 6$; $s(0) = -2$; $v(0) = 5$

     ii $a(t) = t + 1$; $s(0) = 1$; $v(0) = 2$

     iii $a(t) = t^2$; $s(0) = 0$; $v(0) = -1$

8 A piece of rock breaks away from the White Cliffs of Dover.

Its acceleration towards the shore, 90 m below, is 5 m s^{-2}.

Let t seconds be the time since the rock broke free. Let $v(t)$ m s^{-1} be its velocity at time t. Note that when $t = 0$, the velocity of the rock is 0

How long will it take for the rock to hit the shore?

Challenge

9 The velocity of a particle is given by $v(t) = 4 + 3t$ where distance is measured in metres and time in seconds. After one second the particle is 6 m to the right of the origin.

   a Where was the particle initially?

   b What is its acceleration at $t = 5$?

   c How far has it travelled in the fifth second?

   d Work out an expression in n for the distance travelled in the nth second.

Fluency and skills

You can use integration to find the area between a curve and the x-axis. To do this, you perform a calculation using a **definite integral**.

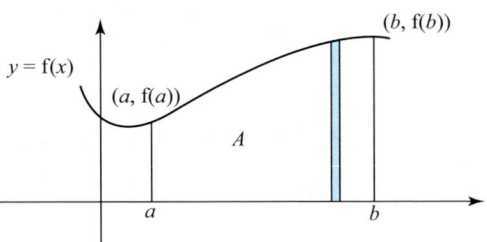

> **Key point**
>
> A definite integral is denoted by $\int_a^b f(x)\,dx$
>
> b is called the **upper limit**, and a the **lower limit**.

Consider a continuous function $y = f(x)$ over some interval and where all points on the curve in that interval lie on the same side of the x-axis. The area, A, is bound by $y = f(x)$, the x-axis and the lines $x = a$ and $x = b$ where $a < b$

Consider a small change in x and the change in area, δA, that results from this change.

Use Leibniz notation where δx represents a small change in x and δy represents the corresponding change in y

ICT Resource online

To experiment with numerical integration using rectangles, click this link in the digital book.

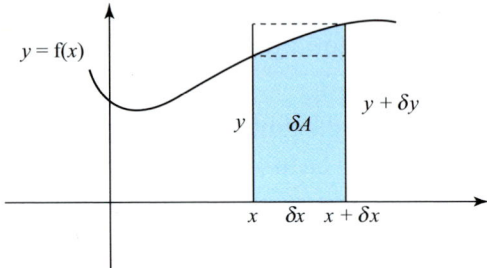

Zooming in you can see a small section of the area, trapped between vertical lines at x and at $x + \delta x$

The small area between the vertical lines at x and $x + \delta x$ (shaded) is denoted by δA

$$y\delta x \le \delta A \le (y + \delta y)\delta x$$

Dividing by δx:
$$y \le \frac{\delta A}{\delta x} \le (y + \delta y)$$

As you let δx tend to zero:

$$y \le \lim_{\delta x \to 0} \frac{\delta A}{\delta x} \le \lim_{\delta x \to 0}(y + \delta y) \Rightarrow y \le \frac{dA}{dx} \le y$$

> **Remember**
>
> $\lim_{\delta x \to 0} \delta y = 0$ and $\lim_{\delta x \to 0} \frac{\delta y}{\delta x} = \frac{dy}{dx}$

So
$$y = \frac{dA}{dx}$$

Integrating this will give you a formula for the area from the origin up to the upper bound, namely $A = \int y\,dx$. Note that, when calculating this small area, the lines $x = a$ and $x = b$ were not used so this has given us a *general* formula for calculating the area. A further calculation is needed to obtain the area between $x = a$ and $x = b$

If you wish to calculate the area between two vertical lines, $x = a$ and $x = b$, then you need only integrate to get the formula for area and substitute a and b for x. The difference between your results is the required area.

If $\int f(x)dx = F(x) + c$ then $(F(b) + c) - (F(a) + c) = F(b) - F(a)$

> **Key point**
>
> The area under a curve between the x-axis, $x = a$, $x = b$ and $y = f(x)$, is given by $A = \int_a^b f(x)dx = F(b) - F(a)$

> The constants of integration sum to zero. This means you don't need to worry about the constant of integration when calculating a definite integral.

Working with areas below the x-axis will produce negative results. As area is a positive quantity you should use the magnitude of the answer only (ignore the negative sign).

Example 1

Evaluate the definite integral $\int_1^4 (3x^2 + 4x + 1)dx$. You must show your working.

$$\int_1^4 (3x^2 + 4x + 1)dx$$
$$= \left[x^3 + 2x^2 + x \right]_1^4$$ — Integrate.
$$= \left(4^3 + 2 \times 4^2 + 4 \right) - \left(1^3 + 2 \times 1^2 + 1 \right)$$ — Substitute the two values of x
$$= 100 - 4$$
$$= 96$$

Example 2

A parabola cuts the axes as shown.

Show that the area A is $10\frac{2}{3}$ units.

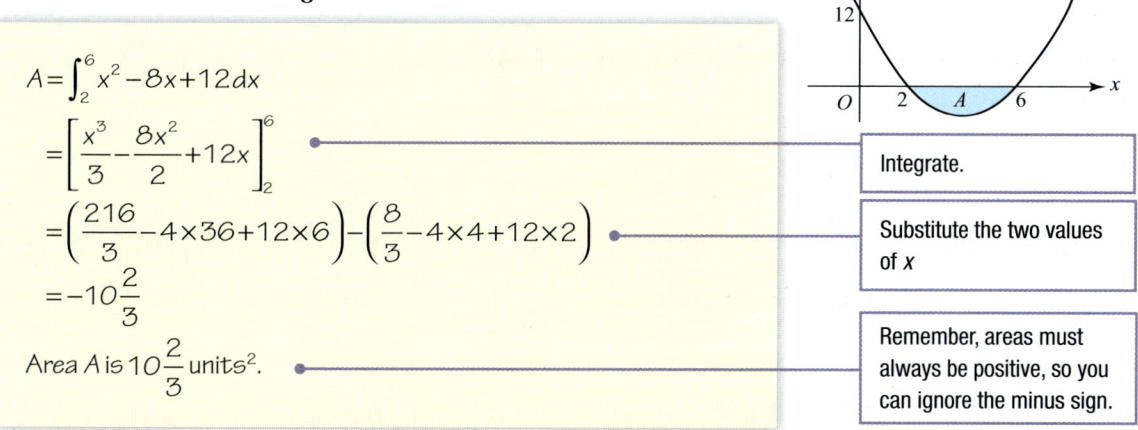

$$A = \int_2^6 x^2 - 8x + 12\,dx$$
$$= \left[\frac{x^3}{3} - \frac{8x^2}{2} + 12x \right]_2^6$$ — Integrate.
$$= \left(\frac{216}{3} - 4 \times 36 + 12 \times 6 \right) - \left(\frac{8}{3} - 4 \times 4 + 12 \times 2 \right)$$ — Substitute the two values of x
$$= -10\frac{2}{3}$$

Area A is $10\frac{2}{3}$ units^2. — Remember, areas must always be positive, so you can ignore the minus sign.

> If you had calculated the definite integral between 0 and 6, the answer would be zero. The positive and negative sections would cancel each other out.

Try it on your calculator

You can use a calculator to evaluate a definite integral.

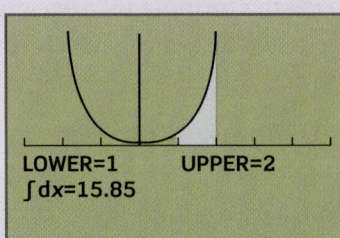

LOWER=1   UPPER=2
∫dx=15.85

Activity

Find out how to evaluate $\int_1^2 3x^4 - x^3 + 1\,dx$ on *your* calculator.

Exercise 4.7A Fluency and skills

1 Evaluate these definite integrals. Show your working in each case.

a $\int_2^4 x+1\,dx$ **b** $\int_1^5 2x-1\,dx$

c $\int_0^3 4-x\,dx$ **d** $\int_{-2}^6 1-3x\,dx$

e $\int_{-2}^3 7\,dx$ **f** $\int_1^7 \pi\,dx$

g $\int_1^3 \pi+1\,dx$ **h** $\int_0^3 3x^2+4x+1\,dx$

i $\int_{-3}^3 x^2-6x-1\,dx$ **j** $\int_{-1}^1 1-x-x^2\,dx$

k $\int_{\sqrt{2}}^{\sqrt{3}} x^3\,dx$ **l** $\int_1^{1.5} x+\frac{1}{x^2}\,dx$

m $\int_{-1}^0 x^3+2x+3\,dx$ **n** $\int_4^{25} \frac{1}{\sqrt{x}}\,dx$

o $\int_2^5 4\pi x^2\,dx$ **p** $\int_{\sqrt{3}}^{\sqrt{5}} x^3+x\,dx$

q $\int_{-3}^3 x^3-2x\,dx$ **r** $\int_{\frac{1}{4}}^{\frac{1}{2}} \frac{\pi}{x^2}\,dx$

2 Work out the total shaded area for each of these graphs. Show your working in each case.

a
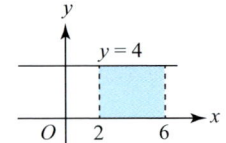
$y = 4$

b
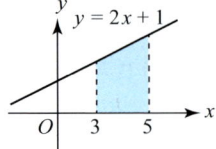
$y = 2x + 1$

c
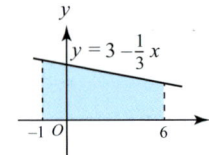
$y = 3 - \frac{1}{3}x$

d
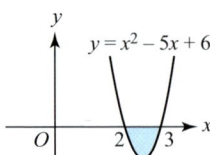
$y = x^2 - 5x + 6$

e
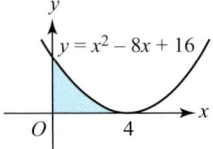
$y = x^2 - 8x + 16$

f
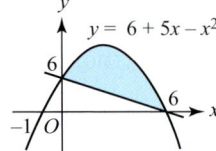
$y = 6 + 5x - x^2$

g
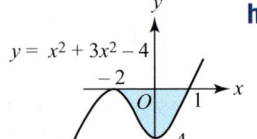
$y = x^2 + 3x^2 - 4$

h
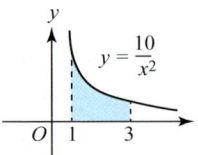
$y = \frac{10}{x^2}$

i
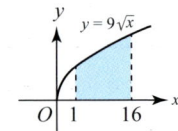
$y = 9\sqrt{x}$

Reasoning and problem-solving

To calculate the area under a curve

1. Make a sketch of the function, if there isn't one provided.
2. Identify the area that has to be calculated.
3. Write down the definite integral associated with the area.
4. Evaluate the definite integral and remember that the area is always positive.

Exercise 4.7B Reasoning and problem-solving

1. An area of 29 units² is bound by the line $y = x$, the x-axis and the lines $x = -3$ and $x = a$, $a < 0$. Use integration to help you calculate the value of a

2. Integrate the function $f(x) = x^2 - 6x + 8$ with respect to x and thus calculate the area bound by the curve $y = f(x)$, the x-axis, and the lines $x = 2$ and $x = 4$

3. A parabola has equation $y = 10 + 3x - x^2$

   Calculate the area of the dome trapped between the curve and the x-axis.

4. The equation $y = \sqrt{x}$ is one part of a parabola with $y = 0$ as the axis of symmetry.

   a Calculate $\int y \, dx$

   b Calculate the area trapped between the curve, the x-axis and the lines $x = 0$ and $x = 1$

   c P is the area below the curve between $x = 1$ and $x = 4$. Q is the area between $x = 4$ and $x = 9$. Calculate the ratio P : Q

   d Calculate the area trapped between the curve $y = 1 + \sqrt{x}$, the x-axis and the lines $x = 0$ and $x = 1$ and explain the difference between this and the answer to part **b**.

5. **a** Area A is defined by $A(a) = \int_{1}^{a} 3x^2 \, dx$

   For what value of a is the area 999 units²?

   b Area B is defined by $B(a) = \int_{a}^{2a} 3x^2 \, dx$

   For what value of a is the area 875 units²?

   c Area C is defined by $C(a) = \int_{a}^{a+1} 3x^2 \, dx$

   For what values of a is the area 19 units²?

6. The area trapped between the x-axis and the parabola $y = (x+3)(x-9)$ is split in two parts by the y-axis. What is the ratio of the larger part to the smaller?

7. An area is bounded by the curve $y = 4x^3$, the x-axis, the line $x = 1$ and the line $x = a$, $a > 1$. What is the value of a if the area is 2400 units²?

8. Sketch a graph of $y = \sin x$ for $0 \le x \le 360°$ and, using this, explain why the definite integral $\int_{0}^{360} \sin x \, dx$ equals zero.

9. On a velocity–time curve, the distance travelled can be obtained by calculating the area under the curve.

   An object is thrown straight up. Its velocity, in m s⁻¹, after t seconds can be calculated using $v(t) = 24t - 5t^2$

   What distance did the object travel between $t = 1$ and $t = 4$?

Challenge

10. Space debris is detected falling into the Earth's atmosphere. Its velocity in kilometres per second is modelled by $v(t) = 5 + 0.01t$ where t is the time in seconds measured from where the debris was detected. It completely burned up after 4 seconds. How far did the debris travel in the atmosphere?

4 Summary and review

Chapter summary

- The gradient of the tangent to a curve at the point P can be approximated by the gradient of the chord PQ, where Q is a point close to P on the curve.
- The derivative at the point P on the curve can be calculated from first principles by considering the limiting value of the gradient of the chord PQ as Q tends to P
- The derivative is denoted by $f'(x)$

$$f'(x) = \lim_{h \to 0} \frac{f(x+h) - f(x)}{h}$$

- If $y = f(x)$ then $\dfrac{dy}{dx} = f'(x)$
- The derivative gives the instantaneous rate of change of y with respect to x.
- The derivative of $y = ax^n$, where a is a constant, is $\dfrac{dy}{dx} = nax^{n-1}$
- $f(x) = ag(x) + bh(x) \Rightarrow f'(x) = ag'(x) + bh'(x)$, where a, b are constants.
- $f'(x)$ gives the rate of change of function f with respect to x. So,

  $f'(x) > 0$ means the function is increasing,
  $f'(x) < 0$ means the function is decreasing,
  $f'(x) = 0$ means the function is stationary.

- By sketching the gradient function, you can determine whether a function is increasing, decreasing, or stationary.
- At a turning point, the tangent is horizontal, so $f'(x) = 0$
- Where the function changes from being an increasing to a decreasing function, the turning point is referred to as a maximum.
- Where the function changes from being a decreasing to an increasing function, the turning point is referred to as a minimum.
- The derivative is itself a function that can be differentiated. The result is called the second derivative and is denoted by $f''(x)$ or $\dfrac{d^2 y}{dx^2}$
- $\dfrac{d^2 y}{dx^2}$ gives the rate of change of the gradient with respect to x. Assuming $\dfrac{dy}{dx} = 0$, then $f''(x) > 0$ means the gradient is increasing—it is a minimum turning point.

  $f''(x) < 0$ means the gradient is decreasing—it is a maximum turning point.
- Since the derivative at a point $(a, f(a))$ gives the gradient of the curve at that point and the gradient of the tangent at that point then $m_{tangent} = f'(a)$ and the equation of the tangent is $y - f(a) = f'(a)(x - a)$
- The normal at $(a, f(a))$ has a gradient $m_{normal} = -\dfrac{1}{f'(a)}$ so the equation of the normal is $y - f(a) = -\dfrac{1}{f'(a)}(x - a)$

- The process of differentiation can be reversed by a process called integration.

$$F'(x) = f(x) \Rightarrow F(x) = \int f(x)\,dx + c$$

where c is the constant of integration.

- $\int ax^n\,dx = \dfrac{ax^{n+1}}{n+1}$ ... read as "the integral of ax^n with respect to x"

- $\int (af(x) + bg(x))\,dx = a\int f(x)\,dx + b\int g(x)\,dx$

- $\int_a^b f(x)\,dx$ is called a definite integral with upper limit b and lower limit a

- $\int f(x)\,dx = F(x) + c \Rightarrow \int_a^b f(x)\,dx = F(b) - F(a)$

- If $f(x)$ is continuous in the interval $a \leq x \leq b$ and all points on the curve in this interval are on the same side of the x-axis, the area bounded by the curve, the x-axis and the lines $x = a$ and $x = b$ is given by the positive value of $\int_a^b f(x)\,dx$

Check and review

You should now be able to...	Try Questions
✔ Differentiate from first principles.	1
✔ Differentiate functions composed of terms of the form ax^n	2, 3
✔ Use differentiation to calculate rates of change.	4
✔ Work out equations, tangents and normals.	5
✔ Work out turning points and determine their nature.	6
✔ Work out and interpret the second derivative.	7
✔ Work out the integral of a function.	8
✔ Understand and calculate definite integrals.	9
✔ Use definite integrals to calculate the area under a curve.	10

1 Differentiate each function from first principles.

   a $y = 3x^2$ at $(1, 3)$ **b** $y = x^3 + 1$ when $x = -1$ **c** $y = x^4 - x - 5$ at $(2, 9)$

   d $y = 1 - x^2$ **e** $y = x - x^2$ **f** $y = \pi x^2$

2 Work out the derivative of

   a $f(x) = x^3 + 2x^2 + 3x + 1$ **b** $y = 4\sqrt{x} + x$ **c** $y = x + \dfrac{1}{x} + \dfrac{1}{x^2}$

   d $f(x) = \sqrt[4]{x} - \sqrt[3]{x}$ **e** $y = \dfrac{x+3}{x^2}$ **f** $y = \dfrac{4}{x} + \dfrac{2}{\sqrt{x}}$

   g $y = 1 - \dfrac{1}{x^2} + \dfrac{1}{\sqrt[3]{x}}$ **h** $y = \dfrac{x^2 + 2x + 3}{x}$ **i** $y = \dfrac{x + 2\sqrt{x}}{\sqrt{x}}$

   j $y = \dfrac{(x^2 + x - 5)}{2x^{\frac{1}{2}}}$

3 Work out the value of the following functions. Show your working.

   a $f'(2)$ when $f(x) = 2x^2 - 5$ **b** $\dfrac{dy}{dx}$ when $x = 3$ given $y = \dfrac{x+3}{x}$

   c $f'(4)$ when $f(x) = 1 + \dfrac{1}{x} - \dfrac{33}{\sqrt{x}}$ **d** $\dfrac{dy}{dx}$ when $x = 9$ given $y = \dfrac{x-9}{2\sqrt{x}}$

4 **a** Work out the rate of change of y with respect to x when $x = 4$ given that $y = x(2x^2 - 5x)$ Show your working.

   b A roll of paper is being unravelled. The volume of the roll is a function of the changing radius, r cm. $V = 25\pi r^2$ [V cm³ is the volume].

      Calculate the rate of change of the volume when the radius is 3 cm. Show your working.

   c A particle moves along the x-axis so that its distance, D cm, from the origin at time t seconds is given by $D(t) = t^2 - 5t + 1$

      i Work out the particle's velocity at $t = 3$ (the rate of change of distance with respect to time). Show your working.

      ii Work out the particle's acceleration at this time (the rate of change of velocity with time). Show your working.

   d A tourist ascends the outside of a tall building in a scenic elevator.

      As she ascends, she can see further. The distance to her horizon, K km, can be calculated by the formula $K = \sqrt{\dfrac{hD}{1000}}$ where h is her height in metres and D is the diameter of the planet in kilometres. On Earth, $D = 12\,742$ km

      i Calculate the rate of change of the distance to her horizon with respect to her height
        1 When she is 50 m up,
        2 When she is 100 m up.

      ii How high up will she be when the distance to the horizon is changing at a rate of 0.4 km per m of height?

      iii Suppose the tourist were on a building on the Moon. The Moon has a diameter of 3474 km. Calculate the rate of change of the distance to her horizon with respect to her height when she is 50 m up.

5 Work out the equations of the tangent and normal to the given curves at the given points. Show your working.

    a $y = x^2 + 4x + 3$ at $(1, 8)$

    b $y = 2x - \dfrac{3}{x}$ at $(3, 5)$

    c $y = 5 + 4x - x^2$ at $(2, 9)$

    d $y = 200\sqrt{x}$ at $(25, 1000)$

6 Work out the turning points on each curve and determine their nature. Show your working.

    a $y = x^2 + 4x - 5$

    b $y = 3x^3 - 4x$

    c $y = 10x^4$

    d $y = ax + \dfrac{a}{x}$ where a is a positive constant.

    e $y = ax^2 + bx + c$ where a, b and c are positive constants.

7 Work out the second derivative of

    a $y = x^2$

    b $y = x^3 + x^2$

    c $y = \dfrac{1}{x}$

    d $y = 1 + x + \dfrac{x^2}{2} + \dfrac{x^3}{6} + \dfrac{x^4}{24} + \dfrac{x^5}{120} + \dfrac{x^6}{720}$

8 Calculate

    a $\displaystyle\int 1 - x - x^3 \, dx$

    b $\displaystyle\int \dfrac{1}{x^2} - 4x^2 \, dx$

    c $\displaystyle\int \sqrt{x} - \dfrac{1}{\sqrt{x}} - \sqrt[3]{x} \, dx$

    d $\displaystyle\int \dfrac{(x+3)^2}{x^{\frac{1}{2}}} \, dx$

9 Calculate the following definite integrals. Show your working.

    a $\displaystyle\int_{1}^{4} x + 6 \, dx$

    b $\displaystyle\int_{1}^{10} 2\pi \, dx$

    c $\displaystyle\int_{-2}^{4} x^2 - 4x - 5 \, dx$

    d $\displaystyle\int_{\sqrt{2}}^{\sqrt{5}} 2x^3 \, dx$

    e $\displaystyle\int_{1}^{9} \dfrac{x+1}{2\sqrt{x}} \, dx$

10 a Calculate the area bounded by $y = x^2 - 7x + 10$, the x-axis, $x = 2$ and $x = 5$. Show your working.

    b Calculate the area bounded by $y = x^2 - 5x + 6$, the x-axis, $x = 2$ and $x = 3$. Show your working.

    c Calculate the area bounded by $y = 27 - x^3$, the x-axis, and the y-axis. Show your working.

What next?

Score		
0 – 5	Your knowledge of this topic is still developing. To improve, search in MyMaths for the codes: 2028–2030, 2054–2056, 2269–2270, 2273	
6 – 8	You're gaining a secure knowledge of this topic. To improve, look at the InvisiPen videos for Fluency and skills (04A)	
9 – 10	You've mastered these skills. Well done, you're ready to progress! Now try looking at the InvisiPen videos for Reasoning and problem-solving (04B)	

Click these links in the digital book

History

Differentiation and integration belong to a branch of maths called calculus, which is the study of change. It took hundreds of years and the work of many mathematicians to develop calculus to the state in which we know it today.

The first person to write a book about both differentiation and integration, and the first woman to write any book about mathematics, was **Gaetana Maria Agnesi**. One of 21 children, Maria displayed incredible ability in a number of disciplines at an early age, and was an early campaigner for women's right to be educated – giving a speech on the topic when she was 9 years old.

Agnesi's book, **Istituzioni analitiche ad uso della gioventu italiana** (analytical institutions for the use of the Italian youth), was published in 1748 when she was just 30 years old.

Research

Before Maria, in the 17th century, the fundamental theorem of calculus was independently discovered by **Isaac Newton** and **Gottfried Leibniz**.

These two were both well-regarded by the mathematical community, and both published very similar works on the subject within only a few years of each other. This led to suspicions of copying and a bitter feud between the two lasted until Leibniz's death in 1716.

Find out the contributions that Newton and Leibniz made to the theorem of calculus and use your research to write a summary. In your summary, discuss any similarities and differences between the works of each mathematician, and how their results added to or built on what was previously established in the field of calculus.

Isaak Newton.

Gottfried Wilhelm Leibniz.

4   Assessment

In questions that tell you to show your working, you shouldn't depend solely on a calculator. For these questions, solutions based entirely on graphical or numerical methods are not acceptable.

1   Differentiate these polynomials with respect to x.

    a $x^4 - 5x^2 + 1$ **[3 marks]**   **b** $x^{-3} + 7x^3$ **[3]**   **c** $\dfrac{3}{x^2} - \sqrt{x}$ **[4]**

2   The radius (in cm) of a circle at time t seconds is given by $r = 20 - 2\sqrt{t}$

    a   Work out an expression for the rate of change of the radius.   **[3]**

    b   Calculate the rate of change of the radius at time 25 s. State the units of your answer.   **[2]**

3 $y = 2x^2 - x^{-3}$

    a   Calculate the gradient of the curve at the point $(1, 1)$. Show your working.   **[3]**

    b   Work out the equation of the normal to the curve at the point $(1, 1)$ **[3]**

4 $f(x) = \dfrac{1}{5}x^5 - x^2$

    a   Work out an expression for $f'(x)$ **[2]**

    b   Calculate $f'(2)$ **[2]**

    c   Work out the equation of the tangent to $y = f(x)$ at the point where $x = 2$

       Give your equation in the form $ax + by + c = 0$ where a, b and c are integers.   **[4]**

5   The curve C has equation $y = 6x^3 - 3x^2 - 12x + 5$

    a   Use calculus to show that C has a turning point when $x = -\dfrac{2}{3}$ **[4]**

    b   Work out the coordinates of the other turning point on C. Show your working.   **[2]**

    c   Is this point is a maximum or a minimum? Explain your reasoning.   **[2]**

6   Work out the range of values of x for which $y = x^3 + 5x^2 - 8x + 4$ is decreasing.   **[5]**

7   Work out these integrals.

    a $\displaystyle\int 2x + 3x^5 \, dx$ **[2]**   **b** $\displaystyle\int x^{-4} - 4x^3 \, dx$ **[3]**   **c** $\displaystyle\int 3\sqrt{x} + \dfrac{1}{\sqrt{x}} \, dx$ **[3]**

8   Calculate the exact values of these definite integrals. You must show your working.

    a $\displaystyle\int_{1}^{2} 5x^4 - 2 \, dx$ **[4]**   **b** $\displaystyle\int_{0}^{3} 2\sqrt{x} \, dx$ **[4]**   **c** $\displaystyle\int_{1}^{2} \dfrac{2}{x^3} - 3x \, dx$ **[4]**

9   Find an expression for $f(x)$ when

    a $f'(x) = 4x^2 + 5x - 1$ **[3]**   **b** $f'(x) = 7x^{-3} - x + \sqrt{x}$ **[5]**

10   The region shown is bounded by the x-axis and the curve $y = -x^2 + 4x - 3$

     Show that the area of the shaded region is $1\dfrac{1}{3}$ square units.   **[4]**

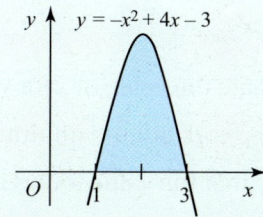

11 $y = 3x^2$

    a   Work out $\dfrac{\mathrm{d}y}{\mathrm{d}x}$ from first principles.   **[5]**

    b   Calculate the gradient of the tangent where $x = 5$ **[2]**

12 $y = x^3 - 2x$

    a   Work out $\dfrac{\mathrm{d}y}{\mathrm{d}x}$ from first principles.   **[5]**

    b   Calculate the gradient of the tangent where $x = 2$ **[2]**

13 Work out $\dfrac{\mathrm{d}y}{\mathrm{d}x}$ when

    a $y = x(x+3)^2$ **[4]**   **b** $y = \dfrac{x+2}{\sqrt{x}}$ **[4]**

14 The volume (in litres) of water in a container at time t minutes is given by

$$V = \frac{t^3 - 8t}{3t}$$

    Calculate the rate of change of the volume after 4 minutes. Show your working.   **[6]**

15 Find $\dfrac{\mathrm{d}y}{\mathrm{d}x}$ and work out the equation of the normal to $y = x^2(2x+1)(x-3)$

    at the point where $x = 2$ **[10]**

16 The curve C has equation $y = \dfrac{3x^3 + \sqrt{x}}{2x}$

    a   Find $\dfrac{\mathrm{d}y}{\mathrm{d}x}$ and work out the equation of the tangent to C at the point where $x = 1$ **[9]**

    The tangent to C at $x = 1$ crosses the x-axis at the point A and the y-axis at the point B

    b   Calculate the exact area of the triangle AOB **[4]**

17 Show that the function $\mathrm{f}(x) = (1+2x)^3$ is increasing for all values of x **[6]**

18 Work out the range of values of x for which $\mathrm{f}(x) = 5\sqrt{x} + \dfrac{3}{\sqrt{x}}$ is a decreasing function.   **[4]**

19 Work out the range of values of x for which $y = \dfrac{1}{3}x(x-1)(5-x)$ is an increasing function.   **[7]**

20 Given that $\mathrm{f}(x) = x^2(2x - \sqrt{x})$, work out expressions for

    a $\mathrm{f}'(x)$ **[4]**   **b** $\mathrm{f}''(x)$ **[2]**

21 Given that $y = 3x^2 - 4\sqrt{x}$, work out $\dfrac{\mathrm{d}^2 y}{\mathrm{d}x^2}$ **[4]**

22 Showing your working, calculate the coordinates of the stationary point on
the curve with equation

$$y = 32x + \frac{2}{x^2} - 15, \ x > 0$$

    Show that this point is a minimum.   **[8]**

23 A cylindrical tin is closed at both ends and has a volume of $200\,\mathrm{cm}^3$.

    a   Express the height, h in terms of the radius, x **[3]**

    b   Show that the surface area, A of the tin is given by

$$A = 2\pi x^2 + \frac{400}{x}$$

       [3]

    c   Calculate the value of x for which A is a minimum.   **[4]**

    d   Hence, work out the minimum value of A **[2]**

    e   Justify that the value found in part **d** is a minimum.   **[2]**

24 A box has a square base of side length x

The volume of the box is $3000\,\text{cm}^3$.

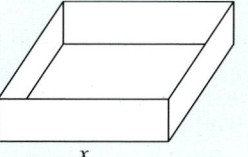

   a Show that the surface area, A, of the box (not including the lid) is given by $A = x^2 + \dfrac{12000}{x}$ **[6]**

   b Calculate the value of x for which A is a minimum.   **[3]**

   c Hence, work out the minimum value of A **[2]**

   d Justify that the value found in part **c** is a minimum.   **[2]**

25 Work out these integrals.

   a $\displaystyle\int (2x+3)^2\,\mathrm{d}x$ **[3]**   **b** $\displaystyle\int \sqrt{x}(5x-1)\,\mathrm{d}x$ **[4]**   **c** $\displaystyle\int \dfrac{2+x}{2\sqrt{x}}\,\mathrm{d}x$ **[4]**

26 Calculate the exact values of these definite integrals. You must show your working.

   a $\displaystyle\int_0^1 (3x-1)^3\,\mathrm{d}x$ **[5]**   **b** $\displaystyle\int_{-1}^1 x^2(x-4)(x-5)\,\mathrm{d}x$ **[5]**

27 Calculate the area of the region bounded by the x-axis and the curve $y = x^2 - 3x - 10$
You must show your working.   **[8]**

28 The shaded region shown is bounded by the x-axis, the line $x = \dfrac{1}{2}$ and the curve with equation $y = \dfrac{5}{x^2} - 3x^2 - 2\sqrt{x}$, $x > 0$

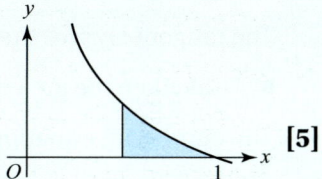

   Calculate the area of the shaded region. You must show your working.   **[5]**

29 The curve with equation $y = f(x)$ passes through the point $(1, 1)$

   Given that $f'(x) = 5x^4 - \dfrac{2}{x^3}$

   a Calculate $f(x)$ **[4]**

   b Work out the equation of the normal to the curve at the point $(1, 1)$ **[4]**

30 **a** Differentiate with respect to x, where k is a constant.

     i $kx + x^k, k \neq -1$ **ii** $\dfrac{1}{x^k} - k, k \neq 1$ **[4]**

   b Integrate the functions in part **a** with respect to x **[5]**

31 $f(x) = x^3 - 2x$

   The tangent to $y = f(x)$ through the point where $x = 2$ meets the normal through the point $x = -1$ at the point P. Calculate the coordinates of P **[13]**

32 The function $f(x)$ is given by $f(x) = x^2 + kx$ where k is a positive constant.

   The tangent to $y = f(x)$ at the point where $x = k$ meets the x-axis at the point A and the y-axis at the point B

   Given that the area of the triangle ABO is 36 square units, work out the value of k **[10]**

33 The normal to the curve $y = 2x^2 - x + 2$ at the point where $x = 1$ intersects the curve again at the point Q. Calculate the coordinates of Q **[9]**

34 Work out and classify all the stationary points of the curve $y = x^4 - 2x^2 + 1$.
Show your working.   **[8]**

35 Work out and classify all the stationary points of the curve with equation
$y = 3x^4 + 4x^3 - 12x^2 + 20$. You must show your working.   **[8]**

36 Calculate the range of values of x for which $\mathrm{f}(x)=4x^4-2x^3$ is an increasing function. **[5]**

37 Calculate the range of values of x for which $\mathrm{f}(x)=-3x^4+8x^3+90x^2+12$ is a decreasing function. **[6]**

38 A triangular prism has a cross-section with base twice its height.

The volume of the prism is $250\,\mathrm{cm}^3$.

Calculate the minimum possible surface area of the prism given that it is closed at both ends. **[12]**

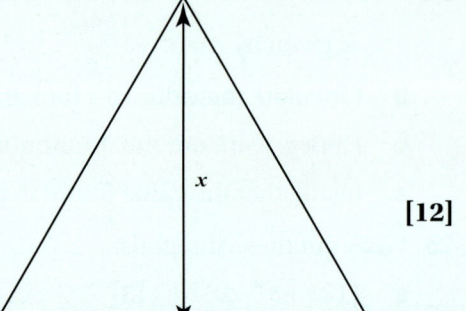

39 A cylinder of radius, $r\,\mathrm{cm}$ is open at one end. The surface area of the cylinder is $700\,\mathrm{cm}^2$.

Calculate the maximum possible volume of the cylinder. **[11]**

40 The curve with equation $y=\mathrm{f}(x)$ passes through the point $A(1,4)$

Given that $\mathrm{f}'(x)=3x^2-\dfrac{2}{x^3}$

  a Work out the equation of the tangent to the curve at the point when $x=-1$ **[7]**

  The tangent crosses the y-axis at the point B

  b Calculate the area of the triangle ABO **[3]**

41 The curve with equation $y=\mathrm{f}(x)$ passes through the point $(0,0)$

Given that $\mathrm{f}'(x)=4x-3x^2$, work out the area enclosed by the curve $y=\mathrm{f}(x)$ and the x-axis. You must show your working. **[8]**

42 The shaded region is bounded by the curve with equation

$y=12-7x-x^2$ and the line with equation $y=4$

Calculate the area of the shaded region. Show your working. **[9]**

43 Calculate the area of the region bounded by the x-axis and the curve with equation $y=x(x+1)(x-2)$ Show your working. **[8]**

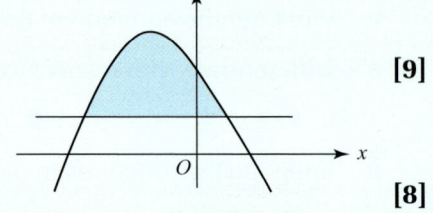

44 The shaded region is bounded by the curve with equation $y=3x^2+1$ and the lines $y=4$ and $y=2$

Calculate the area of the shaded region. Show your working. **[11]**

45 The region R is bounded by the x-axis and the curve with equation

$y=x^2(k-x)$, where k is a positive constant.

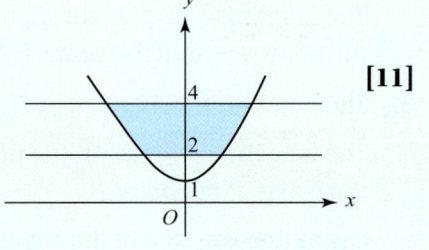

Given that the area of R is 108 square units, calculate the value of k **[7]**

46 The region R is bounded by the curve with equation

$y=13-2x-x^2$ and the line $y=11-x$

Showing your working, calculate the area of R **[9]**

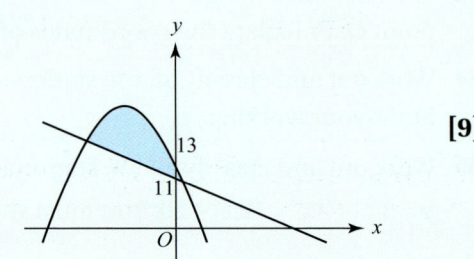

5 Exponentials and logarithms

Animal populations grow at increasing rates when there is adequate food supply, optimal living conditions and the absence of predators. The growth of the population is proportional to its size, leading to faster and faster growth as time progresses. This type of growth is called 'exponential'.

Exponential functions and their inverses, logarithmic functions, appear in many areas of life as well as in many areas of mathematics. They are used to model growth and decline and can be used to calculate populations, radioactive decay and increasing bank balances as a result of interest. This means that these functions are an invaluable part of mathematical modelling.

Orientation

What you need to know	What you will learn	What this leads to
KS4 • Recognise, sketch and interpret graphs of exponential functions $y = k^x$ for positive values of k	• To convert between powers and logarithms. • To manipulate and solve equations involving powers and logarithms. • To use exponential functions and their graphs. • To verify and use mathematical models and consider limitations of these models.	**Ch15 Differentiation 2** Derivation for a^n and ln x and applications to problems.
Ch2.4 Curve sketching • Graph transformations. • Proportionality.		**Ch16 Integration and differential equations** Integration of exponential functions.
Ch4.1 Differentiation • Differentiation from first principles.		**Careers** Archeology. Investment banking. Wildlife biology.

p.58

p.86

 **MyMaths** Practise before you start   🔍 1070, 2024, 2028, 2258

Fluency and skills

Logarithms allow you to perform certain calculations with a large number of digits more efficiently.

You read $10^4 = 10\,000$ as '10 to the **power** 4 equals $10\,000$'.

Similarly, you read $x = a^n$ as 'x is equal to a to the power n'.

Logarithms are a different way of working with powers. Another way of reading $x = a^n$ is to say 'n is the logarithm of x to **base** a'. The following three statements are equivalent.

> **Key point**
>
> $x = a^n$
>
> $n = \log_a x$
>
> n is the log of x to base a

For example, $10^2 = 100$ and $\log_{10} 100 = 2$ are equivalent statements.

The most common base to use is 10, but you can use any positive number.

You need to know the following three cases for $x = a^n$. They are true when $a > 0$ and $a \neq 1$

> **Key point**
>
> When $n = 1$ $\Rightarrow$ $a^1 = a$ $\Rightarrow$ $\log_a a = 1$
>
> When $n = 0$ $\Rightarrow$ $a^0 = 1$ $\Rightarrow$ $\log_a 1 = 0$
>
> When $n = -1$ $\Rightarrow$ $a^{-1} = \dfrac{1}{a}$ $\Rightarrow$ $\log_a\left(\dfrac{1}{a}\right) = -1$

There are three **laws of logarithms**:

> **Key point**
>
> Law 1 $\log_a(xy) = \log_a x + \log_a y$
>
> Law 2 $\log_a\left(\dfrac{x}{y}\right) = \log_a x - \log_a y$
>
> Law 3 $\log_a(x^k) = k \log_a x$

See Ch1.2
For a reminder on index laws.

These laws are derived from the rules of indices.

For example, for Law 1, if $p = \log_a x$ and $q = \log_a y$, then $x = a^p$ and $y = a^q$
You know $xy = a^p \times a^q = a^{p+q}$ so $\log_a(xy) = p + q = \log_a x + \log_a y$

If the base of a logarithm is not given, then always use base 10

> **Key point**
>
> You can write $\log_{10} x$ as simply $\log x$

Example 1

Show that $\log_5 125 = 3$

$\log_5 125 = \log_5 5^3$ — Write 125 using powers.

$= 3 \times \log_5 5$ — Use Law 3 to simplify the log.

$= 3 \times 1 = 3$

Example 2

Write as a single logarithm $5 \log_{10} 2 - \log_{10} 4 + \dfrac{1}{2} \log_{10} x$. Show your working.

$5 \log 2 - \log 4 + \dfrac{1}{2} \log x = \log 2^5 - \log 4 + \log x^{\frac{1}{2}}$ — Use Law 3 to write $5 \log 2$ as $\log 2^5$ and $\dfrac{1}{2} \log x$ as $\log x^{\frac{1}{2}}$

$= \log \left(\dfrac{2^5}{4} \times x^{\frac{1}{2}} \right)$ — Use Laws 1 and 2 to combine logs.

$= \log \left(8\sqrt{x} \right)$ — Simplify the expression.

Exercise 5.1A Fluency and skills

1 Write each of the following using logarithms.

a $2^3 = 8$ **b** $3^2 = 9$

c $10^3 = 1000$ **d** $2 = 16^{\frac{1}{4}}$

e $0.001 = 10^{-3}$ **f** $\dfrac{1}{4} = 4^{-1}$

g $\dfrac{1}{8} = 2^{-3}$ **h** $\dfrac{1}{4} = 8^{-\frac{2}{3}}$

2 Write each of the following in index notation.

a $\log_2 32 = 5$ **b** $\log_2 16 = 4$

c $\log_3 81 = 4$ **d** $\log_5 1 = 0$

e $\log_3 \left(\dfrac{1}{9} \right) = -2$ **f** $1 = \log_6 6$

g $\log_{16} 2 = \dfrac{1}{4}$ **h** $\log_2 \left(\dfrac{1}{64} \right) = -6$

3 Find the value of each of these expressions. Show your working.

a $\log_3 9$ **b** $\log_2 16$

c $\log_4 16$ **d** $\log_{16} 16$

e $\log_5 125 + \log_5 5$ **f** $\log_3 81 - \log_3 27$

g $\log_2 8 + \log_2 2$ **h** $\log_3 9 - \log_3 27$

4 Write each of these as a single logarithm.

a $\log 4 + \log 3$

b $\log 12 - \log 2$

c $\log 2 + \log 6 - \log 3$

d $3 \log 2 + \log 4$

e $3 \log 3 + 2 \log 2$

f $3 \log 6 - 2 \log 3 + \dfrac{1}{2} \log 9$

g $4 \log 2 - 5 \log 1$

h $2 \log 4 + \log \dfrac{1}{2}$

i $\dfrac{1}{2} \log 9 + \dfrac{1}{3} \log 8$

j $\dfrac{1}{2} \log 4 + \dfrac{2}{3} \log 27$

k $4 \log x + 2 \log y$

l $2 \log x + \log y$

5 Write each of these in terms of $\log a$, $\log b$ and $\log c$, where a, b and c are greater than zero.

a $\log(a^2 b)$

b $\log\left(\dfrac{a}{b}\right)$

c $\log\left(\dfrac{a^2}{b^3}\right)$

d $\log(a\sqrt{b})$

e $\log\left(\dfrac{\sqrt{ab}}{c}\right)$

f $\log(\sqrt{abc})$

g $\log\left(a\sqrt{\dfrac{b}{c}}\right)$

6 Write each of these logarithms in terms of $\log 2$ and $\log 3$. By taking $\log 2 \approx 0.301$ and $\log 3 \approx 0.477$, find their approximate values.

a $\log 12$ **b** $\log 18$

c $\log 4.5$ **d** $\log 13.5$

e $\log 5$ **f** $\log 0.125$

7 Prove that

a $\dfrac{\log 125}{\log 25} = 1.5$ **b** $\dfrac{\log 27}{\log 243} = 0.6$

8 Write $\log 40$ in terms of $\log 5$. If $\log 5 = 0.698\,970\,00\ldots$ find the value of $\log 40$ correct to 6 decimal places.

Reasoning and problem-solving

To solve problems with logarithms

(1) Convert between index notation and logarithmic notation.

(2) Apply the laws of logarithms if necessary and any results for special cases.

(3) Manipulate and solve the equation. Check your solution by substituting back into the original equation.

Example 3

Solve the equation $2^{3x+1} = 36$. Show your working.

$3x + 1 = \log_2 36$

$3x = \log_2 36 - 1$

$x = \dfrac{1}{3}(\log_2 36 - 1) = 1.39$ (to 3 sf)

Check: $2^{3x+1} = 2^{4.17+1} = 2^{5.17} = 36.0$

1 Convert to logarithmic notation using the equivalent statements $x = a^n$ and $n = \log_a x$

3 Manipulate and solve for x. Use the original equation to check your solution.

Calculator

Try it on your calculator

You can use a calculator to solve equations with exponents.

Eq: $3^{2x-1} = 10$
  X = 1.547951637
 Lft = 10
 Rgt = 10

REPT

Activity

Find out how to solve $3^{2x-1} = 10$ on *your* calculator.

1 Solve each of these equations. Give your answers to 3 sf and show your working.

a $2^x = 5$ **b** $3^x = 10$

c $5^x = 4$ **d** $2^{x+1} = 7$

e $3^{x-1} = 80$ **f** $2^x \times 2^{x-1} = 20$

g $\dfrac{1}{2^x} = 9$ **h** $\left(\dfrac{3}{4}\right)^x = 2$

i $2^{x+1} = 3^{2x}$ **j** $5^{3x-1} = 2^{x+3}$

2 Solve each of these equations.

Show your working.

a $\log_{10}(2x-40) = 3$

b $\log_5(3x+4) = 2$

c $\log_3(x+2) - \log_3 x = \log_3 8$

d $\log_3(x+2) + \log_3 x = 1$

e $\log_5(x^2 - 10) = \log_5(3x)$

f $\log_4 x + \log_4(x - 12) = 3$

g $2\log_2 x = \log_2(2x - 1)$

3 Show that $y = 316$ and $x = 3.16$ is the solution to these simultaneous equations to 3 sf.

$\log x + \log y = 3$
$\log x - \log y = 2$

4 Let $y = a^x$ and use the fact that $a^{2x} = (a^x)^2$ to solve each of these quadratic equations. Give your answers to 3 sf where appropriate.

a $2^{2x} - 3 \times 2^x + 2 = 0$

b $3^{2x} - 12 \times 3^x + 27 = 0$

c $2^{2x} + 6 = 5 \times 2^x$

d $2^{2x} - 2^{x+1} = 8$

e $5^{2x} + 5^{x+1} - 50 = 0$

f $3^{2x+1} - 26 \times 3^x = 9$

g $2^{2x+1} - 13 \times 2^x + 20 = 0$

h $9^x + 8 = 2(3^{x+1})$

i $25^x + 4 = 5^{x+1}$

j $2^x \times 2^{x+1} = 10$

5 a Solve $3^{2x-1} - 5 \times 3^{x-1} + 2 = 0$

b Hence, find the point of intersection of the graphs of $y = 3^{2x-1} + 2$ and $y = 5 \times 3^{x-1}$

6 Find the point of intersection of the graphs of $y = 2^{2x} - 5$ and $y = 4 \times 2^x$. Show your working.

7 a Solve

     i $3^x > 10\,000$ **ii** $0.2^x < 0.001$

b Calculate the smallest integer for which $2^n > 1\,000\,000$

c Calculate the largest integer for which $0.2^n > 0.000\,005$

8 Calculate the smallest positive integer value of x such that

a $\left(1 + \dfrac{x}{100}\right)^6 > 1.25$ **b** $1 - 0.8^x > 0.95$

9 Given that $x = a^p$ and $y = a^q$, prove that

a $\log_a\left(\dfrac{x}{y}\right) = \log_a x - \log_a y$

b $\log_a(x^k) = k\log_a x$

10 a Prove that $\log_a b = \dfrac{1}{\log_b a}$

b Solve

     i $\log_5 x = 4\log_x 5$

     ii $\log_3 x + 8\log_x 3 = 6$

Challenge

11 a Starting with $x = a^y$, prove that
$$\log_a x = \dfrac{\log_b x}{\log_b a}$$

b Express $\log_2 12$ as a logarithm in base 10, so that $\log_2 12 = k \times \log_{10} 12$ and find the value of k

c Write these as logarithms in base 10 and use a calculator to find their values.

     i $\log_6 37$ **ii** $\log_4 6$ **iii** $\log_3 25$

12 If $x = \log_y z$, $y = \log_z x$ and $z = \log_x y$, prove that $xyz = 1$

Fluency and skills

Key point

The general equation of an **exponential function** is $y = a^x$ where a is a positive constant.

The table shows values for the graphs of $y = 2^x$ and $y = \left(\dfrac{1}{2}\right)^x$

Note that $a^{-x} = \left(\dfrac{1}{a}\right)^x$

x	-2	-1	0	1	2
$y = 2^x$	$2^{-2} = \dfrac{1}{4}$	$2^{-1} = \dfrac{1}{2}$	$2^0 = 1$	$2^1 = 2$	$2^2 = 4$
$y = \left(\dfrac{1}{2}\right)^x$	$\left(\dfrac{1}{2}\right)^{-2} = 4$	$\left(\dfrac{1}{2}\right)^{-1} = 2$	$\left(\dfrac{1}{2}\right)^0 = 1$	$\left(\dfrac{1}{2}\right)^1 = \dfrac{1}{2}$	$\left(\dfrac{1}{2}\right)^2 = \dfrac{1}{4}$

The diagram shows how the shape of the graph varies for different values of a. When a is greater than 1, the graph curves up from left to right. When a is less than 1, the graph curves down from left to right and is a reflection in the y-axis.

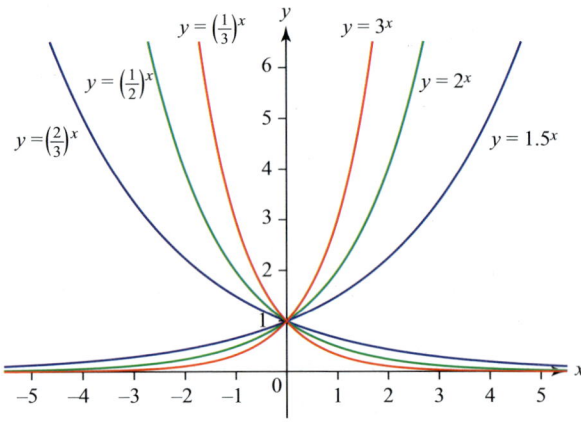

The equations with $a < 1$ can also be written as $y = 1.5^{-x}$, $y = 2^{-x}$ and 3^{-x}, so the diagram also shows the relationship between $y = a^x$ and $y = a^{-x}$

There is one special value of a where the gradient of the curve $y = a^x$ is equal to the value of a^x for all values of x

This is called **the exponential function** and is written $y = e^x$, where **e = 2.71828** (to 5 dp).

e is irrational (like π) and e^x is sometimes written $\exp(x)$

Key point

The graph of $y = e^x$ has a gradient of e^x at any point (x, y)

See Ch2.4

For a reminder on transformations of graphs and proportionality.

You can transform the graph of $y = e^x$ in various ways. In particular, a stretch parallel to the x-axis by a scale factor $\dfrac{1}{k}$ results in a curve with the equation $y = e^{kx}$

The tangent at P_2 is steeper by a factor k than the tangent at P_1

The graph of $y = e^{kx}$ has a gradient of $k \times e^{kx}$ at the point (x, y)

This means that the gradient of $y = e^{kx}$ is proportional to y at the point (x, y), where the constant of proportionality is k

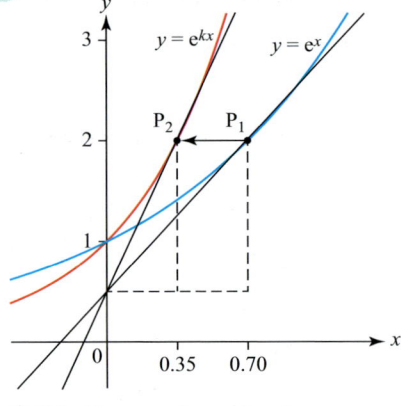

▲ This diagram shows $k = 2$

The inverses of exponential functions are logarithmic functions. You can find the inverse of an exponential function, $y = a^x$, using the following method.

$y = a^x$

$\log_a y = \log_a a^x$

$\log_a y = x\log_a a$

$\log_a y = x$

$y = \log_a x$

> **To find the inverse:**
> Take logs to base a on both sides.
> Use log laws to find x in terms of y
> Remember $\log_a a = 1$
> Then interchange x and y

> **Key point**
>
> The inverse of $y = a^x$ is the logarithmic function, $y = \log_a x$

The diagram shows the shape of, and the relationship between, some exponential functions and their inverses. In particular

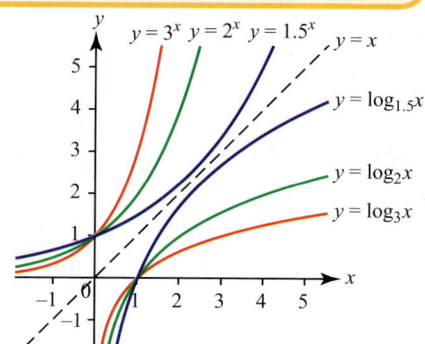

- a^x is positive for all values of x
- $\log_a x$ does not exist for negative values of x
- all the exponential graphs pass through $(0, 1)$ because $a^0 = 1$ for all a $(a \neq 0)$
- all the logarithmic graphs pass through $(1, 0)$ because $\log_a 1 = 0$ for all $a > 0$

ICT Resource online

To experiment with the graph of $y = a^x$ and its inverse, click this link in the digital book.

> **Key point**
>
> The inverse of $y = e^x$ is $y = \log_e x$ which can be written $y = \ln x$
>
> $\ln x$ is called the **natural (or Naperian) logarithm**.

> **See p.614**
> For a list of mathematical notation.

Note that $y = e^x$ passes through the point $(0, 1)$ and its inverse $y = \ln x$ passes through the point $(1, 0)$
The x-axis is an asymptote for the graph $y = e^x$
The y-axis is an asymptote for the graph $y = \ln x$

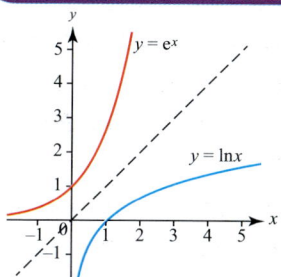

Example 1

a This is a table of values for the function $y = a^x$
Find the value of a. Hence, find the values of p and q

x	0	1	2	3
y	1	p	9	q

b The graph of the function $y = e^{kx}$ contains the point $(3, e^6)$
Find the value of k and the gradient of the graph at this point.

a When $x = 2$, $a^2 = 9$, so $a = 3$ •——— Substitute into $y = a^x$

$p = a^1 = 3^1 = 3$ and $q = a^3 = 3^3 = 27$

b When $x = 3$, $y = e^{3k} = e^6$, so $k = 2$ •——— Substitute into $y = e^{kx}$

The gradient of $y = e^{kx}$ is $k \times e^{kx}$ •——— Use the known fact about $y = e^{kx}$

So the gradient at the point $(3, e^6)$ is $2 \times e^{2 \times 3} = 2e^6$

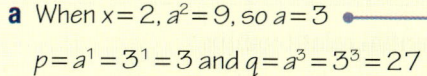

You can solve equations which involve the exponential and natural logarithm functions using the same methods as in Section 5.1

Example 2

Solve the equations. Give your answers to 3 sf.

a $e^{3x+7} = 12$ **b** $\ln(4x - 1) = 6$

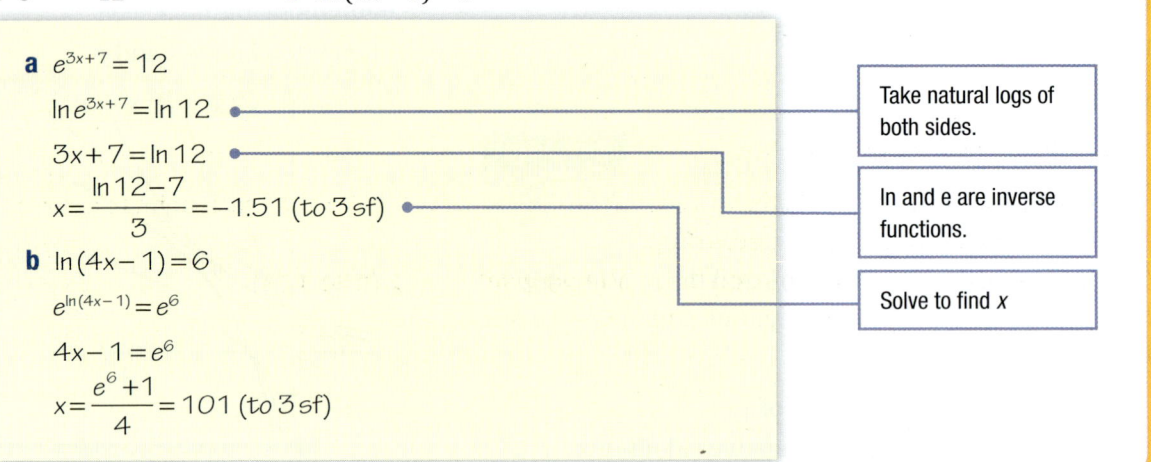

a $e^{3x+7} = 12$

$\ln e^{3x+7} = \ln 12$ ⟶ Take natural logs of both sides.

$3x + 7 = \ln 12$ ⟶ ln and e are inverse functions.

$x = \dfrac{\ln 12 - 7}{3} = -1.51$ (to 3 sf) ⟶ Solve to find x

b $\ln(4x - 1) = 6$

$e^{\ln(4x-1)} = e^6$

$4x - 1 = e^6$

$x = \dfrac{e^6 + 1}{4} = 101$ (to 3 sf)

Exercise 5.2A Fluency and skills

1 Draw the graphs of $y = 3^x$ and $y = 3^{-x}$ for $-2 \le x \le 2$ on the same axes.

Explain why the graph of $y = \left(\dfrac{1}{3}\right)^x$ is identical to one of the graphs you have drawn.

2 Solve these equations. Give your answers to 3 sf.

    a $e^{2x-3} = 8$ **b** $e^{5-x} = 100$

    c $\ln(3x + 6) = 1.5$ **d** $\ln(18x - 55) = 2$

3 These two tables give the values (x, y) for three relationships. Values are given to 4 sf where appropriate. For each table

    i Write the equation for each of the three relationships,

    ii State which one of the three relationships is exponential,

    iii Write the three y-values when $x = 5$

a

x	1	2	3	4
y_1	2	4	6	8
y_2	1	4	9	16
y_3	2	4	8	16

b

x	1	2	3	4
y_1	12	6	4	3
y_2	1	1.414	1.732	2
y_3	$\dfrac{1}{2}$	$\dfrac{1}{4}$	$\dfrac{1}{8}$	$\dfrac{1}{16}$

4 These two tables give the values of x and y for two exponential relationships.

Copy and complete the tables and write the equations for the relationships.

a

x	−2	−1	1	2	3	4
y			3	9	27	

b

x	−2	−1	1	2	3	4
y			$\dfrac{1}{5}$	$\dfrac{1}{25}$	$\dfrac{1}{125}$	

5 a Copy and complete this table for the curve $y = e^x$. Give your answers to 3 sf.

x	y	Gradient at (x, y)
0		
1		
2		
3		
4		

b Repeat for the curve $y = e^{3x}$

6 Each of the points W, X, Y, Z lies on one of the curves A, B, C or D. Match each point to a curve. The coordinates are given correct to 2 dp.

$W(2.10, 0.23)$ $X(0.80, 0.33)$ $Y(1.20, 2.30)$ $Z(0.50, 1.73)$

$A \ y = 2^x$ $B \ y = 3^x$ $C \ y = 2^{-x}$ $D \ y = 4^{-x}$

7 a Write down the gradient of the graph of

   i $y = e^x$ at the point $(2, e^2)$

   ii $y = e^{-x}$ at the point $(2, e^{-2})$

b Find the equation of the tangent to the curves at the given points.

8 Determine the equation of the tangent to the curve $y = e^x$ at the point where

   a $x = 3$ **b** $x = \dfrac{1}{2}$

9 Describe the transformation that maps the curve $y = e^x$ onto each of these curves. Sketch the graph of $y = e^x$ and its image in each case. You can check your sketches on a graphics calculator.

   a $y = e^{-x}$ **b** $y = -e^x$ **c** $y = 2e^x$ **d** $y = e^{2x}$

   e $y = e^x + 1$ **f** $y = e^x - 1$ **g** $y = e^{x+1}$ **h** $y = e^{x-1}$

10 a Draw the graph of $y = 3^x$ on graph paper with both axes labelled from -2 to 9. On the same axes, draw the graph of its inverse by reflecting in the line $y = x$

   Answer parts **b** to **d** using your graphs. You can use your calculator to check your answers.

   b Write the coordinates of the image of the point $(1, 3)$ under the reflection.

   c Write the coordinates of the point where $x = 1.2$ and also of its image point. Hence write the value of $\log_3 3.74$

   d Write the values of

   i $3^{1.5}$

   ii $\log_3 5.20$

   iii $3^{0.5}$

   iv $\log_3 1.73$

   v $\log_3 6$

   vi $\log_3 7$

11 a The graph of $y = e^x$ passes through the points $(3, p)$, (q, e^2) and $(r, 9)$. Write the values of p, q and r

b The graph of $y = e^{-x}$ passes through the points $\left(\dfrac{1}{2}, a\right)$, (b, e^3) and $(c, 9)$. Write the values of a, b and c

Reasoning and problem-solving

To solve problems involving exponential functions

1. Draw or sketch a graph if it is helpful.

2. Use what you know about the gradients of $y = e^x$ and $y = e^{kx}$

3. Use the relationship between an exponential function and its inverse.

Example 3

The tangent to the curve $y = e^x$ at point P, where $x = 2$, intersects the y-axis at point Q
Find point Q

The gradient of $y = e^x$ at $P(2, e^2)$ is e^2

At P, the equation of the tangent is $\dfrac{y - e^2}{x - 2} = e^2$

giving $\qquad y = e^2 x - e^2$

When $x = 0 \qquad y = 0 - e^2 = -e^2$

The point Q is $(0, -e^2)$

> **2** The gradient of $y = e^x$ is e^x

> **2** Use the equation of a straight line to find the equation of the tangent.

> The tangent intersects the y-axis at Q, so $x = 0$

Example 4

State the inverse of the function $y = \ln x$. Find the equation of the normal to this inverse at $x = 3$, giving values to 2 dp.

$y = e^x$ is the inverse

The gradient of the tangent is e^3

so the gradient of the normal is $-\dfrac{1}{e^3}$

When $x = 3$, $y = e^3$

Equation of normal $\dfrac{y - e^3}{x - 3} = -\dfrac{1}{e^3}$

$e^3 y - e^6 = -x + 3$

$y = -e^{-3}x + 3e^{-3} + e^3$

$y = -0.05x + 20.23$

> **3** $\ln x$ and e^x are inverse functions.

> Remember to give your answer in the form asked for in the question.

Exercise 5.2B Reasoning and problem-solving

1 a Draw the graphs of the curve $y = 2^x$ and the line $y = 6$ on the same axes for $0 \le x \le 3$
   Write an estimate of their point of intersection and hence deduce an approximate solution to the equation $2^x = 6$

  b Draw the graphs of $y = 3^x$ and $y = x + 3$ on the same axes for $0 \le x \le 2$. Find an approximate solution to the equation $3^x - x = 3$ from your diagram.

2 a Sketch the graphs of $y = 2^x$ and $y = x^2$ on the same axes and explain why there is more than one possible solution to the equation $2^x = x^2$

  b Draw the two graphs for $-2 \le x \le 4$ and find solutions to the equation in part **a**, giving answers to 2 dp where appropriate. You can check your sketches on a graphics calculator.

3 Use an algebraic method to find the point of intersection for each of these pairs of curves.

   a $y = 4^{3x}$ and $y = 4^{x+6}$

   b $y = 4^x$ and $y = 2^{x+1}$

   c $y = 3^x$ and $y = 9^{x-2}$

   d $y = 4 \times 2^x$ and $y = 2^{-x}$

4 The curves of $y = e^x$ and $y = e^{-x}$ intersect the straight line $x = 2$ at points P and Q respectively. Calculate the distance PQ to 3 sf.

5 Prove that the tangent to the curve $y = e^{3x}$ at the point where $x = 1$ passes through the point $\left(\dfrac{2}{3}, 0\right)$

6 Find the point where the tangent to the curve $y = e^{\frac{1}{2}x}$ at the point $(4, e^2)$ intersects the straight line $x = 6$. Show your working.

7 Find the equation of the normal to the curve $y = e^{-2x}$ when $x = 1$, giving values to 3 dp.

8 Find the x-intercept of the normal to the curve $y = e^{2x}$ at the point where $x = 1$

9 a Write the equation of the graph of the inverse of the function $y = a^x$

   b Show that the inverse of $y = 2^x$ is approximately $y = 1.44 \times \ln x$. Hence, find the value of x if $2^x = 17$

Challenge

10 The diagram shows how to investigate the gradient of a curve $y = a^x$ at P(0, 1) using differentiation from first principles.

See Ch 4.1

For a reminder on differentiation from first principles.

   a Taking $\delta x = 0.0001$, use a spreadsheet to copy and complete the table of values for the gradient of the chord $PQ = \dfrac{a^{\delta x} - 1}{\delta x}$ for curves with different values of a

	A	B	C	D	E	F	G
1	a	2.0	2.2	2.4	2.6	2.8	3.0
2	$\dfrac{a^{\delta x} - 1}{\delta x}$						

   b Between which pair of values does e lie? Give a reason for your answer.

   c Create further tables for narrower ranges of a to find the value of e correct to 2 decimal places.

   d Why will this method only give an approximate value of e?

   (You can take increasingly small values for δx to convince yourself that e does lie between the values you have found.)

11 The graph of $y = e^x$ has a gradient of e^x at the point (x, y)

   a Justify this statement by copying and completing this table of values. Take $\delta x = 0.0001$ and choose your own values of x for $x > 3$. You could use a computer spreadsheet.

	A	B	C	D	E
1	x	1	2	3	
2	e^x	2.71828			
3	$\dfrac{e^{x+\delta x} - e^x}{\delta x}$				

   b Why is this method not a proof of the statement?

Exponential processes

Fluency and skills

Mathematical models are used to describe and make predictions about real-life events using mathematical language and symbols.

The gradient of $y = e^{kx}$ is proportional to y at the point (x, y), where the constant of proportionality is k. This property of proportionality occurs frequently in the natural world and allows you to create mathematical models of events such as radioactive decay and population growth.

For example, if the rate of increase of a population of bacteria is directly proportional to the number of bacteria, y, then $\dfrac{dy}{dt} = ky$. The rate of change of $y = e^{kt}$ is proportional to y, so an exponential function is a good model for this situation.

> **Key point**
>
> An equation of the form $y = Ae^{kt}$ gives an exponential model where A and k are constants.

Example 1

The population, P hundreds of cells, of an organism grows exponentially over time, t hours, according to $P = Ae^{\frac{1}{20}t}$

t	0	5	10	15
P	4			

a Find the value of A, copy and complete the table and draw a graph to represent the data.

b What is the rate of increase in the population when $t = 5$? Show your working.

> $y = e^{kx}$ has a gradient of $k \times e^{kx}$ at the point (x, y)

a When $t = 0$, $P = A \times e^0 = A \times 1 = A$, so $A = 4$

t	0	5	10	15
P	4	$4e^{0.25} = 5.14$	$4e^{0.5} = 6.59$	$4e^{0.75} = 8.47$

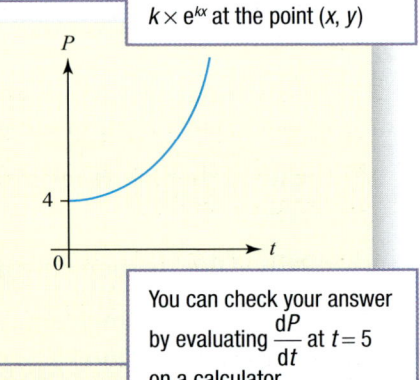

b When $t = 5$, the rate of change of P is

$$A \times \frac{1}{20}e^{\frac{5}{20}} = 4 \times \frac{1}{20}e^{\frac{1}{4}} = 0.257 \text{ hundreds}$$

$$= 25.7 \text{ cells per hour}$$

> You can check your answer by evaluating $\dfrac{dP}{dt}$ at $t = 5$ on a calculator.

Exercise 5.3A Fluency and skills

1 The value of n, after t hours, is given by $n = 2^t$

a Copy and complete this table.

t	0	1	2	3	4
n					

b Calculate the value of n after 6 hours.

2 A patient is injected with 10 units of insulin. The drug content in the body decreases exponentially and the number of units, n, left in the patient after t minutes is modelled by $n = A \times 0.95^t$

The actual number of units in the patient is measured at the times given in this table.

t	0	2	4	6	8
Actual units	10.0	9.1	8.2	7.4	6.7
n					

a State the value of A

b Copy and complete the table. Does the model predict reasonably accurate values of n for $t \leq 8$?

c Does the model indicate that less than half the initial insulin is present in the patient after 13 minutes?

3 The variables x, y and z change with time t. Calculate the rate of change of each variable at the instant when   **i** $t = 0$ **ii** $t = 2$

a $x = e^{2t}$ b $y = 5e^{3t}$ c $z = 100e^{-\frac{t}{20}}$

Show your working.

Do x, y and z grow or decay over time?

4 a The height h cm of a bean shoot t hours after germination is given by $h = 0.3e^{0.1t}$

What is the rate of growth of the bean shoot when $t = 5$? Show your working.

b A radioactive chemical decays so that, after t hours, its mass m is given by $m = 2 \times e^{-0.01t}$ kg. Calculate its rate of decay after 100 hours.

5 The temperature $\theta °C$ of water in a boiler rises so that $\theta = \frac{1}{5}e^{t}$ after t minutes.

a What is the temperature

   i Initially,   **ii** After 2 minutes?

b What is the average rate of change of temperature over the first 2 minutes?

c What is the instantaneous rate of change of temperature when $t = 2$? Show your working.

d The boiler switches off after 6 minutes. What is the temperature at this time?

6 The population of a country at the start of a given year, P millions, is growing exponentially so that $P = 15e^{0.06t}$ where t is the time in years after 2000. Calculate

a The size of the population at the start of 2006

b The rate of increase in the population at the start of 2006

c The average rate of increase in the population from the start of 2000 to the start of 2006

7 An injected drug decays exponentially. Its concentration, C mg per ml, after t hours is given by $C = 2 \times e^{-0.45t}$

a What is the initial concentration?

b Calculate the concentration after 4 hours.

c What is the rate of decay of the drug when   **i** $t = 0$ **ii** $t = 4$?

Show your working.

8 The cost of living in Exruria increased by 5% each year from 2000 to 2010. The weekly food bill per family was £P at the start of 2000

a Show that by the start of 2002 this bill was £$P \times 1.05^2$

b If $P = 102$, find the weekly food bill at the start of 2010

9 £P invested for n years at r% p.a. increases to £A where $A = P\left(1 + \dfrac{r}{100}\right)^n$

Two friends have £1000 each. They place their money in different accounts and make no withdrawals for 4 years. Both accounts have annual fixed rates; one at 6% p.a., the other at 4% p.a. What is the difference in their savings at the end of the 4 years?

10 The mass m of a radioactive material at a time t is given by $m = m_0 e^{-kt}$ where k and m_0 are constants. If $m = \dfrac{9}{10}m_0$ when $t = 10$, find the value of k. Also find the time taken for the initial mass of material to decay to half that mass.

Strategy

To solve modelling problems involving rates of change

1. Calculate data using the model.

2. Consider sketching or using a graphical calculator to graph the model.

3. Use your knowledge of exponential functions and logarithms to find rates of change and solve equations.

4. Compare actual data with your model and, where necessary, comment on any limitations.

Example 2

An area of fungus, $A\,\mathrm{cm}^2$, grows over t days such that $A = 2 + 6e^{0.1t}$

a Sketch a graph of A against t by calculating A for $t = 0, 10, 20$ and 30

b How long does it take for the area of the fungus to double?

c What is the initial rate of change of A?

d Why might this model not be realistic for large values of t?

a

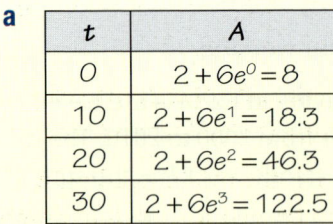

t	A
0	$2 + 6e^0 = 8$
10	$2 + 6e^1 = 18.3$
20	$2 + 6e^2 = 46.3$
30	$2 + 6e^3 = 122.5$

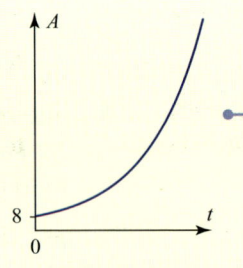

> **1** Substitute the given values of t to calculate A over time.

> **2** Sketch a graph, starting with $t = 0$

> The table shows that t is less than 10

b Initially $A = 8$, so area doubles when $A = 16$

$$16 = 2 + 6e^{0.1t}$$

$$\text{so } e^{0.1t} = \frac{7}{3}$$

$$\ln(e^{0.1t}) = \ln\frac{7}{3}$$

$$0.1t \times 1 = \ln\frac{7}{3}$$

$$t = 8.47 \text{ to 3sf}$$

It takes approximately 8 days for the fungus to double its area.

> Form an equation using the model.

> **3** Take natural logs of both sides and simplify using log laws. Remember $\ln e = 1$

c Rate of change is $0 + 6 \times 0.1e^{0.1t}$

When $t = 0$, initial rate of change $= 0.6\,\mathrm{cm}^2$ per day.

> **3** The rate of change of $y = e^{kx}$ is ke^{kx}. The constant 2 has zero rate of change.

d There is no limit placed on the area of fungus—it would increase to infinity according to the model. The model ignores factors such as limited space and changes in conditions.

> **4** Comment on limitations of the model.

1 A population of insects, n, increases over t days, and can be modelled by $n = 100 - 80e^{-\frac{1}{5}t}$

    a What was the initial number of insects?

    b If there are 72 insects after 5 days, does this data fit the model?

    c How many insects are there after 9 days?

    d Does the model predict a limiting number of insects? If so, what is it?

2 A population of bacteria grows so that the actual number of bacteria is n after t hours. The number, p, predicted by a model of the growth, is given by $p = 2 + Ae^{\frac{1}{20}t}$

t	0	20	40	60	80	100
n	5	11	25	60	140	265
p	5					

    a Calculate A and copy and complete the table.

    b According to the model, how long does it take for the initial number of bacteria to triple?

    c Draw graphs of n and p against t on the same axes. Is the model accurate at predicting the actual numbers? What limitations would you place on the model?

    d What rate of growth is predicted at $t = 40$?

3 For each of the following contexts, select the most appropriate model. Explain your answer.

    a A population of deer on a remote island, D, is modelled over t years.
    Model A: $D = 10 + 6e^{0.5t}$ $t \geq 0$
    Model B: $D = -7 + 6e^{0.5t}$ $t \geq 0$

    b The concentration of a drug, C mg per ml, decays exponentially over t days.
    Model A: $C = 5e^{-3t}$ Model B: $C = 5e^{3t}$

    c An area of bacteria, A cm², grows in a circular petri dish with radius 4 cm over t hours.
    Model A: $A = 2 + 3e^{2t}$ $0 < t < 1.4$
    Model B: $A = 2 + 3e^{2t}$ $0 < t < 2.4$

4 Trees in a local wood are infected by disease. The number of unhealthy trees, N, was observed over t years and modelled by $N = 200 - Ae^{-\frac{1}{20}t}$

    a If there are 91 unhealthy trees after 10 years, calculate the value of A

    b What is the initial number of unhealthy trees and the initial rate of change?

    c How long does it takes for the initial number of unhealthy trees to triple?

    d Explain why the model predicts a limit to N and state its value.

5 A block of steel leaves a furnace and cools. Its temperature, $\theta°C$, is given by $\theta = c + 500e^{-0.005t}$ after it has been cooling for t minutes.

    a If the steel leaves the furnace at 530 °C, how long does it take for the temperature to drop to 100 °C?

    b Calculate the rate of decrease of temperature (to 3 sf) when $t = 20$

    c Explain why the model predicts a minimum temperature below which the steel does not cool. State its value.

    d Why might this model not be appropriate?

Challenge

6 A rare plant has been monitored over time. Its population, n, after t years of monitoring has been modelled as $n = \dfrac{Ae^{\frac{1}{4}t}}{e^{\frac{1}{4}t} + k}$, where A and k are positive constants.

    a There were 250 plants initially and 528 plants after 4 years. Calculate the values of A and k to 2 sf.

    b Show that, according to this model, the population cannot exceed a certain number and calculate this number.

Curve fitting

Fluency and skills

Two common, non-linear relationships take the form $y = ax^n$ and $y = kb^x$, where a, n, b and k are constants. These are called **polynomial** and **exponential relationships** respectively. Given values for x and y, we can use logs to determine the form of the relationship and find values for the constants.

When you plot these relationships on axes where the scales are linear, the range between plotted y-values gets larger and larger, and the graph becomes harder to read accurately. By using logs, you can scale down the y-values and convert the relationships into linear ones. This means that their graphs are transformed into straight line graphs.

> **Key point**
>
> $y = ax^n$ becomes $Y = nX + c$, where $Y = \log y$, $X = \log x$ and $c = \log a$
>
> $y = kb^x$ becomes $Y = mx + c$, where $Y = \log y$, $m = \log b$ and $c = \log k$

These straight lines are **lines of best fit** through the data points.

Example 1

The relationship between x and y is given by $y = ax^n$ where a and n are constants.

x	10	20	30	50	100
y	88	232	409	837	2208

Use the data in this table to find the values of a and n, and write the relationship.

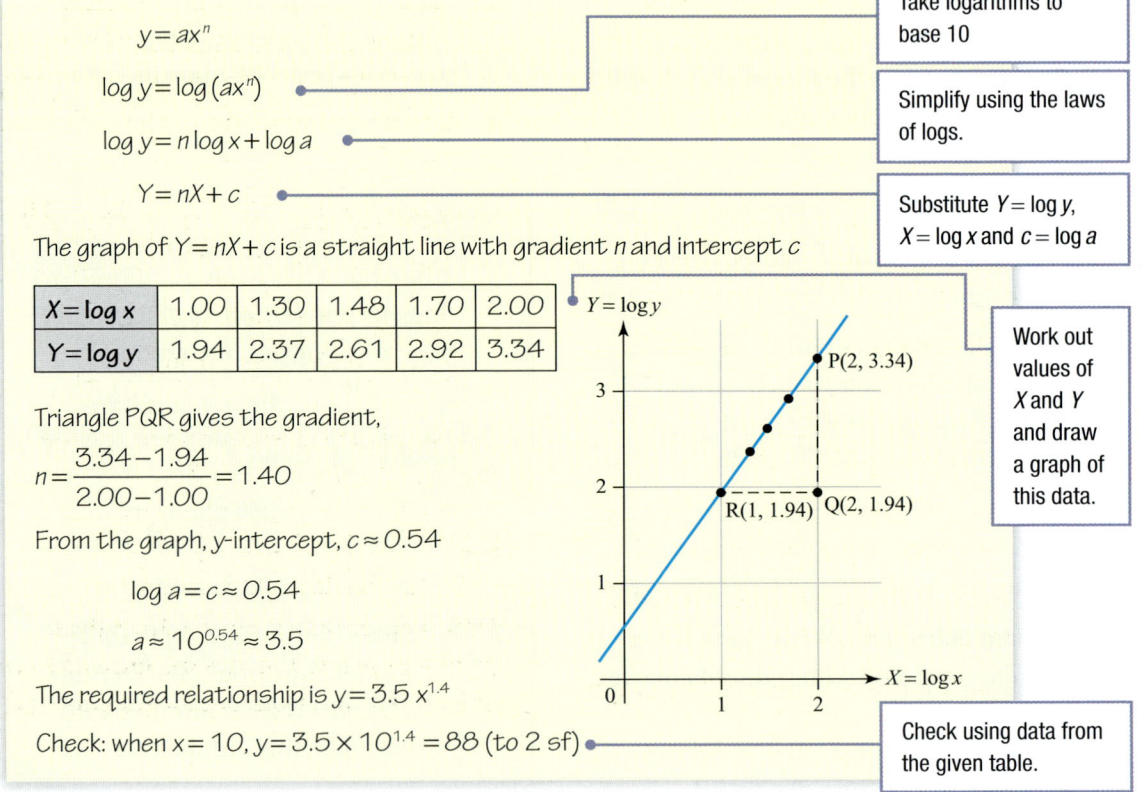

$y = ax^n$ Take logarithms to base 10

$\log y = \log(ax^n)$ Simplify using the laws of logs.

$\log y = n \log x + \log a$

$Y = nX + c$ Substitute $Y = \log y$, $X = \log x$ and $c = \log a$

The graph of $Y = nX + c$ is a straight line with gradient n and intercept c

$X = \log x$	1.00	1.30	1.48	1.70	2.00
$Y = \log y$	1.94	2.37	2.61	2.92	3.34

Triangle PQR gives the gradient,

$n = \dfrac{3.34 - 1.94}{2.00 - 1.00} = 1.40$

From the graph, y-intercept, $c \approx 0.54$

$\log a = c \approx 0.54$

$a \approx 10^{0.54} \approx 3.5$

The required relationship is $y = 3.5\,x^{1.4}$

Check: when $x = 10$, $y = 3.5 \times 10^{1.4} = 88$ (to 2 sf)   Check using data from the given table.

Work out values of X and Y and draw a graph of this data.

1 The relationship $y = ax^n$ is graphed as $Y = nX + c$ where $Y = \log y$, $X = \log x$ and $c = \log a$ as shown. Take readings from the graph to find the values of a and n. Write the relationship between x and y

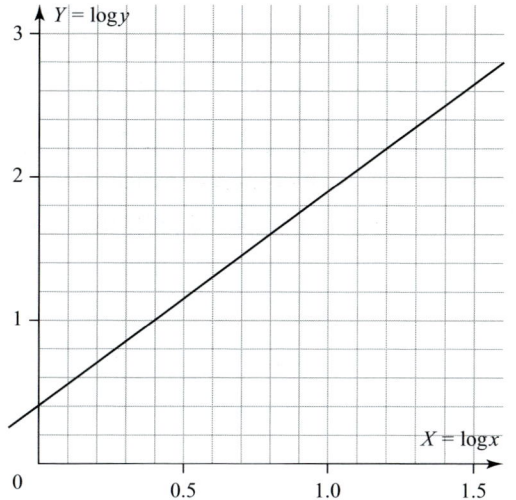

2 The relationship $y = kb^x$ is graphed as $Y = mx + c$ where $Y = \log y$, $m = \log b$ and $c = \log k$ as shown. Use the graph to find the values of b and k. Write the relationship between x and y

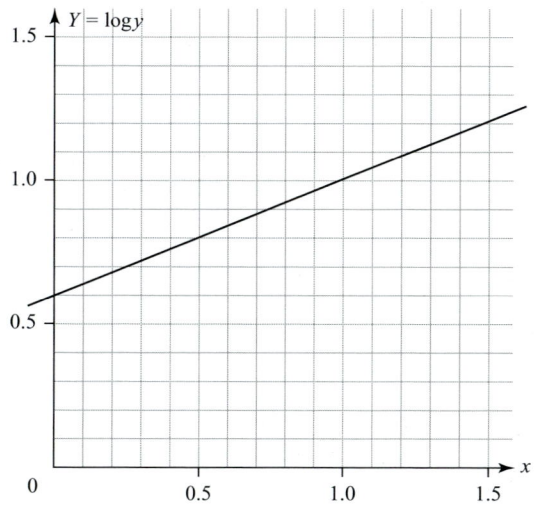

3 For each set of data, $y = ax^n$ where a and n are constants. For each table

   i Draw an appropriate straight-line graph,

   ii Find a and n and write the relationship.

   Check your answers using the data.

a

x	2	3	4	5	6
y	17	40	74	117	172

b

x	1	2	3	4	5
y	1.6	2.3	2.9	3.4	3.8

c

x	10	15	20	25	30
y	376	333	305	286	270

d

x	2	4	6	8
y	1.72	0.57	0.30	0.19

4 For each set of data, $y = kb^x$ where k and b are constants. For each table

   i Draw an appropriate straight-line graph,

   ii Find k and b and write the relationship.

   Check your answers using the data.

a

x	3	5	7	9
y	40	176	774	3415

b

x	0.2	0.3	0.4	0.5	0.6
y	8.2	9.2	10.4	11.6	13.1

c

x	2.0	2.5	3.0	3.5	4.0
y	176	139	110	87	69

d

x	0.3	0.4	0.5	0.6
y	0.33	0.39	0.46	0.54

5 This set of data gives four points on the graph of the function $y = f(x)$

x	0.5	1.0	2.0	3.0
y	0.66	1.08	2.92	7.87

Draw appropriate graphs to investigate whether the function has the form $f(x) = ax^n$ or $f(x) = kb^x$ and work out the function $f(x)$

Strategy

To solve problems involving polynomial and exponential relationships

(1) Transform the non-linear functions $y = ax^n$ and $y = kb^x$ to linear functions using logarithms.

(2) Use the transformed data to draw a straight-line graph, using a line of best fit when necessary.

(3) Use your graph to calculate the constants and work out the relationship between x and y

When experimental data is gathered, it is often not exact. Data points plotted on a scatter diagram may not lie exactly on a straight line. You have to judge by eye where to draw the **line of best fit**.

ICT Resource online

ICT resource

To investigate logarithmic graphs and lines of best fit, click this link in the digital book.

Example 2

A patch of algae grows so that its area is y cm^2 after x days.

x	2	4	6	8
y	4.9	7.4	12.6	18.7

a   Use the data in this table to find the relationship between x and y in the form $y = kb^x$

b   What was the initial area of algae and what would you expect the area to be after 10 days?

a $\log y = \log (kb^x)$ — **Take logarithms to base 10** (1)

$\log y = x \log b + \log k$ — **Simplify using the laws of logs.**

$Y = (\log b) \times x + \log k$ — **Substitute $Y = \log y$** (1)

This is a straight line with gradient $\log b$ and intercept $\log k$

x	$Y = \log y$
2	0.69
4	0.87
6	1.10
8	1.27

Compare with $y = mx + c$

$Y = \log y$

1.4
1.2
1.0
0.8
0.6
0.4
0.2
0   1 2 3 4 5 6 7 8 x

Work out values of Y. Plot the data and draw a line of best fit by eye. (2)

Gradient $= \log b \approx \dfrac{1.27 - 0.69}{8 - 2} = 0.0966\ldots$

$b \approx 10^{0.0966\ldots} \approx 1.25$

y-intercept $= \log k \approx 0.49$

$k = 10^{0.49} \approx 3.1$ — **Use the graph to find the gradient and y-intercept.** (3)

The required relationship is $y = 3.1 \times 1.25^x$

b   When $x = 0$, the initial area $= 3.1 \times 1.25^0 = 3.1$ cm^2

After 10 days, the expected area $= 3.1 \times 1.25^{10} = 28.9$ cm^2

1 A pipe empties water into a river. The table shows the volume of water, $y\,\text{m}^3$, leaving the pipe each second, for different depths of water, $x\,\text{m}$, in the pipe. Engineers expect x and y to be connected by the relationship $y = ax^n$

x	1.2	1.3	1.4	1.5
y	5.8	7.6	8.7	10.0

a Show, by drawing a line of best fit on appropriate axes, that the engineers are right and find approximate values of a and n

b What volume of water (in m^3) would you expect to leave the pipe each second when the depth of water is

  i 1.35 metres   **ii** 3.2 metres?

c Which of your two answers in part **b** is the more reliable? Explain why.

2 When an oil droplet is placed on the surface of water it forms a circle. As time, t seconds, increases, the area, A, of the oil increases.

t	2	3	4	5
A	17	40	73	117

a Show that there is a relationship between A and t given by $A = k \times t^n$ and find the values of the constants k and n

b Explain why this algebraic model is better at estimating the area of the oil slick after 1 second than after 1 minute.

3 A small town has had a growing population, p, for many years, x, since 1980

Year	1980	1985	1995	2000
x	0	5	15	20
p	4140	5000	8400	10200

a Show that the population can be approximately related to the number of years since 1980 using $p = k \times b^x$. State the value of k and find the value of b using an appropriate straight-line graph.

b Estimate the size of the population in 1990 and in 2010. Which of your two answers is the better estimate? Explain why.

4 When $200\,\text{cm}^3$ of gas being held under high pressure suddenly expands, its pressure p (measured in thousands of pascals) decreases rapidly as its volume, $v\,\text{cm}^3$, increases. The values are given in the table.

v	200	250	300	350	400
p	78.0	56.5	43.7	36.0	29.7

a Show that $p = k \times v^n$ is a good approximation for this data and calculate the values of the constants k and n to 3 sf.

b What is the volume of the gas when its pressure has fallen to

  i Half its initial pressure,

  ii A tenth of its initial pressure?

c Which of your two answers to part **b** is more reliable? Explain why.

Challenge

5 The mass, m grams, of a radioactive material decreases with time, t hours. The mass predicted to remain radioactive after various times is given in the table.

t	5	10	15	30
m	47.6	45.3	43.1	38.0

a Draw an appropriate straight-line graph and find the relationship between m and t

b According to this model, what mass of radioactive material is present when $t = 0$?

c How long does the model predict it will take for the radioactive substance to decay to half its original mass? This is its half-life.

Chapter summary

- $x = a^n$ and $n = \log_a x$ are equivalent statements.
- The inverse of $y = a^x$ is $y = \log_a x$
- Logarithmic expressions can be manipulated using the following three laws of logarithms.

  Law 1: $\log_a(xy) = \log_a x + \log_a y$

  Law 2: $\log_a\left(\dfrac{x}{y}\right) = \log_a x - \log_a y$

  Law 3: $\log_a(x^k) = k\log_a x$

- Logs can have different bases. Bases 10 and e are the most common.
- The value of e is 2.71828 (to 5 dp).
- The general exponential function is a^x. The exponential function is e^x
- The inverse of $y = e^x$ is $y = \log_e x$ which can be written as $y = \ln x$
- The gradient of $y = e^x$ is e^x and the gradient of $y = e^{kx}$ is ke^{kx}
- The graph of $y = e^x$ can be transformed in a variety of ways.
- Mathematical models using exponential functions can be used to describe real-life events. A common model is $y = Ae^{kt}$. Contextual limitations of these models should be considered.
- Two common, non-linear relationships take the form $y = ax^n$ and $y = kb^x$
- The graphs of these relationships can be transformed into straight lines using logs.

  $y = ax^n$ becomes $Y = nX + c$, where $Y = \log y$, $X = \log x$ and $c = \log a$

  $y = kb^x$ becomes $Y = mx + c$, where $Y = \log y$, $m = \log b$ and $c = \log k$
- These straight lines are lines of best fit through the data points.

Check and review

You should now be able to…	Try Questions
✔ Convert between powers and logarithms.	1–2
✔ Manipulate expressions and solve equations involving powers and logarithms.	3–7
✔ Use the exponential functions $y = a^x$, $y = e^x$, $y = e^{kx}$ and their graphs.	8–14
✔ Verify and use mathematical models, including those of the form $y = ax^n$ and $y = kb^x$	15–17
✔ Consider limitations of exponential models.	18

1 Express each of these in logarithmic form.

   a $2^5 = 32$ **b** $0.0001 = 10^{-4}$

2 Express each of these in index notation.

   a $\log_2 16 = 4$ **b** $\log_4\left(\dfrac{1}{16}\right) = -2$

3 Evaluate

   a $\log_2 16 + \log_2 2$ **b** $\log_2 16 - \log_2 2$

4 Write $2\log 9 + \log\left(\dfrac{1}{3}\right)$ as a single logarithm.

5 Write $\log\left(\dfrac{a^2}{\sqrt{b}}\right)$ in terms of $\log a$ and $\log b$

6 Solve these equations, giving your answers to 3 sf.

    a $2^x \times 2^{x+1} = 24$

    b $2^{2x} - 9 \times 2^x + 8 = 0$

    c $\log_{10}(2x - 15) = 2$

7 Calculate the smallest positive integer x such that

    a $\left(1 + \dfrac{x}{100}\right)^4 > 2$ **b** $1 - 0.9^x > 0.8$

8 **a** Find the equation of the tangent to the curve $y = e^x$ at the point $(3, e^3)$

    b Find the equation of the tangent to $y = e^{\frac{1}{2}x}$ at the point where $x = 2$. What are the coordinates of the point where this tangent intersects the line $x = 1$?

9 Describe the transformation that maps $y = e^x$ onto $y = e^{2x}$

10 The graph of $y = e^x$ passes through the points $(3, a)$ and $(b, 4)$
Find the exact values of a and b

11 Calculate the value of k (to 3 sf) where the inverse of $y = 3^x$ is $y = k \ln x$

12 Find the equation of the normal to the graph of $y = \dfrac{1}{2}e^x + 2x^2$ at the point $\left(0, \dfrac{1}{2}\right)$

13 Prove that the normal to the curve $y = 1 - x + e^x$ at the point $(1, e)$ passes through the point $(e^2, -1)$

14 Sketch the graph of $y = x^2 + e^x$
Find the minimum point.

15 A bank account has £250 invested in it at a compound rate of $r\%$ p.a.

    a Prove that, after n years, the account holds £A where $A = 250\left(1 + \dfrac{r}{100}\right)^n$

    b Find A when $r = 5$ and $n = 4$

    c How long does it take for the amount invested to double in value when $r = 6$?

16 Two variables are measured. The relationship between them is likely to be $y = k \times b^x$

x	1.0	2.0	3.0	4.0
y	15.0	23.7	43.4	70.2

    a Plot a suitable straight-line graph and estimate the values of the constants b and k

    b Further measurements give the values $x = 7.5$, $y = 372$. Show whether the model predicts these results sufficiently accurately or not.

17 The variables x and y satisfy the relationship $y = a \times x^b$, where a and b are constants. By drawing a line of best fit, show that this is approximately correct and estimate the values of a and b

x	4.0	4.8	5.6	6.4
y	90	140	198	270

18 A scientist proposes that the population of deer, D, on an island can be modelled by the formula $D = 25e^{0.6t}$. Why might this model not be appropriate?

What next?

Score			
	0 – 9	Your knowledge of this topic is still developing. To improve, search in MyMaths for the codes: 2061–2063, 2133, 2134, 2136, 2257, 2268	🔗
	10 – 14	You're gaining a secure knowledge of this topic. To improve, look at the InvisiPen videos for Fluency and skills (05A)	🎞
	15 – 18	You've mastered these skills. Well done, you're ready to progress! Now try looking at the InvisiPen videos for Reasoning and problem-solving (05B)	🎞

Click these links in the digital book

Exploration
Going beyond the exams

5

History

Towards the end of 16th century, rapid advances in science demanded calculations to be carried out on a scale never seen before. This put pressure on mathematicians to devise methods for simplifying the processes involved in multiplication, division and finding powers and roots.

In 1614 the Scottish mathematician **John Napier** published his work on **logarithms**, which scientists, navigators and engineers quickly adopted in their calculations.

Swiss mathematician **Leonhard Euler** did further work on logarithms in the 18th century, creating the system that we still use today.

John Napier

Have a go

Try the following calculations by hand.

$$5438 \times 2149$$

$$4769604 \div 567$$

Now try these calculations using logarithms.

"For the sake of brevity, we will always represent this number 2.718281828459... by the letter e."
- Leonhard Euler

Research

Find the dates for each of the following events and plot them all on a single timeline.

1. Today
2. Your date of birth
3. The moon landing
4. The first powered flight
5. The first use of a printing press
6. The first use of coins
7. The end of the Neolithic age
8. The extinction of dinosaurs

Explain the need for a logarithmic timescale. Have a go at plotting a timeline of the same events with a logarithmic scale.

Research

Find a method for 'extracting square roots' by hand.

Use the method to find the square root of 126 736

In questions that tell you to show your working, you shouldn't depend solely on a calculator. For these questions, solutions based entirely on graphical or numerical methods are not acceptable.

1 Sketch, on the same axes, the graphs of $y = 2^x$, $y = 5^x$ and $y = 0.5^x$ **[4 marks]**

2 Given $f(x) = \log_3 x$

   a Sketch the graph of $y = f(x)$ and write down the equation of any asymptote. **[3]**

   b Solve the equations

      i $f(x) = 3$ **ii** $f(x) = -2$ **iii** $f(x) = 0.5$ **[3]**

3 Write in the form $\log_5 b$. Show your working.

   a $\log_5 7 + \log_5 8$ **[1]**   **b** $\log_5 24 - \log_5 8$ **[1]**   **c** $3\log_5 2 - \log_5 10$ **[2]**

4 **a** Write down the value of

      i $\log_3 81$ **ii** $\log_3\left(\dfrac{1}{3}\right)$ **[2]**

   b Express as a single logarithm

      i $2\log_n 5 + \log_n 3$ **ii** $1 - \log_n 5$ **[4]**

5 Solve these equations, giving your answers to 3 significant figures. Show your working.

   a $6^x = 13$ **[2]**   **b** $e^x = 5$ **[2]**

6 Solve these equations for $x > 0$. Show your working.

   a $\log_2 x - \log_2 3 = \log_2 4$ **[2]**   **b** $\log_2 9 + 2\log_2 x = \log_2 x^4$ **[2]**

7 The number of bacteria on a dish is given by the equation $B = 200e^{3t}$ where t is the time in hours.

   a How many bacteria are there originally? **[1]**

   b How long will it be until there are 10 000 bacteria? **[3]**

   c Give one reason why this model might not be appropriate. **[1]**

8 The value, £V, of an investment after t years is given by the formula $V = V_0 e^{0.07t}$ where V_0 is the amount originally invested.

   How long will it take for the investment to double in value? **[3]**

9 The graph of $y = k^x$ passes through $(2, 15)$

   Find an exact value of k **[2]**

10 The graph of $y = ax^n$ passes through the points $(1, 2)$ and $(-2, 32)$

   Find exact values of a and n **[3]**

11 **a** Sketch the graph of $y = 2 + e^x$, and write down the equation of the asymptote. **[3]**

   b Solve the equation $e^{2x} = 5$, giving your answer in terms of logarithms. Show your working. **[2]**

12 Given that $\log_n 3 = p$

    a Write in terms of p

        i $\log_n 9$ **ii** $\log_n \dfrac{1}{3}$ **iii** $\log_n 3n$ **[6]**

    b Find the value of n given that $\log_n 3n - \log_n 9 = 2$ **[3]**

13 Solve these equations. Show your working.

    a $\log_3 x + \log_3(x-1) = \log_3 12$ **[5]**

    b $\log_3 x - \log_3(x-1) = 2$ **[3]**

    c $3\ln x - \ln 2x = 5$ **[4]**

14 The population of a particular species on an island t years after a study began is modelled as

$$P = \frac{1500a^t}{2+a^t}, \text{ where } a \text{ is a positive constant.}$$

    a What was the population at the beginning of the study?   **[2]**

    b Given that the population after two years was 600

        i Find the value of a

        ii Calculate, to the nearest whole year, how long it takes for the population to double its initial size.   **[8]**

    c Explain why, according to this model, the population cannot exceed 1500   **[2]**

    d Give one reason why this model might not be appropriate.   **[1]**

15 A radioactive isotope has mass, M grams, at time t days given by the equation

$$M = 50\mathrm{e}^{-0.3t}$$

    a What is the initial mass of the isotope?   **[1]**

    b What is the half-life of the isotope?   **[3]**

16 The graph of $y = a(b)^x$ passes through the points $(0,5)$ and $(2, 1.25)$

    a Find exact values for a and b **[3]**

    b Sketch the curve.   **[2]**

17 Find the rate of change of the function $\mathrm{f}(t) = \mathrm{e}^{2t}$ at time $t = 1.5$. Show your working.   **[3]**

18 Solve this equation. You must show your working.

$$2^{2x} + 7\left(2^x\right) - 18 = 0 \qquad \textbf{[3]}$$

19 The concentration of a drug, C mg per litre, in the blood of a patient at time t hours is modelled by the equation

$$C = C_0 \mathrm{e}^{-rt}$$

where C_0 is the initial concentration and r is the removal rate.

The concentration after 1 hour is 9.2 mg/litre and after 2 hours is 8.5 mg/litre.

    a Calculate the initial concentration and the value of r. Give your answers to 2 significant figures.   **[7]**

The drug becomes ineffective when the concentration falls below 3.6 mg/litre.

    b What is the maximum time that can elapse before a second dose should be given? Give your answer to the nearest hour.   **[3]**

6 Vectors

The current pushing on a boat, as well as the wind if there is any, will have an effect on the motion of the boat. All the forces acting on the boat are vectors with a magnitude and direction, and a kayaker needs to consider all of these components to work out how hard, and in which direction, they must paddle: it's not as simple as just paddling in the direction they want to travel.

Vectors can be used to describe anything that has both magnitude and direction, and they are used in many fields, especially those involved with the motion of objects (such as physics and engineering). Velocity, acceleration, force and momentum are all vector quantities. The concept of vectors is relatively simple, but it's fundamental to a lot of maths.

Orientation

What you need to know	What you will learn	What this leads to
Ch1 Algebra 1 • Simultaneous equations. p.28 **Ch3 Trigonometry** • Sine and cosine rules and the area formula. p.76	• To identify vector and scalar quantities. • To solve geometric problems in 2D using vector methods. • To solve problems involving displacements, velocities and forces. • To find the magnitude and direction of a vector and use the components. • To use the position vectors to find displacements and distances.	**Ch7 Units and kinematics** Motion in a straight line. Distance and speed as scalar quanitites. Position, displacement and velocity as vector quantites. **Ch8 Forces and Newton's laws** Resolving in two directions. Magnitude and direction of a resultant force.

 MyMaths Practise before you start 🔍 2005, 2018, 2045, 2046

Fluency and skills

Key point

A **scalar** quantity has **magnitude** (size) *only*.
Examples include distance (100 m) and speed (10 m s⁻¹).
A **vector** quantity has *both* magnitude and **direction**. Examples include displacement (100 m north) and velocity (10 m s⁻¹ on a bearing of 217°).

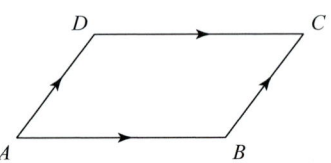

▲ Vector **a** represents the directed line segment from A to B.

You can represent any vector between points A and B by the magnitude and direction of a directed line segment; written $\overrightarrow{AB}$, **a**, a or $\underline{a}$. When you write a vector by hand, using only a single letter, make sure to underline it.

See p.614 You write the magnitude as $|\overrightarrow{AB}|$ or AB or $|\mathbf{a}|$ or a

For a list of mathematical notation.

Key point

Vectors are **equal** if they have the same magnitude and direction.

▲ Two pairs of equal vectors are shown.
$\overrightarrow{AB} = \overrightarrow{DC}$ and $\overrightarrow{AD} = \overrightarrow{BC}$

Multiplying a vector by a number (a **scalar**) changes its magnitude but not its direction.

Key point

$k\mathbf{a}$ is a vector parallel to **a** and with magnitude $k|\mathbf{a}|$
a is parallel to **b** only if $\mathbf{a} = k\mathbf{b}$

Drawing triangles and parallelograms like this is sometimes known as using 'the triangle and parallelogram laws of addition'.

Combining vectors $\overrightarrow{AB}$ and $\overrightarrow{BC}$ gives the vector $\overrightarrow{AC}$
You add vectors by placing them 'nose to tail'.

Key point

The vector sum of two or more vectors is called their **resultant**.

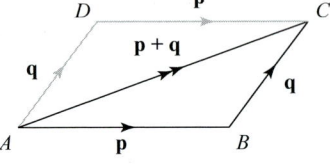

Two or more points are **collinear** if a single vector, or multiple parallel vectors, can pass through the points to form a single, straight line segment.

▲ $\overrightarrow{AC}$ is the resultant of $\overrightarrow{AB}$ and $\overrightarrow{BC}$
$\overrightarrow{AB} + \overrightarrow{BC} = \overrightarrow{AC}$

Key point

A vector with a magnitude of 1 is called a **unit vector.**

The unit vector in the direction of **a** is $\hat{\mathbf{a}} = \dfrac{\mathbf{a}}{|\mathbf{a}|}$

$\overrightarrow{AB} + \overrightarrow{BA} = \mathbf{0}$ (the **zero vector**. It has zero magnitude and undefined direction). This means that $\overrightarrow{BA} = -\overrightarrow{AB}$

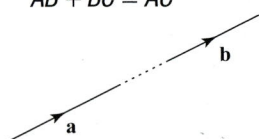

▲ Vectors **a** and **b** are collinear

Key point

a and −**a** have the same magnitude but opposite directions.

In the diagram $\overrightarrow{DB} = \overrightarrow{DA} + \overrightarrow{AB} = -\mathbf{q} + \mathbf{p} = \mathbf{p} - \mathbf{q}$

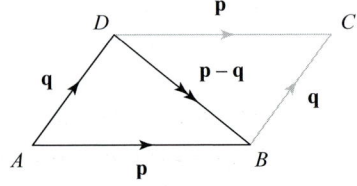

Key point

Subtracting a vector is the same as adding its negative.

Example 1

Vector **p** has magnitude 10 and direction due east. Vector **q** has magnitude 8 and direction 055°. Find the resultant of **p** and **q**

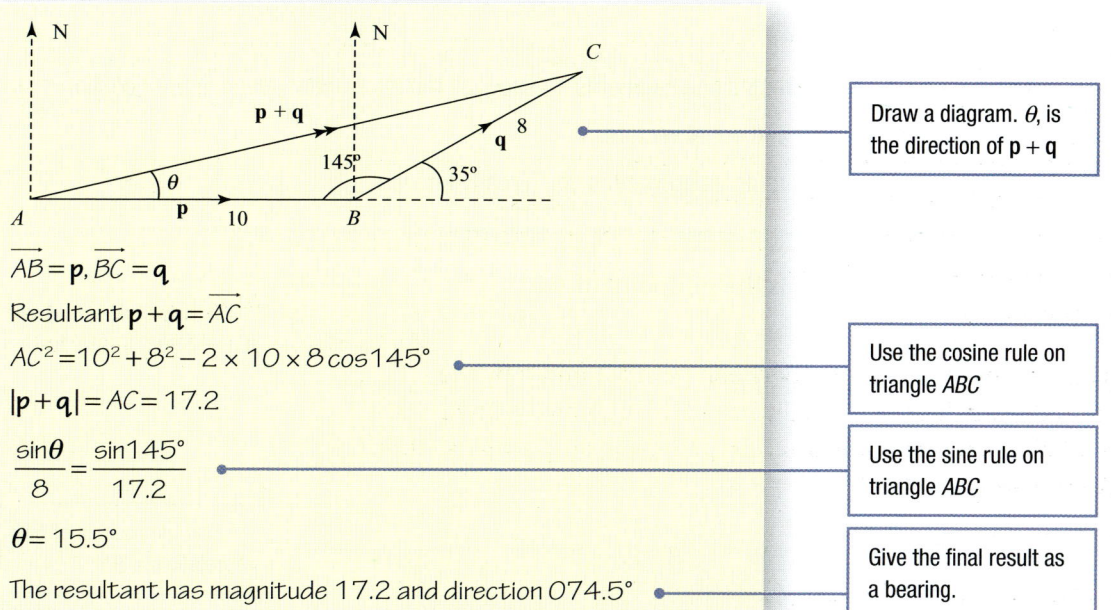

Draw a diagram. θ, is the direction of **p** + **q**

$\overrightarrow{AB} = \mathbf{p}, \overrightarrow{BC} = \mathbf{q}$

Resultant $\mathbf{p} + \mathbf{q} = \overrightarrow{AC}$

$AC^2 = 10^2 + 8^2 - 2 \times 10 \times 8 \cos 145°$

$|\mathbf{p} + \mathbf{q}| = AC = 17.2$

Use the cosine rule on triangle ABC

$\dfrac{\sin\theta}{8} = \dfrac{\sin 145°}{17.2}$

Use the sine rule on triangle ABC

See Ch3.2

For a reminder of the sine and cosine rules.

$\theta = 15.5°$

The resultant has magnitude 17.2 and direction 074.5°

Give the final result as a bearing.

Exercise 6.1A Fluency and skills

1 The diagram shows parallelogram $ABCD$, where $\overrightarrow{AB} = \mathbf{p}$ and $\overrightarrow{AD} = \mathbf{q}$. Express these vectors in terms of **p** and **q**

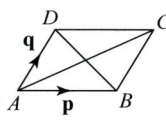

a $\overrightarrow{BC}$ **b** $\overrightarrow{DC}$ **c** $\overrightarrow{BA}$ **d** $\overrightarrow{CB}$
e $\overrightarrow{AC}$ **f** $\overrightarrow{BD}$ **g** $\overrightarrow{DB}$

2 The diagram shows two squares $ABEF$ and $BCDE$, where $\overrightarrow{AB} = \mathbf{p}$ and $\overrightarrow{AF} = \mathbf{q}$. Express these vectors in terms of **p** and **q**

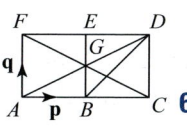

a $\overrightarrow{BD}$ **b** $\overrightarrow{AD}$ **c** $\overrightarrow{CF}$ **d** $\overrightarrow{AG}$

3 The diagram shows a trapezium $ABCD$ in which BC is parallel to AD and twice as long. Express these vectors in terms of **p** and **q**

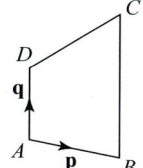

a $\overrightarrow{BC}$ **b** $\overrightarrow{DC}$
c $\overrightarrow{AC}$ **d** $\overrightarrow{BD}$

4 In each part of this question you are given the magnitude and direction (bearing) of two vectors **p** and **q**. In each case, work out the magnitude and direction of their resultant.

a **p**: 16, east **q**: 20, north
b **p**: 5, 100° **q**: 6, 060°
c **p**: 12, north **q**: 18, 200°

5 Vectors **p** and **q** have magnitudes 8 and 5 respectively and the angle between their directions is 45°. Find the magnitude and direction from **p** of these vectors.

a $\mathbf{p} + \mathbf{q}$ **b** $\mathbf{p} - \mathbf{q}$

6 Find the magnitude and direction of the resultant of each pair of given vectors. Show your working.

a A displacement of 3.5 km on a bearing of 050° and a displacement of 5.4 km on a bearing of 128°

b A displacement of 26 km on a bearing of 175° and a displacement of 18 km on a bearing of 294°

c Velocities of 15 km h⁻¹ due north and 23 km h⁻¹ on a bearing of 253°

d Forces of 355 N on a bearing of 320° and 270 N on a bearing of 025°

Reasoning and problem-solving

To solve problems involving vectors

1. Sketch a diagram using directed line segments, to show all the information given in the question.
2. Look for parallel, collinear and equal vectors.
3. Break down vectors into a route using vectors you already know.

Example 2

$ABCD$ is a parallelogram. E is the midpoint of AC

Vector $AB = \mathbf{p}$ and vector $AD = \mathbf{q}$

Prove that E is the midpoint of BD

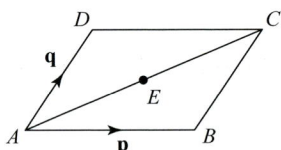

$\overrightarrow{AC} = \mathbf{p} + \mathbf{q}$ so $\overrightarrow{AE} = \dfrac{1}{2}(\mathbf{p} + \mathbf{q})$

$\overrightarrow{BE} = \overrightarrow{BA} + \overrightarrow{AE} = -\mathbf{p} + \dfrac{1}{2}(\mathbf{p} + \mathbf{q}) = \dfrac{1}{2}(\mathbf{q} - \mathbf{p})$

> **3** Find the position of E in relation to B

But $\overrightarrow{BD} = \mathbf{q} - \mathbf{p}$

So $\overrightarrow{BE} = \dfrac{1}{2}\overrightarrow{BD}$

> **2** $\overrightarrow{BE}$ and $\overrightarrow{BD}$ are collinear and BE is half as long as BD

Hence BED is a straight line and E is the midpoint of BD

Example 3

The diagram shows parallelogram $ABCD$. E lies on DC, and $DE : EC = 1 : 3$

AE and BC, when extended (produced), meet at F

$AB = \mathbf{p}$ and $AD = \mathbf{q}$
$AF = \lambda AE$ and $BF = \mu BC$

a   Express AF in terms of λ, $\mathbf{p}$ and $\mathbf{q}$

b   Express AF in terms of μ, $\mathbf{p}$ and $\mathbf{q}$

c   Hence find the values of λ and μ

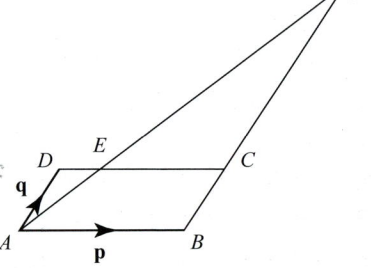

a $\overrightarrow{AE} = \overrightarrow{AD} + \overrightarrow{DE} = \mathbf{q} + \dfrac{1}{4}\mathbf{p}$

> **3** Break down vectors and apply the rules you know.

$\overrightarrow{AF} = \lambda \overrightarrow{AE} = \lambda\left(\mathbf{q} + \dfrac{1}{4}\mathbf{p}\right)$

b $\overrightarrow{AF} = \overrightarrow{AB} + \overrightarrow{BF} = \overrightarrow{AB} + \mu\overrightarrow{BC} = \mathbf{p} + \mu\mathbf{q}$

> Use the answers from parts **a** and **b**.

c $\lambda\left(\mathbf{q} + \dfrac{1}{4}\mathbf{p}\right) = \mathbf{p} + \mu\mathbf{q}$ gives $\left(\dfrac{1}{4}\lambda - 1\right)\mathbf{p} = (\mu - \lambda)\mathbf{q}$

$\dfrac{1}{4}\lambda - 1 = 0$ and $\mu - \lambda = 0$

> $\mathbf{p}$ and $\mathbf{q}$ are not parallel, so the equation is only possible if both sides are the zero vector.

$\lambda = 4$ and $\mu = 4$

1 In triangle ABC, D and E are the midpoints of AB and AC
$\overrightarrow{BC} = \mathbf{p}$ and $\overrightarrow{BD} = \mathbf{q}$
Express these vectors in terms of $\mathbf{p}$ and $\mathbf{q}$

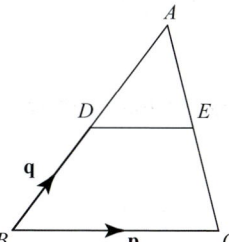

a $\overrightarrow{AC}$

b $\overrightarrow{DE}$

c What does this tell you about DE and BC? Explain why.

2 In the triangle shown, D and E are the midpoints of BC and AC respectively. The point G lies on AD and AG is twice GD

Let $\overrightarrow{AB} = \mathbf{p}$ and $\overrightarrow{AC} = \mathbf{q}$

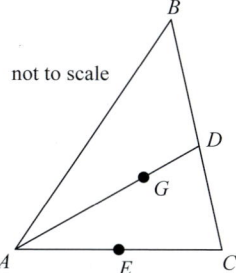

not to scale

Show that BGE is a straight line and find the ratio $BG : GE$

3 In a triangle ABC, the points D, E and F are the midpoints of AB, BC and CA respectively. Point P lies in the plane of the triangle. By setting $\overrightarrow{PA} = \mathbf{a}$, $\overrightarrow{PB} = \mathbf{b}$ and $\overrightarrow{PC} = \mathbf{c}$, show that $\overrightarrow{PD} + \overrightarrow{PE} + \overrightarrow{PF} = \overrightarrow{PA} + \overrightarrow{PB} + \overrightarrow{PC}$

4 $ABCD$ is a quadrilateral. E, F, G and H are the midpoints of AB, BC, CD and DA respectively. By setting $\overrightarrow{AB} = \mathbf{p}$, $\overrightarrow{BC} = \mathbf{q}$ and $\overrightarrow{CD} = \mathbf{r}$, show that $EFGH$ is a parallelogram.

5 In the diagram, D is the midpoint of AE and $AB : BC = 1 : 2$
$\overrightarrow{AB} = \mathbf{p}$ and $\overrightarrow{AD} = \mathbf{q}$
$\overrightarrow{CF} = \lambda \overrightarrow{CD}$ and $\overrightarrow{EF} = \mu \overrightarrow{EB}$

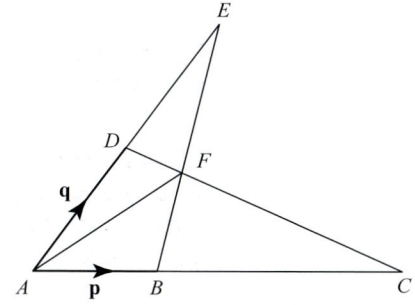

a Express $\overrightarrow{AF}$ in terms of λ, $\mathbf{p}$ and $\mathbf{q}$

b Express $\overrightarrow{AF}$ in terms of μ, $\mathbf{p}$ and $\mathbf{q}$

c Hence determine the values of λ and μ

6 A ship is being steered due east. A current flows from north to south causing the ship to travel at 12 km h^{-1} on a bearing of 120° Work out

a The speed of the current,

b The still-water speed of the ship.

7 Points A and B are directly opposite each other across a river that is 100 m wide and flowing at 2 m s^{-1}. A boat, which can travel at 4 m s^{-1} in still water, leaves A to cross the river.

a If the boat is steered directly across the river, how far downstream of B will it reach the other bank?

b In what direction should it be steered so that it travels directly to B?

Challenge

8 An aircraft has a speed in still air of 300 km h^{-1}. A wind is blowing from the south at 80 km h^{-1}. The pilot must fly to a point south-east of his present position.

a On what bearing should the pilot steer the aircraft?

b At what speed will the aircraft travel?

9 A ship, which can travel at 12 km h^{-1} in still water, is steered due north. A current of 9 km h^{-1} from west to east pushes the ship off course.

a Find the ship's resultant velocity.

b The ship is turned around with the intention of returning to its starting point. On what bearing should it be steered?

 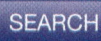

Fluency and skills

Vectors **i** and **j** are unit vectors in the x and y directions.

The vector $\overrightarrow{OP}$ shown has an **x-component** of 3 and a **y-component** of 2. You can write it in **component form**

in terms of **i** and **j**, or as a **column vector**: $\overrightarrow{OP} = 3\mathbf{i} + 2\mathbf{j} = \begin{pmatrix} 3 \\ 2 \end{pmatrix}$

Suppose that a vector $\overrightarrow{OP}$ has magnitude r, direction θ and components x and y, as shown.

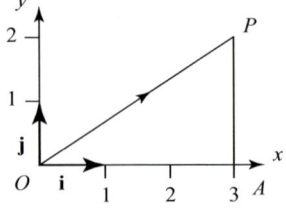

▲ **i** and **j** are unit vectors, $|\mathbf{i}| = 1 = |\mathbf{j}|$

If you know r and θ, you can **resolve** the vector into components.

If you are given x and y, you can find the magnitude and direction.

Key point

$x = r\cos\theta$ and $y = r\sin\theta$

$\overrightarrow{OP} = r\cos\theta\mathbf{i} + r\sin\theta\mathbf{j} = \begin{pmatrix} r\cos\theta \\ r\sin\theta \end{pmatrix}$

Key point

$r = \sqrt{x^2 + y^2}$ and $\tan\theta = \dfrac{y}{x}$

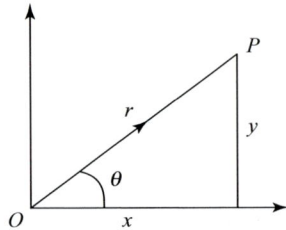

It is a good idea to sketch a diagram when finding θ because, for example, $(3\mathbf{i} - 2\mathbf{j})$ and $(-3\mathbf{i} + 2\mathbf{j})$ would give the same value of $\tan\theta$

Example 1

Express these vectors in component form.

a

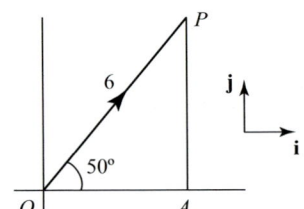

b

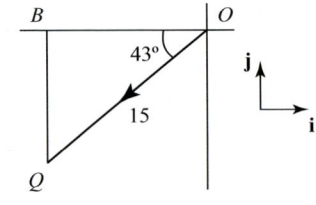

a $OA = 6\cos 50° = 3.86$ and $AP = 6\sin 50° = 4.60$

$\overrightarrow{OP} = 3.86\mathbf{i} + 4.60\mathbf{j}$ or $\begin{pmatrix} 3.86 \\ 4.60 \end{pmatrix}$

b $OB = 15\cos 43° = 11.0$ and $BQ = 15\sin 43° = 10.2$

$\overrightarrow{OQ} = -11.0\mathbf{i} - 10.2\mathbf{j}$ or $\begin{pmatrix} -11.0 \\ -10.2 \end{pmatrix}$

Example 2

Work out the magnitude and direction of these vectors.

a $\mathbf{p} = 5\mathbf{i} + 2\mathbf{j}$
b $\mathbf{q} = -\mathbf{i} - 2\mathbf{j}$

a 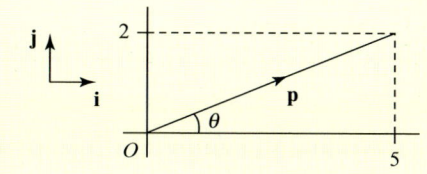 b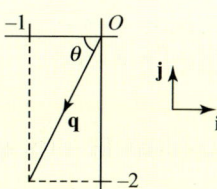

$|\mathbf{p}| = \sqrt{5^2 + 2^2} = 5.39$

$\tan\theta = \dfrac{2}{5}$ given $\theta = 21.80°$

$|\mathbf{q}| = \sqrt{(-1)^2 + (-2)^2} = 2.24$

$\tan\theta = 2$ given $\theta = 63.4°$

You should always make the direction clear – in this case by marking the angle in a diagram. More formally, you can state the direction as a rotation θ from the positive x-direction, where $-180° < \theta \leq 180°$

The answer to part **b** would then be given as $-116.6°$

When vectors are expressed in component form
- to add (or subtract) two vectors, you add (or subtract) the two **i**-components and add (or subtract) the two **j**-components,
- to multiply a vector by a scalar, you multiply both components by that scalar.

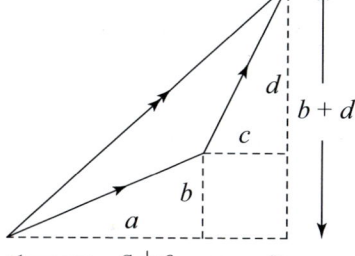

Key point

$(a\mathbf{i} + b\mathbf{j}) + (c\mathbf{i} + d\mathbf{j}) = (a + c)\mathbf{i} + (b + d)\mathbf{j}$

$(a\mathbf{i} + b\mathbf{j}) - (c\mathbf{i} + d\mathbf{j}) = (a - c)\mathbf{i} + (b - d)\mathbf{j}$

$k(a\mathbf{i} + b\mathbf{j}) = ka\mathbf{i} + kb\mathbf{j}$ where k is a scalar.

Example 3

Given $\mathbf{p} = 12\mathbf{i} + 5\mathbf{j}$ and $\mathbf{q} = 3\mathbf{i} - 4\mathbf{j}$, work out

a i $\mathbf{p} - \mathbf{q}$ ii $2\mathbf{p} + 3\mathbf{q}$ b A vector parallel to **p** and with magnitude 39

c The unit vector $\hat{\mathbf{q}}$

a i $\mathbf{p} - \mathbf{q} = 12\mathbf{i} + 5\mathbf{j} - (3\mathbf{i} - 4\mathbf{j}) = (12 - 3)\mathbf{i} + (5 - (-4))\mathbf{j} = 9\mathbf{i} + 9\mathbf{j}$

ii $2\mathbf{p} + 3\mathbf{q} = 2(12\mathbf{i} + 5\mathbf{j}) + 3(3\mathbf{i} - 4\mathbf{j}) = (24\mathbf{i} + 10\mathbf{j}) + (9\mathbf{i} - 12\mathbf{j}) = 33\mathbf{i} - 2\mathbf{j}$

b $|\mathbf{p}| = \sqrt{12^2 + 5^2} = 13$

A parallel vector with magnitude 39 is $3\mathbf{p} = 36\mathbf{i} + 15\mathbf{j}$

Vector $k\mathbf{p}$ is parallel to **p** and has k times the magnitude.

c $|\mathbf{q}| = \sqrt{3^2 + (-4)^2} = 5$

$\hat{\mathbf{q}} = \dfrac{1}{5}\mathbf{q} = 0.6\mathbf{i} - 0.8\mathbf{j}$

You can separate a two-dimensional vector equation into two equations, one for each of the x- and y-components.

Two vectors are **equal** if and only if *both* their **i**- and **j**-components are equal.

If $a\mathbf{i} + b\mathbf{j} = c\mathbf{i} + d\mathbf{j}$ then $a = c$ and $b = d$

Example 4

Given that $\mathbf{p} = x\mathbf{i} + y\mathbf{j}$ and $\mathbf{q} = (2y + 5)\mathbf{i} + (1 - x)\mathbf{j}$, work out the values of x and y for which $\mathbf{p} = \mathbf{q}$

$xi + yj = (2y + 5)i + (1 - x)j$

So $x = 2y + 5$ [1]

and $y = 1 - x$ [2]   Equate components.

$x = 2\dfrac{1}{3}$ and $y = -1\dfrac{1}{3}$ Solve [1] and [2].

To describe the position of a point A relative to an origin O, you use the vector $\overrightarrow{OA}$. This is called the **position vector** of A. It is often labelled $\mathbf{a}$ or $\mathbf{r}_A$

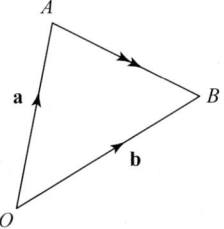

If points A and B have position vectors $\mathbf{a}$ and $\mathbf{b}$ then

vector $\overrightarrow{AB} = \mathbf{b} - \mathbf{a}$ distance $AB = |\mathbf{b} - \mathbf{a}|$

As A has position vector $(2\mathbf{i} + \mathbf{j})$, you can say that its coordinates are $(2, 1)$

Example 5

Points A and B have position vectors $\mathbf{a} = 2\mathbf{i} + \mathbf{j}$ and $\mathbf{b} = 5\mathbf{i} - 6\mathbf{j}$
Evaluate the distance AB

$\overrightarrow{AB} = (5i - 6j) - (2i + j) = 3i - 7j$ Vector $\overrightarrow{AB} = \mathbf{b} - \mathbf{a}$

$|\overrightarrow{AB}| = \sqrt{3^2 + (-7)^2} = 7.62$ Distance $AB = |\mathbf{b} - \mathbf{a}|$

Example 6

Points A, B and C lie on a straight line, with C between A and B, and $AC : CB = 2 : 3$

A and B have position vectors $\mathbf{a} = 2\mathbf{i} + 3\mathbf{j}$ and $\mathbf{b} = 12\mathbf{i} + 5\mathbf{j}$ respectively. Find the position vector $\mathbf{c}$ of C

$\overrightarrow{AB} = \mathbf{b} - \mathbf{a} = (12i + 5j) - (2i + 3j) = 10i + 2j$ Sketch a diagram to help you answer the question.

$\overrightarrow{AC} = \dfrac{2}{5}\overrightarrow{AB} = 4i + 0.8j$

$\mathbf{c} = \mathbf{a} + \overrightarrow{AC} = (2i + 3j) + (4i + 0.8j)$

$= 6i + 3.8j$ This comes from $AC : CB = 2 : 3$

Exercise 6.2A Fluency and skills

1 Write these vectors in the form $x\mathbf{i} + y\mathbf{j}$. Show your working.

a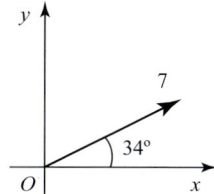

b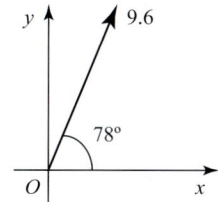

c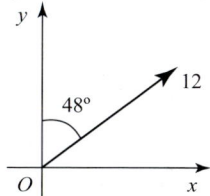

d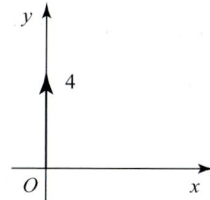

e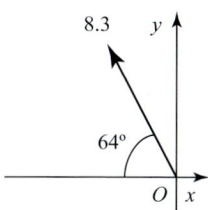

f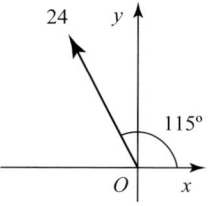

g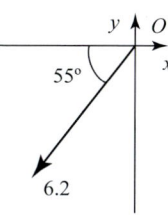

h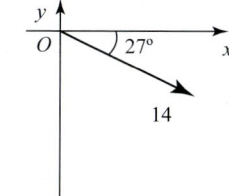

2 Evaluate the magnitude, r, and the direction, θ, of these vectors, where θ is the anticlockwise rotation from the positive x-direction and $-180° < \theta \le 180°$
Show your working.

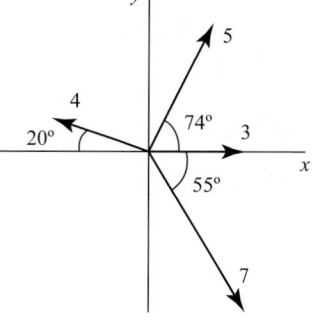

  a $5\mathbf{i} + 2\mathbf{j}$ **b** $7\mathbf{i} + 9\mathbf{j}$ **c** $-5\mathbf{j}$

  d $-2\mathbf{i} + 3\mathbf{j}$ **e** $3\mathbf{i} - 5\mathbf{j}$ **f** $-6\mathbf{i} - 5\mathbf{j}$ **g** $-2\mathbf{i}$

3 The diagram shows the magnitude and direction of four vectors.
By writing each of these vectors in component form, find the
magnitude and direction of their resultant.

4 Given vectors $\mathbf{p} = 2\mathbf{i} - \mathbf{j}$, $\mathbf{q} = -2\mathbf{i} + 3\mathbf{j}$ and $\mathbf{r} = 4\mathbf{i} + \mathbf{j}$, calculate each
of these vectors.

  a $\mathbf{p} + \mathbf{q}$ **b** $\mathbf{p} - \mathbf{r}$ **c** $2\mathbf{q} - \mathbf{p}$ **d** $2\mathbf{p} + 3\mathbf{r}$ **e** $|\mathbf{p}|$ **f** $|\mathbf{q} + \mathbf{r}|$

5 Given vectors $\mathbf{p} = 3\mathbf{i} + u\mathbf{j}$, $\mathbf{q} = v\mathbf{i} - 4\mathbf{j}$ and $\mathbf{r} = 4\mathbf{i} - 6\mathbf{j}$, work out

  a The values of u and v if $\mathbf{p} - \mathbf{q} = \mathbf{r}$

  b The value of u if $\mathbf{p}$ and $\mathbf{r}$ are parallel.

6 Given $\mathbf{p} = -3\mathbf{i} + 4\mathbf{j}$, write down

  a A vector parallel to $\mathbf{p}$ with magnitude 20

  b The unit vector $\hat{\mathbf{p}}$ in the direction of $\mathbf{p}$

7 Evaluate the values of a and b which satisfy these equations.

  a $(2a + 1)\mathbf{i} + (a - 2)\mathbf{j} = (b - 1)\mathbf{i} + (2 - b)\mathbf{j}$

  b $(a^2 - b^2)\mathbf{i} + 2ab\mathbf{j} = 3\mathbf{i} + 4\mathbf{j}$

8 Point A has position vector $(8\mathbf{i} - 15\mathbf{j})$ m.

Work out the distance of point A from the origin. Show your working.

9 Points A and B have position vectors $(-2\mathbf{i} + 3\mathbf{j})$ m and $(4\mathbf{i} + 7\mathbf{j})$ m respectively. Evaluate

    a The length AB,

    b The angle made by AB with the x-direction.

10 A particle is at point A, position vector $(3\mathbf{i} + 7\mathbf{j})$ m. It undergoes a displacement of $(6\mathbf{i} - 3\mathbf{j})$ m, ending up at B

    a Calculate the particle's final distance from the origin O

    b Find the angle AOB

11 $ABCD$ is a parallelogram. The vertices A, B and C have position vectors $\mathbf{a} = \mathbf{i} + 2\mathbf{j}$, $\mathbf{b} = 3\mathbf{i} + 6\mathbf{j}$ and $\mathbf{c} = -4\mathbf{i} + 7\mathbf{j}$ respectively.

    a Find the vector $\overrightarrow{BC}$.

    b Hence find the position vector $\mathbf{d}$ of D

12 $ABCD$ is a parallelogram. A, B and D have position vectors $\mathbf{a} = \begin{pmatrix} -10 \\ -4 \end{pmatrix}$, $\mathbf{b} = \begin{pmatrix} 15 \\ 6 \end{pmatrix}$ and $\mathbf{d} = \begin{pmatrix} -4 \\ 12 \end{pmatrix}$ Find the position vector $\mathbf{c}$ of C

13 Points A, B and C lie on a straight line in that order, with $AB : BC = 7 : 3$

A and C have position vectors $\mathbf{a} = -5\mathbf{i} + 4\mathbf{j}$ and $\mathbf{c} = 7\mathbf{i} + 12\mathbf{j}$ respectively. Find the position vector $\mathbf{b}$ of B

Reasoning and problem-solving

To solve a problem involving vector components

1 Draw a diagram if appropriate.

2 Convert vectors to components if they're not already in component form.

3 When solving vector equations remember that you can equate x- and y-components separately.

Show that the points A (0, 2), B (2, 5) and C (6, 11) are collinear.

$A = \begin{pmatrix} 0 \\ 2 \end{pmatrix}$ $B = \begin{pmatrix} 2 \\ 5 \end{pmatrix}$ $C = \begin{pmatrix} 6 \\ 11 \end{pmatrix}$

> Express each point as a vector from the origin.

$\overrightarrow{AB} = \begin{pmatrix} 2 \\ 5 \end{pmatrix} - \begin{pmatrix} 0 \\ 2 \end{pmatrix} = \begin{pmatrix} 2 \\ 3 \end{pmatrix}$

$\overrightarrow{BC} = \begin{pmatrix} 6 \\ 11 \end{pmatrix} - \begin{pmatrix} 2 \\ 5 \end{pmatrix} = \begin{pmatrix} 4 \\ 6 \end{pmatrix}$

> Determine two vectors that join a common point (B).

$\overrightarrow{BC} = 2\overrightarrow{AB}$

> Show that the vectors are parallel (vectors $\mathbf{a}$ and $\mathbf{b}$ are parallel if $\mathbf{a} = k\mathbf{b}$).

So the points A, B and C must be collinear.

Exercise 6.2B Reasoning and problem-solving

1 A canoeist takes part in a race across a lake. They must pass through checkpoints, whose positions on a grid map are given by the x- and y-coordinates $(1, 11)$, $(7, 6)$ and $(13, 1)$ respectively. Show that the canoeist will pass through all three checkpoints if they paddle in a straight line.

2 Particles A and B have position vectors $\mathbf{a} = (2\mathbf{i} + 5\mathbf{j})$ m and $\mathbf{b} = (6\mathbf{i} + 3\mathbf{j})$ m respectively. Particle A undergoes a displacement of $(2\mathbf{i} - 3\mathbf{j})$ m and particle B moves in the opposite direction and three times as far. Calculate the distance between the particles after these displacements.

3 Points A and B have position vectors $\mathbf{a} = 3\mathbf{i} + \mathbf{j}$ and $\mathbf{b} = 11\mathbf{i} + 6\mathbf{j}$ respectively. Point C lies on the same straight line as A and B. The lengths AC and BC are in the ratio $3 : 2$. Show that there are two possible positions for point C, and find the position vector of each.

4 A town contains four shops A, B, C and D. Shop B is 200 m west of A. Shop C is 100 m north of A. Shop D is 283 m north-east of A. Show that the positions of shops B, C and D are collinear, given that the distances are rounded.

5 Particle A is stationary at the point $(2\mathbf{i} + 3\mathbf{j})$ m, particle B is stationary at the point $(3\mathbf{i} - \mathbf{j})$ m and particle C is stationary at the point $(4\mathbf{i} + 13\mathbf{j})$. Particle B undergoes a displacement of $k(\mathbf{i} + 2\mathbf{j})$ and particle C undergoes a displacement of $k(4\mathbf{i} - \mathbf{j})$ so that all three particles are aligned in a straight line. Determine the possible values of k

6 Particle A starts at the point $(3\mathbf{i} + \mathbf{j})$ m and travels for 2 s along a track, finishing at the point $(7\mathbf{i} + 4\mathbf{j})$ m. A second particle, B, starts at the same time from the point $2\mathbf{i}$ m. It travels along a parallel track for a distance of d m.

  a Work out the final position vector of B in component form, in terms of d

  b If $d = 15$, evaluate the final distance from A to B

7 The road from P to Q makes a detour round a mountain. It first goes 6 km from P on a bearing of 080°, then 7 km on a bearing of 020° and finally 5 km on a bearing of 295° to reach Q. There is a plan to bore a tunnel through the mountain from P to Q. It will be considered cost effective if it reduces the journey by more than 10 km. Determine whether the tunnel should be built based on this information.

8 If a particle moves with constant velocity $\mathbf{v}$, its displacement after time t is $\mathbf{v}t$. Particle A is initially at the point with position vector $(\mathbf{i} + 4\mathbf{j})$ m and moving with constant velocity $(3\mathbf{i} + 3\mathbf{j})$ m s⁻¹. At the same time, particle B is at the point $(5\mathbf{i} + 2\mathbf{j})$ m and moving with constant velocity $(2\mathbf{i} + 3.5\mathbf{j})$ m s⁻¹. After t seconds the particles are at A' and B' respectively.

  a Find the position vector of A' in terms of t

  b Find the vector $\overrightarrow{A'B'}$ in terms of t

  c Decide if the particles will collide, and, if so, find the position vector of the point of collision

9 If a particle moves with constant velocity $\mathbf{v}$, its displacement after time t is $\mathbf{v}t$. A particle starts from the point with position vector $4\mathbf{j}$ m and moves with constant velocity $(2\mathbf{i} - \mathbf{j})$ m s⁻¹. At the same time a second particle starts from the point $(6\mathbf{i} + 8\mathbf{j})$ m and moves with constant velocity $(-\mathbf{i} - 3\mathbf{j})$ m s⁻¹. Show that the particles collide, and find the time at which they do so.

Challenge

10 Particle A starts from the point with position vector $5\mathbf{j}$ m and moves with constant velocity $(2\mathbf{i} + \mathbf{j})$ m s⁻¹. Five seconds later a second particle, B, leaves the origin, moving with constant velocity, and collides with A after a further 2 s. Find the velocity of B

Chapter summary

- A vector quantity has both magnitude and direction. A scalar quantity has magnitude only.
- Equal vectors have the same magnitude *and* direction.
- $k\mathbf{a}$ is parallel to $\mathbf{a}$ and has magnitude $k|\mathbf{a}|$
- Two or more points are collinear if a single vector, or multiple parallel vectors, pass through those points.
- The unit vector has a magnitude of 1 in the direction of $\mathbf{a}$ is $\hat{\mathbf{a}} = \dfrac{\mathbf{a}}{|\mathbf{a}|}$
- $\overrightarrow{AC} = \overrightarrow{AB} + \overrightarrow{BC}$, $\overrightarrow{AC}$ is the resultant of $\overrightarrow{AB}$ and $\overrightarrow{BC}$
- Making a vector negative reverses its direction, so $\overrightarrow{BA} = -\overrightarrow{AB}$
- A vector is written in component form as $x\mathbf{i} + y\mathbf{j}$ or $\begin{pmatrix} x \\ y \end{pmatrix}$
- For magnitude r and direction θ (the positive rotation from the x-direction), $x = r\cos\theta$ and $y = r\sin\theta$
- For components x and y, $r = \sqrt{x^2 + y^2}$ and $\tan\theta = \dfrac{y}{x}$
- Treat components separately in calculations and equations
  $(a\mathbf{i} + b\mathbf{j}) \pm (c\mathbf{i} + d\mathbf{j}) = (a \pm c)\mathbf{i} + (b \pm d)\mathbf{j}$
  $k(a\mathbf{i} + b\mathbf{j}) = ka\mathbf{i} + kb\mathbf{j}$
  if $a\mathbf{i} + b\mathbf{j} = c\mathbf{i} + d\mathbf{j}$ then $a = c$ and $b = d$
- The position vector of a point A relative to an origin O is $\overrightarrow{OA}$, often labelled $\mathbf{a}$ or $\mathbf{r}_A$
- If points A and B have position vectors $\mathbf{a}$ and $\mathbf{b}$ then vector $\overrightarrow{AB} = \mathbf{b} - \mathbf{a}$ and distance $AB = |\mathbf{b} - \mathbf{a}|$

Check and review

You should now be able to...	Try Questions
✔ Identify vector quantities and scalar quantities.	1
✔ Solve geometric problems in two dimensions using vector methods.	2–4
✔ Solve problems involving displacements, velocities and forces.	5, 6
✔ Find and use the components of a vector.	6
✔ Find the magnitude and direction of a vector expressed in component form.	6
✔ Use position vectors to find displacements and distances.	7

1   State whether each quantity is vector or scalar.

    a Speed   **b** Displacement   **c** Force   **d** Velocity   **e** Distance

2 In the diagram, *C* is the midpoint of *BD*, *F* is the midpoint of *AC* and *EB* = 2*AE*. The vectors $\overrightarrow{AE}$ and $\overrightarrow{EF}$ are **p** and **q** respectively.

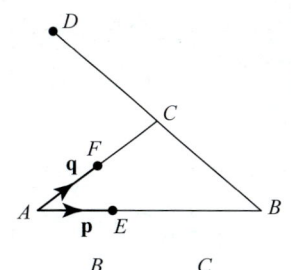

a Express these vectors in terms of **p** and **q**

   i $\overrightarrow{AB}$ **ii** $\overrightarrow{BC}$ **iii** $\overrightarrow{DF}$

b Show that *E*, *F* and *D* are collinear.

3 The diagram shows a regular hexagon *ABCDEF*. $\overrightarrow{AB}$ = **p** and $\overrightarrow{AF}$ = **q**

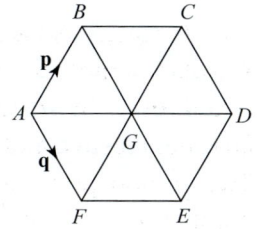

Express these vectors in terms of **p** and **q**

   a $\overrightarrow{BC}$ **b** $\overrightarrow{BE}$ **c** $\overrightarrow{FD}$ **d** $\overrightarrow{CE}$

4 The diagram shows a quadrilateral *ABCD*

$\overrightarrow{AB}$ = **p**, $\overrightarrow{AC}$ = **q** and $\overrightarrow{AD}$ = **r**. *E* is the midpoint of *CD*

The point *F* lies on *BD* and *BF* : *FD* = 3 : 1

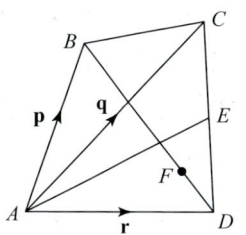

Express these vectors in terms of **p**, **q** and **r**

   a $\overrightarrow{DC}$ **b** $\overrightarrow{DE}$ **c** $\overrightarrow{AE}$ **d** $\overrightarrow{BD}$ **e** $\overrightarrow{BF}$ **f** $\overrightarrow{AF}$

5 **V** is the resultant of two velocities, $\mathbf{v_1}$ and $\mathbf{v_2}$
$\mathbf{v_1}$ has magnitude $3\,\text{m s}^{-1}$ and direction 150° (anticlockwise rotation from the *x*-direction). $\mathbf{v_2}$ has magnitude $10\,\text{m s}^{-1}$ and direction θ. Given that the direction of **V** is 90°, calculate

   a The value of θ **b** The magnitude of **V**

6 Forces **P**, **Q** and **R** have directions 50°, 100° and −20° respectively (measured from the positive *x*-direction) |**P**| = 8 N, |**Q**| = 10 N and |**R**| = 6 N

   a Express **P**, **Q** and **R** in component form.

   b Calculate the magnitude and direction of the resultant of the three forces.

7 In relation to origin *O*, points *A* and *B* have position vectors **a** = 2**i** + 5**j** and **b** = 6**i** − 2**j** respectively. Find

   a The distance *OA*   **b** The displacement $\overrightarrow{AB}$ **c** The distance *AB*

What next?

Score			
	0 – 3	Your knowledge of this topic is still developing. To improve, search in MyMaths for the codes: 2206–2207	Click these links in the digital book
	4 – 5	You're gaining a secure knowledge of this topic. To improve, look at the InvisiPen videos for Fluency and skills (06A)	
	6 – 7	You've mastered these skills. Well done, you're ready to progress! Now try looking at the InvisiPen videos for Reasoning and problem-solving (06B)	

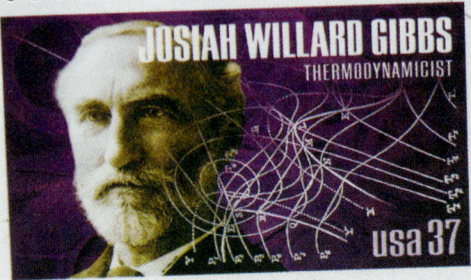

JOSIAH WILLARD GIBBS
THERMODYNAMICIST
usa 37
2005

History

J Willard Gibbs was an American mathematical physicist. Between 1881 and 1884, while teaching at Yale University, he produced lecture notes to help his students understand the electromagnetic theory of light. In his notes he replaced something called **quaternions** with a simpler representation that we now know as **vectors**.

These vector analysis notes proved to be a great success and were later adapted and published as a textbook by one of Gibbs' students, **Edwin Bidwell Wilson**, in 1901. This means that vectors are a relatively recent part of mathematics.

"Mathematics is a language."
– J Gibbs

Research

The theory of quaternions was developed by **Sir William Rowan Hamilton**. What is the connection with Broom Bridge in Dublin?

A British mathematician also developed a theory of vector calculus independently around the same time as Gibbs. Who was it?

Did you know?

Vectors are used extensively in video game development. They are required in order to control or determine the position of objects on the screen.

Programmers combine the language of mathematics with the computer programming language to create the game.

Most people would think that computer programming is a modern invention that didn't truly come about until the 2nd half of the 20th century.

However, the title of 'first computer programmer' is often given to **Ada Lovelace**, a mathematician born in 1815.

Lovelace met **Charles Babbage**, a mathematician and engineer who was working on an 'analytical engine', which was essentially what we would now describe as a programmable computer. She took interest in his work and developed what we now consider to be the first ever computer program.

The analytical engine never did get built, but Lovelace laid the foundations for the type of computer programming that we encounter in so many aspects of our lives today.

Ada Lovelace

1 Given vectors $\mathbf{p} = \begin{pmatrix} 6 \\ -1 \end{pmatrix}$ and $\mathbf{q} = \begin{pmatrix} -3 \\ 4 \end{pmatrix}$

   a Evaluate $3\mathbf{p} + 5\mathbf{q}$ **[2 marks]**

   b Write down the unit vector, $\hat{\mathbf{q}}$, in the direction of $\mathbf{q}$ **[3]**

2 Four vectors, $\mathbf{a}$, $\mathbf{b}$, $\mathbf{c}$ and $\mathbf{d}$ have a resultant of 0. Given $\mathbf{a} = 2\mathbf{i} + 7\mathbf{j}$, $\mathbf{b} = 3\mathbf{i} - 10\mathbf{j}$ and $c = 5\mathbf{i} - 21\mathbf{j}$, evaluate

   a $\mathbf{d}$ **[3]**

   b $|\mathbf{d}|$ **[2]**

3 A sailing boat starts from buoy A and sails 800 metres on a bearing of 032° to buoy B
   It then sails 1200 metres on a bearing of 294° to buoy C. Work out

   a The distance AC **[6]**

   b The bearing of C from A **[4]**

4 Given vectors $\mathbf{p} = x\mathbf{i} + 2\mathbf{j}$, $\mathbf{q} = 3\mathbf{i} + y\mathbf{j}$ and $\mathbf{r} = 4\mathbf{i} + 6\mathbf{j}$, evaluate

   a x and y if $\mathbf{r} = \mathbf{p} - \mathbf{q}$ **[3]**

   b x if $\mathbf{p}$ and $\mathbf{r}$ are parallel.   **[2]**

5 A river flows at 0.3 m s^{-1}. Peter and Mary, who can each swim at 0.5 m s^{-1} in still water, wish to cross the river. The river is 20 metres wide.

   a Peter aims straight across the river, and is carried downstream by the current.   **[3]**

   i How long does he take to reach the other side?

   ii Work out his actual speed of travel.

   b Mary swims to the closest point on the opposite bank.

   i In which direction does she aim?

   ii Work out her actual speed of travel.

   iii How long does it take her to reach the other side?   **[6]**

6 The diagram shows two vectors, $\mathbf{a}$ and $\mathbf{b}$, where $\overrightarrow{OA} = \mathbf{a}$ and $\overrightarrow{OB} = \mathbf{b}$. The point X lies on OA where $OX:XA = 2:1$. The point Y lies on OB where $OY:YB = 3:1$. M is the midpoint of XB

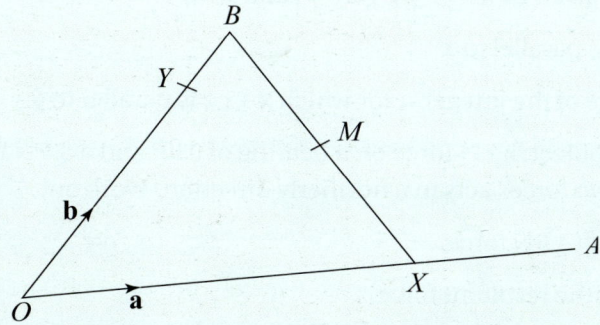

a Express these vectors in terms of **a** and **b** [11]

    i $\overrightarrow{OX}$ **ii** $\overrightarrow{OY}$ **iii** $\overrightarrow{OM}$ **iv** $\overrightarrow{AM}$ **v** $\overrightarrow{MY}$

b Use your answer to **a** to prove that A, M and Y are collinear, and work out the ratio $AM:MY$ [2]

7 Vectors $\mathbf{a} = \begin{pmatrix} 2 \\ -5 \end{pmatrix}$ and vector $\mathbf{b} = \begin{pmatrix} 4 \\ 2 \end{pmatrix}$

Work out the magnitude and direction of the resultant of **a** and **b** [6]

8 A is the point $(-2, 5)$, B is the point $(1, 3)$ and C is the point $(10, -3)$

    a Write down   **i** $\overrightarrow{AB}$ **ii** $\overrightarrow{BC}$ [2]

    b Prove that A, B and C are collinear. [2]

9 Two vectors, **a** and **b**, are given by $\mathbf{a} = \begin{pmatrix} -3 \\ 8 \end{pmatrix}$ and $\mathbf{b} = \begin{pmatrix} 2 \\ 1 \end{pmatrix}$

Given $|\mathbf{a} + \lambda\mathbf{b}| = 13$, work out the possible values of the scalar λ [6]

10 The diagram shows a trapezium $OABC$ where $\overrightarrow{OA} = \mathbf{a}$, $\overrightarrow{OC} = \mathbf{c}$, and $\overrightarrow{CB} = 2\mathbf{a}$
M is the midpoint of AB, and N is the midpoint of BC

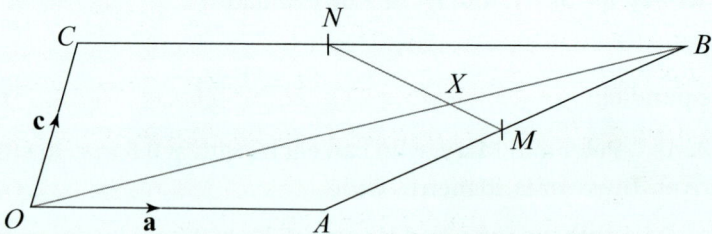

a Express these vectors in terms of **a** and **c** [7]

    i $\overrightarrow{OB}$ **ii** $\overrightarrow{ON}$ **iii** $\overrightarrow{AB}$ **iv** $\overrightarrow{OM}$ **v** $\overrightarrow{MN}$

The line OB meets the line MN at the point X

b Evaluate $\overrightarrow{OX}$ [8]

11 Points P, Q, R and S have position vectors $\mathbf{p} = \begin{pmatrix} 6 \\ 3 \end{pmatrix}$, $\mathbf{q} = \begin{pmatrix} -3 \\ -5 \end{pmatrix}$, $\mathbf{r} = \begin{pmatrix} 1 \\ -3 \end{pmatrix}$ and $\mathbf{s} = \begin{pmatrix} 10 \\ 5 \end{pmatrix}$ [7]

Prove that the quadrilateral $PQRS$ is a parallelogram.

12 Vectors **x**, **y** and **z** are given by $\mathbf{x} = \begin{pmatrix} 1 \\ -1 \end{pmatrix}$, $\mathbf{y} = \begin{pmatrix} 5 \\ 3 \end{pmatrix}$ and $\mathbf{z} = \begin{pmatrix} 2 \\ 1 \end{pmatrix}$

    a Prove that $\mathbf{x} + 3\mathbf{y}$ is parallel to **z** [3]

    b Work out the value of the integer c, for which $\mathbf{x} + c\mathbf{z}$ is parallel to **y** [4]

13 Two forces act on an object: a 2 N force on a bearing of 030° and a 3.5 N force.
The resultant of the two forces acts in a northerly direction. Work out

    a The direction of the 3.5 N force, [6]

    b The magnitude of the resultant force. [2]

In questions that tell you to show your working, you shouldn't depend solely on a calculator. For these questions, solutions based entirely on graphical or numerical methods are not acceptable.

1 **a** Simplify these expressions.

      i $(5^n)^m$ **ii** $\dfrac{3^m \times 3^n}{3}$ **iii** $\cdot \sqrt{7^m}$ **[4 marks]**

  b Solve the equation $2^{x+1} = 4^3$. You must show your working.   **[2]**

2 Solve these simultaneous equations. You must show your working.

    $x + 12y = 5$ $x^2 + 2xy - 5 = 0$ **[5]**

3 $f(x) = x^2 + 3x + k - 4$

  a Show that when $k = 10$ the equation $f(x) = 0$ has no real solutions.   **[2]**

  b Solve the equation $f(x) = 0$ given that

    i $k = 6$ **ii** $k = 5$ **[4]**

  c Find the value of k for which the equation $f(x) = 0$ has precisely one solution.   **[2]**

4 **a** Write $x^2 - 4x + 12$ in the form $(x + p)^2 + q$ where p and q are integers to be found.   **[2]**

  b Show algebraically that the equation $x^2 - 4x + 12 = 0$ has no real solutions.   **[2]**

  c Sketch the graph of $y = x^2 - 4x + 12$ and write down the coordinates of the minimum point.   **[3]**

5 The diagram shows the triangle ABC where $\overrightarrow{AB} = \mathbf{p}$ and $\overrightarrow{AC} = \mathbf{q}$

  The point D is the midpoint of BC

  a Express these vectors in terms of $\mathbf{p}$ and $\mathbf{q}$

    i $\overrightarrow{BC}$ **ii** $\overrightarrow{BD}$ **iii** $\overrightarrow{DA}$ **[5]**

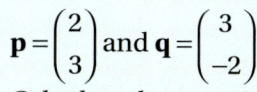

    $\mathbf{p} = \begin{pmatrix} 2 \\ 3 \end{pmatrix}$ and $\mathbf{q} = \begin{pmatrix} 3 \\ -2 \end{pmatrix}$

  b Calculate the magnitude of these vectors.

    i $\overrightarrow{AB}$ **ii** $\overrightarrow{BD}$ **[5]**

  c Show that vectors $\mathbf{p}$ and $\mathbf{q}$ are perpendicular.   **[3]**

6 **a** Express $\dfrac{2 + \sqrt{3}}{1 - \sqrt{3}}$ in the form $p + q\sqrt{3}$ and write down the values of p and q

    You must show your working.   **[4]**

  b Solve the equation $x\sqrt{8} = 5\sqrt{2} - \sqrt{32}$

    Show your working and give your answer in its simplest form.   **[3]**

7 A circle has equation $x^2 + y^2 - 6x + 14y + 33 = 0$

  a Work out the centre and the radius of the circle. Show your working.   **[4]**

  b Sketch the circle, labelling its points of intersection with the coordinate axes.   **[3]**

8 The line p_1 has equation $2x + 8y + 14 = 0$

    a   Find the gradient of p_1 **[2]**

    The line p_2 is perpendicular to p_1 and passes through the point $(3, 6)$

    b   Find the equation of the line p_2 in the form $ax + by + c = 0$ **[3]**

    The lines p_1 and p_2 intersect at the point A

    c   Find the coordinates of A using algebra.   **[3]**

    d   Calculate the length OA **[2]**

9 $\mathbf{a} = 2\mathbf{i} - 3\mathbf{j}$ and $\mathbf{b} = 8\mathbf{i} + \mathbf{j}$ where $\mathbf{i}$ and $\mathbf{j}$ are unit vectors in a due east and due north direction respectively.

    a   Find the vector $\mathbf{c} = \mathbf{a} - 3\mathbf{b}$ **[2]**

    b   Calculate the magnitude and direction of vector $\mathbf{c}$. Show your working.   **[4]**

    c   Describe the geometric relationship between the vector $\mathbf{c}$ and the vector $11\mathbf{i} + 3\mathbf{j}$

        Explain how you determined this relationship.   **[2]**

10 Prove by counter-example that the following statement is false.

    "The product of two consecutive integers is odd."   **[2]**

11 a   Write down the value of $^9C_5$ **[1]**

    b   Find the binomial expansion in ascending powers of x of

        i $(1 - x)^6$ **ii** $(2x + 1)^4$ **[6]**

12 A boat is sailing with a velocity of $\begin{pmatrix} -1 \\ 4 \end{pmatrix}$ m s^{-1}

    a   Calculate the speed of the boat to 1 decimal place.   **[2]**

    b   Find the direction the boat is sailing as a bearing.   **[2]**

13 Given $p(x) = x^3 - 2x^2 - 5x + 6$

    a   Show that $(x - 3)$ is a factor of $p(x)$ **[2]**

    b   Find all the solutions to the equation $p(x) = 0$. Show your working.   **[3]**

    c   Sketch the curve of $y = p(x)$

        Label the coordinates where the curve crosses the x-axis and y-axis.   **[3]**

14 a   Sketch these graphs on the same axes, giving the coordinates of their intersections with the x and y-axis.

        i $y = \dfrac{1}{x+3}$ **ii** $y = 3^x$ **[5]**

    b   State how many solutions there are to the equation $\dfrac{1}{x+3} = 3^x$

        Explain your answer.   **[1]**

15 A triangle has sides 9 cm and 17 cm and angle 65° as shown.

    a   Calculate the length of side x **[2]**

b Work out the size of the acute angle y [3]

c Calculate the area of the triangle. [2]

16 a Solve these equations for $0 \leq x \leq 180°$

Show your working and give your answers to 1 dp where appropriate.

  i $\sin x = 0.5$ **ii** $\cos 2x = 0.75$ [6]

b Sketch the graph of $y = \sin(x + 40)$ for $0 \leq x \leq 360°$

   Label the x-intercepts. [2]

17 A triangle has sides of 12.3 cm and 13.7 cm and angles of 34° and θ as shown.

a Calculate the possible values of θ [4]

b If θ is acute, calculate the area of the triangle. [3]

18 Differentiate the following expressions with respect to x

a $x^3 - 2x^5 + 3$ **b** $\sqrt{x} + \dfrac{1}{x}$ [5]

19 a Find the gradient of the curve $y = x^3 - 2x$ at the point $(2, 4)$. Show your working. [3]

b Find the equation of the normal to the curve $y = x^3 - 2x$ at this point. Give your answer in the form $ax + by = c$ where a, b and c are integers. [3]

20 Work out these integrals.

a $\int (2 - 3x)^2 \, \mathrm{d}x$ **b** $\int \sqrt{x}(3x + 4) \, \mathrm{d}x$ **c** $\int \dfrac{1 - 5x^2}{\sqrt{x}} \, \mathrm{d}x$ [11]

21 The shaded area is enclosed by the x-axis, the line $x = 5$ and the curve with equation $y = -2 + \sqrt{x}$

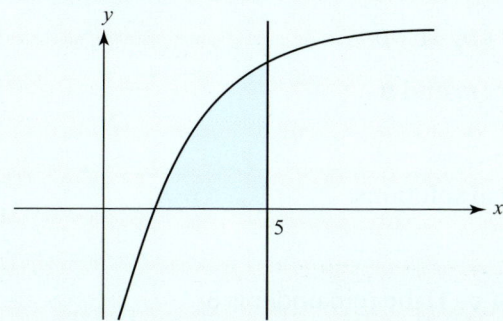

a Find the coordinates of the point where the curve crosses the x-axis. Show your working. [2]

b Calculate the shaded area, giving your answer in surd form. You must show your working. [4]

22 A curve has equation $y = x^3 - 2x^2 + \dfrac{1}{x}$

a Find $\dfrac{\mathrm{d}y}{\mathrm{d}x}$ [2]

b Find the equation of the tangent to the curve through the point where $x = -1$ [3]

   This tangent crosses the x-axis at the point A and the y-axis at the point B

c Calculate the area of triangle OAB [3]

23 The graph of $y = a(x^b)$ passes through the points $(3, 16)$ and $(2, 4)$

Find an approximate equation for the curve. [4]

24 a Write these expressions in the form $\log_2 a$. Show your working.

    i $\log_2 15 - \log_2 3$ **ii** $\log_2 5 + 2\log_2 3$ **iii** $1 + \log_2 5$ **[5]**

    b Solve the equation $3^x = 17$

         Show your working and give your answer to 3 significant figures. **[2]**

25 a Sketch the graph of $y = \log_4 x$

         Label any points of intersection with the coordinate axes and give the equation of any asymptotes. **[3]**

    b Solve these equations, giving your answers as exact fractions. You must show your working.

        i $\log_4 x = -3$ **ii** $\ln x + \ln 5 = \ln(x + 1)$ **[4]**

26 a Solve these inequalities. You must show your working.

        i $10 - 2x \geq 4 - 5x$ **ii** $x^2 + x < 6$ **[5]**

    b Give the range of values of x that satisfies both $10 - 2x \geq 4 - 5x$ and $x^2 + x < 6$ **[2]**

27 The points $A(3, 7)$ and $B(5, -3)$ are points on the circumference of a circle.

     AB is a diameter of the circle.

    a Find the equation of the circle. **[4]**

    b Find an equation of the tangent to the circle at the point A **[4]**

28 a Work out the points of intersection, A and B, between the line with equation $x + 2y = 5$ and the curve with equation $x^2 - xy - y^2 = 5$. Show your working. **[5]**

    b Find an equation of the chord AB in the form $ax + by + c = 0$ **[3]**

29 Prove that $n^2 - m^2$ is odd for any consecutive integers m and n **[3]**

30 You are given $f(x) = 2x^2 + kx - 4k$

     Find the range of values of k for which $f(x) = 0$ has real solutions. **[4]**

31 $f(x) = x^3 - ax^2 + bx + 10$

     $(x - 2)$ is a factor of $f(x)$ and when $f(x)$ is divided by $(x + 1)$ the remainder is 6

     Find the values of the constants a and b **[4]**

32 $f(x) = 2x^3 + 3x^2 - 23x - 12$

     $x = -4$ is a solution of $f(x) = 0$

    a Fully factorise $f(x)$ **[2]**

    b Sketch the graph of $y = f(x)$, giving the coordinates of the x and y-intercepts. **[3]**

33 $\overrightarrow{OA} = \begin{pmatrix} 3 \\ -1 \end{pmatrix}$ and $\overrightarrow{OB} = \begin{pmatrix} -2 \\ 5 \end{pmatrix}$

Calculate the area of triangle OAB to 1 decimal place. **[6]**

34 A ball is kicked from the ground at an angle of 30° to the horizontal at a speed of 15 m s^{-1}

Write down the initial velocity of the ball as a vector. **[3]**

35 Use the binomial expansion to expand then fully simplify these expressions.
Show your working.

a $\left(1+\sqrt{5}\right)^5$ **b** $\left(1-\sqrt{3}\right)\left(2-\sqrt{3}\right)^4$ **[7]**

36 a Find the expansion of $(2+x)^6$ in ascending powers of x **[3]**

b Use your expansion to find the value of 2.001^6 correct to 3 decimal places. **[2]**

37 a Show that the equation $\sin 3x + 2\cos 3x = 0$ can be written $\tan 3x = -2$ **[2]**

b Solve the equation $\sin 3x + 2\cos 3x = 0$ for $0 \le x \le 180°$

Show your working and give your solutions correct to the nearest degree. **[4]**

38 a Sketch, for $-180° \le x \le 180°$, the following graphs. In each case, give the coordinates of intersection with the x- and y-axes, the equations of any asymptotes, and the coordinates of any maximum or minimum points.

i $y = 2\sin x$ **ii** $y = \tan 2x$ **[8]**

b Solve the equation $2\sin(x+30) = \sqrt{3}$ for values of x in the range $-180° < x < 180°$
Show your working. **[4]**

39 Given that $\cos x = \dfrac{1}{3}$ for an acute angle x, find the exact value of the following expressions.

Give your answers in the form $a\sqrt{2}$

a $\sin x$ **b** $\tan x$ **[5]**

40 A triangle has side lengths 19 mm, 24 mm and 31 mm.

a Calculate the size of the smallest angle in the triangle. **[3]**

b Calculate the area of the triangle. **[3]**

41 A triangle has side lengths 12 cm and 17 cm with an angle of 30° as shown.

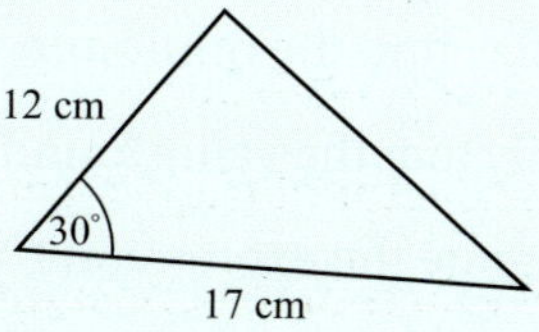

a Find the perimeter of the triangle. Give your
answer to 1 dp. **[3]**

b Calculate the area of the triangle. **[2]**

A second triangle has the same area but is equilateral.

c Find the perimeter of this new triangle. Give your answer to 1 dp. **[3]**

42 A curve has equation $y = \dfrac{\left(x^3+2\right)^2}{x^5}$

a Show that $y = x + 4x^{-2} + 4x^{-5}$ **[2]**

b Find an expression for

i $\dfrac{dy}{dx}$ **ii** $\dfrac{d^2y}{dx^2}$ **[4]**

c Calculate the value of $\displaystyle\int_{-1}^{1} y\, dx$. Show your working. **[4]**

43 A river has parallel banks and is 15 m wide as shown. The water in the river is flowing at a velocity of 3 m s^{-1} parallel to the banks. A boat crosses a river at an angle of θ to the bank at a velocity relative to the water of 4 m s^{-1} perpendicular to the bank.

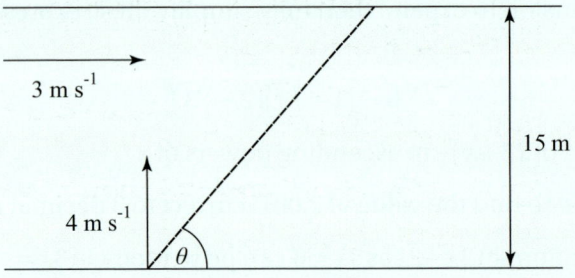

  a   Find the resultant speed of the boat.   **[2]**

  b   Find the angle θ **[2]**

44 $y = 4x^3$

  a   Work out $\dfrac{dy}{dx}$ from first principles.   **[5]**

  b   Calculate the gradient of the tangent to the curve when $x = -2$ **[2]**

45 Find the range of values of x for which $f(x) = -\dfrac{16}{\sqrt{x}} - \dfrac{1}{2}x^2$ is an increasing function.   **[4]**

46 Find and classify all the stationary points of the curve with equation

$$y = \frac{x^4}{2} - x^3 - 4x^2 - 3x + 1$$

You must show your working.   **[8]**

47 A box has a circular base with radius x cm and height h cm.

The volume of the box is 12 cm^3

  a   Show that the surface area, A, of the box (not including the lid) is given by $A = \pi x^2 + \dfrac{24}{x}$ **[6]**

  b   Calculate the value of x for which A is a minimum.   **[3]**

  c   Hence, find the minimum value of A **[2]**

  d   Justify that the value found in part **b** is a minimum.   **[2]**

48 a   Integrate the expression $10x^4 - 12x^2 + 1$ with respect to x **[3]**

     A curve has equation $y = f(x)$

     Given that $\dfrac{dy}{dx} = 10x^4 - 12x^2 + 1$, and that the curve passes through the point (2, 9),

  b   Find the equation of the curve,   **[2]**

  c   Verify that the curve passes through the point $(-1, -24)$ **[2]**

49 A jet ski travels 200 m in a straight line on a bearing of 200°, then 600 m in a straight line on a bearing of 060°. Calculate the distance the jet ski must travel, to the nearest m, and the bearing on which it needs to travel to return directly to its start point. **[6]**

50 Use the data in the table to approximate the equation of the curve, given that it is in the form $y = ax^b$

Give your values of a and b to one decimal place. **[4]**

x	5	10	15	20
y	25	70	130	200

51 The price in £ of a car that is t years old is modelled by the formula $P = 12000e^{-\frac{t}{5}} - 1000$

a When will the car first be worth less than £3000? Give your answer in years and months to the nearest month. **[3]**

b Comment on the appropriateness of this model. **[1]**

52 a Sketch on the same axes the graphs of

    i $y = \log_5 x$ **ii** $y = \ln x$ **[3]**

b Solve the equation $\ln x - 2\ln 3 = 5$

Show your working and give your answer in terms of e **[3]**

53 a Use a suitable method of proof to prove or disprove the following statements.

    i "$2^n + 1$ is a prime number for all positive integers n"

    ii "$3^n + 1$ is even for all integers n in the range $1 \le n \le 4$"

    iii "$5^n + 10$ is a multiple of 5 for all positive integers n" **[6]**

b For each proof in part **a**, state the name of the method of proof you have used. **[3]**

54 The curve shown has equation $y = A + \dfrac{B}{x+C}$

Find the values of A, B and C **[4]**

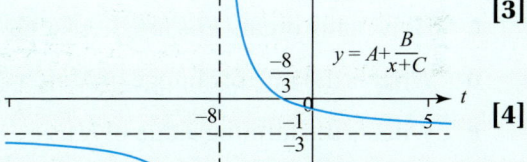

55 The first three terms in the expansion of $(1 + ax)^n$ are $1 - 8x + 30x^2$

Given that $n \in \mathbb{Z}^+$, find the value of

a n **[4]**   **b** a **[1]**

56 From the expansion of $(4 + 3x)^3 (x - 2)^5$, work out

a The constant term,   **[2]**   **b** The term in x^2 **[4]**

57 Solve these equations for $-360 \le x \le 360°$

Show your working and give your answers to 3 significant figures.

a $\sin^2 x = 0.6$ **[4]**   **b** $\cos^2 x - \cos x - 2 = 0$ **[3]**

58 a Prove that $3\sin^2 x - 4\cos^2 x \equiv 7\sin^2 x - 4$ **[3]**

    b Solve the equation $3\sin^2 x - 4\cos^2 x = 1$ for $-180° \leq x \leq 180°$. You must show your working. **[5]**

59 Solve the quadratic equation $\cos x + 3\sin^2 x = 2$ for x in the range $0 \leq x \leq 360°$.
You must show your working. **[6]**

60 $f(x) = x^4 + 5x^3 - 2$

    a The line l_1 is the tangent to $y = f(x)$ at the point where $x = -1$. Find the equation of l_1 **[4]**

    The line l_2 is the normal to $y = f(x)$ at the point where $x = 1$

    b Find the point of intersection of l_1 and l_2. You must show your working. **[6]**

61 Given that $g(x) = x^3 - 9x^2 + 11x + 21$

    a Solve the equation $g(x) = 0$. Show your working. **[3]**

    b Calculate the area bounded by the curve with equation $y = g(x)$ and the x-axis.
Show your working. **[6]**

62 A closed cylinder is such that its surface area is 50π cm^2

    a Calculate the radius of the cylinder that gives the maximum volume. **[7]**

    b Find the maximum volume and prove it is a maximum. **[4]**

63 A population of an organism grows such that after t hours the number of organisms is N
thousand, where N is given by the equation $N = A - 8e^{-kt}$

    Initially there are 3000 organisms and this number doubles after 5 hours.

    a Find the value of

        i A **ii** k **[5]**

    b Sketch the graph of N against t for $t > 0$ **[2]**

    c How many organisms are there after 3 hours? **[2]**

    d What is the rate of change of N after 3 hours? **[3]**

    e What is the limiting value of N as $t \to \infty$? **[2]**

64 Solve the equation $2e^{2x} - 13e^x + 15 = 0$. You must show your working. **[4]**

65 Solve these equations. Show your working.

    a $\log_4 (x+6) - \dfrac{1}{2} = 2\log_4 x$ **[5]**

    b $\log_3 y + \log_9 3y = 2$ **[4]**

    c $(\ln x)^2 - 3\ln x = 1$ **[4]**

66 Solve these simultaneous equations. Show your working.
$$xy = 24$$
$$\ln y + 2\ln x = \frac{1}{2}\ln\left(\frac{64}{9}\right)$$ **[6]**

67 What vector describes the translation of the curve $y = \sqrt{x}$ onto the curve $y = \sqrt{x+3}$? **[1]**

7 Units and kinematics

Racing cars must be designed to cope with the variable acceleration provided by the engine, and to best utilise the constant downwards acceleration of gravity. The velocity and acceleration of a car, or the time it takes to go a certain distance along the track, can be calculated using a number of different mathematical methods.

The equations for constant acceleration—sometimes called the *suvat* equations—can be used to model or calculate the motion of objects in a variety of situations. The time it takes a lift to go up 40 floors in a skyscraper; the speed at which a pebble thrown into a pond hits the water; the acceleration of a rocket as it takes off from Earth – all of these situations can be modelled with a simple set of equations.

Orientation

What you need to know

KS4
- Change freely between related standard units and compound units.
- Plot and interpret graphs to solve simple kinematic problems.

Ch4.6 Integration
- Area under a curve.
- Velocity-time graphs.

p.116

Ch6 Vectors
- Scalar and vector definitions.

p.154

What you will learn

- To understand SI units and convert between them and other metric units.
- Calculate average speed and average velocity.
- Draw and interpret graphs of displacement and velocity against time.
- Derive and use the formulae for motion.
- Use calculus to solve problems involving variable acceleration.

What this leads to

Ch8 Forces and Newton's laws
Systems of forces. Dynamics and applications of $F = ma$ Motion under gravity.

Careers
Road traffic investigators. Aeronautical engineering.

  **MyMaths** Practise before you start 🔍 1322, 1970, 2056, 2206, 2269

Fluency and skills

All quantities in mechanics are defined in terms of three **fundamental quantities** or **dimensions**: mass, length and time. Quantities, or dimensions, are measured in units.

Some SI (Système International d'Unités) base units you'll have come across before are kilogram / kg (mass), metre / m (length), and second / s (time).

Kinematics is the study of motion. In kinematics, you will meet distance, displacement, speed, velocity and time. These are derived quantities that you can describe in terms of the fundamental quantities (mass, length and time).

	Vector	Scalar	Fundamental quantities	SI Units
See Ch6.1 For a reminder on vectors and scalars.	Displacement	Distance	length	metres (m)
	Velocity	Speed	$\dfrac{\text{length}}{\text{time}}$	metres per second (m s^{-1} or m/s)
	Acceleration		$\left(\dfrac{\text{velocity}}{\text{time}}\right) = \dfrac{\text{length}}{(\text{time})^2}$	metres per second squared (m s^{-2} or m/s²)

Mechanics also involves the derived quantities force and weight.

> Force = mass × acceleration. The SI unit is the newton (N).   **Key point**

Weight is the force of gravity on an object. An object with mass m kg has weight mg N, where g is the acceleration due to gravity. On Earth, this is 9.81 m s^{-1} to 3 sf. If you were on the Moon, your mass would be the same but your weight would be less. In common speech, you might use mass and weight to mean the same thing, but make sure you don't do this in Maths.

Correct formulae are dimensionally consistent. If, for example, $a = b + c$ and a is a velocity, then b and c also have the dimensions of a velocity. You must also use the same units throughout, and so you may need to convert some units before carrying out any calculations for a formula to work.

Example 1

Express a speed of 15 km h^{-1} in m s^{-1}

$15 \text{ km h}^{-1} = 15\,000 \text{ m h}^{-1}$ ⟶ $1 \text{ km} = 1000 \text{ m}$

$15\,000 \text{ m h}^{-1} = \dfrac{15\,000}{3600} \text{ m s}^{-1} = 4\dfrac{1}{6} \text{ m s}^{-1}$ ⟶ $1 \text{ h} = 60 \times 60 = 3600 \text{ s}$

Exercise 7.1A Fluency and skills

1   State the quantity described by these units.

   a   Newtons, N   **b**   Kilograms, kg

   c   Metres per second, m s^{-1}

   d   Metres per second squared, m s^{-2}

2   Convert

   a   8.5 km to m   **b**   2.3 m to mm

   c   482 cm to m   **d**   1650 m to km

   e 72 km h^{-1} to m s^{-1} **f** 14 m s^{-1} to km h^{-1}

   g 25 cm s^{-1} to km h^{-1} **h**   2.4 m² to cm²

   i   1.4 kg to g   **j**   1.6 tonnes to kg

3 A car travels 70 km in 35 minutes. Evaluate its speed in

    a $\mathrm{km\,h^{-1}}$ **b** $\mathrm{m\,s^{-1}}$

4 Work out the distance, in km, travelled in a quarter of an hour by a car that has a constant speed of $20\,\mathrm{m\,s^{-1}}$

5 A particle has an acceleration of $200\,\mathrm{km\,h^{-2}}$. Express this in $\mathrm{m\,s^{-2}}$

6 The force F (in N) on an object is related to its mass m (in kg) and its acceleration a (in $\mathrm{m\,s^{-2}}$) by $F = ma$. Work out, in kg, the mass of an object if a force of 0.25 kN (kilonewtons) accelerates it at $20\,\mathrm{km\,min^{-2}}$

Reasoning and problem-solving

Strategy

When answering a question involving units

  (1) Convert units if they're inconsistent and perform any necessary calculations.

  (2) Check that dimensions have been conserved and that your final answer is in the correct units.

Example 2

u and v are velocities, a is acceleration and s is displacement. Use the formula $v^2 = u^2 + 2as$ to work out s if $u = 24\,\mathrm{km\,h^{-1}}$, $v = 32\,\mathrm{km\,h^{-1}}$ and $a = 0.005\,\mathrm{m\,s^{-2}}$. Give your answer in kilometres.

$$u = \frac{24 \times 1000}{3600} = 6.67\,\mathrm{m\,s^{-1}} \text{ and } v = \frac{32 \times 1000}{3600} = 8.89\,\mathrm{m\,s^{-1}}$$

  (1) Convert velocities to $\mathrm{m\,s^{-1}}$

$$8.89^2 = 6.67^2 + 2 \times 0.005s$$

  (2) Substitute values and solve for s

$$s = \frac{8.89^2 - 6.67^2}{0.01} = 3456.8\,\mathrm{m}$$

$$s = 3.46\,\mathrm{km} \text{ (to 3 sf)}$$

  (2) Convert your answer to km.

Exercise 7.1B Reasoning and problem-solving

1 A runner travels 3900 m at $8\,\mathrm{km\,h^{-1}}$
Find, in minutes, the time she takes.

2 In the formula $s = ut + \frac{1}{2}at^2$, u is velocity, a is acceleration, t is time and s is displacement. Find the value of s if $u = 4\,\mathrm{km\,h^{-1}}$, $a = 0.01\,\mathrm{m\,s^{-2}}$ and $t = 40$ minutes. Give your answer in km.

3 A station platform is 180 m long. A train of length 120 m passes it at $30\,\mathrm{km\,h^{-1}}$. How long will it take for the train to pass completely?

Challenge

4 A liquid of density $1.2\,\mathrm{g\,cm^{-3}}$ is flowing at $3\,\mathrm{km\,h^{-1}}$ through a cylindrical pipe of radius 5 cm. Given that $\text{density} = \dfrac{\text{mass}}{\text{volume}}$, and that for a cylinder with radius r and height h its volume is given by $\pi r^2 h$, calculate the mass, in kg, of the liquid emerging from the pipe in 30 seconds.

Fluency and skills

You use these terms to describe location and movement.

Position is a vector: the distance and direction from the origin O

Displacement is a vector: the change of position.

Distance is a scalar: the magnitude of displacement.

Velocity is a vector: the rate of change of displacement.

Speed is a scalar: the magnitude of velocity.

The diagram shows displacement $\overrightarrow{PQ} = -6$ from position 4 to position -2, then displacement $\overrightarrow{QO} = 2$ from position -2 to position 0 (the origin).

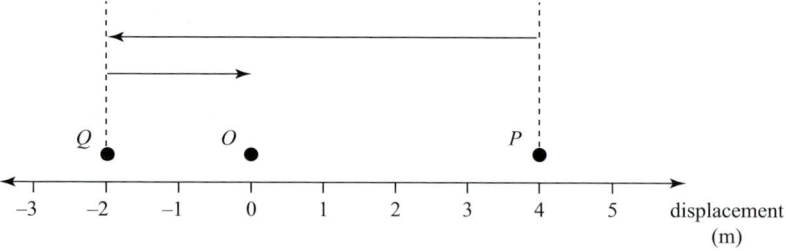

The resultant displacement $\overrightarrow{PO}$ is -4 m but the total distance moved is 8 m.

You can see from this that it's important to distinguish between displacement and distance. Similarly, it's important to distinguish between the average velocity during motion and the average speed during motion. Average speed will not take into account the direction of the motion.

$$\text{Average velocity} = \frac{\text{resultant displacement}}{\text{total time}}$$

$$\text{Average speed} = \frac{\text{total distance}}{\text{total time}}$$

You can illustrate motion with a **displacement–time (or s–t) graph.**

For an s–t graph, $\text{gradient} = \dfrac{\text{change of displacement}}{\text{change of time}}$, which is velocity.

> Displacement is usually represented by the letter s

The **gradient** of a displacement–time graph is the velocity.

For straight-line s–t graphs, you should assume that any changes of gradient are instantaneous. This makes calculations easier, but in reality the velocity would change over a given period of time.

Example 1

The graph shows the motion of a particle along a straight line between 0 and 11 seconds.

a Find the displacement and velocity for the first 6 seconds and for the final 5 seconds.

b Find   **i**   The resultant displacement,

            ii   The average velocity,

           iii   The average speed.

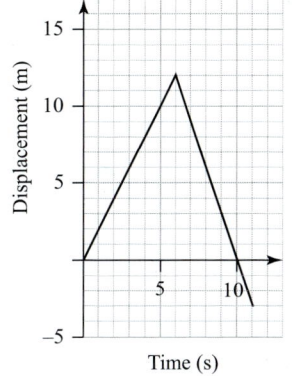

a First 6 seconds:

displacement = 12 m

$\text{Velocity} = \dfrac{12 - 0}{6 - 0} = 2 \text{ m s}^{-1}$ •—————— Velocity = gradient

Final 5 seconds:

displacement = (−3) − 12 = −15 m

$\text{Velocity} = \dfrac{(-3) - 12}{11 - 6} = -3 \text{ m s}^{-1}$

Average velocity = $\dfrac{\text{Resultant displacement}}{\text{Total time}}$

b i   Resultant displacement = 12 + (−15) = −3 m

   ii   Average velocity = $\dfrac{-3}{11} \text{ m s}^{-1}$ •——————

   iii   Average speed = $\dfrac{12 + 15}{11} = 2\dfrac{5}{11} \text{ m s}^{-1}$ •——————

Average speed = $\dfrac{\text{Total distance}}{\text{Total time}}$

Speed is a scalar, so all motion is positive.

You can also draw a **velocity–time graph** (a *v–t* graph).

If velocity changes from u m s^{-1} to v m s^{-1} in t s, as shown, then

$\text{gradient} = \dfrac{v - u}{t}$ m s^{-2} = rate of change of velocity = acceleration.

Key point

Acceleration is the rate of change of velocity.

The gradient of a velocity–time graph is the acceleration.

Key point

$\text{Acceleration} = \dfrac{\text{change in velocity}}{\text{time}} = \dfrac{v - u}{t}$

The shaded area = $\dfrac{1}{2}(u + v)t$ = average velocity × time = displacement

Key point

The area between the *v–t* graph and the *t*-axis is the displacement.

 Q 2183   SEARCH

Example 2

Use this velocity–time graph to answer the questions.

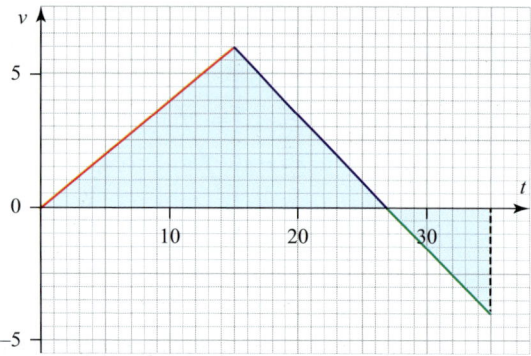

a Calculate the acceleration during

    i The first 15 s   **ii** The remaining 20 s

b Calculate   **i** The resultant displacement,   **ii** The total distance travelled.

c Sketch the corresponding *s*–*t* graph.

a **i** $\text{Acceleration} = \dfrac{6-0}{15-0} = 0.4 \text{ m s}^{-2}$ | Acceleration = gradient

    ii $\text{Acceleration} = \dfrac{(-4)-6}{35-15} = -0.5 \text{ m s}^{-2}$

b **i** Object moves forward for the first 27 s | The velocity is positive.

    $\text{Displacement} = \dfrac{1}{2} \times 27 \times 6 = 81 \text{ m}$ | Displacement = shaded area

    In final 8 s, the object moves backwards:

    $\text{displacement} = \dfrac{1}{2} \times 8 \times (-4) = -16 \text{ m}$

    Resultant displacement $= 81 - 16 = 65 \text{ m}$

    ii Total distance $= 81 + 16 = 97 \text{ m}$

Gradient (velocity) increases (in red) to be the steepest at $t = 15$ s, then decreases (in blue) to zero at $t = 27$ s, when $s = 81$ m

c

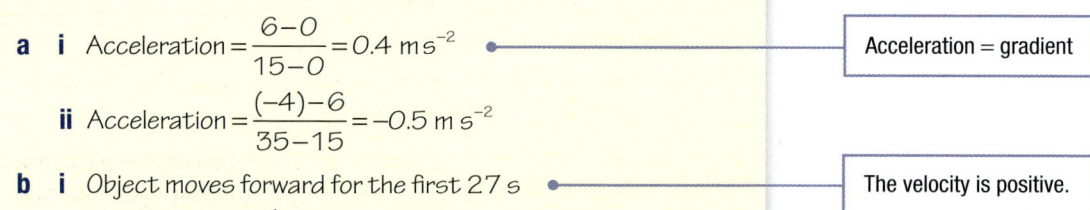

Gradient then becomes negative (in green) and increasingly steep, ending at $s = 65$ m

Negative acceleration is sometimes called **deceleration**. Though it often means slowing down, in the final stage of Example 2 the acceleration is negative but the *speed* is increasing.

1 A ball rolls at a constant speed of $4\,\text{m}\,\text{s}^{-1}$. After 6 m, it hits a wall and rebounds along the same line at a constant speed of $2.5\,\text{m}\,\text{s}^{-1}$. A further 3 s later the ball is stopped.

  a Draw a displacement–time graph.

  b Work out
    i The average speed of the ball,
    ii The average velocity of the ball.

2 The graph shows the motion of a cat along the top of a fence.

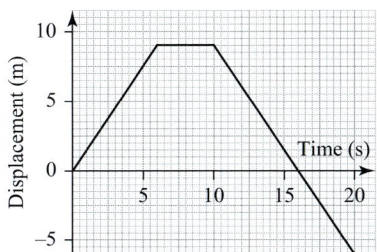

  a Describe the cat's motion during this period of 20 s

  b Calculate the cat's velocity during each phase.

  c Work out
    i The average speed of the cat,
    ii The average velocity of the cat during this period of 20 s

  d Explain why, in reality, the motion of a cat could not be represented by a straight-line graph like the one shown.

3 A car accelerates from rest at point O for 6 s at $2.5\,\text{m}\,\text{s}^{-2}$, then brakes uniformly to rest in 4 s. It immediately reverses, accelerating uniformly and passing point O at a speed of $6\,\text{m}\,\text{s}^{-1}$

  a Sketch a velocity–time graph.

  b Calculate the car's greatest positive displacement from O

  c At what time was the car back at O?

  d Calculate the car's acceleration whilst reversing.

4 A runner and a cyclist start together from rest at the same point. The runner accelerates at $0.6\,\text{m}\,\text{s}^{-2}$ for 10 s, then continues at uniform velocity. The cyclist accelerates uniformly for 8 s then immediately slows uniformly to rest after a further 22 s. At the moment the cyclist stops they have both gone the same distance.

  a Draw a velocity–time graph for the runner and for the cyclist.

  b Work out
    i How far each travels,
    ii The cyclist's greatest speed.

5 Car P is at rest when car Q passes it at a constant speed of $20\,\text{m}\,\text{s}^{-1}$. Immediately, P sets off in pursuit, accelerating at $2\,\text{m}\,\text{s}^{-2}$ until it reaches $30\,\text{m}\,\text{s}^{-1}$ and continues at that speed.

  a Sketch a velocity–time graph for this situation.

  b How far is P behind Q when it reaches full speed?

  c How much longer after this point does it take for P to draw level with Q?

  d In reality, the graph showing this motion would not be made up of perfect straight lines. Suggest why this is.

Challenge

6 A car can accelerate and decelerate at $2.5\,\text{m}\,\text{s}^{-2}$ and has a cruising speed of $90\,\text{km}\,\text{h}^{-1}$. It travels 8 km along a road, but 2 km of road is affected by road works with a speed limit of $36\,\text{km}\,\text{h}^{-1}$. The total journey time will depend on where the road-works occur. By sketching suitable graphs, work out the maximum and minimum journey times. Assume the car starts and finishes at rest.

MECH

Fluency and skills

All motion problems involve some or all of these quantities.

See p.614

For a list of mathematical notation.

> **Key point**
>
> s = displacement u = initial velocity v = final velocity
>
> a = acceleration t = time

For constant acceleration, these quantities satisfy five equations, sometimes called the **suvat equations**. You need to memorise these equations and know how to derive them.

Each one involves four of the five variables, and when solving a problem you should list the variables you know and the variables you want to find. This will let you choose the correct equation to use—find the one with 3 variables that you know and 1 that you don't.

Acceleration = gradient of v-t graph, so $a = \dfrac{v-u}{t}$

> **Key point**
>
> $\Rightarrow v = u + at$

Displacement = area under graph

> **Key point**
>
> $\Rightarrow s = \dfrac{1}{2}(u+v)t$

Substituting for v from ① into ②, $s = \dfrac{1}{2}(u+u+at)t$

> **Key point**
>
> $\Rightarrow s = ut + \dfrac{1}{2}at^2$

Rearranging ① you get $u = v - at$. Substituting this into ②, $s = \dfrac{1}{2}(v-at+v)t$

> **Key point**
>
> $\Rightarrow s = vt - \dfrac{1}{2}at^2$

Rearranging ①, you get $t = \dfrac{v-u}{a}$. Substituting this into ②, $s = \dfrac{(u+v)(v-u)}{2a}$

> **Key point**
>
> $\Rightarrow v^2 = u^2 + 2as$

In problems involving these equations, the moving objects are usually treated as 'particles', with a small or irrelevant size.

In most cases this won't affect the results. However, when you're dealing with a large object moving across a relatively short distance, such as a bus moving between street junctions, the true distance may be shorter than that used in calculations—you might need to consider this when discussing the limitations of mathematical models.

> **ICT Resource online**
>
> To practise choosing which equation of motion to use, click this link in the digital book.

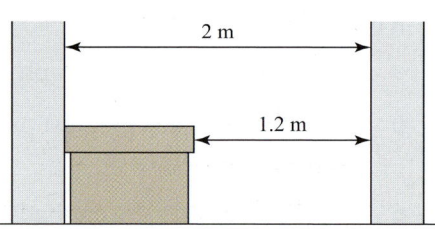

▲ If this box were treated as a particle with irrelevant size, it would move 2 m to travel between the walls of the corridor. In reality, it will only move 1.2 m

Example 1

An object travelling at $15\,\text{m}\,\text{s}^{-1}$ accelerates at $3\,\text{m}\,\text{s}^{-2}$ for 5 s. Work out

a Its final velocity, **b** Its displacement during this period.

a $u = 15, a = 3, t = 5$, find v ●――――――――――――――● List what you know and your 'target' variable.

$v = 15 + 3 \times 5 = 30$ ●

∴ final velocity is $30\,\text{m}\,\text{s}^{-1}$ Use $v = u + at$

b $u = 15, a = 3, t = 5$, find s

$s = 15 \times 5 + \dfrac{1}{2} \times 3 \times 5^2 = 112.5$ ●   Use $s = ut + \dfrac{1}{2}at^2$

∴ displacement $= 112.5\,\text{m}$

In part **b** you could use $s = \dfrac{1}{2}(u+v)t$ with the value of v found in part **a**.

Exercise 7.3A Fluency and skills

1 You are given the initial velocity v and the uniform acceleration a of a particle during time t. Write down the equation you could use to find its displacement during this time.

2 **a** Evaluate v if $u = 4\,\text{m}\,\text{s}^{-1}, a = 2\,\text{m}\,\text{s}^{-2}, t = 7\,\text{s}$

  b Evaluate u if $v = 12\,\text{m}\,\text{s}^{-1}, a = -3\,\text{m}\,\text{s}^{-2}, t = 4\,\text{s}$

  c Evaluate t if $u = 8\,\text{m}\,\text{s}^{-1}, v = 35\,\text{m}\,\text{s}^{-1}, a = 3\,\text{m}\,\text{s}^{-2}$

3 **a** Evaluate s if $u = 5\,\text{m}\,\text{s}^{-1}, a = 2\,\text{m}\,\text{s}^{-2}, t = 4\,\text{s}$

  b Evaluate a if $s = 30\,\text{m}, u = -4\,\text{m}\,\text{s}^{-1}, t = 6\,\text{s}$

  c Evaluate t if $u = 12\,\text{m}\,\text{s}^{-1}, a = -6\,\text{m}\,\text{s}^{-2}, s = 9\,\text{m}$

4 **a** Evaluate s if $u = 4\,\text{m}\,\text{s}^{-1}, v = 6\,\text{m}\,\text{s}^{-1}, a = 5\,\text{m}\,\text{s}^{-2}$

  b Evaluate t if $u = 3\,\text{m}\,\text{s}^{-1}, v = 5\,\text{m}\,\text{s}^{-1}, s = 20\,\text{m}$

  c Evaluate s if $v = 8\,\text{m}\,\text{s}^{-1}, a = 2\,\text{m}\,\text{s}^{-2}, t = 3\,\text{s}$

5 $v = 12.1\,\text{m}\,\text{s}^{-1}, a = -1.20\,\text{m}\,\text{s}^{-1}, s = 73.6\,\text{m}$

  a Calculate u

  b Calculate t

6 A particle starts from rest and accelerates uniformly for 10 s. It moves 150 m in that time. Work out

  a Its acceleration,

  b Its speed at the end of the period.

7 A particle accelerates uniformly for 8.1 s. After this time, it has a displacement of 80 cm and its speed is $14\,\text{cm}\,\text{s}^{-1}$

  a Calculate its acceleration during this time in $\text{m}\,\text{s}^{-2}$

  b Calculate its speed before it underwent the acceleration in $\text{m}\,\text{s}^{-1}$

8 A particle travelling at $144\,\text{km}\,\text{h}^{-1}$ is brought uniformly to rest in a distance of 200 m. Work out

  a Its acceleration,

  b The time it takes to stop.

9 A particle, accelerating at $4\,\text{m}\,\text{s}^{-2}$, takes 5 s to travel from A to B. Its speed at B is $24\,\text{m}\,\text{s}^{-1}$ Work out

  a The distance AB **b** Its speed at A

10 A driver notices the speed limit change and applies the brakes suddenly, causing a uniform deceleration. The car's speed after braking is $60\,\text{km}\,\text{h}^{-1}$, the car's deceleration while braking is $(-)10\,\text{m}\,\text{s}^{-2}$ and the distance it covers while braking is 26 m.

  a How fast, in $\text{km}\,\text{h}^{-1}$, was the car moving before applying the brakes?

  b For how long did the driver apply the brakes?

Reasoning and problem-solving

Strategy

When solving a problem with the equations of motion for constant acceleration

1. Use the information in the question to list the known values and the variable you need to find. Be careful to distinguish between displacement and distance, and between velocity and speed.

2. Choose the correct equation to use.

3. Apply the equation to find the numerical value and use it to answer the original question.

Example 2

A hot-air balloon is drifting in a straight line at a constant velocity of $3\,\text{m s}^{-1}$. A sudden head-wind gives it an acceleration of $-0.5\,\text{m s}^{-2}$ for a period of $16\,\text{s}$. For this period, calculate

a The resultant displacement of the balloon,   **b** The distance travelled.

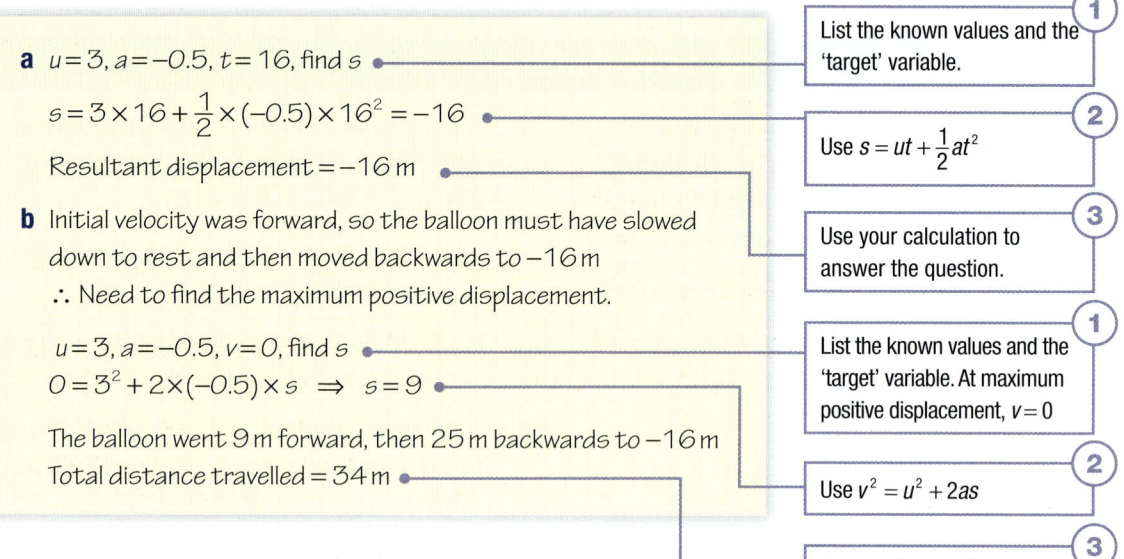

a $u = 3$, $a = -0.5$, $t = 16$, find s

    ① List the known values and the 'target' variable.

$s = 3 \times 16 + \frac{1}{2} \times (-0.5) \times 16^2 = -16$

    ② Use $s = ut + \frac{1}{2}at^2$

Resultant displacement $= -16\,\text{m}$

b Initial velocity was forward, so the balloon must have slowed down to rest and then moved backwards to $-16\,\text{m}$

    ③ Use your calculation to answer the question.

$\therefore$ Need to find the maximum positive displacement.

$u = 3$, $a = -0.5$, $v = 0$, find s

    ① List the known values and the 'target' variable. At maximum positive displacement, $v = 0$

$0 = 3^2 + 2 \times (-0.5) \times s \Rightarrow s = 9$

The balloon went $9\,\text{m}$ forward, then $25\,\text{m}$ backwards to $-16\,\text{m}$

    ② Use $v^2 = u^2 + 2as$

Total distance travelled $= 34\,\text{m}$

    ③ Answer the question.

Example 3

Car P is accelerating at $2\,\text{m s}^{-2}$. When its velocity is $10\,\text{m s}^{-1}$, it is overtaken by car Q, which is travelling at $16\,\text{m s}^{-1}$ and accelerating at $1\,\text{m s}^{-2}$. How long will P take to catch up with Q?

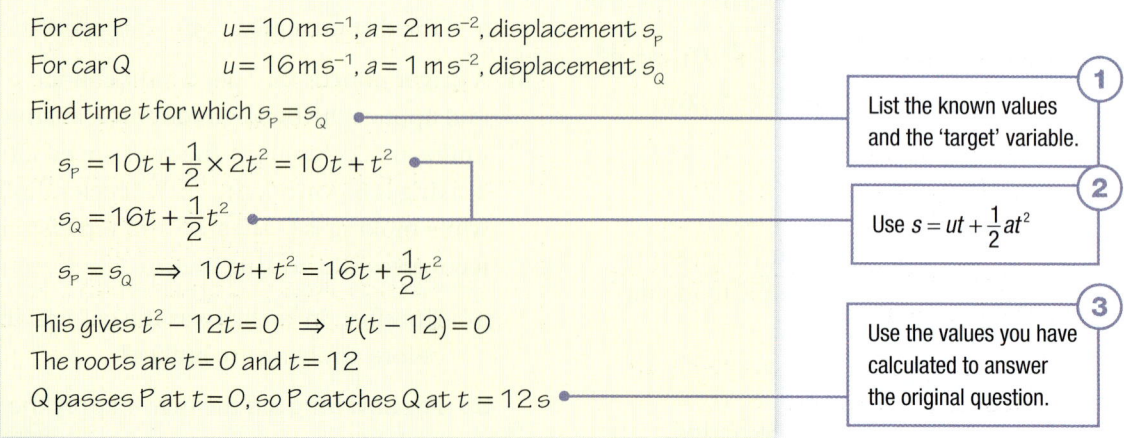

For car P $u = 10\,\text{m s}^{-1}$, $a = 2\,\text{m s}^{-2}$, displacement s_P
For car Q $u = 16\,\text{m s}^{-1}$, $a = 1\,\text{m s}^{-2}$, displacement s_Q

Find time t for which $s_P = s_Q$

    ① List the known values and the 'target' variable.

$s_P = 10t + \frac{1}{2} \times 2t^2 = 10t + t^2$

$s_Q = 16t + \frac{1}{2}t^2$

    ② Use $s = ut + \frac{1}{2}at^2$

$s_P = s_Q \Rightarrow 10t + t^2 = 16t + \frac{1}{2}t^2$

This gives $t^2 - 12t = 0 \Rightarrow t(t - 12) = 0$
The roots are $t = 0$ and $t = 12$
Q passes P at $t = 0$, so P catches Q at $t = 12\,\text{s}$

    ③ Use the values you have calculated to answer the original question.

1 A lorry starts from rest and accelerates uniformly at $3\,\text{m s}^{-2}$. It passes point *A* after 20 s and point *B* after a further 10 s. Calculate the distance *AB*

2 A train leaves station *A* from rest with constant acceleration $0.2\,\text{m s}^{-2}$. It reaches maximum speed after 2 minutes, maintains this speed for 4 minutes, then slows down to stop at station *B* with acceleration $-1.5\,\text{m s}^{-2}$. Calculate the distance *AB*

3 A ferry carries passengers between banks of a river, which are 20 m apart. After setting off, the ferry accelerates at $0.2\,\text{m s}^{-2}$ for 12 seconds before turning off the engine and decelerating at a constant rate and coming to a stop at the opposite bank.

   a Calculate the speed of the ferry after the first 12 seconds.

   b Calculate the distance the ferry travels during these 12 seconds.

   c Calculate the value of the ferry's deceleration after the engines are turned off.

   d State any assumptions you made in part **c** and explain why, in reality, this value would be higher.

4 A train accelerates uniformly from rest for 1 minute, at which time its velocity is $30\,\text{km h}^{-1}$. It maintains this speed until it is 500 m from the next station. It then decelerates uniformly and stops at the station. Calculate the train's acceleration during the first and last phases of this journey.

5 A boat is travelling at $4\,\text{m s}^{-1}$. Its propeller is then put into reverse, giving it an acceleration of $-0.4\,\text{m s}^{-2}$ for 25 seconds.

   a Work out the displacement of the boat during this period.

   b Work out the distance travelled by the boat during this period.

6 An object, starting from rest, travels 10 m during one second and 15 m during the next second. Work out the acceleration of the object, assuming it to be constant.

7 Lorry A is travelling along a straight road, with lorry B following 40 m behind. They are both travelling at a constant $25\,\text{m s}^{-1}$. Lorry A then brakes to a halt, with acceleration $-5\,\text{m s}^{-2}$. The driver of lorry B takes 0.2 seconds to react, then brakes with acceleration $-4\,\text{m s}^{-2}$

   a Do the lorries collide?

   b State any assumptions you made, and explain how this may affect the answer.

MECH

Fluency and skills

You know that velocity is equal to the gradient of a displacement–time (s–t) graph. If the graph is not linear, the gradient changes, but at a particular point on the graph you can say

> **Key point**
> gradient of s–t graph = velocity *at that instant*

Tangent

Displacement

Gradient of tangent here gives velocity at that instant

Time

Similarly, for a non-linear velocity–time (v–t) graph

> **Key point**
> gradient of v–t graph = acceleration *at that instant*

See Ch 4.3 For a reminder of finding gradients with differentiation.

If you know s as a function of t, you can work out the gradient (velocity) by differentiation. The velocity is the rate of change of displacement.

> **Key point**
> $v = \dfrac{ds}{dt}$ or $v = \dot{s}$

Similarly, acceleration is the rate of change of velocity. You can obtain acceleration by differentiating velocity with respect to time, which is the same as differentiating displacement twice.

> **Key point**
> $a = \dfrac{dv}{dt} = \dfrac{d^2 s}{dt^2}$ or $a = \dot{v} = \ddot{s}$

> A dot above a variable denotes the derivative with respect to time. Two dots show the second derivative.

Example 1

A particle moves in a straight line so that at time t seconds its displacement s, in metres, from an origin O is given by $s = t^4 - 32t$. Evaluate its velocity and acceleration when $t = 2$. Show your working.

$$v = \frac{ds}{dt} = 4t^3 - 32$$

When $t = 2$, $v = 4 \times 2^3 - 32 = 0$ so the particle is at rest when $t = 2$

$$a = \frac{dv}{dt} = 12t^2$$

When $t = 2$, $a = 12 \times 2^2 = 48$ so $a = 48\,\text{m s}^{-2}$ when $t = 2$

> Differentiate displacement to find velocity.

> Differentiate velocity to find acceleration.

> You can check your answer by finding $\dfrac{d^2 s}{dt^2}$ at the point where $t = 2$ on a calculator.

See Ch 4.7 For a reminder of integration and areas under graphs.

You can also write the relationship between s, v and a using integrals.

> **Key point**
> $v = \dfrac{ds}{dt} \Rightarrow s = \int v\,dt$ and $a = \dfrac{dv}{dt} \Rightarrow v = \int a\,dt$

> Remember to include the constant of integration.

You know that integration corresponds to area under a graph, and so the first of these relationships confirms that displacement is the area under the v–t graph.

Example 2

A particle starts with velocity $2\,\text{m s}^{-1}$ and moves with acceleration $a = (6t + 4)\,\text{m s}^{-2}$

a Calculate its velocity after $5\,\text{s}$ **b** Calculate the distance it travels in that time.

a $v = \int 6t + 4\ dt = 3t^2 + 4t + c$ ●————— Integrate acceleration to find velocity. Remember to add c

$v = 2$ when $t = 0$, so $c = 2$ ●

$v = 3t^2 + 4t + 2$

When $t = 5$, $v = 3 \times 5^2 + 4 \times 5 + 2 = 97\,\text{m s}^{-1}$ ————— Use initial velocity $= 2$ to find c

b $s = \int 3t^2 + 4t + 2\ dt = t^3 + 2t^2 + 2t + d$ ●————— Integrate velocity to find displacement.

$s = 0$ when $t = 0$, so $d = 0$

$s = t^3 + 2t^2 + 2t$

When $t = 5$, $s = 5^3 + 2 \times 5^2 + 2 \times 5 = 185\,\text{m}$

$\therefore$ The particle travels $185\,\text{m}$ ●————— In this case, distance = displacement

Exercise 7.4A Fluency and skills

Show your working for these questions.

1 Given $s = 4t^2 - t^3$

  a Write an expression for v

  b Evaluate v when $t = 2\,\text{s}$

2 Given $s = (t + 2)(t - 6)$

  a Write an expression for v

  b Evaluate v when $t = 10\,\text{s}$

  c Evaluate v when $t = 100\,\text{s}$

3 Given $v = 6t^2 - 8t$

  a Write an expression for a

  b Evaluate a when $t = 3\,\text{s}$

4 Given $v = t(t^2 + 7)$

  a Write an expression for a

  b Evaluate a when $t = 0\,\text{s}$

5 Given $s = t^4 + 5t^2$, evaluate

  a v when $t = 1\,\text{s}$

  b a when $t = 0.5\,\text{s}$

6 Given $s = 3t^2 - 12\,t + 5$

  a Write an expression for v,

  b Evaluate t when $v = 0\,\text{m s}^{-1}$

7 Given $v = 14t - t^2$

  a Write an expression for a,

  b Evaluate t when $a = 6\,\text{m s}^{-2}$

8 If $v = 24 - 6t$ and $s = 0\,\text{m}$ when $t = 0\,\text{s}$

  a Write an expression for s,

  b Evaluate s when $t = 2\,\text{s}$

9 If $a = 6t^2 + 3$ and $v = 4\,\text{m s}^{-1}$ when $t = 0\,\text{s}$

  a Write an expression for v

  b Evaluate v when $t = 2\,\text{s}$

10 If $v = 9 - t^2$ and $s = 4\,\text{m}$ when $t = 0\,\text{s}$, evaluate

  a s when $t = 1\,\text{s}$

  b s when $v = 0\,\text{m s}^{-1}$

11 If $a = 6t - 12$ and $v = 2\,\text{m s}^{-1}$ when $t = 0\,\text{s}$, evaluate

  a v when $t = 3\,\text{m}$,

  b v when $a = 0\,\text{m s}^{-2}$

12 $a = 24t^2 + 6t - 10$

  a Given that $v = 2\,\text{m s}^{-1}$ when $t = 0\,\text{s}$, work out v when $t = 2\,\text{s}$

  b Given that $s = 5\,\text{m}$ when $t = 0\,\text{s}$, work out s when $t = 1\,\text{s}$

 SEARCH

When solving a motion problem with differentiation or integration

1. Identify what dimension(s) you're dealing with (speed, velocity, distance, displacement) and differentiate or integrate as appropriate.

2. (When integrating) include the constant of integration and calculate its value.

3. Use the result of your differentiation or integration to answer the original question.

Example 3

A particle moves with acceleration $(2t - 3)\,\text{m s}^{-2}$. It is initially at a point O and is travelling with velocity $2\,\text{m s}^{-1}$. Show that its direction of travel changes twice and find the distance between the points where this occurs.

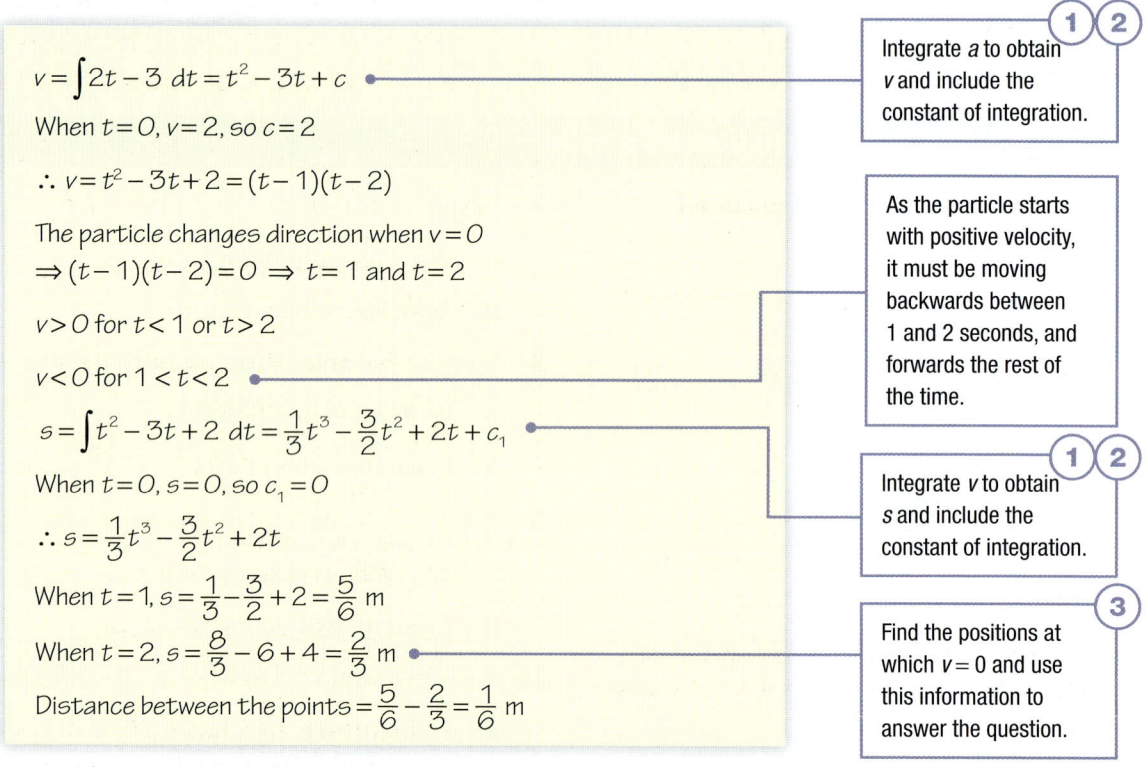

$v = \int 2t - 3\ dt = t^2 - 3t + c$

When $t = 0$, $v = 2$, so $c = 2$

$\therefore v = t^2 - 3t + 2 = (t-1)(t-2)$

The particle changes direction when $v = 0$

$\Rightarrow (t-1)(t-2) = 0 \Rightarrow t = 1$ and $t = 2$

$v > 0$ for $t < 1$ or $t > 2$

$v < 0$ for $1 < t < 2$

$s = \int t^2 - 3t + 2\ dt = \frac{1}{3}t^3 - \frac{3}{2}t^2 + 2t + c_1$

When $t = 0$, $s = 0$, so $c_1 = 0$

$\therefore s = \frac{1}{3}t^3 - \frac{3}{2}t^2 + 2t$

When $t = 1$, $s = \frac{1}{3} - \frac{3}{2} + 2 = \frac{5}{6}$ m

When $t = 2$, $s = \frac{8}{3} - 6 + 4 = \frac{2}{3}$ m

Distance between the points $= \frac{5}{6} - \frac{2}{3} = \frac{1}{6}$ m

(1)(2) Integrate a to obtain v and include the constant of integration.

As the particle starts with positive velocity, it must be moving backwards between 1 and 2 seconds, and forwards the rest of the time.

(1)(2) Integrate v to obtain s and include the constant of integration.

(3) Find the positions at which $v = 0$ and use this information to answer the question.

Exercise 7.4B Reasoning and problem-solving

1. A particle, moving in a straight line, starts from rest at O. Its acceleration (in m s^{-2}) at time t is given by $a = 30 - 6t$

   a Calculate its velocity and position at time t

   b How long does the particle take to return to O?

   c What is its greatest positive displacement from O?

2. A particle is initially traveling at $18\,\text{m s}^{-1}$ It is acted upon by a braking force that brings it to rest. At time t s during braking, its acceleration has magnitude $t\,\text{m s}^{-2}$ Calculate

   a the length of time it takes to stop,

   b the distance it travels before coming to rest.

3 A particle, initially at rest, is acted on for $3\,\text{s}$ by a variable force that gives it acceleration of $(8t+2)\,\text{m}\,\text{s}^{-2}$. Work out the distance it travels while the force is acting.

4 A ball is thrown straight up from an open window. Its height, at time $t\,\text{s}$, is $h\,\text{m}$ above the ground, where $h = 4 + 8t - 5t^2$

   a How high is the window above the ground?

   b For how long is the ball in the air?

   c At what speed does it hit the ground?

   d Evaluate the maximum height it reaches above the ground.

5 A particle starts from point O and moves so that its displacement s, in metres, at time t seconds is given by $s = t^4 - 32t$. Work out

   a   **i** Its position at the moment it is instantaneously at rest,

      ii Its acceleration at the moment it is instantaneously at rest,

   b Its speed at the moment that it returns to O,

   c The distance it travels in the first $3\,\text{s}$

6 A particle travels in a straight line through a point O with constant acceleration. Its velocity is given by $v = 2t - 3\,\text{m}\,\text{s}^{-1}$

   If its displacement from O is $-4\,\text{m}$ at $t = 0\,\text{s}$, work out how long before the particle passes point O

7 A particle moving along a straight line through a point O has acceleration given by $a = (2t - 5)\,\text{m}\,\text{s}^{-2}$. When $t = 4\,\text{s}$, the particle has velocity $2\,\text{m}\,\text{s}^{-1}$ and displacement from O of $+8\,\text{m}$. Work out the two positions where the particle is at rest.

8 A particle is moving in a straight line. Its velocity at time $t\,\text{s}$ is given by $v = 6t - 3t^2\,\text{m}\,\text{s}^{-1}$

   a Evaluate its velocity at times $t = 1$ and $t = 3$

   b What distance does it travel from time $t = 1$ to time $t = 3$?

Challenge

9 A car travels from rest at a set of traffic lights until it stops at the next set of lights. The car's displacement $x\,\text{m}$ from its starting position at time $t\,\text{s}$ is given by $x = \dfrac{1}{13500}t^3(t^2 - 75t + 1500)$

   a Write an expression for the car's velocity.

   b Work out the time it takes for the car to travel between the two sets of lights.

   c What is the distance between the two sets of lights?

   d Work out the car's maximum speed.

10 The *suvat* equations can be derived using differentiation and integration.

   Acceleration is the derivative of velocity, so $a = \dfrac{\mathrm{d}v}{\mathrm{d}t}$

   For a constant acceleration between $t = 0$ and $t = t$, this gives $\displaystyle\int_0^t a\,\mathrm{d}t = \int_u^v \mathrm{d}v$

   a Use this information to derive the *suvat* equation that uses a, t, u and v

   Similarly, $v = \dfrac{\mathrm{d}s}{\mathrm{d}t}$ so, between displacement $s = 0$ and $s = s$, $\displaystyle\int_0^s a\,\mathrm{d}s = \int_0^t \mathrm{d}t$

   b Use this information to derive the suvat equation that uses s, u, a and t

 MyMaths   Q   2289    SEARCH

Chapter summary

- The SI base units used in this chapter are kilograms / kg (mass), metres / m (length) and seconds / s (time).
- In kinematics, the quantities in the table are used.

Vector Quantity	Scalar Quantity	SI Units
Displacement	Distance	m
Velocity	Speed	$m\,s^{-1}$ or m/s
Acceleration	—	$m\,s^{-2}$ or m/s²

- Formulae should be dimensionally consistent and you must use the same units throughout, converting if necessary.
- Position is a vector – the distance and direction from the origin.
  Displacement is a vector – the change of position.
  Distance is a scalar – the magnitude of displacement.
  Velocity is a vector – the rate of change of displacement.
  Speed is a scalar – the magnitude of velocity.
  Acceleration is a vector – the rate of change of velocity.
- Average velocity $= \dfrac{\text{resultant displacement}}{\text{total time}}$ and average speed $= \dfrac{\text{total distance}}{\text{total time}}$
- The gradient of a displacement–time (s–t) graph is the velocity.
- The gradient of a velocity–time (v–t) graph is the acceleration.
- The area between a v–t graph and the t-axis is the displacement.
- In straight-line graphs, changes in motion are assumed to be instantaneous. In reality, this is usually not possible.
- $s =$ displacement, $u =$ initial velocity, $v =$ final velocity, $a =$ acceleration, $t =$ time
  For constant acceleration: $v = u + at$; $s = ut + \dfrac{1}{2}at^2$; $v^2 = u^2 + 2as$; $s = \dfrac{1}{2}(u+v)t$; $s = vt - \dfrac{1}{2}at^2$
- The equations of motion for constant acceleration assume objects to be particles with tiny or irrelevant size. In reality, the size of an object may affect calculations.
- For variable acceleration, use calculus: $v = \dfrac{\mathrm{d}s}{\mathrm{d}t}$ and $a = \dfrac{\mathrm{d}v}{\mathrm{d}t} = \dfrac{\mathrm{d}^2 s}{\mathrm{d}t^2}$; $s = \int v\ \mathrm{d}t$ and $v = \int a\ \mathrm{d}t$

Check and review

You should now be able to…	Try Questions
✔ Use standard SI units and convert between them and other metric units.	1
✔ Calculate average speed and average velocity.	2, 4
✔ Draw and interpret graphs of displacement and velocity against time.	3, 4
✔ Derive and use the formulae for motion in a straight line with constant acceleration.	5
✔ Use calculus to solve problems involving variable acceleration.	6, 7

1 Using the formula $s = vt - \frac{1}{2}at^2$, calculate s if $v = 5\,\text{km}\,\text{h}^{-1}$, $a = 0.002\,\text{m}\,\text{s}^{-2}$ and $t = 10$ minutes. Give your answer in kilometres to 3 sf.

2 An object travels for 10 s at $8\,\text{m}\,\text{s}^{-1}$ and then for a further distance of 120 m at $18\,\text{m}\,\text{s}^{-1}$. Work out its average speed for the whole journey.

3 A log flume climbs a slope for 20 s at a constant $1\,\text{m}\,\text{s}^{-1}$. It then pauses for 5 s before moving rapidly downhill at $10\,\text{m}\,\text{s}^{-1}$ for 5 s

   a Sketch a distance–time graph for this motion.

   b Explain why it is unrealistic that a log flume would be able to move with the motion described by your graph.

4 The velocity–time graph shows the motion of a particle for a period of 10 s

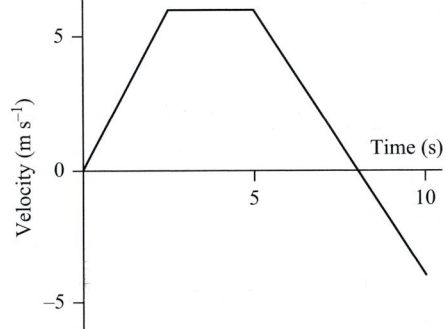

a Work out the acceleration of the object for each of the three stages of the motion.

b Calculate i The resultant displacement,
            ii The distance travelled.

c Calculate the average velocity of the particle.

5 A particle moves with constant acceleration along a straight line. It passes the origin, O, at $2\,\text{m}\,\text{s}^{-1}$ and travels 15 m in the next 5 s

   a Calculate its acceleration.

   b Work out its position when its velocity is $8\,\text{m}\,\text{s}^{-1}$

6 A particle, P, is moving along a straight line. At time t seconds, its displacement, s metres from the origin, O, is given by $s = t(t^2 - 16)$

   a Write down an expression for the velocity of P at time t

   b Work out the acceleration of P when $t = 5$

7 A particle moves along a straight line with acceleration $a = (4t - 3)\,\text{m}\,\text{s}^{-2}$ at time t s. Initially it has a velocity of $5\,\text{m}\,\text{s}^{-1}$

   a Write down an expression for its velocity at time t

   b If the particle started at the origin, O, work out its displacement from O after 4 s

What next?

History

If you have ever struggled with mechanics, you may take some comfort from the fact that the laws and equations that are used today took many centuries to establish.

Galileo (1564 - 1642) has often been credited with establishing the principles governing uniformly accelerated motion. Much of the work, however, had already been done almost three centuries earlier, by a group known as The Oxford Calculators. **The Oxford Calculators** were a group of 14th century thinkers associated with Oxford University's Merton College.

The group distinguished between **kinematics** and **dynamics** and they were the first to formulate the **mean speed theorem**.

This theorem states that an object which uniformly accelerates from rest will cover the same distance as an object travelling at a constant velocity if this velocity is half the final speed of the accelerated object.

Investigation

Explain how this diagram illustrates the mean speed theorem.

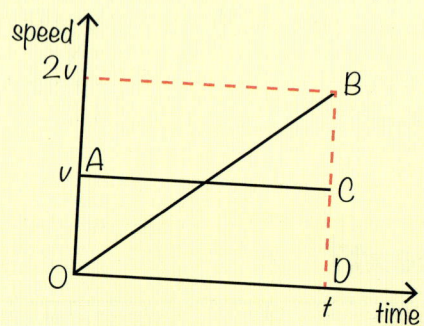

Applications

The theory of kinematics continues to be developed all the time. **Ferdinand Freudenstein** was a 20th century American physicist and engineer, regarded by many as the father of modern kinematics.

The **Freudenstein equation**, which he presented in his 1954 doctoral thesis, can be applied to a number of machines used in daily life such as the braking system of a vehicle. Designers of modern mechanical systems test them using apps that build on Freudenstein's ideas.

The derivation of Freudenstein's equation relies on the theory of vectors and trigonometry that you have already studied.

1   A train leaves a station, P, and accelerates from rest with a constant acceleration of $0.4\,\text{m s}^{-1}$ until it reaches a speed of $24\,\text{m s}^{-1}$. It maintains this speed for 6 minutes. It then decelerates uniformly with a deceleration of $0.2\,\text{m s}^{-1}$, until it comes to rest at station Q.

   a   Draw a velocity–time graph of the journey from P to Q, labelling all the relevant times.   **[5 marks]**

   b   Calculate the distance PQ.   **[2]**

2   At time $t = 0\,\text{s}$, a body passes the origin with a velocity of $60\,\text{m s}^{-1}$, and decelerates uniformly at a rate of $4\,\text{m s}^{-2}$.

   a   Determine the time at which the body is at rest.   **[2]**

   b   Determine the times at which the body is 400 metres from the origin.   **[4]**

3   Points A, B and C lie on a straight line. The distances AB and BC are 80 m and 96 m respectively. A particle moves in a straight line from A to C with an acceleration of $4\,\text{m s}^{-2}$. It takes 5 seconds to travel from A to B. Work out

   a   The speed of the particle at A,   **[3]**

   b   The time taken for the particle to travel from B to C **[6]**

4   The displacement, s metres, of a particle, at a time t seconds, is given by the formula $s = t^3 - 9t^2 + 24t$

   a   Write an expression for the velocity of the particle.   **[2]**

   b   Calculate the times at which the particle is at rest.   **[3]**

   c   Work out the distance travelled by the particle between $t = 0\,\text{s}$ and $t = 5\,\text{s}$ **[5]**

5   A speeding van passes a police car. The van is travelling at $27\,\text{m s}^{-1}$, and the police car is travelling at $15\,\text{m s}^{-1}$. From the instant when the van is level with the police car, the police car accelerates uniformly at $3\,\text{m s}^{-2}$ in order to catch the van. Work out

   a   The time taken until the police car is level with the van,   **[5]**

   b   The speed of the police car at this time.   **[2]**

6   A man on a bicycle accelerates uniformly from rest to a velocity of $10\,\text{m s}^{-1}$ in 5 seconds. He maintains this speed for 20 seconds, and then decelerates uniformly to rest. His journey takes a total of T seconds.

   a   Draw a velocity–time graph of his journey.   **[3]**

   Given that he cycles a total of 265 metres, calculate

   b   The value of T,   **[3]**

   c   The acceleration for the final stage of his journey.   **[2]**

7   A car moving along a straight road with constant acceleration passes points A and B with velocities $10\,\text{m s}^{-1}$ and $40\,\text{m s}^{-1}$ respectively. Work out the velocity of the car at the instant when it passes M, the midpoint of AB **[6]**

8   A bus travels on a straight road with a constant acceleration of $0.8\,\text{m s}^{-2}$. A and B are two points on the road, a distance of 390 metres apart. The bus increases its velocity by $12\,\text{m s}^{-1}$ in travelling from A to B

    a   What is the speed of the bus at A?   **[5]**

    b   Work out the time taken for the bus to travel from A to B **[2]**

9

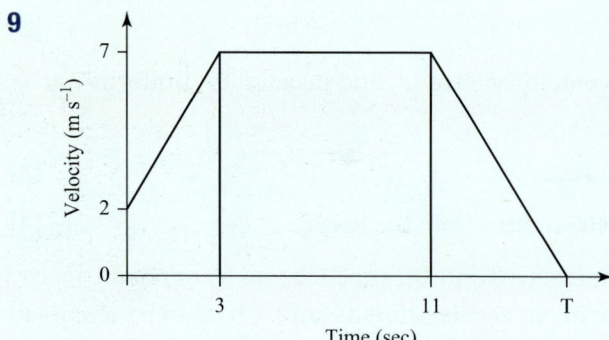

The diagram shows the velocity–time graph of the motion of a runner over a time period of T seconds. During that time, the runner travels a distance of 76 metres.

    a   Write down the initial speed of the runner.   **[1]**

    b   Work out the value of her initial acceleration.   **[2]**

    c   Describe her motion from $t=3$ to $t=11$ **[1]**

    d   Calculate the value of T **[6]**

10 The acceleration, $a\,\text{m s}^{-2}$, of a particle moving in a straight line is given by the formula $a = 2t - 6$

At time $t=0$, the particle is moving through the origin with a velocity of $10\,\text{m s}^{-1}$

    a   Write an expression for the velocity of the particle at time t **[4]**

    b   At which times is the particle moving with a velocity of $2\,\text{m s}^{-1}$?   **[4]**

    c   Write an expression for the displacement of the particle at time t **[4]**

    d   Work out the displacement of the particle when $t=6$ **[2]**

11 A jogger is running along a straight road with a velocity of $4\,\text{m s}^{-1}$ when she passes her friend who is stationary with a bicycle. Three seconds after the jogger is level with her friend, her friend sets off in pursuit. Her friend accelerates from rest with a constant acceleration of $2\,\text{m s}^{-2}$. When the cyclist has been riding for T seconds, the cyclist and her friend are level.

    a   Draw a velocity–time graph for $t=0$ to $t=T+3$ **[3]**

    b   Write down an equation for T **[4]**

    c   Solve this equation to find the value of T **[3]**

    d   How fast is the cyclist travelling when they draw level?   **[2]**

8 Forces and Newton's laws

The abseiler in this picture is in equilibrium. The rope is taut but the abseiler is not moving, which means that all forces acting on the person are balanced. The weight of the abseiler balances with the tension in the rope. The force applied by the abseiler against the rock face balances with the normal reaction from the rock face. This simple example illustrates that systems of forces surround us every day.

Drawing diagrams and labelling forces is a useful way to visualise these systems. It is important to remember that any object that is not accelerating is in a state of equilibrium. This means that the forces can be balanced even if the object is moving. There are many different types of forces, and an understanding of how they work is vital for modelling.

Orientation

What you need to know	What you will learn	What this leads to
Ch6 Vectors (p.153) • Scalar and vector definitions. • Components of a vector.	• To resolve forces in two perpendicular directions. • To calculate the magnitude and direction of a resultant force. • To resolve for particles with constant acceleration, including those which are connected by string over pulleys and 'connected' particles. • Understand mass and weight.	**Ch18 Motion in two dimensions** SUVAT equations in 2D. The significance of g
Ch7 Units and kinematics (p.177) • Formulae for motion. • Acceleration under gravity.		**Ch19 Forces 2** Modelling friction. Inclined planes. Moments.

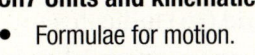

 MyMaths — Practise before you start — 🔍 2206, 2183, 2184

199

Fluency and skills

When you hit a tennis ball, you are applying a **force** to it. When you pick up a book, you are applying a force to it.

See Ch6

For a reminder on vectors.

To describe a force you give both its **magnitude** and the **direction** in which it acts, so force is a vector quantity. The magnitude of a force is measured in newtons (N). The direction can be given as an angle or bearing.

All objects accelerate downwards towards the ground due to the gravitational pull of the Earth, also known as an object's **weight**.

Objects stop accelerating downwards when a contact force pushes upwards to counteract their weight. The contact force is called the **normal reaction**. It is always perpendicular (or normal) to the surface and if the object is at rest then the forces must balance each other.

> Examples of forces include:
>
> Kicking a ball with a force of magnitude 200 N in the easterly direction.
>
> A force $\mathbf{F} = (3\mathbf{i} + 4\mathbf{j})\,\text{N}$ acting on a particle.

Example 1

A box, with weight W N, is at rest on a horizontal table. Explain why it does not move even though its weight is pulling it directly down.

The weight of the box is balanced by the reaction from the table, which pushes directly upwards with the same strength as the weight.

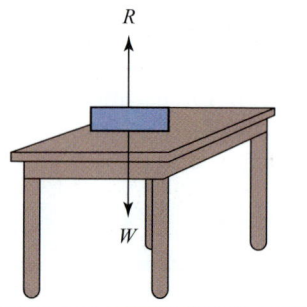

The box in Example 1 is said to be in **equilibrium**.

> **Key point**
>
> An object that is at rest or moving with constant velocity is in equilibrium.

In these situations you can **resolve** the forces in a particular direction. This means you find the overall force acting in that direction.

Resolve vertically $\qquad R - W = 0$

> **Key point**
>
> When you resolve in any direction for an object that is in equilibrium, the overall force in that direction will be zero.

If you apply a horizontal force P to a box laying on a rough surface, a **resistance force** acts in the opposite direction. If the box doesn't move, the resistance force must equal P

> Bracketed arrows are sometimes used to denote the direction in which you are resolving. $(\uparrow)$ indicates that you are resolving vertically and $(\rightarrow)$ indicates that you are resolving horizontally.

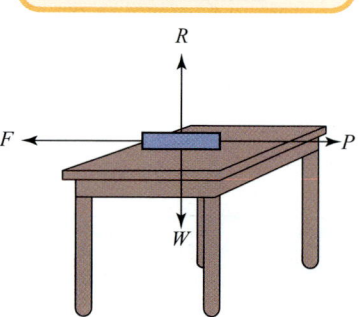

Resolve horizontally $P - F = 0$
Resolve vertically $R - W = 0$

The resistance force always acts in a direction which opposes motion. There is no resistance to motion on smooth surfaces. Resistance to motion appears when rough surfaces try to move relative to each other.

The resistance to motion due to a rough surface is sometimes called **friction**. This is why it is given the letter F

An object does not have to be at rest to be in equilibrium. It could be moving at a constant velocity.

Newton's first law of motion states that an object will remain at rest or continue to move with constant velocity unless an external force is applied to it, i.e. a moving object remains in equilibrium as long as the resultant force on it is zero.

A string is attached to a box and pulled vertically so that the box moves upwards with a constant speed of $1\,\mathrm{m\,s^{-1}}$

Resolve vertically $T - W = 0$

The force T in the string is a **tension**.

The box is pushed upwards at a constant speed of $5\,\mathrm{m\,s^{-1}}$ by two rods.

Resolve vertically $2T - W = 0$

The force T in each rod is called **thrust**.

When a rod or string is being pulled, the force is a tension force. When a rod is in **compression** (being squashed), the force is a thrust force. You cannot compress a string or it just goes slack, so you cannot have a thrust force in a string.

Example 2

A bag of weight W rests on a table. It is pulled by two horizontal strings with tensions T and S, both pulling in the same direction, so that the box moves at a constant speed of $1\,\mathrm{m\,s^{-1}}$. The resistance force is F and the normal reaction of the table on the box is R

a Draw a diagram showing all the forces acting on the box.

b Resolve horizontally and vertically to form two equations.

c The resistance to motion is $5\,\mathrm{N}$ and the tension in string T is $1.5\,\mathrm{N}$. Find the tension S

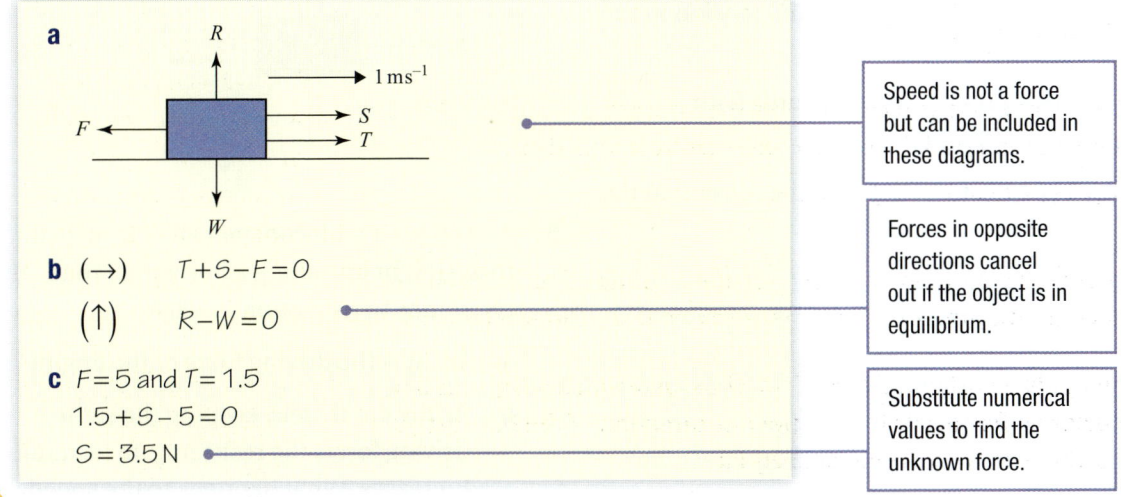

a

b ($\rightarrow$) $T + S - F = 0$

($\uparrow$) $R - W = 0$

c $F = 5$ and $T = 1.5$

$1.5 + S - 5 = 0$

$S = 3.5\,\mathrm{N}$

Speed is not a force but can be included in these diagrams.

Forces in opposite directions cancel out if the object is in equilibrium.

Substitute numerical values to find the unknown force.

1 Draw diagrams showing all the forces acting on the blocks in the following situations. (Use F for the resistance force and R for the normal reaction.)

    a A block of weight W is placed on a smooth horizontal table.

    b A block of weight $4W$ is placed on a rough horizontal table and is pushed with a horizontal force of P which causes it to move at $2\,\mathrm{m\,s}^{-1}$

    c A block of weight W is placed on a rough horizontal table and a string is attached to the block which pulls horizontally, at constant velocity, with tension T

    d A block of weight W is placed on a rough horizontal table and a string, attached to the block, pulls the block along the table with tension T at an angle of $40°$ to the horizontal.

2 **a** Resolve forces vertically in question **1 a**.

    b Resolve horizontally and vertically in question **1 b**.

    c Resolve horizontally and vertically in question **1 c**.

3 A girl holds a string, which is attached to a box of weight $40\,\mathrm{N}$. The box hangs vertically below the girl's hand.

    a Find the magnitude of the tension in the string.

She then pulls the box upwards with constant velocity.

    b Find the magnitude of the tension in the string.

4 Resolve horizontally and vertically for the following objects, which are in equilibrium.

    a
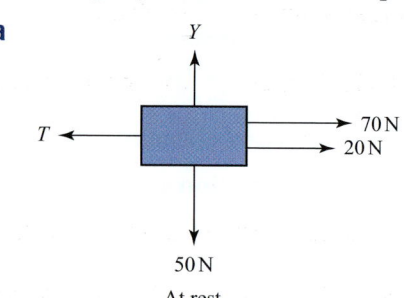

    b
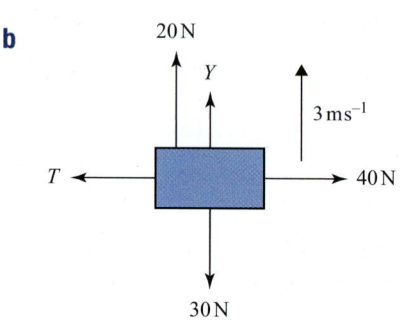

    c
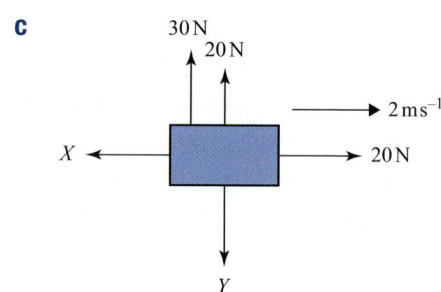

    d
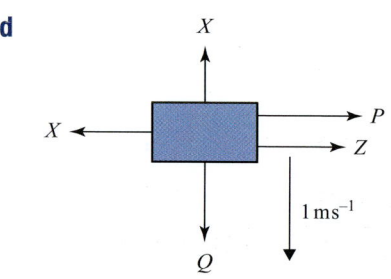

5 A car travels with constant velocity of $8\,\mathrm{m\,s}^{-1}$ in a straight line along a horizontal road. The resistance to its motion is $900\,\mathrm{N}$.

    a What is the driving force of the engine?

    b As the car travels, keeping the same driving force, the resistance to its motion increases. What happens to the speed of the car?

Reasoning and problem-solving

When more than one force acts on an object, the **resultant force** is the single force that is equivalent to all the forces acting on the object.

> **Key point**
>
> If forces $\mathbf{F}_1, \mathbf{F}_2, \ldots, \mathbf{F}_n$ act on an object then the resultant force is $\mathbf{R} = \mathbf{F}_1 + \mathbf{F}_2 + \ldots \mathbf{F}_n$

Strategy

To solve questions involving the resultant force

1. Resolve in two perpendicular directions (always in the direction of one of the forces) to find the sum of the components of all the forces in these two directions. Label the components P and Q

2. Draw a right-angled triangle with P and Q as the two shorter sides.

3. To calculate the resultant $R = \sqrt{P^2 + Q^2}$ is the magnitude and α gives the direction.

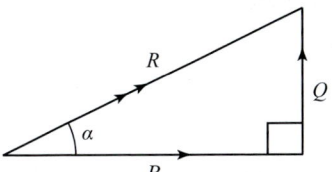

If the resultant R is known, use $P = R\cos\alpha$ and $Q = R\sin\alpha$ to find the components.

Example 3

The forces $(3\mathbf{i} + 14\mathbf{j})\,\text{N}$, $(5\mathbf{i} - 2\mathbf{j})\,\text{N}$ and $(7\mathbf{i} - \mathbf{j})\,\text{N}$ act on an object.

a Calculate the resultant force in the form $(a\mathbf{i} + b\mathbf{j})\,\text{N}$.

b Find the magnitude and direction of the resultant force. Give the direction as the angle between the resultant force and the unit vector $\mathbf{i}$.

> **①** Work out the sum of forces in both directions. See Ch6.2 for a reminder on components of vectors.

a $(3\mathbf{i} + 14\mathbf{j})\,\text{N} + (5\mathbf{i} - 2\mathbf{j})\,\text{N} + (7\mathbf{i} - \mathbf{j})\,\text{N} = (15\mathbf{i} + 11\mathbf{j})\,\text{N}$

b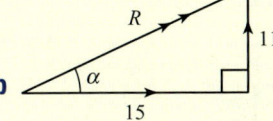

> **②** Draw a sketch.

Magnitude:

$\sqrt{15^2 + 11^2} = 18.6\,\text{N}$ (to 3 sf)

> **③** Use Pythagoras' theorem to find the magnitude.

Direction:

$\tan\alpha = \dfrac{11}{15}$

$\alpha = \tan^{-1}\left(\dfrac{11}{15}\right) = 36.3$ (to 1 dp)

> **③** Use trigonometry to find the angle.

The angle the resultant makes with the horizontal is $36.3°$

Example 4

Work out X and Y if the resultant force on this object has magnitude 24 N and makes an angle of 30° above the rightward horizontal.

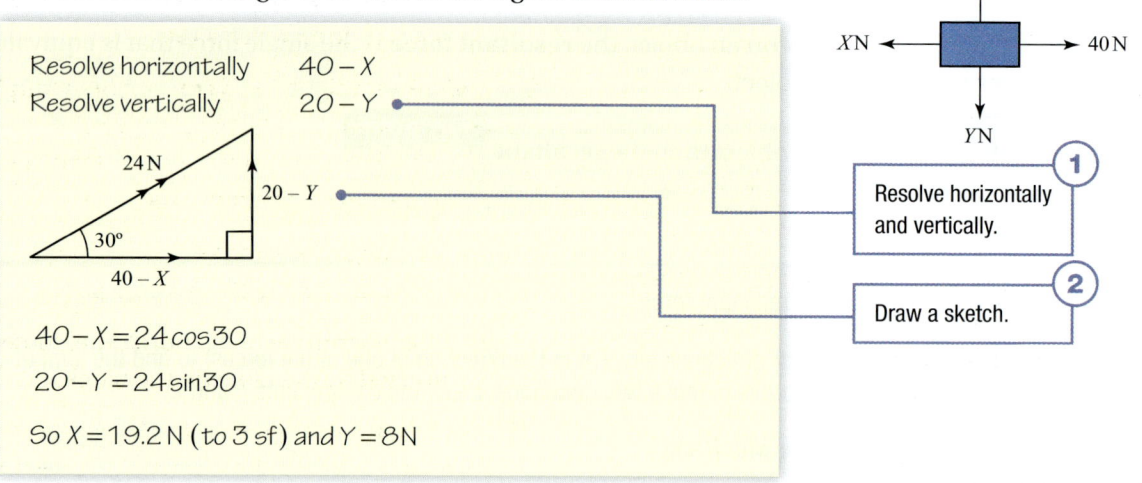

Resolve horizontally $\quad 40 - X$

Resolve vertically $\quad\quad 20 - Y$

① Resolve horizontally and vertically.

② Draw a sketch.

$40 - X = 24\cos 30$

$20 - Y = 24\sin 30$

So $X = 19.2$ N (to 3 sf) and $Y = 8$ N

Exercise 8.1B Reasoning and problem-solving

1 The forces $(10\mathbf{i} + 15\mathbf{j})$ N, $(25\mathbf{i} + 7\mathbf{j})$ N and $(13\mathbf{i} - 4\mathbf{j})$ N act on an object.

   a Work out the resultant force in the form $(a\mathbf{i} + b\mathbf{j})$ N.

   b Find the magnitude and direction of the resultant force. Show your working.

2 The forces $(200\mathbf{i} + 350\mathbf{j})$ N, $(125\mathbf{i} + 75\mathbf{j})$ N and $(-200\mathbf{i} - 300\mathbf{j})$ N act on an object.

   a Work out the resultant force in the form $(a\mathbf{i} + b\mathbf{j})$ N.

   b Find the magnitude and direction of the resultant force. Show your working.

3 The forces $\begin{pmatrix} 90 \\ -25 \end{pmatrix}$ N, $\begin{pmatrix} -75 \\ 60 \end{pmatrix}$ N and $\begin{pmatrix} a \\ b \end{pmatrix}$ N act on an object.

   a Calculate the values of a and b if the resultant force is $\begin{pmatrix} -50 \\ 120 \end{pmatrix}$ N.

   b Find the magnitude and direction of the resultant force. Show your working.

4 These objects are all in equilibrium. Find the values of the lettered forces.

a

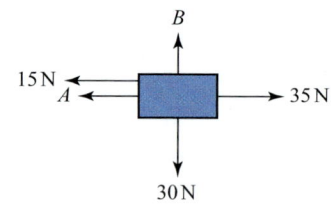

b

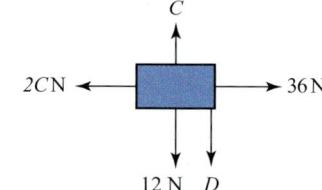

c

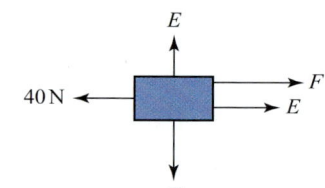

d

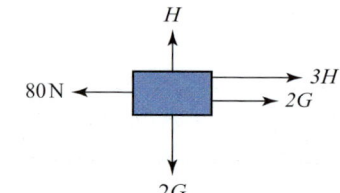

5 Find the magnitude and direction of the resultant force on this object. Show your working.

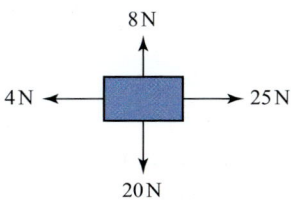

6 Find the magnitude and direction of the resultant force on the following objects. Show your working.

a

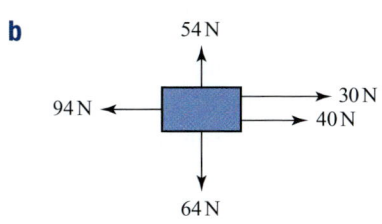

b

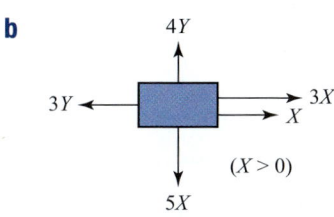

7 State why the following objects cannot be in equilibrium.

a

b

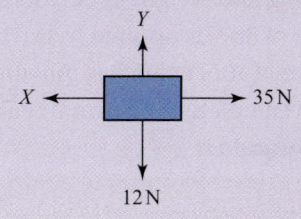

$(X > 0)$

8 A bicycle travels with a constant velocity of $6\,\mathrm{m\,s^{-1}}$ in a straight line along a horizontal road. The resistance to motion is $300\,\mathrm{N}$.

a What is the magnitude of the force applied by the cyclist?

b At a certain point in the journey the cyclist accelerates without increasing the force she applies to the bicycle. Give a possible reason for this.

9 A block of weight $50\,\mathrm{N}$ is placed on a rough, horizontal table. A horizontal force of magnitude $5\,\mathrm{N}$ is applied to the block but the block does not move. Find

a The magnitude of the vertical force exerted by the table on the block,

b The magnitude of the horizontal force exerted by the table on the block.

10 A train moves over rough tracks at a constant velocity. One carriage follows the engine, attached to the engine by a light, inextensible tow bar.

a Is the tow bar in thrust or tension? Give a reason for your answer.

b The train then travels in the opposite direction. Does your answer to part **a** change? Give a reason for your answer.

11 The underside of a block of weight $50\,\mathrm{N}$ is attached to a vertical rod which applies an upward vertical force of $60\,\mathrm{N}$ to the block. State whether the rod is in tension or thrust.

Challenge

12 Work out the positive values of X and Y given that the resultant force has magnitude $30\,\mathrm{N}$ on a $315°$ bearing and Y acts in a northerly direction.

Dynamics 1

Fluency and skills

When you give a large push to a sledge on ice, the sledge begins to accelerate. The bigger the force that you apply, the bigger the acceleration. Acceleration is directly proportional to the force that you apply.

If you had a sledge of twice the mass, you would need a force twice the size to create the same acceleration. The force you apply is directly proportional to the mass.

See Ch2.4
For a reminder on proportionality.

Newton's second law of motion states that the resultant force acting on a particle is proportional to the product of the mass of the particle and its acceleration, $\mathbf{F} \propto m\mathbf{a}$

You measure force in newtons; one newton is the force required to give a 1 kg mass an acceleration of $1\,\text{m s}^{-1}$. Using these units, Newton's second law is $\mathbf{F} = m\mathbf{a}$

> **Key point**
>
> If a resultant force $\mathbf{F}$ N acts on an object of mass m kg giving it an acceleration $\mathbf{a}\,\text{m s}^{-2}$ then $\mathbf{F} = m\mathbf{a}$

Example 1

Calculate the acceleration if the forces acting on an object of mass 25 kg are $(40\mathbf{i}+15\mathbf{j})$ N, $(20\mathbf{i} - 7\mathbf{j})$ N and $(31\mathbf{i} + 23\mathbf{j})$ N.

Remember that acceleration is a vector not a scalar.

The resultant force
$$\mathbf{F} = (40\mathbf{i}+15\mathbf{j}) + (20\mathbf{i}-7\mathbf{j}) + (31\mathbf{i}+23\mathbf{j}) = 91\mathbf{i}+31\mathbf{j}\,\text{N}$$

$$91\mathbf{i}+31\mathbf{j} = 25\mathbf{a}$$ — Use $\mathbf{F} = m\mathbf{a}$

$$\therefore \mathbf{a} = (3.64\mathbf{i}+1.24\mathbf{j})\,\text{ms}^{-2}$$

> **Key point**
>
> The equation $F = ma$ can be used in any direction where F is the overall force in that direction and a is the acceleration in that direction.

Example 2

A box of mass 5 kg and weight 49 N rests on a horizontal floor. A horizontal force of 30 N is applied to the box. The box is subject to a resistance force of 10 N when it is moving. Resolve horizontally and vertically to calculate the normal reaction R and the acceleration a

$$R - 49 = 0, \text{ so } R = 49\,\text{N}$$ — Resolve vertically.

$$30 - 10 = 5a$$

$$\therefore a = 4\,\text{ms}^{-2}$$ — Apply Newton's second law horizontally and solve.

1 A box of mass 5 kg lies on a horizontal table. A horizontal force of magnitude 20 N is applied to the box. Find the magnitude of the resistive force if the acceleration is $1\,\text{m s}^{-2}$.

2 A crate of mass 60 kg lies on a horizontal floor. A horizontal force of magnitude 300 N is applied to the box. Find the acceleration of the crate if the resistive force has magnitude 210 N.

3 A car of mass 1000 kg is travelling along a horizontal road. The total resistance to motion is 400 N and the driving force is 1600 N. Calculate the acceleration of the car.

4 A rope is attached to a block of mass 250 kg, which lies on horizontal ground. The rope is pulled horizontally with tension T. The magnitude of the resistive force is 650 N. Find T if the block accelerates at $0.25\,\text{m s}^{-2}$.

5 Calculate the acceleration acting on an object if

a The resultant force is $(9\mathbf{i} + 18\mathbf{j})\,\text{N}$ and the mass is 5 kg,

b The resultant force is $\begin{pmatrix} 3 \\ 5 \end{pmatrix}\,\text{N}$ and the mass is 2 kg,

c The resultant force is 7 N and the mass is 25 kg,

d The forces acting on the object are $(5\mathbf{i} - \mathbf{j})\,\text{N}$ and $(3\mathbf{i} - 4\mathbf{j})\,\text{N}$ and the mass is 4 kg,

e The forces acting on the object are $(3\mathbf{i} + 8\mathbf{j})\,\text{N}$, $(9\mathbf{i} + 11\mathbf{j})\,\text{N}$, $(-\mathbf{i} - 7\mathbf{j})\,\text{N}$ and $(4\mathbf{i} + 9\mathbf{j})\,\text{N}$ and the mass is 0.25 kg.

6 A car of mass 1200 kg is at rest on a horizontal road. Work out the force needed to give the car an acceleration of $3\,\text{m s}^{-2}$ if the total resistance to motion is 300 N.

7 Work out the missing values in the following diagrams.

a

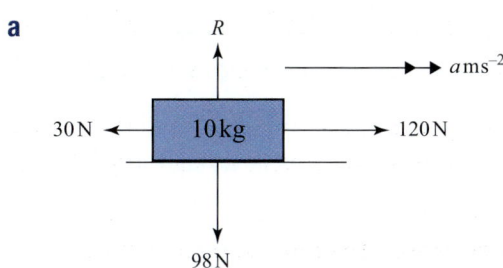

b

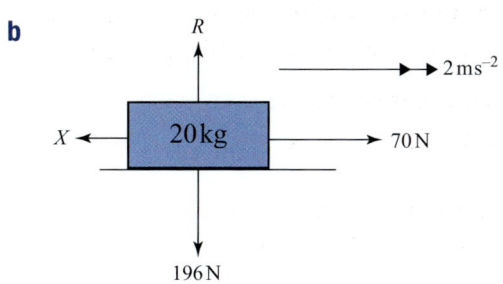

8 A truck of mass 2000 kg is travelling on a horizontal road. The total resistance to motion is 500 N. A horizontal braking force of magnitude 900 N is applied to the truck. Work out the deceleration of the truck.

9 A box of mass m kg rests on a horizontal floor. A horizontal force 40 N is applied to the box which gives it an acceleration of $2\,\text{m s}^{-2}$. Calculate the value of m if the total resistance to motion is 12 N.

10 A car of mass 800 kg is moving along a straight level road with a velocity of $30\,\text{m s}^{-1}$, when the driver spots an obstacle ahead. The driver immediately applies the brakes, providing a net braking force of 3000 N. Calculate

a The deceleration,

b The time taken for the car to come to rest,

c The distance travelled by the car in coming to rest.

Reasoning and problem-solving

You saw earlier that the equation $F = ma$ can be used in any direction where F is the overall force in that direction and a is the acceleration in that direction.

Example 3

A box of mass 10 kg and weight 98 N is pulled upwards by a vertical string. The block is decelerating at a rate of $2\,\mathrm{m\,s^{-2}}$.

a Find the tension in the string.

b State any assumptions you made in part **a**.

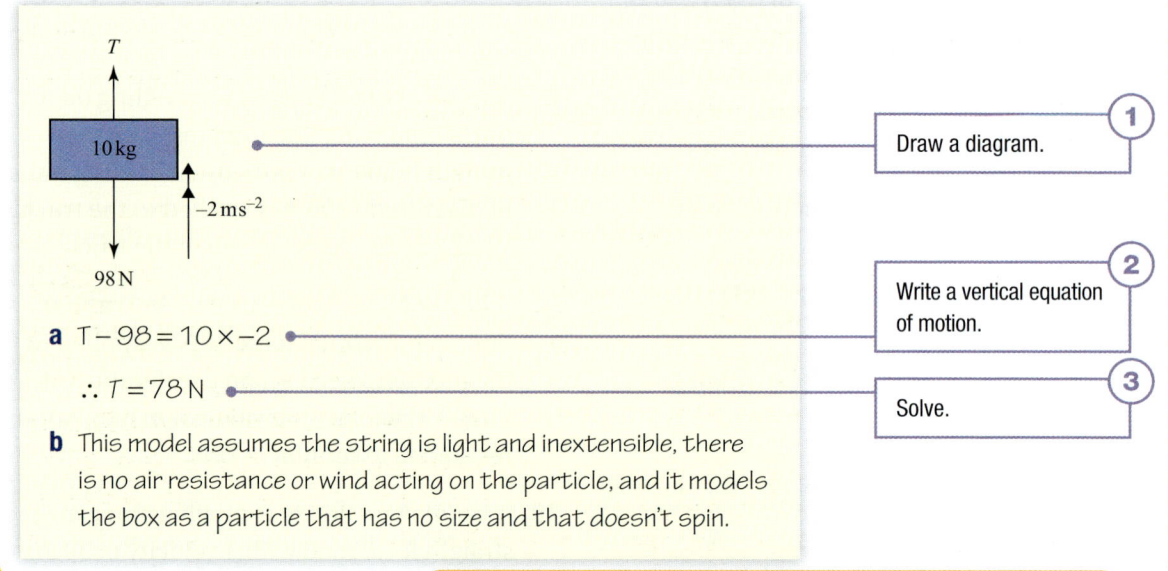

a $T - 98 = 10 \times -2$

$\therefore T = 78\,\mathrm{N}$

b This model assumes the string is light and inextensible, there is no air resistance or wind acting on the particle, and it models the box as a particle that has no size and that doesn't spin.

1 Draw a diagram.

2 Write a vertical equation of motion.

3 Solve.

In mechanics, you often model strings as inextensible which means they do not stretch under tension.

Exercise 8.2B Reasoning and problem-solving

1 Work out the mass of a car if a resultant force of magnitude 700 N causes a constant acceleration that brings the car from rest to $10\,\mathrm{m\,s^{-1}}$ in 200 m.

2 Work out the magnitude of the resultant force on a bike of mass 100 kg that goes from $15\,\mathrm{m\,s^{-1}}$ to $12\,\mathrm{m\,s^{-1}}$ in 20 m.

3 Work out how far an initially stationary object of mass 0.5 kg will travel in 1 second if a resultant force of 30 N is applied to it for that one second.

4 A train of mass 50 tonnes accelerates from rest to a speed of $10\,\mathrm{m\,s^{-1}}$ over a distance of 50 m. If the total resistance to motion is 3000 N then work out the driving force acting on the train.

5 A cyclist and her bike have a combined mass of 80 kg. She travels on a horizontal road at $12\,\mathrm{m\,s^{-1}}$ and the total resistance to motion is 25 N.

   a What is the magnitude of the force that she is applying?

The cyclist sees a problem ahead so immediately stops pedalling and applies the brake. Her braking distance is 10 m.

   b Assuming that the resistance to motion stays at 25 N, find the braking force that she applies.

   c State, with a reason, what will actually happen to the resistance to motion and what effect this will have on your answer to part **b**.

6 A box of mass m kg has the following forces acting on it: $(2\mathbf{i} + 7\mathbf{j})\,\mathrm{N}$, $(3\mathbf{i} - 2\mathbf{j})\,\mathrm{N}$, $(11\mathbf{i} - 2\mathbf{j})\,\mathrm{N}$, $(11\mathbf{i} + 3\mathbf{j})\,\mathrm{N}$ and $(5\mathbf{i} + p\mathbf{j})\,\mathrm{N}$. The resultant acceleration is $(7\mathbf{i} + p\mathbf{j})$. Find the values of m and p

7 A parachutist of mass 80 kg and weight 800 N is falling to the ground. Her speed changes from $50\,\mathrm{m\,s^{-1}}$ to $10\,\mathrm{m\,s^{-1}}$ in 2 seconds.

   a Assuming constant acceleration over these two seconds, find the resistive force of the parachute.

   b Comment on the assumption that the acceleration will be constant over these two seconds.

8 A lorry of mass m kg accelerates from rest to a speed of $20\,\mathrm{m\,s^{-2}}$ over 16 seconds. The total resistance to motion of the lorry is 1200 N and the driving force of the lorry is 3700 N. Find m

9 A string is attached to the top of a box of mass 5 kg and weight 49 N. The box is at rest on the ground. The string is held vertically above the box and the tension in the string is slowly increased until the box is just about to lift off the ground.

   a Find the tension in the string at this point.

The tension in the string is then increased to 60 N.

   b Find how long the box takes to reach a height of 1 m.

When the box is at a height of 1 m above the ground the tension is reduced to 40 N.

   c Find the speed at which the box hits the ground.

10 A force $\begin{pmatrix} 3 \\ 4 \end{pmatrix}\,\mathrm{N}$ is applied to a box of mass 4 kg. Another force $\begin{pmatrix} a \\ b \end{pmatrix}\,\mathrm{N}$ of magnitude 10 N is to be applied in order to give the box the greatest possible acceleration. Find a and b and find the magnitude of the acceleration of the box.

Challenge

11 Calculate the two possible values of x if the forces $\begin{pmatrix} x \\ 1 \end{pmatrix}\,\mathrm{N}$ and $\begin{pmatrix} 8 \\ x \end{pmatrix}\,\mathrm{N}$ applied to a box of mass 5 kg cause an acceleration of magnitude $2.6\,\mathrm{m\,s^{-2}}$. Assume there is no resistance to motion. Assume both forces act only in directions parallel to the ground.

12 Work out R and m in this diagram.

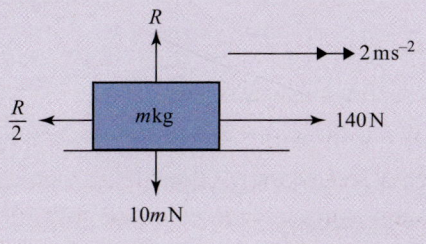

Fluency and skills

If a ball of mass m kg is dropped and air resistance is ignored then the only force acting on the ball is the gravitational pull of the Earth. The acceleration of the ball is entirely due to the weight of the ball.

The weight of the ball depends on where the ball is. If the ball is in space its weight is much less than if it is near the surface of the Earth. Note however that even if the ball's weight changes based on where it is, its mass never changes.

Using Newton's second law,
$F = ma$ gives $W = mg$

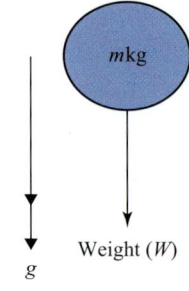

g

Weight (W)

> **Key point**
>
> An object of mass m kg has weight mg newtons.

In questions involving motion on Earth, you should assume that $g = 9.8 \, \text{m s}^{-2}$ unless you are told to use a different value in the question.

> You may be told to use $g = 9.81 \, \text{m s}^{-2}$ or $g = 10 \, \text{m s}^{-2}$. In your answers you should use the same number of significant figures as the value of g you are given.

Example 1

Work out the following quantities.

a   The weight of a car of mass 1230 kg.

b   The weight of a piece of paper of mass 4.54 g.

c   The mass of a cup of water of weight 3.2 N. You may assume that g is $10 \, \text{m s}^{-2}$.

a $W = 1230 \times 9.8 = 12054 \, \text{N}$

So $W = 12000 \, \text{N}$ (to 2 sf)

b $W = 0.00454 \times 9.8 = 0.0445374 \, \text{N}$

$\therefore W = 0.045 \, \text{N}$ (to 2 sf)

c $3.2 = m \times 10$

$m = 0.32 \, \text{kg} = 320 \, \text{g}$

$m = 300 \, \text{g}$ (to 1 sf)

> Assuming $g = 9.8 \, \text{m s}^{-2}$, give your answer to 2 sf.

> Convert the mass to kilograms.

Example 2

A block is pulled along a horizontal surface by a horizontal string. The tension in the string is 40 N. The frictional force is 30 N and the normal reaction is 20 N. Work out the mass and the acceleration of the block. You may assume that g is $10\,\text{m}\,\text{s}^{-2}$.

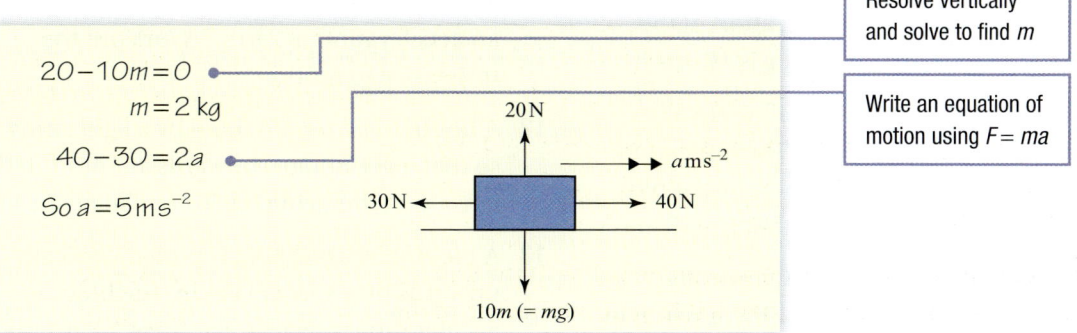

$$20 - 10m = 0$$
$$m = 2\,\text{kg}$$
$$40 - 30 = 2a$$
$$\text{So } a = 5\,\text{m}\,\text{s}^{-2}$$

Resolve vertically and solve to find m

Write an equation of motion using $F = ma$

Example 3

A stone of mass 50 g is dropped from the top of a cliff, which is 20 m above the sea.

a Find the speed at which the stone hits the sea below.

b State any assumptions you have made.

a The acceleration of the stone is $9.8\,\text{m}\,\text{s}^{-2}$.

$$u = 0 \qquad s = 20 \qquad a = 9.8$$

$$v^2 = u^2 + 2as$$
$$v^2 = 2 \times 20 \times 9.8$$
$$v = 20\,\text{m}\,\text{s}^{-1}\text{ (to 2 sf)}$$

b The model assumes that there is no air resistance or wind and it models the stone as a particle that has no size and does not spin.

Write down the quantities that you know. The acceleration is equal to g

Select an appropriate equation and solve for v. See Ch7.3 for a reminder on equations of motion for constant acceleration.

Exercise 8.3A Fluency and skills

1 Work out

   a The weight of a 10 g piece of paper,

   b The weight of a 10 tonne lorry,

   c The mass of a 100 N weight.

2 A block of mass 75 kg is pulled up by a vertical rope. The tension in the rope is 1200 N. Calculate the acceleration of the block.

3 A ball is thrown vertically upwards. It returns to its starting point after 3 seconds. Find the speed with which it was thrown. You may assume that g is $9.81\,\text{m}\,\text{s}^{-2}$.

4 A particle is thrown vertically downwards with a speed of $18\,\text{m}\,\text{s}^{-2}$ from the top of a building, which is 5 m high. Find the speed with which the particle hits the ground. You may assume that g is $10\,\text{m}\,\text{s}^{-2}$.

5 A block is pulled along a horizontal surface by a horizontal string. The tension in the string is 450 N. The resistance force is 300 N and the normal reaction force is 250 N. Work out the mass and the acceleration of the block. You may assume that g is $10\,\text{m}\,\text{s}^{-2}$.

6 A block is pulled along a horizontal surface by a horizontal string. The tension in the string is 70 N. The resistance force is 50 N and the normal reaction force is 40 N. Work out the mass and the acceleration of the block. You may assume that g is $10\,\text{m}\,\text{s}^{-2}$.

7 A crate of mass 10 kg is pulled vertically upwards by a vertical rope. The tension in the rope is 130 N. Find the acceleration of the crate. You may assume that g is $10\,\text{m}\,\text{s}^{-2}$.

8 A box of mass m kg is pulled vertically upwards by a vertical string. The tension in the string is 80 N and the acceleration of the box is $0.2\,\text{m}\,\text{s}^{-2}$. Find m

9 Work out the tension in the cable attached to the top of a lift of mass 400 kg when the lift is

  a Stationary,

  b Accelerating at $1\,\text{m}\,\text{s}^{-2}$ vertically upwards,

  c Accelerating at $2\,\text{m}\,\text{s}^{-2}$ vertically downwards.

10 A box of mass 10 kg is lifted by a light string so that it accelerates upwards at $2.0\,\text{m}\,\text{s}^{-2}$. Work out the tension in the string.

11 A box of mass 2 kg sinks through water with an acceleration of $2\,\text{m}\,\text{s}^{-2}$. Work out the resistance to motion.

12 A box of mass m kg is lifted by a light string so that it accelerates upwards at $3\,\text{m}\,\text{s}^{-2}$. If the tension in the string is 256 N calculate m

13 A block of mass 200 kg is pulled vertically upwards by a vertical cable. The block is accelerating upwards at $0.5\,\text{m}\,\text{s}^{-2}$. Find the magnitude of the tension in the cable. You may assume that g is $9.81\,\text{m}\,\text{s}^{-2}$.

14 A container of mass 60 kg is being lowered vertically downwards by a vertical cable. The container is accelerating downwards at $0.1\,\text{m}\,\text{s}^{-2}$. Find the magnitude of the tension in the cable. You may assume that g is $9.81\,\text{m}\,\text{s}^{-2}$.

Reasoning and problem-solving

Example 4

A bag of mass $10\,b$ kg is lowered by a light, inextensible string so that it accelerates downwards at $3\,b\,\text{m}\,\text{s}^{-2}$. Calculate the possible values of b if the tension is 41.5 N.

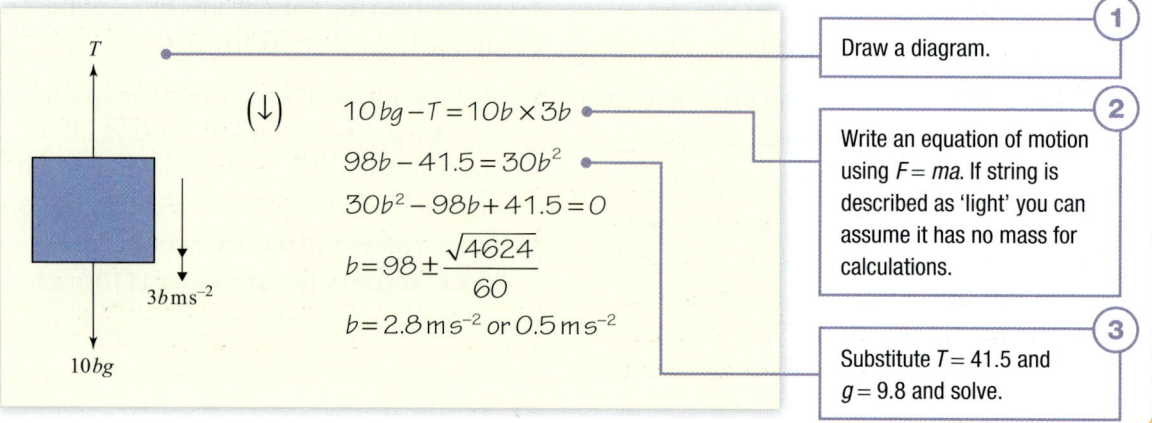

1 Draw a diagram.

$(\downarrow)\quad 10bg - T = 10b \times 3b$

$98b - 41.5 = 30b^2$

$30b^2 - 98b + 41.5 = 0$

$b = 98 \pm \dfrac{\sqrt{4624}}{60}$

$b = 2.8\,\text{m}\,\text{s}^{-2}\ \text{or}\ 0.5\,\text{m}\,\text{s}^{-2}$

2 Write an equation of motion using $F = ma$. If string is described as 'light' you can assume it has no mass for calculations.

3 Substitute $T = 41.5$ and $g = 9.8$ and solve.

1   A crane lifts up a crate of bricks with mass 1200 kg from the ground. The acceleration is constant, and after 1 second the crate is 0.5 m off the ground. Work out the acceleration of the crate and the tension in the crane's cable.

2   A block of mass 5 kg is pulled along a horizontal surface by a horizontal string. The tension in the string is 50 N. The normal reaction force is R and the resistance force is $\frac{R}{2}$. Work out R and the acceleration of the block. You may assume that g is $10\,\text{m}\,\text{s}^{-2}$.

3   A man who weighs 780 N on the surface of the Earth weighs only 130 N on the surface of the moon. Work out the ratio of the value of g on the moon to the value of g on the surface of the Earth.

4   A crate of mass 20 kg is pulled upwards from rest by a light rope to a speed of $4\,\text{m}\,\text{s}^{-1}$. If the tension in the rope is 260 N then work out how far the crate travelled before getting to $4\,\text{m}\,\text{s}^{-1}$.

5   A ball is thrown vertically upwards from the ground. Its highest point is 20 m above the ground. Find how long the ball is in the air.

6   A box of mass 8.5 kg is pulled vertically upwards by a rope.

    a   Calculate the tension in the rope, in terms of g, when

        i   The basket is stationary,

        ii The basket is accelerating at $2\,\text{m}\,\text{s}^{-2}$ upwards.

    b   What assumptions have you made about the rope?

7   A box of mass 25 kg is being lifted. The box starts from rest and then a rope attached to the top of the box pulls it vertically upwards. The tension in the rope is 400 N. Another rope is attached to the bottom of the box and a man pulls lightly down to stop the box from swaying to the side. The tension in this rope is 50 N. Find the acceleration of the box and the height it will be at after 5 seconds. You may assume that g is $10\,\text{m}\,\text{s}^{-2}$.

8   A man on the surface of the moon lifts a box of mass 10 kg by pulling on a string. The tension in the string is 30 N. Find how long it takes to lift the box to a height of 1 m. You may assume that g is $1.6\,\text{m}\,\text{s}^{-2}$.

9   A box of mass 3 kg rests on a tray of mass 2 kg. They are lifted up by two vertical strings each attached to the tray so that the tension in each string is the same. The box and the tray accelerate upwards at $1\,\text{m}\,\text{s}^{-2}$.

    a   Find the tension in each of the strings.

    b   The 3 kg box falls off the tray but the tension in the string stays the same. Find the new acceleration of the tray.

Challenge

10 A block of mass m kg is pulled along a horizontal surface by a horizontal string. The normal reaction force is R. The tension in the string is half the normal reaction force and the resistance force is one fifth of the normal reaction force. You may assume $g = 10\,\text{m}\,\text{s}^{-2}$

    a   Calculate the acceleration of the block.

    b   Calculate the acceleration of the block assuming there is no friction.

11 A bag of mass 4 kg hangs stationary from a taut string at a height of 2 m. It is lowered to the ground at constant acceleration, hitting the ground after twice the time it would have taken if it had been in freefall. Work out the tension in the string in terms of g

MECH

8.4 Systems of forces

Fluency and skills

A person standing in a lift exerts a force, acting vertically downwards, on the floor of the lift. The lift exerts a force, acting vertically upwards, on the person. As the lift moves up and down, the size of this force changes but the force that the person exerts on the lift is always equal and opposite to the force the lift exerts on the person.

Newton's third law states that for every action there is an equal and opposite reaction.

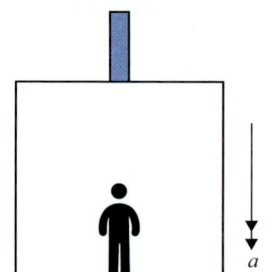

> **Key point**
>
> Newton's third law states that when an object A exerts a force on an object B, object B exerts an equal and opposite force on object A.

If a red ball hits a yellow ball, the force that the red ball exerts on the yellow ball has equal magnitude to the force that the yellow ball exerts on the red ball. These forces act in opposite directions.

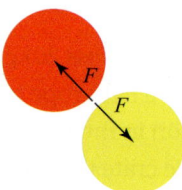

Example 1

A man of mass 80 kg is in a lift of mass 300 kg, which is accelerating upwards at $4.1\,\mathrm{m\,s^{-2}}$.

a Work out the tension in the cable pulling the lift by

   i Considering the set of forces acting on the man and the set of forces acting on the lift,

   ii Considering the man and the lift as one object.

b Use Newton's third law to explain why you can consider the man and the lift as one object.

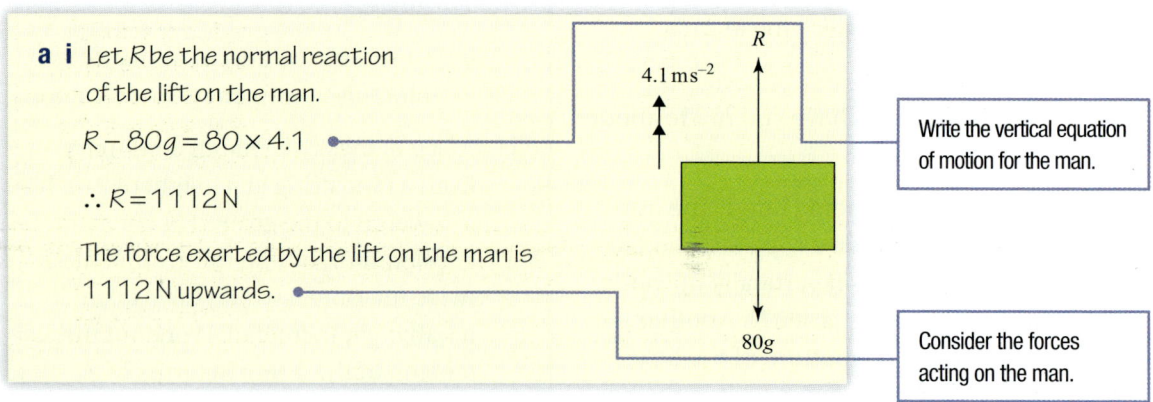

a i Let R be the normal reaction of the lift on the man.

$$R - 80g = 80 \times 4.1$$

$$\therefore R = 1112\,\mathrm{N}$$

The force exerted by the lift on the man is 1112 N upwards.

Write the vertical equation of motion for the man.

Consider the forces acting on the man.

(Continued on the next page)

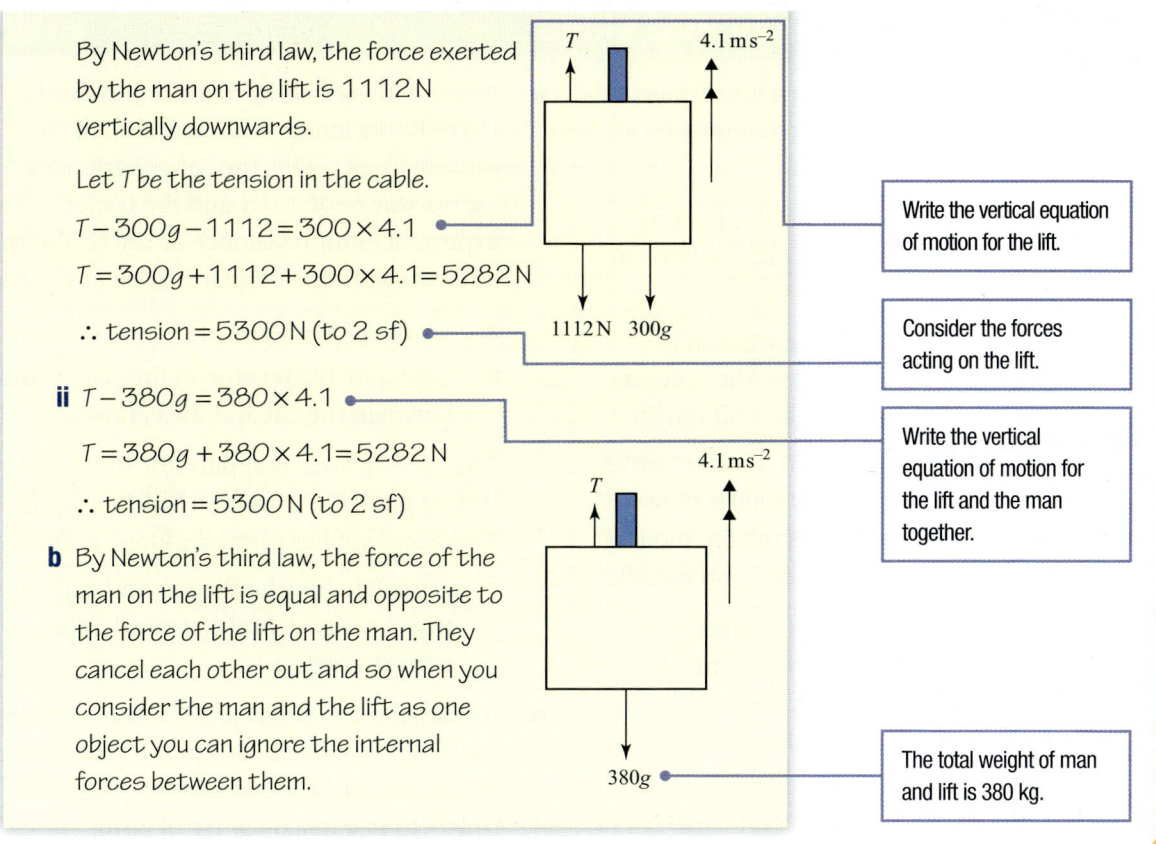

By Newton's third law, the force exerted by the man on the lift is 1112 N vertically downwards.

Let T be the tension in the cable.

$T - 300g - 1112 = 300 \times 4.1$

$T = 300g + 1112 + 300 \times 4.1 = 5282 \, N$

$\therefore$ tension = 5300 N (to 2 sf)

ii $T - 380g = 380 \times 4.1$

$T = 380g + 380 \times 4.1 = 5282 \, N$

$\therefore$ tension = 5300 N (to 2 sf)

b By Newton's third law, the force of the man on the lift is equal and opposite to the force of the lift on the man. They cancel each other out and so when you consider the man and the lift as one object you can ignore the internal forces between them.

> Write the vertical equation of motion for the lift.

> Consider the forces acting on the lift.

> Write the vertical equation of motion for the lift and the man together.

> The total weight of man and lift is 380 kg.

Example 2

A van of mass 1250 kg tows a trailer of mass 250 kg along a horizontal road. The driving force from the engine is D N, the van experiences air resistance of 300 N and the trailer experiences air resistance of 100 N.

> The main source of resistance to motion for a moving vehicle is usually air resistance.

If the acceleration of the van and trailer is $2.4 \, \text{m s}^{-2}$, calculate the tension in the connection between the van and the trailer and calculate the driving force D

Let T be the tension in the connection between the van and the trailer

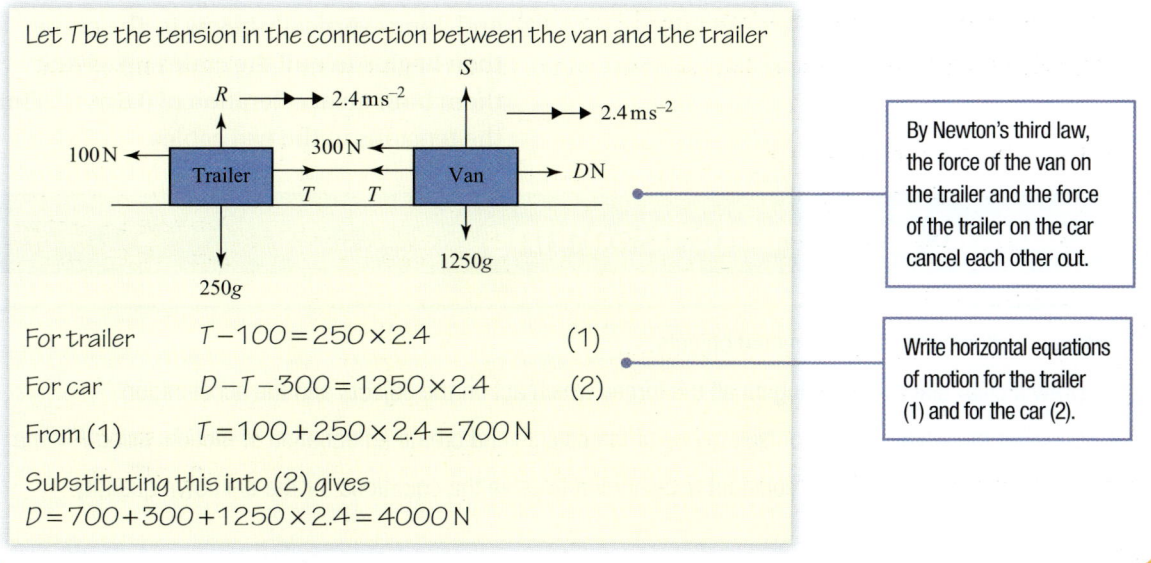

> By Newton's third law, the force of the van on the trailer and the force of the trailer on the car cancel each other out.

For trailer $T - 100 = 250 \times 2.4$ (1)

For car $D - T - 300 = 1250 \times 2.4$ (2)

From (1) $T = 100 + 250 \times 2.4 = 700 \, N$

> Write horizontal equations of motion for the trailer (1) and for the car (2).

Substituting this into (2) gives
$D = 700 + 300 + 1250 \times 2.4 = 4000 \, N$

1 A woman of mass 60 kg is in a lift of mass 250 kg which is accelerating downwards at $3.2\,\text{m\,s}^{-2}$.

   a Resolve vertically for the woman and the lift together to work out the tension in the cable.

   b Resolve vertically for the woman to work out the magnitude of the reaction force between the woman and the lift.

2 A box of mass 20 kg is in a lift of mass 200 kg. If the tension in the cable is 2800 N then work out the acceleration of the lift and the reaction force between the box and the lift.

3 A car of mass 1200 kg tows a caravan of mass 800 kg along a horizontal road. The driving force is 4000 N, the car experiences air resistance of 150 N and the caravan experiences air resistance of 100 N. Calculate

   a The acceleration of the car,

   b The force transmitted through the tow bar.

4 A tray of mass 500 g has a box of mass 750 g placed on it. A string is attached to the tray and the string is pulled upwards to cause the tray and box to accelerate. The tension in the string is 15 N. Find the acceleration of the tray and the box. Find also the force exerted by the tray on the box. You may assume that g is $10\,\text{m\,s}^{-2}$.

5 A car of mass 1500 kg tows a caravan of mass 400 kg along a horizontal road. The driving force is D N, the car experiences air resistance of 250 N and the trailer experiences air resistance of 120 N. The car and caravan accelerate from rest at $1.8\,\text{m\,s}^{-2}$.

   a Work out the value of D

   b Calculate the tension in the connection between the car and the caravan.

This driving force is applied for 10 seconds. At that point the driver puts his foot on the brake and applies a braking force of 2000 N.

   c Calculate how far the car and caravan travel with the brake applied before they come to rest.

6 A man of mass 80 kg carries a bag of mass 2 kg in a lift of mass 500 kg. The lift is moving upwards and decelerates from a speed of $5\,\text{m\,s}^{-1}$ to rest in a distance of 20 m.

   a Calculate the tension in the cable and the normal reaction force between the man and the lift.

   b Calculate the force that the man feels from carrying the bag.

7 The cable from a crane is attached to a crate of mass 220 kg. Another crate of mass 150 kg is connected to the 220 kg crate by a cable and hangs vertically below it. The crane then begins to pull the crates up, giving them both an acceleration of $0.6\,\text{m\,s}^{-2}$. Find the tensions in the two cables.

Reasoning and problem-solving

Strategy

To solve questions involving connected objects

① Draw a clear diagram marking on all the forces which act on the objects and the acceleration.

② Consider the whole system or isolate one of the objects and create an equation of motion, using $F = ma$

③ Use equations of motion for constant acceleration to solve the equations for the unknown quantity.

Example 3

MECH

The two ends of a light, inextensible string are attached to two objects of mass 4 kg and 9 kg. The string passes over a smooth, fixed pulley. The 9 kg mass is initially 0.2 m above the ground and the 4 kg mass is initially 0.6 m above the ground. They are released from rest.

a i Write an equation of motion for each mass.

  ii Use your equations of motion to work out the acceleration of the two masses and the tension in the string.

b Work out the speed at which the 9 kg mass hits the ground.

c Work out the greatest height of the 4 kg mass above the ground in subsequent motion.

---

a i 4 kg mass: $(\uparrow)$ $T - 4g = 4a$ (1)

  9 kg mass: $(\downarrow)$ $9g - T = 9a$ (2)

 ii Adding up (1) and (2)

  gives $5g = 13a$

  $\therefore a = 3.8 \, \text{m s}^{-2}$ (to 2 sf)

  Using (1) $T = 54 \, \text{N}$ (to 2 sf)

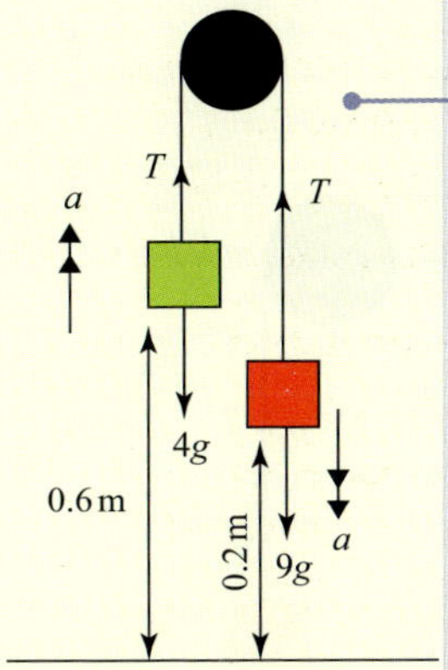

b $u = 0$ $a = 3.8$ $s = 0.2$

  $v^2 = u^2 + 2as = 1.52$

  So $v = 1.2 \, \text{m s}^{-1}$ (to 2 sf)

c For upward motion,

  $a = -9.8 \, \text{m s}^{-2}$ (to 2 sf)

  At highest point $v = 0$

  $u = 1.23$ $v = 0$ $a = -9.8$

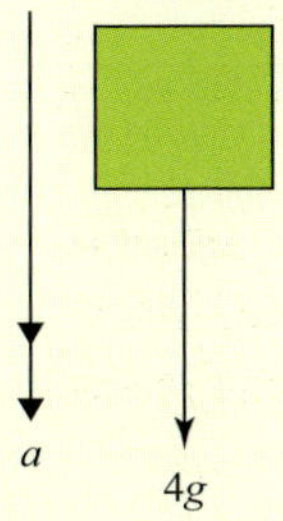

  $v^2 = u^2 + 2as$ gives

  $s = \dfrac{0^2 - 1.23^2}{2 \times (-9.8)} = 0.077$

  So greatest height of 9 kg mass above the ground is $0.6 + 0.2 + 0.077 = 0.88 \, \text{m}$ (to 2 sf)

① Draw a diagram.

T ↑   *T* ↑   *a*

0.6 m   0.2 m   4g   9g   *a*

When the 9 kg mass hits the ground, the string becomes slack. There is no upwards force on the 4 kg mass and it begins to decelerate due to gravity.

② Consider the 4 kg mass.

a   4g

③ Use values from your answer from part **b** and identify the equation of motion needed.

The particle is initially 0.6 m above the ground, travels 0.2 m as the 9 kg mass falls and then travels a further 0.077 m.

See Ch 7.3
For a reminder of the equations of motion.

ICT Resource online

ICT resource

To experiment with the motion of connected masses, click this link in the digital book.

Example 4

A 2 kg mass rests on a rough horizontal table. It is attached to a long string that passes over a smooth pulley at the end of the table, and is tied to a 3 kg mass held still in the air, 1 m above the ground. The 3 kg mass is released. The 2 kg mass experiences a constant resistance to motion of magnitude 10 N.

a Work out the acceleration of the two masses and the tension in the string.

b Work out the speed at which the 3 kg mass hits the ground.

c Work out how far the 2 kg mass travels before it comes to rest.

d Work out the magnitude of the force exerted by the string on the pulley before the 3 kg mass hits the floor.

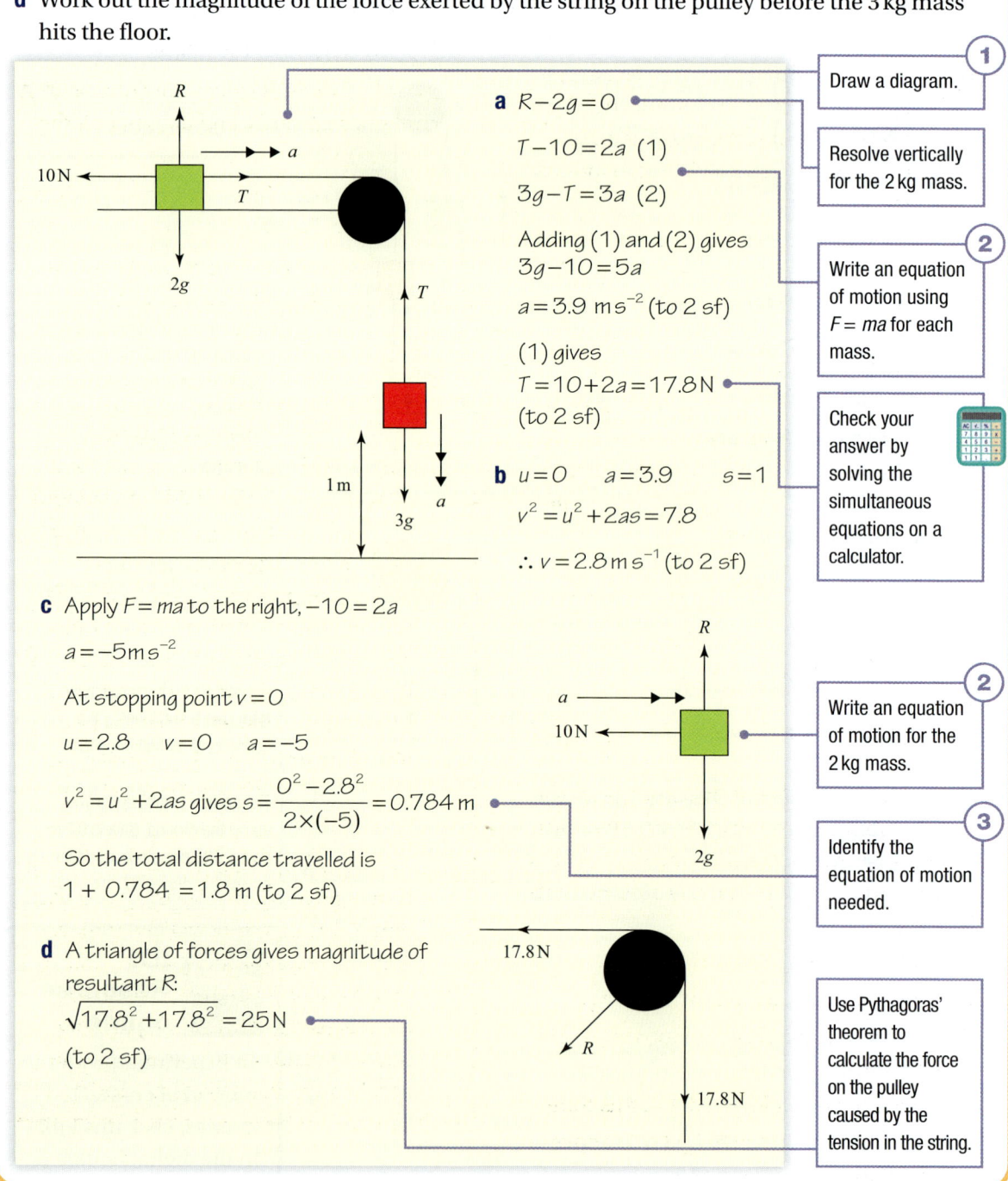

a $R - 2g = 0$

$T - 10 = 2a$ (1)

$3g - T = 3a$ (2)

Adding (1) and (2) gives
$3g - 10 = 5a$

$a = 3.9 \text{ m s}^{-2}$ (to 2 sf)

(1) gives
$T = 10 + 2a = 17.8 \text{ N}$
(to 2 sf)

b $u = 0 \qquad a = 3.9 \qquad s = 1$

$v^2 = u^2 + 2as = 7.8$

$\therefore v = 2.8 \text{ m s}^{-1}$ (to 2 sf)

c Apply $F = ma$ to the right, $-10 = 2a$

$a = -5 \text{ m s}^{-2}$

At stopping point $v = 0$

$u = 2.8 \qquad v = 0 \qquad a = -5$

$v^2 = u^2 + 2as$ gives $s = \dfrac{0^2 - 2.8^2}{2 \times (-5)} = 0.784 \text{ m}$

So the total distance travelled is
$1 + 0.784 = 1.8 \text{ m}$ (to 2 sf)

d A triangle of forces gives magnitude of resultant R:

$\sqrt{17.8^2 + 17.8^2} = 25 \text{ N}$

(to 2 sf)

1 Draw a diagram.

Resolve vertically for the 2 kg mass.

2 Write an equation of motion using $F = ma$ for each mass.

Check your answer by solving the simultaneous equations on a calculator.

2 Write an equation of motion for the 2 kg mass.

3 Identify the equation of motion needed.

Use Pythagoras' theorem to calculate the force on the pulley caused by the tension in the string.

1 A 2 kg mass and a 3 kg mass are connected by a light inextensible string and hang either side of a smooth, fixed pulley. Calculate the tension in the string and the acceleration of the particles.

2 A 3 kg mass rests on a rough horizontal table. It is attached to a long string that passes over a smooth pulley at the end of the table and is tied to a mass of 5 kg, which is held at rest in the air, 0.2 m above the ground. The 5 kg mass is released from rest. The 3 kg mass experiences a constant resistance to motion of magnitude 12 N.

  a Work out the tension in the string and the magnitude of the acceleration of the two masses.

  b Work out the speed at which the 5 kg mass hits the ground.

  c Work out how far the 3 kg mass travels before it comes to rest.

  d Work out the magnitude of the force exerted by the string on the pulley while the 5 kg mass is falling.

3 Two blocks, A and B, of masses 5 kg and 10 kg respectively are connected by a light, inextensible string that passes over a smooth pulley, P. Initially A is at rest on a smooth horizontal table, and B hangs freely, as shown in the diagram. The system is released from rest. You may assume that $g = 10\,\text{m s}^{-2}$

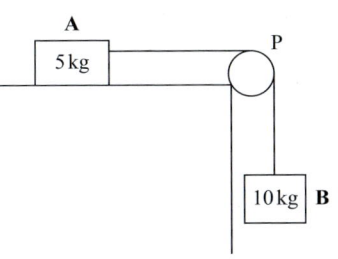

  a Calculate the acceleration of B.

  b Calculate the tension in the string.

After three seconds, A is still moving freely when B hits the floor.

  c Calculate the velocity of A at this time.

  d Calculate the initial height of B above the floor.

4 Two blocks, A and B, of mass 3 kg and 4 kg respectively, are connected by a light, inextensible string, passing over a fixed smooth, light pulley. The blocks are released from rest with the string taut, and the hanging parts vertical. Find

  a The acceleration of B,

  b The tension in the string.

After 5 seconds, B strikes the floor. Block A continues upwards, and does not hit the pulley. Find

  c The velocity of A at the instant when B strikes the floor,

  d The greatest height above its initial starting height reached by A.

Challenge

5 Two trays, each of mass 1 kg, are connected by a long light, inextensible string and hang either side of a smooth, fixed pulley. Both trays are 1 m above the horizontal ground. A mass of 2 kg and 3 kg respectively is placed in each tray and the system is released from rest. When the tray with the 3 kg mass is 0.5 m above the ground, the 3 kg mass slips out. Find the time from the system initially being released from rest and the tray with the 2 kg mass hitting the ground.

6 An x kg and a y kg mass are connected by a light, inextensible string and hang either side of a smooth fixed pulley. The masses are initially both s metres above the horizontal ground. They are released from rest. If $y > x$, work out the tension in the string and the acceleration in terms of x, y and g

Chapter summary

- Newton's first law of motion states that an object will remain at rest or continue to move with constant velocity unless an external force is applied to it.
- Force is a vector. It has both magnitude and direction.
- An object is in equilibrium if it is at rest or moving at constant velocity.
- The resultant force is the single force equivalent to all the forces acting on the object.
- If an object is in equilibrium, the resultant force is zero.
- You can summarise Newton's second law of motion as:
  If a resultant force $\mathbf{F}$ N acts on an object of mass m kg, giving it an acceleration $\mathbf{a}$, then $\mathbf{F} = m\mathbf{a}$
- You can use the equation $F = ma$ in any direction, where F is the overall force in that direction and a is the acceleration in that direction.
- Resistance forces always oppose motion.
- Deceleration of $a\,\text{m s}^{-2}$ in one direction is acceleration of $-a\,\text{m s}^{-2}$ in the opposite direction.
- An object of mass m kg has weight mg N. g is approximately $9.8\,\text{m s}^{-2}$ on the Earth's surface. It decreases as the object moves further from the Earth's surface.
- Newton's third law states that for every action there is an equal and opposite reaction. So when an object A exerts a force on an object B, object B exerts an equal and opposite force on object A.
- If two objects are connected, the internal forces between them can be ignored when the two objects are considered as a whole. E.g. a man standing in a lift or a van towing a trailer.
- A number of assumptions are often made in questions involving forces and Newton's laws:
  - Objects are particles. There is no turning effect and mass acts at one point.
  - Strings are light and inextensible.
  - The acceleration is constant through the string.
  - The tension is constant through the string.
  - The tension in the string is the same on both sides of a pulley.
  - Pulleys and smooth surfaces are perfectly smooth. There is no resistance force acting.

Check and review

You should now be able to...	Try Questions
✔ Resolve in two perpendicular directions for a particle in equilibrium.	3
✔ Calculate the magnitude and direction of the resultant force acting on a particle.	1, 2
✔ Resolve for a particle moving with constant acceleration. Work out acceleration or forces.	4
✔ Understand the connection between the mass and the weight of an object. Know that weight changes depending on where the object is.	5
✔ Resolve for "connected" particles, such as an object in a lift.	6
✔ Resolve for particles moving with constant acceleration connected by string over pulleys.	7

1   Express, as a vector, the resultant force acting on this box.

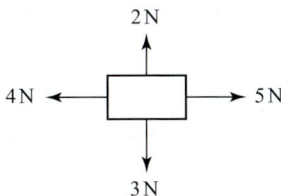

2   The forces $(13\mathbf{i} - 5\mathbf{j})\,$N, $(-5\mathbf{i} + 7\mathbf{j})\,$N and $(12\mathbf{i} + \mathbf{j})\,$N act on an object.
Work out the magnitude and direction of the resultant force.

3   A bird is held in equilibrium so that it is not moving in any direction. Its mass is 3 kg and it is flying with a forwards driving force parallel to the ground of 20 N. If the only external force exerted on the bird is caused by the wind, find the vector that describes the force exerted on the bird by the wind.

4   A car of mass 1250 kg is travelling at $40\,\mathrm{m\,s^{-1}}$ on a horizontal road. There is no resistance force. Work out the braking force needed to bring the car to rest in 100 m.

5   A man lifts up a bucket of mass 200 g. Inside the bucket is a brick of mass 500 g. The acceleration of the bucket and the brick is $0.3\,\mathrm{m\,s^{-2}}$. Find the force exerted by the man on the bucket and the force exerted by the bucket on the brick if

   a   The man is on the surface of the Earth. You may assume that g is $10\,\mathrm{m\,s^{-2}}$.

   b   The man is on the surface of the Moon. You may assume that g is $1.6\,\mathrm{m\,s^{-2}}$.

6   A woman of mass 40 kg is in a lift of mass 450 kg which is moving upwards but decelerating at $0.9\,\mathrm{m\,s^{-2}}$. Work out the tension in the cable that holds up the lift and the magnitude of reaction force between the woman and the lift.

7   Two masses A and B of 5 kg and 10 kg respectively are attached to the ends of a light, inextensible string. Mass A lies on a rough, horizontal platform. The string passes over a small smooth pulley fixed on the edge of the platform. The pulley is initially 3 m from A. Mass B hangs freely below the pulley. The masses are released from rest with the string taut. There is a constant resistive force of 20 N on mass A as it is moving.

   a   Work out the tension in the string and the magnitude of the acceleration of the masses.

After 0.5 seconds, a 9 kg section of the 10 kg mass falls off.

   b   Work out how long it takes for the 10 kg mass to come to rest after the 9 kg section falls off.

   c   Work out the closest distance between A and the pulley.

   d   Calculate the magnitude of the force on the pulley from the strings after the 9 kg section falls off.

MECH

What next?

Score		
0 – 3	Your knowledge of this topic is still developing. To improve, search in MyMaths for the codes: 2185–2188, 2293	🔗
4 – 5	You're gaining a secure knowledge of this topic. To improve, look at the InvisiPen videos for Fluency and skills (08A)	🎞
6 – 7	You've mastered these skills. Well done, you're ready to progress! Now try looking at the InvisiPen videos for Reasoning and problem-solving (08B)	🎞

Click these links in the digital book

Galileo Galilei

History

Galileo Galilei was born in Italy in 1564 and throughout his lifetime he made contributions to astronomy, physics, engineering, philosophy and mathematics.

Before Galileo, the description of **forces** and their influence, given by **Aristotle**, had remained unchallenged for almost two thousand years. Galileo exposed the errors in Aristotle's work through experiment and logic.

Galileo's work in mechanics paved the way for **Newton** to define his **three laws of motion** in 1687

Investigation

"A ship is sailing on a calm sea when a cannon ball is dropped from the crow's nest.
Where does the cannon ball land?"

* Describe the path of the cannon ball as seen by someone on the ship.
* Describe the path of the cannon ball as seen by someone on the shore, as the ship passes by.
* Why is the path of the cannonball different in each case?

Find out how Galileo used the dropping of a cannon ball to demonstrate a flaw in Aristotle's understanding of the way forces behave.

"All truths are easy to understand, once discovered; the point is to discover them."
- Galileo

Research

The work of Galileo and Newton stood the test of time until the 20th century and is now referred to as **classical mechanics**.

In the late 1800s and early 1900s, **Einstein** and other scientists showed that the laws of classical mechanics don't always hold. They needed a more general theory, and this gave rise to a new field of study called **quantum mechanics**.

In 1905, Einstein introduced the concept of the speed of light, and stated that nothing can travel faster this speed. What is the value of the speed of light, and do we still believe his theory to be true?

1 A car of mass 750 kg moves along a level straight road at a constant velocity of 20 m s^{-1}. The engine produces a driving force of 3000 N.

    a Write the magnitude of the resisting force. **[1 mark]**

    The car increases the driving force to 6000 N. Assuming that the resisting force remains constant,

    b Find the acceleration of the car, **[2]**

    c Calculate the distance travelled by the car as it increases its speed from 20 m s^{-1} to 30 m s^{-1}. **[2]**

2 Two particles, P and Q, of mass 20 kg and 30 kg respectively, are connected by a light inextensible string, passing over a fixed smooth light pulley. The particles are released from rest with the string taut, and the hanging parts vertical. Find

    a The acceleration of P, **[6]**   **b** The tension in the string. **[1]**

3 A small block of mass 5 kg is released from rest at the surface of a lake of still water. The water offers a constant resisting force of 29 N.

    a Calculate the acceleration of the block. **[3]**

    After 8 seconds the block hits the bottom of the lake.

    b How fast is the block moving when it hits the bottom of the lake? **[2]**

    c How deep is the lake at that point? **[2]**

4 A car of mass 1200 kg tows a caravan of mass 800 kg along a horizontal road. The car and the caravan experience resistances of 500 N and 300 N respectively. The constant horizontal force driving the car forwards is 1500 N.

    Set up equations of motion for the car and the caravan and solve to find

    a The acceleration of the car and the caravan, **[6]**

    b The tension in the tow bar connecting the car and the caravan. **[1]**

5 The upwards motion of a lift between two floors is in three stages. Firstly the lift accelerates from rest at 2 m s^{-2} until it reaches a velocity of 6 m s^{-1}. It maintains this velocity for 5 seconds, after which it slows to rest with a deceleration of 3 m s^{-2}.

    a Draw a velocity-time graph for the motion of the lift between the two floors. **[3]**

    b Calculate the reaction force between a man of mass 100 kg and the floor of the lift during each of the three stages of the motion. **[5]**

6 The diagram shows the velocity-time graph for the motion of a lift moving up between two floors in a tall building. A parcel of mass 40 kg rests on the floor of the lift. Calculate the vertical force exerted by the floor of the lift on the parcel between

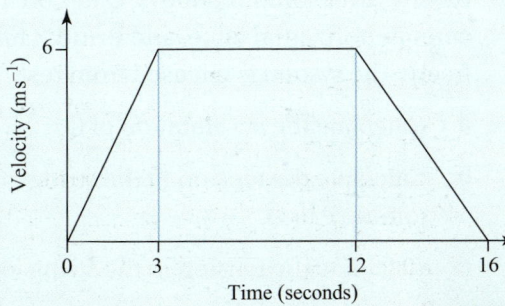

    a t = 0 and t = 3 **[3]**

    b t = 3 and t = 12 **[2]**

    c t = 12 and t = 15 **[3]**

7 A lorry of mass 1900 kg tows a trailer of mass 800 kg along a straight horizontal road. The lorry and the trailer are connected by a light horizontal tow bar. The lorry and the trailer experience resistances to motion of 700 N and 400 N respectively. The constant horizontal driving force on the lorry is 2900 N.

    a   Set up equations of motion for the lorry and the trailer.   **[4]**

    b   Use your equations to work out

        i The acceleration of the lorry and the trailer,   **ii** The tension in the tow bar.   **[3]**

When the speed of the vehicles is $12 \, \text{m s}^{-1}$ the tow bar breaks. The resistance to the motion of the trailer remains 400 N.

    c   Find the distance moved by the trailer from the moment the tow bar breaks to the moment the trailer comes to rest.   **[4]**

8 Two boxes, A and B, of masses 0.2 kg and 0.3 kg respectively are connected by a light, inextensible string that passes over a smooth pulley, P. Initially A is at rest on a rough horizontal platform, a distance 4 m from the pulley, and B hangs freely. The system is released from rest. A experiences a constant resisting force of $0.15 \, g$. In this question give your answers in terms of g

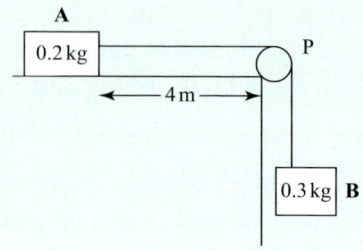

    a   Calculate the acceleration of A.   **[5]**

When A is 1 metre from the pulley, the string breaks.

    b   Calculate the velocity of A at this instant.   **[2]**

    c   Calculate the deceleration of A after the string has broken.   **[2]**

    d   Show that A is moving at a speed of $\sqrt{\dfrac{3g}{10}} \, \text{m s}^{-1}$ when it hits the pulley.   **[2]**

9 Two particles, A and B, of masses 2 kg and 3 kg respectively are attached to the ends of a light inextensible string. The string passes over a smooth fixed pulley. The system is released from rest with both masses a height of 72 cm above a horizontal table. Calculate

    a   The speed with which B hits the table,   **[8]**

    b   How long it takes for B to hit the table.   **[2]**

When B has hit the table, particle A continues upwards without hitting the pulley.

    c   Calculate the greatest height above the table reached by A.   **[3]**

10 The diagram shows three bodies, P, Q and R, connected by two light inextensible strings, passing over smooth pulleys. Q lies on a smooth horizontal table, and P and R hang freely. The system is released from rest.

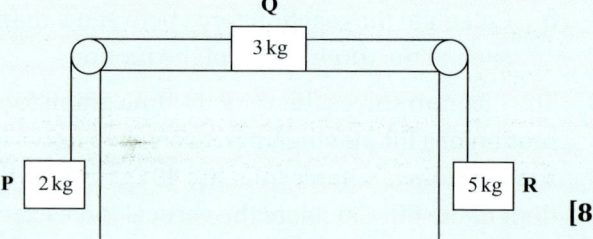

    a   Calculate the acceleration of Q.   **[8]**

    b   Calculate the tension in the string joining P to Q.   **[2]**

    c   Calculate the tension in the string joining Q to R.   **[2]**

11 A lift of mass 820 kg transports a woman of mass 80 kg. The lift is accelerating upwards at $4\,\mathrm{m\,s^{-2}}$.

   a Calculate the tension in the lift cable. [2]

   b Calculate the vertical force exerted on the woman by the floor of the lift. [2]

   Some time later the tension in the lift cable is 8640 N.

   c Calculate the acceleration of the lift. [2]

   d Calculate the vertical force exerted on the woman by the floor of the lift. [2]

12 A tug of mass 8000 kg is pulling a barge of mass 6000 kg along a canal. The tug and the barge are connected by an inextensible horizontal tow rope. The tug and the barge experience resistances to motion of 1200 N and 600 N respectively. The tug is accelerating at $0.2\,\mathrm{m\,s^{-2}}$. Find

   a The force in the tow rope, [2]

   b The tractive force of the tug. [2]

   The tow rope can operate safely up to a maximum force of 2100 N.

   c Calculate the maximum safe tractive force for the tug. [4]

13 A car of mass 1000 kg tows a trailer of mass 400 kg along a horizontal road. The engine of the car exerts a forward force of 4.9 kN. The car and the trailer experience resisting forces that are each proportional to their masses. Given that the car accelerates at $3\,\mathrm{m\,s^{-2}}$ find

   a The tension in the tow bar, [7]

   b The resisting force on the trailer. [1]

14 A body, B, of mass 20 kg, hangs below a mass, A, of 5 kg, connected by a light inextensible string. The system is lifted by a vertical force of 295 N, applied to A

   a Calculate the acceleration of A. [6]

   b Calculate the tension in the string between A and B. [2]

15 A locomotive of mass 10 tonnes pushes a carriage of mass 5 tonnes along straight, horizontal rails. The locomotive and the carriage are joined by a horizontal coupling. The locomotive and the carriage experience resisting forces of 3 kN and 2 kN respectively. They accelerate at $0.3\,\mathrm{m\,s^{-2}}$. Find

   a The force in the coupling, [2]

   b The force of the engine on the locomotive. [3]

When the locomotive and the carriage are travelling at $20\,\mathrm{m\,s^{-1}}$, the locomotive turns off its engine.

   c Calculate the new force in the coupling. [8]

   d Calculate the time until the locomotive and the carriage come to rest. [2]

1 A train travels at a constant speed of 50 km h^{-1} for 5 minutes before decelerating at a constant rate for one minute until it comes to a stop. After waiting for 2 minutes the train goes in the opposite direction, accelerating at a constant rate for 1 minute to reach a speed of 30 km h^{-1}

    a Sketch a velocity-time graph for the journey described. **[3]**

    b Find the total distance travelled. **[2]**

    c Calculate the resultant displacement of the train after 9 minutes. **[2]**

2 An object starts from rest and accelerates at a constant rate for 3 seconds until it is moving at 12 m s^{-1}

    a Calculate

      i The acceleration of the particle,   **ii** The distance taken to reach 12 m s^{-1} **[3]**

    The acceleration of the object is caused by a force of 5 N

    b Calculate the mass of the object. State the unit of your answer. **[3]**

3 A book falls from a shelf and takes 0.6 seconds to hit the ground.

    a Calculate

      i The height of the shelf to the nearest cm,

      ii The speed at which the book hits the ground. **[4]**

    b What assumptions did you make in your answers to part **a**? **[2]**

4 The displacement, s, in metres of a particle from its start point at time t seconds is given by

$$s = 2t^3 - t^2 + \frac{t}{3}$$

Calculate

    a The velocity after 3 seconds,   **[3]**   **b** The acceleration after 3 seconds.   **[3]**

5 A toy train moves back and forwards along a straight section of track.

The displacement–time graph shows the motion of the toy train.

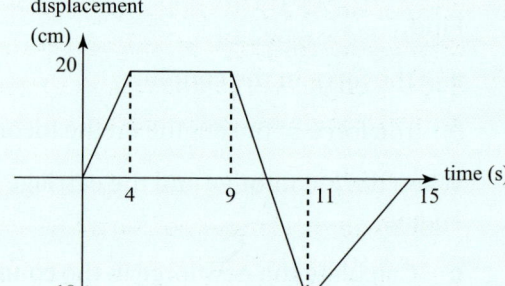

    a Calculate

      i The time spent stationary,

      ii The velocity during the first 3 seconds,

      iii The total distance travelled. **[4]**

    b Find the average speed of the train during the final 6 seconds. State the units of your answer. **[3]**

6 The displacement, s m of a particle at time t s is given by the formula $s = t^3 - 3t + 2$

    a Calculate the displacement of the particle after 5 seconds. **[2]**

    b Find an expression for the velocity of the particle after t seconds. **[2]**

    c Calculate the acceleration of the particle after 5 seconds. **[3]**

    d Sketch a velocity–time graph for this situation. **[3]**

7 A motorbike starts from rest and moves along a straight horizontal road.

It accelerates at 3 m s^{-2} until it reaches a speed of 24 m s^{-1}, then it decelerates at 2 m s^{-2} until it comes to a stop.

    a Find the length of the journey. **[4]**

    b Calculate the average speed over the whole journey. **[4]**

8 A family of total mass 180 kg are in a lift of mass 500 kg. The lift is accelerating upwards at 2.4 m s^{-2}

Use 9.81 m s^{-2} as an approximation of g and give your answers to an appropriate degree of accuracy.

    a Find the tension in the cable. **[3]**

The least massive member of the family has a mass of 22 kg

    b Calculate the reaction force between the child and the lift. **[3]**

    c Without further calculation, state how the reaction force will vary for the heavier members of the family. **[1]**

9 A box of mass 0.5 kg starts from rest and is pulled along a horizontal table by a string. The resistance force is 2 N and the tension in the string is 3.5 N

    a Calculate the distance travelled by the box in the first 2 seconds. **[4]**

After 4 seconds the string breaks.

    b Calculate the length of time until the box comes to rest. **[3]**

10 A train consists of a carriage of mass 3000 kg pulled by an engine of mass 4000 kg with a driving force of 12 000 N. The train moves along a straight, horizontal track. The carriage and the engine are joined by a horizontal light bar and their total resistances to motion are 600 N and 900 N respectively.

    a Calculate

        i The acceleration of the train,   **ii** The tension in the bar. **[4]**

When the speed of the train is 13 m s^{-1} the carriage becomes disconnected from the engine.

    b Calculate the distance travelled by the carriage until it comes to rest. **[3]**

11 This graph shows the acceleration of a particle from rest.

    a Sketch a velocity–time graph. **[4]**

    b Work out the displacement of the particle after 10 seconds. **[3]**

    c Calculate the average speed over the 10 seconds. **[4]**

12 The acceleration (in $m\,s^{-2}$) of a particle at time t s is given by the formula $a = 2t - 4$

Find the distance travelled by the particle in the first 7 seconds given that it is initially travelling at $5\,m\,s^{-1}$ **[5]**

13 A diver of mass 65 kg jumps vertically upwards with a speed of $3.5\,m\,s^{-1}$ from a board 3 m above a swimming pool. By modelling the diver as a particle moving vertically only (but not hitting the board on the way down), calculate

    a The time taken for the diver to reach the water, **[4]**

    b The speed of the diver when they reach the water. **[2]**

Give your answers to an appropriate degree of accuracy.

14 A particle has acceleration $(4t - 5)\,m\,s^{-2}$ and is initially travelling at $3\,m\,s^{-1}$

    a Find the times at which the particle changes direction. **[5]**

    b Calculate the displacement of the particle after 2 seconds. **[4]**

    c Calculate the distance travelled in the first 2 seconds. **[4]**

15 Two remote-control cars are on a smooth, horizontal surface.

One of the cars passes the other at point A whilst travelling at a constant speed of $3\,m\,s^{-1}$

Two seconds after the first car passes, the second car accelerates from rest at a rate of $1.8\,m\,s^{-2}$ until it catches up with the first car at point B

    a Calculate the time the second car has been moving when they meet at point B **[5]**

    b Calculate the distance from A to B **[2]**

16 A ball is dropped from a height of 2 m and at the same time a second ball is projected vertically from the ground at a speed of $6\,m\,s^{-1}$

When and where will the two balls collide? Give your answers to 2 decimal places. **[6]**

17 Two boxes of mass 3 kg and 2.5 kg are connected by a light inextensible string that passes over a smooth pulley. The boxes both hang 2 m above the ground before the system is released from rest.

    a Calculate

        i The initial acceleration of the boxes,   **ii** The tension in the string. **[5]**

    b Assuming it does not reach the pulley, work out the greatest height reached by the 2.5 kg box. **[4]**

Collecting, representing and interpreting data

If a drinks vendor wants to investigate whether temperature affects their sales, they can make simple recordings of mean temperature and sales over a given time period. The data can then be analysed to determine whether a correlation exists. This is an example of bivariate data — temperature and sales are two independent variables that may or may not affect each other. It is important to recognise, however, that correlation does not always imply causation. This means that two sets of data can show correlation without one affecting the other.

Data collection and analysis is the foundation of many different kinds of research. Being able to collect relevant data accurately and without bias, effectively represent it, and then interpret the results in a meaningful way is very important when undertaking investigations and testing hypotheses.

Orientation

What you need to know

KS4
- Apply statistics to describe a population.
- Construct and interpret tables, charts and diagrams for numerical data.
- Recognise appropriate measures of central tendency and spread.
- Use and interpret scatter graphs of bivariate data.
- Recognise correlation.

What you will learn

- To distinguish a population and its parameters from a sample and its statistics.
- To identify and name sampling methods and highlight sources of bias.
- Read discrete and continuous data from a variety of diagrams.
- Plot and use scatter diagrams.
- Summarise raw data.

What this leads to

Ch10 Probability and discrete random variables
Binomial distribution.

Ch11 Hypothesis testing 1
Formulating a test.
The critical region.

  **MyMaths** Practise before you start   🔍 1192, 1194, 1195, 1213, 1248

Fluency and skills

Sampling from a population can provide extremely useful information, if done effectively.

> **Key point**
>
> The **population** is the set of things you are interested in.
> A **sample** is a subset of the population.

The population may be finite, like the current top-selling pop artists, or infinite, such as the range of locations at which an archer's arrow might land.

To find parameters for a population you often need to use every piece of data in the population. This requires a **census**, which is a collection of every data point in the population. Sometimes a census is impractical or impossible, so you use a **sample** to say something about the entire population.

> **Key point**
>
> A **parameter** is a number that describes the entire population. A **statistic** is a number taken from a single sample—you can use one or more of these to estimate the parameter.

> You can use a statistic to estimate a parameter. For example, the mean of a sample is an estimate of the population mean.

	Advantage	Disadvantage
Census	• Guarantees an accurate view of the population.	• Not possible for an infinite population. • Time-consuming and expensive. • Unrealistic when it damages the thing being investigated, e.g. testing battery lifetimes.
Sample	• Can be used for a large or infinite population. • Quicker and cheaper than a census.	• Can give a misleading view of the population.

Example 1

A student wants to know the mean number of sweets in a packet of their favourite snack. They are interested in packets made within the last year.

They open 10 packets and count the number of sweets in each. They find the mean of these totals.

a   Identify the population, the parameter, the sample(s) and the statistic(s) in this example, and say which statistic can be used to estimate the parameter.

b   Explain why it is not sensible to use a census in this situation.

a   The population is all packets of the snack made within the last year. The mean number of sweets per packet is the parameter. The 10 packets of sweets opened by the student is the sample. The statistics are the number of sweets in each of these 10 packets and the sample mean. The sample mean is an estimator for the parameter.

b   It would be difficult to gather all data points as, unless the factory records this data, most packets will have been disposed of or sold to unknown destinations.

When deciding on the best way to produce a sample, it is useful to know if you would, at least in principle, be able to list every single member of the population.

The table shows some typical methods of sampling that you can use if you *are* able to list every member of the population.

Sampling method	Description
Simple random sampling	Every member of the population is equally likely to be chosen. For example, allocate each member of the population a number. Then use random numbers to choose a sample of the desired size.
Systematic sampling	Find a sample of size n from a population of size N by taking one member from the first k members of the population at random, and then selecting every k^{th} member after that, where $k = \dfrac{N}{n}$
Stratified sampling	When you know you want distinct groups to be represented in your sample, split the population into these distinct groups and then sample within each group in proportion to its size.

Often, you're not able to list every member of a population. In this case, you have to generate a sample to represent the population in the best way you can.

Sampling method	Description
Opportunity sampling	Take samples from members of the population you have access to until you have a sample of the desired size.
Quota sampling	When you know you want distinct groups to be represented in your sample, decide how many members of each group you wish to sample in advance and use opportunity sampling until you have a large enough sample for each group.

Exercise 9.1A Fluency and skills

1   A child wants to know the average height of an adult in their family. They measure the heights of all their adult relatives who live in their city. Identify the population, the parameter, the sample(s) and the statistic(s) in this situation.

2   A conservationist is interested to know the maximum height that various species of trees reach in their country's forests. They choose a random forest and measure the heights of all the trees there. Identify the population, the parameter, the sample(s) and the statistic(s) in this situation.

3   You wish to find out the favourite bands of students in a school. There are 1000 students in the school.

    a   Describe how to take a census.

    b   Describe how to take a sample of size 40 using simple random sampling.

    c   Give one argument for using a sample and one argument for using a census.

4   Systematic sampling is used to find a sample from a population of 3783 people. A random number between 1 and 13 is generated, then that numbered member of the population and every 13th member thereafter is chosen for the sample. How large is the sample?

5 A school wishes to know how popular its new after-school club is with the 1000 students in the school. The school has three year-groups of 150 students each and two year-groups of 275 students each. It comes up with three methods of sampling a group of 40 students to get an idea.

State the name of each sampling method.

a 40 students are chosen at random from the list of 1000 students using random numbers between 1 and 1000

b The school lists the students by year and by class and gives each student a number. The school randomly chooses one student from the first 25 listed and then every 25th student from the list, until they've picked 40 students.

c The school lists the students by year and by class and gives each student a number. Six students are chosen at random from each of the three smaller year groups and 11 students are chosen at random from each of the two larger year groups.

6 A manufacturer of lightbulbs produces 2772 coloured lightbulbs in a day, amongst which are 1001 red bulbs, 1309 blue bulbs and 462 green bulbs. 36 bulbs are chosen to test their lifetimes.

a Calculate how many of each colour bulb are chosen if a stratified sample is taken.

b Explain why it would not be appropriate to use a census in this situation.

Reasoning and problem-solving

Strategy

When deciding on a sampling method

1. Consider whether or not you can list every member of a population.
2. Identify any sources of bias and any difficulties you might face in taking certain samples.
3. Compare the different sampling methods you have available and choose the one that best suits your needs and limitations.

Taking a sample that accurately reflects the population is not a simple job. It is all too easy to bias your sample or get results that may not accurately reflect the population.

Key point

A sampling method is **biased** if it creates a sample that does not represent the population.

When deciding on a sampling method, you should aim to produce as unbiased a sample as possible, but you may need to factor in the difficulty and cost of any sampling method chosen.

Example 2

A researcher looks at whether rainfall in eastern China is greater in 2015 than 1987. She gathers 100 measurements in Beijing in 2015 and compares them with measurements for Beijing for the same times of year in 1987.

a Suggest why any conclusions she draws might not be valid.

b Suggest and explain a sampling method that she could use to get a better representation of rainfall in eastern China.

a She has only looked at one location in eastern China. Other locations might have higher results or lower results.

> Think about the possible sources of bias.

b If she has appropriate data from 1987 she could instead use quota sampling, in which she selects several diverse locations in eastern China and takes a predetermined number of measurements from each of them.

> Think about which sampling method best suits this situation.

Exercise 9.1B Reasoning and problem–solving

1 A meteorologist wants to know the average temperature in their town. The values of temperature, taken hourly over a period of three months, give an average result of 14 °C, but the values taken hourly over a period of three years give a result of 21 °C.

a How reliable are the two methods and what might you conclude about the actual average temperature?

The meteorologist takes monthly average values and orders them by time, from first to last. They then use systematic sampling to generate a random number between 1 and 12 for the first monthly value used for a sample, and every 12th monthly value is used after that.

b Say why this wouldn't give a fair representation of the average temperature.

2 A teacher is organising a conference in their city but is not sure where best to hold it so that everyone can attend. There are 293 students spread across 12 schools. The teacher wants to obtain a sample of 36 students to take into account their views. The teacher comes up with two ideas:

Idea 1. The teacher could list all 293 students and number them from 000 to 292. The teacher randomly generates 36 different 3-digit numbers in this range and the corresponding students are included in the sample.

Idea 2. Two schools are selected at random and 18 students from each school are selected at random. If there aren't enough

students then the teacher randomly chooses another school and draws random students until enough are chosen.

a For each idea, decide if every possible sample is equally likely to be chosen.

b State, with reason, which idea is best from a statistical point of view.

c Suggest another method the teacher could use based on stratified sampling.

Challenge

3 Weather data is taken from five different locations in Britain between May and October in both 1987 and 2015

Large data set

a Why might five different locations be chosen and what is the name of this sampling technique?

b Give one reason why the same locations and one why the same months are chosen for both years.

4 The Met Office collects data from approximately 300 weather stations around Britain, an average of 40 km apart.

Large data set

a Why might it be important to have lots of stations and why should they not be too close together?

A website has been set up which accepts data submission from anyone with their own weather station.

b Give one reason to accept any submissions and one reason they need to be cautious about this data.

Central tendency and spread

Fluency and skills

> **Key point**
>
> **Discrete data** can take any one of a finite set of categories (non-numeric) or values (numeric), but nothing in between those values. Often, the values are different categories.
>
> **Continuous data** is always numeric, and it can take any value between two points on a number line.

Statistical investigations generate large quantities of **raw data**. It is useful to reduce this data to some key values, called **summary statistics**. These can be categorised as **measures of central tendency** (also known as 'averages') and **measures of dispersion** (also known as 'spread').

> **Key point**
>
> There are three measures of central tendency: **mode**, **median** and **mean**.

The mode of a set of data is the value or category that occurs most often or has the largest frequency. For grouped data, the **modal interval** or **modal group** is normally given.

See p.614
For a list of mathematical notation.

> **Key point**
>
> To work out the **mean** $\bar{x}$ of a set of n observations, calculate their sum and divide the result by n
>
> $$\bar{x} = \frac{\sum x}{n}$$

The symbol 'Σ' means 'the sum of' whatever follows it. The bar on top of the x indicates the mean of the x-values.

> **Key point**
>
> The mean of a set of data given in the form of a frequency distribution is given by
>
> $$\bar{x} = \frac{\sum fx}{\sum f}$$

f denotes the frequency, so $\sum f$ is the number of observations and $\sum fx$ is the sum of the x-values.

For a grouped frequency distribution, you can only calculate an estimate for the mean rather than the exact value. In this case, x is the middle value of each group.

The **median** of a set of data is the middle value of data listed in order of size.

To calculate the position of the median of a set of n observations, work out the value of $\frac{n+1}{2}$

If the value of $\frac{n+1}{2}$ is a whole number, then the median is the value in that position. If the answer is not a whole number, then the median is the mean of the two values in the positions on either side of $\frac{n+1}{2}$

Calculator

Try it on your calculator

You can calculate the mode, mean ($\bar{x}$) and median for a set of data using your calculator.

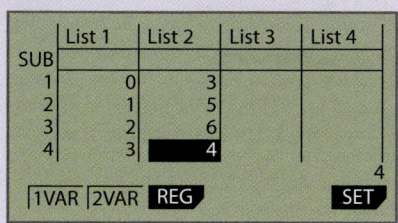

```
1–Variable
x̄     =1.61111111
Σx    =29
Σx²   =65
σx    =1.0076865
sx    =1.03690086
n     =18          ↓
```

```
1–Variable
Q1    =1            ↑
Med   =2
Q3    =2
maxX  =3
Mod   =2
Mod:n =1            ↓
```

Activity

Find out how to calculate the mean, median and mode of the data set

x	0	1	2	3
f	3	5	6	4

on *your* calculator.

Example 1

a Work out the mode and median of this data set.

b Show that the mean of this data set is 1.79

x	0	1	2	3	4
f	2	6	6	4	1

a Modes = 1 and 2 ●————————

   Median = 2 ●————————

b Mean = $\dfrac{0 \times 2 + 1 \times 6 + 2 \times 6 + 3 \times 4 + 4 \times 1}{2 + 6 + 6 + 4 + 1} = 1.79$ ●

> Two values have frequency 6 so this data is bi-modal.

> $\sum f = 19$ so the median is the $\dfrac{19+1}{2} = $ 10th value.

> Use mean $= \dfrac{\sum fx}{\sum f}$

Example 2

Write the modal interval, the interval containing the median and an estimate of the mean for this data set.

x	$0 \leq x < 4$	$4 \leq x < 8$	$8 \leq x < 12$	$12 \leq x < 16$	$16 \leq x < 20$
f	2	3	7	10	1

Modal interval: $12 \leq x < 16$ ●————————

Median lies in $8 \leq x < 12$

Midpoints: 2, 6, 10, 14, 18 ●————————

Estimate for mean = 10.9 ●————————

> $\sum f = 23$ so the median is the $\dfrac{23+1}{2} = $ 12th value.

> Use the midpoint of each interval to calculate your estimate for the mean.

> Use your calculator to get the mean. Check the value is sensible.

 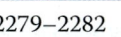 

In order to summarise data, you can also use measures of **spread** or **dispersion**.

> **Key point**
> You should be familiar with four measures of dispersion: range, interquartile range, variance and standard deviation.

> **Key point**
> The **range** of a set of data is the largest value minus the smallest value.

The median, i.e. the middle value of ordered data, is sometimes called the **second quartile**. The **first quartile** (or **lower quartile**) is the middle value between the lowest value and the median, and the **third quartile** (or **upper quartile**) is the middle value between the median and the largest value.

> **Quartiles** divide the ordered data into four groups, with equal numbers of observations in each.

> **Key point**
> To evaluate the first and third quartiles for a set of n ungrouped observations, work out the values of $\frac{n+1}{4}$ and $\frac{3(n+1)}{4}$ respectively.

If the values of $\frac{n+1}{4}$ or $\frac{3(n+1)}{4}$ is a whole number, then the quartile is the value in that position.

If the answer is not a whole number, then the quartile is the mean of the two values in the positions on either side of $\frac{n+1}{4}$ or $\frac{3(n+1)}{4}$

> **Key point**
> If the first and third quartiles of a set of observations are denoted by Q_1 and Q_3, the interquartile range is $IQR = Q_3 - Q_1$

Example 3

Find the interquartile range for this set of data:

1, 6, 9, 23, 4, 19, 2, 7, 13, 24, 2, 7, 14, 19, 14, 20

1, 2, 2, 4, 6, 7, 7, 9, 13, 14, 14, 19, 19, 20, 23, 24 ● —— Write the data in ascending order.

$n = 16, \dfrac{n+1}{4} = 4.25$

So $Q_1 = \dfrac{4+6}{2} = 5$ ● —— Calculate Q_1

$\dfrac{3(n+1)}{4} = 12.75$

So $Q_3 = \dfrac{19+19}{2} = 19$ ● —— You can check the values of Q_1 and Q_3 on your calculator.

$IQR = Q_3 - Q_1 = 19 - 5 = 14$

As well as the quartiles and interquartile range, you can also work out percentiles and the interpercentile range of data.

Example 4

The masses, in grams, of 80 apples are shown in the table.

Calculate the 30% to 80% interpercentile range.

Mass, m	frequency
$60 \leq m < 70$	7
$70 \leq m < 80$	26
$80 \leq m < 90$	29
$90 \leq m < 100$	18

30th percentile $= \dfrac{30}{100} \times 80 = 24$th apple

80th percentile $= \dfrac{80}{100} \times 80 = 64$th apple

Imagine the apples arranged in order of mass and work out which is at the 30th and 80th percentile.

Mass, m	frequency	cumulative frequency
$60 \leq m < 70$	7	7
$70 \leq m < 80$	26	33
$80 \leq m < 90$	29	62
$90 \leq m < 100$	18	80

Work out the cumulative frequencies.

The 24th apple is 17 into the group $70 \leq m < 80$

Imagine the apples are arranged in order of mass.

The 7th apple has mass 70 g and 33rd has mass 80 g

Mass of 24th apple $= 70 + 17 \times \dfrac{80 - 70}{26}$

$= 78.9$ g (to 3 sf)

$\dfrac{80 - 70}{26} = 0.385$ is the difference in the mass between apples in the $70 \leq m < 80$ group. This process is called linear interpolation. It assumes the masses of apples are evenly distributed within each group.

The 64th apple in in the group $90 \leq m < 100$

Mass of 64th apple $= 90 + 2 \times \dfrac{100 - 90}{18} = 91.7$ g (to 3 sf)

So the 30% to 80% interpercentile range $= 91.7 - 78.9 = 12.8$ g

The **variance** of a set of data measures how spread-out the values are from the mean. To find the variance of a population you calculate the difference between each data value (x) and the mean ($\bar{x}$). You then find the mean of the squares of these values. This is the mean of $(x - \bar{x})^2$, and is called the variance, σ^2. The square root of the variance, σ, is the **standard deviation**.

The values are squared to remove minuses.

Key point

The variance of n observations with mean $\bar{x}$ is defined as $\sigma^2 = \dfrac{\sum(x - \bar{x})^2}{n} = \dfrac{\sum x^2}{n} - \bar{x}^2$

The standard deviation of n values with mean $\bar{x}$ is defined as $\sigma = \sqrt{\dfrac{\sum(x - \bar{x})^2}{n}} = \sqrt{\dfrac{\sum x^2}{n} - \bar{x}^2}$

When a population is large it may be more practical to use a sample from the population to estimate the variance of the whole population. An **unbiased estimate of the population variance** using a sample of n observations with sample mean $\bar{x}$ is given by the **sample variance**, s^2

$$s^2 = \dfrac{\sum(x - \bar{x})^2}{n-1} = \dfrac{\sum x^2}{n-1} - \dfrac{\left(\sum x\right)^2}{n(n-1)}$$

The divisor in the sample variance is $n - 1$

The sample standard deviation of a set of n values is defined as

$$s = \sqrt{\dfrac{\sum(x - \bar{x})^2}{n-1}} = \sqrt{\dfrac{\sum x^2}{n-1} - \dfrac{\left(\sum x\right)^2}{n(n-1)}}$$

  2279–2282

In this course you can use σ^2 and σ. You may see s on your calculator so you should know what it means.

You may also see the variance defined as $\dfrac{S_{xx}}{n}$ and the standard deviation defined as $\sqrt{\dfrac{S_{xx}}{n}}$ where $S_{xx} = \sum (x - \bar{x})^2$

ICT Resource online

To investigate standard deviation and mean, click this link in the digital book.

Large data set

Calculator

Try it on your calculator

You can calculate the standard deviation of a population, or estimate the standard deviation using a sample, on your calculator.

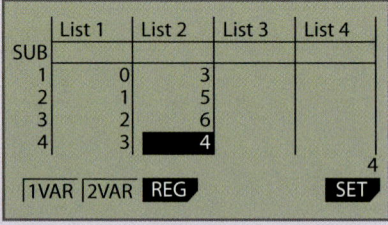

```
        List 1   List 2   List 3   List 4
SUB
  1         0        3
  2         1        5
  3         2        6
  4         3        4
                                        4
 1VAR  2VAR  REG              SET
```

```
   1-Variable
   x̄      =1.61111111
   Σx     =29
   Σx²    =65
   σx     =1.0076865
   sx     =1.03690086
   n      =18              ↓
```

Activity

Find out how to calculate the standard deviation of the data set

x	0	1	2	3
f	3	5	6	4

on *your* calculator.

Example 5

Calculate the standard deviation, σ, of each of these sets of data.

a $n = 49$, $\sum (x - \bar{x})^2 = 68.4$

b $n = 1000$, $\sum x = 47\,000$, $\sum x^2 = 2\,900\,000$

c 6, 7, 3, 4, 4, 5, 6, 6, 8, 9, 4, 5, 6, 7, 8

d

x	$7 \leq x < 8$	$8 \leq x < 9$	$9 \leq x < 10$	$10 \leq x < 11$	$11 \leq x < 12$
f	3	10	11	7	3

a $\sigma = \sqrt{\dfrac{68.4}{49}} = 1.18$

Use the formula with brackets.

b $\sigma = \sqrt{\dfrac{2\,900\,000}{1000} - \left(\dfrac{47\,000}{1000}\right)^2} = 26.3$

Use Σx and n to find the mean, then calculate σ

c $\sigma = 1.67$

Find σ using your calculator.

d Midpoints: 7.5, 8.5, 9.5, 10.5, 11.5

$\sigma = 1.09$

The data is grouped so use midpoints and find σ using your calculator.

Data can be coded before the mean and variance are calculated. For example, masses can be changed from x kilograms to y grams by multiplying each value by 1000, so $y = 1000x$

If you are given the mean and variance of the coded data, you have to 'decode' them to find the mean and variance of the original data.

Coding affects the mean and variance in different ways:

If $y = ax + b$ then $\bar{y} = a\bar{x} + b$ and $\sigma_y^2 = a^2\sigma_x^2 \Rightarrow \sigma_y = a\sigma_x$

The mean and variance of coded y-values are $\bar{y} = 54.6$ and $\sigma_y^2 = 1.04$

Calculate the mean and variance of the original x-values given that $y = 4x - 60$

$\bar{y} = a\bar{x} + b$	$\sigma_y^2 = a^2\sigma_x^2$ so $\sigma_y^2 = 4^2\sigma_x^2$	Substitute values into the coding equation.
$54.6 = 4\bar{x} - 60$	$1.04 = 16\sigma_x$	
$\bar{x} = \dfrac{54.6 + 60}{4} = 28.65$	$\sigma_x = 0.065$	Rearrange and solve for the original mean and standard deviation.

STATS

Exercise 9.2A Fluency and skills

1 For these sets of data, give the

   i Mode(s), **ii** Mean, **iii** Median, **iv** Range, **v** Standard deviation.

   a 6, 8, 9, 2, 5, 6, 10, 8, 5, 7, 4, 8, 11

   b 68, 71, 72, 75, 68, 65, 69, 70, 71, 68, 62, 64, 71

2 For this data, calculate

   a The modal interval, **b** An estimate for the mean.

x	4–9	10–15	16–21	22–27	28–33
f	1	3	7	4	2

3 Work out the 1st and 3rd quartiles and the interquartile range for these observations.

   a 12, 14, 15, 19, 21, 25, 26, 29, 32, 36

   b 1.2, 1.7, 1.9, 2.3, 2.5, 2.9, 3.1, 3.6, 3.7, 3.9, 4.0, 4.1

4 Work out the mean of this data set.

x	4	5	6	7	8
f	1	5	7	3	2

5 50 milk samples, each of mass 100 g, were analysed as part of a food quality inspection and the potassium content, in mg, was recorded. The results are shown in the table.

Potassium (x mg)	Frequency
$140 \leq x < 144$	5
$144 \leq x < 148$	12
$148 \leq x < 152$	16
$152 \leq x < 156$	11
$156 \leq x < 160$	6

    a Estimate the mean of this data.

    b Calculate the 40% to 70% interpercentile range.

6 Work out the mean and variance of this set of x-values: 7.1, 6.4, 8.5, 7.5, 7.8, 7.2, 6.9, 8.1, 8.0, 6.6

7 For these sets of data, work out the standard deviation.

    a $\sum(x - \bar{x})^2 = 141.4$, $n = 10$ (values taken from the entire population)

    b $n = 9$, $\sum x = 295$, $\sum x^2 = 9779$ (values taken from a sample of the population)

8 The heights, in centimetres, of 40 seedlings are given in the table.

    Use the data from this sample to find the standard deviation. Give your answer to 2 dp.

x	$16 \leq x < 18$	$18 \leq x < 20$	$20 \leq x < 22$	$22 \leq x < 24$	$24 \leq x < 26$
f	8	13	11	6	2

9 A data set has been coded using $y = 5x - 3$. The mean of coded y-values is is 76.8 and the standard deviation is 5.324

    Find the mean and standard deviation of the original x-values.

Reasoning and problem-solving

Strategy

To solve a problem about summary statistics

(1) Identify the summary statistics appropriate to the problem.

(2) Calculate values of the required statistics, using a calculator where appropriate.

(3) Use the statistics to describe key features of the data set and make comparisons.

(4) If not already done, identify any outliers and remove them, then see how this affects the calculations.

Key point

Outliers are values that lie significantly outside the typical set of values of the variable.

An outlier can be defined in several ways. One rule says that an outlier can be considered as any value that lies outside the interval $(Q_1 - 1.5 \times IQR, Q_3 + 1.5 \times IQR)$

Outliers may indicate natural variation in the data set or may be the result of errors in measuring or recording data. If an outlier is due to an error, such as reading a value incorrectly or measuring an item from the wrong population, it should be removed. This is one way in which you can clean data.

Example 7

In this question, define an outlier as a value more than three standard deviations above or below the mean.

A group of 15 students complete a timed test for their homework. Their times (in minutes) are recorded:

32, 34, 33, 37, 39, 39, 42, 45, 41, 40, 40, 44, 13, 36, 36

a Calculate the mode, median and mean of the data.

b Show that there is exactly one outlier in the data.

c Give one possible reason for

    i Removing the outlier,   **ii** *Not* removing the outlier.

d A teacher investigates the outlier and decides to remove it. Without further calculation, explain how removing this value would affect your answers to part **a**.

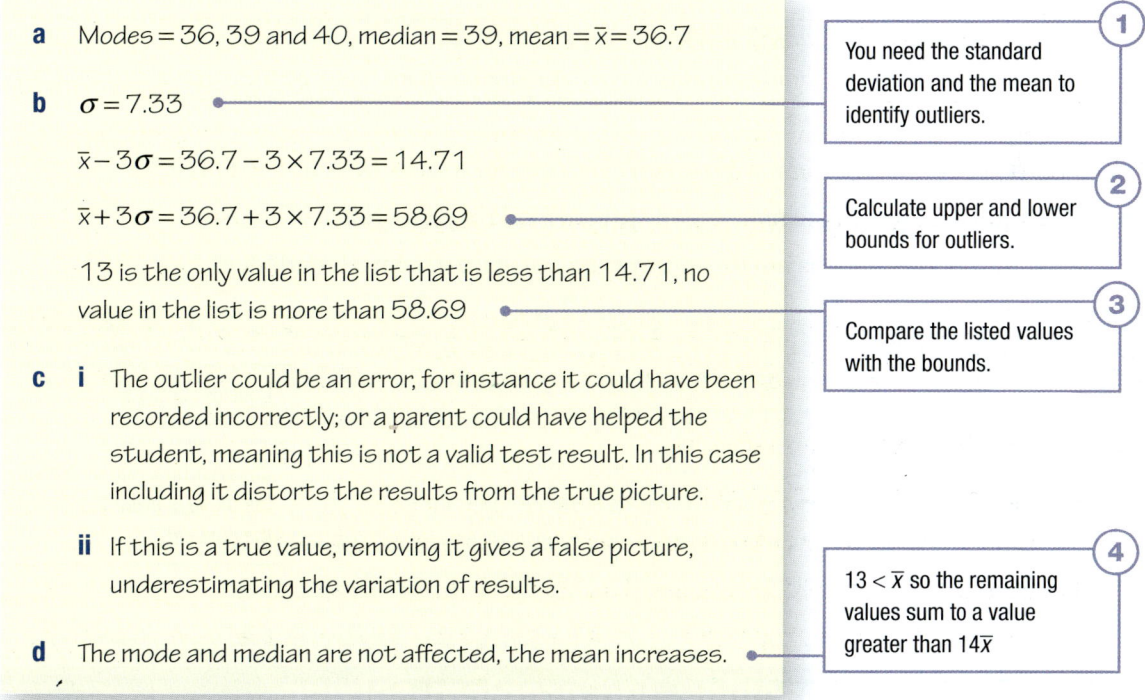

a Modes = 36, 39 and 40, median = 39, mean = $\bar{x}$ = 36.7

b $\sigma = 7.33$

    $\bar{x} - 3\sigma = 36.7 - 3 \times 7.33 = 14.71$

    $\bar{x} + 3\sigma = 36.7 + 3 \times 7.33 = 58.69$

    13 is the only value in the list that is less than 14.71, no value in the list is more than 58.69

c **i** The outlier could be an error, for instance it could have been recorded incorrectly; or a parent could have helped the student, meaning this is not a valid test result. In this case including it distorts the results from the true picture.

    ii If this is a true value, removing it gives a false picture, underestimating the variation of results.

d The mode and median are not affected, the mean increases.

1 You need the standard deviation and the mean to identify outliers.

2 Calculate upper and lower bounds for outliers.

3 Compare the listed values with the bounds.

4 $13 < \bar{x}$ so the remaining values sum to a value greater than $14\bar{x}$

You should choose appropriate statistics for central tendency and spread. You can determine which statistics to use by considering the properties of the data.

Statistic		Pros	Cons
Measure of central tendency	Mode	Useful for non-numerical data. Not usually affected by outliers. Not usually affected by errors or omissions. Is always an observed data point.	Doesn't make use of all the data. May not be representative if it has a low frequency. May not be representative if other values have a similar frequency.
	Median	Not affected by outliers. Not significantly affected by errors in the data.	Doesn't make use of all the data.
	Mean	When the data set is very large a few extreme values have negligible impact.	When the data set is small a few extreme values or errors have a big impact.
Measure of spread	Range	Reflects the full data set.	Distorted by outliers.
	Interquartile range	Not distorted by outliers.	Does not reflect the full data set.
	Standard deviation	When the data set is very large a few outliers have negligible impact.	When the data set is small a few outliers have a big impact.

Example 8

A village records its daily mean windspeed every day for a week (kn): 3, 4, 6, 6, 1, 0, 25

a Calculate an appropriate measure of spread. Explain your choice of statistic.

b Explain why the mode is not an appropriate measure of central tendency.

a $IQR = 6 - 1 = 5$

The range and standard deviation are both distorted by the outlier 25 but the interquartile range is not significantly affected. •——— Consider outliers in the data.

b This data set is small with only two instances of the mode, and thus is not representative of the data. There is enough data to calculate the median, which is more representative of the data. •——— Consider how representative the mode is given the size of the data set.

Exercise 9.2B Reasoning and problem-solving

1 The lifetimes of a batch of 100 batteries are measured and have the following distribution, where x is the lifetime measured to the nearest half-hour.

x	15.5	16.0	16.5	17.0	17.5	18.0	18.5	19.0	19.5
f	13	21	28	22	15	0	0	0	1

a Work out the median lifetime, the range and the interquartile range.

b Explain why the range is *not* an appropriate measure of spread.

c Outliers are defined for this data as being outside the interval $(Q_1 - 2 \times IQR, Q_3 + 2 \times IQR)$. Calculate the limits of this interval.

d To test the quality of another batch of batteries, one is chosen at random and its lifetime is measured. If its lifetime is outside the interval found in part **b**, the whole batch is tested. Otherwise the batch is accepted. If the lifetime of the selected battery was 14.5 hours, what decision should be taken about a test of the whole batch?

2 The number of emails received by a teacher per day was recorded over a period of 16 days:
13, 15, 19, 17, 14, 33, 19, 9, 10, 17, 18, 14, 20, 18, 15, 10

   a Work out the mean, $\bar{x}$, and standard deviation, σ, for this data.

   b An outlier is defined as any observation less than $\bar{x} - 3\sigma$ or more than $\bar{x} + 3\sigma$.
   Should any of the observations be classified as outliers? Show your working.

3 The daily total number of hours of sunshine for Heathrow for the first 14 days of May 2015 are as follows: 4.4   0.7   3.3   6.9   4.7   5.4   5.5   0.1   5.7   7.5   5.6   9.3   14   0

Large data set

   a Find the mean and standard deviation of these values.

   b The mean for the first 15 days of May was 5.25. Using your answer to part **a** or otherwise, find the standard deviation for these days.

4 The maximum daily wind gust in knots over three weeks in July in Leeming 1987 is shown in the following table:

Large data set

Speed	$10 \leq v < 15$	$15 \leq v < 18$	$18 \leq v < 21$	$21 \leq v < 24$	$24 \leq v < 29$
Frequency	4	2	8	2	5

   a Estimate the mean and standard deviation for these data.

   b The corresponding mean and standard deviation for Leeming for the same days in 2015 are 21.57 and 5.46 Write a brief statement comparing maximum gust speeds for the two years.

5 The mean daily windspeed in knots for 8 consecutive days in Leuchars were recorded as

   8   10   6   9   18   36   20   15

   a Find the median and interquartile range (IQR) for these data.

   b Using the rule that outliers are values less than $Q_1 - 1.5 \times IQR$ or more than $Q_3 + 1.5 \times IQR$, where Q_1 is the lower quartile and Q_3 is the upper quartile, identify any outliers within the data.

   c State how the median will change after any outliers have been removed.

Challenge

6 11 members of a golf team record their scores on an 18-hole course:

108, 110, 114, 101, 99, 98, 107, 103, 109, 145, 105

   a Calculate the mode, median and mean of the scores.

   b Explain which of the measures you calculated in part **a** is the most representative measure of central tendency in this case.

A competing team has the same mean score. The team has seven members and their scores are: 110, 112, 115, 108, 111, a, 105

   c Find the value of a and compare the scores of the two teams. Justify your choice of statistics.

STATS

Fluency and skills

Data is often summarised by five main statistics.

> **Key point**
>
> The **five-number summary** gives the **minimum value, lower quartile, median, upper quartile and maximum** values.

You can use a **box-and-whisker plot** to display these values. The values are marked along a linear scale and the points are joined to form a central box and two whiskers. One quarter of the data values in the sample lie between each consecutive pair of vertical lines on the diagram. Lines placed further apart, i.e. longer whiskers and box, show a greater spread of the data but do not show more data values.

You display outliers on a box plot as crosses (×), and they are not included in whiskers.

If you have sufficient information you should use the most extreme value that is not an outlier as the end of the whisker. Otherwise, use the boundary for outliers as the end of the whisker.

Suppose you define an outlier as a value less than $Q_1 - 1.5 \times IQR$ or more than $Q_3 + 1.5 \times IQR$. This box plot represents the set of data 20, 22, 23, 25, 25, 26, 27, 27, 27, 28, 39

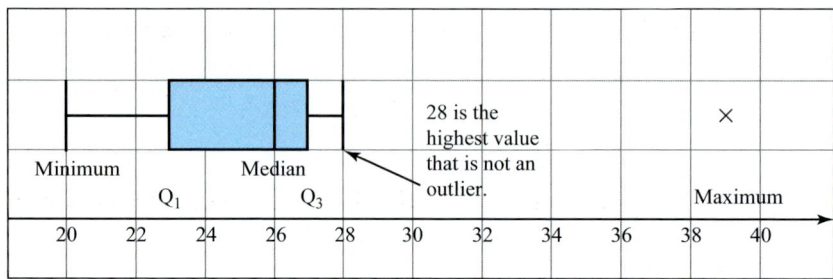

This box plot represents the set of data summarised by minimum = 43, $Q_1 = 46$, median = 49, $Q_3 = 50$, maximum = 65

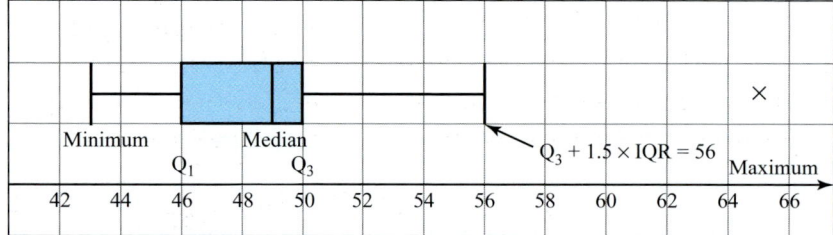

Note that, if you don't know the values of the quartiles or interquartile range, you should also be able to identify outliers by direct observation. In the diagrams above, it should be clear even without calculation that the maximum value is set apart from the rest.

> **Key point**
>
> Box-and-whisker plots are useful for comparing sets of data.

Example 1

A building company works with two plumbers. Over a period of time, they assess how long it takes each plumber to fix leaking pipes. This data is displayed in box-and-whisker plots.

An outlier is defined as a value less than $Q_1 - 1.5 \times IQR$ or more than $Q_3 + 1.5 \times IQR$.

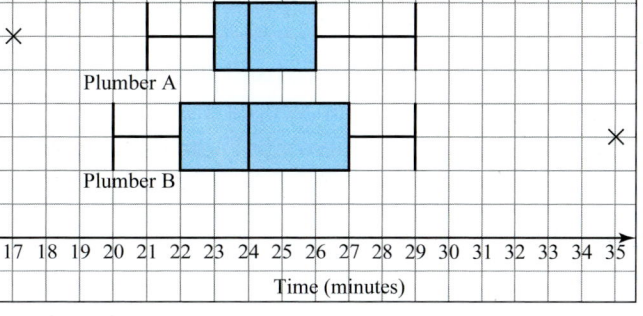

a Write down the minimum, lower quartile, median, upper quartile and maximum for each set of data.

b Recommend a choice of plumber given that no outliers are deleted.

a Plumber A's times: minimum = 17 minutes, Q_1 = 23 minutes, median = 24 minutes, Q_3 = 26 minutes, maximum = 29 minutes

Plumber B's times: minimum = 20 minutes, Q_1 = 22 minutes, median = 24 minutes, Q_3 = 27 minutes, maximum = 35 minutes

b It would be most sensible to choose Plumber A. Although both plumbers have a median time of 24 minutes. Plumber A's data shows less variation – it has a smaller IQR of 3 compared with 5 for Plumber B's data: plumber A's data has one outlier representing a quick time and Plumber B's data has one outlier representing a slow time.

> Use a measure of central tendency and a measure of spread.

Sometimes your interest is in the way in which the data is distributed rather than the individual values.

> **Key point**
> The **distribution** of data is how often each outcome occurs. Each outcome occurs with a given **frequency**.

When representing grouped data, the groups must be consecutive, non-overlapping ranges and do not have to be equal in width.

Distributions for continuous variables can be complicated because data values may be rounded measurements of true values. You may need to apply a continuity correction to ensure that the intervals meet but don't overlap.

> Two people who give their height as 1.50 m are unlikely to be exactly the same height: they are usually rounding up or down.

> **Key point**
> A continuity correction involves altering the endpoints of an interval of rounded data to include values which would fall in the interval when rounded.

> **Key point**
> You can use frequency polygons to display discrete or continuous data.

ICT Resource online
Large data set

To investigate single variable data using graphs, click this link in the digital book.

Points showing the frequencies are plotted and joined with straight lines. If the data is grouped, the frequency is plotted at the mid-point of each group.

Example 2

A scientist measures the lengths of 40 runner beans in a trial. She groups the results and records the frequencies, as shown in the table. Draw a frequency polygon for the given data.

Length (l cm)	Frequency
$10 < l \leq 15$	4
$15 < l \leq 20$	14
$20 < l \leq 25$	12
$25 < l \leq 30$	7
$30 < l \leq 35$	3

The mid-points of the groups are 12.5, 17.5, 22.5, 27.5 and 32.5

Plot the frequencies at the mid-points and join the points with straight lines.

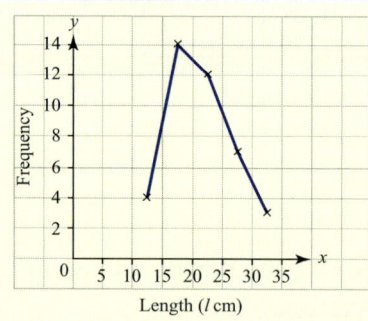

Length (l cm)

Key point

You can use a **histogram** to display continuous data.

A histogram consists of rectangles whose areas are proportional to the frequencies of the groups. The width of each rectangle is the size of the interval. A histogram often displays **frequency density** on its vertical axis.

$$\text{Frequency density} = \frac{\text{Frequency}}{\text{Class width}}$$

When working with grouped data you can make estimates by assuming that data points are distributed equally throughout a group.

A histogram looks similar to a bar chart, but a bar chart displays discrete variables and doesn't use interval width or frequency density.

Example 3

24 students in a gym class balance on one leg for as long as they can, and their times are recorded to the nearest second.

a i Find the missing frequency.

ii Give the width and height of the missing bar on the histogram.

b Estimate the number of students who could stand on one leg for

i Less than 5.5 seconds,

ii More than 10 seconds.

Time (seconds)	Continuity correction	Frequency
0–4	$0 \leq t < 4.5$	
5–7	$4.5 \leq t < 7.5$	3
8–11	$7.5 \leq t < 11.5$	6
12–16	$11.5 \leq t < 16.5$	6
≥ 17	$16.5 \leq t$	0

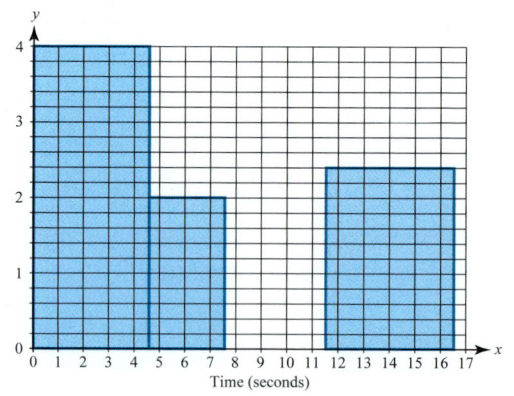

Time (seconds)

Collecting, representing and interpreting data Single-variable data

a **i** When area $= 2(7.5 - 4.5) = 6$, frequency $= 3$

Find the constant of proportionality using a known value.

So area $= 4.5 \times 4 = 18$ means frequency $= 9$

In this diagram area $= 2 \times$ frequency

ii Frequency $= 6$ so area $= 12$

Use the constant of proportionality, 2

Width $= 11.5 - 7.5 = 4$

Use the known area and width of the missing bar to find the height.

Height $= 12 \div 4 = 3$

b **i** $\dfrac{5.5 - 4.5}{7.5 - 4.5} = \dfrac{1}{3}$

Assume the data in the second bar is equally distributed throughout the interval. Find the fraction of the second bar that lies to the left of 5.5

$\dfrac{1}{3} \times 3 = 1$

$1 + 9 = 10$ people

Multiply by the frequency of the group.

ii $\dfrac{11.5 - 10}{11.5 - 7.5} \times 6 = 2.25$

Add the frequency of the first group to $\dfrac{1}{3}$ of the frequency of the second group.

$2.25 + 6 = 8.25$

8 people

Round your answer to a sensible degree of accuracy.

Find the fraction of the third bar that lies to the right of 10 and multiply by frequency.

Another way to display continuous data is a **cumulative frequency diagram** (or graph, or curve). You can use a cumulative frequency curve to estimate values, including the five-number summary.

A cumulative frequency diagram consists of points whose x-coordinates are the upper boundary of each interval, and whose y-coordinates are the sums of the frequencies up to those points. Or, in other words, the y-coordinates are the cumulative frequencies.

You join the points by a smooth curve, which is always increasing from zero to the size of the sample. Drawing a dotted line from the y-axis to the curve at the appropriate point allows you to estimate the value of any required quartile.

For data given in intervals, you can use $\dfrac{n}{2}, \dfrac{n}{4}$ and $\dfrac{3n}{4}$ for the median, lower quartile and upper quartile. There is no need to use $n + 1$ in each case.

Example 4

Work out the median, the 1ˢᵗ and 3ʳᵈ quartiles, and the interquartile range for this data set.
A cumulative frequency curve of the data is shown.

x	$0 \le x < 4$	$4 \le x < 8$	$8 \le x < 12$	$12 \le x < 16$
f	14	21	35	10

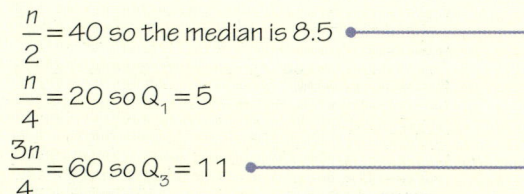

$\dfrac{n}{2} = 40$ so the median is 8.5

$\dfrac{n}{4} = 20$ so $Q_1 = 5$

To find the median, draw a dotted line from $\dfrac{n}{2}$ on the y-axis, and trace this down to read the x-value when it hits the curve.

$\dfrac{3n}{4} = 60$ so $Q_3 = 11$

$\text{IQR} = 11 - 5 = 6$

Repeat this method to find the 1st and 3rd quartiles, using $\dfrac{n}{4}$ and $\dfrac{3n}{4}$

1 A teacher creates box-and-whisker plots to compare the percentage marks gained by two classes in a test. Compare their scores.

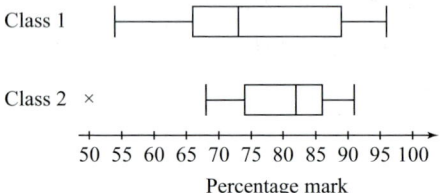

2 An estate agent collects data on the houses sold in various price brackets. 100 houses sold for more than £500 000 each.

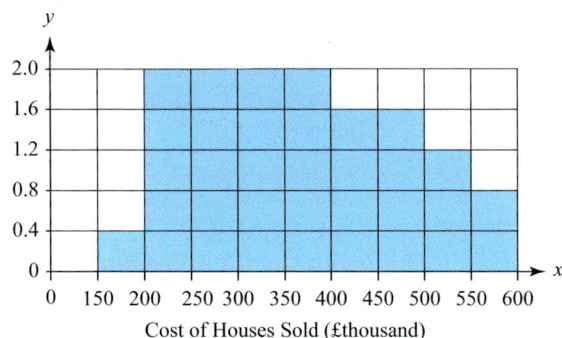

a How many houses sold for less than £300 000?

b Estimate the number of houses that sold for between £360 000 and £480 000

c Draw a frequency curve for the data.

3 The IQ scores of 30 people are measured and a cumulative frequency graph is plotted. Use the graph to estimate the upper quartile and lower quartile.

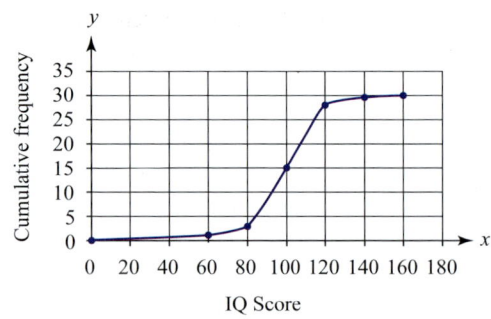

4 A cafe owner is interested in how much his staff are receiving in tips. They gather the data given in the table.

Tips	Frequency
£0.00–£0.24	3
£0.25–£0.49	7
£0.50–£0.99	12
£1.00–£1.99	9
£2.00–£2.99	5
£3.00–£4.99	1
£5.00+	0

a Plot a cumulative frequency diagram for this data.

b Estimate the median tip.

5 A set of continuous data is recorded to one decimal place. The results are summarised in a histogram.

x	Frequency
0–0.4	18
0.5–0.7	
0.8–1.0	
1.1–1.4	15
1.5–1.8	12

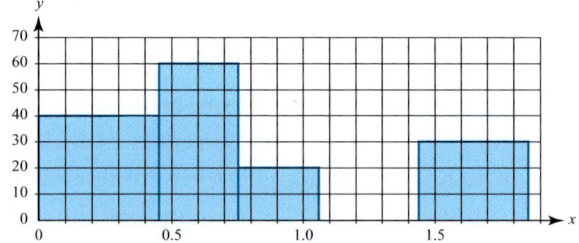

a Find the missing frequencies.

b Calculate the width and height of the missing bar.

c Estimate the percentage of values that are

i Below 0.6   **ii** Above 1.2

Reasoning and problem-solving

When interpreting a diagram displaying data

① Consider what is being represented and whether your data is discrete or continuous.

② If necessary, identify any outliers or missing/incorrect data, and consider the effects of removing them.

③ Read what is being asked for in the question and use the diagram to answer it.

You can use statistical diagrams to find probabilities of events.

If an experiment can have any one of N equally likely outcomes and n of those outcomes result in event A, then the theoretical probability of event A happening is $P(A) = \dfrac{n}{N}$

You can estimate probabilities of certain events happening by using diagrams that show grouped data. This often involves assuming data is equally distributed on each interval.

See Ch10.1
For information about probability.

STATS

Example 5

This box plot shows the masses (in grams) of 52 eggs from a certain species of bird.

An egg is picked at random from this set.

a Estimate the probability that its mass is

   i   More than 22 g

   ii Less than 18 g

   iii Less than 20 g

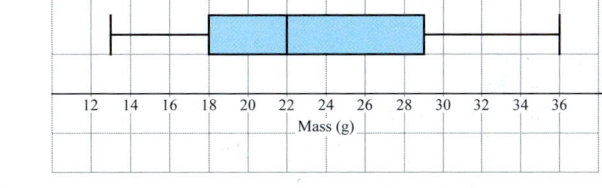

b 40% of the eggs weigh less than a g. Find a

a i 50% •

  ii 25% •

  iii $25\% + 0.5 \times 25\% = 37.5\%$ •

b 25% of the values lie below $18 \Rightarrow (40 - 25) = 15\%$
of the values lie between 18 and a

$18 + \dfrac{15}{25}(22 - 18) = 20.4$ •

$a = 20.4\,g$

> Assume 25% of the data is equally distributed between 18 and 22

> 22 g is the median so 26 of the 52 eggs weigh more than 22 g

> One quarter of the eggs weigh less than the lower quartile value of 18 g

> 25% of the eggs weigh between 18 g and 22 g. 20 g is halfway through this interval so you assume half of the eggs between 18 g and 22 g weigh lass than 20 g

It is important to use an appropriate diagram to represent your data.

	Advantages	Disadvantages
Box plot	Highlights outliers. Makes it easy to compare data sets.	Data is grouped into only four categories so some detailed analysis is not possible.
Histogram	Clearly shows shape of distribution.	Doesn't always highlight outliers. It is possible but not easy to estimate Q_1, Q_2 and Q_3
Cumulative frequency curve	Makes it easy to find the five number summary.	Doesn't always highlight outliers. If interval boundaries are not shown the degree of detail is not clear.

Example 6

The daily maximum relative humidity (%) is measured in 23 locations on a certain day.

40, 41, 41, 44, 48, 51, 53, 53, 54, 54, a, 59, 61, 62, 62, 63, 64, 65, 65, 66, 66, b, 90

a and b are unknown values but the list is known to be in ascending order. An outlier is defined as a value less than $Q_1 - 1.5 \times \text{IQR}$ or greater than $Q_3 + 1.5 \times \text{IQR}$

a Explain why a box-and-whisker plot is not an appropriate choice of diagram to represent this data.

b Describe a more appropriate diagram to represent this data.

a $Q_3 + 1.5 \times \text{IQR} = 65 + 1.5 \times (65 - 51) = 86$, so 90 is known to be an outlier. Hence b may or may not be an outlier so it would be impossible to plot the upper limit of the data accurately.

> The upper and lower quartiles are 65 and 51

b A histogram would better represent the data. As the relative values of a and b are known in relation to values either side, they would not affect the shape of the histogram as long as they are not used as boundaries of categories. Dividing the data into several groups would display the shape of the data clearly.

Exercise 9.3B Reasoning and problem-solving

1 a An art student wants to know how much money she could make from selling her work. Over a one-year period, she collects data on how much money pieces of art sold for at auction. Would a box-and-whisker plot or a histogram be better to display this data?

b A data set is bi-modal. State a graph that would *not* be suitable to display this data.

2 The heights of a sample of a species of plant are recorded. Copy the frequency table and use the cumulative frequency graph to fill in the missing values.

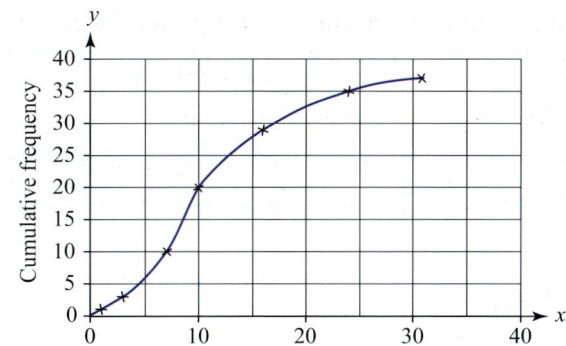

Length in cm	$0 \le x < 1$	$1 \le x < 3$	$3 \le x < 7$	$7 \le x < 10$	$10 \le x < 16$	$16 \le x < 24$	$24 \le x < 31$	$31 \le x$
Frequency	1	2			9			0

3 A doctor is investigating obesity in his local area. He records the masses (in kg) of 1000 patients and creates a box-and-whisker plot. The lightest patient weighs 5 kg.

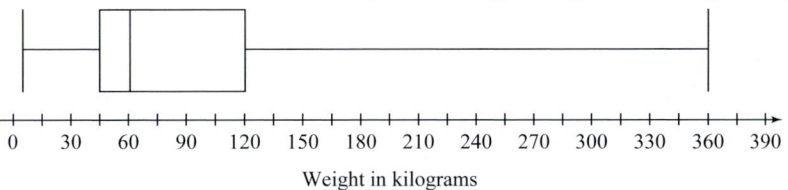

Weight in kilograms

a A patient is selected at random. Estimate the probability that the patient weighs

   i More than 45 kg   **ii** Less than 90 kg

b The two largest masses measured are 360 kg and 165 kg, but it was later decided that the 360 kg mass should be considered an outlier. If it is removed, how will the box-and-whisker plot change?

4 The table shows the number of days for which the daily maximum temperature in Cambourne was in the given range between June and August 2015

Large data set

Temperature range (°C)	Number of days
$T < 13$	0
$13 \le T < 15$	6
$15 \le T < 17$	27
$17 \le T < 20$	
$20 \le T < 21$	
$21 \le T < 25$	1
$25 \le T$	0

Daily maximum temperature (°C)

a **i** Use the histogram to work out the missing values in the table.

   ii Find the width and height of the missing bars in the histogram.

b Estimate the number of days in which the daily maximum temperature was

   i More than 14 °C   **ii** Less than 18 °C

9.4 Bivariate data

Fluency and skills

It is often useful to look for connections between two variables. Data relating to *pairs* of variables is called **bivariate data**.

> **Key point**
>
> Variables that are statistically related are described as **correlated**.

If the variables increase together, they have **positive correlation**. If one variable increases as the other decreases, they have **negative correlation**. Two variables can also be uncorrelated and the data is then said to have **zero correlation**.

You can **identify correlation** by plotting a **scatter diagram,** which shows each pair of data values as a point on a graph. A scatter diagram shows both the type and strength of the relationship between two variables.

The **explanatory** or **independent variable** goes on the *x*-axis. This is the variable that doesn't depend on other things—the values for it are usually selected by the person gathering data. The **response** or **dependent variable** goes on the *y*-axis. This is the variable that is expected to change in response to a change in the other variable.

> Data that lies exactly on a straight line has perfect correlation. Otherwise, the correlation may be described as strong, moderate or weak.

Types and strength of correlation

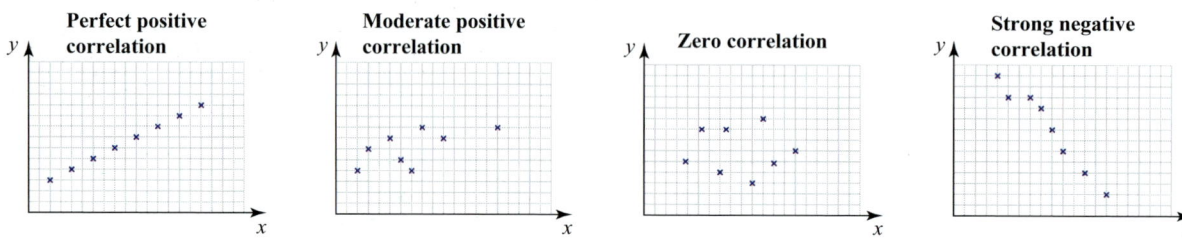

Perfect positive correlation · Moderate positive correlation · Zero correlation · Strong negative correlation

Example 1

The data in the table relates to the width and length of the petals of 9 roses. Plot this data on a scatter diagram and describe the nature of the correlation.

Width, cm	4.9	4.7	4.6	5	5.4	4.6	5	4.4	4.9
Length, cm	3	3.2	3.1	3.6	3.9	3.4	3.4	2.9	3.1

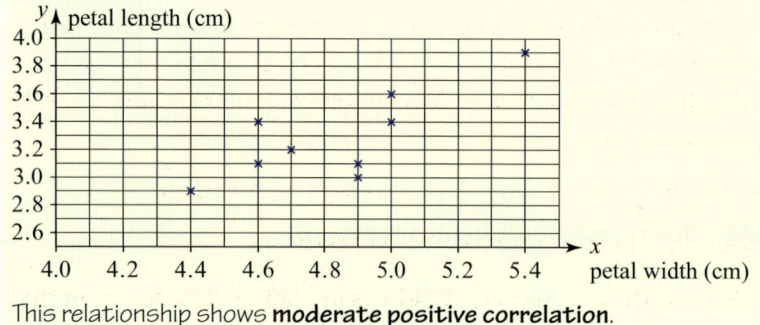

This relationship shows **moderate positive correlation**.

ICT Resource online

To investigate correlation, click this link in the digital book.

Exercise 9.4A Fluency and skills

1   Give the most likely type and strength of correlation (if any) for these pairs of variables.

   a   A town's annual income and crime rates.

   b   The heights of fathers and their adult sons.

   c   The cooking time for a chicken and the weight of the chicken.

   d   The shoe size of 30 adults and their annual salaries.

2   This diagram relates two variables, x and y

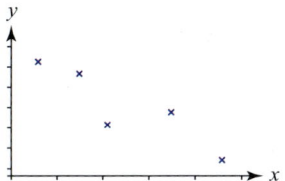

   Describe the correlation.

3   For each of these data sets, draw a scatter diagram and use it to describe the type and degree of correlation.

   a

x	2	4	4	7	8	10	12	14
y	3	4	7	7	9	10	12	12

   b

x	−2	0	2	4	3	5	7	7
y	9	8	5	6	1	1	2	6

4   The table shows the length of service in a company (L years to the nearest year) with corresponding annual salary (S thousand pounds to the nearest thousand) for 10 workers. Plot a scatter diagram for this data and describe the correlation.

L	5	15	2	12	11	1	3	6	4	12
S	22	26	16	21	22	18	23	23	18	27

5   This table gives data for a sample of 10 individuals from a bivariate population. Draw a scatter diagram for this data and describe the correlation.

x	51	62	47	53	71	65	55	69	57	61
y	68	55	68	64	50	58	65	50	61	54

Reasoning and problem-solving

Strategy

To solve a problem about bivariate data and correlation

① Draw a scatter diagram to identify any correlation between two variables.

② Identify data points that don't fit the general pattern shown by the data.

③ Interpret the data in context.

You should not assume that, because two variables correlate, changes in one variable are *causing* changes in the other. For example, rates of diabetes and annual income correlate for certain groups, but this is because they both relate to dietary intake. It's important that you remember the difference between **correlation** and **causation**.

> When a change in one variable *does* affect the other, they have a **causal connection**.   **Key point**
>
> Correlation without a causal connection is known as **spurious correlation**.

When data is correlated, a scatter diagram shows a pattern. **Outliers** are points that do not fit the pattern. They are easy to identify on a scatter diagram.

STATS

In order to identify a trend, it is often helpful to use a **regression line**, that is, a line which fits as well as possible with the points on the scatter graph. You can use the regression buttons on your calculator to find the equation of this line, also called a **line of best fit**.

Once the line has been drawn on the scatter graph, it is possible to use it to estimate values of the response variable (y) for given values of the explanatory variable (x). You can use the line of best fit to make predictions for a value that falls within the range of the observed data. This is called **interpolation**. This is generally reliable, particularly if the variables show a strong correlation. Using the line of best fit to make predictions outside the range of observed data is called **extrapolation**. This is not reliable as there is no evidence the pattern continues beyond the observed range.

Know your data set

Large data set One okta is one eighth of the celestial dome. It is used as a unit of measure for cloud cover. Click this link in the digital book for more information about the Large data set.

Example 2

The following data gives daily total rainfall in mm, y, against daily mean total cloud in oktas, x, on six consecutive Saturdays.

x	2.2	3.5	4.7	5.2	6.6	7.8
y	7.2	9.3	13.8	8.1	4.1	13.1

a Plot a scatter diagram to represent this data.

b The y-value of one of the points was incorrectly recorded. Write the most likely coordinates of this point.

c The regression line $y = 0.9x + 5.9$ is calculated using the other five points. Plot this line on your diagram.

d For the day identified in part **b**, use your sample regression line to estimate the true daily total rainfall.

e The weather forecast predicts daily mean total cloud of 0.5 oktas on the following Saturday. Using the regression line, a researcher expects approximately 6.4 mm of rain that day. Explain why this assumption is not trustworthy.

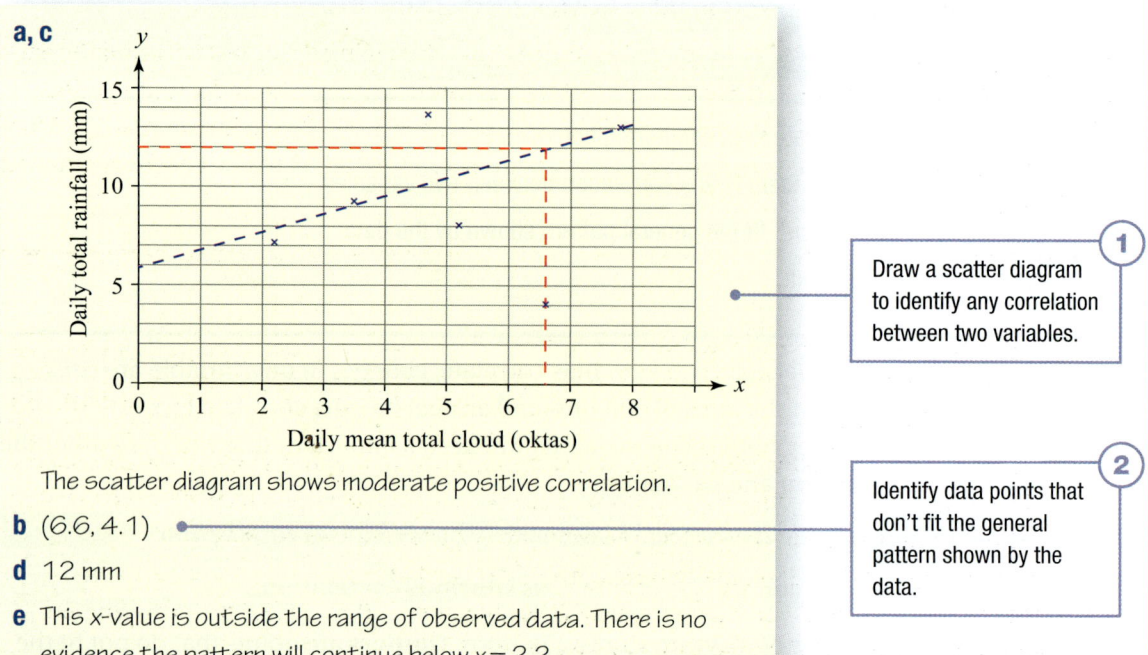

a, c

The scatter diagram shows moderate positive correlation.

Draw a scatter diagram to identify any correlation between two variables.

b (6.6, 4.1)

Identify data points that don't fit the general pattern shown by the data.

d 12 mm

e This x-value is outside the range of observed data. There is no evidence the pattern will continue below $x = 2.2$

Exercise 9.4B Reasoning and problem-solving

1. For each example of bivariate data, state whether the correlation is likely to be positive, negative or zero. If non-zero, state whether you think there is a causal relationship between the variables.
   a Daytime temperature at a seaside resort and number of deckchairs hired out.
   b Plant growth and amount of fertilizer applied.   c A person's annual income and weight.
   d Unemployment rates and measures of the standard of living.

2. This data shows the number of customer arrivals, x, in a five-minute interval, and queue length, y, at the end of the interval, in a shop. One mistake was made in recording the data.

x	16	5	12	8	4	9	17
y	4	0	5	3	1	2	1

   a Draw a scatter diagram and identify which reading was most likely recorded incorrectly.
   b The regression line is given by the equation $y = 0.35x - 0.65$. Plot this on your diagram.
   c Estimate the true queue length for the point you identified in part **a**.
   d During another five-minute interval 25 customers arrive. An observer predicts the queue length will be 8. Explain why this is not a reliable prediction.

3. The maximum daily temperature ($T°C$) and maximum daily relative humidity ($h\%$) for May 1st–14th 2015 in Hurn were recorded as follows:

T	12.6	13.2	16.1	14.7	15.2	14.5	15.6	16.7	18.2	15.8	17.4	17.0	16.7	12.5
h	79	96	99	96	97	91	90	95	91	92	95	95	94	95

   a Draw a scatter diagram for this data, using T as the explanatory variable.
   b Calculate the median and interquartile range for the humidity values.
   c Using the rule that outliers are values less than $Q_1 - 1.5 \times IQR$ or more than $Q_3 + 1.5 \times IQR$, where Q_1 is the lower quartile and Q_3 is the upper quartile, identify any outliers for humidity.
   d Describe the correlation between temperature and humidity and state how it would change if the outlier in **c** were removed.
   e Using the original set of 14 points, the regression line of h on T has equation $h = 0.84T + 80.3$ Estimate the value of h corresponding to a temperature of 14°C.

4. At a weather station in England between March and August 2015, rainfall and relative humidity each decreased month on month and average monthly temperature increased over the same period.
   a State the sign of the correlation between rainfall and **i** Relative humidity, **ii** Temperature, **iii** Month.
   b Do you think the correlation between rainfall and humidity is likely to be causal or spurious? Give reasons.

Challenge

5. A sample of 7 data items is taken from a bivariate population and the following values of the variables, X and Y, are obtained:

X	3	9	10	12	13	15	18
Y	2	9	9	11	11	19	13

   a Plot a scatter diagram for this data and hence describe the type and strength of the correlation. The researcher leading this investigation expected there to be near-perfect positive correlation between X and Y. She found that one y-value had been recorded incorrectly.
   b Identify the most likely value to be incorrectly recorded.
   c This error is corrected and a line of best fit is given by $y = 0.8x + 0.9$. Draw this line on your graph and estimate the value of y when $x = 7$

Chapter summary

- You can estimate a population and its parameters using statistics taken from a sample.
- There are many sampling methods, including simple random sampling, systematic sampling, stratified sampling, opportunity sampling and quota sampling.
- Sampling methods can produce biased samples in some situations.
- Outliers are points that don't fit the pattern of the data.
- The mode of a set of observations is the value that occurs with the largest frequency.
- The median of a set of data is the middle value once the data is in order of size.
- To find the mean of a set of n observations, calculate their sum and divide the result by n
- The range of a set of observations is the largest observation minus the smallest.
- The interquartile range is the difference between the first and third quartiles.
- Linear interpolation can be used to calculate quartiles and percentiles in grouped data.
- The variance of a set of data measures the degree of spread.
- If $y = ax + b$ then $\bar{y} = a + b\bar{x}$ and $\sigma_y^2 = b^2\sigma_x^2$
- Data is either discrete or continuous.
- Continuous data needs summarising for representation.
  - You use the five-number summary to form a box-and-whisker plot.
  - You group data to draw a histogram or a cumulative frequency diagram.
- Variables whose values are linearly related are said to be correlated. The type of correlation can be positive, negative or zero. If non-zero, it can be weak, moderate or strong.
- A scatter diagram shows both the type and strength of the relationship between two variables.
- Variables that have a non-zero correlation are not necessarily causally connected.

Check and review

You should now be able to...	Try Questions
✔ Distinguish a population and its parameters from a sample and its statistics.	1
✔ Identify and name sampling methods.	2
✔ Highlight sources of bias in a sampling method.	2
✔ Read continuous data given in box-and-whisker plots, frequency polygons, histograms and cumulative frequency diagrams.	3
✔ Plot scatter diagrams and use them to identify types and strength of correlation.	4
✔ Use scatter diagrams and rules using quartiles to identify outliers.	5
✔ Summarise raw data using appropriate measures of location and spread.	6, 7

1 Identify the population, sample, parameter and statistics in these situations.

    a 23 LED light bulbs made in a certain factory are tested to see how warm they get when lit.

    b Nine matured casks of whiskey from a certain brewery have their alcohol levels measured.

2   A sample of 15 families is to be taken to determine the average number of children per household in a given city. Name the sampling methods in parts **a** and **b**.

    a   A list of all house or flat numbers and names is found. The numbers and names are ordered and labelled 1, 2, 3, ... , n. 15 random numbers are generated between 1 and n. Those labels give the numbers or names of the households that are sampled.

    b   A surveyor spends three hours on a Saturday in the high street asking passers-by until he records enough responses.

    c   Which of the sampling methods in parts **a** and **b** could be biased?

3   Create a table of frequencies that could produce this histogram.

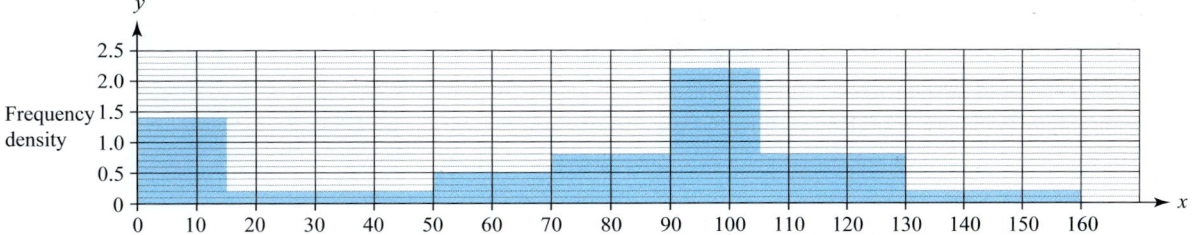

4   **a**   Draw a scatter diagram for this data set and use it to describe the type and degree of correlation.

x	12	10	7	8	8	7	5	6
y	2	5	7	6	9	11	12	13

    b   The equation of the regression line is given by $y = -1.5x + 20$. Plot this on your scatter diagram.

    c   Estimate a y-value corresponding to an x-value of 9. Answer to the nearest whole number.

5   **a**   Draw a scatter diagram for this data and comment on the correlation between the variables.

x	3	9	5	14	10	7
y	12	13	9	1	8	8

    b   The y-value of one point was recorded incorrectly. Given that there is strong negative correlation between the variables, identify the most likely incorrect point and state a more plausible y-value for this point.

6   For this data set, find the modal interval and estimate the median, mean and variance.

x	$0 \leq x < 4$	$4 \leq x < 8$	$8 \leq x < 12$	$12 \leq x < 16$	$16 \leq x < 20$
f	1	4	8	4	3

7   **a**   For this set of observations, find the interquartile range (IQR).

       14, 16, 15, 26, 10, 9, 12, 17, 15, 18, 11, 10

    b   Using the rule that outliers are values less than $Q_1 - 1.5 \times IQR$ or more than $Q_3 + 1.5 \times IQR$, where Q_1 is the first quartile and Q_3 is the third quartile, identify any outliers within the data.

What next?

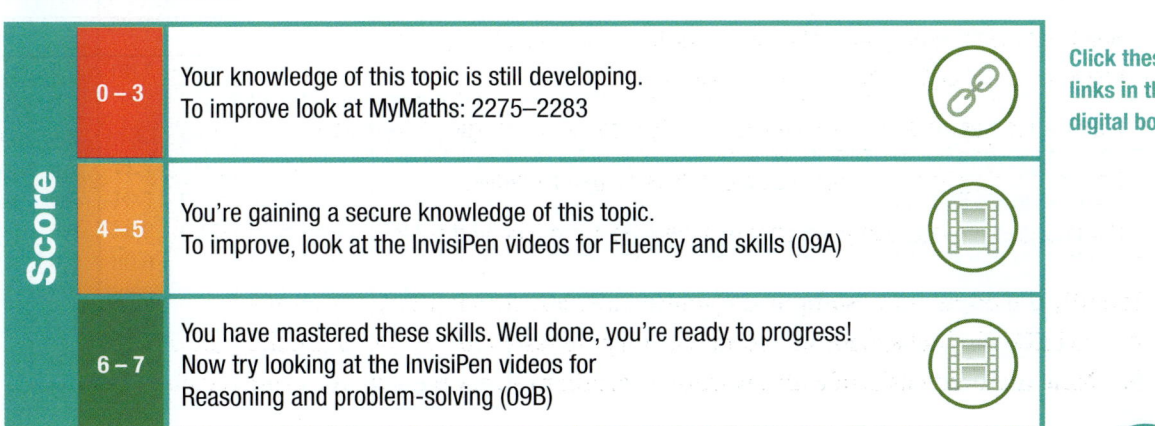

Score		
0 – 3	Your knowledge of this topic is still developing. To improve look at MyMaths: 2275–2283	
4 – 5	You're gaining a secure knowledge of this topic. To improve, look at the InvisiPen videos for Fluency and skills (09A)	
6 – 7	You have mastered these skills. Well done, you're ready to progress! Now try looking at the InvisiPen videos for Reasoning and problem-solving (09B)	

Click these links in the digital book

Florence Nightingale (1820 - 1910) is best known for her contributions to modern nursing, but she was also a very influential statistician.

Nightingale worked to make statistical data more accessible by developing innovative graphs and charts. She used them to present information about soldier mortality in order to convince governments that changes to hospital conditions were needed.

Nightingale's work transformed the way we can represent statistics and highlighted the importance of statistics in government.

Research

Sir Francis Galton (1822 – 1911) is credited with inventing the concepts of **standard deviation** and **correlation**, and was the first to recognise the phenomenon of **regression towards the mean**.

What is regression towards the mean?

Why does it need to be taken into account when designing experiments?

Have a go

In 1973, the English statistician **Francis Anscombe** constructed the four sets shown, known as Anscombe's quartet, in order to make a point about the value of statistical diagrams.

Calculate for each data set
1. The mean value of x
2. The mean value of y
3. The variance of x
4. The variance of y

The correlation between x and y is 0.816 for each data set, and in each case the line of best fit is $y = 0.5x + 3$

Comment on what all of the summary statistics appear to tell us about the data sets.

Plot the data sets on four diagrams, using the same scales.

What do you observe? What point do you think Anscombe wanted to make?

Did you know?

Sir Francis Galton used information from weather stations in England to produce the world's first weather map.

I		II		III		IV	
x	y	x	y	x	y	x	y
10.0	8.04	10.0	9.14	10.0	7.46	8.0	6.58
8.0	6.95	8.0	8.14	8.0	.77	8.0	5.76
13.0	7.58	13.0	8.74	13.0	12.74	8.0	7.11
9.0	8.81	9.0	8.77	9.0	7.11	8.0	8.84
11.0	8.33	11.0	9.26	11.0	7.81	8.0	8.47
14.0	9.96	14.0	8.10	14.0	8.84	8.0	7.04
6.0	7.24	6.0	6.13	6.0	6.08	8.0	5.25
4.0	4.26	4.0	3.10	4.0	5.399	19.0	12.50
12.0	10.84	12.0	9.13	12.0	8.15	8.0	5.56
7.0	4.82	7.0	7.26	7.0	6.42	8.0	7.91
5.0	5.68	5.0	4.74	5.0	5.73	8.0	6.89

1 A council wants to sample households in a town in order to find out what proportion of their waste is recycled. There are 20 000 houses in the town and the council want to take a sample of 10% of these houses.

    a Describe how a systematic random sample could be used. **[2 marks]**

The council intends to contact the selected households by telephone and request permission to track how much material they recycle over a 3-month period.

    b Suggest a practical difficulty that may be encountered when contacting households. **[1]**

    c Give a reason why the results from the sampled households may not be representative of the population. **[1]**

2 A battery manufacturer needs to test the average life of its batteries.

    a What is the population in this case? **[1]**

    b Aside from cost, give a reason why only a sample should be tested. **[1]**

    c What would be a sensible statistic to calculate? **[1]**

3 The daily total sunshine, in hours, recorded over a 12-day period during the summer at Hurn weather station are given.

| 0.3 | 6.2 | 10.5 | 12.4 | 12.3 | 14.3 |
| 15 | 14.9 | 10.3 | 22 | 14.5 | 10.1 |

The mean hours of daily total sunshine over this period was 10.95

    a One of the measurements has been written down incorrectly.

Identify the error and calculate the correct value. **[3]**

    b Calculate the variance of the correct data. **[2]**

    c Calculate the median and explain whether the mean or the median would be the best average to use in this case. **[3]**

4 The box-and-whisker plots show the maximum daily temperature in August 2015 for two different weather stations.

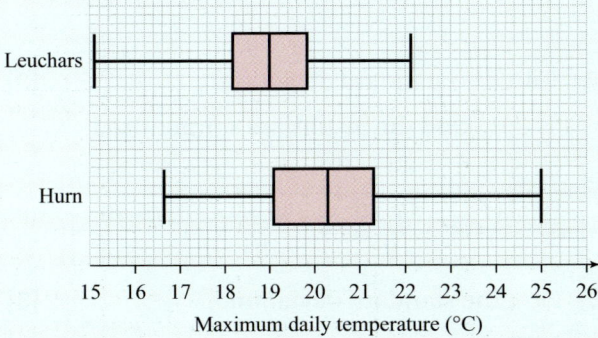

Maximum daily temperature (°C)

    a Estimate the interquartile range at each station. **[3]**

    b Write down the median temperature at each station. **[2]**

    c Compare the maximum daily temperatures during August at each weather station. **[3]**

An outlier is defined as a value outside of the range $[Q_1 - 1.5(Q_3 - Q_1), Q_3 + 1.5(Q_3 - Q_1)]$

d Work out whether or not these values are outliers, showing your working clearly.

   i 15.1°C at Leuchars   **ii** 22.1°C at Leuchars   **iii** 25°C at Hurn   **[6]**

5 The scatter diagram shows the mean windspeed and maximum gust speed in Camborne in October 1987

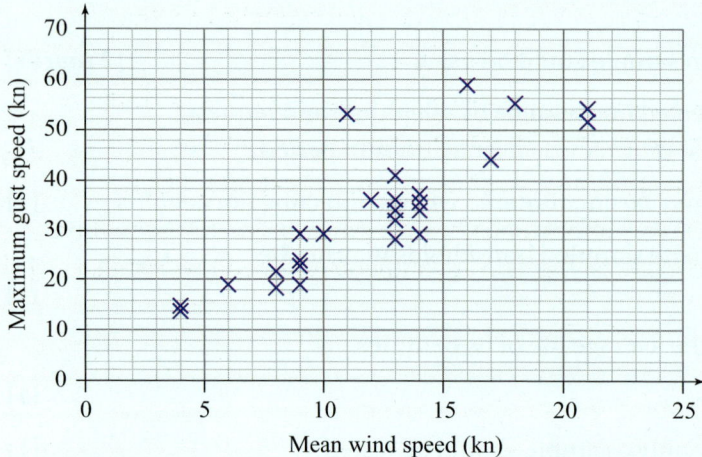

Describe the relationship between the mean windspeed and maximum gust speed. **[2]**

6 A teacher wants to find out the average number of homework assignments that students have been given each week. He takes a sample of 30 students. The results are shown in this graph.

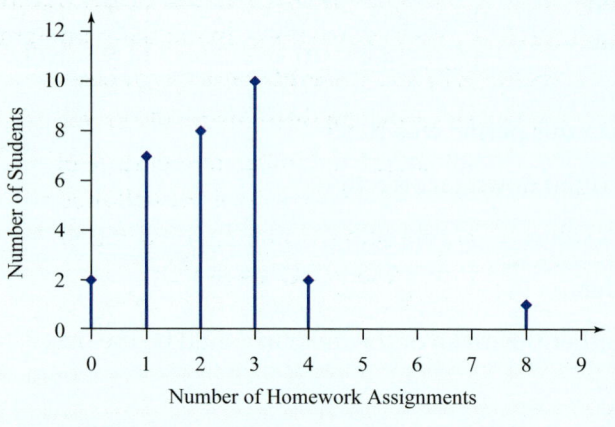

a What is the modal number of assignments? **[1]**

b Calculate the mean and standard deviation of the sample. **[4]**

c State the median and first and third quartiles. **[2]**

d Advise which measure of average would be best to use, stating your reasons clearly. **[1]**

7 The maximum daily relative humidity, h, was measured at Heathrow weather station every day in May 1987

You are given $\sum h = 2824$ and $\sum h^2 = 258304$

a Calculate **i** The mean relative humidity,   **ii** The standard deviation. **[3]**

The same measurement was taken every day in May 2015 and the results are summarised in the table.

Relative Humidity (%)	70–74	75–79	80–84	85–89	90–94	95–99
Number of days	1	2	6	9	8	5

b Estimate **i** The mean relative humidity,   **ii** The standard deviation. **[5]**

c Compare the humidity in May 2015 with May 1987 **[2]**

d Draw a frequency polygon for the 2015 results. **[3]**

8 A sample of students in a sixth form is to be taken and surveyed regarding their use of the library. The numbers of boys and girls in each year is given in the table.

	Year 12	Year 13
Girls	60	51
Boys	39	30

a It is suggested that a sample of size 60 should be taken by randomly selecting 15 boys and 15 girls from each year. State a disadvantage of taking a sample in this way. **[1]**

b It is instead decided to use a stratified sample of 60 students. Calculate the number of boys and the number of girls that should be sampled in each year. **[4]**

c In order to select individuals for the survey, an interviewer will randomly choose students as they leave their common room. Explain why the results of the survey could be biased. **[1]**

9 An outlier is an observation greater than $Q_3 + 1.5(Q_3 - Q_1)$ or less than $Q_1 - 1.5(Q_3 - Q_1)$

A box-and-whisker plot is drawn for a large volume of data. Four extra observations are then recorded. Which of the extra observations A, B, C and D are outliers? Show your working. **[3]**

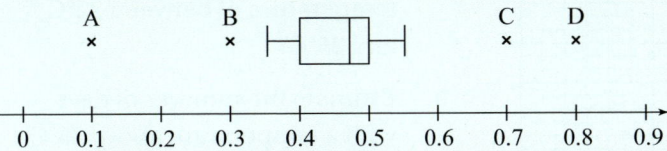

10 An experiment was carried out using tomato plant seeds. Trays of seeds were planted and each tray was placed in a controlled environment with a different temperature for each tray. All other variables, such as light and water, were the same for each of the trays. After 10 days, the number of seeds that had germinated in each was counted. The results are shown in the scatter diagram.

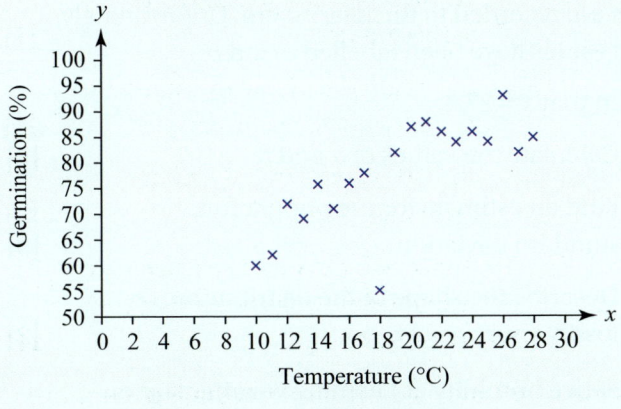

a Which is the explanatory variable? **[1]**

b Describe the relationship observed. **[2]**

c It is suggested that a temperature of 35°C would result in almost all seeds germinating. Comment on whether this is a sensible suggestion. **[2]**

11

Rainfall, r	Number of days
$0 \leq r < 0.1$	84
$0.1 \leq r < 0.5$	27
$0.5 \leq r < 1$	19
$1 \leq r < 2$	14
$2 \leq r < 5$	17
$5 \leq r < 30$	23

The rainfall is measured (in mm) at Leeming weather station every day from May to October 2015. The results are summarised in this table.

Large data set

A histogram is to be drawn to show this data. The bar for $0.5 \leq r < 1$ is 5 mm wide and 76 mm tall.

a Calculate the height and width of the $1 \leq r < 2$ bar. **[4]**

b Estimate to 2 dp

i The median,

ii The interquartile range. **[6]**

12 The police in a town wish to survey members of a number of "Neighbourhood Watch" schemes. They wish to survey people from each scheme and the number selected is to be in proportion to the size of the scheme. Two possible methods of selecting the sample are suggested.

Method A: 10 people are randomly selected from each scheme from a list of all members.

Method B: The person who manages each scheme is asked to choose a proportionally sized sample from their population at random.

a State the name of each of these methods of sampling. **[2]**

b Which method is preferable? Clearly explain why this method is better. **[2]**

13 This histogram shows the maximum daily temperature at Heathrow weather station in July 2015

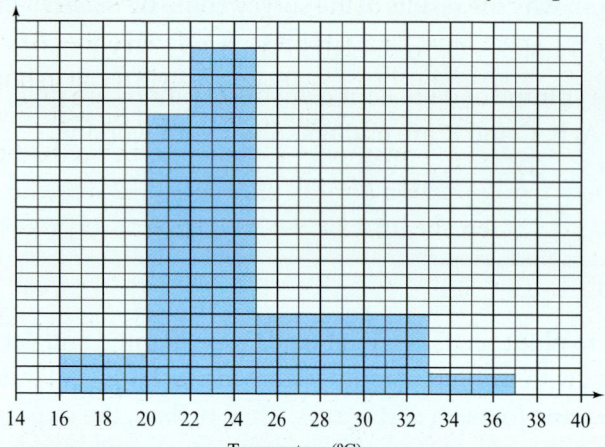

Temperature (°C)

The maximum temperature was below 20°C on just two days.

a How many days had a maximum temperature of between 22°C and 25°C? **[3]**

b Estimate the number of days with a temperature above 26°C. **[4]**

c Use the histogram to estimate the mean maximum daily temperature. **[4]**

14 The lengths of 40 fish caught in a competition are recorded to the nearest cm. Unfortunately, some of the numbers are now illegible. These values have been labelled *a* and *b*

Length (cm)	Frequency
18–22	3
23–25	10
26–28	14
29–31	*a*
32–40	*b*

Given that $\bar{x} = 27.3$

a Calculate the values of *a* and *b* **[5]**

b Find an estimate for the population standard deviation. **[3]**

c Describe the shape of the distribution. Justify your answer. **[2]**

15 The cumulative frequency curve shows the relative humidity at Leeming weather station in May 2015

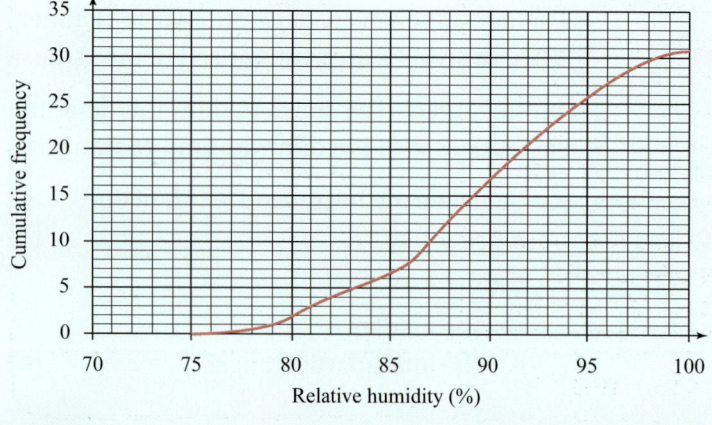

Relative humidity (%)

a What is the median relative humidity? **[1]**

b Calculate the interquartile range. **[2]**

A day with relative humidity over 95% is likely to have been foggy or misty.

c Estimate the percentage of days that were foggy or misty. **[3]**

10 Probability and discrete random variables

Probability is a big part of genetics. Consider a child's chance of inheriting a specific eye colour. The basic model involves two possibilities: *B* (brown) and *b* (blue), where brown is dominant over blue. Each parent passes down either *B* or *b* and only the combination *bb* will result in blue eyes. The combinations *Bb*, *bB* and *BB* will all result in brown eyes. So if both parents have the combination *Bb*, they both have brown eyes and the probability of their child having brown eyes is 75%.

The example above treats eye colour as a discrete variable: either blue or brown in this case. Age and blood type are also typically treated as discrete variables. Probability is important in subjects such as statistics, research, medicine, weather forecasting and business, and this chapter provides an introduction to the terminology and techniques used in calculating and expressing probabilities.

Orientation

What you need to know	What you will learn	What this leads to
KS4 • Record, describe and analyse the frequency of experiment outcomes. • Understand and draw probability tree diagrams. • Construct and interpret diagrams for discrete data.	• To use the vocabulary of probability theory. • To solve problems involving mutually exclusive and independent events. • To use a probability function or given context to find a probability distribution and probabilities for particular events. • To recognise and solve problems related to the binomial distribution.	**Ch20 Probability and continuous random variables** Conditional probability. Modelling with probability. Connecting the binomial distribution with the normal distribution. **Careers** Weather forecasting. Actuarial science. Genetics.

 **MyMaths** | Practise before you start   🔍 1193, 1211, 1935

263

Fluency and skills

A **random experiment** is any repeatable action with a collection of clearly defined outcomes that cannot be predicted with certainty. For example, the experiment 'an ordinary dice is thrown once'.

The **sample space** for an experiment is the collection of all possible outcomes of the experiment. **Events** are groups of outcomes within the sample space. In the diagram, S is the sample space for the experiment 'an ordinary dice is thrown once'. A is the event 'getting an odd number'.

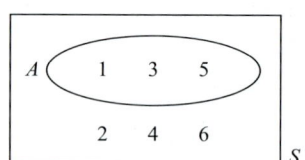

▲ Sample space S with event A

> **Key point**
> The total probability associated with a sample space is 1

If, in a trial where 60 dice are thrown, twelve 6s and seven 5s are thrown, then an estimate of the probability of a 5 or a 6 is $\dfrac{12+7}{60} = \dfrac{19}{60}$

See p.614
For a list of mathematical notation.

> **Key point**
> The probability of an even happening is written 'P(event)'.

Probabilities can be added like this when events are **mutually exclusive**. This means they cannot both happen in one trial.

> Events are mutually exclusive if they cannot occur together. You cannot roll a dice once and get 5 *and* 6

> **Key point**
> If A and B are mutually exclusive, $P(A \text{ or } B) = P(A) + P(B)$

Two events, A and not-A, are known as **complementary events**. They are mutually exclusive and **exhaustive**. Exhaustive means that no other outcome exists.

> The event not-A is written as A'

> **Key point**
> $P(A') = 1 - P(A)$

Example 1

Show that the probability of obtaining a score greater than 4 when a fair dice is thrown is $\dfrac{1}{3}$

$P(1) = P(2) = \ldots = P(6) = \dfrac{1}{6}$

$P(\text{score} > 4) = P(5 \text{ or } 6) = P(5) + P(6) = \dfrac{1}{6} + \dfrac{1}{6} = \dfrac{1}{3}$

> The dice is fair and the outcomes are mutually exclusive, so all outcomes are equally likely.

For a sample space with N equally likely outcomes, the probability of any one occurring is $\dfrac{1}{N}$

> **Key point**
> If event A occurs in $n(A)$ of the equally probable outcomes, the probability of A is given by $P(A) = \dfrac{n(A)}{N}$

> A sample space where there are N equally likely outcomes is known as the **discrete uniform distribution.**

Two events, A and B, are **independent** if the fact that A has occurred does not affect the probability of B occurring.

The $\cup$ and $\cap$ symbols are called the union and intersection symbols respectively.

> **Key point**
>
> If A and B are independent, $P(A \text{ and } B) = P(A) \times P(B)$

$P(A \text{ or } B)$ is sometimes written $P(A \cup B)$. $P(A \text{ and } B)$ is sometimes written $P(A \cap B)$.

Example 2

Two boxes contain counters. Box 1 contains four red and five white counters; box 2 contains two red and three white counters. One counter is taken from each box. Find the probability that one is red and the other is white.

Let R_1, R_2, W_1, W_2 be the events 'red from box 1, …'

$P(\text{red and white}) = P((R_1 \text{ and } W_2) \text{ or } (W_1 \text{ and } R_2))$

$\qquad\qquad\qquad = P(R_1) \times P(W_2) + P(W_1) \times P(R_2)$

$\qquad\qquad\qquad = \dfrac{4}{9} \times \dfrac{3}{5} + \dfrac{5}{9} \times \dfrac{2}{5} = \dfrac{22}{45}$

Find the probability of picking red then white, *or* the probability of picking white then red.

STATS

You can use a **tree diagram** to work out the probabilities in Example 2

To calculate P(two reds), multiply the probabilities along the two 'red' branches.

To calculate P(red and white), multiply along the branches for 'red' and 'white' and also for 'white' and 'red' and add them together.

Venn diagrams can also be used to show probabilities.

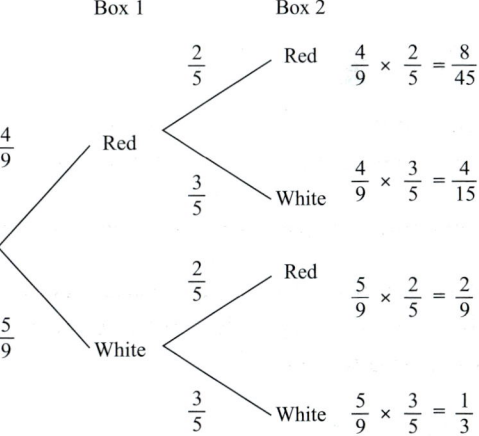

Example 3

The Venn diagram shows the probabilities for the types of films a group of students enjoy watching.

Each letter represents the likelihood of a student enjoying the following types of film.

T: Thrillers, A: Action, C: Comedies,

The probability that a student chosen at random likes one or both of action films or comedies is 0.8. Calculate the values of x and y

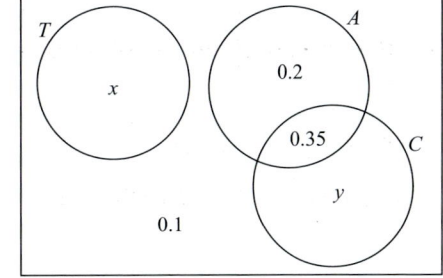

$P(A \text{ or } C) = 0.2 + 0.35 + y = 0.8$

$\qquad\qquad\qquad y = 0.25$

$x + 0.1 + 0.2 + 0.35 + 0.25 = 1$

$x = 0.1$

Use the Venn diagram to write down $P(A \text{ or } C)$

Total probability $= 1$

A **probability distribution** for a random experiment shows how the total probability of 1 is distributed between all the possible outcomes.

A **discrete random variable** X takes values x_i with probabilities $P(X = x_i)$.

Example 4

A fair coin is tossed 3 times and the number of heads noted. By calculating the probabilities of the possible outcomes, copy and complete the table.

Number of heads	0	1	2	3
Probability				

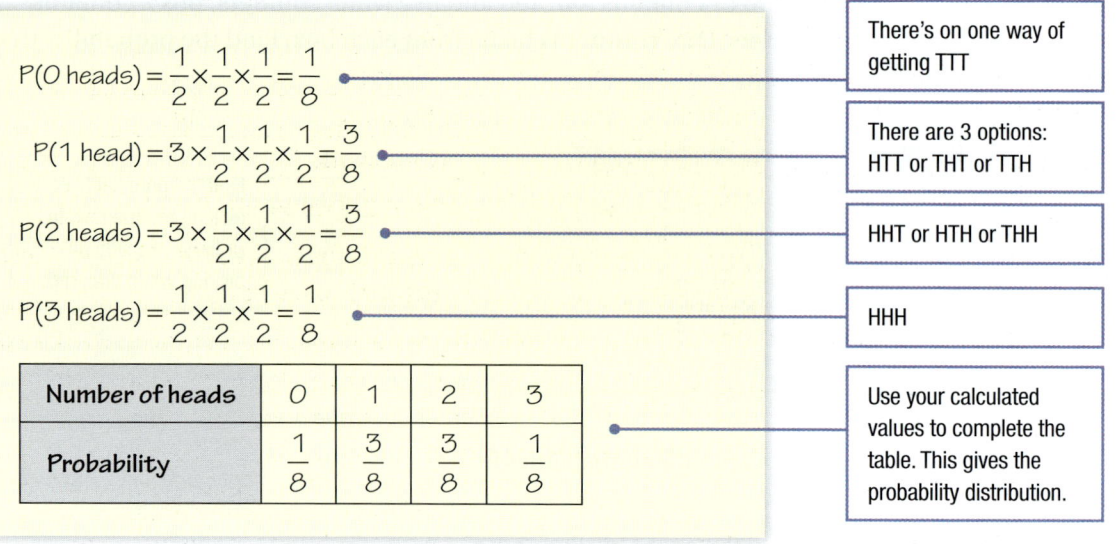

$$P(0 \text{ heads}) = \frac{1}{2} \times \frac{1}{2} \times \frac{1}{2} = \frac{1}{8}$$

There's on one way of getting TTT

$$P(1 \text{ head}) = 3 \times \frac{1}{2} \times \frac{1}{2} \times \frac{1}{2} = \frac{3}{8}$$

There are 3 options: HTT or THT or TTH

$$P(2 \text{ heads}) = 3 \times \frac{1}{2} \times \frac{1}{2} \times \frac{1}{2} = \frac{3}{8}$$

HHT or HTH or THH

$$P(3 \text{ heads}) = \frac{1}{2} \times \frac{1}{2} \times \frac{1}{2} = \frac{1}{8}$$

HHH

Number of heads	0	1	2	3
Probability	$\frac{1}{8}$	$\frac{3}{8}$	$\frac{3}{8}$	$\frac{1}{8}$

Use your calculated values to complete the table. This gives the probability distribution.

A probability distribution can be shown on a graph. The discrete distribution in Example 4 can be shown using a bar chart, where the height of each bar represents the probability of that event. The total of the heights of all the bars is 1 because the total probability associated with any sample space is 1

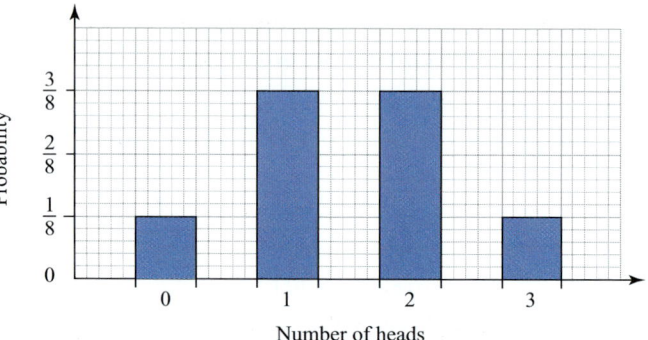

The probability of a continuous distribution is described by a function $f(x)$. The probability of X taking a value between a and b is given by the area under $f(x)$ between $x = a$ and $x = b$. This area is shown shaded on the diagram.

The total area under $f(x)$ equals 1 as this is the total probability associated with the sample space.

In a continuous distribution, $P(X = a) = 0$ because the area under a single point on a curve is zero.

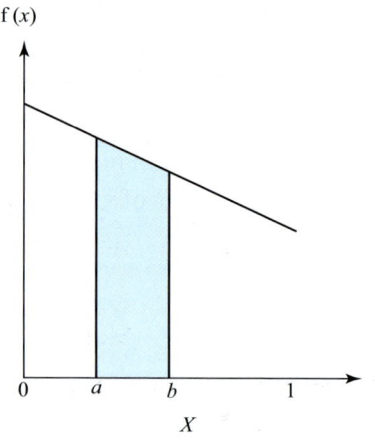

1 A bag contains four counters numbered 1 to 4. A counter is chosen at random, not replaced, and then another counter is chosen. List all the pairs of numbers that make the sample space for this experiment.

2 A small health and fitness club offers gym facilities, personal training and Pilates classes. Its 45 members do only one of these activities each, as shown in the table.

Activity	Gym	Training	Pilates	Total
Male	7	a	1	20
Female	10	b	c	d
Total	17	20	8	e

a Find the values of a to e

b Find the probability that a randomly chosen member

   i Does training,

   ii Is female and does Pilates,

   iii Is male.

3 An employee records the number of emails she receives per day throughout March. The results are shown in the bar chart.

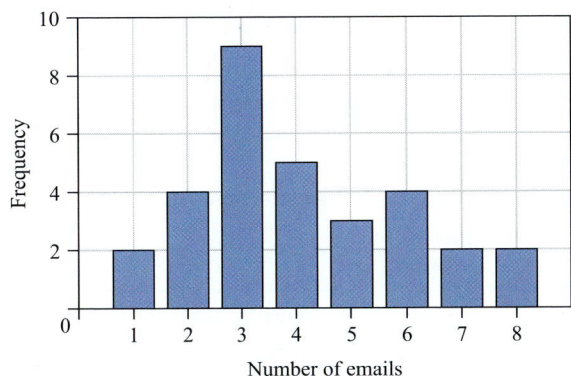

Find the probability that, on a randomly chosen day, the number of emails she receives is

a 5   **b** More than 5

c Between 1 and 4 inclusive.

4 One Saturday afternoon, a student decides to either go to the cinema, visit her friends, go shopping or stay in to do her homework. The probabilities of each of these activities are shown in the table.

Activity	Probability
Cinema	$2h$
Friends	$4h$
Shopping	$3h$
Homework	h

a Find the value of h

b What is the probability that she doesn't do her homework?

5 The three mutually exclusive and exhaustive events, P, Q and R are associated with a random experiment.
$P(Q) = 0.2$ and $P(R) = 0.3$. Find

a $P(P)$ **b** $P(P')$

c $P(P \text{ or } Q)$ **d** $P(P \text{ and } R)$

6 A fair dodecahedral dice has integers 1 to 6, each occurring twice, on its 12 faces. The dice is rolled once. What is the probability that the uppermost face shows a square number or a prime number?

7 You throw two fair six-sided dice.

a What is the probability that one score is less than 4 and the other score is greater than 4?

b X is a variable for the difference (biggest – smallest) of the scores. By drawing a two-way table showing all possible outcomes, write down the probability distribution of X

STATS

To solve a probability problem

1. Identify mutually exclusive events and use the addition rule.
2. Identify independent events and use the multiplication rule.
3. For unknown probabilities, consider using the "probabilities total 1" result.

Example 5

a A and B are two events associated with a random experiment. Find the probability of B if $P(A) = \dfrac{1}{3}$, $P(A \text{ and } B) = 0$ and $P(A \text{ or } B) = \dfrac{3}{5}$

b R and S are two events associated with a random experiment. Given that $P(R) = 0.4$, $P(S) = 0.7$ and $P(R \text{ and } S) = 0.3$, show that R and S are *not* independent.

a $P(A \text{ or } B) = P(A) + P(B)$

$P(B) = P(A \text{ or } B) - P(A) = \dfrac{3}{5} - \dfrac{1}{3} = \dfrac{4}{15}$

> **1** $P(A \text{ and } B) = 0$, so A and B are mutually exclusive.

b $P(R \text{ and } S) = 0.3 \qquad P(R) \times P(S) = 0.4 \times 0.7 = 0.28$

$0.3 \neq 0.28$ so R and S are not independent.

> **2** $P(R \text{ and } S) \neq P(R) \times P(S)$

Example 6

The daily mean windspeed (DMW) and daily mean pressure (DMP) in Jacksonville were investigated between May and October 1987. The number of days where the DMW is at least 5 kn, and the number of days where the DMP is at least 1020 Pa, were recorded. A day is chosen at random from the set, and the probability that the DMW was at least 5 kn is 0.41

	DMP ≥ 1020	DMP < 1020
DMW ≥ 5	27	55
DMW < 5	20	a

Know your data set

Large data set

Windspeed is often measured in knots. 1 knot = 1.15 mph. Click this link in the digital book for more information about the Large data set.

a Find the value of a

b Two students each select a day at random from the set (they can choose the same day). What is the probability that, on both days, DMP ≥ 1020 Pa?

a $\dfrac{27}{27+55+20+a} + \dfrac{55}{27+55+20+a} = 0.41$

$\dfrac{82}{102+a} = 0.41$

$a = \dfrac{82}{0.41} - 102 = 98$

> **1** ≥ 1020 Pa and < 1020 Pa are mutually exclusive.

b $P(\geq 1020 \text{ Pa on both days}) = \dfrac{47}{200} \times \dfrac{47}{200} = 0.06 \ (1 \text{ sf})$

> **2** Events are independent.

1 In a 2-player game, a calculator generates random numbers from 0 to 9. 0–2 scores one point, 3–8 scores two and 9 scores three.

a Copy and complete the following probability distribution table.

Score	1	2	3
Probability			

b Find the probability that the score is

i Less than 3 **ii** 1 or 3

2 X is a continuous random variable which can take any value between 0 and 1 The probability distribution is shown by the straight line graph.

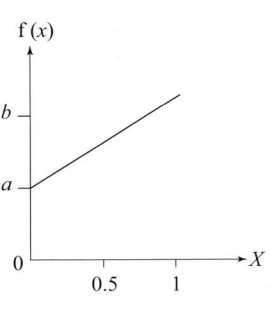

a Write down $P(X = 1)$

b Write down $P(0 \leq X \leq 1)$

c Given that $b = 3a$, calculate a and b

d Show that $P(0.25 \leq X \leq 0.75) < P(0.5 \leq X \leq 1)$

3 The rainfall and temperature at a weather station during May and June 2015 were recorded and the following data obtained.

Temperature / Rainfall	Low	Medium	High	Total
Low	3	6	8	17
Medium	a	20	c	d
High	7	4	3	e
Total	15	b	16	f

a Find the values of constants a–f

b Find the probability that, on a randomly chosen day in May or June 2015,

i Rainfall was high,

ii Temperature was medium or low,

iii Temperature was low and rainfall high.

c Considering only days with low rainfall, what is the probability that on a random day the temperature was high?

4 Two six-sided dice are each thrown once and their scores added. Draw a two-way table to show all possible outcomes. The probability that the sum of the scores is greater than 10 is $\frac{5}{36}$. Are both dice fair? You must give reasons for your answer.

5 For any family of three children, A is the event 'there is at least one boy and girl' and B is the event 'there are more girls than boys'. Assuming all combinations are equally likely, are the events A and B independent of each other? Show your working.

6 A box contains blue, yellow and black beads. There are four blue beads and twice as many yellow as black beads. When a bead is chosen at random, the probability that it is black is $\frac{3}{11}$. How many yellow beads were initially in the box?

Challenge

7 Two events, E and F, are associated with a random experiment. $P(E \text{ and } F') = 0.4$, $P(E' \text{ and } F) = 0.3$, $P(E \text{ and } F) = 0.1$

a Find **i** $P(E)$, $P(F)$, $P(E \text{ or } F)$

ii $P(\text{neither } E \text{ nor } F)$

b Verify the equation $P(E \text{ or } F) = P(E) + P(F) - P(E \text{ and } F)$

STATS

Fluency and skills

Probability often involves sets of identical, independent trials. When trials have two outcomes: 'success' and 'failure', you can use a **binomial probability distribution** to model the situation.

> **Key point**
>
> Conditions for a binomial probability distribution:
>
> - Two possible outcomes in each trial.
> - Fixed number of trials.
> - Independent trials.
> - Identical trials (p is the same for each trial).

You can see from the tree diagram that if a biased coin is thrown three times and the probability of a head in any throw is 0.4, there are three ways to obtain exactly two heads.

$P(\text{two heads}) = 3 \times 0.4^2 \times 0.6$

The trials are independent of each other and there are three ways to obtain exactly two heads.

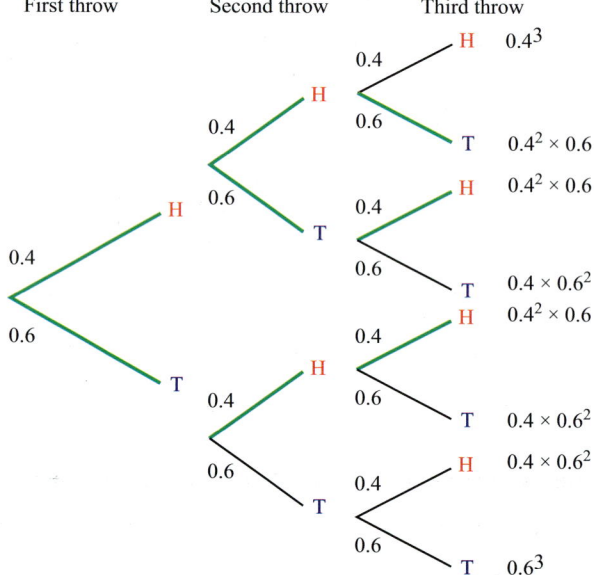

Tree diagrams can be a helpful way to record the possible outcomes of a small number of trials. However, when the number of trials is greater than three this becomes unwieldy in practice and the ^nC_r formula is more appropriate.

In general, for a binomial probability distribution where X is a **random variable** for the number of successes, the probability of x successes is given by:

See Ch2.2
For a reminder on the ^nC_r function.

> **Key point**
>
> $P(X = x) = {}^nC_x\, p^x (1-p)^{n-x}$
>
> where n is the number of trials and p is the probability of success in any given trial.

^nC_x is the number of ways of getting x successes in n trials.

A random variable, X, with this distribution function is said to follow a binomial distribution with parameters n and p. This is written $X \sim B(n, p)$.

Individual probabilities, $P(X = x)$, and cumulative probabilities, $P(X \leq x)$, can be found directly on a calculator. Related probabilities can also be calculated.

ICT Resource online

To practise calculating binomial distribution values, click this link in the digital book.

Key point

If X can only take integer values, $P(X < x) = P(X \leq x - 1)$
and $P(X > x) = 1 - P(X \leq x)$

Try it on your calculator

You can calculate binomial probabilities on a calculator.

```
Binomial    C.D
Data      : Variable
x         : 5
Numtrial  : 30
P         : 0.2
Save Res  : None
Execute
Calc

Binomial    C.D
           p=0.42751243
```

Activity

Find out how to calculate $P(X \leq 5)$ for a random variable $X \sim \text{Bin}(30, 0.2)$ on *your* calculator.

Unless told otherwise, you should work out probabilities using a calculator.

Example 1

Two fair six-sided dice are thrown 24 times. X represents the number of double sixes.

a Write down the probability distribution of X and its distribution function.

b Using the distribution function, find $P(X = 1)$

c Find the value of $P(X < 5)$

d Find the probability of at least three double sixes.

24 independent and identical trials.

a $X \sim B\left(24, \dfrac{1}{36}\right)$

$n = 24$, $p = \dfrac{1}{6} \times \dfrac{1}{6} = \dfrac{1}{36}$

$P(X = x) = {}^{24}C_x \left(\dfrac{1}{36}\right)^x \left(\dfrac{35}{36}\right)^{24-x}$

Check this using the binomial probability distribution function option on your calculator.

b $P(X = 1) = 0.3488 \ (4 \, dp)$

c $P(X < 5) = P(X \leq 4) = 0.9995 \ (4 \, dp)$

d $P(X \geq 3) = 1 - P(X \leq 2) = 0.0281$

Change to $P(X \leq 5 - 1)$ and use your calculator.

The shape of a binomial distribution has some interesting features.

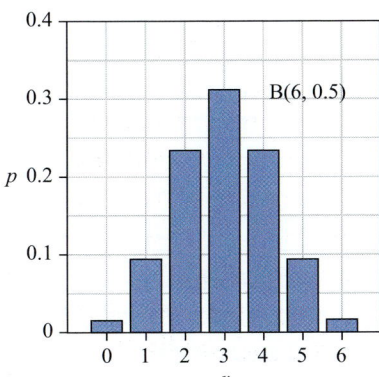

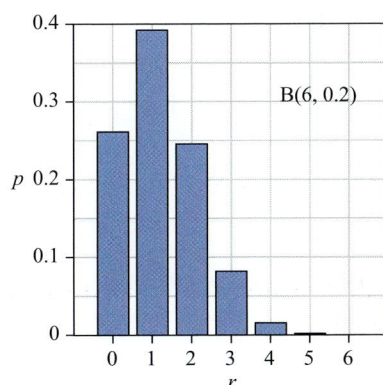

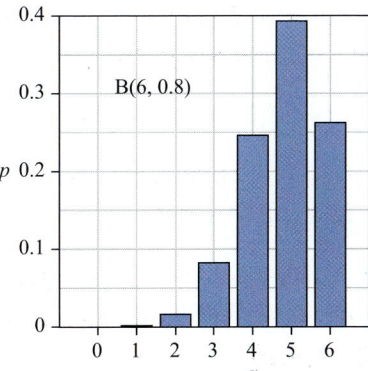

If $p = \dfrac{1}{2}$, the distribution shows symmetry. This is because the binomial coefficients have symmetry.

The distribution of $B(n, p)$ shows reflective symmetry with $B(n, 1 - p)$. For example, $P(X = 4)$ for distribution $X \sim B(6, 0.2)$ is equal to $P(X = 2)$ for $X \sim B(6, 0.8)$

MyMaths 🔍 2110, 2111 SEARCH 🔗

1 Given that $X \sim B(5, 0.3)$, find

    a $P(X = 3)$ **b** $P(X \leq 2)$ **c** $P(X \neq 0)$

2 $X \sim B(8, 0.6)$. Find, to 2 sf

    a $P(x \leq 0)$ **b** $P(x \leq 3)$

    c $P(x < 5)$ **d** $P(x > 2)$

3 The random variable T has a binomial distribution, $n = 8$, $p = \dfrac{1}{4}$. Find, to 2 sf

    a $P(T = 4)$ **b** $P(T \geq 7)$ **c** $P(3 \leq T < 5)$

4 Given that $X \sim B(5, 0.4)$,

    a Write an expression for $P(X = x)$

    b Copy and complete the probability distribution table.

x	0	1	2	3	4	5
$P(X = x)$	0.078			0.230	0.077	0.010

5 A fair six-sided dice is thrown 4 times and the random variable X denotes the number of 6s obtained.

    a Give the distribution of X.

    b Find, giving your answers to 3 dp

    i $P(X = 4)$ **ii** $P(X > 2)$ **iii** $P(1 \leq X < 3)$

6 A bag contains 12 counters. Three are red and the rest are black. A sample of five counters is taken, with each counter being returned before the next is chosen. Find the probability that the sample contains more than 3 red counters.

Reasoning and problem-solving

Strategy

To solve a probability problem involving the binomial distribution

  1 Check the conditions for a binomial distribution are met. List any assumptions.

  2 Identify the random variable and the corresponding values of n and p

  3 Calculate probabilities using the addition and multiplication rules if necessary.

Example 2

In Camborne from May to October 2015 the daily mean windspeed was given using the Beaufort scale. 12 days are picked at random from this set and could be described as either light (probability 0.6), fresh (probability 0.1) or moderate (probability 0.3). Stating any assumptions you make, calculate the probability of

    a 4 days being light,   **b** At least 6 days being either fresh or moderate.

  a Let X be the number of days which are light. Assume the daily mean windspeeds are independent of each other.

    $X \sim B(12, 0.6)$

    $P(X = 4) = 0.0420$

  b Let Y be the number of days which are fresh or moderate.

    $p = P(\text{any day is fresh or moderate})$

    $= 0.1 + 0.3 = 0.4$

    $Y \sim B(12, 0.4)$

    $P(Y \geq 6) = 1 - P(Y \leq 5) = 1 - 0.6652 = 0.3348$

(1) 12 identical, independent trials with two outcomes.

(2) 12 trials. $P(\text{light}) = 0.6$

(3) Calculate $P(X = 4)$

The outcomes are mutually exclusive.

(2) 12 trials, $p = 0.4$

1 **a** In a repeated set of trials, X is a random variable for the total number of 'successes'. State three conditions required to be able to model X by a binomial distribution.

b A weather caution is issued on any day when the wind gust exceeds 24 mph. In Leeming, between 1st May and 31st October 1987, this occurred 33 times out of the 148 days where data was available. **i** Calculate the probability that, in a set of 5 randomly chosen days during this period, fewer than 2 days had wind gusts exceeding 24 mph. **ii** Explain why a binomial model may not be suitable if the five days were consecutive.

2 For each of the following random variables, state whether the binomial distribution can be used as a good probability model. If it can, state the values of n and p; if it can't, or if its use is questionable, give reasons.

a The number of black counters obtained when 4 counters are chosen, with each being returned before the next is chosen, from a bag containing 6 black and 8 white counters.

b The number of patients in an independent random sample of size 8 at a GP practice who are prescribed antibiotics. You are given that 12% of patients are prescribed antibiotics.

c The number of heads in 5 throws of a biased coin where the probability of a head is 0.6

d The number of throws of a fair coin up to and including the first head.

3 A calculator claims it can randomly and independently generate a digit from 0–9. For any four digits generated, the probability of 2 zeros is 0.03. Is the calculator's claim correct? Show your working.

4 The West Cornwall Tourist Office stated that in Camborne half of all days in July 2015 had over $6\frac{1}{2}$ hours of sunshine.

a Use a binomial probability model to find the probability that, in a random sample of 8 days in July 2016, fewer than 4 days had over 6½ hours of sunshine. State any assumptions you make.

b Write down the probability that **i** More than 3 **ii** More than 4 days had over 6½ hours of sunshine.

5 You claim you can get at least one six in four throws of a fair dice. Your friend says you won't succeed. Who is more likely to be right? Show your working.

6 A pair of fair six-sided dice is thrown eight times. Find the probability that a score greater than 7 is scored no more than five times.

7 In the summer 1987 in Hurn, the median of the daily windspeed is 7 kn. Use the binomial distribution to find the probability that, in a random selection of 5 days, more than three had a daily windspeed of less than 7 kn.

8 Somebody claims they can tell the difference between two different brands, A and B, of tea. They are given 5 pairs of cups, where in each pair 1 cup contains brand A and 1 contains brand B. Assuming that they are guessing, find the probability that they correctly identify at least 3 pairs.

Challenge

9 For any family of 5 children, A is the event 'there is at least 1 boy and 1 girl' and B is the event 'there are more girls than boys'. A symmetrical binomial probability distribution can model X, the number of girls in a family of 5 children.

Are the events A and B independent of each other? Show your working.

STATS

Chapter summary

- $P(A \text{ or } B) = P(A) + P(B)$ if A and B are mutually exclusive.
- The total probability of all mutually exclusive events is 1
- If A is an event associated with a random experiment, then $P(A') = 1 - P(A)$
- For any sample space with N equally probable outcomes, if an event A can occur in n(A) of these outcomes, the probability of A is given by $P(A) = \dfrac{n(A)}{N}$
- If A and B are independent, $P(A \text{ and } B) = P(A) \times P(B)$
- If A and B are mutually exclusive, $P(A \text{ and } B) = 0$
- A probability distribution for a random experiment shows how the total probability of 1 is distributed between all the possible outcomes.
- Venn diagrams can be used to show the probabilities of various possible outcomes whose total probability adds to 1
- The conditions for a binomial probability distribution are:
  - Two possible outcomes in each trial.
  - Fixed number of trials.
  - Independent trials.
  - Identical trials (p is the same for each trial).
- If all the conditions are met, the binomial probability distribution can be used to calculate the probabilities of events expressed in terms of the number of 'successes' in a set of trials.
- If $X \sim B(n, p)$ then $P(X = x) = {}^{n}C_{x} p^{x} (1-p)^{n-x}$ where n is the number of trials and p is the probability of success in any given trial.
- Individual probabilities, $P(X = x)$, and cumulative probabilities, $P(X \le x)$, can be found directly on a calculator.

Check and review

You should now be able to...	Try Questions
✔ Use the vocabulary of probability theory, including the terms random experiment, sample space, independent events and mutually exclusive events.	1
✔ Solve problems involving mutually exclusive and independent events using the addition and multiplication rules.	1, 6, 8
✔ Find the probability distribution and probabilities for particular events.	1
✔ Recognise and solve problems relating to experiments which can be modelled by the binomial distribution.	3–9

1 You throw a coin until a head or three tails has occurred. If the probability of a head is p, find the probability distribution of the number of throws made. Check that the sum of the probabilities equals 1

2 Given that $X \sim B(16, 0.4)$, find

    a $P(X = 12)$ **b** $P(X > 6)$

    Give your answers to 3 dp.

3 A fair six-sided dice is thrown 5 times and the random variable X denotes the number of times an odd value is obtained.

    a   Give the distribution of X

    b   Find, giving your answers to 2 dp,

        i $P(X = 3)$ **ii** $P(X > 0)$

        iii $P(1 \leq X < 3)$

4 Telephone calls to an online bank are held in a queue until an advisor is available. Over a long period the bank has found that 8% of callers have to wait more than 4 minutes for a response. In a random sample of 20 callers, find the probability that fewer than 3 have to wait more than 4 minutes.

5 A pair of fair six-sided dice is thrown five times. Find the probability that a double six is scored no more than once.

6 A box contains 15 coloured beads. Three of the beads are red, 2 are black and the rest are white. A sample of 6 beads is taken, with each bead being returned to the box before the next is chosen. Find the probability that the sample contains more than 3 beads which are not white.

7 In a game of dice, you commit to getting at least one double six in 24 throws of two fair six-sided dice. Your friend says that you won't succeed. Who is more likely to be right? You must show your working.

8 A large packet of mixed flower seeds produce flowers which are either red, blue or white. These colours occur in the ratio $2 : 3 : 5$. Find the probability that, when 12 plants are grown, at least 6 of them are white.

9 At a postal sorting office, 32% of letters are classified as large and the rest are standard.

A random sample of 12 letters is taken. The random variable X is the number of standard letters in the sample.

    a   Write down the probability distribution of X

    b   Find the probability that more than two-thirds of the letters are standard.

What next?

Score		
0 – 4	Your knowledge of this topic is still developing. To improve, search in MyMaths for the codes: 2091 – 2094, 2110, 2111, 2114	🔗
5 – 7	You're gaining a secure knowledge of this topic. To improve, look at the InvisiPen videos for Fluency and skills (10A)	🎞
8 – 9	You have mastered these skills. Well done, you're ready to progress! Now try looking at the InvisiPen videos for Reasoning and problem-solving (10B)	🎞

Click these links in the digital book

Exploration

Going beyond the exams

10

History

A popular gambling game in 17th century France was to wager that there would be at least one 6 in every four throws of a standard dice.

Antoine Gombaud, a gambler, reasoned that an equivalent wager was that there would be at least one double 6 in every 24 throws of a pair of standard dice.

Over a period of time, however, Gombaud lost money on the wager. He decided to ask his friend **Blaise Pascal** for help. Pascal, in turn, contacted **Pierre de Fermat** and between them they formulated the fundamental principles of **probability** for the first time.

Blaise Pascal

Pierre de Fermat

Have a go

Find the probability of rolling at least one 6 in four throws of a standard dice.

Find the probability of rolling at least one double 6 in 24 throws of a standard pair of dice.

Explain why Gombaud won money on the first wager but lost money on the second.

Did you know?

Assuming the probability of having a birthday on any given day is the same, you only need 23 people in a room to make it more likely than not that at least two of them share a birthday. If there are 70 people in a room, the chance of two or more of them sharing a birthday is 100% if rounded to three significant figures.

This feels counterintuitive because 23 is such a small number compared to the number of days in a year (365, or 366 on a leap year).

The maths, however, is actually fairly simple. There are $^{23}C_2$ (or 253) ways of choosing a pair of random people out of a group of 23. To find the chances of finding *at least one* match, you just need to subtract the chance of no matches from 1.

$$P(\text{2 or more shared birthdays}) = 1 - \left(\frac{364}{365}\right)^{253} = 0.5004... \, (>50\%)$$

Birthdays actually come in clusters, so in reality this number may actually be lower than 23

1 Data, shown in the table, is collected on the number of students nationally who like two given brands of fast food.

Brand A	Brand B		
		Like	Dislike
	Like	11 713	19 981
	Dislike	9061	15 457

a How many students are polled in total? **[1 mark]**

b What is the probability a randomly chosen student likes

   i Brand A,   ii Brand B,   iii Both brands? **[9]**

c Are opinions of the two brands independent? **[2]**

2 On any given day, the probability that a commuter misses their bus to work is $\frac{1}{10}$ and the probability that they miss the bus home is $\frac{1}{12}$. The probability that they accidentally overcook their dinner is $\frac{1}{7}$. These events are independent.

a What does it mean for events to be independent? **[2]**

b Calculate the probability that the commuter misses their bus home and accidentally overcooks their dinner. **[2]**

c Calculate the probability that the commuter misses both buses but doesn't overcook their dinner. **[3]**

3 Two bags contain balls of various colours. A ball is drawn at random from a bag. The probabilities of drawing a specific colour from each bag are given in the table.

Event	Probability for first bag	Probability for second bag
White	$3k$	?
Blue	$6k$	0.15
Black	$4k$	0.1
Red	$2k$	0.25
Green	$5k$	0.15

a Calculate the value of k **[3]**

b Calculate the probability of drawing a white ball from the second bag. **[2]**

c Tia wants to maximise the probability of drawing a white or a blue ball. Which bag should she choose? **[5]**

4 A satsuma must meet a minimum size requirement in order to be suitable for packaging. Each packet contains 8 satsumas. The grower finds that the probability of a randomly chosen satsuma not being large enough is 0.01

a Find the probability that a random set of 8 satsumas contains at least one that is not suitable for packaging. **[4]**

b Find the probability that a random set of 8 satsumas contains at most one that is not suitable for packaging. **[2]**

A batch is accidentally sent out without being checked for the minimum size. A supermarket receives 60 packets.

c Find the probability that the supermarket has received at least one packet that contains at least one undersized satsuma. **[5]**

5 At a factory, sweets are automatically discarded if they are misshapen. An inspector picks five discarded sweets at random to check that the right decisions are being made. If at least four of the discarded sweets are misshapen, then the inspector is satisfied.

   a What conditions must be true for the binomial distribution to be a suitable model for this situation? [2]

   On average, 84 out of 360 discarded sweets are not misshapen.

   b Find the probability that the first four inspected sweets are misshapen but then the fifth is fine. [2]

   c Find the probability that exactly one sweet is not misshapen. [2]

   d Find the probability that the inspector is satisfied. [4]

6 In a football tournament, only two teams have a chance of winning. Team B will only win the tournament if they win all three of their remaining matches *and* Team A fails to win any of its four remaining matches to win the league. All match results are independent. The probability that Team A wins any of their matches is 0.56 per match and the probability that Team B wins any of their matches is 0.61 per match. Find the probability that Team A wins the tournament. [4]

7 Data is collected on the number of days between May and October 2015 in which the daily maximum temperature and relative humidity are each above or below the average in Hurn.

		Daily maximum relative humidity	
		Above average	Below average
Daily maximum temperature	Above average	25	61
	Below average	40	58

   a How many days are included in the data? [1]

   b What is the probability a randomly chosen day has

      i Above-average daily maximum temperature,

      ii Above-average daily maximum relative humidity,

      iii Both measures above average? [3]

   c Are daily maximum temperature and relative humidity independent in this sample? Justify your answer. [2]

8 An outdoor park has to be closed if it is raining too much. The probability it rains too much on any day is 0.09

   Assuming that the probability it rains on any day is independent of rain on any other day

   a Find the probability that the park is closed at least once in a week, [4]

   b Find the probability that the park is closed at most once in a week. [4]

   Over a year, the weather is tracked to see how often the park has to be closed.

   c Find the probability that the park is closed in at least 20 weeks over the year. [2]

   d What is the expected number of weeks that the park is closed for at least one day? [2]

9 Three weather stations in a town measure the daily total sunshine. The probability that stations A, B and C measure the most sunshine are 0.41, 0.36 and 0.23 respectively. Find the probability that over a five-day period, station A records the most sunshine on more days than the other stations. [5]

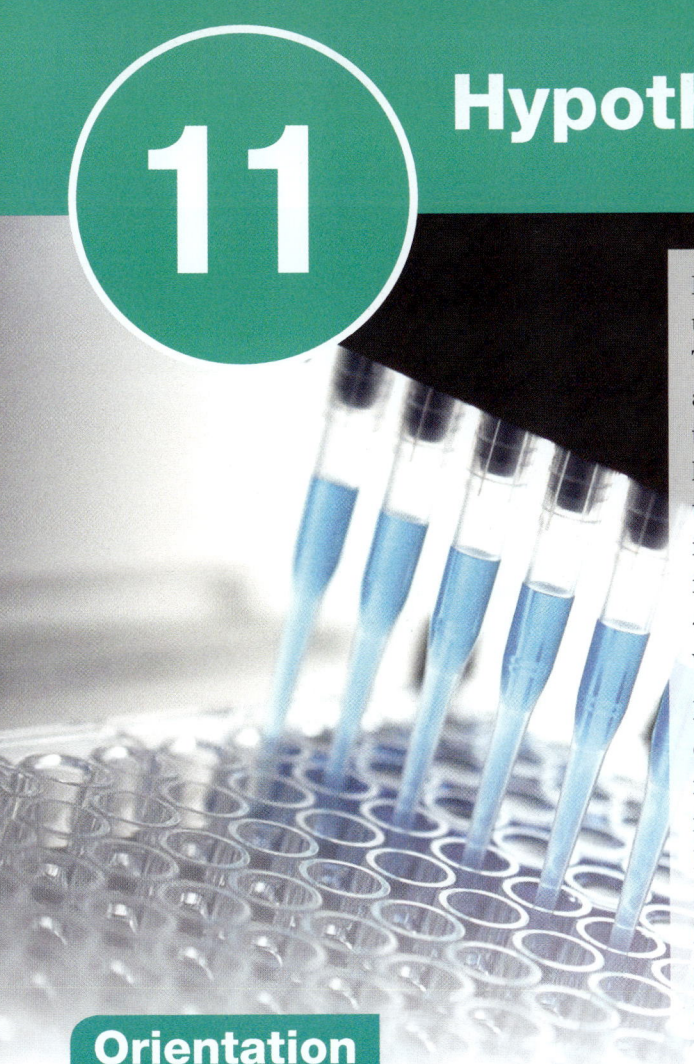

11 Hypothesis testing 1

Before a new medicine can be released for public use, it has to go through a rigorous testing process. These tests include medical trials on people who are candidates for the treatment. This allows for the monitoring of side effects, the observation of how many trial patients actually respond to the treatment, and consideration for what level of improvement is deemed sufficient. For example, if a drug significantly improves the condition of 40% of trial patients, but has no effect on the rest, would it be considered successful?

In order to test a premise, you must have a clearly defined hypothesis as well as unbiased methods for collecting and analysing data. Does the data follow any kind of pattern, or fit a known model? How much error is there in the measurements and how much error is acceptable? What would lead you to reject your hypothesis? This is all part of the scientific method for investigating phenomena and making new discoveries.

Orientation

What you need to know

KS4
- Interpret, analyse and compare the distributions of data sets.
- Recognise appropriate measures of central tendency and spread.

Ch10 Probability and discrete random variables
- Binomial distributions.

p.260

What you will learn

- To understand the terms null hypothesis, alternative hypothesis, critical value, critical region and significance level.
- To calculate the critical region and the p-value.
- To decide whether to reject or accept the null hypothesis and make conclusions.

What this leads to

Ch21 Hypothesis testing 2
Testing correlation.
Testing a normal distribution.

Careers
Scientific research.
Quality control.

 **MyMaths** Practise before you start $\mathbb{Q}$ 2111

Fluency and skills

A shop claims that 85% of its customers are satisfied with its service. You think this figure is too high and take a random sample of 10 customers. If only 8 of the sample are satisfied, would you conclude that 85% is too high? What if only 6 or 7 are satisfied?

You use a **hypothesis test** to determine whether to accept or reject the shop's claim.

> **Key point**
>
> The **null hypothesis**, H_0, is a statistical statement representing your basic assumption.

Don't reject the null hypothesis until there is sufficient evidence to do so.

> **Key point**
>
> The **alternative hypothesis**, H_1, is a statement that contradicts the null hypothesis.

Example 1

A shop makes this claim: '85% of our customers are satisfied with our service.'

Let p be the probability that a customer, chosen at random, is satisfied.

Write the null hypothesis and the alternative hypothesis in these cases.

a The claim is believed to be an overestimate.
b The claim is believed to be an underestimate.
c The claim is believed to be incorrect.

a $H_0: p = 0.85$ and $H_1: p < 0.85$

b $H_0: p = 0.85$ and $H_1: p > 0.85$

c $H_0: p = 0.85$ and $H_1: p \neq 0.85$

H_1 expresses the alternative suggestion.

If H_0 is an underestimate the true value must be higher.

The true value of p could be more or less than 0.85

> **Key point**
>
> The null hypothesis always includes the equality sign.

In parts **a** and **b** of Example 1, you can test the null hypothesis against the alternative hypothesis with a **one-tailed test**. This only tests either *below* the value stated in H_0 or only *above* the value stated in H_0. In part **c**, you can test the hypothesis with a **two-tailed test**. This tests both below and above the value stated in H_0 ($p < 0.85$ and $p > 0.85$)

Let X be the random variable representing the number of customers, in the chosen group of 10, who are satisfied. X is called the **test statistic**.

The value of X determines whether you accept or reject the null hypothesis.

Ch 10.2
Assuming H_0 is true, X is binomially distributed: $X \sim B(10, 0.85)$

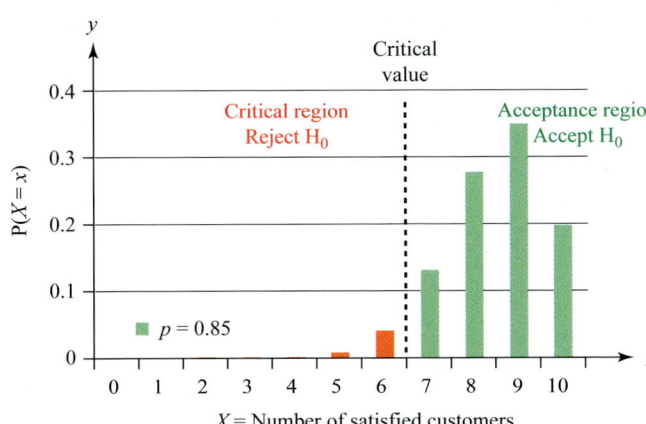

$X =$ Number of satisfied customers

For every hypothesis test, there is a set of values of X for which you accept H_0 and a set for which you reject H_0

> **Key point**
>
> The **critical region** is the set of values that leads you to *reject* the null hypothesis.
> The **acceptance region** is the set of values that leads you to *accept* the null hypothesis.

The **critical value** lies on the border of the critical region. It depends on the **significance level** of the test. The critical region includes the **critical value** and all values that are more extreme than that.

> **Key point**
>
> Every hypothesis test has a significance level. This is equal to the probability of *incorrectly* rejecting the null hypothesis.

If you use a *low* significance level, e.g. 1%, you can be fairly certain of your result if you reject the null hypothesis. If you reject the null hypothesis when using a *high* significance level, e.g. 20%, you may need to carry out further tests to be sure of your result. Significance levels of 10%, 5% and 1% are often used, but the significance level could be any value.

> In a discrete distribution such as the binomial distribution, the probability of incorrectly rejecting H_0 is the probability represented in the critical region and is therefore **less than or equal to** the significance level.

After carrying out a hypothesis test, you must state whether you accept H_0 or reject H_0

You must also state a **conclusion** that relates directly to the problem.

There is always a chance that your decision to accept H_0 or reject H_0 is wrong. Your conclusion must use language that reflects this.

> The lower the significance level, the smaller the critical region, and vice versa.

A different significance level may produce a different critical region and so may lead to a different conclusion.

 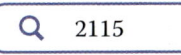 

STATS

Example 2

Let X be the number of satisfied customers in a sample of 10 customers at a shop. Let p be the probability that a customer, chosen at random, is satisfied. A hypothesis test is carried out to assess the shop's claim $p = 0.85$ against the alternative hypothesis $p < 0.85$

At a significance level of 5%, the critical region is $X \leq 6$

If the significance level is changed to a%, the critical region is $X \leq 4$

a Write an inequality for a

b A sample is taken and 6 customers out of 10 are satisfied. Write down the conclusion if the significance level is

    i 5%   **ii** a%

> **a** $a < 5$

> **b** **i** Reject H_0, as there is enough evidence to suggest that the shop's claim is too high.

>   **ii** Accept H_0, as there is not enough evidence to suggest that the shop's claim is too high.

The size of the critical region is reduced so, assuming H_0 is correct, the probability of getting a value in the critical region is reduced.

6 falls into the critical region $X \leq 6$. State your answer in context.

6 falls into the acceptance region $X \geq 5$

Key point

As you lower the significance level, you need more evidence to reject the null hypothesis and you lower the chance of making an incorrect conclusion.

Example 3

A driving instructor claims that 70% of his candidates pass first time. An inspector thinks that this is inaccurate, so he does a survey of 25 former candidates and records the number who passed first time. The significance level of his test is 10% and the critical values are 14 and 21

The null hypothesis is that the driving instructor's claim is correct, so H_0: $p = 0.7$ where p is the probability that a candidate passes first time.

The alternative hypothesis is that the driving instructor's claim is wrong, so H_1: $p \neq 0.7$

a State the critical region and the acceptance region for the test.

b State whether the inspector would accept or reject the null hypothesis if he found that

    i 14 of the former candidates passed first time,

    ii 20 of the former candidates passed first time.

> X is the number of the 25 former candidates who passed first time.

> **a** Critical region is $X \leq 14$ and $X \geq 21$

>   Acceptance region is $15 \leq X \leq 20$

> **b** **i** Reject the null hypothesis.

>   **ii** Accept the null hypothesis.

The critical region consists of the critical value, 14, and more extreme values. Here, that is values less than 14

This is everything not in the critical region. X can only be an integer in this situation.

14 is the critical value and so lies in the critical region.

20 lies in the acceptance region.

In questions **1** to **3**, p is the probability of an event occurring.

1 State the critical region and acceptance region for the following tests.

   a $H_0: p = 0.6$ and $H_1: p < 0.6$
     The critical value is 5

   b $H_0: p = 0.6$ and $H_1: p > 0.6$
     The critical value is 5

   c $H_0: p = 0.6$ and $H_1: p \neq 0.6$
     The critical values are 5 and 12

2 State whether to accept or reject the null hypothesis for the following tests.

   a $H_0: p = 0.4$ and $H_1: p < 0.4$
     The critical value is 3 and in the sample taken, X takes a value of 4

   b $H_0: p = 0.3$ and $H_1: p > 0.3$
     The critical value is 17 and in the sample taken, X takes a value of 18

   c $H_0: p = 0.7$ and $H_1: p \neq 0.7$
     The critical values are 2 and 11 and in the sample taken, X takes a value of 10

3 a $H_0: p = 0.4$ and $H_1: p < 0.4$
     In the sample taken, X takes a value of 18.
     H_0 is rejected at the 10% significance level. State whether you accept or reject the null hypothesis at a 20% significance level.

   b $H_0: p = 0.3$ and $H_1: p > 0.3$
     In the sample taken, X takes a value of 18.
     H_0 is accepted at the 10% significance level. State whether you accept or reject the null hypothesis at a 5% significance level.

4 A football coach claimed that he lost only 15% of his games. One of his players thinks that this claim is inaccurate and decides to test it at the 5% significance level.

   A random sample of 50 games is taken. The critical values for the number of losses are 2 and 14

   a Write a null and alternative hypothesis to represent this situation.

   b State the critical region and the acceptance region for the test.

   c State whether you accept or reject the null hypothesis if in the sample

     i 3 of the games were lost,

     ii 15 of the games were lost,

     iii 14 of the games were lost.

5 A courier company states on its website that 90% of its parcels arrive within 24 hours. An investigator thinks that this claim is too high, so he tests it by taking a random sample of 40 parcels.

   The critical value at the 10% significance level is 33

   a State the critical region and the acceptance region for the test.

   b State whether you accept or reject the null hypothesis if

     i 32 of the parcels arrive within 24 hours,

     ii 35 of the parcels arrive within 24 hours.

6 A factory worker estimates that 1 in 20 products on the production line are faulty. A supervisor thinks that his estimate is too low, so she tests it at the 10% significance level.

   A random sample of 100 products is taken. The critical value is 9

   a State the critical region and the acceptance region for the test.

   b The supervisor says that the critical value for the 20% significance level is more than 9. Is she correct?

STATS

7 A vehicle inspector thinks that 1 in 25 cars do not have the correct tyre pressure. A colleague believes that the proportion suggested by the inspector may be higher, so she tests it at the 10% significance level.

A random sample of 60 cars is taken. She measures tyre pressure and notes the number of cars which do not have the correct pressure. The critical value is 5

a State the critical region and the acceptance region for the test.

b The critical value for the $p\%$ significance level is more than 5. Find an inequality for p

Reasoning and problem-solving

Strategy

To interpret the result of a hypothesis test

(1) Identify the critical region and draw a diagram if it helps.

(2) If the value from the sample lies in the critical region then you reject the null hypothesis.
If the value does not lie in the critical region then you accept the null hypothesis.

(3) End with a conclusion that relates back to the situation described in the question.

Example 4

A researcher claims that the daily mean air temperature exceeds 20 °C on 3 out of every 5 days in Beijing between May and October 2015. A supervisor thinks that this claim may be inaccurate. She tests the claim by choosing a random sample of 50 days during this time period and gathering data values for the daily mean air temperature in Beijing. The critical values at the 5% significance level are 22 and 38

a State the hypotheses needed to investigate this claim.

b If 40 data values were above 20°C, state whether you would accept or reject the null hypothesis. State your conclusion clearly.

The supervisor concludes that there is not enough evidence at the 5% significance level that the researcher's claim is inaccurate.

c Write an inequality for n, the number of data values in the sample that do *not* exceed 20 °C.

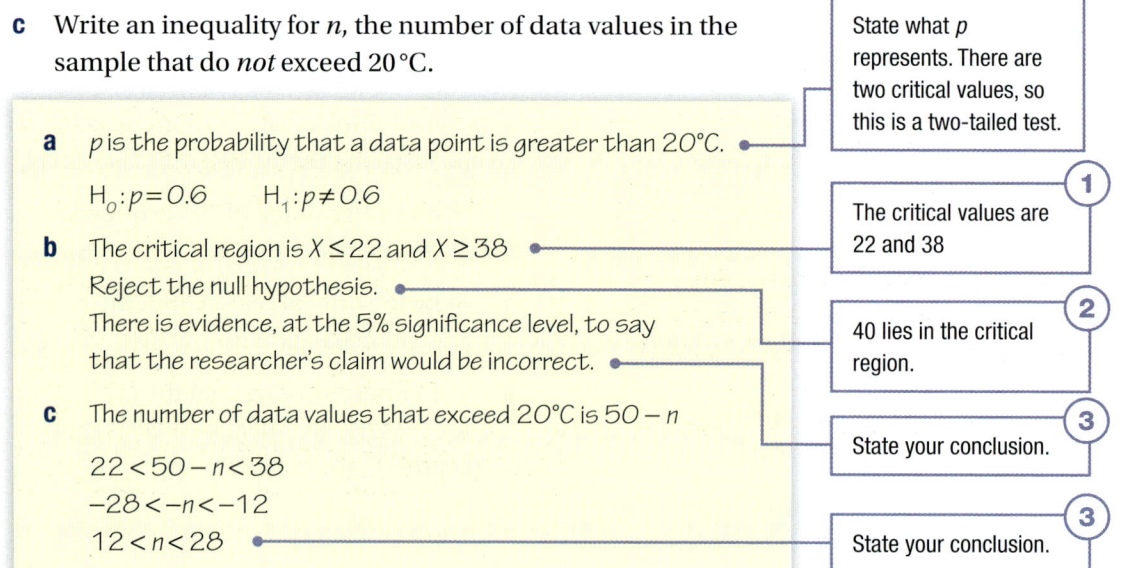

a p is the probability that a data point is greater than 20°C.

$H_0: p = 0.6 \qquad H_1: p \neq 0.6$

> State what p represents. There are two critical values, so this is a two-tailed test.

b The critical region is $X \leq 22$ and $X \geq 38$
Reject the null hypothesis.
There is evidence, at the 5% significance level, to say that the researcher's claim would be incorrect.

> **(1)** The critical values are 22 and 38

> **(2)** 40 lies in the critical region.

> **(3)** State your conclusion.

c The number of data values that exceed 20°C is $50 - n$

$22 < 50 - n < 38$
$-28 < -n < -12$
$12 < n < 28$

> **(3)** State your conclusion.

1 State H_0 and H_1 in each of the following cases.

  a The probability of an event occurring is stated as 0.6 but this is thought to be too low.

  b The probability of an event occurring is stated as 0.7 but this is thought to be too high.

  c The probability of an event occurring is stated as 0.4 but this is thought to be inaccurate.

2 A resident estimates that, in Jacksonville, on 6 in 20 days in 2015, the mean windspeed is over 7 knots. The critical values at the 5% significance level are 1 and 11

  a State the hypotheses needed to investigate this claim.

  b If the null hypothesis is accepted then state the inequality for n, the number of days in 2015 in Jacksonville with a mean windspeed of over 7 knots.

3 A politician believes that 35% of her constituency support her. A researcher tests the claim to see if she is underestimating her support. The critical value at the 15% significance level is 41

  a State the hypotheses needed to investigate this claim.

  b If 39 voters say that they support the politician, decide if you would accept or reject the null hypothesis and state your conclusion.

  c If the conclusion is 'there is evidence that the politician is underestimating her support' then state the lowest possible value of n, the number of voters in the sample who said they support the politician.

4 James estimates that, on 25% of days in a month, Leeming gets fewer than 4 hours of sunshine. His friend thinks that the percentage of days with fewer than 4 hours of sunshine is lower than that. The critical value at the 10% significance level is 6

  a State the hypotheses needed to investigate the claim.

  b If James observes that Leeming gets fewer than 4 hours of sunshine on 5 days in a month, state whether you would accept or reject the null hypothesis. State the conclusion.

  c James observes that Leeming gets fewer than 4 hours of sunshine on n days and the conclusion was 'there is not sufficient evidence, at the 10% significance level, to say that the percentage of days in a month on which Leeming gets fewer than 4 hours of sunshine is less than 25%'. State an inequality for n

  d The critical value at the x% significance level is 5. State an inequality for x

---

Challenge

5 $H_0: p = 0.2$ $H_1: p \neq 0.2$

A random sample of 80 is taken.

The critical values at the 5% significance level are 8 and 24

  a Find inequalities for a and b where a and b are the critical values at the 10% significance level and $a < b$

  b Find an inequality for x if 7 and 26 are the critical values at the x% significance level.

STATS

Fluency and skills

Suppose somebody claims that 40% of people prefer apples to pears. You disagree with this claim, and believe the true value to be lower than 40%. You plan to investigate this claim by asking a sample of n people whether they prefer apples to pears. The hypotheses are then:

$H_0: p = 0.4$ $H_1: p < 0.4$

where p is the probability of a person, chosen at random, saying they prefer apples to pears.

In this kind of hypothesis test, you can find the critical values or critical region using the binomial distribution, provided certain conditions are true: samples are chosen at random, there are two possible outcomes (in this case, yes or no), the outcomes are independent of one another, and the same chance of each outcome can be assumed each time.

You can therefore assume that the number of people, X, who say they prefer apples to pears follows a binomial distribution with n trials. Thus $X \sim B(n, p)$

Say you ask a random sample of 50 people. Assuming the null hypothesis to be true, you would expect $x \approx 20$, where x is the number who say yes. The probabilities for any observed value of x are binomially distributed around $x = 20$, as shown in the graph for $X \sim B(50, 0.4)$

> Always assume H_0 to be true in your calculations.

The alternative hypothesis is $p < 0.4$. That means this is a one-tailed test and the critical region involves values equal to or lower than the critical value. Say you are testing at a significance level of 5%: that gives a critical value such that $P(X \leq x) = 0.05$. Using your calculator, you can see that $P(X \leq 14) = 0.054$ and $P(X \leq 13) = 0.028$

See Ch 10.2

To check how to calculate probabilities using the binomial theorem.

Because the number of people can only take an integer value, the critical value must also be an integer. So you choose the first value where the probability is more extreme than the significance level, that is, $X = 13$. So if 13 or fewer people say they prefer apples to pears, you would reject H_0

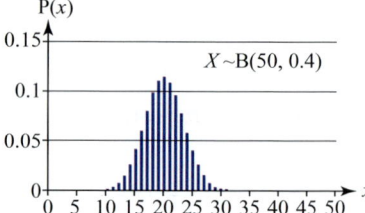

▲ If you assume p to be 0.4, then the most likely outcome is for 20 of the 50 people to prefer apples. But that doesn't mean other numbers are impossible: as you can see, the probability of exactly 20 people is only just over 0.1

> **Key point**
>
> If H_0 is assumed to be true, for discrete random variables, a value lies within the critical region if the probability of X being *equal to or more extreme* than that value is *equal to or less than* the significance level.

In many cases, instead of finding the critical region, it is easier to simply calculate the probability that X is *equal to or more extreme* than the observed value. If the probability of any observed value is *less than or equal to* the significance level, you have grounds to reject the null hypothesis.

> **Key point**
>
> The **p-value** is the probability that X is *equal to or more extreme* than an observed value. If the p-value is *greater* than the significance level, you accept H_0. If the p-value is *less than or equal to* the significance level, you reject H_0

So, going back to the case of people's preference for apples and pears: the p-value for 13 people preferring apples is 0.028 (=2.8%) and the p-value for 14 people preferring apples is 0.054 (=5.4%). At a significance level of 5%, you would reject the null hypothesis for $x = 13$ and accept the null hypothesis for $x = 14$

> If you had been testing at a higher significance level, say 10%, you would reject the null hypothesis in both cases, because the p-value for both is lower than 10%

Example 1

You are told that $X \sim B(40, p)$ and $H_0: p = 0.25$, $H_1: p \neq 0.25$

For each of the following values, **i** State the p-value to 4 decimal places and **ii** State, with clear reason, your conclusion for the hypothesis test at a 10% significance level.

a $X = 7$ **b** $X = 16$

> 10 is the most likely outcome. This is a two-tailed test, so you must consider values much higher and much lower than 10

$0.25 \times 40 = 10$

a i $P(X \leq 7) = 0.1820$

    ii The p-value is greater than 0.05, so accept H_0

b i $P(X \geq 16) = 1 - P(X \leq 15)$
$$= 1 - 0.9738$$
$$= 0.0262$$

    ii The p-value is less than 0.05, so reject H_0

> Use your calculator to find the p-value for $X = 7$

Exercise 11.2A Fluency and skills

1 $X \sim B(30, p)$ and
$H_0: p = 0.2$ $H_1: p < 0.2$
  a Find the critical region for X if the significance level is 2%
  b State the conclusion if $X = 1$

2 $X \sim B(20, p)$ and
$H_0: p = 0.4$ $H_1: p \neq 0.4$
Find the critical region for X if the significance level is 10%

3 $X \sim B(30, p)$ and
$H_0: p = 0.15$ $H_1: p < 0.15$
The significance level is 10%
Find the p-value (to 4 decimal places) and state the conclusion if
  a $X = 1$ **b** $X = 2$

4 $X \sim B(25, p)$ and
$H_0: p = 0.2$, $H_1: p \neq 0.2$
The significance level is 5%

Find the p-value (to 4 decimal places) and state the conclusion if
  a $X = 1$ **b** $X = 9$

5 $X \sim B(60, p)$ and
$H_0: p = 0.45$ $H_1: p \neq 0.45$
The significance level is 10%.
Find the p-value (to 4 decimal places) and state the conclusion if
  a $X = 34$ **b** $X = 21$

6 $X \sim B(30, p)$ and
$H_0: p = 0.15$, $H_1: p > 0.15$
Find the critical region for X if the significance level is 15%

7 $X \sim B(40, p)$ and
$H_0: p = 0.35$, $H_1: p \neq 0.35$
A test has significance level 1%
Without finding the critical region, state the conclusion if
  a $X = 6$ **b** $X = 22$

8 $X \sim B(50, p)$ and $H_0: p = 0.3$, $H_1: p > 0.3$

   a Find the critical region for X if the significance level is

   i 0.2%   **ii** 2%   **iii** 20%

   b State the conclusion in each case if $X = 22$

   c The critical value for the p% significance level is 21. Find an inequality for p

9 $X \sim B(80, p)$ and $H_0: p = 0.7$, $H_1: p > 0.7$

   a Find the critical region for X if the significance level is

   i 1%   **ii** 5%   **iii** 10%

   b State the conclusion in each case if $X = 64$

   c The null hypothesis is rejected at the p% significance level when $X = 64$. Find the smallest possible whole number value of p

10 $X \sim B(40, p)$ and $H_0: p = 0.4$, $H_1: p < 0.4$

   The critical region for X is $X \le 8$ at the y% significance level. Find the greatest possible whole number value of y

11 $X \sim B(50, p)$ and $H_0: p = 0.05$, $H_1: p < 0.05$

   a Find the critical region for X if the significance level is 10%

   b Find the smallest integer value for x such that the critical region at the x% significance level is different to your answer to **a**.

12 $X \sim B(60, p)$ and $H_0: p = 0.2$, $H_1: p \ne 0.2$

   The critical region for X is $X < 6$, $X \ge 18$ at the $2a$% significance level.
   Find the two possible integer values of a

Reasoning and problem-solving

To find a critical region and interpret the result when X is binomially distributed

1 Define X, state its distribution and write down H_0 and H_1

2 Assume that H_0 is true and either find the critical region or calculate the p-value.

3 Decide whether to accept or reject H_0 and interpret the result in the context of the question.

Example 2

A farmer in Camborne claims that it rains on 80% of the days in February. A farmworker thinks that the farmer's claim is inaccurate and so records whether or not it rains on each of the 28 days in one particular February. He uses a hypothesis test at the 10% significance level to test whether or not the farmer's claim is accurate. State the farmworker's conclusion if he found that it rained on 26 out of the 28 days in February.

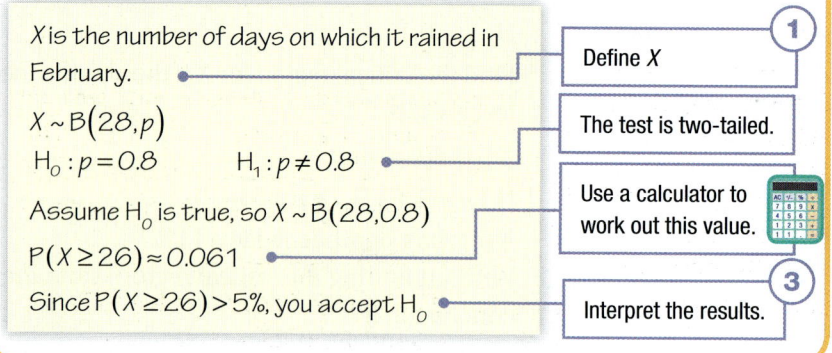

X is the number of days on which it rained in February.

$X \sim B(28, p)$

$H_0: p = 0.8$ $H_1: p \ne 0.8$

Assume H_0 is true, so $X \sim B(28, 0.8)$

$P(X \ge 26) \approx 0.061$

Since $P(X \ge 26) > 5\%$, you accept H_0

Define X ①

The test is two-tailed.

Use a calculator to work out this value.

③

Interpret the results.

Know your data set

Large data set Daily total rainfall includes snow or hail which is measured in its liquid state.
Click this link in the digital book for more information about the Large data set.

ICT Resource online

Large data set

To experiment with hypothesis testing, click this link in the digital book.

1 A dog-food company claimed that 8 of 10 owners say their dogs prefer their product to other brands. 40 owners were surveyed and a hypothesis test was carried out at the 5% significance level to determine whether there was any evidence that the company was overestimating the popularity of its product.

    a State the hypotheses clearly and find the critical region.

    b State the conclusion if 31 owners said that their dogs preferred the product.

2 A meteorologist estimates that, on 60% of summer days in Beijing, the mean air temperature exceeds 24°C. A resident in Beijing believes this claim to be inaccurate, so she chooses 75 summer days at random and carries out a hypothesis test at the 2% significance level. She found that, on 56 of the days, the mean air temperature exceeded 24°C.

    a State the hypotheses clearly and calculate the p-value (to 4 decimal places).

    b State the conclusion that the resident should make.

3 In a large container of sweets, 15% are blackcurrant-flavoured. After a group of children have eaten a lot of the sweets, one of the children wants to see whether the proportion of blackcurrant sweets in the container has changed. She selects a random sample of 60 sweets and finds that 4 of them are blackcurrant-flavoured.

    a Stating your hypotheses clearly, test at the 10% level of significance whether or not there is evidence that the proportion of blackcurrant-flavoured sweets has changed.

    b If the child discovers that she miscounted and that actually there are 5 blackcurrant-flavoured sweets, would your conclusion change?

4 When a coin is tossed a number of times, more than twice as many heads as tails are recorded.

    a Use the p-value to test, at the 5% level, whether a coin is biased towards heads if

      i The coin is tossed 4 times,

      ii The coin is tossed 24 times.

    b If the conclusion is 'there is sufficient evidence to say that the coin is biased towards heads', find the smallest number of times that the coin could have been tossed.

5 The discrete random variable X has the following probability distribution.

x	0	1	2	3	4	5	6
$P(X = x)$	$5k^2$	$2k$	0.1	$3k$	$10k^2$	0.2	0.05

    a State the value of k

    b If the critical values are 0 and 6 then find a lower bound for the significance level.

6 A phone repairer found that 1 in 10 phones brought in for repair had cracked screens. She suspects that over time this proportion has reduced. She carries out a hypothesis test at the 10% significance level on the next 40 phones that are brought in.

    a State clearly H_0 and H_1

She initially thought that n of these phones had cracked screens and concluded that she should reject H_0. She then found that one more screen was cracked and concluded that she should accept H_0

    b Find the value of n

Challenge

7 A coin is tossed n times. X represents the number of heads recorded. It is tested at the 2% level to see whether the coin is fair. The critical region is $X \leq 12$ and $X \geq 28$

    a Find the value of n

    b Find the critical region for a 1% significance level.

STATS

 2115  SEARCH

Chapter summary

- The null hypothesis, H_0, is the basic assumption in a statistical test. Do not reject the null hypothesis until there is sufficient evidence to do so.
- If H_1 has the form $p < ....$ only look at the left tail of the distribution (green in the diagram).
- If H_1 has the form $p > ....$ only look at the right tail of the distribution (red in the diagram).
- If H_1 has the form $p \neq ....$ look at both tails of the distribution (green and red in the diagram).
- In a two-tailed test you always consider half of the significance level at the left tail and half of the significance level at the right tail.

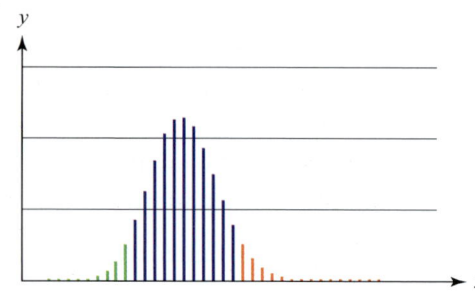

- The critical value lies in the critical region and is the cut-off point for whether values are in the critical region or not. The lower to significance level, the smaller the critical region.
- If you are given a value and asked to test it then it is often quicker to use the p-value method.
- In every hypothesis test you must state whether you accept or reject the null hypothesis.
- If the context is given for a hypothesis test then you must state a conclusion that relates directly to the context of the question.

Check and review

You should now be able to...	Try Questions
✔ Understand the terms null hypothesis and alternative hypothesis.	1–6
✔ Understand the terms critical value, critical region and significance level.	1–6
✔ Calculate the critical region.	2, 3, 5, 6
✔ Calculate the p-value.	4, 6
✔ Decide whether to reject or accept the null hypothesis.	4, 6
✔ Make a conclusion based on whether you reject or accept the null hypothesis.	4, 6

1   A proportion, p, of individuals in a large population have a characteristic, C. An independent random sample of size n is taken from the population and the number, X, of individuals with C is noted. Write down the distribution of X

2   A factory makes security keypads which have 25 buttons on them. 20% of the buttons are thought to be defective after the initial manufacturing process. A manager thinks that the figure of 20% may be wrong so she selects a keypad at random and carries out a hypothesis test at the 5% significance level.

   a   State the hypotheses clearly.

   b   Find the critical region for the test.

   c   Using this critical region, state the probability of concluding the figure of 20% is incorrect when in fact it is correct.

   d   Explain why the answer to part **c** is not equal to 5%.

3 Lydia flips a coin to predict whether or not her school netball team will win their game. Olivia says that her analysis is better, so she records the results of the next 18 games and carries out a hypothesis test at the 10% significance level.

a State the hypotheses clearly.

b What is the lowest number of predictions that Olivia would have to get right in order to justify her claim?

4 Students took a multiple choice maths exam. Each of the 50 questions had 5 possible answers. One student did the test very quickly, didn't seem to read the questions at all and got 15 answers right. The teacher thought that this student had simply guessed all the answers. She conducts a hypothesis test at the 5% significance level to see if there is any evidence to suggest that the student performed better than if he had simply been guessing.

a State the hypotheses clearly.

b Use the p-value method to carry out the test and state the conclusion.

5 a State the conditions under which the binomial distribution provides a good model for a statistical experiment and state the probability distribution function.

b Using the binomial distribution function, copy and complete the following table for binomial probabilities with $n = 6$, $p = 0.34$

x	0	1	2	3	4	5	6
$P(X = x)$	0.0827	0.2555	0.3290				0.0015

c A random variable X is known to have a binomial distribution with $n = 6$. A test with a significance level of 5% is performed to investigate whether the parameter p equals 0.34 against the hypothesis that it is greater than this value. Use your answer to part **b** to find the set of x-values which would suggest that $p > 0.34$ and give the probability of incorrectly rejecting H_0

6 A scratch-card company claims that 10% of cards win prizes. A customer conducts a hypothesis test at the 10% significance level to see if the claim is inaccurate. He collects 45 cards and only wins with one card.

a State the hypotheses clearly.

b The customer says that the probability of winning with only one card is less than 5%, so he concludes that the company's claim is too high. Is he correct?

c The customer then carried out another hypothesis test on 90 cards. He won with two cards. He said that as both the total number of cards and the number of winning cards had doubled he should make the same conclusion as before. Is he right?

What next?

Score			
	0 – 3	Your knowledge of this topic is still developing. To improve, search in MyMaths for the code: 2115	Click these links in the digital book
	4 – 5	You're gaining a secure knowledge of this topic. To improve, look at the InvisiPen videos for Fluency and skills (11A)	
	6	You've mastered these skills. Well done, you're ready to progress! Now try looking at the InvisiPen videos for Reasoning and problem-solving (11B)	

STATS

Ronald Fisher

History

Ronald Fisher (1890 - 1962) was born in England and studied at Cambridge University.

In 1925 he published a book on statistical methods, in which he defined the **statistical significance test**. The book was well received by the mathematical and scientific community and significance testing became an established method of analysis in experimental science.

Two other statisticians, **Jerzy Neyman** and **Egon Pearson**, also published a number of papers on **hypothesis testing**. The papers introduced much of the terminology used today, though their approach was different to Fisher's. Neyman and Fisher argued over the relative merits of their work for many years.

> "Statistics is the servant to all sciences."
> - Jerzy Neyman

Did you know?

Fisher applied his methods to test the claim, made by **Muriel Bristol**, that she could tell whether the tea or the milk was added to a cup first. This became known as the **lady tasting tea** experiment.

Have a go

In the lady tasting tea experiment, the lady was offered 8 cups of tea at the same time. 4 were prepared with milk first and 4 were prepared with tea first.

The lady was then asked to choose the 4 cups which had been prepared with tea first.

How many possible combinations of 4 cups could she give?

(You could use the nCr button on your calculator.)

In the test, Fisher would only consider the lady's claim as valid if she identified all 4 cups correctly. What was the significance level of this test?

How many cups would Fisher have had to use if he wanted to test the lady's claim at a significance level of less than 0.1%?

Muriel Bristol successfully identified all 4 cups, passing Fisher's test.

1. A certain variety of sweet pea produces flowers of various colours. Plants with yellow flowers are particularly prized. A random sample of n plants is chosen to test, at the 5% significance level, whether or not the proportion of plants with yellow flowers is $\frac{1}{2}$

   a. If n is the sample size and p is the proportion of plants producing yellow flowers, write down the null and alternative hypotheses for this test. **[1 mark]**

   b. If $n = 6$, find the critical region, giving the probability of incorrectly rejecting H_0 **[5]**

2. A random variable X has a binomial distribution with parameters $n = 10$ and p, a constant. The value of p was known to be 0.4 but is now believed to have increased.

   a. Write down the null and alternative hypotheses in a test of this belief. **[1]**

   b. Use a significance level of 5% to determine the values of X that would suggest the belief is incorrect. **[4]**

   40% of days in Leuchars during the summer of 1987 had over 6 hours of sunshine. In 2015, a random sample of 10 summer days was taken and 8 of them were found to have over 6 hours of sunshine.

   c. Use your answer to parts **a** and **b** to state whether you believe that the proportion of days with over 6 hours of sunshine has changed during that time period. **[1]**

3. You wish to investigate whether a coin is biased towards heads. You toss the coin 5 times and note the number of heads showing.

   a. Given that the number of heads is 4, perform a 5% significance test, stating clearly your null and alternative hypotheses. **[5]**

   b. Would your conclusion change if the number of heads showing was 5? **[1]**

4. It is estimated that 40% of cars on the road have a mechanical defect which breaks current road traffic regulations. A sample of 20 cars were examined and 6 were found to have such defects.

   a. State a condition on the method of choosing the sample so that a binomial probability model can be used to test the estimate. **[1]**

   b. Assuming that the condition in part **a** is met, test at a 5% significance level whether the data suggests that 40% is an overestimate. You should state clearly your null and alternative hypotheses. **[4]**

5. Over a long period, 6 out of every 10 adults who were asked agreed with the statement 'annual snowfall has decreased over the last 10 years'. This year, in an independent random sample of 12 adults, 10 agreed with the statement.

   Is there evidence that the proportion of adults holding this view has increased? You should use a 10% significance level and describe the critical region. **[5]**

6 The summer maximum daily temperature in 2015 in Leeming was above 17.5 °C on 82% of all days. A random sample of 14 summer days were considered in 1987 and 13 of these had temperatures above 17.5 °C. It is suggested that this data shows that temperatures were higher in Leeming in 1987 than in 2015

    a Explain why the binomial probability distribution provides a good model for this experiment. **[2]**

    b Copy and complete the following table for X, a random variable for the number of summer days with a maximum temperature above 17.5 °C *in 2015* **[2]**

x	11	12	13	14
$P(X=x)$	0.2393	0.2725		

    c Test, using a 10% significance level, whether the suggestion is justified. **[3]**

7 A survey found that 60% of documents printed in an office were printed single-sided. Employees were asked not to print single-sided in order to save paper. A fortnight later the manager of the office wanted to see if there had been any reduction in the rate of single-sided printing. He tested 40 documents and found that 18 of them had been printed single-sided.

    a State the hypotheses clearly. **[1]**

    b If he concluded that there had been an improvement, using a significance level of a%, what is the lowest possible whole number value of a? **[3]**

8 It is known that 35% of all days at Heathrow airport have a mean daily windspeed above 23 mph. Somebody claims that this proportion is lower at Gatwick airport. A sample of 30 days found that there was a mean daily windspeed above 23 mph on 8 of the days. A statistical test is carried out to determine whether windspeeds are lower at Gatwick airport than at Heathrow.

    a State a condition for the sample to be suitable for use in the test and state why the condition is necessary. **[2]**

Let X be a random variable for the number of days with a mean daily windspeed above 23 mph in a sample of size 30 at Gatwick airport.

    b State the null and alternative hypotheses to be used in the test and, assuming that the null hypothesis is true, give the distribution of X **[2]**

    c Perform this test at a significance level of 5% **[3]**

9 $X \sim B(n, p)$ and

$$H_0: p = k \qquad H_1: p > k$$

The critical value is $n-1$

    a Find the probability that X is in the critical region in terms of k and n **[4]**

    b If $n = 2$ and the significance level is at least 19% then find an inequality for k **[5]**

1 A gardener wishes to research the average rainfall in a year. She considers the following methods for finding a sample of approximately 50 days.

State the name of the method of sampling in each case.

   a Numbering each day of the year, then using a list of random numbers to select 50 days. **[1]**

   b Selecting every Saturday during that year. **[1]**

   c Taking a measurement whenever she has time until she has 50 measurements. **[1]**

2 The blood glucose levels of a group of adults is recorded immediately before and 2 hours after eating a meal. The results are summarised in these box and whisker plots.

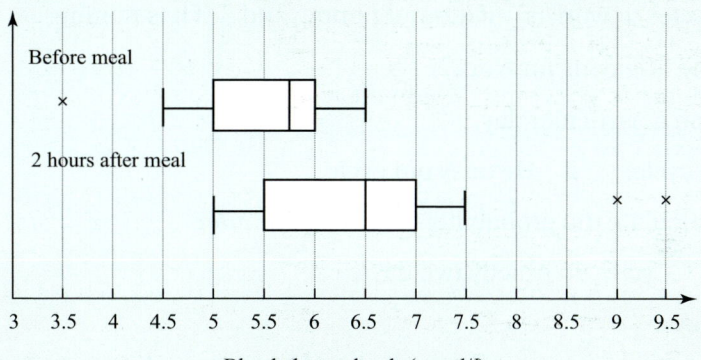

Blood glucose levels (mmol/*l*)

   a What is the interquartile range 2 hours after eating the meal? **[2]**

   b What is the median blood glucose level 2 hours after the meal? **[1]**

   c Compare the blood glucose levels immediately before eating with those 2 hours after eating. **[2]**

A blood glucose level of below 6 mmol/l before eating is considered normal.

   d What percentage of the group have a normal blood glucose level before the meal? **[1]**

3 This cumulative frequency graph shows the results of a general knowledge test taken by 80 people.

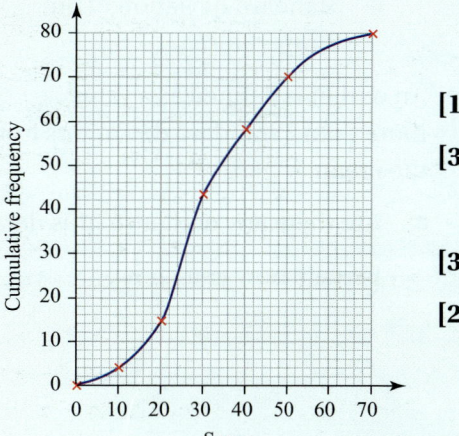

   a What is the median score? **[1]**

   b Calculate the interquartile range. **[3]**

The test was out of 80 marks.

   c How many people scored over 60%? **[3]**

   d What score did 90% of people exceed? **[2]**

4 The number of days with rain and whether the temperature is warm (≥ 18°C) or cool is recorded for 6 months at Hurn weather station.

	No rain	Rain	Total
Warm	59		116
Cool			
Total		97	184

The results are shown in the table, with some values missing.

  a Copy and complete the table. **[2]**

  b Write down the probability that a day was dry and warm. **[1]**

  c Are the events "a day is warm" and "a day has no rain" independent?

You must use probabilities to justify your answer. **[3]**

5 Use the fact that $X \sim B(20, 0.4)$ to find these probabilities.

  a $P(X = 13)$ **[1]**   **b** $P(X \leq 5)$ **[1]**   **c** $P(X > 10)$ **[2]**

6 The probability that a commuter cycles to work is $\frac{3}{5}$ if it is not raining and $\frac{2}{7}$ if it is raining. The probability of rain at the time he is leaving for work is $\frac{1}{20}$

  a Calculate the probability that, on a particular day,

    i It is raining and he does not cycle,   **ii** He does not cycle. **[4]**

  b In a 5-day week with no rain, calculate the probability that the commuter

    i Cycles every day,   **ii** Cycles on exactly two days,

    iii Cycles on more than two days. **[5]**

7 The maximum daily temperature at Leeming weather station was recorded every day in 2015 and the data is shown in the frequency polygon.

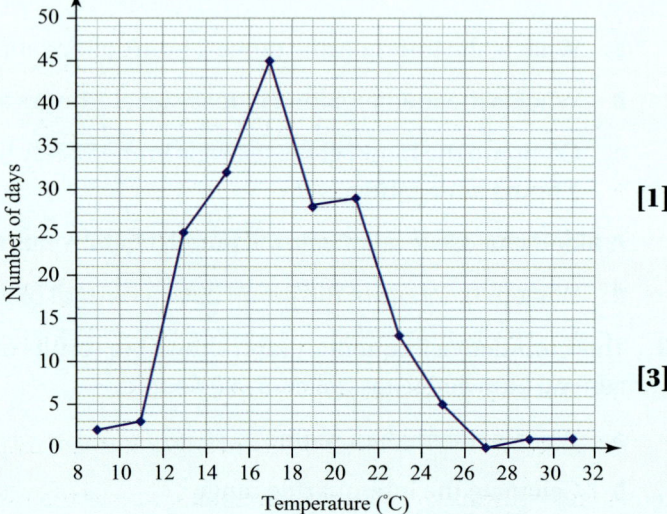

  a What was the modal temperature range? **[1]**

  b Estimate the

    i Mean of the temperature,

    ii Standard deviation of the temperature. **[3]**

An outlier is defined as a value more than 2 standard deviations from the mean.

  c Are there any outliers in this data?

Explain how you know. **[4]**

8   The resting pulse rates (bpm) of a group of people were measured. The mean for the group is 71 bpm.

71	67	22	99	68	67
63	56	76	65	68	75

The data is given in the box but one of the numbers has been entered incorrectly.

    a   Identify the error and calculate the correct pulse rate.   **[3]**

    b   Calculate the variance of the correct data.   **[1]**

    c   Calculate the median and explain why it might be a better average to use than the mean.   **[2]**

9   A smallholder records the mass of all eggs laid by her hens over a week.

The table shows her results.

Size	Mass e, of egg	Frequency
Small	$43 \leq e < 53$	42
Medium	$53 \leq e < 63$	59
Large	$63 \leq e < 73$	24
Very large	$73 \leq e < 83$	18

    a   What is the modal size of egg?   **[1]**

    b   Estimate the mean and standard deviation of the mass of the eggs.   **[3]**

    c   Estimate the median and interquartile range of the mass of the eggs.   **[6]**

The smallholder decides to keep all the small eggs to use at home or give to friends.

    d   Without further calculation, state the effect this will have on the mean and standard deviation of the remaining eggs.   **[2]**

10 $P(A) = 0.3$, $P(B) = 0.75$ and $P(A \cap B) = 0.15$

    a   Calculate

       i $P(A \cup B)$ **ii** $P(A' \cap B)$ **[3]**

Event C is mutually exclusive to event A and to event B

    b   Given that $P(C) \leq x$, calculate the value of x **[2]**

11   The probability of rain on any given day in June at Heathrow is assumed to be 30%.

    a   Assuming a binomial distribution, find the probability that during a ten-day period in June

       i   There is no rain,   **ii**   It rains on fewer than half of the days,

       iii   It rains on more than 7 days.   **[6]**

    b   Comment on the suitability of a binomial distribution in this case.   **[2]**

12   A factory produces thermometers that record the maximum daily outdoor temperature.

The probability of a thermometer being faulty is 1%

One day, a sample of 30 thermometers is taken and 2 are found to be faulty.

    a   Test, at the 5% significance level, whether there is any evidence that the probability of being faulty has increased.   **[6]**

    b   What is the actual significance level in this case?   **[1]**

    c   State the probability of incorrectly rejecting the null hypothesis in this case.   **[1]**

13 A meteorologist is researching the weather in the UK. They require data from a sample of 30 weather stations. Several methods of selecting the sample are suggested.

    a For each method, state what it is called and explain its disadvantages.

        i Using the 30 nearest weather stations to their location.

        ii Using an alphabetical list of the weather stations and selecting every 14th station. **[4]**

Instead, the meteorologist decides to take a stratified random sample based on where the weather station is by country.

The table shows how many weather stations are in each country.

    b Work out how many weather stations should be selected from each country.

Country	No. of stations
England	250
Wales	50
Scotland	90
Northern Ireland	30

[3]

14 In 2016, the population of countries in the European Union, to the nearest million, is summarised in the table.

A histogram is drawn to illustrate the data. The bar for 6–15 million is 1 cm wide and 13.5 cm tall.

    a Calculate the height and width of the 61–85 bar. **[5]**

    b Estimate how many counties have a population, to the nearest million, of under 10 million. **[4]**

    c Estimate the populations that the middle 50% of countries lie between. **[4]**

Population (millions)	Number of countries
0–5	11
6–15	9
16–60	4
61–85	4

15 The probability distribution of windspeeds (using the Beaufort scale) is given in the table.

x	0	2	4	6
$P(X=x)$	0.03	0.72		0.02

    a Calculate

        i $P(X=4)$ **ii** $P(2 \leq X < 6)$ **[4]**

A Beaufort force 4 is described as 'moderate' wind.

    b 50 days are selected at random. Calculate the probability of the windspeed being lower than 'moderate' on more than 30 days. **[3]**

    c What assumptions did you make in part **b**? **[2]**

16 You are given that $X \sim B(25, 0.45)$

Find the largest value of x such that $P(X \geq x) > 0.8$ **[4]**

17 Somebody thinks that the probability of it raining on a day in the summer holiday is 40%

One summer holiday, she recorded that it rained on 10 of the 40 days.

    a Use a hypothesis test with 2.5% significance level to test whether or not she is correct. **[5]**

    b Work out the acceptance region for this test. **[5]**

    c Comment on any assumptions you have made. **[2]**

12 Algebra 2

The secure transmission and storage of digital information requires strong forms of encryption. Public-key encryption is a vital part of online security today. It relies on trapdoor functions – these are functions which have no clear or easily-computable inverse. For example, calculating the prime factors of a number is notoriously difficult when the prime factors are very large, even for computers. The multiplication of two large prime numbers is therefore a trapdoor function: simple to perform, very hard to reverse. This forms the basis of RSA encryption, a widely used form of public-key encryption.

An understanding of functions and how to use them is vital in many areas of science, and they are a cornerstone of mathematics and, in particular, algebra. Functions are applicable to many topics, including parametric equations, partial fractions and other useful mathematical concepts covered in this chapter.

Orientation

What you need to know	What you will learn	What this leads to
Ch1 Algebra 1 • Direct proof, proof by exhaustion and counter examples.	• To make logical deductions and prove statements directly.	**Ch13 Sequences** Finding a binomial series by expanding an algebraic fraction.
Ch2 Polynomials and the binomial theorem • Algebraic division and using the factor theorem. • Transformations of graphs.	• To understand and use functions, parametric equations and algebraic fractions. • To decompose partial fractions.	**Ch15 Differentiation** Differentiation of parametric functions. **Ch16 Integration and differential equations** Integration using partial fractions.
Ch3 Trigonometry • Sine, cosine and tangent.	• To manipulate vectors in 3D and hence solve geometric problems.	

p.4
p.54
p.58
p.70

  **MyMaths** | Practise before you start

🔍 2024, 2042, 2043, 2047, 2048, 2252, 2253, 2284

12.1 Further mathematical proof

Fluency and skills

See Ch1.1
For a reminder of proof.

In chapter 1 you learned how to prove a statement using **direct proof** and **proof by exhaustion** and also how to disprove a statement using a **counter example**.

You can also use **proof by contradiction** to show that, if some statement were true, a logical contradiction occurs, and so the statement must be false.

Example 1

Prove by contradiction that, for every real number $0° < x < 90°$, $\tan x - \sin x > 0$

Assume that $\tan x - \sin x \leq 0$ for some value of $0° < x < 90°$.

$\Rightarrow \dfrac{\sin x}{\cos x} - \sin x \leq 0$

$\Rightarrow \sin x \left(\dfrac{1}{\cos x} - 1 \right) \leq 0$

$\Rightarrow$ Either $\sin x \leq 0$ or $\dfrac{1}{\cos x} \leq 1$ but not both.

$\sin x \leq 0 \Rightarrow 0° \not< x \not< 90°$ which contradicts the original statement.

$\dfrac{1}{\cos x} \leq 1 \Rightarrow 1 \leq \cos x \Rightarrow 0° \not< x \not< 90°$ which contradicts the original statement.

So, for every real number x between $0°$ and $90°$, $\tan x - \sin x > 0$

Write a statement that contradicts the given statement.

Substitute $\tan x = \dfrac{\sin x}{\cos x}$

Write the expression as a product.

Example 2

Prove that, if n^2 is even for $n \in \mathbb{Z}$, then n is even.

$\mathbb{Z}$ is the symbol for the set of integers.

Assume that there is an odd integer n such that n^2 is even.

Let $n = 2a + 1$ where a is an integer

$(2a + 1)^2 = 4a^2 + 4a + 1 = 2(2a^2 + 2a) + 1$

$2(2a^2 + 2a)$ is even.

So $2(2a^2 + 2a) + 1$ must be odd.

This means that n^2 must be odd which contradicts the original statement.

So, if n^2 is even, then n is even.

Write the opposite of the given statement.

You can write any odd number as $2a + 1$

Write the expression in the form $2k + 1$

This is a contradiction so the opposing statement is impossible.

Examples 3 and 4 show two important proofs that you need to know in your course of study.

Example 3

Prove by contradiction that $\sqrt{2}$ is irrational.

> An **irrational number** is one that you cannot express in the form $\frac{a}{b}$ where a and b are integers with no common factors.

Assume that there exist integers a and b, with no common factors, such that $\sqrt{2} = \dfrac{a}{b}$

$\Rightarrow 2 = \dfrac{a^2}{b^2}$ and so $2b^2 = a^2$

a^2 is a multiple of 2 so a must be even.

Let $a = 2m$ for an integer m, and so $a^2 = 4m^2$

So $2b^2 = 4m^2$, or $b^2 = 2m^2$

b^2 is a multiple of 2 so b must be even.

So $\sqrt{2} = \dfrac{even}{even}$

Both a and b have 2 as a factor. This contradicts the original statement that a and b have no common factors.

So the statement is false and $\sqrt{2}$ cannot be rational.

- Write the opposing statement.
- Square both sides.
- Use the result a^2 even $\Rightarrow a$ even.
- The numerator and denominator have a common factor of 2.

Example 4

Prove by contradiction that the number of prime numbers is infinite.

Assume that there are only n primes.

In this case the complete list of primes is

$2, 3, 5, 7, 11, 13, 17, 19, ..., p_n$

Let N be the product of all these primes + 1

So $N = (2 \times 3 \times 5 \times 7 \times ... \times p_n) + 1$

N is not divisible by 2 since it has a remainder of 1 when divided by 2

So 2 cannot be a factor of N

This also applies to all the other primes up to p_n

So none of the known primes is a factor of N and N must be a prime number.

This statement contradicts the original statement that the list of primes, $2, 3, 5, 7, 11, 13, 17, 19, ..., p_n$ was complete.

So the number of primes must be infinite.

- Write the opposing statement.
- Any number is either prime or divisible by a prime number.

Exercise 12.1A Fluency and skills

Use direct proof in questions 1 and 2

1. Prove that, for any integer n,
   $(5n + 1)^4 - (5n - 1)^4$ is divisible by 40

2. Show that for a 3-digit integer, if the sum of the digits is divisible by 9, then the number itself is divisible by 9

Use the method of exhaustion in questions 3 and 4

3. If p is an integer, prove that
   $p! \le 2^p$ for $0 \le p \le 3$

4. Given that an integer can be even or odd, prove that the difference between the cubes of two consecutive integers is odd.

Use counter examples to disprove statements 5 to 7

5 If $a > b$ and $b > c$ then $ab > bc$

6 If p^2 is rational, then p is rational.

7 Cube numbers can end in any digit except 9

Use contradiction to prove statements 8 to 12

8 For any integer n, if n^2 is odd, then n is odd.

9 There are no values of a and b such that $a^2 - 4b = 2$. (Use the alternative statement $a^2 = 4b + 2$ and the fact that a^2 even $\Rightarrow a$ even.)

10 For every real number $0° < x < 90°$, $\sin x + \cos x \geq 1$. HINT: write down the opposing inequality and square both sides.

11 There are no integers m and n such that $\dfrac{m^2}{n^2} = 2$, where $\dfrac{m^2}{n^2}$ is a fully simplified fraction. You may assume that for an integer a, if a^2 is even then a is even.

12 Any integer greater than 1 has at least one prime factor.

Reasoning and problem-solving

Strategy

To prove a statement, P, using the method of contradiction

(**1**) Assume P is not true.

(**2**) Write the statement P′ which is the opposite of P.

(**3**) Show that P′ leads to a contradiction.

> **Natural numbers** are integers greater than 0.

Example 5

State why this proof by contradiction is flawed.

For any natural number n, the sum of all natural numbers less than n is not equal to n.

Opposing statement: For any natural number n, the sum of all smaller natural numbers is equal to n.

However, when $n = 5$, $1 + 2 + 3 + 4 = 10$ which is not equal to 5.

This is a contradiction.

So the assumption was false and the theorem must be true.

> The contradiction of the universal statement "for all x, P(x) is true" is *not* the same as "for all x, P(x) is false".
> The contradiction of the statement "for all x, P(x) is true" is the opposing statement "there exists an x such that P(x) is false".
> So the statement "for any natural number n, the sum of all natural numbers less than n is not equal to n", has the opposing statement "there exists a natural number n such that the sum of all natural numbers smaller than n is equal to n".
> In fact, for $n = 3$, $1 + 2 = 3$ so the statement can be *disproved* by counter example.

Write the opposing statement.

(**2**)

Example 6

Saqib says that the curve $y = e^{2x}$ lies above the curve $y = e^x$ for all $x > 0$

Is he correct? Construct a proof to justify your answer.

Saqib is correct.

Assume the opposite: $y = e^{2x}$ lies below $y = e^x$ for some $x > 0$

$e^{2x} < e^x$ for some $x > 0$

$e^{2x} = (e^x)^2$

$e^x \times e^x < e^x \Rightarrow e^x < 1$

$\Rightarrow x < 0$ which contradicts the initial statement that $x > 0$

Therefore the curve $y = e^{2x}$ lies above the curve $y = e^x$ for all $x > 0$

Use your knowledge of the two graphs to choose a proof rather than a counterproof.	
The statement uses 'for *all* $x > 0$' so the opposite statement uses 'for *some* $x > 0$'	
Divide both sides of the inequality by e^x	
Solve the inequality for x and state the contradiction.	
State the conclusion clearly.	

Exercise 12.1B Reasoning and problem-solving

1 Prove that, if m and n are both integers and mn is odd then m and n must both be odd.

2 Prove that, if $m^2 < 2m$ has any solutions, then $0 < m < 2$

3 Prove that if n is any odd integer, then $(-1)^n = -1$

4 Prove that the statement '$\sin(A - B) = \sin A - \sin B$, for all A and B' is false.

5 Prove that there is no integer k such that $(5m + 3)(5n + 3) = 5k$, where m and n are integers.

6 A student says that the sum of the cubes of any two consecutive numbers always leaves a remainder of 1 when divided by 2. Is she correct? Construct a proof to support your answer.

7 Prove that, for $m < -\dfrac{1}{2}$, $1 - \dfrac{1}{m} < 3$

8 Prove that there is no smallest positive number.

9 A rule for multiplying by 11 in your head is shown.

$(3\ 4\ 5) \times 11$

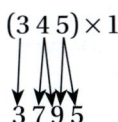

$3\ 7\ 9\ 5$

Working from the right:

1 Write down the 5

2 Add $5 + 4$, add $4 + 3$

3 Write down the 3

Prove algebraically that this rule works but state a 'catch'.

10 Prove that for an integer a, $(a^2 + 2) \div 4$ cannot be an integer.

11 Prove that there is no greatest odd integer.

12 A student is factorising integers. He thinks that, if b is a factor of a and c is a factor of b then c is a factor of a. Is he right? Use proof to justify your answer.

13 Prove that any positive rational number can be expressed as the product of two irrational numbers.

14 Prove that every cube number can be expressed in the form $9k$ or $9k \pm 1$ where k is an integer.

15 Prove that, for $a, b \in \mathbb{Z}$, there are no positive solutions to the equation $a^2 - b^2 = 1$

16 Prove that if a is an integer and is not a square number then $\sqrt{a}$ is irrational.

Challenge

17 State the error in this 'proof'.

Let $a = b$

So $a^2 = ab$

So $a^2 - b^2 = ab - b^2$

So $(a - b)(a + b) = b(a - b)$

So $\cancel{(a - b)}(a + b) = b\cancel{(a - b)}$

So $(a + b) = b$

So, since $a = b$, $2b = b$ and so $2 = 1$

Fluency and skills

A **function**, f, is a mapping from a **domain**, set X, to a **range**, set Y, where each input value, $x \in X$, generates one and only one output value, $f(x) \in Y$. The range, Y, is the image of the domain, X, under the function, f.

> A mapping from set X to set Y is a rule which associates each element of X with one element of Y.

> **Key point**
>
> For a function f: $X \rightarrow Y$, which maps elements of set X to elements of set Y
> The set of all possible input values, X, is called the domain.
> The set of all possible output values, Y, is called the range.

In a **one-to-one** function, each element in the range corresponds to exactly one element in the domain.
For example, $f(x) = 3x - 2$, $x \in \mathbb{R}$ is a one-to-one function.
$x = 3$ generates $f(x) = 7$
No other x value generates $f(x) = 7$
Its domain is the real numbers, $\mathbb{R}$. Its range is $\mathbb{R}$

In a **many-to-one** function, an element in the range can be generated by more than one element in the domain.
For example, $f(x) = x^2$, $x \in \mathbb{R}$ is a many-to-one function.
Both $x = 3$ and $x = -3$ generate the value $f(x) = 9$
Its domain is $\{x: x \in \mathbb{R}\}$. Its range is $\{f(x): f(x) \in \mathbb{R}, f(x) \geq 0\}$

See p.614
For a list of mathematical notation.

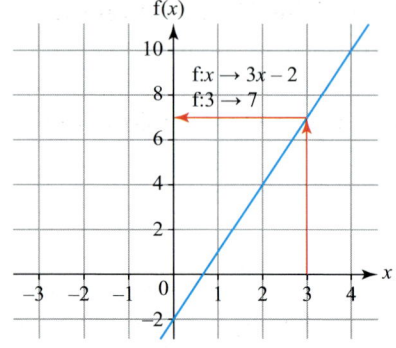

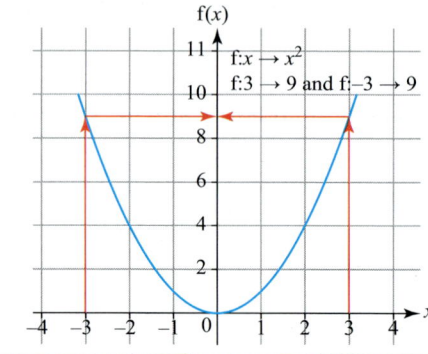

> If any element in the domain generates more than one element in the range, then the relationship is not a function.

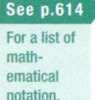
Example 1

For the function $f(x) = \dfrac{1}{\sqrt{x+6}}$

a State the maximum possible domain and the corresponding range.

b Sketch its graph

a Maximum possible domain is $\{x: x \geq -6\}$, range is $\{f(x): f(x) > 0\}$

The denominator is strictly positive.

b
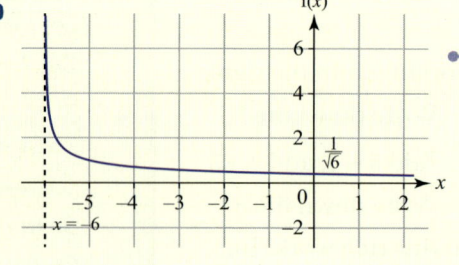

You can check your sketch using a graphical calculator.

A **composite function** is formed when you combine two or more functions. Given two functions $f(x)$ and $g(x)$:

- the composite function $fg(x)$ means 'apply f to the results of $g(x)$'
- the composite function $gf(x)$ means 'apply g to the results of $f(x)$'

For $fg(x)$, the results, or range, of $g(x)$ become the inputs, or members of the domain, of f. So the range of $g(x)$ must be a subset of the domain of f.

The order is important. You always work out the 'inside' function first.

Composite functions are not commutative. $fg(x)$ is not necessarily the same as $gf(x)$.

Example 2

For the functions $f(x) = x^2$, $x \in \mathbb{R}$ and $g(x) = 2x + 1$, $x \in \mathbb{R}$

a write the composite functions $fg(x)$, $gf(x)$, $ff(x)$ and $gg(x)$

b work out the values of $fg(2)$, $gf(2)$, $ff(2)$ and $gg(2)$

For $fg(x)$, apply g first and then apply f to the result. For $gf(x)$, apply f first and then apply g to the result.

a $fg(x) = (2x + 1)^2$ $gf(x) = 2x^2 + 1$
$ff(x) = (x^2)^2 = x^4$ $gg(x) = (2(2x + 1) + 1) = 4x + 3$

For $ff(x)$, apply f and then apply f to the result. For $gg(x)$, apply g first and then apply g to the result.

b $fg(2) = (2 \times 2 + 1)^2 = 25$ $gf(2) = 2 \times 2^2 + 1 = 9$
$ff(2) = (2)^4 = 16$ $gg(2) = 4(2) + 3 = 11$

You can check your answers by evaluating the composite functions at 2 on your calculator.

You can use composite functions in geometry to describe a transformation or a series of transformations.

See Ch2.4
For a reminder on transformations.

Example 3

a Describe the two transformations needed to transform the graph of $y = x^3$ into the graph of $y = 3 - x^3$

b The sequence of transformations applied to $y = x^3$ in part **a** can be described by the composite function $gf(x^3)$, $x \in \mathbb{R}$. Write down
   i $f(a)$ **ii** $g(a)$

c Sketch the curve of $fg(x^3)$ and describe the series of transformations needed to transform the graph of $y = x^3$ into the graph $y = fg(x^3)$

a

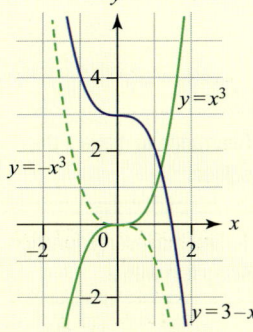

Reflection in the x-axis followed by translation by vector $\begin{pmatrix} 0 \\ 3 \end{pmatrix}$

b i $f(a) = -a$

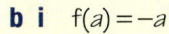

ii $g(a) = a + 3$

c

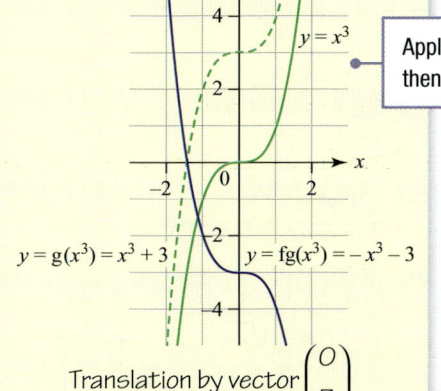

Apply g first, then apply f

Translation by vector $\begin{pmatrix} 0 \\ 3 \end{pmatrix}$ followed by reflection in the x-axis.

The **inverse**, $f^{-1}(x)$, of a function, $f(x)$, reverses the effect of the original function.

You can find the inverse of a function by first rearranging the equation to make x the subject. You then replace the subject x with $f^{-1}(x)$ and replace $f(x)$ with x

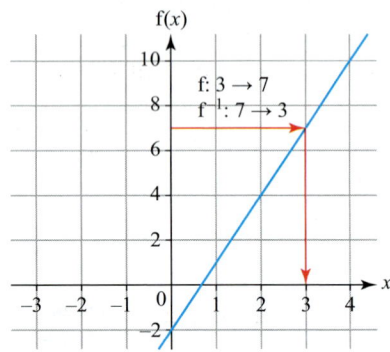

f: 3 → 7
f⁻¹: 7 → 3

f: 3 → 9 and f: −3 → 9
f⁻¹: 9 → 3 and −3

> When inverted, many-to-one functions become one-to-many relationships.

$f(x) = 3x - 2, x \in \mathbb{R}$ maps elements from the domain $\mathbb{R}$ to the range $\mathbb{R}$

This can be rearranged to $x = \dfrac{f(x) + 2}{3}$

$f^{-1}(x) = \dfrac{x+2}{3}, x \in \mathbb{R}$ is a one-to-one function which maps elements from the domain $\mathbb{R}$ to the range $\mathbb{R}$

$f(x) = x^2, x \in \mathbb{R}$ maps elements from the domain $\mathbb{R}$ to the range $f(x) \in \mathbb{R}, f(x) \geq 0$

This can be rearranged to $x = \pm\sqrt{f(x)}$

$f^{-1}(x) = \pm\sqrt{x}, x \in \mathbb{R}, x \geq 0$ is NOT a function because it maps some elements of the domain to more than one element in the range.

> **Key point**
>
> Only one-to-one functions have inverse functions.

> **Key point**
>
> If a function has an inverse, then $ff^{-1}(x) = f^{-1}f(x) = x$
> The domain of $f(x)$ is the range of $f^{-1}(x)$.

> An inverse function, $f^{-1}(x)$, maps the elements in the range of $f(x)$ back onto the elements in the domain of $f(x)$.

Example 4

a Find the inverse of the function $f(x) = 8x^3 + 1, x \in \mathbb{R}$

b On one set of axes
  i sketch the graph of $y = f(x)$ **ii** sketch the line $y = x$
  iii reflect the graph of $y = f(x)$ in the line $y = x$

c Sketch the graph of $y = f^{-1}(x)$ on the same axes. What do you notice?

a $f(x) = 8x^3 + 1$

$f(x) - 1 = 8x^3$

$\dfrac{f(x) - 1}{8} = x^3$

$\dfrac{\sqrt[3]{f(x) - 1}}{2} = x$

The inverse of $f(x) = 8x^3 + 1$ is

$f^{-1}(x) = \dfrac{1}{2}\sqrt[3]{x - 1}$

> Rearrange to make x the subject.

b and **c**

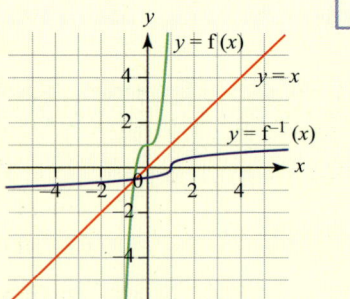

$y = f(x)$
$y = x$
$y = f^{-1}(x)$

The line $y = x$ is a mirror line of the function and its inverse.

> To find the inverse function, swap $f(x)$ with x

Algebra 2 Functions

Calculator

Try it on your calculator

You can sketch the inverse of a function on a graphics calculator.

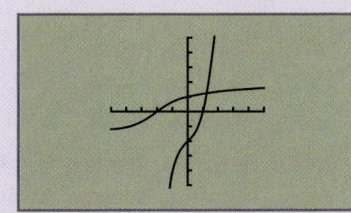

Activity

Find out how to sketch $f(x) = x^3 + x - 2$ and $f^{-1}(x)$ on *your* graphics calculator.

The graph of $f^{-1}(x)$ is a reflection of the graph of $f(x)$ in the straight line $y = x$

$|x|$ is called the **modulus** of x. It is also known as the **absolute value** of x.

The modulus of a real number is always positive. You can think of it as its distance from the origin.

For example, if $x = -3$, then $|x| = 3$

To sketch the graph of $y = |f(x)|$, you first sketch the graph of $y = f(x)$. You then take any part of the graph that lies below the x-axis and reflect it in the x-axis.

You can use your graphical calculator or computer software to sketch modulus functions.

Example 5

a On one set of axes **i** Sketch the graph of $y = 2x - 3$ **ii** Sketch the graph of $y = |2x - 3|$

b Sketch the graph of $y = 5 - |2x - 3|$

a

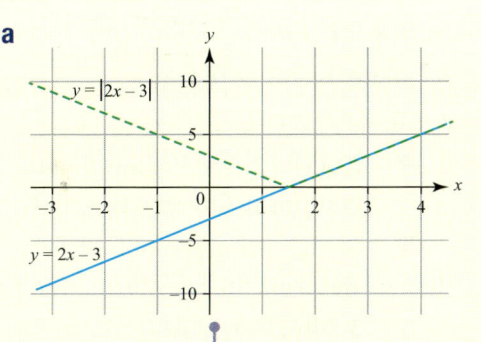

b

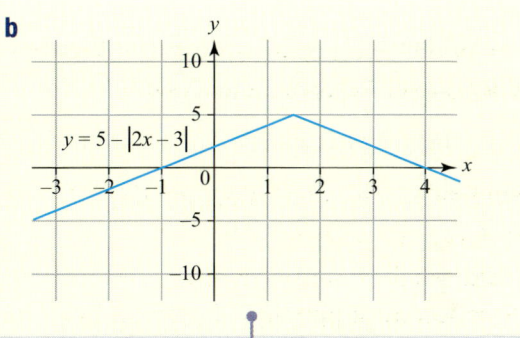

Only the absolute value of $2x - 3$ is taken. The negative portion of the graph is reflected in the line $y = 0$ and the minimum value of y is 0.

Reflect in the x-axis to get $-|2x - 3|$ then translate by $\binom{0}{5}$ to get $5 - |2x - 3|$

Exercise 12.2A Fluency and skills

1   State the range of these functions.

   a $f(x) = 5x - 2, x \in \mathbb{R}$

   b $f(x) = 5x - 2, -6 < x < 6$

   c $f(x) = (2x - 5)^2, x \in \mathbb{R}$

   d $f(x) = (2x - 5)^2, 0 \le x \le 10$

2   State the maximum possible domain of these functions.

   a $f(x) = 3x^3, f \in (-24, 81]$

   b $f(x) = \ln x, f \in \mathbb{R}$

3   State the maximum possible domain and corresponding range of these functions.

   a $f(x) = 4^x$ **b** $f(x) = \dfrac{1}{(x+3)^2}$

4 State if each of these functions is one-to-one or many-to-one. Justify your answers.

   a $f(x) = 2x^2,\ x \in \mathbb{R}$

   b $f(x) = 3^{-x},\ x \in \mathbb{R}$

   c $f(x) = x^4,\ x \in \mathbb{R}$

   d $f(x) = \sin^2 x,\ 90° \le x \le 270°$

   e $f(x) = \dfrac{1}{x^2},\ x \in \mathbb{R},\ x \ne 0$

   f $f(x) = -3x^3,\ x \in \mathbb{R}$

   g $f(x) = \dfrac{1}{x-3},\ x \in \mathbb{R},\ x \ne 3$

   h $f(x) = \cos x,\ 0° \le x \le 360°$

   i $f(x) = \cos x,\ 0° \le x \le 180°$

   j $f(x) = \cos 2x,\ 0° \le x \le 180°$

5 For each function

   i Sketch the function

   ii Find the point(s) of intersection between the function and $y = 3x, x \in \mathbb{R}$

     a $f(x) = x^2,\ x \in [0,3)$

     b $g(x) = x^2,\ x \in [0,5]$

     c $h(x) = x^2,\ x \in (0,7]$

6 **a** Evaluate $fg(1)$, $gf(-2)$, $ff\left(\dfrac{-2}{3}\right)$ for these functions. Show your working.

     i $f(x) = 3x + 7,\ x \in \mathbb{R}$

        $g(x) = 3x^2 - 9,\ x \in \mathbb{R}$

     ii $f(x) = \dfrac{1}{x},\ x \in \mathbb{R},\ x \ne 0$

        $g(x) = \dfrac{-2}{x^2 + 1},\ x \in \mathbb{R}$

   b $f(x) = (x+1)^2,\ x \in \mathbb{R}$ and $g(x) = 1 - x$, $x \in \mathbb{R}$. Work out the values of $fg(2)$ and $gf(2)$, $fg(-4)$ and $gf(-4)$.

7 For each of these functions

   a $f(x) = x^3$ **b** $f(x) = (x-4)^2 - 10$

   i State the maximum possible domain and corresponding range.

   ii Evaluate

     $f(0)$ $f(-4)$ $f(4)$

8 Given that $|t| = 5$ work out all possible values of $|3t + 2|$.

9 In each case, the graph of $y = f(x)$ is a straight line passing through the points given. Sketch the graph

   i $y = f(x)$ **ii** $y = |f(x)|$

   iii $|f(x-2)|$ **iv** $|f(x)| - 2$

   a $(-3, 0)$ and $(0, 6)$

   b $(-3, -2)$ and $(2, -7)$

10 Use the following functions to answer parts **i–iv**.

   a $f(x) = x^2,\ 0 \le x < 8$

   b $f(x) = (x-2)^3,\ x \in (2, 10)$

   c $f(x) = 2^x,\ x < 0$

   i State the range of $f(x)$.

   ii Find the inverse function $f^{-1}(x)$, stating its domain.

   iii Use a graphical calculator or a computer with appropriate software to sketch, on the same grid

     1 $y = f(x)$

     2 $y = f^{-1}(x)$

     3 $y = x$

     Check on your graph that the inverse is the reflection of the original function in the line $y = x$

   iv State the range of $f^{-1}(x)$

11 **a** $f(x) = x^2,\ x \in \mathbb{R}$

     i If $g(x) = 2x, x \in \mathbb{R}$ show that the composite function $fg(x)$ is $4x^2, x \in \mathbb{R}$

     ii Sketch the graph of $y = fg(x)$ and describe the transformation from $y = f(x)$ to $y = fg(x)$

   b $f(x) = 4x^2,\ x \in \mathbb{R}$

     i If $h(x) = x + 3, x \in \mathbb{R}$ show that the composite function $fh(x)$ is $4x^2 + 24x + 36$

     ii Sketch the graphs of $f(x)$ and $fh(x)$ and describe the transformation from $f(x)$ to $fh(x)$.

12 The curve $y = x^2, x \in \mathbb{R}$ is translated by $\begin{pmatrix} -3 \\ 0 \end{pmatrix}$ to create a new function, $f(x)$. $f(x)$ is then stretched parallel to the y-axis by scale factor 4 to create the composite function $gf(x)$

a Write an expression for f(x)

b Write an expression for g(x)

c Sketch the function gf(x)

13 a By writing each expression as a composition of simple functions, explain how you would transform the graph of $y = x^2$ into

    i $x^2 - 6x + 13$ **ii** $4x^2 + 12x + 8$

b In the same way explain how you would transform the graph of $y = x^3$ into

    i $(x + 2)^3 - 7$ **ii** $(3x - 5)^3 + 6$

14 Find an expression for f(x) when

    a $fg(x) = e^{x^2}$, $x \in \mathbb{R}$ and $g(x) = x^2$, $x \in \mathbb{R}$

    b $fg(x) = 3\log(x + 1)$, $x \in \mathbb{R}$ and $g(x) = x + 2$, $x \in \mathbb{R}$

15 For each pair of functions, sketch the graph of the composite function gf(x), stating its maximum possible domain.

    a $f(x) = -x$, $x \in \mathbb{R}$; $g(x) = x^2$, $x \in \mathbb{R}$

    b $f(x) = x^2$, $x \in \mathbb{R}$; $g(x) = x^{\frac{3}{2}}$, $x \in \mathbb{R}$

Reasoning and problem-solving

Strategy 1

To find the inverse of a function

(1) Make sure the function is a one-to-one function.

(2) Rearrange the function to make x the subject.

(3) Interchange the variables: change x to $f^{-1}(x)$ and change $f(x)$ to x

(4) Check that the graphs of $f(x)$ and $f^{-1}(x)$ are reflected in the line $y = x$

Example 6

Two functions exist only for $x \geq 0$

$f: x \mapsto 7x - 2$ and $g: x \mapsto ax^2 + b$ and $fg: x \mapsto 28x^2 - 9$

> $f: x \mapsto 7x - 2$ is another way of writing $f(x) = 7x - 2$

a Explain why these functions are all one-to-one functions.

b Evaluate the values of a and b.   **c** Work out the inverse of fg(x).

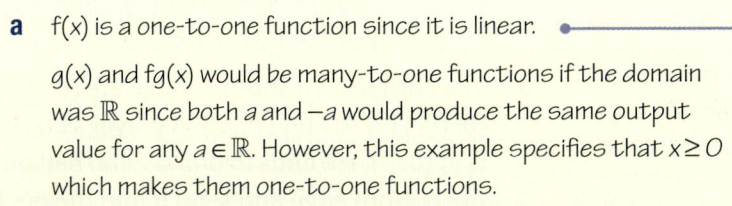

a f(x) is a one-to-one function since it is linear.

    g(x) and fg(x) would be many-to-one functions if the domain was $\mathbb{R}$ since both a and $-a$ would produce the same output value for any $a \in \mathbb{R}$. However, this example specifies that $x \geq 0$ which makes them one-to-one functions.

> **(1)** All linear functions are one-to-one.

b fg: $x \mapsto 7(ax^2 + b) - 2$ and fg: $x \mapsto 28x^2 - 9$
    So $7ax^2 + 7b - 2 = 28x^2 - 9$
    so $7a = 28 \Rightarrow a = 4$
    and $7b - 2 = -9 \Rightarrow b = -1$

> Combine f(x) and g(x) to give an expression for fg(x) and equate this to the expression given for fg(x).

c Rewrite fg: $x \mapsto 28x^2 - 9$ as $y = 28x^2 - 9$
    Interchange x and y: $x = 28y^2 - 9$
    $y = \sqrt{\dfrac{x + 9}{28}}$
    So $(fg)^{-1}: x \mapsto \sqrt{\dfrac{x + 9}{28}}$

> **(2)** Interchange the variables x and y.

> **(3)** Rearrange to make y the subject.

To find the modulus, or absolute values, of a function for particular values of x

1. Substitute the values of x in the function f(x)

2. For any of the values of the function that are negative, replace the $-$ sign with a $+$ sign.

3. To sketch a modulus graph of a function reflect in the x-axis any parts of the graph which would be below the x-axis.

Find, graphically, the set of values of x for which $|4x - 3| < 2x$

2 3

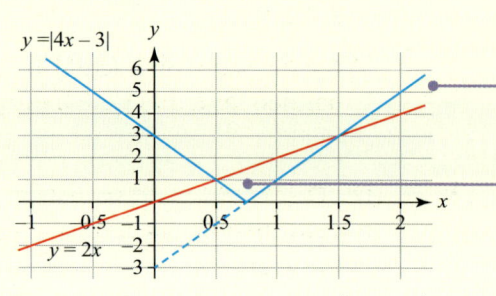

$y = |4x - 3|$

$y = 2x$

Sketch the graph $y = |4x - 3|$ by sketching $y = 4x - 3$ and reflecting the section with negative y-values in the x-axis.

Find the points of intersection.

From the graphs the solution of $|4x - 3| < 2x$ is $0.5 < x < 1.5$

Exercise 12.2B Reasoning and problem-solving

1 The graph of $y = f(x)$, $-1 < x < 5$ is shown.

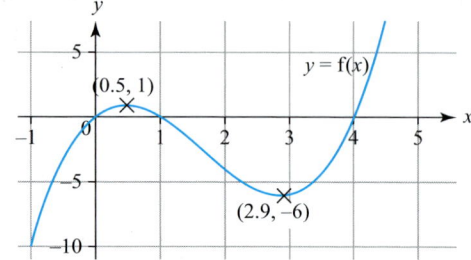

(0.5, 1)

$y = f(x)$

(2.9, −6)

a Sketch the graph of $y = 3f(x) + 12$, $-1 < x < 5$ and indicate the coordinates of any turning points.

b The domain of $y = 3f(x) + 12$ is restricted to make the function invertible. State two different ways this could be done.

2 A company introduces a new product onto the market. Sales, S, (in thousands of pounds) initially increase steadily but then decrease, given by the function

$S = -2|w - 15| + 30$ where w is the time (in weeks).

a Describe a sequence of transformations that maps $S = w$ onto $S = -2|w - 15| + 30$

b Calculate the values of S for values of w every two weeks from 0 to 30

c Sketch the graph of the function.

d What was the maximum amount of sales in any one week?

3 The corners of a snooker table are (0, 0), (0, 6), (12, 0) and (12, 6). A player's cue ball is at (8, 4). He aims to bounce the ball off the bottom edge and send it into the pocket at (6, 6). Work out the equation he needs to pocket the ball and sketch the graph.

4 The shape of the roof of an art gallery is given by the equation

$$f(x) = \frac{-4}{3}|x - 16.5| + 22$$

where f(x) is the distance above ground level and x is horizontal distance in metres.

a Sketch the graph of this function.

b Write down the largest possible domain and corresponding range of this function.

c How high is the roof apex above its base?

d What is the length of the side of the square base on which the roof is mounted?

5 For each function, explain whether the inverse function exists and write an expression for the inverse if it exists.

    a $f(x)=(x+2)^2$, $x\in\mathbb{R}$

    b $g(x)=(x+2)^2$, $x\le 0$

    c $h(x)=(x+2)^2$, $x\le -2$

6 **a** Work out the maximum possible domain and corresponding range of the function $\dfrac{1}{x-2}$

    b Work out the inverse of the function $\dfrac{1}{x-2}$ and write down its domain and range.

    c Compare your answers for the domain and range in parts **a** and **b**. What do you notice?

    d Sketch a graph of the two functions on the same axes.

7 Use a graphical calculator or computer to check your answers in this question. Sketch the graphs of the functions in parts **a** to **d** on the same set of axes, for $-180°\le x\le 180°$

    a $e(x)=\cos x$

    b $f(x)=2\cos x$

    c $g(x)=2\cos(x+90)°$

    d $h(x)=2\cos(x+90)°-3$

    e Describe the sequence of transformations that maps $e(x)$ to $f(x)$ to $g(x)$ to $h(x)$.

8 Describe a sequence of two transformations that maps the graph of $y=\sin x$ to $y=\sin(30-x)$

9 **a** Sketch the graph $y=\log_2 x$

b Transform the graph from part **a** by first translating it by $\begin{pmatrix} -2 \\ 0 \end{pmatrix}$ and then stretching it parallel to the y-axis by scale factor $\dfrac{1}{2}$

c Write the equation of the graph you drew in part **b**.

10 $f(x)=3x$, $x\in\mathbb{R}$ and $g(x)=x-4$, $x\in\mathbb{R}$

    a **i** Describe a sequence of two transformations that maps $y=e^x$ to $y=fg(e^x)$

        ii Write the expression $fg(e^x)$ in expanded form and sketch the graph $y=fg(e^x)$

    b **i** Describe a sequence of two transformations that maps $y=e^x$ to $y=gf(e^x)$

        ii Write the expression $gf(e^x)$ in expanded form and sketch the graph $y=gf(e^x)$

11 Calliste attempts to answer the following question. Explain and correct her mistakes.

$f(x)=4-x$, $x\in\mathbb{R}$. Sketch the graph of $y=|f(x)|$ and find the values of x for which $|f(x)|=\dfrac{1}{2}x$

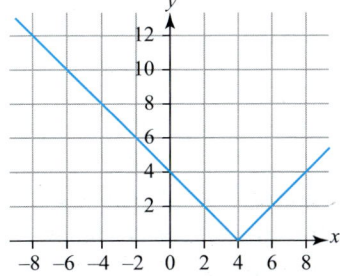

$|f(x)|=4+x$

$4+x=\dfrac{1}{2}x$

$\dfrac{1}{2}x=-4$

$x=-8$

Challenge

12 $f(x)=x^3$, $x\in\mathbb{R}$ and $g(x)=\ln x$

    a State the maximum domain and corresponding range for $g(x)$

    b Saqib says that the composition $fg(x)$ is a function but the composition $gf(x)$ is not a function. Is he right? Explain your answer.

Parametric equations

Fluency and skills

A **Cartesian equation** can be written in the form $y = f(x)$.
A **parametric equation** expresses a relationship between two
variables, x and y, in terms of a third variable, often named θ or t,
which is known as the **parameter**.

> A Cartesian
> equation uses the
> variables x and y

x and y are given as functions of θ or t.

Consider a circle with centre $(3, 0)$ and radius 2. Any point, P, on the
circle can be identified using the angle between the positive x-axis
and the radius to the point P. The angle, θ, is the parameter in this
case.

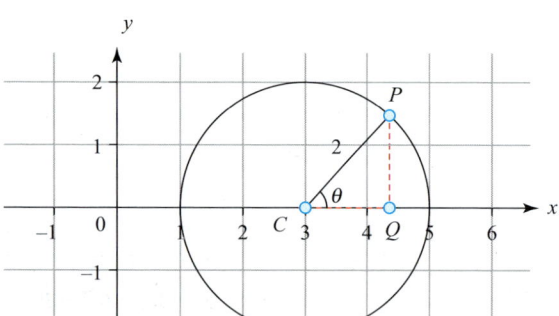

The point P can be expressed in terms of θ as $(3 + CQ, PQ)$

$$\cos\theta = \frac{CQ}{2} \Rightarrow CQ = 2\cos\theta$$

> **See Ch3.1**
> For a
> reminder of
> trigono-
> metric
> functions.

$$\sin\theta = \frac{PQ}{2} \Rightarrow PQ = 2\sin\theta$$

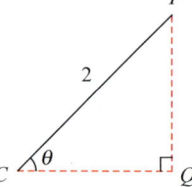

Point P is $(3 + 2\cos\theta, 2\sin\theta)$

The parametric equations for the circle are

$x = 3 + 2\cos\theta$ and $y = 2\sin\theta$, for $0° \le \theta < 360°$

You can use the identity $\sin^2\theta + \cos^2\theta \equiv 1$ to write this as a Cartesian
equation.

$$x - 3 = 2\cos\theta$$
$$(x-3)^2 = 4\cos^2\theta$$
$$y^2 = 4\sin^2\theta$$
$$(x-3)^2 + y^2 = 4\cos^2\theta + 4\sin^2\theta$$
$$= 4(\cos^2\theta + \sin^2\theta)$$
$$= 4$$

The Cartesian equation of the circle is $(x-3)^2 + y^2 = 4$

Example 1

A curve is given by the parametric equations $x = t + 4$ and $y = t^2 - 10$

a Sketch the curve for $-4 \leq t \leq 4$

b Write the Cartesian equation for the curve.

a

t	−4	−3	−2	−1	0	1	2	3	4
x	0	1	2	3	4	5	6	7	8
y	6	−1	−6	−9	−10	−9	−6	−1	6

> You can check your curve by drawing it on a graphics calculator.

b $x = t + 4$

$t = x - 4$

$y = (x - 4)^2 - 10$

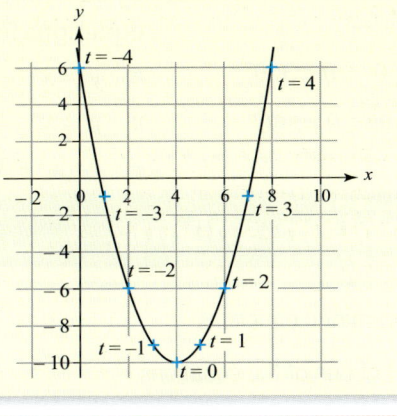

Try it on your calculator

You can sketch a curve given by parametric equations on a graphics calculator.

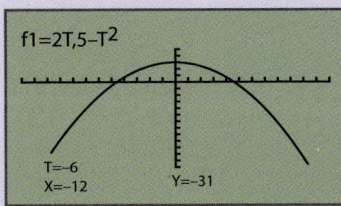

f1=2T,5−T²

T=−6
X=−12   Y=−31

Activity

Find out how to sketch the curve given by $x = 2t$; $y = 5 - t^2$ on *your* graphics calculator.

Exercise 12.3A Fluency and skills

1 Work out the coordinates of the points on these parametric curves where $t = 5$, 2 and −3

  a $x = t; y = \dfrac{4}{t}$ **b** $x = \dfrac{3}{t^2}; y = -2t$

  c $x = \dfrac{1+t}{1-t}; y = \dfrac{2-t}{2+t}$ **d** $x = \dfrac{2t+5}{t+1}; y = \dfrac{t^3+4}{3}$

2 Work out the Cartesian equations given by these parametric equations.

  a $x = t^2 + 2; y = 2t + 1$ **b** $x = 2t; y = -4t^3$

  c $x = 3t; y = \dfrac{1}{2t}$ **d** $x = \dfrac{-2}{t}; y = 3t^4$

  e $x = \dfrac{1+t}{1-t}; y = 2t$ **f** $x = 2\sin\theta; y = 2\cos\theta$

3 Use a table of values for t, x and y to sketch the graphs described by these parametric equations. Use a graphical calculator or computer software to check your sketches.

  a $x = t; y = t^2$ **b** $x = t; y = 2t^3$

  c $x = t^2 + 1; y = t + 2$ **d** $x = t^3; y = t^2$

  e $x = 1 - t; y = \dfrac{1}{t}$ **f** $x = t; y = \dfrac{1}{t^2}$

4 A curve C is defined by parametric equations $x = \dfrac{4}{t-2}, y = \sqrt{t-2}$

  a Theo says the largest possible domain is $t \geq 2$, Akeem says it is $t > 2$. Who is correct? Explain your answer.

  b Write the Cartesian equation of the curve, stating the domain and range.

5 Show that any point on the parabola $y^2 = 4ax$ can be described by $(at^2, 2at)$ for a parameter t.

6 A curve C is defined by parametric equations

$x = e^{2t}$, $y = 5e^{-t}$, $t \in \mathbb{R}$

Write the Cartesian equation of the curve, stating the domain and range.

c $x = \dfrac{1}{2}\cos\theta - 4$, $y = \dfrac{1}{2}\sin\theta + 1$, $0° \leq \theta < 360°$

d $x = \sqrt{2}\cos\theta - 4$, $y = \sqrt{2}\sin\theta - 3$, $0° \leq \theta < 360°$

7 Show that the curve with parametric equations $x = 5\sin\theta$; $y = 5\cos\theta$ is a circle, centre O with radius 5

9 Work out the coordinates of the points on the curve $x = \dfrac{3+2t}{1-t}$, $y = \dfrac{2-t}{1+3t}$ where

a $t = -1$ **b** $t = 8$

8 Find the Cartesian equation of the curve given by the parametric equations

a $x = 7\cos\theta + 8$, $y = 7\sin\theta + 6$, $0° \leq \theta < 360°$

b $x = 5\cos\theta + 3$, $y = 5\sin\theta - 1$, $0° \leq \theta < 360°$

10 By substituting $y = tx$, work out parametric equations for these curves.

a $y^3 = x^4$ **b** $y = x^2 - 3x$

c $y^2 = x^2 - 5x$ **d** $y^4 = x^3 + x^2$

e $y = x^3 + 3xy$

Reasoning and problem-solving

To find a Cartesian equation from parametric equations

(**1**) Make the parameter the subject of one of the equations.

(**2**) Substitute the value for the parameter in the other equation and simplify.

Example 2

Point A moves across a coordinate grid in a straight line with speed $\begin{pmatrix} 6 \\ 8 \end{pmatrix}$ cms^{-1}. Let t be the time in seconds. When $t = 0$, A is at $(12, 0)$.

a Write down parametric equations in t for the position of A

b Find the Cartesian coordinates of the point where A crosses the line $y = x$

a $x = 12 + 6t$ $y = 8t$ •

b $12 + 6t = 8t \Rightarrow t = 6$ s, $(12 + 6 \times 6, 8 \times 6) = (8, 48)$

> Every second, the x-coordinate increases by 6 and the y-coordinate increases by 8.

Exercise 12.3B Reasoning and problem-solving

1 A bullet is fired horizontally out to sea at 750 ms^{-1} from the top of a cliff 50 m high. Its position in relation to the foot of the cliff (in metres) is given by parametric equations $x = 750t$, $y = 50 - 5t^2$. Showing your working, work out

a When the bullet hits the sea

b How far from the base of the cliff it is at this time

c How far from the cliff the bullet is when it is 25 m above the sea.

2 A fairground roundabout has a radius of 10 m, with centre at the origin. A child gets on at the point $(0, -10)$ and moves clockwise. Write parametric equations for the position of the child where the parameter is the angle $\theta°$ between the radius at any time and the negative direction of the y-axis. Give the coordinates of the child when θ is 90°, 135°, 180° and 270°.

3 An ellipse has the parametric equations $x = 5\cos t$, $y = \sin t$. Use a table of values to sketch this graph. Check your sketch using a graphical calculator or computer software.

4 A 'human cannonball' is fired from a cannon at a circus at a speed of $35\,\mathrm{m\,s^{-1}}$ and an angle of 30° to the horizontal. His motion is described by $x = 35t\cos 30°$; $y = 35t\sin 30° - 5t^2$. He aims for a large safety platform with closest edge 90 m away horizontally and 7.5 m high.

Does he succeed?

5 A projectile passes through the air. Its passage can be modelled by the parametric equations

$y = -5t^2 + 20t + 105$, $x = 5t$, $t \geq 0$

where t is time (seconds), x is horizontal displacement (metres) and y is vertical displacement from the ground (metres). Show your working in parts **a** to **c**.

  a What is the greatest height reached by the projectile?

  b After how many seconds does the projectile hit the ground?

  c How far does the projectile travel horizontally before hitting the ground?

6 Find the Cartesian equation of the locus given by $x = \sin\theta$, $y = 2\sin\theta\cos\theta$

7 A curve is given by parametric equations $x = \sin\theta$, $y = 3\cos\theta$, $0° < \theta \leq 360°$

The curve is first translated by the vector $\begin{pmatrix} 6 \\ 0 \end{pmatrix}$

and then stretched parallel to the y-axis by scale factor $\dfrac{1}{2}$

  a Write parametric equations to describe the transformed curve.

  b Write a Cartesian equation to describe the transformed curve.

8 Two curves are defined by parametric equations

Curve A: $x = 5 \times 3^{2t}$, $y = 3^{1-2t}$, $t \in \mathbb{R}$

Curve B: $x = \dfrac{3}{t}$, $y = 5t$, $t > 0$

  a Show that curve A is identical to curve B.

  b Write the Cartesian equation of the curve, stating its domain.

9 Find the point(s) of intersection of the curve given by $x = 5 - 2t$, $y = t^2 - 1$, $t \geq 0$ and the line given by $y = x - 2$, $x \in \mathbb{R}$. Show your working.

10 Find the points of intersection between the parabola $y = 2t$, $x = t^2$ and the circle $y^2 + x^2 - 3x - 90 = 0$. Show your working.

11 Work out the points of intersection of the curve $x = 3t^2 + 2$, $y = 4(t - 2)$ and the line $2x + 3y + 2 = 0$. Show your working.

12

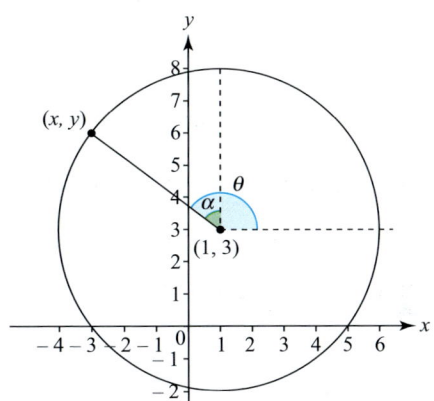

The diagram shows a circle with centre $(1, 3)$ and radius 5. $0 < \theta \leq 360°$ is the angle measured anti-clockwise from the radius to $(6, 3)$ and the radius to (x, y). $0 < \alpha \leq 360°$ is the angle measured anti-clockwise from the radius to $(1, 8)$ and the radius to (x, y).

  a Find parametric equations to represent the circle which use the parameter $0° < \theta \leq 360°$

  b Find parametric equations to represent the circle which use the parameter $0° < \alpha \leq 360°$

Challenge

13 Show that the circle defined by $x = 3\cos\theta$, $y = 4 + 3\sin\theta$, $0° \leq \theta < 360°$ and the parabola defined by $x = -5t$, $y = 2t^2$, $t \in \mathbb{R}$ do not intersect.

12.4 Algebraic fractions

Fluency and skills

You can manipulate algebraic fractions using the same methods you use for arithmetical fractions.

Example 1

Simplify $\dfrac{x^2+x-6}{x^2-x-2}\times\dfrac{x^2-6x-7}{x^2+6x+9}$

$\dfrac{x^2+x-6}{x^2-x-2}\times\dfrac{x^2-6x-7}{x^2+6x+9}$

$=\dfrac{(x-2)(x+3)}{(x-2)(x+1)}\times\dfrac{(x-7)(x+1)}{(x+3)(x+3)}$ — Factorise the quadratic expressions first.

$=\dfrac{(x-2)(x+3)}{(x-2)(x+1)}\times\dfrac{(x-7)(x+1)}{(x+3)(x+3)}$ — Cancel common factors.

$=\dfrac{(x-7)}{(x+3)}$

See Ch2.3 For an introduction to algebraic division.

You can use algebraic long division for a divisor of the form $(ax+b)$ in the same way that you would for a divisor of the form $(x+b)$

Example 2

Calculate $(4x^3+3x-2)\div(2x-1)$

$$
\begin{array}{r}
2x^2+x+2 \\
(2x-1)\overline{)4x^3+0x^2+3x-2} \\
\underline{4x^3-2x^2} \\
2x^2+3x \\
\underline{2x^2-x} \\
4x-2 \\
\underline{4x-2} \\
0
\end{array}
$$

$4x^3\div2x=2x^2$

Use $0x^2$ to fill the place value for x^2

In Example 2 there is no remainder so $(2x-1)$ is a factor of $(4x^3+3x-2)$. You can find factors using a variant of the factor theorem.

In general, for a polynomial f(x) of degree $n\geq1$ and any constants a and b

f$(x)\equiv(ax-b)$g$(x)+R$

where g(x) is a polynomial of order $n-1$ and R is a constant.

When $x = \dfrac{b}{a}$ this gives

$$\mathrm{f}\!\left(\dfrac{b}{a}\right) \equiv \left(a\!\left(\dfrac{b}{a}\right) - b\right)\mathrm{g}(x) + R$$

$$\mathrm{f}\!\left(\dfrac{b}{a}\right) \equiv R$$

If $\mathrm{f}\!\left(\dfrac{b}{a}\right) = 0$ there is no remainder when $\mathrm{f}(x)$ is divided by $(ax - b)$ so $(ax - b)$ is a factor of $\mathrm{f}(x)$

Key point

The factor theorem states that
$$\mathrm{f}\!\left(\dfrac{b}{a}\right) = 0 \text{ if and only if } (ax - b) \text{ is a factor of } \mathrm{f}(x)$$

Example 3

Show that $(3x - 5)$ is a factor of $\mathrm{f}(x) = (6x^3 - 7x^2 - 8x + 5)$

$(3x - 5)$ is a factor of $\mathrm{f}(x)$ if $\mathrm{f}\!\left(\dfrac{5}{3}\right) = 0$ ⟵ $3x - 5 = 0 \Rightarrow x = \dfrac{5}{3}$

$\mathrm{f}\!\left(\dfrac{5}{3}\right) = 6\!\left(\dfrac{5}{3}\right)^3 - 7\!\left(\dfrac{5}{3}\right)^2 - 8\!\left(\dfrac{5}{3}\right) + 5 = 0$ ⟵ Substitute $x = \dfrac{5}{3}$

So $(3x - 5)$ is a factor of $\mathrm{f}(x) = (6x^3 - 7x^2 - 8x + 5)$

Exercise 12.4A Fluency and skills

1 Simplify these fractions

  a $\dfrac{x}{x-3} + \dfrac{2x}{x-5}$ **b** $\dfrac{2y}{y+8} - \dfrac{4y}{y-2}$

  c $\dfrac{2x}{y-3} + \dfrac{3x}{2y-7}$ **d** $\dfrac{3z}{2z+9} - \dfrac{5z}{3z-4}$

2 a Simplify   **i** $\dfrac{2x^2 - 5x + 2}{x^2 - 4}$ **ii** $\dfrac{(x+2)^2}{2x^2 + 3x - 2}$

  b Use your answers from part **a** to write down the value of
$$\dfrac{2x^2 - 5x + 2}{x^2 - 4} \times \dfrac{(x+2)^2}{2x^2 - 3x - 2}$$

3 Simplify these fractions

  a $\dfrac{5z^2 - 10z}{8z + 24} \times \dfrac{15z + 45 - 10z}{12z^2 - 24z}$

  b $\dfrac{3w + 12}{w^2 - 7w} \div \dfrac{4w^2 + 16w}{w - 7}$

  c $\dfrac{2n^2 - 11n - 6}{8n^2 + 22n + 15} \div \dfrac{2n^2 - 11n - 6}{12n^2 - 13n - 35}$

  d $\dfrac{4m^3 - 2m^2 - 12m}{9m^2 + 18m + 5} \times \dfrac{3m^2 - m - 10}{2m^3 - m^2 - 6m}$

4 Divide

  a $8x^2 - 26x - 70$ by $(x - 5)$

  b $3x^2 + 45x + 168$ by $(x + 8)$

5 Divide

  a $x^3 + 6x^2 + 4x - 1$ by $(x + 1)$

  b $2x^3 + x^2 - x - 63$ by $(2x - 3)$

  c $3x^3 + 17x^2 - 11x - 33$ by $(3x + 2)$

  d $6x^3 + x^2 - 19x - 12$ by $(3x + 1)$

  e $x^3 + 27$ by $(x + 3)$

6 Work out the remainder when

  a $2x^3 - 2x^2 + 7x + 14$ is divided by $(2x - 3)$

  b $3x^4 - 13x^3 + 2x^2 - 5x - 10$ is divided by $(3x + 3)$

  c $9x^3 + 5x^2 + 6x + 7$ is divided by $(3x - 1)$

  d $4x^4 - x^3 - 12x^2 + 3x + 4$ is divided by $(2x + 11)$

7 Factorise $x^4 - 4x^2 + 3$

8 a Show that $(3x - 4)$ is a factor of $6x^3 - 5x^2 - 16x + 16$

  b Show that $(2x + 1)$ is a factor of $2x^3 - x^2 + 7x + 4$

  c Show that $(2x - 8)$ is a factor of $8x^3 - 36x^2 + 18x - 8$

9 a Use the factor theorem to show that $(3x+1)$ is a factor of $6x^3+5x^2+13x+4$

  b Write $6x^3+5x^2+13x+4$ in the form $(3x+1)(Ax^2+Bx+C)$ where A, B and C are constants to be found.

10 The polynomial f(x) is given by $3x^3+10x^2-23x+10$

  a Show that $(3x-2)$ is a factor of f(x)

  b Factorise f(x) fully.

c Simplify the fraction $\dfrac{3x^3+10x^2-23x+10}{3x^2+4x-4}$

11 The expressions in parts **a** and **b** include a factor of the form $(x+a)$

  a Factorise fully $6x^3+5x^2-12x+4$

  b Factorise fully $4x^3-9x^2-x+6$

  c Factorise fully $18x^4-27x^3-2x^2+10x$ This expression includes a factor of the form $(6x-b)$

Reasoning and problem-solving

To solve a problem using the factor theorem

1 If the factor under consideration is of the form $(ax+b)$, solve the equation $ax+b=0$ for x

2 Substitute your solution into the expression f(x)

3 Interpret your solution. If the expression is equal to zero, $(ax+b)$ is a factor of f(x)

Work out the values of p and q when f$(x)=6x^4+px^3-24x^2+qx+4$ is divisible by $(2x+1)$ and $(3x-1)$

1 3

$(2x+1)$ is a factor of f$(x) \Rightarrow$ f$\left(-\dfrac{1}{2}\right)=0$

$2x+1=0 \Rightarrow x=-\dfrac{1}{2}$

2

f$\left(-\dfrac{1}{2}\right)=6\left(-\dfrac{1}{2}\right)^4+p\left(-\dfrac{1}{2}\right)^3-24\left(-\dfrac{1}{2}\right)^2+q\left(-\dfrac{1}{2}\right)+4$

Substitute $x=-\dfrac{1}{2}$ into the expression

$=-\dfrac{p}{8}-\dfrac{q}{2}-\dfrac{13}{8}$

$=0$

Simplify the equation by multiplying both sides by -8

$p+4q+13=0$ (1)

1 3

$(3x-1)$ is a factor of f$(x) \Rightarrow$ f$\left(\dfrac{1}{3}\right)=0$

$3x-1=0 \Rightarrow x=\dfrac{1}{3}$

2

f$\left(\dfrac{1}{3}\right)=6\left(\dfrac{1}{3}\right)^4+p\left(\dfrac{1}{3}\right)^3-24\left(\dfrac{1}{3}\right)^2+q\left(\dfrac{1}{3}\right)+4$

Substitute $x=\dfrac{1}{3}$ into the expression

$=\dfrac{p}{27}+\dfrac{q}{3}+\dfrac{38}{27}$

$=0$

Simplify the equation by multiplying both sides by 27

$p+9q+38=0$ (2)

$9q-4q+38-13=0 \Rightarrow q=-5$

Equation (2) – equation (1)

$p+4(-5)+13=0 \Rightarrow p=7$

You can check your answer by solving these equations on your calculator.

1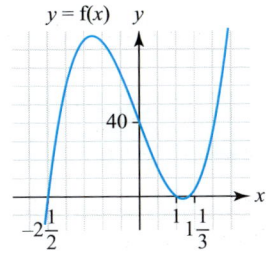

The graph shows the curve $y = f(x)$ where $f(x)$, $x \in \mathbb{R}$ is a cubic function. Write an expression for $f(x)$ in the form $Ax^3 + Bx^2 + Cx + D$

2 Work out the values of p and q when $12x^4 + 4x^3 + px^2 + qx + 8$ is divisible by $(3x - 2)$ and $(x + 1)$

3 Show there is only one vertical asymptote to the curve given by $y = \dfrac{x}{x-1} + \dfrac{1}{(x-1)(x-2)}$ and give its equation.

4 Solve for x, $\dfrac{1}{x} - \dfrac{1}{x-2} = \dfrac{1}{a} - \dfrac{1}{a-2}$

5 Prove that $(x^2 - 9)$ is a factor of the expression $x^4 + 6x^3 - 4x^2 - 54x - 45$

6 $x^4 - 13x^2 + 36$, $2x^3 + 3x^2 - 11x - 6$ and $3x^3 + x^2 - 20x + 12$ have a common quadratic factor. What is it?

7 Work out the HCF of $x^3 + 5x^2 + 2x - 8$ and $2x^3 + 3x^2 - 17x + 12$ in factorised form.

8 Show that $(x - 2)^2$ is a factor of $x^4 - 10x^3 + 37x^2 - 60x + 36$ and evaluate the other factors.

9 a Show that $(5x + 2)$ is a factor of $30x^3 + 7x^2 - 12x - 4$

b Fully simplify the expression $\dfrac{30x^3 + 7x^2 - 12x - 4}{2x^2 - x} \times \dfrac{2x^2 - 9x + 4}{2x^2 - 7x - 4}$

c i State the maximum possible domain and corresponding range for the function
$f(x) = \dfrac{30x^3 + 7x^2 - 12x - 4}{2x^2 - x} \times \dfrac{2x^2 - 9x + 4}{2x^2 - 7x - 4}$

ii Show that the gradient $f'(x)$ is positive at all points on the curve.

10 $(2x - 5)$ is a factor of $f(x) = ax^4 + bx^2 - 75$

a Use the factor theorem to show that $(2x + 5)$ is a factor of $f(x)$

b The third and final factor of $f(x)$ takes the form $(x^2 + c)$. Find the values of a, b and c

11 James attempts to complete the calculation $(4x^4 + 6x^3 + x - 8) \div (2x + 1)$. Explain and correct his mistakes.

$$
\begin{array}{r}
2x^3 + 2x^2 - 0.5 \\
2x+1\overline{)\ 4x^4 + 6x^3 + \ x\ -8} \\
-(4x^4 + 2x^3) \\
\hline
4x^3 + \ x \\
-(4x^3 + 2x) \\
\hline
-x\ -8 \\
-(-x\ -0.5) \\
\hline
-7.5
\end{array}
$$

$(4x^4 + 6x^3 + x - 8) \div (2x + 1) = 2x^3 + 2x^2 - 0.5 - 7.5$
$= 2x^3 + 2x^2 - 8$

12 For what values of b is $(x - b)$ a factor of $3x^3 + (b + 3)x^2 - (4b^2 + b - 7)x - 4$?

Challenge

13 a Prove that, if two polynomials $f(x)$ and $g(x)$ have a common linear factor $(x - a)$, then $(x - a)$ is a factor of the polynomial $f(x) - g(x)$

b Use the result from part **a** to prove that, if the equations $kx^3 + 3x^2 + x + 4 = 0$ and $kx^3 + 2x^2 + 9x - 8 = 0$ have a common linear factor, then $k = \dfrac{-9}{4}$ or $\dfrac{-59}{108}$

Partial fractions

Fluency and skills

The reverse process of adding algebraic fractions is splitting an algebraic fraction into its component parts. This technique is known as decomposing an algebraic fraction into **partial fractions**.

When adding fractions you find the lowest common multiple of the individual denominators. When decomposing a fraction you need to reverse this process, breaking the denominator into factors.

Example 1

Express $\dfrac{5x+2}{(x+4)(x-5)}$ in the form $\dfrac{A}{(x+4)}+\dfrac{B}{(x-5)}$ where A and B are integers.

Let $\dfrac{5x+2}{(x+4)(x-5)} \equiv \dfrac{A}{(x+4)}+\dfrac{B}{(x-5)}$

$(5x+2) \equiv \dfrac{A(x+4)(x-5)}{(x+4)}+\dfrac{B(x+4)(x-5)}{(x-5)}$

Multiply through by the denominator.

$(5x+2) \equiv A(x-5)+B(x+4)$

$5x+2 \equiv (A+B)x+(4B-5A)$

Look at both sides of the equation to compare the coefficient of the x term and the constant.

$A+B=5 \Rightarrow A=5-B$ (1)

$4B-5A=2$ (2)

$4B-5(5-B)=2 \Rightarrow B=3$

$A=5-3=2$

Solve the simultaneous equations by substituting the value of A from equation (1) into equation (2)

$\dfrac{(5x+2)}{(x+4)(x-5)} \equiv \dfrac{2}{(x+4)}+\dfrac{3}{(x-5)}$

If a denominator contains a squared factor, you need to consider the squared factor as a possible denominator, as well as all linear factors.

Example 2

a Find the sum of $\dfrac{3}{(x-5)}+\dfrac{7}{(x+4)}+\dfrac{10}{(x+4)^2}$

b Split $\dfrac{97x+1}{(x-5)(x+4)^2}$ into partial fractions.

a $\dfrac{3}{(x-5)}+\dfrac{7}{(x+4)}+\dfrac{10}{(x+4)^2}$

$\equiv \dfrac{3(x+4)^2+7(x-5)(x+4)+10(x-5)}{(x-5)(x+4)^2}$

$\equiv \dfrac{10x^2+27x-142}{(x-5)(x+4)^2}$

The distinct denominators $(x+4)$ and $(x+4)^2$ cannot be discerned from the final fraction.

(*Continued on the next page*)

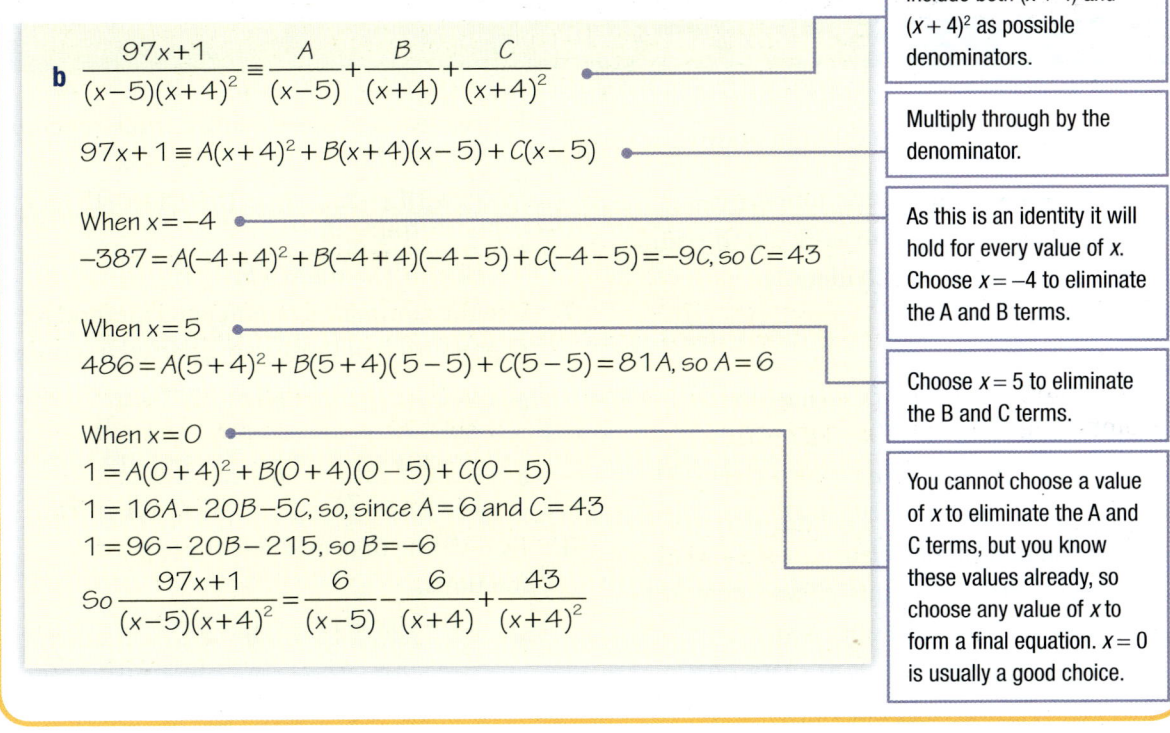

b $\dfrac{97x+1}{(x-5)(x+4)^2} \equiv \dfrac{A}{(x-5)} + \dfrac{B}{(x+4)} + \dfrac{C}{(x+4)^2}$

> Include both $(x+4)$ and $(x+4)^2$ as possible denominators.

$97x+1 \equiv A(x+4)^2 + B(x+4)(x-5) + C(x-5)$

> Multiply through by the denominator.

When $x = -4$

$-387 = A(-4+4)^2 + B(-4+4)(-4-5) + C(-4-5) = -9C$, so $C = 43$

> As this is an identity it will hold for every value of x. Choose $x = -4$ to eliminate the A and B terms.

When $x = 5$

$486 = A(5+4)^2 + B(5+4)(5-5) + C(5-5) = 81A$, so $A = 6$

> Choose $x = 5$ to eliminate the B and C terms.

When $x = 0$

$1 = A(0+4)^2 + B(0+4)(0-5) + C(0-5)$

$1 = 16A - 20B - 5C$, so, since $A = 6$ and $C = 43$

$1 = 96 - 20B - 215$, so $B = -6$

So $\dfrac{97x+1}{(x-5)(x+4)^2} = \dfrac{6}{(x-5)} - \dfrac{6}{(x+4)} + \dfrac{43}{(x+4)^2}$

> You cannot choose a value of x to eliminate the A and C terms, but you know these values already, so choose any value of x to form a final equation. $x = 0$ is usually a good choice.

If the order of the numerator is equal to or greater than the order of the denominator, you divide the numerator by the denominator first and then express the remainder in partial fractions.

Example 3

Express $\dfrac{2x^2 - 3x - 39}{(x+2)(x-3)}$ in the form $C + \dfrac{D}{(x+2)} + \dfrac{E}{(x-3)}$ where C, D and E are integers.

$\dfrac{2x^2 - 3x - 39}{(x+2)(x-3)} \equiv C + \dfrac{D}{(x+2)} + \dfrac{E}{(x-3)}$

$2x^2 - 3x - 39 \equiv C(x+2)(x-3) + D(x-3) + E(x+2)$

> Multiply both sides by the denominator.

Let $x = -2$

$2(-2)^2 - 3(-2) - 39 \equiv C(0)(-5) + D(-5) + E(0)$

$-25 \equiv -5D$

$D = 5$

> An identity is true for all values of x. Substitute a value which removes two unknowns.

Let $x = 3$

$2(3)^2 - 3(3) - 39 \equiv C(5)(0) + D(0) + E(5)$

$-30 \equiv 5E$

$E = -6$

> Repeat the process to find E

Let $x = 0$

$2(0)^2 - 3(0) - 39 \equiv C(2)(-3) + 5(-3) - 6(2)$

$-39 \equiv -6C - 27$

$C = 2$

> Use the values of D and E and any value of x to find the third unknown.

Hence $\dfrac{2x^2 - 3x - 39}{(x+2)(x-3)} \equiv 2 + \dfrac{5}{(x+2)} - \dfrac{6}{(x-3)}$

1 Write $\dfrac{18x+26}{(3x+2)(x-4)}$ as the sum of two fractions with linear denominators.

2 Compare coefficients to work out the values of the constants in each identity.

    a $12x-48 \equiv Ax(x-2)+B(x-3)(x+4)$

    b $x^2-22x-35 \equiv C(x+3)(x-2)$
$$+ D(x+1)(x-2)$$
$$+ E(x+1)(x+3)$$

    c $3x^3+11x^2-20 \equiv Fx(x+2)^2+Gx(x+2)$
$$- H(x+2)^2$$

3 By substituting appropriate values of x, work out the values of the constants in these identities.

    a $7x-15 \equiv A(x-1)+B(x-3)$

    b $4x^2+24x+15 \equiv C(x+2)+D(x+2)^2$
$$+ E(x+1)$$

    c $24x-24 \equiv Fx(x+4)+G(x+4)(x-2)$
$$+ Hx(x-2)$$

4 Substitute $x=2$ to find the value of A in the identity $5x^2-5x-1 \equiv A(x^2-1)+(Bx+C)(x-2)$
Then substitute other values of x to find the values of B and C.

5 Express $\dfrac{1}{(1-x)^2}$ and $\dfrac{x}{(1-x)^2}$ in partial fractions. Compare your results.

6 Express each of these using partial fractions.

    a $\dfrac{x-17}{(x+3)(x-2)}$ **b** $\dfrac{10x-2}{(x-1)(x+1)^2}$

    c $\dfrac{146-38x}{(2x-5)(x+6)(2x+1)}$

7 Use the comparing coefficients method to express each of these using partial fractions.

    a $\dfrac{8}{x(x+4)}$ **b** $\dfrac{4}{(x+1)(x-3)}$

    c $\dfrac{-36x+4}{(x+5)(x-7)^2}$ **d** $\dfrac{18x-12}{x(2x-3)(x+4)}$

8 By dividing first, evaluate these partial fractions.

    a $\dfrac{3x^2-10x+11}{(x-1)(x-3)}$ **b** $\dfrac{5x^2+27x+26}{(x+1)(x+5)}$

9 Express using partial fractions

    a $\dfrac{4x+2\sqrt{3}}{x^2-3}$ **b** $\dfrac{\sqrt{6}x+9\sqrt{5}}{6x^2-5}$

10 Express $\dfrac{x}{(x-a)(x-b)}$ in partial fractions.

11 Given that a, b, c, P and Q are constants, write expressions for P and Q in terms of a, b and c if $\dfrac{ax+b}{(x+c)^2} = \dfrac{P}{x+c} + \dfrac{Q}{(x+c)^2}$

Reasoning and problem-solving

Strategy

To find the partial fractions of an expression

  1 Write a partial fraction for each linear or squared factor in the denominator.

  2 Multiply through by the denominator and create an identity.

  3 Either substitute suitable values or compare coefficients to find the constants and write the expression using partial fractions.

Example 4

a Express $\dfrac{1}{r(r+1)}$ in partial fractions.

b Deduce that $\dfrac{1}{1\times2}+\dfrac{1}{2\times3}+\dfrac{1}{3\times4}+...+\dfrac{1}{n(n+1)}=\dfrac{n}{n(n+1)}$

(Continued on the next page)

a $\dfrac{1}{r(r+1)} \equiv \dfrac{A}{r} + \dfrac{B}{r+1}$

> **1** Write partial fractions with denominators r and $r+1$

$1 \equiv A(r+1) + Br \Rightarrow 1 \equiv (A+B)r + A$

> **2** Multiply through by the denominator and rearrange to compare coefficients.

$A = 1$ and $A + B = 0 \Rightarrow B = -1 \Rightarrow \dfrac{1}{r(r+1)} \equiv \dfrac{1}{r} - \dfrac{1}{r+1}$

b $\dfrac{1}{1\times 2} + \dfrac{1}{2\times 3} + \dfrac{1}{3\times 4} + \dots + \dfrac{1}{n(n+1)}$

$= \left(\dfrac{1}{1} - \dfrac{1}{2}\right) + \left(\dfrac{1}{2} - \dfrac{1}{3}\right) + \left(\dfrac{1}{3} - \dfrac{1}{4}\right) + \dots + \left(\dfrac{1}{n-1} - \dfrac{1}{n}\right) + \left(\dfrac{1}{n} - \dfrac{1}{n+1}\right)$

> Write each fraction as a pair of partial fractions. Include the first few terms and the last few terms to see the pattern.

$= \dfrac{1}{1} + \left(-\dfrac{1}{2} + \dfrac{1}{2}\right) + \left(-\dfrac{1}{3} + \dfrac{1}{3}\right) - \dfrac{1}{4} + \dots + \dfrac{1}{n-1} + \left(-\dfrac{1}{n} + \dfrac{1}{n}\right) - \dfrac{1}{n+1}$

> Use the pattern to cancel zero terms.

$= 1 - \dfrac{1}{n+1} = \dfrac{n+1}{n+1} - \dfrac{1}{n+1} = \dfrac{n}{n+1}$

> Write 1 as $\dfrac{n+1}{n+1}$ and subtract to get one fraction.

Exercise 12.5B Reasoning and problem-solving

1 An athlete ran the first lap of a race in $\dfrac{3}{t-3}$ minutes and the second lap in $\dfrac{2}{t-7}$ minutes. Write a single fraction in t for her total time.

2 An arithmetic progression is of the form a, $a+d$, $a+2d$ etc. where a is the first term and d the common difference. The sixth term, $(a+5d)$, is $\dfrac{12x+4}{4x^2-1}$

Find d when $a = \dfrac{1}{2x+1}$

3 A lens has a focal length, $f = \dfrac{(p+2)(p-7)}{8p-11}$.

The object is at distance, $u = \dfrac{p+2}{A}$ and the image is at distance, $v = \dfrac{p-7}{B}$

Use the formula for a lens, $\dfrac{1}{u} + \dfrac{1}{v} = \dfrac{1}{f}$, to work out A and B.

4 A student says that, for real constants a and b,

$$\dfrac{a+b}{(x+c)(x+d)} \equiv \dfrac{a}{(x+c)} + \dfrac{b}{(x+d)} \quad \text{OR} \quad \dfrac{b}{(x+c)} + \dfrac{a}{(x+d)}$$

a Show that this is not true for all real values of a, b, c and d.

b Find expressions for a and b for which the first identity holds.

5 A student tries to decompose $\dfrac{14}{(x-5)(x+1)^2}$ into partial fractions. Explain and correct her mistakes.

$$\dfrac{14}{(x-5)(x+1)^2} \equiv \dfrac{A}{(x-5)} + \dfrac{B}{(x+1)^2}$$

$$14 \equiv A(x+1)^2 + B(x-5)$$

Let $x = -1 \Rightarrow 14 \equiv 0A - 6B$ so $B = -\dfrac{7}{3}$

Let $x = 5 \Rightarrow 14 \equiv 36A + 0B$ so $A = \dfrac{7}{18}$

$$\dfrac{14}{(x-5)(x+1)^2} \equiv \dfrac{7}{18(x-5)} - \dfrac{7}{3(x+1)^2}$$

Challenge

6 a Express $\dfrac{1}{(r+1)(r+2)}$ in partial fractions.

b Use your partial fractions to show that
$$\dfrac{1}{1(2)} + \dfrac{1}{2(3)} + \dfrac{1}{3(4)} + \dfrac{1}{4(5)} + \dots + \dfrac{1}{(n+1)(n+2)} = 1 - \dfrac{1}{(n+2)}$$

c Explain what happens to the sum if n approaches infinity.

Fluency and skills

To use vectors for three-dimensional problems, you use three perpendicular axes: the x-, y- and z-axes.

If you draw the x- and y-axes as usual, the convention is to draw the z-axis coming "out of" the page. So if the page is lying on your desk, the positive z-axis points towards the ceiling. Axes like these are a **right-hand set** because you can position your thumb, forefinger and second finger in the direction of the x-, y- and z-axis respectively.

Left-hand set   **Right-hand set**

A vector **r** in 3D has three components. You can write it as a column vector, or use **i**, **j** and **k**, which are the unit vectors in the x-, y- and z-directions. For example

> **See Ch6.2**
> For a reminder of vector notation.

$$\mathbf{r} = \begin{pmatrix} a \\ b \\ c \end{pmatrix} = a\mathbf{i} + b\mathbf{j} + c\mathbf{k}$$

In this diagram a right-hand set of axes is used.

The **magnitude** (or **modulus**) of **r** is |**r**|, the length of OP in the diagram.

From triangle PDC you have $PC^2 = a^2 + b^2$

From triangle OPC you have $OP^2 = PC^2 + c^2 = a^2 + b^2 + c^2$

> **Key point**
> The magnitude of vector $\mathbf{r} = a\mathbf{i} + b\mathbf{j} + c\mathbf{k}$ is
> $$|\mathbf{r}| = \sqrt{a^2 + b^2 + c^2}$$

The direction of **r** makes angle α with the positive x-axis, β with the positive y-axis and γ with the positive z-axis, where

> **See Ch6.2**
> For a reminder of magnitude/ direction form.

> **Key point**
> $$\cos\alpha = \frac{a}{|\mathbf{r}|}, \quad \cos\beta = \frac{b}{|\mathbf{r}|} \text{ and } \cos\gamma = \frac{c}{|\mathbf{r}|}$$

It follows that

> **Key point**
> $$\cos^2\alpha + \cos^2\beta + \cos^2\gamma = \frac{a^2 + b^2 + c^2}{|\mathbf{r}|^2} = 1$$

You combine or equate 3D vectors in component form by combining or equating their components.

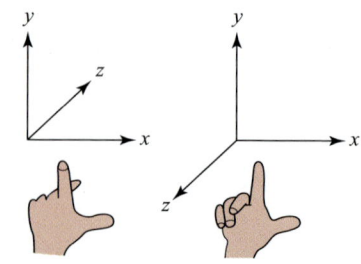

Try it on your calculator

You can find the angle between two vectors on a calculator.

Angle(VctA, VctB)
            43.83017926

Activity

Find out how to calculate the angle between $\begin{pmatrix} 6 \\ 3 \\ 11 \\ -3 \\ 1 \end{pmatrix}$ and $\begin{pmatrix} \\ -1 \\ \end{pmatrix}$ on *your* calculator.

The position vector of a point A defines its position in relation to an origin O.

> **Key point**
>
> The position vector of A is $\overrightarrow{OA}$, you usually write this as **a**.
>
> For points A and B, $\overrightarrow{AB} = \overrightarrow{AO} + \overrightarrow{OB} = \overrightarrow{OB} - \overrightarrow{OA} = \mathbf{b} - \mathbf{a}$

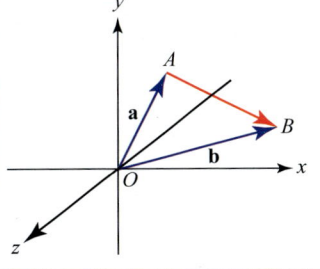

Example 1

Points A and B have position vectors $\mathbf{a} = 3\mathbf{i} + 2\mathbf{j} - 5\mathbf{k}$ and $\mathbf{b} = 4\mathbf{i} - 7\mathbf{k}$ respectively.

Work out **a** the length of AB **b** the angle between $\overrightarrow{AB}$ and the y-direction

c the unit vector in the direction of $\overrightarrow{AB}$

a $\overrightarrow{AB} = \mathbf{b} - \mathbf{a} = (4-3)\mathbf{i} - 2\mathbf{j} + (-7+5)\mathbf{k} = \mathbf{i} - 2\mathbf{j} - 2\mathbf{k}$

$AB = |\overrightarrow{AB}| = \sqrt{1^2 + (-2)^2 + (-2)^2} = 3$

Simplify the question by moving $\overrightarrow{AB}$ to the origin without changing its direction.

b $\overrightarrow{AB}$ makes an angle of β with the positive y-direction

$\cos(180° - \beta) = \dfrac{2}{|\overrightarrow{AB}|} = \dfrac{2}{3}$

$\cos\beta = -\cos(180° - \beta) = \dfrac{-2}{3}$

$\beta = \cos^{-1}\left(-\dfrac{2}{3}\right) = 131.8°$

Draw a simplified sketch.

The $\mathbf{j}$ component is -2

This is equal to $\dfrac{\mathbf{j}\text{-component of } \overrightarrow{AB}}{|\overrightarrow{AB}|}$

You can check this using your calculator.

c The unit vector in the direction of $\overrightarrow{AB} = \dfrac{\overrightarrow{AB}}{|\overrightarrow{AB}|} = \dfrac{1}{3}\overrightarrow{AB} = \dfrac{1}{3}\mathbf{i} - \dfrac{2}{3}\mathbf{j} - \dfrac{2}{3}\mathbf{k}$

> You could state this problem in terms of coordinates $A(3, 2, -5)$ and $B(4, 0, -7)$.

Example 2

Points P and Q have position vectors $\mathbf{p} = -3\mathbf{i} - 3\mathbf{j} + \mathbf{k}$ and $\mathbf{q} = 7\mathbf{i} + 4\mathbf{j} - 4\mathbf{k}$ respectively.

The point R lies on PQ such that $PR : RQ = 2 : 3$. Work out the position vector, $\mathbf{r}$, of R.

$\overrightarrow{PQ} = \mathbf{q} - \mathbf{p} = (7+3)\mathbf{i} + (4+3)\mathbf{j} + (-4-1)\mathbf{k} = 10\mathbf{i} + 7\mathbf{j} - 5\mathbf{k}$

$\overrightarrow{PR} = \dfrac{2}{5}\overrightarrow{PQ} = 4\mathbf{i} + 2.8\mathbf{j} - 2\mathbf{k}$

$\mathbf{r} = \mathbf{p} + \overrightarrow{PR} = (-3+4)\mathbf{i} + (-3+2.8)\mathbf{j} + (1-2)\mathbf{k} = \mathbf{i} - 0.2\mathbf{j} - \mathbf{k}$

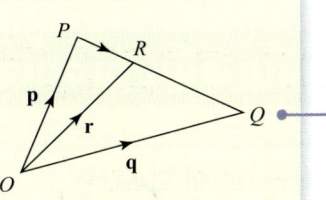

Sketch a diagram.

You can generalise the technique used in Example 2 to get the **ratio formula**.

> **Key point**
>
> For points P, Q and R and scalars μ and λ, if $PR : RQ = \lambda : \mu$ then
>
> $$\mathbf{r} = \dfrac{\mu\mathbf{p} + \lambda\mathbf{q}}{\lambda + \mu}$$

1 Given vectors $\mathbf{a} = 3\mathbf{i} - \mathbf{j} + 2\mathbf{k}$, $\mathbf{b} = 6\mathbf{i} - 3\mathbf{j} - 2\mathbf{k}$ and $\mathbf{c} = \mathbf{i} + \mathbf{j} - 3\mathbf{k}$, work out

   a **i** $\mathbf{a} - \mathbf{b}$ **ii** $2\mathbf{a} + 5\mathbf{c}$ **iii** $|\mathbf{b}|$

      iv $|\mathbf{c} - \mathbf{a}|$ **v** unit vector $\hat{\mathbf{b}}$

      vi the angle between $\mathbf{b}$ and the positive x-, y- and z-directions. Show your working.

   b a vector parallel to $\mathbf{b}$ with magnitude 28

   c The values of p, q and r if $p\mathbf{a} + q\mathbf{c} = 3\mathbf{i} - 5\mathbf{j} + r\mathbf{k}$

2 The vector $\mathbf{r}$ has magnitude 8 and makes angles of $27°$, $85°$ and $63.5°$ with the positive x-, y- and z-directions respectively. Express $\mathbf{r}$ in component form.

3 A vector $\mathbf{p}$ has magnitude 12 and makes angles of $68°$ and $75°$ with the positive y- and z-directions respectively.

   a Work out the two possible angles between $\mathbf{p}$ and the positive x-direction.

   b Use your answer to part **a** to express the two possible vectors $\mathbf{p}$ in component form.

4 Points A, B and C have position vectors
$$\mathbf{a} = \begin{pmatrix} 2 \\ 1 \\ 1 \end{pmatrix}, \mathbf{b} = \begin{pmatrix} 3 \\ 2 \\ 5 \end{pmatrix} \text{ and } \mathbf{c} = \begin{pmatrix} 6 \\ -1 \\ 5 \end{pmatrix} \text{ respectively.}$$

   a Work out the lengths of the sides of triangle ABC

   b Deduce that the triangle is right-angled.

   c State one other fact about the triangle.

5 Points A and B have position vectors $\mathbf{a} = 7\mathbf{i} + 2\mathbf{j} + 5\mathbf{k}$ and $\mathbf{b} = 5\mathbf{i} - 6\mathbf{j} + \mathbf{k}$ respectively. C is the midpoint of AB. Work out

   a The position vector, $\mathbf{c}$, of C,

   b The distance of C from the origin,

   c The unit vector, $\hat{\mathbf{c}}$, in the direction of $\mathbf{c}$.

6 Points A and B have position vectors $\mathbf{a} = 9\mathbf{i} + 2\mathbf{j} - 4\mathbf{k}$ and $\mathbf{b} = 3\mathbf{i} + 5\mathbf{j} + \mathbf{k}$ respectively. The point C lies on AB, and $AC : CB = 7 : 3$. Work out the position vector of C

7 Given vectors
$$\mathbf{p} = \begin{pmatrix} 2 \\ 3 \\ -4 \end{pmatrix}, \mathbf{q} = \begin{pmatrix} 1 \\ 1 \\ -2 \end{pmatrix} \text{ and } \mathbf{r} = \begin{pmatrix} -2 \\ 2 \\ 1 \end{pmatrix}, \text{ work out}$$

   a $\mathbf{p} + \mathbf{q} - 2\mathbf{r}$,

   b $|\mathbf{p} - \mathbf{q}|$,

   c A vector of magnitude 15 in the direction of $\mathbf{r}$,

   d The angle between $\mathbf{r}$ and the positive x-direction,

   e The values of λ and μ if $\lambda\mathbf{p} + \mu\mathbf{q} = \begin{pmatrix} 1 \\ 3 \\ -2 \end{pmatrix}$

Reasoning and problem-solving

Strategy

To solve a problem using 3D vectors

 1 Sketch a diagram. It does not need to be to scale, and it does not always need to be on 3D axes.

 2 Express the required vectors in terms of the given ones.

 3 Write a clear conclusion.

A, B, C and D have position vectors $\mathbf{a} = \mathbf{i} + \mathbf{j} - 2\mathbf{k}$, $\mathbf{b} = 3\mathbf{i} + 5\mathbf{j} + 4\mathbf{k}$, $\mathbf{c} = -\mathbf{i} + 7\mathbf{j} - 2\mathbf{k}$ and $\mathbf{d} = 7\mathbf{i} - 3\mathbf{j} + 6\mathbf{k}$ respectively.

P, Q, R and S are the midpoints of AB, BD, CD and AC respectively.
Show that $PQRS$ is a parallelogram.

To get from the origin to P, use $\overrightarrow{OA} = \mathbf{a}$, then go halfway along $\overrightarrow{AB} = \mathbf{b} - \mathbf{a}$

$$\mathbf{p} = \mathbf{a} + \frac{1}{2}(\mathbf{b} - \mathbf{a}) = \frac{1}{2}(\mathbf{a} + \mathbf{b})$$

So $\quad \mathbf{p} = \frac{1}{2}(\mathbf{i} + \mathbf{j} - 2\mathbf{k} + 3\mathbf{i} + 5\mathbf{j} + 4\mathbf{k}) = 2\mathbf{i} + 3\mathbf{j} + \mathbf{k}$

② Express $\mathbf{p}$, $\mathbf{q}$, $\mathbf{r}$ and $\mathbf{s}$ in terms of $\mathbf{a}$, $\mathbf{b}$, $\mathbf{c}$ and $\mathbf{d}$

Similarly $\quad \mathbf{q} = \frac{1}{2}(\mathbf{b} + \mathbf{d}) = 5\mathbf{i} + \mathbf{j} + 5\mathbf{k}$

$\mathbf{r} = \frac{1}{2}(\mathbf{c} + \mathbf{d}) = 3\mathbf{i} + 2\mathbf{j} + 2\mathbf{k}$

① Sketch a diagram

$\mathbf{s} = \frac{1}{2}(\mathbf{a} + \mathbf{c}) = 4\mathbf{j} - 2\mathbf{k}$

Hence $\overrightarrow{PQ} = \mathbf{q} - \mathbf{p} = 3\mathbf{i} - 2\mathbf{j} + 4\mathbf{k}$ and
$\overrightarrow{SR} = \mathbf{r} - \mathbf{s} = 3\mathbf{i} - 2\mathbf{j} + 4\mathbf{k}$

③ These conditions are necessary and sufficient to prove $PQRS$ is a parallelogram.

So PQ and SR are parallel and equal in length. Hence $PQRS$ is a parallelogram.

Exercise 12.6B Reasoning and problem-solving

1 Points A, B and C have positions vectors $2\mathbf{i} + 3\mathbf{j} - 3\mathbf{k}$, $\mathbf{i} - 4\mathbf{j} + 2\mathbf{k}$ and $3\mathbf{i} - 5\mathbf{j} + \mathbf{k}$ respectively. Prove that ABC is a right-angled triangle.

2 The points A, B and C with position vectors $\mathbf{a} = \mathbf{i} + 2\mathbf{j}$, $\mathbf{b} = 2\mathbf{i} + \mathbf{j} - 2\mathbf{k}$ and $\mathbf{c} = 3\mathbf{i} - \mathbf{j} + \mathbf{k}$ are three vertices of a parallelogram. Work out all possible positions of the fourth vertex, D.

3 The vector $\mathbf{V}$ has magnitude 6 and makes the same angle with each of the positive x-, y- and z-directions. Evaluate the possible values of $\mathbf{V}$

4 In this question east, north and upwards are the positive x-, y- and z-directions respectively. A child, standing at the origin O, flies a toy drone. She first sends it to A, 25 m north and at a height of 15 m, then for 35 m in the direction of the vector $6\mathbf{i} + 3\mathbf{j} + 2\mathbf{k}$ to B

  a Work out the angle of elevation of B from O

  b At B the drone's battery runs out and it falls to the ground. How far does she have to walk to retrieve it?

5 $ABCD$ is a tetrahedron. The position vectors of its vertices are $\mathbf{a}$, $\mathbf{b}$, $\mathbf{c}$ and $\mathbf{d}$ respectively.

P, Q and R are the respective midpoints of AB, AD and BC. S divides PC in the ratio $1:2$. T is the midpoint of QR

  a Show that D, T and S are collinear.

  b Work out the ratio $DT:TS$

Challenge

6 The points A, B, C and D lie on a straight line, as shown.

$$AC:CB = AD:BD = \lambda:\mu, \text{ where } \lambda > \mu$$

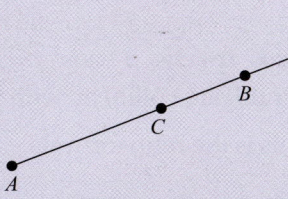

  a If $\lambda = 3$, $\mu = 2$ and A and B have position vectors $2\mathbf{i} + \mathbf{j} + 3\mathbf{k}$ and $7\mathbf{i} + 3\mathbf{j} + 9\mathbf{k}$ respectively, work out the length of CD.

  b Show that in general $CD:AB = 2\lambda\mu:(\lambda^2 - \mu^2)$

Chapter summary

- You can use proof by contradiction to prove important results.
- A function is a relationship between two variables (usually x and y) where each input number (x) must generate only one output number (y). A function can be one-to-one or many-to-one.
- The sets of all possible values of x and y, are called the domain and the range respectively.
- A composite function is formed when two (or more) functions are combined.
- The inverse of a function is found by interchanging the variables x and y (or f(x)) and rearranging the equation to make y the subject. Only one-to-one functions have inverses.
- $|x|$ means the 'modulus of x' or the 'absolute value of x'
- The modulus of a real number is always positive. You can think of it as its distance from the origin.
- A parameter is a single variable that can be used to express other variables.
- To obtain the Cartesian equation from the parametric equations you need to eliminate the parameter.
- Algebraic fractions use the same rules and techniques as arithmetical fractions.
- For a polynomial f(x), $\text{f}\left(\dfrac{b}{a}\right) = 0$ if and only if $(ax - b)$ is a factor of f(x)
- The process of splitting a fraction into its component parts is called decomposing a fraction into partial fractions.
- When solving a partial fractions problem you can either compare coefficients or substitute appropriate values into the identity, or a mixture of the two.
- To split an expression in partial fractions, the degree of the numerator must be at least one lower than the degree of the denominator.
- Axes in 3D form a right-hand set if conventional x- and y-axes are drawn on a horizontal surface and the positive z-axis points towards the ceiling.
- 3D vectors can be expressed in component form $\mathbf{r} = \begin{pmatrix} a \\ b \\ c \end{pmatrix} = a\mathbf{i} + b\mathbf{j} + c\mathbf{k}$
- The magnitude of a 3D vector $\mathbf{r}$ is $|\mathbf{r}| = \sqrt{a^2 + b^2 + c^2}$
- If $\mathbf{r}$ makes angles α, β and γ with the x, y and z directions respectively, then
$$\cos\alpha = \frac{a}{|\mathbf{r}|}, \ \cos\beta = \frac{b}{|\mathbf{r}|} \text{ and } \cos\gamma = \frac{c}{|\mathbf{r}|}$$
It follows that $\cos^2\alpha + \cos^2\beta + \cos^2\gamma = 1$
- If points P, Q and R are collinear, with position vectors $\mathbf{p}$, $\mathbf{q}$ and $\mathbf{r}$, and $PR : RQ = \lambda : \mu$, then $\mathbf{r} = \dfrac{\mu\mathbf{p} + \lambda\mathbf{q}}{\lambda + \mu}$. This is the ratio formula.

Check and review

You should now be able to...	Try Questions
Make logical deductions and prove statements directly by exhaustion, by counter example and by contradiction.	1–3
Understand and use functions.	4–6

✔ Understand and use parametric equations.		7, 8
✔ Understand and use algebraic fractions in all their forms.		9–11
✔ Decompose fractions into partial fractions.		12
✔ Manipulate vectors in three dimensions.		13
✔ Solve geometrical problems in three dimensions by vector methods.		13

1 Prove directly that, apart from 1! and 0!, all factorials are even.

2 Prove by exhaustion that, whichever way you factorise 385 you end up with the same factors.

3 Prove by contradiction that, if x is a rational number and y an irrational number, then $x - y$ is irrational.

4 $f(x) = 3x + 2, x \in \mathbb{R}$ and $g(x) = x^3, x \in \mathbb{R}$

Work out

   a f(4)   **b** g(−5)   **c** fg(2)

5 **a** State a sequence of two transformations which map the graph of $y = x$ onto the graph of $y = 5x - 4$

   b Sketch, on the same axes, the graphs of $y = 5x - 4$ and $y = |5x - 4|$

   c Given that $|x| = 3$, work out all the possible values of $|5x - 4|$. Use your sketch to confirm your solution.

6 Work out the inverse functions of

   a $\dfrac{5}{x^2}, x > 1$ **b** $\dfrac{3x + 7}{4}, x \in \mathbb{R}$ **c** $\dfrac{x - 2}{x - 4}, x \neq 4$

7 Work out the Cartesian equation represented by $x = 5t$; $y = \dfrac{-4}{t}$

8 Work out the points of intersection of the curve $x = 2t + 1$, $y = 3 - t^2$ and the straight line $x - y = 6$. Show your working.

9 Divide $2x^4 + x^3 - 5x^2 - 5x - 3$ by $2x + 3$

10 Simplify $\dfrac{n^2 - n - 6}{8n^2 + 4n + 3} \div \dfrac{n^2 - n - 6}{n^2 - 4n - 5}$

11 Fully factorise $4x^3 - 12x^2 - x + 3$

12 Evaluate the partial fractions of $\dfrac{72}{(x + 1)(x - 5)^2}$

13 Points A and B have position vectors $\mathbf{a} = 2\mathbf{i} + \mathbf{j} - 3\mathbf{k}$ and $\mathbf{b} = 5\mathbf{i} - 2\mathbf{j} + 3\mathbf{k}$. The point C, position vector $\mathbf{c}$, lies between A and B

   a Evaluate the vector $\overrightarrow{AB}$

   b Work out the length of AB

   c Work out the angle between $\overrightarrow{AB}$ and the positive z-direction. Show your working.

   d Work out $\mathbf{c}$, given that $AC : CB = 2 : 1$

What next?

Score			
	0 – 7	Your knowledge of this topic is still developing. To improve, search in MyMaths for the codes: 2049, 2135, 2138, 2139, 2142, 2200, 2208, 2224, 2254, 2259–2262	🔗
	8 – 10	You're gaining a secure knowledge of this topic. To improve, look at the InvisiPen videos for Fluency and skills (12A)	🎞
	11 – 13	You've mastered these skills. Well done, you're ready to progress! To develop your techniques, look at the InvisiPen videos for Reasoning and problem-solving (12B)	🎞

Click these links in the digital book

ICT

Some of the most interesting curves can be produced from **parametric equations** which use a combination of sines and cosines.

$x = t\cos(2.34t),$
$y = t\sin(2.34t)$
$-4.7 \leq t \leq 4.7$

A small change to even just one of the functions can make a significant difference.

$x = t + t\cos(2.34t),$
$y = t\sin(2.34t)$
$-4.7 \leq t \leq 4.7$

Other types of function can make different, even more interesting curves.

$x = \cos(2t) - \cos^3(10t)$
$y = \sin(2t) - \sin^3(10t)$
$-2 \leq t \leq 2$

Use a graph plotter or graphing software to plot different parametric curves.
Try the equations above and vary the numbers – see what happens.

Applications

Lissajous curves (also known as **Bowditch curves**) represent complex harmonic motion. They take the form
$x = A\sin(at + d),\ y = B\sin(bt)$

For example: $x = 6\sin(3t + \frac{\pi}{2})$
$y = 4\sin(2t)$
$-3.5 \leq t \leq 3.5$

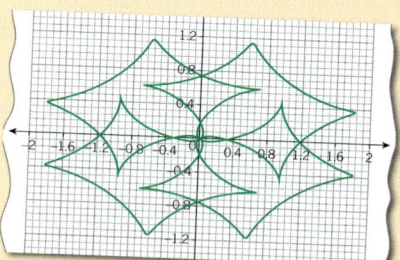

In 2009 the European Space Agency launched the Herschel and Planck observatories. Their orbits were made to follow a Lissajous curve in order to save fuel.

Challenge

A ball is thrown at $20\,\text{m s}^{-1}$ at an angle of $60°$ to the ground.
Write parametric equations to show the trajectory of the ball. Where does it land?

1 If $\dfrac{3x^4+x^3-8x^2+3x+1}{x+2} \equiv Ax^3+Bx^2+Cx+D+\dfrac{E}{x+2}$ find the constants A, B, C, D and E **[6 marks]**

2 A function is defined by $f(x)=x^2-2$, $x\in\mathbb{R}$, $x\geq 0$

   a State the range of $f(x)$ **[1]**

   b Write an expression for the inverse function $f^{-1}(x)$, stating its domain. **[3]**

   c Sketch the graphs of $f(x)$ and $f^{-1}(x)$ on the same set of axes. **[4]**

   d Find the value of x for which $f(x)=f^{-1}(x)$ **[3]**

3 Functions $f(x)$ and $g(x)$ are defined by

$$f(x)=\frac{x}{x-3},\ x\in\mathbb{R},\ x\neq 3,\ \text{and}\ g(x)=\frac{5x-2}{x},\ x\in\mathbb{R},\ x\neq 0$$

   a Work out an expression for the inverse function $f^{-1}(x)$ **[2]**

   b Work out an expression for the composite function $gf(x)$ **[3]**

   c Solve the equation $f^{-1}(x)=gf(x)$. Show your working. **[3]**

4 **a** Sketch the graph of $y=|2x-15|$ **[2]**

   b Solve the equation $|2x-15|=3$ **[2]**

   c Solve the inequality $|2x-15|\leq 3$ **[2]**

5 Determine which of the following functions are one-to-one, and which are many-to-one. Justify your answers.

   A $y=3x+2$, $x\in\mathbb{R}$ **B** $y=x^2-5$, $x\in\mathbb{R}$

   C $y=\dfrac{1}{x-3}$, $x\in\mathbb{R}$, $x\neq 3$ **D** $y=\sin x$, $x\in\mathbb{R}$ **[8]**

6 Use proof by contradiction to prove that, if n is an integer, and n^n is odd, then n is odd. **[5]**

7 The function $f(x)$ is defined by $f(x)=\dfrac{3x-7}{x^2-3x-4}-\dfrac{1}{x-4}$

   a Show that $f(x)=\dfrac{2}{x+1}$ **[4]**

   b What is the largest possible domain of $f(x)$? **[1]**

   c Work out an expression for the inverse function $f^{-1}(x)$, stating its domain. **[2]**

   d Solve the equation $f(x)=f^{-1}(x)$. Show your working. **[3]**

8 Decide which of the following statements are true and which are false. For those that are true, prove that they are true. For those that are false, give a counter example in each case.

   A For $x\neq -1$, $\dfrac{4x}{(x+1)^2}\leq 1$

   B $n!+1$ is prime for all positive integers, n

   C The product of three consecutive odd integers is always a multiple of 15

   D n^3-n is divisible by 6 for all positive integers, n **[13]**

9 Solve the equation $\dfrac{3}{x-2}-\dfrac{4}{x+1}=2$. Show your working. **[3]**

10 Write the Cartesian equation of the curve that is given parametrically

by $x = \dfrac{1}{2t+1}$; $y = \dfrac{2}{3-t}$, $t > 3$. **[7]**

11 $f(x) = |3x|$, $x \in \mathbb{R}$, $g(x) = 2x-1$, $x \in \mathbb{R}$

    a   Sketch the graph of $y = f(x)$ **[2]**   **b**   Sketch the graph of $y = gf(x)$ **[2]**

    c   Describe the transformation from $f(x)$ to $gf(x)$ **[2]**

12 If $\dfrac{6x^4 + 5x^3 - 4x^2 - 3x + 1}{2x+3} \equiv Ax^3 + Bx^2 + Cx + D + \dfrac{E}{2x+3}$, find the constants A, B, C, D and E **[6]**

13 Functions $f(x)$ and $g(x)$ are defined by

$$f(x) = e^{2x}, \ x \in \mathbb{R}, \quad \text{and} \quad g(x) = \ln(3x-2), \ x \in \mathbb{R}, \ x > \frac{2}{3}$$

    a   Write an expression for $fg(x)$ **[3]**

    b   Solve the equation $fg(x) = x^2$. Show your working.   **[4]**

    c   Work out an expression for $f^{-1}(x)$ **[2]**

    d   Solve the equation $f(x) = 5$. Show your working.   **[2]**

14 a   Work out the values of the constants A and B for which $\dfrac{3x-5}{(x-2)(x-1)} \equiv \dfrac{A}{x-2} + \dfrac{B}{x-1}$ **[4]**

    b   Hence show that $\dfrac{3x-5}{(x-2)(x-1)}$ is a decreasing function for $x > 2$ **[2]**

15 a   Prove that if a is an integer, and a^2 is a multiple of three, then a is also a multiple of three.   **[4]**

    b   Use the method of proof by contradiction to prove that $\sqrt{3}$ is irrational.   **[7]**

16 A curve, C, is given parametrically by $x = \sqrt{\sin t}$, $y = 3\sin t \cos t$, $0° \leq t \leq 90°$

    a   Show that a Cartesian equation for C is $y = 3x^2\sqrt{1-x^4}$ **[4]**

    b   Explain why there is no point on the curve for which $y = 2$ **[5]**

17 a   Express each of these in partial fractions.

      i $\dfrac{4x+1}{(x+1)(x-2)}$ **ii** $\dfrac{15-9x}{(x-1)(x-2)}$ **[4]**

    b   Hence solve the equation $\dfrac{4x+1}{(x+1)(x-2)} + \dfrac{15-9x}{(x-1)(x-2)} = 1$ **[3]**

18 Three vectors are given by $\mathbf{p} = \begin{pmatrix} 2 \\ -1 \\ 3 \end{pmatrix}$, $\mathbf{q} = \begin{pmatrix} 5 \\ 0 \\ 2 \end{pmatrix}$ and $\mathbf{r} = \begin{pmatrix} 0 \\ -7 \\ 6 \end{pmatrix}$

    a   Find $5\mathbf{p} - 3\mathbf{q} + \mathbf{r}$ **[2]**

    b   Find a vector of magnitude 3, in the direction of the vector $5\mathbf{p} - 3\mathbf{q} + \mathbf{r}$ **[2]**

19 $A(-1, 3, 0)$, $B(5, 0, 9)$ and $C(7, -1, 12)$ are points on a set of 3D axes.

    a   Write down the vector   **i** $\overrightarrow{AB}$ **ii** $\overrightarrow{BC}$ **[4]**

    b   Prove that A, B and C are collinear.   **[2]**

20 Points A, B and C have position vectors $\overrightarrow{OA} = \begin{pmatrix} -2 \\ -5 \\ 1 \end{pmatrix}$, $\overrightarrow{OB} = \begin{pmatrix} x \\ y \\ 2 \end{pmatrix}$ and $\overrightarrow{OC} = \begin{pmatrix} 7 \\ y^2 \\ 4 \end{pmatrix}$,

where x and y are constants to be determined. A, B and C are collinear.

    a   Find the value of x **[4]**

    b   Find the possible values of y **[3]**

13 Sequences

The Fibonacci sequence (1 1 2 3 5 8 13...) is one of the most well known in mathematics, and it shows up time and time again in nature. The number of petals that a flower will have is frequently a Fibonacci number, and so is the number of spirals in pineapples, sunflower heads and pinecones. The ratio of any two adjacent terms tends towards the golden ratio as the values get larger. This is a ratio found in seashells, hurricanes, spiral galaxies, art and architecture.

Identifying and defining sequences is valuable for modelling situations which are expected to follow a pattern, rather than be random. They are applicable to financial analysis, research, computer science and engineering in determining the likelihood of different scenarios, and in analysing and predicting these scenarios. This chapter covers the groundwork for understanding different types of sequence and being able to describe them.

Orientation

What you need to know

Ch2 Polynomials and the binomial theorem
- The expansion of $(1 + x)^n$
- ^nC_r notation.

p.48

Ch12 Algebra 2
- Algebraic fractions.
- Partial fractions.

p.316
p.320

What you will learn

- To use the binomial expansion and recognise the range of validity.
- To use the binomial expansion to estimate the value of a surd.
- To understand if a sequence is increasing or decreasing.
- To find the order of a periodic sequence.
- To find the nth term and the sum of an arithmetic series and a geometric series.
- To evaluate a series given in sigma notation.

What this leads to

Ch17 Numerical methods
Iterative root finding.

Further Core Pure Maths
Summing powers.
Summing series and the method of differences.

 MyMaths Practise before you start

Q 2041, 2200, 2260

The binomial series

Fluency and skills

See Ch2.2
For a reminder on the binomial theorem.

Key point

You know by the binomial theorem that the expansion of $(1+x)^n$, when n is a positive integer, is

$$(1+x)^n = 1 + {}^nC_1 x + {}^nC_2 x^2 + {}^nC_3 x^3 + \ldots + x^n$$

You write $n \in \mathbb{Z}^+$ when n is a positive integer.

You can find the coefficients ${}^nC_1$, ${}^nC_2$, ${}^nC_3$ in this expansion by using Pascal's triangle or the $\boxed{nCr}$ button on your calculator.

For example, $(1+x)^3 = 1 + 3x + 3x^2 + x^3$

The sum on the right-hand side is a **series** of ascending powers of x. It is the result obtained when you expand $(1+x)(1+x)(1+x)$ and collect like terms.

Pascal's triangle
1
1 1
1 2 1
1 3 3 1

$${}^3C_0 = \mathbf{1}, {}^3C_1 = \mathbf{3}, {}^3C_2 = \mathbf{3},$$
$${}^3C_3 = \mathbf{1}$$

When n is not a positive integer

- You cannot find $(1+x)^n$ by multiplying out $(1+x)$ n times,
- You cannot use the $\boxed{nCr}$ button to calculate coefficients. For example, when $n = -2$ ${}^{-2}C_1 = \text{Math ERROR}$

You write $n \notin \mathbb{Z}^+$ when n is *not* a positive integer.

Therefore, you must use a more general form of the binomial theorem.

Key point

When $n \notin \mathbb{Z}^+$, the binomial expansion of $(1+x)^n$ in ascending powers of x is

$$(1+x)^n = 1 + nx + \frac{n(n-1)}{2!}x^2 + \frac{n(n-1)(n-2)}{3!}x^3 + \ldots$$

The three dots mean the series has an **infinite** number of terms. You can also use this formula if n is a positive integer.

Remember that $3! = 3 \times 2 \times 1 = 6$

Example 1

Work out the binomial expansion of $(1+x)^{-2}$ up to and including the term in x^3

$$(1+x)^n = 1 + nx + \frac{n(n-1)}{2!}x^2 + \frac{n(n-1)(n-2)}{3!}x^3 + \ldots$$

$$n = -2, \quad (1+x)^{-2} = 1 + (-2)x + \frac{(-2)(-3)}{2!}x^2 + \frac{(-2)(-3)(-4)}{3!}x^3 + \ldots$$

$$= 1 - 2x + \frac{6}{2!}x^2 - \frac{24}{3!}x^3 + \ldots = 1 - 2x + 3x^2 - 4x^3 + \ldots$$

Simplify coefficients.

You stop expanding at the x^3 term even though the series is infinite.

You can only use the binomial theorem in this way if the bracket is in the form $(1+x)^n$. If it is not, you need to re-write it in this form.

Example 2

Work out the binomial expansion of these expressions up to and including the term in x^2

a $(1+4x)^{\frac{1}{2}}$ **b** $(2+x)^{-1}$

a $(1+4x)^{\frac{1}{2}} = (1+y)^{\frac{1}{2}}$ where $y = 4x$

    $= 1 + \dfrac{1}{2}y + \dfrac{(\frac{1}{2})(-\frac{1}{2})}{2!}y^2 + \ldots = 1 + \dfrac{1}{2}(4x) + \dfrac{(\frac{1}{2})(-\frac{1}{2})}{2!}(4x)^2 + \ldots$

    $= 1 + \dfrac{1}{2} \times 4x - \dfrac{1}{8} \times 16x^2 + \ldots = 1 + 2x - 2x^2 + \ldots$

> Rewrite the expression in the form $(1 + y)^n$. To do this, use $y = 4x$

> $(4x)^2 = 16x^2$

b $(2+x)^{-1} = \left[2\left(1+\dfrac{x}{2}\right)\right]^{-1} = 2^{-1}\left[1+\left(\dfrac{x}{2}\right)\right]^{-1} = 2^{-1}\left[1+y\right]^{-1}$ where $y = \dfrac{x}{2}$

    $= \dfrac{1}{2}\left[1 + (-1)y + \dfrac{(-1)(-2)}{2!}y^2 + \ldots\right]$

    $= \dfrac{1}{2}\left[1 + (-1)\left(\dfrac{x}{2}\right) + \dfrac{(-1)(-2)}{2!}\left(\dfrac{x}{2}\right)^2 + \ldots\right]$

    $= \dfrac{1}{2}\left[1 - \dfrac{x}{2} + \dfrac{x^2}{4} + \ldots\right] = \dfrac{1}{2} - \dfrac{1}{4}x + \dfrac{1}{8}x^2 + \ldots$

> Take out a factor of 2 so that the bracket starts with a 1

> Keep the index -1 on the 2 when removed from the brackets.

The infinite sum $1 + nx + \dfrac{n(n-1)}{2!}x^2 + \dfrac{n(n-1)(n-2)}{3!}x^3 + \ldots$ only makes sense if its terms get progressively smaller. To ensure this, you require $-1 < x < 1$

> **Key point**   See Ch 12.2
>
> When $n \notin \mathbb{Z}^+$, the expansion $(1+x)^n = 1 + nx + \dfrac{n(n-1)}{2!}x^2 + \dfrac{n(n-1)(n-2)}{3!}x^3 + \ldots$ is only
>
> **valid** if $-1 < x < 1$ (i.e. $|x| < 1$)

For a reminder on domains and the modulus function.

This restriction is called the **range of validity** of the expansion. For values of x outside this range, the series has no value and the expansion is not valid.

Example 3

Evaluate the range of values of x for which the binomial expansion of $\dfrac{3}{1-2x}$ is valid.

$\dfrac{3}{1-2x} = 3(1+y)^{-1}$ where $y = -2x$

    $= 3\left[1 - y + y^2 + \ldots\right]$

The expansion of $3(1+y)^{-1}$ is valid provided $-1 < y < 1$

In terms of x, the expansion of $\dfrac{3}{1-2x}$ is valid
provided $-1 < -2x < 1$

$-1 < -2x < 1$ means $\dfrac{-1}{-2} > x > \dfrac{1}{-2}$

The range of validity is $-\dfrac{1}{2} < x < \dfrac{1}{2}$ or, equivalently, $|x| < \dfrac{1}{2}$

> The multiplier 3 does not affect the range of validity.

> Use $y = -2x$

> See Ch 1.7
>
> For a reminder on inequalities.

> You reverse an inequality sign when you divide through by a negative number.

Using similar methods to those used in Example 3, you can find the range of validity for the expansion of $(a+bx)^n$

$$(a+bx)^n = \left(a\left(1+\frac{b}{a}x\right)\right)^n$$

$$= (a(1+y))^n \qquad \text{where } y = \frac{b}{a}x$$

The expansion of $(a(1+y))^n$ is valid provided $-1 < y < 1$

So, in terms of x, the expansion of $(a+bx)^n$ is valid provided $-1 < \frac{b}{a}x < 1$

> **Key point**
>
> For $n \notin \mathbb{Z}^+$, the binomial expansion of $(a+bx)^n$ is only valid for $\left|\frac{b}{a}x\right| < 1$

Exercise 13.1A Fluency and skills

1 Work out the binomial expansions of these expressions up to and including the term in x^3

  a $(1+x)^{-3}$ **b** $(1+x)^{\frac{1}{2}}$

  c $(1+x)^{\frac{2}{3}}$ **d** $(1+4x)^{-1}$

  e $(1-3x)^{-2}$ **f** $\left(1+\frac{1}{2}x\right)^{\frac{1}{3}}$

2 Work out the binomial expansions of these expressions up to and including the term in x^2

  a $(2+x)^{-4}$ **b** $(3+2x)^{-1}$

  c $(4-3x)^{\frac{3}{2}}$ **d** $\dfrac{3}{(3+x)^2}$

  e $\sqrt{9+x}$ **f** $\dfrac{6}{3-4x}$

  g $\dfrac{8}{\sqrt{4+9x}}$ **h** $\dfrac{27}{2(3-2x)^2}$

3 Work out the binomial expansions of these expressions, up to and including the term in x^2. Simplify coefficients in terms of the positive constant k

  a $(1+kx)^{-3}$ **b** $\sqrt{(4+kx)}$

  c $\left(\dfrac{k}{k+x}\right)^2$ **d** $\left(\sqrt{k}+x\right)^{-2}$

4 Evaluate the range of values of x for which the binomial expansion of these expressions is valid.

  a $(1+4x)^{-3}$ **b** $(3+x)^{-2}$

  c $(4-5x)^{\frac{3}{2}}$ **d** $\left(\dfrac{1}{2}+\dfrac{1}{3}x\right)^{\frac{1}{4}}$

5 Given that $(1+x)^{-2} = 1-2x+3x^2+...$, work out the expansion of these expressions up to and including the term in x^2

  a $x(1+x)^{-2}$

  b $(2+x)(1+x)^{-2}$

  c $(3-2x)(1+x)^{-2}$

  d $(1-3x^2)(1+x)^{-2}$

6 By simplifying these expressions, or otherwise, work out their binomial expansions up to and including the term in x^2

  a $\dfrac{1}{4+4x+x^2}$

  b $\dfrac{4-6x}{4-9x^2}$

  c $\dfrac{3}{\sqrt{1+4x+4x^2}}$

  d $\dfrac{2}{\left(2-\sqrt{x}\right)} \times \dfrac{6}{\left(2+\sqrt{x}\right)}$

7 Given that the binomial expansion of $(a+4x)^{-3}$, where $a>0$, is only valid for $-\dfrac{1}{2} < x < \dfrac{1}{2}$

  a Calculate the value of a

  b Hence evaluate the coefficient of x^3 in this expansion.

Reasoning and problem-solving

To find an approximate value of an expression

1. Expand using the binomial expansion formula.
2. Equate coefficients if necessary.
3. Substitute the required value of x into the first few terms of the series.

Recall that, for $n \notin \mathbb{Z}^+$, $(1+x)^n = 1 + nx + \dfrac{n(n-1)}{2!}x^2 + \dfrac{n(n-1)(n-2)}{3!}x^3 + \ldots$

You can use the first few terms of this series to approximate $(1+x)^n$

Key point

If x is small enough that x^3 and higher powers can be ignored, then $(1+x)^n \approx 1 + nx + \dfrac{n(n-1)}{2!}x^2$

$\approx$ is the approximation symbol.

For example, $(1+x)^{-2} = 1 - 2x + 3x^2 - 4x^3 + \ldots$ so $(1+x)^{-2} \approx 1 - 2x + 3x^2$ for suitably small values of x

If you substitute a low value of x, such as 0.1, then $1 - 2(0.1) + 3(0.1)^2 = 0.83$

This is a reasonable approximation for 1.1^{-2} which is $0.8264\ldots$

Similarly, if you substitute $x = 0.01$, then $1 - 2(0.01) + 3(0.01)^2$, which is a very good approximation for 1.01^{-2}

The greater number of terms you use, and the closer your value of x is to zero, the better the approximation. You can also use a binomial expansion to find estimates for surds.

Example 4

The first three terms in the binomial expansion of $\sqrt{1+px}$, where p is a negative constant, are $1 - x - \dfrac{1}{2}x^2$

a Work out the value of p **b** Use this series approximation and $x = 0.04$ to estimate $\sqrt{23}$

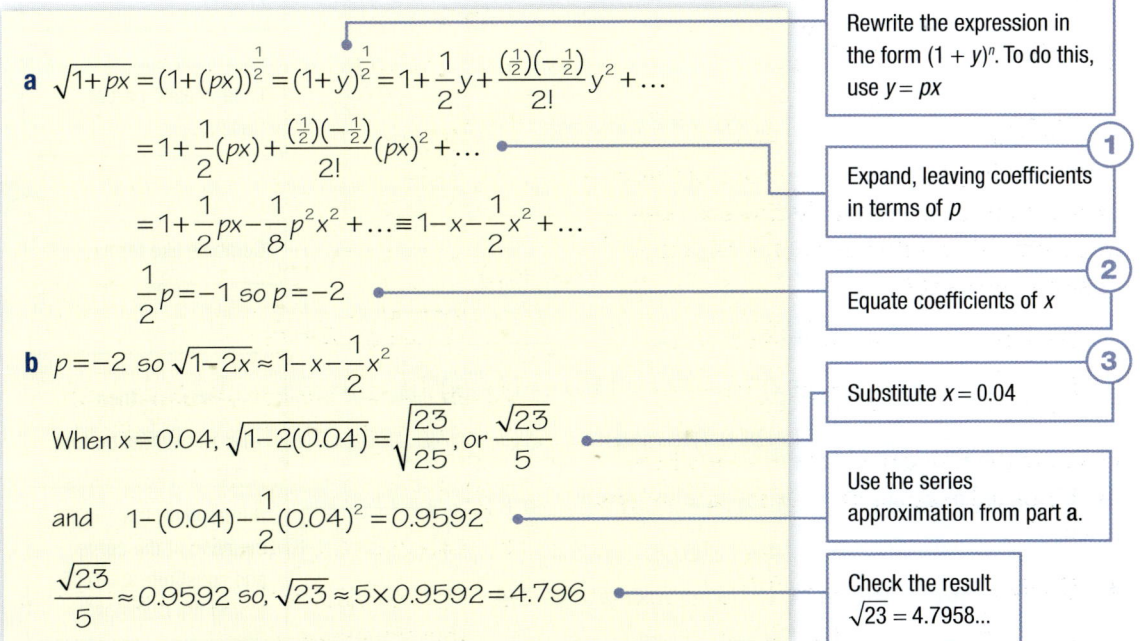

a $\sqrt{1+px} = (1+(px))^{\frac{1}{2}} = (1+y)^{\frac{1}{2}} = 1 + \dfrac{1}{2}y + \dfrac{\left(\frac{1}{2}\right)\left(-\frac{1}{2}\right)}{2!}y^2 + \ldots$

Rewrite the expression in the form $(1+y)^n$. To do this, use $y = px$

$= 1 + \dfrac{1}{2}(px) + \dfrac{\left(\frac{1}{2}\right)\left(-\frac{1}{2}\right)}{2!}(px)^2 + \ldots$

1 Expand, leaving coefficients in terms of p

$= 1 + \dfrac{1}{2}px - \dfrac{1}{8}p^2x^2 + \ldots \equiv 1 - x - \dfrac{1}{2}x^2 + \ldots$

$\dfrac{1}{2}p = -1$ so $p = -2$

2 Equate coefficients of x

b $p = -2$ so $\sqrt{1-2x} \approx 1 - x - \dfrac{1}{2}x^2$

3 Substitute $x = 0.04$

When $x = 0.04$, $\sqrt{1 - 2(0.04)} = \sqrt{\dfrac{23}{25}}$, or $\dfrac{\sqrt{23}}{5}$

and $1 - (0.04) - \dfrac{1}{2}(0.04)^2 = 0.9592$

Use the series approximation from part **a**.

$\dfrac{\sqrt{23}}{5} \approx 0.9592$ so, $\sqrt{23} \approx 5 \times 0.9592 = 4.796$

Check the result $\sqrt{23} = 4.7958\ldots$

You can use partial fractions to find the expansion of an algebraic fraction.

See Ch 12
For a reminder on partial fractions.

To find the expansion of an algebraic function involving fractions

1. Split the function into partial fractions.
2. Write each partial fraction in the form $(1+x)^n$
3. Work out the binomial expansion of each partial fraction.
4. Combine like terms.

The curve C has equation $y = f(x)$, where $f(x) = \dfrac{3+5x}{(1+x)(1+2x)}$

a Work out the expansion of $f(x)$ up to and including the term in x^2

b State the range of values of x for which the full expansion of $f(x)$ is valid.

c Estimate the gradient of C at the point where $x = 0.05$

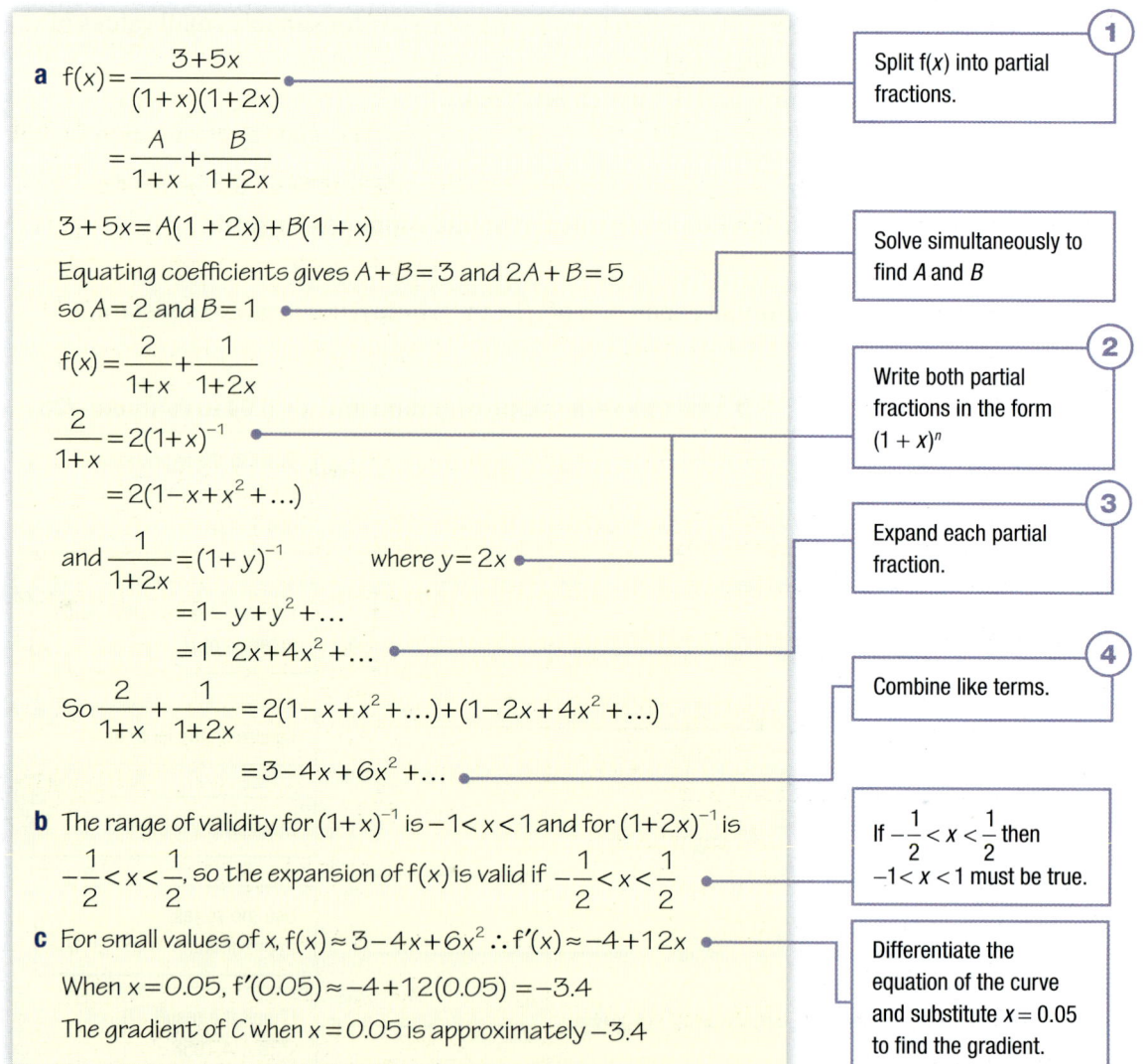

a $f(x) = \dfrac{3+5x}{(1+x)(1+2x)}$

$= \dfrac{A}{1+x} + \dfrac{B}{1+2x}$

$3 + 5x = A(1+2x) + B(1+x)$

Equating coefficients gives $A + B = 3$ and $2A + B = 5$

so $A = 2$ and $B = 1$

$f(x) = \dfrac{2}{1+x} + \dfrac{1}{1+2x}$

$\dfrac{2}{1+x} = 2(1+x)^{-1}$

$= 2(1 - x + x^2 + \ldots)$

and $\dfrac{1}{1+2x} = (1+y)^{-1}$ where $y = 2x$

$= 1 - y + y^2 + \ldots$

$= 1 - 2x + 4x^2 + \ldots$

So $\dfrac{2}{1+x} + \dfrac{1}{1+2x} = 2(1 - x + x^2 + \ldots) + (1 - 2x + 4x^2 + \ldots)$

$= 3 - 4x + 6x^2 + \ldots$

b The range of validity for $(1+x)^{-1}$ is $-1 < x < 1$ and for $(1+2x)^{-1}$ is

$-\dfrac{1}{2} < x < \dfrac{1}{2}$, so the expansion of $f(x)$ is valid if $-\dfrac{1}{2} < x < \dfrac{1}{2}$

c For small values of x, $f(x) \approx 3 - 4x + 6x^2 \therefore f'(x) \approx -4 + 12x$

When $x = 0.05$, $f'(0.05) \approx -4 + 12(0.05) = -3.4$

The gradient of C when $x = 0.05$ is approximately -3.4

Callout boxes:
1. Split $f(x)$ into partial fractions.
— Solve simultaneously to find A and B
2. Write both partial fractions in the form $(1+x)^n$
3. Expand each partial fraction.
4. Combine like terms.
— If $-\dfrac{1}{2} < x < \dfrac{1}{2}$ then $-1 < x < 1$ must be true.
— Differentiate the equation of the curve and substitute $x = 0.05$ to find the gradient.

See Ch 4.2
For a reminder on differentiation.

Use the quotient rule to work out the exact value of $f'(0.05)$ and compare your answer with this estimate.

Sequences The binomial series

1 a Work out the first three terms, in ascending powers of x, in the binomial expansion of $\sqrt{1+5x}$

b Use this series approximation and $x = 0.05$ to show that $\sqrt{5} \approx \dfrac{143}{64}$

c Explain why substituting $x = 0.8$ into the full expansion of $\sqrt{1+5x}$ does not give an answer equal to $\sqrt{5}$

2 a Work out the expansion of $(4+x)^{\frac{1}{2}}$, up to and including the term in x^2. State the range of values of x for which the full expansion of $(4+x)^{\frac{1}{2}}$ is valid.

b Use the series approximation found in part **a** and

  i $x = 0.25$ to show that $\sqrt{17} \approx \dfrac{2111}{512}$

  ii $x = -0.25$ to calculate an estimate for $\sqrt{15}$

c Hence show that $\sqrt{17} - \sqrt{15} \approx \dfrac{1}{4}$

3 The first three terms in the binomial expansion of $(1+3x)^n$ are $1 + kx - x^2$, where n and k are constants, $n > \dfrac{1}{2}$

a Work out the value of n and the value of k

b Use this series approximation and a suitable value of x to calculate an estimate for $\sqrt[3]{1.69}$

4 The first three terms in the binomial expansion of $(1+px)^n$ are $1 + 12x + 24x^2$, where p and n are constants.

Using the series approximation and $x = 0.01$, show that $\sqrt{3} \approx \dfrac{1403}{810}$

5 a Work out the expansion of $\dfrac{3+7x}{(1+x)(1+3x)}$ up to and including the term in x^3

b State the range of values of x for which this full expansion is valid.

6 The curve C has equation $y = f(x)$, where $f(x) = \dfrac{2-3x}{2-3x-2x^2}$

a Work out the expansion of $f(x)$ up to and including the term in x^2

b Sketch a graph which approximates C for values of x close to zero.

7 $f(x) = \dfrac{x+4}{(1+x)(2+x)^2}$

The curve C has equation $y = f(x)$ and passes through point P with x-coordinate -0.1

a Calculate the y-coordinate of point P, giving your answer to 1 decimal place.

b Work out the first three terms in the expansion of $f(x)$, giving your answer in the form $p + qx + rx^2$, stating the value of each constant p, q and r

c Use your answers to parts **a** and **b** to show that the tangent to this curve at point P can be approximated by the line with equation $y = 0.98 - 2.2x$

Challenge

8 a By using $\cos^2 x + \sin^2 x \equiv 1$, show that
$$\frac{1}{\cos^2 x} = 1 + \sin^2 x + \sin^4 x + \sin^6 x + \ldots$$
and explain why this expansion is valid for $-90° < x < 90°$

It is given that
$$\frac{1}{\sin^2 x} = 1 + \cos^2 x + \cos^4 x + \cos^6 x + \ldots$$
where $0° < x < 90°$

b Read through this argument and explain where the error has occurred.

For $0° < x < 90°$, $\dfrac{1}{\cos^2 x} \approx 1 + \sin^2 x$

and $\dfrac{1}{\sin^2 x} \approx 1 + \cos^2 x$

Adding these expressions
$$\frac{1}{\cos^2 x} + \frac{1}{\sin^2 x} \approx (1 + \cos^2 x) + (1 + \sin^2 x)$$

so $\dfrac{1}{\cos^2 x} + \dfrac{1}{\sin^2 x} \approx 3$

(as $\cos^2 x + \sin^2 x \equiv 1$)

But when $x = 15°$,
$$\frac{1}{\cos^2 15°} + \frac{1}{\sin^2 15°} = 16$$

which is not close to 3

Fluency and skills

A **sequence** is a list of numbers, called **terms**. You can refer to each term by its position in the list. For example, for the sequence 3 , 6 , 9 , 12 , ..., 3 is the first term, 6 is the second term and so on. You can use the notation u_n for the nth term of a sequence, where n is a positive integer. So for the sequence 3 , 6 , 9 , 12 , ... , you know that u_1 = 1st term = 3, u_2 = 2nd term = 6, and u_n = nth term.

Example 1

Work out the value of the first three terms of the sequence with nth term u_n where
$u_{n+1} = 2u_n + 3$, $u_1 = 5$

$u_1 = 5$ —— Write down the first term.

$n = \mathbf{1}$, $u_{1+1} = 2u_1 + 3$ —— Apply the given formula to find the second term.

$\quad u_2 = 2(5) + 3 = 13$

$n = \mathbf{2}$, $u_{2+1} = 2u_2 + 3$

$\quad u_3 = 2(13) + 3 = 29$ —— Use u_2 to find u_3

The first three terms are 5, 13 and 29

Example 1 uses a **recurrence relation** (or iterative formula), which can be simpler than a direct formula.

Calculator

Try it on your calculator

You can calculate terms of a sequence defined by a recurrence relation on a calculator.

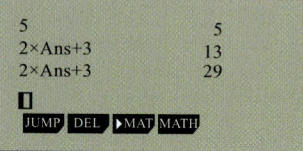

```
5                    5
2×Ans+3             13
2×Ans+3             29

JUMP  DEL  ▸MAT MATH
```

Activity

Find out how to calculate the first three terms of the sequence defined by $u_{n+1} = 2u_n + 3$, $u_1 = 5$ on *your* calculator.

Key point

A sequence with nth term u_n is

- **Increasing** if $u_{n+1} > u_n$ for all $n \in \mathbb{Z}^+$
- **Decreasing** if $u_{n+1} < u_n$ for all $n \in \mathbb{Z}^+$
- **Periodic** if the sequence consists of repeated terms.

Periodic sequences are neither increasing nor decreasing. For example, in the sequence 4, 3, 4, 3, 4, 3 ..., $u_{n+1} > u_n$ and $u_{n+1} < u_n$ are only sometimes true.

Example 2

Describe the list u_1, u_2, u_3, u_4 as either an increasing, decreasing or periodic sequence, where

a $u_n = n^2 + 2n + 3$ **b** $u_n = \cos(180n)^\circ$ **c** $u_n = 0.5^n + 2$

a The terms 6, 11, 18, 27 form an increasing sequence. —— e.g. $u_3 = (3)^2 + 2(3) + 3$ $= 18$

b The terms −1, 1, −1, 1 form a periodic sequence. —— e.g. $u_2 = \cos(180 \times 2)^\circ$ $= 1$

c The terms 2.5, 2.25, 2.125, 2.0625 form a decreasing sequence. —— e.g. $u_1 = 0.5^1 + 2 = 2.5$

> **Key point**
>
> A periodic sequence has **period** (or **order**) k, where k is the smallest number of terms that repeat as a block of numbers. For example, the periodic sequence 4, 1, 7, 4, 1, 7, 4, 1, 7, ... has period 3 since the first three terms 4, 1 and 7 repeat in blocks.

> **Key point**
>
> A sequence u_1, u_2, u_3, ... **converges** to a number L if the terms get ever closer to L. L is called the **limit** of the sequence.

For example, if $u_n = \dfrac{n+1}{n}$, then $u_1 = 2$, $u_{10} = 1.1$, $u_{100} = 1.01$ and $u_{1000} = 1.001$

The terms of this sequence are getting ever closer to 1, which suggests that 1 is the limit of this sequence. You write $u_n \to 1$ as $n \to \infty$

> The symbol $\to$ means 'tends to'. So $n \to \infty$ means 'n tends to infinity'.

You can also find the limit of a sequence by solving an equation.

Example 3

A sequence is defined by $u_{n+1} = 0.2u_n + 2$, $u_1 = 3$. The limit of u_n as $n \to \infty$ is L. Find the value of L

As $n \to \infty$, the terms u_n and u_{n+1} are approximately equal to L

Since $u_{n+1} = 0.2u_n + 2$, L must satisfy the equation $L = 0.2L + 2$

$0.8L = 2$ so $L = 2.5$

The limiting value of this sequence is 2.5

> Replace u_n and u_{n+1} with L in the equation $u_{n+1} = 0.2u_n + 2$

> Solve the equation $L = 0.2L + 2$

> Use your calculator to verify that $u_n \to 2.5$ as $n \to \infty$

Exercise 13.2A Fluency and skills

In these questions, u_n represents the nth term of a sequence where $n \in \mathbb{Z}^+$

1 Work out the values of the first four terms of these sequences.

   a $u_n = 5n + 3$ **b** $u_n = n^2 - 3$

   c $u_n = 2n^2(n-1)$ **d** $u_n = \dfrac{n}{n^2 - 6}$

2 Work out the values of u_2, u_3 and u_4 for these sequences.

   a $u_{n+1} = 2u_n + 5$, $u_1 = 1$

   b $u_{n+1} = u_n^2 + 2$, $u_1 = 1$

   c $u_{n+1} = \dfrac{1}{u_n + 1}$, $u_1 = 3$

   d $u_{n+1} = (u_n - 3)^2$, $u_1 = 5$

3 Describe the list u_1, u_2, u_3, u_4 as either an increasing sequence, a decreasing sequence or neither where

   a $u_n = \dfrac{6}{n+1}$ **b** $u_n = n^2 + 4n - 3$

   c $u_n = n^2 - 6n + 1$ **d** $u_n = \sqrt[n]{n}$

   e $u_n = \log_{n+1}(n+2)$

   f $u_n = \sin(30n)° \cos(30n)°$

4 Evaluate the order of these periodic sequences.

   a $u_n = \tan(n \times 60°)$

   b $u_{n+1} = 3 - \dfrac{9}{u_n}$, $u_1 = 1$

   c $u_{n+1} = (-1)^n u_n$, $u_1 = 2$

   d $u_{n+1} = \dfrac{2}{9}\left(3 - \dfrac{1}{u_n}\right)$, $u_1 = 3$

5 The nth term of a sequence is given by $u_n = 3n - 2$

   a Calculate the value of n such that $u_n = 229$

   b Work out the values of n for which $u_n > 1000$

6 The nth term of a sequence is given by $u_n = n^2 - 8n + 18$

   a Calculate the value of n such that $u_n = 83$

   b Work out the value of the smallest term in this sequence.

7 By solving an equation, find the limit L of these sequences as $n \to \infty$. Where appropriate, give answers in simplified surd form.

a $u_{n+1} = 0.4u_n + 3$ **b** $u_{n+1} = 0.6u_n - 3$

c $u_{n+1} = 3 - 0.25u_n$ **d** $u_{n+1} = \dfrac{3}{5}u_n - \dfrac{1}{2}$

e $u_{n+1} = 0.1(7u_n + 2)$ **f** $u_{n+1} = \dfrac{\sqrt{5}u_n + 4}{3}$

g $u_{n+1} = \left(\sqrt{3} - 2\right)u_n + 2\sqrt{3}$

h $u_{n+1} = \dfrac{1}{4}u_n^2 + 1$

Use your calculator or a spreadsheet with starting value $u_1 = 1$ to verify each answer.

Reasoning and problem-solving

You can describe, or **model**, real-life situations using sequences.

Example 4

The total number of people, p_n, who became infected with a virus n weeks after it started to spread can be modelled by the equation $p_{n+1} = 1.05p_n + c$, where c is a constant.

After week 1, 20 people were infected. During week 2, 41 people became infected.

Use this model to estimate the number of people who became infected during week 7

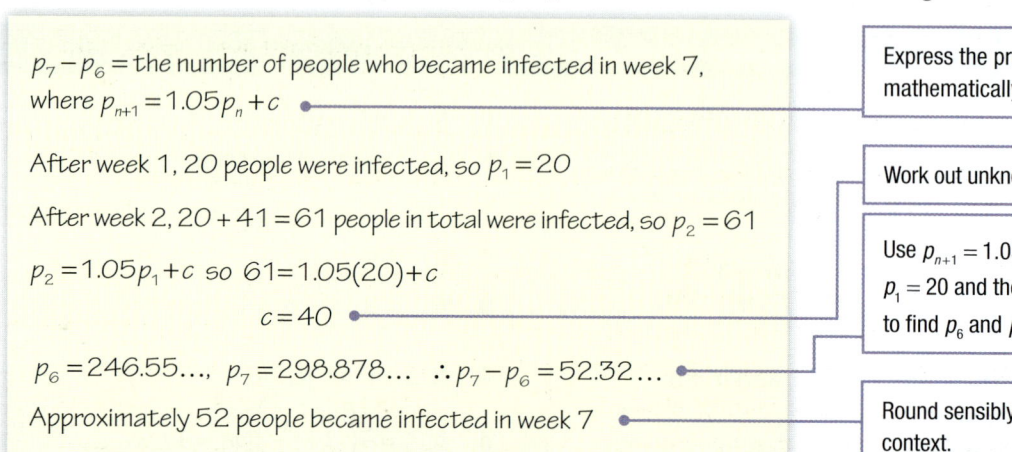

$p_7 - p_6 =$ the number of people who became infected in week 7, where $p_{n+1} = 1.05p_n + c$

After week 1, 20 people were infected, so $p_1 = 20$

After week 2, $20 + 41 = 61$ people in total were infected, so $p_2 = 61$

$p_2 = 1.05p_1 + c$ so $61 = 1.05(20) + c$

$c = 40$

$p_6 = 246.55\ldots$, $p_7 = 298.878\ldots$ $\therefore p_7 - p_6 = 52.32\ldots$

Approximately 52 people became infected in week 7

1 Express the problem mathematically.

2 Work out unknown values.

Use $p_{n+1} = 1.05p_n + 40$, $p_1 = 20$ and the $\boxed{ANS}$ key to find p_6 and p_7

3 Round sensibly given the context.

Exercise 13.2B Reasoning and problem-solving

In all questions, assume $n \in \mathbb{Z}^+$

1 A sequence is defined by the equation $u_{n+1} = 4u_n + k$, $u_1 = 1$, where k is a constant. Given that $u_3 = 31$

  a Work out the value of u_4

  b Evaluate the sum of the first four terms of this sequence.

2 A sequence is defined by the equation $u_{n+1} = au_n + 7$, $u_1 = -1$, where a is a constant. Given that $u_3 = 19$

  a Show that $a^2 - 7a + 12 = 0$

  b Work out the possible values of u_4

3 A sequence is defined by $u_{n+1} = au_n + b$ for constants a and b. The first three terms of this sequence are 15, 24 and 42, respectively.

a Calculate the values of a and b

Given that $u_{m+1} = 1.96u_m$, where $m \geq 2$

b Work out the value of u_{m+2}

4 A periodic sequence satisfies the equation
$u_{n+1} = a - \dfrac{2a}{u_n}$, where $a > 0$ is a constant.
The sequence has period 3 and $u_3 = -2$

  a Show that

  i $u_2 = a - 1$ **ii** $a^2 - a - 2 = 0$

  b Hence find the sum of the first 300 terms of this sequence.

5 Monthly observations at a wildlife sanctuary suggest that its population of otters is declining. The number of otters, p_n, seen during the $(n+1)$th observation was modelled by the equation $p_{n+1} = 0.75(p_n + 8)$. During the first observation 56 otters were seen.

  a Show that there was a 12.5% decrease in the number of otters seen in the third observation compared to the second.

In one particular observation, at least 24 otters were observed.

  b Show that the number of otters seen in the next observation was also at least 24

  c Explain why, according to this model, the population of otters will never die out.

6 A code breaking competition consists of 10 rounds, each more difficult than the previous one. A round starts when the code is issued and contestants must break the code within two hours before being allowed to progress to the next round. It takes one of the contestants, Sam, m_n minutes to break the code in round n where $m_{n+1} = a(m_n - 1)$ and a is a positive constant. Sam takes 4 minutes to break the code in round 2 and 10 minutes to break the code in round 4

  a Show that $3a^2 - a - 10 = 0$

  b Work out the time, in minutes, that it takes Sam to break the code in round 1

  c How many rounds of this competition does Sam successfully take part in?

7 In preparation to run a race, Paula undertakes weekly training sessions. In the nth session she runs e_n miles due East from her house, turns due South and runs s_n miles and then runs directly back to her house, so that the path she takes in each session is a right-angled triangle.

In the first session she runs 1.5 miles due East and 2 miles due South.

  a Calculate the total distance she runs in session 1

For $n \geq 1$, it is given that $e_{n+1} = \dfrac{2}{3}e_n + 2$ and that $s_{n+1} = \dfrac{1}{2}s_n + k$ where k is a constant. In session 2, Paula runs 12 miles in total.

  b Show that $\sqrt{10 + 2k + k^2} = 8 - k$ and hence evaluate the value of k

Paula correctly calculates that, to the nearest mile, the distance she runs in session 8 equals the length of the race.

  c Calculate, to the nearest mile, the length of the race.

8 A sequence is defined by $u_{n+1} = pu_n + 6$, $u_1 = 5$, where p is a positive constant.

Given that $u_3 = 9.2$

  a Show that $5p^2 + 6p - 3.2 = 0$

The limit of u_n as $n \to \infty$ is L

  b Find the value of L

Challenge

9 A sequence is defined by
$u_{n+1} = pu_n + q$, $u_1 = 12$, where p and q are constants. The second term of this sequence is 10 and the limit as $n \to \infty$ is 2

  a Find the value of p and the value of q

  b For these values of p and q, find the limit as $n \to \infty$ of the sequence defined by $v_{n+1} = qv_n + p$, $v_1 = 12$

Fluency and skills

Key point

A sequence is **arithmetic** if adding a constant to any term gives the next term.

For example, 4, 7, 10, 13, ... is an arithmetic sequence. You add 3 to each term to get the next term.

In this example, the constant 3 is called the **common difference** of the sequence. The difference $u_3 - u_2 = 10 - 7 = 3$

This means the difference $u_{n+1} - u_n = 3$ for all values of n

Key point

If an arithmetic sequence has first term a and common difference d, then

- For all n, $d = u_{n+1} - u_n$
- The terms are $a, a+d, a+2d, a+3d, \ldots$
- The nth term of the sequence is $u_n = a + (n-1)d$. For example the **4**th term is $a + (4-1)d$ or $a + 3d$
- If $d > 0$, the sequence is increasing and if $d < 0$, the sequence is decreasing.

Example 1

a Work out the nth term of the arithmetic sequence 3, 8, 13, 18, ...

b Hence evaluate the value of the 50th term of this sequence.

a First term a is 3. Hence, common difference d is $8 - 3 = 5$

n^{th} term is $u_n = a + (n-1)d = 3 + (n-1) \times 5 = 5n - 2$

b When $n = 50$, $u_{50} = 5(50) - 2 = 248$

$d = u_2 - u_1$

Put a and d into the nth term formula.

Substitute $n = 50$

Key point

An **arithmetic series** is the sum of the terms of an arithmetic sequence.

For the arithmetic sequence $a, a + d, a + 2d, a + 3d, \ldots$

the corresponding arithmetic series is $a + (a + d) + (a + 2d) + (a + 3d) + \ldots$

An arithmetic sequence and its series share the same nth term and the same common difference.

You can use the notation S_n for the sum of the first n terms of an arithmetic series.

For example, for the arithmetic series $4 + 7 + 10 + 13 + \ldots$

$S_2 = 4 + 7 = 11$ and $S_3 = 4 + 7 + 10 = 21$

You can work out a formula for the sum S_n. If you write out the series twice—once in order and once in reverse order—then a shortcut for finding the sum becomes clear.

$$S_n = u_1 + u_2 + u_3 + \ldots + u_{n-1} + u_n$$

$$S_n = u_n + u_{n-1} + u_{n-2} + \ldots + u_2 + u_1$$

$$\downarrow \quad \downarrow \quad \downarrow \quad \downarrow \quad \downarrow \quad \downarrow$$

$$\therefore S_n + S_n = (u_1 + u_n) + (u_2 + u_{n-1}) + (u_3 + u_{n-2}) + \ldots + (u_{n-1} + u_2) + (u_n + u_1)$$

You can see that each of the bracketed terms in this sum are equivalent to $u_1 + u_n$. For example
$u_2 + u_{n-1} = (u_1 + d) + (u_n - d) = u_1 + u_n$ and $u_3 + u_{n-2} = (u_1 + 2d) + (u_n - 2d) = u_1 + u_n$

So $2S_n = \underbrace{(u_1 + u_n) + (u_1 + u_n) + (u_1 + u_n) + \ldots + (u_1 + u_n) + (u_1 + u_n)}_{n \text{ brackets}}$

$$= n(u_1 + u_n)$$

Therefore, $S_n = \dfrac{1}{2}n(u_1 + u_n)$

Key point

If an arithmetic series has first term a and common difference d, then $S_n = \dfrac{1}{2}n(a+l)$, where $l = u_n$, the last term of the sum.

You can also write $S_n = \dfrac{1}{2}n[2a + (n-1)d]$, since $u_1 + u_n = a + (a + (n-1)d) = 2a + (n-1)d$. You can use this formula if you do not know the last term.

'Evaluate' means 'find the value of'.

Example 2

The arithmetic series $7 + 12 + 17 + \ldots + 102$ has 20 terms.

a Evaluate this series. **b** Calculate the sum of the first 10 terms of this series.

a The sum has $n = 20$ terms, the first is $a = 7$ and the last is $l = 102$
$$S_{20} = \frac{1}{2}(20)(7 + 102)$$
$$= 1090$$

b The first term is $a = 7$ and the common difference, $d = 5$
$$S_{10} = \frac{1}{2}(10)\left[2 \times 7 + (10-1) \times 5\right]$$
$$= 295$$

Use $S_n = \dfrac{1}{2}n(a+l)$

Use $S_n = \dfrac{1}{2}n[2a + (n-1)d]$

You can use the symbol $\sum$ for a series. In the notation $\displaystyle\sum_{k=1}^{3} k^2$, the **index** k varies, taking in turn the values 1, 2, and 3

So $\displaystyle\sum_{k=1}^{3} k^2 = 1^2 + 2^2 + 3^2 = 14$

Similarly, $\displaystyle\sum_{k=1}^{3} (4k - 3) = (4 \times 1 - 3) + (4 \times 2 - 3) + (4 \times 3 - 3) = 1 + 5 + 9 = 15$

You read $\displaystyle\sum_{k=1}^{3} k^2$ as 'the sum of k^2 as the integer k varies from 1 to 3'.

Key point

For any function f, $\displaystyle\sum_{k=1}^{n} f(k) = f(1) + f(2) + \ldots + f(n)$

In particular, $\displaystyle\sum_{k=1}^{n} 1 = n$

You may want to find the sum of a series where the value of the index k does not start at 1

e.g. $\displaystyle\sum_{k=3}^{6} k^2 = \underbrace{3^2 + 4^2 + 5^2 + 6^2}_{4\,\text{terms}}$

In these instances, you can use the rule $\displaystyle\sum_{k=a}^{n} u_k = \sum_{k=1}^{n} u_k - \sum_{k=1}^{a-1} u_k$

Try it on your calculator

You can calculate the sum of a series on a calculator.

$\displaystyle\sum_{x=2}^{7} (8x-3)$

198

Activity

Find out how to calculate $\displaystyle\sum_{x=2}^{7}(8x-3)$ on *your* calculator.

Example 3

Use the formula to evaluate the arithmetic series $\displaystyle\sum_{k=1}^{11}(6k+1)$

$\displaystyle\sum_{k=1}^{11}(6k+1) = (6\times1+1)+(6\times2+1)+(6\times3+1)+\ldots+(6\times11+1)$

$= \underbrace{7+13+19+\ldots+67}_{11\,\text{terms}}$

There are 11 terms in this arithmetic series. The first is 7 and the last is 67

$S_{11} = \dfrac{1}{2}(11)(7+67) = 407$

> Write out the first few terms and the last term of the series.

> Use $S_n = \dfrac{1}{2}n(a+l)$

Exercise 13.3A Fluency and skills

In these questions u_n represents the nth term of a sequence where $n \geq 1$

1 Work out the values of the first four terms of these arithmetic sequences.

  a $u_n = 3+2n$ **b** $u_n = 12-3n$

  c $u_n = 7n-4$ **d** $u_n = \dfrac{3}{2}+\dfrac{5}{2}n$

2 Work out expressions for the nth terms of these arithmetic sequences, simplifying each answer as far as possible.

  a $7, 11, 15, \ldots$ **b** $2, -1, -4, \ldots$

  c $4, 5.5, 7, \ldots$ **d** $2, \dfrac{8}{3}, \dfrac{10}{3}, \ldots$

  e $\dfrac{3}{5}, 1, \dfrac{7}{5}, \ldots$ **f** $\sqrt{2}, \sqrt{8}, \sqrt{18}, \ldots$

3 Work out the value of the 50th term in each of these sequences.

  a $3, 6, 9, \ldots$ **b** $8, 5, 2, \ldots$

  c $\dfrac{5}{4}, \dfrac{5}{2}, \dfrac{15}{4}, \ldots$ **d** $5.4, 3.2, 1.0, \ldots$

4 Evaluate these arithmetic series. The number of terms in each sum is shown in brackets.

  a $60+57+\ldots+18$ (15 terms)

  b $10+6+2+\ldots-34$ (12 terms)

  c $2+2.5+3+\ldots$ (99 terms)

  d $\sqrt{3}+\sqrt{27}+\ldots$ (10 terms)

5 Work out the number of terms in each of these arithmetic series.

  a $4+\ldots+151$ when $S_n = 3875$

  b $7+\ldots+52$ when $S_n = 295$

  c $4+\ldots+44$ when $S_n = 2400$

  d $\dfrac{1}{5}+\ldots+\dfrac{28}{5}$ when $S_n = 29$

6 Use the formula to evaluate these arithmetic series.

  a $\displaystyle\sum_{k=1}^{20}(2k+3)$ **b** $\displaystyle\sum_{k=1}^{12}(45-3k)$

  c $\displaystyle\sum_{k=3}^{9}(3k-1)$ **d** $\displaystyle\sum_{k=5}^{40}(1.2k+3)$

Reasoning and problem-solving

Strategy

To solve a problem using an arithmetic sequence or series

①　Define variables such as the first term a, the last term l and the common difference d using information given in the question.

②　Work out any unknown values.

③　Use appropriate formulae and give your answer in context, rounding appropriately.

Example 4

The third term of an arithmetic series is 15 and the seventh term is 31.
Calculate the sum of the first 10 terms of this series.

nth term $u_n = a + (n-1)d$

$u_3 = a + 2d$ so $a + 2d = 15$ [1]

$u_7 = a + 6d$ so $a + 6d = 31$ [2]

$d = 4,\ a = 7$

$S_{10} = \dfrac{1}{2}(10)\left[2 \times 7 + (10-1) \times 4\right] = 250$

① Set up equations to find the first term a and common difference d

② Solve [1] and [2] simultaneously.

③ Use $S_n = \dfrac{1}{2}n\left[2a + (n-1)d\right]$

You can create a mathematical model to approximate real-life situations involving an arithmetic sequence or series.

In this question you need to work with S_n, the **total** amount repaid after n working years.

Example 5

After graduating, Jane found a job and started to repay her student loan of £10 500

In her first working year, she repaid £550. In her second working year, she repaid £700.

The amount she repaid in each working year is modelled by an arithmetic sequence.

a　Use this model to

  i　Show that, by the end of her fourth working year, Jane owed £7400

  ii　Work out the number of whole working years during which Jane owed more than £500

b　Comment on the suitability of this model.

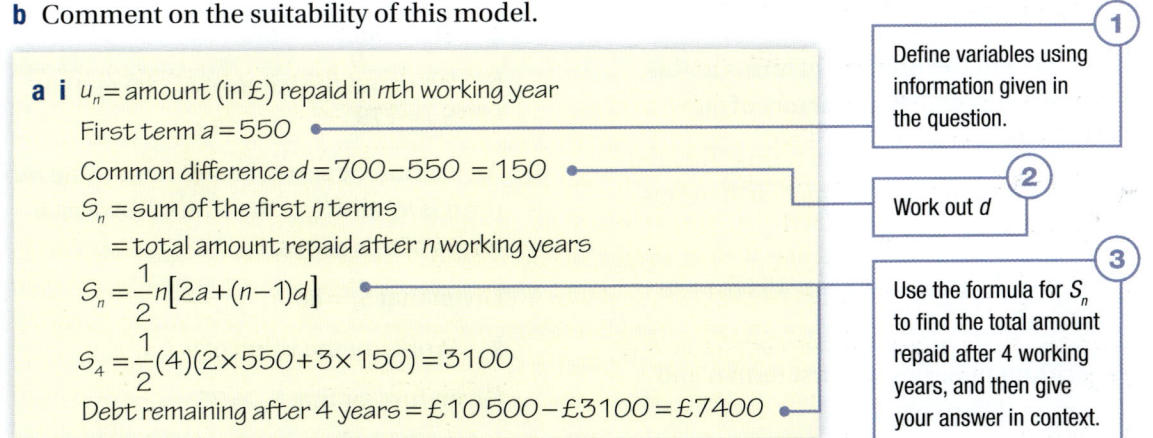

a i u_n = amount (in £) repaid in nth working year

First term $a = 550$

Common difference $d = 700 - 550 = 150$

S_n = sum of the first n terms

    = total amount repaid after n working years

$S_n = \dfrac{1}{2}n\left[2a + (n-1)d\right]$

$S_4 = \dfrac{1}{2}(4)(2 \times 550 + 3 \times 150) = 3100$

Debt remaining after 4 years = £10 500 − £3100 = £7400

① Define variables using information given in the question.

② Work out d

③ Use the formula for S_n to find the total amount repaid after 4 working years, and then give your answer in context.

(Continued on the next page)

 MyMaths   Q   2039   SEARCH

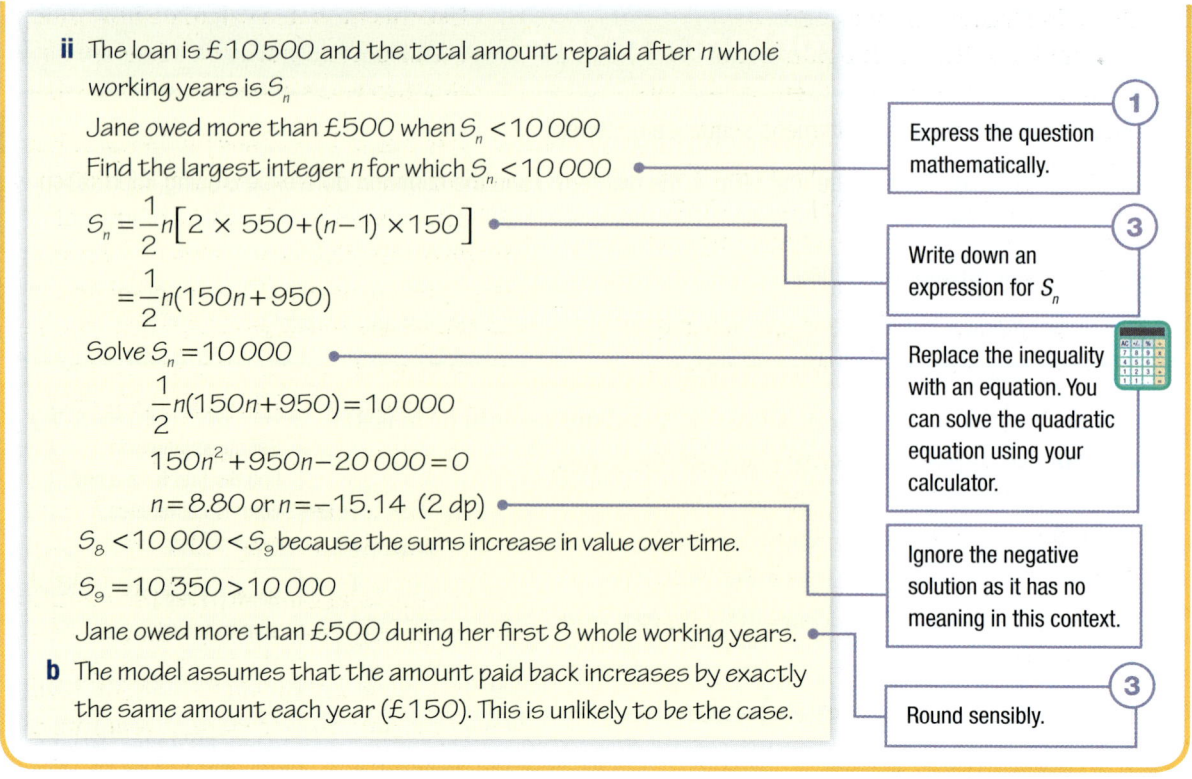

ii The loan is £10 500 and the total amount repaid after n whole working years is S_n

Jane owed more than £500 when $S_n < 10\,000$

Find the largest integer n for which $S_n < 10\,000$

$S_n = \dfrac{1}{2}n[2 \times 550 + (n-1) \times 150]$

$= \dfrac{1}{2}n(150n + 950)$

Solve $S_n = 10\,000$

$\dfrac{1}{2}n(150n + 950) = 10\,000$

$150n^2 + 950n - 20\,000 = 0$

$n = 8.80$ or $n = -15.14$ (2 dp)

$S_8 < 10\,000 < S_9$ because the sums increase in value over time.

$S_9 = 10\,350 > 10\,000$

Jane owed more than £500 during her first 8 whole working years.

b The model assumes that the amount paid back increases by exactly the same amount each year (£150). This is unlikely to be the case.

1 Express the question mathematically.

3 Write down an expression for S_n

Replace the inequality with an equation. You can solve the quadratic equation using your calculator.

Ignore the negative solution as it has no meaning in this context.

3 Round sensibly.

Exercise 13.3B Reasoning and problem-solving

1 The nth term of an arithmetic sequence is u_n, where $u_4 = 21$ and $u_7 = 36$

    a Evaluate the first term a and the common difference d of this sequence.

    b Calculate the value of N such that $u_N = 6 \times u_{10}$

2 The nth term of an arithmetic sequence is u_n, where $u_5 = 13$ and $u_{10} = 7u_2$

    a Show that $u_{18} = 52$

    b Work out the number of terms in this sequence which are factors of u_{18}

3 An arithmetic series has 20 terms. The first term is 4 and the sum of the first 10 terms of this series is 175

Calculate the sum of the last 10 terms of this series.

4 An arithmetic series has first term a and common difference d

The nth term is u_n and S_n is the sum of the first n terms of this series.

Given that $u_8 = 26$ and $S_5 = 205$

    a Calculate the value of the smallest positive term of this series,

    b Work out the greatest value of S_n

5 The first term of an arithmetic series is 5 and the common difference is 4

The nth term is u_n and S_n is the sum of the first n terms of this series.

    a Show that $S_n = n(2n+3)$

    b Work out the value of u_N, given that $S_N = 779$

6 An arithmetic series has 20 terms. The nth term is u_n and S_n is the sum of the first n terms of this series.

Given that $S_5 = 85$

    a Work out the value of u_3

Given further that $S_{17} = 35 \times u_3$

    b Evaluate the sum of all the terms of this series.

7 Quickline trains has invested in new timetabling software. The number of complaints received from their passengers each month in the following year was modelled using an arithmetic sequence. In the first month, 152 complaints were received. This decreased to 140 complaints in the second month.

a Use this model to show that a total of 1032 complaints were received during the year.

The train company claimed that there was roughly a 60% reduction in the number of complaints received between the first six months and the last six months of that year.

b Use this model to determine whether this claim is accurate.

c Comment on the suitability of this model in the following year.

8 The number of departures from a new airport each month was modelled using an arithmetic sequence. There were 450 departures in its first month of being operational. In the first three months of the airport being operational, there were 1470 departures.

a Use this model to work out the total number of departures in the first six months of the airport being operational.

If the number of departures in any complete month exceeded 1000 then the airport authority paid an environmental fine of £5000 for that month.

b Use this model to calculate how much the airport authority was fined in its first two years of being operational.

9 Jim has trained for a new job in sales. Each year he earns a basic salary of £24 000 plus 5% commission on the value of any sales made during that year. In his first working year, Jim earned £27 500. In his second working year, he made sales worth £80 000. The amount Jim earned each year was modelled using an arithmetic sequence.

a Use this model to work out

i The total amount Jim earned in his first five working years,

ii The total value of sales Jim made in his first five working years.

At the end of each working year, Jim paid 8% of the amount earned that year into a pension fund. The amount paid in each year formed an arithmetic sequence.

b Calculate the minimum number of whole years Jim will need to work for these pension contributions to have a total value of more than £30 000

c Give one criticism of using an arithmetic sequence to model the amount Jim earned each year.

Challenge

10 Without using a calculator, work out a relationship between the sums

$1+2+3+...+n$ and $1^3+2^3+3^3+...+n^3$

and hence find the value of the positive integer n for which

$1^3+2^3+3^3+...+n^3 = 90\,000$

Geometric sequences

Fluency and skills

Key point

A sequence is **geometric** if multiplying any term by a fixed non-zero number gives the next term in the sequence.

For example, 3, 6, 12, 24, ... is a geometric sequence. You multiply each term by 2 to get the next term.

In this example, the multiplier 2 is called the **common ratio** of the sequence. The ratio $\dfrac{u_3}{u_2} = \dfrac{12}{6} = 2$

This means $\dfrac{u_{n+1}}{u_n} = 2$ for all values of n

Key point

If a geometric sequence has first term a and common ratio r (both non-zero), then

- For all n, $r = \dfrac{u_{n+1}}{u_n}$

- The terms are $a, a \times r, a \times r^2, a \times r^3, ...$

- The nth term of the sequence is $u_n = a \times r^{n-1}$. For example, the **3**rd term is $a \times r^{3-1}$, or ar^2

Example 1

a Work out the nth term of the geometric sequence 2, 6, 18, 54, ...

b Calculate the value of the 10th term of this sequence.

a The first term $a = 2$, the common ratio $r = \dfrac{u_2}{u_1} = \dfrac{6}{2} = 3$

nth term is $u_n = a \times r^{n-1} = 2 \times 3^{n-1}$

b When $n = 10$, $u_{10} = 2 \times 3^{10-1} = 39\,366$

Key point

A **geometric series** is the sum of the terms of a geometric sequence.

For the geometric sequence $a, a \times r, a \times r^2, a \times r^3, ...$
the corresponding geometric series is $a + (a \times r) + (a \times r^2) + (a \times r^3) + ...$
You say the series has nth term $a \times r^{n-1}$

A geometric sequence and its series share the same nth term and the same common ratio.

You can use the notation S_n for the sum of the first n terms of a geometric series.

For example, for the geometric series $3 + 6 + 12 + 24 + ...$

$S_2 = 3 + 6 = 9$ and $S_3 = 3 + 6 + 12 = 21$

You can work out a formula for S_n

$$S_n = a + ar + ar^2 + \ldots + ar^{n-2} + ar^{n-1} \qquad [1]$$

$$\therefore S_n \times r = (a + ar + ar^2 + \ldots + ar^{n-2} + ar^{n-1}) \times r$$

$$S_n \times r = ar + ar^2 + ar^3 + \ldots + ar^{n-1} + ar^n \qquad [2]$$

[2]−[1]:

$$(S_n \times r) - S_n = (ar + ar^2 + ar^3 + \ldots + ar^{n-1} + ar^n) - (a + ar + ar^2 + \ldots + ar^{n-1})$$

$$\therefore S_n(r-1) = ar^n - a \qquad \text{so } S_n = \frac{a(r^n - 1)}{r - 1} \qquad \text{(provided } r \neq 1\text{)}$$

ICT Resource online

To investigate sequences, click this link in the digital book.

> **Key point**
>
> If a geometric series has first term a and common ratio r $(r \neq 1)$, then $S_n = \dfrac{a(r^n - 1)}{r - 1}$ or, equivalently, $S_n = \dfrac{a(1 - r^n)}{1 - r}$.

> When $r < 1$ use $S_n = \dfrac{a(1 - r^n)}{1 - r}$ to avoid unnecessary negative signs in your working.

Example 2

Calculate the sum of the first 12 terms of the geometric series $4 + 6 + 9 + \ldots$, to 3 sf.

$$r = \frac{u_2}{u_1} = \frac{6}{4} = 1.5$$

$$S_n = \frac{a(r^n - 1)}{r - 1} \text{ so, } S_{12} = \frac{4(1.5^{12} - 1)}{1.5 - 1}$$

$a = 4$, $r = 1.5$ and $n = 12$

$$= 1029.970 \ldots = 1030 \text{ (3 sf)}$$

You can check your answer using the Σ function on your calculator.

You can evaluate a geometric series that is written using sigma notation.

Example 3

Use a formula to evaluate the geometric series $\displaystyle\sum_{k=1}^{8} (10 \times 0.2^k)$. Give your answer to 2 decimal places.

$$\sum_{k=1}^{8} (10 \times 0.2^k) = \overbrace{(10 \times 0.2^1) + (10 \times 0.2^2) + \ldots + (10 \times 0.2^8)}^{8 \text{ terms}}$$

$$= 2 + 0.4 + \ldots + (10 \times 0.2^8)$$

Evaluate the first two terms to work out a and r

$$a = 2, \ r = \frac{0.4}{2} = 0.2 \quad \text{and } n = 8$$

$$S_n = \frac{a(1 - r^n)}{1 - r}$$

$r < 1$ so use this version of the equation.

$$S_8 = \frac{2(1 - 0.2^8)}{1 - 0.2} = 2.499 \ldots = 2.50 \text{ (2 dp)}$$

PURE

Exercise 13.4A Fluency and skills

1 Work out the values of the first four terms of the geometric sequences defined by

   a $u_n = 4 \times 3^{n-1}$ **b** $u_n = 3 \times 4^{n-1}$

   c $u_n = 8 \times 0.5^n$ **d** $u_n = 6 \times 3^{-n}$

   e $u_n = 5 \times 2^{-n-1}$ **f** $u_n = 2 \times 0.5^{-(n-3)}$

2 Work out an expression for the nth term of these geometric sequences.

   a $5, 10, 20, \dots$ **b** $36, 24, 16, \dots$

   c $2, -6, 18, \dots$ **d** $-8, 6, -4.5, \dots$

   e $7, -24.5, 85.75, \dots$ **f** $1, \dfrac{1}{6}, \dfrac{1}{36}, \dots$

3 Evaluate these geometric series. Give non-integer answers to 3 significant figures.

   a $6 + 12 + 24 + \dots$ (10 terms)

   b $64 + 96 + 144 + \dots$ (7 terms)

   c $50 + 40 + 32 + \dots$ (15 terms)

   d $4 - 6 + 9 + \dots$ (20 terms)

4 Use a formula to evaluate these geometric series.

   a $\displaystyle\sum_{k=1}^{12}(2 \times 3^k)$ **b** $\displaystyle\sum_{k=1}^{6}(192 \times 0.5^k)$

   c $\displaystyle\sum_{k=3}^{10}(5 \times 2^k)$ **d** $\displaystyle\sum_{k=5}^{15}3 \times (-2)^{k-1}$

Reasoning and problem-solving

Strategy

To solve a problem using a geometric sequence or series

(1) Define variables such as the first term a, the common ratio r and the nth term u_n using information given in the question.

(2) Work out any unknown values.

(3) Use appropriate formulae and give your answer in context, rounding appropriately.

You can use simultaneous equations to solve a problem involving a geometric sequence or series.

Example 4

A geometric sequence has first term a, and common ratio r. The second term is 0.5 and the fifth term is 108. Work out the value of the ninth term of this sequence.

The nth term is u_n, where $u_n = a \times r^{n-1}$ •————— Define suitable variables. **(1)**

$u_2 = a \times r$ so $ar = 0.5$ [1]

$u_5 = a \times r^4$ so $ar^4 = 108$ [2] •————— You need to find a and r **(2)**

$\dfrac{\cancel{a}r^4}{\cancel{a}r} = r^{4-1} = r^3$ •————— Divide [2] by [1] to find r

$\dfrac{ar^4}{ar} = \dfrac{108}{0.5}$ so $r^3 = 216$

$r = \sqrt[3]{216} = 6$

$ar = 0.5$ so $a \times 6 = 0.5$ and $a = \dfrac{1}{12}$ •————— Use [1] and $r = 6$ to find a

$u_n = \dfrac{1}{12} \times 6^{n-1}$ so when $n = 9$, $u_9 = \dfrac{1}{12} \times 6^{9-1} = 139\,968$

The geometric series $a + a \times r + a \times r^2 + a \times r^3 + \ldots$ has an **infinite** number of terms.

You know that the first n terms of this series has sum $S_n = \dfrac{a(1-r^n)}{1-r}$

If $-1 < r < 1$, then, as n increases, the term r^n in this formula approaches zero, which means S_n approaches $\dfrac{a(1-0)}{1-r}$, or simply $\dfrac{a}{1-r}$

> **Key point**
>
> $\dfrac{a}{1-r}$ is the **sum to infinity** of the series. You can write S_∞ for this value.
>
> If $|r| < 1$ then $S_\infty = \dfrac{a}{1-r}$

$|r| < 1$ means
$-1 < r < 1$

Example 5

a Evaluate the sum to infinity of the geometric series $48 + 12 + 3 + \ldots$

b Another geometric series has first term 36 and sum to infinity 20. Calculate the sum of the first six terms of this series. Give the answer to 1 decimal place.

a $a = 48, r = \dfrac{u_2}{u_1} = 0.25$

Substitute values of a and r

$S_\infty = \dfrac{a}{1-r}$ so $S_\infty = \dfrac{48}{1-0.25} = 64$

2 You need to find r

b $20 = \dfrac{36}{1-r}$ so $r = -0.8$

3 Use the appropriate formula for $r < 1$

$S_n = \dfrac{a(1-r^n)}{1-r}$

$a = 36, r = -0.8$ and $n = 6$ so $S_6 = \dfrac{36(1-(-0.8)^6)}{1-(-0.8)}$

3 Round appropriately.

$= 14.7571 = 14.8 \ (1 \ dp)$

You can describe, or **model**, a real-life situation using a geometric sequence or series.

Example 6

At the start of year 1, £3000 was invested in a savings account.

The account paid 3% compound interest per year and no further deposits or withdrawals were allowed.

The balance of the account at the start of year n was modelled using a geometric sequence with common ratio r

a Work out the balance at the start of year 2 and state the value of r

b At the start of which year did the balance first exceed £3600?

(*Continued on the next page*)

MyMaths 🔍 2040 SEARCH 🔗

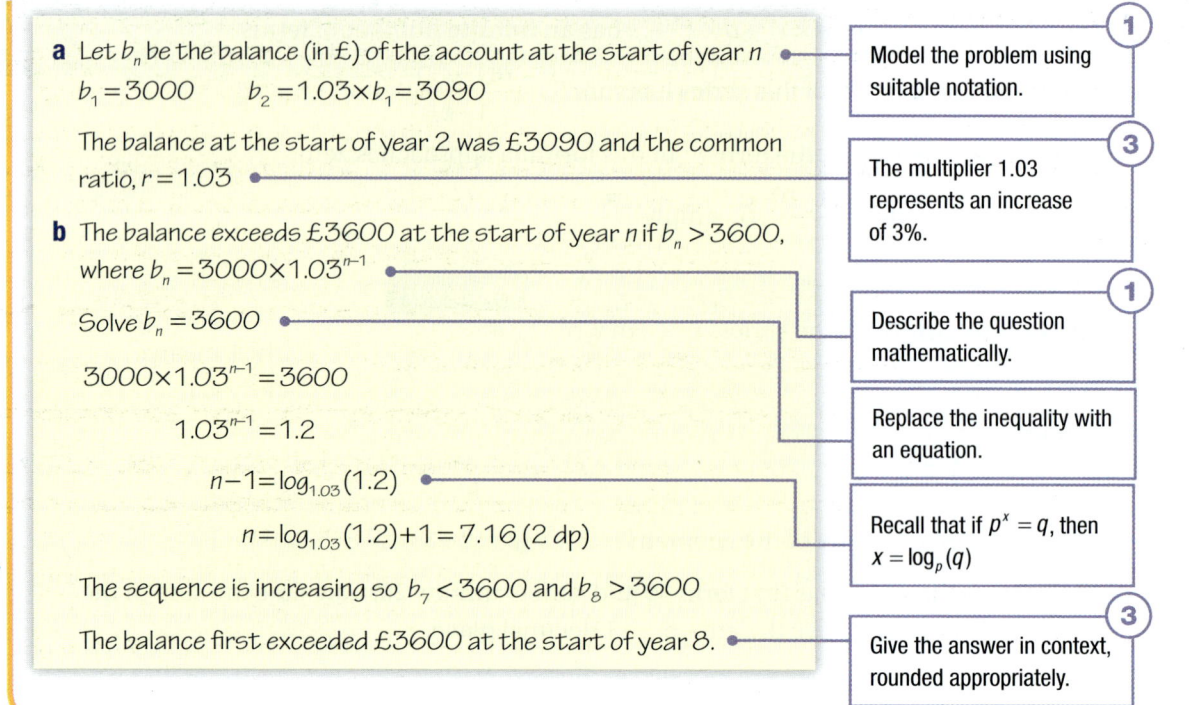

a Let b_n be the balance (in £) of the account at the start of year n

$b_1 = 3000 \qquad b_2 = 1.03 \times b_1 = 3090$

The balance at the start of year 2 was £3090 and the common ratio, $r = 1.03$

b The balance exceeds £3600 at the start of year n if $b_n > 3600$, where $b_n = 3000 \times 1.03^{n-1}$

Solve $b_n = 3600$

$3000 \times 1.03^{n-1} = 3600$

$1.03^{n-1} = 1.2$

$n-1 = \log_{1.03}(1.2)$

$n = \log_{1.03}(1.2) + 1 = 7.16 \ (2 \ dp)$

The sequence is increasing so $b_7 < 3600$ and $b_8 > 3600$

The balance first exceeded £3600 at the start of year 8.

Annotations:

① Model the problem using suitable notation.

③ The multiplier 1.03 represents an increase of 3%.

① Describe the question mathematically.

Replace the inequality with an equation.

Recall that if $p^x = q$, then $x = \log_p(q)$

③ Give the answer in context, rounded appropriately.

See Ch5.2
For a reminder on logarithms and exponential functions.

Exercise 13.4B Reasoning and problem-solving

1 The nth term of a geometric sequence is u_n, where $u_3 = 2$ and $u_6 = 128$

   a Work out the common ratio r and the first term a of this sequence.

   b Calculate the value of the eighth term of this sequence.

   c Express u_n in the form 2^{pn+q} stating the value of each constant p and q

2 The nth term of a geometric series is u_n, where $u_3 = 45$ and $u_5 = 405$. The series has common ratio r, where $r > 0$

   a Calculate the sum of the first six terms of this series.

   The sum of the first n terms of this series exceeds 1 000 000

   b Evaluate the least possible value of this sum.

3 A finite geometric series has first term 6 and common ratio r, where $r > 0$. The fifth term is 96 and the last term of this series has value 24 576

   a Work out the value of r

   b Evaluate the sum of all the terms of this series.

4 The nth term of a geometric series is u_n with first term a and common ratio 1.5

   Given that $S_2 = 40$, where S_n is the sum of the first n terms of this series

   a Calculate, to 3 significant figures, the value of S_n for which $9000 < S_n < 10000$

   b Evaluate the sum to infinity of the geometric series $\dfrac{1}{u_1} + \dfrac{1}{u_2} + \dfrac{1}{u_3} + ...$

5 A geometric series has first term 50 and common ratio r, where $0 < r < 1$. The nth term is u_n and the sum of the first n terms of this series is S_n. Given that $S_3 = 98$

   a Work out the value of S_∞

   b **i** Evaluate the largest integer N for which $S_\infty - S_N > 0.5$

   ii For this N evaluate $\displaystyle\sum_{k=N}^{\infty} u_k$, giving your answer to 3 decimal places.

6 At the start of year 1, £2500 is invested in a savings account. The account paid 2% compound interest per year and no further deposits or withdrawals were allowed. The balance (in £) of the account at the start of year n was modelled using a geometric sequence with common ratio r

   a Write down the value of r

   Use this model to calculate, to the nearest £

   b The balance of the account at the start of year 4

   c The amount of interest earned on the account in year 7

7 Toni bought a car for £24 000 which decreased in value by 25% each year after its purchase. The value of the car n years after its purchase was modelled by a geometric sequence with common ratio r

   a Work out the value of r

   b Use this model to show that three years after its purchase the value of the car was £10 125

   Toni sold the car in the year its value fell below £2000

   c Use this model to find the number of whole years Toni owned the car.

8 In his first year of driving, Tom drove 3125 miles. In his first two years of driving he drove 5625 miles. The distance (in miles) driven in Tom's nth year of driving was modelled using a geometric sequence.

   a Use this model to

   i Work out the distance Tom drove in his fourth year of driving,

   ii Calculate the total distance Tom drove in his first six years of driving,

   iii Show that the total distance Tom can drive in his lifetime is less than 15 625 miles.

   b Comment on the suitability of this model in the long-term.

9 A rectangular piece of foil has length 50 cm, width 40 cm and is 0.4 mm thick. The foil was folded in half, and then in half again, and so on. The thickness (in mm) of the shape formed when the foil was folded n times was modelled using a geometric sequence.

   a Calculate the thickness, in mm, of the shape when the foil is folded four times.

   b Calculate the volume, in cm^3, of the shape when the foil is folded four times. You may assume after all folds the shape formed was a cuboid.

   c Calculate the thickness of the shape if the foil were folded 40 times. Compare your answer with the distance from the Earth to the Moon (approximately 384 000 km).

10 At the end of the first month of opening, the electricity bill for a small business was £450. The amount due increased by 5% each month. The bills were analysed and it was correctly stated that, to the nearest £10, the average electricity bill per month for the first year was £600

   Use a suitable model to determine which average (mean or median, or both) was used in this statement. Justify your answer.

Challenge

11 At 1 pm on a particular day, three people were told a rumour. At 2 pm that day, these three people had told three other people the rumour, who, at 3 pm that day, had each told three other people the rumour, and so on. Nobody was told the rumour more than once.

Given that the current estimate for the world's population is 7.4×10^9, how many hours will it take for the world's population to have heard this rumour?

Chapter summary

- When $n \notin \mathbb{Z}^+$, $1 + nx + \dfrac{n(n-1)}{2!}x^2 + \dfrac{n(n-1)(n-2)}{3!}x^3 + \ldots$ is the binomial expansion of $(1+x)^n$

- The expansion of $(a+bx)^n$ when $n \notin \mathbb{Z}^+$ is only valid for $\left|\dfrac{b}{a}x\right| < 1$

- A sequence with nth term u_n is increasing if $u_{n+1} > u_n$ for all $n \in \mathbb{Z}^+$, decreasing if $u_{n+1} < u_n$ for all $n \in \mathbb{Z}^+$, and periodic if the sequence consists of one block of terms repeated throughout.

- The period (or order) of a periodic sequence is the smallest number of terms that repeat as a block of numbers.

- If a sequence defined by $u_{n+1} = au_n + b$, where a and b are constants, has a limit L as $n \to \infty$ then L satisfies the equation $L = aL + b$. More generally, if the sequence defined by $u_{n+1} = \mathrm{f}(u_n)$ has limit L then $L = \mathrm{f}(L)$

- A sequence is arithmetic if for some constant d, $u_{n+1} - u_n = d$ for all $n \in \mathbb{Z}^+$
  d is the common difference of the sequence.

- $u_n = a + (n-1)d$ is the nth term of an arithmetic sequence with first term a

- A series is formed by adding the terms of a sequence together.

- $S_n = \dfrac{1}{2}n(a+l)$ is the sum of the first n terms of an arithmetic series, with first term a and last term l. This is also given by $S_n = \dfrac{1}{2}n[2a + (n-1)d]$

- Sigma notation can be used for series $\displaystyle\sum_{k=1}^{n} \mathrm{f}(k) = \mathrm{f}(1) + \mathrm{f}(2) + \ldots + \mathrm{f}(n)$

- A sequence is geometric if for some non-zero constant r, $\dfrac{u_{n+1}}{u_n} = r$ for all $n \in \mathbb{Z}^+$
  r is the common ratio of the sequence.

- $u_n = a \times r^{n-1}$ is the nth term of a geometric sequence with first term a

- $S_n = \dfrac{a(1-r^n)}{1-r}$, $r \neq 1$ is the sum of the first n terms of a geometric series.

- $S_\infty = \dfrac{a}{1-r}$, $|r| < 1$ is the sum to infinity of a geometric series.

Check and review

You should now be able to...	Try Questions
✔ Use the binomial expansion and recognise the range of validity.	1
✔ Use the binomial expansion to estimate the value of a surd.	2
✔ Understand if a sequence is increasing or decreasing.	3
✔ Work out the order of a periodic sequence.	4
✔ Work out the nth term and the sum of an arithmetic series.	5
✔ Work out the nth term and the sum of a geometric series.	6
✔ Evaluate a series given in sigma notation.	7

1 Work out the binomial expansion of these expressions up to and including the term in x^3. State the range of validity of each full expansion.

   a $(1+x)^{\frac{4}{3}}$

   b $\dfrac{1}{(1+2x)^3}$

   c $(9-4x)^{\frac{3}{2}}$

   d $\sqrt{\dfrac{64}{(4+x)^3}}$

2 Work out and use the first three terms in the binomial expansion of these expressions to estimate $\sqrt{3}$

   a $(1+x)^{\frac{1}{2}}$ $x=0.08$

   b $(4+3x)^{\frac{1}{2}}$ $x=-\dfrac{1}{3}$

3 Work out u_1, u_2, u_3 and u_4 for each of these sequences and describe as increasing, decreasing or neither.

   a $u_{n+1}=3u_n-5$, $u_1=4$

   b $u_{n+1}=\dfrac{1}{2}u_n^2-1$, $u_1=2$

   c $u_{n+1}=u_n+\dfrac{1}{3}$, $u_1=\dfrac{1}{2}$

   d $u_{n+1}=5u_n-0.5$, $u_1=-1$

   e $u_{n+1}=\dfrac{4}{u_n}-1$, $u_1=3$

   f $u_{n+1}=\sqrt{2u_n+3}$, $u_1=3$

4 Work out the order of these periodic sequences.

   a $u_n=4(-1)^n+2(-1)^{n+1}$

   b $u_n=2\sin(n\times90)°+3\cos(n\times90)°$

   c $u_n=\dfrac{3}{2}-\dfrac{9}{4u_n}$, $u_1=\dfrac{1}{2}$

   d $u_{n+1}=4\left(\dfrac{u_n-2}{u_n}\right)$, $u_1=3$

5 Work out the nth term and the sum of these arithmetic series.

   a $7+13+19+...+73$ (12 terms)

   b $15+11+7+...-81$ (25 terms)

   c $9+14+19+...$ (30 terms)

   d $\dfrac{1}{2},1,\dfrac{3}{2},\,...$ (50 terms)

   e $18.5, 24, 29.5, ...$ (10 terms)

   f $\dfrac{1}{2}+\dfrac{1}{3}+...$ (19 terms)

6 Work out the nth term and the sum of these geometric series. Give non-integer answers to 3 significant figures.

   a $3+18+108+...$ (8 terms)

   b $100+10+1+...$ (10 terms)

   c $200-190+180.5+...$ (15 terms)

   d $\log_2(3)+\log_2(9)+\log_2(81)+...$ (9 terms)

7 Use formulas to evaluate these series, each of which is either arithmetic or geometric. Give non-integer answers to 3 significant figures.

   a $\displaystyle\sum_{k=1}^{30}(5k+3)$

   b $\displaystyle\sum_{k=1}^{8}(25\times1.8^k)$

   c $\displaystyle\sum_{k=5}^{20}(40-3k)$

   d $\displaystyle\sum_{k=1}^{\infty}(250\times0.6^k)$

What next?

Score				
	0 – 4	Your knowledge of this topic is still developing. To improve, search in MyMaths for the codes: 2038–2040, 2204, 2205, 2264		**Click these links in the digital book**
	5 – 6	You're gaining a secure knowledge of this topic. To improve, look at the InvisiPen videos for Fluency and skills (13A)		
	7	You've mastered these skills. Well done, you're ready to progress! To develop your exam techniques, look at the InvisiPen videos for Reasoning and problem-solving (13B)		

ICT

An **iterative** formula can be used to produce a sequence of numbers.

By using a spreadsheet, investigate the iterative formula $x_{n+1} = \dfrac{x_n}{a} + b$ for different values of a and b and for a range of starting values, x_0

In terms of a and b, what happens to these sequences as n increases?

	B5			fx	=B4/\$B\$1+\$B\$2
	A	B	C	D	E
1	a = 2				
2	b = 2				
3					
4	x0	654			
5	x1	332			
6	x2	171			
7	x3	90.5			
8	x4	50.25			

Have a go

Using methods of calculus, **Isaac Newton** found an iterative formula for a sequence which will converge to the square root of any positive number, a

The iterative formula is given by

$$x_{n+1} = \frac{1}{2}\left(\frac{a}{x_n} + x_n\right)$$

Try using the formula to calculate $\sqrt{13}$

Note

A **convergent sequence** is a sequence for which the terms get closer and closer to a particular value.

For example, the sequence

$1, \dfrac{1}{2}, \dfrac{1}{3}, \dfrac{1}{4}, \dfrac{1}{5}, \ldots$ converges to 0

Information

Sequences can converge to their limits at different rates. Consider these sequences.

a $x_{n+1} = \dfrac{x_n}{10}$ this is equivalent to the sequence with nth term $\left(\dfrac{1}{10}\right)^n$

b $y_{n+1} = \dfrac{y_n}{10^{2n+1}}$ this is equivalent to the sequence with nth term $\left(\dfrac{1}{10}\right)^{n^2}$

Each of these sequences converges to 0, however sequence **b** has a higher rate of convergence. Despite the fact that there exists no value of n for which either sequence ever equals the limit, 0, you can think of sequence **b** as converging faster.

To get a sense of this, look at the size of the terms of each sequence for different values of n
For example, when $n = 3$
$x_3 = 1 \times 10^{-3} = 0.001$
$y_3 = 1 \times 10^{-9} = 0.000\,000\,001$

Even for small values of n, sequence **b** is already much closer to 0

Research

Find out what is meant by the terms **linear convergence**, **sublinear convergence** and **superlinear convergence**.

1 A sequence is defined by the recurrence relation $u_{n+1} = 1 - \dfrac{1}{u_n}$, where $u_1 = 2$

   a Write down the values of

   i u_2

   ii u_3

   iii u_4 **[6 marks]**

   b Deduce the value of u_{50} **[2]**

2 Write the first four terms of each sequence, then describe the sequences as either increasing, decreasing or periodic.

   a $u_n = 2\cos(180n)°$ **[3]**

   b $u_n = 0.2^n + 4$ **[3]**

   c $u_n = n^2 + 4n - 2$ **[3]**

3 Write down the first four terms in the binomial expansion, in ascending powers of x, of $(1 - 2x)^{-2}$, stating the values of x for which the expansion is valid.   **[6]**

4 A car costs £30 000. Its value depreciates by 20% per annum. Work out

   a Its value after 1 year,   **[1]**

   b Its value after 4 years,   **[2]**

   c The year in which it will be worth less than £5000   **[4]**

5 A sequence of terms is defined by the recurrence relation $u_{n+1} = 4 - ku_n$, where k is a constant.

   Given that $u_1 = 3$

   a Work out an expression in terms of k for u_2 **[2]**

   b Work out an expression in terms of k for u_3 **[2]**

   Given also that $u_1 + u_2 + u_3 = 9$

   c Calculate the possible values of k **[4]**

6 The sum to infinity of a geometric series is 20. The first term is 4

   a Calculate the common ratio of the series.   **[3]**

   b Evaluate the third term of the series.   **[2]**

7 Adam plans to pay money into a savings scheme each year for 20 years. He will pay £800 in the first year, and every year he will increase the amount that he pays into the scheme by £100

   a Show that he will pay £1000 into the scheme in year 3   **[1]**

b Calculate the total amount of money that he will pay into the scheme over the 20 years. **[2]**

c Over the same 20 years, Ben will also pay money into a savings scheme. He will pay £610 in the first year, and every year he will increase the amount that he pays into the scheme by £d. Given that Adam and Ben will pay in exactly the same total amounts over the 20 years, calculate the value of d **[3]**

8 When $(1+ax)^n$ is expanded the coefficients of x and x^2 are -4 and 20 respectively.

a Work out the value of a and the value of n **[8]**

b Evaluate the coefficient of x^3 **[2]**

9 The second term of a geometric series is 120 and the fifth term is 15. Work out

a The common ratio of the series, **[4]**

b The first term of the series, **[1]**

c The sum to infinity of the series. **[2]**

10 a Use a formula to evaluate $\displaystyle\sum_{r=1}^{40}(3r+1)$ **[3]**

b Calculate the value of n for which $\displaystyle\sum_{r=1}^{n}(3r+1)=9800$ **[4]**

11 a Write down the first three terms in the binomial expansion of $(1-2x)^{\frac{1}{2}}$, in ascending powers of x **[3]**

b Write down the first three terms in the binomial expansion of $(1+x)^{-\frac{1}{2}}$, in ascending powers of x **[3]**

c Use your answers to **a** and **b** to prove that $\sqrt{\dfrac{1-2x}{1+x}}=1-\dfrac{3}{2}x+\dfrac{3}{8}x^2+\ldots.$ **[4]**

12 The fourth term of an arithmetic series is 11 and the sum of the first three terms is -3

a Write down the first term of the series. **[4]**

b Work out the common difference of the series. **[1]**

c Given that the sum of the first n terms of the series is greater than 500, calculate the least possible value of n **[5]**

13 The first three terms of a geometric series are $(3p-1)$, $(p-3)$ and $(2p)$ respectively.

a Use algebra to work out the possible values of p **[5]**

b For the negative value of p, calculate the sum to infinity of the series. **[3]**

c For the positive value of p, evaluate the sum of the first 999 terms of the series. **[2]**

14 a Write down the first four terms in the binomial expansion $\sqrt{1-x}$, in ascending powers of x **[6]**

b By substituting $x=\dfrac{1}{4}$, work out a fraction that is an approximation to $\sqrt{3}$ **[4]**

15 A salesman sells vacuum cleaners for £120 each. In one week, he receives 2% commission on the first vacuum cleaner he sells, 4% commission on the second vacuum cleaner he sells, with commission increasing in steps of 2% , so that he receives commission of 30% on the sale of his fifteenth vacuum cleaner. Commission stays fixed at 30% for the sale of all vacuum cleaners, after the sale of his fifteenth vacuum cleaner in that week.

a Calculate how much commission he receives in a week for the sale of

    i His first vacuum cleaner,

    ii His fifth vacuum cleaner,

    iii His twentieth vacuum cleaner. [6]

In one week he sells 40 vacuum cleaners.

b How much commission does he receive in total that week? [5]

16 The sum to infinity of a geometric series is 48, and the sum of the first two terms of the series is 45

The common ratio of the series is r

a Prove that r satisfies the equation $1 - 16r^2 = 0$ [4]

b Calculate the sum of the first four terms of the series. [4]

17 The training programme of a cyclist requires her to cycle 3 km on the first day of training.

Then, on each day that follows, she cycles 2 km more than she cycled on the day before.

a Calculate how far she cycles on the seventh day. [1]

b Calculate the total distance she has cycled by the end of the tenth day. [2]

c On which day of training will she cycle more than 100 km? [3]

d On which day of training will the total distance that she has cycled exceed 1000 km? [5]

18 a Write down the first three terms in the binomial expansion of $\sqrt{4-x}$, in ascending powers of x [7]

b Deduce an approximate value of $\sqrt{399}$, giving your answer to 3 decimal places. [5]

19 An investment scheme pays 3% compound interest per annum. The interest is paid annually.

A deposit of £1000 is invested in this scheme at the start of each year.

The initial investment of £1000 is made at the start of year 1

a Explain why the value of the investment at the start of year 2 is £2030 [2]

b Calculate the value of the investment at the start of year 3 [2]

c Work out the year in which the total value of the investment exceeds £50 000 [4]

20 The sum of the first two terms of an arithmetic series is 2. The sum of the first ten terms of the series is 330

    a Work out the common difference of the series. **[5]**

    b Write down the first term of the series. **[1]**

    c Given that the sum of the first n terms of the series is equal to 1170, find the value of n **[4]**

21 Given that $f(x) = \dfrac{5x}{(2+x)(1-2x)} \equiv \dfrac{A}{2+x} + \dfrac{B}{1-2x}$

    a Work out the values of the constants, A and B **[5]**

    b Write down the series expansion of $f(x)$, in ascending powers of x, up to and including the term in x^3 **[11]**

    c State the values of x for which the expansion is valid. **[1]**

22 Given that $f(x) = \dfrac{13x-33}{(5-x)(1+3x)}$

    a Work out the expansion of $f(x)$ up to and including the term in x^3 **[14]**

    b State the values of x for which the expansion is valid. **[1]**

23 When a ball is dropped from a height of h metres above a hard floor it rebounds to a height of $\dfrac{3}{4}h$

A ball is dropped from an initial height of 2 metres. Calculate

    a The height to which the ball rises after the first bounce, **[2]**

    b The total distance the ball has travelled when it hits the floor for the second time, **[2]**

    c The total distance that the ball travels. **[3]**

24 Given that x, 15 and y are consecutive terms of an arithmetic series, and 1, x and y are consecutive terms of a geometric series, work out the possible values of x and y **[9]**

25 By solving an equation, find the limit of these sequences as $n \to \infty$. Where appropriate, give answers in simplified surd form.

    a $u_{n+1} = 0.2u_n + 4$ **[2]**

    b $u_{n+1} = 9 - 0.2u_n$ **[2]**

    c $u_{n+1} = \dfrac{1}{2}\left(\dfrac{1}{3}u_n - 10\right)$ **[2]**

    d $u_{n+1} = \left(\sqrt{2}-1\right)u_n + 4$ **[2]**

    e $u_{n+1} = \dfrac{1}{\sqrt{2}}u_n + \sqrt{2}$ **[2]**

    f $u_{n+1} = 0.5u_n^2 + 0.5$ **[2]**

14 Trigonometric identities

Tides are affected by the gravity of both the Moon and the Sun, and follow a predictable oscillatory pattern. Since tides are periodic, the sine and cosine functions can be used to model them, with components governed by the Sun and Moon. Predicting and modelling the tides has been essential to marine life for centuries, and now also plays an important role in calculating the effect and extent of climate change.

Trigonometric functions are suitable for modelling anything known to follow a periodic pattern, including tides, seasonal temperature fluctuations hours of daylight, the motion of a Ferris wheel, the orbits of planets and satellites and music waves. The sine and cosine functions are very useful mathematical tools, and there are many trigonometric identities that aid in the manipulation and application of trigonometric expressions.

Orientation

What you need to know	What you will learn	What this leads to
Ch3 Trigonometry • Sine, cosine and tangent. • The sine and cosine rules.	• To convert between degrees and radians and use radians in problems. • To use reciprocal and inverse trigonometric functions. • To use trigonometric formulae for compound angles, double angles and half angles. • To find and use equivalent forms for $a\cos\theta + b\sin\theta$ • To simplify and solve equations using trigonometric formulae.	**Ch15 Differentiation 2** Derivatives of trigonometric functions. **Ch18 Motion in two dimensions** Resolving forces. Projectiles.
Ch12 Algebra 2 • Transformations of functions. • Inverse functions.		

p.70   p.76   p.304

  **MyMaths**   Practise before you start   🔍 2045–2048, 2138, 2142, 2284

Fluency and Skills

As well as degrees, you can measure angles in **radians**. If you draw a circle with centre O and radius r, then 1 radian is the angle subtended at O by an arc that is equal in length to the radius.

As the circumference is $2\pi \times r$, you can fit 2π of these arcs around the circumference. So, the total angle at O is 2π radians or $360°$

> **Key point**
>
> 2π radians $= 360°$ π radians $= 180°$
>
> 1 radian $= \dfrac{180°}{\pi} \approx 57.3°$ 1 degree $= \dfrac{\pi}{180}$ radians

The symbol for radians is rad or, more rarely, ^{c} or ^{r}, as in 2^c and 2^r, where c means circular measure and rad is short for radians.

For sector POQ, the arc length, s, is a fraction of the circumference, and the sector area, A, is a fraction of the circle's area.

> **Key point**
>
> $s = \dfrac{\theta}{2\pi} \times 2\pi r \Rightarrow s = r\theta$
>
> $A = \dfrac{\theta}{2\pi} \times \pi r^2 \Rightarrow A = \dfrac{1}{2}r^2\theta$
>
> where θ is in radians.

Compare the formulae where θ is in radians with those where θ is in degrees: $s = \dfrac{\theta}{360°} \times 2\pi r$

and $A = \dfrac{\theta}{360°} \times \pi r^2$

Example 1

a Convert an angle of $30°$ into radians, writing it in terms of π

b Calculate the exact values of arc length PQ and sector area POQ

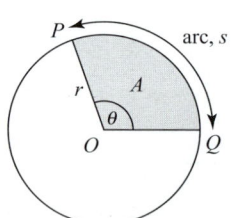

a $30° = 30 \times \dfrac{\pi}{180} = \dfrac{\pi}{6}$ radians

b Arc length $= r\theta = 9 \times \dfrac{\pi}{6} = \dfrac{3\pi}{2}$ cm

Sector area $= \dfrac{1}{2}r^2\theta = \dfrac{1}{2} \times 9^2 \times \dfrac{\pi}{6} = \dfrac{27\pi}{4}$ cm^2

1 degree $= \dfrac{\pi}{180}$ radians

When x is in radians, the graph of $y = x$ almost coincides with the graphs of $y = \sin x$ and $y = \tan x$ for small values of x. Similarly, the graph of $y = 1 - \dfrac{1}{2}x^2$ gives a close approximation to $y = \cos x$ for small x

Compare the graphs by plotting them on your graphics calculator, and check the results.

> **Key point**
>
> For small values of x in radians,
>
> $\sin x \approx x$ $\tan x \approx x$ $\cos x \approx 1 - \dfrac{1}{2}x^2$

Example 2

Calculate the percentage error if 8° in radians is used as an approximation for $\sin 8°$

$\sin 8° = 0.13917$ (to 5 dp)

$8° = \dfrac{8}{180} \times \pi = 0.13963$ radians (to 5 dp)

Percentage error $= \dfrac{\text{absolute error}}{\text{correct value}} \times 100\% = \dfrac{0.13963 - 0.13917}{0.13917} \times 100\%$

$= 0.3\%$ (to 1 sf)

These two special triangles can be used to find trigonometric ratios for angles in radians. It is useful to know these.

For example, $60° = \dfrac{1}{3}$ of $180°$, which is $\dfrac{\pi}{3}$ radians

So, using the triangle, $\sin\dfrac{\pi}{3} = \sin 60° = \dfrac{\sqrt{3}}{2}$ and $\cos\left(\dfrac{\pi}{3}\right) = \cos 60° = \dfrac{1}{2}$

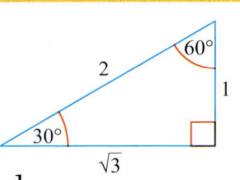 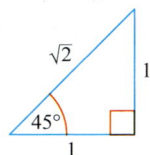

You can still use the sine rule, cosine rule and area of a triangle formula for triangles where angles are measured in radians.

You can change the set-up of your calculator from degrees to radians. Note that $\dfrac{\pi}{3} = 60$, so you use the special triangle here.

See Ch3.2
For a reminder on the sine and cosine rules.

Example 3

Calculate the length x and the area of the triangle, giving your answers as surds.

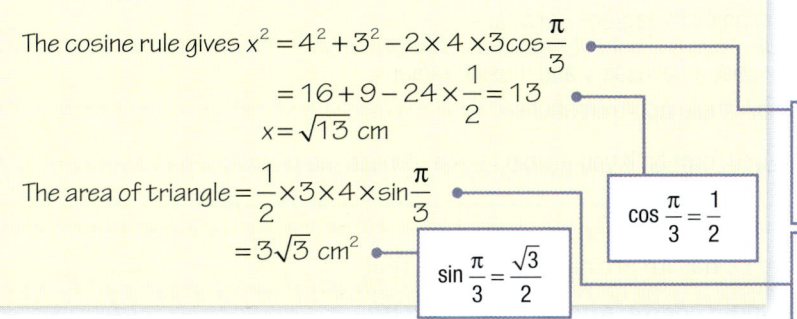

The cosine rule gives $x^2 = 4^2 + 3^2 - 2 \times 4 \times 3\cos\dfrac{\pi}{3}$

$= 16 + 9 - 24 \times \dfrac{1}{2} = 13$

$x = \sqrt{13}$ cm

The area of triangle $= \dfrac{1}{2} \times 3 \times 4 \times \sin\dfrac{\pi}{3}$

$= 3\sqrt{3}$ cm^2

$\sin\dfrac{\pi}{3} = \dfrac{\sqrt{3}}{2}$

$\cos\dfrac{\pi}{3} = \dfrac{1}{2}$

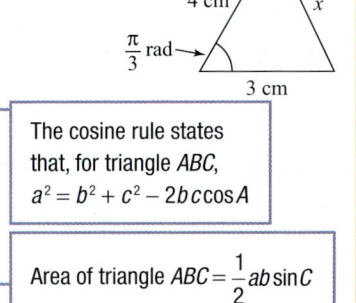

The cosine rule states that, for triangle ABC,
$a^2 = b^2 + c^2 - 2bc\cos A$

Area of triangle $ABC = \dfrac{1}{2}ab\sin C$

Exercise 14.1A Fluency and skills

1 Convert these angles into degrees, correct to 1 dp.

  a 2 rad   **b** 3 rad   **c** 0.5 rad

2 Convert these angles into radians. Give your answers in terms of π and also as decimals to 2 dp.

  a 90°   **b** 60°   **c** 270°

3 Find the arc length and the sector area of a circle with radius 4 cm and an angle of 2 radians at the centre.

4 Copy and complete this table, giving radians in terms of π and exact values for the trigonometric ratios. Use the special triangles below Example 2 to do this.

θ	θ (radians)	$\sin\theta$	$\cos\theta$	$\tan\theta$
45°				
120°				
135°				
270°				
360°				

5 Calculate the unknown sides and angles in $\triangle DEF$ where $d = 72.4\,\text{cm}$, $e = 43.2\,\text{cm}$ and angle $F = \dfrac{\pi}{4}$ rad

6 Giving your answers in terms of π, find the arc length and the sector area of a circle of radius $6\,\text{cm}$ with an angle of $\dfrac{3}{4}\pi$ radians at the centre.

7 a Copy and complete this table, correct to 4 dp.

θ	θ (radians)	$\sin \theta$	$\tan \theta$
$10°$			
$5°$			
$2°$			

b At what value of $\theta°$ do $\sin \theta$, $\tan \theta$ and θ in radians have the same value correct to 4 dp?

8 If $\theta = 12°$, use the value of θ in radians as an approximation for $\tan \theta$ and calculate the percentage error in this approximation.

9 In this triangle, $\theta = 1°$ and $x = 2.5\,\text{km}$

Calculate the height of the triangle y, in metres, using an approximation for $\tan \theta$

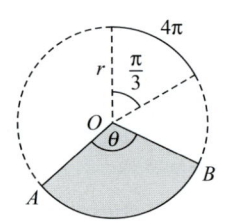

10 Find approximations for these expressions when θ and 2θ (in radians) are both small.

a $\dfrac{\sin 2\theta}{2\theta}$ **b** $\dfrac{1 - \cos \theta}{2\theta^2}$ **c** $\dfrac{\theta \sin \theta}{1 - \cos 2\theta}$

11 Solve these equations for $-\pi \le \theta \le \pi$. Show your working.

a $2\sin \theta - 1 = 0$ **b** $\sqrt{3} + 2\cos \theta = 0$

c $\sin \theta = \cos \theta$

Reasoning and problem-solving

Strategy

To solve problems that involve calculations with angles

1 Decide whether to work with angles in degrees or radians.

2 Choose appropriate formulae and, if necessary and if using radians, use values involving π or small-angle approximations.

3 Give your answer in the required form or, if you have a choice, choose the most appropriate form.

Example 4

A sector within a circle, centre O, has an arc of length $4\pi\,\text{cm}$ that makes an angle of $\dfrac{\pi}{3}$ radians at the centre. Another sector, AOB, has area $45\pi\,\text{cm}^2$

Calculate

a The radius, r, of the circle,

b The angle θ, in radians, in the sector AOB

a $s = r\theta$

$4\pi = r \times \dfrac{\pi}{3}$

$r = \dfrac{4\pi \times 3}{\pi} = 12$

Radius $r = 12\,\text{cm}$

b $A = \dfrac{1}{2}r^2\theta$

$45\pi = \dfrac{1}{2} \times 12^2 \times \theta$

$\theta = \dfrac{45\pi \times 2}{144} = \dfrac{5\pi}{8}$

Angle $\theta = \dfrac{5\pi}{8}$ radians

1 2 Choose the appropriate formula considering that you're working in radians.

2 Substitute known values.

3 Give your answer in radians.

1 Through how many radians does the minute hand of a clock turn in 20 minutes?

2 Copy and complete the table showing values for the angle θ, arc length s, and area A, of sectors with radius r

r cm	θ radians	s cm	A cm²
4		12	
5		2	
4			10
5			30

3 An aeroplane climbs at an angle of 5° for the first 2 km of its flight. Using an approximation for the angle, how high is it (in metres) after flying 2 km?

4 A hill is known to be 160 m high. The angle of elevation from a point x km away is 0.3°. Use an approximation to calculate the value of x

5 A comet is 10 million kilometres away from Earth, and its sides subtend an angle of 0.04° when viewed from Earth. Estimate the diameter of the comet.

6 A triangle has sides of length 5 cm, 7 cm and 12 cm. Calculate the size of the largest angle in the triangle. Give your answer in radians.

7 A wheel of radius 2 m rolls 0.5 m along the ground. How many radians has it turned through?

8

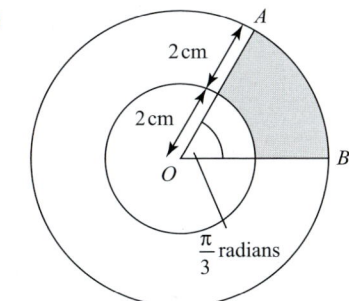

Find the perimeter and area of the shaded region in terms of π when $\angle AOB = \dfrac{\pi}{3}$

9 Solve these equations for $-\dfrac{\pi}{2} \leq \theta \leq \dfrac{\pi}{2}$. Show your working.

a $2\sin 2\theta = \sqrt{3}$

b $\tan\left(\theta + \dfrac{\pi}{6}\right) = 1$

c $\sqrt{2}\sin 2\theta = 2\cos\theta$

d $2\tan\theta - 1 = \dfrac{1}{\tan\theta}$

10 A chord AB subtends an angle of 0.6 radians at the centre O of a circle of radius 8 cm. Calculate the area of the segment cut off by AB

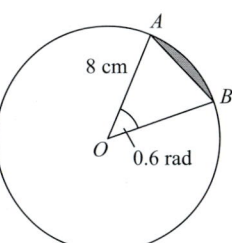

Challenge

11 Two equal circles of radius 7 cm have their centres 10 cm apart. Calculate the area of the overlap of the two circles.

12

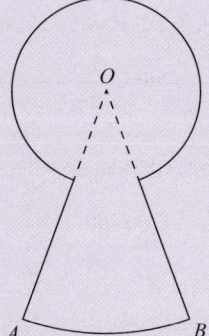

A keyhole is made from two sectors of two circles with a common centre O and radii 1 cm and 3 cm. If angle $AOB = \dfrac{\pi}{6}$ radians, calculate

a The perimeter of the keyhole,

b Its area,

giving answers in terms of π

13 Three equal circles of radius r are placed in the same plane so each one touches the other two. Find the area between them in terms of r

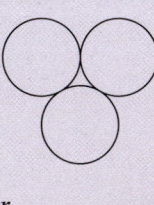

Fluency and Skills

Reciprocal functions

You know the **reciprocal** of x is $\dfrac{1}{x}$. The reciprocals of sine, cosine and tangent are cosecant (or cosec), secant (or sec) and cotangent (or cot).

> **Key point**
>
> $$\operatorname{cosec}\theta = \frac{1}{\sin\theta} \qquad \sec\theta = \frac{1}{\cos\theta} \qquad \cot\theta = \frac{1}{\tan\theta}$$

> Be careful: cosec is related to sine (not cos) and sec is related to cosine.

Your calculator may have buttons for cosec, sec and cot, but if not then you can calculate their values using the equivalent sin, cos and tan functions.

For example, $\sec 40° = \dfrac{1}{\cos 40°} = 1.305$

Example 1

Using your knowledge of trigonometric identities, and showing your working, find the exact values of

a $\operatorname{cosec} 225°$ **b** $\cot\dfrac{7\pi}{6}$

a $\operatorname{cosec} 225° = \dfrac{1}{\sin 225°} = \dfrac{1}{\sin 45°} = -\sqrt{2}$

> $225°$ is in the 3rd quadrant in the CAST diagram, so sin and cosec are negative.

See Ch3.1
For a reminder on the CAST diagram.

b $\dfrac{7\pi}{6}$ radians $= \dfrac{7 \times 180°}{6} = 210°$

$\cot\dfrac{7\pi}{6} = \cot 210° = \dfrac{1}{\tan 210°} = \dfrac{1}{\tan 30°} = \sqrt{3}$

> $210°$ is in the 3rd quadrant so tan and cot are positive.

You can use the graphs of $\sin\theta$, $\cos\theta$ and $\tan\theta$ to sketch the graphs of their reciprocal functions.

Sine and cosec

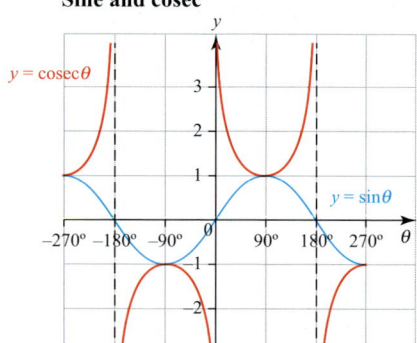

Cos and sec

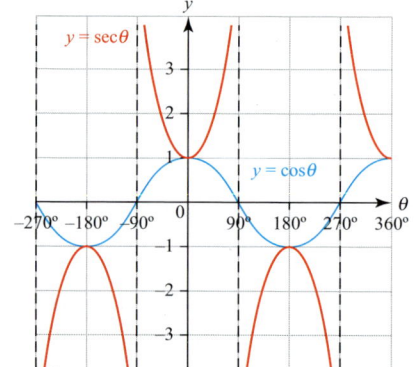

Tan and cot

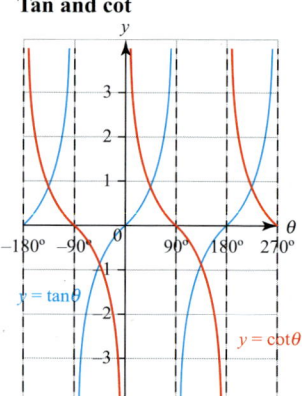

These graphs show the relationships between the original trigonometric functions and their reciprocals. They also show the **domains** and **ranges** of the reciprocal functions.

	Domain	Range
$y = \operatorname{cosec} \theta$	$\theta \in \mathbb{R}, \theta \neq 0°, \pm 180°, \pm 360°, \dots$	$y \in \mathbb{R}, y \geq 1, y \leq -1$
$y = \sec \theta$	$\theta \in \mathbb{R}, \theta \neq \pm 90°, \pm 270°, \dots$	$y \in \mathbb{R}, y \geq 1, y \leq -1$
$y = \cot \theta$	$\theta \in \mathbb{R}, \theta \neq 0°, \pm 180°, \pm 360°, \dots$	$y \in \mathbb{R}$

In Section 3.1 you derived $\sin^2 \theta + \cos^2 \theta \equiv 1$ by dividing Pythagoras' theorem by c^2

In a similar way, you can divide Pythagoras' theorem by a^2 or b^2 to derive results involving cosec, sec and cot.

> **Key point**
>
> $$\sec^2 \theta \equiv 1 + \tan^2 \theta \qquad \text{and} \qquad \operatorname{cosec}^2 \theta \equiv 1 + \cot^2 \theta$$

See Ch 12.2
For a reminder on domains and ranges.

PURE

Example 2

Prove the identity $\dfrac{1}{\cos^2 x} - \dfrac{\sec^2 x}{\operatorname{cosec}^2 x} \equiv 1$

$$\frac{1}{\cos^2 x} - \frac{\sec^2 x}{\operatorname{cosec}^2 x} \equiv \sec^2 x - \frac{\left(\dfrac{1}{\cos^2 x}\right)}{\left(\dfrac{1}{\sin^2 x}\right)}$$

Use the definitions of sec x and cosec x to rearrange the expression and get rid of fractions.

$$\equiv \sec^2 x - \frac{\sin^2 x}{\cos^2 x} \equiv \sec^2 x - \frac{1}{\cos^2 x}\sin^2 x$$

$$\equiv \sec^2 x - \sec^2 x \sin^2 x$$

$$\equiv \sec^2 x (1 - \sin^2 x)$$

$$\equiv \sec^2 x \cos^2 x$$

Use $\sin^2 x + \cos^2 x \equiv 1$

$$\equiv \frac{1}{\cos^2 x} \times \cos^2 x \equiv 1$$

Use the definition of sec x.

The identity is proved.

Inverse functions

The trigonometric functions have **inverse functions**. They can be written in two ways:

> **Key point**
>
> $\arcsin x$ or $\sin^{-1} x$ $\quad$ $\arccos x$ or $\cos^{-1} x$ $\quad$ $\arctan x$ or $\tan^{-1} x$

For example, $\sin \dfrac{\pi}{6} = 0.5$, so $\arcsin 0.5 = \dfrac{\pi}{6}$

Another way of thinking about arcsin 0.5 is 'the angle whose sine is 0.5'

Do not confuse the inverse $\sin^{-1} x$ with the reciprocal $(\sin x)^{-1}$

See Ch12.2

For a reminder on inverse functions.

The functions $\sin x$, $\cos x$ and $\tan x$ are many-to-one functions, so their inverses have many possible values (for example, arcsin 0.5 = 30°, 150°, 330°, ...). However, inverse functions must have one-to-one mapping. This means that the domain of the original function must be restricted to ensure that it is one-to-one.

> For the reflections below to work, axes must have the same scale and the angles must be in radians.

The graphs of the inverse functions are reflections of the original trigonometric graphs in the line $y = x$

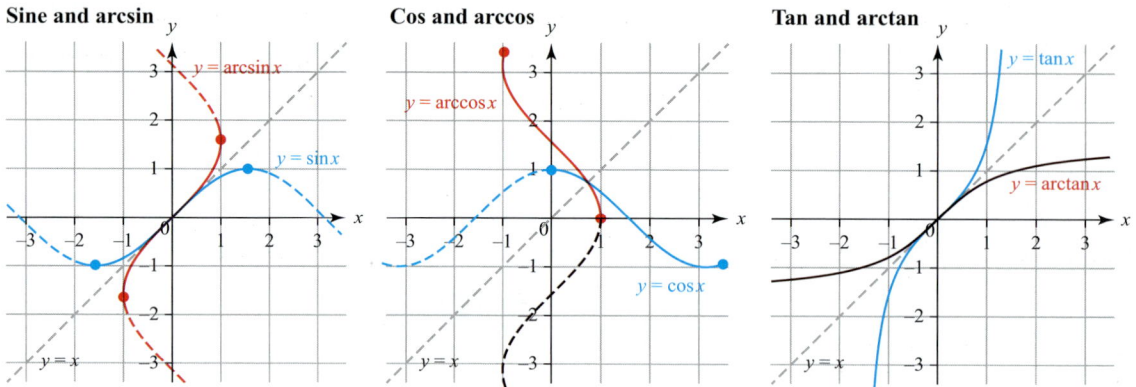

Sine and arcsin   Cos and arccos   Tan and arctan

The graphs show the restricted domains and ranges as follows.

	$y = \sin x$	$y = \arcsin x$	$y = \cos x$	$y = \arccos x$	$y = \tan x$	$y = \arctan x$
Domain	$-\dfrac{\pi}{2} \leq x \leq \dfrac{\pi}{2}$	$-1 \leq x \leq 1$	$0 \leq x \leq \pi$	$-1 \leq x \leq 1$	$-\dfrac{\pi}{2} \leq x \leq \dfrac{\pi}{2}$	$x \in \mathbb{R}$
Range	$-1 \leq y \leq 1$	$-\dfrac{\pi}{2} \leq y \leq \dfrac{\pi}{2}$	$-1 \leq y \leq 1$	$0 \leq y \leq \pi$	$y \in \mathbb{R}$	$-\dfrac{\pi}{2} < y < \dfrac{\pi}{2}$

The **principal value** of arcsin x, for a given value of x, is the unique value of arcsin x within the allowed range $-\dfrac{\pi}{2} \leq y \leq \dfrac{\pi}{2}$

Similarly, the principal values of arccos x are in the range $0 \leq y \leq \pi$ and the principal values of arctan x are in the range $-\dfrac{\pi}{2} \leq y \leq \dfrac{\pi}{2}$

Example 3

Prove that $\arcsin\left(-\dfrac{1}{\sqrt{2}}\right) = -\dfrac{\pi}{4}$

Let $\arcsin\left(-\dfrac{1}{\sqrt{2}}\right) = \alpha$

So $\sin\alpha = \left(-\dfrac{1}{\sqrt{2}}\right)$

> You need to find the angle whose sine is $-\dfrac{1}{\sqrt{2}}$

Angle α is a principle value of arcsin and so it lies between $-\dfrac{\pi}{2}$ and $\dfrac{\pi}{2}$ in the 1st or 4th quadrants.

> See the graphs and table shown just before this example.

See Ch3.1

For a reminder on quadrants.

However, as $\sin\alpha$ is negative, α must lie in the 4th quadrant, not the 1st quadrant.

Using the special triangle, $\alpha = -\dfrac{\pi}{4}$ radians

So, $\arcsin\left(-\dfrac{1}{\sqrt{2}}\right) = -\dfrac{\pi}{4}$

$\sqrt{2}$ 1 $45° = \dfrac{\pi}{4}$ rad   1

> Recall this special triangle.

1 Use the graphs following Example 1 to help you answer this question.

When the original trigonometric functions, $\sin\theta$, $\cos\theta$ and $\tan\theta$, have values of 0, +1 and −1, what can you say about the graphs of their reciprocal functions?

2 Use your calculator to find, to 3 sf, the values of

a $\sec 200°$ b $\cot 130°$ c $\csc 340°$

d $\sec\dfrac{3\pi}{5}$ e $\cot\dfrac{5\pi}{6}$ f $\csc\dfrac{2\pi}{9}$

3 Copy and complete this table, giving answers in exact form.

x, degrees	x, radians	cosec x	sec x	cot x
30°				
60°				
90°				

4 Without using a calculator, write down the exact values of

a $\cot 135°$ b $\sec 120°$ c $\csc 210°$

d $\cot\dfrac{4\pi}{3}$ e $\sec\dfrac{7\pi}{4}$ f $\csc\dfrac{3\pi}{2}$

5 Solve for $-180° \le x \le 180°$, showing your working.

a $\csc x = 2$ b $\sec x = 4$

c $\cot x = -1$ d $\sec(x-10°) = 3$

e $\csc(x+10°) = -4$ f $\cot(x+30°) = -2$

g $\sec 3x = \sqrt{2}$ h $\csc 2x = 2$

6 Simplify these expressions.

a $\cot\theta\tan\theta$ b $\cot\theta\sin\theta\tan\theta$

c $\cot^2\theta\sec\theta\sin\theta$ d $\csc\theta\sec\theta\sin^2\theta$

7 Use your calculator to evaluate

a $\tan^{-1}3$ b $\cos^{-1}0.25$

c $\sin^{-1}\left(\dfrac{3}{4}\right)$ d $\cos^{-1}(-0.4)$

e $\tan^{-1}(-2)$ f $\sin^{-1}\left(-\dfrac{1}{4}\right)$

8 Giving your answers in terms of π, find the exact values of

a $\arcsin\dfrac{1}{\sqrt{2}}$ b $\arctan\sqrt{3}$

c $\arccos 1$ d $\arctan(-1)$

e $\arccos\dfrac{\sqrt{3}}{2}$ f $\arcsin 0$

g $\arccos\left(-\dfrac{1}{\sqrt{2}}\right)$ h $\arctan\left(-\dfrac{1}{\sqrt{3}}\right)$

9 Sketch the curves on two sets of axes and find algebraically the points of intersection of these pairs of curves for $0 < x < \pi$

a $y = 1 + 2\tan x$, $y = 1 + \sec x$

b $y = \sec x$, $y = 2 - \cos x$

Check your answers on a graphics calculator.

10 Solve these equations for $0 \le x \le 180°$, showing your working.

a $\sin^{-1}\left(\dfrac{1}{2}\right) = x$

b $\sin^{-1}\left(\dfrac{\sqrt{3}}{2}\right) = x - 30°$

c $\cos^{-1}0.4 = x$

d $\cos^{-1}0.7 = x + 10°$

e $\tan^{-1}1 = x$

f $\tan^{-1}(-1) = x - 50°$

g $\sin^{-1}\left(\dfrac{1}{\sqrt{2}}\right) = x + 45°$

h $\cos^{-1}\left(\dfrac{1}{2}\right) = x - 45°$

11 Find the values, in degrees, of

a $\arccos\dfrac{\sqrt{3}}{2} + \arcsin\left(-\dfrac{1}{2}\right)$

b $\arctan 1 - \arctan(-1)$

12 Describe the transformations which map

a The graph of $y = \sec x$ onto

i $y = \sec(x - 45°)$ ii $y = 2\sec\left(\dfrac{1}{3}x\right)$

b The graph of $y = \csc x$ onto

i $y = 2\csc(x + 45°)$ ii $y = \csc\left(\dfrac{1}{2}x\right)$

c The graph of $y = \cot x$ onto

i $y = -\dfrac{1}{4}\cot x$ ii $y = 2 + \cot x$

13 Given that y is in radians, sketch the graphs of

a $y = \sin^{-1}\left(\dfrac{1}{2}x\right)$ b $y = \cos^{-1}(3x)$

c $y = 2\tan^{-1}\left(\dfrac{1}{3}x\right)$ d $y = 2 + \sin^{-1}(2x)$

In each case, write the exact coordinates of the end points of each graph.

Strategy

To solve problems involving reciprocal or inverse trigonometric functions

(1) Look for opportunities to rearrange or simplify functions.

(2) Use the appropriate definitions and trigonometric identities.

(3) Find values using your calculator, graphs or special triangles.

Example 4

Solve the equation $2\tan^2\theta + 3\sec\theta = 0$ for $-\pi < \theta < \pi$. Show your working.

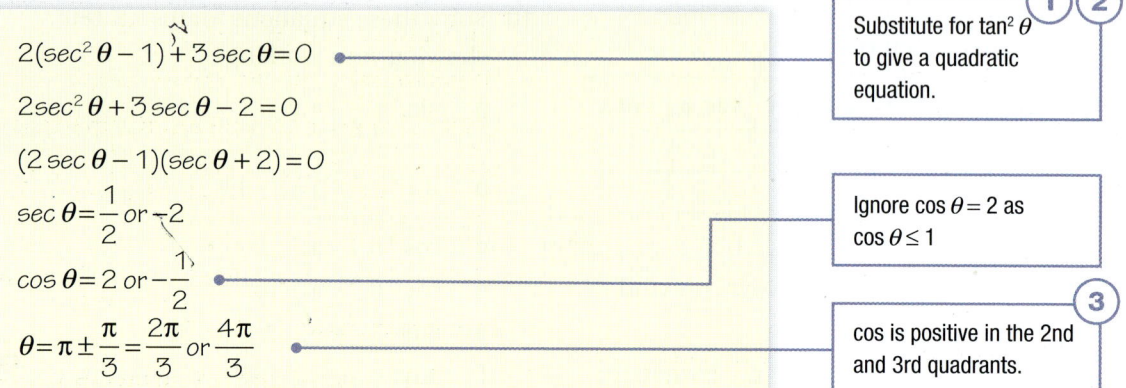

$2(\sec^2\theta - 1) + 3\sec\theta = 0$

$2\sec^2\theta + 3\sec\theta - 2 = 0$

$(2\sec\theta - 1)(\sec\theta + 2) = 0$

$\sec\theta = \dfrac{1}{2}$ or -2

$\cos\theta = 2$ or $-\dfrac{1}{2}$

$\theta = \pi \pm \dfrac{\pi}{3} = \dfrac{2\pi}{3}$ or $\dfrac{4\pi}{3}$

> (1)(2) Substitute for $\tan^2\theta$ to give a quadratic equation.

> Ignore $\cos\theta = 2$ as $\cos\theta \leq 1$

> (3) cos is positive in the 2nd and 3rd quadrants.

Example 5

Find the exact value of $\tan\left(\arcsin\dfrac{3}{4}\right)$

Find $\tan\theta$ given that $\sin\theta = \dfrac{3}{4}$

The principal value of $\arcsin\dfrac{3}{4}$ is positive, so θ is in the 1st quadrant.

$x^2 + 3^2 = 4^2$

$x = \sqrt{16-9} = \sqrt{7}$

$\tan\theta = \dfrac{3}{\sqrt{7}}$ and $\tan\left(\arcsin\dfrac{3}{4}\right) = \dfrac{3}{\sqrt{7}}$

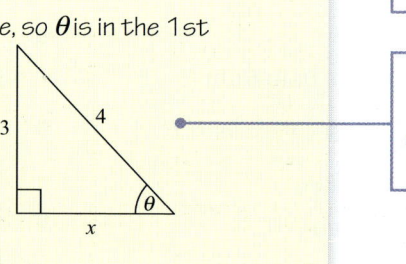

> (2) Find the tangent of the angle whose sine is $\dfrac{3}{4}$

> (3) Sketch a triangle with $\sin\theta = \dfrac{3}{4}$ and use Pythagoras' theorem.

For all questions in this exercise you must show your working.

1 Solve these equations for $-\pi < \theta < \pi$

  a $\sec^2\theta = 2$

  b $\cot^2\theta = 3$

  c $4\sin\theta = 3\operatorname{cosec}\theta$

  d $\tan\theta = 4\sin\theta\cos\theta$

  e $3\cos\theta = \sec\theta$

  f $4\cot\theta = 3\tan\theta$

  g $\cot 2\theta = \tan 2\theta$

  h $\sec 3\theta - \operatorname{cosec} 3\theta = 0$

2 Solve these equations for $0 \le \theta \le 360°$

  a $2 + \sec^2\theta = 4\tan\theta$

  b $2\cot\theta = \tan\theta + 1$

  c $3\sin\theta - 2\operatorname{cosec}\theta = 1$

  d $2\sec\theta - 1 = \tan^2\theta$

3 Eliminate the trigonometric functions from these pairs of equations.

  a $x = 4\sec\alpha,\ y = 2\tan\alpha$

  b $x = 3\operatorname{cosec}\alpha,\ y = 2\cot\alpha$

  c $x = 4\cos\alpha,\ y = 3\tan\alpha$

  d $x = 1 - \cot\alpha,\ y = 1 + \operatorname{cosec}\alpha$

4 **a** Angle α is acute and $\sin\alpha = \dfrac{4}{5}$

    Find the value of **i** $\operatorname{cosec}\alpha$ **ii** $\cot\alpha$

  b Angle θ is obtuse and $\tan\theta = -\dfrac{8}{15}$

    Find the value of **i** $\sec\theta$ **ii** $\cot\theta$

  c Angle β is reflex and $\cos\beta = \dfrac{9}{41}$

    Find the value of **i** $\operatorname{cosec}\beta$ **ii** $\cot\beta$

5 Write each of these expressions as a power of $\sec\alpha$, $\operatorname{cosec}\alpha$ or $\cot\alpha$

  a $\dfrac{1}{\tan^2\alpha}$ **b** $\dfrac{\sec\alpha}{\cos^2\alpha}$ **c** $\dfrac{1-\cos^2\alpha}{\sin^4\alpha}$

6 Find the value of $\cot\theta$ when

  a $2\sin\theta = 3\cos\theta$ **b** $4\tan\theta = 1$

7 Prove these identities.

  a $\tan x\sin x + \cos x \equiv \sec x$

  b $\dfrac{\cos x\tan x}{\operatorname{cosec}^2 x} \equiv \dfrac{\cos^3 x}{\cot^3 x}$

  c $\tan\theta + \cot\theta \equiv \sec\theta\operatorname{cosec}\theta$

  d $\sec\theta - \tan\theta \equiv \dfrac{1}{\sec\theta + \tan\theta}$

  e $(1 + \sec\theta)(1 - \cos\theta) \equiv \tan\theta\sin\theta$

8 The force $F = 20\sin(3t - 1.5)$ N acts on an oscillating object where t is the time in seconds. Make t the subject of the formula and find its value when $F = 15\,\text{N}$.

9 If $\theta = \arcsin x$, find $\tan\theta$ in terms of x

10 Given that $x = \tan\theta$, find $\operatorname{arccot} x$ in terms of θ

11 Show that, if $f(\theta) = \dfrac{1}{1+\sin\theta} + \dfrac{1}{1-\sin\theta} = 8$,

then $\sec^2\theta = 4$

Hence, solve the equation $f\left(x - \dfrac{\pi}{6}\right) = 8$ where x is in radians and $0 < x < \pi$

12 **a** If $\arcsin x = \dfrac{\pi}{5}$, prove that $\arccos x = \dfrac{3\pi}{10}$

  b If $2\arcsin x = \arccos x$, find the value of x

Challenge

13 Prove these identities.

  a $\dfrac{\sin\theta}{1+\cos\theta} + \dfrac{1+\cos\theta}{\sin\theta} \equiv 2\operatorname{cosec}\theta$

  b $\dfrac{1}{\cot\theta + \operatorname{cosec}\theta} \equiv \dfrac{1-\cos\theta}{\sin\theta}$

  c $\dfrac{\tan^2\theta + \cos^2\theta}{\sin\theta + \sec\theta} \equiv \sec\theta - \sin\theta$

14 Given that $f(\theta) = 4\sec^2\theta - \tan^2\theta = k$, show that $\operatorname{cosec}^2\theta = \dfrac{k-1}{k-4}$

Hence, solve the equation $f\left(x + \dfrac{\pi}{4}\right) = 7$ where x is in radians and $0 < x < \pi$

Fluency and Skills

A **compound angle** is the result of adding or subtracting two (or more) angles.

There are several identities that use compound angles.

You can prove that $\sin(A+B) \neq \sin A + \sin B$ by counter-example.

Substituting $A = B = 45°$ gives $\sin(45° + 45°) = \sin 90° = 1$

whereas $\sin 45° + \sin 45° = 2 \times \dfrac{1}{\sqrt{2}} = \sqrt{2}$ and $1 \neq \sqrt{2}$

To find the correct expansion of $\sin(A+B)$, you can use this diagram where $\angle ROP$ is the compound angle $A + B$

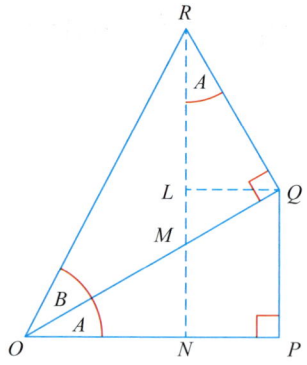

Triangles MON and MRQ are similar, so $\angle MRQ = \angle MON =$ angle A

In $\triangle ORN$, $\sin(A+B) = \dfrac{NR}{OR} = \dfrac{NL + LR}{OR}$

$$= \dfrac{PQ}{OR} + \dfrac{LR}{OR}$$

$$= \dfrac{PQ}{OQ} \times \dfrac{OQ}{OR} + \dfrac{LR}{RQ} \times \dfrac{RQ}{OR}$$

$$= \sin A \cos B + \cos A \sin B$$

> This step introduces OQ and RQ to use in $\triangle OQP$ and $\triangle RQL$

Although proved here for A and B as acute angles, this formula is true for all values of A and B

Replacing B by $-B$ gives $\sin(A-B) = \sin A \cos B - \cos A \sin B$

There are also **compound angle formulae** for $\cos(A \pm B)$ and $\tan(A \pm B)$

> Angle $-B$ is in the 4th quadrant where sine is $-$ve and cosine is $+$ve, so $\sin(-B) = -\sin B$ and $\cos(-B) = \cos B$

Key point

$\sin(A \pm B) \equiv \sin A \cos B \pm \cos A \sin B$

$\cos(A \pm B) \equiv \cos A \cos B \mp \sin A \sin B$

$\tan(A \pm B) \equiv \dfrac{\tan A \pm \tan B}{1 \mp \tan A \tan B}$

> Take care with the signs, especially the $\mp$ signs.

Example 1

Using special triangles and showing your working, find the value of cos 15° as a surd.

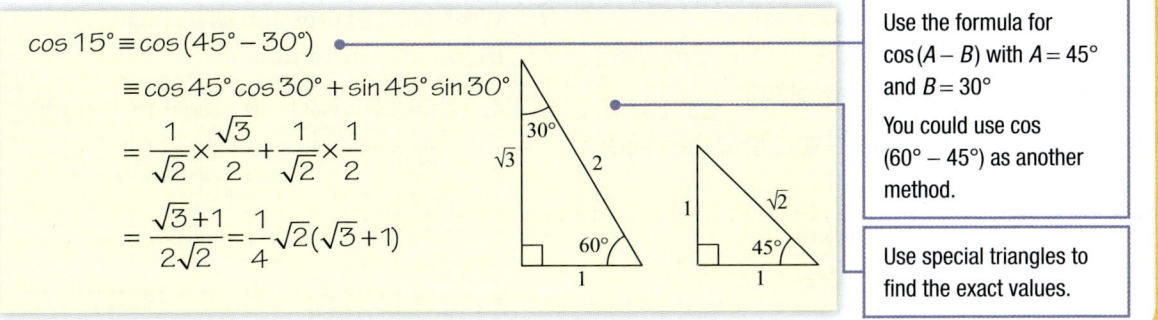

$\cos 15° \equiv \cos(45° - 30°)$

$\equiv \cos 45° \cos 30° + \sin 45° \sin 30°$

$= \dfrac{1}{\sqrt{2}} \times \dfrac{\sqrt{3}}{2} + \dfrac{1}{\sqrt{2}} \times \dfrac{1}{2}$

$= \dfrac{\sqrt{3}+1}{2\sqrt{2}} = \dfrac{1}{4}\sqrt{2}(\sqrt{3}+1)$

Use the formula for $\cos(A - B)$ with $A = 45°$ and $B = 30°$

You could use cos (60° − 45°) as another method.

Use special triangles to find the exact values.

Letting $A = B$ in the compound angle formulae creates the **double angle formulae**. $A + A = 2A$ is called a double angle.

Key point

$$\sin 2A \equiv 2 \sin A \cos A \qquad \cos 2A \equiv \cos^2 A - \sin^2 A \qquad \tan 2A \equiv \dfrac{2 \tan A}{1 - \tan^2 A}$$

You can derive three versions of the formula for $\cos 2A$, by substituting the identity $\sin^2 A + \cos^2 A \equiv 1$

Key point

$$\cos 2A = \qquad \cos^2 A - \sin^2 A \qquad \textbf{or} \qquad 2\cos^2 A - 1 \qquad \textbf{or} \qquad 1 - 2\sin^2 A$$

You can rearrange the above identities for $\cos 2A$ to give expressions for $\cos^2 A$ and $\sin^2 A$

Key point

$$\cos^2 A \equiv \dfrac{1}{2}(1 + \cos 2A) \qquad \sin^2 A \equiv \dfrac{1}{2}(1 - \cos 2A)$$

By replacing $2A$ by A, you can write **half-angle formulae** such as

Key point

$$\sin A \equiv 2 \sin \dfrac{A}{2} \cos \dfrac{A}{2} \qquad \cos A \equiv \cos^2 \dfrac{A}{2} - \sin^2 \dfrac{A}{2}$$

Example 2

For the acute angle α, find the value of

a $\sin \alpha$ when $\cos 2\alpha = \dfrac{17}{25}$ **b** $\tan 2\alpha$ when $\sin \alpha = \dfrac{1}{3}$

a $\sin^2 \alpha \equiv \dfrac{1}{2}(1 - \cos 2\alpha)$

$= \dfrac{1}{2}\left(1 - \dfrac{17}{25}\right) = \dfrac{4}{25}$

$\sin \alpha = \dfrac{2}{5}$

b $\tan 2\alpha \equiv \dfrac{2 \tan \alpha}{1 - \tan^2 \alpha} = \dfrac{2 \times \dfrac{1}{2\sqrt{2}}}{1 - \dfrac{1}{8}}$

$= \dfrac{1}{\sqrt{2}} \times \dfrac{8}{7} = \dfrac{4}{7}\sqrt{2}$

$\sqrt{8} = 2\sqrt{2}$

Choose the correct formula.

Substitute and simplify.

α is acute, so ignore $-\dfrac{2}{5}$

$\sin \alpha = \dfrac{1}{3}$ so use a right-angled triangle and Pythagoras' theorem to work out $\tan \alpha$

For questions that ask for exact values, remember to show your working. You can check your final answer using a calculator.

1 **a** By writing $15° = 60° - 45°$, find the exact value of $\sin 15°$

   b Find the exact values of

   i $\cos 75°$ **ii** $\tan 75°$ **iii** $\tan 105°$

2 Write each expression as a single trigonometric ratio and find the exact value.

   a $\sin 35° \cos 10° + \cos 35° \sin 10°$

   b $\sin 70° \cos 10° - \cos 70° \sin 10°$

   c $\cos 40° \cos 10° + \sin 40° \sin 10°$

   d $\dfrac{\tan 75° - \tan 45°}{1 + \tan 75° \tan 45°}$

3 Simplify

   a $\sin 2\theta \cos \theta + \cos 2\theta \sin \theta$

   b $\cos 3\theta \cos \theta - \sin 3\theta \sin \theta$

   c $\dfrac{\tan 2x + \tan x}{1 - \tan 2x \tan x}$ **d** $\dfrac{1 + \tan 3x \tan x}{\tan 3x - \tan x}$

4 Write each expression as a single trigonometric ratio.

   a $\dfrac{\sqrt{3}}{2}\cos x + \dfrac{1}{2}\sin x$ **b** $\dfrac{\sqrt{3} + \tan x}{1 - \sqrt{3}\tan x}$

   c $\dfrac{1}{\sqrt{2}}\cos x - \dfrac{1}{\sqrt{2}}\sin x$ **d** $\dfrac{1 + \tan x}{1 - \tan x}$

   e $\dfrac{\cot x - \sqrt{3}}{1 + \sqrt{3}\cot x}$

5 Use compound angle formulae to show that

   a $\sin(90° - A) = \cos A$

   b $\sin(180° - A) = \sin A$

   c $\cos(90° - A) = \sin A$

   d $\cos(180° - A) = -\cos A$

6 Given acute angles α and β such that $\sin \alpha = \dfrac{12}{13}$ and $\tan \beta = \dfrac{3}{4}$, use trigonometric formulae to show that

   a $\sin(\alpha + \beta) = \dfrac{63}{65}$ **b** $\tan(\alpha - \beta) = \dfrac{33}{56}$

7 Write each expression as a single trigonometric ratio.

   a $2 \sin 23° \cos 23°$ **b** $\cos^2 42° - \sin^2 42°$

   c $\dfrac{2\tan 70°}{1 - \tan^2 70°}$ **d** $2\cos^2 50° - 1$

   e $2 \sin 3\theta \cos 3\theta$ **f** $\sin \theta \cos \theta$

   g $\dfrac{1}{2}(1 + \cos 40°)$ **h** $1 - 2\sin^2 4\theta$

   i $1 + \cos 2\theta$ **j** $\dfrac{1}{2}(1 - \cos 50°)$

   k $\cos^2 \dfrac{\pi}{5} - \sin^2 \dfrac{\pi}{5}$ **l** $\sec \theta \operatorname{cosec} \theta$

   m $\cot \theta - \tan \theta$ **n** $\dfrac{1 - \tan^2 4\theta}{2\tan 4\theta}$

8 Write each expression as a single trigonometric ratio and find the exact value.

   a $2\sin \dfrac{\pi}{12}\cos \dfrac{\pi}{12}$ **b** $2\cos^2 \dfrac{\pi}{8} - 1$

   c $1 - 2\sin^2 \dfrac{\pi}{4}$ **d** $\dfrac{1 - \tan^2 22\frac{1}{2}°}{2\tan 22\frac{1}{2}°}$

9 Use trigonometric formulae to find exact values of $\sin 2\theta$ and $\sin \dfrac{1}{2}\theta$ when θ is acute and

   a $\cos \theta = \dfrac{3}{5}$ **b** $\tan \theta = \dfrac{5}{12}$

10 Find, in surd form, the values of $\sin \dfrac{x}{2}$, $\cos \dfrac{x}{2}$ and $\tan \dfrac{x}{2}$ when x is acute and

   a $\cos x = \dfrac{1}{9}$ **b** $\sin x = \dfrac{3}{5}$

11 **a** Find the exact value of $\sin \theta$ when θ is acute and $\cos 2\theta = \dfrac{39}{49}$

   b Find $\cos \theta$ when θ is acute and $\tan 2\theta = \dfrac{3}{4}$

12 **a** If $\tan(\alpha + \beta) = 4$ and $\tan \alpha = 3$, find $\tan \beta$

   b If $\sin(\alpha + \beta) = \cos \beta$ and $\sin \alpha = \dfrac{3}{5}$, find $\tan \beta$

   c If $\tan(\alpha - \beta) = 5$, find $\tan \alpha$ in terms of $\tan \beta$

Reasoning and problem-solving

To solve problems involving compound trigonometric angles

(1) Decide on the method you will use for a particular problem.

(2) Use the appropriate angle formulae.

(3) Simplify and manipulate the equation or expression.

Example 3

Solve the equation $\sin(\theta - 30°) = 3\cos\theta$ for $0 \le \theta \le 360°$. Show your working.

$\sin\theta\cos 30° - \cos\theta\sin 30° = 3\cos\theta$ ●—— (2) Use $\sin(A - B) \equiv \sin A\cos B - \cos A\sin B$

$\sin\theta \times \dfrac{\sqrt{3}}{2} - \cos\theta \times \dfrac{1}{2} = 3\cos\theta$ ●—— $\cos 30° = \dfrac{\sqrt{3}}{2}$ and $\sin 30° = \dfrac{1}{2}$

$\dfrac{\sqrt{3}}{2}\sin\theta = \dfrac{7}{2}\cos\theta \quad \Rightarrow \quad \dfrac{\sin\theta}{\cos\theta} = \dfrac{\frac{7}{2}}{\frac{\sqrt{3}}{2}}$

$\tan\theta = \dfrac{7}{\sqrt{3}}$ ●—— (3) Simplify the equation.

$\theta = 76.1°$ or $256.1°$ ●—— $\tan\theta$ is +ve, so θ is in the 1st and 3rd quadrants.

Example 4

Prove the identity $\cot\theta - \tan\theta \equiv 2\cot 2\theta$

$\text{LHS} \equiv \dfrac{\cos\theta}{\sin\theta} - \dfrac{\sin\theta}{\cos\theta} \equiv \dfrac{\cos^2\theta - \sin^2\theta}{\sin\theta\cos\theta}$ ●—— (2) Use $\cos^2 A - \sin^2 A \equiv \cos 2A$

$\equiv \dfrac{\cos 2\theta}{\sin\theta\cos\theta}$

$\equiv \dfrac{2\cos 2\theta}{\sin 2\theta}$ ●—— (2) Use $2\sin A\cos A \equiv \sin 2A$

$\equiv 2\cot 2\theta \equiv \text{RHS}$ ●—— Recognise $\dfrac{\cos 2\theta}{\sin 2\theta} \equiv \cot 2\theta$

The identity is proved.

Example 5

Use the identity $\cos 2\theta \equiv \cos^2\theta - \sin^2\theta$ to find a Cartesian equation for the curve given by parametric equatins $x - \sin\theta°$, $y = \cos 2\theta°$

$\cos 2\theta \equiv \cos^2\theta - \sin^2\theta \equiv (1 - \sin^2\theta) - \sin^2\theta$ ●—— Use the identity $\sin^2\theta + \cos^2\theta \equiv 1$

$\equiv 1 - 2\sin^2\theta$ ●—— Write the identity in terms of $\sin\theta$ and $\cos 2\theta$

$so\ y = 1 - 2x^2$

1 Solve these equations for $0 \le \theta \le 180°$. Show your working.

    a $\quad 3\sin\theta = \sin(\theta + 45°)$

    b $\quad 2\cos\theta = \cos(\theta + 30°)$

    c $\quad 2\sin\theta + \sin(\theta + 60°) = 0$

2

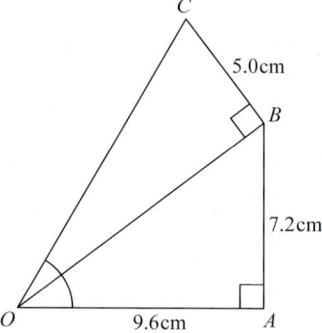

    a Calculate lengths OB and OC in quadrilateral $OABC$

    b Hence use compound angle formulae to calculate the sine, cosine and tangent of angle AOC

3 Prove these identities.

    a $\quad \tan A + \cot A \equiv 2\operatorname{cosec} 2A$

    b $\quad \operatorname{cosec} 2A + \cot 2A \equiv \cot A$

    c $\quad 2\operatorname{cosec} 2A \equiv \sec A \operatorname{cosec} A$

    d $\quad \operatorname{cosec} A - \cot A \equiv \tan\dfrac{A}{2}$

4 If $\sin(\theta + \phi) = \cos\phi$, show that $\tan\theta + \tan\phi = \sec\theta$

5 Prove that
$$\arcsin\frac{1}{2} + \arcsin\frac{1}{3} = \arcsin\left(\frac{2\sqrt{2} + \sqrt{3}}{6}\right)$$

6 **a** Angle A is obtuse and angle B is acute such that $\tan A = -2$ and $\tan B = \sqrt{5}$. Use trigonometric formulae to find the values, in surd form, of

      i $\cot(A + B)$ **ii** $\sin(A + B)$

    b If $\tan\left(\theta + \dfrac{\pi}{3}\right) = \dfrac{1}{3}$, show that
$$\tan\theta = 2 - \frac{5}{3}\sqrt{3}$$

7 Solve these equations for $0 \le \theta \le 2\pi$. Show your working.

    a $\tan 2\theta = 3\tan\theta$ **b** $\cos 2\theta + 3\sin\theta = -1$

    c $\sin 2\theta - 1 = \cos 2\theta$ **d** $\sin 2\theta + \sin\theta = \tan\theta$

    e $\cos 2\theta = \tan 2\theta$ **f** $2\sin 2\theta = \tan\theta$

8 Solve these equations for $0 \le \theta \le 360°$. Show your working.

    a $\sin\theta = \sin\dfrac{\theta}{2}$ **b** $\tan\theta = 6\tan\dfrac{\theta}{2}$

    c $3\cos\dfrac{\theta}{2} = 2 + \cos\theta$ **d** $\sin\theta = \cot\dfrac{\theta}{2}$

9 Find the Cartesian equation for the curve that has the following parametric equations.
$$x = 4\cos\theta \qquad y = 2\cos 2\theta$$

10 A curve is defined by the parametric equations $x = 3\sec\theta$, $y = 6\tan\theta$. Find a Cartesian equation for the curve.

11 **a** Prove that $\tan(A + B) \equiv \dfrac{\tan A + \tan B}{1 - \tan A\tan B}$

    b Find a similar formula for $\tan(A - B)$

12 Prove these identities.

    a $\dfrac{\sin(\theta + \phi)}{\cos\theta\cos\phi} \equiv \tan\theta + \tan\phi$

    b $\cot\theta \equiv \dfrac{\sin 2\theta + \cos\theta}{1 + \sin\theta - \cos 2\theta}$

    c $\dfrac{\cos\theta + \sin\theta}{\cos\theta - \sin\theta} \equiv \sec 2\theta + \tan 2\theta$

    d $\dfrac{1 + \sin 2\theta}{1 - \sin 2\theta} \equiv \tan^2\left(\theta + \dfrac{\pi}{4}\right)$

    e $\dfrac{\cos^3\theta + \sin^3\theta}{\cos\theta + \sin\theta} \equiv 1 - \dfrac{1}{2}\sin 2\theta$

    f $\dfrac{\sin 2\theta}{1 - \cos 2\theta} \equiv \cot\theta$

    g $\sin 2\theta \equiv \dfrac{2\tan\theta}{1 + \tan^2\theta}$

    h $\dfrac{\sin 2\theta + \sin\theta}{\cos 2\theta + \cos\theta + 1} \equiv \tan\theta$

13 Prove that $\cos^4\theta + 2\sin^2\theta - \sin^4\theta \equiv 1$

14 Prove that $\dfrac{1 - \tan 15°}{1 + \tan 15°} = \dfrac{1}{\sqrt{3}}$

15 Given that $\sin\theta\tan\theta = 2 - \cos\theta$, prove that $\cos\theta = 0.5$

Hence, solve the equation $\sin\phi\tan\phi = 2 - \cos\phi$ for $0 \le \phi < \pi$

16 Given that $3\cot\theta = 8\sec\theta$, prove that $3\sin^2\theta + 8\sin\theta - 3 = 0$

Hence solve the equation $3\cot\theta = 8\sec\theta$ for $0 < \theta < 180°$. Show your working.

17 a Prove that
$$(\cos x + \cos y)^2 + (\sin x + \sin y)^2 = 4\cos^2\left(\frac{x-y}{2}\right)$$

b Use this result to prove that
$$\cos 15° = \frac{\sqrt{2}+\sqrt{6}}{4}$$

18 a Use double angle formulae to prove that
$$\sin A + \sin B = 2\sin\left(\frac{A+B}{2}\right)\cos\left(\frac{A-B}{2}\right)$$

b Solve the equation $\sin x + \sin y = 0$ for $0 \le x < \pi$ and $0 \le y < \pi$. Show your working.

19 Find all the solutions of the equation $\operatorname{cosec}^2\theta - \frac{1}{2}\cot\theta = 3$ for $0 \le \theta \le 2\pi$. Show your working.

20 a Show that the equation
$$\frac{1}{1-\sin x} + \frac{1}{1+\sin x} = 8 \text{ can be rearranged}$$
as $\sec^2 x = 4$

b Use your result to solve the equation
$$\frac{1}{1-\sin(2\theta+10°)} + \frac{1}{1+\sin(2\theta+10°)} = 8 \text{ for}$$
$0 \le \theta \le 180°$. Show your working.

21 The curve with the parametric equations $x = 4k\sin t$, $y = 3 + k\cos 2t$ intersects the y-axis at the point $(0, 5)$. If k is a non-zero constant, find the value of k.

22 A curve has the Cartesian equation $y = \frac{4\sqrt{9-x^2}}{x}$. Two parametric equations also define this curve. If one of these equations is $x = 3\sin t$, find the other equation

23 a Given that $3\operatorname{cosec}^2 x - \cot^2 x = n$, use the identity $1 + \cot^2 x = \operatorname{cosec}^2 x$ to show that $\sec^2 x = \frac{n-1}{n-3}$, provided that $n \ne 3$

b Use your answer to **a** to solve the equation $3\operatorname{cosec}^2(\theta+30°) - \cot^2(\theta+30°) = 5$, giving your answers in degrees where $0 \le \theta \le 180°$. Show your working.

24 Two seas, with different tidal patterns, meet in a strait. The depth of each sea is given by $d_1 = 6 + 4\sin\left(t \times \frac{\pi}{6}\right)$ and $d_1 = 6 + 2\sin\left[(t-1) \times \frac{\pi}{6}\right]$, respectively, where t is the number of hours after midnight. The tides clash except when their depths are equal. At what times between 00:00 and 12:00 is there no clash? Show your working.

Challenge

25 a Prove these identities.

  i $\sin 3A \equiv 3\sin A - 4\sin^3 A$

  ii $\cos 3A \equiv 4\cos^3 A - 3\cos A$

b Express $\tan 3A$ in terms of $\tan A$

c Find identities for $\sin 4A$ and $\cos 4A$ in terms of $\sin A$ and $\cos A$

26 a Given that $2\arctan 3 = \operatorname{arccot} x$, substitute $\tan\alpha = 3$ and $\cot\beta = x$ to explain why $\tan 2\alpha = \tan\beta$

Hence use trigonometric formulae to show that $x = -\frac{4}{3}$

b Use similar substitutions to show the solution of
$$\arcsin x + \arccos\frac{12}{13} = \arcsin\frac{4}{5}$$
is $x = \frac{33}{65}$

c Solve this equation for x
$$x = \arcsin k + \arcsin\sqrt{1-k^2}$$

Equivalent forms for $a \cos \pi + b \sin \pi$

Fluency and Skills

Compare the graphs of $y = 3 \cos \theta$, $y = 4 \sin \theta$ and $y = 3 \cos \theta + 4 \sin \theta$

For each θ-value on the graph, you can add the two y-values on the blue and green curves to give the y-value on the black curve. Check this result by drawing these three graphs on a graphics calculator.

The graph of $y = 3 \cos \theta + 4 \sin \theta$ can also be created from the graph of $y = \sin \theta$ by

(a) Stretching, scale factor 5, parallel to the y-axis, then

(b) Translating by about $37°$ to the left. (You will work out the exact value of the translation in Example 1.)

So, $3 \cos \theta + 4 \sin \theta = 5 \sin(\theta + 37°)$

You can check this result using a graphics calculator.

Similarly, you can use a graphics calculator to show that
$3 \cos \theta + 4 \sin \theta = 5 \cos(\theta - 53°)$

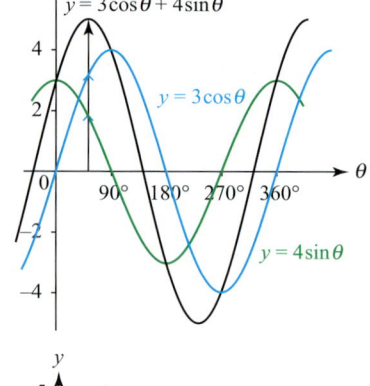

> **Key point**
>
> $a \cos \theta \pm b \sin \theta$ can be written as $r \sin(\theta \pm \alpha)$ or $r \cos(\theta \pm \alpha)$, where r is positive and angle α is acute.

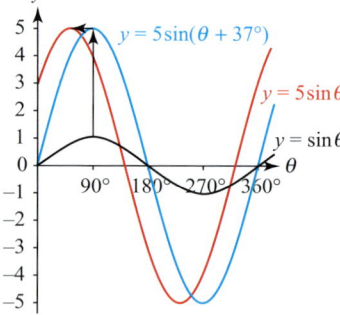

Example 1

If $4 \sin \theta + 3 \cos \theta = r \sin(\theta + \alpha)$, find r and α such that $r > 0$ and α is acute.

$4 \sin \theta + 3 \cos \theta = r \sin \theta \cos \alpha + r \cos \theta \sin \alpha$

$\qquad 4 = r \cos \alpha \quad [1]$

$\qquad 3 = r \sin \alpha \quad [2]$

$\qquad \dfrac{3}{4} = \dfrac{r \sin \alpha}{r \cos \alpha} = \tan \alpha$

$\qquad \alpha = \arctan\left(\dfrac{3}{4}\right) = 36.9°$

$3^2 + 4^2 = r^2(\sin^2 \alpha + \cos^2 \alpha)$

$\qquad = r^2$

$r = \sqrt{25} = 5$

Hence, $4 \sin \theta + 3 \cos \theta = 5 \sin(\theta + 36.9°)$

Use a compound angle formula to expand the right-hand side.

Compare coefficients of $\sin \theta$ and $\cos \theta$

Divide [2] by [1] and use $\tan \alpha \equiv \dfrac{\sin \alpha}{\cos \alpha}$

α is acute.

Square [1] and [2] and add to find r

Use $\sin^2 \alpha + \cos^2 \alpha \equiv 1$

Ignore $r = -5$ as r is +ve.

PURE

1 In each case, find the values of r and α where $r > 0$ and α is acute.

Give r as a surd where appropriate and give α in degrees.

a $5\sin\theta + 12\cos\theta = r\sin(\theta+\alpha)$

b $4\cos\theta + 3\sin\theta = r\cos(\theta-\alpha)$

c $\cos\theta - 2\sin\theta = r\cos(\theta+\alpha)$

d $4\sin\theta - 2\cos\theta = r\sin(\theta-\alpha)$

e $3\sin\theta - 4\cos\theta = r\sin(\theta+\alpha)$

f $8\cos\theta + 15\sin\theta = r\cos(\theta+\alpha)$

g $2\cos 3\theta + 5\sin 3\theta = r\sin(3\theta+\alpha)$

2 Find the value of r so that

a $\cos\theta + \sin\theta = r\sin\left(\theta + \dfrac{\pi}{4}\right)$

b $\cos\theta - \sin\theta = r\cos\left(\theta + \dfrac{\pi}{4}\right)$

3 Find the value of α so that

a $\sqrt{3}\cos\theta + \sin\theta = 2\sin(\theta+\alpha)$

b $\cos\theta - \sqrt{3}\sin\theta = 2\cos(\theta+\alpha)$

4 **a** Use a graphics calculator or otherwise to draw the graph of $y = 12\sin\theta + 5\cos\theta$

If $y = r\sin(\theta+\alpha)$, write the values of r and α from your calculator display.

Check your answers algebraically.

b Repeat for $y = 4\sin\theta - 5\cos\theta$

5 The graph of $y = \sin\theta$ is stretched by a scale factor r parallel to the y-axis, and then translated by the vector $\alpha\mathbf{i} + 0\mathbf{j}$ to create the graph of $y = r\sin(\theta-\alpha)$, where α is in degrees. Describe the two translations which, when applied to the graph of $y = \sin\theta$, transform it into the graph of $y = 5\cos\theta + 12\sin\theta$

Reasoning and problem-solving

Strategy

To solve problems of the form $a\cos\theta + b\sin\theta = c$

(1) Choose from $r\sin(\theta\pm\alpha)$ and $r\cos(\theta\pm\alpha)$ by matching the signs.

(2) Expand using the compound angle formula.

(3) Evaluate the values of r and α by matching coefficients.

(4) Use $r\sin(\theta\pm\alpha)$ or $r\cos(\theta\pm\alpha)$ to solve equations or find stationary points.

You have four choices when converting $a\cos\theta + b\sin\theta$ to either $r\sin(\theta\pm\alpha)$ or $r\cos(\theta\pm\alpha)$

The algebra is easier if you match the signs. For example, since $2\sin\theta - 3\cos\theta$ has a negative sign, you should use $r\sin(\theta-\alpha)$ or $r\cos(\theta+\alpha)$ as they too have a negative sign when expanded. If instead you choose $r\sin(\theta+\alpha)$ or $r\cos(\theta-\alpha)$ for $2\sin\theta - 3\cos\theta$, you get a different sign when you expand and α will be negative.

Example 2

a Solve the equation $2\cos\theta + 3\sin\theta = 1$ for θ between 0 and 360°. Show your working.

b Find the maximum value of $2\cos\theta + 3\sin\theta$ and the smallest positive value of θ at which the maximum occurs.

a $2\cos\theta + 3\sin\theta = r\cos(\theta - \alpha)$ where α is acute

$= r\cos\theta\cos\alpha + r\sin\theta\sin\alpha$

$r\cos\alpha = 2$ [1]

$r\sin\alpha = 3$ [2]

$r^2(\cos^2\alpha + \sin^2\alpha) = 2^2 + 3^2 = 13$

$r = \sqrt{13}$

$\dfrac{r\sin\alpha}{r\cos\alpha} = \tan\alpha = \dfrac{3}{2}$

$\alpha = \arctan\dfrac{3}{2} = 56.3°$

The equation is now $\sqrt{13}\cos(\theta - 56.3°) = 1$

$\cos(\theta - 56.3°) = \dfrac{1}{\sqrt{13}} = 0.2773\ldots$

$\theta - 56.3° = 73.9°$ or $360° - 73.9° = 286.1°$

$\theta = 73.9° + 56.3°$ or $286.1° + 56.3°$

$\theta = 130.2°$ or $342.4°$

b $2\cos\theta + 3\sin\theta = \sqrt{13}\cos(\theta - 56.3°)$

The maximum value of $2\cos\theta + 3\sin\theta$ is $\sqrt{13} \times 1 = \sqrt{13}$

Maximum occurs when $\theta - 56.3° = 0°, 360°$

The smallest positive value of θ is 56.3°

1 Choose $r\cos(\theta - \alpha)$ so that both sides have + signs. Alternatively, you could use $r\sin(\theta + \alpha)$

2 Expand using the compound angle formula.

3 Compare coefficients of $\sin\theta$ and $\cos\theta$

Square and add [1] and [2] to find r

Divide [2] by [1] and solve to find α

4 cos is +ve in 1st and 4th quadrants.

4 The maximum value of $\cos(\theta - 56.3°)$ is 1 at 0° and 360°

Exercise 14.4B Reasoning and problem-solving

1 Solve these equations for $0 \leq \theta \leq 360°$. Show your working.

a $5\cos\theta + 12\sin\theta = 6$

b $2\cos\theta - 3\sin\theta = 1$

c $8\sin\theta + 15\cos\theta = 10$

d $3\sin\theta - 5\cos\theta = 4$

e $\sin\theta - \sqrt{3}\cos\theta = 0$

f $5\sin\theta + 8\cos\theta = 0$

2 Solve these equations for $-\pi \leq \theta \leq \pi$. Show your working.

a $\cos\theta + \sqrt{3}\sin\theta = 2$

b $\cos\theta + \sin\theta = \dfrac{1}{\sqrt{2}}$

c $\sin\theta + \sqrt{3}\cos\theta = 1$

d $\sqrt{3}\cos\theta - \sin\theta = \sqrt{2}$

3 Solve these equations for $0 \leq \theta \leq 180°$. Show your working.

a $2\sin2\theta + \cos2\theta = 1$

b $2\cos3\theta - 6\sin3\theta = 5$

c $6\cos\dfrac{\theta}{2}+8\sin\dfrac{\theta}{2}=9$

d $\sin\dfrac{\theta}{2}-4\cos\dfrac{\theta}{2}=1$

4 a Use a graphics calculator or otherwise to draw the graphs of $y=\sqrt{2}\sin(\theta+45°)$ and $y=\cos(\theta-30°)$ on the same axes. Find the points of intersection in the range $-180°\le\theta\le180°$

b Use an algebraic method to calculate the same points of intersection.

5 Use a graphics calculator or otherwise to draw the graph of $y=3\sin x+4\cos x$

Find algebraically the values of x in the range $-90°\le x\le90°$ for which $y\ge3$. Show your working.

6 Use a graphics calculator or otherwise to draw the graph of $y=\sin2x+2\cos2x$. Showing your working, find algebraically the values of x in the range $0<x<180°$ for which

a $y>2$ **b** $y<-2$

7 An alternating electrical current i amps at a time t seconds (where $t\ge0$) is given by $i=12\cos3t-5\sin3t$

a What is the initial value of the current?

b How many seconds does it take for the current to fall to 5 amps?

8 a Show that $\sqrt{2}\cos\theta+\sqrt{3}\sin\theta$ can be written in the form $r\cos(\theta-\alpha)$ where $r>0$ and α is acute. Hence, find the maximum value of $\sqrt{2}\cos\theta+\sqrt{3}\sin\theta$ and the values of θ between 0 and 360° at which the maximum value occurs.

b Find the minimum value of $\dfrac{1}{\sqrt{2}\cos\theta+\sqrt{3}\sin\theta}$ and the smallest value of θ at which it occurs. You can check your answer on a graphics calculator.

9 Use the compound angle formula to find the maximum and minimum values of each expression, giving your answers in surd form if necessary. In each case, state the smallest positive value of θ at which each maximum and minimum occurs.

a $\cos\theta-\sqrt{3}\sin\theta$ **b** $24\sin\theta-7\cos\theta$

c $3\sin\theta-2\cos\theta$ **d** $8\cos2\theta-6\sin2\theta$

Check your answers using a graphics calculator.

10 a Express $7\cos\theta-24\sin\theta$ in the form $r\cos(\theta-\alpha)$, giving the values of r and θ

Show that $7\cos\theta-24\sin\theta+3\le28$ and find the minimum value of $7\cos\theta-24\sin\theta+3$

b Use a graphics calculator or otherwise to draw the graph of $\mathrm{f}(\theta)=\dfrac{1}{7\cos\theta-24\sin\theta}$ for $0\le\theta\le2\pi$

Describe the transformations required to map the graph of $y=\sec\theta$ onto the graph of $y=\mathrm{f}(\theta)$

Challenge

11 Use the compound angle formula to find the requested stationary values of these expressions and, in each case, state the smallest positive value of θ at which these values occur.

a The maximum value of

   i $20+5\sin\theta-12\cos\theta$

   ii $20-(5\sin\theta-12\cos\theta)$

b The minimum value of $\dfrac{65}{5\sin\theta-12\cos\theta}$

You can check your results on a graphics calculator.

12 Two alternating electrical currents are combined so that the resultant current I is given by $I=2\cos\omega t-4\sin\omega t$, where the constant $\omega=4$, and where $t\ge0$ is the time in seconds.

a Use the compound angle formula to find the maximum value of I and the first time at which it occurs.

b For how many seconds in each cycle is the value of I more than half its maximum value?

Chapter summary

- Angles are measured in degrees or radians. $180° = \pi$ radians.
- Arc length $= r\theta$ and sector area $= \dfrac{1}{2}r^2\theta$, with θ in radians.
- For small θ in radians, $\sin\theta \approx \theta$, $\tan\theta \approx \theta$ and $\cos\theta \approx 1 - \dfrac{1}{2}\theta^2$
- The reciprocal trigonometric functions are $\operatorname{cosec}\theta = \dfrac{1}{\sin\theta}$, $\sec\theta = \dfrac{1}{\cos\theta}$ and $\cot\theta = \dfrac{1}{\tan\theta}$
- The inverse trigonometric functions are $\arcsin x$, $\arccos x$ and $\arctan x$, also known as $\sin^{-1} x$, $\cos^{-1} x$ and $\tan^{-1} x$
- The compound angle formulae give expansions for $\sin(A \pm B)$, $\cos(A \pm B)$ and $\tan(A \pm B)$
  - $\sin(A \pm B) \equiv \sin A \cos B \pm \cos A \sin B$
  - $\cos(A \pm B) \equiv \cos A \cos B \mp \sin A \sin B$
  - $\tan(A \pm B) \equiv \dfrac{\tan A \pm \tan B}{1 \mp \tan A \tan B}$
- The double angle formulae convert trigonometric functions for $2A$ into trigonometric functions for A
  - $\sin 2A \equiv 2\sin A \cos A$
  - $\cos 2A \equiv \cos^2 A - \sin^2 A$
  - $\tan 2A \equiv \dfrac{2\tan A}{1 - \tan^2 A}$
- The expression $a\cos\theta \pm b\sin\theta$ can be rewritten in the form $r\sin(\theta \pm \alpha)$ or $r\cos(\theta \pm \alpha)$ where $r > 0$ and angle α is acute.

Check and review

You should now be able to...	Try Questions
✔ Convert between degrees and radians and use radians in problems.	1, 2, 6, 14
✔ Use reciprocal and inverse trigonometric functions.	3, 7, 9, 14, 15
✔ Use trigonometric formulae for compound angles, double angles and half angles.	4, 8, 13
✔ Find and use equivalent forms for $a\cos\theta + b\sin\theta$	11, 12
✔ Solve equations using trigonometric formulae to simplify expressions.	5, 10, 15

1 a Convert these angles to degrees, giving your answers to 1 dp.

   i 1 radian   **ii** 2.5 radians

   iii 3.5 radians

b Convert these angles to radians, giving your answers in terms of π

   i 15°   **ii** 72°   **iii** 210°

2 Find approximations for these expressions when θ is a small angle.

   a $\dfrac{\tan\theta}{\sin\theta}$ **b** $\dfrac{\sin 2\theta}{4\theta}$ **c** $\dfrac{1 - \cos 2\theta}{\theta}$

3 Use special triangles, and showing any working, write the exact values of

   a $\sec\dfrac{\pi}{3}$ **b** $\cot\dfrac{\pi}{6}$

   c $\operatorname{cosec}\dfrac{3\pi}{4}$ **d** $\operatorname{cosec}\dfrac{3\pi}{2}$

e $\arccos \dfrac{1}{\sqrt{2}}$ **f** $\arctan \dfrac{1}{\sqrt{3}}$

g $\arccos 0$ **h** $\arcsin\left(-\dfrac{1}{\sqrt{2}}\right)$

4 Write each expression as a single trigonometric ratio.

   a $\cos 40° \cos 10° - \sin 40° \sin 10°$

   b $\dfrac{1 - \tan 100° \tan 35°}{\tan 100° + \tan 35°}$

5 Solve for $0 \le x \le 180°$. Show your working.

   a $\cot 2x = \dfrac{1}{2}$ **b** $\sec(2x + 40°) = 2$

   c $3\cos x - \cot x = 0$ **d** $2\cot x = \csc x$

   e $4\cos x - \csc x = 0$

   f $\csc^2 x = 4\cot x - 3$

   g $\tan x + \cot x = 2$

6 A circle, of radius 4 cm, has a sector AOB subtending and angle of $\dfrac{\pi}{4}$ radians at the centre O

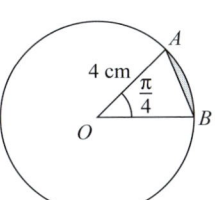

   Calculate the exact area of

   a The sector AOB

   b The shaded segment.

7 Simplify

   a $\sin x\left(\dfrac{\cot x}{\sec x} + \sin x\right)$ **b** $\dfrac{1}{\sin^2 \theta} - \dfrac{\cot \theta}{\tan \theta}$

8 Find the values of $\tan 15°$ and $\sec 75°$, writing your answers as surds.

9 Prove the identity $\arctan x \equiv \dfrac{\pi}{2} - \text{arccot}\, x$

10 Solve these equations for $0 \le x \le 180°$. Show your working.

   a $\sin(\theta - 60°) = 3\cos(\theta - 30°)$

   b $4\sin 2\theta = \sin \theta$

   c $\sin 2\theta + \sin \theta = \tan \theta$

   d $\cos 2\theta = 3 - 5\sin \theta$

11 **a** Express $7\sin \theta - 24\cos \theta$ in the form $r\sin(\theta - \alpha)$, where $r > 0$ and α is acute.

   b Hence, solve the equation $7\sin \theta - 24\cos \theta = 15$ for $0 \le \theta \le 360°$. Show your working.

12 **a** Write $7\cos \theta + 6\sin \theta$ in the form $r\sin(\theta + \alpha)$, where $r > 0$ and α is acute.

   b Hence, find the maximum value of $7\cos \theta + 6\sin \theta$ and the values of θ at which it occurs in the interval $0 \le \theta \le 360°$

13 Prove these identities.

   a $1 - \cos 2A \equiv \tan A \sin 2A$

   b $\sin 2A(1 + \tan^2 A) \equiv 2\tan A$

   c $\cot A - \tan A \equiv 2\cot 2A$

14 Sketch the graph of $y = \cos^{-1}\left(\dfrac{x}{2}\right)$ with y in radians. Write the exact coordinates of the end points of the graph.

15 If $f(x) = \dfrac{1 - \sin x}{\cos x} + \dfrac{\cos x}{1 - \sin x}$, show that $f(x) = 2\sec x$. Hence, solve the equation $f(x) = \tan^2 x - 2$ for $0 \le x < 360°$. Show your working.

What next?

	Score		
	0 – 7	Your knowledge of this topic is still developing. To improve, search in MyMaths for the codes: 2050, 2155–2160, 2262, 2266	**Click these links in the digital book**
	8 – 11	You're gaining a secure knowledge of this topic. To improve, look at the InvisiPen videos for Fluency and skills (14A)	
	12 – 15	You've mastered these skills. Well done, you're ready to progress! Now try looking at the InvisiPen videos for Reasoning and Problem-solving (14B)	

History

The demands of navigation and astronomy in 16th century Europe required the solution of problems from a branch of geometry called **spherical trigonometry**, the study of sides and angles of **spherical polygons**.

The process demanded many calculations involving multiplication and division, and so a method was needed in order to make them easier.

Prior to the invention of **logarithms** in 1614, the most common method involved the use of trigonometric tables with an identity such as

$$\cos A \cos B \equiv \frac{1}{2} \left(\cos (A + B) + \cos (A - B) \right)$$

Tycho Brahe (1546–1601) made great use of this method over a twenty year period of observations and calculations, taking great pains to achieve accurate results. On his death, he left his entire work to **Johannes Kepler**, who used it to develop his theory of planetary motion.

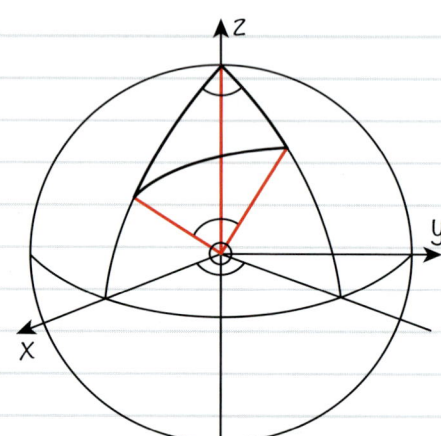

Kepler's work influenced both **Galileo** and **Newton**.

Have a go

Suppose that you need to calculate
0.4379×0.9768

By considering the identity above, let
$\cos A = 0.4379$ and $\cos B = 0.9768$

Use arccos (inverse cosine) to find A and B.
$A = 64.0300$ $B = 12.3659$

$A + B = 76.3959$ $A - B = 51.6641$

$\cos (76.3959) = 0.2352$
$\cos (51.6641) = 0.6203$

$\frac{1}{2} (0.2352 + 0.6203) = ?$

(Work this out without a calculator.)

The exact answer to the original question is
0.42774072 – how does this compare?

Try calculating 87.14×3.519 using this method.

Note

The values of the six trigonometric ratios had, by this time, been worked out *by hand* to at least 10 decimal places.

"Geometry is the archetype of the beauty of the world."
    - Johannes Kepler

Note

To work out 43.79×976.8, for example, the calculation is exactly the same, but the position of the decimal point is adjusted at the end.

In questions that tell you to show your working, you shouldn't depend solely on a calculator. For these questions, solutions based entirely on graphical or numerical methods are not acceptable.

1 a Find the exact values of x for which $\tan x = \sqrt{3}$ in the interval $-180° \leq x \leq 180°$ **[3 marks]**

   b Sketch the graph of $y = \tan 2x$ for $-90° \leq x \leq 90°$. State the equations of the asymptotes. **[3]**

2 A sector of a circle of radius 5 cm has an angle of $\dfrac{2\pi}{3}$ at its centre. Calculate the exact value of

   a The area of the sector, **[2]**

   b The arc length of the sector. **[2]**

3 A sector of a circle with radius 3 cm has an area of 5.4 cm².

   a Calculate the angle subtended at the centre of the circle. **[2]**

   b Calculate the perimeter of the sector. **[3]**

4 a State the exact value of **i** $\sec\left(\dfrac{\pi}{4}\right)$ **ii** $\cot\left(\dfrac{5\pi}{6}\right)$ **[4]**

   b Sketch the graph of $y = \operatorname{cosec} x$ for the interval $-\pi \leq x \leq \pi$
     Write the equations of the asymptotes. **[3]**

   c State the range of $y = \operatorname{cosec} x$ **[1]**

5 Solve these equations for $0 \leq x \leq 360°$. Show your working.

   a $\operatorname{cosec} x = 2$ **[3]**   b $\sec x = -\sqrt{2}$ **[3]**

6 a On the same set of axes, sketch the graphs of $y = \sin x$ for $-\dfrac{\pi}{2} \leq x \leq \dfrac{\pi}{2}$ and $y = \arcsin x$ for $-1 \leq x \leq 1$ **[3]**

   b Describe the geometric relationship between the two curves. **[1]**

   c Work out the value of these expressions giving your answers as exact multiples of π

     i $\arcsin\left(-\dfrac{\sqrt{3}}{2}\right)$ **ii** $\arcsin\left(\cos\dfrac{\pi}{3}\right)$ **[2]**

7 The velocity of a particle at time t seconds is given by $v = 5\cos\left(\dfrac{t}{2} - \dfrac{\pi}{3}\right)\,\mathrm{m\,s^{-1}}$.

   a State the maximum speed of the particle. **[1]**

   b Give the times at which this maximum speed occurs for $0 \leq t < 15$ **[3]**

8 Given that $\arccos\left(x + \dfrac{1}{2}\right) = \dfrac{\pi}{4}$, find the exact value of x **[3]**

9 a Use the identity $\sin^2\theta + \cos^2\theta \equiv 1$ to show that $\sec^2\theta - \tan^2\theta \equiv 1$ **[3]**

   b Given that $\tan\theta = \sqrt{5}$, find the exact value of

     i $\sec\theta$ **ii** $\cos\theta$ **[5]**

10 By using the formula $\cos(A \pm B) \equiv \cos A \cos B \mp \sin A \sin B$, find the exact value of

   a $\cos 75°$ **[3]**   b $\cos 15°$ **[2]**

11 Given that $\sin(A+B) \equiv \sin A \cos B + \sin B \cos A$, show that $\sin 2x \equiv 2 \sin x \cos x$ **[2]**

12 Write $6 \sin\theta + 8 \cos\theta$ in the form $r \sin(\theta + \alpha)$, where $r > 0$ and $0 < \alpha < 90°$ **[4]**

13 For $\theta = 0.05$ radians, state the approximate value of

    a $\sin\theta$ **[1]**   **b** $\cos\theta$ **[1]**   **c** $\tan\theta$ **[1]**

14 **a** Sketch the graphs of these equations for $0 \le x \le 2\pi$. Label the x- and y-intercepts with their exact values.

    i $y = \sin 3x$ **ii** $y = \cos\left(x + \dfrac{\pi}{3}\right)$ **[6]**

    b Solve the equation $\sin 3x = \dfrac{\sqrt{2}}{2}$ for $0 \le x \le \pi$. Show your working and give your answers in terms of π **[5]**

15 A triangle ABC has $AB = 10\,\text{cm}$, $BC = 13\,\text{cm}$ and $\angle BCA = \dfrac{\pi}{4}$

    a Calculate the possible lengths of AC **[5]**

    b What is the minimum possible area of the triangle? **[3]**

16 The area of an isosceles triangle is $100\,\text{cm}^2$. Calculate the perimeter of the triangle, given that one of the angles is $\dfrac{\pi}{6}$ rad. **[7]**

17 The population (in thousands) of a particular species of insect around a lake t weeks after a predator is released is modelled by

$P = 6.5 - 4.1 \sin\left(\dfrac{\pi t}{2.3}\right)$

    a What was the initial population? **[2]**

    b State the maximum possible population of the insect. **[1]**

    c When does this maximum first occur? Give your answer to the nearest day. **[3]**

18 AB is the arc of a circle with radius $3.5\,\text{cm}$ and centre C as shown.

    a Calculate the

        i Area of the sector,   **ii** Perimeter of the sector. **[5]**

    The segment S is bounded by the arc and the chord AB

    b Calculate the

        i Area of S **ii** Perimeter of S **[6]**

19 **a** Sketch these graphs for $0 \le x \le 2\pi$. Label the x- and y-intercepts and any asymptotes in terms of π

    i $y = 2\operatorname{cosec} x$ **ii** $y = \sec\dfrac{x}{2}$ **[6]**

    b State the range of each of the graphs in part **a**. **[3]**

20 $f(x) = \cot(x - 30°)$

    a Sketch the graph of $y = f(x)$ for x in the interval $-180° \le x \le 180°$
    Label the x- and y-intercepts and give the equations of any asymptotes. **[4]**

    b Solve the equation $\cot(x - 30°) = 0.2$ for $-180° \le x \le 180°$. Show your working. **[3]**

21 Solve these equations for $0 \le x \le 2\pi$. Show your working and give your solutions as exact multiples of π

    i $\sec(x+\pi)=2$ **ii** $\operatorname{cosec}\left(x-\dfrac{\pi}{8}\right)=\sqrt{2}$ **[8]**

22 $f(x)=\arccos(x-1)$

    a   Sketch the graph of $y=f(x)$ for $0 \le x \le 2$ **[3]**

    b   State the range of $f(x)$ **[2]**

    c   Work out the inverse of $f(x)$ and state its domain and range.   **[4]**

23 Find the exact value of x for which

    a $\arctan(2x-1)=\dfrac{\pi}{3}$ **[2]**   **b** $\operatorname{arccot}(x-5)=\dfrac{2\pi}{3}$ **[2]**

24 **a**   Show that the equation $2\tan^2 x=\sec x-1$ can be written as $2\sec^2 x-\sec x-1=0$ **[3]**

    b   Hence solve the equation $2\tan^2 x=\sec x-1$ for x in the interval $0 \le x \le 2\pi$, showing your working.   **[4]**

25 **a**   Given that $\sin^2\theta+\cos^2\theta \equiv 1$, show that $\operatorname{cosec}^2\theta-\cot^2\theta \equiv 1$ **[3]**

    b   Solve the equation $\operatorname{cosec}^2\theta-3\cot\theta-1=0$ for $0 \le \theta < 360°$, showing your working.   **[6]**

26 **a**   Show that $\sec^4 x-\tan^4 x \equiv \sec^2 x+\tan^2 x$ **[2]**

    b   Find the values of x in the range $-\pi \le x \le \pi$ that satisfy $\sec^4 x-\tan^4 x=5+\tan^2 x$. Show your working.   **[4]**

27 **a**   By writing $\cos 3x$ as $\cos(2x+x)$, show that $\cos 3x=4\cos^3 x-3\cos x$ **[5]**

    b   Hence solve the equation $8\cos^3 x-6\cos x=\sqrt{3}$ for x in the interval $0 \le x \le 2\pi$

        Show your working and give your answers as exact multiples of π **[5]**

28 **a**   Use the identity $\cos(A+B) \equiv \cos A\cos B-\sin A\sin B$ to show that $\cos 2x \equiv 2\cos^2 x-1$ **[3]**

    b   Hence solve the equation $\cos 2x+3\cos x+2=0$ for $0 \le \theta < 360°$, showing your working.   **[5]**

29 Prove by counter-example that $\cos(A+B) \ne \cos A+\cos B$ **[2]**

30 $f(x)=8\cos x+4\sin x$

    a   Write $f(x)$ in the form $r\cos(x-\alpha)$, where $r>0$ and $0<\alpha<\dfrac{\pi}{2}$ **[4]**

    b   Hence solve the equation $8\cos x+4\sin x=\sqrt{5}$ for $0<x \le 2\pi$. Show your working.   **[4]**

31 Sketch these graphs for $0 \le x \le 360°$

    a $y=2\cos(x+60°)$ **[4]**   **b** $y=-\sin\left(\dfrac{x}{2}\right)$ **[2]**

32 $g(x)=\tan\left(\dfrac{\pi}{3}-x\right)$, for $0 \le x \le 2\pi$

    a   Sketch the graph of $y=g(x)$, clearly labelling the x- and y-intercepts, and state the equations of the asymptotes.   **[4]**

    b   Solve the equation $g(x)=\dfrac{\sqrt{3}}{3}$. Show your working and give your solutions in terms of π **[4]**

33 The area of sector ABC is $56.7\,\text{cm}^2$ and the length of the arc AB is $12.6\,\text{cm}$

    Calculate the area of the shaded segment.   **[8]**

34 The shape ABC is formed from a sector and a triangle as shown in the diagram.

    a Find the length of AC **[2]**

    b Calculate

       i The area of ABC **ii** The perimeter of ABC **[7]**

35 **a** Sketch the graphs of

       i $y = \cot(2x - 60)$ for $0 \leq x \leq 180°$

       ii $y = 1 - \operatorname{cosec}\left(\dfrac{x}{2}\right)$ for $0 \leq x \leq 360°$

    Give the equations of the asymptotes.   **[7]**

    b Solve the equation $\sec(3x + 20) = \sqrt{2}$ for x in the interval $0 \leq x \leq 180°$. Show your working.   **[5]**

36 **a** Solve the equation $\cot^2\theta = 5$ for $0 \leq \theta \leq 2\pi$ **[4]**

    b Find the exact solutions of $\sec^2\left(\theta + \dfrac{\pi}{6}\right) = 2$ for $0 \leq \theta \leq 2\pi$. Show your working.   **[5]**

37 Find the coordinates of the points of intersection of the curves $y = \operatorname{cosec} x$ and $y = \sin x$ in the interval $0 \leq x \leq 360°$ **[4]**

38 **a** Sketch the graph of $y = |\arcsin x|$ for $-1 \leq x \leq 1$ **[2]**

    b Sketch the graph of $y = 1 + 2\arccos x$ for $-1 \leq x \leq 1$ and label the y-intercept with its exact value.   **[3]**

39 Show that the curve with Cartesian equation $\dfrac{x^2}{25} - \dfrac{y^2}{9} = 1$ has parametric equations $x = 5\sec\theta$, $y = 3\tan\theta$ **[2]**

40 Solve the equation $3\operatorname{cosec}^2 x - \cot x = 7$ for x in the interval $0 \leq x \leq 360°$. Show your working.   **[8]**

41 Find all the solutions of $2\tan^2 2\theta + \sec 2\theta - 4 = 0$ for $0 \leq \theta \leq \pi$. Show your working.   **[9]**

42 Prove that $\cos\theta\cot\theta - \sin\theta \equiv \operatorname{cosec}\theta - 2\sin\theta$ **[4]**

43 Solve the inequality $\cot^2 x > 1 + \operatorname{cosec} x$ for x in the interval $0 \leq x \leq \pi$. Show your working.   **[6]**

44 **a** **i** Prove that $\dfrac{\cos x}{\sin x} - \dfrac{\sin x}{1 - \cos x} \equiv -\operatorname{cosec} x$

       ii For what values of x is this identity valid?   **[5]**

    b Solve the equation $\dfrac{\cos x}{\sin x} - \dfrac{\sin x}{1 - \cos x} = 3$ for $0 \leq x \leq 2\pi$. Show your working.   **[4]**

45 **a** Prove that $\cos x \equiv 1 - 2\sin^2\left(\dfrac{x}{2}\right)$ **[3]**

    b Find all the solutions of $\cos x + 5\sin^2\left(\dfrac{x}{2}\right) = 3$ for $0 < x < 360°$. Show your working.   **[6]**

46 The height of a tide (in metres) in a harbour t hours after midnight is given approximately by the equation $h = 2.8 + \sqrt{3}\sin\left(\dfrac{t}{2}\right) - 3\cos\left(\dfrac{t}{2}\right)$

A particular boat requires a depth of at least 3.5 m in order to safely leave or enter the harbour. The owners of the boat wish to depart the harbour in the afternoon.   **[8]**

47 Solve the equation $\sin 2x + \sin x = 0$ for x in the interval $0 < x < 360°$. Show your working.   **[7]**

48 Given that $\sin\theta + 2\cos\theta$ can be written in the form $r\sin(\theta + \alpha)$ where $r > 0$ and $0 < \alpha < 90°$,

    a Find the value of r and the value of α **[4]**

    b Calculate the minimum value of $\dfrac{1}{(\sin\theta + 2\cos\theta)^2}$ and the smallest positive value of θ for which this minimum occurs,   **[4]**

    c Find the maximum value of $\dfrac{1}{3 + \sin\theta + 2\cos\theta}$ and express it in the form $a + b\sqrt{c}$.

    Find also the smallest positive value of θ for which this maximum occurs.   **[5]**

Dido's problem refers to a conundrum that faced Queen Dido of Carthage in around 850 BC. According to legend, she was offered as much land as she could cover with an animal hide, and using intuition she cut it into thin strips and created a large circular area. She may not have used calculus at the time, but differentiation can be used to find solutions to problems like this: as it happens, the largest area that can be bound by a perimeter of given length is indeed a circle.

Differentiation is an incredibly powerful tool that crops up throughout mathematics, and has applications in subjects such as physics, engineering, biology and economics. It can be applied to simple examples, such as the one above, as well as more complex ones involving trigonometric functions, logarithms and so on. This chapter builds on what you studied in Chapter 4, and your method will depend on the expression you need to differentiate.

Orientation

What you need to know

Ch4 Differentiation and integration
- Introduction to differentiation. **p.86**
- Turning points and the second derivative. **p.106**

Ch5 Exponentials and logarithms
- The laws of logarithms. **p.130**
- The natural logarithm. **p.134**

Ch12 Algebra 2
- Parametric equations. **p.312**

Ch14 Trigonometric identities
- Small angle approximations. **p.364**
- Double angle formula. **p.374**

What you will learn

- To find points of inflection and determine when a curve is convex or concave.

- To understand and use limits.

- To differentiate $\sin x$, $\cos x$, e^x, a^x, and $\ln x$

- To use the product, quotient and chain rules.

- To find the derivative of a function which is defined implicitly and of a function which is defined parametrically.

- To find the derivative of an inverse function.

What this leads to

Ch16 Integration and differential equations
Integration by parts.

Further Core Pure Maths
- Differentiate inverse trigonometric functions.
- Differentiate hyperbolic functions.

Fluency and skills

See Ch4.1

For an introduction to differentiation.

The first differential, or derivative, of a function gives you the gradient of that function, which tells you about the shape of its graph.

> **Key point**
>
> When $\dfrac{dy}{dx} > 0$, the function is increasing and the curve is **rising** in the positive x-direction.
>
> When $\dfrac{dy}{dx} < 0$, the function is decreasing and the curve is **falling** in the positive x-direction.
>
> When $\dfrac{dy}{dx} = 0$, the function is stationary (neither rising nor falling).

> Remember that Leibniz notation for the first derivative is $\dfrac{dy}{dx}$, and $\dfrac{d^2 y}{dx^2}$ for the second derivative. $\dfrac{dy}{dx}$ gives the rate of change of a function (i.e. the gradient), and $\dfrac{d^2 y}{dx^2}$ gives the rate of change of the gradient.

See Ch4.5

For a reminder on turning points.

The graph shows the shape of a cubic function.

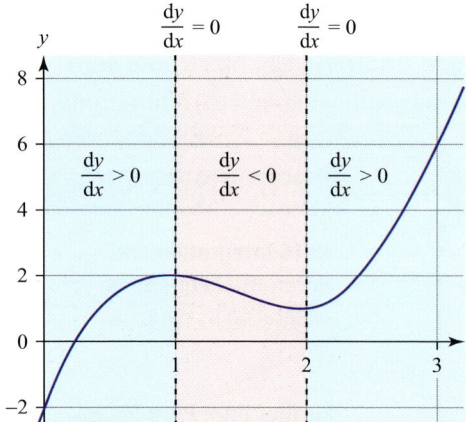

$\dfrac{dy}{dx} > 0$ in the regions where $x < 1$ or $x > 2$, so the curve is rising.

$\dfrac{dy}{dx} < 0$ in the regions where $1 < x < 2$, so the curve is falling.

$\dfrac{dy}{dx} = 0$ when $x = 1$ and $x = 2$, so the curve is neither rising nor falling.

The second derivative of a function, $\dfrac{d^2 y}{dx^2}$, gives you further information about the shape of the curve.

In the graph, at $x = 1$, there is a turning point where the gradient changes from positive to negative. So the second derivative is negative (it's decreasing). This point is a **maximum**, and the curve is described as **concave**. A curve is concave if any line segment joining two points on the curve stays below the curve.

> **Key point**
>
> If $\dfrac{dy}{dx} = 0$ and $\dfrac{d^2 y}{dx^2} < 0$, the point is a maximum and the curve is concave.

At $x = 2$, there is a turning point where the gradient changes from negative to positive. So the second derivative is positive (it's increasing). This point is a **minimum**, and the curve is described as **convex**. A curve is convex if any line segment joining two points on the curve stays above the curve.

> **Key point**
>
> If $\dfrac{dy}{dx} = 0$ and $\dfrac{d^2 y}{dx^2} > 0$, the point is a minimum and the curve is convex.

Between $x = 1$ and $x = 2$, the curve changes from concave to convex, that is, it changes **concavity**. The point at which a curve changes concavity is known as a **point of inflection**. At a point of inflection, the first derivative can take any value (the function does not need to be stationary), and the second derivative is *always* zero.

However, the reverse isn't true: a second derivative of zero does *not* necessarily show a point of inflection: you need to inspect the gradient at that point. If the gradient is non-zero, you have a point of inflection. If the gradient is zero, you must check that it has the same sign either side of the point for it to be a point of inflection. Alternatively, the second derivative must switch sign either side of the point to indicate a change in concavity.

> **Key point**
>
> At a point of inflection, it is always the case that $\dfrac{d^2y}{dx^2} = 0$
>
> But if $\dfrac{d^2y}{dx^2} = 0$, further inspection is needed to determine the nature of the point.

The table shows what can be concluded from the value of the first and second derivatives.

Example of a curve	Lower x value	At turning point	Higher x value
Minimum	$\dfrac{dy}{dx} < 0, \dfrac{dy}{dx} > 0$ Convex	$\dfrac{dy}{dx} = 0, \dfrac{d^2y}{dx^2} > 0$ Convex	$\dfrac{dy}{dx} > 0, \dfrac{d^2y}{dx^2} > 0$ Convex
Maximum	$\dfrac{dy}{dx} > 0, \dfrac{d^2y}{dx^2} < 0$ Concave	$\dfrac{dy}{dx} = 0, \dfrac{d^2y}{dx^2} < 0$ Concave	$\dfrac{dy}{dx} < 0, \dfrac{d^2y}{dx^2} < 0$ Concave
	$\dfrac{dy}{dx} > 0, \dfrac{d^2y}{dx^2} < 0$ Concave	$\dfrac{dy}{dx} \geq 0, \dfrac{d^2y}{dx^2} = 0$ Point of inflection	$\dfrac{dy}{dx} > 0, \dfrac{d^2y}{dx^2} > 0$ Convex
	$\dfrac{dy}{dx} < 0, \dfrac{d^2y}{dx^2} > 0$ Convex	$\dfrac{dy}{dx} \leq 0, \dfrac{d^2y}{dx^2} = 0$ Point of inflection	$\dfrac{dy}{dx} < 0, \dfrac{d^2y}{dx^2} < 0$ Concave

Example 1

Use algebra to describe the shape of the curve $y = x^3 - 3x^2 + 2x$ at the following points.
a $(0, 0)$ **b** $(1, 0)$ **c** $(2, 0)$

$y = x^3 - 3x^2 + 2x$ $\dfrac{dy}{dx} = 3x^2 - 6x + 2$ $\dfrac{d^2y}{dx^2} = 6x - 6$

> Work out the first and second derivatives.

a At $x = 0$, $\dfrac{d^2y}{dx^2} = 0 - 6 = -6$

> Evaluate the second derivative at that point.

$\dfrac{d^2y}{dx^2} < 0$, so the curve is concave.

b At $x = 1$, $\dfrac{d^2y}{dx^2} = 6 - 6 = 0$ and $\dfrac{dy}{dx} = 3 - 6 + 2 = -1$

So this is a point of inflection, on a decreasing section of the curve.

> Given that $\dfrac{d^2y}{d^2x} = 0$, also evaluate the first derivative. As the first derivative is non-zero, you can conclude it is a point of inflection.

(Continued on the next page)

c At $x = 2$, $\dfrac{d^2y}{dx^2} = 12 - 6 = 6$

$\dfrac{d^2y}{dx^2} > 0$, so the curve is convex.

You can check your answers by drawing the curve on a graphics calculator.

Example 2

Use algebra to identify the points of inflection in each of these curves.

a $y = x^3$ **b** $y = x^3 - 3x$

a $y = x^3$ $\dfrac{dy}{dx} = 3x^2$ $\dfrac{d^2y}{dx^2} = 6x$

Work out the first and second derivatives.

When $\dfrac{d^2y}{dx^2} = 0$, $x = 0$. When $x = 0$, $\dfrac{dy}{dx} = 0$

Use the condition $\dfrac{d^2y}{d^2x} = 0$ to find the possible points of inflection. Also evaluate the first derivative. As the first derivative is zero, further inspection is required.

x	-0.1	0	0.1
$\dfrac{dy}{dx} = 3x^2$	$+$	0	$+$
tangent	/	—	/
$\dfrac{d^2y}{dx^2} = 6x$	$-$	0	$+$
convex or concave	concave	neither	convex

Use a table to inspect what is happening either side of the point on the curve where $x = 0$

So, the tangent is horizontal at $x = 0$ and the function is increasing either side of $x = 0$

Examine the first derivative.

The shape of the curve changes from concave to convex at $x = 0$

Examine the second derivative.

So, at $x = 0$, there is a horizontal point of inflection on the curve $y = x^3$

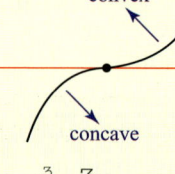
convex
concave
tangent gradient 0

Draw a sketch of the curve at this point to check your conclusions.

b $y = x^3 - 3x$

$\dfrac{dy}{dx} = 3x^2 - 3$

$\dfrac{d^2y}{dx^2} = 6x$

Work out the first and second derivatives.

When $\dfrac{d^2y}{dx^2} = 0$, $x = 0$. When $x = 0$, $\dfrac{dy}{dx} = -3$

At $x = 0$, there is a point of inflection on a decreasing part of the curve $y = x^3 - 3x$. At the point of inflection, the gradient of the curve is -3

Given that $\dfrac{d^2y}{d^2x} = 0$, also evaluate the first derivative. As the first derivative is non-zero, you can conclude it is a point of inflection.

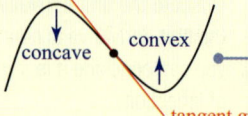
concave   convex
tangent gradient -3

Use a graphics calculator to draw the curve and check your conclusions.

1 For each of the following graphs, identify the regions in which the curve is concave and the regions in which the curve is convex.

a

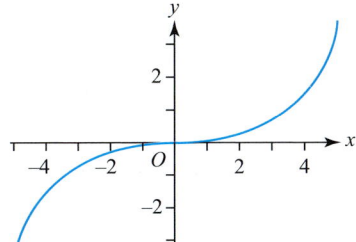

b

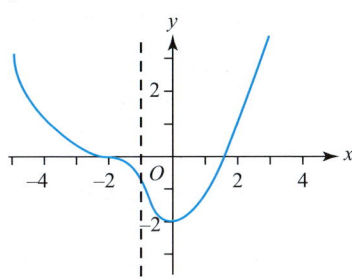

iv

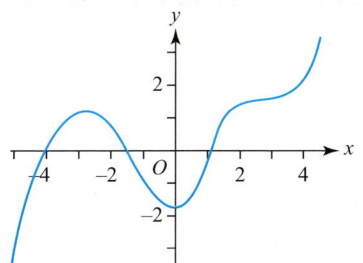

2 For each of the following graphs

  a Identify the number of points of inflection,

  b Describe the change of concavity at each point of inflection.

i

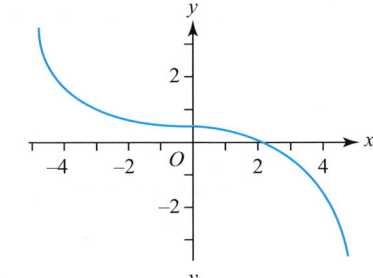

ii

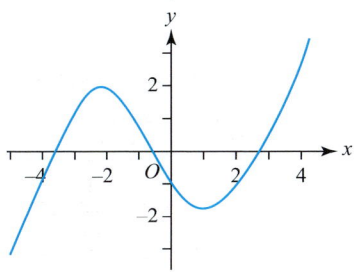

iii

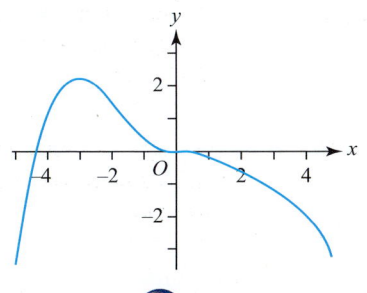

3 Use algebra to describe the shape of each curve at the given point. Show your working.

  a $y = x^2 + 2x - 1$ at $(1, 2)$

  b $y = \dfrac{1}{x}$ at $(-1, -1)$

  c $y = \dfrac{x+1}{x}$ at $(1, 2)$

  d $y = 1 + x - x^2 - x^3$ at $(1, 0)$

4 The curve $y = \dfrac{1}{3}x^3 + 2x^2 + 4x$ has a horizontal point of inflection.

  a Determine the x-coordinate of the point.

  b Show that the gradient of the curve at this point is zero.

  c Determine whether the curve is concave or convex

    i To the left of the point,

    ii To the right of the point.

5 The curve $y = x^4 - 2x^3$ has two points of inflection. Identify them both and describe them.

6 Determine and describe the point of inflection on the curve $y = x^2 + \dfrac{8}{x}$

7 Determine and describe the points of inflection on the curve $y = 3x^5 - 10x^3$

8 a Use the second derivative to show that the curve $y = 1 + 2x - x^2$ is always concave.

  b Show that the curve $y = 1 - \sqrt{x}$, $x > 0$ is always convex.

To identify the key features of a curve

(1) Work out both $\dfrac{dy}{dx}$ and $\dfrac{d^2y}{dx^2}$

(2) Solve $\dfrac{d^2y}{dx^2} = 0$ and, if necessary, $\dfrac{dy}{dx} = 0$

(3) Examine the possible values of x and interpret your findings.

(4) If required, use the results to sketch the curve.

Example 3

$y = x^2(x+3)$

a Show that the curve has one point of inflection.

b Identify and classify any other turning points.

c Sketch the curve.

a $y = x^3 + 3x^2$ so $\dfrac{dy}{dx} = 3x^2 + 6x = 3x(x+2)$ and $\dfrac{d^2y}{dx^2} = 6x + 6 = 6(x+1)$

When $\dfrac{d^2y}{dx^2} = 0$, $x = -1$

When $x = -1$, $\dfrac{dy}{dx} = (3 \times (-1)^2) + (6 \times -1) = -3$

So a point of inflection occurs when $x = -1$, on a decreasing part of the curve.

The curve is concave before $x = -1$ and convex after.

b When $\dfrac{dy}{dx} = 0$, $x = 0$ or $x = -2$

Therefore there are stationary points at $(0, 0)$ and $(-2, 4)$

When $x = 0$, $\dfrac{d^2y}{dx^2} = 6$

so there is a minimum turning point at $(0, 0)$

When $x = -2$, $\dfrac{d^2y}{dx^2} = -6$

so there is a maximum turning point at $(-2, 4)$

c

(1) Work out the first and second derivatives.

(2) Solve the second derivative equal to zero to find points of inflection.

(3) Evaluate the gradient at this point. Check your answer by evaluating the gradient at $x = -1$ on your calculator.

(2) Solve the first derivative equal to zero to find stationary points.

Substitute back into $y = x^2(x+3)$ to find the y-values.

(4) Use the results to sketch the graph. You could check your sketch using a graphics calculator.

Example 4

Show that the graph of the function $y = x^4 - 4x^3 + 7x^2 - 12x - 1$ is convex for all values of x

$y = x^4 - 4x^3 + 7x^2 - 12x - 1$

So $\dfrac{dy}{dx} = 4x^3 - 12x^2 + 14x - 12$ and $\dfrac{d^2y}{dx^2} = 12x^2 - 24x + 14$

(1) Work out the first and second derivatives.

(Continued on the next page)

$$\frac{d^2 y}{dx^2} = 12(x^2 - 2x) + 14$$
$$= 12((x-1)^2 - 1^2) + 14$$
$$= 12(x-1)^2 + 2$$

The smallest value this expression can take is 2, when $x = 1$

③ Examine the possible values of x

So $\frac{d^2 y}{dx^2} > 0$ for all x, therefore the curve is always convex.

③ Interpret findings.

Exercise 15.1B Reasoning and problem-solving

1 Sketch the following curves. Identify clearly any stationary points, points of inflection and intersections with the axes.

   a $y = (x-4)(x-25)(x+20)$

   b $y = x^4 - 24x^3 - 540$

   c $y = x^3 - 9x^2 + 24x - 16$

2 Prove, using the second derivative, that the general quadratic $y = ax^2 + bx + c$, $a \neq 0$ is

   a   Always convex when $a > 0$

   b   Always concave when $a < 0$

3 **a**   Show that the graph of the quartic function $y = x^4 + 8x^3 + 25x^2 - 5x + 10$ is convex for all values of x

   b   Show that the graph of the quartic function $y = 1 - 2x - 10x^2 + 4x^3 - x^4$ is concave for all values of x

   c   Show that the graph of the quartic function $y = 1 - 2x - 6x^2 + 4x^3 - x^4$ is never convex for any value of x

4 The general equation of the cubic function whose roots are a, b and c is $y = k(x-a)(x-b)(x-c)$, where k is a constant. Show that the point of inflection of the curve has an x-coordinate equal to the mean value of the roots.

5 The curve $y = 3x^5 - 5x^3$ has three stationary points. Identify the roots, the stationary points and the points of inflection, showing your working. Hence sketch the curve.

6 The curve $y = x^4 - 2x^3 - 12x^2$ has two points of inflection. Find the equation of the line that passes through both.

7 For what values of x is the graph of $y = x^3 - ax^2 - ax$ convex? Give your answer in terms of a

Challenge

8 The general form of a cubic function is $f(x) = ax^3 + bx^2 + cx + d$ where a, b, c and d are constants and $a \neq 0$

   a   What conditions must be placed on the constants a, b and c so that the graph of $y = f(x)$ has

      i   No stationary points,

      ii   Exactly one stationary point,

      iii   Two distinct stationary points?

   b   In terms of a, b, c and d, for what values of x is the graph of $y = f(x)$

      i   Concave,

      ii   Convex,

      iii   At a point of inflection?

9 The general form of a quartic function is $f(x) = ax^4 + bx^3 + cx^2 + dx + e$ where a, b, c, d and e are constants and $a \neq 0$. What conditions must be placed on these constants so that there are exactly two changes of concavity on the curve $y = f(x)$?

Fluency and skills

See Ch14.1
For more on small angle estimations.

You may remember that, when working in radians, for small values of θ, $\sin\theta \approx \theta$ and $\cos\theta \approx 1 - \frac{1}{2}\theta^2$

These are known as small angle estimations. Hence, you can derive the following two results.

Key point

$$\lim_{\theta\to 0}\frac{\sin\theta}{\theta}=1 \quad \text{and} \quad \lim_{\theta\to 0}\frac{1-\cos\theta}{\theta}=0$$

See Ch14.3
To revise the double angle formula.

Using these results, and the compound angle formula for trigonometric functions, you can derive the first derivatives of $\sin x$ and $\cos x$ from first principles.

Proof

$$f(x)=\sin x \text{ so } f'(x)=\lim_{h\to 0}\frac{\sin(x+h)-\sin x}{h}$$

Applying the compound angle formula,

$$f'(x)=\lim_{h\to 0}\frac{\sin x\cos h+\cos x\sin h-\sin x}{h}=\lim_{h\to 0}\frac{\sin x(\cos h-1)+\cos x\sin h}{h}$$

$$=\lim_{h\to 0}\sin x\frac{(\cos h-1)}{h}+\lim_{h\to 0}\cos x\frac{\sin h}{h}=\sin x\lim_{h\to 0}\frac{(\cos h-1)}{h}+\cos x\lim_{h\to 0}\frac{\sin h}{h}$$

Using the results derived from the small angle estimations, $f'(x)=(\sin x\times 0)+(\cos x\times 1)$

Therefore, $f'(x)=\cos x$

Key point

If $y=\sin x$, then $\dfrac{dy}{dx}=\cos x$ If $y=\cos x$, then $\dfrac{dy}{dx}=-\sin x$

You need to work in radians if you're differentiating trig functions.

Example 1

Determine the first and second derivatives of the function $y=2x^3+3\sin x$

Apply $\dfrac{d}{dx}\sin x=\cos x$

$$y=2x^3+3\sin x \qquad \frac{dy}{dx}=6x^2+3\cos x \qquad \frac{d^2y}{dx^2}=12x-3\sin x$$

To find the second derivative, you must differentiate again, so apply $\dfrac{d}{dx}\cos x=-\sin x$

Example 2

Given that $f(x)=\sin x+2\cos x$, find the exact value of $f'\left(\dfrac{\pi}{3}\right)$, showing your working.

$$f(x)=\sin x+2\cos x \qquad f'(x)=\cos x-2\sin x$$

$$f'\left(\frac{\pi}{3}\right)=\cos\frac{\pi}{3}-2\sin\frac{\pi}{3} \qquad =\frac{1}{2}-2\times\frac{\sqrt{3}}{2} \qquad =\frac{1}{2}-\sqrt{3}$$

See Ch14.1 for a reminder on the special triangles which are useful when dealing with angles in radians.

Exercise 15.2A Fluency and skills

1 Work out the derivative of each of these functions.

   a $3\sin x$

   b $\dfrac{\cos x}{3}$

   c $3-\cos x$

   d $x+\sin x$

   e $x^2+\cos x$

   f $\sin x+\cos x$

   g $3\sin x-4\cos x$

2 Find $\dfrac{dy}{dx}$ and $\dfrac{d^2y}{dx^2}$ for each of these functions.

   a $y=2\sin x$

   b $y=2x^2-\cos x$

   c $y=2\sin x-3\cos x$

   d $y=\sin x+\cos x+x^3$

   e $y=\dfrac{\cos x-\sin x}{2}$

3 Show your working in parts **a–e**.

   a   Given that $f(x)=3\sin x$, calculate

      i $f'\left(\dfrac{\pi}{3}\right)$ ii $f''\left(\dfrac{\pi}{3}\right)$

   b   Given that $f(x)=2\cos x$, calculate

      i $f'\left(\dfrac{\pi}{6}\right)$ ii $f''\left(\dfrac{\pi}{6}\right)$

   c   Given that $f(x)=\sin x+\cos x$, calculate

      i $f'\left(\dfrac{\pi}{4}\right)$ ii $f''\left(\dfrac{\pi}{4}\right)$

   d   Given that $f(x)=1-\cos x$, calculate

      i $f'\left(\dfrac{\pi}{2}\right)$ ii $f''\left(\dfrac{\pi}{2}\right)$

   e   Given that $f(x)=3\cos x-4\sin x$, calculate

      i $f'\left(\dfrac{3\pi}{2}\right)$ ii $f''\left(\dfrac{3\pi}{2}\right)$

4   a   By considering when $\dfrac{dy}{dx}=0$, find the turning points on the curve $y=1+\sin x$ in the interval $0\le x\le 2\pi$. Show your working.

   b   By considering when $\dfrac{d^2y}{dx^2}=0$, find the points of inflection on the curve $y=1-\cos x$ in the interval $0\le x\le 2\pi$

5 Find the gradient of the tangent to the curve $y=5\sin x-\sqrt{3}\cos x$ when $x=\dfrac{\pi}{3}$. Show your working.

6 Is the curve $y=x+\sin x$ convex or concave when $x=\dfrac{\pi}{3}$? Show your working.

7 Determine, from first principles, the derivative of

   a $2\sin x$

   b $-\cos x$

   c $x-\sin x$

Reasoning and problem-solving

Strategy

To identify the features of a trigonometric function

1) Manipulate the function until it is the sum of terms of the form ax^n, $a\sin x$ or $a\cos x$, where a is a constant.

2) Find the first derivative, $\dfrac{dy}{dx}$, and/or the second derivative, $\dfrac{d^2y}{dx^2}$

3) Interpret their meaning in terms of the gradient or shape of the curve respectively.

PURE

:::: MyMaths   Q 2165   SEARCH

Example 3

a Show that the function $f(x) = x + \sin x$ is never decreasing.

b Show that its concavity changes at $x = 0, \pi, 2\pi, 3\pi, \ldots, n\pi$

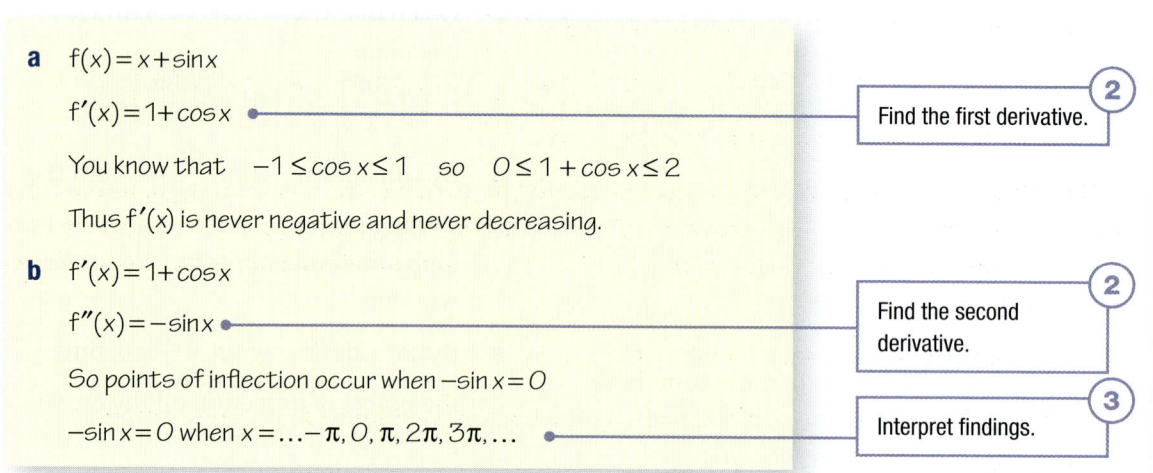

a $f(x) = x + \sin x$

$f'(x) = 1 + \cos x$ •————— Find the first derivative. ②

You know that $-1 \leq \cos x \leq 1$ so $0 \leq 1 + \cos x \leq 2$

Thus $f'(x)$ is never negative and never decreasing.

b $f'(x) = 1 + \cos x$

$f''(x) = -\sin x$ •————— Find the second derivative. ②

So points of inflection occur when $-\sin x = 0$

$-\sin x = 0$ when $x = \ldots -\pi, 0, \pi, 2\pi, 3\pi, \ldots$ •————— Interpret findings. ③

Exercise 15.2B Reasoning and problem-solving

1 Find the first derivative of the following functions.

a $\dfrac{x \sin x - x}{x}$

b $\dfrac{\sin x}{\tan x}$

c $\dfrac{\sin x - \tan x}{\tan x}$

d $2 \tan x \left(\cos x - \dfrac{\sin x}{2 \tan x} \right)$

2 a Find the equation of the tangent to the curve $y = \sin x$ at the point where $x = \dfrac{\pi}{3}$ expressing your answer in the form $ax + by = c$ and leaving c as an exact value.

b Verify that the curve is concave at this point.

3 Differentiate the following functions.

a $\dfrac{x^2 - x \sin x}{2x}$

b $\dfrac{4 - x^3 \cos x}{x^3}$

c $\cos x (\tan x + 1)$

d $\dfrac{\cos^2 x - \sin^2 x}{\cos x - \sin x}$

4 A curve has equation $y = \dfrac{x^2}{2\pi} - 2\sin x$

    a Showing your working, find its gradient when x is

        i $\;0$ **ii** $\;\dfrac{\pi}{2}$ **iii** $\;\pi$ **iv** $\;\dfrac{3\pi}{2}$ **v** $\;2\pi$

    b Show that the function is always increasing when $x > 2\pi$

    c A related family of functions has equation $y = \dfrac{ax^2}{2\pi} - 2\sin x$ where a is a constant.

       For what values of a will the graph always be convex?

5 A Big Wheel is set up in a city as a tourist attraction. It takes just over 6 minutes (2π minutes) to make a complete turn. The vertical height of a passenger, y metres above the ground, is modelled by the function $y = 21 - 20\cos t$, where t is the time in minutes since the passenger was at the bottom of the wheel.

    a How high up is the passenger when $t = \dfrac{\pi}{3}$ minutes?

    b What is the vertical speed of the passenger when $t = \dfrac{\pi}{6}$ minutes? Show your working.

    c What is the vertical acceleration of the passenger when $t = \dfrac{\pi}{2}$ minutes?

6 It can be convenient to break the year into 2π units rather than 365 days.

The time of sunrise in Liverpool over a year can be modelled by $y = 12 - 5\cos x$, where sunrise occurs at y hours after midnight, on the date given as x units.

    a Find the rate at which y is changing when the date $x = 0.5$ units.

    b When is the rate of change of the time of sunrise equal to zero? Show your working.

    c There are two points of inflection on the curve $y = 12 - 5\cos x$ in a single year.

       Find the values of x at which they occur.

7 The depth of water in a harbour basin over a day can be modelled by $y = 2\sin\left(x + \dfrac{\pi}{3}\right) + 5$, where y metres is the depth and x hours is the time since midnight.

    a Find $\dfrac{dy}{dx}$, the rate of change of the depth with time.

    b Find this rate at   **i** 4 am   **ii** 4 pm

    c Is the water rising or falling at noon ($x = 12$)?

Challenge

8 **a** Can you find the derivative of $\sin 2x$ from first principles? Hint: Remember that $\dfrac{1}{h} = \dfrac{2}{2h}$

    b Can you find the derivative of $\sin ax$ from first principles, where a is a constant?

       Hint: Remember that $\dfrac{1}{h} = \dfrac{a}{ah}$

15.3 Exponential and logarithmic functions

Fluency and skills

See Ch5.1–5.3 For more about exponentials.

The mathematical constant e was first discovered by the mathematician Jacob Bernoulli whilst researching compound interest. Its value, to 6 significant figures, is 2.71828, although, like the value of π, it is irrational.

e is defined as $e = \lim\limits_{n \to \infty} \left(1 + \dfrac{1}{n}\right)^n$

Key point

It can be shown that if $y = e^x$ then $\dfrac{dy}{dx} = e^x$

And more generally that if $y = e^{ax}$ then $\dfrac{dy}{dx} = ae^{ax}$

Example 1

Find the gradient of the curve $y = x + 3e^{2x}$ where the curve crosses the y-axis. Show your working.

$y = x + 3e^{2x}$ $\dfrac{dy}{dx} = 1 + 3 \times 2e^{2x} = 1 + 6e^{2x}$ — Find the first derivative.

When $x = 0$, $\dfrac{dy}{dx} = 1 + 6e^0 = 7$ — Check your answer by evaluating the derivative at $x = 0$ on your calculator.

The curve crosses the y-axis when $x = 0$

See Ch5.2 For more about the natural log.

$\log_a x$ is known as the 'natural log' and is more commonly referred to as **ln x**

Key point

It can be shown that if

$y = \ln x$ OR $y = \ln ax$ then $\dfrac{dy}{dx} = \dfrac{1}{x}$

$\ln ax = \ln a + \ln x$. As $\ln a$ is a constant, it differentiates to zero and so the derivative of $\ln x$ is equal to the derivative of $\ln ax$

Example 2

Given that $y = \ln 2x^3$, find $\dfrac{dy}{dx}$

$y = \ln 2x^3 = \ln 2 + 3\ln x$ $\dfrac{dy}{dx} = 3 \times \dfrac{1}{x} = \dfrac{3}{x}$

Use the laws of logarithms to rearrange y including $\log_a (x^k) = k \log_a (x)$ and then differentiate.

See Ch5.1 For a reminder on the laws of logarithms.

You can differentiate the function $f(x) = a^x$ if you first express it in base e, rather than a

Let $y = a^x$

Take natural logs on both sides $\ln y = \ln a^x$

Use the log law $\log_a (x^k) = k \log_a x$ $\ln y = x \ln a$

Take exponentials on both sides $y = e^{x \ln a}$ $(a^x = e^{x \ln a})$

Differentiate $\dfrac{dy}{dx} = \ln a \times e^{x \ln a}$

Since $e^{x \ln a} = a^x$, you have $\dfrac{dy}{dx} = a^x \ln a$

Differentiation Exponential and logarithmic functions

Key point

If $y = a^x$, then $\dfrac{dy}{dx} = a^x \ln a$

Example 3

Differentiate 3^x

$y = 3^x$ $\qquad \dfrac{dy}{dx} = 3^x \ln 3$ ●————————————— y is in the form a^x where $a = 3$

Exercise 15.3A Fluency and skills

1 Differentiate each of these functions.

   a $4e^x$ **b** $x - 2e^x$

   c $5e^x - 3x^2$ **d** $x^{-1} + e^x$

   e $\dfrac{1 + xe^x}{x}$ **f** $e^x - \dfrac{1}{x^2}$

   g $x^3 + e^x - \cos x$ **h** $\dfrac{1}{3} - \dfrac{e^x}{4}$

2 Find $\dfrac{dy}{dx}$ and $\dfrac{d^2 y}{dx^2}$ for each of these functions.

   a e^{3x} **b** $2e^{-4x}$

   c $e^x - e^{2x}$ **d** $6e^{0.5x}$

   e $e^x + e^{-x}$ **f** $e^x - \dfrac{1}{e^x}$

   g $e^{1.25x} + e^{0.5x}$ **h** $e^x(1 + e^{-3x})$

3 Find the first and second derivatives for each of these functions.

   a $f(x) = 3\ln x$ **b** $f(x) = 1 - 2\ln x$

   c $f(x) = \ln 2x$ **d** $f(x) = 3\ln x + \ln 3x$

   e $f(x) = \ln x^5$ **f** $f(x) = \ln x^{-1}$

   g $f(x) = \ln 3x^2$ **h** $f(x) = \ln \dfrac{3}{x}$

   i $f(x) = \ln \sqrt{x}$

4 Find the derivative of each of these functions.

   a 5^x **b** 2^x **c** 6^x

   d $\dfrac{3^x}{5}$ **e** $8^x - 7^x$ **f** $5x - 5^x$

   g 3×4^x

5 Showing your working, given that

   a $f(x) = 3e^x$, find $f'(0)$

   b $f(x) = x - 2e^x$, find $f'(1)$

   c $f(x) = x^2 + 3e^{2x}$, find $f'(0.5)$

   d $f(x) = \ln x$, find $f'(0.5)$

   e $f(x) = \ln x^2$, find $f'(0.25)$

   f $f(x) = \ln 4x$, find $f'(2)$

6 Find the gradient of each of these curves at the given point. Show your working.

   a $y = 3e^x$ at $(0, 3)$ **b** $y = 5e^x$ at $(1, 5e)$

   c $y = 4^x$ at $(0, 1)$ **d** $y = 5^x$ at $(1, 5)$

   e $y = 2^x$ at $(-1, 0.5)$ **f** $y = x + 6^x$ at $(2, 38)$

   g $y = 3\ln x$ at $(1, 0)$

   h $y = x - 2\ln x$ at $(e, e - 2)$

Reasoning and problem-solving

Strategy

To construct a binomial expansion

  (1) Find the first and/or second derivative.

  (2) Use these to find the gradient of the curve or rate of change, as appropriate.

  (3) Interpret the solution within the context of the problem.

Example 4

A car initially has a value of £20 000

Its value after x years can be modelled by $y = 20000 \times e^{-0.357x}$ $(x \geq 0)$

Showing your working, find the annual rate of change of the car's value after

a 3 years, **b** 10 years.

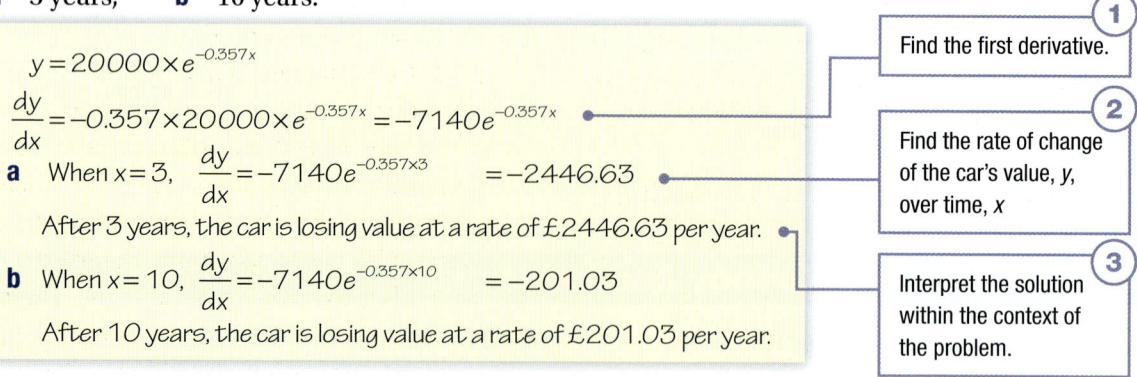

$y = 20000 \times e^{-0.357x}$

$\dfrac{dy}{dx} = -0.357 \times 20000 \times e^{-0.357x} = -7140e^{-0.357x}$

a When $x = 3$, $\dfrac{dy}{dx} = -7140e^{-0.357 \times 3}$ $= -2446.63$

After 3 years, the car is losing value at a rate of £2446.63 per year.

b When $x = 10$, $\dfrac{dy}{dx} = -7140e^{-0.357 \times 10}$ $= -201.03$

After 10 years, the car is losing value at a rate of £201.03 per year.

1 Find the first derivative.

2 Find the rate of change of the car's value, y, over time, x

3 Interpret the solution within the context of the problem.

Example 5

£100 is placed in an investment bond and its value increases.

The value of the bond is £x after y years. The time, y, and the value, x, are related by
$y = 17.2 \ln x - 79.2$

a After how many years would the value be £106?

b Find the rate of change of the value of the bond. Showing your working, calculate the time it takes to make £1 if there is

  i £110   **ii** £200   in the bond.

a $y = 17.2 \ln x - 79.2$

When $x = 106$, $= 17.2 \ln 106 - 79.2$

$= 1.0111$

It will take 1.01 years (to 3 sf)

b $y = 17.2 \ln x - 79.2$

$\dfrac{dy}{dx} = \dfrac{17.2}{x}$

i When $x = 110$, $\dfrac{dy}{dx} = \dfrac{17.2}{110}$

$= 0.156$

So with £110 in the bond, it will take 0.156 years to make £1

ii When $x = 200$, $\dfrac{dy}{dx} = \dfrac{17.2}{200}$

$= 0.086$

So with £200 in the bond, it will take 0.086 years to make £1

1 Find the first derivative.

2 Find the rate of change of the bond's value, x, over time, y

3 Interpret the solution within the context of the problem.

Exercise 15.3B Reasoning and problem-solving

1 In a managed moorland, the number of breeding pairs of pheasants is modelled by $P = 100e^{0.095t} - 50$, where P is the number of breeding pairs at the start of year t. At the beginning, $t = 0$

a How long will it take for the population to double?

b At what rate is the population changing at this time? Show your working.

c The population of pheasant pairs is plotted on a graph against time. Use the second derivative to prove that the curve is always convex.

2 A new dishwasher costs £170. Due to depreciation and other factors, its value drops. Its actual value, £y, in year t can be modelled by $y = 230e^{-0.134t} - 60$. Show your working in parts **a–d**.

  a Verify that the dishwasher costs £170 at $t = 0$

  b At what rate was the value changing when $t = 5$?

  c In what year was the dishwasher dropping £12 a year in value?

  d When the dishwasher reaches the end of its useful life, its value is zero.

   i When does the model predict this will happen?

   ii At what rate is it losing value at this point?

  e The value of the dishwasher is plotted on a graph against time. Discuss the concavity of the curve.

3 A cup of tea is made from boiling water. Left alone, it will cool down according to the model $X = Ae^{-kt} + R$, where $X°$C is the temperature t minutes after the tea is made. $R°$C is the surrounding room temperature.
  A and k are constants. Show your working in parts **a–d**.

  a Calculate the values of the constants A and k given that, initially, the temperature of the tea is 100 °C and that, after 3 minutes, it had cooled to 80 °C. The room temperature was 18 °C.

  b At what rate is the tea cooling

   i The instant that it is poured,

   ii After 10 minutes?

  c After how many minutes will the rate of cooling drop below 1 °C per minute?

  d At what rate is the rate of cooling changing when $t = 6$?

4 An altimeter works on the principle that altitude is a function of atmospheric pressure. When the altitude is a metres above sea level, the air pressure is P units, and $a = 54\,098 - 8155.5 \ln P$

  a What should the altimeter read at the top of Snowdon where the pressure is 665.3 units?

  b At what rate is the altitude changing as the pressure changes (in metres per unit) when the pressure is 700 units? Show your working.

  c What is the rate of change of altitude with pressure, in metres per unit, when a walker is 1000 m up? Show your working.

5 A parasitic tick has infected a flock of sheep. The area is sprayed with an experimental insecticide. A control group is examined on day one and again on day seven after the spraying. On day one, 250 ticks were discovered in the group. On day seven, there were 100 ticks found. The number of ticks, P, is related to the day, t, by a relationship of the form $P(t) = k \ln t + c$, where k and c are constants.

  a Use this information to find the values of the constants.

  b At what rate is the number of ticks changing on day four? Show your working.

  c The flock will be deemed clear of ticks when P falls below one. Showing your working,

   i Find the value of t when this occurs,

   ii Work out the rate of change of the number of ticks at this time.

15.4 The product and quotient rules

Fluency and skills

The **product rule** is used to differentiate two functions, u and v, that have been multiplied together.

> **Key point**
>
> In Leibniz notation, the product rule is written
> $$\frac{d}{dx}(uv) = v\frac{du}{dx} + u\frac{dv}{dx}$$

The **quotient rule** is used to differentiate two functions, u and v, one of which has been divided by the other.

> **Key point**
>
> In Leibniz notation, the quotient rule is written
> $$\frac{d}{dx}\left(\frac{u}{v}\right) = \frac{v\dfrac{du}{dx} - u\dfrac{dv}{dx}}{v^2}$$

A *quotient* is a quantity or function divided by another.

Example 1

Find $\dfrac{dy}{dx}$ when $y = x^4 \ln x$

Let $u = x^4$ and $v = \ln x$ — Define u and v where y is in the form uv

So $\dfrac{du}{dx} = 4x^3$ and $\dfrac{dv}{dx} = \dfrac{1}{x}$ — Differentiate u and v separately.

Therefore, $\dfrac{dy}{dx} = \ln x \times 4x^3 + x^4 \times \dfrac{1}{x} = 4x^3 \ln x + x^3$ — Substitute into the product rule.

Example 2

Find $\dfrac{dy}{dx}$ when $y = \tan x$

$y = \tan x = \dfrac{\sin x}{\cos x}$ — Write $\tan x$ as the quotient of $\sin x$ and $\cos x$

Let $u = \sin x$ and $v = \cos x$

So $\dfrac{du}{dx} = \cos x$ and $\dfrac{dv}{dx} = -\sin x$ — Define u and v and differentiate.

$\dfrac{dy}{dx} = \dfrac{\cos x \times \cos x - \sin x \times (-\sin x)}{\cos^2 x}$ — Substitute into the quotient rule.

$\dfrac{dy}{dx} = \dfrac{\cos^2 x + \sin^2 x}{\cos^2 x} = \dfrac{1}{\cos^2 x} = \sec^2 x$ — Simplify using the trigonometric identity $\sin^2 x + \cos^2 x = 1$

Therefore, $\dfrac{d}{dx}(\tan x) = \sec^2 x$

See Ch3.1

For a reminder of the basic trigonometric identities.

1 Differentiate each of these functions.

a $x\sin x$ b xe^x

c $e^x\cos x$ d $3x\ln x$

e $\sin x\ln x$ f $e^{3x}(x^2+1)$

g $\sqrt{x}\ln 3x$ h $\dfrac{1}{x}\cos x$

i $\left(x+\dfrac{1}{x}\right)\left(x-\dfrac{1}{x}\right)$ j $e^{3x}\ln x$

k $(2x^3+2x)(x-x^2)$ l $x^2\ln x^2$

m $(3-4x^2)\dfrac{1}{\sqrt{x}}$ n $\sqrt{x}\ln 3x$

o $(1+\sin x)(1-\sin x)$ p $e^{2x+1}(2x+1)$

2 a By writing $\sin^2 x = \sin x\sin x$, find the derivative of $\sin^2 x$

b Find the derivative of $\cos^2 x$

c Using the fact that $\sin 2x = 2\sin x\cos x$, find the derivative of $\sin 2x$

d Using the fact that $\cos 2x = \cos^2 x - \sin^2 x$, find the derivative of $\cos 2x$

3 Show your working in parts **a–c**.

a $f(x)=x\sin x$

find   i $f'\left(\dfrac{\pi}{2}\right)$ ii $f'\left(\dfrac{\pi}{6}\right)$

b $f(x)=x^3 e^x$

find   i $f'(0)$ ii $f'(1)$

c $f(x)=(x^2+2x+1)(1-3x-x^2)$

find   i $f'(0)$ ii $f'(-1)$

4 Find the derivative of each of these functions.

a $\dfrac{x+1}{x-1}$ b $\dfrac{\sin x}{x}$

c $\dfrac{\sin x}{e^x}$ d $\dfrac{(x^2+1)}{x+1}$

e $\dfrac{\cos x}{\sin x}$ f $\dfrac{\ln x}{x}$

g $\dfrac{3}{\sin x}$ h $\dfrac{x}{\cos x}$

i $\dfrac{x^2}{1-\sqrt{x}}$ j $\dfrac{x^2-3x+1}{\sqrt{x}}$

k $\dfrac{e^{2x+3}}{x^3}$ l $\dfrac{1+\sin x}{\cos x}$

5 Show your working in parts **a–c**.

a Given that $f(x)=\dfrac{x}{\cos x}$

find   i $f'(\pi)$ ii $f'(0)$

b Given that $f(x)=\dfrac{\ln x}{e^x}$

find   i $f'(e)$ ii $f'(1)$

c Given that $f(x)=\dfrac{x^2+2x-1}{1-2x}$

find   i $f'(0)$ ii $f'(-1)$ iii $f'(1)$

Reasoning and problem-solving

Strategy

To solve problems involving the differentiation of functions expressed as products or quotients

(1) Separate the function into two appropriate parts, u and v, and differentiate these.

(2) Apply the product rule or the quotient rule as appropriate.

(3) Simplify the answer as much as possible.

(4) Give the answer within the context of the problem.

Example 3

Work out the equation of the tangent to $y = \dfrac{(3x^2+1)}{(x+1)}$ **at the point (1, 2). Show all your working.**

$y = \dfrac{(3x^2+1)}{(x+1)}$

Let $u = 3x^2 + 1$ and $v = x + 1$

so $\dfrac{du}{dx} = 6x$ and $\dfrac{dv}{dx} = 1$

$\dfrac{dy}{dx} = \dfrac{(x+1)\times 6x - (3x^2+1)\times 1}{(x+1)^2}$

$\dfrac{dy}{dx} = \dfrac{6x(x+1) - (3x^2+1)}{(x+1)^2}$

$= \dfrac{3x^2 + 6x - 1}{(x+1)^2}$

When $x = 1$, $\dfrac{dy}{dx} = \dfrac{3(1)^2 + 6(1) - 1}{(1+1)^2}$

$= \dfrac{8}{4} = 2$

So the equation of the tangent is $y - 2 = 2(x - 1)$
which simplifies to $y = 2x$

1 Define u and v and differentiate.

2 Apply the quotient rule.

3 Simplify the expression.

Substitute in the values of the given point to find the gradient.

4 From Ch1.7, use the equation of a straight line $y - y_1 = m(x - x_1)$, where (x_1, y_1) is a point on the line and m is the gradient, to find the equation of the tangent.

Example 4

A flu patient's temperature will quickly rise then slowly come back to normal over a week. The situation can be modelled by $X = \dfrac{12t}{e^t}$, $0 \le t \le 7$ where $X°C$ is the temperature above the patient's normal and t is the day since the flu began ($t = 0$).

When will the fever peak? Show your working.

Let $u = 12t$ and $v = e^t$

$\dfrac{du}{dt} = 12$ and $\dfrac{dv}{dt} = e^t$

$\dfrac{dX}{dt} = \dfrac{(e^t \times 12) - (12t \times e^t)}{e^{2t}} = \dfrac{12(1-t)}{e^t}$

At a stationary point, $\dfrac{12(1-t)}{e^t} = 0$, so $t = 1$

When $t = 1$, $X = \dfrac{12 \times 1}{e^1} \approx 4.4$

$\dfrac{dX}{dt} = \dfrac{12(1-t)}{e^t}$

Let $u = 12 - 12t$ and $v = e^t$

$\dfrac{du}{dt} = -12$ and $\dfrac{dv}{dt} = e^t$

1 Define u and v and differentiate.

2 Apply the quotient rule and simplify.

Identify when the first derivative equals zero to find any stationary points.

1 To determine the nature of this stationary point, you need to inspect the sign of the second derivative, so you must use the quotient rule again. Separate the gradient function, $\dfrac{dx}{dt}$, into two appropriate parts.

(*Continued on the next page*)

$$\frac{d^2X}{dt^2} = \frac{e^t \times (-12) - (12 - 12t) \times e^t}{e^{2t}} = \frac{-12(2-t)}{e^t}$$

② Apply the quotient rule again to work out the second derivative, and simplify.

When $t = 1$, $\dfrac{d^2X}{dt^2} = \dfrac{-12}{e^1}$

Evaluate for $t = 1$, when the function is stationary.

The second derivative is negative at $t = 1$, so $(1, 4.4)$ is a maximum turning point.

The fever will peak on day 1 when it will be 4.4 °C above normal.

④ Give the answer in context.

Exercise 15.4B Reasoning and problem-solving

1 Work out the gradient of the tangent to the curve $y = \dfrac{2x+1}{x^2 - 1}$ at the point $(0, -2)$. Show your working.

2 Find the equation of the tangent to the curve $y = xe^x + 1$ at the point where it crosses the y-axis. Show your working.

3 Find the gradient of the tangent to the curve $y = x\cot x$ at the point where $x = \dfrac{\pi}{4}$ Show your working and leave your answer in terms of π

4 The sketch shows the function

$$f(x) = \frac{\ln x}{e^x}, \quad 0 < x \le 8$$

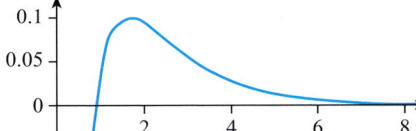

a Where does it cross the x-axis? Show your working.

b Show that there is a stationary point at $x^x = e$ and that this occurs when $x \approx 1.763$

c Show that a point of inflection occurs at $x = e^{\left(\frac{1+2x}{x^2}\right)}$ and that this occurs when $x \approx 2.55245$

5 Use the definitions

$$\sec x = \frac{1}{\cos x}; \quad \operatorname{cosec} x = \frac{1}{\sin x}; \quad \cot x = \frac{1}{\tan x}$$

to prove that

a $\dfrac{d}{dx}\sec x = \sec x \tan x$ **b** $\dfrac{d}{dx}\operatorname{cosec} x = -\operatorname{cosec} x \cot x$ **c** $\dfrac{d}{dx}\cot x = -\operatorname{cosec}^2 x$

Challenge

6 Differentiate   **a** $\sin^3 x$ **b** $xe^x \sin x$

7 A river runs with a current of x miles per hour. A boat, which can reach 10 mph in still water, travels up-river for one mile, and then down-river for one mile, in T hours.

T is a function of x, the speed of the current, and can be expressed by the equation

$$T(x) = \frac{20}{(10-x)(10+x)}, \quad 0 \le x < 10$$

a Prove that, in the defined domain, T is an increasing function.

b What is the rate of change of T with respect to x when the current is 5 mph? Show your working.

c Showing your working, find the value of   **i** $T(0)$ **ii** $T'(0)$

d In context, what happens as x approaches 10?

The chain rule

Fluency and skills

You will need to know how to differentiate a function of a function, known as a **composite function**. You can do this using the **chain rule**.

Take the function $y = \sin x^2$. You can say that y is a function of x, but, if you let $u = x^2$, then you can also say that y is a function of u, and u is a function of $x

> In other words, $y = f(u)$ and $u = g(x)$

> **Key point**
>
> The chain rule states that, for composite functions, $\dfrac{dy}{dx} = \dfrac{dy}{du} \times \dfrac{du}{dx}$

As you will see in Example 2, in certain contexts the chain rule is a significantly more efficient way of calculating a derivative.

Example 1

a Differentiate $y = \sin x^2$

b Differentiate $y = \sqrt{\cos x}$

a $y = \sin u$ $u = x^2$

Define y as a function of u, and u as a function of x

$\dfrac{dy}{du} = \cos u = \cos x^2$ $\dfrac{du}{dx} = 2x$

Work out $\dfrac{dy}{du}$ and $\dfrac{du}{dx}$, and express both in terms of x

$\dfrac{dy}{dx} = \dfrac{dy}{du} \times \dfrac{du}{dx} = \cos x^2 \times 2x = 2x \cos x^2$

Apply the chain rule and simplify.

b $y = \sqrt{u} = u^{\frac{1}{2}}$ $u = \cos x$

Define each function and differentiate.

$\dfrac{dy}{du} = \dfrac{1}{2} u^{-\frac{1}{2}} = \dfrac{1}{\sqrt{\cos x}}$ $\dfrac{du}{dx} = -\sin x$

$\dfrac{dy}{dx} = \dfrac{dy}{du} \times \dfrac{du}{dx} = \dfrac{1}{\sqrt{\cos x}} \times -\sin x = -\dfrac{\sin x}{\sqrt{\cos x}}$

Apply the chain rule and simplify.

You can also write the chain rule using a slightly different notation.

> **Key point**
>
> If you define a composite function as $y = f(g(x))$, then $\dfrac{dy}{dx} = f'(g(x)) \times g'(x)$

> Note that you could achieve the same result here by first expanding the brackets and differentiating each term.

Example 2

Differentiate $(3x^2 + 1)^3$

$y = f(u)$ and $u = g(x)$ where $f(u) = u^3$ and $g(x) = 3x^2 + 1$

This is a composite function, so define $g(x)$

$f'(u) = 3u^2$ and $g'(x) = 6x$

$\dfrac{dy}{dx} = f'(g(x)) \times g'(x)$

$= 3(3x^2 + 1)^2 \times 6x = 18x(3x^2 + 1)^2$

Differentiate each function then multiply them together.

Exercise 15.5A Fluency and skills

1 For each of these functions, define y as a function of u and u as a function of x

a $y = \sin^2 x$

b $y = \tan 2x$

c $y = \sqrt{7x}$

d $y = \ln \cos x$

e $y = e^{7x}$

f $y = (2x+1)^4$

g $y = (3x-2)(3x-2)(3x-2)$

h $y = \dfrac{1}{\sqrt{\sin x}}$

i $y = \dfrac{2}{\sqrt{x^3}}$

j $y = \sqrt{x + \ln x}$

k $y = (\sin x + \ln x)^5$

l $y = \sqrt[3]{\cos x - \ln x}$

2 Differentiate each of these functions.

a $(3x+4)^5$

b $(2x-1)^7$

c $(x^2+1)^6$

d $(1-2x-3x^2)^3$

e $\sqrt{2x+1}$

f $\sqrt{3-5x}$

g $\sqrt{3x^2+4}$

h $\sqrt[3]{1-2x}$

i $\sqrt{x^2+3x+4}$

j $\dfrac{1}{(2x+3)^2}$

k $\dfrac{1}{\sqrt{1-3x}}$

l $\dfrac{3}{(x^2-2x+5)}$

m $\cos^2 x$

n $\sqrt{\cos x}$

o $\sin(3x+2)$

p $\tan(5x-1)$

q $\cos\sqrt{x+1}$

r $\sin(\cos x)$

s $e^{\sin x}$

t $e^{\sqrt{2x-1}}$

u $e^{(e^x)}$

v $\ln(\sin x)$

w $\ln\left(\sqrt{2x+3}\right)$

x $\ln(\ln x)$

y $\sin(\ln x)$

z $\dfrac{1}{\ln x}$

3 Find the gradient of each of these curves at the given point. Show your working.

a $y = (2x+1)^3$ at $x=0$

b $y = \sin^3 x$ at $x = \dfrac{\pi}{3}$

c $y = \sqrt{3x+1}$ at $x=5$

d $y = \ln(9-4x)$ at $x=2$

e $y = e^{\sqrt{10-3x}}$ at $x=2$

f $y = \ln(\cos x)$ at $x = \dfrac{\pi}{4}$

g $y = \dfrac{1}{(3x+1)^4}$ at $x=0$

h $y = \sin(x^2+x-2)$ at $x=1$

4 Find the derived function given that

a $f(x) = 2^{\sin x}$

b $f(x) = \log_{10}(2x+1)$

    Hint: $\log_a b = \dfrac{\ln b}{\ln a}$

c $f(x) = 3^{x^2+2x-1}$

5 Find the equation of the tangent to each of these curves at $x=1$. Show your working.

a $y = (x^2-3x+1)^5$

b $y = e^{3x^2-2x}$

c $y = \sqrt{\sqrt{x}+3}$

d $y = \sin(x^2-1)$

e $y = \tan\left(2x^2-2x+\dfrac{\pi}{4}\right)$

f $y = e\ln(x^3-x+e)$

6 A semicircle with centre $(4, 0)$ and radius 5 units has equation $y = \sqrt{9+8x-x^2}$, $-1 \le x \le 9$

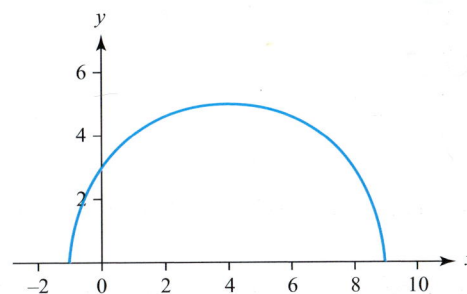

a Find the rate of change of y with respect to x when $x=7$. Show your working.

b Find the equation of the tangent to the circle when $x=1$ by using $\dfrac{dy}{dx}$. Show your working.

c What is the gradient of the curve as it crosses the y-axis? Show your working.

d What is the value of $\dfrac{d^2y}{dx^2}$ at this point?

Strategy

To solve problems involving differentiation of a composite function

1. Identify each function within the function and differentiate. Note that there may be more than one function within a function, and some of the functions may be products or quotients.

2. Apply the chain rule, and other necessary rules, to find the derivative of the composite function.

3. Simplify your answer and apply it to the problem.

Sometimes a composite function contains more than two distinct functions. You can still use the chain rule in these cases.

Key point

If $y = f(u)$ where $u = g(v)$ and $v = h(x)$, then

$$\frac{dy}{dx} = \frac{dy}{du} \times \frac{du}{dv} \times \frac{dv}{dx}$$

You can use this principle for any number of functions within a composite function.

Example 3

Differentiate $\sin\sqrt{x^2 + 1}$

$$y = \sin u, \ u = v^{\frac{1}{2}}, \ v = x^2 + 1$$

This is a function, within a function, within a function. First, define y, u and v ①

$$\frac{dy}{du} = \cos u = \cos v^{\frac{1}{2}} = \cos\sqrt{x^2 + 1}$$

$$\frac{du}{dv} = \frac{1}{2\sqrt{v}} = \frac{1}{2\sqrt{x^2 + 1}} \qquad \frac{dv}{dx} = 2x$$

Calculate $\frac{dy}{du}, \frac{du}{dv}$ and $\frac{dv}{dx}$ and express in terms of x ①

$$\frac{dy}{dx} = \frac{dy}{du} \times \frac{du}{dv} \times \frac{dv}{dx} = \cos\sqrt{x^2 + 1} \times \frac{1}{2\sqrt{x^2 + 1}} \times 2x = \frac{x\cos\sqrt{x^2 + 1}}{\sqrt{x^2 + 1}}$$

Apply the chain rule and simplify your answer. ② ③

Example 4

A cylinder is placed in a hydraulic press and compressed, so that its height, h, decreases at a rate of $0.1\,\text{m s}^{-1}$ and its radius squared, r^2, increases at a rate of $0.05\,\text{m s}^{-1}$. At what rate is its volume changing when $h = 2\,\text{m}$ and $r^2 = 0.25\,\text{m}^2$?

V is a function of r^2 and h, both of which are functions of t ①

$$\text{Volume } V = \pi r^2 h = \pi u \qquad \text{where } u = r^2 h$$

$$\frac{dV}{dt} = \frac{dV}{du} \times \frac{du}{dt}$$

You'll need the chain rule to calculate $\frac{dV}{dt}$ ②

$$\frac{dV}{du} = \pi \qquad \frac{du}{dt} = r^2\frac{dh}{dt} + h\frac{dr^2}{dt}$$

u is a product of two functions, r^2 and h, so use the product rule. ②

$$= (-0.1)r^2 + 0.05h$$

You're given the values of $\frac{dh}{dt}$ and $\frac{dr^2}{dt}$ ②

$$\frac{dV}{dt} = \frac{dV}{du} \times \frac{du}{dt} = \pi \times ((-0.1)r^2 + 0.05h) = 0.05\pi(h - 2r^2)$$

$$= 0.05\pi(2 - 2 \times 0.25) = 0.075\pi$$

Apply the chain rule and substitute in the values for h and r^2 given in the question. ③

1 Find the derivative of

a $\sin(\cos 2x)$ **b** $e^{\sin 3x}$

c $(\sin x + \cos 2x)^5$ **d** $\sqrt{\sin(4x+1)}$

e $\sqrt{(x-1)(x+2)}$ **f** $\sin(xe^x)$

g $e^{x(3x+4)}$ **h** $\cos(x^2 \sin x)$

i $\left(\dfrac{x+1}{x-1}\right)^{\frac{3}{2}}$ **j** $\sin\left(\dfrac{x}{x+1}\right)$

k $\ln\left(\dfrac{2x}{1-x}\right)$ **l** $\ln(xe^x)$

2 'Lighting up time' is defined as the time during which cars must use their headlights on public roads. Lighting up times can be calculated from the formula

$T = 2.5\cos\left(\dfrac{\pi D}{180}\right) + 19$, where T hours is the lighting up time on day D after the longest day of the year.

a Find $\dfrac{dT}{dD}$, the rate at which lighting up time changes as the days pass.

b Use this to find the minimum value of T and the value of D when this occurs.

c Find $\dfrac{d^2 T}{dD^2}$ and use it to verify that the answer to **b** is indeed the minimum.

3 A square, with side length x cm, is increasing in size such that its side length changes at a rate of 2 cm s^{-1}. At what rate is the area increasing when the side length is 10 cm?

4 Quadratic equations of the form $x^2 + 2bx + 1 = 0$, where $b > 1$, have two roots, one of which is $x = \sqrt{b^2 - 1} - b$

a Find $\dfrac{dx}{db}$

b Show that the graph of the function $x = \sqrt{b^2 - 1} - b$ is always increasing when $b > 1$

c Find $\dfrac{d^2 x}{db^2}$ and show that the curve of the function is always concave.

5 £1000 is deposited in a special account. The amount of money in the account, £P, can be calculated from the formula $P = 1000e^{0.001y}$, where y is the number of years the money has been deposited.

a Find $\dfrac{dP}{dy}$ **b** At what rate is the money growing when $y = 5$?

c Find $\dfrac{d^2 P}{dy^2}$

6 The growth of a particular tree is modelled by $h = 15\left(1 - e^{-0.22t}\right)$, where h metres is the height of the tree after t years.

a Find $\dfrac{dh}{dt}$

b Show that $\dfrac{dh}{dt} = 0.22(15 - h)$

7 A boy is collecting stickers. There are 200 stickers to collect and he starts with 10

Each week he buys a new pack of stickers and discards duplicates. His number of stickers, N, at the end of week t is modelled by

$N = \dfrac{200}{19e^{-0.7t} + 1}$

a Find $\dfrac{dN}{dt}$

b Express $19e^{-0.7t}$ in terms of N

c Show that $\dfrac{dN}{dt} = \dfrac{7}{2000}N(200 - N)$

Challenge

8 Differentiate these functions.

a $\tan 3x$ **b** $\sqrt{\sin x \cos x}$

c $\sqrt{\sin^2 x \cos 7x}$

9 A weather balloon can be modelled as a cube. As the balloon rises, the length of each side in metres is given by $x = 2 + 1.5t$, where t is the time in minutes after the balloon is released. At what rate does the volume of the balloon increase when $x = 8$ metres?

Fluency and skills

You will need to know how to find $\dfrac{dy}{dx}$ when x is expressed as a function of y, i.e. $x = f(y)$. Inverse functions often make the differentiation of equations easier, as you will see in Example 1.

Given that $x = f(y)$, then $y = f^{-1}(x)$ and therefore $x = f(f^{-1}(x))$

Using the chain rule to differentiate $x = f(f^{-1}(x))$ with respect to x on both sides gives $\quad x\left(\dfrac{d}{dx}\right) = f'(f^{-1}(x)) \times \dfrac{d}{dx}(f^{-1}(x))$

Substituting $y = f^{-1}(x)$ gives $\quad 1 = f'(x) \times \dfrac{dx}{dy} \Rightarrow 1 = \dfrac{dy}{dx} \times \dfrac{dx}{dy} \Rightarrow \dfrac{dy}{dx} = \dfrac{1}{\frac{dx}{dy}}$

> **Key point**
>
> For $x = f(y)$, $\quad \dfrac{dy}{dx} = \dfrac{1}{\frac{dx}{dy}}$

Remember that inverse trigonometric functions can be written in two ways.

$\sin^{-1} x = \arcsin x$
$\cos^{-1} x = \arccos x$
$\tan^{-1} x = \arctan x$

Example 1

Use the fact that $\dfrac{dy}{dx} = \dfrac{1}{\frac{dx}{dy}}$ to show that the derivative of $y = \sin^{-1} x$ is $\dfrac{1}{\sqrt{1+x^2}}$

$y = \sin^{-1} x, \quad so \quad x = \sin y$

$\dfrac{dx}{dy} = \cos y \quad and \quad \dfrac{dy}{dx} = \dfrac{1}{\cos y}$

$-\dfrac{\pi}{2} \le \sin^{-1} x \le \dfrac{\pi}{2} \Rightarrow -\dfrac{\pi}{2} \le y \le \dfrac{\pi}{2}$

$\cos y \ge 0$ for $-\dfrac{\pi}{2} \le y \le \dfrac{\pi}{2}$

$\dfrac{dy}{dx} = \dfrac{1}{\sqrt{1 - \sin^2 y}}$

$\dfrac{dy}{dx} = \dfrac{1}{\sqrt{1 - x^2}}$

Therefore, $\dfrac{d}{dx}\left(\sin^{-1} x\right) = \dfrac{1}{\sqrt{1-x^2}}$

Use the fact that $\sin^2 y + \cos^2 y = 1$

$x = \sin y$ so substitute x for $\sin y$

Exercise 15.6A Fluency and skills

1 For each of these functions
   i Find its derivative and state its inverse,
   ii State the derivative of the inverse.
   You can assume each function is defined in a suitable domain so that its inverse exists.
   a $f(x) = x^6$ **b** $f(x) = x^{\frac{2}{3}}$
   c $f(x) = x^2 + 2$ **d** $f(x) = (x+4)^2 + 1$
   e $f(x) = \sqrt{3x+2}$ **f** $f(x) = \sqrt[3]{2x+1}$

2 Use the method shown in Example 1 to prove that

   a $\dfrac{d}{dx}(\cos^{-1} x) = -\dfrac{1}{\sqrt{1-x^2}}$

   b $\dfrac{d}{dx}(\tan^{-1} x) = \dfrac{1}{1+x^2}$

   c $\dfrac{d}{dx}(\sec^{-1} x) = \dfrac{1}{x\sqrt{x^2-1}}$

   d $\dfrac{d}{dx}(\cot^{-1} x) = -\dfrac{1}{1+x^2}$

3 For each of these functions
   i Find its derivative,   **ii** Find its inverse,
   iii Find the derivative of the inverse.
   a $f(x) = 3x+1$ **b** $f(x) = 1-5x$
   c $f(x) = 3x^2$ **d** $f(x) = \dfrac{x+1}{2}$
   e $f(x) = \dfrac{1}{x+1}$ **f** $f(x) = \dfrac{\frac{2}{x}}{x+1}$
   g $f(x) = \dfrac{1}{x^2}$ **h** $f(x) = \sqrt{x}$
   i $f(x) = \sqrt{x} - 1$ **j** $f(x) = x^{\frac{1}{3}}$
   k $f(x) = e^{2x+1}$ **l** $f(x) = \ln 4x$

Reasoning and problem-solving

Strategy

To solve problems involving the differentiation of inverse functions

① Differentiate with respect to y and find $\frac{dy}{dx}$ by using the result $\frac{dy}{dx} = \frac{1}{\frac{dx}{dy}}$

② Find the equation of the tangent or normal by substituting into the equation of a straight line $y - y_1 = m(x - x_1)$

Example 2

A function has the inverse $x = y^3 + 2y + 4$. Find the equation of the tangent of the function at the point $(7, 1)$.

$x = y^3 + 2y + 4$

$\frac{dx}{dy} = 3y^2 + 2$ so, $\frac{dy}{dx} = \frac{1}{3y^2 + 2}$

When $y = 1$, $\frac{dy}{dx} = \frac{1}{3 \times 1^2 + 2} = \frac{1}{5}$

Equation of the tangent: $y - 1 = \frac{1}{5}(x - 7)$

① Differentiate with respect to y and then use $\frac{dy}{dx} = \frac{1}{\frac{dx}{dy}}$

② From Ch1.7, use the equation of a straight line $y - y_1 = m(x - x_1)$, where (x_1, y_1) is a point on the line and m is the gradient, to find the equation of the tangent.

Exercise 15.6B Reasoning and problem-solving

1 A function is defined by $f(x) = \dfrac{4}{x^3 + 1}$

Find the equation of the normal to the function $y = f^{-1}(x)$ at the point $(2, 1)$

2 Find the equation of the tangent to the curve $y = \sin^{-1}\left(\dfrac{2x}{2x+1}\right)$, $x \geq 0$, at the point where $x = \dfrac{1}{2}$. Show your working.

3 a Find the equation of the tangent to the curve $y = \sin^{-1}(x+1)$; $-2 \leq x \leq 0$ at the point where $x = -\dfrac{1}{2}$. Show your working.

b What can be said about the tangent to the curve at the point where $x = -2$? State its equation.

4 A door is 1.5 m wide. As the door opens, it sweeps over an area, $A\,\text{m}^2$, making an angle of θ rad at the hinge. The leading edge of the door is h metres from this line.

a Show that $A = \dfrac{9}{8}\sin^{-1}\left(\dfrac{2h}{3}\right)$

b Find $\dfrac{dA}{dh}$

c Calculate the rate of increase of the area when $h = 1.2$

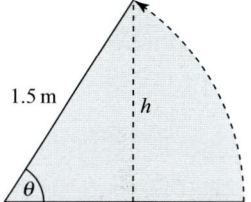

1.5 m h θ

Challenge

5 As the tide comes into a harbour, the time passed since low tide, t hours, can be calculated from the depth of water using the formula

$t = \dfrac{6}{\pi}\cos^{-1}(2 - 0.2D)$, where D is the depth in feet.

a Find an expression for $\dfrac{dt}{dD}$

b Find the rate of change of time passed with respect to depth when the water is 10 feet deep.

Implicit differentiation

Fluency and skills

In an equation of the form $y = f(x)$, y is called the dependent variable and x the independent variable. When the dependent variable is the subject of the equation, the function is said to be **explicit**.

However, equations are sometimes expressed in a form where the dependent variable is not explicitly given as the subject.

For example, the equation $2y + 3x = 5$ is the equation of a straight line and the fact that y is a function of x is only implied. This is an example of an **implicit function**.

The equation $(x-2)^2 + (y-1)^2 = 4$ is the equation of a circle and a second example of an implicit function.

> **Key point**
>
> When a function cannot be easily rearranged into the form $y = f(x)$, you can differentiate it implicitly.
>
> If a term is expressed as a function of y, you must use the chain rule $\dfrac{d}{dx}f(y) = f'(y) \times \dfrac{dy}{dx}$

Example 1

A function is defined by $x^2 + 3xy + 4y^3 = 0$. Express $\dfrac{dy}{dx}$ in terms of x and y

Differentiating each term •————— This is an implicit function, so differentiate throughout with respect to x

Differentiating x^2 gives $\quad 2x$

Differentiating $3xy$ gives $\quad \left(3x \times \dfrac{dy}{dx}\right) + (3y \times x^0) = 3x\dfrac{dy}{dx} + 3y$ •——— The second term is a product of x and y, so you must use the product rule to differentiate. See Ch15.4 for a reminder.

Differentiating $4y^3$ gives $\quad (4 \times 3y^2) \times \dfrac{dy}{dx} = 12y^2\dfrac{dy}{dx}$ •——— The third term is a function of y, so you must use the chain rule to differentiate. See Ch15.5 for a reminder.

Putting these terms together gives $2x + 3x\dfrac{dy}{dx} + 3y + 12y^2\dfrac{dy}{dx} = 0$

$\dfrac{dy}{dx}(3x + 12y^2) + (2x + 3y) = 0$ •——— Collect terms multiplied by $\dfrac{dy}{dx}$

$\dfrac{dy}{dx} = -\dfrac{2x + 3y}{3x + 12y^2}$

1 For each expression, find $\dfrac{dy}{dx}$ in terms of x and y

   a $x^2 = y^3$

   b $(x+1)^2 + (y-3)^2 = 4$

   c $2x^2 + y^2 = 4$

   d $\dfrac{1}{x} + \dfrac{1}{y} = 1$

   e $x^2 + 2xy + 3y = 0$

   f $y + \dfrac{1}{y} = x^2$

   g $x^2 + 2xy + y^2 = 6$

   h $3e^x = \sqrt{y}$

   i $\sin x = \cos 2y$

   j $\ln(y+1) = x^2 + 1$

   k $x \cos y = \tan x$

   l $x^{\frac{3}{4}} + y^{\frac{3}{4}} = \pi$

   m $x + \sin^{-1} x = y + \cos^{-1} y$

   n $\ln(x+y) = x + xy$

   o $e^{y+1} = x^2 + 2xy + 1$

   p $\tan^{-1} y + xy = 0$

   q $\dfrac{1}{x^2} + \dfrac{1}{y^2} = 144$

2 A hyperbola has equation $xy + 2x^2y^2 = 1$
   Find the gradient of the curve at the point $(1, 0.5)$. Show your working.

3 A circle has equation $x^2 + y^2 = 25$

   a Find $\dfrac{dy}{dx}$ in terms of x and y

   b Find the gradients of the tangents to the circle at the points where $x = 3$

   c Find the gradient of the tangent to the circle at the point where $x = 0$ and $y = 5$

4 Find the stationary points on each of the following curves.

   a $x^2 + 5y^2 = 20x$

   b $x^2 + 2y^2 - 4x + 7y + 9 = 0$

   c $x^3 + 3y^3 = 81$

5 A curve has equation $e^x + e^{2y} + 3x = 2 - 4y$

   a Show that the point $(0, 0)$ lies on the curve.

   b Find $\dfrac{dy}{dx}$

   c Find the gradient of the tangent at the point $(0, 0)$

6 A curve has equation $\ln(y-1) = x \ln x$

   a Express $\dfrac{dy}{dx}$ in terms of x and y

   b Show that the point $(1, 2)$ lies on the curve.

   c Find the gradient of the tangent to the curve at this point.

7 A circle has equation $x^2 + y^2 - 2x - 4y + 1 = 0$

   a Find $\dfrac{dy}{dx}$ in terms of x and y

   b The point $(5, -1)$ lies on the circle. Determine the gradient of the tangent to the circle at this point.

8 St Valentine's Equation is $(x^2 + y^2 - 1)^3 = x^2 y^3$

When the points which satisfy this equation are plotted, a heart is produced.

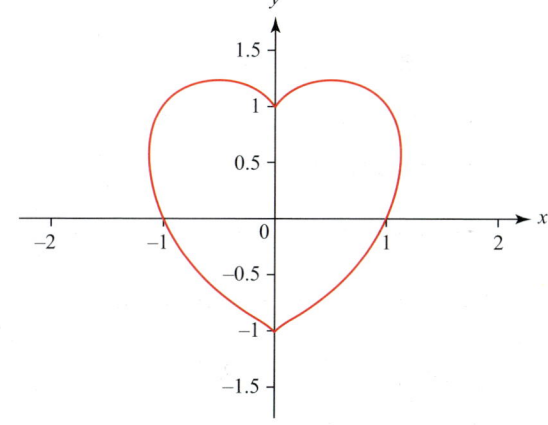

   a Find an expression in x and y for $\dfrac{dy}{dx}$

   b Find the gradient of the curve at the point $(1, 1)$

PURE

Reasoning and problem-solving

To find the point where a tangent of given gradient touches the curve of an implicit function

1. Find the derivative of the implicit function.

2. Use the given conditions to express y in terms of x and obtain an equation for the tangent.

3. Substitute the expression into the implicit function and solve to find the x-values of the points of intersection.

4. Substitute the x-values into the equation of the tangent to find the corresponding y-values.

Example 2

The equation of a circle is $x^2 + y^2 + 2x - 7 = 0$

Find where the tangents to the circle with gradient 1 touch the circle.

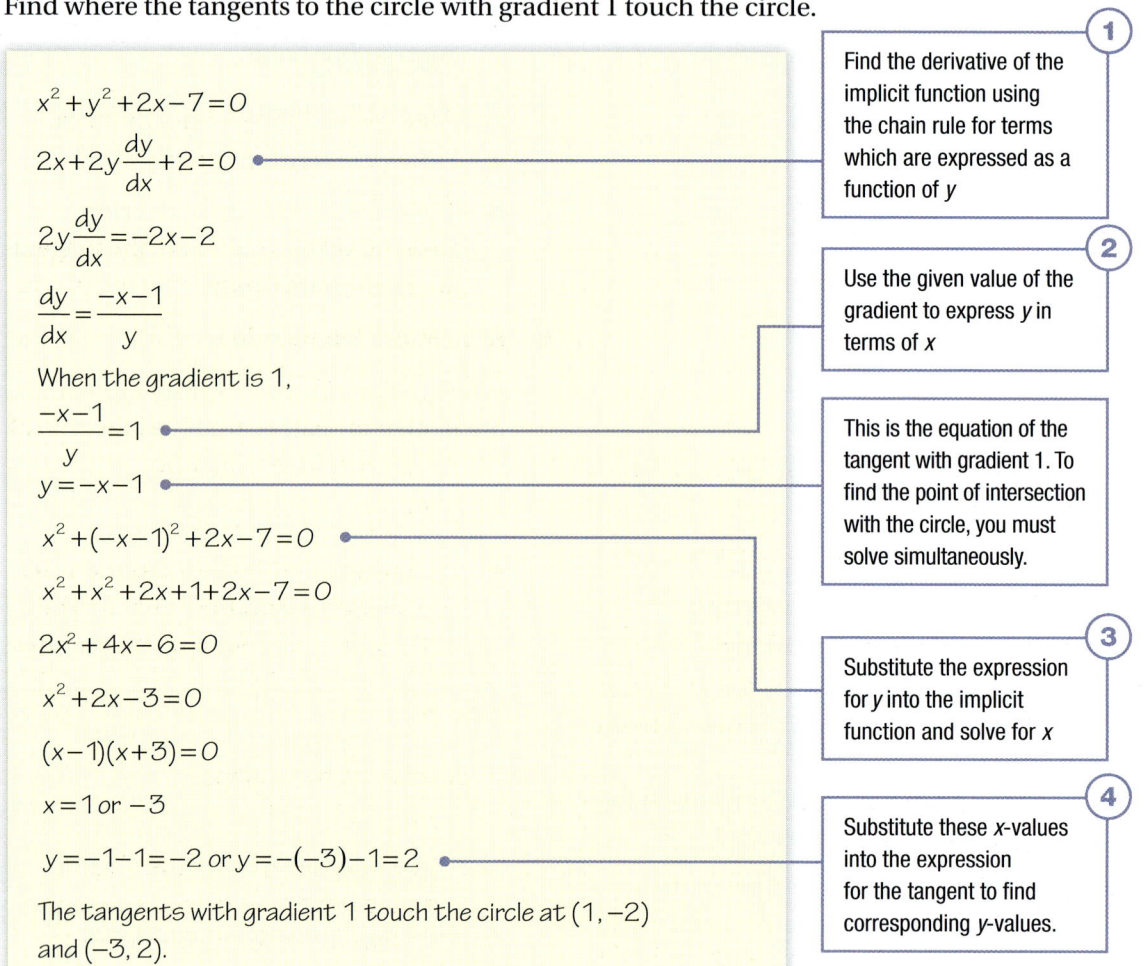

$x^2 + y^2 + 2x - 7 = 0$

$2x + 2y\dfrac{dy}{dx} + 2 = 0$

> **1** Find the derivative of the implicit function using the chain rule for terms which are expressed as a function of y

$2y\dfrac{dy}{dx} = -2x - 2$

$\dfrac{dy}{dx} = \dfrac{-x-1}{y}$

> **2** Use the given value of the gradient to express y in terms of x

When the gradient is 1,

$\dfrac{-x-1}{y} = 1$

$y = -x - 1$

> This is the equation of the tangent with gradient 1. To find the point of intersection with the circle, you must solve simultaneously.

$x^2 + (-x-1)^2 + 2x - 7 = 0$

$x^2 + x^2 + 2x + 1 + 2x - 7 = 0$

$2x^2 + 4x - 6 = 0$

$x^2 + 2x - 3 = 0$

> **3** Substitute the expression for y into the implicit function and solve for x

$(x-1)(x+3) = 0$

$x = 1 \text{ or } -3$

$y = -1 - 1 = -2 \text{ or } y = -(-3) - 1 = 2$

> **4** Substitute these x-values into the expression for the tangent to find corresponding y-values.

The tangents with gradient 1 touch the circle at $(1, -2)$ and $(-3, 2)$.

1 An asteroid has the equation $x^{\frac{2}{3}} + y^{\frac{2}{3}} = 5$

The asteroid passes through the point $(1, 8)$. What is the equation of the tangent to the asteroid at this point?

2 A circle has the equation
$x^2 + y^2 - 8x - 4y - 5 = 0$

Find where the tangents to the circle with gradient $\dfrac{3}{4}$ touch the circle.

3 An ellipse has the equation $\dfrac{x^2}{a^2} + \dfrac{y^2}{b^2} = 1$ where a and b are constants.

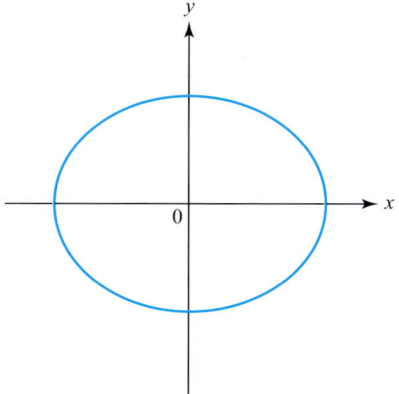

a Show that the horizontal width of the ellipse is $2a$

b Find the rate of change of y with respect to x

c Find the stationary points of the curve.

d What is the relationship between x and y when the gradient of the tangent to the ellipse is 1?

4 A hyperbola has the equation
$(y+1)^2 = (x+3)(2x+1)$

a Find $\dfrac{\mathrm{d}y}{\mathrm{d}x}$ in terms of x and y

b Find the equations of the tangents to the curve when $x = 2$

c Find the point where the two tangents intersect. Show your working.

5 The ellipse shown has the equation
$x^2 + xy + y^2 - 9 = 0$

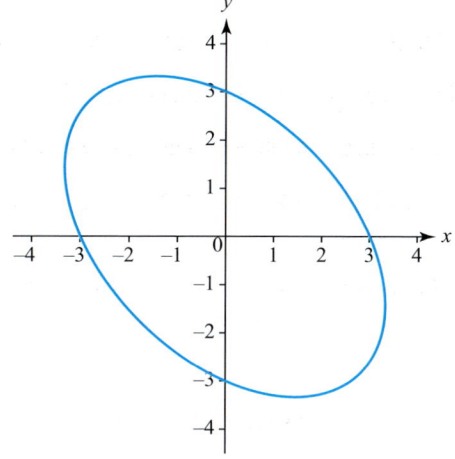

a Find the points at which the gradient of the ellipse is 1

b Find the maximum and minimum values of y, leaving your answer in surd form.

c Find the values of x where the tangents to the ellipse are parallel to the y-axis.

6 A circle has the equation
$x^2 + y^2 - 2x - 10y + 1 = 0$

a By differentiating, find the equation of the normal to the circle at

i $(4, 9)$ **ii** $(5, 8)$

b Find where the two normals intersect, showing your working. What do you notice about this point of intersection?

Challenge

7 A curve has the equation
$x^2 - xy + 2y^2 = 144$

a Find the equation of the tangent to the curve at the point where the curve intersects the line $y = 3x$ in the first quadrant.

b Find the stationary points of the curve. Use graph plotting software to draw the curve and classify the stationary points.

Parametric functions

Fluency and skills

See Ch12.3

For a reminder on parametric equations.

When two variables, x and y, are expressed as functions of a parameter, the derivative, $\dfrac{dy}{dx}$, can be found using the chain rule.

Key point

You can find $\dfrac{dy}{dx}$ from the parametric equations $x = f(t)$ and $y = g(t)$ by using the chain rule $\dfrac{dy}{dx} = \dfrac{dy}{dt} \times \dfrac{dt}{dx}$

Parametric equations are useful because they allow you to relate two variables with respect to a third variable. For example, in kinematics, you can represent the trajectory of an object using parametric equations which depend on time as the parameter.

Example 1

The parametric equations of a parabola are $x = 4t$; $y = 30t - 5t^2$

a Find $\dfrac{dy}{dx}$

b Find the turning point of the parabola. Show your working.

a $x = 4t$

$\dfrac{dx}{dt} = 4$

$y = 30t - 5t^2$

$\dfrac{dy}{dt} = 30 - 10t$ ← Differentiate both parametric equations with respect to t

$\dfrac{dy}{dx} = (30 - 10t) \times \dfrac{1}{4}$ ← Substitute into $\dfrac{dy}{dx} = \dfrac{dy}{dt} \times \dfrac{dt}{dx}$

$\dfrac{dy}{dx} = \dfrac{5}{2}(3 - t)$ ← Simplify.

b At the turning point,

$\dfrac{dy}{dx} = 0$

$\dfrac{5}{2}(3 - t) = 0$ ← Equate the first derivative to zero.

$\Rightarrow t = 3$ ← Solve to find t

When $t = 3$,

$x = 4 \times 3 = 12$

$y = (30 \times 3) - (5 \times 3^2) = 45$ ← Substitute $t = 3$ into the parametric equations to find the coordinates of the turning point.

So the turning point is at $(12, 45)$

Example 2

The curve shown has parametric equations $x = \sin 2\theta$; $y = 2\cos\theta$

Find an expression for its derivative in terms of the parameter θ

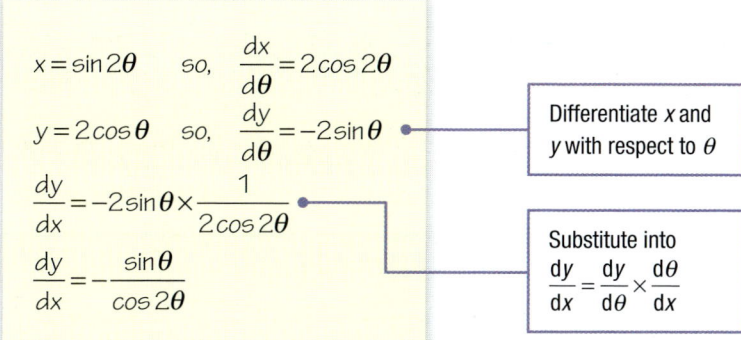

$x = \sin 2\theta$ so, $\dfrac{dx}{d\theta} = 2\cos 2\theta$

$y = 2\cos\theta$ so, $\dfrac{dy}{d\theta} = -2\sin\theta$

Differentiate x and y with respect to θ

$\dfrac{dy}{dx} = -2\sin\theta \times \dfrac{1}{2\cos 2\theta}$

$\dfrac{dy}{dx} = -\dfrac{\sin\theta}{\cos 2\theta}$

Substitute into $\dfrac{dy}{dx} = \dfrac{dy}{d\theta} \times \dfrac{d\theta}{dx}$

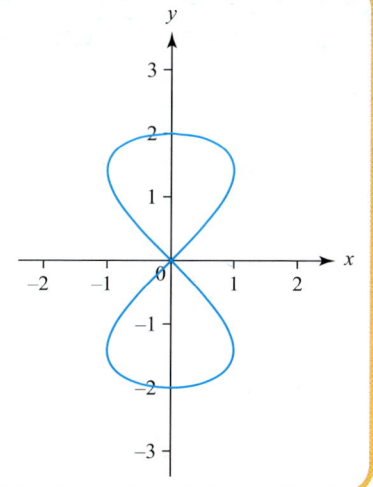

Exercise 15.8A Fluency and skills

1 Find $\dfrac{dy}{dx}$ for each pair of parametric equations.

a $x = 2t$; $y = t + 1$

b $x = t + 4$; $y = 2t - 1$

c $x = t^2$; $y = t$

d $x = \sin\theta$; $y = \cos\theta$

e $x = \sin(\theta + 3)$; $y = \cos(\theta + 2)$

f $x = 3 + \sin\theta$; $y = 1 + \cos\theta$

g $x = \sin(2\theta + 1)$; $y = \cos(2\theta + 1)$

h $x = e^t + e^{-t}$; $y = e^t - e^{-t}$

i $x = 10\ln t$; $y = 10\,e^{-t}$

j $x = 2t^3$; $y = 3t^2$

k $x = \sin t$; $y = t^2$

2 Find $\dfrac{dy}{dx}$ for each pair of parametric equations.

a $x = 9\cos\theta + 3\cos 3\theta$; $y = 9\sin\theta - 3\sin 3\theta$

b $x = 3\sin 2\theta$; $y = 4\cos 2\theta$

c $x = \sin^{-1}\theta$; $y = \cos^{-1}\theta$, $-1 \leq \theta \leq 1$

d $x = \sin^{-1}\theta$; $y = \tan^{-1}\theta$, $-1 \leq \theta \leq 1$

e $x = \sin^2\theta$; $y = \cos^2\theta$

f $x = \sqrt{3t}$; $y = \dfrac{1}{\sqrt{t}}$

g $x = \sqrt{\sin t}$; $y = \sqrt{\cos t}$

h $x = \cos 2\theta$; $y = 4\sin\theta$

i $x = e^{\sin t}$; $y = e^{\cos t}$

j $x = \dfrac{1}{\sin t}$; $y = \dfrac{\cos t}{\sin t}$

k $x = 1 - 5t$; $y = 1 + 20t - 5t^2$

3 A spiral has parametric equations
$x = \theta\cos\theta$, $y = \theta\sin\theta$, $0 \leq \theta \leq 2\pi$

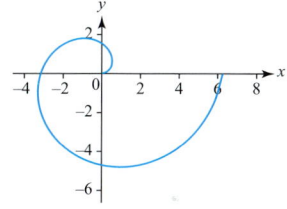

a Find the Cartesian coordinates of the point corresponding to a parameter value $\theta = \pi$

b Find the gradient of the tangent at this point. Show your working.

4 For each curve, find the coordinates of the point corresponding to the given parameter value. Find the gradient at that point, showing your working.

a $x = t^3$; $y = t$; when $t = 2$

b $x = \dfrac{t}{t+1}$; $y = \dfrac{t^2}{t+1}$, $t \neq -1$; when $t = 3$

c $x = \sqrt{2}\sin t$; $y = 2\sqrt{2}\cos t$; when $t = \dfrac{\pi}{4}$

d $x = \sqrt{2}\cos 3t$; $y = 3\sin 2t$; when $t = \dfrac{\pi}{12}$

e $x = 3\sec t$; $y = 4\tan t$; when $t = \dfrac{\pi}{3}$

f $x = 2t - 1$; $y = 4 + 3t$; when $t = 4$

Reasoning and problem-solving

To solve problems involving differentiating parametric equations

(1) Differentiate the equations with respect to the parameter.

(2) Using the chain rule, substitute the results into $\dfrac{dy}{dx} = \dfrac{dy}{dt} \times \dfrac{dt}{dx}$

(3) Find an expression relating only y and x

(4) Use the given conditions to answer the question.

Example 3

An ellipse has parametric equations $x = 2\sin\theta$; $y = 3\cos\theta$

a Find the rate of change of y with respect to x when $x = 1$. Show your working.

b State the relationship between the variables y and x alone.

a $x = 2\sin\theta$ so, $\dfrac{dx}{d\theta} = 2\cos\theta$

$y = 3\cos\theta$ so, $\dfrac{dy}{d\theta} = -3\sin\theta$

> Differentiate with respect to θ ①

$\dfrac{dy}{dx} = \dfrac{-3\sin\theta}{2\cos\theta} = -\dfrac{3}{2}\tan\theta$

> Substitute into $\dfrac{dy}{dx} = \dfrac{dy}{d\theta} \times \dfrac{d\theta}{dx}$ ②

When $x = 1$, $\sin\theta = \dfrac{1}{2}$ so, $\theta = \dfrac{\pi}{6}$

$\dfrac{dy}{dx} = -\dfrac{3}{2}\tan\dfrac{\pi}{6}$

> Find the value of θ when $x = 1$ ④

$= -\dfrac{3}{2} \times \dfrac{1}{\sqrt{3}}$

$= -\dfrac{\sqrt{3}}{2}$

> Substitute $\theta = \dfrac{\pi}{6}$ into the equation to find the rate of change. ④

So the rate of change is $-\dfrac{\sqrt{3}}{2}$ when $x = 1$

b $x = 2\sin\theta$ so, $\dfrac{x}{2} = \sin\theta$

$y = 3\cos\theta$ so, $\dfrac{y}{3} = \cos\theta$

Squaring both equations: $\dfrac{x^2}{4} = \sin^2\theta$; $\dfrac{y^2}{9} = \cos^2\theta$

Adding both equations: $\dfrac{x^2}{4} + \dfrac{y^2}{9} = \sin^2\theta + \cos^2\theta$

The required relationship is: $\dfrac{x^2}{4} + \dfrac{y^2}{9} = 1$

> Manipulate the equations so that you can eliminate θ by using the identity $\sin^2\theta + \cos^2\theta = 1$. This allows you to find an expression relating only y and x ③

1 A function is defined by the parametric equations $x = \dfrac{t}{2}$; $y = 2t^3 - t^2 + 1$

    a Find the rate of change of y with respect to x when $x = 4$. Show your working.

    b Express y in terms of x alone.

2 An ellipse has parametric equations $x = \sin\theta$; $y = 5\cos\theta$

    a Find an expression relating only the variables y and x

    b Find the equations of the two tangents to the ellipse which are parallel to the x-axis.

3 The sketch shows the function with parametric equations $x = 2 - 4t^2$; $y = 3t^3 - t + 1$

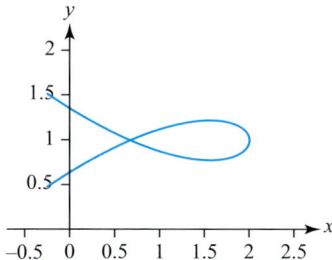

    a Find $\dfrac{dy}{dx}$

    b At what values of the parameter, t, will the curve be stationary?

    c Find the stationary points and use the sketch to identify their nature.

    d **i** Find the three values of the parameter, t, that correspond to $y = 1$

       ii With reference to the sketch, comment on what's happening at these three points.

4 The parametric equations $x = 1 + 5\cos\theta$; $y = 2 + 5\sin\theta$ represent a circle.

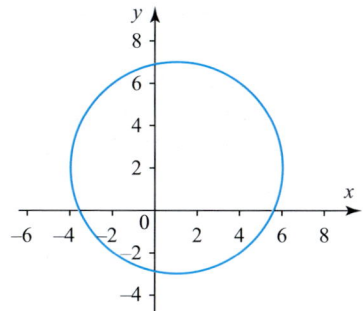

    a Find $\dfrac{dy}{dx}$

    b Find the gradient of the normal to the circle at the point that corresponds to a parameter value of $\theta = \dfrac{\pi}{3}$

    c Using the parametric equations, show that the Cartesian equation of the circle is $(x-1)^2 + (y-2)^2 = 25$

5 A ball is thrown in the air. The flight-path follows a parabolic trajectory.

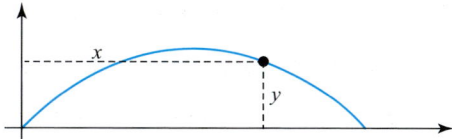

    t seconds after release, the position of the ball (x, y) can be expressed by the parametric equations

    $x = 10t$; $y = 10t - 5t^2$

    a Find the rate of change of y with respect to x

    b **i** What are the t-values that correspond to a y-value of zero?

       ii What are the corresponding x-values?

    c Using the parametric equations, show that the Cartesian equation of the flight-path is a quadratic of the form $y = ax^2 + bx + c$, where a, b and c are constants to be found.

Challenge

6 An ellipse has parametric equations $x = 2\cos\theta$; $y = 6\sin\theta$. Point A has coordinates $(\sqrt{2}, 3\sqrt{2})$ and lies on the ellipse. Find the point at which the normal to the ellipse at point A intersects the x-axis.

15 Summary and review

Chapter summary

- You can partially determine the shape of a curve by considering the derivative and second derivative.

  If $\frac{dy}{dx} > 0$, the function is increasing. If $\frac{dy}{dx} < 0$, the function is decreasing.

  If $\frac{d^2y}{dx^2} > 0$, the curve is convex. If $\frac{d^2y}{dx^2} < 0$, the curve is concave.

- A **point of inflection** is where the graph of the curve changes concavity, i.e. where the second derivative changes sign.

- At a point of inflection, $\frac{d^2y}{dx^2} = 0$, but $\frac{d^2y}{dx^2} = 0$ does not always imply a point of inflection.

- If $\frac{dy}{dx} = 0$ and there is a change in concavity, then you have a horizontal point of inflection.

- You can use small-angle approximations for $\sin\theta$ and $\cos\theta$ to derive the following results.

  $$\lim_{\theta \to 0} \frac{\sin\theta}{\theta} = 1 \text{ and } \lim_{\theta \to 0} \frac{1-\cos\theta}{\theta} = 0$$

  Using these results, you can derive, from first principles, the derivatives of $\sin x$ and $\cos x$

  $$\frac{d}{dx}(\sin x) = \cos x \,;\, \frac{d}{dx}(\cos x) = -\sin x$$

- You need to know the following derivatives.

  - $\frac{d}{dx}(e^{kx}) = ke^{kx}$

  - $\frac{d}{dx}(\ln x) = \frac{1}{x}$

  - $\frac{d}{dx}(a^{kx}) = ka^{kx}\ln a$

- The product rule is used to differentiate two functions, u and v, that have been multiplied together.

  In Leibniz notation, it is written $\frac{d}{dx}(uv) = v\frac{du}{dx} + u\frac{dv}{dx}$

- The quotient rule is used to differentiate two functions, u and v, one of which has been divided by the other.

  In Leibniz notation, it is written $\frac{d}{dx}\left(\frac{u}{v}\right) = \frac{v\frac{du}{dx} - u\frac{dv}{dx}}{v^2}$

- If x is expressed as a function of y, then $\frac{dy}{dx} = \frac{1}{\frac{dx}{dy}}$

- The chain rule is used to differentiate a function of a function, known as a composite function. Consider the composite function $y = f(g(x))$. Let $g(x) = u$ so that $y = f(u)$.

  Then $\dfrac{dy}{dx} = \dfrac{dy}{du} \times \dfrac{du}{dx}$

  The chain rule allows you to solve problems involving connected rates of change.

- Equations presented in a form where the dependent variable is not explicitly given as the subject are known as implicit functions.

- To differentiate implicit functions, you must use the chain rule to differentiate terms expressed as a function of y and the product rule to differentiate terms which involve a product of x and y

- If x and y are defined parametrically, $x = f(t)$ and $y = g(t)$, you can use the chain rule to find $\dfrac{dy}{dx}$

  $\dfrac{dy}{dx} = \dfrac{dy}{dt} \times \dfrac{dt}{dx}$

Check and review

You should now be able to...	Try Questions
✔ Determine when a curve is convex or concave.	1, 2
✔ Find points of inflection on a curve.	1, 2
✔ Use $\lim\limits_{\theta \to 0} \dfrac{\sin \theta}{\theta} = 1$ and $\lim\limits_{\theta \to 0} \dfrac{1 - \cos \theta}{\theta} = 0$	3
✔ Differentiate $\sin x$ and $\cos x$	3, 4, 6, 7, 9, 10, 15
✔ Differentiate e^x, a^x, and $\ln x$	5–7, 9–11, 14
✔ Use the product and quotient rules for differentiation.	6, 7, 11
✔ Use the chain rule for differentiation.	4, 5, 8–11, 15
✔ Find the derivative of a function which is defined implicitly.	11–13
✔ Find the derivative of an inverse function.	14, 15
✔ Find the derivative of a function which is defined parametrically.	16–18

1 The curve $y = \dfrac{x^3}{3} + x^2 - 3x + 1$ has two turning points and a point of inflection.

Find each point and describe each fully. Show your working.

2 **a** Find the stationary points on the curve $y = x^4 + 8x^3 + 18x^2 + 1$. Show your working.

   b Determine the nature of each of the points.

3 **a** Differentiate $y = \sin 2x$ from first principles.

   b $f(x) = 2\sin x + 3\cos x$. Showing your working, calculate   **i** $f'\left(\dfrac{\pi}{4}\right)$ **ii** $f''\left(\dfrac{\pi}{4}\right)$

4 A toy attached to a spring bobs up and down. Its position relative to its starting position is given by $y = 4\sin\left(3x + \dfrac{\pi}{4}\right)$ where the toy is y metres from its starting position after x seconds.

When $y < 0$, the toy is below its starting position.

   a Find the rate of change of the distance of the toy from its starting position at $x = 10$. Show your working.

   b Is the toy rising or falling after 3 seconds? Show your working.

   c When is *the rate of change of the rate of change* of y first equal to zero?

5 Find the derivative of each of the following.

   a $4e^{2x+1}$ **b** $\ln(3x+1)$

   c 2×3^x **d** 2^{-x}

6 Differentiate each of the following functions with respect to x

   a $e^x \sin^2 x$ **b** $\dfrac{3\ln x}{2x}$

   c $\sqrt{x}\,\tan x$ **d** $\dfrac{x^3 + 6x + 11}{\cos x}$

7 Use both the product rule and the quotient rule to differentiate these functions.

   a $\dfrac{xe^x}{\ln x}$ **b** $\dfrac{\cos x}{x\sin x}$ **c** $\dfrac{x\cos x}{2x+1}$

8 $y = (2x^2 - 3)^5$

   a Calculate $\dfrac{dy}{dx}$

   b Hence, state the range of values of x for which y is increasing.

9 Find the gradient of the curve at the point given. Show your working.

   a $y = (3x-1)^{\frac{1}{3}}$ at $x = 3$ **b** $y = \cos\left(2x^2 + \dfrac{3\pi}{2}\right)$ at $x = 0$

   c $y = \ln(x^2 - 8)$ at $x = -3$ **d** $y = e^{\sqrt{2x+1}}$ at $x = 4$

10 Find the derived function when

   a $f(x) = 3^{\cos x}$ **b** $f(x) = 4^{\ln x}$

11 Find $\dfrac{dy}{dx}$ when

   a $\dfrac{3}{x^2} + \dfrac{1}{y^2} = 6$ **b** $\cos x \sin y = 5$

   c $xe^y + ye^x = x$ **d** $e^y \ln x = x + 2y$

12 A curve has equation $3x + 2y^2 = 6y$

   a Find an expression for $\dfrac{dy}{dx}$

   b Find the gradient of the curve at the point $(-12, -3)$.

13 A parabola has equation $x^2 + 6xy + 9y^2 + x + 3y + 1 = 0$. Find $\dfrac{dy}{dx}$

14 Given that $x = e^y + 2y$

   a Find an expression for $\dfrac{dy}{dx}$ in terms of y

   b Calculate the gradient at the point $(1, 0)$

15 Differentiate each of the following functions with respect to x

   a $\sin^{-1}\left(\sqrt{1-x^4}\right)$ **b** $\cos^{-1}\left(1 - e^{-\frac{x}{10}}\right)$ **c** $\tan^{-1}\sqrt{x^2-1}$

16 Determine $\dfrac{dy}{dx}$ for each pair of parametric equations.

   a $x = 3t - 4;\ y = 6t + 1$ **b** $x = t^2 + 1;\ y = t - 1$

   c $x = 3\sin\theta + 1;\ y = 4\cos\theta - 5$ **d** $x = \ln t;\ y = t^2$

17 A curve has parametric equations

   $x = 2t,\qquad y = \dfrac{3}{t^2},\qquad t \neq 0$

   a Find an expression for $\dfrac{dy}{dx}$

   b Calculate the gradient of the curve at the point where $t = 4$

18 The sketch shows the function $x = \cos t - \sqrt{3}\sin t;\ y = 2\sin t - \sqrt{3}\cos t$

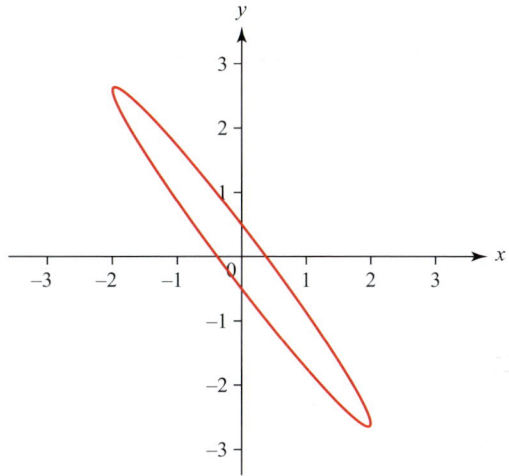

   a Find where the curve cuts the x-axis. Show your working.

   b Find the gradient of the ellipse at these points. Show your working.

What next?

Score		
0–9	Your knowledge of this topic is still developing. To improve, search in MyMaths for the codes: 2161–2166, 2222, 2223, 2271, 2272	🔗
10–14	You're gaining a secure knowledge of this topic. To improve, look at the InvisiPen videos for Fluency and skills (15A)	🎞
15–18	You've mastered these skills. Well done, you're ready to progress! To develop your exam techniques, look at the InvisiPen videos for Reasoning and problem-solving (15B)	🎞

Click these links in the digital book

History

Gottfried Leibniz (1640–1716) is generally credited with the product rule for differentiation. In **Leibniz notation**

$$\frac{d(uv)}{dx} = u\,\frac{dv}{dx} + v\,\frac{du}{dx}$$

Isaac Barrow (1630 – 1677) also contributed to the early development of calculus and may have independently discovered the product rule. **Isaac Newton**, who was once a student of Barrow's, then went on to develop the theory of calculus, causing a bitter dispute over who got there first between him and Leibniz.

Note

The brackets are used here to show that we are finding the nth derivative rather than the nth power.

This may be written as
$(f.g)^{(1)} = f.g^{(1)} + f^{(1)}g$

The dots indicate that we are taking the product of the functions rather than finding one as a function of the other.

Investigation

Suppose that f and g are functions of x
Using $f(n)$ to represent the nth derivative of f with respect to x, we can write the rule for the derivative of a product as
$(fg)^{(1)} = fg^{(1)} + f^{(1)}g$

Differentiating a second time gives
$(fg)^{(2)} = fg^{(2)} + f^{(1)}g^{(1)} + f^{(1)}g^{(1)} + f^{(2)}g$
$\qquad\;\; = fg^{(2)} + 2f^{(1)}g^{(1)} + f^{(2)}g$

The process may be continued.
Differentiating a third time gives
$(fg)^{(3)} = fg^{(3)} + f^{(1)}g^{(2)} + 2f^{(1)}g^{(2)} + 2f^{(2)}g^{(1)} + f^{(2)}g^{(1)} + gf^{(3)}$
$(fg)^{(3)} = fg^{(3)} + 3f^{(1)}g^{(2)} + 3f^{(2)}g^{(1)} + gf^{(3)}$

Does this look familiar? Compare it to the binomial expansion formula
$(a+b)^n = \sum_{r=0}^{n} {}^nC_r\, a^{n-r}b^r$

Try to write a rule for the derivative of $(fg)^{(n)}$. This rule is known as **Leibniz's rule**.

Challenge

Try using **Leibniz's rule** in reverse to integrate the expression

$x^4e^x + 16x^3e^x + 72x^2e^x + 96xe^x + 24e^x$

"Nothing of worth or weight can be achieved with half a mind, with a faint heart, and with a lame endeavour."
- Isaac Barrow

In questions that tell you to show your working, you shouldn't depend solely on a calculator. For these questions, solutions based entirely on graphical or numerical methods are not acceptable.

1 $y = 3x^2 - \dfrac{2}{\sqrt{x}}$

   a Find $\dfrac{\mathrm{d}y}{\mathrm{d}x}$ **[2 marks]**

   b Calculate the rate of change of y with respect to x at the point where $x = 4$ **[2]**

   c Find the equation of the normal to the curve at the point where $x = 1$ **[5]**

2 The volume (in m³) of water in a tank at time t seconds is given by

$$V = t - \frac{4}{t^2}$$

At what time will the rate of change of the volume be $2\,\mathrm{m^3\,s^{-1}}$? Show your working.   **[4]**

3 **a** Differentiate the following expression with respect to x

     i $5\cos x$ **ii** $\ln x$ **iii** $\dfrac{1}{4}\mathrm{e}^x$ **[3]**

   b Find the equation of the normal to $y = 2\ln x$ at the point where $x = 1$

     Give your answer in the form $ax + by + c = 0$ where a, b and c are integers.   **[5]**

4 **a** Given that $y = x\mathrm{e}^x$ find expressions for   **i** $\dfrac{\mathrm{d}y}{\mathrm{d}x}$ **ii** $\dfrac{\mathrm{d}^2 y}{\mathrm{d}x^2}$ **[4]**

   b Write down an expression for the kth derivative of $y = x\mathrm{e}^x$ **[1]**

5 $\mathrm{f}(x) = 3x\sin x$

   a Calculate $\mathrm{f}'(x)$ **[2]**

   b Find an equation of the tangent to $y = \mathrm{f}(x)$ at the point where $x = \pi$ **[3]**

6 **a** Differentiate these expressions with respect to x

     i $\dfrac{x}{x+2}$ **ii** $\dfrac{3x^2}{\cos x}$ **iii** $(3x^3 + 5)\mathrm{e}^x$ **[6]**

   b Show that the derivative of $\dfrac{x^2 + 3x}{x - 5}$ can be written as $\dfrac{ax^2 + bx + c}{(x-5)^2}$, where a, b and c are constants to be found.   **[3]**

7 **a** Differentiate these expressions with respect to x

     i $\cos 3x$ **ii** $2\mathrm{e}^{3x}$ **iii** $\sin(2x - 5)$ **[3]**

   b Calculate the exact gradient of the curve of $y = \sin^2 x$ at the point where $x = \dfrac{\pi}{3}$.

     Show your working.   **[3]**

8 Given that $x = 2y^2 - 8\sqrt{y}$

   a Find   **i** $\dfrac{\mathrm{d}x}{\mathrm{d}y}$ **ii** $\dfrac{\mathrm{d}y}{\mathrm{d}x}$ in terms of y **[4]**

   b Work out the equation of the normal at the point where $y = 4$ **[4]**

9 Find $\dfrac{\mathrm{d}y}{\mathrm{d}x}$ given that $5xy - y^3 = 7$ **[4]**

10 A curve is defined by the parametric equations $\quad x=t^2, \qquad y=6t, \qquad t>0$

    a Calculate the gradient of the curve when $y=18$. Show your working.   **[4]**

    b Find the equation of the tangent at the point where $x=4$ **[4]**

11 $y=2x^2\,\mathrm{e}^x$

    a Use calculus to find the stationary points of the curve. Show your working.   **[6]**

    b Classify each stationary point.   **[4]**

12 A curve C has the equation $y=x^3-4x+3$

    a Identify and describe any points of inflection on the curve.   **[4]**

    b Sketch the curve C **[2]**

    c For what values of x is the curve concave?   **[1]**

13 A curve C has the equation $y=x^2\mathrm{e}^x+\mathrm{e}^x$

    a Showing your working, find the stationary point on the curve and show that it is a point of inflection.   **[5]**

    b By considering $\dfrac{\mathrm{d}^2y}{\mathrm{d}x^2}$, show that the curve has another point of inflection.   **[4]**

14 A cuboid has length twice its width as shown.

    The volume of the cuboid is $192\,\mathrm{cm}^3$

    a Show that the surface area of the cuboid, S, is given by

       $S=4x^2+\dfrac{k}{x}$, where k is a constant to be found.   **[5]**

    b Find the minimum value of S, showing your working.   **[5]**

    c Use calculus to justify this is a minimum.   **[3]**

15 Find and classify the stationary point on the curve $y=x^3+3x^2+3x-5$. Show your working.   **[5]**

16 Use differentiation from first principles to find the derivative of

    a $\sin x$ **[4]**   **b** $\cos 2x$ **[4]**

17 $\mathrm{f}(x)=\tan x$

    a Write down $\mathrm{f}'(x)$ **[1]**

    b Show that $\mathrm{f}''(x)=2\tan x\sec^2 x$ **[3]**

18 You are given that $y=4^x$

    a Write down $\dfrac{\mathrm{d}y}{\mathrm{d}x}$ **[1]**

    b The tangent to the curve at the point $(-1, 0.25)$ cuts the x-axis at point A

       Find the exact x-coordinate of A **[4]**

19 Given that $y=\mathrm{e}^x\sin 5x$

    a Calculate   **i** $\dfrac{\mathrm{d}y}{\mathrm{d}x}$ **ii** $\dfrac{\mathrm{d}^2y}{\mathrm{d}x^2}$ **[6]**

    b Find the smallest positive value of x to give a maximum value of y and prove it is a maximum. Show your working.   **[5]**

20 **a** Differentiate the following with respect to t

    i $\dfrac{e^{3t}}{t^2+1}$ **ii** $3t\ln t$ **iii** $e^{-t}\sin 4t$ **[8]**

  b Find $g'(x)$, when $g(x)=2^x\tan x$ **[2]**

21 Given that $y=\sec x$

  a Prove that $\dfrac{dy}{dx}=\sec x\tan x$ **[4]**

  b Find $\dfrac{d^2y}{dx^2}$, giving your answer in terms of $\sec x$ only.   **[3]**

22 The value of money in an account after t years is approximated by the formula

    $V=ke^{0.03t}$, $k>0$ is a constant.

  Given that £5000 is invested originally,

  a Work out the value of the account after 10 years,   **[3]**

  b Calculate the rate in increase in value when $t=5$. Show your working.   **[3]**

23 A curve C has equation $x=y\ln y$

  a Calculate $\dfrac{dy}{dx}$ at the point where $y=2$. Show your working.   **[4]**

  The normal to C at the point where $y=2$ intersects the y-axis at point A

  b Find the exact y-coordinate of A. Show your working.   **[5]**

24 Given that $x=\cos y$, show that $\dfrac{dy}{dx}=-\dfrac{1}{\sqrt{1-x^2}}$ **[4]**

25 Find an expression for the gradient in each case.

  a $xe^y-\ln x=12$ **[4]**   **b** $x\sin y-2x^2y=0$ **[4]**

26 A curve C has equation $\sin x+\dfrac{1}{2}\tan y=\sqrt{3}$, $0\le x, y\le \pi$

  Find $\dfrac{dy}{dx}$ when $x=\dfrac{\pi}{3}$ **[7]**

27 A curve has parametric equations $x=3t+7$, $y=2+\dfrac{3}{t}$, $t\ne 0$

  a Find the gradient of the curve at the point where $x=10$. Show your working.   **[4]**

  b Find a Cartesian equation of the curve in the form $y=f(x)$ where $f(x)$ is expressed as a single fraction.   **[3]**

28 A curve is defined by the parametric equations $x=2\sin t$, $y=3\cos t$

  a Find an expression for $\dfrac{dy}{dx}$ **[3]**

  b Work out the equation of the tangent at the point when $t=\dfrac{2\pi}{3}$

    Give your answer in the form $a\sqrt{3}x+by+c=0$, where a, b and c are integers.   **[4]**

29 Find and classify the stationary point of the curve $y=x^4-3$. Show your working.   **[5]**

30 Use calculus to determine if the following functions are convex or concave.

  a $f(x)=3x^2-7x+8$ **[3]**   **b** $g(x)=(2-x)^4$ **[4]**

31 Find the range of values of x for which the function $f(x)=\ln(x^2+1)$ is concave. Show your working.   **[5]**

32 A sector of a circle has radius r and angle θ as shown.

Given that the arc length is 6 cm.

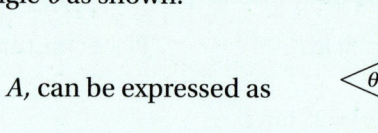

a Show that the area of the triangle, A, can be expressed as

$A = \dfrac{18}{\theta^2}\sin\theta$ **[5]**

b Show that the stationary point occurs when $\tan\theta = \dfrac{\theta}{2}$ **[4]**

33 $f(x) = x - \ln(2x-3), \quad x > \dfrac{3}{2}$

The tangent to the curve $y = f(x)$ at $x = 3$ intersects the x-axis at point P

Find the x-coordinate of P, giving your answer in the form $a + \ln b$. Show your working. **[7]**

34 Given that $y = \operatorname{cosec}3x$

a Show that $\dfrac{dy}{dx} = A\cot 3x\operatorname{cosec}3x$, where A is a constant to be found, **[4]**

b Find $\dfrac{d^2y}{dx^2}$ **[4]**

35 Given that $y = \arctan x$

a Find $\dfrac{dy}{dx}$ **[4]**

b Find an equation of the tangent to $y = \arctan x$ at the point where $x = 1$ **[3]**

The tangent intersects the x-axis at point A and the y-axis at point B

c Show that the area of triangle OAB is $\dfrac{1}{16}(\pi^2 - 4\pi + 4)$ **[5]**

36 A cube has side length x

The volume of the cube is increasing at a rate of $12\ \text{cm}^3\text{s}^{-1}$

Find the rate at which x is increasing when the volume is $216\ \text{cm}^3$ **[6]**

37 The volume of a spherical balloon, $V\ \text{cm}^3$, is increasing at a constant rate of $6\ \text{cm}^3\,\text{s}^{-1}$

Find the rate at which the radius of the sphere is increasing when the volume is $36\pi\ \text{cm}^3$

Leave your answer in exact form. $\left[V = \dfrac{4}{3}\pi r^3\right]$ **[5]**

38 Prove that the derivative of $\arcsin 2x$ is $\dfrac{2}{\sqrt{1-4x^2}}$ **[6]**

39 $xy^2 + 2y = 3x^2$

a Find an expression in terms of x and y for $\dfrac{dy}{dx}$ **[4]**

b Calculate the possible rates of change of y with respect to x when $y = 1$ **[5]**

40 Use implicit differentiation to prove that the derivative of a^x is $a^x\ln a$ **[4]**

41 Given that $\quad x = \dfrac{2}{3-t}, \qquad y = \dfrac{t^2}{3-t}, \qquad t \neq 3,$

a Show that $\dfrac{dy}{dx} = \dfrac{6t-t^2}{2}$ **[5]**

b Find a Cartesian equation in the form $y = f(x)$. Simplify your answer. **[3]**

42 A curve C is defined by the parametric equations $\quad x = \sec(\theta-4), \qquad y = \tan(\theta-4)$

a Show that $\dfrac{dy}{dx} = \operatorname{cosec}(\theta-4)$ **[5]**

b Find a Cartesian equation of C **[2]**

c Show that the equation of the tangent to C at the point where $x = 3$ and y is positive is given by $3x - 2y\sqrt{2} = 1$ **[6]**

16 Integration and differential equations

The study of the motion of objects such as planets, moons and stars is referred to as celestial mechanics. This motion is modelled using differential equations constructed by applying Newton's second law to the inverse square law of gravity. The orbit of Uranus was observed for decades until an accurate theory of motion could be constructed. However, the observed motion did not match the motion predicted by the differential equations when they were compared. This was because the motion of Uranus was being affected by an unknown planet, later to be discovered as Neptune. Once Neptune's motion had been observed, differential equations could be used to make an accurate model of Uranus's orbit.

Differential equations enable the modelling of complex systems of motion. Integration is the reverse of differentiation, and is a tool required to solve differential equations. This chapter introduces several techniques which can be applied to many different scenarios involving integration.

Orientation

What you need to know	What you will learn	What this leads to
Ch4 Differentiation and integration • Introduction to integration. • Area under a curve.	• To integrate a set of standard functions, f(x) and the related functions, f(ax + b) • To find the area between curves.	**Applications** Financial modelling. Demographic modelling. Epidemiology.
Ch12 Algebra 2 • Partial fractions.	• To integrate by substitution, by parts and by using partial fractions.	**Further Core Pure Maths** Volumes of revolution.
Ch15 Differentiation • The product rule for differentiation.	• To understand the expression 'differential equation'. • To use integration where the variables are separable.	

p.112
p.116
p.320
p.406

 **MyMaths** Practise before you start

🔍 2054, 2056, 2163, 2260

Fluency and skills

See Ch4.7
For a reminder on finding the area under a curve.

You can use definite integration to find the area under a curve.

Let A be the area between the curve $y = f(x)$, the x-axis, and the lines $x = a$ and $x = b$. The area of this region can be approximated by dividing it up into rectangles, finding the area of each rectangle, and adding the areas together.

See Ch4.2
For a reminder on Leibniz notation.

Suppose that the area is divided into n rectangles, each of width δx. The height of each rectangle is given by the value of y at the left-hand side of the rectangle.

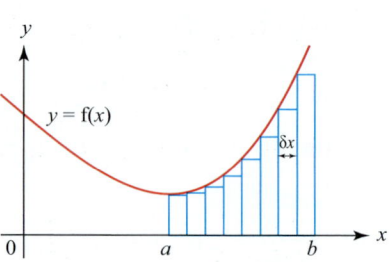

Summing these areas, you get

$$A \approx y_1 \delta x + y_2 \delta x + \ldots + y_n \delta x$$

This can be written using sigma notation $\quad A \approx \sum_{i=1}^{n} y_i \delta x$

To make this approximation more accurate, you must increase the number of rectangles between a and b. In other words, in order to find an accurate value of A, you let the number of rectangles, n, tend to infinity.

Letting $n \to \infty, \quad A = \lim_{n \to \infty} \sum_{i=1}^{n} y_i \delta x$

As the number of rectangles is increased, the limit is equal to the definite integral.

Key point

$$\lim_{n \to \infty} \sum_{i=1}^{n} y_i \delta x = \int_{a}^{b} y \, dx$$

ICT Resource online

To investigate the integrals of trigonometric functions, click this link in the digital book.

Example 1

Consider the curve $y = 3x^2 + 2x + 1$. The area under the curve and between the lines $x = 1$ and $x = 3$ can be estimated by calculating and adding the area of 4 rectangles of equal width, δx

a What is the value of δx in this case?

b Use the total area of the four rectangles, $\sum_{i=1}^{4} y_i \delta x$, to find an estimate for this area.

c Calculate the exact area under the curve, showing your working.

a $\delta x = \dfrac{3-1}{4} = \dfrac{1}{2}$

b When $x = 1, y = 6$

When $x = 1.5, y = 10.75$

When $x = 2, y = 17$

When $x = 2.5, y = 24.75$

$$\sum_{i=1}^{4} y_i \delta x = \left(6 \times \dfrac{1}{2}\right) + \left(10.75 \times \dfrac{1}{2}\right) + \left(17 \times \dfrac{1}{2}\right) + \left(24.75 \times \dfrac{1}{2}\right)$$

$$= 29.25$$

To find δx, divide the total width of the interval by the number of rectangles.

To find each rectangle height, calculate the y-value at the left-hand side of each rectangle.

Integration and differential equations Standard integrals

(*Continued on the next page*)

c $\displaystyle\int_{1}^{3}(3x^2+2x+1)dx=[x^3+x^2+x]_1^3$

$= (3^3+3^2+3)-(1^3+1^2+1)$

$= 27+9+3-3$

$= 36$

Find the exact area using integration. Remember that you don't need to worry about the constant of integration when calculating a definite integral.

Substitute the limits and evaluate.

Integration is the reverse of differentiation. If the derivative of a function F(x) is f(x), then you can say that an indefinite integral of f(x) with respect to x is F(x)

Key point

$F'(x)=f(x) \Rightarrow \int f(x)dx = F(x)+c$

where c is the constant of integration.

An function to be integrated, i.e. $\int f(x)dx$, is sometimes called an integrand.

Using this definition, you can immediately work out a number of **standard integrals**.

Key point

1. $\dfrac{d}{dx}\left(\dfrac{1}{n+1}x^{n+1}\right)=x^n \Rightarrow \int x^n dx = \dfrac{1}{n+1}x^{n+1}+c,\ n\neq-1$

2. $\dfrac{d}{dx}(e^x)=e^x \Rightarrow \int e^x dx = e^x+c$

3. $\dfrac{d}{dx}(\ln x)=\dfrac{1}{x} \Rightarrow \int \dfrac{1}{x}dx = \ln|x|+c$

4. $\dfrac{d}{dx}(\sin x)=\cos x \Rightarrow \int \cos x\,dx = \sin x+c$

5. $\dfrac{d}{dx}(-\cos x)=\sin x \Rightarrow \int \sin x\,dx = -\cos x+c$

6. $\dfrac{d}{dx}(\tan x)=\sec^2 x \Rightarrow \int \sec^2 x\,dx = \tan x+c$

Example 2

Integrate $3\sin x + 2\cos x$ with respect to x

$\int 3\sin x + 2\cos x\,dx = 3\int \sin x\,dx + 2\int \cos x\,dx = -3\cos x + 2\sin x + c$

Example 3

Find   **a** $\int\dfrac{1}{x}dx$ **b** $\int e^x + \sec^2 x\,dx$ **c** $\int\left(\cos x+\dfrac{3}{2x}\right)dx$

a $\int\dfrac{1}{x}dx = \ln|x|+c$

b $\int e^x + \sec^2 x\,dx = \int e^x dx + \int \sec^2 x\,dx$

$= e^x + \tan x + c$

c $\int\left(\cos x+\dfrac{3}{2x}\right)dx = \int\cos x\,dx + \dfrac{3}{2}\int\dfrac{1}{x}dx$

$= \sin x + \dfrac{3}{2}\ln|x|+c$

Use standard integral 3 from above. See Ch12.2 for a reminder on modulus notation.

Use standard integrals 2 and 6 from above.

Use standard integrals 3 and 4 from above.

1 Find

a $\int 24x^5\,dx$ **b** $\int 7\sin x\,dx$

c $\int 4\cos x\,dx$ **d** $\int \dfrac{2}{3}\sec^2 x\,dx$

e $\int 5e^x\,dx$ **f** $\int 2x+\sqrt{x}\,dx$

g $\int \dfrac{3}{x^2}\,dx$ **h** $\int \dfrac{3}{x}\,dx$

i $\int \sin x+\sec^2 x\,dx$ **j** $\int 4\cos x-3\sin x\,dx$

k $\int 3\sec^2 x+2\cos x\,dx$ **l** $\int 5e^x\,dx$

2 Find

a $\int 3-4e^x\,dx$ **b** $\int \dfrac{1}{2x}\,dx$

c $\int \dfrac{1}{3x}+\dfrac{2}{5x}\,dx$ **d** $\int 1-3e^x\,dx$

e $\int 7-6\sin x\,dx$ **f** $\int 4x^3-\dfrac{5}{x}\,dx$

g $\int 1-2\cos x\,dx$ **h** $\int x-\dfrac{1}{x^3}\,dx$

i $\int \dfrac{1}{e}+\dfrac{e^x}{3}\,dx$ **j** $\int \dfrac{3}{4}+\dfrac{2}{x^3}-\dfrac{\sin x}{3}\,dx$

k $\int 4x^{-2}-5\sqrt{x}\,dx$ **l** $\int 1+x^{\frac{2}{3}}\,dx$

3 Evaluate the following integrals. Show your working.

a $\displaystyle\int_4^9 \dfrac{1}{\sqrt{x}}\,dx$ **b** $\displaystyle\int_{\frac{\pi}{2}}^{\pi} \sin x\,dx$

c $\displaystyle\int_{\frac{\pi}{6}}^{\frac{\pi}{3}} \cos x\,dx$ **d** $\displaystyle\int_{\frac{\pi}{4}}^{\frac{\pi}{4}} \sec^2 x\,dx$

e $\displaystyle\int_0^1 e^x\,dx$ **f** $\displaystyle\int_e^{e^3} \dfrac{1}{x}\,dx$

g $\displaystyle\int_2^4 x+\dfrac{8}{x^2}\,dx$ **h** $\displaystyle\int_1^4 1-2x\,dx$

4 Evaluate the following integrals. Show your working.

a $\displaystyle\int_0^5 x^2+2x+1\,dx$ **b** $\displaystyle\int_{\frac{\pi}{2}}^{\pi} 1-\cos x\,dx$

c $\displaystyle\int_{\frac{\pi}{6}}^{\frac{\pi}{3}} 2\sin x+3\,dx$ **d** $\displaystyle\int_{\frac{\pi}{4}}^{\frac{\pi}{4}} \sin x+\sec^2 x\,dx$

e $\displaystyle\int_0^1 x+e^x\,dx$ **f** $\displaystyle\int_{e^2}^{e^4} \dfrac{2}{3x}\,dx$

5 For each of the two following curves

i Split the area under the curve into four rectangles of equal width and find an approximation for each shaded area.

ii Write down a definite integral which, when evaluated, will give the exact shaded area.

iii Use integration to calculate the exact area. Show your working.

a   **b**

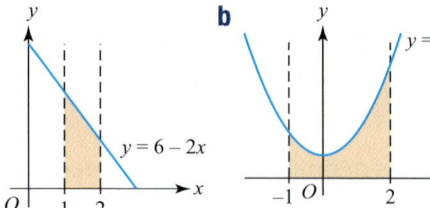

6 For each of the curves below

i Split the area under the curve into six rectangles of equal width and find an approximation for each shaded area.

ii Write down a definite integral which, when evaluated, will give the exact shaded area.

iii Use integration to calculate the exact area. Show your working.

a   **b**

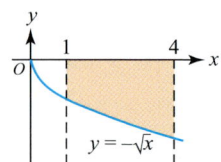

7 Find the shaded area between the curve $y=1+2\sin x$, the x-axis and the lines $x=\dfrac{\pi}{3}$ and $x=\pi$ as shaded in the diagram.

8 The graph shows the function $y=\dfrac{1}{2x}$ in the domain $1\le x\le 9$

Calculate the shaded area between the curve, the x-axis and the lines $x=3$ and $x=3e$. Show your working.

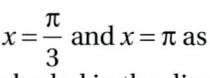

Reasoning and problem-solving

To find the area trapped between two curves, you subtract the area below the bottom curve from the area below the top curve.

> **Key point**
>
> If $f(x) \geq g(x)$ or $f(x) \leq g(x)$ for all x in the interval $a \leq x \leq b$, then the area between the two curves $y = f(x)$ and $y = g(x)$ and the lines $x = a$ and $x = b$ is given by
> $$A = \left| \int_a^b f(x) - g(x) \, dx \right|$$

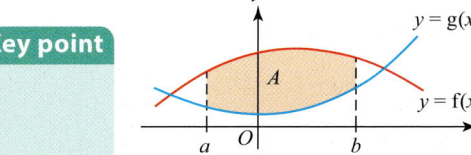

The formula for the area between two curves holds true even when the trapped area straddles the x-axis.

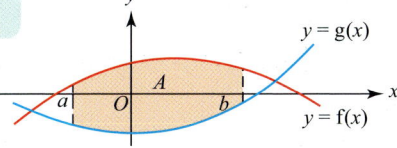

If the curves intersect in the interval, then each trapped area must be considered separately.

The total area is $A + B = \left| \int_a^c f(x) - g(x) \, dx \right| + \left| \int_c^b f(x) - g(x) \, dx \right|$

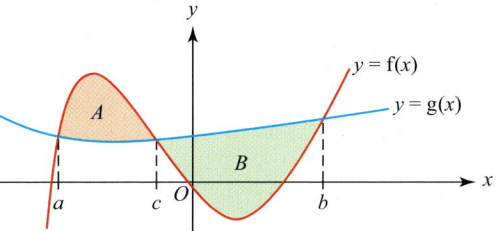

Strategy

To find the area trapped between two curves in an interval

(1) Identify where the curves intersect. This will determine the limits of integration.

(2) Draw a sketch to help you plan which integrals will need to be evaluated.

(3) Integrate and substitute limits.

(4) Where necessary, find the total area by addition.

Example 4

The curve $y = x^2 + x + 4$ intersects the curve $y = -x^2 + 2x + 5$ at the points $\left(-\dfrac{1}{2}, 3\dfrac{3}{4}\right)$ and $(1, 6)$

Find the exact area between the two curves. Show your working.

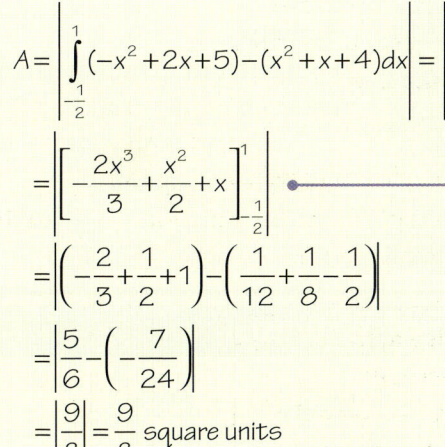

$$A = \left| \int_{-\frac{1}{2}}^{1} (-x^2 + 2x + 5) - (x^2 + x + 4) \, dx \right| = \left| \int_{-\frac{1}{2}}^{1} -2x^2 + x + 1 \, dx \right|$$

$$= \left| \left[-\frac{2x^3}{3} + \frac{x^2}{2} + x \right]_{-\frac{1}{2}}^{1} \right|$$

$$= \left| \left(-\frac{2}{3} + \frac{1}{2} + 1 \right) - \left(\frac{1}{12} + \frac{1}{8} - \frac{1}{2} \right) \right|$$

$$= \left| \frac{5}{6} - \left(-\frac{7}{24} \right) \right|$$

$$= \left| \frac{9}{8} \right| = \frac{9}{8} \text{ square units}$$

① Use the x-values of the points of intersection as the limits of integration.

③ Integrate and substitute the limits.

Note that the same result is same regardless of which area is subtracted from which.

Example 5

Find the area trapped between the curves $y = x^3 - 2x^2 + 2x - 2$ and $y = x^2 - 2$.
Show each step of working.

$x^3 - 2x^2 + 2x - 2 = x^2 - 2$ ● ——— Equate the curves.

$x^3 - 3x^2 + 2x = 0$

$x(x-1)(x-2) = 0$

$x = 0$ or $x = 1$ or $x = 2$

$y = -2$ or $y = -1$ or $y = 2$ respectively. ●

1 Identify where the curves intersect. These x-values will be used as the limits for integration.

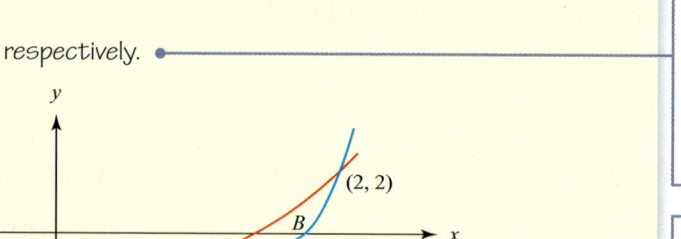

$y = x^2 - 2$

$(2, 2)$

B

O x

A $(1, -1)$

$(0, -2)$

$y = x^3 - 2x^2 + 2x - 2$

2 Draw a sketch. You could check your sketch using a graphics calculator.

$A + B = \left| \int_0^1 (x^3 - 2x^2 + 2x - 2) - (x^2 - 2)\,dx \right| + \left| \int_1^2 (x^3 - 2x^2 + 2x - 2) - (x^2 - 2)\,dx \right|$

$= \left| \int_0^1 x^3 - 3x^2 + 2x\,dx \right| + \left| \int_1^2 x^3 - 3x^2 + 2x\,dx \right|$ ●

3 Integrate and substitute the limits.

$= \left| \left[\frac{1}{4}x^4 - x^3 + x^2 \right]_0^1 \right| + \left| \left[\frac{1}{4}x^4 - x^3 + x^2 \right]_1^2 \right|$

$= \left| \left(\frac{1}{4} - 1 + 1 \right) - 0 \right| + \left| (4 - 8 + 4) - \left(\frac{1}{4} - 1 + 1 \right) \right|$

$= \left| \frac{1}{4} \right| + \left| -\frac{1}{4} \right| = \frac{1}{2}$ ●

4 Find the total area by addition.

Trapped area = 0.5 square units

 **Try it on your calculator**

You can find the area between two curves on a graphics calculator.

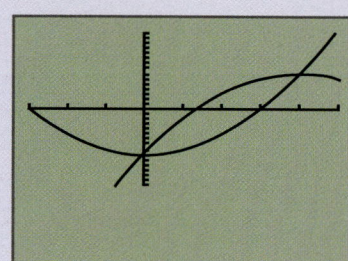

Activity
Find out how to find the area between $y = x^2 - 9$ and $y = -x^2 + 8x - 9$ on *your* graphics calculator.

Exercise 16.1B Reasoning and problem-solving

1 Find

a $\displaystyle\int \frac{x^2}{\sqrt{x}}\,dx$

b $\displaystyle\int \frac{x+1}{3x}\,dx$

c $\displaystyle\int x^2(2x+3)\,dx$

d $\displaystyle\int \sqrt{x}(x+2\sqrt{x})\,dx$

e $\displaystyle\int \frac{x^2 + 2x + 1}{x}\,dx$

f $\displaystyle\int \frac{x+\sqrt{x}}{\sqrt{x}}\,dx$

g $\displaystyle\int \frac{1+\cos^3 x}{\cos^2 x}\,dx$

h $\displaystyle\int \frac{1+\sin x\cos^2 x}{\cos^2 x}\,dx$

i $\int e^x(e^x+1)\,dx$ **j** $\int(e^x+4)(e^x-1)\,dx$

k $\int(e^x+e^{-x})(e^x-e^{-x})\,dx$

l $\int \cos x\tan x\,dx$ **m** $\int \sin^2 x+\cos^2 x\,dx$

2 Each pair of curves traps a single region. Calculate the area of each region, showing your working.

a $y=x^2+2x+1$ and $y=x+3$

b $y=2x^2+12x-11$ and $y=x+10$

c $y=x^2+5x+6$ and $y=1-6x-x^2$

d $y=x^2$ and $y=\sqrt{x}$

3 a The curves $y=\sin x$ and $y=\cos x$ intersect at two points in the domain $0\le x\le 2\pi$. Find the two points, showing your working.

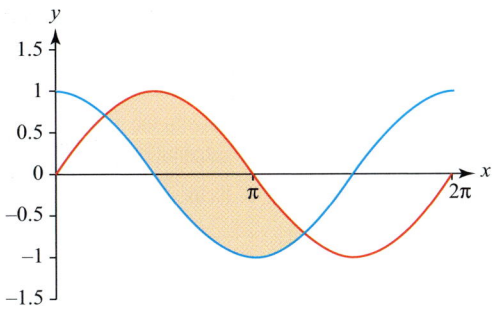

b Find the area trapped between the two curves between these two points, showing your working.

c What is the area of the region trapped between the curves and the y-axis?

4 Each pair of curves traps two distinct regions. Find the total area trapped in each case. Show your working.

a $y=x^3$ and $y=2x^2+x-2$

b $y=x^3+2x^2-3x$ and $y=x^2-x$

c $y=x^4+x^3-x^2$ and $y=2x^2-x^3$

d $y=x^2-6x+11$ and $y=\dfrac{6}{x}$

5 The cross-section of a riverbed can be represented by the equation $y=-2x(-x+4)$, for $0\le x\le 4$. The river flows at a rate of $0.4\,\mathrm{m\,s^{-1}}$. Assuming the river water reaches the top of the bed, calculate the volume of water that flows past a given point in 1 minute. Show your working.

6 Two cars, A and B, enter a race to see which can travel the furthest in a straight line over 4 seconds. The speed, v, with respect to time, of each car during these 4 seconds is given by the equations

$$v_A=4\sqrt{t}$$

$$v_B=t^2-\left(\frac{t}{2}\right)^3$$

a Show that the velocity of each car is the same after 4 seconds.

b Show that car A wins the race by a distance of 8 metres. (Hint: the area under a velocity–time graph gives the distance travelled)

PURE

Integration by substitution

Fluency and skills

Say you are presented with following integral.

$$\int (2x+1)^5 \, dx$$

Without expanding out 5 sets of brackets, the function would be difficult to integrate with the techniques you've learnt so far. But if you substitute the function inside brackets with a u, then integrating u^5 is actually fairly simple. This technique is known as **integration by substitution**.

Example 1

Calculate $\int (2x+1)^5 \, dx$

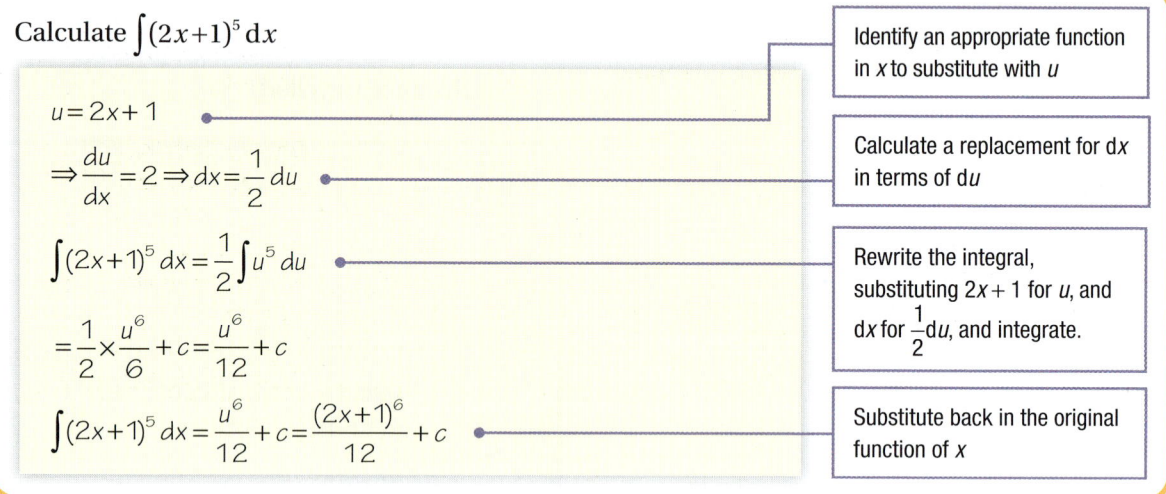

Identify an appropriate function in x to substitute with u

$$u = 2x + 1$$

Calculate a replacement for dx in terms of du

$$\Rightarrow \frac{du}{dx} = 2 \Rightarrow dx = \frac{1}{2} du$$

Rewrite the integral, substituting $2x + 1$ for u, and dx for $\frac{1}{2} du$, and integrate.

$$\int (2x+1)^5 \, dx = \frac{1}{2} \int u^5 \, du$$

$$= \frac{1}{2} \times \frac{u^6}{6} + c = \frac{u^6}{12} + c$$

Substitute back in the original function of x

$$\int (2x+1)^5 \, dx = \frac{u^6}{12} + c = \frac{(2x+1)^6}{12} + c$$

See Ch15.5
For a reminder on the chain rule.

You may notice that this technique is essentially just the chain rule in reverse, so it's useful when you can spot a function and its derivative within the integral.

If the derivative of the function is, or contains, a constant, you can compensate by multiplying the integral by the reciprocal of that constant: in Example 1, the function is $2x + 1$ and the derivative of that is 2, which is why you need to multiply the substituted integral by $\frac{1}{2}$

Try using the chain rule, where $u = 2x + 1$, to differentiate $y = \frac{(2x+1)^6}{12}$:

$$\frac{dy}{dx} = \frac{dy}{du} \times \frac{du}{dx} = \frac{6u^5}{12} \times 2 = \frac{12}{12} u^5 = (2x+1)^5$$

This calculation shows Example 1 in reverse.

Some functions can be integrated quickly without the need for much working out. For example, the following result can be deduced using the integration by substitution technique.

Key point

In general, when $\int f(x) \, dx = F(x) + c$

$$\int f(ax+b) \, dx = \frac{1}{a} F(ax+b) + c$$

Example 2

Find   **a** $\int \sin(3x+4)\,dx$ **b** $\int \dfrac{1}{4x+3}\,dx$ **c** $\int e^{1-5x}\,dx$

a $\int \sin(3x+4)\,dx = -\dfrac{1}{3}\cos(3x+4)+c$ ●────────────────────

b $\int \dfrac{1}{4x+3}\,dx = \dfrac{1}{4}\ln|4x+3|+c$ ●────

c $\int e^{1-5x}\,dx = -\dfrac{1}{5}e^{1-5x}+c$ ●────────

> Use $\int f(ax+b)\,dx = \dfrac{1}{a}F(ax+b)+c$

When using substitution to find a **definite integral**, you substitute the corresponding values of u for the values of x in the limits. By doing this, you do not need to make the substitution back into the original variable after integration.

Example 3

Evaluate $\displaystyle\int_0^4 x\sqrt{x^2+9}\,dx$ using the substitution $u = x^2+9$

> Separate the operators du and dx

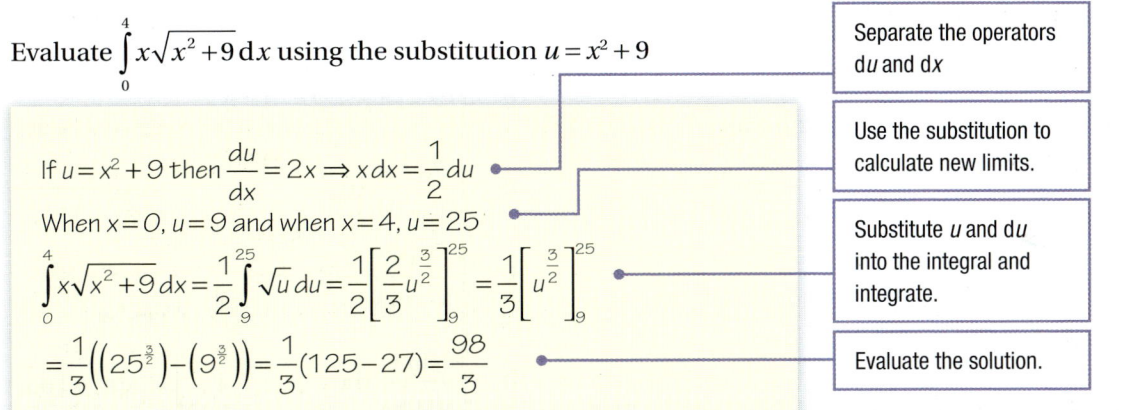

If $u = x^2 + 9$ then $\dfrac{du}{dx} = 2x \Rightarrow x\,dx = \dfrac{1}{2}du$ ●────

When $x = 0$, $u = 9$ and when $x = 4$, $u = 25$

$$\int_0^4 x\sqrt{x^2+9}\,dx = \frac{1}{2}\int_9^{25}\sqrt{u}\,du = \frac{1}{2}\left[\frac{2}{3}u^{\frac{3}{2}}\right]_9^{25} = \frac{1}{3}\left[u^{\frac{3}{2}}\right]_9^{25}$$

$$= \frac{1}{3}\left(\left(25^{\frac{3}{2}}\right)-\left(9^{\frac{3}{2}}\right)\right) = \frac{1}{3}(125-27) = \frac{98}{3}$$

> Use the substitution to calculate new limits.

> Substitute u and du into the integral and integrate.

> Evaluate the solution.

Exercise 16.2A Fluency and skills

1 Find

   a   **i** $\int \dfrac{1}{\sqrt{2x}}\,dx$ **ii** $\int \dfrac{2}{\sqrt{4x-1}}\,dx$

   b   **i** $\int \sin 4x\,dx$ **ii** $\int \sin(5x-2)\,dx$

   c   **i** $\int \cos 3x\,dx$ **ii** $\int 3\cos(2-3x)\,dx$

   d   **i** $\int \sec^2 5x\,dx$ **ii** $\int \sec^2(4+3x)\,dx$

   e   **i** $\int e^{7x}\,dx$ **ii** $\int 4e^{-2x}\,dx$

   f   **i** $\int \dfrac{1}{2x}\,dx$ **ii** $\int \dfrac{2}{4x+5}\,dx$

2 State the integral of

   a $(3x+2)^{10}$ **b** $(5x-1)^8$

   c $(7x-3)^{100}$ **d** $(3x-8)^{-8}$

   e $(1-3x)^9$ **f** $(6-x)^7$

   g $\dfrac{3}{(2x-1)^5}$ **h** $\dfrac{1}{(10-x)^5}$

   i $\sin(3-5x)$ **j** $2\cos(4x-1)$

   k $\cos(2x)$ **l** $\sec^2(4x+3)$

   m $3\sec^2(2x+1)$ **n** e^{5x+2}

   o $\dfrac{7}{3x+9}$ **p** $\dfrac{4}{8-x}$

3 Find each integral. A suitable substitution has been suggested.

   a $\int 2x(x^2+2)^4\,dx$; let $u = x^2+3$

   b $\int (2x+1)(x^2+x-1)^3\,dx$; let $u = x^2+x-1$

   c $\int 6x^2\sqrt{2x^3-1}\,dx$; let $u = 2x^3-1$

   d $\int x(2x^2-5)^3\,dx$; let $u = 2x^2-5$

   e $\int \dfrac{x}{(x^2-7)^4}\,dx$; let $u = x^2-7$

   f $\int \dfrac{2x+3}{(x^2+3x-1)^2}\,dx$; let $u = x^2+3x-1$

g $\displaystyle\int \frac{3x}{\sqrt{1-x^2}}\,dx$; let $u=1-x^2$

h $\displaystyle\int \frac{1}{\sqrt[3]{x+4}}\,dx$; let $u=x+4$

i $\displaystyle\int x\sin(3x^2+1)\,dx$; let $u=3x^2+1$

j $\displaystyle\int \sec^2 x\tan^3 x\,dx$; let $u=\tan x$

k $\displaystyle\int \frac{\cos x}{\sin x-1}\,dx$; let $u=\sin x$

l $\displaystyle\int \frac{\sin x}{3-2\cos x}\,dx$; let $u=3-2\cos x$

m $\displaystyle\int xe^{2x^2+1}\,dx$; let $u=2x^2+1$

n $\displaystyle\int \frac{e^{\sqrt{x}}}{\sqrt{x}}\,dx$; let $u=\sqrt{x}$

o $\displaystyle\int \sin x\,e^{\cos x}\,dx$; let $u=\cos x$

p $\displaystyle\int \frac{\ln x}{x}\,dx$; let $u=\ln x$

4 Find each integral using a suitable substitution.

a $\displaystyle\int x(2x^2-1)^3\,dx$

b $\displaystyle\int (x^2+x)\left(\frac{x^3}{3}+\frac{x^2}{2}\right)^{\frac{1}{2}}\,dx$

c $\displaystyle\int \frac{x^2-2x}{\sqrt{x^3-3x^2}}\,dx$

d $\displaystyle\int (x-1)\sin(x^2-2x+3)\,dx$

e $\displaystyle\int \frac{x^2+2x+3}{x^3+3x^2+9x+1}\,dx$

f $\displaystyle\int \frac{\sin\sqrt{x}}{\sqrt{x}}\,dx$

g $\displaystyle\int \frac{x+\cos\sqrt{x+1}}{\sqrt{x+1}}\,dx$

h $\displaystyle\int \frac{1}{\sqrt[3]{2x-1}}\,dx$

i $\displaystyle\int \sin x(1-\cos^3 x)\,dx$

j $\displaystyle\int \sec^2 x(1+\tan^2 x)\,dx$

k $\displaystyle\int \frac{\cos x}{\sqrt{1+\sin x}}\,dx$

l $\displaystyle\int \frac{e^x}{1+e^x}\,dx$

m $\displaystyle\int \sin 2x\,e^{\cos 2x}\,dx$

n $\displaystyle\int \frac{\sqrt{\ln(2x+3)}}{2x+3}\,dx$

5 Evaluate the following integrals. Show your working.

a $\displaystyle\int_1^7 2x\sqrt{x^2+15}\,dx$

b $\displaystyle\int_1^4 \frac{(\sqrt{x}-1)^2}{2\sqrt{x}}\,dx$

c $\displaystyle\int_0^1 (x-1)(x^2-2x-1)^3\,dx$

d $\displaystyle\int_1^2 \frac{2x+3}{x^2+3x}\,dx$

e $\displaystyle\int_5^{11} \frac{x}{\sqrt{x^2-21}}\,dx$

f $\displaystyle\int_1^2 \frac{x-2}{x^2-4x+1}\,dx$

g $\displaystyle\int_0^{\frac{\pi}{6}} \frac{\sin x}{\cos^2 x}\,dx$

h $\displaystyle\int_0^{\frac{\pi}{6}} \frac{\sin x}{\sqrt{\cos x}}\,dx$

i $\displaystyle\int_{\ln 3}^{\ln 8} e^x\sqrt{e^x+1}\,dx$

Reasoning and problem-solving

To integrate functions with terms involving $\sin^2 x$ and $\cos^2 x$

(1) Use the basic identities to express $\sin^2 x$ or $\cos^2 x$ in terms of $\cos 2x$

(2) Integrate using the substitution method.

Example 4

Find $\displaystyle\int (1-3\sin^2 x)\,dx$

Recall that $\sin^2 x=\dfrac{1}{2}(1-\cos 2x)$

$\displaystyle\int (1-3\sin^2 x)\,dx=\int\left(1-3\times\frac{1}{2}(1-\cos 2x)\right)dx$

$\displaystyle =\int\left(-\frac{1}{2}+\frac{3}{2}\cos 2x\right)dx$

$\displaystyle =-\frac{1}{2}x+\frac{3}{2}\times\frac{1}{2}\sin 2x+c$

$\displaystyle =-\frac{1}{2}x+\frac{3}{4}\sin 2x+c$

(1) Express $\sin^2 x$ in terms of $\cos 2x$

(2) Integrate using substitution. Remember that the integral of $f(ax+b)$ is $\dfrac{1}{a}F(ax+b)$, where in this case the function is cos, $a=2$ and $b=0$

Example 5

Find $\int \cos^5 x \, dx$

$$\int \cos^5 x \, dx = \int \cos^4 x \cos x \, dx$$

$$= \int (\cos^2 x)^2 \cos x \, dx$$

$$= \int (1 - \sin^2 x)^2 \cos x \, dx$$

Let $u = \sin x \Rightarrow \dfrac{du}{dx} = \cos x \Rightarrow du = \cos x \, dx$

$$\int (1 - \sin^2 x)^2 \cos x \, dx = \int (1 - u^2)^2 \, du$$

$$= \int (1 - 2u^2 + u^4) \, du$$

$$= u - \frac{2}{3}u^3 + \frac{1}{5}u^5 + c$$

$$= \sin x - \frac{2}{3}\sin^3 x + \frac{1}{5}\sin^5 x + c$$

1 Substitute $\cos^2 x = 1 - \sin^2 x$

2 Look for a function and its derivative within the integrand to choose an appropriate substitution. Here, the function is sin x and its derivative is cos x

2 Integrate.

Substitute the value of u back into the answer.

Exercise 16.2B Reasoning and problem-solving

1 a Find

   i $\int \sin^2 x \, dx$ **ii** $\int \cos^2 x \, dx$

  b Hence state

   i $\int \sin^2 2x \, dx$ **ii** $\int \cos^2 2x \, dx$

   iii $\int \sin^2 (2x-1) \, dx$ **iv** $\int \cos^2 (2x+3) \, dx$

   v $\int \sin^2 (1-x) \, dx$ **vi** $\int \cos^2 (1-2x) \, dx$

2 Make use of trigonometric identities to find

  a $\int 2\sin x \cos x \, dx$ **b** $\int 2\sin x (\cos x + 1) \, dx$

  c $\int (\cos x - \sin x)(\cos x - \sin x) \, dx$

3 a Transform the integral $\int \dfrac{1}{4+x^2} \, dx$ using the substitution $x = 2\tan u$

  b Use the fact that $1 + \tan^2 x = \sec^2 x$ to simplify the integral.

  c Integrate with respect to u

  d Substitute $u = \tan^{-1}\left(\dfrac{x}{2}\right)$ to complete the integration with respect to x

  e Similarly, find

   i $\int \dfrac{1}{9+x^2} \, dx$ **ii** $\int \dfrac{1}{1+4x^2} \, dx$

4 a Transform the integral $\int f(x) f'(x) \, dx$ using the substitution $u = f(x)$

  b Perform the integration with respect to u

  c Substitute $f(x) = u$ to the integral with respect to x

d Use your result to find the following integrals.

   i $\int (2x+1)(x^2+x+5) \, dx$

   ii $\int \sin x \cos x \, dx$

   iii $\int \dfrac{\ln x}{x} \, dx$ **iv** $\int 2e^{2x}(e^{2x}+1) \, dx$

e Prove $\int \dfrac{f'(x)}{f(x)} \, dx = \ln|f(x)| + c$

f Find each of the following integrals.

   i $\int \dfrac{\cos x}{\sin x} \, dx$ **ii** $\int \dfrac{\sec^2 x}{\tan x} \, dx$

   iii $\int \dfrac{e^x}{e^x+3} \, dx$ **iv** $\int \dfrac{1}{x\ln x} \, dx$

Challenge

5 What is the integral of a^x?

  a Use the identity that $a = e^{\ln a}$ and the rule that $\ln a^b = b \ln a$ to change the base of the exponential function from a to e

  b Use the substitution $u = x \ln a$ to simplify the integral.

  c Integrate and change the base back to a

  d Use the technique to find

   i $\int 4^x \, dx$ **ii** $\int 5^{2x} \, dx$ **iii** $\int 3^{x+1} \, dx$

16.3 Integration by parts

Fluency and skills

See Ch15.4

For a reminder of the product rule for differentiation.

Integration by parts is the technique used when you want to integrate the product of two functions.

You can derive a formula for integration by parts using the product rule, which is the technique used to *differentiate* the product of two functions.

Using the product rule for differentiation, $\dfrac{d}{dx}(uv) = v\dfrac{du}{dx} + u\dfrac{dv}{dx}$, where u and v are functions of x

Integrating throughout with respect to x gives $\displaystyle\int \dfrac{d}{dx}(uv)dx = \int v\dfrac{du}{dx}dx + \int u\dfrac{dv}{dx}dx$

Rearranging this gives $\displaystyle\int u\dfrac{dv}{dx}dx = uv - \int v\dfrac{du}{dx}dx$

> **Key point**
>
> To integrate by parts, consider the integral as being made up of two parts, u and $\dfrac{dv}{dx}$, and use
> $$\int u\frac{dv}{dx}dx = uv - \int v\frac{du}{dx}dx$$

> You usually select u as the function that becomes simpler when differentiated.

Example 1

Find $\quad$ **a** $\displaystyle\int 2x\cos x\,dx$ $\quad$ **b** $\displaystyle\int x^2 \ln x\, dx$

a Let $u = 2x$ $\qquad$ Let $\dfrac{dv}{dx} = \cos x$

$\Rightarrow \dfrac{du}{dx} = 2$ $\qquad \Rightarrow v = \displaystyle\int \cos x\,dx = \sin x$

> Here the two functions are $2x$ and $\cos x$. You know $2x$ becomes simpler when differentiated, so use this for u

> Differentiate to find $\dfrac{du}{dx}$ and integrate to find v

$\displaystyle\int u\dfrac{dv}{dx}dx = uv - \int v\dfrac{du}{dx}dx$

$\displaystyle\int 2x\cos x\,dx = 2x\sin x - \int \sin x \times 2\,dx$

$\qquad = 2x\sin x - 2\displaystyle\int \sin x\,dx$

$\qquad = 2x\sin x + 2\cos x + c$

> Substitute $u = 2x$, $\dfrac{du}{dx} = 2$, $v = \sin x$ and $\dfrac{dv}{dx} = \cos x$ into the equation and integrate.

b Let $u = \ln x$ $\qquad$ Let $\dfrac{dv}{dx} = x^2$

$\Rightarrow \dfrac{du}{dx} = \dfrac{1}{x}$ $\qquad \Rightarrow v = \displaystyle\int x^2\,dx = \dfrac{1}{3}x^3$

> Although x^2 becomes simpler on differentiating, you don't yet know how to integrate $\ln x$, so use $u = \ln x$

$\displaystyle\int u\dfrac{dv}{dx}dx = uv - \int v\dfrac{du}{dx}dx$

$\displaystyle\int \ln x \times x^2\,dx = \left(\ln x \times \dfrac{1}{3}x^3\right) - \int\left(\dfrac{1}{3}x^3 \times \dfrac{1}{x}\right)dx$

$\qquad = \dfrac{1}{3}x^3 \ln x - \dfrac{1}{3}\displaystyle\int x^2\,dx$

$\qquad = \dfrac{1}{3}x^3 \ln x - \dfrac{1}{9}x^3 + c$

> Differentiate to find $\dfrac{du}{dx}$ and integrate to find v

> Substitute $u = \ln x$, $\dfrac{du}{dx} = \dfrac{1}{x}$, $v = \dfrac{1}{3}x^3$ and $\dfrac{dv}{dx} = x^2$ into the equation and integrate.

Example 2

Evaluate the integral $\int_{-1}^{1}(x+1)e^x\,dx$. Show your working and give your answer in exact form.

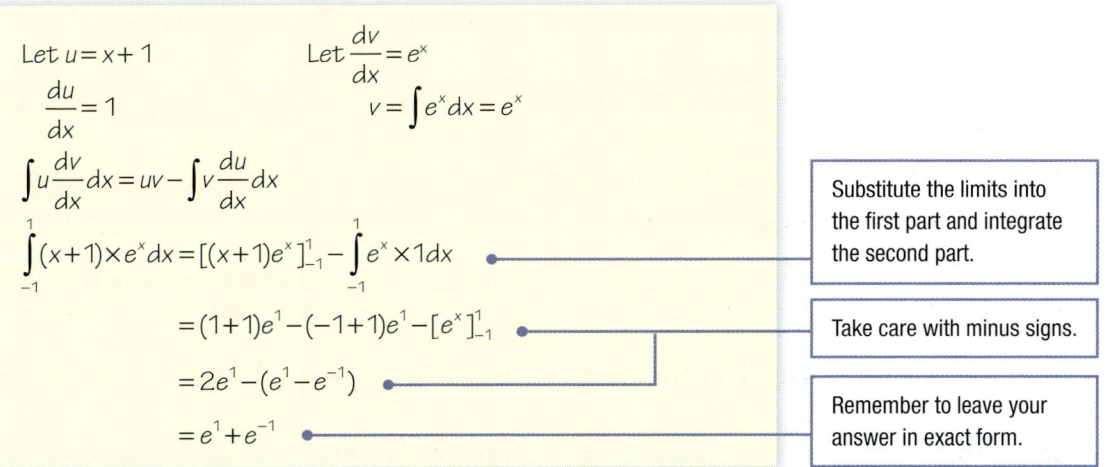

Let $u = x+1$ Let $\dfrac{dv}{dx} = e^x$

$\dfrac{du}{dx} = 1$ $v = \int e^x\,dx = e^x$

$\int u\dfrac{dv}{dx}\,dx = uv - \int v\dfrac{du}{dx}\,dx$

$\int_{-1}^{1}(x+1)\times e^x\,dx = [(x+1)e^x]_{-1}^{1} - \int_{-1}^{1}e^x \times 1\,dx$ Substitute the limits into the first part and integrate the second part.

$= (1+1)e^1 - (-1+1)e^1 - [e^x]_{-1}^{1}$ Take care with minus signs.

$= 2e^1 - (e^1 - e^{-1})$

$= e^1 + e^{-1}$ Remember to leave your answer in exact form.

Exercise 16.3A Fluency and skills

1 Use integration by parts to find

   a $\int 3x\cos x\,dx$ **b** $\int 2x\sin x\,dx$

   c $\int \dfrac{1}{2}x\cos x\,dx$ **d** $\int 3x\sin 2x\,dx$

   e $\int x\cos(3x+1)\,dx$ **f** $\int \dfrac{x}{5}\cos(1-4x)\,dx$

   g $\int (2x+3)\sin x\,dx$ **h** $\int (2x+1)\cos(x+1)\,dx$

   i $\int (1-3x)\cos(4x)\,dx$

2 Use integration by parts to find

   a $\int 3xe^x\,dx$ **b** $\int 2xe^{2x}\,dx$

   c $\int \dfrac{1}{2}xe^{2x+1}\,dx$ **d** $\int (x+2)e^{2x}\,dx$

   e $\int (3-5x)e^{1+2x}\,dx$ **f** $\int (3x-2)e^{1-3x}\,dx$

3 Use integration by parts to find

   a $\int 2x\ln x\,dx$ **b** $\int 5x\ln x\,dx$

   c $\int (x+1)\ln x\,dx$ **d** $\int (2x-1)\ln 2x\,dx$

   e $\int \dfrac{1}{x^2}\ln(3x)\,dx$ **f** $\int (x+1)^2\ln x\,dx$

   g $\int \dfrac{1}{\sqrt{x}}\ln(2x)\,dx$ **h** $\int x\sqrt{x}\,dx$

   i $\int 2x\sqrt{x+1}\,dx$ **j** $\int x\sec^2 x\,dx$

4 Use integration by parts to evaluate the following integrals. Show your working.

   a $\int_{0}^{1}xe^x\,dx$ **b** $\int_{1}^{3}(x^2-2x)\ln x\,dx$

   c $\int_{0}^{\frac{\pi}{4}}(x+1)\cos x\,dx$ **d** $\int_{1}^{5}\dfrac{\ln x}{x^2}\,dx$

   e $\int_{0}^{1}5xe^{-x}\,dx$ **f** $\int_{\frac{\pi}{2}}^{\frac{\pi}{2}}x\cos x\,dx$

   g $\int_{3}^{8}x\sqrt{x+1}\,dx$ **h** $\int_{0}^{\frac{\pi}{2}}x\sin\left(x+\dfrac{\pi}{2}\right)dx$

5 The graph shows $y = x\sin x$ in the interval $0 \le x \le 2\pi$

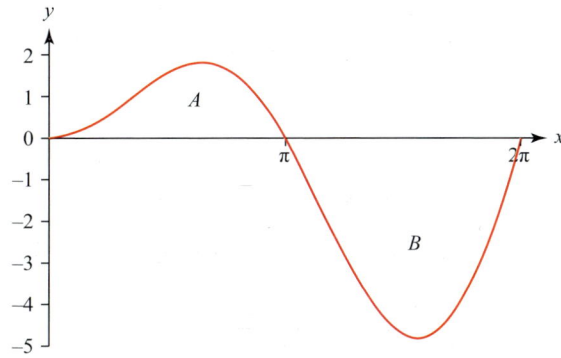

   a Verify that it cuts the x-axis at 0, π and 2π

   b Showing your working, calculate the area   **i** A **ii** B

   c Hence calculate the area trapped between the curve and the x-axis in the interval $0 \le x \le 2\pi$

Reasoning and problem-solving

Strategy

To integrate by parts when there appears to be only one function, $f(x)$, and its integral is not known

1. Rewrite $\int f(x)\,dx$ as $\int f(x) \times 1\,dx$
2. Let $u = f(x)$ and let $\dfrac{dv}{dx} = 1$
3. Use integration by parts.

Example 3

Find $\int \ln x\,dx$

$$\int \ln x\,dx = \int \ln x \times 1\,dx$$

Let $u = \ln x$ Let $\dfrac{dv}{dx} = 1$

$\Rightarrow \dfrac{du}{dx} = \dfrac{1}{x}$ $\Rightarrow v = x$

$$\int u\frac{dv}{dx}dx = uv - \int v\frac{du}{dx}dx$$

$$\int \ln x\,dx = (\ln x \times x) - \int \left(x \times \frac{1}{x}\right)dx$$

$$= x\ln x - \int 1\,dx$$

$$= x\ln x - x + c$$

> 1. Rewrite the function in the expanded form.
>
> 2. Let $u = f(x)$ and let $\dfrac{dv}{dx} = 1$
>
> 3. Substitute $u = \ln x$, $\dfrac{du}{dx} = \dfrac{1}{x}$, $v = x$ and $\dfrac{dv}{dx} = 1\,dx$

Strategy

To use integration by parts more than once

1. Use integration by parts to get $uv - \int v\dfrac{du}{dx}\,dx$
2. Use integration by parts on $\int v\dfrac{du}{dx}\,dx$

Example 4

Find $\int x^2 \sin x\,dx$

Let $u = x^2$ Let $\dfrac{dv}{dx} = \sin x$

$\Rightarrow \dfrac{du}{dx} = 2x$ $\Rightarrow v = -\cos x$

$$\int x^2 \sin x\,dx = (x^2 \times -\cos x) - \int(-\cos x \times 2x)\,dx$$

$$= -x^2 \cos x + 2\int x\cos x\,dx$$

To integrate $\int x\cos x\,dx$

Let $u = x$ Let $\dfrac{dv}{dx} = \cos x$

$\Rightarrow \dfrac{du}{dx} = 1$ $\Rightarrow v = \sin x$

$$\int x\cos x\,dx = x\sin x - \int \sin x\,dx$$

$$\int x^2 \sin x\,dx = -x^2\cos x + 2\int x\cos x\,dx = -x^2\cos x + 2\left(x\sin x - \int \sin x\,dx\right)$$

$$= -x^2\cos x + 2x\sin x - 2\int \sin x\,dx$$

$$= -x^2\cos x + 2x\sin x + 2\cos x + c$$

> 1. Use integration by parts to get $uv - \int v\dfrac{du}{dx}\,dx$
>
> 2. Use integration by parts a second time to get $\int x\cos x\,dx$
>
> Replace $\int x\cos x\,dx$ with $x\sin x - \int \sin x\,dx$
>
> 2. Integrate the final term.

1 Use integration by parts to find each of the following.

a $\int \ln(4x)\,dx$ **b** $\int \ln(3x+1)\,dx$

c $\int x^7 \ln(x^3)\,dx$ **d** $\int \sin x \ln(\cos x)\,dx$

e $\int \ln(1-5x)\,dx$ **f** $\int \ln\left(\dfrac{1}{x}\right)dx$

g $\int (\ln x)^2\,dx$ **h** $\int (\ln x^2)\,dx$

i $\int x \sin^2 x\,dx$

2 Apply integration by parts to find

a $\int x^2 e^x\,dx$ **b** $\int x^2 \sin x\,dx$

c $\int x^2 \cos x\,dx$ **d** $\int (x+1)^2 \sin x\,dx$

e $\int (x^2+2x)\cos x\,dx$ **f** $\int (1-3x)^2 e^x\,dx$

g $\int (x^2+x+1)e^{-x}\,dx$ **h** $\int x^2(x+1)^7\,dx$

i $\int x^2(x+1)^{-2}\,dx$ **j** $\int (x+1)^2(x+3)^5\,dx$

k $\int x(x+1)\sin x\,dx$ **l** $\int x^2 \sin 2x\,dx$

m $\int x^2 \cos 3x\,dx$

3 Apply integration by parts twice to evaluate each of the following integrals. Show your working and give your answers in exact form.

a $\int_0^1 (x^2+5)e^{\frac{1}{3}x}\,dx$ **b** $\int_0^{\frac{\pi}{4}} x^2 \cos 2x\,dx$

c $\int_0^{\pi} (x-\pi)^2 \cos x\,dx$

4 A function is such that $\dfrac{dy}{dx} = x^2 e^x$

a Express y in terms of x and c, the constant of integration.

b When $x=0$, $y=2$. Express y in terms of x alone.

c By considering the discriminant of a quadratic, prove that y is always positive.

5 **a** Find $\int x \ln x\,dx$

b Hence find $\int 4x(\ln x)^2\,dx$

6 The curve $y = (\pi - x)^2 \sin x$, $0 \le x \le 2\pi$, exhibits half-turn symmetry about the point $(\pi, 0)$

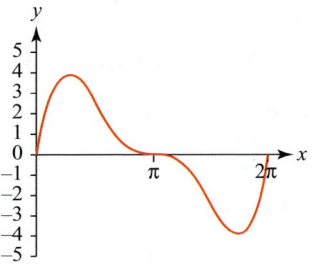

a Check that when $x=0$, π, and 2π, the curve cuts the x-axis.

b Find the area trapped between the curve and the x-axis in the interval $0 \le x \le 2\pi$. Show your working.

7 The sketch shows the area below the curve $y = \dfrac{x^2}{\sqrt{2x-1}}$, bounded by $x=1$ and $x=2.5$

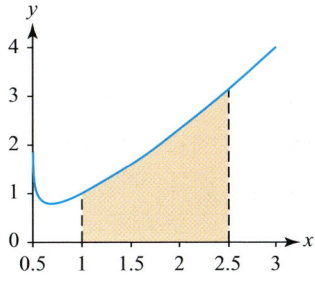

Use integration by parts twice to calculate the exact area. Show your working.

Challenge

8 Consider the integral $P = \int e^x \sin x\,dx$

a Let $u = e^x$ and $\dfrac{dv}{dx} = \sin x$ and perform integration by parts to get

$P = $ (an expression which includes $\int e^x \cos x\,dx$)

b Let $u = e^x$ and $\dfrac{dv}{dx} = \cos x$ and perform integration by parts again.

You should get

$P = $ (an expression which includes $\int e^x \sin x\,dx$)

c Replacing this integral by P, you get

$P = $ (an expression which includes P)

d Make P the subject of the equation to find $\int e^x \sin x\,dx$

Fluency and skills

See Ch12.5

For a reminder on partial fractions.

You already know how to decompose a rational function of the form $\dfrac{px+q}{(ax+b)(cx+d)}$ into partial fractions of the form $\dfrac{A}{ax+b}+\dfrac{B}{cx+d}$ by a suitable choice of A and B. This technique often enables you to integrate rational functions. In the first form, no standard integral is evident. However, once resolved into partial fractions, each part is a standard integral.

> Using partial fractions can also make differentiation easier. You'll use this in question **3** of Exercise 16.4A

Example 1

Find $\displaystyle\int \frac{x+1}{(x+2)(x+3)}\,\mathrm{d}x$

Let $\dfrac{x+1}{(x+2)(x+3)}=\dfrac{A}{x+2}+\dfrac{B}{x+3}$

$x+1=A(x+3)+B(x+2)$

Comparing coefficients for x: $Ax+Bx=x$ so $A+B=1$
Comparing coefficients for the constant: $3A+2B=1$
$A=-1$ and $B=2$

$\displaystyle\int\frac{x+1}{(x+2)(x+3)}dx=-\int\frac{1}{x+2}dx+\int\frac{2}{x+3}dx$

$=-\ln|x+2|+2\ln|x+3|+c$

$=\ln k+2\ln|x+3|-\ln|x+2|$

$=\ln\left(k\dfrac{|x+3|^{2}}{|x+2|}\right)$

- Express the integral using partial fractions and then multiply throughout by $(x+2)(x+3)$
- Solve the simultaneous equations.
- Substitute your values for A and B to resolve into partial fractions and integrate.
- Sometimes it can be convenient to express the constant of integration as a log.

Example 2

Find $\displaystyle\int \frac{2x+1}{x(x-1)(3x+1)}\,\mathrm{d}x$

$\dfrac{2x+1}{x(x-1)(3x+1)}=\dfrac{A}{x}+\dfrac{B}{x-1}+\dfrac{C}{3x+1}$

$2x+1=A(x-1)(3x+1)+Bx(3x+1)+Cx(x-1)$

Comparing coefficients for x^{2}: $3A+3B+C=0$

Comparing coefficients for x: $-2A+B-C=2$

Comparing coefficients for the constant: $A=-1$

$A=-1$ so $3B+C=3$ and $B=C$

$A=-1$, $B=\dfrac{3}{4}$ and $C=\dfrac{3}{4}$

- Express the integral using partial fractions and then multiply throughout by $x(x-1)(3x+1)$
- Substitute $A=-1$ into the simultaneous equations.
- Substitute $B=C$ into $3B+C=3$ to solve the simultaneous equations.

(*Continued on the next page*)

$$\int \frac{2x+1}{x(x-1)(3x+1)} = -\int \frac{1}{x}\,dx + \frac{3}{4}\int \frac{1}{x-1}\,dx + \frac{3}{4}\int \frac{1}{3x+1}\,dx$$

Substitute your values for A, B and C to resolve into partial fractions and integrate.

$$= -\ln|x| + \frac{3}{4}\ln|x-1| + \frac{3}{4} \times \frac{1}{3}\ln|3x+1| + \ln k$$

Remember the constant of integration.

$$= \ln\left(k \frac{|x-1|^{\frac{3}{4}}|3x+1|^{\frac{1}{4}}}{|x|} \right)$$

Exercise 16.4A Fluency and skills

1 Find

a $\int \dfrac{1}{x(x+1)}\,dx$ **b** $\int \dfrac{6}{x(x+3)}\,dx$

c $\int \dfrac{1}{x(3x+1)}\,dx$ **d** $\int \dfrac{x+1}{x(x+2)}\,dx$

e $\int \dfrac{8}{(x-2)(x+2)}\,dx$ **f** $\int \dfrac{5}{(x+1)(x-4)}\,dx$

g $\int \dfrac{2}{(x+2)(x+1)}\,dx$ **h** $\int \dfrac{21}{(x-4)(x+3)}\,dx$

f $\int \dfrac{x^2+2}{x(x-1)(x-2)}\,dx$

g $\int \dfrac{20x^2}{(x+3)(x+4)(x-1)}\,dx$

h $\int \dfrac{x^2+x-2}{(x+1)(x-3)(2x-1)}\,dx$

i $\int \dfrac{x^2+x+1}{x(x-1)(x+2)}\,dx$

j $\int \dfrac{x^2+x+1}{x(1-x)(x+2)}\,dx$

2 Find

a $\int \dfrac{4}{(2x+1)(2x-3)}\,dx$ **b** $\int \dfrac{5}{(x+2)(3x+1)}\,dx$

c $\int \dfrac{7}{(x+5)(3x+1)}\,dx$ **d** $\int \dfrac{x+1}{(x-1)(x+3)}\,dx$

e $\int \dfrac{4x-2}{(x+4)(x-2)}\,dx$ **f** $\int \dfrac{5}{(2x+1)(3x-1)}\,dx$

g $\int \dfrac{6x}{(x-2)(x+1)}\,dx$ **h** $\int \dfrac{2x+1}{(2x+3)(x-1)}\,dx$

i $\int \dfrac{4x-1}{(x-1)(2x+1)}\,dx$ **j** $\int \dfrac{3x}{(2x+1)(2x-1)}\,dx$

3 Express each integrand as the sum of three rational functions, each of which has a linear denominator, and then integrate.

a $\int \dfrac{1}{x(x-1)(x+1)}\,dx$

b $\int \dfrac{10}{x(3x+2)(x-1)}\,dx$

c $\int \dfrac{x+3}{x(x-1)(4x-3)}\,dx$

d $\int \dfrac{6}{(x-1)(x-2)(x+1)}\,dx$

e $\int \dfrac{3x}{(2x-1)(x-2)(x-1)}\,dx$

4 Express each f$'(x)$ using partial fractions and then

   i Integrate to find f(x),

   ii Differentiate to find f$''(x)$

a $f'(x) = \dfrac{2x-1}{x(5-x)}$ **b** $f'(x) = \dfrac{x+1}{(1-x)(3x-1)}$

c $f'(x) = \dfrac{1+2x}{(1-2x)(x-1)}$

d $f'(x) = \dfrac{2x+3}{(2-3x)(5x+1)}$

5 Evaluate each of the following integrals. Show your working and give your answers in exact form.

a $\int_1^3 \dfrac{5}{x(x-5)}\,dx$ **b** $\int_3^6 \dfrac{4}{x(x-2)}\,dx$

c $\int_2^4 \dfrac{2}{x(x+1)}\,dx$ **d** $\int_1^2 \dfrac{1}{x(4-x)}\,dx$

e $\int_4^5 \dfrac{2}{(x-3)(x-1)}\,dx$ **f** $\int_1^3 \dfrac{6}{(x-5)(x-2)}\,dx$

MyMaths   2217   SEARCH

To solve integration problems when dealing with improper rational functions

(1) If necessary, expand the denominator.

(2) Perform algebraic division to resolve the function into a polynomial and a proper rational function.

(3) Resolve the rational function into partial fractions.

(4) Integrate.

(5) Simplify.

Example 3

Integrate $\dfrac{x^3+5x^2+1}{(x+1)(x+3)}$

This example uses the law of logarithms $k\log_a x = \log_a x_n$, which was covered in Section 5.1

$f(x)=\dfrac{x^3+5x^2+1}{(x+1)(x+3)}=\dfrac{x^3+5x^2+1}{x^2+4x+3}$

(1) Expand the denominator.

$$
\begin{array}{r}
x+1 \\
x^2+4x+3\overline{\smash{)}\,x^3+5x^2+1} \\
\underline{x^3+4x^2+3x} \\
x^2-3x+1 \\
\underline{x^2+4x+3} \\
-7x-2
\end{array}
$$

(2) Perform algebraic division.

See Ch2.3
For a reminder on algebraic division.

$f(x)=x+1+\left(\dfrac{-7x-2}{x^2+4x+3}\right)=x+1-\left(\dfrac{7x+2}{(x+1)(x+3)}\right)$

(2) Resolve the function into a polynomial and a proper rational function.

$\dfrac{7x+2}{(x+1)(x+3)}=\dfrac{A}{x+1}+\dfrac{B}{x+3}$

$7x+2=A(x+3)+B(x+1)$

Multiply through by $(x+1)(x+3)$

Comparing coefficents for x: $A+B=7$

Comparing coefficients for the constant: $3A+B=2$

$A=-\dfrac{5}{2}$ and $B=\dfrac{19}{2}$

Solve the simultaneous equations.

$\dfrac{7x+2}{(x+1)(x+3)}=-\dfrac{5}{2(x+1)}+\dfrac{19}{2(x+3)}$

(3) Resolve into partial fractions.

$\displaystyle\int\dfrac{x^3+5x^2+1}{(x+1)(x+3)}dx=\int(x+1)dx+\dfrac{5}{2}\int\dfrac{1}{x+1}dx-\dfrac{19}{2}\int\dfrac{1}{x+3}dx$

$=\dfrac{1}{2}x^2+x+\dfrac{5}{2}\ln|x+1|-\dfrac{19}{2}\ln|x+3|+c$

(4) Integrate.

$=\dfrac{\ln|x+1|^{\frac{5}{2}}}{\ln|x+3|^{\frac{19}{2}}}+\dfrac{1}{2}x^2+x+c$

(5) Simplify.

Exercise 16.4B Reasoning and problem-solving

1 By first factorising the denominator, find

a $\int \dfrac{1}{x^2+2x}\,\mathrm{d}x$ **b** $\int \dfrac{2}{x^2+5x}\,\mathrm{d}x$

c $\int \dfrac{1}{2x^2+x}\,\mathrm{d}x$ **d** $\int \dfrac{x+2}{x^2+x}\,\mathrm{d}x$

e $\int \dfrac{1}{x^2-1}\,\mathrm{d}x$ **f** $\int \dfrac{12}{x^2-9}\,\mathrm{d}x$

2 By first factorising the denominator, find

a $\int \dfrac{2}{4x^2-1}\,\mathrm{d}x$ **b** $\int \dfrac{3x}{9x^2-4}\,\mathrm{d}x$

c $\int \dfrac{1}{x^2-4x+3}\,\mathrm{d}x$ **d** $\int \dfrac{3}{x^2-5x+4}\,\mathrm{d}x$

e $\int \dfrac{x}{2x^2+3x-2}\,\mathrm{d}x$ **f** $\int \dfrac{2x}{x^2-7x+10}\,\mathrm{d}x$

g $\int \dfrac{x+1}{x^2+3x-10}\,\mathrm{d}x$ **h** $\int \dfrac{3x-1}{2x^2+7x-4}\,\mathrm{d}x$

3 Find

a $\int \dfrac{x^2+x-1}{(x-1)(x+2)}\,\mathrm{d}x$ **b** $\int \dfrac{x^2+3x+6}{(x+1)(x+2)}\,\mathrm{d}x$

c $\int \dfrac{x^2+2x}{(x-1)(x+3)}\,\mathrm{d}x$ **d** $\int \dfrac{2x^2-5}{(x-2)(x+2)}\,\mathrm{d}x$

e $\int \dfrac{2x^2+8x+7}{(x+1)(x+3)}\,\mathrm{d}x$ **f** $\int \dfrac{3x^2-8x+6}{(x-1)(x-2)}\,\mathrm{d}x$

4 Evaluate the following integrals. Show your working and give your answers in exact form.

a $\int_{0}^{2} \dfrac{4x-10}{(x+2)(x-4)}\,\mathrm{d}x$ **b** $\int_{4}^{5} \dfrac{2x-3}{(x-1)(x-2)}\,\mathrm{d}x$

c $\int_{3}^{4} \dfrac{1-2x}{(x-2)(x-5)}\,\mathrm{d}x$ **d** $\int_{0}^{2} \dfrac{-x-1}{(x-1)(x-3)}\,\mathrm{d}x$

e $\int_{1}^{3} \dfrac{x-8}{x(x-4)}\,\mathrm{d}x$ **f** $\int_{2}^{4} \dfrac{2x-14}{(x-1)(x-5)}\,\mathrm{d}x$

5 This is a sketch of $y = \dfrac{2x-4}{(x-1)(x-3)}$

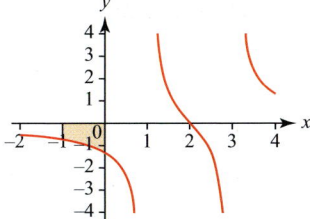

a Calculate the area bounded by the curve, the x-axis, $x=-1$ and $x=0$. Show your working.

b Calculate the area bounded by the curve, the x-axis, $x=4$ and $x=5$. Show your working.

c Why would it not be sensible to look for the area bounded by the curve, the x-axis, $x=0$ and $x=2$?

6 The curve shows sketches of $y = \dfrac{-8}{x(x-4)}$ and $y = \dfrac{-9}{x(x-3)}$

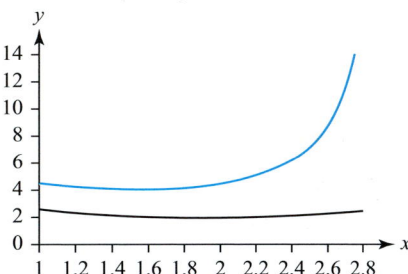

a Identify which curve goes with which equation.

b Showing your working, find the area bounded by

i $y = \dfrac{-8}{x(x-4)}$, the x-axis and $x=1$ and $x=2$

ii $y = \dfrac{-9}{x(x-3)}$, the x-axis and $x=1$ and $x=2$

c Hence find the area between the two curves in this interval, giving your answer in the form $a\ln 3 + b\ln 2$

Challenge

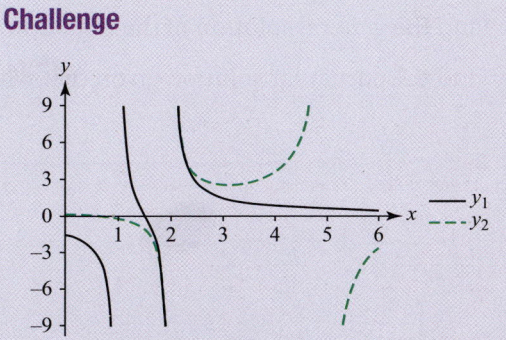

7 The sketch shows the two curves
$y_1 = \dfrac{2x-3}{(x-1)(x-2)}$ and $y_2 = \dfrac{1-2x}{(x-2)(x-5)}$

a Calculate the area between the curves in the interval $3 \le x \le 4$. Show your working.

b Describe the intervals over which it would not be meaningful to consider the area between the curves.

Fluency and skills

A **differential equation** is an equation that expresses a relationship between functions and their derivatives. In order to find the **solution** of a differential equation you must **separate the variables**. That is, you must express the left-hand side purely in terms of y and the right-hand side purely in terms of x

> **Key point**
>
> Let y be a function of x, let f be a function of y, and let g be a function of x
>
> If $f(y)\dfrac{dy}{dx} = g(x)$
>
> Integrating both sides with respect to x gives $\displaystyle\int f(y)\dfrac{dy}{dx}dx = \int g(x)dx$
>
> The chain rule then gives the result $\displaystyle\int f(y)dy = \int g(x)dx$

Integrating will result in a **general solution** that involves a constant of integration. The general solution represents a **family of solutions**. If you know at least one point (x, y) in the solution, then you can find the constant of integration and hence a **particular solution**. The point (x, y) is usually referred to as **the initial condition**.

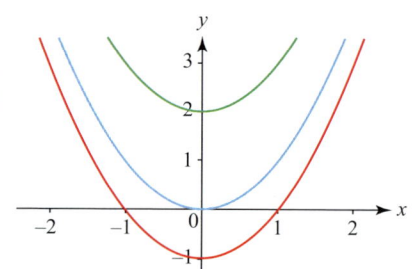

The differential equation $\dfrac{dy}{dx} = x$ has the general solution $y = x^2 + c$

The diagram shows the particular solutions for $c = -1$, 0 and 2

Example 1

a Find the general solution of the equation $\dfrac{dy}{dx} = 2x(y+1); \; y > -1$

b Find the particular solution given that when $x = 3$, $y = 0$

a $\dfrac{1}{y+1}\dfrac{dy}{dx} = 2x$

    You can rewrite the equation in this form. This step is known as 'separating the variables'.

$\displaystyle\int\dfrac{1}{y+1}\dfrac{dy}{dx}dx = \int 2x\,dx$

    Integrate both sides with respect to x to find the general solution written as an implicit function.

$\Rightarrow \displaystyle\int\dfrac{1}{y+1}dy = \int 2x\,dx \Rightarrow \ln|y+1| = x^2 + c$

$\Rightarrow e^{\ln|y+1|} = e^{x^2+c} \Rightarrow y+1 = e^{x^2+c}$

    To get y in terms of x, take the exponential of each side of the equation.

$\Rightarrow \quad y = e^{x^2+c} - 1$

    This gives the general solution written as an explicit function.

b When $x = 3$ and $y = 0$, $\ln|0+1| = 3^2 + c$

$\Rightarrow c = -9$

    Substitute the initial conditions to find c

So the particular solution is given by $y = e^{x^2-9} - 1$

You also may be required to draw upon other integration techniques in order to solve a differential equation.

Example 2

Solve $\dfrac{dy}{dx} = \dfrac{2x}{y\sqrt{y^2-1}}$ given that when $x = 2$, $y = 1$

$\dfrac{dy}{dx} = \dfrac{2x}{y\sqrt{y^2-1}}$ so $y\sqrt{y^2-1}\dfrac{dy}{dx} = 2x$

> Separate the variables.

$\displaystyle\int y\sqrt{y^2-1}\dfrac{dy}{dx}dx = \int 2x\,dx \Rightarrow \int y\sqrt{y^2-1}\,dy = \int 2x\,dx$

Let $u = y^2 - 1 \Rightarrow \dfrac{du}{dy} = 2y \Rightarrow dy = \dfrac{du}{2y}$

So, $\displaystyle\int y\sqrt{u}\dfrac{du}{2y} = \int 2x\,dx \Rightarrow \int \dfrac{1}{2}\sqrt{u}\,du = \int 2x\,dx$

> Decide on the method of integration. In this case, use substitution.

$\dfrac{u^{\frac{3}{2}}}{3} = x^2 + c$

> Integrate to find the general solution.

$\dfrac{(y^2-1)^{\frac{3}{2}}}{3} = x^2 + c$

> Replace u with $(y^2 - 1)$

When $x = 2$, $y = 1$ $\Rightarrow$ $\dfrac{(1^2-1)^{\frac{3}{2}}}{3} = 2^2 + c$ $\Rightarrow$ $c = -4$

> Use the initial conditions to find the particular solution.

Therefore, $\dfrac{(y^2-1)^{\frac{3}{2}}}{3} = x^2 - 4$

Exercise 16.5A Fluency and skills

1 Find the general solution to $\dfrac{dy}{dx} = x - 2$ and on the same axes sketch the graphs of the members of the family of solutions corresponding to $c = -1$, 0 and 2

2 Find the general solution to each of the following differential equations.

   a $\dfrac{dy}{dx} - 2x + 1 = 0$ b $\dfrac{dy}{dx} = 1 - x^2$ c $\dfrac{dy}{dx} = \sqrt{2x+1}$ d $\dfrac{dy}{dx} = \dfrac{3}{x^2}$

   e $\dfrac{dy}{dx} = \dfrac{2}{x}$ f $\dfrac{dy}{dx} = 3y$ g $\dfrac{dy}{dx} = 5xy$ h $\dfrac{dy}{dx} = \dfrac{x}{y}$

   i $\dfrac{dy}{dx} = \dfrac{x+1}{y-3}$ j $\dfrac{dy}{dx} = x(1+y)$ k $\dfrac{dy}{dx} + y = 5$ l $\dfrac{dy}{dx} = (x+2)(2y+1)$

   m $2y\dfrac{dy}{dx} - 3x^2 = 0$ n $\dfrac{dy}{dx} = \dfrac{\sin x}{\cos y}$ o $\dfrac{dy}{dx} = e^{2x}\cos^2 y$ p $\dfrac{dy}{dx} = e^{x-y}$ q $\dfrac{dy}{dx} = e^{2x-3y+1}$

3 Find the particular solution to the differential equation that corresponds to the given initial conditions.

   a $\dfrac{dy}{dx} - 4x + 1 = 0$; $(1, 4)$ b $\dfrac{dy}{dx} + 9x^2 - 2 = 0$; $(2, -10)$ c $\dfrac{dy}{dx} = \dfrac{1}{x}$; $(1, 3)$

   d $\dfrac{dy}{dx} = 6y$; $\left(\dfrac{1}{3}, 1\right)$ e $\dfrac{dy}{dx} = x(y+3)$; $(2, -2)$ f $\dfrac{dy}{dx} = xy + 5y$; $(0, e)$

   g $\dfrac{dy}{dx} = \dfrac{x+2}{2y-1}$; $(2, 1)$ h $\dfrac{dy}{dx} = \dfrac{\cos x}{\sin y}$; $\left(\dfrac{\pi}{5}, \dfrac{\pi}{3}\right)$ i $\dfrac{dy}{dx} = e^{2x-y}$; $\left(\dfrac{1}{2}, 0\right)$

   j $2x\dfrac{dy}{dx} - y = 0$; $(1, 2)$

 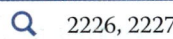 

4 Find the general solution, stated explicitly if possible.

a **i** $\dfrac{dy}{dx} = \dfrac{(y^2+1)^5}{y}$ **ii** $\dfrac{dy}{dx} = x\dfrac{\sin y}{\cos y}$ **iii** $e^{\sqrt{y}}\dfrac{dy}{dx} - 2\sqrt{y} = 0$

b **i** $\dfrac{dy}{dx} = \dfrac{\sin x}{y\sin y}$ **ii** $\dfrac{dy}{dx} = \dfrac{xe^x}{y}$ **iii** $\dfrac{dy}{dx} = \dfrac{9}{y^2\ln y}$

c **i** $\dfrac{dy}{dx} = \dfrac{6y}{x(2x-3)}$ **ii** $\dfrac{dy}{dx} = \dfrac{4y^2}{(x+1)(x-1)}$ **iii** $\dfrac{dy}{dx} = \dfrac{\sqrt{y}}{(x-2)(x-3)}$

5 The gradient of a curve at the point (x, y) is given by $\dfrac{\cos x}{e^y}$. The curve passes through the point $(\pi, 0)$. Find an expression for y in terms of x

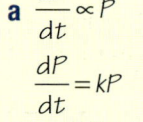

Reasoning and problem-solving

Strategy

To solve modelling problems involving differential equations

1. Form a differential equation using the information provided.

2. Separate the variables.

3. Integrate to find the general solution.

4. Use the initial conditions to find the constant and hence the particular solution.

5. Use the model to make predictions.

6. Consider any assumptions and any limitations of the model.

Example 3

In 1820, Thomas Malthus suggested that the rate of growth of a population, P, over time t, is proportional to the size of the population.

a Express this model using a differential equation.

b Find a general solution to your differential equation.

c Consider any limitations of the model.

a $\dfrac{dP}{dt} \propto P$

$\dfrac{dP}{dt} = kP$

> Form a differential equation. The rate of change of P is proportional to P. See Ch2.4 for a reminder on proportionality. **(1)**

b $\dfrac{1}{P}\dfrac{dP}{dt} = k$

$\displaystyle\int \dfrac{1}{P}\dfrac{dP}{dt}\,dt = \int k\,dt \Rightarrow \int \dfrac{1}{P}\,dP = \int k\,dt$

$\ln P = kt + c$

$P = e^{kt+c} = e^{kt}e^{c} = Ae^{kt}$ where $A = e^{c}$, the population at time zero.

> Separate the variables and integrate to find the general solution both sides with respect to t. **(2)(3)**

c Most natural populations do not grow exponentially fast since they are constrained by food supply, habitat or other external factors.

> Consider any limitations of the model. **(5)**

Example 4

A species of tree is expected to grow to 10 m in height.

The rate of growth in metres per year is directly proportional to the height it still has to grow to reach its potential. The rate of growth can be modelled by $\dfrac{dh}{dt} = k(10-h)$ where h is the height of the tree in metres, t is the time in years and k is a constant.

a Solve the differential equation to find the general solution.

b After 2 years, the tree is 1 m tall. Find the particular solution to the differential equation.

c Use the model to predict the age, to the nearest year, of a tree that is 6 m tall.

a $\dfrac{dh}{dt} = k(10-h) \Rightarrow \dfrac{1}{(10-h)}\dfrac{dh}{dt} = k$

$\displaystyle\int \dfrac{1}{(10-h)}\dfrac{dh}{dt}\,dt = \int k\,dt \Rightarrow \int \dfrac{1}{(10-h)}\,dh = \int k\,dt$

$\ln|c| - \ln|10-h| = kt$

$\ln\dfrac{|c|}{|10-h|} = kt$

$\dfrac{c}{10-h} = e^{kt} \Rightarrow h = 10 - ce^{-kt}$

b At $t=0$, $h=0 \Rightarrow c=10$

This gives $h = 10(1-e^{-kt})$

At $t=2$, $h=1 \Rightarrow 1 = 10(1-e^{-2k})$

$e^{-2k} = 0.9$

$-2k = \ln 0.9$

$k = 0.053$ (to 3 dp)

$h = 10(1-e^{-0.053t})$

c Using $h=6$, $6 = 10(1-e^{-0.053t})$

$e^{-0.053t} = 0.4$

$-0.053t = \ln 0.4$

$t = 17.288$ (to 3 dp)

So the tree will be approximately 17 years old.

Step	Description
2	Separate the variables and integrate.
	For convenience, express the constant of integration as $\ln|c|$
	Take the exponential of both sides to find the general solution.
3	Use the initial conditions to find c and substitute into the general solution.
3	Use the initial conditions to find k.
3	Substitute the value of k back in to get the particular solution.
4	Substitute $h=6$ into the particular solution and solve to find t

Exercise 16.5B Reasoning and problem-solving

1 Express each sentence as a differential equation.

a When buying gold, the rate of change of the cost with respect to its weight is a constant. Let w grams cost £C

b A ball rolls down an inclined plane. The rate at which its distance from the top changes with respect to time is directly proportional to the time it has been rolling. Let S cm be the distance after t seconds.

c Under compound interest conditions, the rate at which the amount in an account grows is directly proportional to the amount in the account. There are £A in the account after T years.

d The rate at which a cup of tea cools is directly proportional to the difference between the temperature of the tea and the temperature of its surroundings. After t minutes, with the room at a constant $20°$C, the tea is at a temperature of $T°$C

e When selling luxury items the rate of change of demand, D, with respect to the price, P, is inversely proportional to the price.

f Earthquakes are measured using the Richter Scale. The rate of change of an earthquake's intensity, as its Richter scale number changes, is directly proportional to 10^R.
R is the Richter scale number when the intensity is I units.

g A square metal plate expands when heated. The rate at which the length of the side increases with respect to the area is inversely proportional to the square root of the length of the side. A square of area x cm^2 has a side of y cm.

h The growth of a particular plant is such that the rate at which its height is changing with respect to time is inversely proportional to $\sec^2 h$, where h cm is the height after t days.

2 Water is poured into a cistern which can hold 50 litres. The rate at which it fills can be modelled by $\dfrac{dV}{dt} = 2 + 0.6t$, where there are V litres in the cistern after t minutes.

a Solve the differential equation to express V in terms of t and a constant.

b Initially the tank is empty. Express V in terms of t

c The flow cuts off when the cistern is full. At what time will this occur?

3 Over a month, the rate at which the percentage of the moon which is visible changes with time can be modelled by $\dfrac{dP}{dt} = \dfrac{25\pi}{7}\cos\left(\dfrac{\pi t}{14}\right)$ where P is the percentage visible on day t of the month.

a Solve the differential equation to express P in terms of t and a constant.

b The month began with a half-moon. [Initially, $t = 0$, $P = 50$.] Express P in terms of t

c What percentage of the moon was visible on day 4?

d On which day was there a full moon?

4 A 10 metre ladder is held vertically against a wall. The foot of the ladder starts to slip away from the wall. The rate at which the height of the midpoint of the ladder changes with respect to its distance from the wall is modelled by $\dfrac{dy}{dx} = k\dfrac{x}{y}$ where the midpoint is y metres high when it is x metres from the wall and k is a constant.

a Solve the differential equation to express y in terms of x, k and c (the constant of integration).

b Initially the midpoint is 5 metres high and touching the wall. Find the value of c

When it is 4 metres high, it is 3 metres from the wall.

c Express the relationship between x and y as an implicit equation and describe the path traced out by the midpoint of the ladder as it falls.

5 Bacteria are growing in a petri dish. The rate at which they multiply is directly proportional to the number of bacteria. Initially there were 500 bacteria in the dish. On day t there are N bacteria.

a Form a differential equation.

b Find the general solution of the differential equation.

c Use the initial conditions to find the constant of integration.

d At the end of day 3, there were 4000 bacteria in the dish. Find constant of proportion and the particular solution to the equation, expressing N as an explicit function of t

e When will the number of bacteria exceed 50 000?

f Why might this model not be appropriate?

6 An osprey can be expected to reach an adult weight of 2000 g.

On day zero, a chick will weigh 50 g on hatching. It fledges after 60 days when its weight is 1990 g.

Its rate of growth is directly proportional to the difference between its weight and its expected adult weight. On day t, its weight is w grams.

a Form a differential equation to model the development of the osprey chick,

b Evaluate the constant of integration if at $t = 0$, $w = 50$

c Find the constant of proportion,

d Express w explicitly as a function of t

e When is the chick's weight expected to exceed 1500 grams?

f Discuss any assumptions and any limitations of the model.

7 On a small island, it is estimated that there are enough resources to support a population of 500 breeding pairs of rabbits. Initially, 10 pairs are introduced on the island. By the end of year 1, there are 50 pairs. The rate of growth is found to be jointly proportional to both the population size and to the difference between the population size and the island's capacity for breeding pairs. Let P be the number of pairs on the island at the end of year t

a Form a differential equation to model the rabbit population in year t

b Express P explicitly in terms of t

Challenge

8 Use a spreadsheet to explore the models introduced in questions **3** and **4**

Draw graphs of each and see how they have been chosen to model each context.

How closely do they model the situation

a At the beginning,

b As time goes on?

Which one might best model the rate of uptake of a new idea in Britain (for example—the membership of a social media site)?

You'll need to get initial conditions from the web.

Chapter summary

- If the derivative of a function $F(x)$ is $f(x)$, then you can say that an indefinite integral of $f(x)$ with respect to x is $F(x)$. That is, $F'(x) = f(x) \Rightarrow \int f(x)\,dx = F(x) + c$ where c is the constant of integration.
  This leads to the following set of standard integrals.

  1. $\int x^n\,dx = \dfrac{1}{n+1}x^{n+1} + c; \quad n \neq -1$ **2.** $\int e^{kx}\,dx = \dfrac{1}{k}e^{kx} + c$ **3.** $\int \dfrac{1}{x}\,dx = \ln|x| + c$

  4. $\int \cos kx\,dx = \dfrac{1}{k}\sin kx + c$ **5.** $\int \sin kx\,dx = -\dfrac{1}{k}\cos kx + c$ **6.** $\int \sec^2 kx\,dx = \dfrac{1}{k}\tan kx + c$

- The area between the curves $y = f(x)$ and $y = g(x)$ is given by $A = \left| \int_a^b f(x) - g(x)\,dx \right|$ if the curves do not intersect in the interval $a < x < b$. If the curves do intersect in the interval, then each trapped area must be considered separately.

- To integrate the product of a function and a derivate function, you can use integration by substitution.
  If $\int f(g(x))g'(x)\,dx$, let $u = g(x)$ and $du = g'(x)\,dx$. This gives $\int f(u)\,du$

- To integrate a product of two functions, you can use integration by parts. To do this, consider the integral as being made up of two parts, u and dv, and use $\int u \dfrac{dv}{dx}\,dx = uv - \int v \dfrac{du}{dx}\,dx$

- You can reduce a rational function of the form $\dfrac{px+q}{(ax+b)(cx+d)}$ into partial fractions of the form $\dfrac{A}{ax+b} + \dfrac{B}{cx+d}$ by a suitable choice of A and B. This technique often allows you to integrate rational functions more easily.

- An equation which relates functions to their derivatives is called a **differential equation**.

- The solution to a differential equation is the set of functions which satisfy the equation.

- If a differential equation can be expressed in the form $f(y)\dfrac{dy}{dx} = g(x)$ then you say the variables are separable.

- Integrating $\int f(y)\,dy = \int g(x)\,dx$ yields the *general solution* of the differential equation.

- If you know one point (x, y) in the solution, you can find a *particular solution*.

Check and review

You should now be able to...	Try Questions
✔ Integrate a set of standard functions, $f(x)$ and the related functions, $f(ax + b)$	1,2,3
✔ Find the area between two curves.	3
✔ Simplify an integral by changing the variable, referred to as *substitution*.	4,5
✔ Use integration by parts to integrate the product of two functions.	6,7,8
✔ Simplify an integral by decomposing a rational function into partial fractions.	9,10
✔ Understand the meaning of the expression 'differential equation'.	11,12
✔ Use integration where the variables are separable.	11,12

1 Find

a $\int 4x^6\,dx$ **b** $\int 2\sin(3x+1)\,dx$

c $\int \cos 2x\,dx$ **d** $\int x+\sec^2 x\,dx$

e $\int (e^x + e^{-x})\,dx$

2 Evaluate the following integrals. Show your working.

a $\int_1^4 \dfrac{2}{\sqrt{x}}\,dx$ **b** $\int_{\frac{\pi}{6}}^{\frac{\pi}{2}} \sin x\,dx$

c $\int_{-\frac{\pi}{4}}^{\frac{\pi}{4}} \cos 2x\,dx$ **d** $\int_0^{\frac{\pi}{4}} 1-\sec^2 x\,dx$

e $\int_{\ln 2}^{\ln 5} e^x\,dx$ **f** $\int_{e^2}^{e^4} \dfrac{3}{x}\,dx$

3 In each case the two functions trap a single region between them. Find the area of the region, showing your working.

a $y = x^2 + x$ and $y = 2x + 2$

b $y = 1 + 5x - x^2$ and $y = x^2 - 4x + 5$

4 Find each integral. A suitable substitution has been suggested.

a $\int x(x^2+4)^5\,dx$; let $u = x^2 + 4$

b $\int 8x^3(x^4-2)^3\,dx$; let $u = x^4 - 2$

c $\int (x-2)(x^2-4x+1)\,dx$; let $u = x^2 - 4x + 1$

d $\int \cos x\sqrt{\sin x}\,dx$; let $u = \sin x$

5 Evaluate the following integrals. Show your working.

a $\int_0^{\sqrt{3}} \dfrac{x}{\sqrt{x^2+1}}\,dx$ **b** $\int_0^{\frac{\pi}{3}} \dfrac{\sin x}{2\cos^2 x}\,dx$

6 Use integration by parts to find

a $\int 2x\sin x\,dx$ **b** $\int (3x+1)e^{2x}\,dx$

c $\int x^2\cos 3x\,dx$

7 Evaluate each of the following, showing your working.

a $\int_0^1 (x+3)e^{x+3}\,dx$ **b** $\int_1^e 9x^2\ln x\,dx$

8 Apply integration by parts twice to find

a $\int (x^2+2x-1)e^x\,dx$ **b** $\int 2x(\ln x)^2\,dx$

9 Find

a $\int \dfrac{8}{x(x-2)}\,dx$ **b** $\int \dfrac{2}{(x+2)(x+3)}\,dx$

10 Evaluate $\int_4^5 \dfrac{3x+1}{(x-1)(x-3)}\,dx$ expressing your answer in the form $a\ln b$ where a and b are integers. Show your working.

11 Find the particular solution to each differential equation.

a $\dfrac{dy}{dx} = y(x+1)$ given that when $x = 1$, $y = 1$

b $\dfrac{dy}{dx} = \dfrac{2x}{\cos y}$ given that when $x = 1$, $y = \dfrac{\pi}{6}$

12 The rate at which a car loses value is directly proportional to the value of the car. The car is worth £V after T years.

Initially the car was worth £20 000
After 3 years it was worth £14 580

a Form a differential equation.

b Use the initial conditions to find the constants of proportion and integration.

c Express V as an explicit function of T

What next?

Score			
	0 – 6	Your knowledge of this topic is still developing. To improve, search in MyMaths for the codes: 2057, 2167–2171, 2216–2221, 2226–2227, 2274	
	7–9	You're gaining a secure knowledge of this topic. To improve, look at the InvisiPen videos for Fluency and skills (16A)	
	10–12	You've mastered these skills. Well done, you're ready to progress! Now try looking at the InvisiPen videos for Reasoning and Problem-solving (16B)	

Click these links in the digital book

Have a go

Not all differential equations have **separable variables**.

Consider, for example

$$\frac{dy}{dx} + \frac{2y}{x} = \frac{1}{x^2}\cos x$$

Start by multiplying both sides of the equation by x^2

$$x^2\frac{dy}{dx} + 2xy = \cos x$$

Recognise that the left-hand side of this equation is the derivative of x^2y with respect to x

$$\frac{d(x^2y)}{dx} = \cos x$$

Complete the solution to give y in terms of x

Note

No amount of rearranging this equation will leave all of the x terms on one side and the y on the other.

The trick here is to recognise that the left-hand side is the **derivative of a product**.

Information

The solution of the differential equation above relied on multiplying both sides by a factor that turned the left-hand side into the derivative of a product. Such a factor is called an **integrating factor**.

Use the chain rule to differentiate $e^{\int P dx}\, y$ with respect to x:

$$\frac{d\left(e^{\int P dx}\, y\right)}{dx} = e^{\int P dx}\frac{dy}{dx} + Pe^{\int P dx}y$$

$$= e^{\int P dx}\left(\frac{dy}{dx}dx + Py\right)$$

This result shows that a differential equation of the form $\dfrac{dy}{dx} + Py = Q$,

where P and Q are functions of x, has integrating factor $I = e^{\int P dx}$

Challenge

Use the integrating factor method to solve the differential equation

$$\frac{dy}{dx} + \frac{3}{x}\,y = \sqrt{x}$$

Given that $y = 7$ when $x = 1$, find the value of y when $x = 4$

Challenge

Find the integrating factor for the equation

$$\frac{dy}{dx} + \frac{2y}{x} = \frac{1}{x^2}\cos x$$

Note

When working out $\int P dx$ it is not necessary to include a constant.

In questions that tell you to show your working, you shouldn't depend solely on a calculator. For these questions, solutions based entirely on graphical or numerical methods are not acceptable.

1 Integrate each of these expressions.

    a $\dfrac{1}{x}$ **[1]**   **b** $\sin(x-3)$ **[2]**   **c** e^{2x} **[2]**   **d** $\cos 2x$ **[2]**

2 $f'(x) = 2 + 3\sin 6x$

    Find the equation of the curve $y = f(x)$, which passes through the point $(0, 1)$ **[4]**

3 a Find these integrals.

    i $\displaystyle\int \dfrac{1}{x+3}\,dx$ **ii** $\displaystyle\int \sin x \cos^2 x\,dx$ **iii** $\displaystyle\int 2x(x^2+4)^3\,dx$ **[5]**

  b Find the exact value of $\displaystyle\int_0^1 \dfrac{x}{x^2+1}\,dx$. Show your working.   **[4]**

4 a Use integration by parts to calculate each of these integrals.

    i $\displaystyle\int xe^x\,dx$

    ii $\displaystyle\int x\sin x\,dx$ **[5]**

  b Show that the integral $\displaystyle\int_0^{\frac{\pi}{6}} x\sin 2x\,dx$ can be written in the form $\beta\sqrt{3} - \alpha\pi$

    where α and β are constants to be found.   **[4]**

5 a Find the general solution to the differential equation $\dfrac{dy}{dx} = \dfrac{y}{x+1}$

    Give your answer in the form $y = f(x)$ **[4]**

  b Find the particular solution for curve $y = f(x)$ which passes through the point $(1, 8)$ **[2]**

6 a Express $\dfrac{10-13x}{(3+x)(1-2x)}$ in the form $\dfrac{A}{3+x} + \dfrac{B}{1-2x}$, where A and B are integers.   **[3]**

  b Hence find the area bounded by the curve $y = \dfrac{10-13x}{(3+x)(1-2x)}$, the x-axis and the lines

    $x = -1$ and $x = 0$. Show your working.   **[4]**

7 a Use the substitution $u = 1+2x$ to find $\displaystyle\int \dfrac{2x}{1+2x}\,dx$. Give your answer in terms of x **[5]**

  b Find the exact area of the region bounded by the curve $y = \dfrac{2x}{1+2x}$, the x-axis and

    the lines $x = 0$ and $x = 1$. Show your working.   **[2]**

8 Find the value of $\displaystyle\int_1^2 \dfrac{4x+3}{2x^2+3x-2}\,dx$. Show your working.   **[6]**

9 $f(x) = \dfrac{6x^3+14x^2+11x-1}{3x^2+7x+2}$

  a Given that $f(x)$ can be expressed as $Ax + \dfrac{B}{x+2} + \dfrac{C}{3x+1}$, find the values of A, B and C **[4]**

  The gradient of a curve is given by $\dfrac{dy}{dx} = f(x)$

  b Find the equation of the curve given that it passes through the point $(0, 0)$ **[4]**

10 Find the general solution to the differential equation $\dfrac{dy}{dx} = \dfrac{5y}{2-3x-2x^2}$

  Give your answer in the form $y = f(x)$ **[7]**

11 A radioactive material decays such that the rate of change of the number, N, of particles is proportional to the number of particles at time t days.

    a   Write down a differential equation in N and t **[1]**

Initially there are N_0 particles and it takes T days for N to halve.

    b   Solve your differential equation, giving the solution in terms of N_0, t and T, with N as the subject.   **[6]**

12 a   Sketch the graphs of $y = \sin 2x$ and $y = \cos x$ on the same axes for $0 \le x \le \pi$
         Give the points of intersection with the coordinate axes.   **[4]**

    b   Find the values of x in the interval $[0, \pi]$ of the points of intersection of the two curves. Show your working.   **[4]**

    c   Calculate the total area enclosed between the two curves in the interval $[0, \pi]$. Show your working.   **[4]**

13 The curve C has a tangent at the point P as shown.

The equation of C is $y = x\mathrm{e}^{\frac{x}{3}}$ and P has x-coordinate 3

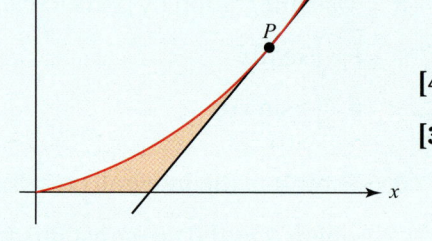

    a   Find the equation of the tangent to the curve at P.
         Show all your working.   **[4]**

    b   Work out $\int x\mathrm{e}^{\frac{x}{3}}\,\mathrm{d}x$ **[3]**

The area bounded by the curve C, the tangent to the curve at P and the coordinate axes is shaded.

    c   Calculate the exact value of the shaded area. Show your working.   **[5]**

14 $\mathrm{f}(x) = \dfrac{4}{\sqrt{x}(x-4)}$

Use the substitution $u = \sqrt{x}$ to find $\int \mathrm{f}(x)\,\mathrm{d}x$

Give your answer as a single logarithm in terms of x **[8]**

15 Use an appropriate substitution to find $\int x(2x-5)^4\,\mathrm{d}x$, give your answer in terms of x **[5]**

16 a   Calculate $\int x^2 \sin x\,\mathrm{d}x$ **[4]**

    b   The area R is bounded by the curve $y = x^2 \sin x$, the x-axis and the lines $x = 0$ and $x = 2\pi$

         Calculate the area of R. Show your working and give your answer in terms of π **[4]**

17 At time t minutes, the volume of water in a cylindrical tank is $V\,\mathrm{m}^3$. Water flows out of the tank at a rate proportional to the square root of V

    a   Show that the height of water in the tank satisfies the differential equation

$$\frac{\mathrm{d}h}{\mathrm{d}t} = -k\sqrt{h}$$ **[4]**

    b   Find the general solution of this differential equation in the form $h = \mathrm{f}(t)$ **[3]**

The tank is 2 m tall and is initially full. It then takes 2 minutes to fully empty.

    c   Show that the particular solution to the differential equation is $h = 2\left(1 - \dfrac{1}{2}t\right)^2$ **[4]**

17 Numerical methods

Atmospheric modelling is important for both understanding climate change and improving weather forecasting accuracy. Numerical analysis is an important tool for this kind of modelling. Partial differential equations, which express relationships between velocity, pressure and temperature, as well as laws for momentum, mass and energy, are needed. Each model will use a slightly different set of equations to place focus on different geographical locations and different atmospheric processes.

Equations like these can often be too complicated to solve algebraically, so mathematicians have developed a range of numerical techniques to find approximate solutions to problems. It is important to determine if these techniques are useful and if the solutions they provide are accurate enough for the purpose. As modern computing has advanced and computers have become more and more powerful, finding and evaluating these numerical methods has been made easier, and hence become a significant area of study.

Orientation

What you need to know	What you will learn	What this leads to
Ch1 Algebra 1 (p.16) • Finding roots of an equation.	• To use the change of sign method to find and estimate the root(s) of an equation. • To use an iterative formula to estimate the root of an equation. • To recognise the conditions that cause an iterative sequence to converge. • To use the Newton-Raphson method to estimate the root of an equation. • To use the trapezium rule to find the area under a curve.	**Applications** Weather forecasting. Computer-aided engineering. Software development.
Ch4 Differentiation and integration (p.116) • Area under a curve.		
Ch13 Sequences (p.340) • Increasing, decreasing or periodic sequences.		
Ch15 Differentiation 2 (p.398) • Differentiation of trigonometric functions.		

 **MyMaths** Practise before you start 🔍 2026, 2056, 2165, 2264

463

17.1 Simple root finding

Fluency and skills

See Ch1.4

For a reminder on finding the roots of an equation.

When you solve an equation you are finding its **roots**.

The roots of the equation $f(x) = 0$ are the x-coordinates of the points where the graph of $y = f(x)$ crosses the x-axis.

For example, the exact roots of the equation $x^2 - 2x - 2 = 0$ are $1 \pm \sqrt{3}$ and are shown on the diagram.

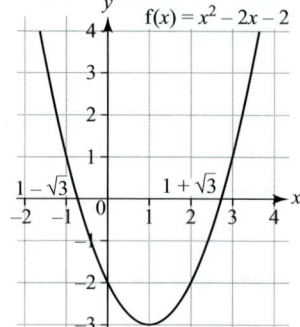

Often, it is difficult, or even impossible, to find the exact value of a root. When this happens you can use a **numerical method** to estimate its value. The simplest numerical method for detecting a root is the **change of sign** method.

> **Key point**
> If two real numbers c and d are such that $f(c)$ and $f(d)$ have opposite signs, you say $f(x)$ changes sign between $x = c$ and $x = d$

The diagram shows the graph of $y = f(x)$, which passes through points P and Q, with x-coordinates c and d, respectively. $f(x)$ changes sign between $x = c$ and $x = d$

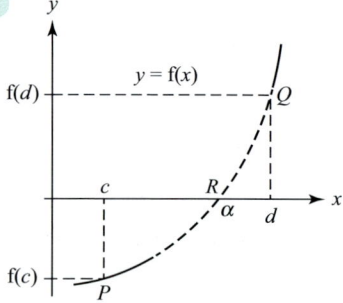

Since the curve joins P (which lies below the x-axis) to Q (which lies above it), the curve must cross the x-axis somewhere between $x = c$ and $x = d$. This is the point $R(\alpha, 0)$. Since $f(\alpha) = 0$, α is a root of the equation $f(x) = 0$

> **Key point**
> If $f(x)$ is **continuous** and changes sign between $x = c$ and $x = d$, then the equation $f(x) = 0$ has a root α, where $c < \alpha < d$

Example 1

> There is a cubic formula that gives exact roots but it can be difficult to use.

$f(x) = x^3 - 4x - 4$ is a continuous function.

The equation $x^3 - 4x - 4 = 0$ has exactly one real root α

a Show that $2.35 < \alpha < 2.45$ **b** State the value of α, correct to 1 dp.

a $f(x) = x^3 - 4x - 4$

$f(2.35) = (2.35)^3 - 4(2.35) - 4$ and $f(2.45) = (2.45)^3 - 4(2.45) - 4$

$= -0.422125$ $= 0.906125$

> Show that $f(2.35)$ and $f(2.45)$ have opposite signs.

$f(2.35) < 0$ and $f(2.45) > 0$ so $f(x)$ changes sign between $x = 2.35$ and $x = 2.45$

> By the change of sign method, the equation $f(x) = 0$ has a root between 2.35 and 2.45

So $2.35 < \alpha < 2.45$ (as α is the only real root of this equation)

b Every number between 2.35 and 2.45 equals 2.4 when rounded to 1 dp, so $\alpha = 2.4$ to 1 dp.

2.35   **2.4**   2.45

> Make a conclusion about the root.

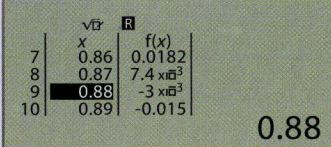
Exercise 17.1A Fluency and skills

For this exercise you can assume that all functions are continuous. Show your working in each question.

1 Use the change of sign method to show that these equations have a root between the given values c and d

a $7 - 3x - x^3 = 0$, $c = 1$, $d = 2$

b $x^2 - \dfrac{1}{x} - 4 = 0$, $c = 2.1$, $d = 2.2$

c $\sin(2x^c) - x^2 + 3 = 0$, $c = \dfrac{1}{2}\pi$, $d = \dfrac{2}{3}\pi$

d $e^x \ln x - x^2 = 0$, $c = 1.69$, $d = 1.71$

2 a Show that $e^x - x^3 = 0$ has a root

    i α between 1.85 and 1.95

    ii β between 4.535 and 4.545

  b Hence write down the value of α correct to 1 dp and the value of β correct to 2 dp.

3 i Show that these equations have a root between the given values of c and d. Work in radians where appropriate.

  ii Write down the value of this root to 1 dp.

    a $\dfrac{2}{x^3} - \dfrac{1}{x} - 2 = 0$, $c = 0.85$, $d = 0.95$

    b $e^{-x} + 2x - 1 = 0$, $c = -1.35$, $d = -1.25$

    c $x^2 \sin x - 0.5 = 0$, $c = 3.05$, $d = \pi$

4 i Show that these equations have a root between the given values of c and d. Work in radians where appropriate.

  ii Write down the value of this root to as many decimal places as can be justified.

    a $x^4 - 3x^3 + 1 = 0$, $c = 2.955$, $d = 2.965$

    b $e^{\frac{1}{x}} - x^2 = 0$, $c = 1.414$, $d = 1.424$

    c $x^2 - \sqrt{x} - 2 = 0$, $c = 1.8305$, $d = 1.8315$

d $2\ln x - \sec x = 0$, $c = \dfrac{8}{5}\pi$, $d = 5.02725$

e $e^{\cos x} - \cos(e^x) = 0$, $c = -\dfrac{3}{5}e$, $d = -\dfrac{1}{2}\pi$

5 Each of these equations has exactly one real root between 0 and 2

  i For which equations does this root lie between 1.75 and 1.85?

    a $x^2 - \sin x^c - 2 = 0$ **b** $x^3 - \cos x^c - 6 = 0$

    c $\dfrac{1}{1 + \tan x^c} - x + 2 = 0$ **d** $2 - x \operatorname{cosec} x = 0$

  ii For which of these equations is this root equal to 1.76 to 2 dp?

6 Each of the equations **i**, **ii** and **iii** can be paired, in some order, with exactly one of **A**, **B** and **C** to make a true statement.

  a Find these pairings.

  b Use these pairings to write down the value of each root to as many decimal places as can be justified.

    i $x^2 - \dfrac{1}{x} - 2 = 0$

    ii $e^{-x} - 3\sin^2 x^c + 1.8 = 0$

    iii $x - \dfrac{1}{x^2} - 2 = 0$

    A Has a root between 2.205 and 2.215

    B Has a root between 2.21 and 2.22

    C Has a root between $\dfrac{1}{2}\pi$ and $\dfrac{3}{5}e$

7 Show that each of these equations has a root between 0 and the positive constant a

  a $x^2 + 2x - 2a = 0$ **b** $ax^2 + x - a^3 = 0$

  c $\cos\left(\dfrac{\pi}{a}x\right) - \dfrac{a}{\pi}x = 0$ **d** $x^3 + (a+1)x^2 - 2a^3 = 0$

See Ch14.1

For a reminder on radian notation. A superscript c shows that x is in radians.

Strategy

To solve problems involving finding roots of equations

(1) Sketch a graph to determine the number of real solutions in the interval.

(2) Rearrange the equation into the form $f(x) = 0$

(3) Use a suitable interval to test for any possible roots.

(4) Test using the change of sign method and give any necessary conclusions.

Example 2

a Use a sketch to show that the equation $2^x = 4 - x$ has exactly one solution, α

> The change of sign method works only for continuous functions – not when there is an asymptote.

b Show that $\alpha = 1.4$ to 1 dp.

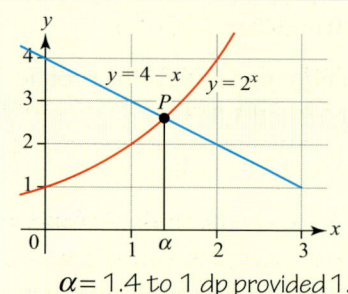

a The graphs have only one point of intersection, P, so there is only one solution.

> (1) Sketch the graphs of $y = 2^x$ and $y = 4 - x$
> You can check your sketch using a graphics calculator.

b $2^x = 4 - x \implies 2^x - 4 + x = 0$

α is the root of the equation $f(x) = 0$ where $f(x) = 2^x - 4 + x$

> (2) Rearrange the equation into the form $f(x) = 0$

$\alpha = 1.4$ to 1 dp provided $1.35 < \alpha < 1.45$

$f(1.35) = -0.100$ and $f(1.45) = 0.182$

> (3) Use a suitable interval.

By the change of sign method, the equation has a root between 1.35 and 1.45. Since α is the only root, $\alpha = 1.4$ to 1 dp.

> (4) Test using the change of sign method.

You can describe a range of numbers using **interval** notation. For $c, d \in \mathbb{R}$, the **open interval** (c, d) is the set of real numbers x such that $c < x < d$. $f(x)$ is **continuous** on an interval (c, d) if you can draw the graph of $y = f(x)$ without taking your pen off the paper.

Example 3

If $f(x) = \dfrac{1}{x-2}$, $x \neq 2$, **a** Show that $f(x)$ changes sign across the interval $(1, 3)$

b Use a sketch to show that the equation $f(x) = 0$ has no real roots. Comment on this result.

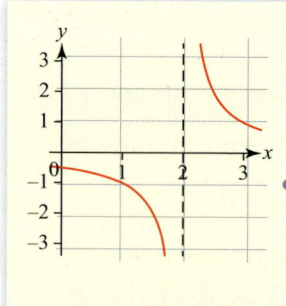

a $f(1) = -1 < 0$; $f(3) = 1 > 0$; the sign changes across the interval.

> (4) Test using the change of sign method.

b The curve never touches the x-axis. So the equation has no real roots.

The vertical asymptote at $x = 2$ means that $f(x)$ is not continuous on the interval $(1, 3)$ and so the change of sign method should not be applied.

> (1) Sketch the graph of $y = \dfrac{1}{x-2}$

> (1) Determine the number of real roots.

Key point

A continuous function $f(x)$ does not change sign in an interval which contains an even number of roots (counting repetitions) of the equation $f(x) = 0$

1 a Show that the equation $x^3 - x^2 = 1$ has a solution α in the interval $(1.4, 1.5)$

b By sketching on a single diagram the graph of $y = x^3 - x^2$ and the graph of $y = 1$, show that α is the only real solution of the equation $x^3 - x^2 = 1$

c Let β be the solution to the equation $8x^3 - 4x^2 = 1$. Use the result of part **a** to find the value of β to 1 dp.

2 a Find the coordinates of the stationary point on the curve $y = 2x - x^2$. Show your working.

b Show, by sketching a graph, that $2x - x^2 = 0.5^x$ has exactly two solutions.

c Given that these two solutions are α and β, where $\beta > 0.5$, show that

 i α lies in the interval $(0.44, 0.48)$

 ii β lies in the interval $(1.84, 1.88)$

d Hence find the value of $\beta - \alpha$ to 1 dp.

3 $f(x) = \tan(x - 1) - 1$, here x is in radians.

a Solve $f(x) = 0$ for $0 \le x \le \pi$, to 1 dp. Show your working.

b Show that $f(x)$ changes sign across the interval $(2, 3)$

c Hence explain why $f(x)$ cannot be continuous in the interval $(2, 3)$

d Find, to 1 dp, the x-value at which $f(x)$ is not continuous for $2 \le x \le 3$

4 The diagram shows the curve of $y = \sqrt{x}$, $x \ge 0$, which is an increasing function.

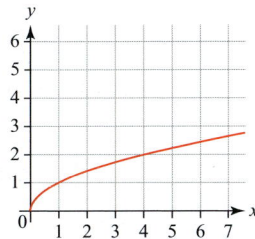

a On a copy of this diagram, draw a suitable straight line to show that the equation $2\sqrt{x} = 6 - x$ has exactly one real solution, α

b By applying the change of sign method to a suitable function and interval, show that $\alpha = 2.7$ correct to 1 dp.

c i Find the exact solution of the equation $2\sqrt{x} = 6 - x$, giving your answer in the form $a + b\sqrt{7}$. Show your working.

 ii Hence show that $\sqrt{7} \approx \dfrac{53}{20}$

5 Given the continuous function $f(x) = 4\sin(\pi x) - 6x - 1$, where x is in radians,

a Show that $f(x)$ does not change sign across the interval $(0, 1)$,

b Evaluate $f\left(\dfrac{1}{6}\right)$ and hence explain why the equation $f(x) = 0$ must have at least two roots in the interval $(0, 1)$,

c Find all the roots of $4\sin(\pi x) - 6x - 1 = 0$. Justify that you have found all the roots by sketching two suitable graphs on a single diagram.

6 The continuous and differentiable function $f(x)$ changes sign across the intervals (a, b) and (b, c). Which of these statements are true? Support your answers with a reason.

a $f(x)$ does not change sign across (a, c)

b The equation $f(x) = 0$ has no roots in (a, c)

c The equation $f'(x) = 0$ has at least one root in the interval (a, c)

Challenge

7 Let $f(x) = x^3 - 2x - 1$

The equation $f(x) = 0$ has exactly one real root α in the interval $(1, 2)$ as shown in the diagram.

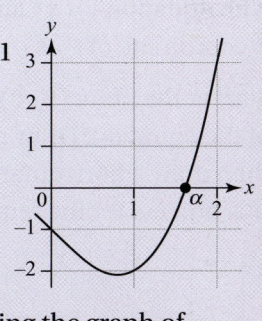

a By approximating the graph of $y = f(x)$ as a straight line across the interval $(1, 2)$, show that an approximate value of α is 1.4

b By repeating this method across $(1.4, 2)$, find another approximation for α and show it is accurate to 1 dp.

c Find the exact value of α

Fluency and skills

If an equation $x = g(x)$ has a solution α, then you can use an **iterative formula** written as $x_{n+1} = g(x_n)$ to solve the equation numerically. If the starting point x_1 is close to α, then the iterative formula produces a sequence $x_2, x_3, \ldots$ which can **converge** to α

See Ch 13.2 For a reminder on sequences.

Key point

A sequence $x_1, x_2, x_3 \ldots$ converges to α if, as n increases, x_n gets ever closer to α

Example 1

A convergent sequence is defined by $x_{n+1} = 0.5x_n + 2$ with $x_1 = 3$

a Find the values of x_2, x_3 and x_4 **b** To what value does the sequence converge?

a $x_2 = 0.5x_1 + 2 \qquad x_3 = 0.5x_2 + 2 \qquad x_4 = 0.5x_3 + 2$

$\quad = 0.5(3) + 2 \qquad = 0.5(3.5) + 2 \qquad = 0.5(3.75) + 2$

$\quad = 3.5 \qquad\qquad = 3.75 \qquad\qquad = 3.875$

The terms $x_2, x_3 \ldots$ are the iterates of the formula.

Substitute the values of x_1, x_2 and x_3

b Further calculations give $x_{10} = 3.998\ldots$ and $x_{20} = 3.999998\ldots$
The sequence appears to converge to 4. This can be confirmed by substituting 4 into the iterative formula and checking it is a solution.

As n increases, x_n gets ever closer to 4

Staircase or **cobweb** diagrams can be used to display iterates given by $x_{n+1} = g(x_n)$

To do this, draw the curve C with equation $y = g(x)$ and the straight line L with equation $y = x$ Label the point of intersection P and mark the x-coordinate of P as α. This is a solution to the equation $x = g(x)$

Start at the point $(x_1, 0)$, given that x_1 is your initial guess for the root of the equation. Draw a vertical line until you reach the curve C, and then draw a horizontal line until you reach the line L. Repeat this process for the required number of iterations.

Go to curve C first – C comes before L in the alphabet.

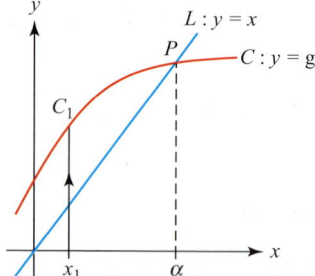
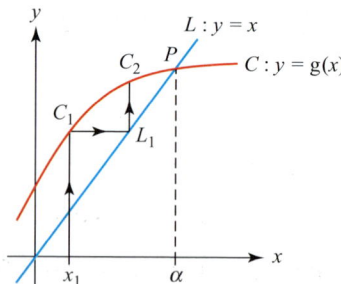
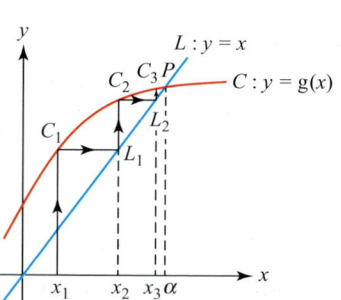

As the points $C_1, C_2, C_3, \ldots$ on the curve C converge to the point P, the sequence $x_1, x_2, x_3, \ldots$ converges to the solution α

This is an example of a **staircase** diagram.

Try it on your calculator

You can draw a staircase diagram on a graphics calculator.

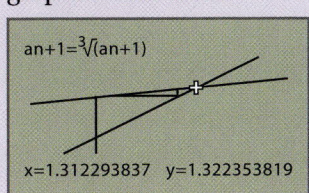

an+1=³√(an+1)

x=1.312293837   y=1.322353819

Activity

Find out how to draw a staircase diagram to display the iteration given by $x_{n+1} = \sqrt[3]{x_n + 1}$ with $x_0 = 1$ on *your* graphics calculator.

ICT Resource online

To investigate iterative root finding, click this link in the digital book.

Example 2

The diagram shows the curve with equation $y = g(x)$ and the line $y = x$. Also shown is a solution α to the equation $x = g(x)$ and an approximation to α, x_1

Display the iterates x_2, x_3 and x_4 found using the iterative formula $x_{n+1} = g(x_n)$

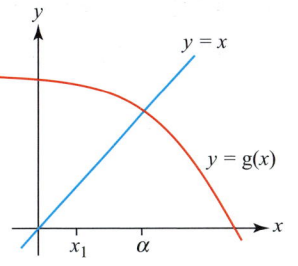

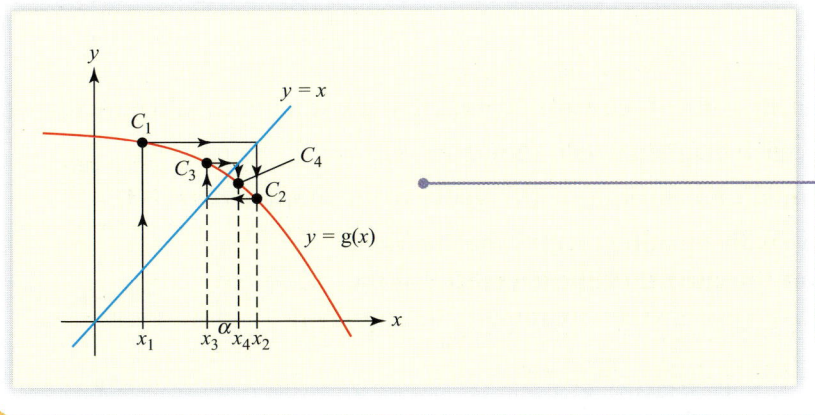

Start at the point $(x_1, 0)$. Draw a vertical line until you reach the curve $y = g(x)$, and then draw a horizontal line until you reach the line $y = x$. As you repeat this process, you obtain a more and more accurate approximation to the root.

This is a converging **cobweb** diagram.

Exercise 17.2A Fluency and skills

1 Each of these equations $x = g(x)$ has a solution α given to 3 dp.

  i Use the iterative formula $x_{n+1} = g(x_n)$ and starting value $x_1 = 1$ to calculate x_2, x_3 and x_4, giving answers to 3 dp where appropriate.

  ii Find the first iterate x_n such that $x_n = \alpha$ when both of these values are rounded to 3 dp.

  a $x = 1 - 0.1x^3, \alpha = 0.922$ **b** $x = 2 + \sqrt[3]{x}, \alpha = 3.521$

  c $x = \dfrac{2}{x^3 - 3}, \alpha = -0.618$ **d** $x = \cos(x - 1) + 0.6$, x in radians, $\alpha = 1.485$

2 For each equation $x = g(x)$, the sequence defined by the iterative formula $x_{n+1} = g(x_n)$ with the given starting value, converges to a solution α

  i Use this iterative formula and starting value x_1 to find α correct to 2 dp, working in radians where appropriate.

 MyMaths   2174    SEARCH

ii Write down an interval of width 0.01 with a mid-value of α

a $x = e^{-x} + 2$, $x_1 = 1$ **b** $x = \dfrac{3x^2 + 2}{x^2 - 2}$, $x_1 = 3$ **c** $x = \sqrt{2x + \ln x}$, $x_1 = 2$

d $x = \left(4x^2 - 1\right)^{\frac{1}{3}}$, $x_1 = 3$ **e** $x = 2\sin x - \cos x$, $x_1 = 3$ **f** $x = \sqrt{x + e^{-x}}$, $x_1 = 0$

3 The equation $x = \sin(\cos x) - \cos(\sin x) + 1$, where x is in radians, has exactly one real root $\alpha = 0.878$ to 3 dp. It is given that the sequence defined by the iterative formula $x_{n+1} = \sin(\cos x_n) - \cos(\sin x_n) + 1$, with starting value $x_1 = 0$, converges to α

a Find the smallest number, N, of iterations required to produce an iterate which equals α when both numbers are rounded to 3 dp.

b Show that using this iterative formula with starting value $x_1 = 100$ requires *fewer* than N iterations to achieve this level of accuracy.

4 A student is asked to draw either a staircase or cobweb diagram on each of these sketches. Starting at x_1, the student draws three lines on each sketch, but in each case they make a mistake.

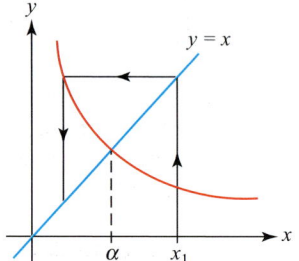 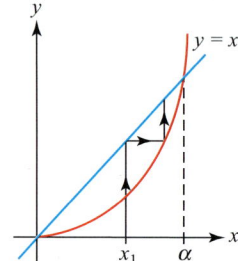 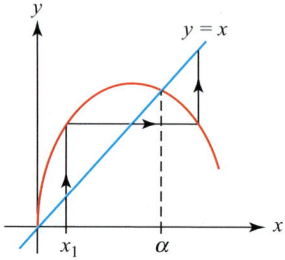

a On a copy of each sketch, draw the three arrowed lines that the student should have drawn.

It is given that, if continued, each diagram would show convergence.

b Which diagram will definitely not illustrate convergence to the solution α?

5 The diagram shows the curve with equation $y = g(x)$, the line $y = x$ and the points P and Q where the curve and line intersect. P is the stationary point on the curve $y = g(x)$ and the solutions of the equation $x = g(x)$ are α and β

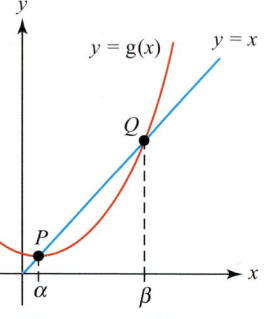

Using a copy of the diagram and the iterative formula $x_{n+1} = g(x_n)$

a Determine which of α or β these iterates converge to when x_1 is just less than β,

b Illustrate and describe the behaviour of these iterates when x_1 is just greater than β

Reasoning and problem-solving

To solve problems that involve finding an approximation to the root of an equation

1 Select an appropriate way to rearrange the equation $f(x) = 0$ into the required form $x = g(x)$

2 Apply the iterative formula $x_{n+1} = g(x_n)$ with a suitable starting value x_1

3 Use the change of sign method to prove accuracy levels have been achieved.

4 Examine the value of $g'(x)$ near a solution to determine if convergence to that root will occur.

5 Give your conclusion.

Example 3

The diagram shows the graph of $y = f(x)$, where $f(x) = x^3 - 6x + 2$, $x \geq 0$

The equation $f(x) = 0$ has exactly two positive roots α and β

a i Show that the equation $f(x) = 0$ can be rearranged into the

form $x = \dfrac{x^3 + 2}{6}$

ii Use the iterative formula $x_{n+1} = \dfrac{x_n^3 + 2}{6}$ with $x_1 = 0.5$ to find α

to 2 dp, justifying your answer.

b i Show that the equation $f(x) = 0$ can be rearranged into the

form $x = \sqrt{\dfrac{6x - 2}{x}}$

ii Comment on the suitability of the iterative formula

$x_{n+1} = \sqrt{\dfrac{6x_n - 2}{x_n}}$ with $x_1 = 0.5$ for estimating α

a i $x^3 - 6x + 2 = 0$

$6x = x^3 + 2$

$x = \dfrac{x^3 + 2}{6}$

ii $x_{n+1} = \dfrac{x_n^3 + 2}{6}$, $x_1 = 0.5$ so $x_2 = 0.35$, $x_3 = 0.34$, $x_4 = 0.34$,

$x_5 = 0.34$ (2 dp)

The interval is $(0.335, 0.345)$

$f(0.335) = 0.028$ and $f(0.345) = -0.029$ so $f(x)$ changes

sign across $(0.335, 0.345)$

To 2 dp, 0.34 is a root of the equation $f(x) = 0$

$\alpha = 0.34$ to 2 dp.

b i $x^3 - 6x + 2 = 0$

$x^3 = 6x - 2$

$\dfrac{x^3}{x} = \dfrac{6x - 2}{x}$

$x = \sqrt{\dfrac{6x - 2}{x}}$

ii $x_{n+1} = \sqrt{\dfrac{6x_n - 2}{x_n}}$, $x_1 = 0.5$ so $x_2 = 1.41$, $x_3 = 2.14$,

$x_4 = 2.25$, $x_5 = 2.26$, $x_6 = 2.26$ (2 dp).

This sequence converges to β so this formula is not suitable for

estimating α

① Rearrange $x^3 - 6x + 2 = 0$ to make x the subject.

② Apply the iterative formula. Continue calculating iterates until their values agree to 2 dp.

All numbers in this interval round to 0.34 to 2 dp.

③ Use the change of sign method to check accuracy.

⑤ Give your conclusion.

① Rearrange $x^3 - 6x + 2 = 0$ to make x the subject.

② Apply the iterative formula.

⑤ Give your conclusion.

You can use the gradient of the curve to determine if $x_{n+1} = g(x_n)$ converges to the solution α

Key point

If $-1 < g'(x) < 1$ for all x in an interval which contains
α and the starting value x_1, then $x_{n+1} = g(x_n)$ converges.

Example 4

The equation $x = x^3 - 2$ has a solution α near 1.5

a Show that the sequence produced by $x_{n+1} = x_n^3 - 2$ with $x_1 = 1$ fails to converge to α

b Explain the cause of this failure.

a $x_{n+1} = x_n^3 - 2$, $x_1 = 1$ so $x_2 = -1$, $x_3 = -3$, $x_4 = -29$, $x_5 = -24\,391$

This sequence does not converge to α

b $g(x) = x^3 - 2 \Rightarrow g'(x) = 3x^2$

$g'(\alpha) \approx g'(1.5) = 3(1.5)^2 = 6.75$

The gradient is greater than 1 for values of x close to α

> **2** Apply the iterative formula.
>
> Differentiate $g(x)$
>
> **4** Evaluate $g'(x)$ near the root α
>
> **5** Give your conclusion.

Exercise 17.2B Reasoning and problem-solving

1 The diagram shows part of the graph of $y = f(x)$ where $f(x) = x^4 - 2x - 1$

The equation $f(x) = 0$ has exactly two real roots α and β

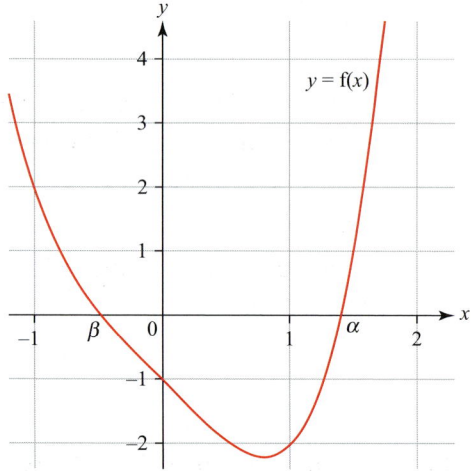

a Show that the equation $x^4 - 2x - 1 = 0$ can be rearranged into the form
$$x = \frac{\sqrt{2x+1}}{x}$$

b Use the iterative formula
$$x_{n+1} = \frac{\sqrt{2x_n + 1}}{x_n}$$ with $x_1 = 1$ to find α to 1 dp, justifying your answer.

c Show that $x_1 = -0.5$ is not an appropriate starting value when using the iterative formula $x_{n+1} = \frac{\sqrt{2x_n + 1}}{x_n}$ to estimate β

2 Given that $f(x) = x^3 - 3x^2 - 4$ and that the equation $f(x) = 0$ has exactly one real root, α

a Use the change of sign method to show that $\alpha = 3.355$ to 3 dp.

Two iterative formulae (I) and (II), found by rearranging the equation $f(x) = 0$, are used to estimate α

(I) $x_{n+1} = \sqrt[3]{3x_n^2 + 4}$, $x_1 = 3$

(II) $x_{n+1} = \sqrt{\dfrac{3x_n^2 + 4}{x_n}}$, $x_1 = 3$

b Compare how quickly each of these formulae produces an estimate for α which is correct to 3 dp.

c **i** Show that the equation $f(x) = 0$ can be rearranged into the form $x_{n+1} = g(x_n)$, where $g(x) = \dfrac{2}{\sqrt{x-3}}$

ii By finding $g'(x)$ determine whether or not this is a suitable rearrangement for estimating α. Support your answer by evaluating iterates.

3 At the start of an experiment substance A is being heated whilst substance B is cooling down. All temperatures are measured in °C. The equation $T_A = 10e^{0.1t}$ models the temperature T_A of substance A and the equation $T_B = 16e^{-0.2t} + 25$ models the temperature T_B of substance B, t minutes from the start.

a Show that the time t at which the two substances have equal temperatures satisfies the equation $t = 10\ln(1.6e^{-0.2t} + 2.5)$

b Use the iterative formula $t_{n+1} = 10\ln(1.6e^{-0.2t_n} + 2.5)$ with $t_1 = 0$ to find this time, giving your answer to the nearest minute.

All the logarithm keys on a student's calculator have stopped working.

c By letting $x = e^{0.1t}$ use a suitable iterative formula of the form $x_{n+1} = g(x_n)$ that will enable the student to find the approximate time at which the two substances have equal temperatures.

4 a By sketching a pair of graphs on a single diagram, show that the equation $\theta = 2\sin\theta$, where θ is in radians, has exactly one solution between $\frac{1}{2}\pi$ and π

The diagram (not drawn to scale) shows a sector OAB of a circle, radius 4 cm. The angle subtended by the arc AB at the centre O is α radians.

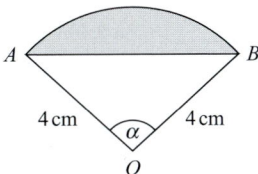

Given that the area of triangle OAB equals the area of the shaded segment

b Show that $\alpha = 2\sin\alpha$

c Use the iterative formula $\theta_{n+1} = 2\sin\theta_n$ and a suitable starting value to find angle α to 2 dp, justifying your answer.

5 Two points A and B on a circle, centre O, radius r cm are such that the length of arc AB is twice the length of the chord AB

a Show that $\alpha = \sqrt{8 - 8\cos\alpha}$, where α is the angle, in radians, subtended by this arc at the centre of this circle.

The equation $\theta = \sqrt{8 - 8\cos\theta}$ has exactly one positive solution.

b Use the iterative formula $\theta_{n+1} = \sqrt{8 - 8\cos\theta_n}$ with starting value $\theta_1 = 3$ radians to find angle α to 1 dp, justifying your answer.

c Given that $r = 2$ cm, calculate the area of triangle OAB to 1 decimal place.

Two students are investigating the convergent sequence defined by $\theta_{n+1} = \sqrt{2 - 2\cos\theta_n}$ with $\theta_1 = 1$ radian. One claims this sequence converges to a positive value, the other claims that it converges to zero. To determine who is correct, they are given this diagram which shows a circle with centre R and radius 1. The angle subtended by the arc PQ at the centre O is θ radians.

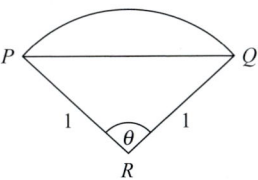

d i What does the equation $\theta = \sqrt{2 - 2\cos\theta}$ imply about the relationship between the arc length PQ and the length of the chord PQ?

ii Determine which student made the correct claim.

Fluency and skills

See Ch13.2
For a reminder on sequences.

See Ch15
For a reminder on differentiation.

The **Newton-Raphson** method is another way to estimate a root of an equation.

> **Key point**
>
> If α is a root of the equation $f(x) = 0$, then the iterative sequence given by $x_{n+1} = x_n - \dfrac{f(x_n)}{f'(x_n)}$ converges to α, if it converges.

The starting value x_1 is a **first approximation** to α. In general, x_{n+1} is a better approximation to α than x_n

Example 1

The equation $x^3 - 2x - 3 = 0$ has exactly one real root α, where $1 < \alpha < 2$

a Taking $x_1 = 2$ as a first approximation to α, use the Newton-Raphson method to find the second and the third approximations to α. Give answers to 3 decimal places where appropriate.

b Use a change of sign to show that the iterates converge to 1.893 (3 dp).

---

a $f(x) = x^3 - 2x - 3$ — Define a suitable function $f(x)$ which has a root α

$f'(x) = 3x^2 - 2$ — Differentiate.

$x_2 = x_1 - \dfrac{f(x_1)}{f'(x_1)} = 2 - \dfrac{f(2)}{f'(2)}$ — Substitute $x_1 = 2$

$f(2) = 2^3 - 2 \times 2 - 3 = 1$ and $f'(2) = 3 \times 2^2 - 2 = 10$ — Work out f(2) and f'(2)

$x_2 = 2 - \dfrac{1}{10} = 1.9$

A second approximation for α is $x_2 = 1.9$ — Substitute and evaluate x_2

$x_3 = x_2 - \dfrac{f(x_2)}{f'(x_2)}$ — Use x_2 to find x_3

$= 1.9 - \dfrac{f(1.9)}{f'(1.9)}$ — Substitute $x_2 = 1.9$

$f(1.9) = 0.059$ and $f'(1.9) = 8.83$ — Substitute and evaluate x_3

$x_3 = 1.8933...$

A third approximation for α is $x_3 = 1.893$ (3 dp)

b $f(x) = x^3 - 2x - 3$

$f(1.8925) = -0.0069...$ and $f(1.8935) = 0.0018...$ — Apply the change of sign method across a suitable interval.

This change of sign shows that $\alpha = 1.893$ (3 dp)

> Since $x_3 = 1.893$, the Newton-Raphson method has found α correct to 3 dp in just two iterations.

In practice, you can use the $\boxed{\text{Ans}}$ key to calculate the iterates $x_2, x_3, x_4, \ldots$ To do this for the previous example you press $\boxed{2}$ $\boxed{=}$ to enter the starting value $x_1 = 2$, then key in the sequence

$\boxed{\text{Ans}} - \left(\dfrac{\boxed{\text{Ans}}^3 - 2\boxed{\text{Ans}} - 3}{3\boxed{\text{Ans}}^2 - 2} \right)$ to enter the iterative formula and then repeatedly press $\boxed{=}$ to display each iterate.

1   Use the Newton-Raphson method to find second and third approximations to a root α of the given equation, where x_1 is a first approximation to this root. Give answers to 2 dp where appropriate.

a $x^3 - 2x^2 + x - 3 = 0$, $2 < \alpha < 3$, $x_1 = 2$

b $2x - x^3 + 5 = 0$, $2 < \alpha < 3$, $x_1 = 2$

c $x^4 - 2x^3 - 1 = 0$, $-1 < \alpha < 0$, $x_1 = -1$

d $x^2 - 2x^{\frac{1}{2}} - 8 = 0$, $3 < \alpha < 4$, $x_1 = 3$

e $x^3 + \dfrac{4}{x} - 6 = 0$, $1 < \alpha < 2$, $x_1 = 2$

f $3\sqrt[3]{x} + \dfrac{1}{2x^2} - 5 = 0$, $4 < \alpha < 5$, $x_1 = 4$

2   Each of these equations has exactly one real root, α. Use the Newton-Raphson method with the given first approximation x_1 to find α to 3 dp. Justify that this level of accuracy has been achieved by using the change of sign method.

a $e^x + 3x - 4 = 0$, $x_1 = 1$

b $x^2 - 3e^{2x} = 0$, $x_1 = 0$

c $x^2 + 3\ln x = 0$, $x_1 = 1$

d $\sin x + x - 3 = 0$, $x_1 = 0$ radians

e $x - \cos^2 x - 3 = 0$, $x_1 = 0$ radians

f $x^2 \ln x - 2 = 0$, $x_1 = 0.7$

3   The equation $x^3 - 2x^2 - 7 = 0$ has exactly one real root α

a   Show that two iterations of the Newton-Raphson method with first approximation $x_1 = 3$ are sufficient to produce an estimate for α which is accurate to 3 dp.

b   Determine whether using the Newton-Raphson method with first approximation $x_1 = 1$ produces a reliable estimate for α

4   The equation $(x + \sin x)^2 - 1 = 0$ has exactly one positive root α

a   Working in radians, show that two iterations of the Newton-Raphson method with first approximation $x_1 = 1$ produces an estimate for α which is

i   Accurate to 2 dp,

ii   Not accurate to 3 dp.

b   Determine whether using the Newton-Raphson method with first approximation $x_1 = 2$ produces a reliable estimate for α

5   Given that $f(x) = \sin x + e^{-x}$, use the Newton-Raphson method (working in radians) with $x_1 = 2$ to find, correct to 3 dp, an approximate solution to the equation

a $f(x) = 0$ b $f'(x) = 0$

For each equation, justify that this level of accuracy has been achieved.

6   The Newton-Raphson method, with the given first approximation x_1, is used to solve these equations.

i   For each equation, show that x_3 cannot be calculated, giving the reason.

ii   Use the Newton-Raphson method to find all the solutions to each equation shown in the diagrams. Give your answers to 3 dp, justifying that this level of accuracy has been achieved.

a $2\sqrt{x} - \dfrac{1}{x} + 1 = 0$, $x_1 = 1$

b $x^3 - 8\sqrt{x} + 2 = 0$, $x_1 = 1$

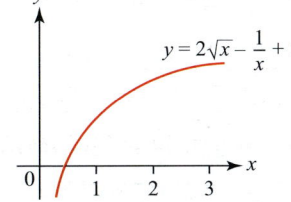

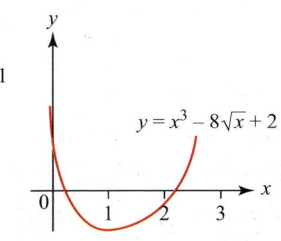

Strategy

To use the Newton-Raphson method to find solutions to practical problems

(1) Rearrange the equation you are trying to solve into the form f(x) = 0

(2) Use your calculator to efficiently calculate iterates.

(3) Use the change of sign method to prove the accuracy of your estimate.

Example 2

The diagram shows the design for a sports field in the shape of a sector of a circle ABC, centre A, radius r. The acute angle $BAC = \alpha$ radians, and point D on AB is such that the line CD is perpendicular to AB

The length CB must be 30% longer than CD

Use the Newton-Raphson method to find α to 1 dp.

See Ch14.1

For a reminder on radians and the arc length formula.

Arc length $CB = r\alpha$

Using trigonometry, $\sin \alpha = \dfrac{CD}{AC}$ Length $AC = r$, so length $CD = r\sin \alpha$

Length $CB = 1.3 \times$ length CD, so $r\alpha = 1.3 r \sin \alpha \Rightarrow \alpha = 1.3 \sin \alpha$

α is a solution to the equation $\theta = 1.3\sin\theta \Rightarrow \theta - 1.3\sin\theta = 0$

$f(\theta) = \theta - 1.3\sin\theta$ so, $f'(\theta) = 1 - 1.3\cos\theta$

See Ch15.2

For a reminder on differentiating trigonometric functions.

$\theta_{n+1} = \theta_n - \dfrac{f(\theta_n)}{f'(\theta_n)} = \theta_n - \dfrac{(\theta_n - 1.3\sin\theta_n)}{(1 - 1.3\cos\theta_n)}$, with $\theta_1 = 1$ radian

$\theta_2 = 1.31..., \theta_3 = 1.22..., \theta_4 = 1.22...$

$f(1.15) = -0.03659...$ and $f(1.25) = 0.01632...$

This change of sign proves that $\alpha = 1.2$ radians (to 1 dp)

Find expressions for the lengths of CB and CD

Rearrange the equation. (1)

Differentiate.

Find the iterative formula. (2)
1 radian is a reasonable first approximation.

The iterates converge to $\alpha = 1.2$ radians (to 1 dp).

Use the change of sign method across the interval (1.15, 1.25) (3)

When using the Newton-Raphson method, you must choose a suitable value for the first approximation. In the example, $\theta_1 = 1$ radian was used as a first approximation to α

If, instead, $\theta_1 = 0.6$ radian, then $\theta_2 = -1.237..., \theta_3 = -1.221..., \theta_4 = -1.221...$

Clearly this sequence does not converge to the required root α. The reason for this failure is that the first approximation is close to the θ-coordinate of a stationary point on the curve $y = \theta - 1.3\sin\theta$

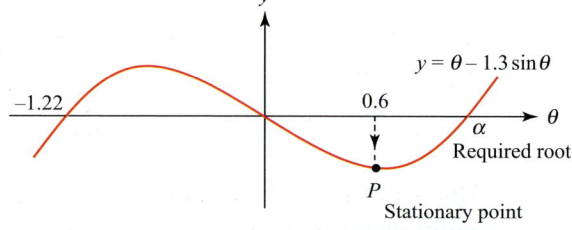

The diagram shows the point P on this curve where $\theta = 0.6$

Key point

The Newton-Raphson method may not converge to the required root if the first approximation is close to the x-coordinate of a stationary point on the curve $y = f(x)$.
The Newton-Raphson method will also fail to converge if the derivative is zero at one of the values for x_n

1 The equation $4x^3 - 12x^2 + 9x - 1 = 0$ has exactly one real root α in the interval $(0,1)$

   a Use the Newton-Raphson method with a first approximation $x_1 = 0$ to find α to 2 dp, justifying your answer.

   b Explain why the Newton-Raphson method fails to find an estimate for α when the first approximation is

     i $x_1 = 0.5$ **ii** $x_1 = 1$

   c Using the factor theorem, find the exact value of α and hence find the number of iterations of the Newton-Raphson formula starting with $x_1 = 0$ that are required to find α correct to 8 dp.

2 The curve C has equation $y = f(x)$ where $f(x) = x^4 + 4x^2 - 6x$. C has exactly two real roots and exactly one stationary point.

   a Write down one of the roots of C

   b Use the Newton-Raphson method on a suitable function with starting value $x_1 = 1$ to find, to 1 dp,

     i The other root of this curve,

     ii The coordinates of the stationary point P on C

   c Sketch the curve C

3 The curve C with equation $y = x\sin x + 2\cos x$, where x is in radians, has exactly one stationary point P in the interval $(\pi, 2\pi)$. The x–coordinate of P is β

   a Show that β is a root of the equation $g(x) = 0$, where $g(x) = \tan x - x$

   b Show that, when applied to $g(x)$, the Newton-Raphson formula can be written as $x_{n+1} = x_n - \left(\dfrac{\tan x_n - x_n}{\tan^2 x_n} \right)$

   c Using this formula with first approximation $x_1 = 4.5$ radians, find the coordinates of point P to 2 dp.

   d Use any appropriate technique to show that P is a minimum point.

4 **a** **i** Show with a sketch that $2e^x = 6x + 40$ has exactly one positive solution.

     ii Apply the Newton-Raphson method to a suitable function to solve this equation to 1 dp, justifying your answer. Use $x_1 = 2$ as a first approximation.

For a particular country, the equation $S = 2e^{0.5t}$ models the total S (£millions) paid in subscriptions for streaming music. The equation $D = 3t + 40$ models the total D (£millions) paid for music downloads t years after 1st January 2010

   b At the start of which year did the total amount paid in subscriptions for streaming first exceed the total amount paid for downloads? Show your working.

   c Find, to the nearest million, the total amount paid in subscriptions and downloads from 1st January 2010 to the start of the year found in part **b**.

Challenge

5 For different values of $k > 0$, investigate the behaviour of the sequence defined by $x_{n+1} = \dfrac{1}{2}\left(x_n + \dfrac{k}{x_n} \right)$, $x_1 = 1$

   a How is the value to which this sequence converges related to the value of k?

   b How can your findings be used to estimate the value of $\sqrt{54321}$ if the square-root button on your calculator is not working?

   c Use the Newton-Raphson method to explain your findings.

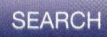

Fluency and skills

See Ch4.6

For a reminder on how to use integration to find the area under a curve.

The definite integral $\int_a^b f(x)\,dx$ can be used to calculate the area of the region **R** bounded by the curve $y = f(x)$ and the x-axis between $x = a$ and $x = b$

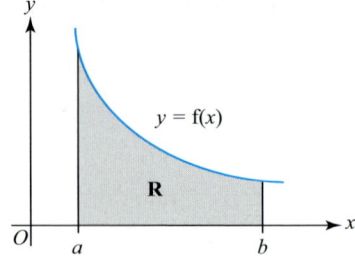

Often, $\int f(x)\,dx$ cannot be found, so you can't calculate the exact value of the definite integral. You can, however, use the trapezium rule to find an approximation to $\int_a^b f(x)\,dx$

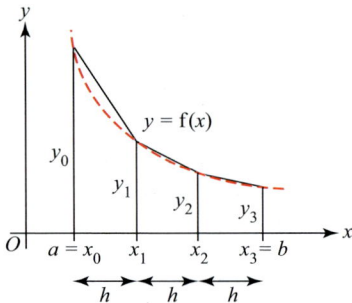

The diagram shows the region **R** split into three trapeziums of equal width. The x-coordinates between a and b are x_0, x_1, x_2 and x_3

Each trapezium has width h and their vertical sides are the y-values (or ordinates, where, $y_0 = f(x_0)$, $y_1 = f(x_1)$ and so on).

The area of the three trapeziums is

$$\frac{1}{2}h(y_0 + y_1) + \frac{1}{2}h(y_1 + y_2) + \frac{1}{2}h(y_2 + y_3)$$

The area of **R** $\approx \frac{1}{2}h\{(y_0 + y_1) + (y_1 + y_2) + (y_2 + y_3)\}$

so $\int_a^b f(x) \approx \frac{1}{2}h\{(y_0 + y_3) + 2(y_1 + y_2)\}$

More generally,

Key point

The trapezium rule is $\int_a^b f(x) \approx \frac{1}{2}h\{(y_0 + y_n) + 2(y_1 + y_2 + \ldots + y_{n-1})\}$

where the interval $a \le x \le b$ is divided into n intervals of equal width $h = \dfrac{b-a}{n}$ defined by the values $a = x_0, x_1, \ldots, x_{n-1}, x_n = b$

The ordinates are given by $y_0 = f(x_0), y_1 = f(x_1)$ etc.

See Ch4.6

For a reminder on how to use a calculator to evaluate a definite integral.

Remember that you can evaluate definite integrals using your calculator. You could use this to check answers found using the trapezium rule.

Recall that the area of a trapezium is found using $\frac{1}{2}h(a+b)$ where a and b are the lengths of the parallel sides and h is the distance between them.

Increasing the number of intervals improves the accuracy of the estimate for $\int_a^b f(x)$

Example 1

PURE

The diagram shows the curve with equation $y = \sqrt{\ln x}$ for $x \geq 1$

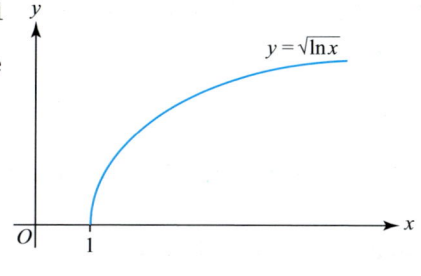

a Use the trapezium rule with four intervals to estimate the value of $\int_{2}^{4} \sqrt{\ln x}\, dx$, giving the answer to 1 dp.

b Explain why the trapezium rule gives an underestimate for the value of $\int_{2}^{4} \sqrt{\ln x}\, dx$

c Explain why increasing the number of intervals when using the trapezium rule gives a better estimate for this integral.

a The number of intervals $n = 4$

The lower and upper limits of $\int_{2}^{4} \sqrt{\ln x}\, dx$ are $a = 2$ and $b = 4$, respectively.

The width of each interval $h = \dfrac{b-a}{n} = \dfrac{4-2}{4} = 0.5$

$x_0 = 2$, $x_1 = 2.5$, $x_2 = 3$, $x_3 = 3.5$, $x_4 = 4$

x_i	2	2.5	3	3.5	4
$y_i = \sqrt{\ln x_i}$	0.83	0.96	1.05	1.12	1.18

> Draw a table to record ordinates to at least one more decimal place than that required in the final answer.

$$\int_{2}^{4} \sqrt{\ln x}\, dx \approx \frac{1}{2}h\big[(y_0 + y_4) + 2(y_1 + y_2 + y_3)\big]$$

$$\approx \frac{1}{2}(0.5)\big[(0.83 + 1.18) + 2(0.96 + 1.05 + 1.12)\big]$$

$$= 2.0675$$

$$\int_{2}^{4} \sqrt{\ln x}\, dx \approx 2.1 \,(\text{to 1 dp})$$

> Substitute $h = 0.5$ and the y-values into the formula.

> Give the answer produced by the trapezium rule to the required level of accuracy.

b The curve is concave so each trapezium lies entirely under the curve $y = \sqrt{\ln x}$ so the total area of the trapeziums is less than the exact area.

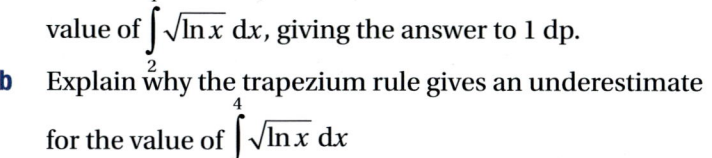

c Increasing the number of intervals decreases the width of each trapezium. The thinner each trapezium, the more accurately they approximate the curve.

Exercise 17.4A Fluency and skills

1 Use the trapezium rule with the stated number of intervals to find an estimate for these integrals. Give each estimate to two decimal places.

a $\displaystyle\int_{1}^{5} e^{\sqrt{x}}\, dx$, 4 intervals

b $\displaystyle\int_{1}^{3} 8^{\frac{1}{x}}\, dx$, 4 intervals

c $\displaystyle\int_{3}^{6} \ln\left(1 + \sqrt{2x}\right) dx$, 5 intervals

d $\displaystyle\int_{0}^{\frac{\pi}{2}} \sin\left(x^2\right) dx$, 3 intervals

e $\displaystyle\int_{-1}^{2} \frac{1}{x^3 + 5}\, dx$, 6 intervals

f $\displaystyle\int_{1}^{2} x^x\, dx$, 5 intervals

2 i Use the trapezium rule with five intervals to find an estimate for the area of the region **R** shown in each diagram. Work in radians where appropriate and give your answer to 3 dp.

ii Is your answer to part **i** an underestimate or an overestimate for the area of **R**? Why?

a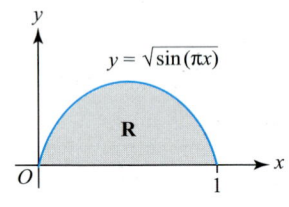
$y = \sqrt{\sin(\pi x)}$

b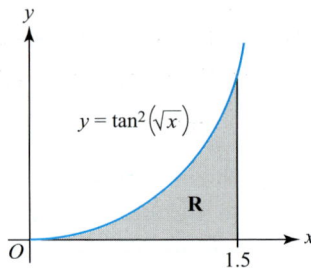
$y = \tan^2\left(\sqrt{x}\right)$

c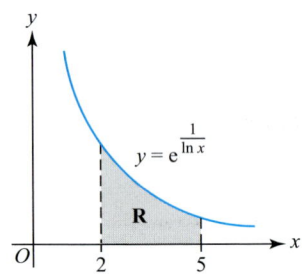
$y = e^{\frac{1}{\ln x}}$

d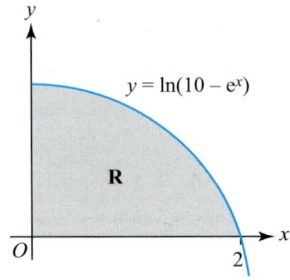
$y = \ln(10 - e^x)$

3 The table shows some values for x and y, where $y = 3^{x^2-1}$

The values of y have been rounded to 2 dp where appropriate.

x_i	1	1.1	1.2	1.3	1.4	1.5	1.6	1.7	1.8	1.9	2
y_i	1	1.26	1.62		2.87		5.55	7.98		17.59	27

a Copy and complete the table.

b Use the trapezium rule with all of the y-values in the completed table to find an estimate for $\int_{1}^{2} 3^{x^2-1} \, dx$. Give your answer to 1 dp.

4 a Use the trapezium rule with the stated number of intervals to find an estimate for these integrals. Give each estimate to 3 dp.

i $\int_{1}^{4} \dfrac{e^x}{x} - 2 \, dx$, 4 intervals   **ii** $\int_{1}^{4} 6 - e^{0.1x^2} \, dx$, 5 intervals

The diagrams below show, in some order, part of the graphs with equations $y = \dfrac{e^x}{x} - 2$ and $y = 6 - e^{0.1x^2}$ for $1 \le x \le 4$

b Match each graph to its equation and hence state, with a reason, whether your answers to part **a** are underestimates or overestimates for each integral.

Graph 1

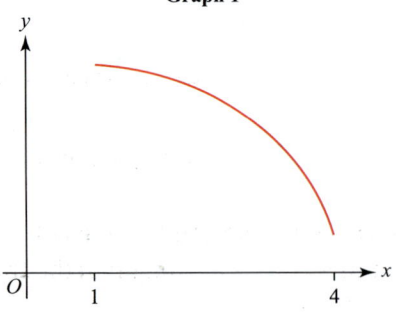

Graph 2

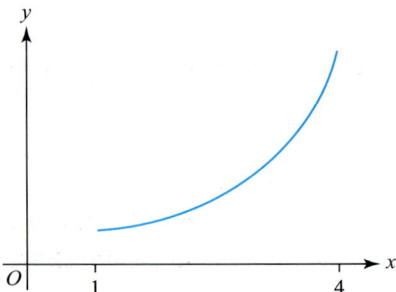

5 For each function $f(x)$

   i   Sketch the graph of $y = f(x)$ for $0 \le x \le 5$,

   ii Use the trapezium rule with the stated number of intervals to estimate $R = \int_{0}^{5} f(x)\, dx$, to 2 dp,

   iii Is this answer an underestimate or an overestimate for the integral? Why?

   iv Verify your answer to part **iii** by using integration to find the exact area of R. Show your working.

   a $f(x) = \dfrac{1}{(x+1)} + 4$, 4 intervals   **b** $f(x) = e^{0.1x} + 2$, 5 intervals

Reasoning and problem-solving

To use the trapezium rule to find numerical solutions to real-life problems

  1 Calculate the width of each trapezium using the number of intervals.

  2 Check that the number of ordinates is one more than the number of intervals used.

  3 Substitute the values of h and the ordinates into the trapezium rule.

  4 Use a sketch of a graph to determine if your answer is an underestimate or overestimate.

  5 Interpret an estimate for the integral in context.

Example 2

The diagram shows the curve with the equation $v = 1.7e^{\sqrt{t}} - 0.5t^2$, where v models the speed (in $\mathrm{ms^{-1}}$) of an athlete t seconds after he starts a warm-up sprint.

It takes the athlete six seconds to run the length of the track.

a Use the trapezium rule with five intervals to find an estimate for the value of $\int_{0}^{6} 1.7e^{\sqrt{t}} - 0.5t^2\, dt$, giving the answer to 1 dp

Is your answer an underestimate or overestimate? Why?

b State the minimum length of the track, to the nearest metre.

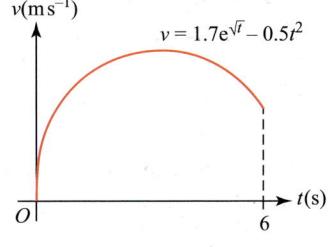

a The interval width $h = \dfrac{6-0}{5} = 1.2$ •   Use $h = \dfrac{b-a}{n}$ ①

t_i	0	1.2	2.4	3.6	4.8	6
$v_i = 1.7e^{\sqrt{t_i}} - 0.5t_i^2$ (2 dp)	1.70	4.36	5.12	4.86	3.68	1.69

There are six ordinates and five intervals. ②

$\int_{0}^{6} 1.7e^{\sqrt{t}} - 0.5t^2\, dt$

$\approx \dfrac{1}{2}(1.2)[1.70 + 1.69 + 2(4.36 + 5.12 + 4.86 + 3.68)]$ •   Substitute. ③

$= 23.7$ (1 dp)

The curve is concave so this is an underestimate for the length of the track. •   Make use of the sketch. ④

b 24 metres. •   Interpret in context. ⑤

Recall that the area under a speed–time graph equals the distance travelled, which in this case, is the length of the track.

1 a Use the trapezium rule with five intervals to estimate the value of
$$\int_0^2 1 - \sqrt{\cos(0.25\pi x)}\ \mathrm{d}x,$$ where x is in radians. Give your answer to 2 sf.

The shaded region in the diagram shows the plan of a large field OAB where O is the origin. Point A has coordinates $(0, 3)$ and point B has coordinates $(2, 3)$. The curve OB has equation $y = 3 - 3\sqrt{\cos(0.25\pi x)}$ and all distances are measured in kilometres.

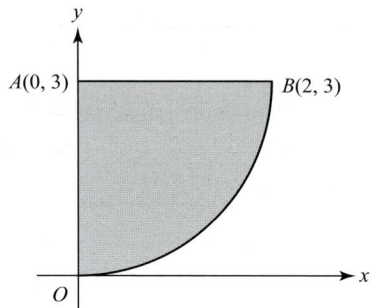

b i Use your answer to part **a** to find an estimate for the area of this field. Give your answer to 3 sf.

ii State, with a reason, whether this answer is an underestimate or an overestimate for the exact area of this field.

2 $I = \int_{4.5}^{6} \dfrac{x}{\sqrt{x^2 - 20}}\ \mathrm{d}x$

a Use the trapezium rule with five intervals to estimate the value of I. Give your answer to 2 dp.

The trapezium rule with 20 intervals is used to estimate I. The sum of all the ordinates used is 52.725 (to 3 dp).

b Find this estimate for I, to 2 dp.

c Use integration to find I exactly. Show your working.

d Show that, as an estimate for I, your answer to part **b** is nine times more accurate than your answer to part **a**.

3 The diagram shows the cross section of a river from one side of its bank (point A) to the other (point B).

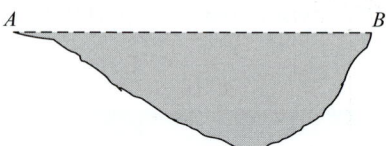

The depth h (in metres) of the river is modelled by the equation $h = 4\sin\left(\dfrac{\pi}{36}x^2\right)$, where x is the distance (in metres) across the river from point A

a Find the width of the river from A to B

b Using the trapezium rule with six intervals, working in radians, find an estimate for the area of this cross section. Give your answer to 1 dp.

c A stick floating between A and B has speed $0.5\ \mathrm{m\,s^{-1}}$. Given that 1 cubic metre = 1000 litres, find an estimate for the flow rate of the river. Give your answer in litres per second, to 1 sf.

4 a i Use the trapezium rule with five intervals to estimate the value of
$$\int_0^8 2^{0.25x}\ \mathrm{d}x \text{ to 1 dp.}$$

ii State, with a reason, whether your answer is an underestimate or an overestimate for this integral.

At the start of a race, runner B has a 10 metre head start on runner A. The equations $v_A = k + 2^{0.25t}$ and $v_B = 0.25t + 1$ model the speeds (in $\mathrm{m\,s^{-1}}$) of A and B, respectively, t seconds from the start, where k is a constant. After four seconds, A is running $1\ \mathrm{m\,s^{-1}}$ faster than B,

b Find the value of k

The diagram shows the speed–time graphs for runners A and B.

It takes B exactly eight seconds to complete the race.

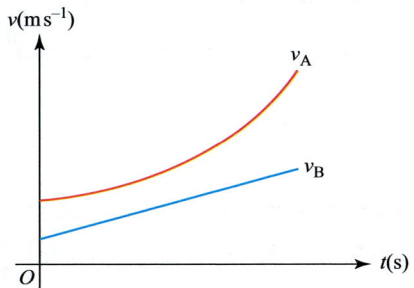

c Determine who wins the race, justifying your answer.

5 a Write down the formula for the sum of the first n terms of a geometric series with first term a and common ratio r

b Show that by using the trapezium rule with n intervals to estimate the value of $\int_0^1 2^{nx}\,\mathrm{d}x$ the answer given is $\dfrac{3}{2n}\left(2^n-1\right)$

For a particular value of n, the estimate for $\int_0^1 2^{nx}\,\mathrm{d}x$ found using the trapezium rule with n intervals is 6144 to the nearest integer.

c i Use a suitable numerical method to find the value of n, justifying your answer.

ii For this n, find an estimate for the value of $\int_0^1 4^{\sqrt{n}x}\,\mathrm{d}x$, giving your answer to 1 dp.

6 $\quad I=\displaystyle\int_2^3 \mathrm{e}^{-\frac{3}{x}}\,\mathrm{d}x$

a Use the trapezium rule with five intervals to estimate I to 3 dp.

b Using the substitution $u=-\dfrac{3}{x}$ show that

$$I=\int_{-1.5}^{-1}\frac{3\mathrm{e}^u}{u^2}\,\mathrm{d}u$$

c Use the trapezium rule with five intervals to estimate the value of

$$\int_{-1.5}^{-1}\frac{3\mathrm{e}^u}{u^2}\,\mathrm{d}u \text{ to 3 dp.}$$

The diagram shows parts of the curves with equations $y=\mathrm{e}^{-\frac{3}{x}}$ and $y=\dfrac{3\mathrm{e}^x}{x^2}$

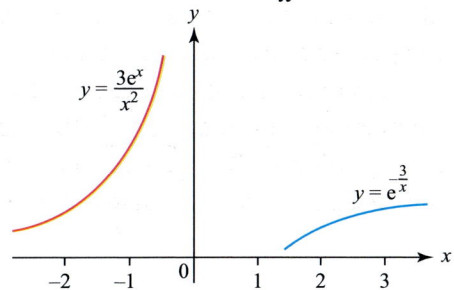

d Use your estimates and the diagram to find the value of I to as many decimal places as can be justified.

7 a Use the trapezium rule to estimate

$$I=\int_0^1\frac{4}{1+x^2}\,\mathrm{d}x \text{ using } n \text{ intervals}$$

where

i $n=4$

ii $n=5$

iii $n=8$

b Which famous value do these answers appear to approach as n increases?

c Use the substitution $x=\tan\theta$ to find the exact value of I

Chapter summary

- An open interval (a,b) is the set of real numbers x such that $a < x < b$
- The solutions to the equation $f(x) = 0$ are called the roots of the equation.
- $f(x)$ changes sign across (a,b) if $f(a)$ and $f(b)$ have opposite signs.
- $f(x)$ is continuous if you can draw the graph $y = f(x)$ without taking your pen off the paper.
- **The change of sign method:** If the continuous function $f(x) = 0$ changes sign in the interval (c,d) then the equation $f(x) = 0$ has a real root α, $c < \alpha < d$
- When the equation $f(x) = 0$ is rearranged into the form $x = g(x)$, the sequence defined by the iterative formula $x_{n+1} = g(x_n)$ can converge to a root α of the equation $f(x) = 0$
- If $|g'(x)| \geq 1$ for values of x close to a root α of the equation $f(x) = 0$ then the sequence $x_1, x_2, x_3, \ldots$ will not converge to this root.
- Convergence and divergence can be displayed on a staircase or cobweb diagram.
- **The Newton-Raphson method:** If α is a root of the equation $f(x) = 0$ then the iterative sequence given by $x_{n+1} = x_n - \dfrac{f(x_n)}{f'(x_n)}$ can converge to α
- If x_1 is close to an x-coordinate of a stationary point of the graph of $y = f(x)$ then the Newton-Raphson method may fail to provide a good estimate for α
- **The trapezium rule:** $\displaystyle\int_a^b f(x) \approx \frac{1}{2} h \left\{ (y_0 + y_n) + 2(y_1 + y_2 + \ldots + y_{n-1}) \right\}$ where the interval $a \leq x \leq b$ is divided into n intervals of equal width $h = \dfrac{b-a}{n}$ defined by the values $a = x_0, x_1, \ldots, x_{n-1}, x_n = b$

  The ordinates $y_0, y_1, \ldots, y_n$ are given by $y_i = f(x_i)$ for $i = 1, 2, \ldots, n$
- Increasing the number of intervals (and therefore trapeziums used) improves the accuracy of the estimate for an integral.
- If a curve is convex, the trapezium rule overestimates an integral.
- If a curve is concave, the trapezium rule underestimates an integral.

Check and review

You should now be able to...	Try Questions
✔ Use the change of sign method to find and estimate the root(s) of an equation.	1, 2
✔ Use an iterative formula to estimate the root of an equation.	3
✔ Recognise the conditions that cause an iterative sequence to converge.	3, 4, 5
✔ Use the Newton-Raphson method to estimate the root of an equation.	4, 5
✔ Use the trapezium rule to find the area under a curve.	6

1 a Show that each equation has a solution in the given interval. Work in radians where appropriate.

i $x^3 + 3 = 5x^2, (0, 1)$

iii $2^x - e^{\sqrt{x}-1} = 3, (2.15, 2.25)$

ii $x^3 + 3 = 5x^2, (-0.8, -0.7)$

iv $x \sin x = \cos(\pi x) + 1, (3, \pi)$

b Using these intervals, which of these solutions are known to 1 dp?

2 $f(x) = \dfrac{1}{x-2} - x$

a Use algebra to solve the equation $f(x) = 0$

b Show that $f(x)$ changes sign between $x = 1.5$ and $x = 2.4$

c What information about the function $f(x)$ do the results of part **a** and part **b** give?

3 a Sketch graphs to show that the equation $2e^x = 4 - x^2$ has exactly one positive solution, α

b Show that the equation $2e^x = 4 - x^2$ can be rearranged into the form $x = \ln\left(2 - 0.5x^2\right)$

c Use the iterative formula $x_{n+1} = \ln\left(2 - 0.5x_n^2\right)$ with $x_1 = 1$ to find α to 3 decimal places, justifying your answer.

4 Use the Newton-Raphson method with first approximation $x_1 = 1$ to find a solution of these equations correct to 3 dp. Work in radians where appropriate.

a $x^3 + 6x = 3$ **b** $x^2 = 2e^{\frac{1}{2}x}$ **c** $\ln(2x+1) = 2x - 1$ **d** $\sin(x^2 + 4) = e^{-x}$

5 $f(x) = x^2 - 4\sqrt{x} - 1$

a Use the Newton-Raphson method with first approximation $x_1 = 2$ to find a root of the equation $f(x) = 0$. Give your answer to 3 dp, justifying your answer.

b Explain why $x_1 = 1$ would not be an appropriate first approximation to use when applying the Newton-Raphson method to solve the equation $f(x) = 0$

6 a Use the trapezium rule with five intervals to estimate the value of $\displaystyle\int_0^2 4^x \, dx$

Give your answer to 2 dp. State, with reason, whether this is an underestimate or overestimate.

b Use your answer to part **a** to find an estimate for the value of $\displaystyle\int_0^2 2^{2x+3} \, dx$

Give this estimate to 3 sf.

What next?

Score		
0 – 3	Your knowledge of this topic is still developing. To improve, search in MyMaths for the codes: 2060, 2173, 2174, 2176	Click these links in the digital book
4 – 5	You're gaining a secure knowledge of this topic. To improve, look at the InvisiPen videos for Fluency and skills (17A)	
6	You've mastered these skills. Well done, you're ready to progress! To develop your exam techniques, look at the InvisiPen videos for Reasoning and problem-solving (17B)	

History

Leonhard Euler (1707–1783) was a Swiss mathematician, physicist, astronomer and engineer.
He made important discoveries in many areas of mathematics and published 886 papers and books,
many of which were in the last two decades of his life, when he was completely blind.
One of the contributions that Euler made to calculus was a method
for finding numerical solutions to differential equations.

Did you know?

Euler's identity, written as

$$e^{i\pi} + 1 = 0$$

is regarded by many as the most beautiful result in mathematics. In one
equality, it links the five most fundamental mathematical constants.

Information

Given a differential equation such as $\dfrac{dy}{dx} + \dfrac{2y}{x} = \dfrac{4}{x}$ where $y = 4$ when $x = 2$, Euler's method allows
you to find an approximate value for y, given a value for x, *without solving the equation*.
This is particularly useful when there is no obvious algebraic method to solve the equation.

When $x = 3$, for example, you would proceed as follows.
First substitute $x = 2$ and $y = 4$ into the equation to find that $\dfrac{dy}{dx} = -2$

Suppose that you now increase the value of x by some small amount, h, then the corresponding change in y is
approximately $h\dfrac{dy}{dx} = -2h$. It follows that the point $(2 + h, 4 - 2h)$ will be close to the solution curve.
For example, if you choose $h = 0.1$, then this gives the point $(2.1, 3.8)$

The whole process may now be repeated, this time starting from the new point $(2.1, 3.8)$ and continuing in
steps of 0.1 until the value of y is found when $x = 3$

Challenge

The numerical solution outlined above involves many calculations that take a while to complete manually.
It is much easier to use a spreadsheet to carry out all the difficult calculations.

	A	B	C	D
1	xn	yn	y'	h
2	2	4	–2	0.1
3	=A2+D2	=B2+C2*D2	=4/A3–2/A3*B3	

Find y when $x = 3$ using $h = 0.1$
Experiment with smaller values of h and compare the results with the theoretical value of 2.89, correct to 2 dp.

In questions that tell you to show your working, you shouldn't depend solely on a calculator. For these questions, solutions based entirely on graphical or numerical methods are not acceptable.

1 Show that the equation $x^3 + 2x^2 - 3x - 2 = 0$ has a root between $x = 1$ and $x = 2$ **[2 marks]**

2 Given that $f(x) = x \sin x$, where x is in radians, show that $f(x) = 0$ has a root in the interval $3 < x < 3.5$ **[2]**

3 **a** Show that the equation $x^3 - 4x - 1 = 0$ has a root in the interval $(2, 2.5)$ **[2]**

  b Use the iterative formula $x_{n+1} = \sqrt{4 + \dfrac{1}{x_n}}$, starting with $x_1 = 2$ to find x_2 and x_3 to 2 dp. **[3]**

4 **a** Use the iterative formula $x_{n+1} = \ln(5 - x_n)$, starting with $x_1 = 1$ to find, to 2 decimal places, a root of the equation $e^x + x - 5 = 0$ **[4]**

  b Prove that your solution is correct to 2 decimal places. **[3]**

5 **a** Show that the equation $x^3 - 3x + 1 = 0$ has a root between 1 and 2 **[2]**

  b Taking 2 as a first approximation, use the Newton-Raphson process twice to find an approximation to the root of $x^3 - 3x + 1 = 0$, to 2 dp. **[4]**

6 You are given that a particle's motion is modelled by $f(x) = 2x^4 - 3x^3 + 4x$

  a Use the Newton-Raphson process twice, taking $x = -1$ as the first approximation to find the negative root of the equation $f(x) = 0$ to 2 decimal places. **[4]**

  b Prove that your solution is correct to 2 dp. **[2]**

7 Use the trapezium rule with four strips to estimate the integral $\displaystyle\int_1^3 \sin^3 x \, dx$, to 3 sf. **[4]**

8 Use the trapezium rule with four ordinates to estimate the integral $\displaystyle\int_0^3 \tan^3 x \, dx$, to 3 sf. **[4]**

9 An object's temperature is modelled by $f(x) = 5x - e^x$

  a Prove that there is a root of $f(x) = 0$ between $x = 0$ and $x = 0.5$ **[2]**

  The graph of $y = f(x)$ is shown.

  b Explain how you know from the graph that there is a root of $f(x) = 0$ between 2 and 3 **[1]**

  c Show that $x = 2.5$ is the root, correct to 1 dp. **[3]**

10 **a** Sketch, on the same axes, the graphs of $y = x + 1$ and $y = \dfrac{4}{x}$ **[2]**

  b Use your graphs to explain how many roots there are to the equation $x + 1 = \dfrac{4}{x}$ **[1]**

  c Show that the equation $x + 1 = \dfrac{4}{x}$ has a root in the interval $(1.5, 1.6)$ **[3]**

  d Find the solutions to the equation $x + 1 = \dfrac{4}{x}$, give your answers to 3 significant figures. **[2]**

11 The graphs of $y = e^x \sin x$ and $y = x + 2$ are shown, where x is in radians.

   a Explain how many solutions there are to the equation $e^x \sin x = x + 2$ **[1]**

   b Show that one of the roots, α, is such that $1.2 < \alpha < 1.3$ **[3]**

   c Find an interval of size 0.1 that the other positive root lies within.   **[3]**

12 a Show that the equation $x^3 - 3x^2 - 5 = 0$ can be written $x = \sqrt{\dfrac{5}{x} + 3x}$ **[2]**

   b Use the iteration formula $x_{n+1} = \sqrt{\dfrac{5}{x_n} + 3x_n}$, starting with $x_1 = 3$ to find x_5

     Give your answer to 2 decimal places.   **[3]**

13 The graph of $y = 2e^x + 3x - 7$ is shown.

   a Use the iteration formula $x_{n+1} = \dfrac{7 - 2e^{x_n}}{3}$ with $x_1 = 0.8$

     to find x_2, x_3, x_4, x_5 to 2 decimal places.   **[3]**

   b Explain what is happening in this case.   **[1]**

   c Now derive a different iteration formula and, again
     using $x_1 = 0.8$. calculate x_2, x_3, x_4, x_5 to 2 dp.   **[4]**

14 a Sketch the graphs of $y = x - 3$ and $y = \sqrt{x}$ on the same axes.   **[2]**

   b Use an appropriate iteration formula with $x_1 = 2$ to find a root of $\sqrt{x} = x - 3$ to 2 dp.   **[4]**

   c **i** Draw a suitable diagram to illustrate the results of the first two iterations.   **[3]**

     ii Write down the name of this diagram.   **[1]**

15 a Sketch the graphs of $y = \ln x$ and $y = e^x - 5$ on the same axes.   **[3]**

   b Explain how many roots the equation $\ln x = e^x - 5$ has.   **[1]**

   c Show that one of the roots occurs between $x = 1.6$ and $x = 1.8$ **[3]**

   d Use the Newton–Raphson process, to find this root correct to 2 decimal places.   **[4]**

16 a Show that the equation $x^3 + 4x - 3 = 0$ can be written $x = a\left(b - x^3\right)$, where a and b are
    constants to be found.   **[2]**

   b Use the iteration formula $x_{n+1} = a\left(b - x_n^3\right)$ for the values of a and b found in part **a** with
    $x_1 = 0.1$ to find x_5 correct to 2 significant figures.   **[3]**

   c **i** Draw a suitable diagram to illustrate the results of the first 3 iterations.   **[2]**

     ii Write down the name of this diagram.   **[1]**

17 Use the Newton–Raphson method to find, to 3 significant figures, the solution of
the equation $x \sin x = 2 \ln x$, where x is in radians, which is near 2   **[5]**

18 Population growth can be modelled by the graph of $y = xe^{2x}$

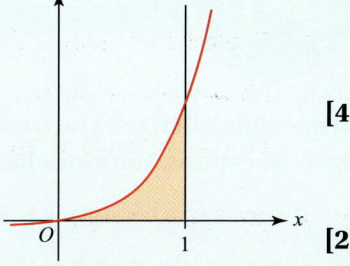

  a Use the trapezium rule with five strips to estimate, to 4 significant figures, the area enclosed by the curve, the x-axis and the line $x = 1$ **[4]**

  b State without further calculation whether this will be an overestimate or an underestimate of the actual area. Justify your answer.   **[2]**

  c **i** Use integration by parts to find the actual area.   **[3]**

     ii Calculate the percentage error in your approximation.   **[2]**

19 The graph of $y = x^3 - 5x - 3$ is shown.

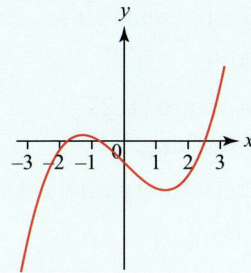

  a How many solutions are there to the equation $x^3 - 5x - 3 = 0$? Justify your answer.   **[1]**

  b Show that $y = x^3 - 5x - 3 = 0$ can be written in the form $x = \pm\sqrt{a + \dfrac{b}{x}}$, where a and b are constants to be found.   **[3]**

  c Use the iteration formula $x_{n+1} = \sqrt{a + \dfrac{b}{x_n}}$ with the values you have found for a and b to calculate the positive root of the equation correct to 3 significant figures.   **[3]**

  d Use Newton-Raphson to find the largest negative root, correct to 2 significant figures.   **[4]**

  e Verify that the smallest negative root is -1.83 to 3 sf.   **[2]**

20 Explain how the change of sign method will fail to find a root, α, to $f(x) = 0$ in these cases

  a $f(x) = \dfrac{1}{x-3}$ for $2.5 < \alpha < 3.5$ **[2]**

  b $f(x) = (3x-2)(2x-1)(x-4)$ for $0 < \alpha < 1$ **[2]**

21 Use the Newton-Raphson method to find a root, correct to 2 decimal places, to the equation $\sin^2 x = e^{-x}$, where x is in radians, using

  a $x_1 = 1$ **[3]**   **b** $x_1 = 3$ **[3]**

22 a Use the trapezium rule with five ordinates to estimate, to 3 significant figures, the area enclosed by the curve with equation $y = \sqrt{\ln x}$, the x-axis and the line $x = 2$ **[4]**

  b Comment on the suggestion that the actual area is close to 0.5   **[2]**

23 $f(x) = x \ln x - 1, x > 0$

  a Find an interval of size 0.2 that contains the solution to $f(x) = 0$ **[3]**

  b Use Newton-Raphson to approximate the root of the equation $f(x) = 0$

    Ensure your answer is correct to 3 decimal places.   **[8]**

In questions that tell you to show your working, you shouldn't depend solely on a calculator. For these questions, solutions based entirely on graphical or numerical methods are not acceptable.

1 The area of a triangle is $\left(-1+3\sqrt{5}\right)$ cm². The height of the triangle is $\left(3+2\sqrt{5}\right)$ cm.

Show that the length of the base of the triangle is $\left(6-2\sqrt{5}\right)$ cm. **[4 marks]**

2 Find the solutions to the equation $2^{2x+1}-7(2^x)+6=0$. Show your working and give your answers to 3 significant figures where appropriate. **[4]**

3 $f(x)=x^2+(k+1)x+2$

    a Find the range of values of k for which the equation $f(x)=0$ has distinct real roots. **[4]**

    b Find the solutions of the equation $x^4-3x^2+2=0$. Show your working. **[3]**

4 Calculate the points of intersection between a circle with radius 5 and centre (1, 2) and a line that passes through the points (1, 3) and (–2, 6). Show your working. **[8]**

5 Find the range of values of x that satisfy both $2-2x-3x^2\ge 0$ and $4x+7>1$. Show your working. **[5]**

6 Solve the equation $5-\sin\theta-6\cos^2\theta=0$ for $0<\theta<360°$. Show your working. **[6]**

7 Solve the simultaneous equations $\quad e^{x+y}=3 \quad 3x+2y=0$

Show your working and give each of your solutions as a single logarithm. **[5]**

8 $g(x)=6x^3-19x^2-12x+45$

    a $y=g(x)$ and $y=0$ intersect at (3, 0) and at two other points. Calculate the remaining points of intersection, showing your working. **[5]**

    b Sketch the curve $y=g(x)$, clearly labelling the points of intersection with the coordinate axes. **[3]**

    c Calculate the area enclosed by the curve and the x-axis. Show your working. **[6]**

9 Find and classify all the stationary points of the curve with equation

$$y=\frac{1}{2}x^4-3x^3+2x^2+15x+1$$ **[8]**

Show your working.

10 The number of cases of a viral infection in a school with 2000 students after t days is given by $N=Ae^{kt}$. There are initially 2 cases of the infection and this number doubles after three days.

    a Calculate the exact values of A and k **[4]**

    b According to this model, how many days until a quarter of the students have been infected? Show your working. **[3]**

The number of cases of a second type of viral infection after t days is given by $M=Br^t$. There are initially 10 cases of this infection and after five days there are 15 cases.

    c After how many days will the number of cases of the first infection overtake the number of the second infection? Show your working. **[7]**

    d Sketch on the same axes the graphs of N and M against t for $t>0$ **[4]**

    e How realistic do you think these models are? Explain your answer. **[3]**

11 $f(x) = 6x^3 - 19x^2 - 51x - 20$

    a   Show that $2x + 1$ is a factor of $f(x)$ **[3]**

    b   Find all the solutions to $f(x) = 0$, showing your working.   **[3]**

12 Write each of these expressions in partial fractions.

    a $\dfrac{3x+1}{(x+3)(2x+1)}$ **[4]**   **b** $\dfrac{3x-5}{x^2-25}$ **[4]**

13 The function f is defined by $f : x \mapsto \dfrac{2x-14}{x^2-2x-3} + \dfrac{2}{x-3}, \, x > 3$

    a   Show that $f(x)$ can be written as $\dfrac{k}{x+1}$, where k is an integer to be found.   **[4]**

    b   Write down the

        i   Domain of $f(x)$ **ii**   Range of $f(x)$ **[3]**

    c   Find the inverse function, $f^{-1}(x)$ and state its domain.   **[4]**

14 Given that $g(x) = x^2 + 3, \, x \in \mathbb{R}$ and $h(x) = \dfrac{3}{x-2}, \, x \neq 2$

    a   Write down $hg(-2)$,   **[3]**

    b   Solve the equation $gh(x) = 12$. Show your working.   **[4]**

    c   Is the range of $gh(x)$ the same as the range of $hg(x)$? Explain how you know.   **[3]**

15 Work out the first four terms of the binomial expansion of $(1+2x)^{\frac{1}{4}}, |x| < \dfrac{1}{2}$, in ascending powers of x **[4]**

16 A sequence is defined by $u_{n+1} = 2u_n + 1, \, u_1 = -2$. Showing your working, calculate

    a u_2 **[2]**   **b** $\displaystyle\sum_{1}^{5} u_r$ **[3]**

17 AB is the arc of a circle of radius 5 cm and centre C as shown.

    The segment S is bounded by the arc and the line AB

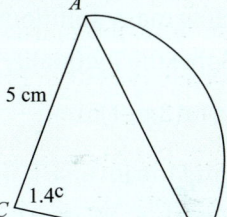

    a   Calculate the area of S **[4]**

    b   Calculate the perimeter of S **[4]**

18 **a**   Sketch each of these graphs on separate axes, for x in the range $0 \leq x \leq 2\pi$

        i $y = 2\sec x$ **ii** $y = \operatorname{cosec} 2x$ **[6]**

    b   Solve the equation $3\cot x + 2 = 4$ for x in the range $0 \leq x \leq 360°$. Show your working.   **[4]**

19 Showing your working, find the exact solutions to the equations.

    a $2\arcsin x = \dfrac{\pi}{2}$ **[3]**   **b** $\arctan 4x = \dfrac{\pi}{3}$ **[2]**

20 **a**   Differentiate with respect to x

        i $3\sin x$ **ii** $x\ln x$ **[3]**

    Given that $f(x) = (2x+1)\cos x$

    b   Find the exact gradient of the curve $y = f(x)$ when $x = \dfrac{\pi}{6}$. Show your working.   **[4]**

21 **a**   Find the coordinates of the minimum point of the curve $y = xe^{2x}$. Show your working.   **[5]**

    b   Explain how you know this is a minimum point.   **[4]**

22 a Work out each of these integrals

    i $\int \sin x \, dx$ **ii** $\int \dfrac{3}{x} \, dx$ **[3]**

  b Calculate the exact value of the integral $\int_{2}^{6} \dfrac{2}{x-1} \, dx$. Show your working and give your answer in its simplest form. **[4]**

23 Calculate the area bounded by the x-axis and the curve $y = \cos x$ for $0 \leq x \leq \pi$. Show your working. **[4]**

24 $f(x) = x^3 - 6x - 12$

  a Show that the equation $f(x) = 0$ has a root in the interval $(3, 3.5)$. **[2]**

  b Use the iterative formula $x_{n+1} = \sqrt{6 + \dfrac{12}{x}}$, starting with $x_1 = 3$ to find x_2 and x_3 to 2 decimal places. **[2]**

  c Prove that your value of x_3 is a solution to $f(x) = 0$, correct to 2 decimal places. **[3]**

25 Use the trapezium rule with four strips to estimate the integral $\int_{0}^{1} \cos^5 x \, dx$ to 3 significant figures. **[5]**

26 Prove by contradiction that if n^2 is odd then n is odd for all integers n **[5]**

27 Show that $\dfrac{2x^2 + 4x + 3}{2x^2 - x - 1}$ can be written $A + \dfrac{B}{x-1} + \dfrac{C}{2x+1}$ where A, B and C are integers to be found. **[5]**

28 The function f is given by $f : x \to |3 - 2x|$

  a Sketch the graph of $y = f(x)$. **[2]**

  b How many solutions will there be to the equation $|3 - 2x| = x$? Explain how you know. **[2]**

  c Solve the inequality $|3 - 2x| \geq x$, showing your working. **[4]**

29 $f(x) = \ln(3x+1)$, $x > -\dfrac{1}{3}$

  a Find the inverse $f^{-1}(x)$. **[3]**

  b Sketch $y = f(x)$ and $y = f^{-1}(x)$ on the same axes. **[5]**

  c Write down the range and domain of $f^{-1}(x)$. **[2]**

30 a Express $\dfrac{6x+10}{(x-1)(x+3)^2}$ in partial fractions. **[5]**

  b Integrate $\dfrac{6x+10}{(x-1)(x+3)^2}$ with respect to x **[4]**

31 The points A and B have position vectors $12\mathbf{i} + 7\mathbf{j} - 5\mathbf{k}$ and $3\mathbf{i} - 2\mathbf{j} - \mathbf{k}$ respectively.

Calculate the magnitude of the vector $\overrightarrow{AB}$. Show your working. **[4]**

32 A sequence is given by $x_{n+1} = (x_n)^2 - 2x_n$ where $x_1 = 1$

  a Write down the value of x_2 and x_3 **[3]**

  b Find an expression in terms of n for $\sum_{1}^{n} x_r$ **[4]**

33 The first term of a geometric series is 36 and the common ratio is $\dfrac{1}{3}$

  a Find the difference between the second and third terms of the sequence. Show your working. **[3]**

  b Calculate the difference between the sum to infinity and the sum of the first five terms of the series. Give your answer as a fraction. **[5]**

34 a Derive a formula for the sum of the first n terms of an arithmetic series with first term a and common difference d **[4]**

An arithmetic series has first term -3 and common difference 1.5.

The sum of the first n terms of this an arithmetic series is 63

b Find the value of n **[4]**

35 a Sketch the graph in part **i** for $-1 \leq x \leq 1$ and the graph in part **ii** for $-2 \leq x \leq 0$

    i $y = \arccos x$ **ii** $y = 2\arcsin(x+1)$ **[6]**

b State the range of each function in part **a**. **[2]**

c Write down the inverse of $f(x) = 2\arcsin(x+1)$, $-2 \leq x \leq 0$ and state its domain. **[4]**

36 a Sketch the graph of $y = 3\ln(x-1)$ and give the equation of any asymptotes. **[3]**

b Calculate the exact gradient of the curve at the point where $x = 3$. Show your working. **[3]**

37 a Using a small angle approximation, show that $\sec 2x \approx \dfrac{1}{(1-2x^2)}$ **[4]**

b Hence, find the first three terms of the binomial expansion for $\sec 2x$ in ascending powers of x **[4]**

c Use your expansion to find an approximate value for $\sec(0.2)$ **[3]**

38 Solve these equations for $0 \leq x \leq 2\pi$. Show your working and give your answers to 3 significant figures.

  a $\sec 2x = 1.5$ **[5]**   **b** $3\operatorname{cosec}\left(\dfrac{x+\pi}{2}\right) = 5$ **[4]**

39 a Prove that $3\sec^2\theta - 7\tan^2\theta \equiv 3 - 4\tan^2\theta$ **[3]**

b Hence solve the equation $1 + 14\tan^2\theta = 6\sec^2\theta$ for $0 \leq \theta \leq 2\pi$ **[5]**

40 a Use the formula for $\cos(A+B)$ to prove that $\cos 2x \equiv 1 - 2\sin^2 x$ **[3]**

b Hence solve the equation $\cos 2x + \sin x = 0$ for $-180° \leq x \leq 180°$ **[5]**

41 a Express $4\sin\theta + \cos\theta$ in the form $R\sin(\theta+\alpha)$, where $R > 0$ and $0 < \alpha < \pi$ **[5]**

b Write down the maximum value of $4\sin\theta + \cos\theta$ and the state the smallest positive value of θ at which this value occurs. **[3]**

42 Find the range of values of x for which the curve with equation $y = \dfrac{1}{12}x^4 + \dfrac{1}{3}x^3 - \dfrac{3}{2}x^2 + 5x$ is concave. **[5]**

43 A curve has equation $y = e^{2x}\cos 3x$

By first finding an expression for $\dfrac{dy}{dx}$, work out the equation of the tangent to the curve when $x = 0$ **[7]**

44 Given that $f(x) = \dfrac{\sin x}{x+2}$,

a Find $f'(x)$, **[2]**

b Prove that the x-coordinate of the minimum points of the curve $y = f(x)$ satisfies the equation $x = \arctan(x+2)$ **[4]**

c Use the iterative formula $x_{n+1} = \arctan(x_n+2)$ with $x_1 = 1.2$ to find the x-coordinate of a minimum point of the curve $y = f(x)$ to 2 decimal places. **[3]**

45 A curve is defined by the parametric equations

$$x = t^2 - 3, \quad y = 1 + 2t$$

  a Find $\dfrac{dy}{dx}$ in terms of t **[3]**

  b Find an equation of the normal to the curve at the point where $t = 2$ **[4]**

  c Find a Cartesian equation of the curve.   **[3]**

46 A curve is defined by the equation $x^2 + 2y^2 - 3xy = 1$

  Find the gradient of the curve at each of the points where $x = 1$ **[7]**

47 Using small angle approximations, differentiate $\cos x$ from first principles.   **[6]**

48 **a** Integrate these expressions.

      i $x \sin x$ **ii** $\ln x$ **[6]**

  b Use the substitution $u = x^2 + 1$ to find $\displaystyle\int \frac{x}{(x^2+1)^3}\, dx$ **[4]**

49 Find the exact area enclosed by the curve $y = \dfrac{4x+5}{x^2+3x+2}$, the coordinate axes and the

  line $x = 1$. Show your working and give your answer as a single logarithm in its simplest form. **[8]**

50 Solve the differential equation $\dfrac{dy}{dx} = y \cos x$, given that $y = 1$ when $x = \dfrac{\pi}{6}$

  Give your answer in the form $y = f(x)$ **[5]**

51 **a** Sketch the graph of $y = 1 + \ln x$ and $y = x^2 - x$ on the same axes for $x > 0$ **[4]**

  b Use an appropriate iteration formula to find, to 3 significant figures, a root
     of the equation $1 + \ln x = x^2 - x$. Show your working.   **[5]**

  c Verify that the solution is correct to 3 significant figures.   **[2]**

52 This graph shows the curve $y = f(x)$

  where $f(x) = x^3 - 2x^2 - 3x + 2$

  a Use the Newton-Raphson process twice,
     taking $x_1 = -1.3$ as the first approximation, to
     find a root of the equation $f(x) = 0$

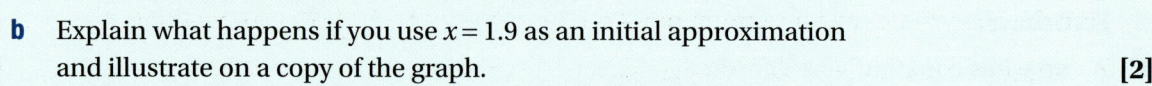

     Give your answer correct to 3 decimal places.   **[4]**

  b Explain what happens if you use $x = 1.9$ as an initial approximation
     and illustrate on a copy of the graph.   **[2]**

53 **a** Explain how many solutions there are to the equation $\sqrt{x} = e^{-x}$ **[3]**

  b Show that there is a solution between 0.4 and 0.5   **[2]**

  c Use the Newton-Raphson method to find this solution to 2 decimal places.
     Demonstrate how you know your answer is correct to this degree of accuracy.   **[7]**

54 $f(x) = a - e^{2x}$ where $a > 0$ is a constant.

  a Sketch the graph of $y = f(x)$. You should label the points where the curve crosses
     the coordinate axes and give the equation of any asymptotes.   **[4]**

  b Find the equation of $f^{-1}(x)$ and state its domain and range, in terms of a if appropriate. **[6]**

55 You are given that $\overrightarrow{OA} = \begin{pmatrix} 2 \\ -1 \\ 5 \end{pmatrix}$ and $\overrightarrow{OB} = \begin{pmatrix} -3 \\ 2 \\ 0 \end{pmatrix}$

Calculate the magnitude of $\overrightarrow{OC}$ where $OACB$ is a rhombus. **[5]**

56 A woman plans to improve her fitness by running a miles in the first week, $a + d$ miles in the second week, and so on, with the number of miles forming an arithmetic sequence.

She runs 11 miles in the 5th week and a total of 208 miles after 13 weeks.

Calculate the values of a and d **[6]**

57 A geometric series has first term a and common ratio r. The 4th term of the series is $-\dfrac{3}{64}$ and the 7th term is $\dfrac{3}{4096}$

Find the sum to infinity of the series. **[7]**

58 a i Describe fully the series given by $u_n = 2\left(\dfrac{1}{3}\right)^r$

    ii Calculate the exact value of $\displaystyle\sum_{1}^{\infty} 2\left(\dfrac{1}{3}\right)^r$ **[4]**

  b State the type of series formed by $u_n = \ln 3^r$ and find an expression for $\displaystyle\sum_{r=1}^{n} \ln 3^r$ **[5]**

59 a Use the binomial expansion to expand $(8 - 5x)^{\frac{1}{3}}$, $|x| < \dfrac{8}{5}$ in ascending powers of x up to the term in x^2 **[4]**

  b Use your expansion to find an approximation to $\sqrt[3]{7950}$ to 6 significant figures. **[5]**

60 Use compound angle formulae to prove that $\cos A - \cos B \equiv K \sin\left(\dfrac{A+B}{2}\right) \sin\left(\dfrac{A-B}{2}\right)$, state the value of K **[5]**

61 Solve the inequality $\operatorname{cosec}^2 x \leq \cot x + 2$ for x in the interval $0 \leq x \leq \pi$

Show your working and give your limits to 3 significant figures. **[6]**

62 Solve the equation $2\cos 2x + 3\sin x + 1 = 0$ for $0 < x < 360°$

Show your working and give your answers to 1 decimal place. **[7]**

63 You are given that $f(x) = \left| \dfrac{1}{3\cos x + 4\sin x} \right|$, $0 < x < 180°$

Find the coordinates of a minimum point on the curve $y = f(x)$. Show your working. **[8]**

64 a Show that $\sin^4 x + \cos^4 x \equiv \dfrac{1}{4}(3 + \cos 4x)$ **[5]**

  b Hence solve $3\sin^4 x + 3\cos^4 x = 2$ for $-\pi < x < \pi$. Show all your working and give your answers to 2 decimal places. **[6]**

65 The graph of $y = f(x)$ is shown.

  a Sketch the following graphs on separate axes, stating the equations of any asymptotes.

    i $y = 2 + f(x - 1)$ **ii** $y = f(3 - x)$ **[5]**

  Given that $f(x) = \dfrac{a}{x + b}$,

b   Find the values of a and b **[3]**

c   Write the equations of the graphs from part **a** with f(x) as a single fraction in its simplest form.   **[5]**

66   The curve, C is defined by the parametric equations

$$x = 4(t-5)^2, \quad y = \frac{t^2}{t-5}$$

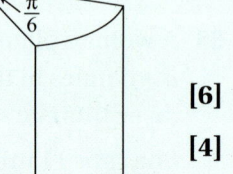

a   Calculate the gradient of C at the point where $t = 2$. Show your working.   **[6]**

b   Find a Cartesian equation in the form $y = $ f(x), simplify your answer.   **[4]**

67   You are given that $3x - 4x^2 y + 3y^2 = 2$

Calculate the possible rates of change of y with respect to x when $y = 1$ **[9]**

68   A closed box of height h is a prism. The cross-section of the prism is a sector of a circle with radius r and angle $\dfrac{\pi}{6}$ as shown.

The volume of the box is 100 cm³

Find the minimum possible surface area of the box and the value of r that gives this surface area.   **[11]**

69   Find the value of $\displaystyle\int_0^{21} \frac{x-5}{3x^2 + 10x + 3}\,\mathrm{d}x$. Show your working and give your

answer in the form $k\ln 2$.   **[9]**

70   Prove that the derivative of $\arccos 3x$ is $-\dfrac{3}{\sqrt{1-9x^2}}$ **[5]**

71   You are given that f$'(\theta) = \tan\theta$ and that f$\left(\dfrac{\pi}{3}\right) = \ln 6$

Prove that f$(\theta) = \ln|\sec\theta| + c$ where c is a constant to be found.   **[7]**

72 a   Find the general solution to the differential equation $\dfrac{\mathrm{d}y}{\mathrm{d}x} = \dfrac{\mathrm{e}^{2x-y}}{y}$ **[4]**

Given that the curve passes through the point (0, 1),

b   Find the particular solution in the form $x = $ f(y).   **[4]**

73   Use an appropriate substitution to find $\displaystyle\int x^2 (3x-1)^5\,\mathrm{d}x$, give your answer in terms of x **[6]**

74   Calculate the area bounded by the curve $y = x^2 \mathrm{e}^x$ and the line $y = 2x\mathrm{e}^x$. Show your working.   **[8]**

75 a   Use the trapezium rule with 5 ordinates to estimate the area enclosed by the curve $y = \operatorname{cosec} x$, the x-axis and the ordinates $x = 1$ and $x = 3$ **[3]**

b   Comment on the suggestion that the actual area is closer to 4   **[3]**

76   Calculate the area enclosed by the circle with equation $(x-5)^2 + (y-3)^2 = 16$ and the line with equation $2x - 5y = 3$. Show your working.   **[10]**

77   Calculate the exact area bounded by the curve $y = \sin^2 x$, the x-axis and the

lines $x = \dfrac{\pi}{3}$ and $x = \dfrac{5\pi}{6}$. Show your working.   **[6]**

18 Motion in two dimensions

Many forces are involved in bungee jumping, the main ones being those due to gravity and the elasticity of the cord. Although a real-life jump takes place in three dimensions, it can be modelled in two dimensions by assuming there is no side-to-side motion. To calculate the motion of the bungee jumper in such a model, you would need to know the vertical and horizontal components of the fall and the force due to the elastic cord, as well as the length of the cord and its 'stretch' factor.

Scenarios which can be modelled in two dimensions include the motion of a snooker ball on a table, the movement of a boat on the ocean, the orbit of the Earth around the Sun and the motion of a thrown ball. The ability to understand and work in two dimensions is very useful for grasping the basic principles involved, and a stepping stone to working in three dimensions.

Orientation

What you need to know

Ch6 Vectors
- Resolving vectors into components.

p.158

Ch7 Units and kinematics
- Motion with variable acceleration.

p.190

Ch8 Forces and Newton's laws
- Motion under gravity.
- Resolving forces horizontally and vertically.

p.210

What you will learn

- To use the constant acceleration equation for motion in two dimensions.
- To use calculus to solve problems in two-dimensional motion with variable acceleration.
- To solve problems involving the motion of a projectile under gravity.
- To analyse the motion of an object in two dimensions under the action of a system of forces.

What this leads to

Ch19 Forces 2
Motion under a resultant force, including motion on a rough surface.
Finding forces and moments acting on a body.

Further Mechanics
Conservation of momentum.
Collisions.
Impulses.

 MyMaths   Practise before you start   Q 2185, 2207, 2289

18.1 Two-dimensional motion with constant acceleration

Fluency and skills

See Ch6.2

For a reminder on components of a vector.

An object moving in two dimensions has position vector **r**, and its displacement, velocity and acceleration have components in the x- and y-directions. You use the following notation.

> **Key point**
>
> displacement $\mathbf{s} = x\mathbf{i} + y\mathbf{j}$ acceleration $\mathbf{a} = a_x\mathbf{i} + a_y\mathbf{j}$
>
> initial velocity $\mathbf{u} = u_x\mathbf{i} + u_y\mathbf{j}$ final velocity $\mathbf{v} = v_x\mathbf{i} + v_y\mathbf{j}$

Provided a_x and a_y are constant, the motion in each direction will obey the equations for constant acceleration. So, for example,

$$v_x = u_x + a_x t \qquad \text{and} \qquad v_y = u_y + a_y t$$

You can combine these to give the vector equation

$$\mathbf{v} = \mathbf{u} + \mathbf{a}t$$

You can also write the other equations in vector form.

See Ch7.3

For a reminder on equations of motion for constant acceleration.

> **Key point**
>
> $\mathbf{v} = \mathbf{u} + \mathbf{a}t$ $\mathbf{s} = \dfrac{1}{2}(\mathbf{u} + \mathbf{v})t$
>
> $\mathbf{s} = \mathbf{u}t + \dfrac{1}{2}\mathbf{a}t^2$ $\mathbf{s} = \mathbf{v}t - \dfrac{1}{2}\mathbf{a}t^2$

Notice that the position vector **r** of a point is its displacement from the origin. You can sometimes write the relationships between displacement, velocity and acceleration with **r** in place of **s**

Expressing the equation, $v^2 = u^2 + 2as$, in vector form uses techniques beyond the scope of this book. You can still use it separately for components in the x- and y-directions.

> **Key point**
>
> $v_x^2 = u_x^2 + 2a_x x$ and $v_y^2 = u_y^2 + 2a_y y$

Example 1

An object is initially at the point with position vector $\mathbf{r} = (6\mathbf{i} + 3\mathbf{j})\,\mathrm{m\,s^{-1}}$ and travelling with velocity $(2\mathbf{i} + 15\mathbf{j})\,\mathrm{m\,s^{-1}}$ when it undergoes an acceleration of $(\mathbf{i} - 2\mathbf{j})\,\mathrm{m\,s^{-2}}$ for a period of four seconds.

Work out

a Its velocity at the end of the four seconds,

b The displacement it undergoes during the four seconds,

c Its position at the end of the four seconds,

d Its speed and direction of motion at the end of the four seconds.

(Continued on the next page)

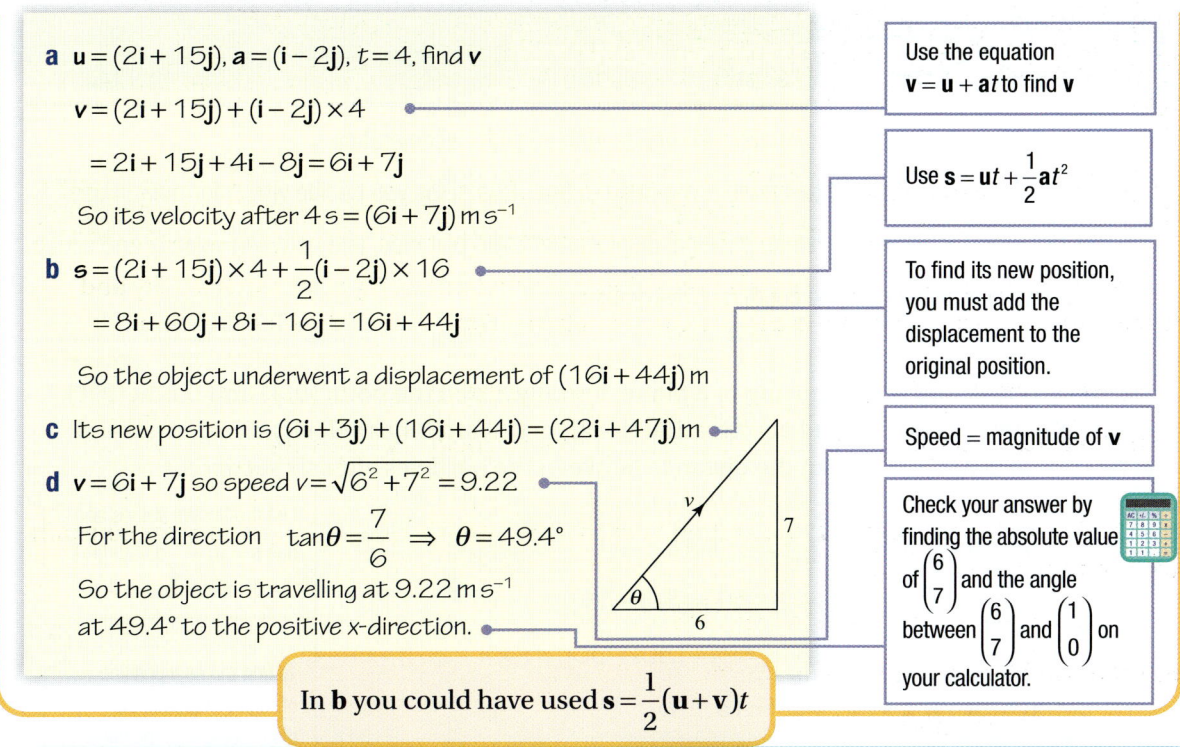

a $\mathbf{u} = (2\mathbf{i} + 15\mathbf{j})$, $\mathbf{a} = (\mathbf{i} - 2\mathbf{j})$, $t = 4$, find $\mathbf{v}$

$\mathbf{v} = (2\mathbf{i} + 15\mathbf{j}) + (\mathbf{i} - 2\mathbf{j}) \times 4$

$= 2\mathbf{i} + 15\mathbf{j} + 4\mathbf{i} - 8\mathbf{j} = 6\mathbf{i} + 7\mathbf{j}$

So its velocity after $4\,\text{s} = (6\mathbf{i} + 7\mathbf{j})\,\text{m s}^{-1}$

> Use the equation $\mathbf{v} = \mathbf{u} + \mathbf{a}t$ to find $\mathbf{v}$

b $\mathbf{s} = (2\mathbf{i} + 15\mathbf{j}) \times 4 + \dfrac{1}{2}(\mathbf{i} - 2\mathbf{j}) \times 16$

$= 8\mathbf{i} + 60\mathbf{j} + 8\mathbf{i} - 16\mathbf{j} = 16\mathbf{i} + 44\mathbf{j}$

So the object underwent a displacement of $(16\mathbf{i} + 44\mathbf{j})\,\text{m}$

> Use $\mathbf{s} = \mathbf{u}t + \dfrac{1}{2}\mathbf{a}t^2$

> To find its new position, you must add the displacement to the original position.

c Its new position is $(6\mathbf{i} + 3\mathbf{j}) + (16\mathbf{i} + 44\mathbf{j}) = (22\mathbf{i} + 47\mathbf{j})\,\text{m}$

d $\mathbf{v} = 6\mathbf{i} + 7\mathbf{j}$ so speed $v = \sqrt{6^2 + 7^2} = 9.22$

For the direction $\tan\theta = \dfrac{7}{6} \Rightarrow \theta = 49.4°$

So the object is travelling at $9.22\,\text{m s}^{-1}$ at $49.4°$ to the positive x-direction.

> Speed = magnitude of $\mathbf{v}$

> Check your answer by finding the absolute value of $\begin{pmatrix} 6 \\ 7 \end{pmatrix}$ and the angle between $\begin{pmatrix} 6 \\ 7 \end{pmatrix}$ and $\begin{pmatrix} 1 \\ 0 \end{pmatrix}$ on your calculator.

In **b** you could have used $\mathbf{s} = \dfrac{1}{2}(\mathbf{u} + \mathbf{v})t$

Exercise 18.1A Fluency and skills

1 A particle has initial velocity $(3\mathbf{i} + 2\mathbf{j})\,\text{m s}^{-1}$ and acceleration $(2\mathbf{i} - \mathbf{j})\,\text{m s}^{-2}$

   a Use $\mathbf{v} = \mathbf{u} + \mathbf{a}t$ to work out its velocity after $5\,\text{s}$

   b Use $\mathbf{s} = \mathbf{u}t + \dfrac{1}{2}\mathbf{a}t^2$ to work out its displacement after $3\,\text{s}$

2 A particle has initial velocity $(\mathbf{i} + 3\mathbf{j})\,\text{m s}^{-1}$ and travels for $3\,\text{s}$

   a Use $\mathbf{v} = \mathbf{u} + \mathbf{a}t$ to work out its final velocity if its acceleration is $(4\mathbf{i} + \mathbf{j})\,\text{m s}^{-2}$

   b Use $\mathbf{v} = \mathbf{u} + \mathbf{a}t$ to work out its acceleration if its final velocity is $(-2\mathbf{i} + 9\mathbf{j})\,\text{m s}^{-1}$

   c Use $\mathbf{s} = \mathbf{u}t + \dfrac{1}{2}\mathbf{a}t^2$ to work out its displacement if its acceleration is $(3\mathbf{i} - 2\mathbf{j})\,\text{m s}^{-2}$

3 A particle starts with velocity $(4\mathbf{i} - 3\mathbf{j})\,\text{m s}^{-1}$ and after $4\,\text{s}$ its velocity is $(8\mathbf{i} + 5\mathbf{j})\,\text{m s}^{-1}$. Use $\mathbf{s} = \dfrac{1}{2}(\mathbf{u} + \mathbf{v})t$ to work out its displacement.

4 A particle accelerates at $(\mathbf{i} + 3\mathbf{j})\,\text{m s}^{-2}$ for $4\,\text{s}$ and undergoes a displacement of $(16\mathbf{i} - 12\mathbf{j})\,\text{m}$. Use $\mathbf{s} = \mathbf{v}t - \dfrac{1}{2}\mathbf{a}t^2$ to work out its final velocity.

5 For each of these situations, choose a suitable equation to work out the required value.

   a $\mathbf{u} = (2\mathbf{i} + 5\mathbf{j})$, $\mathbf{v} = (10\mathbf{i} - 7\mathbf{j})$, $t = 6$, work out $\mathbf{s}$

   b $\mathbf{u} = (-2\mathbf{i} + 3\mathbf{j})$, $\mathbf{a} = (3\mathbf{i} - 4\mathbf{j})$, $t = 8$, work out $\mathbf{s}$

   c $\mathbf{u} = (14\mathbf{i} + 6\mathbf{j})$, $\mathbf{v} = (4\mathbf{i} + 21\mathbf{j})$, $t = 5$, work out $\mathbf{a}$

   d $\mathbf{s} = (-4\mathbf{i} + 52\mathbf{j})$, $\mathbf{a} = (2\mathbf{i} - \mathbf{j})$, $t = 4$, work out $\mathbf{v}$

6 A particle is initially at the point with position vector $\mathbf{r} = (3\mathbf{i} - 2\mathbf{j})\,\text{m}$ and travelling with velocity $(5\mathbf{i} - \mathbf{j})\,\text{m s}^{-1}$ when it undergoes an acceleration of $(-\mathbf{i} + 2\mathbf{j})\,\text{m s}^{-2}$ for a period of $3\,\text{s}$. Work out its position at the end of this period.

7 A particle moving with velocity $(-\mathbf{i} + 2\mathbf{j})\,\text{m s}^{-1}$ undergoes a constant acceleration of $(2\mathbf{i} + \mathbf{j})\,\text{m s}^{-2}$ for $5\,\text{s}$. Work out its speed and direction at the end of this period. Show your working.

8 A particle undergoes a constant acceleration of $(2\mathbf{i} - 3\mathbf{j})\,\text{m s}^{-2}$. After $4\,\text{s}$ it has velocity $(10\mathbf{i} - 4\mathbf{j})\,\text{m s}^{-1}$. If it was initially at the point with position vector $(\mathbf{i} + 5\mathbf{j})\,\text{m}$, work out its position at the end of the period.

Reasoning and problem-solving

Strategy

To solve problems involving two-dimensional motion with constant acceleration

1. List the known values and the variable you need to find. This helps you decide which formula to use.
2. Be careful to distinguish between position, displacement and distance, and between velocity and speed.
3. Interpret your answers in the context of the question.

Example 2

Two particles, A and B, are moving in a plane. Initially A is at the point $(0, 3)$ and B is at $(2, 1)$. A has velocity $(2\mathbf{i} + \mathbf{j})\,\mathrm{m\,s^{-1}}$ and acceleration $(\mathbf{i} - 2\mathbf{j})\,\mathrm{m\,s^{-2}}$, and B has velocity $(3\mathbf{i} - \mathbf{j})\,\mathrm{m\,s^{-1}}$ and acceleration $2\mathbf{i}\,\mathrm{m\,s^{-2}}$. Work out the distance between the particles after six seconds.

For A: $\mathbf{u} = (2\mathbf{i} + \mathbf{j})$, $\mathbf{a} = (\mathbf{i} - 2\mathbf{j})$, $t = 6$, find $\mathbf{s}$

$\mathbf{s} = (2\mathbf{i} + \mathbf{j}) \times 6 + \frac{1}{2}(\mathbf{i} - 2\mathbf{j}) \times 36 = (30\mathbf{i} - 30\mathbf{j})$

After 6 s, A has position vector $\mathbf{r}_A = 3\mathbf{j} + (30\mathbf{i} - 30\mathbf{j}) = (30\mathbf{i} - 27\mathbf{j})$

For B: $\mathbf{u} = (3\mathbf{i} - \mathbf{j})$, $\mathbf{a} = 2\mathbf{i}$, $t = 6$, find $\mathbf{s}$

$\mathbf{s} = (3\mathbf{i} - \mathbf{j}) \times 6 + \frac{1}{2} \times 2\mathbf{i} \times 36 = (54\mathbf{i} - 6\mathbf{j})$

After 6 s, B has position vector $\mathbf{r}_B = (2\mathbf{i} + \mathbf{j}) + (54\mathbf{i} - 6\mathbf{j}) = (56\mathbf{i} - 5\mathbf{j})$

The displacement of B from A is $\overrightarrow{AB} = \mathbf{r}_B - \mathbf{r}_A = (26\mathbf{i} + 22\mathbf{j})$

$|\overrightarrow{AB}| = \sqrt{26^2 + 22^2} = 34.1$

The particles are 34.1 m apart.

1. List the known values and use $\mathbf{s} = \mathbf{u}t + \frac{1}{2}\mathbf{a}t^2$ to find the displacement of A.

2. To find the new position of A, you must add the displacement to the initial position.

2. The distance is the magnitude of the displacement.

3. Use suitable units.

Example 3

A particle, P, is initially at the point with position vector $12\mathbf{j}$ m and moving with velocity $3\mathbf{i}\,\mathrm{m\,s^{-1}}$. It undergoes an acceleration of $(-\mathbf{i} + 2\mathbf{j})\,\mathrm{m\,s^{-2}}$ for a period of 2 s. Show that it is then moving directly away from the origin O, and work out its speed.

See Ch6.1
For a reminder on collinear vectors.

$\mathbf{u} = 3\mathbf{i}$, $\mathbf{a} = (-\mathbf{i} + 2\mathbf{j})$, $t = 2$, find $\mathbf{v}$ and $\mathbf{s}$

$\mathbf{v} = 3\mathbf{i} + (-\mathbf{i} + 2\mathbf{j}) \times 2 = (\mathbf{i} + 4\mathbf{j})\,\mathrm{m\,s^{-1}}$

$\mathbf{s} = 3\mathbf{i} \times 2 + \frac{1}{2}(-\mathbf{i} + 2\mathbf{j}) \times 4 = (4\mathbf{i} + 4\mathbf{j})$

Position $\mathbf{r} = 12\mathbf{j} + (4\mathbf{i} + 4\mathbf{j}) = (4\mathbf{i} + 16\mathbf{j})$ m

$(4\mathbf{i} + 16\mathbf{j}) = 4(\mathbf{i} + 4\mathbf{j})$, so $\mathbf{v}$ is in the same direction as $\mathbf{r}$

The direction $\overrightarrow{OP}$ is as shown, so the particle is moving directly away from O

Speed $v = \sqrt{1^2 + 4^2} = 4.12\,\mathrm{m\,s^{-1}}$

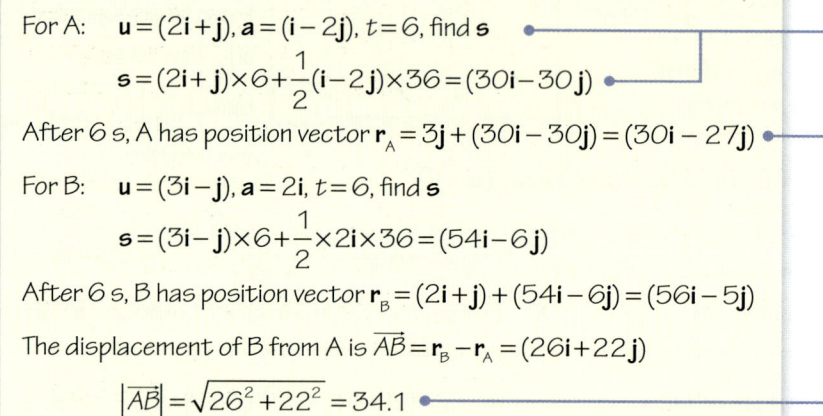

1. To find its velocity and displacement after 2 s, use $\mathbf{v} = \mathbf{u} + \mathbf{a}t$ and $\mathbf{s} = \mathbf{u}t + \frac{1}{2}\mathbf{a}t^2$

2. To find its new position, you must add the displacement to the initial position.

3. Use your answer to show the direction of movement.

The speed is the magnitude of the velocity.

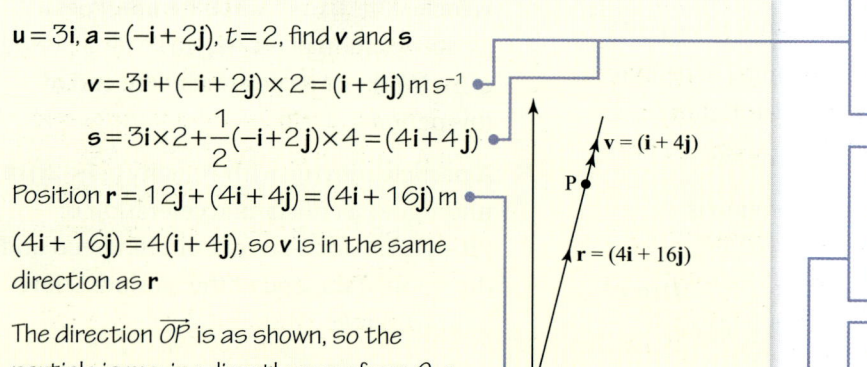

1 A particle is moving in a plane with acceleration $(4\mathbf{i} + 3\mathbf{j})\,\text{m s}^{-2}$. Its initial velocity is $(2\mathbf{i} - 6\mathbf{j})\,\text{m s}^{-1}$. After eight seconds it has velocity $\mathbf{v}$. Work out

   a Its speed and direction at this time, showing your working,

   b The distance between the initial and final positions of the particle.

2 An object is moving in a plane with constant acceleration. Its initial velocity is $(3\mathbf{i} - 2\mathbf{j})\,\text{m s}^{-1}$ and, six seconds later, its velocity is $(9\mathbf{i} + 4\mathbf{j})\,\text{m s}^{-1}$. Its initial position is $(5\mathbf{i} + 2\mathbf{j})\,\text{m}$. Work out its final position.

3 Two particles, A and B, are moving in a plane. Initially A is at the point $(0, 5)$ and B is at $(3, 7)$. A has velocity $(\mathbf{i} + 3\mathbf{j})\,\text{m s}^{-1}$ and acceleration $(2\mathbf{i} + 2\mathbf{j})\,\text{m s}^{-2}$, and B has velocity $2\mathbf{j}\,\text{m s}^{-1}$ and acceleration $(3\mathbf{i} + \mathbf{j})\,\text{m s}^{-2}$. Work out the distance between the particles after four seconds.

4 A boat moving with velocity $(6\mathbf{i} + 2\mathbf{j})\,\text{m s}^{-1}$ undergoes an acceleration of $(-\mathbf{i} - 7\mathbf{j})\,\text{m s}^{-2}$ for two seconds. Show that, at the end of that time, it is travelling in a direction perpendicular to its initial direction, and at twice the speed. Show your working.

5 A snooker ball is initially travelling with velocity $(4\mathbf{i} + 5\mathbf{j})\,\text{m s}^{-1}$. It undergoes an acceleration of magnitude $2.5\,\text{m s}^{-2}$ in a direction given by the vector $(3\mathbf{i} - 4\mathbf{j})$. Work out the velocity and displacement of the snooker ball from its initial position after four seconds.

6 A particle, P, is initially at the point $(2, 6)$ in relation to an origin O. It is moving with velocity $(3\mathbf{i} + \mathbf{j})\,\text{m s}^{-1}$ and constant acceleration $(16\mathbf{i} + 24\mathbf{j})\,\text{m s}^{-2}$. Show that, after two seconds, it is moving directly away from O. Work out its speed at that time. Show your working.

7 An object is moving in a plane. At time $t = 0$ it is at the origin, O, and moving with velocity $\mathbf{u}$. After two seconds, it is at A, where $\overrightarrow{OA} = -2\mathbf{i} - 4\mathbf{j}$. After a further three seconds, it is at B, where $\overrightarrow{AB} = 10\mathbf{i} - 40\mathbf{j}$. Assuming the object has constant acceleration a, work out **a** and **u**

Challenge

8 A particle starts at the origin O, and then moves with constant velocity $(2\mathbf{i} + 3\mathbf{j})\,\text{m s}^{-1}$ for four seconds. It then accelerates with constant acceleration $(\mathbf{i} - 3\mathbf{j})\,\text{m s}^{-2}$ for six seconds, before slowing uniformly to rest in four seconds. Work out

   a Its displacement during the first stage of the journey,

   b Its velocity after the second stage of the journey,

   c Its acceleration during the third stage of the journey,

   d Its position at the end of the journey.

9 An aircraft is travelling horizontally in an easterly direction at a height of $400\,\text{m}$ above level ground and at a speed of $120\,\text{m s}^{-1}$. When it is directly above an observer, O, it releases a package. In relation to the origin O, and to the x- and y- directions of east and up respectively, the package has acceleration $(-4\mathbf{i} - 8\mathbf{j})\,\text{m s}^{-2}$. P is the position of the package at time t Work out

   a An expression for the vector $\overrightarrow{OP}$

   b The time the package takes to reach the ground,

   c How far from the observer the package lands.

Fluency and skills

See Ch7.4

For a reminder on motion with variable acceleration.

For motion in one dimension, the displacement s, velocity v and acceleration a are related by $v = \dfrac{ds}{dt}$ and $a = \dfrac{dv}{dt}$. You can extend these relationships to motion in which the displacement $\mathbf{s}$, velocity $\mathbf{v}$ and acceleration $\mathbf{a}$ are two-dimensional vectors.

See Ch15.8

For a reminder on differentiating parametric equations.

> **Key point**
>
> $$\mathbf{v} = \frac{d\mathbf{s}}{dt} \quad \text{and} \quad \mathbf{a} = \frac{d\mathbf{v}}{dt} = \frac{d^2\mathbf{s}}{dt^2}$$

To differentiate a vector, you differentiate its components, so

> **Key point**
>
> If $\mathbf{s} = x\mathbf{i}, + y\mathbf{j}$, then $\quad \mathbf{v} = \dfrac{dx}{dt}\mathbf{i} + \dfrac{dy}{dt}\mathbf{j} \quad$ or $\quad \mathbf{v} = \dot{x}\mathbf{i} + \dot{y}\mathbf{j}$
>
> $$\mathbf{a} = \frac{d^2x}{dt^2}\mathbf{i} + \frac{d^2y}{dt^2}\mathbf{j} \quad \text{or} \quad \mathbf{a} = \ddot{x}\mathbf{i} + \ddot{y}\mathbf{j}$$

The differentiation of a vector involves parametric equations. The horizontal component, x, and the vertical component, y, of the displacement are expressed in terms of a third parameter. This third parameter is time, t

Example 1

The displacement of a particle at time t s is given by $\mathbf{s} = 4t^2\mathbf{i} + (3t - 5t^3)\mathbf{j}$ metres. Write an expression for its velocity at time t s

$$\mathbf{v} = \frac{d\mathbf{s}}{dt} = 8t\mathbf{i} + (3 - 15t^2)\mathbf{j}\,\text{m s}^{-1}$$

Differentiate displacement to find velocity. You need to differentiate each component with respect to t

Example 2

The velocity of a particle at time t s is given by $\mathbf{v} = \sin 2\pi t\mathbf{i} + t^2\mathbf{j}\,\text{m s}^{-1}$. Work out its acceleration when $t = 1$

$$\mathbf{a} = \frac{d\mathbf{v}}{dt} = 2\pi\cos 2\pi t\mathbf{i} + 2t\mathbf{j}$$

When $t = 1$, $\mathbf{a} = (2\pi\mathbf{i} + 2\mathbf{j})\,\text{m s}^{-2}$

Differentiate velocity to find acceleration. See Ch15.2 for a reminder on differentiating trigonometric functions.

You can also express the relationships between $\mathbf{s}$, $\mathbf{v}$ and $\mathbf{a}$ in terms of integration.

> **Key point**
>
> $$\mathbf{v} = \int \mathbf{a}\,dt \quad \text{and} \quad \mathbf{s} = \int \mathbf{v}\,dt$$

As with differentiation, to integrate a vector, you integrate each of its components.

Example 3

The acceleration of a particle at time t is given by $\mathbf{a} = 6t^2\mathbf{i} + (3t^2 - 8t)\mathbf{j}$. Write an expression for its velocity at time t, given that $\mathbf{v} = 5\mathbf{i} - 6\mathbf{j}$ when $t = 0$

$\mathbf{v} = \int 6t^2\mathbf{i} + (3t^2 - 8t)\mathbf{j} \, dt = 2t^3\mathbf{i} + (t^3 - 4t^2)\mathbf{j} + \mathbf{c}$

When $t = 0$, $\mathbf{v} = 5\mathbf{i} - 6\mathbf{j}$, so $\mathbf{c} = 5\mathbf{i} - 6\mathbf{j}$

$\mathbf{v} = 2t^3\mathbf{i} + (t^3 - 4t^2)\mathbf{j} + 5\mathbf{i} - 6\mathbf{j}$

So $\mathbf{v} = (2t^3 + 5)\mathbf{i} + (t^3 - 4t^2 - 6)\mathbf{j}$

> Integrate acceleration to find velocity. The constant of integration is a vector.

> Use the initial conditions to find $\mathbf{c}$

MECH

Example 4

A particle is initially at the point with position vector $(2\mathbf{i} + \mathbf{j})$ m and has velocity $\mathbf{v} = (4\mathbf{i} + 6t\mathbf{j})$ m s^{-1}. Work out its position four seconds later.

$\mathbf{r} = \int 4\mathbf{i} + 6t\mathbf{j} \, dt = 4t\mathbf{i} + 3t^2\mathbf{j} + \mathbf{c}$

When $t = 0$, $\mathbf{r} = 2\mathbf{i} + \mathbf{j}$, so $\mathbf{c} = 2\mathbf{i} + \mathbf{j}$

$\mathbf{r} = 4t\mathbf{i} + 3t^2\mathbf{j} + 2\mathbf{i} + \mathbf{j} = (4t + 2)\mathbf{i} + (3t^2 + 1)\mathbf{j}$

When $t = 4$, $\mathbf{r} = 18\mathbf{i} + 49\mathbf{j}$, so the position after 4 s is $(18\mathbf{i} + 49\mathbf{j})$ m

> Integrate velocity to find displacement, $\mathbf{r}$, from the origin.

> Substitute $t = 4$ into your expression for the position vector.

Exercise 18.2A Fluency and skills

1 For each of these displacement or position vectors, express the velocity at time t

  a $\mathbf{s} = 4t\mathbf{i} + (8 + 2t^2)\mathbf{j}$

  b $\mathbf{r} = (t^2 - 4t)\mathbf{i} + (t^3 - 2t^2)\mathbf{j}$

  c $\mathbf{s} = 2\cos t\,\mathbf{i} + 4\sin t\,\mathbf{j}$

  d $\mathbf{r} = e^t\mathbf{i} + \ln(t+1)\,\mathbf{j}$

2 For each of these acceleration vectors, write an expression for the velocity vector consistent with the given initial condition.

  a $\mathbf{a} = (3 - 2t)\mathbf{i} + (2t - 6t^3)\mathbf{j}$, $\mathbf{v} = 3\mathbf{i}$ when $t = 0$

  b $\mathbf{a} = 4\cos 2t\,\mathbf{i} + 8\sin 2t\,\mathbf{j}$, $\mathbf{v} = 2\mathbf{i} + \mathbf{j}$ when $t = 0$

3 For each of these velocity vectors, write expressions for the acceleration and displacement or position vectors consistent with the given initial condition.

  a $\mathbf{v} = 15\mathbf{i} + (20 - 10t)\mathbf{j}$, $\mathbf{s} = 4\mathbf{i} - 3\mathbf{j}$ when $t = 0$

  b $\mathbf{v} = -8\sin t\,\mathbf{i} + 4e^{2t}\mathbf{j}$, $\mathbf{r} = \mathbf{i} + 3\mathbf{j}$ when $t = 0$

  c $\mathbf{v} = 2t(1 - 3t)\mathbf{i} + t^2(3 - 4t)\mathbf{j}$, $\mathbf{r} = 2\mathbf{i}$ when $t = 0$

4 A particle has acceleration $\mathbf{a} = (4t\mathbf{i} + 3t^2\mathbf{j})$ m s^{-2} and initial velocity $\mathbf{v} = (-3\mathbf{i} + \mathbf{j})$ m s^{-1}. Work out its velocity when $t = 2$ s

5 A particle has position vector $\mathbf{r} = (\sin \pi t\,\mathbf{i} + \cos \pi t\,\mathbf{j})$ m. Work out its velocity when $t = 3$ s. Show your working.

6 A particle starts from the point whose position vector is $(2\mathbf{i} + 3\mathbf{j})$ m. Its velocity at time t is given by $\mathbf{v} = ((4t - 2)\mathbf{i} + 3t^2\mathbf{j})$ m s^{-1}. Work out its position when $t = 2$ s

7 The acceleration of a particle at time t s is $\mathbf{a} = (6t\mathbf{i} + 2\mathbf{j})$ m s^{-2}. When $t = 0$, its velocity is $(3\mathbf{i} + 4\mathbf{j})$ m s^{-1}. Work out

  a   Its velocity at time t

  b   Its speed after 2 s

  c   Its direction of travel after 2 s, showing your working.

 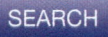

8 At time t the position of a particle is given by $\mathbf{r} = (t^3\,\mathbf{i} + 8\ln(1+t)\mathbf{j})$ m. What is its velocity and acceleration when $t = 2$? Show your working.

9 A particle has velocity at time t given by $\mathbf{v} = 6\cos 3t\,\mathbf{i} + 8\sin 2t\,\mathbf{j}$. It initially has position vector $\mathbf{r} = 2\mathbf{i} - 5\mathbf{j}$. Work out

a Its acceleration at time t

b Its position at time t

Reasoning and problem-solving

To solve problems involving two-dimensional motion with variable acceleration

1 If the problem involves a variable force, use Newton's second law ($\mathbf{F} = m\mathbf{a}$) to find acceleration.

2 Identify whether you need to move from $\mathbf{s} \rightarrow \mathbf{v} \rightarrow \mathbf{a}$ (differentiation) or $\mathbf{a} \rightarrow \mathbf{v} \rightarrow \mathbf{s}$ (integration).

3 Be clear whether you are asked for displacement, position or distance, and velocity or speed.

4 Substitute any known values into your expressions to give the required solution.

Example 5

A particle of mass 4 kg is acted on by a force of $(8\mathbf{i} + 12t\mathbf{j})$ N. Initially, the particle has position vector $(3\mathbf{i} - \mathbf{j})$ m and velocity $(3\mathbf{i} - 5\mathbf{j})$ m s^{-1}. Work out

a Its speed after 4 s   **b** Its position at that time.

a $8\mathbf{i} + 12t\mathbf{j} = 4\mathbf{a}$

$\mathbf{a} = 2\mathbf{i} + 3t\mathbf{j}$ and $\mathbf{v} = \int 2\mathbf{i} + 3t\mathbf{j}\, dt = 2t\mathbf{i} + \dfrac{3}{2}t^2\mathbf{j} + \mathbf{c}_1$

When $t = 0$, $\mathbf{v} = 3\mathbf{i} - 5\mathbf{j}$, so $\mathbf{c}_1 = 3\mathbf{i} - 5\mathbf{j}$

$\mathbf{v} = (2t+3)\mathbf{i} + \left(\dfrac{3}{2}t^2 - 5\right)\mathbf{j}$

When $t = 4$, $\mathbf{v} = 11\mathbf{i} + 19\mathbf{j}$ so speed $= \sqrt{11^2 + 19^2} = 22.0$ m s^{-1}

b $\mathbf{r} = \int (2t+3)\mathbf{i} + \left(\dfrac{3}{2}t^2 - 5\right)\mathbf{j}\, dt$

$= (t^2 + 3t)\mathbf{i} + \left(\dfrac{1}{2}t^3 - 5t\right)\mathbf{j} + \mathbf{c}_2$

When $t = 0$, $\mathbf{r} = 3\mathbf{i} - \mathbf{j}$, so $\mathbf{c}_2 = 3\mathbf{i} - \mathbf{j}$

$\mathbf{r} = (t^2 + 3t + 3)\mathbf{i} + \left(\dfrac{1}{2}t^3 - 5t - 1\right)\mathbf{j}$

When $t = 4$, $\mathbf{r} = (31\mathbf{i} + 11\mathbf{j})$ m

1 Use $\mathbf{F} = m\mathbf{a}$ to find the acceleration.

2 Integrate acceleration to find velocity. Use the initial velocity to find $\mathbf{c}_1$

3 The speed is the magnitude of the velocity.

2 Integrate velocity to find displacement/position.

4 Substitute $t = 4$ into your expression.

Example 6

The displacement of a particle at time t s is given by $\mathbf{s} = 6t^2\mathbf{i} + (3t^2 - 12t)\mathbf{j}$ m. Work out the distance of the particle from its initial position at the instant it is travelling in the x-direction.

$\mathbf{s} = 6t^2\mathbf{i} + (3t^2 - 12t)\mathbf{j}$ and $\mathbf{v} = \dfrac{d\mathbf{s}}{dt} = 12t\mathbf{i} + (6t - 12)\mathbf{j}$

Travels in x-direction when $6t - 12 = 0 \Rightarrow t = 2$

$t = 2$ gives $\mathbf{s} = 24\mathbf{i} - 12\mathbf{j}$

The distance $= \sqrt{24^2 + (-12)^2} = 26.8$ m

2 Differentiate to find velocity.

The y-component of the velocity must be zero.

3 The distance is the magnitude of the displacement.

Exercise 18.2B Reasoning and problem-solving

1 A particle of mass 5 kg is acted on by a force $(10\mathbf{i} + 5t\mathbf{j})$ N. Initially the particle has position vector $(2\mathbf{i} + 4\mathbf{j})$ m. When $t = 2$, $\mathbf{v} = (5\mathbf{i} + 5\mathbf{j})$ m s^{-1}. Work out

    a The initial speed of the particle,

    b The position of the particle at $t = 2$

2 The displacement of a boat at time t s is given by $\mathbf{s} = (5t^3 - 12t)\mathbf{i} + 10t\mathbf{j}$ m. Work out the distance of the boat from its initial position at the instant it is travelling in the y-direction. Leave your answer in surd form.

3 A particle moves on a plane such that its position at time t s is given by $\mathbf{r} = (3t - 2)\mathbf{i} + (4t - 2t^2)\mathbf{j}$ m

    a Write expressions for the velocity and acceleration of the particle at time t

    b Work out the initial speed of the particle.

    c At what time(s) is the particle moving parallel to the x-axis?

    d Is the particle ever stationary? Give a reason for your answer.

4 A particle moves with constant acceleration vector $\mathbf{a}$. Its initial velocity is $\mathbf{u}$ and, at time t, it has velocity $\mathbf{v}$ and displacement $\mathbf{s}$ from its starting position.

    a Show by integration that $\mathbf{v} = \mathbf{u} + \mathbf{a}t$

    b Show by integration that $\mathbf{s} = \mathbf{u}t + \dfrac{1}{2}\mathbf{a}t^2$

5 A ball of mass 5 kg is acted upon by a force of $(20t\mathbf{i} - 15\mathbf{j})$ N. Initially the ball is at the point with position vector $(2\mathbf{i} + 3\mathbf{j})$ m and is travelling with velocity $(-2\mathbf{i} + 12\mathbf{j})$ m s^{-1}. Work out its velocity and position when $t = 6$

6 At time t a particle has position given by $\mathbf{r} = (2t - 1 + \cos t)\mathbf{i} + \sin 2t\mathbf{j}$. The particle starts at the origin.

    a Work out the value of t for which it next touches the x-axis.

    b For that value of t, work out its instantaneous velocity and acceleration. Show your working.

7 A force $\mathbf{F} = (2000t\mathbf{i} - 4000\mathbf{j})$ N acts on a particle of mass 500 kg at time t s. Initially the particle is at the origin and travelling with velocity 10i m s^{-1}. Work out

    a The speed of the particle when $t = 2$

    b The distance of the particle from the origin at that time.

8 An object moves on a plane so that its acceleration at time t s is given by $\mathbf{a} = (-4\cos 2t\mathbf{i} - 4\sin 2t\mathbf{j})$ m s^{-2}. It is initially at the point $(1, 0)$ and travelling at 2 m s^{-1} in the y-direction.

    a Show that the object moves at constant speed.

    b Work out the distance of the object from the origin at time t and hence describe the path of the object.

Challenge

9 Particles P and Q, each of mass 0.5 kg, move on a horizontal plane, with east and north as the $\mathbf{i}$ and $\mathbf{j}$ directions. Initially P has velocity $(2\mathbf{i} - 5\mathbf{j})$ m s^{-1} and Q is travelling north at 2 m s^{-1}. Each particle is acted on by a force of magnitude t N. The force on P acts towards the north-east, while that on Q acts towards the south-east. Work out the value of t for which

    a The two particles have the same speed,

    b The two particles are travelling in the same direction. Show your working.

10 The position of a particle at time t is given by $\mathbf{r} = (2 + 3t + 8t^2)\mathbf{i} + (6 + t + 12t^2)\mathbf{j}$. Work out

    a The times at which the particle is moving directly towards or directly away from the origin,

    b The position of the particle at the time(s) found in part **a**, and identify whether it is moving towards or away from the origin.

Motion under gravity 2

Fluency and skills

See Ch8.3

For a reminder on motion under gravity.

See Ch6.2

For a reminder on components of a vector.

When an object moves vertically under gravity you usually assume that it is a particle (it has no size and does not spin) and that there is no wind or air resistance. Its acceleration is $g\,\text{m s}^{-2}$ downwards.

However, an object thrown into the air is usually a projectile, with displacement and velocity in two dimensions. This means that to investigate the motion of a projectile, you will often need to use trigonometry to resolve velocity into its horizontal and vertical components.

For example, consider a particle which is projected from a point O on a horizontal plane with a speed of $25\,\text{m s}^{-1}$ at an angle of $30°$ to the horizontal.

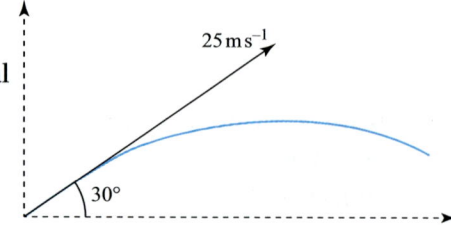

Resolving horizontally: $25\cos 30° \Rightarrow$ The initial horizontal component of velocity is $21.7\,\text{m s}^{-1}$

Resolving vertically: $25\sin 30° \Rightarrow$ The initial vertical component of velocity is $12.5\,\text{m s}^{-1}$

You will also need to use the equations of motion for constant acceleration. Remember that g will always act as the vertical component of acceleration in a downwards direction. Assuming there is no wind or air resistance, there will be no horizontal acceleration, so horizontal velocity is constant.

Example 1

A particle is projected from a point O on a horizontal plane with speed $40\,\text{m s}^{-1}$ at an angle of $30°$ to the horizontal. It lands on the plane at A. Taking $g = 9.81\,\text{m s}^{-2}$, work out

a How long it was in the air, b The distance OA c The greatest height it reached.

a The initial vertical component of velocity is $40\sin 30°$
The vertical acceleration is $-g$

At time t, the height $y = 40t\sin 30° - \dfrac{1}{2}gt^2$

> Use $s = ut + \dfrac{1}{2}at^2$ to find an expression for the height of the particle.

At A, $y = 0$, so $40t\sin 30° - \dfrac{1}{2}gt^2 = 0$

> When the particle lands, the vertical height will be zero.

$$\dfrac{1}{2}t(80\sin 30° - gt) = 0$$

So $y = 0$ when $t = 0$ (the start time) or

when $t = \dfrac{80\sin 30°}{g} = \dfrac{80\sin 30°}{9.81} = 4.08$ (to 2 dp)

The particle is in the air for $4.08\,\text{s}$

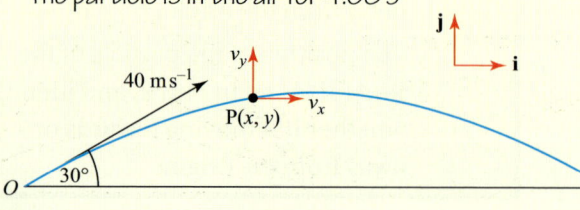

(*Continued on the next page*)

b There is no horizontal acceleration.

The horizontal velocity component is constant, $v_x = 40\cos 30°$

When $t = 4.08$, $x = 40\cos 30° \times 4.08 = 141.3$ So $OA = 141.3\,\text{m}$

> Distance = speed × time

> Use $v = u + at$

c At time t, the vertical velocity component $v_y = 40\sin 30° - gt$

At greatest height, $v_y = 0$, so $t = \dfrac{40\sin 30°}{g} = 2.04$ (to 2 dp)

When $t = 2.04$, $y = 40\sin 30° \times 2.04 - \dfrac{1}{2}g \times 2.04^2 = 20.4$

The greatest height reached is $20.4\,\text{m}$

> At the greatest height, the particle's vertical velocity changes sign, so set v_y equal to zero. This is exactly halfway through the flight.

> Use $s = ut + \dfrac{1}{2}at^2$

You can also write the equations in vector form. You will need to know how to derive formulae for the time of flight, range and greatest height and the equation of the path of a projectile.

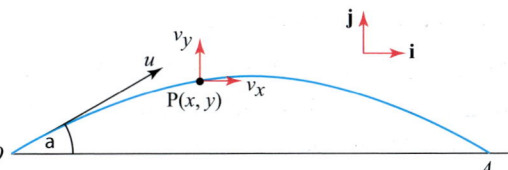

Particle, P, is projected from a point O on a horizontal plane with initial speed u and elevation α. Take **i** and **j** directions, as shown.

The horizontal and vertical components of initial velocity are $u_x = u\cos\alpha$ and $u_y = u\sin\alpha$, so the initial velocity vector is

$$\mathbf{u} = u\cos\alpha\,\mathbf{i} + u\sin\alpha\,\mathbf{j}$$

At time t, the particle is at the point with position vector $\mathbf{r} = x\mathbf{i} + y\mathbf{j}$ and has velocity $\mathbf{v} = v_x\mathbf{i} + v_y\mathbf{j}$. Its acceleration is $\mathbf{a} = -g\mathbf{j}$

From $\mathbf{v} = \mathbf{u} + \mathbf{a}t$: $\mathbf{v} = u\cos\alpha\,\mathbf{i} + (u\sin\alpha - gt)\mathbf{j}$ ....[1]

From $\mathbf{r} = \mathbf{u}t + \dfrac{1}{2}\mathbf{a}t^2$: $\mathbf{r} = ut\cos\alpha\,\mathbf{i} + \left(ut\sin\alpha - \dfrac{1}{2}gt^2\right)\mathbf{j}$ ....[2]

The horizontal distance the particle travels before landing (OA in the diagram above) is the **range**.

When the particle lands, $y = 0$ and so, from [2],

$$ut\sin\alpha - \dfrac{1}{2}gt^2 = 0$$

This gives $t = 0$ (the start time) and $t = \dfrac{2u\sin\alpha}{g}$, the **time of flight**. For this value of t, $x = \dfrac{2u^2\sin\alpha\cos\alpha}{g}$

You can simplify this using the formula $\sin 2\alpha = 2\sin\alpha\cos\alpha$

> **See Ch14.3**
> For a reminder on double angle formulae.

> **Key point**
>
> $\text{Range} = \dfrac{u^2\sin 2\alpha}{g}$

As $\sin 2\alpha = \sin(180° - 2\alpha) = \sin 2(90° - \alpha)$, it follows that elevations of α and $(90° - \alpha)$ give the same range.

The range is greatest when $\sin 2\alpha = 1$. This gives $2\alpha = 90° \Rightarrow \alpha = 45°$

> **Key point**
>
> Elevations of α and $(90° - \alpha)$ give the same range.
>
> The maximum range $= \dfrac{u^2}{g}$ when $\alpha = 45°$

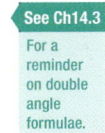

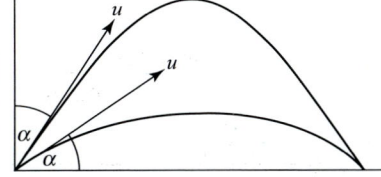

MECH

When the particle is at its **maximum height** the vertical component of velocity is zero, so, from [1],
$$u\sin\alpha - gt = 0$$

This gives $t = \dfrac{u\sin\alpha}{g}$, which is half the time of flight.

At this time, from [2],
$$y = u\left(\frac{u\sin\alpha}{g}\right)\sin\alpha - \frac{1}{2}g\left(\frac{u\sin\alpha}{g}\right)^2 = \frac{u^2\sin^2\alpha}{2g}$$

> **Key point**
>
> Maximum height $= \dfrac{u^2\sin^2\alpha}{2g}$

ICT Resource online

To investigate motion under gravity, click this link in the digital book.

You can find the **equation of the path** followed by the projectile. At time t the particle is at the point (x, y), where, from [2],
$$x = ut\cos\alpha \quad\quad \dots[3]$$
$$y = ut\sin\alpha - \frac{1}{2}gt^2 \quad \dots[4]$$

From [3], $t = \dfrac{x}{u\cos\alpha}$. Substituting this into [4] gives
$$y = u\left(\frac{x}{u\cos\alpha}\right)\sin\alpha - \frac{1}{2}g\left(\frac{x}{u\cos\alpha}\right)^2$$

This gives a quadratic function of x. The path is a parabola.

> **Key point**
>
> The equation of the path is $y = x\tan\alpha - \left(\dfrac{g\sec^2\alpha}{2u^2}\right)x^2$

> Remember that if a value of g is not specified, you should use $g = 9.8\,\text{m s}^{-2}$

Example 2

A particle is projected from a point O on horizontal ground with initial speed $12\,\text{m s}^{-1}$ at an angle of $60°$ to the horizontal. Work out

a Its height above the ground after travelling $7.2\,\text{m}$ horizontally,

b Its speed and direction at that instant.

a Initial velocity is $\mathbf{u} = 12\cos60°\mathbf{i} + 12\sin60°\mathbf{j} = 6\mathbf{i} + 6\sqrt{3}\mathbf{j}$

Acceleration is $\mathbf{a} = -g\mathbf{j} = -9.8\mathbf{j}$

The position vector at time t is $\mathbf{r} = x\mathbf{i} + y\mathbf{j} = 6t\mathbf{i} + \left(6t\sqrt{3} - 4.9t^2\right)\mathbf{j}$

When $x = 7.2$, $6t = 7.2$ and so $t = 1.2$

When $t = 1.2$, $y = 6 \times 1.2\sqrt{3} - 4.9 \times 1.2^2 = 5.41$ so its height above the ground is $5.41\,\text{m}$.

b The velocity at time t is $\mathbf{v} = 6\mathbf{i} + (6\sqrt{3} - 9.8t)\mathbf{j}$

When $t = 1.2$, $\mathbf{v} = 6\mathbf{i} - 1.37\mathbf{j}$ and $v = |\mathbf{v}| = \sqrt{6^2 + (-1.37)^2} = 6.15$

$\tan\theta = \dfrac{1.37}{6} = 0.228$ and so $\theta = 12.8°$

The particle is travelling at $6.15\,\text{m s}^{-1}$ at $12.8°$ below the horizontal.

Take $\mathbf{i}$ and $\mathbf{j}$ as the horizontal and vertical directions and resolve the velocity into these components.

Use $\mathbf{s} = \mathbf{u}t + \dfrac{1}{2}\mathbf{a}t^2$ to find the time when the horizontal distance $x = 7.2$

Use $\mathbf{s} = \mathbf{u}t + \dfrac{1}{2}\mathbf{a}t^2$ to find the vertical height when $t = 1.2$

Use $\mathbf{v} = \mathbf{u} + \mathbf{a}t$

Check your answer by evaluating the angle between $\mathbf{v}$ and $\begin{pmatrix}1\\0\end{pmatrix}$ and the absolute value of $\mathbf{v}$ on your calculator.

1   A projectile is launched from a point O on horizontal ground with speed $15\,\text{m s}^{-1}$ at an angle of 35° to the horizontal.

   a   For how long is the projectile in the air?

   b   What is the horizontal range of the projectile?

   c   Work out the time taken for the projectile to reach its maximum height.

   d   What is the greatest height reached by the projectile?

2   A projectile is launched from a point O on horizontal ground with speed $10\,\text{m s}^{-1}$ at an angle of 60° to the horizontal. Take $g = 9.81\,\text{m s}^{-2}$

   a   For how long is the projectile in the air?

   b   What is the horizontal range of the projectile?

   c   Work out the time taken for the projectile to reach its maximum height.

   d   What is the greatest height reached by the projectile?

3   A particle is projected from a point O on horizontal ground with speed $24\,\text{m s}^{-1}$ at an angle of 50° to the horizontal. Taking $g = 9.81\,\text{m s}^{-2}$, work out

   a   The height of the particle after it has travelled 30 m horizontally,

   b   The speed and direction of the particle at this instant. Show your working.

4   A ball is projected from a point O on horizontal ground. Its initial velocity is $(30\mathbf{i} + 40\mathbf{j})\,\text{m s}^{-1}$. Work out

   a   Its position after

      i   1 s

      ii   2 s

   b   The length of time it is in the air,

   c   Its range,

   d   The equation of its path.

5   A particle is projected at $196\,\text{m s}^{-1}$ at an angle α to the horizontal. It reaches a maximum height of 490 m. Work out the value of α.

6   A particle is projected at 45° to the horizontal from a point O on horizontal ground. It strikes the ground again 100 m away. Take $g = 10\,\text{m s}^{-2}$ and work out the speed with which it was projected.

MECH

Reasoning and problem-solving

Strategy

To solve problems involving motion under gravity

  (1) Sketch a diagram.

  (2) Decide whether to work separately with the horizontal and vertical motion or in terms of vectors.

  (3) State what you know and choose the appropriate constant acceleration equation.

  (4) Interpret your solutions in the context of the question.

Example 3

A girl throws a stone horizontally from the top of a 40 m cliff at a speed which is just sufficient to reach the water's edge 50 m from the base of the cliff. Calculate

a The initial speed of the stone,   **b** The speed and direction of the stone as it hits the water.

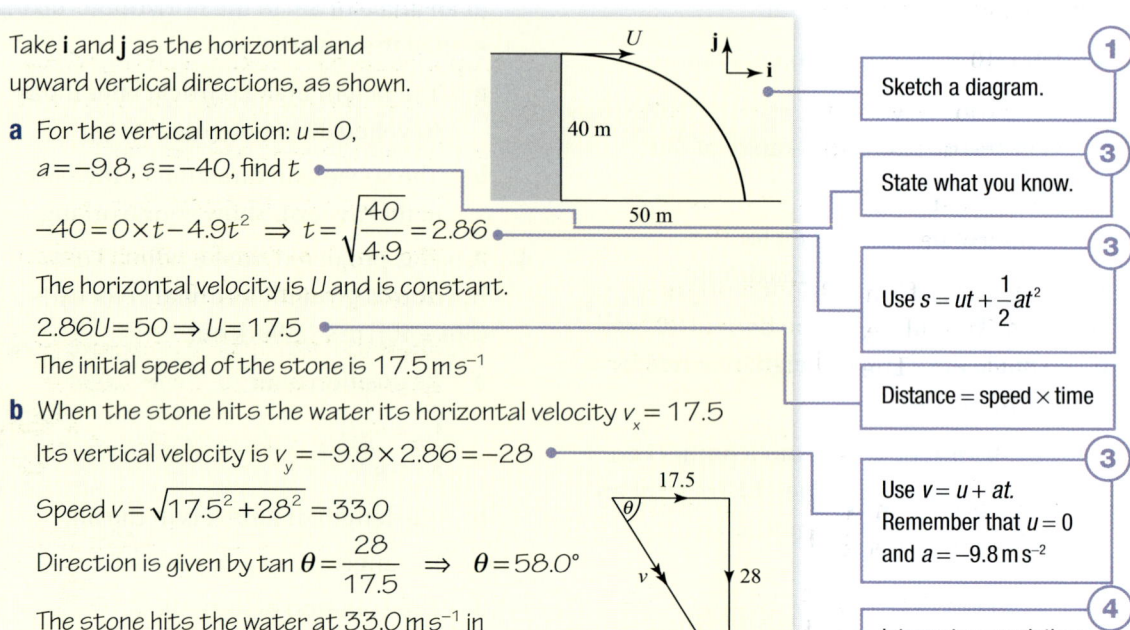

Take **i** and **j** as the horizontal and upward vertical directions, as shown.

a For the vertical motion: $u = 0$, $a = -9.8$, $s = -40$, find t

$$-40 = 0 \times t - 4.9t^2 \Rightarrow t = \sqrt{\frac{40}{4.9}} = 2.86$$

The horizontal velocity is U and is constant.

$$2.86U = 50 \Rightarrow U = 17.5$$

The initial speed of the stone is $17.5\,\text{m s}^{-1}$

b When the stone hits the water its horizontal velocity $v_x = 17.5$

Its vertical velocity is $v_y = -9.8 \times 2.86 = -28$

$$\text{Speed } v = \sqrt{17.5^2 + 28^2} = 33.0$$

$$\text{Direction is given by } \tan\theta = \frac{28}{17.5} \Rightarrow \theta = 58.0°$$

The stone hits the water at $33.0\,\text{m s}^{-1}$ in a direction $58.0°$ below the horizontal.

① Sketch a diagram.

③ State what you know.

③ Use $s = ut + \frac{1}{2}at^2$

Distance = speed × time

③ Use $v = u + at$. Remember that $u = 0$ and $a = -9.8\,\text{m s}^{-2}$

④ Interpret your solutions in the context of the question.

Example 4

A projectile is fired at $50\,\text{m s}^{-1}$ from ground level and strikes a target 100 m away and 20 m above its point of projection. Work out the two possible angles of elevation at which it was fired.

Take $g = 9.81\,\text{m s}^{-2}$

Let the angle of elevation be θ

$$\mathbf{u} = \begin{pmatrix} 50\cos\theta \\ 50\sin\theta \end{pmatrix}, \mathbf{a} = \begin{pmatrix} 0 \\ -9.81 \end{pmatrix} \text{ and, at time } t, \mathbf{r} = \begin{pmatrix} 100 \\ 20 \end{pmatrix}$$

$$\begin{pmatrix} 100 \\ 20 \end{pmatrix} = \begin{pmatrix} 50\cos\theta \\ 50\sin\theta \end{pmatrix} t + \begin{pmatrix} 0 \\ -4.905 \end{pmatrix} t^2$$

$$100 = 50t\cos\theta \Rightarrow t = \frac{2}{\cos\theta} \quad [1]$$

$$20 = 50t\sin\theta - 4.905t^2 \quad [2]$$

$$20 = 100\tan\theta - 19.62\sec^2\theta$$

$$20 = 100\tan\theta - 19.62(1 + \tan^2\theta)$$

$$19.62\tan^2\theta - 100\tan\theta + 19.62 + 20 = 0$$

$$19.62\tan^2\theta - 100\tan\theta + 39.62 = 0$$

So $\tan\theta = 4.66$ or 0.433

The two angles of elevation are $\theta = 77.9°$ or $23.4°$

②③ Use vectors and start by stating what you know.

③ Use $\mathbf{r} = \mathbf{u}t + \frac{1}{2}\mathbf{a}t^2$

Compare x components.

Compare y components.

Substitute [1] into [2]

You could use your calculator to solve this quadratic equation.

510   **Motion in two dimensions** Motion under gravity 2

1 A stone is thrown from the top of a 50 m high vertical cliff at 30 m s^{-1} at an angle of 20° below the horizontal.

Take $g = 10$ m s^{-2}

   a Work out how far from the foot of the cliff the stone hits the sea.

   b What assumptions have you made in your answer?

2 A ball is kicked from the floor along a corridor. The ball leaves the floor at 20 m s^{-1} at an angle of θ. The ceiling of the corridor is 3 m high.

Take $g = 9.81$ m s^{-2}

   a Work out the maximum value of θ if the ball does not touch the ceiling.

   b If you could not ignore the size of the ball, what effect would it have on your answer?

3 A ball is kicked from the floor of a sports hall at 40° to the horizontal and at a speed of V m s^{-1}. The ceiling of the hall is 10 m above the floor.

Take $g = 9.81$ m s^{-2}

   a Work out in terms of V the maximum height to which the ball rises.

   b Use your answer from part **a** to work out the greatest value of V for which the ball does not hit the ceiling.

   c For this value of V work out the distance the ball travels before hitting the floor.

4 A man standing outside a football ground can see the ball if it rises above a height of 10 m. A ball is kicked from ground level at an angle of 50° to the horizontal and with speed 25 m s^{-1}. For how long is it visible?

5 A tower of height h m stands on horizontal ground. An arrow is fired from the top of the tower with speed 40 m s^{-1} at an angle of

25° above the horizontal. It hits the ground 170 m from the base of the tower. Calculate the value of h

6 A projectile has a range of 300 m, which takes it 8 s. Taking $g = 9.81$ m s^{-2}, calculate

   a Its initial speed and direction,

   b Its speed and direction after 2 s

   c The length of time for which it was at a height greater than 40 m.

7 Show that a projectile fired with initial speed u at an angle of elevation α

reaches a maximum height of $\dfrac{u^2 \sin^2 \alpha}{2g}$

Challenge

8 A tennis player serves the ball. At the moment that the racket strikes the ball, it is 11.6 m from the net and 3.2 m above the ground. The racket imparts a horizontal speed u m s^{-1} and a downward vertical speed v m s^{-1}. The ball hits the ground 6.4 m beyond the net after 0.3 s.

   a Calculate the values of u and v

   b By how much does the ball clear the net, which is 0.91 m high?

9 A projectile is fired with initial speed u at an angle of elevation α

   a Show that the equation of its path is $y = x \tan \alpha - \left(\dfrac{g \sec^2 \alpha}{2u^2} \right) x^2$

   b The projectile passes through the points (a, b) and (b, a). Show that its range is $\dfrac{a^2 + ab + b^2}{a + b}$

MECH

Fluency and skills

When motion is in two dimensions it is often necessary to resolve forces into components.

A 40 N force acts horizontally to the right and a 30 N force acts vertically upwards. Using Pythagoras' theorem and trigonometry, you can find a single force which is equivalent to these two, known as the **resultant**, R

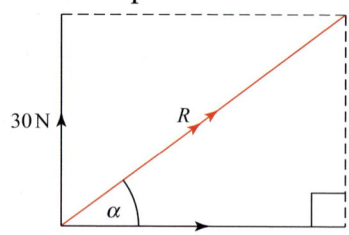

$$R = \sqrt{30^2 + 40^2} = 50\,\text{N} \qquad \alpha = \tan^{-1}\!\left(\frac{30}{40}\right) = 36.9°\ (\text{to 1 dp})$$

So the resultant of the two forces is a single force of 50 N acting at an angle of 36.9° above the horizontal.

You will often need to reverse this process. You can use trigonometry to resolve the resultant 50 N force into its horizontal and vertical **components** X and Y

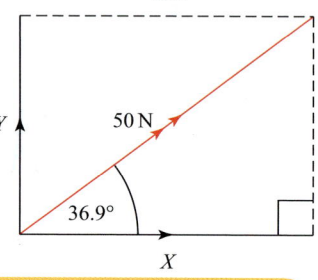

$X = 50 \cos 36.9° = 40\,\text{N}$ (to 2 sf)
$Y = 50 \sin 36.9° = 30\,\text{N}$ (to 2 sf)

Example 1

The diagram shows three forces acting on an object at a point O. The mass of the object is 4 kg

The effect of the forces accelerates the object from rest in the direction $\overrightarrow{OA}$. Work out

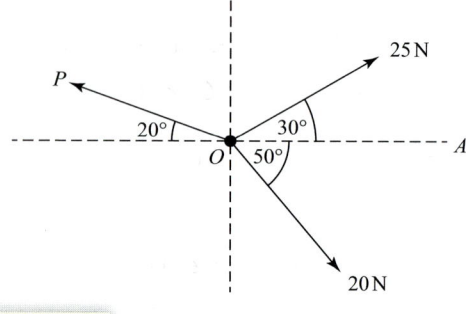

a The magnitude of the unknown force P

b The magnitude of the acceleration,

c How far the object travels in the first five seconds.

a Resolve perpendicular to OA

$P \sin 20° + 25 \sin 30° - 20 \sin 50° = 0$

$P = \dfrac{20 \sin 50° - 25 \sin 30°}{\sin 20°} = 8.25$

P has magnitude 8.25 N

> There is no acceleration in the vertical direction so you can set this equation equal to zero.

b Let the magnitude of the acceleration be a

$25 \cos 30° + 20 \cos 50° - 8.25 \cos 20° = 4a$

$a = \dfrac{25 \cos 30° + 20 \cos 50° - 8.25 \cos 20°}{4} = 6.69$

The acceleration is 6.69 m s⁻²

> Resolve along OA

> Use $F = ma$

c $s = \dfrac{1}{2} \times 6.69 \times 25 = 83.6$ The object travels 83.6 m

> Use $s = ut + \dfrac{1}{2}at^2$. The object starts from rest so take $u = 0$

You can also express your solution in terms of vectors.

Example 2

A block of mass 10 kg rests on a smooth horizontal plane. Forces of 50 N and 20 N, acting respectively at 20° and 40° to the horizontal, pull the block in opposite directions on the plane. Work out

a The acceleration of the block,

b The normal reaction between the plane and the block.

The forces acting are shown in the diagram.

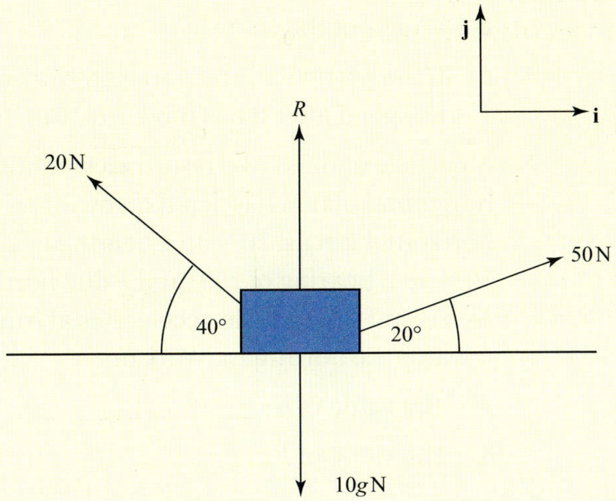

Take **i** and **j** directions as shown.

Sketch a diagram.

$$\begin{pmatrix} 50\cos 20° \\ 50\sin 20° \end{pmatrix} + \begin{pmatrix} -20\cos 40° \\ 20\sin 40° \end{pmatrix} + \begin{pmatrix} 0 \\ R \end{pmatrix} + \begin{pmatrix} 0 \\ -10g \end{pmatrix} = 10\begin{pmatrix} a \\ 0 \end{pmatrix}$$

Use **F** = m**a**. The block accelerates in the **i**-direction.

a $50\cos 20° - 20\cos 40° = 10a$

Comparing x-components.

$a = 3.17$

The block accelerates horizontally at $3.17\,\text{m s}^{-2}$

b $50\sin 20° + 20\sin 40° + R - 10g = 0$

Comparing y-components.

$R = 68.0\,\text{N}$

MECH

Exercise 18.4A Fluency and skills

1 A particle of mass 2 kg moves under the action of three forces: $(9\mathbf{i} + 4\mathbf{j})$ N, $(-2\mathbf{i} + 5\mathbf{j})$ N and $(5\mathbf{i} - \mathbf{j})$ N. Showing your working, work out

    a The magnitude of its acceleration,

    b The angle between its acceleration vector and the **i**-direction.

2 A particle of mass 1 kg moves under the action of three forces: $5\mathbf{i}$ N, $(2\mathbf{i} - 3\mathbf{j})$ N and $(-3\mathbf{i} + 5\mathbf{j})$ N. Showing your working, work out

    a The magnitude of its acceleration,

    b The angle between its acceleration vector and the **i**-direction.

3 A particle of mass 3 kg moves horizontally to the left under the action of its weight and two forces, P and Q, as shown in the diagram.

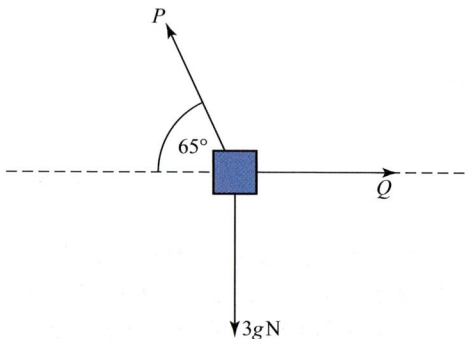

If the acceleration of the particle is $2\,\mathrm{m\,s^{-2}}$, work out the values of P and Q.
Take $g = 9.81\,\mathrm{m\,s^{-2}}$

4 A particle of mass 5 kg rests on a smooth horizontal plane. It is acted upon by forces of 20 N, 12 N and P on bearings 035°, 160° and 295° respectively. As a result the particle accelerates from rest in a northerly direction. Work out

a The value of P

b How far the particle moves in the first three seconds.

5 The diagram shows a particle of mass 8 kg which is accelerating horizontally at $a\,\mathrm{m\,s^{-2}}$

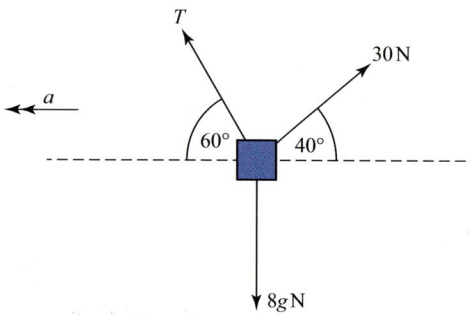

Taking $g = 9.81\,\mathrm{m\,s^{-2}}$, work out

a The force T

b The value of a

6 The diagram shows a particle of mass 4 kg moving on a horizontal plane under the action of three forces. It has acceleration $a\,\mathrm{m\,s^{-2}}$ in the direction shown.

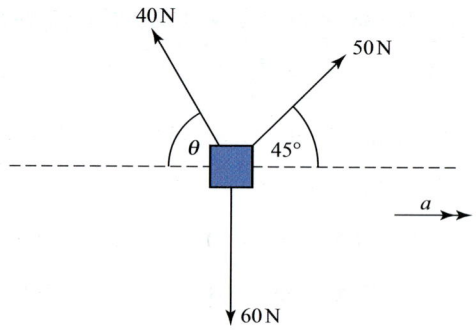

a Work out the value of θ

b Work out the value of a

c If the particle starts from rest, work out its speed after it has travelled 10 m

7 A particle of mass 4 kg rests on a smooth horizontal plane. It is acted on by three horizontal forces: 20 N on a bearing of 120°, 10 N on a bearing of 250° and P due north. As a result the particle accelerates at $a\,\mathrm{m\,s^{-2}}$ towards the north-east. Work out

a The value of a

b The value of P

8 A particle of mass 10 kg is at rest on a smooth horizontal plane. It is acted upon by three horizontal forces: 60 N on a bearing of 065°, 45 N on a bearing of 320° and 22 N on a bearing of 200°. Work out its displacement after 6 s. Show all your working.

9 A body of mass 5 kg lies on a smooth horizontal plane. Forces of 60 N, 40 N and 50 N act in directions 030°, 120° and 200° respectively. Work out the magnitude and direction of the acceleration of the body. Show all your working.

10 A particle of mass 2 kg lies on a smooth horizontal plane. It is acted upon by a force of 20 N on a bearing of 040°, another force of 40 N on a bearing of 300° and a third force P on a bearing of 160°. As a result the particle accelerates in a westerly direction. Work out

a The magnitude of P

b The magnitude of its acceleration.

Reasoning and problem-solving

Strategy

To solve problems involving motion under forces

1. Draw a diagram showing all the forces in the problem.
2. Resolve each force into its two components and use $\mathbf{F} = m\mathbf{a}$
3. Interpret your solutions in the context of the question.

Example 3

A wedge is modelled as a prism of mass 10 kg, the cross-section of which is an isosceles right-angled triangle. It is placed with its largest face on a rough horizontal surface. A force of 60 N is applied to the wedge, perpendicular to one of its sloping faces. The normal reaction of the surface on the wedge is R and the motion is resisted by a frictional force of $0.2R$. Work out the acceleration of the wedge.

Take $g = 9.81\text{ m s}^{-2}$

Let the acceleration be a horizontally.

The forces are shown in the diagram.

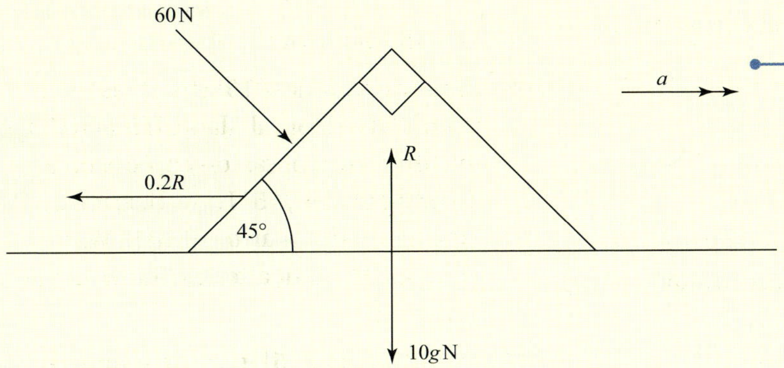

60 N
a
R
0.2R
45°
10gN

Draw a diagram. (1)

$$\begin{pmatrix} 60\cos 45° \\ -60\sin 45° \end{pmatrix} + \begin{pmatrix} 0 \\ R \end{pmatrix} + \begin{pmatrix} -0.2R \\ 0 \end{pmatrix} + \begin{pmatrix} 0 \\ -10g \end{pmatrix} = 10\begin{pmatrix} a \\ 0 \end{pmatrix}$$

Resolve each force. (2)

Equating y-components $\quad -60\sin 45° + R - 10g = 0$

$$R = 140.5$$

Equating x-components $\quad 60\cos 45° - 0.2R = 10a$

$$a = 1.43$$

The acceleration is 1.43 m s^{-2}

The question asks for the acceleration of the wedge. (3)

MECH

MyMaths 2192 SEARCH

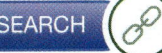

Example 4

A block of mass 5 kg rests on a rough horizontal plane. The block then accelerates uniformly from rest, pulled by a string inclined at 60° to the horizontal. The normal reaction of the plane on the block is R and the motion is resisted by a frictional force of $0.4R$. After 9 m, the block is travelling at $6\,\mathrm{m\,s^{-1}}$. Work out the tension in the string.

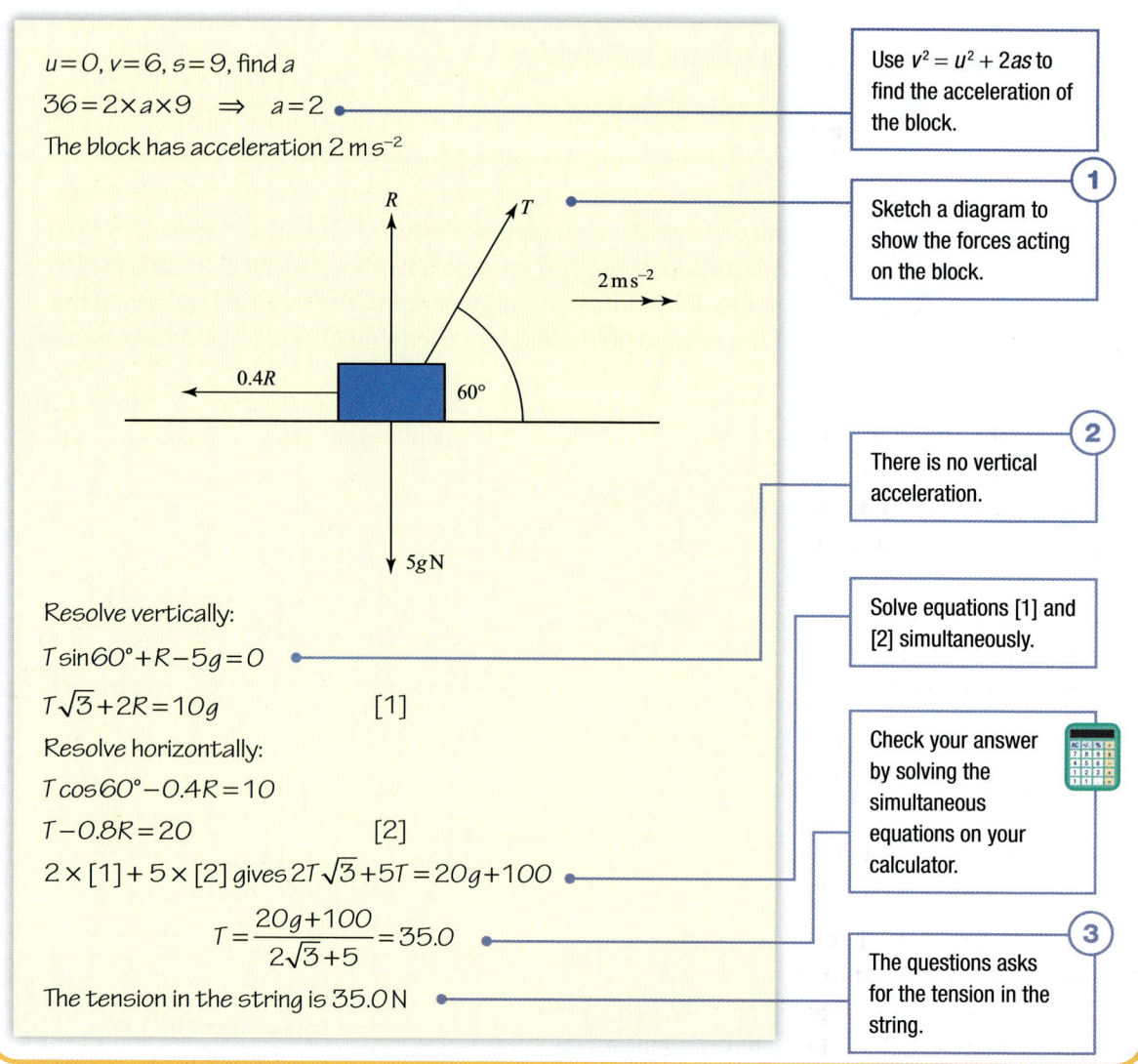

$u = 0$, $v = 6$, $s = 9$, find a

$36 = 2 \times a \times 9 \implies a = 2$

The block has acceleration $2\,\mathrm{m\,s^{-2}}$

Use $v^2 = u^2 + 2as$ to find the acceleration of the block.

Sketch a diagram to show the forces acting on the block. ①

There is no vertical acceleration. ②

Resolve vertically:

$T \sin 60° + R - 5g = 0$

$T\sqrt{3} + 2R = 10g$ [1]

Resolve horizontally:

$T \cos 60° - 0.4R = 10$

$T - 0.8R = 20$ [2]

$2 \times [1] + 5 \times [2]$ gives $2T\sqrt{3} + 5T = 20g + 100$

$$T = \frac{20g + 100}{2\sqrt{3} + 5} = 35.0$$

The tension in the string is 35.0 N

Solve equations [1] and [2] simultaneously.

Check your answer by solving the simultaneous equations on your calculator.

The questions asks for the tension in the string. ③

Exercise 18.4B Reasoning and problem-solving

1 A particle of mass 5 kg is pushed up a smooth slope inclined at 30° to the horizontal by a horizontal force of 40 N. Taking $g = 9.81\,\mathrm{m\,s^{-2}}$, work out

  a The reaction between the particle and the plane,

  b The acceleration of the particle.

2 A rectangular block of mass 2 kg rests on a rough horizontal plane. The block is pulled by a string inclined at 30° to the horizontal. The normal reaction of the surface on the block is R and the motion is resisted by a frictional force of $0.2R$. If the tension in the string is 20 N, work out the acceleration of the block.

3 The diagram shows a crate of mass 600 kg suspended from a helicopter by two cables, each inclined at 60° to the horizontal. The crate is being accelerated horizontally at $0.5\,\mathrm{m\,s^{-2}}$ against air resistance of 400 N. Taking $g = 9.81\,\mathrm{m\,s^{-2}}$, work out T_1 and T_2, the tensions in the two cables.

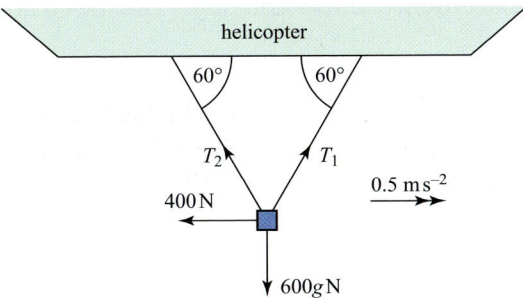

4 A block of mass m rests on a rough horizontal surface. The block is accelerated from rest, pulled by a string inclined at an angle θ to the horizontal. The normal reaction of the surface on the block is R and the motion is resisted by a frictional force of $0.75R$. The tension in the string equals the weight of the block.

  a Show that the acceleration of the block is
  $$a = \frac{g(3\sin\theta + 4\cos\theta - 3)}{4}$$

  b Hence show that for maximum acceleration $\theta = 36.9°$

5 A "kite buggy" is a wheeled cart pulled along by the force of the wind acting on a kite attached to it. A buggy has a mass of 100 kg (including the driver). It is travelling down a 10° slope towed by a kite whose tether makes an angle of 40° with the horizontal. The tension in the tether is 250 N and resistance forces total 50 N. Taking $g = 10\,\mathrm{m\,s^{-2}}$, work out the acceleration of the buggy.

6 A ship of mass 800 tonnes is towed from rest by two tugs. The first exerts a force of 50 kN on a bearing of 020°, the second a force of 70 kN on a bearing of 100°. The resistance of the water is a force of magnitude 20 kN. Work out the distance travelled by the ship in the first 20 seconds.

Challenge

7 A rectangular block of mass 2 kg is at rest at a point A on a rough slope inclined at 30° to the horizontal. It is accelerated up the line of greatest slope, pulled by a string inclined at 20° to the slope. The normal reaction of the surface on the block is R and the motion is resisted by a frictional force of $0.6R$. The tension in the string is 30 N. After 2 s the string breaks and the block slows to rest at B. Work out the distance AB.

8 A cable car and its occupant have a mass of 100 kg. The cradle car is suspended from a trolley and travelling on a cable inclined at 10° to the horizontal. The cradle is supported by two ropes, AC and BC attached at A and B to the trolley as shown, where $AB = 4\,\mathrm{m}$. The ropes have lengths $AC = 4\,\mathrm{m}$ and $BC = 5\,\mathrm{m}$. The car is accelerating up the cable at $0.2\,\mathrm{m\,s^{-2}}$. Work out the tensions, T_1 and T_2, in the ropes. Take $g = 9.81\,\mathrm{m\,s^{-2}}$

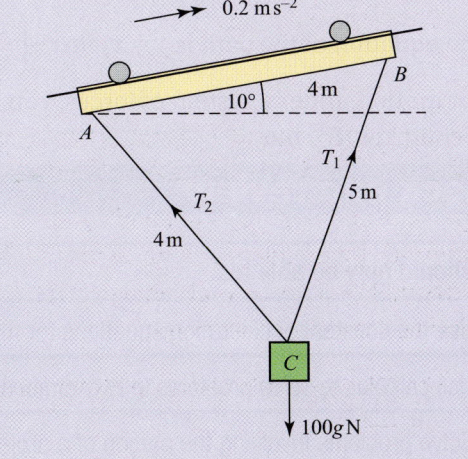

Chapter summary

- For motion in two dimensions the position $\mathbf{r}$, displacement $\mathbf{s}$, initial and final velocities $\mathbf{u}$ and $\mathbf{v}$ and acceleration $\mathbf{a}$ are two-dimensional vectors.

- For constant acceleration: $\mathbf{v} = \mathbf{u} + \mathbf{a}t$ $\mathbf{s} = \frac{1}{2}(\mathbf{u} + \mathbf{v})t$ $\mathbf{s} = \mathbf{u}t + \frac{1}{2}\mathbf{a}t^2$ $\mathbf{s} = \mathbf{v}t - \frac{1}{2}\mathbf{a}t^2$

- You can use the equation, $v^2 = u^2 + 2as$, in component form, that is
$$v_x^{\ 2} = u_x^{\ 2} + 2a_x x \qquad \text{and} \qquad v_y^{\ 2} = u_y^{\ 2} + 2a_y y$$

- For variable acceleration $\mathbf{v} = \dfrac{d\mathbf{s}}{dt}$ and $\mathbf{a} = \dfrac{d\mathbf{v}}{dt} = \dfrac{d^2\mathbf{s}}{dt^2}$

- Differentiate a vector by differentiating components, so if $\mathbf{s} = x\mathbf{i} + y\mathbf{j}$ then $\mathbf{v} = \dfrac{dx}{dt}\mathbf{i} + \dfrac{dy}{dt}\mathbf{j}$, also written as $\mathbf{v} = \dot{x}\mathbf{i} + \dot{y}\mathbf{j}$, and $\mathbf{a} = \dfrac{d^2x}{dt^2}\mathbf{i} + \dfrac{d^2y}{dt^2}\mathbf{j}$, also written as $\mathbf{a} = \ddot{x}\mathbf{i} + \ddot{y}\mathbf{j}$

  You can express these relationships in terms of integration: $\mathbf{v} = \int \mathbf{a}\ dt$ and $\mathbf{s} = \int \mathbf{v}\ dt$

- For a projectile fired from a point on a horizontal plane with speed u and elevation α the initial velocity is $\mathbf{u} = u\cos\alpha\,\mathbf{i} + u\sin\alpha\,\mathbf{j}$ and acceleration is $\mathbf{a} = -g\mathbf{j}$. The velocity and position at time t are $\mathbf{v} = u\cos\alpha\,\mathbf{i} + (u\sin\alpha - gt)\mathbf{j}$ and $\mathbf{r} = ut\cos\alpha\,\mathbf{i} + \left(ut\sin\alpha - \frac{1}{2}gt^2\right)\mathbf{j}$

- You must be able to derive the following formulae:
$$\text{Time of flight} = \frac{2u\sin\alpha}{g} \qquad \text{Range} = \frac{u^2\sin 2\alpha}{g} \qquad \text{Maximum height} = \frac{u^2\sin^2\alpha}{2g}$$
$$y = x\tan\alpha - \left(\frac{g\sec^2\alpha}{2u^2}\right)x^2$$

- Elevations of α and $(90° - \alpha)$ give the same range.

- The maximum range $= \dfrac{u^2}{g}$ when $\alpha = 45°$

- The equation of the path is $y = x\tan\alpha - \left(\dfrac{g\sec^2\alpha}{2u^2}\right)x^2$

- For motion under a system of forces, you resolve forces into components and apply Newton's second law, $\mathbf{F} = m\mathbf{a}$

Check and review

You should now be able to...	Try Questions
✔ Use the constant acceleration equations for motion in two dimensions.	1, 2
✔ Use calculus to solve problems in two-dimensional motion with variable acceleration.	3, 4, 5
✔ Solve problems involving the motion of a projectile under gravity.	6, 7
✔ Analyse the motion of an object in two dimensions under the action of a system of forces.	8

1 For each of these situations choose a suitable equation to work out the required value. Show all your working.

    a $\mathbf{u} = (3\mathbf{i} - 2\mathbf{j})$, $\mathbf{v} = (7\mathbf{i} + 6\mathbf{j})$, $t = 5$, work out $\mathbf{s}$

    b $\mathbf{u} = (2\mathbf{i} - 3\mathbf{j})$, $\mathbf{a} = (\mathbf{i} - 2\mathbf{j})$, $t = 6$, work out $\mathbf{s}$

    c $\mathbf{u} = (12\mathbf{i} + 4\mathbf{j})$, $\mathbf{v} = (2\mathbf{i} + 19\mathbf{j})$, $t = 5$, work out $\mathbf{a}$

    d $\mathbf{s} = (-8\mathbf{i} + 48\mathbf{j})$, $\mathbf{a} = (\mathbf{i} - 2\mathbf{j})$, $t = 10$, work out $\mathbf{v}$

2 A particle is initially at the point with position vector $\mathbf{r} = (5\mathbf{i} - \mathbf{j})$ m and travelling with velocity $(8\mathbf{i} - 3\mathbf{j})$ m s^{-1} when it undergoes an acceleration of $(-2\mathbf{i} + 7\mathbf{j})$ m s^{-2} for a period of 4 s. Work out its position at the end of this period.

3 A particle has position at time t given by $\mathbf{r} = (3t^2 - 2t)\mathbf{i} + (2t^3 - t^2)\mathbf{j}$. Work out its velocity and acceleration when $t = 2$. Show all your working.

4 A particle has velocity at time t given by $\mathbf{v} = t(4 - 9t)\mathbf{i} + t^2(6 - 2t)\mathbf{j}$. Its initial position is $\mathbf{r} = 3\mathbf{i} + 2\mathbf{j}$. Work out its position when $t = 2$

5 A particle has acceleration at time t s given by $\mathbf{a} = 4\sin 4t\,\mathbf{i} + 4\cos 2t\,\mathbf{j}$. At time $t = \dfrac{\pi}{4}$ s it has velocity $\mathbf{v} = \mathbf{i} + 2\mathbf{j}$ and position $\mathbf{r} = 3\mathbf{i} + \mathbf{j}$. Show that at time $t = \dfrac{\pi}{2}$ s its distance from the origin is $\sqrt{13}$

6 A projectile is launched from a point O on horizontal ground with speed 12 m s^{-1} at an angle of 25° to the horizontal. Take $g = 9.8$ m s^{-2}

    a   For how long is the projectile in the air?

    b   What is the horizontal range of the projectile?

    c   Work out the time taken for it to reach maximum height.

    d   What is the greatest height reached by the projectile?

7 A projectile is fired at 30 m s^{-1} from a point on horizontal ground. It is directed so that its range is a maximum.

Taking $g = 9.81$ m s^{-2}, work out

    a   Its range,

    b   The equation of its path.

8 A particle of mass 4 kg rests on a smooth horizontal plane. It is acted upon by forces of 30 N, 18 N and P on bearings 030°, 150° and 305° respectively. As a result the particle accelerates in a northerly direction. Work out

    a   The value of P

    b   How far the particle moves in the first four seconds.

MECH

What next?

Score		
0 – 4	Your knowledge of this topic is still developing. To improve, search in MyMaths for the codes: 2192, 2198, 2199, 2290, 2291	
5 – 6	You're gaining a secure knowledge of this topic. To improve, look at the InvisiPen videos for Fluency and skills (18A)	
7 – 8	You've mastered these skills. Well done, you're ready to progress! To develop your exam techniques, look at the InvisiPen videos for Reasoning and problem-solving (18B)	

Click these links in the digital book

Investigation

The diagram shows a particle, P, moving with constant speed in a circular path of radius r and centre O

Does P have acceleration?

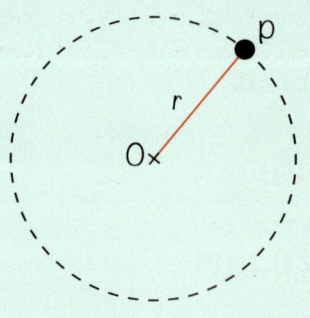

Note

Speed is a scalar quantity – it has magnitude but no direction.

Velocity is a vector quantity – it has both magnitude and direction.

Information

Vectors are a useful tool that can help to describe the motion of P

The first step is to find an expression for the position vector, **r**, of P, relative to O, in terms of the unit vectors **i** and **j**

$$\mathbf{r} = r\cos\theta\,\mathbf{i} + r\sin\theta\,\mathbf{j}$$

Differentiating with respect to time will give us the velocity vector, **v**, of P

$$\mathbf{v} = -r\sin\theta\frac{d\theta}{dt}\,\mathbf{i} + r\cos\theta\frac{d\theta}{dt}\,\mathbf{j}$$

$\frac{d\theta}{dt}$ is the angular velocity, which is constant.

Writing $\omega = \frac{d\theta}{dt}$ gives the velocity vector as

$$\mathbf{v} = -\omega r\sin\theta\,\mathbf{i} + \omega r\cos\theta\,\mathbf{j}$$

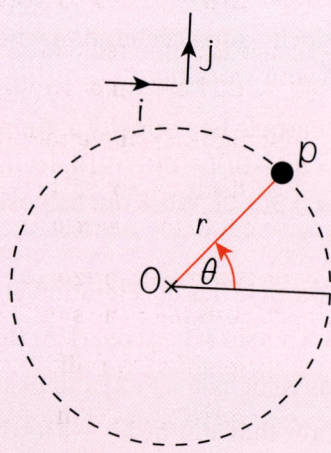

Note

The **chain rule** has been used to differentiate r as θ is also a function of t

Challenge

Differentiate the expression above found for **v** and write the acceleration, **a**, in terms of ω and r

Interpret the result to give the magnitude and direction of the acceleration of P What is the name given to this type of acceleration?

If P has mass m kg, what force is required to maintain the motion?

In questions that tell you to show your working, you shouldn't depend solely on a calculator. For these questions, solutions based entirely on graphical or numerical methods are not acceptable.

1   A particle moves with an initial velocity of $(14\mathbf{i} - 4\mathbf{j})$ m s^{-1} and a constant acceleration of $(-\mathbf{i} + 7\mathbf{j})$ m s^{-2}

     a   Work out its velocity at time $t = 4$ **[3 marks]**

     b   Work out its speed at this time.   **[2]**

2   A box of mass 6 kg moves up a smooth plane, at constant velocity, under action of a horizontal force of magnitude F Newtons. The plane is inclined at an angle of 30° to the horizontal. Work out the value of F **[5]**

3   An object moving with constant acceleration has an initial velocity of $(-2\mathbf{i} + 5\mathbf{j})$ m s^{-1}. Six seconds later it has a velocity of $(4\mathbf{i} - 7\mathbf{j})$ m s^{-1}. Work out

     a   Its acceleration,   **[3]**

     b   The distance of the object from its starting position after these six seconds.   **[4]**

4   A particle is projected with a velocity of 58.8 m s^{-1} at an angle of 30° to the horizontal from a point, X, on horizontal ground. Taking $g = 9.81$ m s^{-2}, work out

     a   The highest point above the ground of the path of the particle,   **[4]**

     b   The distance from X to the point where the particle lands, giving your answer correct to the nearest metre.   **[5]**

5   A stone is thrown horizontally with a velocity of 20 m s^{-1} from the top of a vertical cliff. Five seconds later the stone reaches the sea. Work out

     a   The height of the cliff,   **[3]**

     b   The horizontal distance from the base of the cliff to the point where the stone lands,   **[2]**

     c   The direction in which the stone is travelling at the instant before it hits the sea.   **[4]**

6   Referred to a fixed origin, the position vector, $\mathbf{r}$ metres, of a particle, P, at a time t seconds is given by $\mathbf{r} = (\sin 2t)\mathbf{i} + (\cos 4t)\mathbf{j}$, where $t \geq 0$. At the instant when $t = \dfrac{\pi}{6}$, work out

     a   The velocity of P, showing your working,   **[4]**

     b   The acceleration of P, showing your working.   **[4]**

7   A particle moves through the origin with constant acceleration. It has an initial velocity of $(2\mathbf{i} - 3\mathbf{j})$ m s^{-1}. After six seconds it is moving through the point with position vector $(-6\mathbf{i} + 18\mathbf{j})$ m. Work out

     a   Its velocity at $t = 6$ **[3]**

     b   Its acceleration,   **[2]**

     c   The time at which the particle is moving parallel to the x-axis.   **[2]**

8   A parcel, P, of mass 5 kg, hangs from the horizontal roof of a lift, supported by two light inextensible strings, AP and BP. The lift is accelerating upwards at $\frac{1}{5}g \, \text{m s}^{-2}$. Work out

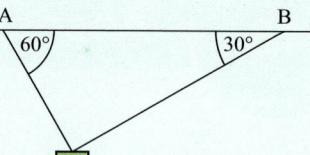

   a   The tension in BP, showing all your working,   [7]

   b   The tension in AP   [1]

9   A particle, P, of mass 0.4 kg moves under the action of a single force,

    F Newtons. At time t, the velocity, **v** m s^{-1} of P is given by $\mathbf{v} = \begin{pmatrix} 3t^2 + 5 \\ 14t - 2 \end{pmatrix} \text{m s}^{-1}$
    At time $t = 0$, P is at the origin. Work out

   a   **F** when $t = 2$. Show all your working.   [5]

   b   The distance of P from the origin when $t = 2$ [7]

10 A ball is hit from a point that is one metre above horizontal ground, with a velocity of 20 m s^{-1} at an angle of elevation of α where $\tan \alpha = \frac{4}{3}$. The ball just clears a vertical wall, which is 12 metres horizontally from the point where the ball was hit.
    Work out (to 3 sf)

   a   The height of the wall,   [7]

   b   The speed of the ball at the instant when it passes over the wall,   [4]

   c   The direction in which the ball is travelling at the instant when it passes over the wall.   [2]

11 At time $t = 0$, a particle, P, is at rest at the point $(2, 0)$. At time t seconds, its acceleration, **a** m s^{-2}
    is given by $\mathbf{a} = \begin{pmatrix} 16 \cos 4t \\ \sin t - 2 \sin 2t \end{pmatrix}$. Work out

   a   The acceleration of P when $t = \frac{\pi}{2}$ [2]

   b   The velocity of P when $t = \frac{\pi}{4}$ [7]

   c   The position of P when $t = \pi$ [7]

12 Two boats, P and Q, are travelling with constant velocities $(3\mathbf{i} - 8\mathbf{j}) \, \text{km h}^{-1}$ and $(-7\mathbf{i} + 12\mathbf{j}) \, \text{km h}^{-1}$ respectively, relative to a fixed origin O. At noon, the position vectors of P and Q are $(4\mathbf{i} + 11\mathbf{j})$ km and $(9\mathbf{i} + 3.5\mathbf{j})$ km respectively. At time t hours after noon, the position vectors of P and Q, relative to O, are $\mathbf{S}_P$ and $\mathbf{S}_Q$. Write

   a   An expression in terms of t for $\mathbf{S}_P$ [2]

   b   An expression in terms of t for $\mathbf{S}_Q$ [2]

    At a time, t hours after noon, the distance between the boats is given by d km

   c   Prove that $d^2 = (-5 + 10t)^2 + (7.5 - 20t)^2$ [4]

   d   Work out the time at which the boats are closest together. Show all your working.   [5]

   e   Work out the distance between the boats at the time when they are closest together.   [2]

13 A particle is projected from a point O, with an initial velocity of u m s^{-1}, at an angle of α to the horizontal. In the vertical plane of projection, taking x and y as the horizontal and vertical axes respectively

   a   Show that $y = x \tan \alpha - \frac{gx^2}{2u^2} \sec^2 \alpha$ [5]

    Given that $u = 42$ and that the particle passes through the point $(60, 70)$

   b   Find the two possible angles of projection. Take $g = 9.8 \, \text{m s}^{-2}$ and show all your working.   [6]

Forces 2

Using a wrench to loosen a bolt is a simple action, but also involves the application of a particular type of force: the *turning force*, also called the *moment*. A turning force is the effect of a force around a fixed point (a pivot). Pushing a door open, removing a bottle cap, steering a car, even bending your leg – these are all examples of turning forces.

This chapter introduces simple systems of forces: static equilibria, motion due to a force and turning forces. Objects are subject to forces all the time and an understanding of how these forces work and how they can be modelled is therefore vital in areas such as physics and engineering.

Orientation

What you need to know	What you will learn	What this leads to
Ch6 Vectors p.158 • Resolving vectors into components.	• To understand that there is a maximum value that the frictional force can take when the object is moving or on the point of moving. • To find unknown forces when a system is at rest or has constant acceleration. • To use constant acceleration formulae. • To solve differential equations arising from $F = ma$ • To take moments about points and resolve to find unknown forces. • To solve ladder problems.	**Applications** Civil engineering. Structural engineering. Mechanical engineering. Nuclear physics. Aeronautical engineering. Rocket science.
Ch7 Units and kinematics p.186 • The constant acceleration formulae. p.190 • Equations of motion for variable acceleration.		
Ch8 Forces and Newton's laws p.200 • Equilibrium requires the resultant force to be zero. • Newton's laws of motion.		

 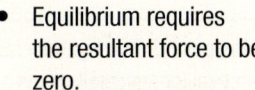 Practise before you start

🔍 2184, 2186–2188, 2207, 2289

Fluency and skills

Statics is the study of forces in equilibrium. In the previous chapter, you learnt to resolve forces in motion. However, if an object is in equilibrium, then the resultant force acting on it is zero.

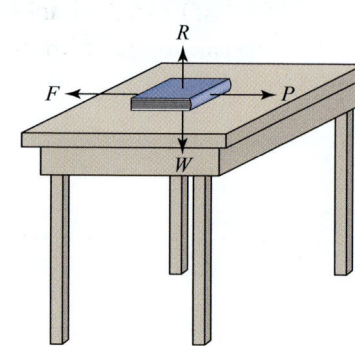

Imagine a book resting in equilibrium on a table. If you apply a horizontal force, P, to the book and it doesn't move, then friction F is acting to oppose P and $F = P$

As you increase P and the book still does not move, F is increasing in line with P, but only up to the point when the book is on the point of moving. At this point, F has its maximum value which is called **limiting friction**. If P increases further, the book moves.

If you repeat this with different books, you'll find the limiting friction changes—it's proportional to the normal reaction, R. So, limiting friction $F = \mu \times R$, where μ is the **coefficient of friction**. The rougher the surface, the greater the value of μ. However, if the book is not on the point of moving, then $F < \mu R$

See Ch8.1

For a reminder on equilibrium.

> **Key point**
>
> Friction acts to oppose any motion.
>
> Friction, F, and normal reaction, R are related by $F \leq \mu R$, where μ is the coefficient of friction.
>
> When the object is about to move, it is in **limiting equilibrium** and limiting friction $F = \mu R$

> Remember to convert the mass to a weight.

Example 1

A crate of mass 50 kg lies on a rough horizontal surface. A string is attached to the crate and is pulled at 20° above the horizontal until the crate is just about to slip. At this point, the tension in the string is 200 N. Find the normal reaction, R, and the coefficient of friction, μ.

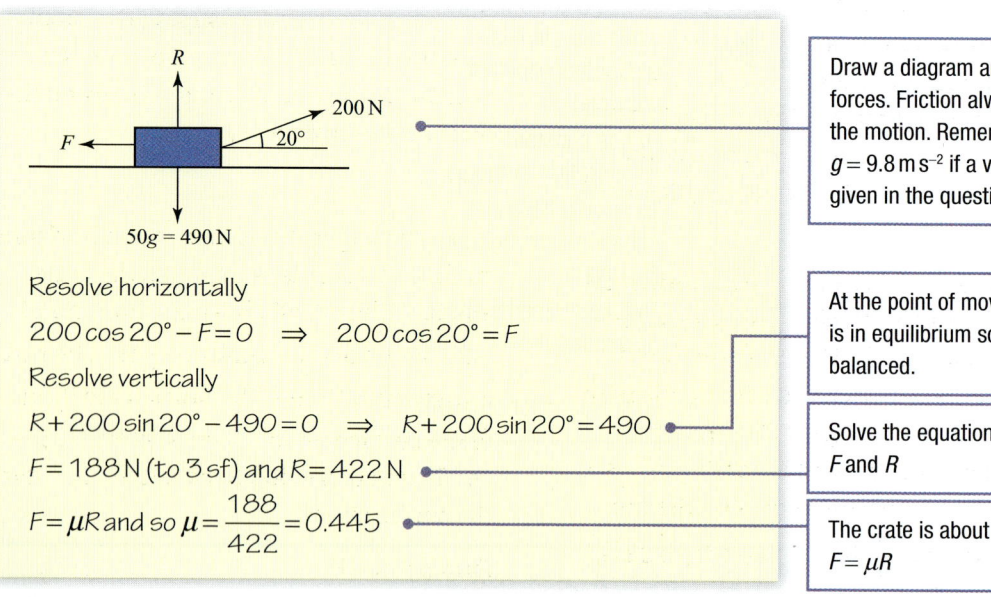

Resolve horizontally

$200 \cos 20° - F = 0 \quad \Rightarrow \quad 200 \cos 20° = F$

Resolve vertically

$R + 200 \sin 20° - 490 = 0 \quad \Rightarrow \quad R + 200 \sin 20° = 490$

$F = 188\,\text{N} \text{ (to 3 sf) and } R = 422\,\text{N}$

$F = \mu R \text{ and so } \mu = \dfrac{188}{422} = 0.445$

Draw a diagram and label all the forces. Friction always opposes the motion. Remember to take $g = 9.8\,\text{m s}^{-2}$ if a value isn't given in the question.

At the point of moving, the crate is in equilibrium so all forces are balanced.

Solve the equations to calculate F and R

The crate is about to slip, so $F = \mu R$

1 For each part **a** and **b**, calculate

   i The horizontal component, X, and the vertical component, Y, (to 3 sf) of the resultant force,

   ii The magnitude of the resultant force (to 2 sf) and the angle it makes with the upward vertical (to the nearest degree).

a
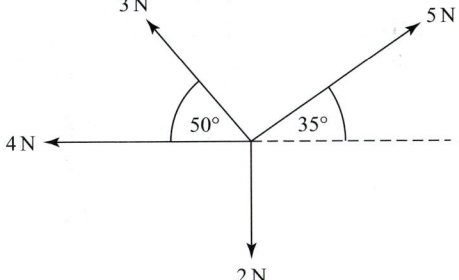

b
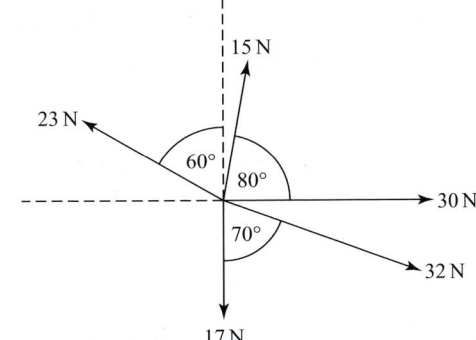

2 A crate of weight 800 N lies on a rough horizontal surface. A string is attached to the crate and is pulled at 30° above the horizontal until the crate is just about to slip. At this point the tension in the string is 350 N. Resolve to find the normal reaction of the surface on the crate, R, and the coefficient of friction, μ

3 A box of weight W lies on a rough horizontal surface. A string is attached to the box and pulled at an angle of 45° to the horizontal until the box is just about to slip. At this point the tension in the string is 600 N and the normal reaction of the surface on the box is R. The coefficient of friction between the box and the surface is $\dfrac{3}{4}$. Find W and R (to 2 sf).

4 A boy sits on a sledge on horizontal ground. The combined weight of the boy and the sledge is 750 N. A friend pulls the rope attached to the sledge so that it makes an angle of 10° to the horizontal.

When the sledge is just about to slip, the friend is pulling with a force T. The normal reaction of the ground on the block is R and the coefficient of friction between the sledge and the ground is $\dfrac{2}{3}$

Show that $T = 450$ N (to 2 sf) and find R (to 2 sf).

5 An object is free to move in a smooth horizontal plane. Four horizontal forces act on the object at the same time. One has magnitude 25 N and is directed due south, one has magnitude 60 N and is directed south-west, one has magnitude 40 N and is directed due west and the final force has magnitude 30 N and is directed north-west.

Find the magnitude of the resultant force and the bearing on which the object moves.

6 A box of weight 400 N lies on a rough horizontal surface. A string is attached to the crate and is pulled at 40° above the horizontal with a tension of magnitude T. The box is just about to slip. The normal reaction of the surface on the crate is 300 N and the coefficient of friction between the box and the surface is μ

Find μ and T

7 A box of weight W lies on a rough horizontal surface. A string is attached to the crate and is pulled at 20° above the horizontal with a tension of magnitude $\dfrac{1}{2}W$. The box is just about to slip. The normal reaction of the surface on the crate is 30 N and the coefficient of friction between the box and the surface is μ

Find μ and W

MECH

Reasoning and problem-solving

When an object is in equilibrium under several forces, you can consider three methods for calculating unknown forces and angles

1. Resolve in two perpendicular directions,

2. Draw a vector diagram and use trigonometry,

3. Use Lami's Theorem.

You have already used the first two of these methods in Chapter 18. You can use them in any problem. Lami's Theorem can only be used when you have just three forces and you know at least one of the angles between them. Although this is a specific case, it can be much faster than using vector diagrams or resolving in perpendicular directions.

Let three forces A, B and C, act on an object so that it is in equilibrium as shown.

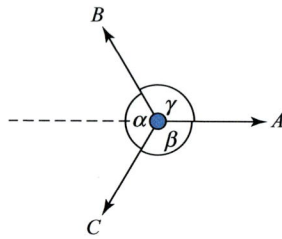

See Ch3.2
For a reminder on the sine rule

You can draw a vector diagram of the forces, nose-to-tail, to have zero resultant.

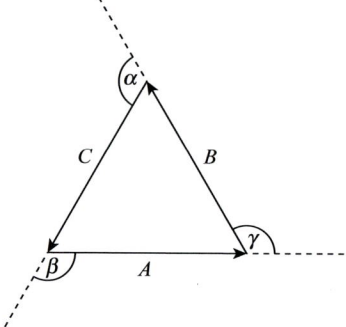

Applying the sine rule to the vector diagram gives

$$\frac{A}{\sin(180°-\alpha)} = \frac{B}{\sin(180°-\beta)} = \frac{C}{\sin(180°-\gamma)}$$

$(180° - \alpha) = \sin\alpha$, and similarly for β and γ, so

$$\frac{A}{\sin\alpha} = \frac{B}{\sin\beta} = \frac{C}{\sin\gamma}$$

> Lami's theorem isn't specifically required in this course, but it can be useful to know when solving equilibrium questions.

> **Key point**
>
> Lami's Theorem states that if there are three forces acting on an object and these forces are in equilibrium, each force is proportional to the sine of the angle between the other two forces.
>
> You can use it when there are three forces acting on an object and you know at least one of the angles between them.

Strategy

To solve problems where objects are in equilibrium

1. Draw a diagram and mark all known and unknown forces and angles.

2. Decide which of these methods to use and form equations.

3. Solve the equations to find unknown forces and angles.

Example 2

A block of weight W is suspended by two ropes. One rope makes an angle of 30° with the vertical and has a tension 60 N. The other rope makes an angle of 50° with the vertical and has a tension T

Find T and W (to 3 sf).

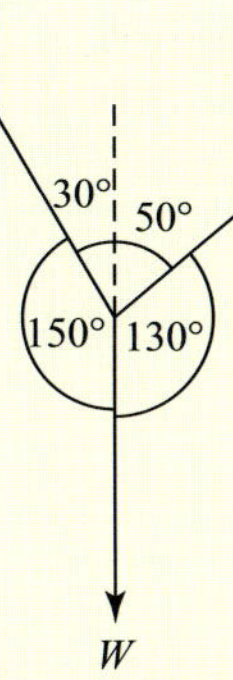

60 N

T

30°   50°

150° 130°

W

① Draw a diagram. Combine the 30° and 50° angles to give a single angle of 80° between the two ropes.

Using Lami's theorem

$$\frac{T}{\sin 150°} = \frac{W}{\sin 80°} = \frac{60}{\sin 130°}$$

② There are three forces in equilibrium and the angles are known, so use Lami's theorem.

To find T, solve $\dfrac{T}{\sin 150°} = \dfrac{60}{\sin 130°}$

$$T = \frac{60 \sin 150°}{\sin 130°} = 39.2\,\text{N (to 3 sf)}$$

To find W, solve $\dfrac{W}{\sin 80°} = \dfrac{60}{\sin 130°}$

$$W = \frac{60 \sin 80°}{\sin 130°} = 77.1\,\text{N (to 3 sf)}$$

Calculate the missing forces T and W. Give your answers to the required degree of accuracy.

Example 3

A block of weight 60 N lies on a rough slope inclined at an angle 40° to the horizontal. A string is attached to the block and passes over a smooth pulley at the top of the slope. It is attached at its other end to a block of weight W which hangs vertically. The block on the slope is about to slip *down* the slope. The normal reaction of the slope on the block is R

The coefficient of friction between the block and the slope is $\dfrac{1}{2}$

Find W (to 3 sf).

(Continued on the next page)

 MyMaths 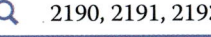 Q   2190, 2191, 2193   SEARCH

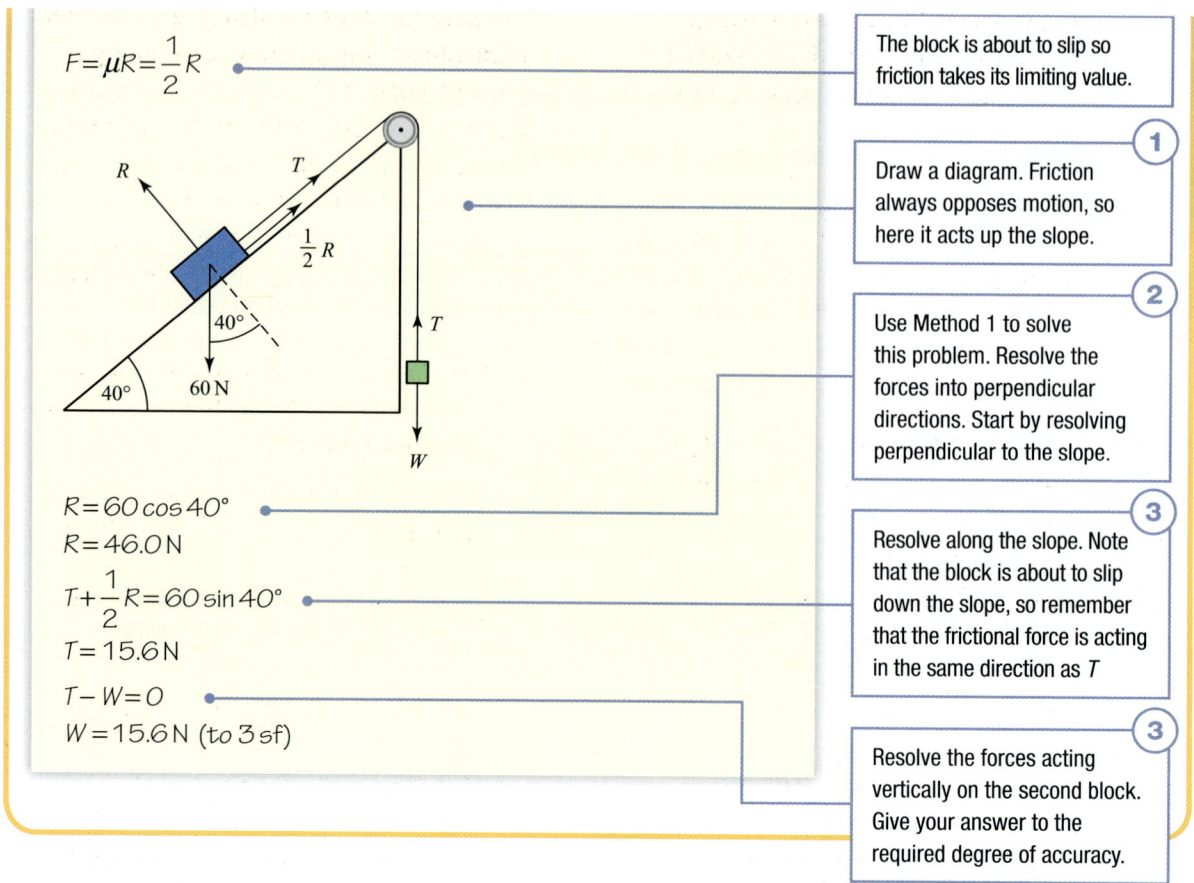

$F = \mu R = \dfrac{1}{2} R$

$R = 60 \cos 40°$

$R = 46.0\,\text{N}$

$T + \dfrac{1}{2} R = 60 \sin 40°$

$T = 15.6\,\text{N}$

$T - W = 0$

$W = 15.6\,\text{N}$ (to 3 sf)

Exercise 19.1B Reasoning and problem-solving

1 Two strings are attached to a box of weight 90 N so that the box is held in mid-air by the strings. Tension R in one string makes an angle of 30° with the upward vertical, whilst tension S in the the other string makes an angle of 20° with the upward vertical. Find R and S

2 A block of weight 120 N lies on a rough slope inclined at an angle of 25° to the horizontal. The coefficient of friction between the block and the slope is $\dfrac{3}{5}$

A string is attached to the block and is passed over a smooth pulley at the top of the slope. Its other end is attached to a block of weight W which hangs vertically. The block on the slope is about to slip *up* the slope. Find W (to 3 sf).

3 A rescue dinghy of weight 250 N is launched into water along a slope, which lies at an angle of 35° to the horizontal. The coefficient of friction between the two is 0.2. Before being launched, the dinghy is held by a light inextensible rope that is parallel to the slope, with tension T

a Find the smallest possible value of T which keeps the dinghy in equilibrium.

b Find the largest possible value of T which keeps the dinghy in equilibrium.

4 Two particles of weights 15 N and 10 N are connected by a light inextensible string which passes over a smooth pulley fixed at the end of a rough horizontal table. The 15 N weight rests on the table and the 10 N weight hangs freely below the pulley. Find the coefficient of friction between the 15 N weight and the table if the system is in limiting equilibrium.

5 A block of weight $6W$ lies on a rough slope inclined at an angle of 30° to the horizontal. The coefficient of friction between the block and the slope is $\frac{1}{3}$

A string is attached to the block and is passed over a smooth pulley at the top of the slope. Its other end is attached to a block of weight kW which hangs vertically. The block on the slope is about to slip *up*. Find k. Give your answer as an exact value using surds.

6 Three children kick a football at the same time. One kick is directed due north with magnitude 40 N. One kick is directed south-east with magnitude T and the third kick is directed due west with magnitude S. Find the magnitude of T and S if the magnitude of the resultant force is zero.

7 Two particles of weights kW and W are connected by a light inextensible string which passes over a smooth pulley fixed at the end of a rough horizontal platform. The kW weight rests on the platform and the W weight hangs freely below the pulley. Find the coefficient of friction, in terms of k, between the kW weight and the table if the system is in limiting equilibrium.

8 A block of weight W lies on a rough slope inclined at an angle of 10° to the horizontal. The coefficient of friction between the block and the slope is μ. A string is attached to the block and is passed over a pulley at the top of the slope. It is attached at its other end to a block of weight $\frac{1}{2}W$ which hangs vertically. The block on the slope is about to slip *up* the slope. Determine the value of the coefficient of friction, μ

9 A block of mass 7 kg lies on a rough horizontal table. The coefficient of friction between the block and the table is $\frac{3}{7}$

Two light inextensible strings are attached to the block. One passes over a pulley at one end of the table and the other passes over a pulley at the other end of the table. The pulleys, the strings and the block are all in a straight line. Masses of m kg and 5 kg respectively are tied to the other ends of the string. These two masses are held with the strings taut.

a Calculate the value of m that will result in the 5 kg mass being on the point of moving downwards.

b Calculate the value of m that will result in the 5 kg mass being on the point of moving upwards.

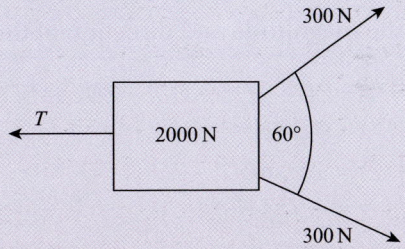

Fluency and skills

See Ch8.2

For a reminder on Newton's second law of motion.

Dynamics is the study of forces in motion. From Newton's second law of motion, you know that, if a force F N acts on an object of mass m kg giving it an acceleration a m s^{-2}, then $F = ma$

You used this equation in Chapter 18 to solve problems where the forces need to be resolved in perpendicular directions. In this chapter, you will recap resolving forces in two dimensions, and then learn to use Newton's second law to solve more complicated problems including the coefficient of friction, smooth pulleys and connected particles, and scenarios where acceleration is not constant.

ICT resource

ICT Resource online

To investigate resolving forces, click this link in the digital book.

Example 1

Three forces, of magnitudes 15 N, 20 N and 30 N, act on a particle of mass 4 kg as shown in the diagram.

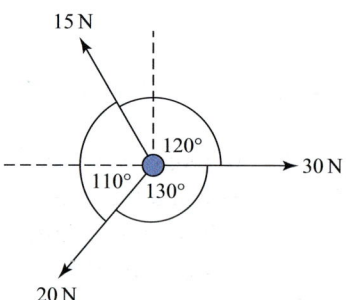

Find the magnitude and direction of the acceleration of the particle.

$(\rightarrow)$ $X = 30 - 15\cos 60° - 20\cos 50°$
  $X = 9.64$ N
$(\uparrow)$ $Y = 15\sin 60° - 20\sin 50°$
  $Y = -2.33$ N

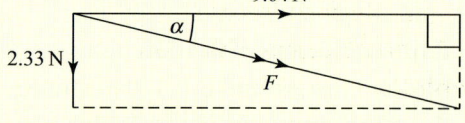

$F = \sqrt{9.64^2 + 2.33^2} = 9.92$

$\alpha = \tan^{-1}\left(\dfrac{2.33}{9.64}\right) = 13.6°$

Using $F = ma$

$a = \dfrac{F}{m}$

$a = \dfrac{9.92}{4} = 2.48\,\text{m s}^{-2}$ (to 3 sf) at an angle of 13.6° below the 30 N force.

Let X be the component of the resultant in the direction of the 30 N force and let Y be the component in the perpendicular direction.

Draw a diagram. Y is in the downwards direction because it has a negative value.

Calculate the magnitude and direction of the resultant force.

Example 2

A box of mass 60 kg is at rest on horizontal ground. A girl pulls a rope attached to the box so that it makes an angle of 35° to the horizontal. The coefficient of friction between the box and the floor is μ. When the girl is pulling with a force of 400 N, the box accelerates at $3\,\mathrm{m\,s^{-2}}$. Find μ (to 2 sf).

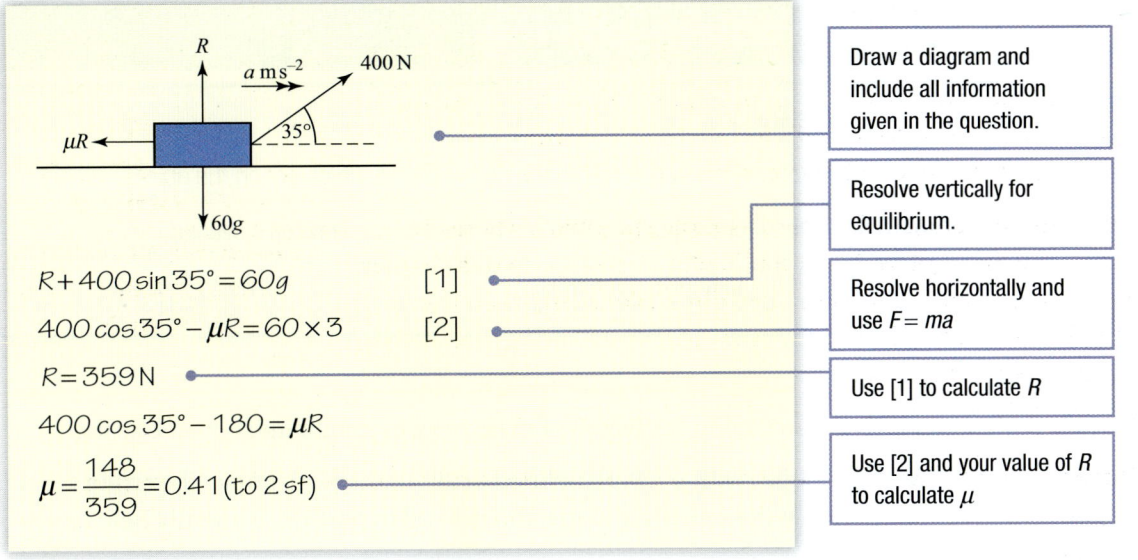

| Draw a diagram and include all information given in the question. |

$R + 400\sin 35° = 60g$ [1] — Resolve vertically for equilibrium.

$400\cos 35° - \mu R = 60 \times 3$ [2] — Resolve horizontally and use $F = ma$

$R = 359\,\mathrm{N}$ — Use [1] to calculate R

$400\cos 35° - 180 = \mu R$

$\mu = \dfrac{148}{359} = 0.41$ (to 2 sf) — Use [2] and your value of R to calculate μ

Exercise 19.2A Fluency and skills

1 Two perpendicular forces of magnitude X and 20 N are applied to an object of mass 2 kg. The mass ends up accelerating in a direction which makes an angle of 30° with the 20 N force. Find the value of X and the magnitude of the acceleration.

2 Three children kick a football of mass 0.45 kg. One kicks it with a magnitude of 30 N in the direction of due north, another kicks it with a magnitude of 40 N in the direction of due east and the third kicks it with a magnitude of 50 N in the direction of north-west. Find the magnitude and direction of the acceleration of the ball.

3 A box of mass 2 kg lies on a rough horizontal table. It is pulled by a tension of 10 N in a string acting at 25° to the horizontal. The coefficient of friction between the box and the table is 0.2. Find the acceleration of the box. Use $g = 10\,\mathrm{m\,s^{-2}}$

4 A book of mass 300 g lies on a rough horizontal desk. A boy pulls it along the table, applying a force of magnitude 2 N at an angle of 50° to the horizontal. The acceleration of the book is $0.8\,\mathrm{m\,s^{-2}}$. Find the coefficient of friction between the book and the desk.

5 Three forces of magnitude 55 N, 40 N and 35 N act on a particle of mass 12 kg in the directions shown in the diagram. Find the magnitude and direction of the acceleration of the particle.

(Diagram: 40 N, 100°, 145°, 115°, 55 N, 35 N)

6 A boy sits on a sledge on rough horizontal ground. The combined mass is 80 kg. A friend pulls the rope attached to the sledge with a force of 500 N at an angle of 25° to the horizontal. The coefficient of friction between the sledge and the floor is $\dfrac{2}{3}$. Find the acceleration of the sledge to 2 sf.

7 A force of 100 N acting at an angle of 40° above the horizontal is applied to a crate of mass 20 kg lying on a rough horizontal surface.

The acceleration of the crate is $2\,\text{m}\,\text{s}^{-2}$. Find μ (to 2 sf).

Reasoning and problem-solving

To solve questions involving acceleration

(1) Draw a clear diagram, marking on the acceleration and all the forces which act on the object.

(2) Resolve in suitable directions to form equations. You might need to use Newton's second law.

(3) Solve these equations to find the unknown quantities.

Example 3

Particle A of mass 7 kg is connected to particle B of mass m kg by a light inextensible string.

Particle A rests on a rough slope (with $\mu = 0.4$) inclined at 30° to the horizontal. The string passes over a smooth pulley at the top of the slope and is fixed to particle B initially held at rest in mid-air.

Particle B is released and accelerates downwards at $2\,\text{m}\,\text{s}^{-2}$. Calculate m

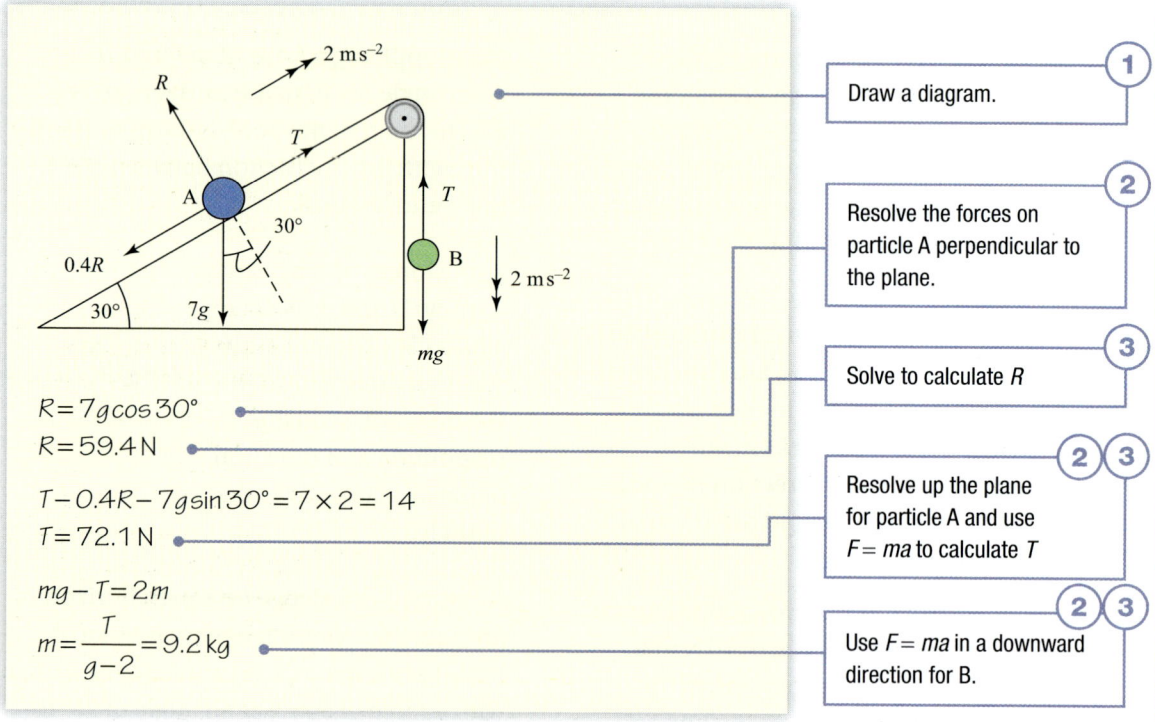

(1) Draw a diagram.

(2) Resolve the forces on particle A perpendicular to the plane.

$R = 7g\cos 30°$
$R = 59.4\,\text{N}$

(3) Solve to calculate R

$T - 0.4R - 7g\sin 30° = 7 \times 2 = 14$
$T = 72.1\,\text{N}$

(2)(3) Resolve up the plane for particle A and use $F = ma$ to calculate T

$mg - T = 2m$
$m = \dfrac{T}{g-2} = 9.2\,\text{kg}$

(2)(3) Use $F = ma$ in a downward direction for B.

In situations when the particles are moving with constant acceleration, you can use the constant acceleration formulae.

See Ch7.3
For a reminder on the constant acceleration formulae.

Example 4

Particle A, of mass 10 kg, sits on a rough table and is attached to one end of a light inextensible string. The string passes over a smooth pulley fixed on the edge of the table 3 m from A. The other end of the string is attached to particle B, of mass 5 kg, held in mid-air.

Mass B is released from rest and, after two seconds, particle A reaches the pulley.

Show that

a The acceleration of the masses is $1.5\,\text{m}\,\text{s}^{-2}$

b The coefficient of friction is 0.27 (to 2 sf).

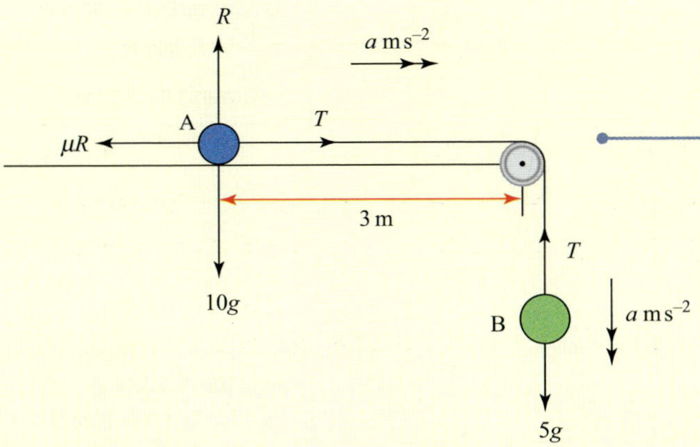

[Diagram: Particle A on table with forces R up, $10g$ down, μR left, T right, acceleration $a\,\text{m}\,\text{s}^{-2}$; pulley 3 m from A; particle B hanging with tension T up, $5g$ down, acceleration $a\,\text{m}\,\text{s}^{-2}$]

1 Draw a diagram.

a $u=0 \quad t=2 \quad s=3$

$3=\dfrac{1}{2}\times a\times 2^2$

$a=1.5\,\text{m}\,\text{s}^{-2}$

2 Use $s=ut+\dfrac{1}{2}at^2$ to calculate a

b $5g-T=5\times 1.5$

$T=41.5\,\text{N}$

2 3 Resolve vertically downwards at B and use $F=ma$ to calculate T

$R=10g$

$R=98\,\text{N}$

2 3 Resolve vertically at A to calculate R. The particle is accelerating horizontally so the vertical forces should sum to zero.

$T-\mu R=10\times 1.5$

$41.5-98\mu=15$

$41.5-15=98\mu$

$\mu=0.27$ (to 2 sf)

2 3 Resolve horizontally at A and use $F=ma$ to calculate μ

In situations where the acceleration is not constant, you can still use the equation $F=ma$ in the form $F=m\dfrac{\text{d}v}{\text{d}t}$

See Ch7.4
For a reminder on equations of motion for variable acceleration.

Example 5

Florence, mass 55 kg, parachutes from a helicopter. She falls vertically at $14.7\,\mathrm{m\,s^{-1}}$. Once Florence opens her parachute, she experiences air resistance of magnitude $275v\,\mathrm{N}$, where $v\,\mathrm{m\,s^{-1}}$ is Florence's speed t seconds after she opens her parachute.

Show that $\dfrac{\mathrm{d}v}{\mathrm{d}t} = -5(v - 1.96)$ and hence find v in terms of t

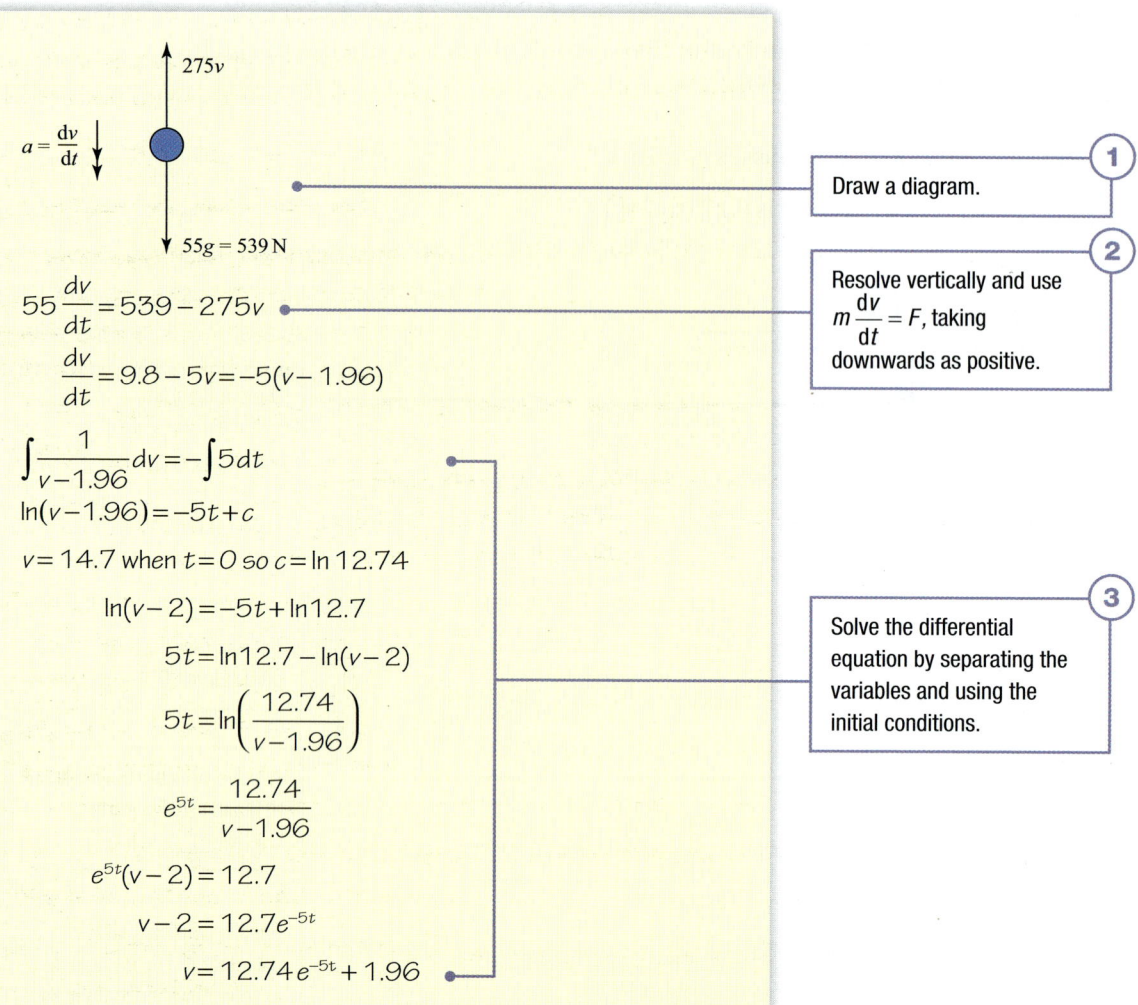

$375v$

$a = \dfrac{\mathrm{d}v}{\mathrm{d}t}$

$55g = 539\,\mathrm{N}$

① **Draw a diagram.**

$55\dfrac{\mathrm{d}v}{\mathrm{d}t} = 539 - 275v$

② **Resolve vertically and use $m\dfrac{\mathrm{d}v}{\mathrm{d}t} = F$, taking downwards as positive.**

$\dfrac{\mathrm{d}v}{\mathrm{d}t} = 9.8 - 5v = -5(v - 1.96)$

See Ch16.5
For a reminder on solving differential equations.

$\displaystyle \int \frac{1}{v - 1.96}\,\mathrm{d}v = -\int 5\,\mathrm{d}t$

$\ln(v - 1.96) = -5t + c$

$v = 14.7$ when $t = 0$ so $c = \ln 12.74$

$\ln(v - 2) = -5t + \ln 12.7$

$5t = \ln 12.7 - \ln(v - 2)$

$5t = \ln\left(\dfrac{12.74}{v - 1.96}\right)$

③ **Solve the differential equation by separating the variables and using the initial conditions.**

$e^{5t} = \dfrac{12.74}{v - 1.96}$

$e^{5t}(v - 2) = 12.7$

$v - 2 = 12.7e^{-5t}$

$v = 12.74\,e^{-5t} + 1.96$

Exercise 19.2B Reasoning and problem-solving

1. A block of mass 7 kg lies on a rough slope inclined at 35° to the horizontal. A string is attached to the block and is pulled with a force of 50 N up the slope. The coefficient of friction between the mass and the slope is 0.1. Find the acceleration of the block (to 2 sf).

2. A box of mass 5 kg lies on a rough slope inclined at 40° to the horizontal. A light inextensible string is attached to the box. The string passes over a smooth pulley fixed

   to the top of the slope. The other end of the string is attached to a box of mass 6 kg which hangs vertically, 1 m above the floor.

   The 6 kg mass is released from rest and after two seconds it hits the floor. Find the coefficient of friction between the 5 kg mass and the slope.

3. A block of mass 75 kg lies on a rough slope inclined at 45° to the horizontal. The coefficient of friction between the block and

the slope is 0.25. A cable is attached to the block and is pulled with a force T parallel to the slope so that the block is accelerating up the slope at $4\,\text{m s}^{-2}$. Find T (to 2 sf).

4 Rachael, mass 50 kg, parachutes from a helicopter. She falls vertically at $12.8\,\text{m s}^{-1}$. Once Rachael opens her parachute, she experiences air resistance of magnitude $175v\,\text{N}$, where $v\,\text{m s}^{-1}$ is her speed t seconds after she opens her parachute.

   Show that $\dfrac{dv}{dt} = -3.5(v - 2.8)$ and hence find v in terms of t.

5 A block of mass 125 kg is at rest on a rough slope inclined at 40° to the horizontal. The coefficient of friction between the block and the slope is $\dfrac{1}{5}$. A cable is attached to the block and is pulled with a force of 1800 N so that the block is accelerating at $a\,\text{m s}^{-2}$ up the slope.

   Show that $a = 6.6$ (to 2 sf).

   Hence find how long it takes to travel 10 m (to 2 sf).

6 Three forces of magnitude 500 N, 600 N and X act on a particle of mass 80 kg as shown in the diagram. All forces act parallel to the ground. The particle accelerates in a direction 10° below the XN force. Find the value of X and the magnitude of the acceleration.

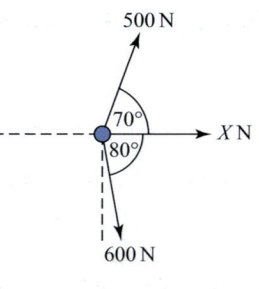

7 A boy kicks a block of mass 2 kg so that it slides up a rough plane inclined at 30° to the horizontal. The block has an initial speed of $3\,\text{m s}^{-1}$. The coefficient of friction between the block and the plane is $\dfrac{1}{2}$

   a Show that, after it has been kicked, the deceleration of the block up the slope is $9.1\,\text{m s}^{-2}$ (to 2 sf).

   b Find the distance travelled by the block before it comes to an instantaneous rest (to 2 sf).

   c Find the acceleration of the block (to 2 sf) as it slides back down the slope.

   d Show that the block will have a speed of $0.8\,\text{m s}^{-1}$ (to 1 sf) as it returns to the point where it was kicked.

8 A car of mass 800 kg drives along a horizontal road. The car's engine provides a driving force of 3600 N and, as the car accelerates from rest, it experiences a resistance force to its motion equal to $120v\,\text{N}$, where v is its speed in m s^{-1}. Form an equation for the car's speed v in terms of t, and hence show that its maximum speed is $30\,\text{m s}^{-1}$ assuming the driving force does not change.

Challenge

9 Two particles, A and B, of mass $2m$ and $3m$ respectively, are attached to the ends of a light inextensible string. A is held at rest on a rough horizontal desk. The string passes over a small smooth pulley fixed on the edge of the desk. B hangs freely below the pulley.

   The particles are released from rest with the string taut. They accelerate at $\dfrac{1}{2}gm\,\text{s}^{-2}$

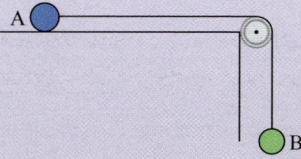

   a Find the tension in the string in terms of m and g and show that $\mu = \dfrac{1}{4}$

   When B has fallen h m it hits the ground and does not rebound.

   In subsequent motion, particle A stops just before the pulley.

   b How far, in terms of h, was A initially away from the pulley?

19.3 Moments

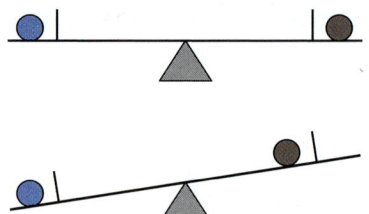

Fluency and skills

If two people with the same weight sit on either end of a see-saw then the see-saw will be balanced. If one person moves forwards, closer to the middle of the see-saw, then the see-saw will lift up at that end. This is because the turning effect of a force is greater the further it acts from the pivot. These turning effects are called **moments**.

> **Key point**
>
> If a force F acts at a *perpendicular* distance d from a point P then the turning effect (or moment) of F about P is the product $F \times d$. Moments are measured in newton-metres (Nm).

You draw the force F as a line and d is the perpendicular distance from this line to the point P. The diagram shows this for one person sitting on a see-saw. Note that d is not simply the length of see-saw between the pivot and the person.

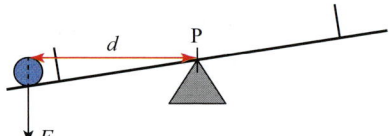

The notation P ⤸ is used when taking moments about P where the clockwise direction is chosen to be positive.

The notation P ⤹ is used when the anti-clockwise direction is chosen to be positive.

Example 1

For each set of forces, calculate the total clockwise moment about the point A.

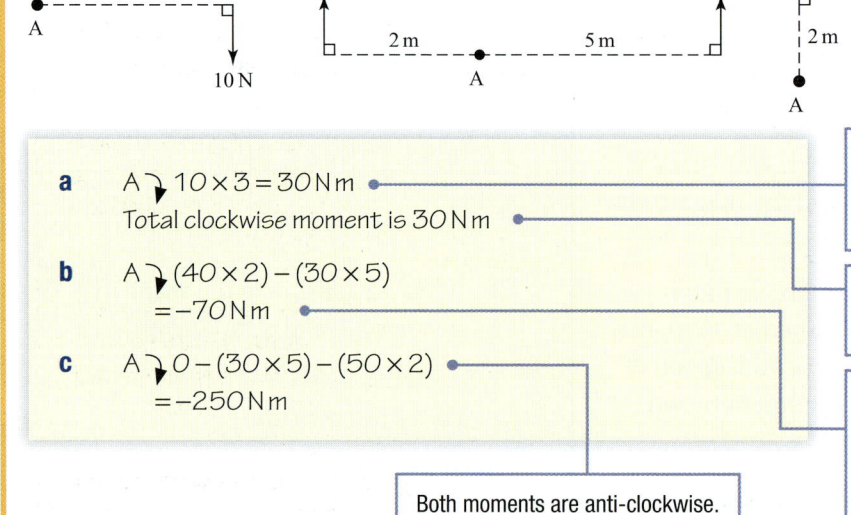

a A⤸ $10 \times 3 = 30\,\text{Nm}$ ●

Total clockwise moment is $30\,\text{Nm}$ ●

b A⤸ $(40 \times 2) - (30 \times 5)$

$= -70\,\text{Nm}$ ●

c A⤸ $0 - (30 \times 5) - (50 \times 2)$ ●

$= -250\,\text{Nm}$

| Choose the clockwise direction to be positive. Take moments using $F \times d$ |
| There is only one moment and it acts clockwise. |
| Since the clockwise direction is taken as positive, any anti-clockwise moments are negative. |

Both moments are anti-clockwise.

Example 2

The force $(2\mathbf{i} - 3\mathbf{j})\,\text{N}$ is applied at point A with position vector $(3\mathbf{i} + \mathbf{j})\,\text{m}$. Calculate the magnitude of the moment of this force about the point B with position vector $(\mathbf{i} - \mathbf{j})\,\text{m}$.

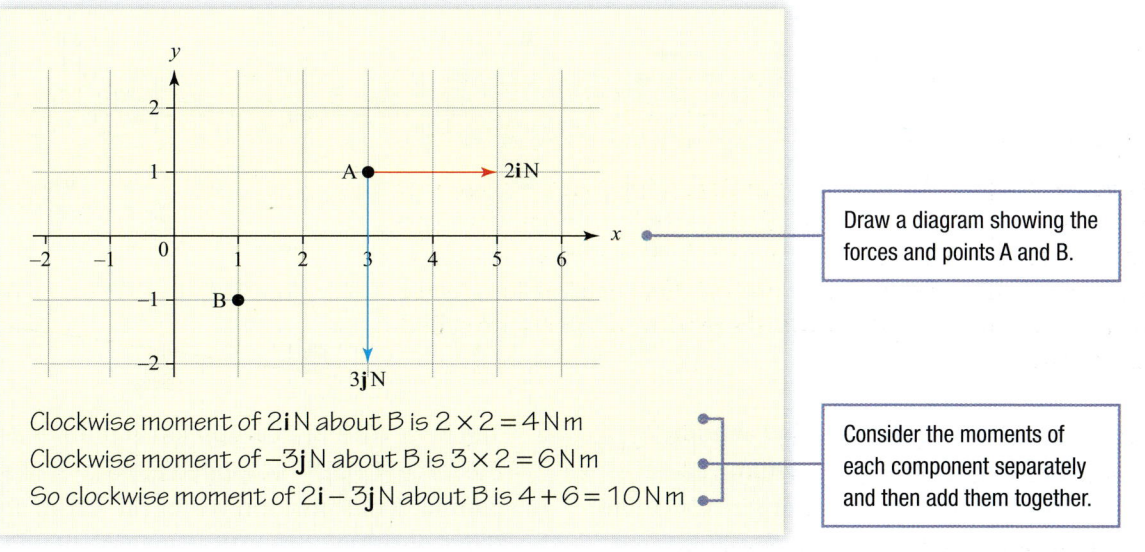

Draw a diagram showing the forces and points A and B.

Clockwise moment of $2\mathbf{i}\,\text{N}$ about B is $2 \times 2 = 4\,\text{N m}$
Clockwise moment of $-3\mathbf{j}\,\text{N}$ about B is $3 \times 2 = 6\,\text{N m}$
So clockwise moment of $2\mathbf{i} - 3\mathbf{j}\,\text{N}$ about B is $4 + 6 = 10\,\text{N m}$

Consider the moments of each component separately and then add them together.

A beam, made from a material that is uniformly spread along its length, is called a **uniform beam**. You can assume that its weight acts at the midpoint of the beam. This point is called the **centre of mass**.

> **Key point**
>
> Any uniform shape (or lamina) that is symmetrical, such as a rectangle, has its centre of mass on every line of symmetry.

If an object is in equilibrium, then

- The resultant force in any direction is zero,
- The resultant moment of all forces about any point is zero, that is, the clockwise and anticlockwise moments balance.

Example 3

A uniform beam AB of weight $300\,\text{N}$ has length $6\,\text{m}$. The beam rests on a support at A and is held horizontally in equilibrium by a vertical string at B. There is a vertical reaction R at A. Calculate the tension, T, in the string.

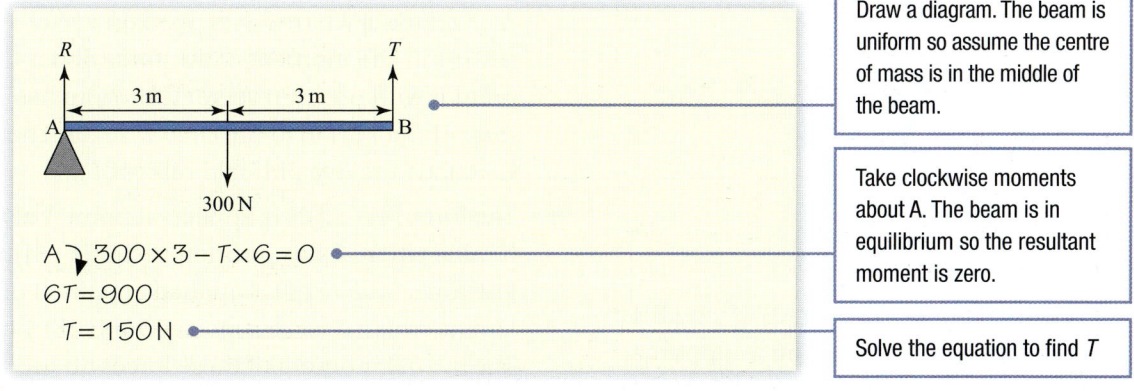

Draw a diagram. The beam is uniform so assume the centre of mass is in the middle of the beam.

$\text{A} \curvearrowright 300 \times 3 - T \times 6 = 0$
$6T = 900$
$T = 150\,\text{N}$

Take clockwise moments about A. The beam is in equilibrium so the resultant moment is zero.

Solve the equation to find T

By taking moments about A, the force R is not in the equation because it is at zero distance from A. This technique is important as it allows you to eliminate an unknown force from an equation.

1 Calculate the total clockwise moment of the forces about the point A.

a

2 m

A

8 N

b
10 N

1 m

A

2 m

20 N

c

2 m

A

5 m

3 N

8 N

d

9 N

3 m

12 N

1 m

A

2 The force $(3\mathbf{i} - 4\mathbf{j})$ N is applied at point A of a lamina where A has position vector $(2\mathbf{i} - \mathbf{j})$ m relative to a fixed origin. Calculate the magnitude of the moment of this force about the point B with position vector $(\mathbf{i} + \mathbf{j})$ m

3 In each diagram, the circle acts as a pivot and the forces acting on the object are shown. Determine whether the object will turn, and if so then state the direction. Assume the objects have zero weight.

a

15 N

2 m

1 m

30 N

b
32 N

1.8 m

18 N

1.2 m

1 m

10 N

4 A horizontal uniform beam AB of length 4 m has weight 120 N. It rests in on a support at A and is held horizontally in equilibrium by a vertical string attached 3 m from A. There is a supporting force P acting at A and the tension in the string is T. Calculate T

5 The force $(5\mathbf{i} + 2\mathbf{j})$N is applied at point A and the force $(\mathbf{i} + 3\mathbf{j})$N is applied at point B of a lamina. A has position vector $(3\mathbf{i} - 2\mathbf{j})$m and B has position vector $(2\mathbf{i} + \mathbf{j})$m. Calculate the resultant moment of these forces about the point C with position vector $(\mathbf{i} - \mathbf{j})$m

6 A horizontal uniform beam with a length of 6 m has a weight of 100 N. The beam rests on a pivot at one end and is held up by a rope at the other. Three rocks, each with a weight of 200 N, are placed at 1.5 m intervals along the beam as shown.

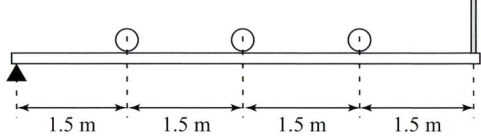

1.5 m   1.5 m   1.5 m   1.5 m

Calculate the tension in the rope.

7 The force $(p\mathbf{i} - 2p\mathbf{j})$N is applied at point A of a lamina where A has position vector $(\mathbf{i} + p\mathbf{j})$m. The moment of this force about the point B with position vector $(2\mathbf{i} - p\mathbf{j})$m has magnitude 12 N m in the clockwise direction. Calculate the two possible values of p

8 The force $(4\mathbf{i} - 2\mathbf{j})$N is applied at point A of a lamina where A has position vector $(\mathbf{i} + 3\mathbf{j})$m. The force $(10\mathbf{i} + x\mathbf{j})$N is applied at point B of a lamina where B has position vector $(-2\mathbf{i} + \mathbf{j})$m. The overall moment of these forces about the point C with position vector $(5\mathbf{i} - 2\mathbf{j})$m is zero. Calculate the value of x

Reasoning and problem-solving

To solve questions involving moments

1. Draw a clear diagram, marking on all the forces that act on the object.
2. Take moments and/or resolve in suitable directions to create equations.
3. Solve the equations to find the unknown quantities.

If a horizontal rod is resting in equilibrium on two supports, and the rod is about to tilt about one of the supports, the reaction at the other support is zero.

A uniform beam AB of length 6 m and weight 300 N rests in equilibrium on supports C and D where AC = 1 m and BD = 2 m

A 100 N weight is attached at A. Find the largest weight W which can be applied at B for the beam to remain in equilibrium.

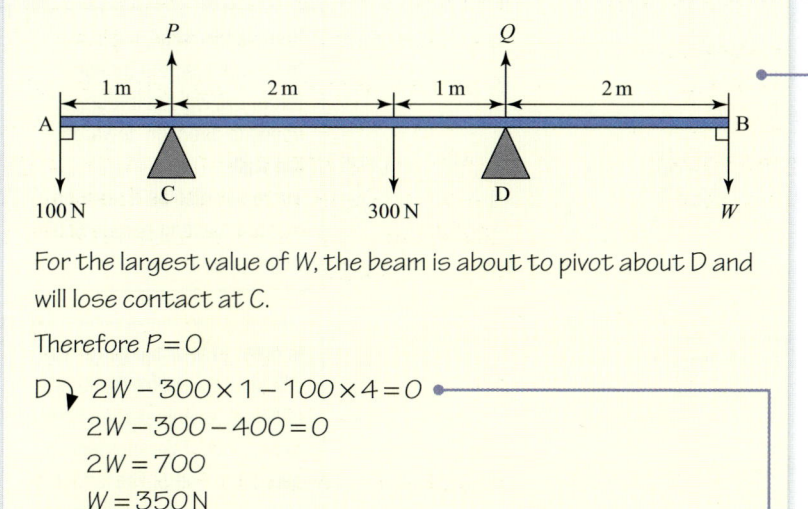

For the largest value of W, the beam is about to pivot about D and will lose contact at C.

Therefore $P = 0$

$$D \quad 2W - 300 \times 1 - 100 \times 4 = 0$$
$$2W - 300 - 400 = 0$$
$$2W = 700$$
$$W = 350\,N$$

1. Draw a diagram showing all the information. You may expect some of the weight of the beam itself to act to the left of C or to the right of D, but as you are assuming that the weight acts only at the midpoint of the beam AB, this is not the case.

The beam remains in equilibrium until W is large enough to lift it off the support at C.

2. 3. Take moments about D. Since the beam is in equilibrium, the moments will sum to zero.

Questions involving a ladder resting against a horizontal floor and a vertical wall can be solved by a combination of resolving and taking moments. It is best to take moments about a point that one or more forces pass through as this allows you to eliminate unknown values from equations.

Example 5

A uniform ladder of length 2 m and mass 15 kg rests against a smooth vertical wall, with its feet on rough horizontal ground. The coefficient of friction between the ladder and the ground is μ. The ladder is inclined at 75° to the horizontal. If a boy of mass 30 kg can climb up to the end of the ladder without it slipping then find the minimum value of μ

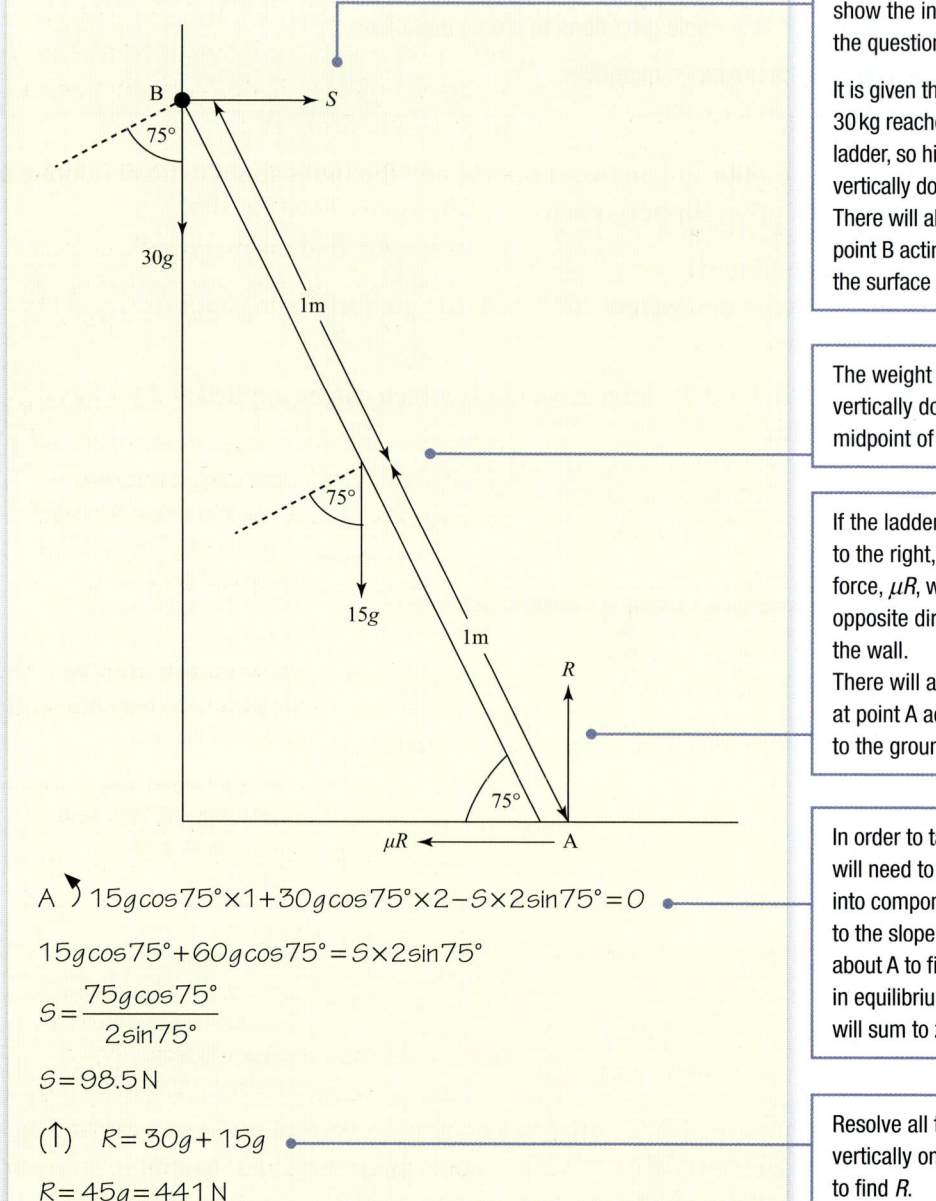

1 Start by sketching a diagram to show the information given in the question.

It is given that a boy of mass 30 kg reaches the top of the ladder, so his weight will act vertically downwards at point B. There will also be a reaction at point B acting perpendicular to the surface of the wall.

1 The weight of the ladder acts vertically downwards at the midpoint of the ladder.

1 If the ladder slips, it will slip to the right, so the frictional force, μR, will act in the opposite direction, towards the wall.
There will also be a reaction at point A acting perpendicular to the ground.

$$A \, \text{⟲} \quad 15g\cos 75° \times 1 + 30g\cos 75° \times 2 - S \times 2\sin 75° = 0$$

2 In order to take moments, you will need to resolve the forces into components perpendicular to the slope. Take moments about A to find S. The ladder is in equilibrium so the moments will sum to zero.

$$15g\cos 75° + 60g\cos 75° = S \times 2\sin 75°$$

$$S = \frac{75g\cos 75°}{2\sin 75°}$$

$$S = 98.5 \, \text{N}$$

$$(\uparrow) \quad R = 30g + 15g$$

3 Resolve all the forces acting vertically on the whole ladder to find R.

$$R = 45g = 441 \, \text{N}$$

$$(\rightarrow) \quad S = \mu R$$

3 Resolve all the forces acting horizontally on the whole ladder to find an equation in μ

$$98.471... = \mu \times 441$$

$$\mu = 0.22$$

3 Substitute your values for S and R into the equation to find μ

1 A uniform rod of length 4 m weighs 50 N. 1 m of the rod lies on a horizontal table and the rest lies over the edge.

   a Where does the reaction act when the rod is about to tilt?

   b What downward force must be applied to the end of the rod to stop it from tilting?

   c What is the reaction of the table on the rod when it is about to tilt?

2 A uniform plank, AB, of mass 28 kg and length 9 m, lies on a horizontal roof in a direction at right angles to the edge of the roof. The end B projects 2 m over the edge. A man of mass 70 kg walks out along the plank.

   a Find how far along the plank he can walk without causing the plank to tip up.

   b Find also the mass which must be placed on the end A so that the man can reach B without upsetting the plank.

3 A uniform beam AB, of length 4 m and weight 200 N, rests in equilibrium on supports C and D as shown in the diagram. A 500 N weight is attached at A and a weight W is attached at B.

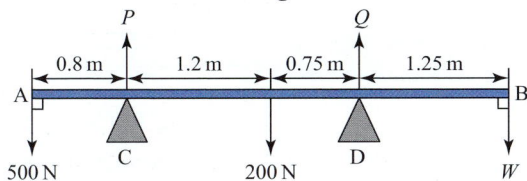

   a Find the largest possible value of W, for which the beam remains in equilibrium.

   b Find the smallest possible value of W for which the beam remains in equilibrium.

4 A uniform beam AB, of length 2.2 m and mass 150 kg is supported in a horizontal position by two supports at C and D where AC = 0.75 m and AD = 1.5 m

   a Find, in terms of g, the force exerted by each support.

   b Find the mass of the heaviest man who can sit at

   i End A   **ii** End B

   of the beam without it tilting.

5 A uniform ladder, of mass 15 kg and length 4m, is placed with its base on rough horizontal ground. The coefficient of friction between the ladder and the ground is $\frac{1}{3}$. The upper end of the ladder rests against a smooth vertical wall, the ladder making an angle of α to the horizontal.

   a Find the minimum value of α such that the ladder does not slip.

   b If $\alpha = 20°$, find the magnitude of the minimum horizontal force that must be applied to the base of the ladder to prevent slipping.

6 A uniform ladder, of length 2.4 m and mass 10 kg, rests against a smooth vertical wall and it lower end is on rough horizontal ground. The ladder is inclined at 70° to the horizontal. A girl of mass 60 kg can climb to a point on the ladder which is 1.8 m away from the foot of the ladder.

   a Find the minimum possible value of the coefficient of friction between the ladder and the ground.

   b A friend of the girl then puts her foot on the base of the ladder and applies a vertical force. Find what force she must she apply to enable her friend to reach the top of the ladder.

Challenge

7 A rectangular uniform drone, when viewed from above, is powered by four propellers with position vectors $(2\mathbf{i} + 3\mathbf{j})$, $(4\mathbf{i} + 2\mathbf{j})$, $(4\mathbf{i} + 4\mathbf{j})$ and $(6\mathbf{i} + 3\mathbf{j})$

The first three propellers, in the order listed, are providing a driving force of $(-7\mathbf{i} + 3\mathbf{j})$ N, $(3\mathbf{i} - 5\mathbf{j})$ N and $(3\mathbf{i} + 5\mathbf{j})$ N

What force must the fourth propeller be providing if the drone is

   a Spinning but otherwise not moving,

   b Moving in the $\mathbf{j}$ direction but not spinning?

MECH

Chapter summary

- Forces can be resolved in any two perpendicular directions. If X and Y are the resolved parts or components of a force F, then F is the resultant of the forces X and Y
- If an object is in contact with a surface, then the frictional force F exerted by the surface on the object satisfies the inequality $F \le \mu R$, where μ is the coefficient of friction between the object and the surface, and R is the normal reaction of the surface on the object.
  - If the object is moving, or on the point of moving, then $F = \mu R$
- If an object in contact with a surface is about to move parallel to the surface, then it is said to be in limiting equilibrium and the friction is the limiting friction.
- $F = ma$ can be used in any direction to find unknown forces acting on an object or the object's acceleration, where F is the component of the resultant force in that direction, and a is the acceleration in that direction.
- If a force F acts at a perpendicular distance d from a point P, then the turning effect or moment of F about P is the product $F \times d$. Moments are measured in Newton-metres (Nm).
- If a body is in equilibrium under a system of forces then
  - The vector sum of the forces is zero, i.e. the sum of the components of the forces in any given direction is zero,
  - The sum of the moments about any given point is zero, that is, the clockwise moments balance the anticlockwise moments.

Check and review

You should now be able to...	Try Questions
Understand that there is a maximum value that the frictional force can take (μR) and that it takes this value when the object is moving or on the point of moving.	1, 4
Resolve in suitable directions to find unknown forces when the system is at rest.	2
Resolve in suitable directions to find unknown forces when the system has constant acceleration.	1, 4
Use constant acceleration formulae for problems involving blocks on slopes or blocks connected by pulleys.	4
Solve differential equations which arise from problems involving $F = ma$	3
Take moments about suitable points and resolve in suitable directions to find unknown forces.	2, 5
Solve ladder problems.	5

1 A particle of weight $5g$ is pushed up a rough plane by a horizontal force of magnitude $3g$, giving it an acceleration of $\frac{g}{13}$ m s^{-2}. The plane is inclined to the horizontal at an angle α, where $\tan \alpha = \frac{5}{12}$

Find the coefficient of friction between the surface and the particle.

2 A uniform beam AB has mass 35 kg and length 6 m. The beam rests in equilibrium in a horizontal position on two smooth supports C and D where AC = 1 m and AD = 4.5 m

a Find, in terms of g, the magnitudes of the reactions on the beam at C and at D.

b Amanda wants to make the reactions at C and D equal and so she adds a mass of m kg at A. Find the value of m

3 A particle of mass 5 kg is travelling north along a horizontal surface with an initial velocity of 12.5 m s^{-1} when it is subject to a resistive force to the south of magnitude $12v$ N, where v m s^{-1} is the speed of the particle t seconds after the resistive force is applied. There is a constant force of 15 N acting on the particle in a northerly direction.

a Find the speed of the particle after two seconds.

b Find the minimum speed of the particle in the subsequent motion.

4 Two particles, A and B, of mass 5 kg and 2 kg respectively, are attached to the ends of a light inextensible string. A is held at rest on a rough horizontal table. The string passes over a small smooth pulley, P, fixed on the edge of the table which is $\frac{14}{15}$ m away from A. B hangs freely below the pulley. The particles are released from rest with the string taut. The coefficient of friction between A and the table is μ

a Find an inequality for μ if particle A moves towards the pulley.

b If $\mu = \frac{1}{3}$, calculate how long it takes for particle A to reach the pulley.

5 A uniform ladder of length 3 m and mass 12 kg rests against a smooth vertical wall and its lower end is on smooth horizontal ground. The ladder is held in equilibrium by a light inextensible string of length 2 m which has one end attached to the bottom of the ladder and the other end fastened to a point at the base of the wall vertically below the top of the ladder.

a Calculate the angle that the ladder makes with the horizontal.

b Find the tension in the string.

What next?

Score		
0 – 2	Your knowledge of this topic is still developing. To improve, search in MyMaths for the codes: 2190, 2191, 2193, 2194, 2197	🔗
3 – 4	You're gaining a secure knowledge of this topic. To improve, look at the InvisiPen videos for Fluency and skills (19A)	🎞
5	You've mastered these skills. Well done, you're ready to progress! To develop your exam techniques, look at the InvisiPen videos for Reasoning and problem-solving (19B)	🎞

Click these links in the digital book

History

Albert Einstein (1879–1955) was born in Ulm, Germany.

His mass-energy equivalence, expressed as $E = mc^2$ has been called the most famous equation in the world. It was proposed as part of his **Special Theory of Relativity** in 1905.

In November 1915, Einstein published the equation of **General Relativity**, a theory of gravitation. The predictions of general relativity differ significantly from the classical mechanics of Newton, particularly in terms of time and space. Despite this, these predictions have been confirmed in all experiments and observations to date.

The one remaining prediction of the theory awaiting confirmation was the existence of ripples in the fabric of the Universe, so called **gravity waves**. It took 100 years, but on the 11th November 2016 it was confirmed that gravity waves had been detected for the first time.

Information

The differences in the predictions of Einstein's general relativity and Newton's classical mechanics only become apparent in extreme conditions. Newton's laws still remain the cornerstone of mechanics in everyday situations.

Newton considered that gravity was a force that would make a moving object deviate from a straight line.

Einstein's view was that gravity is a distortion of space-time.

> "Imagination is more important than knowledge. Knowledge is limited. Imagination encircles the world."
> -Albert Einstein

Applications

Global positioning systems work by using 24 satellites, each with a precise clock. A GPS receiver uses the time at which the signal from each satellite was received in order to determine position.

According to the theory of general relativity, gravity distorts space and time. This means that time on the satellites is moving slightly faster than time on Earth. Without taking into account this difference, a GPS tracker would become completely inaccurate within the course of about 2 minutes.

Research

In 1916, Albert Einstein proposed three tests of general relativity, subsequently called the **classical tests of general relativity**.
What were they? What was the outcome?

1   Four forces, **P**, **Q**, **R** and **S**, act on an object. The object is in equilibrium.
    $\mathbf{P} = 4\mathbf{i} + 5\mathbf{j}$, $\mathbf{Q} = \mathbf{i} - 8\mathbf{j}$ and $\mathbf{R} = 3\mathbf{i} - 12\mathbf{j}$

   a   Calculate **S**   [3 marks]

   b   Calculate |**S**| [2]

2   A particle, P, of mass 20 kg, is attached to one end of a light
    inextensible string. The other end of the string is attached to
    a fixed point, O. A force, F, is applied to P, at right
    angles to the string. The system rests in equilibrium with
    the string at an angle of 30° to the vertical.

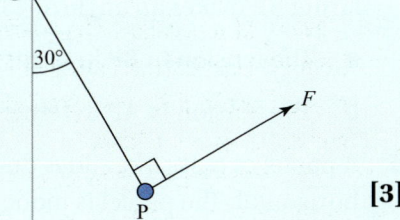

   a   Calculate the value of F in terms of g [3]

   b   Work out the tension in the string in terms of g [3]

3   A box of mass 20 kg is being pulled along a rough horizontal
    floor by a rope. The box is accelerating at $2\,\mathrm{m\,s^{-2}}$. The tension
    in the rope is 148 N, and the rope makes an angle of 30° with
    the horizontal. The coefficient of friction between the box
    and the floor is μ. Modelling the rope as a light inextensible
    string,

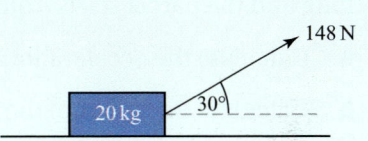

   a   Calculate the normal reaction between the box and the floor,   [3]

   b   Calculate μ [5]

4   Two forces, $\mathbf{F}_1$ and $\mathbf{F}_2$, act on a body. $\mathbf{F}_1 = \begin{pmatrix} 5 \\ -6 \end{pmatrix}$ and $\mathbf{F}_2 = \begin{pmatrix} -1 \\ -2 \end{pmatrix}$. Find the magnitude and
    direction of their resultant.   [6]

5   A uniform rod, AB, has weight 60 N and length 5 m.
    It rests in a horizontal position on two supports
    placed at P and Q, where AP = 1 m, and PQ = 2 m,
    as shown in the diagram. Calculate the magnitude of the force in the support

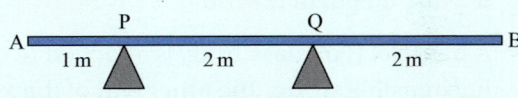

   a   At Q,   [3]

   b   At P.   [2]

6   A box of mass 5 kg rests in equilibrium on a rough plane. The plane is
    inclined at an angle of 30° to the horizontal. The coefficient of friction
    between the box and the plane is $\dfrac{\sqrt{3}}{5}$. The box is acted on by a force
    of P newtons, where P acts up the plane, along the line of greatest
    slope of the plane.

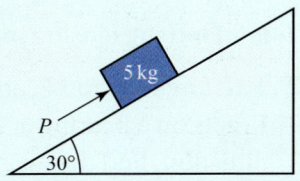

   a   Write an expression, in terms of g, for the normal reaction between the plane and
       the box.   [2]

   b   Calculate, in terms of g, the range of possible values of P [7]

7   Two forces, **P** and **Q**, act on a particle. The force **P** is of magnitude 6 N and acts on a bearing of 035°. The force **Q** is of magnitude 8 N and acts on a bearing of 288°

    a   Draw a diagram and calculate the magnitude of the resultant of **P** and **Q**   **[5]**

    b   Calculate the direction of the resultant of **P** and **Q**   **[3]**

8   A parcel, P, of mass 5 kg, is attached to the ends of two light inextensible strings. The other ends of the strings are attached to two points, A and B, on a horizontal ceiling. The parcel hangs in equilibrium.

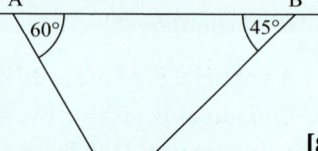

    The string AP makes an angle of 60° with the ceiling, and the string BP makes an angle of 45° with the ceiling. Find, in exact form

    a   The tension in BP in terms of g **[8]**

    b   The tension in AP in terms of g **[1]**

9   A parcel of mass 8 kg slides down a ramp which is inclined at an angle of 40° to the horizontal. The parcel is modelled as a particle and the ramp as a rough plane. The coefficient of friction between the parcel and the ramp is 0.6. The ramp is 9 metres long and the parcel starts from rest at the top of the ramp.

    a   Calculate the acceleration of the parcel.   **[7]**

    b   Calculate the speed of the parcel at the bottom of the ramp.   **[2]**

10

```
        ↑           ↑
P ——————|———————————|—————————— Q
   1.2 m  X   1.8 m   Y
```

A uniform rod, PQ, has weight 200 N. It hangs in a horizontal position, supported by two light inextensible strings attached at X and Y, where PX = 1.2 m, and XY = 1.8 m as in the diagram. The tension in the string at Y is 120 N. Calculate

    a   The tension in the string at X,   **[2]**

    b   The length of the rod.   **[3]**

11   A particle, P, of mass 10 kg, is attached to one end of a light inextensible string. The other end of the string is attached to a fixed point, O. A horizontal force of 50 N is applied to P, so that the system rests in equilibrium with the string at an angle of $x°$ to the vertical.

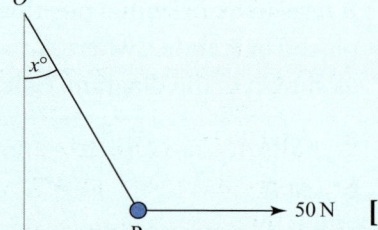

    a   Find the value of x **[7]**

    b   Find the tension in the string.   **[2]**

12   Two forces $\mathbf{F}_1$ and $\mathbf{F}_2$ act on an object. $\mathbf{F}_1$ is of magnitude 5 N, and acts on a bearing of 041°. $\mathbf{F}_2$ acts on a bearing of 289°. Given that the resultant of $\mathbf{F}_1$ and $\mathbf{F}_2$ acts in a north-westerly direction, find the magnitude of $\mathbf{F}_2$ **[5]**

13 Two rings, A and B, each of mass 20 kg, are threaded on a rough horizontal wire. The coefficient of friction between each ring and the wire is μ. The rings are attached to the ends of a light inextensible string.

A smooth ring, C, of mass 30 kg, is threaded on the string, and hangs in equilibrium below the wire. The rings are in limiting equilibrium on the wire, on the point of slipping inwards. The angles between the strings and the wire are each $x°$, where $\tan x = \dfrac{3}{4}$.

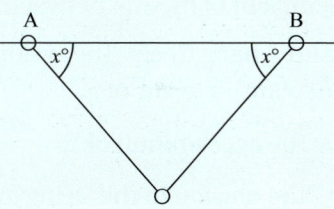

a Calculate the tension in the string in terms of g [4]

b Show that $\mu = \dfrac{4}{7}$ [7]

14 A box of mass 6 kg rests on a rough plane. The plane is inclined at an angle of 30° to the horizontal.

The coefficient of friction between the box and the plane is $\dfrac{\sqrt{3}}{2}$. A horizontal force of P newtons acts on the parcel, and it is in limiting equilibrium, on the point of moving up the plane. Calculate, in terms of g

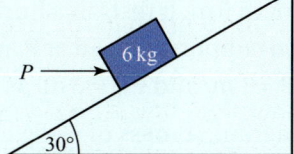

a The force P [5]

b The normal reaction between the plane and the box, [2]

c The frictional force between the plane and the box. [2]

15 A particle is projected up the line of greatest slope of a rough plane with an initial velocity of 4 m s^{-1}. The plane makes an angle of 15° with the horizontal, and the coefficient of friction between the particle and the plane is $\dfrac{1}{10}$

a Calculate the acceleration of the particle. [7]

b Calculate the distance move by the particle before it comes instantaneously to rest. [2]

c Will the particle start to move back down the slope? Justify your answer. [2]

16

A uniform plank, AB, has mass 100 N and length 6 m. It rests in a horizontal position on two supports placed at P and Q, where AP = 1 m, as shown in the diagram. The reaction at Q is 20 N more than the reaction at P.

a Find the magnitude of the reaction at P. [2]

b Find the magnitude of the reaction at Q. [1]

c Find the distance AQ. [3]

17 The diagram shows three bodies, X, Y and Z, connected by two light inextensible strings, passing over smooth pulleys. Y lies on a rough horizontal table, and the coefficient of friction between Y and the table is $\frac{2}{5}$. X and Z hang freely. The system is released from rest. Calculate, in terms of g,

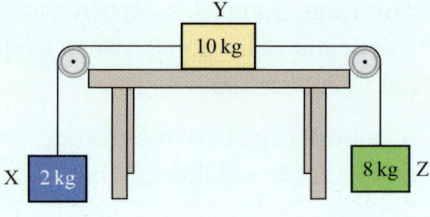

a The acceleration of Y, [10]

b The tension in the string joining X to Y, [2]

c The tension in the string joining Y to Z. [2]

18 A uniform plank, AB, has mass M kg and length 5 m. It rests in a horizontal position on two supports placed at P and Q, where AP = 1 m, and PQ = 3 m, as shown in the diagram. A mass of $\frac{M}{2}$ kg hangs from the rod at B.

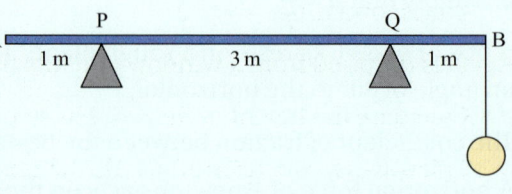

a Find expressions, in terms of M and g, for

  i The magnitude of the force in the support at Q,

  ii The magnitude of the force in the support at P. [5]

The mass at B is now replaced with a new mass of λM kg. The plank is on the point of tipping about Q.

b Find

  i The value of λ

  ii An expression, in terms of M and g, for the magnitude of the force in the support at Q. [4]

19 The diagram shows a particle, A, of mass 5 kg, on a rough plane, connected to a particle, B, of mass 10 kg, by a light inextensible string that passes over a smooth pulley. B hangs freely, and the string is parallel to the line of greatest slope of the plane. The coefficient of friction between A and the plane is $\frac{1}{4}$, and the plane is inclined at an angle of $x°$ to the horizontal, where $\sin x = \frac{3}{5}$.

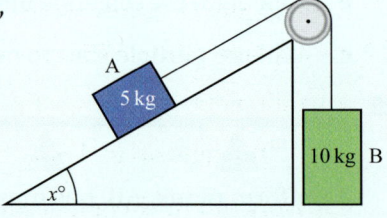

The particles are released from rest.

a Calculate, in terms of g, the acceleration of A. [11]

After B has dropped a distance of 1 metre, it hits the floor, and does not rebound.

b Calculate, in terms of g, the velocity with which B hits the floor. [2]

c Given that A does not hit the pulley, find the further distance travelled by A, until A is instantaneously at rest. [4]

1   A car accelerates at 1.5 m s^{-2} from rest on a straight, horizontal road until it reaches
    a speed of 9 m s^{-1}. It then continues along the road at this constant speed for 3 s before the
    brakes are applied and the car comes to a stop 2 s later. After being stationary for 2 s, it then
    reverses for 2 s with a constant acceleration of 1 m s^{-2}. The car then stops instantly.

   a   Sketch a velocity-time graph.   [4]

   b   Work out the displacement of the car from its starting point when it stops
           the second time.   [3]

2   A ball is dropped from a window and hits the ground 1.5 s later.

   a   Calculate the height of the window in terms of g [2]

   b   Calculate the speed at which the ball hits the ground in terms of g [3]

   At the same time as the ball is dropped, a second ball is thrown vertically upwards
   from the ground. The balls pass each other after one second.

   c   Calculate the possible speeds at which the second ball is thrown upwards in terms of g [4]

3   A particle is held in equilibrium by a force of 3 N acting due east, a force of X
    acting on a bearing of 300° and a force of Y acting on a bearing of 210°

   Calculate the values of X and Y [5]

4   A train of mass 50 tonnes is travelling at 35 m s^{-1} when the brakes are applied,
    causing a resultant braking force of 25 kN

   Find the distance the train travels before coming to rest.   [4]

5   Two particles, A and B, of mass 30 g are 70 g respectively, are connected
    by a light inextensible string that passes over a smooth pulley. The particles
    are released from rest both 0.3 m from the pulley and 1 m from the ground.

   a   Calculate

   i   The tension in the string,   **ii**   The acceleration of the system.   [5]

   b   Work out the height particle A travels to after particle B hits the ground.   [4]

6   An object of mass 1.2 kg rests on a smooth horizontal table. It is attached to a second object of
    mass 0.8 kg that hangs over the edge of the table via a light, inextensible string that passes over a
    smooth pulley as shown in the diagram.

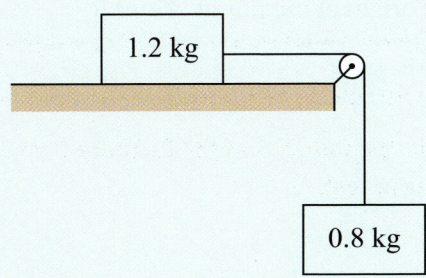

   a   The system is released from rest, calculate

      i   The acceleration of the objects,   **ii**   The tension in the string.   **[5]**

   b   Explain what is meant by each of these modelling assumptions,

      i   The string is light and inextensible,   **ii**   The pulley and table are smooth.   **[3]**

7   A particle passes through the origin at velocity $\begin{pmatrix} 2 \\ -4 \end{pmatrix}$ m s^{-1} then moves with constant

    acceleration for 5 seconds until it reaches a velocity of $\begin{pmatrix} -3 \\ 6 \end{pmatrix}$ m s^{-1}

   a   Find the acceleration over these 5 seconds.   **[3]**

    After 5 seconds, the particle is at point A

   b   Find $|\overrightarrow{OA}|$, showing all your working.   **[4]**

8   A boat starts from rest then moves with acceleration $(2\mathbf{i} - 0.5\mathbf{j})$ m s^{-2}, where $\mathbf{i}$ and $\mathbf{j}$ are unit vectors directed due east and due north respectively.

   a   Find the speed of the boat 10 seconds after it starts moving.   **[3]**

    The boat stops accelerating after travelling 500 m.

   b   Calculate the time taken to travel to this point.   **[4]**

9   A particle moves with velocity $v = t + \cos 2t$

   a   Calculate the acceleration of the particle when $t = \dfrac{\pi}{6}$. Show your working.   **[3]**

   b   Find an expression for the displacement of the particle from the origin at time t given that the particle is at the origin when $t = \pi$ **[4]**

10   The velocity of a particle is given by $\boldsymbol{v} = t^2\mathbf{i} + 2t\mathbf{j}$,

   a   Calculate the acceleration when $t = 3$. Show your working.   **[3]**

    When $t = 0$, $\boldsymbol{r} = 3\mathbf{i} - 5\mathbf{j}$

   b   Find an expression for the displacement at time t **[4]**

11   A cyclist moves on a horizontal road. The position vector of the particle at

    t seconds is given by $\boldsymbol{r} = \begin{pmatrix} \dfrac{1}{3}t^3 - \dfrac{1}{2}t^2 - 3 \\ t^2 - \dfrac{1}{6}t^3 \end{pmatrix}$ m

   a   When $t = 3$, showing each step of working, calculate

      i   The speed of the cyclist,   **ii**   The magnitude of the acceleration of the cyclist.   **[7]**

    The cyclist is due east of the origin at the points A and B

   b   Calculate the distance AB **[6]**

12 An object of mass 2 kg is acted on by a force of 12 N at 30° to the horizontal and a force of P at $\theta°$ to the horizontal as shown in the diagram.

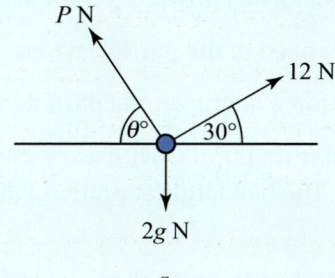

Find the values of P and θ when

a The system is in equilibrium, [5]

b The object is accelerating at 3 m s^{-2} in the direction shown. [5]

13 A block of mass 7 kg rests on a rough horizontal plane, the coefficient of friction between the block and the plane is 0.7.

A horizontal force P acts on the block as shown in the diagram.

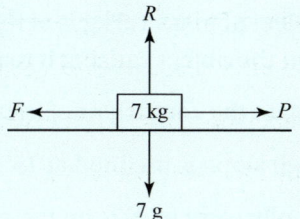

a Calculate the frictional force acting on the block when

    i $P = 30$ N   **ii** $P = 55$ N [5]

b Calculate the acceleration when $P = 60$ N [2]

14 A light beam, AB, of length 7 m is supported at points C and D which are positioned 1 m from the ends of the plank as shown. A mass of 3 kg is placed on the beam, 4 m from point C and the beam remains in equilibrium.

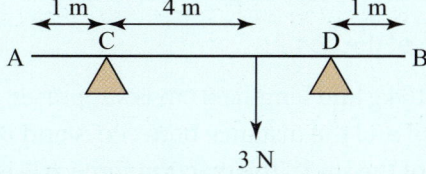

a Calculate the reaction force at points C and D. [4]

b Explain how you have used the assumption that the beam is light. [1]

15 A uniform rod, AB, of length 9 m and mass 30 kg is supported at each end. A mass of 12 kg is placed on the beam, 2 m from A.

a Find the reaction at each of the supports. [4]

b Explain how you have used the assumption that the rod is uniform. [1]

16 A jet ski is initially at rest at position vector $r = \begin{pmatrix} 3 \\ -2 \end{pmatrix}$ m. It then starts moving with constant acceleration for four seconds until it reaches the point A where $\overrightarrow{OA} = \begin{pmatrix} -5 \\ 1 \end{pmatrix}$

a Calculate the acceleration over these four seconds. [3]

b Calculate the speed the jet ski is moving when it reaches point A [4]

c Work out the bearing the jet ski travels on. Show your working. [3]

17 A particle of mass 2 kg has position vector at time t given by $r = \begin{pmatrix} 1 + \cos 2t \\ 3\sin 2t \end{pmatrix}$

   a Calculate the speed of the particle when $t = \dfrac{5\pi}{6}$. Show all your working.   **[5]**

   b Calculate the force acting on the particle when $t = \dfrac{5\pi}{6}$. Show each step of working.   **[5]**

18 A ball is projected from point O with a velocity of 6 m s^{-1} at an angle of 60° to the horizontal. The ball lands at point A which is at the same horizontal level as point O

   a Find the time taken for the ball to reach point A **[3]**

   b Calculate the maximum height reached by the ball.   **[3]**

   c Calculate the distance OA **[3]**

   d What assumptions have you made in modelling this situation?   **[2]**

19 An object of mass 1.5 kg is at rest on a rough horizontal surface. The force $F = \begin{pmatrix} 9 \\ 5 \end{pmatrix}$ N acts on the object causing it to accelerate at 2 m s^{-2} along the surface.

   Calculate the coefficient of friction, μ, between the object and the surface.   **[5]**

20 A rough slope is inclined at 15° to the horizontal.

   A box of weight 60 N is on the slope and held in place by a rope which is parallel to the slope as shown.

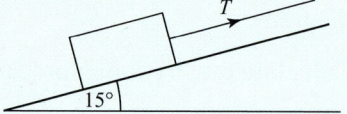

   The coefficient of friction between the box and the slope is 0.13.

   a Calculate the tension in the rope.   **[5]**

   The rope is detached and the box slides down the slope.

   b Find the acceleration of the box.   **[4]**

21 A uniform shelf of mass 16 kg and length 60 cm is supported by brackets at points A and B where B is at one end of the shelf and the distance between A and B is x cm. An object of mass 5 kg is placed on the other end of the shelf. The reaction force at A is double the reaction force at B.

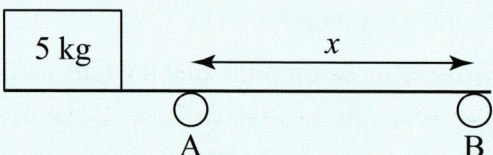

   a By modelling the object as a particle, calculate

     i The reaction forces at A and B,   **ii** The length x **[7]**

   The box is in fact a cube of side length 5 cm

   b Without further calculation, explain the effect this will have on the value of x **[2]**

22 A particle of weight 7 N is hanging from two light, inextensible strings as shown in the diagram.

Find the tension in each string given that the system rests in equilibrium. [5]

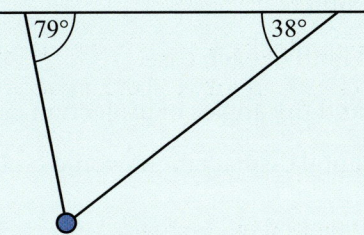

23 A particle of mass 7 kg is initially at rest at the point with position vector $\begin{pmatrix} -1 \\ 3 \end{pmatrix}$ m. It then moves

with acceleration given by $\boldsymbol{a} = \begin{pmatrix} te^t \\ 1-2e^t \end{pmatrix}$ m s^{-2}

Calculate the distance of the particle from the origin after one second. [10]

24 A ball is hit at ground level on a horizontal surface and is caught 1.5 seconds later at a height of 2.1 m above the surface and a horizontal distance of 25 m from where it was projected.

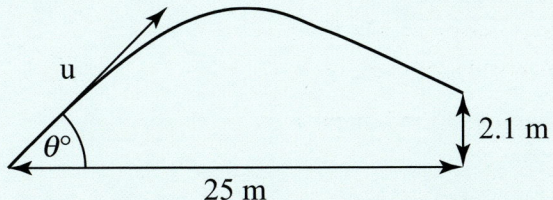

a Calculate

i The initial velocity, ii The angle of projection. [6]

b Calculate the speed and direction of the ball when it is caught. Show your working. [6]

25 A car of mass 2500 kg is towing a caravan of mass 1000 kg up a hill which is inclined at 12° to the horizontal. The caravan is attached to the car using a light, inextensible tow bar.

The coefficient of friction between the vehicles and the road is 0.4. Initially the car and the caravan are at rest, and then they start to move with an acceleration of 4.5 m s^{-2}

a Calculate

i The driving force, ii The tension in the tow bar. [8]

After four seconds, the driving force is removed.

b Calculate the length of time until the car and the caravan come to instantaneous rest. [5]

c Find the driving force required for the car and caravan to be on the point of moving up the hill. [2]

26 A particle is projected with a velocity of 21 m s⁻¹ and lands 50 m away at the same height as it was projected from.

a Calculate

    i The possible angles of projection,

    ii The time taken for the ball to land in each case. **[10]**

Assuming the same initial velocity and the angles of projection calculated in part **a**,

b Explain how and why the horizontal displacement would change if the object being projected was a football. **[2]**

27 A uniform plank has mass 30 kg and length 3.6 m. It is held horizontal by two vertical strings attached at A and B which are 0.5 m and 2 m respectively from one end of the plank.

a Find the magnitudes of the tensions in the strings in terms of g **[5]**

When an object of weight of 40 N is attached x m from the end of the plank as shown, the plank is on the point of rotating about B.

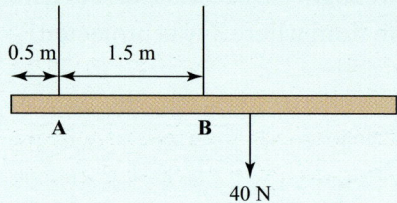

b Calculate the length x **[5]**

28 A particle of mass 700 g is initially travelling at a constant speed of 5 m s⁻¹ and passes through the origin when $t = 0$. It is then acted in on by a horizontal force F, where $F = 3t + 6$, which causes it to accelerate in a straight line.

Calculate the distance of the particle from the origin when $t = 3$ s. **[8]**

29 A tennis player hits the ball directly towards the net at a height of 0.4 m above the ground and a perpendicular distance of 3.1 m from the net. The net is 1 m high and the velocity of the ball is 54 km h⁻¹. The ball passes over the net.

a Calculate the range of possible angles of projection of the ball, giving your answer correct to the nearest tenth of a degree. **[8]**

The total length of the tennis court is 23.8 m and the net is exactly in the middle.

b Calculate the ranges of possible angles of projection of the ball given that it lands within the court on the other side of the net. Give your answer correct to the nearest degree. **[9]**

30 A particle of mass m kg moves in a straight line across a horizontal surface with speed v m s⁻¹. As it moves it experiences a resultant resistive force of magnitude $\frac{1}{5}m\sqrt{v}$. When $t = 0$, the speed of the particle is 9 m s⁻¹

Find the total distance travelled by the particle at its speed decreases from 9 m s⁻¹ to zero. **[10]**

20

Probability and continuous random variables

Although batteries have an expected number of hours that they will last, they rarely last exactly the length of time printed on the box. Their lifespan varies depending on factors such as the environment, manufacture, extent of use and so on, and this can be modelled using a Normal distribution.

The Normal distribution is very useful for modelling outcomes which are influenced by a lot of small contributions, where individual outcomes are varied but the overall pattern is predictable. The distribution shows up in many places, from people's heights and IQs to measurement errors to exam marks. It is a useful tool for modelling the distributions of many random variables. It provides information on the probability of, say, a person being no taller than a certain height. Being able to model probability is very useful in many subject areas, including the social and natural sciences.

Orientation

What you need to know	What you will learn	What this leads to
Ch9 Collecting, representing and interpreting data • Variance and standard deviation. p.234	• To calculate conditional probabilities from data given in different forms. • To apply binomial and Normal probability models in different circumstances. • To use data to assess the validity of probability models. • To solve problems involving both binomial and Normal distributions, including the Normal approximation to the binomial distribution.	**Ch21 Hypothesis testing 2** Testing a Normal distribution. **Applications** Weather forecasting. Actuarial science. Stochastic analysis.
Ch10 Probability and discrete random variables • Binomial distribution. p.270		

 **MyMaths** Practise before you start   🔍 2111, 2281

Fluency and skills

If a fair dice is thrown once, the probability of getting a six is $\frac{1}{6}$

See Ch10.1

For a reminder on the definition of **sample space**.

If you know the outcome of the dice roll is even, the probability it is a six changes to $\frac{1}{3}$ because there are now only 3 possible outcomes. **Conditional probability** is used when information about the outcome is known and it results in a smaller sample space.

Conditional probabilities can be calculated using a Venn diagram, a tree diagram, a two-way table, or formulae. You can also use **set notation** when calculating conditional probabilities.

Let $\varepsilon = \{1, 2, 3, ..., 10\}$, A = prime numbers, B = factors of 20

ε	The **universal set** consists of all elements under consideration	$\{1,2,3,4,5,6,7,8,9,10\}$
$A \cap B$	The **intersection** of A and B consists of elements common to *both* A and B	$\{2,5\}$
$A \cup B$	The **union** of A and B consists of elements that appear in *either* A or B or *both*	$\{1,2,3,4,5,7,10\}$
A'	The **complement** of A consists of elements that do *not* appear in A	$\{1,4,6,8,9,10\}$
$\subset$	A **subset** of ε is a set of elements which are all contained in ε	$A \subset \varepsilon$
$\varnothing$	The **empty set** contains no elements	$A' \cap A = \varnothing$

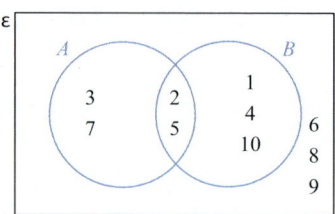

> **Key point**
>
> $P(A \mid B)$ represents the probability of A occurring *given that* B has occurred.

Example 1

25 consumers try three brands of mints, A, B and C, and say which of them, if any, they like. The Venn diagram shows the results.

Find the probability that a randomly chosen consumer

a Likes brand A

b Likes brand A and brand B

c Likes brand B given that they are known to like brand A

d Likes brand B or brand C

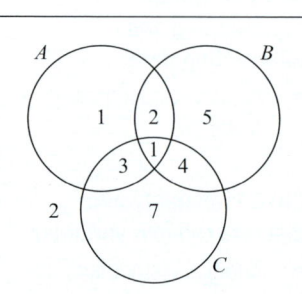

a $P(A) = \dfrac{n(A)}{N} = \dfrac{1+3+1+2}{25} = \dfrac{7}{25}$

b $P(A \cap B) = \dfrac{2+1}{25} = \dfrac{3}{25}$

c $P(B \text{ given } A) = P(B \mid A) = \dfrac{n(A \cap B)}{n(A)} = \dfrac{2+1}{7} = \dfrac{3}{7}$

> Out of the 7 people who like A, only 3 also like B

d $P(B \cup C) = \dfrac{2+5+1+4+3+7}{25} = \dfrac{22}{25}$

If it is 'given' that event A has occurred, the sample space is reduced to just those outcomes involving A

> **Key point**
>
> If A and B are two events associated with a random experiment,
> $$P(B\,|\,A) = \frac{P(A \cap B)}{P(A)} \text{ and, rearranging, } P(A \cap B) = P(A) \times P(B\,|\,A)$$

This result can be used to solve tree diagram problems, including in cases where there is dependence between the stages.

In Example 1, $n(B \cup C)$ is the number of consumers who like at least one of B or C
$n(B \cup C) = 2 + 5 + 1 + 4 + 3 + 7 = 22$
This is the same as $n(B) + n(C) - n(B \cap C)$. You subtract $n(B \cap C)$ to avoid double counting.

> **Key point**
>
> If A and B are two events associated with a random experiment, then
> $$P(A \cup B) = P(A) + P(B) - P(A \cap B)$$

Example 2

Two art experts, S and T, view a painting. The probabilities that they correctly identify the artist are 0.8 and 0.65, respectively. One of the experts is chosen at random and asked to identify the artist.

a Calculate the probability that expert S is chosen and the artist is correctly identified.

b Calculate the probability that the artist is correctly identified.

a Define the events A: S chosen, B: T chosen and
C: artist correctly identified.
$$P(A \text{ and } C) = P(A \cap C) = P(A) \times P(C\,|\,A) = 0.5 \times 0.8 = 0.4$$

> 0.8 is the *conditional* probability of C given A

b $P(C) = P(A \cap C) \text{ or } P(B \cap C)$

> Correct identification is by *either* S or T.

$$= P(A) \times P(C\,|\,A) + P(B) \times P(C\,|\,B)$$
$$= 0.5 \times 0.8 + 0.5 \times 0.65 = 0.725$$

Example 3

In a certain population with equal numbers of males and females, 1 in every 100 males and 1 in every 10 000 females are colour blind. A person is chosen at random from the population. Draw a tree diagram to find the probability that the person is

a Male and colour blind, **b** Colour blind.

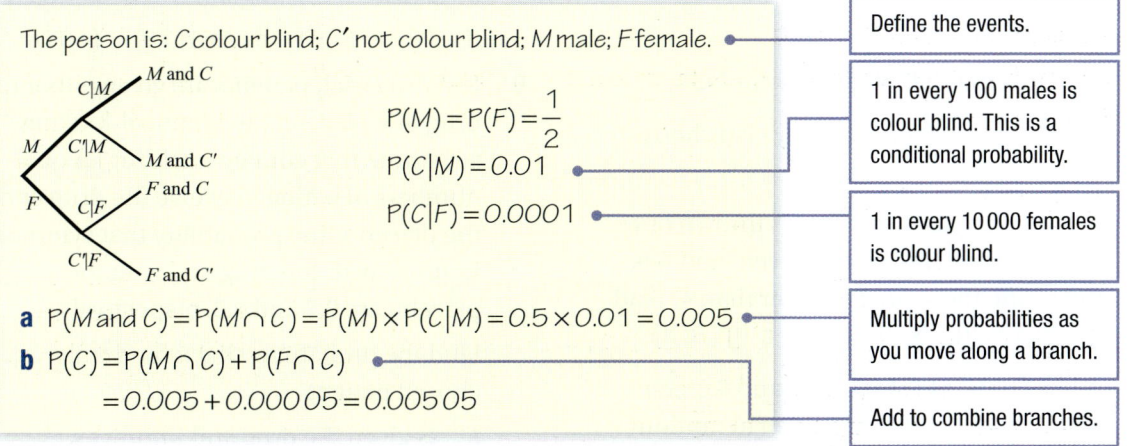

The person is: C colour blind; C' not colour blind; M male; F female.

> Define the events.

$$P(M) = P(F) = \frac{1}{2}$$
$$P(C\,|\,M) = 0.01$$
$$P(C\,|\,F) = 0.0001$$

> 1 in every 100 males is colour blind. This is a conditional probability.

> 1 in every 10 000 females is colour blind.

a $P(M \text{ and } C) = P(M \cap C) = P(M) \times P(C\,|\,M) = 0.5 \times 0.01 = 0.005$

> Multiply probabilities as you move along a branch.

b $P(C) = P(M \cap C) + P(F \cap C)$

$$= 0.005 + 0.00005 = 0.00505$$

> Add to combine branches.

Conditional probability can be used to define independence between events.

STATS

Exercise 20.1A Fluency and skills

1 In a survey, 32 people are asked if they like Indian, Bangladeshi or Thai foods. The Venn diagram shows the number of people who like each type of food.

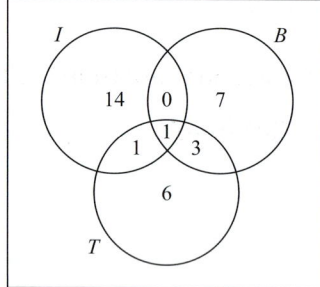

Find the probability that a randomly chosen customer

a Likes Indian food,

b Likes two types of food,

c Likes two types, given that they like Thai food.

2 In a sports club, 28 students choose from football, rugby and athletics. 7 choose only football, 6 choose only rugby and 7 choose only athletics. 6 choose football and rugby, and 2 choose football and athletics. No one chooses all three or rugby and athletics. A student is chosen at random. Use a Venn diagram to calculate the probability that the student chooses

a Rugby,

b Only football, only rugby or both,

c Football, given that they also choose athletics.

3 An ordinary six-sided dice is thrown once. A is the event 'the score is even' and B is the event 'the score is greater than 4'. Find
a $P(A \cap B)$ b $P(A|B)$ c $P(B|A)$

4 A box contains 8 red beads and 6 green beads. Two beads are chosen at random without being replaced and their colours

are recorded. Draw a tree diagram to find the probability that the two chosen beads are different colours.

5 In a hospital survey, 40 patients were classified by diet (vegetarian or not) and blood pressure status (raised or not raised). The results are shown in the table.

BP / Diet	R	R'	Total
V	3	5	8
V'	3	29	32
Total	6	34	40

Find the probability that a patient chosen at random

a Has raised blood pressure,

b Does not have raised blood pressure given that they are vegetarian.

6 If $P(A) = 0.5$, $P(B) = 0.7$ and $P(A \cap B) = 0.4$, find

a $P(A|B)$ b $P(A|B')$ c $P(A'|B)$
d $P(A'|B')$ e $P(A \cup B)$

7 A fair dodecahedral dice has sides numbered 1–12. Event A is rolling more than 9, B is rolling an even number and C is rolling a multiple of 3. Find

a $P(B)$ b $P(A \cap B)$ c $P(A|B)$
d $P((A \cap B)|C)$ e $P(A \cup B)$ f $P(B \cup C)$

8 In a drug trial, patients are given either the drug or a placebo (an identical-looking substitute that contains no drug). Equal numbers of patients receive the drug and the placebo. The probability that symptoms improve with the drug and with the placebo are 0.7 and 0.2, respectively.

Find the probability that a randomly chosen patient

a Is given the drug and improves,

b Improves.

Reasoning and problem-solving

To solve a problem using conditional probability

① Clearly define the events.

② If it's helpful, display the data using a two-way table, Venn diagram or tree diagram.

③ Use formulae to find conditional probabilities.

Example 4

A bag contains the seeds of three different varieties of flower, R, S and T in proportions $1:3:6$

The germination rates for these varieties are 74%, 45% and 60%, respectively.

One randomly-chosen seed is planted.

Given that it germinates, what is the probability that it was of variety S?

R, S and T represent each variety being chosen and G is 'seed germinates'.

Define the events. ①

$P(R) = 0.1 \qquad P(S) = 0.3 \qquad P(T) = 0.6$

Use the ratio to find the probabilities.

$P(G|R) = 0.74 \quad P(G|S) = 0.45 \quad P(G|T) = 0.60$

$$P(S|G) = \frac{P(S \cap G)}{P(G)}$$

For the numerator, use $P(A \cap B) = P(A) \times P(B|A)$ ③

$$= \frac{P(S) \times P(G|S)}{P(G \text{ and } R) + P(G \text{ and } S) + P(G \text{ and } T)}$$

For the denominator, remember that germination can occur with all varieties.

$$= \frac{P(S) \times P(G|S)}{P(R)P(G|R) + P(S)P(G|S) + P(T)P(G|T)}$$

$$= \frac{0.3 \times 0.45}{0.1 \times 0.74 + 0.3 \times 0.45 + 0.6 \times 0.6}$$

$$= \frac{135}{569}$$

Alternatively:

	R	S	T	(total)
G	0.74×0.1 $= 0.074$	0.45×0.3 $= 0.135$	0.6×0.6 $= 0.36$	0.569
Ḡ	$0.1 - 0.074$ $= 0.026$	$0.3 - 0.135$ $= 0.165$	$0.6 - 0.36$ $= 0.24$	0.431
(total)	0.1	0.3	0.6	1

Construct a 2-way table showing all the possible outcomes. ②

$$P(S|G) = \frac{0.135}{0.569}$$

$$= \frac{135}{569}$$

Use the values in the table to calculate the final answer.

STATS

Example 5

S is a sample space for a random experiment where $X \cup Y = S$, $X \cap Y = \varnothing$ and $T \subset S$

$P(X) = P(Y)$, $P(T \mid Y) = \dfrac{1}{4}$ and $P(T \mid X) = \dfrac{2}{3}$

Find   **a** $P(T)$ **b** $P(Y \mid T)$

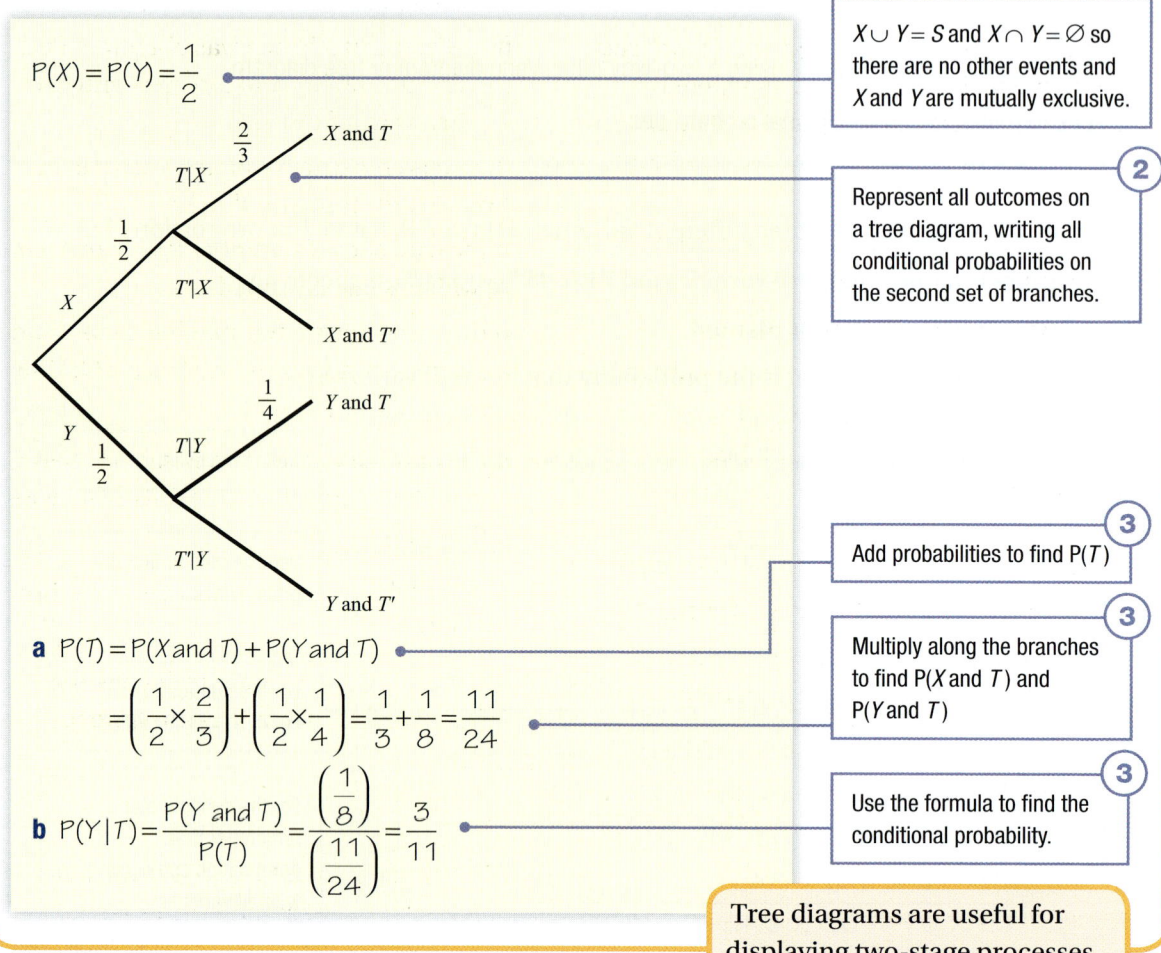

$X \cup Y = S$ and $X \cap Y = \varnothing$ so there are no other events and X and Y are mutually exclusive.

2 Represent all outcomes on a tree diagram, writing all conditional probabilities on the second set of branches.

3 Add probabilities to find P(T)

3 Multiply along the branches to find P(X and T) and P(Y and T)

3 Use the formula to find the conditional probability.

a $P(T) = P(X \text{ and } T) + P(Y \text{ and } T)$

$= \left(\dfrac{1}{2} \times \dfrac{2}{3}\right) + \left(\dfrac{1}{2} \times \dfrac{1}{4}\right) = \dfrac{1}{3} + \dfrac{1}{8} = \dfrac{11}{24}$

b $P(Y \mid T) = \dfrac{P(Y \text{ and } T)}{P(T)} = \dfrac{\left(\dfrac{1}{8}\right)}{\left(\dfrac{11}{24}\right)} = \dfrac{3}{11}$

Tree diagrams are useful for displaying two-stage processes.

Know your data set

Large data set

Jacksonville and Perth had significantly more rain in May–October 2015 than any other locations recorded in the LDS. It's worth knowing the outliers in the LDS. Click this link in the digital book for more information about the Large data set.

Exercise 20.1B Reasoning and problem-solving

1 The sample space for an experiment consists of two mutually exclusive and exhaustive events, P and Q. An event R is a subset of the same sample space. You are given that $P(P) = 0.2$, $P(R \mid P) = 0.4$ and $P(R \mid Q) = 0.7$

   a Use a tree diagram to find $P(R)$

   b Find $P(P \mid R)$

2 A test for a certain disease gives either a positive (disease present) or negative (no disease present) result. It correctly identifies 93% of cases where the disease is present. The proportion of cases where no disease is present but the test result is positive is 7%.

   a Find the proportion of all cases where there is a positive result if 29% of the people being tested actually have the disease.

b Given that a person tests positive for the disease, what is the probability that she has the disease?

3 In a drug trial, patients are given either drug A or drug B. Equal numbers of patients are given each drug. The probability that symptoms improve is 0.71 with drug A and 0.43 with drug B

A patient is chosen at random. Find the probability that the patient

a Receives drug B and improves,

b Received drug B given she has improved.

4 The daily mean air temperature during May 2015 in Beijing had an average value of 21.6°C. On the 18 days when the temperature was above 21.6°C (T), the probability of the daily mean wind speed being less than 5 kn (W) was $\frac{7}{18}$. On the 13 days when the temperature was less than or equal to 21.6°C (T'), the probability of W was $\frac{5}{13}$

For a randomly chosen day from the sample find

a P($T \cap W$)

b P($T \mid W$)

5 In Leeming between 1st June and 31st July 2015 the median daily total sunshine was 5.4 hours. On 30 days the total sunshine was less than 5.4 hours, and the daily maximum temperature was less than 20°C on 19 of these days. The same statistic for days when the total sunshine was greater than or equal to 5.4 hours was 13 out of 31

Find the probability that, on a randomly chosen day

a The temperature was less than 20°C and the total sunshine was less than 5.4 hours

b The temperature was less than 20°C

c The total sunshine was less than 5.4 hours given that the temperature was less than 20°C

6 Three coins are used in a game. Two of them are fair and one has two heads. One coin is chosen at random and flipped.

a Find the probability that a heads is obtained.

b Given that a heads was obtained, find the probability that the coin with two heads was flipped.

7 A card is chosen at random from a set of twelve cards numbered 1–12

If the card shows a number less than 4, coin A, which is fair, is flipped.

If the card shows a number between 4 and 8 inclusive, coin B, for which the probability of a heads is $\frac{2}{3}$, is flipped.

If the number on the card is greater than 8, coin C, for which the probability of a heads is $\frac{1}{3}$, is flipped.

a Find the probability that the coin shows tails.

b If the coin shows tails, calculate the probability that coin B was flipped.

Challenge

8 A box contains 4 black and 8 red balls. A ball is drawn at random, its colour recorded, and the ball is placed back in the box along with *another* ball of the same colour.

a Find the probability that two black balls were drawn.

b If the second ball was black, find the probability that the first was black.

9 Suppose that 1 in every 100 males and 1 in every 10 000 females are colour blind. Assuming that there are equal numbers of males and females in the population, use conditional probability formulae to find the probability that a person chosen at random is female, given that they are not colour blind.

Fluency and skills

To model real-life situations mathematically, you often have to make simplifying assumptions. You can analyse and improve your model by

See Ch 10.2
For a reminder on the binomial distribution.

- comparing predicted results with actual data
- questioning any assumptions that have been made.

To test a binomial model you can use the mean and variance.

Key point

For $X \sim B(n,p)$, the mean (μ) and variance (σ^2) are given by $\mu = np$ and $\sigma^2 = np(1-p)$

Example 1

Jenny says she arrives at school before Claire about 90% of the time. At the end of each week she records how many days, X, out of five she arrives first, and says X follows a binomial distribution. The table shows her results for 40 weeks.

X	0	1	2	3	4	5
frequency	0	1	5	12	15	7

a State any assumptions Jenny has made and use the data to show that $X \sim B(5, 0.9)$ is not a good model for the data.

b Use the mean of the data to suggest a better value for p

Make simplifying assumptions.

a Assume that the probability remains constant every day, and the events are independent.

	mean	variance
Using $X \sim B(n, p)$	$5 \times 0.9 = 4.5$	$5 \times 0.9 \times 0.1 = 0.45$
Using the data	3.55	0.9975

Use your calculator to find the mean and variance.

Calculate and compare the mean and variance using the model and the actual data.

Values are not a close match, so the model is likely not a good one.

b A better value for p is $3.55 \div 5 = 0.71$

Use $\mu = np$

Exercise 20.2A Fluency and skills

1 Red and blue sweets are sold in randomly assorted packets of five. 40 packets are examined and the number of red sweets in each, X, is recorded.

X	0	1	2	3	4	5
frequency	0	3	10	12	13	2

a Calculate the mean and variance of the sample.

b Bo suggests that X can be modelled by a binomial distribution B(5, 0.6). Use the sample to comment on the suitability of this model.

2 A dice is thrown 60 times. The number of scores greater than 2 is half the number of scores less than or equal to 2. Suggest a probability model for the number of outcomes greater than 2 when the dice is thrown 300 times, X. Justify your answer.

3 An experiment is modelled by a random variable X which is thought to have a binomial distribution, $n = 3$, $p = 0.2$. The experiment is performed 100 times.

X	0	1	2	3
frequency	49	40	9	2

Does the data support the suggested probability model?

Reasoning and problem-solving

STATS

Strategy

To solve a probability problem using modelling

1. Consider the assumptions that can be made in the context.
2. Where available, use real data to check the validity of the model.
3. Try to improve the model so it fits the data more closely.

Example 2

A town-centre bus service is scheduled to run six times an hour every weekday. The number of buses due between 7 pm and 8 pm which arrive on time, X, is modelled by $X \sim B(6, 0.72)$

a Give two reasons why a binomial model may not be suitable in this context.

b Explain why this model would not be appropriate for Y, the number of buses due between 4.30 pm and 5.30 pm which arrive on time. How will the model change?

---

a If one bus is late the increased passenger load may cause the following bus to be late, so events are not likely to be independent as assumed by a binomial model.

1 — Think about the conditions for a binomial model in context.

b Traffic is likely to be heavier during this time period because a high proportion of people travel home from work or school at this time. As a result, the chance of delays may be greater than at other times of day so the value of p is likely to be lower.

1 — Challenge the assumption that the risk of delays is constant throughout the day.

3 — Consider how the model could be improved.

Exercise 20.2B Reasoning and problem-solving

1 A coin is tossed 30 times, and yields X heads.

   a Suggest a suitable probability distribution for X, explaining your reasoning.

   b Give a reason why your model may not be suitable.

2 The daily mean temperature was measured throughout summer 2015 at Leuchars. The median and first and third quartiles for this data were 13.9°C, 12.2°C and 15.5°C respectively. During the same year, the daily mean temperature was measured on a random sample of 30 summer days at Town A on the east coast of England.

   a Using the distribution observed in Leuchars, calculate the probability that a randomly chosen day from the Town A sample had temperature

      i less than 13.9°C   ii more than 12.2°C.

      Give one reason why this model might not be appropriate for Town A.

The results of the Town A sample are given in the table.

tempera-ture, t (°C)	$t<8$	$8 \le t <10$	$10 \le t <12$	$12 \le t <14$	$14 \le t <16$	$16 \le t <18$	$18 \le t <20$	$t \ge 20$	Total
frequency	0	1	2	2	9	9	4	3	30

   b Comment on the differences and similarities in temperature distributions at the two locations and explain why these may have come about.

3 Sian plans to revise for 8 hours each day in the 3 weeks leading up to her exams. She reviews the first 15 pages of her notes in 90 minutes, so anticipates that the number of pages she will review each day, X, will have mean 80 and standard deviation 0.05

   a Explain why Sian's model is based on unrealistic assumptions.

   b Describe how more realistic assumptions would change Sian's model.

Fluency and skills

When looking at the probability distribution of a **discrete random variable** (DRV), A, you assign a probability to each value that A could take.

A **continuous random variable** (CRV), X, could take any one of an infinite number of values on a given interval. Instead of assigning probabilities to individual values of X, you assign probabilities to *ranges* of values of X and the probability distribution is represented by a curve, or sequence of curves, called a **probability density function**, $f(x)$.

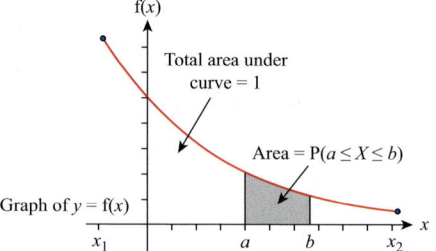

$$P(a \leq X \leq b) = \int_a^b f(x)\, dx$$

For a continuous distribution the probability of an individual value, a, is 0 because the area between a and a is 0. One result of this is that the signs $<$ and $\leq$ become interchangeable.

▲ Graph of $y = f(x)$ where X is a CRV taking values between x_1 and x_2 and $f(x)$ is its probability density function

Key point

If X is a continuous random variable

$$P(a \leq X \leq b) = P(a < X \leq b) = P(a \leq X < b) = P(a < X < b)$$

One of the most important and frequently used probability distributions is the **Normal distribution**. Continuous variables such as height, weight and error measurements are often Normally distributed.

The Normal **probability density function** has a bell-shaped curve. It is a **continuous function** and the area under the curve can be used to calculate probabilities. The total area under the curve equals 1

ICT Resource online

To investigate the Normal distribution, click this link in the digital book.

In a Normal distribution

- Mean $\approx$ median $\approx$ mode
- The distribution is symmetrical
- There are points of inflection one standard deviation (σ) from the mean
- ~68% of values lie within σ of the mean
- ~99.8% of values lie within 3σ of the mean

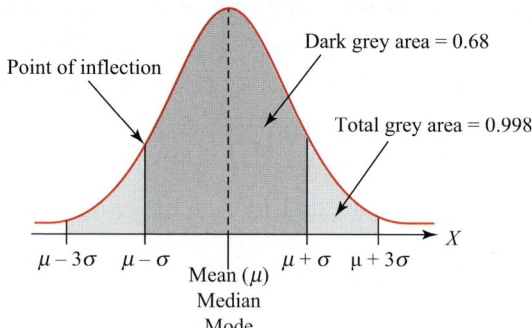

Each Normal distribution is distinguished by its mean, μ, and its variance, σ^2. If a variable X follows a Normal distribution with mean μ and variance σ^2 you write $X \sim N(\mu, \sigma^2)$

Try it on your calculator

You can find probabilities from a Normal distribution on a calculator.

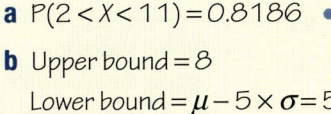

```
Normal   C.D
Data      : Variable
Lower     : 4
Upper     : 5.5
σ         : 1.5
μ         : 3
Save Res  : None      ↓
None LIST
```

```
Normal   C.D
P    =0.20470218
z:Low=0.66666666
z:Up =1.66666667
```

Activity

Find out how to find $P(4 \leq X \leq 5.5)$ where $X \sim N(3, 1.5^2)$ on *your* calculator.

Example 1

If $X \sim N(5, 3^2)$, find the probability that

a $P(2 < X < 11)$　　**b** $P(X < 8)$　　**c** $P(X > 8)$

All calculators work differently, make sure you know how to do this on your own calculator.

a $P(2 < X < 11) = 0.8186$

Use your calculator with $\mu = 5$, $\sigma = 3$

b Upper bound $= 8$

Lower bound $= \mu - 5 \times \sigma = 5 - 5 \times 3 = -10$

$P(X < 8) \approx P(-10 < X < 8) = 0.8413$

The probability of a value lying more than five standard deviations below the mean is negligible, so use $\mu - 5\sigma$ as a default lower bound.

c $P(X > 8) = 1 - P(X < 8)$

$= 1 - 0.8413$

$= 0.1587$

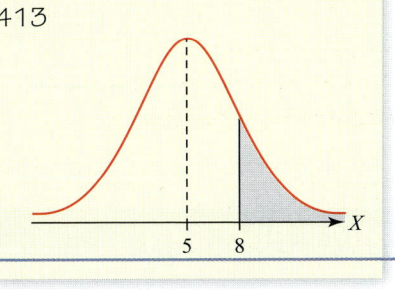

The event $X > 8$ is the **complement** of $X \leq 8$, and $P(X \leq 8) = P(X < 8)$

Since some calculators only give cumulative probabilities for $P(X < x)$ where $x \geq 0$, to calculate a probability you may first have to write it in this form. You can use the symmetry of the graph and the fact that the total area under the graph is 1 to do this.

Key point

$P(X > a) = 1 - P(X < a)$

$P(X < -a) = P(X > a) = 1 - P(X < a)$

$P(a < X < b) = P(X < b) - P(X < a)$

Some calculators allow you to skip this step and find $P(Z > z)$ or $P(Z < -z)$ directly.

To make calculations easier, you can transform any Normal distribution to the **standard Normal distribution**, which has a mean of 0 and a standard deviation of 1. This is a particularly useful instance of the Normal distribution and is usually given the symbol Z, written $Z \sim N(0, 1^2)$

Some calculators don't let you input the values of μ and σ directly. In this case, use the standard Normal distribution.

Key point

For any x-value in a Normal distribution $X \sim N(\mu, \sigma^2)$,

the corresponding z-value in the distribution $Z \sim N(0, 1)$ is $z = \dfrac{x - \mu}{\sigma}$

Rather than finding probabilities corresponding to given values of a random variable, sometimes you need to find the value of a variable corresponding to a given probability. To do this you need to use inverse calculator functions.

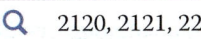 

Try it on your calculator

You can solve problems using the inverse Normal distribution on a calculator.

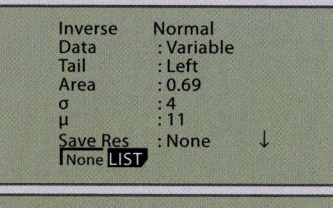

```
Inverse    Normal
Data       : Variable
Tail       : Left
Area       : 0.69
σ          : 4
μ          : 11
Save Res   : None      ↓
None LIST
```

```
Inverse  Normal
 xInv=12.9834014
```

Activity

Find out how to find a when $P(X \le a) = 0.69$ and $X \sim N(11, 4^2)$ on *your* calculator.

Example 2

The random variable, X, is randomly distributed with a mean of 20 and a variance of 8. Find the values of s and t given that

$P(X < s) = 0.7500$ and $P(t < X < s) = 0.5965$

$P(X < s) = 0.75, \sigma = \sqrt{8}, \mu = 20$
$\Rightarrow s = 21.9 \text{ (3 sf)}$

$P(t < X < s) = P(X < s) - P(X < t) = 0.5965$

$0.75 - P(X < t) = 0.5965$

$P(X < t) = 0.1535$

$P(X < t) = 0.1535, \sigma = \sqrt{8}, \mu = 20 \Rightarrow t = 17.1 \text{ (3 sf)}$

Area = 0.5965
Area = 0.25

Draw a sketch to display the information.

Use the inverse Normal function on your calculator.

Exercise 20.3A Fluency and skills

1. If $Z \sim N(0, 1)$, find
   a. $P(Z < 1)$
   b. $P(1 < Z < 1.5)$
   c. $P(-1 < Z < 2)$

2. If $X \sim N(6, 4^2)$, find
   a. $P(X < 9)$
   b. $P(5 < X < 8)$
   c. $P(2 < X < 14)$

3. X is a CRV with $X \sim N(3, \sigma^2)$ and $P(X > 8) = 0.35$
   a. Calculate $P(X < 8)$
   $P(X < a) = 0.35$
   b. Find a
   c. Find $P(X > a)$

4. Find a in each of the following cases, where Z follows the standard Normal distribution $Z \sim N(0, 1)$

   a. $X \sim N(0, 16), P(X \le 5) = P(Z \le a)$
   b. $X \sim N(4, 1), P(X \le 6) = P(Z \le a)$
   c. $X \sim N(2, 9), P(X \le 2) = P(Z \le a)$
   d. $X \sim N(0.5, 1.7^2), P(X \le 3) = P(Z \le a)$
   e. $X \sim N(-1, 2.3^2), P(X \le -4) = P(Z \le a)$
   f. $X \sim N(3, 0.25), P(X \le -2) = P(Z \le a)$

5. The Normal random variable, X, has mean 12 and standard deviation 4. Find the value of d given that $P(X < d) = 0.35$

6. The Normal random variable, X, has mean 21 and variance 10. Find the value of c given that $P(c < X < 21) = 0.40$

7. The quantity of coffee dispensed from a drinks machine is Normally distributed with mean 350 ml and standard deviation 12 ml. Find the probability that a randomly chosen cup of coffee will have a volume between 340 ml and 370 ml.

Reasoning and problem-solving

Strategy

To solve a probability problem using a Normal distribution

1. Identify or find the mean and variance.

2. If necessary, calculate z-values and use the properties of the Normal distribution curve to write probabilities in the form $P(Z < z)$ where z is positive.

3. Use your calculator to find the probability or (using inverse functions) the value of the variable.

4. Compare the results of the model with the actual data.

Example 3

A machine produces metal rods whose lengths are intended to be 1 cm. The machine is set so that, on average, the rods are the correct length and the errors in the length have standard deviation 1 mm.

On inspection, it turns out that 12% of the rods are more than 2 mm from the correct length. Does the Normal distribution provide a good model to describe the length of the rods? Give reasons for your answer.

Let Z be a random variable for the error in the length, in mm, of a randomly chosen rod.

Assume that $Z \sim N(0, 1)$

$P(Z < -2 \text{ or } Z > 2) = 1 - P(-2 < Z < 2)$

$\qquad\qquad = 1 - 0.9545$

$\qquad\qquad = 0.0455$

Using a Normal model, about 4.6% are more than 2 mm from the correct length.

The actual proportion is 12%, so the Normal distribution is not a good model.

① The mean error is 0 and the standard deviation is 1 so the error is modelled by a standard Normal distribution.

② P(rod is more than 2 mm from the correct length) = 1 − P(rod is within 2 mm of the correct length)

③ Use your calculator.

④ Compare the results of the model with the data.

Sometimes you do not know the mean and the variance of a Normal distribution and you need to find them. You can work these out using known probabilities.

Example 4

A council records the waiting times at one of its centres over a one-week period. It finds that 8% of users wait less than 1 minute but 14% wait more than 5 minutes. Council guidelines state that no more than 2.5% of users should wait more than 6 minutes.

a Stating any necessary assumptions, show that this guideline is currently not being met.

b Assuming that the spread of waiting times remains unchanged, find the maximum mean waiting time that would meet the guidelines.

(Continued on next page.)

 **MyMaths** 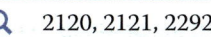 2120, 2121, 2292 SEARCH

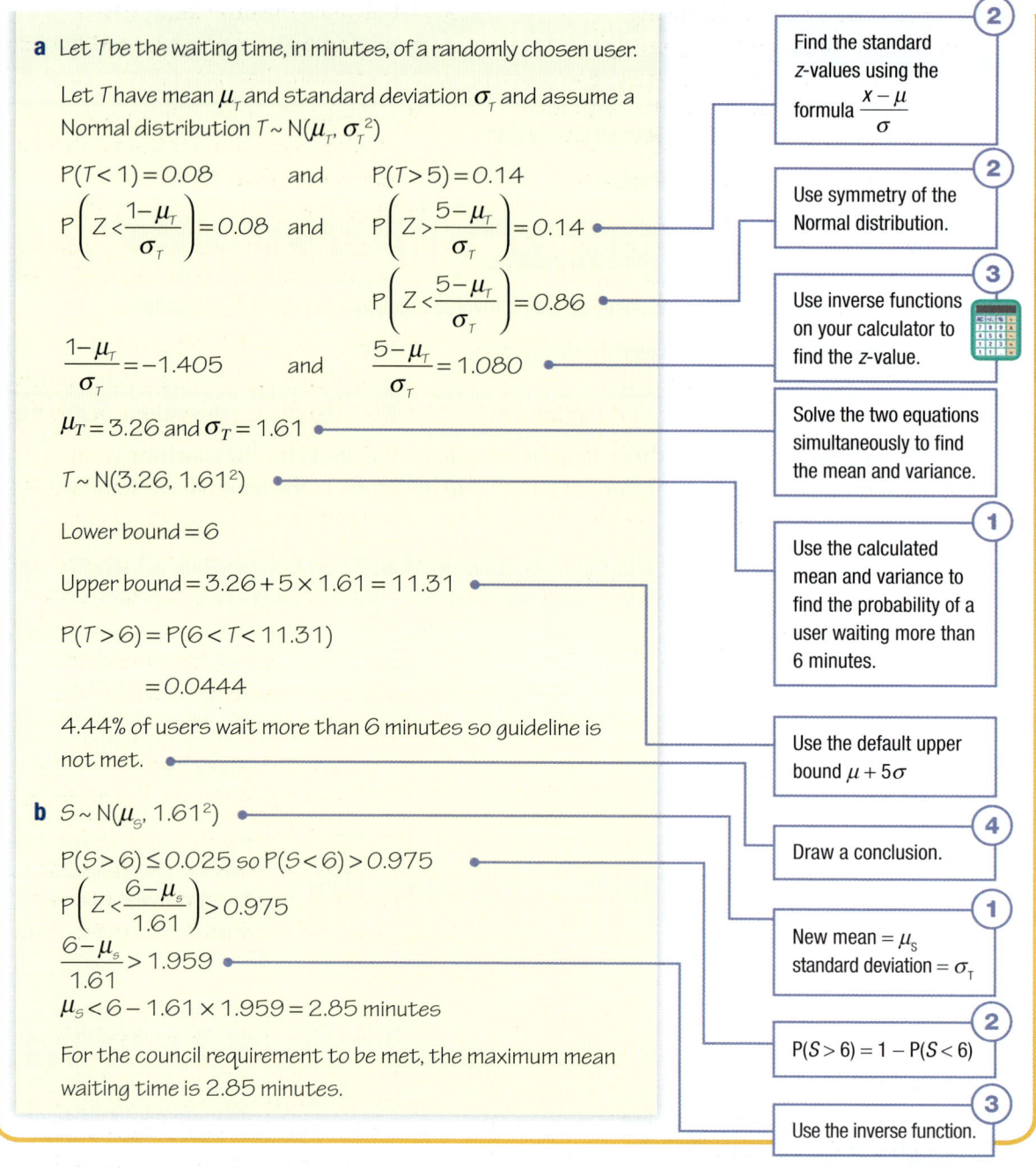

a Let T be the waiting time, in minutes, of a randomly chosen user.

Let T have mean μ_T and standard deviation σ_T and assume a Normal distribution $T \sim N(\mu_T, \sigma_T^2)$

$P(T < 1) = 0.08$ and $P(T > 5) = 0.14$

$P\left(Z < \dfrac{1 - \mu_T}{\sigma_T}\right) = 0.08$ and $P\left(Z > \dfrac{5 - \mu_T}{\sigma_T}\right) = 0.14$

$P\left(Z < \dfrac{5 - \mu_T}{\sigma_T}\right) = 0.86$

$\dfrac{1 - \mu_T}{\sigma_T} = -1.405$ and $\dfrac{5 - \mu_T}{\sigma_T} = 1.080$

$\mu_T = 3.26$ and $\sigma_T = 1.61$

$T \sim N(3.26, 1.61^2)$

Lower bound $= 6$

Upper bound $= 3.26 + 5 \times 1.61 = 11.31$

$P(T > 6) = P(6 < T < 11.31)$

$\qquad = 0.0444$

4.44% of users wait more than 6 minutes so guideline is not met.

b $S \sim N(\mu_S, 1.61^2)$

$P(S > 6) \leq 0.025$ so $P(S < 6) > 0.975$

$P\left(Z < \dfrac{6 - \mu_S}{1.61}\right) > 0.975$

$\dfrac{6 - \mu_S}{1.61} > 1.959$

$\mu_S < 6 - 1.61 \times 1.959 = 2.85$ minutes

For the council requirement to be met, the maximum mean waiting time is 2.85 minutes.

> **2** Find the standard z-values using the formula $\dfrac{x - \mu}{\sigma}$
>
> **2** Use symmetry of the Normal distribution.
>
> **3** Use inverse functions on your calculator to find the z-value.
>
> Solve the two equations simultaneously to find the mean and variance.
>
> **1** Use the calculated mean and variance to find the probability of a user waiting more than 6 minutes.
>
> Use the default upper bound $\mu + 5\sigma$
>
> **4** Draw a conclusion.
>
> **1** New mean $= \mu_S$ standard deviation $= \sigma_T$
>
> **2** $P(S > 6) = 1 - P(S < 6)$
>
> **3** Use the inverse function.

Exercise 20.3B Reasoning and problem-solving

1 Find the unknown parameters in each distribution.

    a $R \sim N(\mu_R, 9)$ given $P(R < 15) = 0.7$

    b $S \sim N(8, \sigma_S^2)$ given $P(S < 6) = 0.4$

    c $T \sim N(\mu_T, \sigma_T^2)$ given $P(T < 15) = 0.7$ and $P(T < 18) = 0.9$

2 a Over several years, a school's cross-country running event was known to be completed in a mean time of 12 minutes 10 seconds with a standard deviation of 1 minute 20 seconds. One year 34 runners took part and a commendation was given to any runner who ran the course in less than 11 minutes. Estimate the number of runners receiving the commendation. State any distributional assumptions made.

b In the same event the following year, organisers wanted to specify a minimum time required to achieve a special commendation which would be awarded to the fastest 10% of runners. What should that time be? Give your answer to the nearest second.

3 a Show that, for a Normal population, approximately 95% of observations are within 2 standard deviations of the mean.

b In a reaction time test, 100 individuals were timed responding to a visual stimulus. The mean, median and modal times were 275 ms, 265 ms and 270 ms, and 93% of observations were within 2 standard deviations of the mean.

Is a Normal distribution a good probability model for these results? Give reasons for your answer.

4 The masses of 28 birds eggs, in grams, are shown in the histogram.

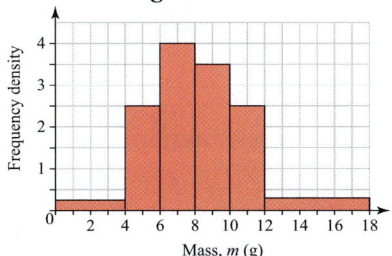

Mass, m (g)

a Estimate the mean and standard deviation of this sample.

b Use linear interpolation to find the proportion of observations within 1 standard deviation of the mean.

c Comment on the distribution of egg weights in the light of your answers to parts **a** and **b**.

5 Following the installation of a new passenger scanner at an airport, the standard deviation of passenger waiting times remains at 3.4 minutes but the mean is reduced from 8. If the probability of a passenger waiting less than 8 minutes is now 0.35, find the new mean waiting time.

6 Two mills produce bags of flour. Mill A produces bags with mass, X kg, $X \sim N(1.2, 0.05^2)$. Mill B produces bags with mass, Y kg, $Y \sim N(1.3, 0.1^2)$

a i Calculate the probability that a randomly chosen bag from Mill A has mass more than 1.25 kg

ii Calculate the probability that a randomly chosen bag from Mill B has mass more than 1.4 kg

iii Show that, for $W \sim N(\mu, \sigma^2)$, the probability of W taking a value more than one standard deviation above the mean is 0.1587

iv Show that, for $W \sim N(\mu, \sigma^2)$, the probability of W taking a value more than n standard deviations below the mean is $P(Z < -n)$

b The two mills are equally likely to produce a bag of mass less than a kg. Find a.

7 The daily mean pressure (hPa) at Heathrow between 1st September and 31st October 2015, X, has median 1020, mean 1018 and standard deviation 10.027. The number of observations between 1008 and 1028 inclusive is 45

Large data set

a Give two reasons why X is likely to be approximately Normally distributed.

b Find the probability that a randomly chosen day at Heathrow in September–October 2015 would have a daily mean pressure between 1015 and 1025 inclusive.

Challenge

8 A waste disposal company recorded waiting times over a one-week period at one of its centres. The results showed that 14% of users waited more than 5 minutes and that 8% of users waited less than 1 minute. The company requirement is that no more than 2.5% of users will wait more than 6 minutes.

a Making any necessary distributional assumptions, show that this requirement is currently not being met.

b Assuming that the spread of waiting times remains unchanged, find the maximum mean waiting time allowed to meet the requirement.

Fluency and skills

You can use a Normal distribution to approximate a binomial distribution *as long as n is large enough*. 'Large enough' is usually taken to be *n* greater than 30 if *p* is not too far from 0.5

See Ch 10.2
For a reminder of the binomial distribution.

The binomial distribution models situations where the random variable takes only discrete values. The Normal distribution models continuous variables. The approximation only works if *n* is quite large.

The random variables *P*, *Q* and *R*, with distributions B(5, 0.4), B(10, 0.4) and B(100, 0.4) respectively, are shown as probability histograms. As the number of trials, *n*, increases, the shape of the distribution becomes increasingly symmetric about its mean value and increasingly resembles a Normal distribution.

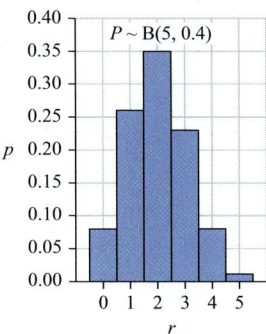
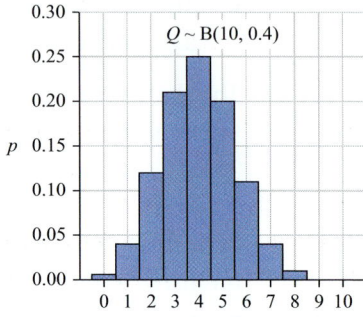
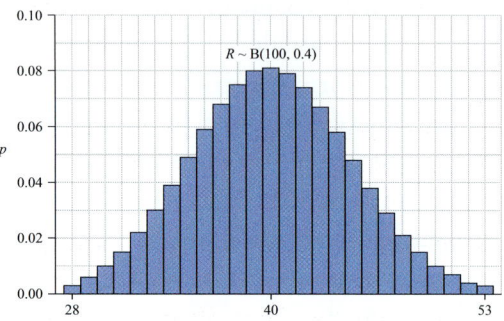

Key point

For $X \sim \text{B}(n, p)$, as *n* increases, the distribution of *X* tends to that of the random variable *Y* where $Y \sim \text{N}(np, np(1 - p))$

The diagram shows the binomial distribution, $n = 30$, $p = 0.5$ in red. The Normal distribution, which is used as an approximation, is shown by a blue line.

The area most closely approximating $\text{P}(X = 18)$ is the area under the curve between 17.5 and 18.5, and the area most closely approximating $\text{P}(X \le 18)$ is the area under the curve to the left of 18.5

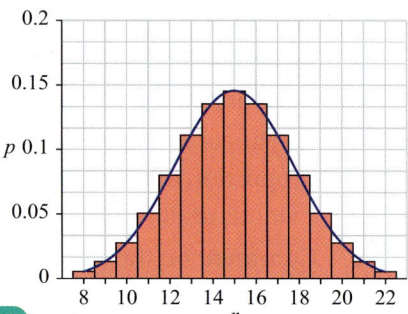

Key point

If *n* is large and *p* is close to 0.5 then $X \sim \text{B}(n, p)$ can be approximated by $Y \sim \text{N}(np, np(1 - p))$

$$\text{P}(X = x) \approx \text{P}\left(x - \frac{1}{2} < Y < x + \frac{1}{2}\right)$$

Inclusion of the $\frac{1}{2}$ increases the accuracy of the approximation. It is known as a **continuity correction**. You should always use it when approximating the binomial by the Normal.

> You use the continuity correction because *X* is discrete whilst *Y* is continuous.

Example 1

If $X \sim B(35, 0.4)$, use the Normal approximation to the binomial distribution to find approximations of

a $P(X = 19)$ **b** $P(X < 19)$ **c** $P(13 < X < 17)$

You should use a continuity correction in each part.

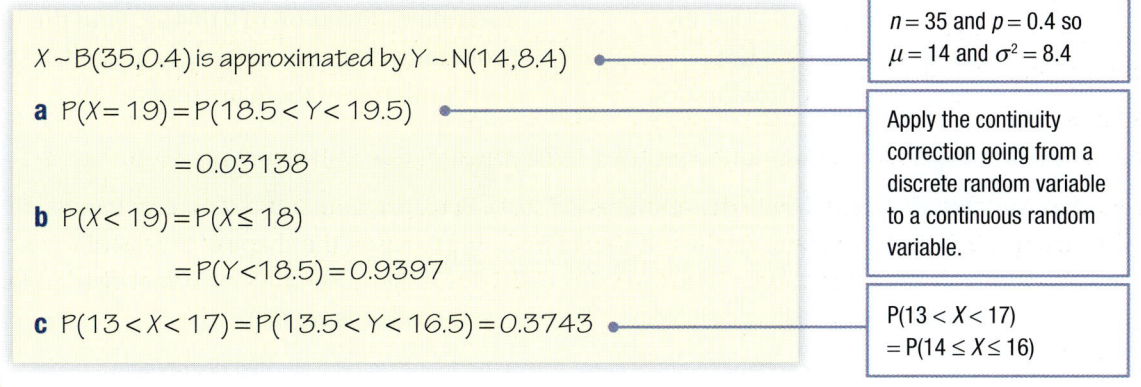

$X \sim B(35, 0.4)$ is approximated by $Y \sim N(14, 8.4)$

$n = 35$ and $p = 0.4$ so $\mu = 14$ and $\sigma^2 = 8.4$

a $P(X = 19) = P(18.5 < Y < 19.5)$

$= 0.03138$

Apply the continuity correction going from a discrete random variable to a continuous random variable.

b $P(X < 19) = P(X \le 18)$

$= P(Y < 18.5) = 0.9397$

c $P(13 < X < 17) = P(13.5 < Y < 16.5) = 0.3743$

$P(13 < X < 17)$
$= P(14 \le X \le 16)$

The Normal distribution can be used to find the value of p, the probability of success, in a binomial distribution.

Example 2

Waiting times at an airport security gate are Normally distributed with mean 10.2 minutes and standard deviation 3.4 minutes. Out of a randomly chosen 12 passengers, find the probability that more than five wait for less than 8 minutes.

Let X be the waiting time in minutes. $X \sim N(10.2, 3.4^2)$

$P(X < 8) = 0.2588$

Use the Normal distribution function on your calculator.

Let Y be the number out of 12 who wait less than 8 minutes.

$Y \sim B(12, 0.2588)$

$P(Y > 5) = 1 - P(Y \le 5)$

$= 1 - 0.93666 = 0.06334$

Use the binomial function on your calculator.

Exercise 20.4A Fluency and skills

1 If $X \sim B(40, 0.45)$ use the Normal approximation to the binomial distribution to find

a $P(X = 21)$ **b** $P(X < 21)$

c $P(16 \le X \le 23)$

Use a continuity correction in each part.

2 The discrete random variable $X \sim B(60, 0.45)$ can be approximated by the continuous random variable $Y \sim N(27, 14.85)$

i Apply a continuity correction to write down the equivalent probability statement for Y.

ii Show that the two distributions yield roughly the same probability in each case.

a $P(X = 23)$ **b** $P(X \le 50)$

c $P(X \ge 12)$ **d** $P(32 \le X \le 51)$

e $P(X > 17)$ **f** $P(X < 40)$

3 A random variable X is Normally distributed with mean 26 and standard deviation 6. An independent random sample of size 6 is taken from the population. Find the probability that more than 3 of the observations are greater than 24

4 A plane carries 300 passengers. Usually, one third of passengers order a hot meal. Use an appropriate approximating distribution to find the probability that more than 115 passengers have a hot meal.

5 The mass shown on packets of red lentils is 1 kg. To satisfy weights and measures legislation, the manufacturer ensures that the mean weight of bags is 1.003 kg with a standard deviation of 0.004 kg. Find the probability that, out of 8 bags checked, less than a quarter of them are under 1 kg.

Reasoning and problem-solving

Strategy

To solve a probability problem by modelling with the binomial and Normal distributions

1 Use an appropriate approximating distribution if the necessary conditions are met.

2 For a binomial problem, the probability of success, p, may be found using a Normal distribution.

3 Use known data to check the validity of any model used.

If you are not told what distribution to use, you must make assumptions about what is appropriate and test against the given data.

Example 3

A machine pours mineral water into bottles. The bottles are labelled '500 ml' and the machine is set so that the mean volume of water in the bottles is 505 ml with a standard deviation of 7 ml.

a Assuming a Normal distribution for volumes, find the probability that a bottle chosen at random from the output will contain within 2 ml of 500 ml.

b Quality inspectors take samples of 5 consecutive bottles every 15 minutes for 10 hours and record how many bottles, Y, in each batch of 5 contain within 2 ml of 500 ml. The results are shown in the table.

Y	0	1	2	3	4	5
f	0	2	8	12	12	6

Find the mean and variance of Y from this data.

c Assuming a binomial model is appropriate for Y, does the result from part **b** suggest the Normal distribution $N(505, 7^2)$ is a suitable model for the distribution of volumes?

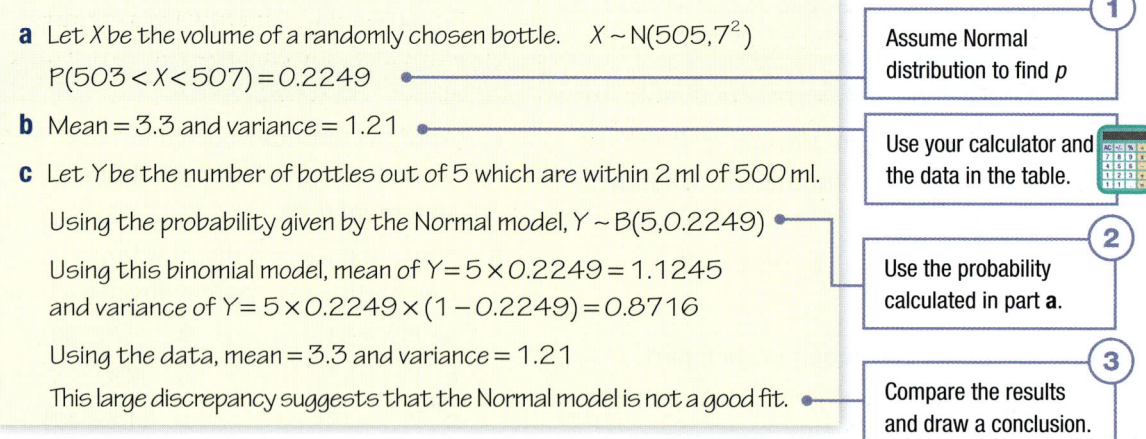

a Let X be the volume of a randomly chosen bottle. $X \sim N(505, 7^2)$

$P(503 < X < 507) = 0.2249$ •——————— Assume Normal distribution to find p ①

b Mean = 3.3 and variance = 1.21 •——————— Use your calculator and the data in the table.

c Let Y be the number of bottles out of 5 which are within 2 ml of 500 ml.

Using the probability given by the Normal model, $Y \sim B(5, 0.2249)$ •———

Using this binomial model, mean of $Y = 5 \times 0.2249 = 1.1245$ and variance of $Y = 5 \times 0.2249 \times (1 - 0.2249) = 0.8716$ ——— Use the probability calculated in part **a**. ②

Using the data, mean = 3.3 and variance = 1.21

This large discrepancy suggests that the Normal model is not a good fit. •——— Compare the results and draw a conclusion. ③

Exercise 20.4B Reasoning and problem-solving

1 A large number of batteries has a mean lifetime of 31 hours and a standard deviation of 21 hours. Half of all lifetimes are less than 20. Give two reasons why a Normal distribution would not provide a good probability model for this population.

2 A random variable X has a binomial distribution with parameters $n = 30$ and $p = 0.4$

  a Using an approximating Normal distribution, find the probability that a randomly chosen value will lie within one standard deviation of the mean.

  b Does your answer to part **a** suggest that the Normal distribution provides a good approximation?

3 a A coin is tossed 10 times and a head appears 7 times. Does this suggest that the coin is biased?

  b The same coin is tossed 10 000 times and a head appears 5450 times. By finding the probability of obtaining at least 5100 heads, decide whether this new evidence changes your view of the coin.

4 a Write down the modal daily rainfall in Beijing during September and October 2015 and explain why this statistic could not be modelled by a Normal distribution.

  b The mean daily mean pressure, measured in hPa, in Beijing during September and October 2015 was 1018 with a standard deviation of 5.17 (2 dp).

   i What is the standard SI unit of pressure and the meaning of the prefix h in hPa?

   ii The daily mean pressure follows a Normal distribution. Find the probability that a randomly chosen day in this period will have a mean pressure below 1015 hPa.

   iii Find the probability that, out of 30 days, more than 8 have a mean pressure below 1015 hPa.

5 The total vegetable content of chicken and mushroom pies has a mean of 98 g and a standard deviation of 10 g. A sample of 10 pies is taken every hour for 24 hours and the value of X, the number of pies with a vegetable content over 98 g, is noted. The mean and standard deviation of these X values are 3.1 and 2.1. Does the Normal distribution provide a good model for the vegetable content of the pies? Give reasons for your answer.

Challenge

6 Airlines sometimes sell more tickets than the number of seats available for the flight. For a particular airline, it is known that on average 6% of passengers fail to turn up for their flight. For a flight with 300 seats available, show that the number of tickets sold should not exceed 309 if the probability that the flight cannot accommodate all passengers who turn up has to be less than 1%.

Chapter summary

- $P(A\,|\,B) = \dfrac{P(A \cap B)}{P(B)}$, and events A and B are independent if $P(A\,|\,B) = P(A)$

- $P(A \cup B) = P(A) + P(B) - P(A \cap B)$

- For the distribution $X \sim B(n, p)$, the mean is $\mu = np$ and the variance is $\sigma^2 = np(1 - p)$

- Normal probability density functions are symmetrical about a clearly defined central mean value so $P(X > \mu + a) = P(X < \mu - a)$ for any constant, a.

- A Normal distribution curve has points of inflection one standard deviation away from the mean.

- For a Normal distribution curve
  - Total area under the curve $= 1$
  - $\sim$68% of values lie within σ of the mean
  - $\sim$99.8% of values lie within 3σ of the mean

- If $X \sim N(\mu, \sigma^2)$, then the variable, Z, given by $Z = \dfrac{X - \mu}{\sigma}$, is also Normal with mean 0, variance 1

- $X \sim B(n, p)$ can be approximated by a Normal distribution $Y \sim N(np, np(1 - p))$ when n is large.

- If a discrete random variable $X \sim B(n, p)$ is approximated by a continuous random variable $Y \sim N(np, np(1 - p))$ then a continuity correction must be applied, so $P(X = x) = P\left(x - \dfrac{1}{2} < Y < x + \dfrac{1}{2}\right)$

Check and review

You should now be able to...	Try Questions
✔ Calculate conditional probabilities from data given in different forms.	1–3
✔ Apply binomial and Normal probability models in different circumstances.	4-6
✔ Use data to assess the validity of probability models.	4, 7
✔ Solve problems involving both binomial and Normal distributions, including the Normal approximation to the binomial distribution.	8

1 You choose a card from a set of thirty cards numbered 1–30. If the card shows a multiple of 4 you flip a fair coin. If it is not a multiple of 4 you flip a coin for which the probability of a heads is $\dfrac{2}{3}$. Find the probability of obtaining

  a A heads given the score was a multiple of 4

  b A heads and not a multiple of 4

2 In a survey of travel arrangements to school, 40 students are questioned about their means of transport. Students who don't walk the whole distance either take the bus or cycle or do both. Of those who don't walk the whole distance, 10 take only the bus, 15 cycle at some point on their journey and 12 cycle the whole way.

  a Show the results on a Venn diagram.

b Find the probability that a randomly chosen student

    i Both cycles and takes the bus,

    ii Takes the bus at some point given that they don't walk.

3 You have two fair dice. Dice 1 is 6-sided and numbered 1–6. Dice 2 is 10-sided and numbered 1–10. You choose a dice at random and roll it. What is the probability that dice 1 was chosen given that the score was 6?

4 A spinner has four sides coloured red, blue, green and yellow. You believe that the outcomes are equally likely. In an experiment, you spin it four times and record the number of times you get yellow.

  a Copy and complete the following table with the theoretical probability of each outcome.

x	0	1	2	3	4
$P(X=x)$	0.3164	0.4219			

The experiment is repeated 80 times giving these results.

x	0	1	2	3	4
f	22	34	17	6	1

  b Do these results support your belief that the spinner is fair? Give reasons for your answer.

5 a Given $X \sim N(12, 3)$, calculate

    i $P(X < 10)$ **ii** $P(9 < X < 14)$

  b The random variable, X, has mean 19 and variance 8. Find the value of a given that $P(a < X < 22) = 0.65$

6 In a packing warehouse, peppers have weights that are Normally distributed with mean 150 g and standard deviation 12 g. Find the probability that a randomly chosen pepper will have a weight between 140 g and 170 g.

7 At a health spa, 75 people were timed to complete a fitness test. The mean and modal times were 281 s and 288 s respectively. Half of the observations were less than 279 s and 69% were within one standard deviation of the mean. Would a Normal distribution be a good probability model for this data? Give reasons for your answer.

8 In a call centre, waiting times are Normally distributed with mean 8.2 minutes and standard deviation 2.7 minutes. Out of a randomly chosen 18 callers, find the probability that fewer than 4 wait for more than 10 minutes.

What next?

Score			
	0–4	Your knowledge of this topic is still developing. To improve, search in MyMaths for the codes: 2095, 2113, 2120, 2121, 2286, 2292	Click these links in the digital book
	5–6	You're gaining a secure knowledge of this topic. To improve, look at the InvisiPen videos for Fluency and skills (20A)	
	7–8	You've mastered these skills. Well done, you're ready to progress! To develop your exam techniques, look at the InvisiPen videos for Reasoning and problem-solving (20B)	

History

Thomas Bayes (1701 – 1761) was an English statistician and Presbyterian minister. He published two books during his lifetime, but it was not until after his death that his most famous work, now referred to as **Bayes' theorem**, was published.

Bayes' theorem provides a formula for calculating the probability of an event based on related, pre-existing conditions. It can be written as

$$P(A|B) = \frac{P(B|A)\,P(A)}{P(B)}$$

$P(A|B)$ is the probability that A occurs given that B is true.
$P(B|A)$ is the probability that B occurs given that A is true.

Investigation

The **Monty Hall problem** is a probability question based on the idea of an American game show, of which Monty Hall was the presenter. This problem became famous when it appeared in the American magazine, 'Parade'.

Suppose that you are presented with three doors. Behind one is a car and behind each of the others is a goat.

You pick a door at random. The presenter, knowing what's behind each door, opens a different door to reveal a goat. You are then offered the chance to switch your choice to the remaining door.

Write a short description of whether you would switch, and why. Note, there is a correct answer.

Note

The solution to the Monty Hall problem, given by columnist **Marilyn vos Savant**, provoked a huge response from readers, including many with PhDs, proclaiming that she was wrong. Marilyn had the highest recorded IQ according to the Guinness Book of Records. Unsurprisingly, she was right!

Research

Joseph Bertrand proposed a similar problem to the Monty Hall problem in 1889, known as **Bertrand's Box Paradox**.
Have a go at solving it. Try using **Kolmogorov's axioms**.

20 Assessment

1 For two events A and B, it is known that $P(A)=0.3$, $P(B|A)=0.4$ and $P(B'|A')=0.35$

    a Represent this information on a tree diagram showing all the individual probabilities. **[6 marks]**

    b Use your tree diagram to calculate $P(B)$ **[3]**

    c Represent the information on a Venn diagram. **[6]**

2 Weather conditions are recorded in Heathrow over six months in 2015. The data shows the number of days on which the temperature exceeded 25°C (heatwaves) and the number of days on which there was at least 1 mm of rain (rainy days).

	Heat-wave	Not heat-wave	
Rainy	2	41	43
Not rainy	20	121	141
	22	162	184

    a Calculate the probability that a randomly-chosen day

       i Had a heatwave.

       ii Was dry given that there was a heatwave.

       iii Was dry.

       iv Had a heatwave given that it was dry. **[6]**

    b If another day in the past is known to be dry, would you assume it had a heatwave? **[1]**

3 Between 05:30 and 22:30 inclusive, 171 number 1 buses arrive at a given bus stop.

    a Calculate the average length of time between bus arrivals. **[3]**

    b Assuming the buses arrive at regular intervals and never run late, what is the probability that a bus arrives between 08:30 and 08:40? **[2]**

    c Discuss whether your answer to part **b** is reasonable for a real bus service, and state any assumptions that are likely to be wrong. **[1]**

4 For each of the following Normal distributions, calculate the probabilities to three significant figures.

    a For $X \sim N(4,3)$ find $P(X<2.3)$ **[2]**

    b For $X \sim N(-4,21)$ find $P(X>-0.5)$ **[2]**

    c For $X \sim N(17,4)$ find $P(X=16)$ **[1]**

5 For $X \sim N(4,4)$, $P(X<2)=0.15866$ and $P(X>7)=0.066807$ to five significant figures.

    a Use this information to calculate $P(2<X<7)$ to three significant figures. **[3]**

    b Express the following probabilities in terms of X and use the information above to calculate their values to 3 significant figures.

       i $P(4<Y)$, where $Y \sim N(8, 16)$

    ii $P(Z < -1)$, where $Z \sim N(0, 1)$

    iii $P(W < -12 \text{ or } W > -2)$, where $W \sim N(-8, 16)$ **[7]**

6 The distance an amateur archer lands an arrow from the centre of the target is modelled by a Normal distribution, with mean 0 cm and variance 100 cm²

  a Find the probability that the arrow lands within 3 cm of centre of the target.   **[1]**

  b The archer shoots ten arrows, one after the other. Assuming the arrows are shot independently, find the probability that at least three arrows land within 3 cm of the centre.   **[3]**

  c Is it reasonable to expect that the arrows have independent probabilities of landing within 3 cm of the centre?   **[1]**

  d In each round of a competition, the archers need to land three out of ten arrows within 3 cm of the centre to score a point for that round. The archer who scores the most points over five rounds wins. Assuming each round is independent of the others, find the expected number of points the archer in this question will score.   **[2]**

7 50 bags of flour are weighed and their masses are recorded in a histogram.

  a Calculate estimates for the mean and variance of the data to 2 decimal places.   **[4]**

  b Let X be the mass of a randomly chosen bag. Show that X can be modelled by a Normal distribution.   **[5]**

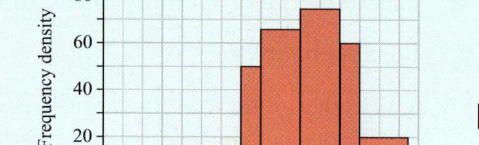

  c Use the Normal model from part **b** to calculate

    i $P(X < 1)$ **ii** $P(X \geq 0.7)$ **iii** $P(X \leq 1.36)$ **[3]**

  d Use the Normal model from part **b** to write the interval, centred at the mean, that 99.8% of the data lies in.   **[1]**

8 A casino uses a dice testing machine to ensure a dice is rolling the right number of sixes.

  a Assuming the dice is fair

    i Calculate μ, the expected number of sixes in ten rolls,

    ii Calculate the variance in the number of sixes over ten rolls,

    iii Calculate $\mu + \dfrac{\sqrt{10}}{2}$ and $\mu - \dfrac{\sqrt{10}}{2}$

    iv Find the probability that in ten rolls the number of sixes rolled is within $\dfrac{\sqrt{10}}{2}$ of the expected number.   **[6]**

  b If the dice is rolled n times, where n is a large number, state a suitable approximate distribution for the number of sixes rolled.   **[2]**

  c Calculate the probability that over n rolls, the number of sixes rolled is within $\dfrac{\sqrt{n}}{2}$ of the expected number, for

    i $n = 900$ **ii** $n = 10\,000$ **iii** $n = 1\,000\,000$ **[9]**

21

Hypothesis testing 2

Imagine a gymnast who wants to improve her success rate at achieving a score of over 12 for her beam routine. She spends three hours every day for a month practising this. Would you expect the number of points she scores for each performance to increase as the month progresses? This is a simple example of correlation: you would expect a clear link between the number of hours of practice and the number of points scored.

Correlation testing is an important part of statistics. While it does not prove causation, the ability to calculate the correlation between two variables helps us determine the likelihood of one having an influence on the other, particularly when the correlation is not as obvious as in the example above. It is a facet of hypothesis testing, along with techniques such as testing for a Normal distribution, and a vital part of statistics and research in many subject areas.

Orientation

What you need to know	What you will learn	What this leads to
Ch9 Collecting, representing and interpreting data • Correlation and causation. *p.252*	• To state the null and alternative hypothesis when testing for correlation or when testing the mean of a Normal distribution. • To compare a given PMCC to a critical value, or its p-value to a significance level, and use this to accept or reject the null hypothesis. • To calculate the test statistic, compare to a critical value, or its p-value to a significance level, and use this to accept or reject the null hypothesis. • To decide what the conclusion means in context about the correlation and about the mean of the distribution.	**Applications** Scientific research. Quality control. Test engineering.
Ch11 Hypothesis testing 1 • Significance and p-value. *p.286*		
Ch20 Probability and continuous random variables • Introduction to the Normal distribution. *p.564*		

 MyMaths   Practise before you start   🔍 2115, 2120, 2283

Fluency and skills

To find out how correlated two variables are, you need a way to test the strength of the **correlation**.

See Ch9.4

For a reminder on correlation and causation.

> **Key point**
>
> The **population correlation coefficient**, ρ, describes how correlated two variables are.
>
> **Pearson's product moment correlation coefficient (PMCC)**, r, is a statistic that estimates ρ

The PMCC is a measure of correlation in a sample, and is used to estimate the population correlation coefficient. The estimate becomes better as the sample size increases, but it is likely to differ from the true value.

> The population correlation coefficient takes values between -1 and 1. 0 indicates zero correlation, 1 and -1 indicate perfect correlation, so the points plotted on a scatter diagram lie on a straight line, with positive and negative gradients respectively.

See Ch11.1

For a reminder on hypothesis testing.

In a hypothesis test, the **null hypothesis** states that the two variables have no correlation, i.e. $H_0: \rho = 0$. To start with, this is assumed to be true.

The **alternative hypothesis** is

- $H_1: \rho \neq 0$ if you think there could be any kind of correlation,
- $H_1: \rho > 0$ if you think there could be positive correlation,
- $H_1: \rho < 0$ if you think there could be negative correlation.

> $H_1: \rho \neq 0$ is a two-tailed test.
>
> $H_1: \rho > 0$ or $H_1: \rho < 0$ is a one-tailed test.

> **Key point**
>
> If the PMCC, r, is further from zero than the **critical value** then you have sufficient evidence to reject the null hypothesis and decide that $\rho \neq 0$.

Example 1

An art auction house measures the size and price of 10 portraits sold during one month, selected at random, to see if there is any correlation between the two. The null hypothesis is $H_0: \rho = 0$ and the alternative hypothesis is $H_1: \rho \neq 0$ with a 5% significance level.

a Explain why the alternative hypothesis takes this form.

The product moment correlation coefficient for the data is $r = 0.532$ and the critical value for this test is ± 0.6319

b State, with a reason, whether H_0 is accepted or rejected.

> **a** The auction house is investigating to see if there is any correlation present. This is a two-tailed test.
>
> **b** Since the PMCC for the sample is closer to zero than the critical value is, there is insufficient evidence to reject H_0

> 'Any correlation' means there could be positive or negative correlation. A two-tailed test considers the possibility that the correlation coefficient is positive and the possibility that it is negative.

The p-value is the probability of getting a result at least as extreme as the one obtained from the sample if the variables are not correlated.

See Ch11.2
For a reminder on p-values.

Key point
Instead of using the critical value, you can compare the **p-value** for the PMCC to the significance level. You accept the null hypothesis if the p-value is greater than the significance level.

ICT Resource online

To investigate correlation using hypothesis tests, click this link in the digital book.

Example 2

The number of hours spent training for a marathon and the number of hours taken to complete a marathon for a random sample of 30 marathon entrants are suspected to have a negative correlation. The hypotheses $H_0 : \rho = 0$ and $H_1 : \rho < 0$ are being considered at the 10% significance level. The PMCC for the sample is -0.3061, which has a p-value of 0.05 for a one-tailed test. State, with a reason, whether H_0 is accepted or rejected.

The p-value is 5%. Since 5% is less than 10%, the result is significant and so the null hypothesis is rejected.

If $\rho = 0$, the probability of this PMCC is less than the significance level.

STATS

Calculator

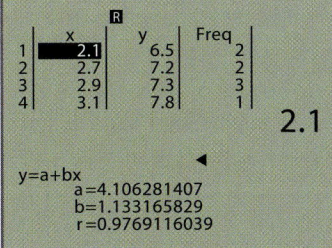

Try it on your calculator

You can calculate the product moment correlation coefficient (PMCC) on a calculator.

Activity

Find out how to calculate the PMCC for the data set (2.1, 6.5), (2.1, 6.5), (2.7, 7.2), (2.7, 7.2), (2.9, 7.3), (2.9, 7.3), (2.9, 7.3), (3.1, 7.8) on *your* calculator.

Exercise 21.1A Fluency and skills

1 The average annual temperature and rainfall in 11 UK towns are measured to investigate whether there is any correlation between the two. The hypotheses $H_0: \rho = 0$ and $H_1: \rho \neq 0$ are being considered at the 5% significance level.

Temp	Rain
18.1	3.06
18.0	2.59
18.4	2.86
18.8	2.62
19.1	2.86
19.1	2.83
19.6	3.41
19.6	3.08
19.7	2.81
20.1	3.13
20.6	3.26

a Use your calculator to find the PMCC.

b The p-value of the PMCC is 0.0446. State, with a reason, whether H_0 is accepted or rejected.

2 A jewellery auction house measures the masses and selling prices of 10 pieces, selected at random, to see if there is any correlation between the two. The null hypothesis is $H_0 : \rho = 0$ and the alternative hypothesis is $H_1 : \rho \neq 0$ with a 5% significance level.

a Explain why the alternative hypothesis takes this form.

The product moment correlation coefficient for the data is $r = -0.148$. The critical value for this test is ± 0.6319

b State, with a reason, whether H_0 is accepted or rejected.

3 The number of caffeinated drinks a person consumes during a day and the number of hours of sleep they get that night are suspected to have a negative correlation. A random sample of 30 people is surveyed to investigate whether any correlation is present. The hypotheses $H_0 : \rho = 0$ and $H_1 : \rho < 0$ are being considered at the 10% significance level. The critical value for the test is -0.2826 and the PMCC for the sample is $r = -0.837$. State, with a reason, whether H_0 is accepted or rejected.

4 A sample is taken across 40 towns to see if limiting alcohol sales after different times in the evening helps to reduce crime levels. The hypotheses $H_0 : \rho = 0$ and $H_1 : \rho < 0$ are tested at the 5% level where ρ measures the correlation between the number of hours before midnight that alcohol is limited and the number of crimes committed that night in the area. The sample is found to have a PMCC of -0.2935. Given that the critical value is -0.2605, state, with a reason, whether H_0 is accepted or rejected.

Reasoning and problem-solving

To solve problems involving correlation

(1) Conduct a test to decide if there is sufficient evidence to reject the null hypothesis.

(2) Reject or accept the null hypothesis.

(3) If the null hypothesis is rejected, then you conclude there is correlation and, if the null hypothesis is accepted, then you conclude there is no correlation.

Example 3

The latitude of a random sample of weather stations and the average daily temperatures recorded in a year are examined to see if there is any correlation. The hypotheses are being considered at the 5% significance level. A sample of 12 weather stations is taken and the PMCC is found to be 0.6581, which has a p-value of 0.02 for a two-tailed test. State the hypotheses of the test and determine the conclusion in context.

The null hypothesis is $H_0: \rho = 0$ and the alternative hypothesis is $H_1: \rho \neq 0$

1 Conduct a test to decide if there is sufficient evidence to reject the null hypothesis.

The p-value is 2% which is lower than the significance level, so the result is more extreme than required. There is sufficient evidence to reject H_0

2 Reject the null hypothesis.

It can be concluded that there is some correlation between the latitude of a weather station and the average daily temperatures recorded in a year.

3 Conclude that there is correlation.

Note that the conclusion doesn't involve any statement about the strength of the correlation.

1 Economists study the relationship between levels of taxation and the amount individuals give to charity to see if there is any correlation. The hypotheses $H_0 : \rho = 0$ and $H_1 : \rho \neq 0$ are being considered at the 5% significance level. 24 countries are measured and the PMCC is found to be -0.1601. The critical value for the test is -0.4044. State, with a reason, whether H_0 is accepted or rejected and determine your conclusion in context.

2 A teacher measures his students' test scores in Maths and English to check for positive correlation. The hypotheses $H_0 : \rho = 0$ and $H_1 : \rho > 0$ are considered at the 10% significance level. All 28 students' scores are recorded and the PMCC is found to be 0.2516, which has a p-value of 0.0983 for a one-tailed test. State, with a reason, whether H_0 is accepted or rejected and give your conclusion in context.

3 Food scientists investigate 20 different flavours of crisps to see if any show signs of being carcinogenic. For each flavour, the amount eaten is measured against a carcinogen index to check for positive correlation. Each flavour is tested at the 5% significance level.

   a What does it mean for the test to have a 5% significance level?

   b How many of the 20 flavours are expected to show positive correlation if in fact there is no correlation for any of the flavours?

It is found that exactly one of the 20 flavours has a PMCC which exceeds the critical value.

   c What might the scientists conclude?

   d Given your answer to part **b**, why would the scientists treat this conclusion suspiciously?

4 The daily mean temperatures in Camborne and Hurn are believed to be strongly positively correlated. The hypotheses $H_0 : \rho = 0$ and $H_1 : \rho > 0$ are considered at the 5% significance level and the critical value is 0.1216. The temperatures are measured every day in May to October 2015 to test this. The test statistic is 0.788. State, with a reason, whether H_0 is accepted or rejected.

5 Links between daily total sunshine and daily mean windspeed are investigated, taking measurements every day in May to October 2015 at a weather station in Heathrow. The hypotheses $H_0 : \rho = 0$ and $H_1 : \rho \neq 0$ are considered at the 10% significance level. The PMCC for the 184 days is -0.0493 and the critical value is ± 0.122

   a State, with a reason, whether H_0 is accepted or rejected and determine the conclusion in context.

   b Explain why a different conclusion might be drawn in Camborne.

6 Education specialists want to find out if there is any correlation between the geographical location of a school and the likelihood that a student attending the school will get at least one A at A-level. A random sample of 46 schools is taken to test the hypotheses $H_0 : \rho = 0$ and $H_1 : \rho \neq 0$ at the 5% level. The PMCC is 0.1319, which has a p-value of 0.382. State, with a reason, whether H_0 is accepted or rejected and determine your conclusion in context.

Challenge

7 A, B and C are properties which are tested for positive correlation at the 10% level. The critical value is 0.365

   a The PMCC between A and B is 0.6 State, with a reason, whether there is reason to say that A and B are positively correlated.

   b The PMCC between B and C is 0.6. State, with a reason, whether there is reason to say that B and C are positively correlated.

   c The PMCC between C and A is 0.262. State, with a reason, whether you can say that C and A are positively correlated.

   d Discuss how your answers to parts **a**, **b** and **c** relate to one another.

STATS

21.2 Testing a Normal distribution

Fluency and skills

See Ch20.3

For an introduction to the Normal distribution.

If a variable is Normally distributed, then you can perform a hypothesis test to identify whether the mean of the sample is sufficiently different to the hypothesised mean of the distribution.

The **null hypothesis** is the assertion that the population the sample is taken from has a particular mean. This is usually based on previous results.

You call this hypothesised mean μ_0, so $H_0 : \mu = \mu_0$

The **alternative hypothesis** is

- $H_1 : \mu \neq \mu_0$ if the mean of the sample is different to the hypothesised mean.
- $H_1 : \mu < \mu_0$ or $H_1 : \mu > \mu_0$ if the mean is lower or higher than the hypothesised mean.

If, $X \sim N(\mu, \sigma^2)$, then $\overline{X} \sim N\left(\mu, \dfrac{\sigma^2}{n}\right)$. To test these types of hypotheses you need a test statistic, z

> **Key point**
>
> $z = \dfrac{\overline{x} - \mu_0}{\dfrac{\sigma}{\sqrt{n}}}$, $\overline{x}$ is the mean of the sample, μ_0 is the hypothesised mean of the distribution, σ^2 is the variance of the distribution and n is the sample size.

> $H_1 : \mu \neq \mu_0$ is a two-tailed test.
>
> $H_1 : \mu < \mu_0$ or $H_1 : \mu > \mu_0$ is a one-tailed test.

You can perform a z-test using your calculator.

You can either compare this test statistic to the critical value, or you can find the p-value and compare that to the significance level. These calculations rely on the standardised Normal distribution $\Phi(z) = P(Z \leq z)$ where $Z \sim N(0, 1)$

> **Key point**
>
> You accept the null hypothesis if the test statistic is smaller in size than the critical value, or if the p-value for the test statistic is greater than the significance level.

Example 1

A machine at a sausage factory is suspected of being faulty. The lengths of the sausages produced follow a Normal distribution. The variance is $16 \, cm^2$, but the mean may have changed from the intended setting of 12 cm.

a State the null and alternative hypotheses.

A random sample of 25 sausages is taken and found to have a mean length of 13.2 cm. The test is carried out at the 10% significance level.

b Calculate the test statistic.

c Calculate the critical value from the significance level.

d State, with a reason, whether the null hypothesis is accepted or rejected.

> **a** $H_0 : \mu = 12$ and $H_1 : \mu \neq 12$

The units of variance are cm^2. Standard deviation is often used in place of variance because it has the same units as the variable.

Even though the observed mean of the sample is higher than μ, a two-tailed test is used here since there is no reason to expect a shift in a particular direction.

(Continued on the next page)

b The test statistic $z = \dfrac{13.2 - 12}{\dfrac{4}{\sqrt{25}}} = 1.5$

> Substitute the values of $\bar{x}, \mu_0, \sigma$ and n into the formula for the test statistic.

c $\Phi^{-1}\left(1 - \dfrac{10}{200}\right) = 1.645$ so the critical value is ± 1.645

d Since the test statistic is less than the critical value, you accept the null hypothesis.

> Compare the test statistic with the critical value.

Exercise 21.2A Fluency and skills

1 A Normal distribution is assumed to have variance 625 but its mean is unknown. A sample of size 49 is taken to investigate the claim that its mean could be 19. State the null and alternative hypotheses for this test.

2 A Normal distribution is known to have variance 196 but has an unknown mean. A sample of size 28 is collected to test whether or not the mean might be 43. State the null and alternative hypotheses for this test.

3 A machine at a spaghetti factory is suspected of being faulty. It is known that the length of each strand of spaghetti can be well-modelled by a Normal distribution with variance $2.25\,\text{cm}^2$, but it is thought that the mean may have changed from the intended setting of 32 cm.

  a State the null and alternative hypotheses.

A random sample of 36 spaghetti strands is taken and is found to have a mean length of 31.5 cm. The test is done at the 10% significance level.

  b Calculate the test statistic.

  c Calculate the critical value at the given significance level.

  d State, with a reason, whether the null hypothesis is accepted or rejected.

4 It is known that lengths of genetically-modified celery stalks follow a Normal distribution and the variance is $9\,\text{cm}^2$. It is believed that the mean may have changed from the original design of 25 cm.

  a State the null and alternative hypotheses.

A random sample of 32 stalks is taken and is

found to have a mean length of 25.9 cm. The test is done at the 10% significance level.

  b Calculate the test statistic.

  c Calculate the critical value from the significance level.

  d State, with a reason, whether the null hypothesis is accepted or rejected.

5 A variable is Normally distributed with variance 196. The mean was known to be 29 at one point, but a sample of 36 is taken to see if the mean has increased. The mean of the sample is 33

  a Calculate the test statistic.

  b Calculate the p-value of the statistic.

  c State, with a reason, whether the null hypothesis is accepted or rejected at the 10% level.

6 The heights of adult men in a large country are well-modelled by a Normal distribution with mean 177 cm and variance $529\,\text{cm}^2$. It is thought that men who live in a poor town may be shorter than those in the general population. The hypotheses $H_0 : \mu = 177$ and $H_1 : \mu < 177$ are tested at the 10% significance level with the assumption that the variance of heights is the same in the town as in the general population. A sample of 25 men is taken from the town and their heights are found to have a mean value of 171 cm.

  a Calculate the test statistic.

  b Calculate the p-value of the statistic.

  c State, with a reason, whether the null hypothesis is accepted or rejected.

STATS

To solve problems involving hypothesis testing

(1) Define the null hypothesis and the alternative hypothesis.

(2) Calculate the test statistic.

(3) Calculate the critical value or the p-value.

(4) Accept or reject the null hypothesis.

(5) Give your conclusion.

Know your dataset

Leuchars, in eastern Scotland, has the highest latitude of any location in the LDS and, in 2015, it had the lowest daily mean temperature of the LDS locations. Click this link in the digital book for more information about the Large data set.

Example 2

A machine that makes copper rods is designed to produce rods with a mean length of 30 cm. An inspector is making sure that the mean is not less than this and assumes, from experience with similar machines, that the variance is 0.16 cm². The inspector measures the lengths of 32 random rods and finds that they have a mean length of 29.9 cm.

a State null and alternative hypotheses for this test.

b Calculate the test statistic and hence calculate the p-value.

c State, with a reason, whether the null hypothesis is accepted or rejected at the 10% significance level. Determine the conclusion of the hypothesis test.

a $H_0 : \mu = 30$ $H_1 : \mu < 30$

b $z = \dfrac{29.9 - 30}{\dfrac{\sqrt{0.16}}{\sqrt{32}}} = -1.414 \ (4 \text{ sf})$

The p-value is $1 - \Phi(1.414) = 7.87\%$

c Since the p-value is less than the significance level, the null hypothesis can be rejected. There is sufficient evidence to suggest that the machine might be producing rods which are too short.

(1) Since the inspector only wants to test if the mean is significantly lower than μ, you use a one-tailed test.

(2)(3) Calculate the test statistic and use this to calculate the p-value.

(4)(5) Reject the null hypothesis and give your conclusion.

When testing a mean, you simply need to show that there is or is not enough evidence to accept or reject the null hypothesis. That is, it is likely that the mean has or has not changed from the hypothesised mean.

Example 3

Two scientists conduct an experiment. The results are modelled by a Normal distribution with variance 53.8, but the scientists think that the mean could be lower than the intended 167. They both separately take samples of size 41 to test the hypotheses $H_0 : \mu = 167$ and $H_1 : \mu < 167$

a One scientist tests at the 5% level and their sample has test statistic −2.53. The critical value is −1.64. State, with a reason, whether the null hypothesis is accepted or rejected.

b The other scientist tests at the 1% level and their sample has test statistic −2.27. The critical value is −2.33. State, with a reason, whether the null hypothesis is accepted or rejected.

c Determine the conclusion the two scientists should reach.

(Continued on the next page)

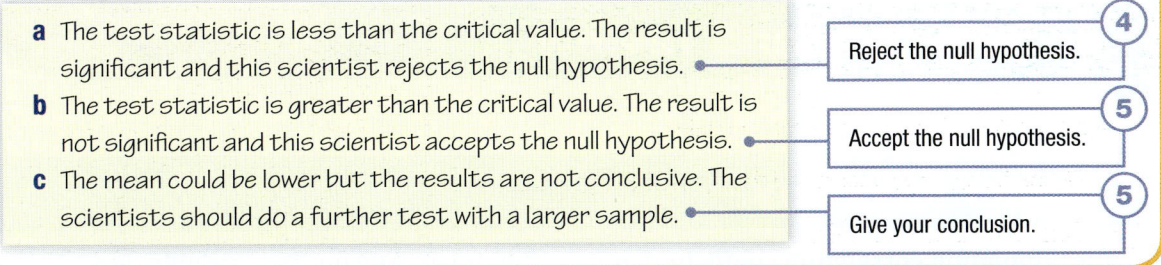

a The test statistic is less than the critical value. The result is significant and this scientist rejects the null hypothesis. ● ——— Reject the null hypothesis. **4**

b The test statistic is greater than the critical value. The result is not significant and this scientist accepts the null hypothesis. ● ——— Accept the null hypothesis. **5**

c The mean could be lower but the results are not conclusive. The scientists should do a further test with a larger sample. ● ——— Give your conclusion. **5**

Exercise 21.2B Reasoning and problem-solving

Large data set

1 Using data from Camborne, Heathrow, Leuchars, Hurn and Leeming in 1987, the average daily total hours of sunshine during June in the UK is modelled by a Normal distribution with mean 4.38 hours and variance 14.0 hours². The sunshine levels are measured in those locations during June 2015 to see whether it was a particularly sunny month by comparison.

a State null and alternative hypotheses for this test.

The 150 daily readings have an average of 6.76 hours.

b Calculate the test statistic and the critical value if the test is at the 5% level.

c State, with a reason, whether the null hypothesis is accepted or rejected. Determine the conclusion of the hypothesis test.

2 A banana farm finds that the banana masses are well-modelled by a Normal distribution with mean 125 g and variance 150 g². During one summer, they are thought to be larger than usual.

a State the null and alternative hypotheses.

A sample of 42 bananas is selected and their mean mass is 131 g.

b Calculate the test statistic and hence calculate the p-value.

c State, with a reason, whether the null hypothesis is accepted or rejected at the 1% significance level.

d Determine the conclusion of the hypothesis test.

e Discuss whether it's reasonable to say that the mean mass of that summer's bananas is 131 g.

3 The average daily maximum relative humidity in Camborne in 1987 is believed to be well-modelled by a Normal distribution with mean 93.59 and variance 35.32. Measurements are taken from a weather station in Heathrow at the same time and they are checked to see if the two locations have the same average.

Large data set

a State the null and alternative hypotheses.

In Heathrow, the average daily maximum relative humidities are recorded and found to have an average of 93.13

b Calculate the test statistic.

c Calculate any critical values at the 10% level.

d State, with a reason, whether the null hypothesis is accepted or rejected.

Challenge

4 A Normal distribution is known to have variance 9 but has an unknown mean. One statistician believes the mean is 16 but another believes that it is 17

a State each person's null and alternative hypotheses.

They take a sample of size 64 to decide at the 10% significance level who is correct. The observed mean of the sample is 16.6

b Determine the conclusion of each statistician's hypothesis test and discuss the results.

STATS

Chapter summary

- The **population correlation coefficient**, ρ, describes how correlated two variables are.
- The product moment correlation coefficient (PMCC), r, is a statistic that estimates ρ. It is calculated from a sample and used to determine the presence of correlation.
- In a test to detect correlation, the null hypothesis is that there is none, i.e. $H_0 : \rho = 0$ and the alternative hypothesis can take any of the following forms.
  - There is some correlation, $H_1 : \rho \neq 0$, a 2-tailed test.
  - There is positive correlation, $H_1 : \rho > 0$, a 1-tailed test.
  - There is negative correlation, $H_1 : \rho < 0$, a 1-tailed test.
  The PMCC is the test statistic.
- A Normal distribution has mean μ and variance σ^2. Given a known or assumed variance you can use a sample to test if a value is a suitable estimate of the mean.
- The null hypothesis when testing your estimated mean for a Normal distribution is that the value you are testing is correct, i.e. $H_0 : \mu = \mu_0$ and the alternative hypothesis can take any of the following forms.
  - Your value is incorrect, $H_1 : \mu \neq \mu_0$, a 2-tailed test.
  - Your value is too small, $H_1 : \mu > \mu_0$, a 1-tailed test.
  - Your value is too large, $H_1 : \mu < \mu_0$, a 1-tailed test.
  For a sample of size n with mean value $\bar{x}$, the test statistic is $z = \dfrac{\bar{x} - \mu}{\frac{\sigma}{\sqrt{n}}}$
- When hypothesis testing:
  - If the test statistic is further from zero than the critical value, or
  - If the p-value associated with the PMCC is smaller than the significance level of the test, then you have sufficient evidence to reject the null hypothesis. Otherwise, you accept the null hypothesis. You should interpret this result in the context of the situation.

Check and review

You should now be able to...	Try Questions
State null and alternative hypotheses when testing for correlation.	1, 2, 3, 4
Compare a given PMCC to a critical value or its p-value to the significance level, and use this comparison to decide whether to accept or reject the null hypothesis.	2, 3, 4, 5
Decide what the conclusion means in context about the correlation.	3, 4, 5
State null and alternative hypotheses when testing the mean of a Normal distribution.	6, 7
Calculate the test statistic, compare it to a critical value or compare its p-value to the significance level, and use this comparison to decide whether to accept or reject the null hypothesis.	7
Decide what the conclusion means in context about the mean of the distribution.	7

1 It's suspected that two variables are correlated but it is not known if the correlation would be positive or negative. State null and alternative hypotheses for a test to identify correlation.

2 The temperature in a seaside town is believed to be positively correlated with the number of ice cream cones sold by a van by the beach.

   a State null and alternative hypotheses for a test with a 5% significance level.

In a sample of 30 days randomly selected over the course of the year, the data is found to have a PMCC of 0.6215. The critical value for this test is 0.3061

   b State, with a reason, whether the null hypothesis is accepted or rejected.

3 A scientist believes there is negative correlation between the amount of a chemical in a petri dish and the number of bacteria present. To test this, she prepares 20 different petri dishes with varying amounts of the chemical in each. The null hypothesis is $H_0 : \rho = 0$ and the alternative hypothesis is $H_1 : \rho < 0$, which are tested at the 1% level. The test statistic is 0.4438, which has a p-value of 2.5%

   a State, with a reason, whether the null hypothesis is accepted or rejected.

   b Determine the scientist's conclusion.

4 Levels of annual household income and the amount spent per year on books are suspected to be positively correlated.

   a State the null and alternative hypotheses.

40 households are used as a sample and the PMCC is found to be 0.198. The critical value at the 5% level is 0.264

   b Determine the conclusion in context.

5 The heights and lengths heptathletes can jump in the high jump and long jump are tested for correlation. The hypotheses $H_0 : \rho = 0$ and $H_1 : \rho \neq 0$ are being considered at the 5% significance level. A sample of 29 competitors is taken and the PMCC is found to be 0.416, which has a p-value of 2.48% for a two-tailed test. State, with a reason, whether H_0 is accepted or rejected and determine the conclusion in context.

6 A Normal distribution is assumed to have variance 18 but has an unknown mean. It is believed that the mean could be −5, but could be larger. A sample is taken to test this. State the null and alternative hypotheses.

7 The lengths of some sticks of rock are well-modelled by a Normal distribution with variance 4 cm². Their mean length is unknown, but is supposed to be 30 cm. A sample of 28 sticks is taken to test their mean length at the 5% significance level.

   a State the null and alternative hypotheses for this test.

The sample has a mean length of 30.45 cm and the test statistic gives a p-value of 0.2338

   b State, with a reason, whether the null hypothesis is accepted or rejected.

   c Determine the conclusion of this hypothesis test in context.

What next?

Score	0 – 3	Your knowledge of this topic is still developing. To improve, search in MyMaths for the codes: 2287, 2288		**Click these links in the digital book**
	4 – 5	You're gaining a secure knowledge of this topic. To improve, look at the InvisiPen videos for Fluency and skills (21A).		
	6 – 7	You've mastered these skills. Well done, you're ready to progress! To develop your exam techniques, look at the InvisiPen videos for Reasoning and problem-solving (21B).		

History

The **Pearson product-moment correlation coefficient** (PMCC) is named after **Karl Pearson** (1857–1936).

Pearson developed the mathematics in partnership with **Sir Francis Galton** (1822–1911), who originally developed the concepts of correlation and regression.

> "All great scientists have, in a certain sense, been great artists; the man with no imagination may collect facts, but he cannot make great discoveries."
>
> –Karl Pearson

ICT

	B5		fx	=CORREL(A2:A8,B2:B8)	
	A	B	C	D	E	F
1	X	Y				
2	32	160				
3	35	158				
4	34	162				
5	36	166				
6	40	172				
7	45	166				
8	48	172				
9						
10		PPMC =	0.762452			

The PMCC can be calculated using spreadsheet software.

Collect data for two variables that you suspect may be connected, for example height and foot length.

Enter the data into two columns on a spreadsheet.

Use the *CORREL* function in Microsoft Excel on the two columns to find the PMCC.

Did you know?

Francis Galton was the half-cousin of **Charles Darwin**. The publication of Darwin's book, On the Origin of Species, in 1859, had a huge impact on Galton and it was Darwin who encouraged him to research inherited traits.

When studying human height, for example, Galton first thought that tall parents would tend to have children that were even taller. He found instead that the children tended to be slightly smaller.

Galton found similar results for other characteristics, which lead him to coin the term **'regression to the mean'**.

Note

Much of Galton's research pointed to the fact that a characteristic may not be the result of a single cause, but affected by multiple different causes, each with varying levels of influence.

This discovery led to the idea of **multiple regression**. Multiple regression is a way to model the data by considering various different factors and the amount of impact they have on the dependent variable.

1 A sample of size 15 was taken from a Normally distributed population with unknown mean and standard deviation 4. If the mean of the sample was 22, test at a 5% significance level the hypothesis that the population mean is 20 **[4 marks]**

2 A Normally distributed population has a variance of 8. A sample of size 25 was taken and had a mean of 38.1. Stating clearly your null and alternative hypotheses, test at the 3% significance level whether the mean is less than 40 **[4]**

3 A sample of size 10 is taken from a Normal population and gives the following values:

18.2, 19.6, 24.1, 19.3, 21.5, 22.6, 23.3, 20.9, 21.7, 20.3

Using a significance level of 3%, test whether the population mean is less than 22
Assume that the standard deviation of the population is 1.8 **[4]**

4 From extensive experience, a manufacturer knows that their halogen light bulbs have a mean lifetime of 1930 hours with a standard deviation of 245 hours. The firm introduces a new type of light bulb. The lifetimes of 20 of the new bulbs are given in the table.

Lifetime (l hours)	Frequency
$1400 \leq l < 1800$	3
$1800 \leq l < 2000$	5
$2000 \leq l < 2100$	6
$2100 \leq l < 2300$	4
$2300 \leq l < 2700$	2

a Find the mean lifetime of these bulbs. **[1]**

b Assuming that the variation in the lifetimes of the bulbs remains unchanged, test at a 5% significance level whether the lifetimes have increased. **[3]**

5 A Normally distributed population has variance 20 and a mean believed to be 12. A sample of size 32 was taken from the population and gave a sample mean of 10.3. By finding the probability of the sample mean taking a value less than 10.3, test the hypothesis that the population mean is 12 against the alternative hypothesis that it is less than 12. You should use a significance level of 5%. **[4]**

6 The average daily maximum temperature in a country can be modelled by a Normal distribution with mean 17.7 °C and variance 11.3 °C². In one particular region, it is thought that the temperature may be higher than the rest of the country. It is assumed that the variance in temperature is the same in the region as in the whole country. A sample of 25 random measurements of the daily maximum temperature is taken for the region and found to have a mean value of 18.6 °C.

a State the null and alternative hypotheses for this test. **[1]**

b Calculate the test statistic and the critical value at the 10% significance level. **[2]**

c State, with a reason, whether the null hypothesis is accepted or rejected and determine the conclusion in context. **[2]**

7 A sample of 8 weather stations is used to see if there is any positive correlation between daily total rainfall and daily maximum relative humidity.

a State null and alternative hypotheses for this test. **[1]**

The hypotheses are being considered at the 10% significance level. The PMCC is found to be 0.5101, which has a p-value of 0.0983 for a one-tailed test.

b State, with a reason, whether H_0 is accepted or rejected and determine the conclusion in context. **[2]**

8 Two scientists have an experiment whose results are modelled by a Normal distribution with variance 17.6 but they think the mean could be different to the intended −13.6. They both

separately take samples of size 72 to test the hypotheses $H_0 : \mu = -13.6$ and $H_1 : \mu \neq -13.6$ at the 5% level. The critical value is ± 1.96

a One scientist's sample has test statistic -2.04. State, with a reason, whether the null hypothesis is accepted or rejected. [1]

b The other scientist's sample has test statistic 2.13. State, with a reason, whether the null hypothesis is accepted or rejected. [1]

c Determine the conclusion the two scientists reach. [2]

9 According to EU legislation, one of the few products that is allowed to be labelled in Imperial measure is milk sold in returnable containers. 30 '1 pint' bottles filled at a dairy farm have the following volumes, given to the nearest ml.

    569 567 568 570 569 568 563 568 572 571
    568 573 570 572 569 571 568 574 567 569
    570 572 564 566 568 572 576 566 571 570

It is required that the mean of the volume in each bottle must be greater than 1 pint (568 ml). Based on this sample, do you believe that this requirement is being met?
You should assume that the sample standard deviation gives a good estimate of the population standard deviation and test at a significance level of 5%. [5]

10 The correlation coefficient for two variables, X and Y, is 0.31 based on a sample size of 19. Given that the critical value is ± 0.468, test at the 5% significance level whether the population correlation coefficient is zero against the alternative hypothesis that it is not zero. [2]

11 At Heathrow, the daily maximum temperature and the daily total sunshine have a sample correlation coefficient of 0.51. You wish to test the suggestion that the population correlation coefficient between the variables is positive.

a Write down the null and alternative hypotheses for the test. [1]

b Given that the critical value is ± 0.367, perform the test at a significance level of 2.5% given that the sample size was 12 [3]

12 Given that the sample product moment correlation coefficient between variables X and Y is -0.17 based on a sample of size 46, and the critical value is -0.248, test the hypothesis that the population correlation coefficient, ρ, is less than zero at the 5% significance level. You should state your null and alternative hypotheses. [3]

13 A high-speed fabric weaving machine increases in temperature as it is operated. The number of flaws per square metre is measured at various temperatures and these variables are found to have a correlation coefficient of -0.42 based on a sample of size 30. The manufacturer claims that the number of flaws is independent of the temperature. Given that the critical value is ± 0.367, test at a significance level of 5% the manufacturer's claim. [3]

14 a A Normal random variable X has an unknown mean μ and known standard deviation σ. A sample of size n is taken from the population and gives a sample mean of $\bar{x}$. A test at significance level 2% is to be carried out on whether the population mean has increased from a value μ_0. Find, in terms of μ_0, σ and n, the set of $\bar{x}$ values which would lead to the belief that the mean had increased. [4]

b Bars of steel of diameter 2 cm are known to have a mean breaking point of 80 kN with a standard deviation of 2.1 kN. An increase in the bars' diameter of 0.2 cm is thought to increase the mean breaking point. A sample of 40 bars with the greater diameter have a mean breaking point of 80.9 kN. Test at a significance level of 2% whether the bars with the greater diameter have a greater mean breaking point. State any assumptions used. [3]

Large
data set

1 Describe the type of correlation you would expect between these pair of variables and explain your answers.

    a Mean daily temperature in Perth and Beijing. **[2 marks]**

    b Daily mean total cloud in Heathrow and in Hurn. **[2]**

2 This two-way table shows the handedness of a group of men and women working in a large office. Not all the cells have been filled in.

Handedness	Men	Women	Total
Left	20		32
Right			
mixed/both	3	2	
Total	210		350

    a Copy and complete the table. **[3]**

    b A person is chosen at random. Find the probability they are

        i Left-handed,   **ii** A woman,

        iii A right-handed man. **[3]**

    c Are the following two events independent? Explain how you know.

        "a randomly chosen person is mixed handed/ambidextrous"

        "a randomly chosen person is a man" **[3]**

3 A town has 20 000 homes and a sample of 100 of these is to be surveyed to find the mean expenditure on clothing.

    a Explain the difference between the parameter and the statistic in this case. **[2]**

    A market researcher proposes to undertake the survey on a Monday morning and has chosen a road with over 100 houses. He plans to knock on every door until he has 100 responses.

    b Explain any problems there might be with this proposal. **[2]**

    A second researcher proposes taking a stratified random sample based on the income of the household.

    c 800 households are in the highest income band. Calculate the number of these households that will be included in the sample. **[2]**

4 In a sample of 100 days, the amount of rain per day is shown in the table.

    a Estimate

        i The mean,

        ii The standard deviation,

Rain, R (mm)	Number of days
$0 \leq R < 0.1$	1
$0.1 \leq R < 0.5$	27
$0.5 \leq R < 1$	38
$1 \leq R < 5$	28
$5 \leq R < 10$	4
$10 \leq R < 20$	0
$20 \leq R < 30$	2

    iii The median,

    iv The interquartile range.   **[10]**

  b Explain whether you think the mean or the median is a better measure of the average in this case.   **[2]**

5 At a college, the probability a student studies Maths is 0.55, the probability they study Physics is 0.3, and the probability they study both is 0.25

  a Calculate the probability that a randomly selected student does not study either Maths or Physics.   **[3]**

  b Draw a Venn diagram to illustrate this information.   **[4]**

  c Calculate the probability that a student studies Maths given that they study Physics.   **[2]**

  d Are the events "a student studies Maths" and "a student studies Physics" independent? Explain how you know.   **[2]**

6 The table shows the number of sunny and rainy days at Camborne weather station in June. A rainy day (R) is defined as one with more than a trace of rain at any time during the day. A sunny day (S) is defined as one with more than 8 hours of bright sunshine.

	R	R'
S	3	13
S'	9	5

  a Find each of these probabilities

    i $P(S)$ **ii** $P(R|S)$ **[3]**

  b Copy and complete the tree diagram with the probabilities on each branch.   **[3]**

7 You are given that $P(B) = \frac{1}{5}$, $P(A|B) = \frac{1}{7}$ and $P(A \cup B) = \frac{5}{7}$.

  a Calculate

    i $P(A \cap B)$ **ii** $P(B \cap A')$ **iii** $P(A|B')$ **[7]**

  b The event C is independent to both A and B and $P(C) = \frac{1}{3}$. Calculate

    i $P(A \cap C)$ **ii** $P(C \cup B)$ **[5]**

8 The tree diagram shows the probability of rain at Heathrow in October, depending on whether or not it is also a windy day (that is, there are gusts of more than 20 km).

  a Calculate

    i $P(W \cap R)$ **ii** $P(R)$ **[5]**

  b Draw a Venn diagram showing the probabilities.   **[4]**

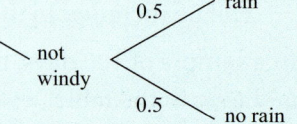

9 The continuous random variable X is modelled by a Normal distribution with mean 2.4 and standard deviation 0.3.

  a Calculate these probabilities

    i $P(X < 3)$ **ii** $P(X > 2.7)$

    iii $P(1.8 < X < 2.2)$ **[6]**

  b Find the value α such that $P(X < \alpha) \approx 0.96$ **[2]**

10 Two sets of times are recorded from 12 people taking part in an experiment.

The product moment correlation coefficient between the times is calculated to be $r = 0.782$ Investigate whether there is a positive correlation between the two times using a 1% significance level. State your hypotheses clearly. **[4]**

11 The mean daily air temperature and the amount of rainfall are recorded for 100 days at a Beijing weather station and the product moment correlation coefficient is found to be $r = -0.261$.

a Investigate whether there is evidence of a negative correlation with a 5% significance level. State your hypotheses clearly. **[4]**

b Write down the critical region if you were to test at the 1% significance level. **[2]**

12 Find the range of test statistics that would lead to the rejection of the null hypothesis in a hypothesis test with significance level of 5% when there is a sample of size 20 and the hypotheses are defined as

$H_0: \rho = 0 \qquad H_1: \rho \neq 0$ **[4]**

13 The probability of a sports team winning a match in any weather is 0.36. If it is raining, the probability of them winning is 0.3. There is a 10% chance of it raining during the match.

a Calculate the probability of the team winning, given that it is not raining. **[4]**

b Calculate the probability that it was raining, given that the team won a match. **[4]**

14 The histogram shows the lengths (not including the tail), L, of a sample of fully-grown meerkats.

Given that there are 6 meerkats with lengths between 18 and 22 cm,

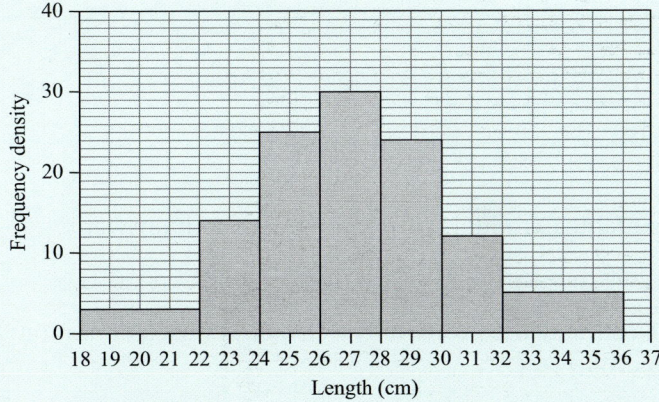

a Calculate the number of meerkats with a length between 22 and 24 cm, **[3]**

b Estimate

   i The mean,   **ii** The standard deviation. **[5]**

c Explain why the Normal distribution can be used to model the lengths, L, of the meerkats. **[2]**

d Using your answers to part **b** as approximations for μ and σ, use the Normal distribution to calculate each of these probabilities.

   i $P(L < 31)$ **ii** $P(21.5 < L < 25.5)$ **[5]**

The product moment correlation coefficient between length and mass is calculated for n of these meerkats and found to be 0.6

A hypothesis test is carried out at the 1% level of significance with hypotheses $H_0: \rho = 0$ and $H_1: \rho > 0$

e What is the range of values of n that will lead to the rejection of the null hypothesis? **[2]**

15 You are given that $X \sim N(\mu, 1.5)$ and that $P(X > 10) = 0.9$

    a Calculate the value of μ to 1 decimal place. **[5]**

    b Find the x-coordinates of the points of inflection of this Normal distribution. **[3]**

16 The probability of more than 2 mm of rain in a day in the summer at a Jacksonville weather station is found to be 0.25. Use the binomial distribution to model the number of days, X with more than 2 mm of rain.

    a Calculate the probability that it doesn't rain for a whole week. **[2]**

    b Calculate the probability that in a ten-day period it rains more than 4 times. **[3]**

    c Use a Normal approximation to estimate the probability that in a 60-day period there is rain on fewer than 10 days. **[4]**

17 The maximum daily temperature, T, during May at Leuchars weather station is shown in the box and whisker diagram.

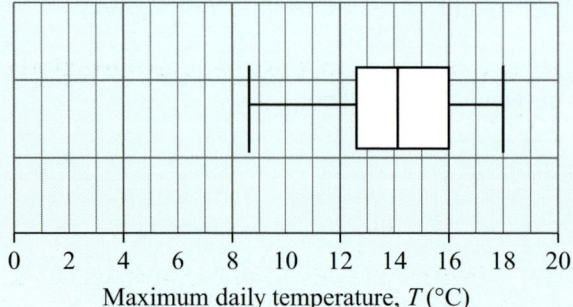

Maximum daily temperature, $T\,(°C)$

    a Would a Normal distribution be suitable to model this data? Explain your answer. **[2]**

T is modelled by a Normal distribution with mean 14°C and standard deviation 2.6°C.

    b **i** Use the Normal distribution to estimate the probability that a day chosen at random has a maximum temperature of less than 16°C.

       ii Compare this result to the data summarised in the box and whisker diagram. **[4]**

    c Calculate these probabilities.

       i $P(T > 15.5)$ **ii** $P(T < 13)$

       iii $P(12.1 < T < 14.9)$ **[6]**

18 For each of these random variables, decide if it could be modelled using the binomial distribution. If you think a binomial distribution is suitable, give the parameters. If not, explain why.

    a A midwife delivers eight babies in a week, and X is the number that are girls.   **[2]**

    b A child rolls a dice repeatedly, and X is the number of throws until they obtain a six.   **[2]**

    c A bag contains 20 blue and 15 green counters, five counters are removed, and X is the number of blue counter removed.   **[2]**

19 A group of 10 volunteers exercise strenuously and their maximum heart-rate is recorded along with their age. The product moment correlation coefficient between age and maximal heart-rate is found to be –0.42. Test whether there is a negative correlation between age and maximal heart-rate using a 10% significance level. State your hypotheses clearly.   **[4]**

20 The daily maximum temperature, T, at a weather station in August is found over several years to have a mean of 22.1 °C and a standard deviation of 3.86 °C. T is modelled by the Normal distribution.

Last year, the maximum daily temperature in a particular week had a mean of 25.6 °C.

    a Investigate whether this week was significantly warmer than usual using a 1% significance level.   **[7]**

    b Write down the acceptance region for the test statistic when using a 10% significance level.   **[2]**

21 A continuous random variable is Normally distributed with mean μ and standard deviation σ

Using the facts that $P(X > 120) = 0.9868$ and $P(X > 140) = 0.0694$

    a Calculate the values of μ and σ to 3 significant figures,   **[9]**

    b Find the value of α such that $P(\mu - \alpha < X < \mu + \alpha) = 0.4582$ **[4]**

22 The average wind speed, W, in a town is Normally distributed with mean 8 kn and standard deviation σ. The probability of the average being less than 6 kn is 0.19

    a Calculate the value of σ **[4]**

    b Calculate the probability the wind speed is more than 12 kn   **[2]**

    c Calculate the probability that during a 5-day period,

        i The average wind speed is never more than 12 kn   **[2]**

        ii The average wind speed is more than 12 kn on fewer than two days.   **[3]**

23 $X \sim B(100, 0.2)$

    a Estimate these probabilities.

        i $P(X < 25)$ **ii** $P(17 \leq X \leq 21)$ **[8]**

    b Explain why the Normal distribution is suitable to use as an approximation.   **[2]**

24 The daily maximum humidity, H, at Leuchars weather station in the month of May is modelled by a Normal distribution with mean μ and standard deviation σ

The probability of the maximum humidity being less than 80% is 0.23 and the probability of it being greater than 99% is 0.29

a Calculate the values of μ and σ **[10]**

b Calculate the probability that a day has a maximum humidity of over 95% **[2]**

c Estimate the probability that, over the month of May, there are at most 10 days with a humidity of over 95% **[5]**

25 The probability of a person in town A being vegetarian is 5%. A random sample of 10 people is taken, and the random variable X is the number in the sample who are vegetarian.

a Explain whether the binomial is suitable to model the distribution of X **[3]**

b Calculate each of these probabilities.

   i $P(X = 2)$ **ii** $P(X > 2)$ **[4]**

A random sample of 200 people is taken from a different town, B, and 4 are found to be vegetarian.

c Use a Normal approximation to test at the 5% significance level whether there is any evidence that the probability of being vegetarian is different in town B to town A. **[7]**

26 The lifetime of certain batteries is known to be Normally distributed with a mean of 30 hours of continuous use and a standard deviation of 5 hours. A customer purchases eight batteries and records their lifetimes, in hours, as shown.

      26.6   25.7   30.5   27.3   20.1   29.5   28.2   25.3

The customer believes they have a faulty batch.

Test, at the 5% level, the customer's claim that the mean is less than 30 hours. **[7]**

27 The average hours of daily sunshine, S, in a particular county in September is Normally distributed with $S \sim N(6.68, 10.24)$

Luke lives in the county and claims that his town is less sunny than the county as a whole.

He records the hours of daily sunshine for n days and finds an average of 5.8 hours of sunshine per day.

a State the hypotheses and the critical value for conducting a hypothesis test, with a 2.5% significance level, about the mean of the Normal distribution in this case. **[3]**

b Find the smallest value of n that will lead to the acceptance of Luke's claim. **[5]**

Mathematical formulae
For A Level Maths

The following mathematical formulae will be provided for you.

Pure Mathematics
Mensuration

Surface area of sphere $= 4\pi r^2$

Area of curved surface of cone $= \pi r \times$ slant height

Arithmetic series

$$S_n = \frac{1}{2}n(a+l) = \frac{1}{2}n\left[2a+(n-1)d\right]$$

Binomial series

$$(a+b)^n = a^n + \binom{n}{1}a^{n-1}b + \binom{n}{2}a^{n-2}b^2 + \dots + \binom{n}{r}a^{n-r}b^r + \dots + b^n \qquad (n \in \mathbb{N})$$

where $\binom{n}{r} = {}^nC_r = \dfrac{n!}{r!(n-r)!}$

$$(1+x)^n = 1 + nx + \frac{n(n-1)}{1\times 2}x^2 + \dots + \frac{n(n-1)\dots(n-r+1)}{1\times 2\times \dots \times r}x^r + \dots \qquad (|x|<1, \ n \in \mathbb{R})$$

Logarithms and exponentials

$$\log_a x = \frac{\log_b x}{\log_b a}$$

$$e^{x\ln a} = a^x$$

Geometric series

$$S_n = \frac{a(1-r^n)}{1-r}$$

$$S_\infty = \frac{a}{1-r} \quad \text{for } |r| < 1$$

Trigonometric identities

$$\sin(A \pm B) = \sin A \cos B \pm \cos A \sin B$$

$$\cos(A \pm B) = \cos A \cos B \mp \sin A \sin B$$

$$\tan(A \pm B) = \frac{\tan A \pm \tan B}{1 \mp \tan A \tan B} \qquad \left(A \pm B \neq \left(k+\frac{1}{2}\right)\pi\right)$$

$$\sin A + \sin B = 2\sin\frac{A+B}{2}\cos\frac{A-B}{2}$$

$$\sin A - \sin B = 2\cos\frac{A+B}{2}\sin\frac{A-B}{2}$$

$$\cos A + \cos B = 2\cos\frac{A+B}{2}\cos\frac{A-B}{2}$$

$$\cos A - \cos B = 2\sin\frac{A+B}{2}\sin\frac{A-B}{2}$$

Differentiation

First principles

$$f'(x) = \lim_{h \to 0} \frac{f(x+h) - f(x)}{h}$$

$f(x)$	$f'(x)$
$\tan kx$	$k\sec^2 kx$
$\sec kx$	$k\sec kx \tan kx$
$\cot kx$	$-k\operatorname{cosec}^2 x$
$\operatorname{cosec} kx$	$-\operatorname{cosec} kx \cot kx$
$\dfrac{f(x)}{g(x)}$	$\dfrac{f'(x)g(x) - f(x)g'(x)}{(g(x))^2}$

Integration (+ constant)

$f(x)$	$\int f(x)\,dx$
$\sec^2 kx$	$\dfrac{1}{k}\tan kx$
$\tan kx$	$\dfrac{1}{k}\ln|\sec kx|$
$\cot kx$	$\dfrac{1}{k}\ln|\sin kx|$
$\operatorname{cosec} kx$	$-\dfrac{1}{k}\ln|\operatorname{cosec} kx + \cot kx|, \quad \dfrac{1}{k}\ln\left|\tan\left(\dfrac{1}{2}kx\right)\right|$
$\sec kx$	$-\dfrac{1}{k}\ln|\sec kx + \tan kx|, \quad \dfrac{1}{k}\ln\left|\tan\left(\dfrac{1}{2}kx + \dfrac{1}{4}\pi\right)\right|$

$$\int u\frac{dv}{dx}\,dx = uv - \int v\frac{du}{dx}\,dx$$

Numerical Methods

The trapezium rule: $\displaystyle\int_a^b y\,dx \approx \frac{1}{2}h\{(y_0 + y_n) + 2(y_1 + y_2 + \ldots + y_{n-1})\}$, where $\quad h = \dfrac{b-a}{n}$

The Newton-Raphson iteration for solving $f(x) = 0$: $x_{n+1} = x_n - \dfrac{f(x_n)}{f'(x_n)}$

Mechanics

Kinematics

For motion in a straight line with constant acceleration:

$$v = u + at$$

$$s = ut + \frac{1}{2}at^2$$

$$s = vt - \frac{1}{2}at^2$$

$$v^2 = u^2 + 2as$$

$$s = \frac{1}{2}(u+v)t$$

Statistics

Probability

$$P(A') = 1 - P(A)$$

$$P(A \cup B) = P(A) + P(B) - P(A \cap B)$$

$$P(A \cap B) = P(A)P(A|B)$$

$$P(A|B) = \frac{P(B|A)P(A)}{P(B|A)P(A) + P(B|A')P(A')}$$

For independent events A and B,

$$P(B|A) = P(B),$$
$$P(A|B) = P(A),$$
$$P(A \cap B) = P(A)P(B)$$

Standard deviation

Standard deviation $= \sqrt{\text{Variance}}$

Interquartile range $= \text{IQR} = Q_3 - Q_1$

For a set of n values $x_1, x_2, ..., x_i, ... x_n$

$$S_{xx} = \sum (x_i - \bar{x})^2 = \sum x_i^2 - \frac{\left(\sum x_i\right)^2}{n}$$

Standard deviation $= \sqrt{\dfrac{S_{xx}}{n}}$ or $\sqrt{\dfrac{\sum x^2}{n} - \bar{x}^2}$

Discrete distributions

Distribution of X	$P(X=x)$	Mean	Variance
Binomial B(n, p)	$\binom{n}{x}p^x(1-p)^x$	np	$np(1-p)$

Sampling distributions

For random sample of n observations from $N(\mu, \sigma^2)$

$$\frac{\bar{X} - \mu}{\sigma/\sqrt{n}} \sim N(0,1)$$

The following statistical tables will be provided for you.

Binomial Cumulative Distribution Function

The tabulated value is $P(X \leq x)$, where X has a binomial distribution with index n and parameter p

$p =$	0.05	0.10	0.15	0.20	0.25	0.30	0.35	0.40	0.45	0.50
$n = 5, x = 0$	0.7738	0.5905	0.4437	0.3277	0.2373	0.1681	0.1160	0.0778	0.0503	0.0313
1	0.9774	0.9185	0.8352	0.7373	0.6328	0.5282	0.4284	0.3370	0.2562	0.1875
2	0.9988	0.9914	0.9734	0.9421	0.8965	0.8369	0.7648	0.6826	0.5931	0.5000
3	1.0000	0.9995	0.9978	0.9933	0.9844	0.9692	0.9460	0.9130	0.8688	0.8125
4	1.0000	1.0000	0.9999	0.9997	0.9990	0.9976	0.9947	0.9898	0.9815	0.9688
$n = 6, x = 0$	0.7351	0.5314	0.3771	0.2621	0.1780	0.1176	0.0754	0.0467	0.0277	0.0156
1	0.9672	0.8857	0.7765	0.6554	0.5339	0.4202	0.3191	0.2333	0.1636	0.1094
2	0.9978	0.9842	0.9527	0.9011	0.8306	0.7443	0.6471	0.5443	0.4415	0.3438
3	0.9999	0.9987	0.9941	0.9830	0.9624	0.9295	0.8826	0.8208	0.7447	0.6563
4	1.0000	0.9999	0.9996	0.9984	0.9954	0.9891	0.9777	0.9590	0.9308	0.8906
5	1.0000	1.0000	1.0000	0.9999	0.9998	0.9993	0.9982	0.9959	0.9917	0.9844
$n = 7, x = 0$	0.6983	0.4783	0.3206	0.2097	0.1335	0.0824	0.0490	0.0280	0.0152	0.0078
1	0.9556	0.8503	0.7166	0.5767	0.4449	0.3294	0.2338	0.1586	0.1024	0.0625
2	0.9962	0.9743	0.9262	0.8520	0.7564	0.6471	0.5323	0.4199	0.3164	0.2266
3	0.9998	0.9973	0.9879	0.9667	0.9294	0.8740	0.8002	0.7102	0.6083	0.5000
4	1.0000	0.9998	0.9988	0.9953	0.9871	0.9712	0.9444	0.9037	0.8471	0.7734
5	1.0000	1.0000	0.9999	0.9996	0.9987	0.9962	0.9910	0.9812	0.9643	0.9375
6	1.0000	1.0000	1.0000	1.0000	0.9999	0.9998	0.9994	0.9984	0.9963	0.9922
$n = 8, x = 0$	0.6634	0.4305	0.2725	0.1678	0.1001	0.0576	0.0319	0.0168	0.0084	0.0039
1	0.9428	0.8131	0.6572	0.5033	0.3671	0.2553	0.1691	0.1064	0.0632	0.0352
2	0.9942	0.9619	0.8948	0.7969	0.6785	0.5518	0.4278	0.3154	0.2201	0.1445
3	0.9996	0.9950	0.9786	0.9437	0.8862	0.8059	0.7064	0.5941	0.4770	0.3633
4	1.0000	0.9996	0.9971	0.9896	0.9727	0.9420	0.8939	0.8263	0.7396	0.6367
5	1.0000	1.0000	0.9998	0.9988	0.9958	0.9887	0.9747	0.9502	0.9115	0.8555
6	1.0000	1.0000	1.0000	0.9999	0.9996	0.9987	0.9964	0.9915	0.9819	0.9648
7	1.0000	1.0000	1.0000	1.0000	1.0000	0.9999	0.9998	0.9993	0.9983	0.9961
$n = 9, x = 0$	0.6302	0.3874	0.2316	0.1342	0.0751	0.0404	0.0207	0.0101	0.0046	0.0020
1	0.9288	0.7748	0.5995	0.4362	0.3003	0.1960	0.1211	0.0705	0.0385	0.0195
2	0.9916	0.9470	0.8591	0.7382	0.6007	0.4628	0.3373	0.2318	0.1495	0.0898
3	0.9994	0.9917	0.9661	0.9144	0.8343	0.7297	0.6089	0.4826	0.3614	0.2539
4	1.0000	0.9991	0.9944	0.9804	0.9511	0.9012	0.8283	0.7334	0.6214	0.5000
5	1.0000	0.9999	0.9994	0.9969	0.9900	0.9747	0.9464	0.9006	0.8342	0.7461
6	1.0000	1.0000	1.0000	0.9997	0.9987	0.9957	0.9888	0.9750	0.9502	0.9102
7	1.0000	1.0000	1.0000	1.0000	0.9999	0.9996	0.9986	0.9962	0.9909	0.9805
8	1.0000	1.0000	1.0000	1.0000	1.0000	1.0000	0.9999	0.9997	0.9992	0.9980

$p=$	0.05	0.10	0.15	0.20	0.25	0.30	0.35	0.40	0.45	0.50
$n=10, x=0$	0.5987	0.3487	0.1969	0.1074	0.0563	0.0282	0.0135	0.0060	0.0025	0.0010
1	0.9139	0.7361	0.5443	0.3758	0.2440	0.1493	0.0860	0.0464	0.0233	0.0107
2	0.9885	0.9298	0.8202	0.6778	0.5256	0.3828	0.2616	0.1673	0.0996	0.0547
3	0.9990	0.9872	0.9500	0.8791	0.7759	0.6496	0.5138	0.3823	0.2660	0.1719
4	0.9999	0.9984	0.9901	0.9672	0.9219	0.8497	0.7515	0.6331	0.5044	0.3770
5	1.0000	0.9999	0.9986	0.9936	0.9803	0.9527	0.9051	0.8338	0.7384	0.6230
6	1.0000	1.0000	0.9999	0.9991	0.9965	0.9894	0.9740	0.9452	0.8980	0.8281
7	1.0000	1.0000	1.0000	0.9999	0.9996	0.9984	0.9952	0.9877	0.9726	0.9453
8	1.0000	1.0000	1.0000	1.0000	1.0000	0.9999	0.9995	0.9983	0.9955	0.9893
9	1.0000	1.0000	1.0000	1.0000	1.0000	1.0000	1.0000	0.9999	0.9997	0.9990
$n=12, x=0$	0.5404	0.2824	0.1422	0.0687	0.0317	0.0138	0.0057	0.0022	0.0008	0.0002
1	0.8816	0.6590	0.4435	0.2749	0.1584	0.0850	0.0424	0.0196	0.0083	0.0032
2	0.9804	0.8891	0.7358	0.5583	0.3907	0.2528	0.1513	0.0834	0.0421	0.0193
3	0.9978	0.9744	0.9078	0.7946	0.6488	0.4925	0.3467	0.2253	0.1345	0.0730
4	0.9998	0.9957	0.9761	0.9274	0.8424	0.7237	0.5833	0.4382	0.3044	0.1938
5	1.0000	0.9995	0.9954	0.9806	0.9456	0.8822	0.7873	0.6652	0.5269	0.3872
6	1.0000	0.9999	0.9993	0.9961	0.9857	0.9614	0.9154	0.8418	0.7393	0.6128
7	1.0000	1.0000	0.9999	0.9994	0.9972	0.9905	0.9745	0.9427	0.8883	0.8062
8	1.0000	1.0000	1.0000	0.9999	0.9996	0.9983	0.9944	0.9847	0.9644	0.9270
9	1.0000	1.0000	1.0000	1.0000	1.0000	0.9998	0.9992	0.9972	0.9921	0.9807
10	1.0000	1.0000	1.0000	1.0000	1.0000	1.0000	0.9999	0.9997	0.9989	0.9968
11	1.0000	1.0000	1.0000	1.0000	1.0000	1.0000	1.0000	1.0000	0.9999	0.9998
$n=15, x=0$	0.4633	0.2059	0.0874	0.0352	0.0134	0.0047	0.0016	0.0005	0.0001	0.0000
1	0.8290	0.5490	0.3186	0.1671	0.0802	0.0353	0.0142	0.0052	0.0017	0.0005
2	0.9638	0.8159	0.6042	0.3980	0.2361	0.1268	0.0617	0.0271	0.0107	0.0037
3	0.9945	0.9444	0.8227	0.6482	0.4613	0.2969	0.1727	0.0905	0.0424	0.0176
4	0.9994	0.9873	0.9383	0.8358	0.6865	0.5155	0.3519	0.2173	0.1204	0.0592
5	0.9999	0.9978	0.9832	0.9389	0.8516	0.7216	0.5643	0.4032	0.2608	0.1509
6	1.0000	0.9997	0.9964	0.9819	0.9434	0.8689	0.7548	0.6098	0.4522	0.3036
7	1.0000	1.0000	0.9994	0.9958	0.9827	0.9500	0.8868	0.7869	0.6535	0.5000
8	1.0000	1.0000	0.9999	0.9992	0.9958	0.9848	0.9578	0.9050	0.8182	0.6964
9	1.0000	1.0000	1.0000	0.9999	0.9992	0.9963	0.9876	0.9662	0.9231	0.8491
10	1.0000	1.0000	1.0000	1.0000	0.9999	0.9993	0.9972	0.9907	0.9745	0.9408
11	1.0000	1.0000	1.0000	1.0000	1.0000	0.9999	0.9995	0.9981	0.9937	0.9824
12	1.0000	1.0000	1.0000	1.0000	1.0000	1.0000	0.9999	0.9997	0.9989	0.9963
13	1.0000	1.0000	1.0000	1.0000	1.0000	1.0000	1.0000	1.0000	0.9999	0.9995
14	1.0000	1.0000	1.0000	1.0000	1.0000	1.0000	1.0000	1.0000	1.0000	1.0000

$p =$	0.05	0.10	0.15	0.20	0.25	0.30	0.35	0.40	0.45	0.50
$n = 20, x = 0$	0.3585	0.1216	0.0388	0.0115	0.0032	0.0008	0.0002	0.0000	0.0000	0.0000
1	0.7358	0.3917	0.1756	0.0692	0.0243	0.0076	0.0021	0.0005	0.0001	0.0000
2	0.9245	0.6769	0.4049	0.2061	0.0913	0.0355	0.0121	0.0036	0.0009	0.0002
3	0.9841	0.8670	0.6477	0.4114	0.2252	0.1071	0.0444	0.0160	0.0049	0.0013
4	0.9974	0.9568	0.8298	0.6296	0.4148	0.2375	0.1182	0.0510	0.0189	0.0059
5	0.9997	0.9887	0.9327	0.8042	0.6172	0.4164	0.2454	0.1256	0.0553	0.0207
6	1.0000	0.9976	0.9781	0.9133	0.7858	0.6080	0.4166	0.2500	0.1299	0.0577
7	1.0000	0.9996	0.9941	0.9679	0.8982	0.7723	0.6010	0.4159	0.2520	0.1316
8	1.0000	0.9999	0.9987	0.9900	0.9591	0.8867	0.7624	0.5956	0.4143	0.2517
9	1.0000	1.0000	0.9998	0.9974	0.9861	0.9520	0.8782	0.7553	0.5914	0.4119
10	1.0000	1.0000	1.0000	0.9994	0.9961	0.9829	0.9468	0.8725	0.7507	0.5881
11	1.0000	1.0000	1.0000	0.9999	0.9991	0.9949	0.9804	0.9435	0.8692	0.7483
12	1.0000	1.0000	1.0000	1.0000	0.9998	0.9987	0.9940	0.9790	0.9420	0.8684
13	1.0000	1.0000	1.0000	1.0000	1.0000	0.9997	0.9985	0.9935	0.9786	0.9423
14	1.0000	1.0000	1.0000	1.0000	1.0000	1.0000	0.9997	0.9984	0.9936	0.9793
15	1.0000	1.0000	1.0000	1.0000	1.0000	1.0000	1.0000	0.9997	0.9985	0.9941
16	1.0000	1.0000	1.0000	1.0000	1.0000	1.0000	1.0000	1.0000	0.9997	0.9987
17	1.0000	1.0000	1.0000	1.0000	1.0000	1.0000	1.0000	1.0000	1.0000	0.9998
18	1.0000	1.0000	1.0000	1.0000	1.0000	1.0000	1.0000	1.0000	1.0000	1.0000
$n = 25, x = 0$	0.2774	0.0718	0.0172	0.0038	0.0008	0.0001	0.0000	0.0000	0.0000	0.0000
1	0.6424	0.2712	0.0931	0.0274	0.0070	0.0016	0.0003	0.0001	0.0000	0.0000
2	0.8729	0.5371	0.2537	0.0982	0.0321	0.0090	0.0021	0.0004	0.0001	0.0000
3	0.9659	0.7636	0.4711	0.2340	0.0962	0.0332	0.0097	0.0024	0.0005	0.0001
4	0.9928	0.9020	0.6821	0.4207	0.2137	0.0905	0.0320	0.0095	0.0023	0.0005
5	0.9988	0.9666	0.8385	0.6167	0.3783	0.1935	0.0826	0.0294	0.0086	0.0020
6	0.9998	0.9905	0.9305	0.7800	0.5611	0.3407	0.1734	0.0736	0.0258	0.0073
7	1.0000	0.9977	0.9745	0.8909	0.7265	0.5118	0.3061	0.1536	0.0639	0.0216
8	1.0000	0.9995	0.9920	0.9532	0.8506	0.6769	0.4668	0.2735	0.1340	0.0539
9	1.0000	0.9999	0.9979	0.9827	0.9287	0.8106	0.6303	0.4246	0.2424	0.1148
10	1.0000	1.0000	0.9995	0.9944	0.9703	0.9022	0.7712	0.5858	0.3843	0.2122
11	1.0000	1.0000	0.9999	0.9985	0.9893	0.9558	0.8746	0.7323	0.5426	0.3450
12	1.0000	1.0000	1.0000	0.9996	0.9966	0.9825	0.9396	0.8462	0.6937	0.5000
13	1.0000	1.0000	1.0000	0.9999	0.9991	0.9940	0.9745	0.9222	0.8173	0.6550
14	1.0000	1.0000	1.0000	1.0000	0.9998	0.9982	0.9907	0.9656	0.9040	0.7878
15	1.0000	1.0000	1.0000	1.0000	1.0000	0.9995	0.9971	0.9868	0.9560	0.8852
16	1.0000	1.0000	1.0000	1.0000	1.0000	0.9999	0.9992	0.9957	0.9826	0.9461
17	1.0000	1.0000	1.0000	1.0000	1.0000	1.0000	0.9998	0.9988	0.9942	0.9784
18	1.0000	1.0000	1.0000	1.0000	1.0000	1.0000	1.0000	0.9997	0.9984	0.9927
19	1.0000	1.0000	1.0000	1.0000	1.0000	1.0000	1.0000	0.9999	0.9996	0.9980
20	1.0000	1.0000	1.0000	1.0000	1.0000	1.0000	1.0000	1.0000	0.9999	0.9995
21	1.0000	1.0000	1.0000	1.0000	1.0000	1.0000	1.0000	1.0000	1.0000	0.9999
22	1.0000	1.0000	1.0000	1.0000	1.0000	1.0000	1.0000	1.0000	1.0000	1.0000

$p =$	0.05	0.10	0.15	0.20	0.25	0.30	0.35	0.40	0.45	0.50
$n = 30, x = 0$	0.2146	0.0424	0.0076	0.0012	0.0002	0.0000	0.0000	0.0000	0.0000	0.0000
1	0.5535	0.1837	0.0480	0.0105	0.0020	0.0003	0.0000	0.0000	0.0000	0.0000
2	0.8122	0.4114	0.1514	0.0442	0.0106	0.0021	0.0003	0.0000	0.0000	0.0000
3	0.9392	0.6474	0.3217	0.1227	0.0374	0.0093	0.0019	0.0003	0.0000	0.0000
4	0.9844	0.8245	0.5245	0.2552	0.0979	0.0302	0.0075	0.0015	0.0002	0.0000
5	0.9967	0.9268	0.7106	0.4275	0.2026	0.0766	0.0233	0.0057	0.0011	0.0002
6	0.9994	0.9742	0.8474	0.6070	0.3481	0.1595	0.0586	0.0172	0.0040	0.0007
7	0.9999	0.9922	0.9302	0.7608	0.5143	0.2814	0.1238	0.0435	0.0121	0.0026
8	1.0000	0.9980	0.9722	0.8713	0.6736	0.4315	0.2247	0.0940	0.0312	0.0081
9	1.0000	0.9995	0.9903	0.9389	0.8034	0.5888	0.3575	0.1763	0.0694	0.0214
10	1.0000	0.9999	0.9971	0.9744	0.8943	0.7304	0.5078	0.2915	0.1350	0.0494
11	1.0000	1.0000	0.9992	0.9905	0.9493	0.8407	0.6548	0.4311	0.2327	0.1002
12	1.0000	1.0000	0.9998	0.9969	0.9784	0.9155	0.7802	0.5785	0.3592	0.1808
13	1.0000	1.0000	1.0000	0.9991	0.9918	0.9599	0.8737	0.7145	0.5025	0.2923
14	1.0000	1.0000	1.0000	0.9998	0.9973	0.9831	0.9348	0.8246	0.6448	0.4278
15	1.0000	1.0000	1.0000	0.9999	0.9992	0.9936	0.9699	0.9029	0.7691	0.5722
16	1.0000	1.0000	1.0000	1.0000	0.9998	0.9979	0.9876	0.9519	0.8644	0.7077
17	1.0000	1.0000	1.0000	1.0000	0.9999	0.9994	0.9955	0.9788	0.9286	0.8192
18	1.0000	1.0000	1.0000	1.0000	1.0000	0.9998	0.9986	0.9917	0.9666	0.8998
19	1.0000	1.0000	1.0000	1.0000	1.0000	1.0000	0.9996	0.9971	0.9862	0.9506
20	1.0000	1.0000	1.0000	1.0000	1.0000	1.0000	0.9999	0.9991	0.9950	0.9786
21	1.0000	1.0000	1.0000	1.0000	1.0000	1.0000	1.0000	0.9998	0.9984	0.9919
22	1.0000	1.0000	1.0000	1.0000	1.0000	1.0000	1.0000	1.0000	0.9996	0.9974
23	1.0000	1.0000	1.0000	1.0000	1.0000	1.0000	1.0000	1.0000	0.9999	0.9993
24	1.0000	1.0000	1.0000	1.0000	1.0000	1.0000	1.0000	1.0000	1.0000	0.9998
25	1.0000	1.0000	1.0000	1.0000	1.0000	1.0000	1.0000	1.0000	1.0000	1.0000

$p =$	0.05	0.10	0.15	0.20	0.25	0.30	0.35	0.40	0.45	0.50
$n = 40, x = 0$	0.1285	0.0148	0.0015	0.0001	0.0000	0.0000	0.0000	0.0000	0.0000	0.0000
1	0.3991	0.0805	0.0121	0.0015	0.0001	0.0000	0.0000	0.0000	0.0000	0.0000
2	0.6767	0.2228	0.0486	0.0079	0.0010	0.0001	0.0000	0.0000	0.0000	0.0000
3	0.8619	0.4231	0.1302	0.0285	0.0047	0.0006	0.0001	0.0000	0.0000	0.0000
4	0.9520	0.6290	0.2633	0.0759	0.0160	0.0026	0.0003	0.0000	0.0000	0.0000
5	0.9861	0.7937	0.4325	0.1613	0.0433	0.0086	0.0013	0.0001	0.0000	0.0000
6	0.9966	0.9005	0.6067	0.2859	0.0962	0.0238	0.0044	0.0006	0.0001	0.0000
7	0.9993	0.9581	0.7559	0.4371	0.1820	0.0553	0.0124	0.0021	0.0002	0.0000
8	0.9999	0.9845	0.8646	0.5931	0.2998	0.1110	0.0303	0.0061	0.0009	0.0001
9	1.0000	0.9949	0.9328	0.7318	0.4395	0.1959	0.0644	0.0156	0.0027	0.0003
10	1.0000	0.9985	0.9701	0.8392	0.5839	0.3087	0.1215	0.0352	0.0074	0.0011
11	1.0000	0.9996	0.9880	0.9125	0.7151	0.4406	0.2053	0.0709	0.0179	0.0032
12	1.0000	0.9999	0.9957	0.9568	0.8209	0.5772	0.3143	0.1285	0.0386	0.0083
13	1.0000	1.0000	0.9986	0.9806	0.8968	0.7032	0.4408	0.2112	0.0751	0.0192
14	1.0000	1.0000	0.9996	0.9921	0.9456	0.8074	0.5721	0.3174	0.1326	0.0403
15	1.0000	1.0000	0.9999	0.9971	0.9738	0.8849	0.6946	0.4402	0.2142	0.0769
16	1.0000	1.0000	1.0000	0.9990	0.9884	0.9367	0.7978	0.5681	0.3185	0.1341
17	1.0000	1.0000	1.0000	0.9997	0.9953	0.9680	0.8761	0.6885	0.4391	0.2148
18	1.0000	1.0000	1.0000	0.9999	0.9983	0.9852	0.9301	0.7911	0.5651	0.3179
19	1.0000	1.0000	1.0000	1.0000	0.9994	0.9937	0.9637	0.8702	0.6844	0.4373
20	1.0000	1.0000	1.0000	1.0000	0.9998	0.9976	0.9827	0.9256	0.7870	0.5627
21	1.0000	1.0000	1.0000	1.0000	1.0000	0.9991	0.9925	0.9608	0.8669	0.6821
22	1.0000	1.0000	1.0000	1.0000	1.0000	0.9997	0.9970	0.9811	0.9233	0.7852
23	1.0000	1.0000	1.0000	1.0000	1.0000	0.9999	0.9989	0.9917	0.9595	0.8659
24	1.0000	1.0000	1.0000	1.0000	1.0000	1.0000	0.9996	0.9966	0.9804	0.9231
25	1.0000	1.0000	1.0000	1.0000	1.0000	1.0000	0.9999	0.9988	0.9914	0.9597
26	1.0000	1.0000	1.0000	1.0000	1.0000	1.0000	1.0000	0.9996	0.9966	0.9808
27	1.0000	1.0000	1.0000	1.0000	1.0000	1.0000	1.0000	0.9999	0.9988	0.9917
28	1.0000	1.0000	1.0000	1.0000	1.0000	1.0000	1.0000	1.0000	0.9996	0.9968
29	1.0000	1.0000	1.0000	1.0000	1.0000	1.0000	1.0000	1.0000	0.9999	0.9989
30	1.0000	1.0000	1.0000	1.0000	1.0000	1.0000	1.0000	1.0000	1.0000	0.9997
31	1.0000	1.0000	1.0000	1.0000	1.0000	1.0000	1.0000	1.0000	1.0000	0.9999
32	1.0000	1.0000	1.0000	1.0000	1.0000	1.0000	1.0000	1.0000	1.0000	1.0000

$p=$	0.05	0.10	0.15	0.20	0.25	0.30	0.35	0.40	0.45	0.50
$n=50, x=0$	0.0769	0.0052	0.0003	0.0000	0.0000	0.0000	0.0000	0.0000	0.0000	0.0000
1	0.2794	0.0338	0.0029	0.0002	0.0000	0.0000	0.0000	0.0000	0.0000	0.0000
2	0.5405	0.1117	0.0142	0.0013	0.0001	0.0000	0.0000	0.0000	0.0000	0.0000
3	0.7604	0.2503	0.0460	0.0057	0.0005	0.0000	0.0000	0.0000	0.0000	0.0000
4	0.8964	0.4312	0.1121	0.0185	0.0021	0.0002	0.0000	0.0000	0.0000	0.0000
5	0.9622	0.6161	0.2194	0.0480	0.0070	0.0007	0.0001	0.0000	0.0000	0.0000
6	0.9882	0.7702	0.3613	0.1034	0.0194	0.0025	0.0002	0.0000	0.0000	0.0000
7	0.9968	0.8779	0.5188	0.1904	0.0453	0.0073	0.0008	0.0001	0.0000	0.0000
8	0.9992	0.9421	0.6681	0.3073	0.0916	0.0183	0.0025	0.0002	0.0000	0.0000
9	0.9998	0.9755	0.7911	0.4437	0.1637	0.0402	0.0067	0.0008	0.0001	0.0000
10	1.0000	0.9906	0.8801	0.5836	0.2622	0.0789	0.0160	0.0022	0.0002	0.0000
11	1.0000	0.9968	0.9372	0.7107	0.3816	0.1390	0.0342	0.0057	0.0006	0.0000
12	1.0000	0.9990	0.9699	0.8139	0.5110	0.2229	0.0661	0.0133	0.0018	0.0002
13	1.0000	0.9997	0.9868	0.8894	0.6370	0.3279	0.1163	0.0280	0.0045	0.0005
14	1.0000	0.9999	0.9947	0.9393	0.7481	0.4468	0.1878	0.0540	0.0104	0.0013
15	1.0000	1.0000	0.9981	0.9692	0.8369	0.5692	0.2801	0.0955	0.0220	0.0033
16	1.0000	1.0000	0.9993	0.9856	0.9017	0.6839	0.3889	0.1561	0.0427	0.0077
17	1.0000	1.0000	0.9998	0.9937	0.9449	0.7822	0.5060	0.2369	0.0765	0.0164
18	1.0000	1.0000	0.9999	0.9975	0.9713	0.8594	0.6216	0.3356	0.1273	0.0325
19	1.0000	1.0000	1.0000	0.9991	0.9861	0.9152	0.7264	0.4465	0.1974	0.0595
20	1.0000	1.0000	1.0000	0.9997	0.9937	0.9522	0.8139	0.5610	0.2862	0.1013
21	1.0000	1.0000	1.0000	0.9999	0.9974	0.9749	0.8813	0.6701	0.3900	0.1611
22	1.0000	1.0000	1.0000	1.0000	0.9990	0.9877	0.9290	0.7660	0.5019	0.2399
23	1.0000	1.0000	1.0000	1.0000	0.9996	0.9944	0.9604	0.8438	0.6134	0.3359
24	1.0000	1.0000	1.0000	1.0000	0.9999	0.9976	0.9793	0.9022	0.7160	0.4439
25	1.0000	1.0000	1.0000	1.0000	1.0000	0.9991	0.9900	0.9427	0.8034	0.5561
26	1.0000	1.0000	1.0000	1.0000	1.0000	0.9997	0.9955	0.9686	0.8721	0.6641
27	1.0000	1.0000	1.0000	1.0000	1.0000	0.9999	0.9981	0.9840	0.9220	0.7601
28	1.0000	1.0000	1.0000	1.0000	1.0000	1.0000	0.9993	0.9924	0.9556	0.8389
29	1.0000	1.0000	1.0000	1.0000	1.0000	1.0000	0.9997	0.9966	0.9765	0.8987
30	1.0000	1.0000	1.0000	1.0000	1.0000	1.0000	0.9999	0.9986	0.9884	0.9405
31	1.0000	1.0000	1.0000	1.0000	1.0000	1.0000	1.0000	0.9995	0.9947	0.9675
32	1.0000	1.0000	1.0000	1.0000	1.0000	1.0000	1.0000	0.9998	0.9978	0.9836
33	1.0000	1.0000	1.0000	1.0000	1.0000	1.0000	1.0000	0.9999	0.9991	0.9923
34	1.0000	1.0000	1.0000	1.0000	1.0000	1.0000	1.0000	1.0000	0.9997	0.9967
35	1.0000	1.0000	1.0000	1.0000	1.0000	1.0000	1.0000	1.0000	0.9999	0.9987
36	1.0000	1.0000	1.0000	1.0000	1.0000	1.0000	1.0000	1.0000	1.0000	0.9995
37	1.0000	1.0000	1.0000	1.0000	1.0000	1.0000	1.0000	1.0000	1.0000	0.9998
38	1.0000	1.0000	1.0000	1.0000	1.0000	1.0000	1.0000	1.0000	1.0000	1.0000

Percentage Points of the Normal Distribution

The values z in the table are those which a random variable $Z \sim N(0, 1)$ exceeds with probability p; that is, $P(Z > z) = 1 - \phi(z) = p$

p	z	p	z
0.5000	0.0000	0.0500	1.6449
0.4000	0.2533	0.0250	1.9600
0.3000	0.5244	0.0100	2.3263
0.2000	0.8416	0.0050	2.5758
0.1500	1.0364	0.0010	3.0902
0.1000	1.2816	0.0005	3.2905

Critical Values for Correlation Coefficients

These tables concern tests of the hypothesis that a population correlation coefficient ρ is 0. The values in the tables are the minimum values which need to be reached by a sample correlation coefficient in order to be significant at the level shown, on a one-tailed test.

Product Moment Coefficient					Sample level	Spearman's Coefficient		
level						Level		
0.100	0.050	0.025	0.010	0.005		0.050	0.025	0.010
0.8000	0.9000	0.9500	0.9800	0.9900	4	1.0000	—	—
0.6870	0.8054	0.8783	0.9343	0.9587	5	0.9000	1.0000	1.0000
0.6084	0.7293	0.8114	0.8822	0.9172	6	0.8286	0.8857	0.9429
0.5509	0.6694	0.7545	0.8329	0.8745	7	0.7143	0.7857	0.8929
0.5067	0.6215	0.7067	0.7887	0.8343	8	0.6429	0.7381	0.8333
0.4716	0.5822	0.6664	0.7498	0.7977	9	0.6000	0.7000	0.7833
0.4428	0.5494	0.6319	0.7155	0.7646	10	0.5636	0.6485	0.7455
0.4187	0.5214	0.6021	0.6851	0.7348	11	0.5364	0.6182	0.7091
0.3981	0.4973	0.5760	0.6581	0.7079	12	0.5035	0.5874	0.6783
0.3802	0.4762	0.5529	0.6339	0.6835	13	0.4835	0.5604	0.6484
0.3646	0.4575	0.5324	0.6120	0.6614	14	0.4637	0.5385	0.6264
0.3507	0.4409	0.5140	0.5923	0.6411	15	0.4464	0.5214	0.6036
0.3383	0.4259	0.4973	0.5742	0.6226	16	0.4294	0.5029	0.5824
0.3271	0.4124	0.4821	0.5577	0.6055	17	0.4142	0.4877	0.5662
0.3170	0.4000	0.4683	0.5425	0.5897	18	0.4014	0.4716	0.5501
0.3077	0.3887	0.4555	0.5285	0.5751	19	0.3912	0.4596	0.5351
0.2992	0.3783	0.4438	0.5155	0.5614	20	0.3805	0.4466	0.5218
0.2914	0.3687	0.4329	0.5034	0.5487	21	0.3701	0.4364	0.5091
0.2841	0.3598	0.4227	0.4921	0.5368	22	0.3608	0.4252	0.4975
0.2774	0.3515	0.4132	0.4815	0.5256	23	0.3528	0.4160	0.4862
0.2711	0.3438	0.4044	0.4716	0.5151	24	0.3443	0.4070	0.4757
0.2653	0.3365	0.3961	0.4622	0.5052	25	0.3369	0.3977	0.4662
0.2598	0.3297	0.3882	0.4534	0.4958	26	0.3306	0.3901	0.4571
0.2546	0.3233	0.3809	0.4451	0.4869	27	0.3242	0.3828	0.4487
0.2497	0.3172	0.3739	0.4372	0.4785	28	0.3180	0.3755	0.4401
0.2451	0.3115	0.3673	0.4297	0.4705	29	0.3118	0.3685	0.4325
0.2407	0.3061	0.3610	0.4226	0.4629	30	0.3063	0.3624	0.4251
0.2070	0.2638	0.3120	0.3665	0.4026	40	0.2640	0.3128	0.3681
0.1843	0.2353	0.2787	0.3281	0.3610	50	0.2353	0.2791	0.3293
0.1678	0.2144	0.2542	0.2997	0.3301	60	0.2144	0.2545	0.3005
0.1550	0.1982	0.2352	0.2776	0.3060	70	0.1982	0.2354	0.2782
0.1448	0.1852	0.2199	0.2597	0.2864	80	0.1852	0.2201	0.2602
0.1364	0.1745	0.2072	0.2449	0.2702	90	0.1745	0.2074	0.2453
0.1292	0.1654	0.1966	0.2324	0.2565	100	0.1654	0.1967	0.2327

Random numbers

86 13	84 10	07 30	39 05	97 96	88 07	37 26	04 89	13 48	19 20
60 78	48 12	99 47	09 46	91 33	17 21	03 94	79 00	08 50	40 16
78 48	06 37	82 26	01 06	64 65	94 41	17 26	74 66	61 93	24 97
80 56	90 79	66 94	18 40	97 79	93 20	41 51	25 04	20 71	76 04
99 09	39 25	66 31	70 56	30 15	52 17	87 55	31 11	10 68	98 23
56 32	32 72	91 65	97 36	56 61	12 79	95 17	57 16	53 58	96 36
66 02	49 93	97 44	99 15	56 86	80 57	11 78	40 23	58 40	86 14
31 77	53 94	05 93	56 14	71 23	60 46	05 33	23 72	93 10	81 23
98 79	72 43	14 76	54 77	66 29	84 09	88 56	75 86	41 67	04 42
50 97	92 15	10 01	57 01	87 33	73 17	70 18	40 21	24 20	66 62
90 51	94 50	12 48	88 95	09 34	09 30	22 27	25 56	40 76	01 59
31 99	52 24	13 43	27 88	11 39	41 65	00 84	13 06	31 79	74 97
22 96	23 34	46 12	67 11	48 06	99 24	14 83	78 37	65 73	39 47
06 84	55 41	27 06	74 59	14 29	20 14	45 75	31 16	05 41	22 96
08 64	89 30	25 25	71 35	33 31	04 56	12 67	03 74	07 16	49 32
86 87	62 43	15 11	76 49	79 13	78 80	93 89	09 57	07 14	40 74
94 44	97 13	77 04	35 02	12 76	60 91	93 40	81 06	85 85	72 84
63 25	55 14	66 47	99 90	02 90	83 43	16 01	19 69	11 78	87 84
11 22	83 98	15 21	18 57	53 42	91 91	26 52	89 13	86 00	47 61
01 70	10 83	94 71	13 67	11 12	36 54	53 32	90 43	79 01	95 15

Mathematical formulae – to learn
For A Level Maths

You are expected to know the following formulae for A Level Mathematics.

Pure Mathematics

Quadratic equations

$ax^2 + bx + c = 0$ has roots $\dfrac{-b \pm \sqrt{b^2 - 4ac}}{2a}$

Laws of indices

$a^x a^y \equiv a^{x+y}$

$a^x \div a^y \equiv a^{x-y}$

$(a^x)^y = a^{xy}$

Laws of logarithms

$x = a^n \iff \log_a x$ for $a > 0$ and $x > 0$

$\log_a x + \log_a y \equiv \log_a xy$

$\log_a x - \log_a y \equiv \log_a \left(\dfrac{x}{y} \right)$

$k \log_a x \equiv \log_a (x)^k$

Coordinate geometry

A straight line graph, gradient m passing through (x_1, y_1) has equation $y - y_1 = m(x - x_1)$

Straight lines with gradients m_1 and m_2 are perpendicular when $m_1 m_2 = -1$

Sequences

General term of an arithmetic progression: $\qquad u_n = a + (n-1)d$

General term of a geometric progression: $\qquad u_n = ar^{n-1}$

Trigonometry

In the triangle ABC

Sine rule $\quad \dfrac{a}{\sin A} = \dfrac{b}{\sin B} = \dfrac{c}{\sin C}$

Cosine rule $\quad a^2 = b^2 + c^2 - 2bc \cos A$

Area $\qquad \dfrac{1}{2} ab \sin C$

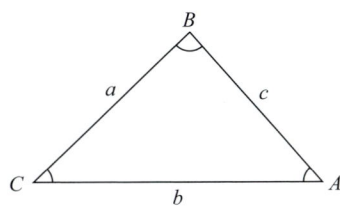

$\cos^2 A + \sin^2 A \equiv 1$

$\sec^2 A \equiv 1 + \tan^2 A$

$\operatorname{cosec}^2 A \equiv 1 + \cot^2 A$

$\sin 2A \equiv 2 \sin A \cos A$

$\cos 2A \equiv \cos^2 A - \sin^2 A$

$\tan 2A \equiv \dfrac{2 \tan A}{1 - \tan^2 A}$

Mathematical formulae - to learn for A Level Maths

Mensuration

Circumference, C, and area, A, of circle with radius r and diameter d:

$C = 2\pi r = \pi d \qquad A = \pi r^2$

Pythagoras' Theorem:

In any right-angled triangle where a, b and c are the lengths of the sides and c is the hypotenuse, $c^2 = a^2 + b^2$

Area of a trapezium $= \frac{1}{2}(a+b)h$, where a and b are the lengths of the parallel sides and h is their perpendicular separation.

Volume of a prism = area of cross section × length

For a circle of radius r, where an angle at the centre of θ radians subtends an arc of length s and encloses an associated sector of area A:

$s = r\theta \qquad A = \frac{1}{2}r^2\theta$

Calculus and differential equations

Differentiation

Function	Derivative
x^n	nx^{n-1}
$\sin kx$	$k\cos kx$
$\cos kx$	$-k\sin x$
e^{kx}	ke^{kx}
$\ln x$	$\frac{1}{x}$
$f(x)+g(x)$	$f'(x)+g'(x)$
$f(x)g(x)$	$f'(x)g(x)+f(x)g'(x)$
$f(g(x))$	$f'(g(x))g'(x)$

Integration

Function	Integral
x^n	$\frac{1}{n+1}x^{n+1}+c, \quad n \neq -1$
$\cos kx$	$\frac{1}{k}\sin kx + c$
$\sin kx$	$-\frac{1}{k}\cos kx + c$
e^{kx}	$\frac{1}{k}e^{kx}+c$
$\frac{1}{x}$	$\ln|x|+c, \quad x \neq 0$
$f'(x)+g'(x)$	$f(x)+g(x)+c$
$f'(g(x))g'(x)$	$f(g(x))+c$

Area under a curve $= \int_a^b y\,dx \qquad (y \geq 0)$

Vectors

$|x\mathbf{i}+y\mathbf{j}+z\mathbf{k}| = \sqrt{x^2+y^2+z^2}$

Statistics

The mean of a set of data: $\bar{x} = \dfrac{\sum x}{n} = \dfrac{\sum \mathrm{f}x}{\sum \mathrm{f}}$

The standard Normal variable: $Z = \dfrac{X - \mu}{\sigma}$ where $X \sim \mathrm{N}(\mu,\, \sigma^2)$

Mechanics

Forces and equilibrium

Weight $=$ mass $\times g$

Friction: $F \leq \mu R$

Newton's second law in the form: $F = ma$

Kinematics

For motion in a straight line with variable acceleration:

$$v = \frac{\mathrm{d}r}{\mathrm{d}t} \qquad a = \frac{\mathrm{d}v}{\mathrm{d}t} = \frac{\mathrm{d}^2 r}{\mathrm{d}t^2}$$

$$r = \int v \,\mathrm{d}t \qquad v = \int a \,\mathrm{d}t$$

Mathematical notation
For AS and A Level Maths

You should understand the following notation without need for further explanation.
Anything highlighted is only used in the full A level, and so will not be needed at AS Level.

Set Notation

$\in$	is an element of
$\notin$	is not an element of
$\subseteq$	is a subset of
$\subset$	is a proper subset of
$\{x_1, x_2, \ldots\}$	the set with elements $x_1, x_2, \ldots$
$\{x: \ldots\}$	the set of all x such that ...
$n(A)$	the number of elements in set A
$\varnothing$	the empty set
ε	the universal set
A'	the complement of the set A
$\mathbb{N}$	the set of natural numbers, $\{1, 2, 3, \ldots\}$
$\mathbb{Z}$	the set of integers, $\{0, \pm 1, \pm 2, \pm 3, \ldots\}$
$\mathbb{Z}^+$	the set of positive integers, $\{1, 2, 3, \ldots\}$
$\mathbb{Z}_0^+$	the set of non-negative integers, $\{0, 1, 2, 3, \ldots\}$
$\mathbb{R}$	the set of real numbers
$\mathbb{Q}$	the set of rational numbers, $\left\{\dfrac{p}{q} : p \in \mathbb{Z},\ q \in \mathbb{Z}^+\right\}$
$\cup$	union
$\cap$	intersection
(x, y)	the ordered pair x, y
$[a, b]$	the closed interval $\{x \in \mathbb{R} : a \le x \le b\}$
$[a, b)$	the interval $\{x \in \mathbb{R} : a \le x < b\}$
$(a, b]$	the interval $\{x \in \mathbb{R} : a < x \le b\}$
(a, b)	the open interval $\{x \in \mathbb{R} : a < x < b\}$

Miscellaneous Symbols

$=$	is equal to
$\ne$	is not equal to
$\equiv$	is identical to or is congruent to
$\approx$	is approximately equal to
∞	infinity
$\propto$	is proportional to
$<$	is less than
$\leqslant, \le$	is less than or equal to, is not greater than
$>$	is greater than
$\geqslant, \ge$	is greater than or equal to, is not less than
$\therefore$	therefore
$\because$	because
$\angle A$	angle A
$p \Rightarrow q$	p implies q (if p then q)
$p \Leftarrow q$	p is implied by q (if q then p)
$p \Leftrightarrow q$	p implies and is implied by q (p is equivalent to q)

a	first term for an arithmetic or geometric sequence
l	last term for an arithmetic sequence
d	common difference for an arithmetic sequence
r	common ratio for a geometric sequence
S_n	sum to n terms of a sequence
S_∞	sum to infinity of a sequence

Operations

$a+b$	a plus b
$a-b$	a minus b
$a \times b,\ ab,\ a \cdot b$	a multiplied by b
$a \div b,\ \dfrac{a}{b}$	a divided by b

$\displaystyle\sum_{i=1}^{n} a_i$	$a_1 + a_2 + \ldots + a_n$
$\displaystyle\prod_{i=1}^{n} a_i$	$a_1 \times a_2 \times \ldots \times a_n$
$\sqrt{a}$	the positive square root of a
$|a|$	the modulus of a
$n!$	n factorial: $n! = n \times (n-1) \times \ldots \times 2 \times 1,\ n \in \mathbb{N};\ 0! = 1$
$\dbinom{n}{r},\ {}^{n}C_r,\ {}_{n}C_r$	the binomial coefficient $\dfrac{n!}{r!(n-r)!}$ for $n,\ r \in \mathbb{Z}_0^+,\ r \le n$ or $\dfrac{n(n-1)\ldots(n-r+1)}{r!}$ for $n \in \mathbb{Q},\ r \in \mathbb{Z}_0^+$

Functions

$\mathrm{f}(x)$	the value of the function f at x
$\mathrm{f} : x \mapsto y$	the function f maps the element x to the element y
f^{-1}	the inverse function of the function f
gf	the composite function of f and g which is defined by $\mathrm{gf}(x) = \mathrm{g}(\mathrm{f}(x))$
$\displaystyle\lim_{x \to a} \mathrm{f}(x)$	the limit of $\mathrm{f}(x)$ as x tends to a
$\Delta x,\ \delta x$	an increment of x
$\dfrac{\mathrm{d}y}{\mathrm{d}x}$	the derivative of y with respect to x
$\dfrac{\mathrm{d}^n y}{\mathrm{d}x^n}$	the nth derivative of y with respect to x
$\mathrm{f}'(x) \ldots,\ \mathrm{f}^{(n)}(x)$	the first, ..., nth derivatives of $\mathrm{f}(x)$ with respect to x
$\dot{x},\ \ddot{x},\ \ldots$	the first, second, ... derivatives of x with respect to t
$\displaystyle\int y\,\mathrm{d}x$	the indefinite integral of y with respect to x
$\displaystyle\int_a^b y\,\mathrm{d}x$	the definite integral of y with respect to x between the limits $x = a$ and $x = b$

Mathematical notation for AS and A Level Maths

Exponential and Logarithmic Functions

e	base of natural logarithms
e^x, $\exp x$	exponential function of x
$\log_a x$	logarithm to the base a of x
$\ln x$, $\log_e x$	natural logarithm of x

Trigonometric Functions

sin, cos, tan, cosec, sec, cot   the trigonometric functions

$\sin^{-1}$, $\cos^{-1}$, $\tan^{-1}$, arcsin, arccos, arctan   the inverse trigonometric functions

$^\circ$	degrees
rad	radians

Vectors

$\mathbf{a}$, $\underline{a}$, $\underset{\sim}{a}$	the vector $\mathbf{a}$, $\underline{a}$, $\underset{\sim}{a}$
$\overrightarrow{AB}$	the vector represented in magnitude and direction by the directed line segment AB
$\hat{\mathbf{a}}$	a unit vector in the direction of $\mathbf{a}$
$\mathbf{i}, \mathbf{j}, \mathbf{k}$	unit vectors in the directions of the Cartesian coordinate axes
$|\mathbf{a}|$, a	the magnitude of $\mathbf{a}$
$|\overrightarrow{AB}|$, AB	the magnitude of $\overrightarrow{AB}$
$\begin{pmatrix} a \\ b \end{pmatrix}$, $a\mathbf{i} + b\mathbf{j}$	column vector and corresponding unit vector notation
$\mathbf{r}$	position vector
$\mathbf{s}$	displacement vector
$\mathbf{v}$	velocity vector
$\mathbf{a}$	acceleration vector

Probability and Statistics

A, B, C, etc.	events
$A \cup B$	union of the events A and B
$A \cap B$	intersection of the events A and B
$P(A)$	probability of the event A
A'	complement of the event A
$P(A \mid B)$	probability of the event A conditional on the event B
X, Y, R, etc.	random variables
x, y, r, etc.	values of the random variables X, Y, R etc.
$x_1, x_2, \ldots$	observations
$f_1, f_2, \ldots$	frequencies with which the observations $x_1, x_2, \ldots$ occur
$p(x)$, $P(X = x)$	probability function of the discrete random variable X
$p_1, p_2, \ldots$	probabilities of the values $x_1, x_2, \ldots$ of the discrete random variable X

$E(X)$	expectation of the random variable X
$\text{Var}(X)$	variance of the random variable X
$\sim$	has the distribution
$B(n, p)$	binomial distribution with parameters n and p, where n is the number of trials and p is the probability of success in a trial
q	$q = 1 - p$ for binomial distribution
$N(\mu, \sigma^2)$	Normal distribution with mean μ and variance σ^2
$Z \sim N(0,1)$	standard Normal distribution
ϕ	probability density function of the standardised Normal variable with distribution $N(0, 1)$
Φ	corresponding cumulative distribution function
μ	population mean
σ^2	population variance
σ	population standard deviation
$\bar{x}$	sample mean
s^2	sample variance
s	sample standard deviation
H_0	null hypothesis
H_1	alternative hypothesis
r	product moment correlation coefficient for a sample
ρ	product moment correlation coefficient for a population

Mechanics

kg	kilograms
m	metres
km	kilometres
m/s, m s^{-1}	metres per second (velocity)
m/s^2, m s^{-2}	metres per second per second (acceleration)
F	Force or resultant force
N	Newton
N m	Newton metre (moment of a force)
t	time
s	displacement
u	initial velocity
v	velocity or final velocity
a	acceleration
g	acceleration due to gravity
μ	coefficient of friction

Answers

Full solutions to all of these questions can be found at the link in the page footer.

Chapter 1
Exercise 1.1A Fluency and skills

1 A prime number, by definition, has exactly two factors: 1 and the number itself. The number 1 has only one factor so is NOT a prime number.

2 Let the numbers be $2m + 1$ and $2n + 1$
$2m + 1 + 2n + 1 = 2m + 2n + 2 = 2(m + n + 1)$
$m + n + 1 = k$, an integer, so the sum is $2k$ which is the definition of an even number.

3 Let the smaller odd number be $2m + 1$
The next one is $2m + 3$
$(2m + 1)(2m + 3) = 4m^2 + 8m + 3 = 4(m^2 + 2m + 1) - 1$
$m^2 + 2m + 1$ is an integer so this is one less than a multiple of 4

4 Let the integers be m, $m + 1$ and $m + 2$
The mean $= \dfrac{m + m + 1 + m + 2}{3} = \dfrac{3m + 3}{3} = m + 1$, which is the middle number.

5 **a** Let the integers be m and $m + 1$
$m^2 + (m + 1)^2 = 2m^2 + 2m + 1$
The first two terms are even and the third is odd so the sum must be odd.
   b Let the integers be $2m$ and $2m + 2$
$(2m)^2 + (2m + 2)^2 = 8m^2 + 8m + 4 = 4(2m^2 + 2m + 1)$
4 is a factor of this expression so the squares of two consecutive even numbers is always a multiple of 4

6 Let the integers be m, $m + 1$, $m + 2$ and $m + 3$
$m + m + 1 + m + 2 + m + 3 = 4m + 6 = 2(2m + 3)$
2 is even; $2m$ is even so $2m + 3$ must be odd. Hence the sum has both odd and even factors.

7 Let the numbers be m and n. $(m + n)^2 = m^2 + 2mn + n^2$ which is $(m^2 + n^2) + 2mn$, which is $2mn$ more than the sum of the squares, since, if m and n are positive, then mn must also be positive.

8 Let the equal side be a. The hypotenuse $= \sqrt{a^2 + a^2} = a\sqrt{2}$
Thus the perimeter is $2a + a\sqrt{2}$
Since $\sqrt{2} > 1$, it follows that the perimeter is always greater than three times the length of one of the equal sides.

9 If the sum = the product then $a + b = ab$
Hence $b - 2 + b = (b - 2)b \Rightarrow 2b - 2 = b^2 - 2b \Rightarrow b^2 - 4b + 2$
This quadratic does not factorise and hence b cannot be an integer. Consequently, since $a = b - 2$, neither can a

10 If $(5y)^2$ is even, then $5y$ is even and, since 5 is odd, y must be even to make $5y$ even.

11 Checking each number, only 25 is a square number and only 27 is a cube.

12 For example: JANUARY, YARN; FEBRUARY, FRAY; AUGUST, STAG; SEPTEMBER, TERM; OCTOBER, BOOT; NOVEMBER, BONE; DECEMBER, BRED

13 There are 5 'cases' and we investigate each one: $(0 + 1)^3 \geq 3^0$
$\Rightarrow 1 = 1$ TRUE; $(1 + 1)^3 \geq 3^1 \Rightarrow 8 > 3$ TRUE; $(2 + 1)^3 \geq 3^2$
$\Rightarrow 27 > 9$ TRUE; $(3 + 1)^3 \geq 3^3 \Rightarrow 64 > 27$ TRUE; $(4 + 1)^3 \geq 3^4$
$\Rightarrow 125 > 81$ TRUE. Thus the statement is true.

14 Let the square number be m^2
$m = 10p + k$ for integers p and k, $0 \leq k \leq p$
If $m = 10p$ then $m^2 = (10p)^2 = 100p^2$ which ends in 0, if $m = 10p + 1$ then $m^2 = (10p + 1)^2 = 100p^2 + 20p + 1$ which ends in 1, if $m = 10p + 2$ then $m^2 = (10p + 2)^2 = 100p^2 + 40p + 4$

which ends in 4, … if $m = 10p + 9$ then
$m^2 = (10p + 9)^2 = 100p^2 + 180p + 81$ which ends in 1
Therefore all square numbers cannot have a last digit 2, 3, 7 or 8

15 $2 \times 3 = 6$ which disproves the statement.

16 $1 + 2 = 3$ which is < 6 and disproves the statement.

17 $3 - (-4) = 3 + 4 = 7$ which disproves the statement.

18 $5 \times -2 = -10$ which disproves the statement.

19 e.g. $a = 4$, $b = 3$; $4 > 3$ but $4^3 = 64 < 3^4 = 81$

20 Try any set with two odd numbers, such as $1 \times 2 \times 3 = 6$, which is not divisible by 4

Exercise 1.1B Reasoning and problem-solving

1 **a** Case 1: P is 2 so even. PQ is even $\times$ odd which is even.
Case 2: P is odd. PQ is odd $\times$ odd which is odd.
Graham is right.
   b Case 1: P is 2 so even and $Q + 1$ is even.
$P(Q + 1)$ is even $\times$ even which is even.
Case 2: P is odd. $P(Q + 1)$ is odd $\times$ even which even.
Sue is right in this case.

2 $99 = 3 \times 33$ and 33 is not prime. The statement is false.

3 $9^1 - 1 = 9 - 1 = 8$ which is 8×1; $9^2 - 1 = 81 - 1 = 80$ which is 8×10; $9^3 - 1 = 729 - 1 = 728$ which is 8×91; $9^4 - 1 = 6561 - 1 = 6560$ which is 8×820; $9^5 - 1 = 59\,049 - 1 = 59\,048$ which is 8×7381; $9^6 - 1 = 531\,441 - 1 = 531\,440$ which is $8 \times 66\,430$
Thus the value of $9^n - 1$ is divisible by 8 for $1 \leq n \leq 6$

4 An equilateral triangle is not obtuse. A right-angled triangle is not obtuse.

5 A convex hexagon can be split into 4 triangles. The sum of the interior angles of a triangle is $180°$. $4 \times 180 = 720$. So the sum of the interior angles of a convex hexagon is $720°$

6 False. A Rhombus has equal sides but is not a square.

7 A convex n-sided polygon can be split into $n - 2$ triangles. The sum of the interior angles of a triangle is $180°$
$(n - 2) \times 180 = 180(n - 2)$. So the sum of the interior angles of a convex n-sided polygon is $180(n - 2)°$

8 False. $5^2 = (-5)^2$ but $5 \neq -5$

9 The square of the remaining side $= (2s + a)^2 - (2s - a)^2$
$= 4as + 4as = 8as$
Therefore the square of the remaining side is a multiple of 8

10 $\dfrac{1}{1 \times 2} = \dfrac{1}{1 + 1}$; $\dfrac{1}{1 \times 2} + \dfrac{1}{2 \times 3} = \dfrac{1}{2} + \dfrac{1}{6} = \dfrac{3 + 1}{6} = \dfrac{4}{6} = \dfrac{2}{3}$;
$\dfrac{1}{1 \times 2} + \dfrac{1}{2 \times 3} + \dfrac{1}{3 \times 4} = \dfrac{1}{2} + \dfrac{1}{6} + \dfrac{1}{12} = \dfrac{6 + 2 + 1}{12} = \dfrac{9}{12} = \dfrac{3}{4}$;
$\dfrac{1}{1 \times 2} + \dfrac{1}{2 \times 3} + \dfrac{1}{3 \times 4} + \dfrac{1}{4 \times 5} = \dfrac{1}{2} + \dfrac{1}{6} + \dfrac{1}{12} + \dfrac{1}{20} = $
$\dfrac{30 + 10 + 5 + 3}{60} = \dfrac{48}{60} = \dfrac{4}{5}$; $\dfrac{1}{1 \times 2} + \dfrac{1}{2 \times 3} + \dfrac{1}{3 \times 4} + \dfrac{1}{4 \times 5} + $
$\dfrac{1}{5 \times 6} = \dfrac{1}{2} + \dfrac{1}{6} + \dfrac{1}{12} + \dfrac{1}{20} + \dfrac{1}{30} = \dfrac{30 + 10 + 5 + 3 + 2}{60} = \dfrac{50}{60} = \dfrac{5}{6}$
Thus $\dfrac{1}{1 \times 2} + \dfrac{1}{2 \times 3} + \dfrac{1}{3 \times 4} + \ldots + \dfrac{1}{n \times (n + 1)} = \dfrac{n}{n + 1}$ for $1 \leq n \leq 5$

11 Let the two-digit number be X with tens digit y and units digit z
Therefore $X = 10y + z = 9y + (y + z)$. Let the sum of the digits be divisible by 3, i.e. $(y + z) = 3P$
Thus $X = 9y + 3P = 3(3y + P)$ and hence X is a multiple of three.

Exercise 1.2A Fluency and skills

1 64

2 -243

3 2401

4 c^{11}

5 p^{12}

6 $-p^{12}$

7 $64c^{-18}$

8 d^{14}

9 $70e^9$

10 $-108f^{12}$

11 $4g^6$

12 $-4k^{55}$

13 $8f^3$

14 $\dfrac{2}{3}e^2$

15 $15ab$

16 $-120wx^2$

17 $24def^2$

18 $-9h^{14}$

19 $5r^8s^{10}$

20 $5r^2s^2$

21 $-g^{10}h^9i^4$

22 $-g^8h^7i^{-4}$

23 $5z^5y^5$

24 $6u^{18}$

25 $6u^{18}$

26 $5t^9$

27 $-5t^9c^4$

28 $\dfrac{1}{5}$

29 5

30 6

31 -50

32 3

33 $\dfrac{1}{16}$

34 $\dfrac{1}{1024}$

35 $\dfrac{1}{9w^2}$

36 $\dfrac{w^4}{9}$

37 512

38 4

39 256

40 8

41 $\dfrac{1}{8}$

42 $\dfrac{6}{7}$

43 $\dfrac{7}{6}$

44 $\dfrac{216}{343}$

45 $n=4$

46 $m=5$

47 $t=1$

48 $b=-2$

Exercise 1.2B Reasoning and problem-solving

1 a $4s^4$ inches^2 **b** $5p^2q^3$ cm

2 a Circumference $=6\pi w^5$ ft; Area $=9\pi w^{10}$ ft^2

   b Surface area $=36\pi w^8$ ft^2; Volume $=36\pi w^{12}$ ft^3

3 $6c^{-4}d^{-4}e^{-5}$

4 $36p^5q^5$

5 a $2y^3z^4$

   b Area $=8$. Therefore the area is independent of m and n

6 $\dfrac{4}{3c^{\frac{1}{2}}}$ mph

7 a $2d^3$ cm

   b Volume $=225\pi s^6t^0=225\pi s^6$m^3. It is independent of t since $t^0=1$

8 $7\pi v^4z^{-4}$ cm^2

9 a $13n^{\frac{1}{2}}$ **b** $30n$

10 a $18m^2n^{-7}$ V **b** $108n^{-10}$ W

11 $\dfrac{81}{2}mx^{\frac{3}{2}}c^{\frac{3}{2}}$

12 $\dfrac{3t^3}{16g}$

13 $\dfrac{15}{8}rs^2$ ohms

14 a $A=\sqrt{s(s-a)(s-b)(s-c)}$

   b $A=30x^2y^2$

   c The sides of the triangle are in the ratio $5:12:13$
   Since $13^2=5^2+12^2$ the triangle is right-angled and so the area is $\dfrac{1}{2}\times 5xy\times 12xy=30x^2y^2$

Exercise 1.3A Fluency and skills

1 Parts **a**, **e** and **f** are rational; parts **b**, **c**, **d** and **g** are irrational

2 a $2\sqrt{21}$ **b** $2\sqrt{14}$ **c** $5\sqrt{3}$ **d** $3\sqrt{3}$

   e $2\sqrt{2}$ **f** $16\sqrt{2}$ **g** $17\sqrt{17}$ **h** $6\sqrt{6}$

   i $210\sqrt{3}$ **j** $288\sqrt{5}$

3 a 8 **b** 5

4 a $\dfrac{4}{10}$ or 0.4 **b** $\dfrac{5}{6}$

5 a $3\sqrt{6}$ **b** $12\sqrt{3}$ **c** $16\sqrt{5}$ **d** $22\sqrt{7}$

   e $2\sqrt{10}$ **f** $6\sqrt{7}$ **g** $3\sqrt{5}$ **h** $2\sqrt{2}$

   i $3\sqrt{6}$ **j** $7\sqrt{3}$ **k** $2\sqrt{3}$ **l** $4\sqrt{5}$

   m $7\sqrt{3}$ **n** $4\sqrt{2}$

6 a $59+6\sqrt{30}$ **b** $14+7\sqrt{2}$ **c** $-14+7\sqrt{2}$

   d $6-10\sqrt{5}$ **e** $120-14\sqrt{6}$

7 a $\dfrac{\sqrt{13}}{13}$ **b** $\dfrac{4\sqrt{6}}{3}$

   c $\dfrac{\sqrt{55}}{10}$ **d** $3(\sqrt{2}+1)$

   e 5 **f** $\dfrac{13\sqrt{5}}{20}-\dfrac{\sqrt{30}}{30}$

   g $\dfrac{5(8+\sqrt{5})}{59}$ **h** $\dfrac{2(2\sqrt{2}-1)}{7}$

   i $11-2\sqrt{30}$ **j** $\dfrac{5-\sqrt{77}}{4}$

8 a $\dfrac{a\sqrt{b}+b}{b}$ **b** $\dfrac{a^2+2a\sqrt{b}+b}{a^2-b}$

   c $\dfrac{\sqrt{ac}+bc}{bc}$ **d** $\dfrac{a-\sqrt{ab}-b\sqrt{ac}+b\sqrt{bc}}{a-b}$

Exercise 1.3B Reasoning and problem-solving

1 $6\sqrt{6}$ cm^2

2 a $\pi(9\sqrt{3})^2=\pi(81\times 3)=243\pi$ cm^2

   b $14\sqrt{5}$

3 a $3\sqrt{5}$ m s^{-1} **b** $8\sqrt{5}$ m

4 $50+19\sqrt{7}$ m^3

5 A is $\dfrac{3\sqrt{21}}{7}$ times bigger than B

6 $5\sqrt{3}$ cm   **7** $10\sqrt{3}$ miles

8 $\dfrac{45}{360}\times 2\pi\left(\dfrac{12}{\sqrt{3}}\right)=\sqrt{3}\,\pi$ m   **9** $2\sqrt{3}$

10 $\dfrac{\left(6\sqrt{8}\right)^3}{\left(4\sqrt{2}\right)^3}=\dfrac{3456\sqrt{2}}{128\sqrt{2}}=27$ **11** $\dfrac{19\sqrt{a}}{3a}$

12 $\dfrac{40\pi}{13}$ m^2 **13** $40(3-\sqrt{6})$

14 Thus coordinates of the centroid are $\left(\dfrac{5\sqrt{6}}{2},\dfrac{5\sqrt{2}}{2}\right)$, so the distance from the origin to the centroid $=5\sqrt{2}$

15 $2s=5+5\sqrt{2}+3+3\sqrt{2}+6+4\sqrt{2}+2+4\sqrt{2}=16+16\sqrt{2}$

   $s=8+8\sqrt{2}$

   $A^2=(3+3\sqrt{2})(5+5\sqrt{2})(2+4\sqrt{2})(6+4\sqrt{2})$

   $=3(1+\sqrt{2})5(1+\sqrt{2})2(1+2\sqrt{2})2(3+2\sqrt{2})$

   $=2^2(1+\sqrt{2})^2 15(11+8\sqrt{2})$

   $A=2(1+\sqrt{2})\sqrt{15(11+8\sqrt{2})}$

Exercise 1.4A Fluency and skills

1 a $x=\pm 3\sqrt{2}$ **b** $x=\sqrt{3}$ or $-\sqrt{3}$

   c $x=0$ or $\dfrac{-5}{4}$ **d** $x=-\sqrt{3}$ twice

   e $x=\dfrac{1}{2}$ or -3 **f** $x=7$ or $\dfrac{2}{3}$

   g $x=\dfrac{3}{4}$ twice **h** $x=\dfrac{9}{4}$ or -2

2 a i $f(x)=x^2+3x+2=(x+1)(x+2)$

      ii

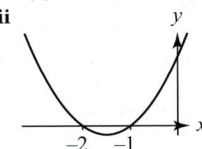

b i $f(x) = x^2 + 6x - 7 = (x - 1)(x + 7)$

ii

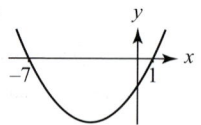

c i $f(x) = -x^2 - x + 2 = (x + 2)(1 - x)$

ii

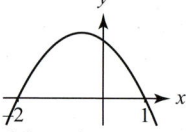

d i $f(x) = -x^2 - 7x - 12 = (x + 4)(-x - 3)$

ii

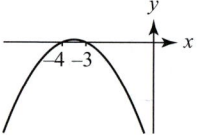

e i $f(x) = 2x^2 - x - 1 = (2x + 1)(x - 1)$

ii

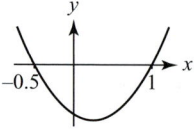

f i $f(x) = -3x^2 + 11x + 20 = (x - 5)(-3x - 4)$

ii

3 a $x = \dfrac{-3}{2} \pm \dfrac{\sqrt{21}}{6}$ or -2.26 or -0.74

b $x = \dfrac{-5}{8} \pm \dfrac{\sqrt{41}}{8}$ or 0.18 or -1.43

c $x = -6 \pm \sqrt{31}$ or -0.43 or -11.57

d $x = -1 \pm \sqrt{29}$ or -6.39 or 4.39

e $x = \dfrac{-15}{2} \pm \dfrac{\sqrt{365}}{2}$ or 2.05 or -17.05

f $x = \dfrac{3}{2} \pm \dfrac{\sqrt{145}}{2}$ or -4.52 or 7.52

g $x = 4.5$ twice

h $x = \dfrac{23}{6} \pm \dfrac{\sqrt{277}}{6}$ or 6.61 or 1.06

i $x = \dfrac{-8}{5} \pm \dfrac{\sqrt{19}}{5}$ or -0.73 or -2.47

j $x = \dfrac{1}{20} \pm \dfrac{\sqrt{41}}{20}$ or 0.37 or -0.27

4 a $f(x) = (x - 7)^2$; minimum $(7, 0)$

b $f(x) = (x + 1)^2 - 6$; minimum $(-1, -6)$

c $f(x) = -1(x + 3)^2 + 4$; maximum $(-3, 4)$

d $f(x) = -1(x - 2)^2 + 7$; maximum $(2, 7)$

e $f(x) = 9\left(x - \dfrac{1}{3}\right)^2 - 6$; minimum $\left(\dfrac{1}{3}, -6\right)$

f $f(x) = -2(x + 7)^2 + 63$; maximum $(-7, 63)$

5 a $x = 1 \pm 1 = 2$ or 0 **b** $x = -2 \pm \sqrt{7}$

c $x = 7 \pm 4 = 11$ or 3 **d** $x = -4 \pm \sqrt{6}$

e $x = 3$ twice **f** $x = -5 \pm 1 = -6$ or -4

g $x = -11 \pm \sqrt{3}$ **h** $x = 8 \pm \sqrt{10}$

i $x = \dfrac{3 \pm \sqrt{7}}{2}$ **j** $x = \dfrac{-2 \pm \sqrt{6}}{3}$

k $x = \dfrac{-11 \pm \sqrt{109}}{2}$ **l** $x = \dfrac{5 \pm \sqrt{57}}{3}$

6 a $x = 0$ or 3 **b** $x = -5$ or 3

c $x = -1$ or 6 **d** $x = -2$ or 4

e $x = -2.5$ or 3 **f** $x = -0.5$ or 1.5

7 a two real distinct roots **b** two real distinct roots

c two real distinct roots **d** no real roots

e two real coincident roots **f** no real roots

g two real coincident roots **h** no real roots

Exercise 1.4B Reasoning and problem-solving

1 a $3x^2 + 25x - 2035 = 0 \Rightarrow x = 22.2$ cm (nearest mm)

b 14.8 litres

2 a $x^2 + 3x - 10 = 0 \Rightarrow x = 2$ years

b 28 years

3 The card has side length 39.7 cm and the photo is 29.8 cm by 19.7 cm.

4 a Area $= 30x^2 - x^4$

b $z^2 - 30z + 85 = 0$

c $x = 5.18$ inches or 1.78 inches

5 a $h = -5\left(t - \dfrac{5}{2}\right)^2 + \dfrac{325}{4}$

b

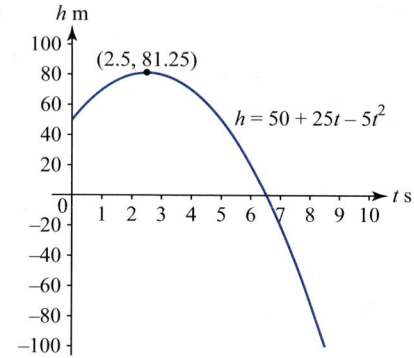

c i $t = 2.5$ s, $h = 81.25$ m **ii** $t = 5$ s **iii** $t = 6.5$ s

6 a

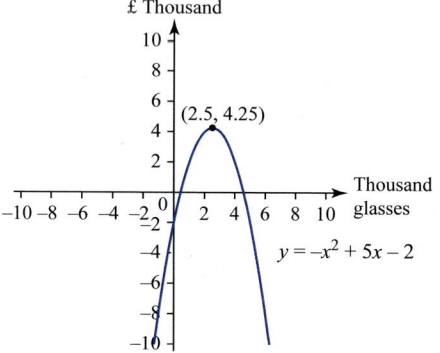

b i 2.5 **ii** 4.6 or 0.4 **iii** 1.5 to 3.5

7 a

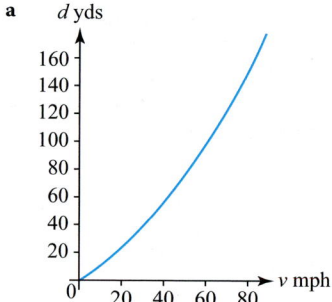

b i 10 m **ii** 55 m **iii** 138 m

c 42 mph

8 $k = \sqrt[3]{16}$ or $-\sqrt[3]{5}$

Exercise 1.5A Fluency and skills

1 a $2x + 3y - 13 = 0$ **b** No.

2 The gradient is $\dfrac{4}{3}$ and the y-intercept is $-\dfrac{8}{3}$

3 a The gradient of $2x - 3y = 4$ is $\dfrac{2}{3}$, the gradient of

$6x + 4y = 7$ is $\dfrac{-6}{4} = \dfrac{-3}{2}$, and $\dfrac{2}{3} \times \dfrac{-3}{2} = -1$ so the lines are perpendicular.

b The gradient of $2x - 3y = 4$ is $\dfrac{2}{3}$ and the gradient of

$8x - 12y = 7$ is $\dfrac{8}{12} = \dfrac{2}{3}$

Both lines have the same gradient, so the lines are parallel.

4 Gradient $= \dfrac{-8}{9}$ and y-intercept $= -\dfrac{7}{6}$

5 $m = \dfrac{-1 - (-6)}{4 - -5} = \dfrac{5}{9}$ and $9y - 5x + 29 = 0$

6 $y = -3x - 25$ or $y + 3x + 25 = 0$

7 a $(4, 6)$, length $4\sqrt{5}$ **b** $\left(\dfrac{-1}{2}, \dfrac{-7}{2}\right)$, length $\sqrt{26}$

c $\left(\dfrac{\sqrt{5}}{2}, \dfrac{2\sqrt{3}}{2}\right)$, length $\sqrt{17}$

8 $y = 4x + 8$ and $3y - 12x = 7$ are parallel; $2x + 3y = 4$ and $6x + 9y = 12$ are parallel; $4x - 5y = 6$ and $10x + 8y = 5$ are perpendicular.

9 a $3y + 2x + 2 = 0$ **b** $y - 2x - 1 = 0$

10 a $x^2 - 2x + y^2 - 16y + 40 = 0$

b $x^2 - 12x + y^2 + 14y + 76 = 0$

c $x^2 - 2\sqrt{5}x + y^2 - 2\sqrt{2}\,y - 4 = 0$

11 a centre $= (-9, 7)$; radius $= 10$

b centre $= (-6, -5)$; radius $= \sqrt{86}$

c centre $= (\sqrt{3}, -\sqrt{7})$; radius $= \sqrt{11}$

12 $m_{AB} = \dfrac{18 + 12}{6 + 10} = \dfrac{30}{16} = \dfrac{15}{8}$; $m_{BC} = \dfrac{-14 - 18}{-2 - 6} = \dfrac{-32}{-8} = 4$;

$m_{CA} = \dfrac{-12 + 14}{-10 + 2} = \dfrac{2}{-8} = \dfrac{-1}{4}$

Since $4 \times \dfrac{-1}{4} = -1$, BC is perpendicular to CA and hence ABC is a right-angled triangle.

Since the angle in a semicircle is a right angle it follows that AB is a diameter and the points form a semicircle.

13 $y = 3x + 5$

14 a $x^2 - 14x + y^2 + 2y - 274 = 0$

b $4\sqrt{31}$

c $\left(-3 + 3\sqrt{62}, 9 + 3\sqrt{62}\right)$ and $\left(-3 - 3\sqrt{62}, 9 - 3\sqrt{62}\right)$

15 a $x^2 + y^2 - 20y = 0$

b $x^2 - 8x + y^2 - 8y + 24 = 0$

c $x^2 - x + y^2 - 14y - 44 = 0$

d $x^2 + (4 + \sqrt{2})x + y^2 + (5 - \sqrt{5})y + (4\sqrt{2} - 5\sqrt{5}) = 0$

Exercise 1.5B Reasoning and problem-solving

1 PQ: $12y - 5x - 63 = 0$, QR: $15y - 8x - 84 = 0$,

RS: $12y - 5x - 84 = 0$, and SP: $15y - 8x - 105 = 0$

Since the quadrilateral contains two pairs of parallel sides of different lengths, it must be a parallelogram.

2 a $(-2, -5)$

b Q is $(-5, -1)$; S is $(1, -9)$

c RP: $4y - 3x + 14 = 0$, QS: $3y + 4x + 23 = 0$

d QP: $7y + x + 12 = 0$, PS: $y - 7x + 16 = 0$,

SR: $3y - x + 28 = 0$, RQ: $9y - 13x - 56 = 0$

e $100\,\text{units}^2$

f $30 + 10\sqrt{10} + 10\sqrt{2}$ units

3 a AB: $y + 2 = -\dfrac{4}{3}(x - 6)$ or $3y + 4x - 18 = 0$,

AC: $y + \dfrac{11}{2} = 7\left(x - \dfrac{11}{2}\right)$ or $2y - 14x + 88 = 0$

---

b Centre $= (6, -2)$, equation of circle is $(x - 6)^2 + (y + 2)^2 = 25$ or $x^2 - 12x + y^2 - 4y + 15 = 0$

$m_{BC} = \dfrac{-6 - 1}{9 - 10} = \dfrac{-7}{-1} = 7$. $m_{AC} \times m_{BC} = \dfrac{-1}{7} \times 7 = -1$ so AC and BC are perpendicular. Hence the triangle is right-angled.

4 a $x^2 + y^2 - 4x - 12y + 15 = 0$; $(x - 2)^2 + (y - 6)^2 = 25$ so centre is $(2, 6)$ and radius 5. If radius is 5, lowest possible point is $(2, 1)$ so the circle does not intersect the x-axis.

b $CP = \sqrt{61} = 7.81...$; radius is 5 so P lies outside the circle.

c $k = -15$ or 35

5 $160\,\text{units}^2$

6 a Centre $= \left(\dfrac{a + b}{2}, \dfrac{c + d}{2}\right)$, radius $= \dfrac{1}{2}\sqrt{(a - b)^2 + (c - d)^2}$

Equation of circle:

$\left(x - \dfrac{(a + b)}{2}\right)^2 + \left(y - \dfrac{(c + d)}{2}\right)^2 = \dfrac{1}{4}((a - b)^2 + (c - d)^2)$

so $(x - a)(x - b) + (y - c)(y - d) = 0$

b $x^2 - 10x + y^2 - 12y - 39 = 0$

Exercise 1.6A Fluency and skills

1 $x = 4, y = 3$ **2** $a = 1, b = 2$

3 $c = 5, d = -9$ **4** $e = 6, f = -1$

5 $x = 12, y = -18$ **6** $n = 5, m = -\dfrac{5}{3}$

7 $c = -1, d = 7$ **8** $f = \dfrac{1}{3}, e = 6$

9 When $x = 0$, $y = 3$ and when $x = 1$, $y = 2$

10 When $g = -3$, $h = 4$ and when $g = 2$, $h = -1$

11 When $g = 3$, $h = 2$ and when $g = -\dfrac{9}{4}$, $h = \dfrac{79}{8}$

12 When $n = -1$, $m = 1$ and when $n = \dfrac{17}{10}$, $m = \dfrac{289}{100}$

13 $(0, 0)$, $(5, 5)$

14 $(-1, 1)$, $(3, 5)$

15 $\left(\dfrac{2}{9}, \dfrac{4}{3}\right)$, $(1, -1)$ **16** $\left(-\dfrac{1}{4}, \dfrac{1}{2}\right)$, $(-2, -3)$

17 $y = \dfrac{20}{x}$ so $\dfrac{20}{x} = 8 + x$

$20 = 8x + x^2$ so $x^2 + 8x - 20 = 0$ so $(x - 2)(x + 10) = 0$

$x = 2$ or -10; $y = 10$ or -2; points of intersection are $(2, 10)$ and $(-10, -2)$

18 $x = 3, y = 4$ or $x = 4, y = 3$

Coordinates are $(3, 4)$ and $(4, 3)$.

19 a $y = -\dfrac{1}{2}x + \dfrac{3}{2}$

b $(x - 2)^2 + (y - 2)^2 = 9$

c $(-1, 2)$ and $\left(\dfrac{19}{5}, -\dfrac{2}{5}\right)$

20 a $y = 2x - 1$

b $(x - 1)^2 + y^2 = \dfrac{289}{4}$

c $\left(\dfrac{3}{5} + \dfrac{\sqrt{1441}}{10}, \dfrac{1}{5} + \dfrac{\sqrt{1441}}{5}\right)$ and $\left(\dfrac{3}{5} - \dfrac{\sqrt{1441}}{10}, \dfrac{1}{5} - \dfrac{\sqrt{1441}}{5}\right)$

Exercise 1.6B Reasoning and problem-solving

1 $x - y = 257$; $x + y = 1619$

The candidates polled 938 and 681 respectively.

2 Maggots cost 11 p each and worms cost 12 p each

3 $m = -3, c = -1$

4 $m = 1\dfrac{1}{2}, n = 3$

5 $q = 25\,\text{cm}, p = 40\,\text{cm}$

6 Florence is 8 years old and Zebedee is 12

7 a Both equations reduce to $y - 2x = 3$ so they are, in fact, the same line. Therefore there are an infinite number of solutions.

b The equations reduce to $y = 2x + 3$ and $y = 2x + 4$ so they are, in fact, parallel lines. Hence they do not meet and so there are no solutions.

8 $x - y + 3 = 0$: $(2, 5)$ or $(8, 11)$; This line intersects the curve in two points.
$y + 11x - 27 = 0$: $(2, 5)$ or $(2, 5)$. This is the tangent.
$y + 2x + 4 = 0$: the line misses the curve.

9 $2x - 9 = x^2 - x - 6 \Rightarrow x^2 - 3x + 3 = 0 \Rightarrow (-3)^2 - 4 \times 1 \times 3$
$= 9 - 12$ which is negative.
Therefore there are no real roots to the quadratic and the line does not intersect the curve.

10 $n = 7$ or -4. However $n = 7$ is the only valid solution since n stands for a number of terms and thus cannot be negative so $n = -4$ is not valid.

11 The field is 225 m long and 75 m wide.

12 $3x + 4y = 25 \Rightarrow y = \dfrac{25 - 3x}{4} \Rightarrow x^2 + \left(\dfrac{25 - 3x}{4}\right)^2 = 25$
$\Rightarrow 25x^2 - 150x + 225 = 0 \Rightarrow x^2 - 6x + 9 = 0$
$\Rightarrow (x - 3)(x - 3) = 0 \Rightarrow x = 3, y = 4$. The line is a tangent because there are two coincident values of x from the solution of the quadratic. The point of intersection is $(3, 4)$

13 $4(2x + 1)^2 + 9x^2 = 36 \Rightarrow 25x^2 + 16x - 32 = 0$

$x = \dfrac{-16 \pm \sqrt{16^2 + 4 \times 25 \times 32}}{50} = \dfrac{-16 \pm \sqrt{3456}}{50} = \dfrac{-16 \pm 24\sqrt{6}}{50}$

$= \dfrac{-8 \pm 12\sqrt{6}}{25}$

Hence $y = 2x + 1 = 2\left(\dfrac{-8 \pm 12\sqrt{6}}{25}\right) + 1 = \dfrac{9 \pm 24\sqrt{6}}{25}$

14 $(10, -80)$

15 a 4   **b** $y = -2x + 20$

16 a A is 2 units from O and B is 65 units from O
b Gradient is negative, hence x (distance) is decreasing.
c For the first 4 seconds, x increases $(2 \rightarrow 9 \rightarrow 14 \rightarrow 17 \rightarrow 18)$. Hence it is moving away; and then decreases $(17 \rightarrow 14 \rightarrow 9 \rightarrow 2 \rightarrow -7)$. Hence it is moving back.
d 18
e 7 seconds
f Back towards O

Exercise 1.7A Fluency and skills

1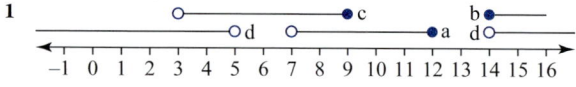

2 a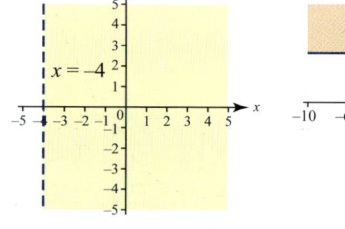
b
c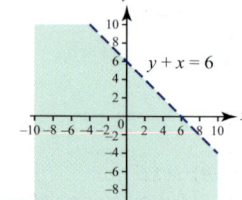
d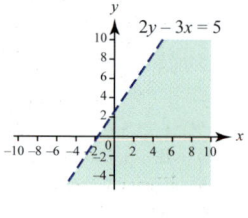

e 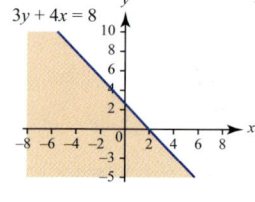   **f**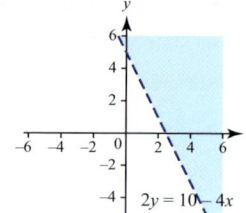

g 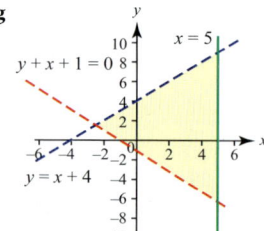   **h**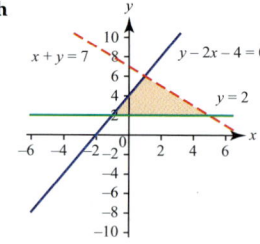

3 a $x > \dfrac{3}{2}$ **b** $x \le -1.9$ **c** $x < 1$

d $x \ge \dfrac{-11}{7}$ **e** $x < \dfrac{-4}{3}$ **f** $x < 0$

4 a $x < -3$ or $x > 2$ **b** $-7 < x < -4$
c $3 \le x \le 8$ **d** $x \le -4$ or $x \ge 6$

5 a $x > 3$ or $x < -1$ **b** $-6 < x < \dfrac{1}{3}$
c $x \le -1$ or $x \ge 3$ **d** $x \ge 0$ or $x \le -3$

6 a $x < -3.83$ or $x > 1.83$ **b** $-5.56 < x < -1.44$
c $1.76 \le x \le 10.24$ **d** $x < -3.32$ or $x > 6.32$
e $x < -0.91$ or $x > 2.57$ **f** $-4.47 < x < 0.22$
g $1.00 \le x \le 2.40$ **h** $x \le -0.38$ or $x \ge 3.05$

7 a i $y = 2x + 3$; $y = x^2$
ii $(-1, 1)$ and $(3, 9)$
iii

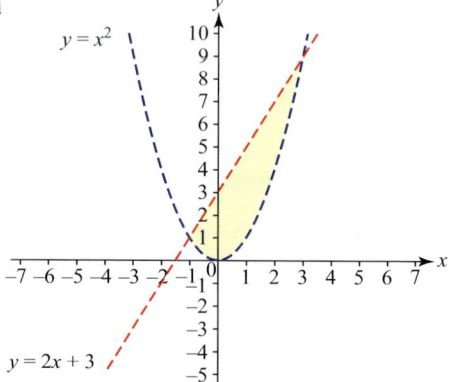

b i $x + y = 4$; $y = x^2 - 5x + 4$
ii $(0, 4)$ and $(4, 0)$
iii

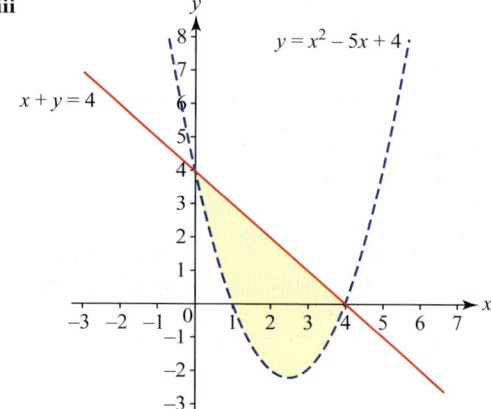

c **i** $x = -2, 4$ and $y = 9, 33$

ii $(-2, 9)$ and $(4, 33)$

iii

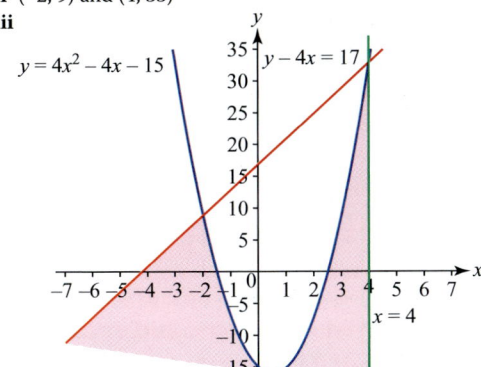

d **i** $x = -5, -4, -\dfrac{7}{3}, 6, 11$ and $y = -18, 12, 26, \dfrac{46}{3}, 42$

ii $(-4, 12)$ and $(11, 42)$, $(-5, 26)$ and $(6, -18)$, $\left(-\dfrac{7}{3}, \dfrac{46}{3}\right)$

iii

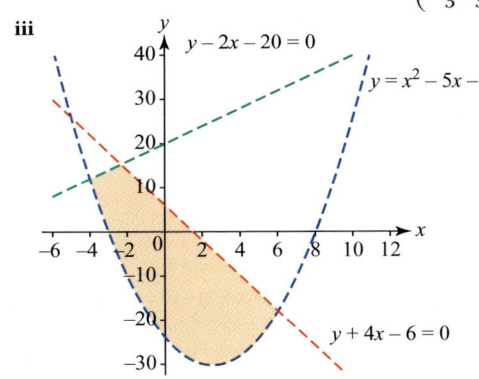

Exercise 1.7B Reasoning and problem-solving

1 **a**

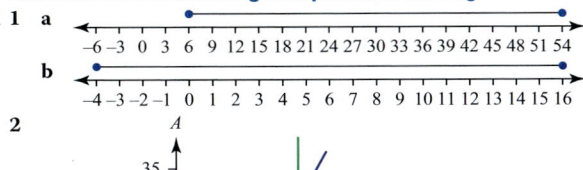

b

2

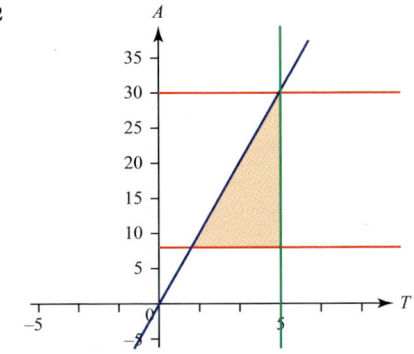

3 **a** $2w + 3p \geq 200$

b **i** $p \geq 17.3\ldots$ i.e. p must be at least 18

ii $w \geq 86.5$ i.e. w must be at least 87

c Yes. She could score 100 in her written paper which would give her a total of 200

4 $m > 3$

5 $99 \leq n \leq 105$

6 $r \leq 8.125$ and $g \geq 4$. Hence we have 8 red, 11 green; 7 red, 10 green; 6 red, 9 green; 5 red, 8 green; 4 red, 7 green; 3 red, 6 green; 2 red, 5 green; 1 red, 4 green.

7 The sister must be more than 10 years old.

8 $\dfrac{10}{3} < b < 5$

9 $6 \leq n \leq 40$

10 $p = 3, q = 7$ or $p = 4, q = 6$ or $p = 5, q = 5$ or $p = 6, q = 4$ or $p = 7, q = 3$

11 **a** $x = 2.16$ or 57.84

b

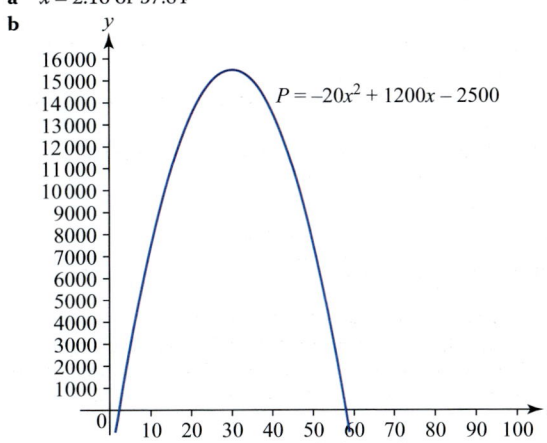

c **i** The firm makes a loss if it sells fewer than 3000 decanters or more than 57 000

ii $13.42 \leq x \leq 46.58$. Thus the firm makes at least £10 000 if they sell between 14 000 and 46 000 decanters.

Review exercise 1

1 Let the numbers be $2m + 1$ and $2n + 1$
Product is $(2m + 1)(2n + 1) = 4mn + 2(m + n) + 1$
$4mn$ and $2(m + n)$ are both even so $4mn + 2(m + n) + 1$
is even + even + odd, which is odd.

2 Find one prime e.g. 41

3 No. $\dfrac{1}{-3} > -3$ OR $\dfrac{1}{\left(\dfrac{1}{4}\right)} > \dfrac{1}{4}$ i.e. $4 > \dfrac{1}{4}$

4 **a** $-s^{12}$ **b** $8c^{32}$ **c** $\dfrac{1}{81}$ **d** $\dfrac{1}{\sqrt{k^3}}$

5 **a** $5\sqrt{11}$ **b** $\dfrac{4\sqrt{a} - a - 3}{a - 1}$

6 $6\sqrt{2}$ cm

7 **a** $c = -5$ or $\dfrac{1}{2}$ **b** $x = -3.43$ or 1.63

8

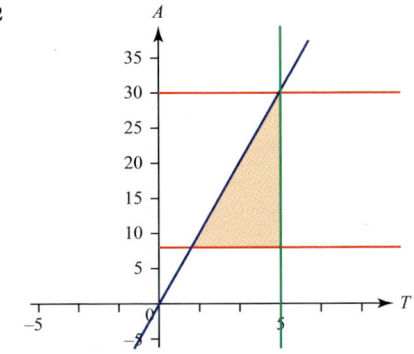

9 **a** **i** $y + 6x = 29$ **ii** $3y = 2x + 18$

b 1st Diagonal: Equation is $14y + 2x = 10$
2nd Diagonal: Equation is $2y - 14x + 20 = 0$
Product of gradients $= -1$ so perpendicular

10 **a** **i** $x^2 - 6x + y^2 - 12y - 19 = 0$

ii $x^2 + 6x + y^2 - 18y + 74 = 0$

iii $x^2 + 4x + y^2 + 14y - 68 = 0$

b $12y = 5x + 273$ **c** $(5, -4)$ and $(11, 14)$

11 **a** $y = -1$ and $x = 3$ **b** $y = -3$ twice and $x = -2$ twice

12 $(-6, -14)$ and $\left(\dfrac{14}{3}, 18\right)$

13 a i $\{x : x \le 1\}$ **ii** $\left\{x : x \ge \dfrac{6}{13}\right\}$

b i $1.26 \le x \le 12.74$ **ii** $x \le -0.67$ or $x \ge 3.27$

14 a

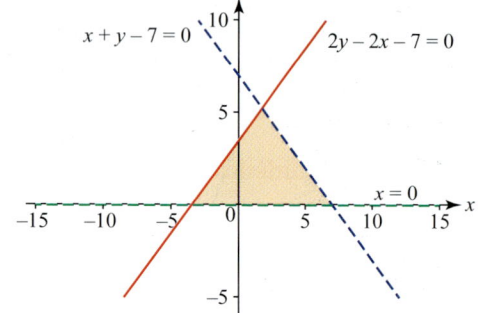

b

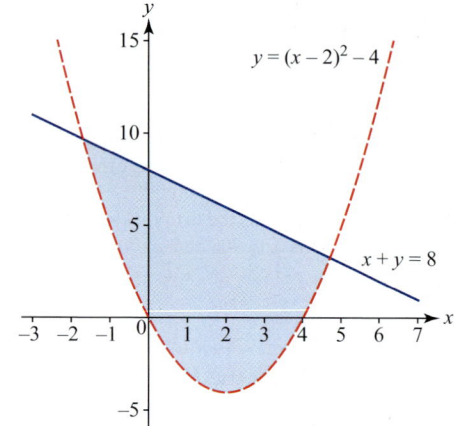

Assessment 1

1 a i 2^{m+n} **ii** 5^{m-2n+1} **iii** $3^{\frac{5m}{2}}$

  b 2^{2p+q}

2 a i $4\sqrt{3}$ **ii** $-1-\sqrt{2}$ **iii** $\dfrac{-12+5\sqrt{7}}{31}$

  b $6+2\sqrt{5}$

3 $y = \dfrac{2}{3}x + \dfrac{7}{3}$ or $2x - 3y + 7 = 0$

4 a $(x+3)^2 + 4$

  b

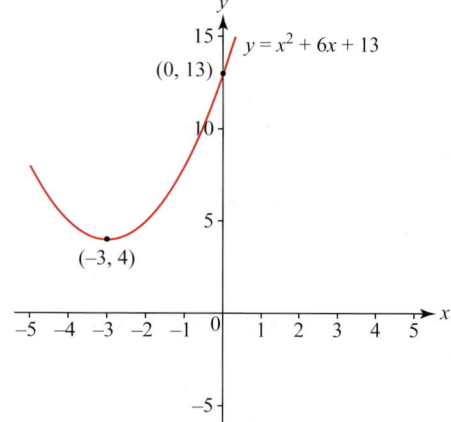

5 $x = 7$, $x = -1$; $y = -11$, $y = 5$

6 $x(x+4) = x+4+2x-5 \Rightarrow x^2 + x + 1 = 0$
$\Rightarrow \Delta = 1^2 - 4 \times 1 \times 1 = -3 < 0 \Rightarrow$ No real solutions

7 a i $x < 4$ **ii** $1 \le x \le 5$

b

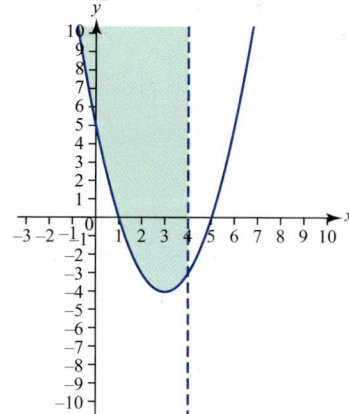

8 $x = \sqrt{5}$

9 a i $(5, -1)$ **ii** 7

  b $\left(1 - \dfrac{\sqrt{34}}{2}, 3 - \dfrac{\sqrt{34}}{2}\right), \left(1 + \dfrac{\sqrt{34}}{2}, 3 + \dfrac{\sqrt{34}}{2}\right)$

10 $k < -\dfrac{3}{4}$ or $k > 3$

11 $\left(\sqrt{a} - \sqrt{b}\right)^2 \ge 0$

$a - 2\sqrt{a}\sqrt{b} + b \ge 0$

$a + b \ge 2\sqrt{ab}$

$\dfrac{a+b}{2} \ge \sqrt{ab}$

12 a $(2u-1)(u-8)$ **b** $x = -1$ or $x = 3$

13 a $x^2 + (mx+2)^2 + 4x - 6(mx+2) + 10 = 0$
$\Rightarrow x^2 + m^2x^2 + 4mx + 4 + 4x - 6mx - 12 + 10 = 0$
$\Rightarrow (m^2+1)x^2 + 2(2-m)x + 2 = 0$

  b $m = -2 \pm \sqrt{6}$

14 a 4 **b** $x = 4\dfrac{2}{3}$, $y = -\dfrac{2}{3}$

15 a False: eg $a = 4$, $b = -5$

  b True: $n^2 + n \equiv n(n+1)$; either n or $(n+1)$ must be even

  c True: $(b-2a)^2 \ge 0 \Rightarrow b^2 - 4ab + 4a^2 \ge 0 \Rightarrow b^2 \ge 4ab - 4a^2$

  d False: eg $n = 4$

16 a $(0, 9)$

  b $(2, 1)$

  c $C(3-\sqrt{2}, 7-4\sqrt{2})$, $D(3+\sqrt{2}, 7+4\sqrt{2})$

17 $(x+3)^2 - 9 + (y-2)^2 - 4 - 2 = 0$; Centre, $C_1 = (-3, 2)$;
Radius, $r_1 = \sqrt{15}$

$(x-1)^2 - 1 + (y-5)^2 - 25 - 55 = 0$; Centre, $C_2(1,5)$;
Radius, $r_2 = 9$

Distance $|C_1C_2| = \sqrt{(1--3)^2 + (5-2)^2} = 5$; $|C_1C_2| + r_1 < r_2$

Chapter 2

Exercise 2.1A Fluency and skills

1 a Degree 2 **b** Degree 4 **c** Degree 3

2 a $6x^2 + 16x$ **b** $6x^3 + 16x^2 - 18x$

  c $12y^2 - 13y - 14$ **d** $12y^3 + 24y^2 - 21y$

  e $t^2 - 10t + 25$ **f** $t^3 - 7t^2 - 5t + 75$

3 a $2x^2 + 32$ or $2(x^2+16)$ **b** $20pq$

4 a $2m^2(2m+3)$ **b** $4n(4n^3-3)$

  c $p(5p^3 - 2p + 6)$ **d** $3y(3y-5x)$

  e $3x(2x - y + 3)$ **f** $7z(y - 3z^2)$

  g $4e(e - 5f)$ **h** $(p-10)(p+10)$

  i $3q(9 - 4q)$ **j** $\dfrac{y(24 - 5y)}{75}$

  k $2(d+1)(d-1)$ **l** $w(2w+3)(5w-2)$

5 a $m(2m+5)(2m-3)$ **b** $n(7n-1)(n-2)$

6 $2(x+2)(2x-1)(2x+3)$

7 a $80pq$ **b** $4x(y+z)$

c $6x^2+32$ **d** $8x^2+(8\sqrt{5}-8\sqrt{3})x+32$

8 a $(x-1)$ **b** $(2x-7)$

c $(2y-1)$ **d** (z^2+3z+4)

e $(3a^2-4a+9)$

f $(2x^3-5x^2+2x+21)\equiv(x^2-4x+7)(2x+3)$

g $(k+7)$ and $(k-1)$

Exercise 2.1B Reasoning and problem-solving

1 $(2a^2-2a+5)-(a^2-2a+4)=a^2+1$, $(a^2+1)-(3a-8)$
$=a^2-3a+9$, $(a^2-2a+4)-(a^2-3a+9)=a-5$

2 $16b^2-56ab+49a^2\,\text{cm}^2$

3 $(6c^3-5c^2-27c+14)\,\text{cm}^3$

4 $(a+3)(3a+7)\text{cm}^2$

5 $2a^2x-6ax^2+4x^3\,\text{cm}^3$

6 a $t=5$ **b** $31.25\,\text{ft}$

7 a $t=0,\ \dfrac{4}{3}$, or 8 seconds

$t=0$: $v=32\,\text{m s}^{-1}$, $a=-56\,\text{m s}^{-2}$

$t=\dfrac{4}{3}$: $v=-\dfrac{80}{3}\,\text{m s}^{-1}$, $a=-32\,\text{m s}^{-2}$

$t=8$: $v=160\,\text{m s}^{-1}$, $a=88\,\text{m s}^{-2}$

b $x=-\dfrac{19\,712}{243}$, $v=-\dfrac{496}{9}\,\text{m s}^{-1}$

c $t=\dfrac{28}{9}+\dfrac{4\sqrt{31}}{9}\,\text{s or }\dfrac{28}{9}-\dfrac{4\sqrt{31}}{9}\,\text{s}$

$a=8\sqrt{31}\,\text{m s}^{-2}\text{ or }a=-8\sqrt{31}\,\text{m s}^{-2}$

8 a $p=-x^2+8x-9$, $q=x^2+x+5$

b Perimeter $\equiv4(x^2+4x-5)$
Area $\equiv x^4+8x^3+5x^2-34x+16$

9 a $66.89^2-33.11^2=(66.89+33.11)(66.89-33.11)=$
$(100)(33.78)=3378$

b $\left(2\sqrt{2}\right)^3-\left(\sqrt{2}\right)^3=8\times2\sqrt{2}-2\sqrt{2}=14\sqrt{2}$

10 $(2h+1)\,\text{cm}$

11 $(s-2)(s-5)\,\text{cm}$

12 $b\equiv[2t^2-3t+7]$

13 $V=\pi(20p^3+9p^2-56p-45)$

Exercise 2.2A Fluency and skills

1 a 120 **b** 5040 **c** 39 916 800

2 a 10 **b** 84 **c** 330 **d** 1287

3 a 10 **b** 10 **c** 1287 **d** 38 760

4 a $1+9x+27x^2+27x^3$

b $1-\dfrac{5}{2}z+\dfrac{5}{2}z^2-\dfrac{5}{4}z^3+\dfrac{5}{16}z^4-\dfrac{1}{32}z^5$

c $1-\dfrac{4m}{3}+\dfrac{2m^2}{3}-\dfrac{4m^3}{27}+\dfrac{m^4}{81}$

d $1+\dfrac{15x}{2}+\dfrac{45x^2}{2}+\dfrac{135x^3}{4}+\dfrac{405x^4}{16}+\dfrac{243x^5}{32}$

5 a $1+8x+28x^2+56x^3+\ldots$

b $1-21x+189x^2-945x^3+\ldots$

c $1+18x+144x^2+672x^3+\ldots$

d $64-576x+2160x^2-4320x^3+\ldots$

e $256-1024x+1792x^2-1792x^3+\ldots$

f $1-20x+180x^2-960x^3+\ldots$

6 a $8-48y+96y^2-64y^3$

b $81b^4+540b^3+1350b^2+1500b+625$

c $1024z^5-\dfrac{1280z^4y}{3}+\dfrac{640z^3y^2}{9}-\dfrac{160z^2y^3}{27}+\dfrac{20zy^4}{81}-\dfrac{y^5}{243}$

7 a $x^6+12x^5+60x^4+\ldots$

b $256x^8-1024x^7+1792x^6+\ldots$

c $-x^9+27x^8-324x^7+\ldots$

d $x^7+28x^6+336x^5+\ldots$

e $1024x^{10}+15\,360x^9+103\,680x^8+\ldots$

f $\dfrac{x^{11}}{2048}+\dfrac{11x^{10}}{256}+\dfrac{55x^9}{32}+\ldots$

8 a $16+96t+216t^2+216t^3+81t^4$

b $81-216p+216p^2-96p^3+16p^4$

c $1024p^5+3840p^4q+5760p^3q^2+4320p^2q^3+1620pq^4+243q^5$

d $243p^5-1620p^4q+4320p^3q^2-5760p^2q^3+3840pq^4-1024q^5$

e $81z^4-216z^3+216z^2-96z+16$

f $64z^6-96z^5+60z^4-20z^3+\dfrac{15}{4}z^2-\dfrac{3}{8}z+\dfrac{1}{64}$

g $8+8x+\dfrac{8}{3}x^2+\dfrac{8}{27}x^3$

h $\dfrac{r^8}{6561}+\dfrac{2r^7s}{2187}+\dfrac{7r^6s^2}{2916}+\dfrac{7r^5s^3}{1944}+\dfrac{35r^4s^4}{10368}+\dfrac{7r^3s^5}{3456}$
$+\dfrac{7r^2s^6}{9216}+\dfrac{rs^7}{6144}+\dfrac{s^8}{65536}$

i $\dfrac{x^3}{8}+\dfrac{x^2y}{4}+\dfrac{xy^2}{6}+\dfrac{y^3}{27}$

9 a $^5C_3\,p^2 5^3=1250p^2$

b $^9C_5(4)^4y^5=32\,256y^5$

c $^{12}C_7(3)^5q^7=192\,456q^7$

d $^5C_3 4^2(-3m)^3=-4320m^3$

e $^{15}C_{11}(2z)^4(-1)^{11}=-21\,840z^4$

f $^8C_2(z)^6\left(\dfrac{3}{2}\right)^2=63z^6$

g $^5C_1(3x)^4(4y)=1620x^4y$

h $^{10}C_5(2a)^5(-3b)^5=-1\,959\,552a^5b^5$ and
$^{10}C_4\,(2a)^6(-3b)^4=1\,088\,640a^6b^4$

i $^3C_1(4p)^2(\tfrac{1}{4})=12p^2$

j $^{11}C_6(4a)^5\left(-\dfrac{3b}{4}\right)^6=\dfrac{168399}{2}a^5b^6$ and

$^{11}C_5(4a)^6\left(-\dfrac{3b}{4}\right)^5=-449\,064a^6b^5$

k $^{11}C_4\left(\dfrac{a}{2}\right)^7\left(-\dfrac{2b}{3}\right)^4=\dfrac{55}{108}a^7b^4$ and

$^{11}C_5\left(\dfrac{a}{2}\right)^6\left(-\dfrac{2b}{3}\right)^5=\dfrac{-77}{81}a^6b^5$

10 a $c^8+4c^6d^2+6c^4d^4+4c^2d^6+d^8$

b $v^{10}-5v^8w^2+10v^6w^4-10v^4w^6+5v^2w^8-w^{10}$

c $8s^6+60s^4t^2+150s^2t^4+125t^6$

d $8s^6-60s^4t^2+150s^2t^4-125t^6$

e $d^3+3d+\dfrac{3}{d}+\dfrac{1}{d^3}$

f $16w^4+96w^2+216+\dfrac{216}{w^2}+\dfrac{81}{w^4}$

11 a $x^3+6x+\dfrac{12}{x}+\dfrac{8}{x^3}$

b $x^8-8x^6+24x^4-32x^2+16$

c $x^{10}-5x^7+10x^4-10x+\dfrac{5}{x^2}-\dfrac{1}{x^5}$

d $\dfrac{1}{x^{12}}+\dfrac{18}{x^9}+\dfrac{135}{x^6}+\dfrac{540}{x^3}+1215+1458x^3+729x^6$

12 a $96x^6-1200x^5+6000x^4-15\,000x^3+18\,750x^2-9375x$

b $16+48x+56x^2+32x^3+9x^4+x^5$

13 a $206+66x+276x^2+88x^3+16x^4$

b $27x+42x^2+334x^3+405x^4+243x^5$

14 a $125+40\sqrt{3}$

b $-31-174\sqrt{5}$

15 a $1+2nx+2n(n-1)x^2+\dfrac{4n(n-1)(n-2)}{3}x^3+\ldots$

b $1-3nx+\dfrac{9n(n-2)}{2}x^2-\dfrac{9n(n-1)(n-2)}{2}x^3+\ldots$

16 a $1+24x+240x^2+\ldots$ **b** 1.264

17 a $1 - 14x + 84x^2 - 280x^3 + \dots$ **b** $0.932\,07$

18 a $7 + 5\sqrt{2}$ **b** $41 - 29\sqrt{2}$ **c** $577 + 408\sqrt{2}$

    d $792 - 560\sqrt{2}$ **e** $\dfrac{5}{2} - \dfrac{7}{4}\left(\sqrt{2}\right)$ **f** $\dfrac{7537}{81} + \dfrac{332}{9}\left(\sqrt{2}\right)$

19 a $28 + 16\sqrt{3}$ **b** $576 - 256\sqrt{5}$

    c $13\,100 - 4924\sqrt{7}$ **d** $198\sqrt{6} + 485$

    e $112 + 64\sqrt{3}$ **f** $485 - 198\sqrt{6}$

Exercise 2.2B Reasoning and problem-solving

1 120 ways   **2** $184\,756$

3 $8s^3 - 36s^2w + 54sw^2 - 27w^3$

4 a $1.340\,096$ (to 6 dp)   **b** 7.5295 (to 4 dp)

5 a 1.0773 (to 4 dp)   **b** 973.94 (to 5 sf)

6 $362.590\,82$ (to 5 dp)

7 $56\,700\sqrt{15}$

8 a 480   **b** 270

9 a $-\dfrac{5}{2}$ **b** $\dfrac{15}{4}$

10 a 210   **b** 120

11 a 5   **b** 7

12 a 8   **b** -3

13 9

14 14

15 a n **b** $\dfrac{1}{6}n^3 - \dfrac{1}{2}n^2 + \dfrac{1}{3}n$ **c** $-n$

16 a $\dfrac{1}{n+1}$ **b** $n + 5 + \dfrac{6}{n}$

17 -2416

18 -168

19 a 15   **b** 0.0198 (to 3 sf)

    c 0.297 (to 3 sf)   **d** 0.132 (to 3 sf)

Exercise 2.3A Fluency and skills

1 a $x - 10$ **b** $3x + 2$ **c** $4x - 3$

2 a $x^2 + 4x - 7$ **b** $x^2 - x - 1$ **c** $x^2 + 7x + 9$

    d $x^2 - 5x + 7$ **e** $x^3 - 4x^2 - 7x + 1$

3 $(3x^3 + 16x^2 + 68x + 264)$ rem 1061

4 $(2x^4 - 5x^3 + 5x^2 - 5x + 3) \div (2x - 3) = (x^3 - x^2 + x - 1)$ rem 0

5 a $x^2 + x - 1$ **b** $x^2 + x + 1$

    c $2x^2 - 7x + 3$ **d** $3x^3 - 5x^2 + 4x - 7$

    e $2x^3 + 7x^2 - 10x - 1$

6 a $f(0) = 0$; $f(1) = 9$; $f(-1) = -13$; $f(2) = 20$; $f(-2) = -36$

    b $f(0) = -2$; $f(1) = -5$; $f(-1) = -3$; $f(2) = -6$; $f(-2) = -14$

    c $f(0) = 2$; $f(1) = 1$; $f(-1) = -3$; $f(2) = 0$; $f(-2) = -20$

    d $f(0) = 2$; $f(1) = 0$; $f(-1) = 6$; $f(2) = 12$; $f(-2) = 0$

    e $f(0) = 4$; $f(1) = 0$; $f(-1) = 6$; $f(2) = 0$; $f(-2) = 0$

7 a $f(-6) = (-6)^3 + 4(-6)^2 - 9(-6) + 18 =$

       $-216 + 144 + 54 + 18 = 0$

    b $f(8) = 2(8)^3 - 13(8)^2 - 20(8) - 32$

       $= 1024 - 832 - 160 - 32 = 0$

    c $f\left(\dfrac{1}{3}\right) = 3\left(\dfrac{1}{3}\right)^3 + 11\left(\dfrac{1}{3}\right)^2 - 25\left(\dfrac{1}{3}\right) + 7 = \dfrac{1}{9} + \dfrac{11}{9} - \dfrac{25}{3} + 7 = 0$

    d $f\left(\dfrac{-2}{5}\right) = 10\left(\dfrac{-2}{5}\right)^3 + 19\left(\dfrac{-2}{5}\right)^2 - 39\left(\dfrac{-2}{5}\right) - 18$

       $= \dfrac{-16}{25} + \dfrac{76}{25} + \dfrac{78}{5} - 18 = 0$

8 $x(4x - 1)(x + 7)$

9 $(x - 1)(x + 1)(2x + 9)$

10 a $(x + 6)(x - 1)(x - 2)$ **b** $(x + 7)(x - 9)(x - 4)$

    c $(x + 3)(2x - 1)(3x + 2)$ **d** $(x + 4)(x - 4)(x^2 + 3)$

Exercise 2.3B Reasoning and problem-solving

1 a $p + q + 2 = 0$ **b** $-9p - q + 38 = 0$ **c** $p = 5$, $q = -7$

2 a $\dfrac{-9}{4}$ **b** $b = 2$ **c** $p = 5$ and $q = -7$

---

3 $(x - 2)$

4 $(x + 1)(x - 2)(x + 3)$

5 $(x - 1)(x + 3)$

6 LCM $(x - 3)(x + 4)(2x + 7)$, HCF $(2x + 7)$

7 LCM $(x + 5)(x - 7)(x + 7)(x + 9)$, HCF $(x + 5)(x + 9)$

8 $(x - 1)$. $f(1) - g(1) = 7x^2 + 24x - 31 \equiv 7(1)^2 + 24(1) - 31 = 0$

    $\Rightarrow (x - 1)$ is a factor of $f(x) - g(x)$

9 $a = 15$ and $b = -36$

10 a $(2x - 3)$ m   **b** $(x - 3)\sqrt{3}$ cm

11 a $t = 2, 3$ and 4.5 secs

    b i When $t = 2$, $a = 5\,\text{ms}^{-1}$ **ii** $t = \dfrac{19 \pm \sqrt{19}}{6}$ s

        When $t = 3$, $a = -3\,\text{ms}^{-1}$

        When $t = 4.5$, $a = 7.5\,\text{ms}^{-1}$

12 a The radius could be $(x - 3)$ m if the height is $(3x + 7)$ m. (Accept alternatives.)

    b $x > 3$

13 a $t = 1, 5$ and 6 seconds

    b When $5 < t < 6$, $f(t)$ is negative; when $1 < t$, $f(t)$ is negative; hence the ride is above ground level when $1 < t < 5$

14 a The dimensions could be $(x + 1)$ cm, $(x + 2)$ cm and $(3x + 12)$ cm (Accept alternatives.)

    b $x > -1$

15 $\left(32\pi x^2 + 64\pi x + \dfrac{608\pi}{3}\right)$ cm^3

Exercise 2.4A Fluency and skills

1 a $(0, -6)$ $(-2, 0)$ and $(3, 0)$ **b** $(0, -35)$, $(\dfrac{-5}{2}, 0)$ and $(7, 0)$

    c $(-2, 0)$ and $(0, 8)$ **d** $(3, 0)$ and $(0, -54)$

    e $(3, 0)$ and $(0, -27)$ **f** $(0, 118)$ and $\left(\dfrac{1}{2}\left(\sqrt[3]{7} - 5\right),\, 0\right)$

2 $x = 1$ is a vertical asymptote; $y = 0$ is a horizontal asymptote

3 a $x = 4$ **b** $x = -2$

    c $y = 3$ **d** $x = \dfrac{1}{2}$

4 a   **b**

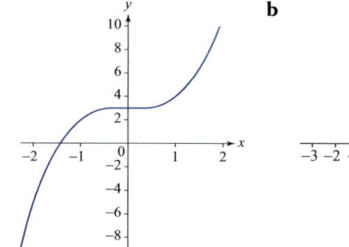 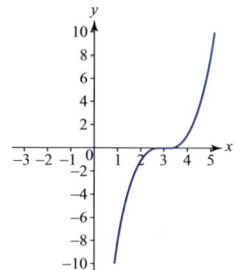

 c   **d**

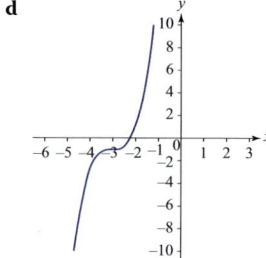

 e   **f**

g
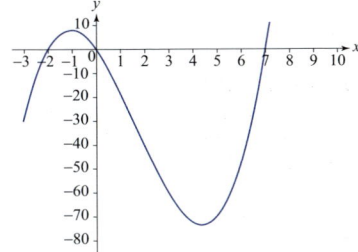

b
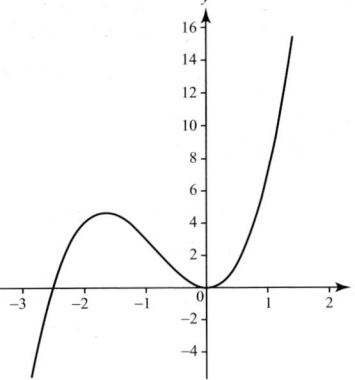

h
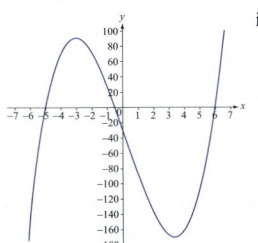

i
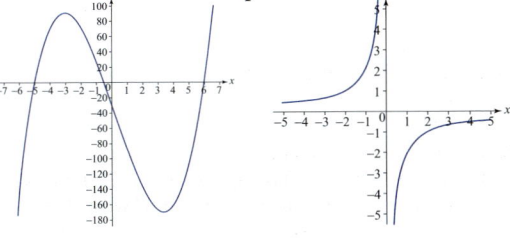

c
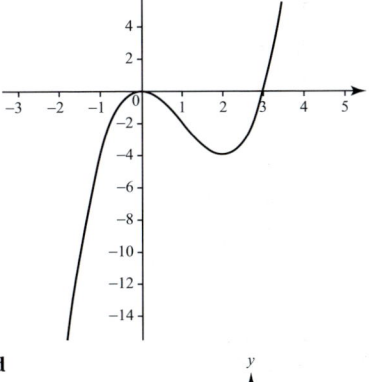

j
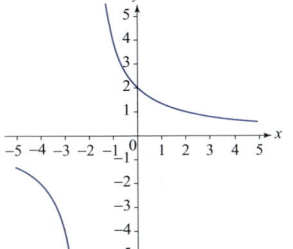

d
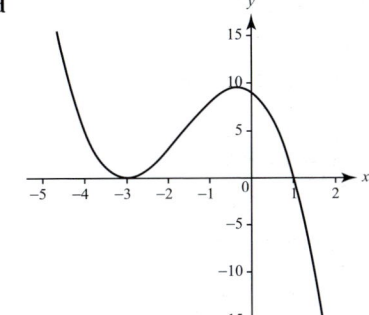

k
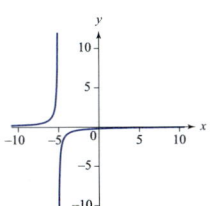

5 a
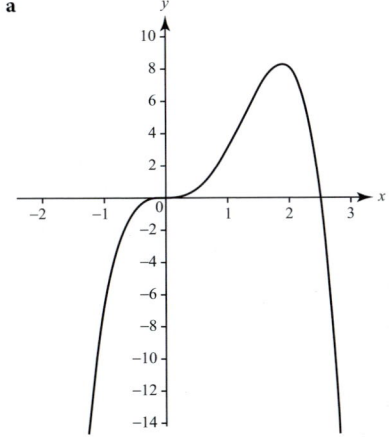

e
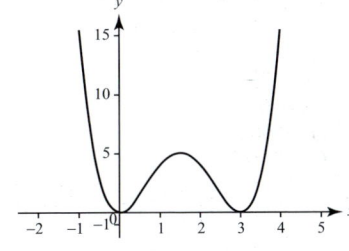

f
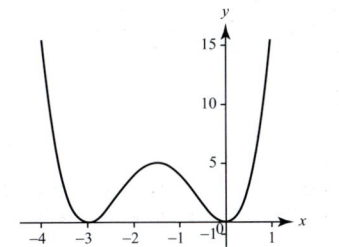

g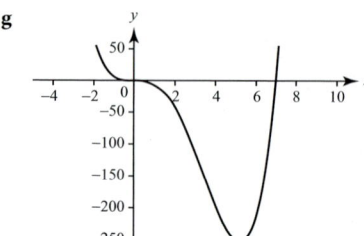

h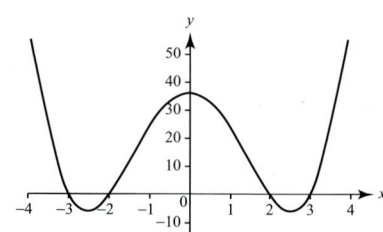

6 a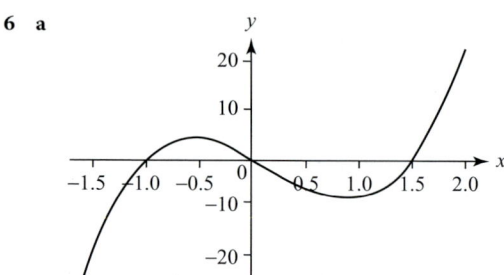

b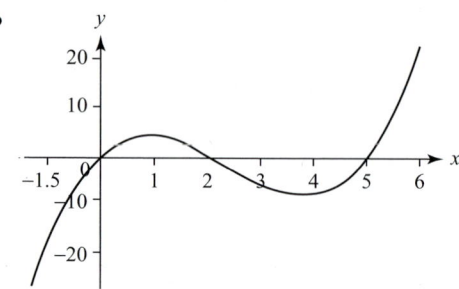

c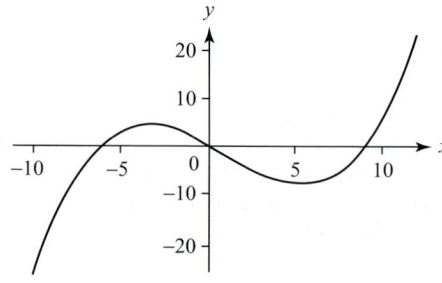

d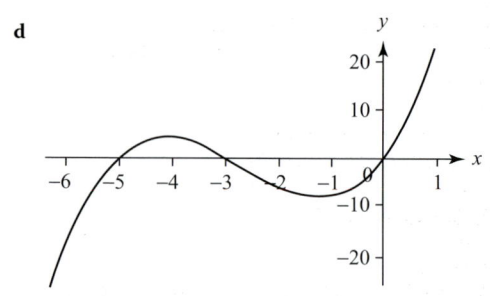

e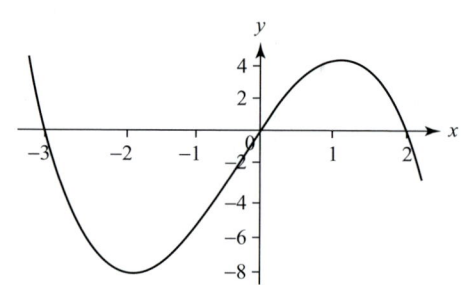

f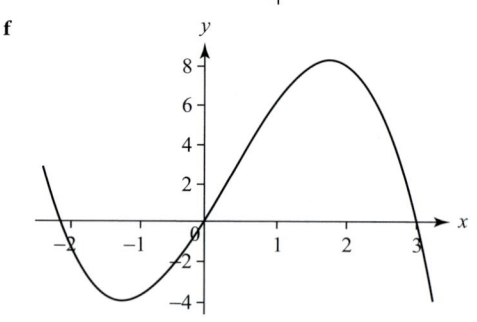

7 a Max (−0.5, 5), Min (2, −4) **b** Max (−2, 15), Min (8, −12)
c Max (−9, 5), Min (1, −4) **d** Max (−2, 9), Min (8, 0)
e Max (−2, 2.5), Min (8, −2) **f** Min (−2, −5), Max (8, 4)
g Max (2, 5), Min (−8, −4) **h** Max (−4, 5), Min (16, −4)

8 a Stretch scale factor ¼ in x-direction
b Stretch scale factor 3 in y-direction
c Translate 7 units left/ by vector $\begin{pmatrix} -7 \\ 0 \end{pmatrix}$
d Translate 4 units up/ by vector $\begin{pmatrix} 0 \\ 4 \end{pmatrix}$
e Stretch scale factor ½ in y-direction
f Reflect in x-axis (or line $y = 0$)
g Reflect in y-axis (or line $x = 0$)
h Stretch scale factor 2 in x-direction

9 a $y = (x+3)^3$ **b** $y = x^3 + 2$ **c** $y = 2x^3$ **d** $y = \left(\frac{x}{3}\right)^3$

10 a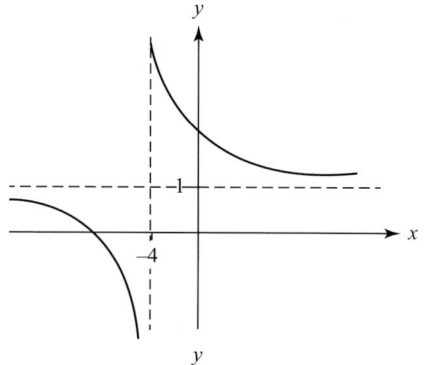

b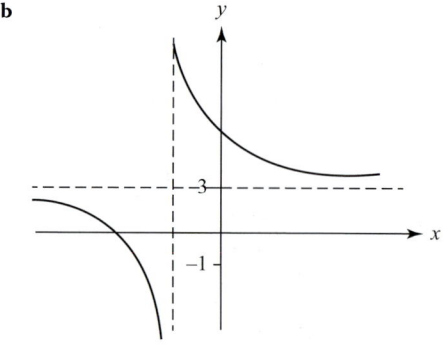

c

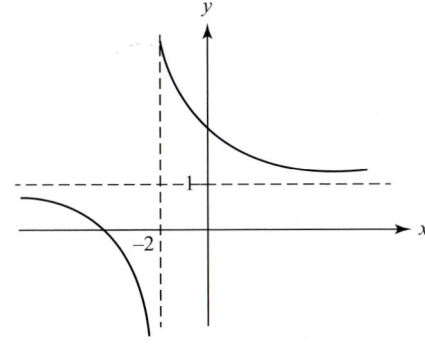

d

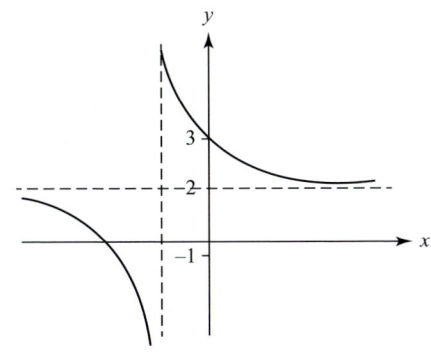

Exercise 2.4B Reasoning and problem-solving

1 a $r = \dfrac{56}{h}$

b

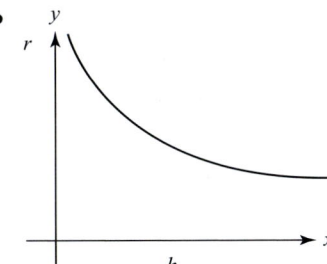

2 a $v = 60\sqrt{t}$

b

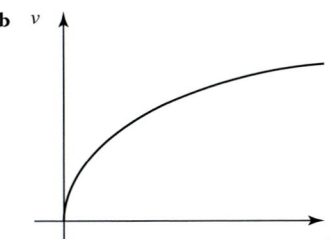

3 a

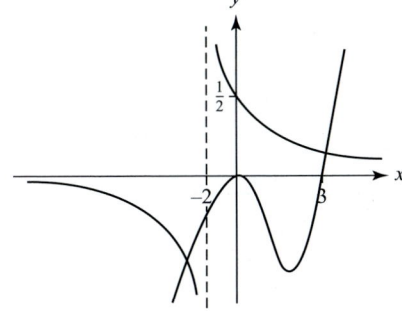

b Two solutions as they intersect twice.

4 a i

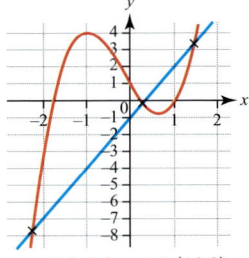

ii $x = -2.3$, 0.3 or 1.4 (± 0.3)

b i

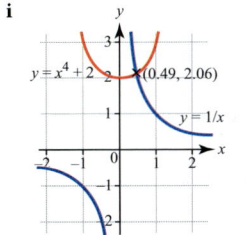

ii $x = 0.5$ (± 0.3)

c i

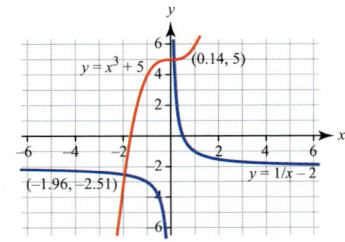

ii $x = -2.0$ (± 0.3) or 0.1 (accept values > 0 and ≤ 0.4)

5 a $y = f(x+1)$ **b** $y = -f(x)$

6 $A = -4$, $B = 4$ and $C = 0$

7 a $A = -5$, $B = -3$ and $C = 17$

  b i Reflection in the x-axis.   **ii** Translation by the vector $\begin{pmatrix} 0 \\ 20 \end{pmatrix}$

8 $A = 4$, $B = -3$ and $C = -3$

9 a $a < 0 \Rightarrow$ maximum, $a > 0 \Rightarrow$ minimum

  b $\left(-\dfrac{b}{2a},\ c - \dfrac{b^2}{4a} \right)$

  c $(0, c)$, $\left(\dfrac{-b + \sqrt{b^2 - 4ac}}{2a},\ 0 \right)$ and $\left(\dfrac{-b - \sqrt{b^2 - 4ac}}{2a},\ 0 \right)$

  d $x = \dfrac{-b}{2a}$

Review exercise 2

1 $2x^5 - 6x^4 - 3x^3 + 15x^2 + 9x - 9$

2 $n(2n - 3)(2n + 5)$

3 a $2y^3 + 9y^2 + 4y - 15$ **b** $2z^3 - 7z^2 + 4z + 4$

4 a $-4(m + 4)$ **b** $(d + 1)(5 - 3d)$

5 $a = -5$, $b = -31$, $c = 84$

6 $2x^2 + 3x - 13$

7 1.4641

8 $16s^8 - 128s^6t + 384s^4t^2 - 512s^2t^3 + 256t^4$

9 a $28 + 16\sqrt{3}$ **b** $-1760\sqrt{5}$

10 $\dfrac{-938223}{16}$

11 a $a = -1$, $b = 4608$, $c = 5376$

   b $512 + 1792x + 2304x^2 + 768x^3 + \ldots$

12 $2x^2 - 9x + 1$

13 $(x + 1)$, $(x - 2)$ and $(x - 3)$

14 $4x^2 - 3x + 7$

15 $x^2 + 3$ rem 10

16 $f\left(\dfrac{3}{2}\right) = 4\left(\dfrac{3}{2}\right)^3 - 8\left(\dfrac{3}{2}\right)^2 + \left(\dfrac{3}{2}\right) + 3 = \dfrac{27}{2} - 18 + \left(\dfrac{3}{2}\right) + 3 = 0$

17 $(x-3)(2x+1)(x+3)$

18 a She has moved the graph one unit left, it should be one unit right.

b

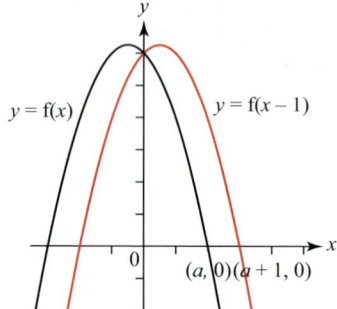

$y = f(x)$ $y = f(x-1)$

$(a, 0)(a+1, 0)$

19

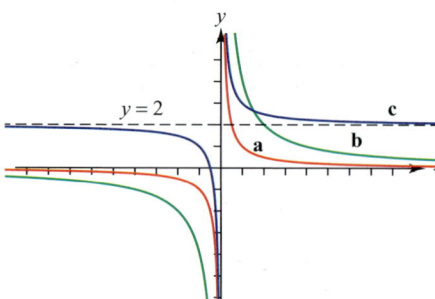

$y = 2$ **c**

a   **b**

20

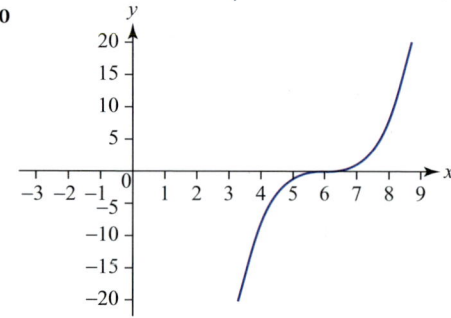

21

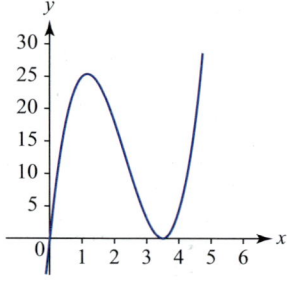

Starts from O at $t = 0$ and accelerates to a distance of about 25.5 m in about 1.2 secs; It then returns to O, which it reaches at 3.5 secs. After that it accelerates continuously away from O

22 a Length $= 16 - 2x$, width $= 10 - 2x$
$\Rightarrow$ Volume $= x(16 - 2x)(10 - 2x)$
$= 160x - 52x^2 + 4x^3$

b Volume

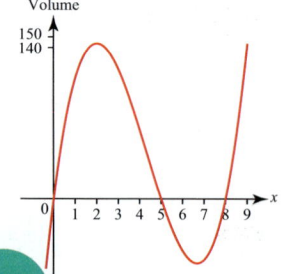

From the graph the maximum value of V is about 145 in³ when x is 2

23 The particle is decelerating away from O for the first $\dfrac{4}{3}$ seconds and then returns back to O, reaching it when $t = 4$ s After that it accelerates away from O indefinitely.

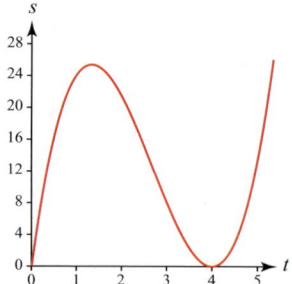

Assessment 2

1 a i $12x^2 - 16x - 3$ **ii** $4a^2 - 12ab + 9b^2$
iii $5x^3 - 13x^2y - 11xy^2 - 2y^3$
b $a = 9, b = 0, c = -16$

2 $1 + 4x + 7x^2 + 7x^3$

3 a $p(p-5)^2$
b $(2x+5)(2x+5-5)^2 = 4x^2(2x+5)$
so $a = 4$

4 $f(3) = 20 \neq 0$

5 a 18 000   **b** 469 000   **c** 15

6 a i $c = 3$ **ii** $A = 189$ **iii** $B = 945$
b 2079

7 $(x-2)(x^2 - x + 3) + 7$

8 a i $x^4 + 4x^3y + 6x^2y^2 + 4xy^3 + y^4$
ii $x^4 - 4x^3y + 6x^2y^2 - 4xy^3 + y^4$
b $2x^4 + 12x^2y^2 + 2y^4 = 2(\sqrt{5})^4 + 12(\sqrt{5})^2(\sqrt{2})^2 + 2(\sqrt{2})^4$
$= 2 \times 25 + 12 \times 5 \times 2 + 2 \times 4 = 178$

9 5376

10 a i $1 + 12x + 60x^2$ **ii** $64 - 192x + 240x^2$
b $64 + 576x + 1776x^2$

11 a $2(2)^3 + (2)^2 - 7(2) - 6 = 16 + 4 - 14 - 6 = 0$, hence $(x-2)$ is a factor
b $2x^3 + x^2 - 7x - 6 = (x-2)(2x^2 + 5x + 3)$
$= (x-2)(2x+3)(x+1)$
$x = 2, -1\dfrac{1}{2}, -1$

12 a $a = 2, b = -1$ **b** $(x+3)(x-1)(2x-5)$

c

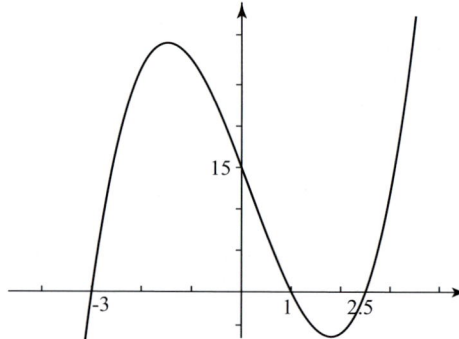

d $-3 \leq x \leq 1$ or $x \geq 2.5$

13 a $x^6 + 6x^4 + 15x^2 + 20 + \dfrac{15}{x^2} + \dfrac{6}{x^4} + \dfrac{1}{x^6}$

b $x^6 - 6x^4 + 15x^2 - 20 + \dfrac{15}{x^2} - \dfrac{6}{x^4} + \dfrac{1}{x^6}$

c $12x^4 + 40 + \dfrac{12}{x^4} = 64 \Rightarrow x^8 - 2x^4 + 1 = 0 \Rightarrow$
$(x^4 - 1)(x^4 - 1) = 0 \Rightarrow (x-1)(x+1)(x^2+1) = 0 \Rightarrow$
$x = \pm 1$

14 a $^{n}C_{r} + {}^{n}C_{r-1} = \dfrac{n!}{r! \times (n-r)!} + \dfrac{n!}{(r-1)! \times (n-(r-1))!}$

$$= \dfrac{n!}{r! \times (n-r)!} + \dfrac{n!}{(r-1)! \times (n+1-r)}$$

$$= \dfrac{n!}{r! \times (n-r)!} \times \dfrac{(n+1-r)}{(n+1-r)} + \dfrac{n!}{(r-1)! \times (n+1-r)!} \times \dfrac{r}{r}$$

$$= \dfrac{(n+1)!}{r! \times (n+1-r)!} = {}^{n+1}C_{r}$$

b $^{n+2}C_{3} - {}^{n}C_{3} = \dfrac{(n+2)!}{3! \times (n+2-3)!} - \dfrac{n!}{3! \times (n-3)!}$

$$= \dfrac{(n+2) \times (n+1) \times n!}{3! \times (n-1)!} - \dfrac{(n-2) \times (n-1) \times n!}{3! \times (n-1)!}$$

$$= \dfrac{n!}{3! \times (n-1)!} \times \left[(n^2 + 3n + 2) - (n^2 - 3n + 2) \right]$$

$$= \dfrac{n \times (n-1)!}{6 \times (n-1)!} \times [6n] = n^2$$

15 a i B   **ii** D   **iii** E   **iv** no graph   **v** C

b $y = \dfrac{1}{x+1}$ (or anything of the form $y = \dfrac{a}{x+b}$ $a > 0$ and $b > 0$)

c

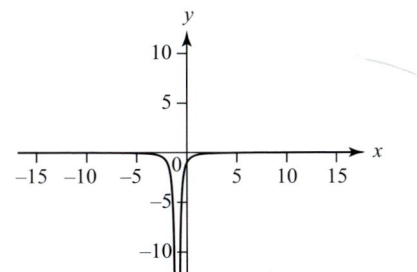

Chapter 3

Exercise 3.1A Fluency and skills

1 a $\sin\theta = \dfrac{4}{5}$; $\tan\theta = \dfrac{4}{3}$

b $\sin\theta = 0.6$; $\tan\theta = 0.75$

c $\sin\theta = \dfrac{5}{13}$; $\tan\theta = \dfrac{5}{12}$

2 a $-\cos 10°$ **b** $-\tan 20°$ **c** $-\sin 20°$

   d $-\cos 22°$ **e** $\tan 35°$ **f** $-\sin 75°$

3

θ	$-90°$	$0°$	$90°$	$180°$	$270°$	$360°$
$\sin\theta$	-1	0	1	0	-1	0
$\cos\theta$	0	1	0	-1	0	1
$\tan\theta$	$-$	0	$-$	0	$-$	0

4 a i, ii

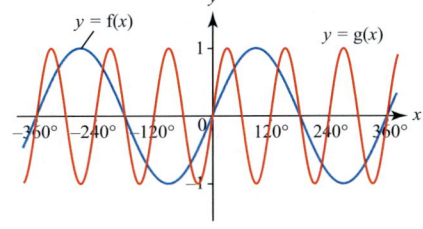

Line symmetry about $x = \pm 90°$, $\pm 270°$, ...
Rotational symmetry (order 2) about every point of intersection with x-axis: $(0°, 0)$, $(\pm 180°, 0)$, ...

iii Horizontal stretch, scale factor $\dfrac{1}{3}$

b i, ii

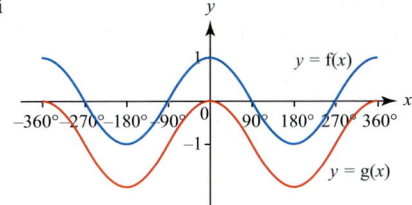

Line symmetry about $x = 0°$, $\pm 180°$, $\pm 360°$, ...
Rotational symmetry (order 2) about every point of intersection with x-axis: $(\pm 90°, 0)$, $(\pm 270°, 0)$, ...

iii Translation by $\begin{pmatrix} 0 \\ -1 \end{pmatrix}$

c i, ii

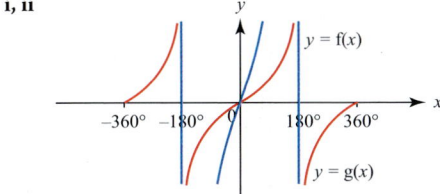

No line symmetry
Rotational symmetry (order 2) about $(0°, 0)$, $(\pm 90°, 0)$, $(\pm 180°, 0)$ ...

iii Horizontal stretch, scale factor 2

5 a $\tan\theta$ **b** $\tan\theta$ **c** 1

   d $\sin\theta$ **e** $\cos\theta$ **f** $\sin^2\theta$

6 a $\sin 20°$ **b** $\tan 30°$ **c** $\cos 20°$

   d $-\tan 42°$ **e** $-\cos 22°$ **f** $\sin 23°$

7 a $\theta = 45°, -135°$ **b** $\theta = -45°, 135°$

8 a $\dfrac{1}{2}$ **b** $\dfrac{1}{2}$ **c** $-\sqrt{3}$

   d $-\dfrac{\sqrt{3}}{2}$ **e** $\dfrac{1}{2}$ **f** $\dfrac{1}{\sqrt{3}}$

9 a $\theta = 23.6°, 156.4°, -203.6°, -336.4°$

   b $\theta = 56.3°, 236.3°, -123.7°, -303.7°$

   c $\theta = 120°, 240°, -120°, -240°$

10 a $\theta = 53.1°$ **b** $\theta = 323.1°$

11 a $\theta = 48.6°$ or $131.4°$ **b** $\theta = 53.1°$ or $233.1°$

   c $\theta = 210°$ or $330°$ **d** $\theta = 131.8°$ or $228.2°$

   e $\theta = 108.4°$ or $288.4°$ **f** $\theta = 224.4°$ or $315.6°$

   g $\theta = 138.6°$ or $221.4°$ **h** $\theta = 156.0°$ or $336.0°$

Exercise 3.1B Reasoning and problem-solving

1 a $\theta = 0°, 120°, 360°$ **b** $\theta = 90°, 330°$

   c $\theta = 25°, 205°$ **d** $\theta = 180°$ or $300°$

2 a $\theta = 15°$ or $75°$ **b** $\theta = 16.8°$ or $106.8°$

   c $\theta = 22.1°, 97.9°$ or $142.1°$ **d** $\theta = 72.3°$ or $107.7°$

   e $\theta = 49.7°$ or $139.7°$ **f** $\theta = 97.2°$

3 Any correct transformation, for example:

   a A translation of $\begin{pmatrix} -90° \\ 0 \end{pmatrix}$

   b A 180° rotation about $(0°, 0)$

4 a $(3\cos\theta, 3\sin\theta)$

   b $(3\cos\theta)^2 + (3\sin\theta)^2 = 9\cos^2\theta + 9\sin^2\theta$

$$= 9(\cos^2\theta + \sin^2\theta)$$
$$= 9 = 3^2$$
$$r = 3$$

   c $\dfrac{5}{2}$

5 $x^2 + y^2 = 100$

6 a $\theta = 26.6°$ or $-153.4°$

   b $\theta = 38.7°$ or $-141.3°$

   c $\theta = 9.2°, 99.2°, -80.8°, -170.8°$

d $\theta = 0°, 90°, 180°, -90°$ or $-180°$ and $\theta = 24.1°, 155.9°, -24.1°$
or $-155.9°$

e $\theta = 0°, 180°$ or $-180°$ and $\theta = \pm 109.5°$

f $\theta = \pm 90°$

7 a $\theta = 90°$

b $\theta = 135°, 315°, 63.4°, 243.4°$

c $\theta = 19.5°, 160.5°, 270°$

d $\theta = 0°, 180°, 360°$

e $\theta = 60°, 180°, 300°$

f $\theta = 48.6°$ or $131.4°$

8 $a = 5, b = 8$

9 a Max = 5, min = 1   **b** 12 hours   **c** 3 am and 3 pm

10 a $\theta = 60°, 300°, 48.2°, 311.8°$ **b** $\theta = 190.1°$ or $349.9°$

11 a

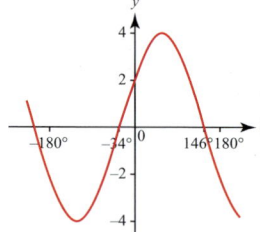

b **i** From graph, $\theta \approx -34°$ and $146°$

    ii $\theta = -33.7°$ or $146.3°$

12 a $\cos^4 x - \sin^4 x = (\cos^2 x - \sin^2 x)(\cos^2 x + \sin^2 x)$
$$= (\cos^2 x - \sin^2 x) \times 1 = \cos^2 x - \sin^2 x$$

b $\tan x + \dfrac{1}{\tan x} = \dfrac{\sin x}{\cos x} + \dfrac{\cos x}{\sin x}$

$$= \dfrac{\sin^2 x + \cos^2 x}{\cos x \sin x} = \dfrac{1}{\sin x \cos x}$$

13 $\dfrac{1 - \tan^2 x}{1 + \tan^2 x} = \dfrac{1 - \frac{\sin^2 x}{\cos^2 x}}{1 + \frac{\sin^2 x}{\cos^2 x}} = \dfrac{\cos^2 x - \sin^2 x}{\cos^2 x + \sin^2 x}$

$$= \dfrac{\cos^2 x - \sin^2 x}{1} = \cos^2 x - \sin^2 x$$

$$= 1 - \sin^2 x - \sin^2 x = 1 - 2\sin^2 x$$
$$x = 30°, 150° \text{ or } 210°, 330°$$

Exercise 3.2A Fluency and skills

1 $BC = 10.7$ cm, $PR = 19.7$ cm

2 a $62.4°$ **b** $115.7°$ **c** $110.1°$

3 a $BC = 6.97$ cm, $PR = 15.0$ cm

b area of triangle $ABC = 26.8$ cm^2
area of triangle $PQR = 29.8$ cm^2

4 a $BC = 7.48$ cm, angle $B = 64.8°$, angle $C = 40.2°$

b $EF = 16.0$ cm, angle $E = 30.8°$, angle $F = 24.2°$

c angle $J = 62°$, $HJ = 10.6$ cm, $HI = 9.82$ cm

d Angle $M = 94°$, $KM = 3.74$ cm, $LM = 5.06$ cm

e Angle $R = 66.8°$ or $113.2°$, angle $Q = 63.2°$ or $16.8°$
Two triangles are possible. $RP = 17.5$ cm or 5.67 cm

5 a Angle $A = 31.5°$, angle $C = 94.0°$, angle $B = 54.5°$

b $DE = 59.2$ cm, angle $D = 81.4°$, angle $E = 47.6°$

c $HI = 40.5$ cm, angle $I = 110.6°$, angle $H = 36.4°$

6 $x = \sqrt{13}$ cm, area $= 3\sqrt{3}$ cm^2

7 a $\dfrac{\sin C}{12} = \dfrac{\sin 55°}{10} \Rightarrow \sin C = \dfrac{12 \times \sin 55°}{10} = 0.982...$

Angle $C = 79.4°$ or $100.6°$
So angle $B = 180° - 55° - 79.4° = 45.6°$
or $= 180° - 55° - 100.6° = 24.4°$
so both versions of angles B and C are possible.

b $\dfrac{\sin Z}{12} = \dfrac{\sin 55°}{13} \Rightarrow \sin Z = \dfrac{12 \times \sin 55°}{13} = 0.756...$

Angle $Z = 49.1°$ or $130.9°$

So angle $Y = 180° - 55° - 49.1° = 75.9°$
or $= 180° - 55° - 130.9° = -5.9°$
which is impossible.

Exercise 3.2B Reasoning and problem-solving

1 11.4 cm^2

2 a $44.4°$ **b** $111.8°$

3 $BC^2 = 12^2 + (4\sqrt{3}\,)^2 - 2 \times 12 \times 4\sqrt{3} \times \cos 30°$
$$= 144 + 48 - 144 \Rightarrow BC = \sqrt{48} = 4\sqrt{3} = AB$$
So ΔABC is isosceles.

4 5.44 cm and 12.2 cm

5 a $\dfrac{9}{4}\sqrt{3}$ **b** $\dfrac{1}{2}$

6 a $h = b\sin A = a\sin B$
Hence,
$$\dfrac{\sin A}{a} = \dfrac{\sin B}{b}$$
Area of $\Delta ACP = \dfrac{1}{2} h \times AP$

Area of $\Delta CBP = \dfrac{1}{2} h \times PB$

So area of $\Delta ACB = \dfrac{1}{2} h \times (AP + PB) = \dfrac{1}{2} h \times c$

$h = b\sin A = a\sin B$ so $\dfrac{1}{2} h \times c = \dfrac{1}{2} bc\sin A = \dfrac{1}{2} ca\sin B$

b $CP = b\sin A$
$AP = b\cos A$
$BP = c - b\cos A$
Pythagoras in ΔCBP gives
$$a^2 = (b\sin A)^2 + (c - b\cos A)^2$$
$$a^2 = b^2(1 - \cos^2 A) + c^2 - 2cb\cos A + b^2\cos^2 A$$
$$a^2 = b^2 + c^2 - 2cb\cos A$$

7 $\dfrac{\sin Z}{2\sqrt{3}} = \dfrac{\sin 30°}{2} \Rightarrow \sin Z = \dfrac{2\sqrt{3}}{2} \times \dfrac{1}{2} = \dfrac{\sqrt{3}}{2}$

$Z = 60°$ or $120°$
Angle sum of triangle gives $X = 90°$ or $30°$
The angles of the triangle are either $30°, 60°, 90°$
or $30°, 30°, 120°$
The triangle is either isosceles or right-angled.

8 $AC = 14.5$ cm, $AB = 18.3$ cm

9 11.4 cm

10 Angle at centre = 2 × angle at circumference
$$B\hat{O}C = 2 \times \hat{A}$$
ΔBOC is isosceles and symmetrical
$$B\hat{O}P = \dfrac{1}{2} B\hat{O}C = \hat{A}$$
In ΔBOP, $\sin A = \dfrac{\frac{1}{2}a}{r} = \dfrac{a}{2r}$
$$2r = \dfrac{a}{\sin A}$$
Similarly with angles B and C at the apex, giving
$$2r = \dfrac{a}{\sin A} = \dfrac{b}{\sin B} = \dfrac{c}{\sin C}$$

11 In ΔABC, $B = 180° - A - C = 45°$

$$\dfrac{a}{\sin 22.5°} = \dfrac{b}{\sin 45°} \Rightarrow a = \dfrac{b\sin 22.5°}{\sin 45°}$$

In ΔBCP, $h = a\sin 67.5° = \dfrac{b\sin 67.5° \times \sin 22.5°}{\sin 45°} = 0.5 \times b$

The height is half the length of the base.

12 The cosine rule gives
$$(n^2 - n + 1)^2 = (n^2 - 2n)^2 + (n^2 - 1)^2 - 2(n^2 - 2n)(n^2 - 1)\cos Y$$
$$n^4 - n^3 + n^2 - n^3 + n^2 - n + n^2 - n + 1$$
$$= n^4 - 4n^3 + 4n^2 + n^4 - 2n^2 + 1 - 2(n^4 - 2n^3 - n^2 + 2n)\cos Y$$
$$2n^3 + n^2 - 2n = n^4 - 2(n^4 - 2n^3 - n^2 + 2n)\cos Y$$
$$\cos Y = \dfrac{n(n^3 - 2n^2 - n + 2)}{2n(n^3 - 2n^2 - n + 2)} = \dfrac{1}{2}$$
Angle $Y = 60°$

Review exercise 3

1 a $\sin\theta = 0.6$, $\tan\theta = 0.75$ **b** $\cos\theta = \dfrac{12}{13}$, $\tan\theta = \dfrac{5}{12}$

2 a $\tan\theta$ **b** $\sin\theta$

3 a $-\sin 10°$ **b** $\tan 80°$ **c** $-\cos 40°$
 d $-\tan 42°$ **e** $\sin 11°$ **f** $-\cos 60°$
 g $\tan 30°$ **h** $-\cos 20°$ **i** $\sin 80°$

4 a Max $= 4$, period $= 360°$ **b** Max $= 5$, period $= 180°$
 c Max $= 6$, period $= 72°$

5 a $x = 45°$ **b** $x = 45°$ for $0° \leq x \leq 180°$

6 a $x^2 + y^2 = 4\cos^2\theta + 4\sin^2\theta = 4(\cos^2\theta + \sin^2\theta) = 4 \times 1 = 4$
 $x^2 + y^2 = 4$ is equation of circle of radius $\sqrt{4} = 2$
 b At Q, $x = 1$ and $y = \sqrt{3}$
 $x^2 + y^2 = 1 + 3 + = 4$
 Hence, Q lies on the circle. $\theta = 60°$

7 a $\dfrac{1}{\sqrt{2}}$ **b** $-\dfrac{1}{\sqrt{2}}$ **c** -1
 d $\dfrac{1}{\sqrt{2}}$ **e** $-\dfrac{1}{\sqrt{2}}$ **f** -1

8 a $\theta = 45.6°, 314.4°, -45.6°$ or $-314.4°$
 b $\theta = 68.2°, 248.2°, -111.8°$ or $-291.8°$
 c $\theta = 210°, 330°, -30°$ or $-150°$

9 a $\theta = 41.8°$ or $138.2°$ **b** $\theta = 74.1°$ or $254.1°$
 c $\theta = 120°, 240°$ **d** $\theta = 210°$ or $330°$
 e $\theta = 28°$ or $332°$

10 a $\theta = 36.9°$ or $-143.1°$ **b** $\theta = 0°, \pm 180°$ or $\pm 41.4°$
 c $\theta = 0°, \pm 180°, 19.5°$ or $160.5°$ **d** $\theta = 80°$ or $140°$
 e $\theta = 30°$ or $-90°$ **f** $\theta = 145°$ or $-35°$

11 a $\theta = 9.7°$ or $80.3°$
 b $\theta = 10.9°$ or $100.9°$
 c $\theta = 3.8°, 56.2°, 123.8°, 176.2°$
 d $\theta = 16.1°, 103.9°, 136.1°$
 e $\theta = 9.2°, 99.2°$
 f $\theta = 53.1°$

12 a $\theta = 210°, 330°, 90°$
 b $\theta = 60°, 300°, 180°$
 c $\theta = 99.6°, 260.4°, 0°, 360°$
 d $\theta = 64.3°, 295.7°, 219.9°, 140.1°$
 e $\theta = 90°$
 f $\theta = 53.6°, 306.4°, 147.5°, 212.5°$

13 a $BC = 6.46\,\text{cm}$, area $= 25.6\,\text{cm}^2$
 b Angle $E = 26.9°$, area $= 16.1\,\text{cm}^2$

Assessment 3

1 a

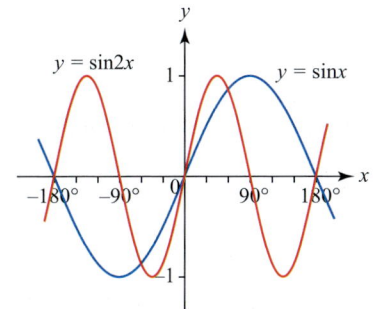

 b

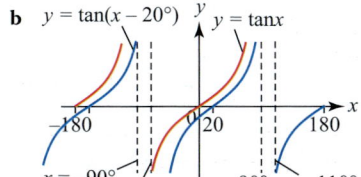

$x = -90°$ $x = -70°$ $x = 90°$ $x = 110°$

2 a $x = 78.5°, 218.5°$ **b** $x = 136.3°, 316.3°$

3 a

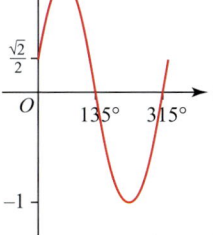

 b Maximum at $(45°, 1)$; Minimum at $(225°, -1)$
 c $x = 117.5°, 332.5°$

4 a $(0°, 1)$; other values possible for x, e.g. $\pm 720°$
 b $(360°, -1)$; other values possible for x, e.g. $-360°$

5 a $57.9°$ **b** $20.3\,\text{m}^2$

6 $15.7\,\text{cm}$

7 2 units

8 a $x = -1.5°, -158.5°$
 b $x = 11.7°, 71.7°, 131.7°, -48.3°, -108.3°, -168.3°$

9 a

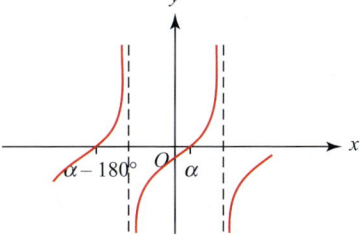

 b $x = \alpha + 90°$, $x = \alpha - 90°$ **c** $x = \alpha + 60°$, $\alpha - 120°$

10 a

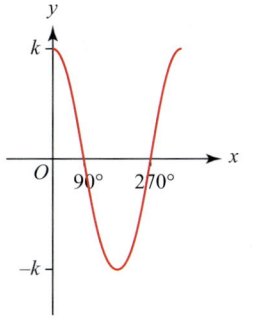

 b $x = 71.6°, 251.6°$

11 a

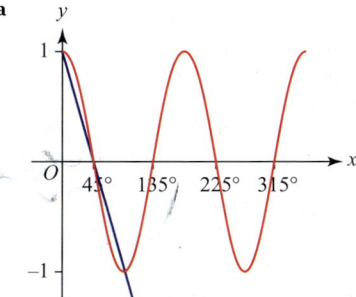

 b 3 as the line intersects the curve 3 times

12 a 19.9 or 7.86 cm **b** $31.4\,\text{cm}^2$

13 a $31.3°$
 b $180° - 31.3° = 148.7°$, but $148.7° + 54° > 180°$, which makes the triangle impossible

14 a 10.1 cm **b** $94.1°$

15 $117.3°$

16 $45° < x < 135°$

17 a $\cos\theta + \tan\theta\sin\theta \equiv \cos\theta + \dfrac{\sin\theta\sin\theta}{\cos\theta}$

$\equiv \dfrac{\cos^2\theta + \sin^2\theta}{\cos\theta} \equiv \dfrac{1}{\cos\theta}$

b $\theta = 66.4°,\ -66.4°$

18 a $x = 53.7°,\ 126.3°,\ 233.7°,\ 306.3°$

b $x = 71.6°,\ 251.6°,\ 135°,\ 315°$

19 $x = 0°,\ 90°,\ 110.9°,\ 159.1°,\ 180°$

20 $\theta = 0°,\ 60°,\ 120°,\ 180°,\ 37.9°,\ 82.1°,\ 157.9°$

21 a $\dfrac{2\sqrt{6}}{5}$ **b** $\dfrac{\sqrt{6}}{12}$

22 a $-8 < k < 0$ **b** $\theta = 30°,\ 150°,\ 270°$

23 $\dfrac{\alpha}{\sqrt{1-\alpha^2}}$

24 $x = 56.9°,\ 123.1°,\ 236.9°,\ 303.1°$

25 $157.3\,\text{cm}$

Chapter 4

Exercise 4.1A Fluency and skills

1 a 6   **b** 8   **c** 8   **d** 10
  e 4   **f** 15   **g** 3   **h** 36

2 a 2   **b** 12   **c** 27   **d** 4
  e 6   **f** 4   **g** 6   **h** 12

3 a 24   **b** 18   **c** 4   **d** 2
  e -8 **f** 3   **g** 13

4 a $4x$ **b** $8x$ **c** $12x$ **d** x
  e $2x$ **f** 2   **g** x^2 **h** $6x^2$

5 a **i** $1 + 2x$ **ii** $1 + 2x$ **iii** $1 + 2x$
    iv $2x + 2$ **v** $2x - 3$ **vi** $4x + 5$
  b **i** 0   **ii** 0   **iii** 0   **iv** 0
  c **i** 1   **ii** -1 **iii** 2   **iv** -3

6 a $6x^2$ **b** $12x^3 - 4x$ **c** $10x^4$
  d $-3x^2$ **e** $2x - 6x^2$ **f** $1 - 2x + 3x^2$

Exercise 4.1B Reasoning and problem-solving

1 a $x = 2$ **b** $x = 10$ **c** $x = 0.05$
  d $(1, 5)$ **e** $(1, 5)$ **f** $(4, 80)$

2 a $(1, 1)$ and $(-1, -1)$ **b** $(2, 8)$ and $(-2, -8)$
  c $(3, 27)$ and $(-3, -27)$ **d** $(0.1, 0.001)$ and $(-0.1, -0.001)$
  e $\left(\dfrac{1}{3}, \dfrac{1}{27}\right), \left(-\dfrac{1}{3}, -\dfrac{1}{27}\right)$
  f $\left(0.7, 0.343\right), \left(-0.7, -0.343\right)$

3 a $x = 1$ **b** $x = 3$
  c $x = 2.5$ **d** $x = \dfrac{b}{2a}$

4 a $x = \dfrac{5}{3}, -1$ **b** $x = 3, -1$
  c $x = \dfrac{1}{3}, 1$ **d** $x = \dfrac{2}{3}, 2$
  e $x = 2$ **f** $x = 2.5$
  g $x = 3$ **h** $x = -2, 1$

5 a nx^{n-1} **b** anx^{n-1} **c** $\dfrac{dy}{dx} = 6x^2; \dfrac{dy}{dx} = 6x$
  d No. No matter how many special cases you show work, you have not shown it works for all n

6 a $f(x) = \dfrac{1}{x} \Rightarrow f'\left(\dfrac{1}{2}\right) = \lim_{h\to0} \dfrac{\frac{1}{\frac{1}{2}+h} - \frac{1}{\frac{1}{2}}}{h}$

$= \lim_{h\to0} \dfrac{\frac{1}{2} - \left(\frac{1}{2}+h\right)}{h\frac{1}{2}\left(\frac{1}{2}+h\right)} = \lim_{h\to0} \dfrac{-h}{h\frac{1}{2}\left(\frac{1}{2}+h\right)}$

$= \lim_{h\to0} \dfrac{-1}{\left(\frac{1}{2}\right)^2 + \frac{1}{2}h} = -\dfrac{1}{\left(\frac{1}{2}\right)^2} = -4$

b $f(x) = \dfrac{1}{x} \Rightarrow f'(x) = \lim_{h\to0} \dfrac{\frac{1}{x+h} - \frac{1}{x}}{h}$

$= \lim_{h\to0} \dfrac{x - (x+h)}{hx(x+h)} = \lim_{h\to0} \dfrac{-h}{hx(x+h)}$

$= \lim_{h\to0} \dfrac{-1}{x^2 + xh} = -\dfrac{1}{x^2}$

Exercise 4.2A Fluency and skills

1 a $21x^6$ **b** $24x^3$ **c** 5
  d 0   **e** 0   **f** 0
  g $-10x^4$ **h** $-7x^6$ **i** $-2x^{-2}$
  j $3x^{-4}$ **k** $\dfrac{1}{2}x^{-\frac{1}{2}}$ **l** $\dfrac{2}{3}x^{-\frac{1}{3}}$
  m $\dfrac{25}{3}x^{\frac{2}{3}}$ **n** $\dfrac{1}{4}x^{-\frac{5}{4}}$ **o** 0

2 a $\dfrac{1}{2}x^{-\frac{1}{2}}$ **b** $\dfrac{1}{3}x^{-\frac{2}{3}}$
  c $\dfrac{1}{5}x^{-\frac{4}{5}}$ **d** $\dfrac{3}{2}x^{\frac{1}{2}}$
  e $\dfrac{2}{3}x^{-\frac{1}{3}}$ **f** $\dfrac{\sqrt[3]{3}}{3}x^{-\frac{2}{3}}$
  g $-x^{-2} = -\dfrac{1}{x^2}$ **h** $-6x^{-3} = -\dfrac{6}{x^3}$
  i $24x^{-5} = \dfrac{24}{x^5}$ **j** $-\dfrac{1}{2}x^{-\frac{3}{2}}$
  k $-\dfrac{9}{2}x^{-\frac{5}{2}}$ **l** 0
  m 0

3 a $2x + 2$ **b** $-2 - 10x$
  c $3x^2 + 4x - 3$ **d** $2x - 4x^3$
  e $1 + \dfrac{1}{x^2}$ **f** $1 - \dfrac{4}{x^3}$
  g $3x^2 + \dfrac{3}{x^4} - \dfrac{3}{x^2}$ **h** $-\dfrac{10}{x^2} - \dfrac{1}{10}$
  i $\dfrac{1}{2\sqrt{x}} - \dfrac{1}{2\sqrt{x^3}}$ **j** $5x^4 - \dfrac{5}{3\sqrt[3]{x^4}}$
  k $-x^{-\frac{3}{2}} + \dfrac{10}{3}x^{-\frac{5}{3}}$ **l** $4x^{-3} - \dfrac{3}{2}x^{\frac{1}{2}}$

4 a **i** $6x + 4$ **ii** -8 **b** **i** $6x^2 - 10x$ **ii** -4
  c **i** $10 - \dfrac{8}{x^2}$ **ii** 8   **d** **i** $-\dfrac{8}{x^2} - \dfrac{8}{x^3}$ **ii** -3
  e **i** $y = x^2 - x$ **ii** 7   **iii** 7

5 a $\dfrac{dy}{dx} = 4x - 5,\ -1$ **b** $\dfrac{dy}{dx} = -5 - \dfrac{10}{x^2},\ -\dfrac{30}{4}$
  c $\dfrac{dy}{dx} = 4x + 1,\ -11$ **d** $\dfrac{dy}{dx} = \dfrac{1}{2}x^{-\frac{1}{2}} - x^{-\frac{3}{2}},\ \dfrac{1}{8}$

6 a $y = x^2$ **b** $y = x^2 + 3x - 8$
  c $y = 2x + 3$ **d** $y = x^2 - 2x + 3$
  e $y = 2x - 15$

7 a True   **b** True   **c** True
  d False   **e** True   **f** False

Exercise 4.2B Reasoning and problem-solving

1 a $2x + 1$ **b** $2x$ **c** $2x - 9x^2$
  d $\dfrac{3}{2}x^{\frac{1}{2}} + \dfrac{3}{2}x^{-\frac{1}{2}}$ **e** $18x^2 + 6x - 9$ **f** $\dfrac{3}{2}x^{\frac{1}{2}} + 1$
  g $-4x^{-2}$ **h** $-x^{-2} - 2$ **i** $\dfrac{3}{8}x^{\frac{1}{2}} + \dfrac{3}{8}x^{-\frac{1}{2}} - \dfrac{1}{8}x^{-\frac{3}{2}}$

2 a 1   **b** $1 + \dfrac{1}{2}x^{-\frac{1}{2}}$
  c $-\dfrac{1}{2}x^{-\frac{3}{2}}$ **d** $-x^{-2} + x^{-\frac{3}{2}}$

Answers For full solutions go to http://www.oxfordsecondary.co.uk/edexcelalevelmaths-answers

e $\frac{1}{2}x^{-\frac{1}{2}} - \frac{3}{2}x^{\frac{1}{2}}$ **f** $-x^{-2} - x^{-\frac{3}{2}}$

g $\frac{1}{2}x^{\frac{1}{2}} + \frac{1}{2}x^{-\frac{1}{2}} + \frac{1}{6}x^{-\frac{3}{2}}$ **h** $-x^{-2} - 2x^{-3}$

i $-\frac{1}{3}x^{-\frac{4}{3}}$

3 a $2x - 2$ **b** $2x - 9$ **c** $-2 - 2x$

d $4x + 3$ **e** $3x^2 - 1$ **f** $\frac{3}{2}x^{\frac{1}{2}} + 1 + \frac{1}{2}x^{-\frac{1}{2}}$

4 a $2x - 8$ **b** $x = 1$ or $x = 7$

c $(4, -9)$ **d i** $x < 4$ **ii** $x > 4$

e i falling **ii** rising

5 a $10 - \dfrac{40}{x^2}$

b i $x = 2$ **ii** $y = 40$

c i $0.5 \le x < 2$ **ii** $2 < x \le 5.5$

6 $f(x) = (x+1)^4 = x^4 + 4x^3 + 6x^2 + 4x + 1 \Rightarrow$
$f'(x) = 4x^3 + 12x^2 + 12x + 4$
$\Rightarrow f'(1) = 32$

7 a $2x + 1$

b 11

c $x = -\dfrac{1}{2}, y = -2\dfrac{1}{4}$

d i $x = \dfrac{1}{2}, y = -1\dfrac{1}{4}$ **ii** $k = -1$

Exercise 4.3A Fluency and skills

1 a 18 **b** 0 **c** 5 **d** $\dfrac{8}{9}$

e 24 **f** 0 **g** $-\dfrac{27}{2}$ **h** 9

i 0 **j** $-\dfrac{3}{4}$ **k** 0 **l** -3

m 2 **n** -3 **o** 0 **p** -2

2 a 18 **b** $\dfrac{5}{2}$ **c** $\dfrac{3}{4}$ **d** 46

e 1 **f** 2 **g** -2 **h** $\dfrac{125}{8}$

i 0 **j** 16 **k** 1280 **l** $-\dfrac{1}{3}$

3 a i $x = -2$ **ii** $x > -2$ **iii** $x < -2$

b i $x = -1.25$ **ii** $x < -1.25$ **iii** $x > -1.25$

c i $x = 0$ **ii** $x > 0$ **iii** $x < 0$

d i $x = -2, 2$ **ii** $x < -2$ or $x > 2$ **iii** $-2 < x < 2$

e i $x = -7$ or $x = 3$ **ii** $-7 < x < 3$

iii $x < -7$ or $x > 3$

f i $x = 1$ and $x = 2$ **ii** $x < 1$ or $x > 2$ **iii** $1 < x < 2$

g i no stationary points

ii increasing for all x in the domain

iii never decreasing

h i no stationary points

ii increasing for all x in the domain

iii never decreasing

i i $x = 0$ **ii** $x < 0$ or $x > 0$ **iii** never decreasing

j i $x = 0$ **ii** $x > 0$ **iii** $x < 0$

4 a $y = -x^2$ **b** $y = 5 - 3x$

c $y = x^2 - 2$ **d** $y = 1 + 10x - x^3$

Exercise 4.3B Reasoning and problem-solving

1 £2.50 per kilometre

2 a -7.5 **b** It is losing mass.

3 a $4\pi r^2$ **b** 36π **c** 24π

4 a i 2 **ii** -6 **iii** -12

b $x = -1$

5 a $(3 - 9.8t)\,\text{m s}^{-1}$

b falling at $46\,\text{m s}^{-1}$

c falling at $95\,\text{m s}^{-1}$

d $-9.8\,\text{m s}^{-2}$ or $9.8\,\text{m s}^{-2}$ towards the ground

6 a $(10 - 3.24t)\,\text{m s}^{-1}$

b 3.09 seconds (to 3 sf)

c $3.24\,\text{m s}^{-2}$ towards lunar surface

7 a $360 - 12t$

b i $240\,\text{ml s}^{-1}$

ii $120\,\text{ml s}^{-1}$

c 30 secs **d** $5400\,\text{ml}$

8 a i $x^2 + 10x + 30$

ii 30

iii $\dfrac{dy}{dx} = (x+5)^2 - 5^2 + 30 = (x+5)^2 + 5$, which is

positive for all x. So function increasing for all x

b i $x^2 - 6x + 12$

ii 12

iii $\dfrac{dy}{dx} = (x-3)^2 - (-3)^2 + 12 = (x-3)^2 + 3$,

which is positive for all x. So function increasing
for all x

c i $x^2 - 5x + 8$

ii 8

iii $\dfrac{dy}{dx} = \left(x - \dfrac{5}{2}\right)^2 - \left(-\dfrac{5}{2}\right)^2 + 8 = \left(x - \dfrac{5}{2}\right)^2 + \dfrac{7}{4}$,

which is positive for all x. So function increasing
for all x

9 $\dfrac{dy}{dx} = -\left(x^2 - 8x + 17\right) = -\left((x-4)^2 - 4^2 + 17\right) = -(x-4)^2 - 1$,

which is negative for all x. So function decreasing for all x

10 a i $4x^3 - 2x^{-3}$

ii $12x^2 + 6x^{-4} = 12x^2 + \dfrac{6}{x^4}$

b The second derivative is positive for all $x \ge 1$ so the
derivative is an increasing function for all x in the
domain.

Exercise 4.4A Fluency and skills

1 a $y - 4 = 7(x - 1)$ **b** $y - 1 = 3(x + 2)$

c $y - 5 = 12(x - 1)$ **d** $y - 4 = -12(x - 2)$

e $y - 1 = -\dfrac{1}{3}(x - 3)$ **f** $y - 1 = -\dfrac{1}{2}(x - 4)$

g $y - 3 = \dfrac{1}{6}(x - 9)$ **h** $y - 5 = -\dfrac{1}{10}(x - 25)$

i $y - 3 = \dfrac{1}{8}(x - 4)$ **j** $y - 2 = -\dfrac{3}{2}(x - 1)$

2 a $y - 1 = -\dfrac{1}{6}(x - 2)$ **b** $y - 10 = -(x + 3)$

c $y + 5 = \dfrac{1}{12}(x - 2)$ **d** $y - 4 = -\dfrac{1}{12}(x - 1)$

e $y - 3 = \dfrac{2}{3}(x - 2)$ **f** $y - 6 = -\dfrac{4}{5}(x - 4)$

g $y - \dfrac{3}{4} = 8(x - 4)$ **h** $y - 2 = \dfrac{2}{7}(x - 1)$

3 a $(3, 27)$ **b** 18 **c** $y - 27 = 18(x - 3)$

4 a $k = 3$ **b** 5 **c** $y - 3 = 5(x - 1)$

5 a $(1, -1)$ **b** 9 **c** $-\dfrac{1}{9}$

d $y + 1 = -\dfrac{1}{9}(x - 1)$

6 a $(2, 5)$ **b** $\dfrac{3}{2}$

c $-\dfrac{2}{3}$ **d** $y - 5 = -\dfrac{2}{3}(x - 2)$

7 a i 0 **ii** $y = -4$

b $x = -3$

c i tangent: $y = -25$, normal: $x = -1$

ii tangent: $y = -25$, normal: $x = -5$

iii tangent: $y = 25$, normal: $x = 2$

Exercise 4.4B Reasoning and problem-solving

1 a $x = 2$ **b** $k = 5$

2 $\left(\dfrac{3}{2}, \dfrac{7}{2}\right)$

3 a $y = x - 2$
  b $(3, 1)$
  c $m_{\tan B}$ at $(3,1) = 2(3) - 3 = 3 \Rightarrow m_{\text{norm } B} = -\dfrac{1}{3}$;

$$m_{\text{norm } B} = -\dfrac{1}{3} \neq m_{\text{norm } A}$$

4 a Consider the line as tangent to first function and normal to second. In 1st function: $m_{\tan} = \dfrac{dy}{dx} = 2x$; in second

function $m_{\text{norm}} = -\dfrac{1}{m_{\tan}} = -\dfrac{1}{\frac{dy}{dx}} = -\dfrac{1}{-x^{-2}} = x^2$

b $\left(2, 4\dfrac{1}{2}\right)$

c $y - 4\dfrac{1}{2} = 4(x - 2)$

5 a $y + 1 = 1(x - 3)$ **b** $y + 1 = -1(x - 3)$ **c** 9 units^2

6 a $y - 1 = 2(x - 0)$ or $y = 2x + 1$ **b** $(-1, -1)$

  c $\left(-\dfrac{2}{3}, 0\right)$ **d** $\dfrac{1}{3}$ units^2

7 a $y - 3 = 5(x - 2)$ **b** $y - 3 = -\dfrac{1}{5}(x - 2)$

8 a $p = 2, q = 9$ **b** $y - 9 = 12(x - 2)$

  c $y - 9 = -\dfrac{1}{12}(x - 2)$

9 a $(3, 9)$ **b** $y - 9 = -(x - 3)$

10 a i $(x + 3)^2 - 10$ **ii** $(-3, -10)$
    iii tangent: $y = -10$; normal: $x = -3$

  b i $(x - 5)^2 - 20$ **ii** $(5, -20)$
    iii tangent: $y = -20$; normal: $x = 5$

  c i $\left(x - \dfrac{3}{2}\right)^2 - \dfrac{37}{4}$ **ii** $\left(\dfrac{3}{2}, -\dfrac{37}{4}\right)$

    iii tangent: $y = -\dfrac{37}{4}$; normal: $x = \dfrac{3}{2}$

  d i $20 - (x + 4)^2$ **ii** $(-4, 20)$
    iii tangent: $y = 20$; normal: $x = -4$

  e i $(x + 0)^2 + 4$ **ii** $(0, 4)$
    iii tangent: $y = 4$; normal: $x = 0$

  f i $2 - (x + 1)^2$ **ii** $(-1, 2)$
    iii tangent: $y = 2$; normal: $x = -1$
    iv In each case at a turning point the tangent was horizontal (gradient = 0) and the normal was vertical (gradient undefined). This might be used to test if a point is a turning point.

11 a i $y - 10\sqrt{p} = -\dfrac{\sqrt{p}}{5}(x - p)$ **ii** $y + 10\sqrt{p} = \dfrac{\sqrt{p}}{5}(x - p)$

  b Normal at A cuts x-axis when $y = 0$:

$$-10\sqrt{p} = -\dfrac{\sqrt{p}}{5}(x - p) \Rightarrow x = 50 + p$$

  Normal at B cuts x-axis when $y = 0$:

$$10\sqrt{p} = \dfrac{\sqrt{p}}{5}(x - p) \Rightarrow x = 50 + p$$

  Thus both cut x-axis at $(50 + p, 0)$

12 a $p = 0, q = 4$

  b 2

  c $y - 4 = -\dfrac{1}{2}(x - 0)$ or $y = 4 - \dfrac{1}{2}x$

d i $(2, 0)$ and $(-1, 0)$ **ii** $y = \dfrac{1}{6}(x - 2)$ and $y = -\dfrac{1}{6}(x + 1)$
e $x = 0$

13 a $y - 12 = -\dfrac{2}{5}(x - 30)$

  b $y - 6 = 10(x - 60)$

  c i $(6, 60)$ **ii** $y - 60 = \dfrac{1}{10}(x - 6)$

  d i $(3, 120)$ **ii** $k = 240$

Exercise 4.5A Fluency and skills

1 a $(-2, -9)$ min   **b** $(-2, -36)$ min
  c $(3, -16)$ min   **d** $(0, 1)$ max

  e $\left(-\dfrac{7}{4}, -\dfrac{1}{8}\right)$ min   **f** $\left(-\dfrac{1}{2}, 20\dfrac{1}{4}\right)$ max

  g $\left(\dfrac{1}{12}, -1\dfrac{1}{24}\right)$ min   **h** $\left(-\dfrac{13}{14}, \dfrac{225}{28}\right)$ max

  i $\left(-\dfrac{1}{4}, 6\dfrac{1}{8}\right)$ max   **j** $(2, 4)$ min

  k $(3, 12)$ min, $(-3, -12)$ max   **l** $(-1, 12)$ min, $(1, 8)$ max
  m $(25, -25)$ min   **n** $(4, -48)$ min

2 a $(0,0), (4, -32)$ **b** $(0,0)$ max, $(4, -32)$ min

  c

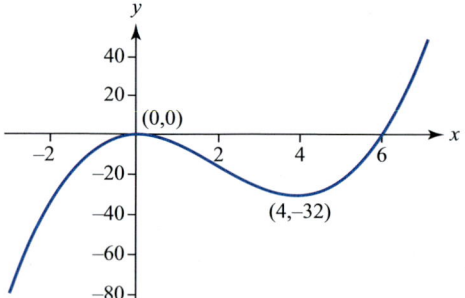

3 a $(0, 1)$ min, $(-10, 1001)$ max
  b

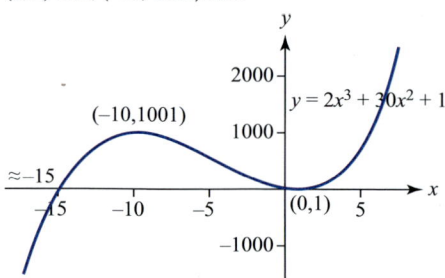

4 a $6x^2 - 18x + 12$
  b $(1, 12)$ max, $(2, 11)$ min
  c i $3x^2 - 6x - 24$; $(-2, 29)$ max, $(4, -79)$ min
    ii $3x^2 + 6x - 45$; $(-5, 130)$ max, $(3, -126)$ min
    iii $-6x^2 - 42x - 36$; $(-1, 18)$ max, $(-6, -107)$ min
    iv $6x^2 - 22x - 8$; $\left(-\dfrac{1}{3}, \dfrac{91}{27}\right)$ max, $(4, -78)$ min
    v $-6x^2 + 10x - 4$; $(1, 2)$ max, $\left(\dfrac{2}{3}, \dfrac{53}{27}\right)$ min
    vi $-12x^2 - 4x + 1$; $\left(\dfrac{1}{6}, \dfrac{275}{54}\right)$ max, $\left(-\dfrac{1}{2}, \dfrac{9}{2}\right)$ min

5 a $f(x) = 3x^4 + 8x^3 - 6x^2 - 24x - 1$

$$\Rightarrow \dfrac{dy}{dx} = 12x^3 + 24x^2 - 12x - 24$$

At turning points

$$\dfrac{dy}{dx} = 0 \Rightarrow 12x^3 + 24x^2 - 12x - 24$$

$$= 0 \Rightarrow x^3 + 2x^2 - x - 2 = 0$$

  Answers   For full solutions go to http://www.oxfordsecondary.co.uk/edexcelalevelmaths-answers

Checking: at $x=1$, $\dfrac{dy}{dx}=1^3+2.1^2-1-2=0$

at $x=-1$, $\dfrac{dy}{dx}=(-1)^3+2.(-1)^2-(-1)-2=0$

at $x=-2$, $\dfrac{dy}{dx}=(-2)^3+2.(-2)^2-(-2)-2=0$

b $f(1)=(1,-20)$, $f(-1)=(-1,12)$, $f(-2)=(-2,7)$

c $(1,-20)$ is a min, $(-1,12)$ is a max, $(-2,7)$ is a min

6 a $f'(x)=12x^3-12x^2-72x$. At stationary points $f'(x)=0$

$12x^3-12x^2-72x=0 \Rightarrow x(x^2-x-6)=0$

$\Rightarrow x(x-3)(x+2)=0$

Hence result.

b minima at $x=-2$ and $x=3$; maximum at $x=0$

7 a $f(x)=8x+\dfrac{72}{x} \Rightarrow f'(x)=8-\dfrac{72}{x^2}$;

At stationary point $f'(x)=0 \Rightarrow 8-\dfrac{72}{x^2}=0 \Rightarrow x=\pm 3$

but in the definition $x>0$. So only $x=3$ is a solution.

b A minimum at $(3,48)$ **c** A maximum.

Exercise 4.5B Reasoning and problem-solving

1 Both numbers would be 500

2 Both numbers would be 60

3 $x=8$, $y=4$

4 a $y=150-3x$ **b** $V=300x^2-6x^3$

c $x=33\dfrac{1}{3}$, $y=50$

5 a $x=4$, $y=8$ **b** $20\,$m

6 a $6x-5x^2$ **b** $\left(\dfrac{3}{5},3\right)$

7 $2\,$cm

8 a $h=\dfrac{1000}{x^2}$ **b** $2x^2+\dfrac{4000}{x}$ **c** $x=10$

9 a $h=\dfrac{440}{\pi x^2}$ **b** $x=4.1$ (to 1 dp)   **c** $x=5.2$ (to 1 dp)

Exercise 4.6A Fluency and skills

1 a $10x+c$ **b** $\dfrac{3}{2}x^2+c$

c $2x^3+c$ **d** $3x^4+c$

e $5x^5+c$ **f** $\dfrac{x^7}{7}+c$

g $\dfrac{3x^2}{2}+x+c$ **h** $5x-2x^2+c$

i x^3+3x^2+2x+c **j** $4x^3+6x+c$

k $3x-2x^2-2x^3+c$ **l** $\dfrac{x^4}{4}+\dfrac{x^3}{3}+\dfrac{x^2}{2}+x+c$

m $3x-\dfrac{x^2}{2}-6x^4+c$ **n** $\dfrac{1}{2}x^4+2x^2+x+c$

o $\dfrac{1}{4}x^4-3x^{-1}+c$ **p** $-x^{-2}+x^{-1}+c$

q $\dfrac{2}{3}x^{\frac{3}{2}}-\dfrac{1}{4}x^4+x+c$

r $\dfrac{2}{3}x^{\frac{3}{2}}-2x^{\frac{1}{2}}+\dfrac{1}{2}x^2+6x+c$ **s** $\dfrac{9}{4}x^{\frac{4}{3}}-\dfrac{3}{5}x^{\frac{5}{3}}+c$

t $\dfrac{4}{5}x^{\frac{5}{4}}-\dfrac{4}{7}x^{\frac{7}{4}}+2x^{\frac{1}{2}}+c$ **u** $x+x^{-1}-3x^{\frac{1}{3}}+c$

v $x^{\frac{1}{3}}-\dfrac{1}{5}x^5+4x+c$ **w** $\dfrac{1}{3}\pi x^4-\dfrac{1}{4}\pi x^{\frac{4}{3}}+\pi x^2+c$

x $\dfrac{3}{8}x^{\frac{8}{3}}+\dfrac{6}{5}x^{\frac{5}{3}}+\dfrac{3}{2}x^{\frac{2}{3}}+c$

2 a $x+c$ **b** $\dfrac{1}{2}x^2+c$

c $3x^2+7x+c$ **d** $3x-x^2+c$

e $\dfrac{1}{3}x^3+\dfrac{1}{2}x^2+x+c$ **f** $x-2x^2-x^3+c$

g x^4+x^2-7x+c **h** $2x+3x^3-3x^4+c$

i $\dfrac{2}{3}x^{\frac{3}{2}}+c$ **j** $\dfrac{3}{4}x^{\frac{4}{3}}+c$

k $2x^{\frac{1}{2}}+c$ **l** $\dfrac{3}{5}x^{\frac{5}{3}}+c$

3 a $\pi x+c$ **b** $3\pi x+\dfrac{x^2}{2}+c$

c $\dfrac{x^3}{6}+c$ **d** $\dfrac{x^3}{3}+3x^2+c$

e $\dfrac{4x^3}{3}+2x^2-28x+c$ **f** $\dfrac{x^3}{3}-x^{-1}+c$

g $\dfrac{2x^{\frac{3}{2}}}{3}-4x^{\frac{1}{2}}+c$ **h** $4x^2+3x^{-1}+c$

i $-\dfrac{x^{-2}}{2}+\dfrac{x^{-3}}{3}+c$ **j** $-x^{-1}-\dfrac{x^2}{2}+\dfrac{x^{-2}}{2}+c$

k $\dfrac{1}{2}x^2+2x^{\frac{1}{2}}+c$ **l** $\dfrac{1}{3}x^3-6x^{\frac{1}{2}}+x+c$

m $-1x^{-1}-\dfrac{3}{2}x^{-2}+c$ **n** $\dfrac{1}{6}x^3+\dfrac{\sqrt{3}}{4}x^2+c$

o $x^3+2x^{-\frac{1}{2}}+c$ **p** $\dfrac{1}{2}x^2-x+c$

q $\dfrac{2}{5}x^{\frac{5}{2}}+\dfrac{8}{3}x^{\frac{3}{2}}+8x^{\frac{1}{2}}+c$ **r** $\dfrac{2}{3}x^{\frac{3}{2}}+2x^{\frac{1}{2}}+c$

4 a $f(x)=2x^3+3$ **b** $f(x)=3x^4-30$

c $f(x)=\dfrac{10}{3}x^{\frac{3}{2}}+10$

5 a $2x^2+3x-10$ **b** $10x+2$

c x^3+x^2+x+7 **d** $2x^2+2x^{\frac{3}{2}}+1$

Exercise 4.6B Reasoning and problem-solving

1 a $P(t)=7t-26$ **b** 9   **c** 6

2 a $f(-2)=16$ **b** $x=-3$ or $x=\dfrac{10}{3}$

3 a $y=r^{\frac{3}{2}}+c$ **b** $y=r^{\frac{3}{2}}$

c 1.84 (to 3 sf)   **d** 9.53 (to 3 sf)

4 a $h=\tfrac{2}{3}t^4-2t^2+4$

b $3.54m$ (to 3 sf)

c $t=\sqrt{6}$ days

5 a $\dfrac{dy}{dx}=6x^2+x-3$ **b** 13

6 19

7 a i $v(t)=t^2+2$; $s(t)=\dfrac{1}{3}t^3+2t+2$

ii $3\,cm\,s^{-1}$; $\dfrac{13}{3}$ cm

b i $v(t)=6t+5$; $s(t)=3t^2+5t-2$; $11\,cm\,s^{-1}$; $6\,cm$

ii $v(t)=\dfrac{1}{2}t^2+2$; $s(t)=\dfrac{1}{6}t^3+2t+1$; $\dfrac{5}{2}\,cm\,s^{-1}$; $\dfrac{19}{6}$ cm

iii $v(t)=\dfrac{1}{3}t^3-1$; $s(t)=\dfrac{1}{12}t^4-t$; $-\dfrac{2}{3}\,cm\,s^{-1}$; $-\dfrac{11}{12}$ cm

8 6 seconds

9 a 0.5 metres to the right of the origin.   **b** $3\,ms^{-2}$

c $17.5\,$m   **d** $3n+\dfrac{5}{2}$

Exercise 4.7A Fluency and skills

1 a 8   **b** 20   **c** $\dfrac{15}{2}$

d -40 **e** 35   **f** 6π

g $2\pi+2$ **h** 48   **i** 12

j $\dfrac{4}{3}$ **k** $\dfrac{5}{4}$ **l** $\dfrac{23}{24}$

m $\dfrac{7}{4}$ **n** 6   **o** 156π

p 5   **q** 0   **r** 2π

2 a 16   **b** 18   **c** $15\dfrac{1}{6}$

   d $\dfrac{1}{6}$ **e** $\dfrac{64}{3}$ **f** 36

   g $6\dfrac{3}{4}$ **h** $6\dfrac{2}{3}$ **i** 378

Exercise 4.7B Reasoning and problem-solving

1 $-\sqrt{67}$

2 $1\dfrac{1}{3}$ square units

3 $57\dfrac{1}{6}$ square units

4 a $\dfrac{2}{3}x^{\frac{3}{2}}+c$ **b** $\dfrac{2}{3}$ square units

   c $P:Q=7:19$ **d** $1\dfrac{2}{3}$ square units

   The difference is a square of 1 square unit: 1 unit along the x-axis by 1 unit translation in the y-direction.

5 a $a=10$ **b** $a=5$ **c** $a=-3,2$

6 27:5

7 7

8 In the domain $0 \le x \le 360$, the area below the x-axis is equal to the area above the axis. Since each will be of opposite sign, their sum is zero.

9 75 m

10 20.08 km

Review exercise 4

1 a 6   **b** 3   **c** 31

   d $-2x$ **e** $1-2x$ **f** $2x\pi$

2 a $3x^2+4x+3$ **b** $\dfrac{2}{\sqrt{x}}+1$

   c $1-\dfrac{1}{x^2}-\dfrac{2}{x^3}$ **d** $\dfrac{1}{4}x^{-\frac{3}{4}}-\dfrac{1}{3}x^{-\frac{2}{3}}$

   e $-x^{-2}-6x^{-3}$ **f** $-4x^{-2}-x^{-\frac{3}{2}}$

   g $2x^{-3}-\dfrac{1}{3}x^{-\frac{4}{3}}$ **h** $1-3x^{-2}$

   i $\dfrac{1}{2}x^{-\frac{1}{2}}$ **j** $\dfrac{3}{4}x^{\frac{1}{2}}+\dfrac{1}{4}x^{-\frac{1}{2}}+\dfrac{5}{4}x^{-\frac{3}{2}}$

3 a 8   **b** $-\dfrac{1}{3}$ **c** 2   **d** $\dfrac{1}{6}$

4 a 56 y units per x unit

   b 150π cm³ per cm

   c i $1\,\text{cm s}^{-1}$ **ii** $2\,\text{cm s}^{-2}$

   d i $0.25\,\text{km m}^{-1}$ (to 2 sf); $0.18\,\text{km m}^{-1}$ (to 2 sf)

   ii 20 m (to 2 sf)

   iii $0.13\,\text{km m}^{-1}$ (to 2 sf)

5 a tangent: $y-8=6(x-1)$; normal $y-8=-\dfrac{1}{6}(x-1)$

   b tangent: $y-5=\dfrac{7}{3}(x-3)$; normal $y-5=-\dfrac{3}{7}(x-3)$

   c tangent: $y=9$; normal $x=2$

   d Equation of tangent: $y-1000=20(x-25)\Rightarrow y=20x+500$

   Equation of normal: $y-1000=-\dfrac{1}{20}(x-25)$

6 a $(-2,-9)$ min

   b $\left(\dfrac{2}{3},-\dfrac{16}{9}\right)$ min, $\left(-\dfrac{2}{3},\dfrac{16}{9}\right)$ max

   c $(0,0)$ min

   d $(1,2a)$ min, $(-1,-2a)$ max

   e At the stationary point $x=-\dfrac{b}{2a}$ the turning point is a minimum.

7 a 2   **b** $6x+2$ **c** $\dfrac{2}{x^3}$

   d $\dfrac{dy}{dx}=1+x+\dfrac{x^2}{2}+\dfrac{x^3}{6}+\dfrac{x^4}{24}+\dfrac{x^5}{120}$

   $\Rightarrow \dfrac{d^2y}{dx^2}=1+x+\dfrac{x^2}{2}+\dfrac{x^3}{6}+\dfrac{x^4}{24}$

8 a $x-\dfrac{x^2}{2}-\dfrac{x^4}{4}+c$ **b** $-\dfrac{1}{x}-\dfrac{4x^3}{3}+c$

   c $\dfrac{2x^{\frac{3}{2}}}{3}-2\sqrt{x}-\dfrac{3x^{\frac{4}{3}}}{4}+c$ **d** $\dfrac{2}{5}x^{\frac{5}{2}}+4x^{\frac{3}{2}}+18x^{\frac{1}{2}}+c$

9 a $25\dfrac{1}{2}$ **b** 18π

   c -30 **d** $10\dfrac{2}{3}$

10 a $4\dfrac{1}{2}$ units²   **b** $\dfrac{1}{6}$ units²   **c** $60\dfrac{3}{4}$ units²

Assessment 4

1 a $4x^3-10x$ **b** $-3x^{-4}+21x^2$ **c** $-6x^{-3}-\dfrac{1}{2}x^{-\frac{1}{2}}$

2 a $\dfrac{dr}{dt}=-\dfrac{1}{\sqrt{t}}$ **b** $-\dfrac{1}{5}\,\text{cm s}^{-1}$

3 a 7   **b** $x+7y-8=0$ or $y=-\dfrac{1}{7}x+\dfrac{8}{7}$

4 a x^4-2x **b** 12   **c** $60x-5y-108=0$

5 a $\dfrac{dy}{dx}=0\Rightarrow 18x^2-6x-12=0\Rightarrow x=-\dfrac{2}{3}$

   b $(1,-4)$

   c Minimum

6 $-4<x<\dfrac{2}{3}$

7 a $x^2+\dfrac{1}{2}x^6+c$ **b** $-\dfrac{1}{3}x^{-3}-x^4+c$ **c** $2x^{\frac{3}{2}}+2x^{\frac{1}{2}}+c$

8 a 29   **b** $4\sqrt{3}$ **c** $-\dfrac{15}{4}$

9 a $f(x)=\dfrac{4}{3}x^3+\dfrac{5}{2}x^2-x(+c)$

   b $f(x)=-\dfrac{7}{2}x^{-2}-\dfrac{1}{2}x^2+\dfrac{2}{3}x^{\frac{3}{2}}(+c)$

10 Integration between $x=1$ and $x=3$ gives the result of $\dfrac{4}{3}$

11 a $\dfrac{dy}{dx}=\lim_{h\to0}\left(\dfrac{3(x+h)^2-3x^2}{(x+h)-x}\right)$

   $=\lim_{h\to0}\left(\dfrac{3x^2+6xh+3h^2-3x^2}{h}\right)=\lim_{h\to0}(6x+3h)=6x$

   b 30

12 a $\dfrac{dy}{dx}=\lim_{h\to0}\left(\dfrac{[(x+h)^3-2(x+h)]-[x^3-2x]}{(x+h)-x}\right)$

   $=\lim_{h\to0}\left(\left[3x^2+3xh+h^2-2\right]\right)=3x^2-2$

   b 10

13 a $3x^2+12x+9$ **b** $\dfrac{1}{2}x^{-\frac{1}{2}}-x^{-\frac{3}{2}}$

14 $\dfrac{8}{3}$ litres per min

15 $x-8y-162=0$ or $y=\dfrac{1}{8}x-\dfrac{81}{4}$

16 a $y-2=\dfrac{11}{4}(x-1)$ **b** $\dfrac{9}{88}$

17 $f(x)=1+6x+12x^2+8x^3\Rightarrow f'(x)=6+24x+24x^2$
   $=6(1+2x)^2\ge 0$

18 $0<x<\dfrac{3}{5}$

19 $1<x<\dfrac{5}{3}$

20 a $6x^2 - \dfrac{5}{2}x^{\frac{3}{2}}$ **b** $12x - \dfrac{15}{4}x^{\frac{1}{2}}$

21 $6 + x^{-\frac{3}{2}}$

22 $\dfrac{dy}{dx} = 32 - 4x^{-3}$; $\dfrac{dy}{dx} = 0 \Rightarrow x^{-3} = 8 \Rightarrow x = \dfrac{1}{2} \Rightarrow y = 9$;

$\dfrac{d^2 y}{dx^2} = 12x^{-4} \Rightarrow$ at $x = \dfrac{1}{2}$, $\dfrac{d^2 y}{dx^2} = 192 > 0$ so a minimum

23 a $h = \dfrac{200}{\pi x^2}$

  b $A = 2 \times \pi x^2 + 2\pi xh = 2\pi x^2 + \dfrac{400}{x}$

  c 3.17 cm (to 3 sf)

  d 189.3 cm^2

  e $\dfrac{d^2 A}{dx^2} = 4\pi + \dfrac{800}{x^3} \Rightarrow$ at $x = 3.17$, $\dfrac{d^2 A}{dx^2} > 0$ so a minimum

24 a $x^2 l = 3000 \Rightarrow l = \dfrac{3000}{x^2}$; $A = 4xl + x^2 = 4x \times \dfrac{3000}{x^2} + x^2$

  $= \dfrac{12000}{x} + x^2$

  b 18.2 cm (to 3 sf)

  c 991 cm^2 (to 3 sf)

  d $\dfrac{d^2 A}{dx^2} = \dfrac{24000}{x^3} + 2 \Rightarrow \dfrac{d^2 A}{dx^2} > 0$ so a minimum

25 a $\dfrac{4}{3}x^3 + 6x^2 + 9x + c$

  b $2x^{\frac{5}{2}} - \dfrac{2}{3}x^{\frac{3}{2}} + c$

  c $2x^{\frac{1}{2}} + \dfrac{1}{3}x^{\frac{3}{2}} + c$

26 a $\dfrac{5}{4}$ **b** $\dfrac{206}{15}$

27 $\dfrac{343}{6}$

28 $\dfrac{67 + 8\sqrt{2}}{24}$ (or 3.26)

29 a $x^5 + x^{-2} - 1$ **b** $y - 1 = -\dfrac{1}{3}(x - 1)$

30 a **i** $k + kx^{k-1}$ **ii** $-kx^{-k-1}$

  b **i** $\dfrac{k}{2}x^2 + \dfrac{x^{k+1}}{k+1}(+c)$ **ii** $\dfrac{x^{1-k}}{1-k} - kx(+c)$

31 $P\left(\dfrac{16}{11}, -\dfrac{16}{11}\right)$ **32** 6 **33** $Q\left(-\dfrac{2}{3}, \dfrac{32}{9}\right)$

34 $(0, 1)$ maximum, $(-1, 0)$ and $(1, 0)$ minimums

35 $(0, 20)$ maximum, $(1, 15)$ and $(-2, -12)$ minimums

36 $x > \dfrac{3}{8}$ **37** $-3 < x < 0$ or $x > 5$

38 270 cm^2 (to 3 sf) **39** 2010 cm^3 (to 3 sf)

40 a $y - 2 = 5(x + 1)$ **b** $\dfrac{7}{2}$

41 $\dfrac{4}{3}$ **42** $\dfrac{243}{2}$

43 $\dfrac{37}{12}$ **44** $\dfrac{36 - 4\sqrt{3}}{9}$ (or 3.23)

45 6 **46** $\dfrac{9}{2}$

Chapter 5
Exercise 5.1A Fluency and skills

1 a $\log_2 8 = 3$ **b** $\log_3 9 = 2$

  c $\log_{10} 1000 = 3$ **d** $\log_{16} 2 = \dfrac{1}{4}$

  e $\log_{10} 0.001 = -3$ **f** $\log_4 \dfrac{1}{4} = -1$

g $\log_2 \dfrac{1}{8} = -3$ **h** $\log_8 \dfrac{1}{4} = -\dfrac{2}{3}$

2 a $2^5 = 32$ **b** $2^4 = 16$ **c** $3^4 = 81$ **d** $5^0 = 1$

  e $3^{-2} = \dfrac{1}{9}$ **f** $6^1 = 6$ **g** $16^{\frac{1}{4}} = 2$ **h** $2^{-6} = \dfrac{1}{64}$

3 a 2 **b** 4 **c** 2 **d** 1

  e 4 **f** 1 **g** 4 **h** -1

4 a $\log 12$ **b** $\log 6$ **c** $\log 4$

  d $\log 32$ **e** $\log 108$ **f** $\log 72$

  g $\log 16$ **h** $\log 8$ **i** $\log 6$

  j $\log 18$ **k** $\log(x^4 y^2)$ **l** $\log(x^2 y)$

5 a $2\log a + \log b$ **b** $\log a - \log b$

  c $2\log a - 3\log b$ **d** $\log a + \dfrac{1}{2}\log b$

  e $\dfrac{1}{2}\log a + \dfrac{1}{2}\log b - \log c$ **f** $\dfrac{1}{2}(\log a + \log b + \log c)$

  g $\log a + \dfrac{1}{2}\log b - \dfrac{1}{2}\log c$

6 a $\log 3 + 2\log 2 \approx 1.079$ **b** $\log 2 + 2\log 3 \approx 1.255$
  c $2\log 3 - \log 2 \approx 0.653$ **d** $3\log 3 - \log 2 \approx 1.130$
  e $1 - \log 2 \approx 0.699$ **f** $0 - 3\log 2 \approx -0.903$

7 a $\dfrac{3\log 5}{2\log 5} = \dfrac{3}{2} = 1.5$ **b** $\dfrac{3\log 3}{5\log 3} = \dfrac{3}{5} = 0.6$

8 $\log 40 = 3 - 2\log 5 = 1.602060$

Exercise 5.1B Reasoning and problem-solving

1 a 2.32 **b** 2.10 **c** 0.861
  d 1.81 **e** 4.99 **f** 2.66
  g -3.17 **h** -2.41 **i** 0.461
  j 0.892

2 a 520 **b** 7 **c** $\dfrac{2}{7}$ **d** 1
  e 5 **f** 64 **g** 1

3 Adding gives $2\log x = 5$

  $\log x = \dfrac{5}{2} = 2.5$

  $x = 10^{2.5} = 316$ (3 sf)

  $y = 10^{0.5} = 3.16$ (3 sf)

4 a 0, 1 **b** 1, 2 **c** 1, 1.58
  d 2 **e** 1 **f** 2
  g 1.32, 2 **h** 0.631, 1.26 **i** 0, 0.861
  j 1.16

5 a 1 or 0.631 **b** $(1, 5)$ and $\left(0.631, \dfrac{10}{3}\right)$

6 $(2.32, 20)$

7 a **i** $x > 8.38$ **ii** $x > 4.29$
  b 20
  c 7

8 a 4 **b** 14

9 a $\dfrac{x}{y} = \dfrac{a^p}{a^q} = a^{p-q}$

  So $\log_a\left(\dfrac{x}{y}\right) = p - q$

  $= \log_a x - \log_a y$

  b $x^k = (a^p)^k = a^{pk}$

  So $\log_a(x^k) = pk$

  $= k \times \log_a x$

10 a Let $y = \log_a b$ so $b = a^y$

  Take logs base b on both sides

  $\log_b b = \log_b(a^y)$

  $1 = y\log_b a$

  $1 = \log_a b \times \log_b a$

  So $\log_a b = \dfrac{1}{\log_b a}$

  b **i** $25, \dfrac{1}{25}$ **ii** $9, 81$

11 a Let $x = a^y$

Take logs base b on both side

$\log_b x = \log_b(a^y)$

$\log_b x = y \log_b a$

$\dfrac{\log_b x}{\log_b a} = y$

So $\dfrac{\log_b x}{\log_b a} = \log_a x$

b $\log_2 12 = 3.32 \times \log_{10} 2$

c i $\dfrac{\log_{10} 37}{\log_{10} 6} = 2.015$

ii $\dfrac{\log_{10} 6}{\log_{10} 4} = 1.292$

iii $\dfrac{\log_{10} 25}{\log_{10} 3} = 2.930$

12 $xyz = \log_y z \times \log_z x \times \log_x y$

$= \log_y z \times \log_z x \times \dfrac{1}{\log_y x}$ (From Q10a)

$= \dfrac{\log_y z}{\log_y x} \times \log_z x$ (Rearranging)

$= \log_x z \times \dfrac{1}{\log_x z}$ (From Q11a and Q10a)

$= 1$

Exercise 5.2A Fluency and skills

1

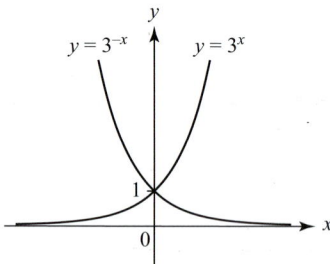

$\left(\dfrac{1}{3}\right)^x = (3^{-1})^x = 3^{-x}$

2 a 2.54 (to 3 sf)   **b** 0.395 (to 3 sf)

c −0.506 (to 3 sf)   **d** 3.47 (to 3 sf)

3 a i $y_1 = 2x$

$y_2 = x^2$

$y_3 = 2^x$

ii y_3 is exponential

iii 10, 25, 32

b i $y_1 = \dfrac{12}{x}$

$y_2 = \sqrt{x}$

$y_3 = 2^{-x}$ or $\left(\dfrac{1}{2}\right)^x$

ii y_3 is exponential

iii 2.4, 2.236, $\dfrac{1}{32}$

4 a

x	−2	−1	1	2	3	4
y	$\dfrac{1}{9}$	$\dfrac{1}{3}$	3	9	27	81

$y = 3^x$

b

x	−2	−1	1	2	3	4
y	25	5	$\dfrac{1}{5}$	$\dfrac{1}{25}$	$\dfrac{1}{125}$	$\dfrac{1}{625}$

$y = \left(\dfrac{1}{5}\right)^x$ or 5^{-x}

5 a

x	y	gradient at (x, y)
0	1	1
1	2.72	2.72
2	7.39	7.39
3	20.1	20.1
4	54.6	54.6

b

x	y	gradient at (x, y)
0	1	3
1	20.1	60.3
2	403	1210
3	8100	24300
4	163000	488000

6 WC, XD, YA, ZB

7 a i $e^2 = 7.39$ **ii** $-e^{-2} = -0.14$

b i $y = e^2 x - e^2$ **ii** $y = \dfrac{3-x}{e^2}$

8 a $y = e^3(x-2)$ **b** $y = \sqrt{e}\left(x + \dfrac{1}{2}\right)$

9 a Reflection in y-axis   **b** Reflection in x-axis

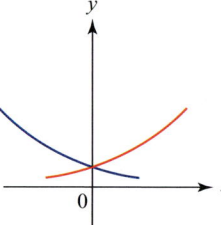 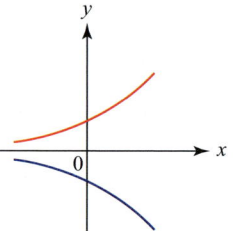

c Stretch (sf = 2) parallel to y-axis   **d** Stretch (sf = $\dfrac{1}{2}$) parallel to x-axis

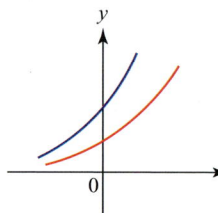 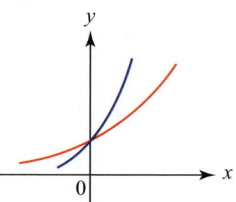

e Translation of $\begin{pmatrix} 0 \\ 1 \end{pmatrix}$ **f** Translation of $\begin{pmatrix} 0 \\ -1 \end{pmatrix}$

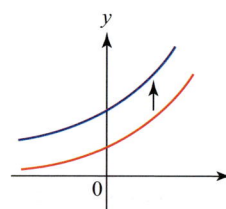 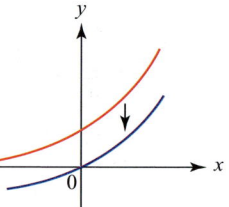

g Translation of $\begin{pmatrix} -1 \\ 0 \end{pmatrix}$ **h** Translation of $\begin{pmatrix} 1 \\ 0 \end{pmatrix}$

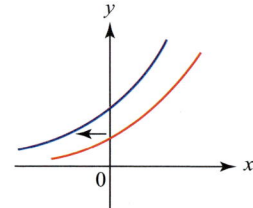 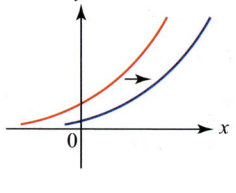

Answers For full solutions go to http://www.oxfordsecondary.co.uk/edexcelalevelmaths-answers

10 a

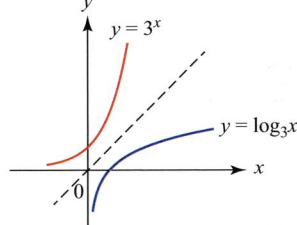

$y = 3^x$

$y = \log_3 x$

b $(3, 1)$

c $(1.2, 3.74)$ and $(3.74, 1.2)$
$\log_3 3.74 = 1.2$

d i 5.20 **ii** 1.50 **iii** 1.73
iv 0.50 **v** 1.63 **vi** 1.77

11 a $p = e^3$, $q = 2$, $r = \ln 9$

b $a = e^{-\frac{1}{2}}$, $b = -3$, $c = -\ln 9$

Exercise 5.2B Reasoning and problem-solving

1 a

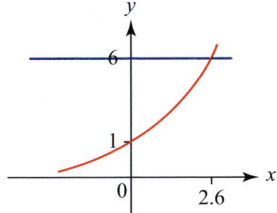

$(2.6, 6)$ so approximately 2.6

b

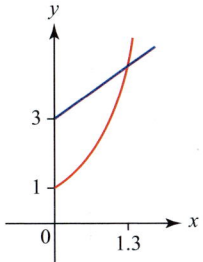

$(1.3, 4.3)$ so approximately 1.3

2 a

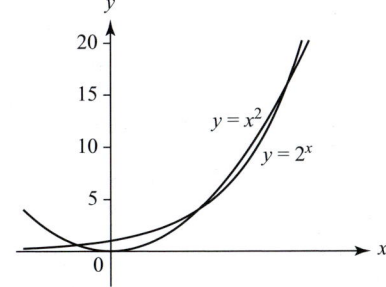

$y = x^2$

$y = 2^x$

There are 3 points of intersection.

b $x = -0.75, 2, 4$

3 a $(3, 4^9)$ **b** $(1, 4)$ **c** $(4, 81)$ **d** $(-1, 2)$

4 7.25

5 $\dfrac{y - e^3}{x - 1} = 3e^3$

$y - e^3 = 3e^3 x - 3e^3$

$y = 3e^3 x - 2e^3$

When $y = 0$, $x = \dfrac{2}{3}$

Tangent passes through $\left(\dfrac{2}{3}, 0\right)$

6 $(6, 2e^2)$

7 $y = 3.695x - 3.559$

8 $(2e^4 + 1, 0)$

9 a $y = \log_a x$

b
$$y = 2^x$$
$$\ln y = \ln 2^x$$
$$\ln y = x \ln 2$$
$$x = \frac{\ln 2}{\ln y}$$
$$y = \frac{\ln 2}{\ln x}$$
$$y = \frac{1}{\ln 2} \times \ln x$$
$$\frac{1}{\ln 2} \approx 1.44$$

So if $2^x = 17$ then $x = 1.44 \times \ln 17 = 4.08$

10 a

a	2.0	2.2	2.4	**2.6**	**2.8**	3
f(a)	0.693	0.788	0.876	**0.956**	**1.030**	1.099

b Between 2.6 and 2.8, because 1 is in between 0.956 and 1.030

c

a	2.6	**2.7**	**2.8**
f(a)	0.956	**0.993**	**1.030**

a	2.7	**2.71**	**2.72**	2.73	2.74	2.75
f(a)	0.993	**0.997**	**1.001**	1.004	1.008	1.012

a	**2.715**
f(a)	**0.9988**

$e = 2.72$ to 2 dp

d $\dfrac{a^{\delta x} - 1}{\delta x}$ is the gradient of a chord and only approximates the gradient of the tangent.

11 a

x	1	2	3	e.g. 12
$y = e^x$	2.7183	7.3891	20.086	162755
Gradient	2.7184	7.3894	20.087	162763

b As in Q10, the gradient is of a chord not a tangent.

Exercise 5.3A Fluency and skills

1 a

t	0	1	2	3	4
n	1	2	4	8	16

b 64

2 a $A = 10$

b

t	0	2	4	6	8
Actual units	10.0	9.1	8.2	7.4	6.7
n	10	9.03	8.15	7.35	6.63

Yes, predictions are reasonably accurate.

c Yes, about 44% of the initial insulin is still present.

3 a i 2 **ii** 109

b i 15 **ii** 6050

c i −5 **ii** −4.52

x and y grow over time. z decays over time.

4 a $0.049\,\text{cm}\,\text{h}^{-1}$ **b** $0.0074\,\text{kg}\,\text{h}^{-1}$

5 a i $\dfrac{1}{5}\,°\text{C}$ **ii** $1.48\,°\text{C}$

b $0.639\,°\text{C}\,\text{min}^{-1}$

c $1.48\,°\text{C}\,\text{min}^{-1}$

d $80.7\,°\text{C}$

6 a 21.5 millions

b 1.29 millions per year

c 1.08 millions per year

7 a $2\,\text{mg ml}^{-1}$

b $0.331\,\text{mg ml}^{-1}$

c i 0.9 (mg/ml)/h   **ii** 0.149 (mg/ml)/h

8 a After 1 year, cost $= £P\left(1 + \dfrac{5}{100}\right)$

$= £P(1.05)$

After 2 years, cost $= £P(1.05) \times 1.05$

$= £P \times (1.05)^2$

b £166.15

9 £92.62

10 $k = 0.0105$, $t = 65.8$

Exercise 5.3B Reasoning and problem-solving

1 a 20

b Model gives $100 - 80\mathrm{e}^{-1}$
$= 70.6$, which is reasonably accurate (roughly 2% error). Yes, model fits data.

c 87 insects

d As $t \to \infty$, $\mathrm{e}^{-\frac{t}{5}} \to 0$
So yes, the model does predict a limiting number and the predicted limiting number $= 100$

2 a $A = 5 - 2 = 3$

t	0	20	40	60	80	100
n	5	11	25	60	140	265
p	5	10	24	62	166	447

b 29.3 hours

c Accurate for $t < 60$
Fairly accurate for $t = 60$
Not accurate for $t = 100$
No limit is placed on the number of bacteria. It would increase to infinity according to the model.

d 1.1 bacteria per hour

3 a Model B. The population of deer would never be a negative number.

b Model A. This model correctly show exponential decay. Model B exhibits exponential growth.

c Model A. When $t = 1.4$, the area of bacteria reaches the area of the petri dish, and so the area of bacteria would not continue to grow exponentially after this time.

4 a 180

b 20, 9 trees/yr

c 5 years

d As $t \to \infty$, $\mathrm{e}^{-\frac{t}{20}} \to 0$
So $N \to 200$

5 a 393 minutes

b $2.26\,^\circ\text{C min}^{-1}$

c As $t \to \infty$, $\mathrm{e}^{-0.005t} \to 0$
So $\theta \to 30$
Minimum temperature $= 30\,^\circ\text{C}$

d Any valid reason e.g. the minimum temperature to which the block of steel cools may change depending on environmental factors. The block of steel will cool to the ambient temperature which may be less than 30°C.

6 a $k = 5.0$, $A = 1500$

b $n = \dfrac{A\mathrm{e}^{\frac{t}{4}}}{\mathrm{e}^{\frac{t}{4}} + k} = \dfrac{A}{1 + k\mathrm{e}^{-\frac{t}{4}}}$

$1 + k\mathrm{e}^{-\frac{t}{4}} > 1$ so $\dfrac{A}{1 + k\mathrm{e}^{-\frac{t}{4}}} < A$ so the population cannot exceed 1500

Exercise 5.4A Fluency and skills

1 $y \approx 2.5x^{1.5}$

2 $y \approx 4 \times 2.5^x$

3 a i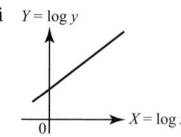

$X = \log x$	0.301	0.477	0.602	0.699	0.778
$Y = \log y$	1.23	1.60	1.87	2.07	2.24

ii $a = 4$, $n = 2.1$, $y = 4 \times x^{2.1}$

b i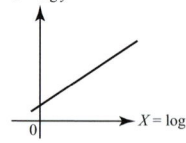

$X = \log x$	0	0.301	0.477	0.602	0.699
$Y = \log y$	0.204	0.362	0.462	0.531	0.580

ii $a = 1.6$, $n = 0.54$, $y = 1.6 \times x^{0.54}$

c i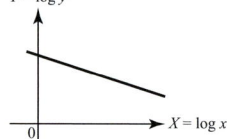

$X = \log x$	1	1.176	1.301	1.398	1.477
$Y = \log y$	2.58	2.52	2.48	2.46	2.43

ii $a = 750$, $n = -0.3$, $y = 750 \times x^{-0.3}$

d i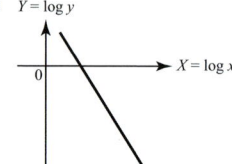

$X = \log x$	0.301	0.602	0.778	0.903
$Y = \log y$	0.236	−0.244	−0.523	−0.721

ii $a = 5.2$, $n = -1.6$, $y = 5.2 \times x^{-1.6}$

4 a i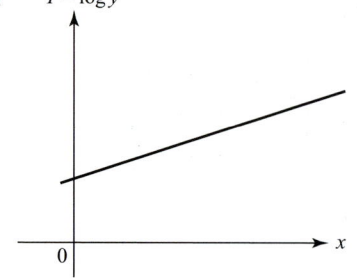

x	3	5	7	9
$Y = \log y$	1.60	2.25	2.89	3.53

ii $b = 2.1$, $k = 4.3$, $y = 4.3 \times 2.1^x$

b i $Y = \log y$

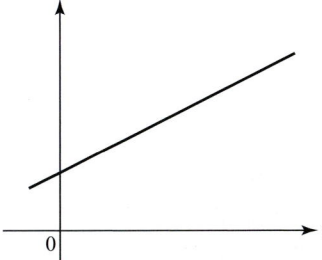

x	0.2	0.3	0.4	0.5	0.6
$Y = \log y$	0.914	0.964	1.017	1.064	1.117

   ii $b = 3.2$, $k = 6.5$, $y = 6.5 \times 3.2^x$

c i $Y = \log y$

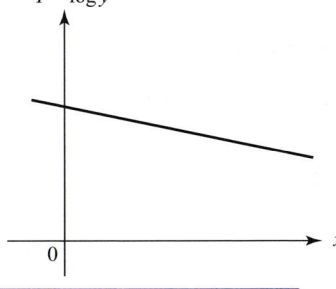

x	2	2.5	3	3.5	4
$Y = \log y$	2.24	2.14	2.04	1.94	1.84

   ii $b = 0.625$, $k = 450$, $y = 450 \times 0.625^x$

d i $Y = \log y$

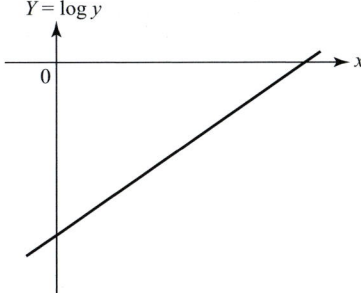

x	0.3	0.4	0.5	0.6
$Y = \log y$	−0.481	−0.409	−0.337	−0.268

   ii $b = 5.2$, $k = 0.2$, $y = 0.2 \times 5.2^x$

5

x	0.5	1	2	3
y	0.66	1.08	2.92	7.87
$X = \log x$	−0.301	0	0.301	0.477
$Y = \log y$	−0.180	0.033	0.465	0.896

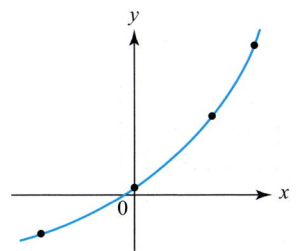

$Y = \log y$

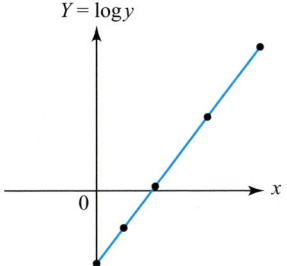

Graph of $Y = \log y$ against x gives straight line so function has the form $y = kb^x$

$y = 0.4 \times 2.7^x$

Exercise 5.4B Reasoning and problem-solving

Some numbers here are subjective so answers might vary.

1 a

$X = \log x$	0.079	0.114	0.146	0.176
$Y = \log y$	0.763	0.881	0.940	1.000

$Y = \log y$

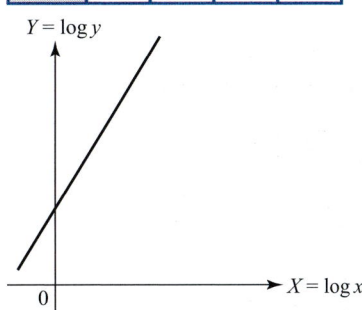

$n = 2.4$, $a = 3.9$, $y = 3.9 \times x^{2.4}$

b i $8\,\text{m}^3$ **ii** $64\,\text{m}^3$

c (i) is more reliable, because 1.35 is within the range of the task 1.2 to 1.5 (interpolation).
3.2 is well outside 1.2 to 1.5 and liable to inaccuracy (extrapolation).

2

$Y = \log t$	0.301	0.477	0.602	0.699
$Y = \log A$	1.23	1.60	1.86	2.07

a $n = 2.1$, $a = 4.0$ to 2 sf, $A = 4.0 \times t^{2.1}$

b 1 second is only slightly outside the data range 2 to 5 seconds. 1 min (60 sec) is well outside the data range 2 to 5 seconds.

3

x	0	5	15	20
p	4140	5000	8400	10200
$Y = \log p$	3.617	3.699	3.924	4.009

a $Y = \log p$

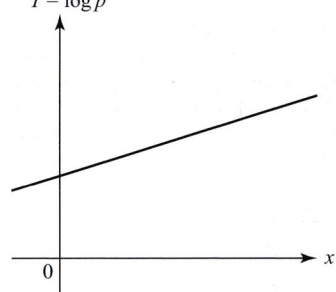

$b = 1.05$, $k = 4140$, $p = 4140 \times 1.05^x$

b $x = 10, p = 6740$

$x = 30, p = 17\,900$

6740 is better as 10 lies in the range 0 to 20 (interpolation).

17 900 is less good as 30 lies outside the range 0 to 20 (extrapolation).

4

v	200	250	300	350	400
p	78.0	56.5	43.7	36.0	29.7
$X = \log v$	2.301	2.398	2.477	2.544	2.602
$Y = \log p$	1.892	1.752	1.640	1.556	1.473

a $n = -1.40, k = 119000, p = 119000 \times v^{-1.4}$

b **i** $v = 308$ **ii** $v = 973$

c (i) is more reliable as it is found by interpolation, whereas (ii) is found by extrapolation.

5 $M = \log m$

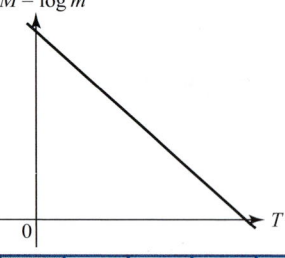

T	5	10	15	30
M	47.6	45.3	43.1	38.0
$\log m$	1.677	1.656	1.634	1.580

a $m = 50 \times 0.991^t$

b $m = 50\,\text{grams}$

c 77 hours

Review exercise 5

1 **a** $\log_2 32 = 5$

   b $\log_{10} 0.0001 = -4$

2 **a** $2^4 = 16$ **b** $4^{-2} = \dfrac{1}{16}$

3 **a** 5   **b** 3

4 $\log 27$

5 $2\log a - \dfrac{1}{2}\log b$

6 **a** $x = 1.79$ **b** $x = 3$ or 0 **c** $x = 57.5$

7 **a** 19   **b** 16

8 **a** $y = e^3 x - 2e^3 = e^3(x - 2)$ **b** $y = \dfrac{1}{2}ex, \left(1, \dfrac{e}{2}\right)$

9 Stretch sf $\dfrac{1}{2}$ parallel to x-axis

10 $a = e^3, b = \ln 4$

11 $k = 0.910$

12 $y = \dfrac{1}{2} - 2x$

13 $y = 1 - x + e^x$

Equation of normal is $y = -\dfrac{x - 1}{e - 1} + e$

When $x = e^2$, $y = -(e + 1) + e = -1$

So normal passes through the point $(e^2, -1)$

14

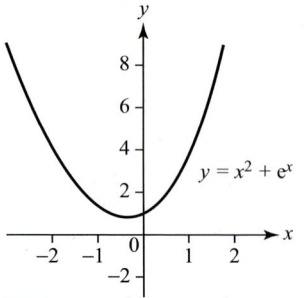

Minimum at $(-0.35, 0.83)$

15 **a** After 1 year, $A = \left(1 + \dfrac{r}{100}\right) \times 250$

After 2 years,

$$A = 250\left(1 + \dfrac{r}{100}\right) + \dfrac{r}{100} \times 250\left(1 + \dfrac{r}{100}\right)$$

$$= 250\left(1 + \dfrac{r}{100}\right)\left(1 + \dfrac{r}{100}\right)$$

$$= 250\left(1 + \dfrac{r}{100}\right)^2$$

etc. until after n years $A = 250\left(1 + \dfrac{r}{100}\right)^n$

b $A = 303.88$

c 12 years

16 **a**

x	1.0	2.0	3.0	4.0
y	15.0	23.7	43.4	70.2
$Y = \log y$	1.176	1.375	1.637	1.846

$b = 1.67, k = 9.0, y = 9.0 \times 1.67^x$

These numbers are subjective so answer might vary.

b Model gives $9.0 \times 1.67^{7.5} = 421$

This is not sufficiently close to 372, so the model does not predict the results accurately enough.

17 Line of best fit is $y = 3.58 \times x^{2.33}$

x	4	4.8	5.6	6.4
y	90	140	198	270
Model	90.6	138	198	271

$a = 3.58, b = 2.33$

These numbers are subjective so answer might vary.

18 No constraint on the number of deer – it would increase to infinity according to the model.

Continuous model but number of deer is discrete.

Model ignores external factors such as disease, predation, and limited food supply.

Assessment 5

1

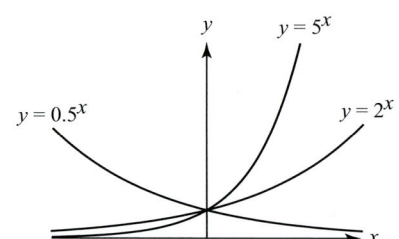

2 **a**

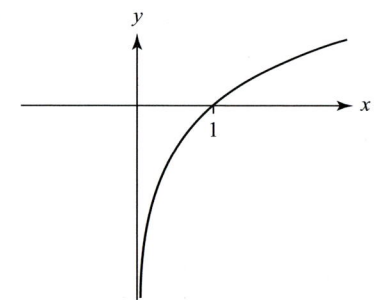

Asymptote is $x = 0$

b **i** 27   **ii** $\dfrac{1}{9}$ **iii** $\sqrt{3}$

3 **a** $\log_5 56$ **b** $\log_5 3$ **c** $\log_5 0.8$

4 **a** **i** 4   **ii** -1

   b **i** $\log_n 75$ **ii** $\log_n\left(\dfrac{n}{5}\right)$

5 **a** $x = 1.43$ **b** $x = 1.61$

Answers For full solutions go to http://www.oxfordsecondary.co.uk/edexcelalevelmaths-answers

6 a $x = 12$　　**b** $x = 3$
7 a $B = 200$
　b $t = 1.3$ hours
　c E.g. No constraint on the number of bacteria (i.e. would go off to infinity according to the model).
　　Continuous model but number of bacteria is discrete.
　　Ignores extraneous factors such as limited dish size.
8 $t = 9.9$ years　　**9** $k = \sqrt{15}$
10 $a = 2$, $n = 4$
11 a

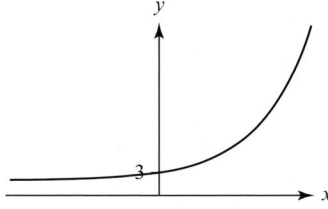

　　Asymptote is $y = 2$
　b $x = \frac{1}{2}\ln 5$ or $\ln\sqrt{5}$
12 a i $2p$
　　ii $-p$
　　iii $p + 1$
　b $n = \frac{1}{3}$
13 a $x = 4$　　**b** $x = \frac{9}{8}$　　**c** $x = \sqrt{2e^{2.5}}$ or 17.2
14 a 500
　b i $a = \frac{2}{\sqrt{3}}$ or $\frac{2\sqrt{3}}{3}$ or 1.15　　**ii** $t = 9.64 \approx 10$ years
　c $P = \dfrac{1500 a^t}{2 + a^t} = \dfrac{1500}{\frac{2}{a^t} + 1}$
　　But $\dfrac{2}{a^t} + 1 > 1$ so $\dfrac{1500}{\frac{2}{a^t} + 1} < 1500$
　　So the population can never exceed 1500
　d E.g. Model is continuous but population of a species is discrete.
15 a $50\,\text{g}$　　**b** $t = 2.31$ days
16 a $a = 5$, $b = 0.5$
　b

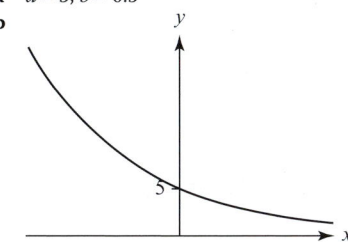

17 40.2
18 $x = 1$
19 a $r = 0.079$ per hr, $C_0 = 10$ mg per litre
　b $t = 13$ hours

Chapter 6

Exercise 6.1A Fluency and skills

1 a $\mathbf{q}$　　**b** $\mathbf{p}$　　**c** $-\mathbf{p}$　　**d** $-\mathbf{q}$
　e $\mathbf{p} + \mathbf{q}$　　**f** $\mathbf{q} - \mathbf{p}$　　**g** $\mathbf{p} - \mathbf{q}$
2 a $\mathbf{p} + \mathbf{q}$　　**b** $2\mathbf{p} + \mathbf{q}$　　**c** $\mathbf{q} - 2\mathbf{p}$　　**d** $\mathbf{p} + \frac{1}{2}\mathbf{q}$
3 a $2\mathbf{q}$　　**b** $\mathbf{p} + \mathbf{q}$　　**c** $\mathbf{p} + 2\mathbf{q}$　　**d** $\mathbf{q} - \mathbf{p}$
4 a Magnitude = 25.6, direction = 038.7°
　b Magnitude = 10.3, direction = 078.1°
　c Magnitude = 7.88, direction = 231.4°

5 a Magnitude = 12.1, direction = 17.0° to $\mathbf{p}$
　b Magnitude = 5.69, direction = 38.4° to $\mathbf{p}$
6 a Resultant is 7.02 km on a bearing of 098.8°
　b Resultant is 23.4 km on a bearing of 217.3°
　c Resultant is 23.5 km h^{-1} on a bearing of 290.6°
　d Resultant is 529 N on a bearing of 347.5°

Exercise 6.1B Reasoning and problem-solving

1 a $\mathbf{p} - 2\mathbf{q}$　　**b** $\frac{1}{2}\mathbf{p}$
　c DE is parallel to BC and half as long.
2 $\overrightarrow{BC} = \mathbf{q} - \mathbf{p}$ so $\overrightarrow{BD} = \frac{1}{2}(\mathbf{q} - \mathbf{p})$
　$\overrightarrow{AD} = \overrightarrow{AB} + \overrightarrow{BD} = \mathbf{p} + \frac{1}{2}(\mathbf{q} - \mathbf{p}) = \frac{1}{2}(\mathbf{p} + \mathbf{q})$
　$\overrightarrow{AG} = \frac{2}{3}\overrightarrow{AD} = \frac{1}{3}(\mathbf{p} + \mathbf{q})$
　$\overrightarrow{BG} = \overrightarrow{BA} + \overrightarrow{AG} = -\mathbf{p} + \frac{1}{3}(\mathbf{p} + \mathbf{q}) = \frac{1}{3}(\mathbf{q} - 2\mathbf{p})$
　But　$\overrightarrow{BE} = \frac{1}{2}\mathbf{q} - \mathbf{p} = \frac{1}{2}(\mathbf{q} - 2\mathbf{p})$
　So　$\overrightarrow{BG} = \frac{2}{3}\overrightarrow{BE}$
　Hence $\overrightarrow{BG}$ and $\overrightarrow{BE}$ are collinear, and $BG{:}GE = 2{:}1$
3 $\overrightarrow{AB} = \mathbf{b} - \mathbf{a}$ so $\overrightarrow{AD} = \frac{1}{2}(\mathbf{b} - \mathbf{a})$
　$\overrightarrow{PD} = \overrightarrow{PA} + \overrightarrow{AD} = \mathbf{a} + \frac{1}{2}(\mathbf{b} - \mathbf{a}) = \frac{1}{2}(\mathbf{a} + \mathbf{b})$
　Similarly $\overrightarrow{PE} = \frac{1}{2}(\mathbf{b} + \mathbf{c})$ and $\overrightarrow{PF} = \frac{1}{2}(\mathbf{c} + \mathbf{a})$
　$\overrightarrow{PD} + \overrightarrow{PE} + \overrightarrow{PF} = \frac{1}{2}(\mathbf{a} + \mathbf{b}) + \frac{1}{2}(\mathbf{b} + \mathbf{c}) + \frac{1}{2}(\mathbf{c} + \mathbf{a}) = \mathbf{a} + \mathbf{b} + \mathbf{c}$
　Hence $\overrightarrow{PD} + \overrightarrow{PE} + \overrightarrow{PF} = \overrightarrow{PA} + \overrightarrow{PB} + \overrightarrow{PC}$
4 $\overrightarrow{EF} = \overrightarrow{EB} + \overrightarrow{BF} = \frac{1}{2}\mathbf{p} + \frac{1}{2}\mathbf{q}$
　$\overrightarrow{AD} = \mathbf{p} + \mathbf{q} + \mathbf{r}$
　$\overrightarrow{HG} = \overrightarrow{HD} + \overrightarrow{DG} = \frac{1}{2}(\mathbf{p} + \mathbf{q} + \mathbf{r}) - \frac{1}{2}\mathbf{r} = \frac{1}{2}\mathbf{p} + \frac{1}{2}\mathbf{q}$
　Hence $\overrightarrow{EF} = \overrightarrow{HG}$, so $\overrightarrow{EF}$ and $\overrightarrow{HG}$ are parallel.
　$\overrightarrow{FG} = \frac{1}{2}\mathbf{q} + \frac{1}{2}\mathbf{r}$
　$\overrightarrow{EH} = \overrightarrow{EA} + \overrightarrow{AH} = -\frac{1}{2}\mathbf{p} + \frac{1}{2}(\mathbf{p} + \mathbf{q} + \mathbf{r}) = \frac{1}{2}\mathbf{q} + \frac{1}{2}\mathbf{r}$
　Hence $\overrightarrow{FG} = \overrightarrow{EH}$, so FG and EH are parallel.
　Hence $EFGH$ is a parallelogram.
5 a $\overrightarrow{AF} = 3\mathbf{p} + \lambda(-3\mathbf{p} + \mathbf{q})$
　b $\overrightarrow{AF} = 2\mathbf{q} + \mu(-2\mathbf{q} + \mathbf{p})$
　c $\lambda = 0.8$ and $\mu = 0.6$
6 a 6 km h^{-1}　　**b** 10.4 km h^{-1}
7 a It will travels 50 m downstream.
　b The boat should steer upstream at 60° to the bank.
8 a 145.9°　　**b** 238 km h^{-1}
9 a The velocity is 15 km h^{-1} on a bearing of 036.9°
　b Steer on a bearing of 253.7°

Exercise 6.2A Fluency and skills

1 a $5.80\mathbf{i} + 3.91\mathbf{j}$　　　　**b** $2.00\mathbf{i} + 9.39\mathbf{j}$
　c $8.92\mathbf{i} + 8.03\mathbf{j}$　　　　**d** $4\mathbf{j}$
　e $-3.64\mathbf{i} + 7.46\mathbf{j}$　　　**f** $-10.1\mathbf{i} + 21.8\mathbf{j}$
　g $-3.56\mathbf{i} - 5.08\mathbf{j}$　　　**h** $12.5\mathbf{i} - 6.36\mathbf{j}$
2 a 5.39, $21.8°$　　**b** 11.4, $52.1°$　　**c** 5, $-90°$
　d 3.61, $124°$　　**e** 5.83, $-59.0°$　　**f** 7.81, $-140°$
　g 2, $180°$

3 Magnitude is 4.66 and direction is 5.43° to positive x-direction.

4 **a** $2\mathbf{j}$ **b** $-2\mathbf{i} - 2\mathbf{j}$ **c** $-6\mathbf{i} + 7\mathbf{j}$
  d $16\mathbf{i} + \mathbf{j}$ **e** $\sqrt{5} = 2.24$ **f** $\sqrt{20} = 4.47$

5 **a** $u = -10, v = -1$ **b** $u = -4.5$

6 **a** $-12\mathbf{i} + 16\mathbf{j}$ **b** $-0.6\mathbf{i} + 0.8\mathbf{j}$

7 **a** $a = \dfrac{2}{3}, b = 3\dfrac{1}{3}$ **b** $a = 2, b = 1$ or $a = -2, b = -1$

8 17 m (by Pythagoras)

9 $\overrightarrow{AB} = 6\mathbf{i} + 4\mathbf{j}$

  a $AB = \sqrt{52} = 7.21$ m **b** $\tan^{-1}\left(\dfrac{4}{6}\right) = 33.7°$

10 a 9.85 m **b** 42.8°

11 a $-7\mathbf{i} + \mathbf{j}$ **b** $-6\mathbf{i} + 3\mathbf{j}$

12 $\mathbf{c} = \mathbf{b} + \overrightarrow{BC} = \mathbf{b} + \overrightarrow{AD} = \mathbf{b} + \mathbf{d} - \mathbf{a} = \begin{pmatrix} 21 \\ 22 \end{pmatrix}$

13 $\mathbf{b} = \mathbf{a} + 0.7\overrightarrow{AC} = \mathbf{a} + 0.7(\mathbf{c} - \mathbf{a}) = 3.4\mathbf{i} + 9.6\mathbf{j}$

Exercise 6.2B Reasoning and problem-solving

1 Vectors between each successive checkpoint can be shown to be parallel and share a common point, so they are collinear.

2 Distance = 10.8 m

3 The diagram shows the two possible situations.
  C_1 lies between A and B
  $\mathbf{c}_1 = \mathbf{a} + \dfrac{3}{5}(\mathbf{b} - \mathbf{a}) = 7.8\mathbf{i} + 4\mathbf{j}$
  C_2 lies on AB produced, with $AC_2 = 3AB$
  $\mathbf{c}_2 = \mathbf{a} + 3(\mathbf{b} - \mathbf{a}) = 27\mathbf{i} + 16\mathbf{j}$

4 Vectors $\overrightarrow{BC}$ and $\overrightarrow{CD}$ can be shown to be parallel and share a common point, so they are collinear.

5 $k = 3$ or $-\dfrac{2}{3}$

6 **a** $((2 + 0.8d)\mathbf{i} + 0.6d\mathbf{j})$ m **b** 8.60 m

7 Journey would reduce by 7.6 km. The tunnel should not be built.

8 **a** $\mathbf{a}' = (1 + 3t)\mathbf{i} + (4 + 3t)\mathbf{j}$
  b $\overrightarrow{A'B'} = (4 - t)\mathbf{i} + (0.5t - 2)\mathbf{j}$
  c They collide after 4 s at $\mathbf{a}' = 13\mathbf{i} + 16\mathbf{j}$ m

9 After t s the first particle is at $2t\mathbf{i} + (4 - t)\mathbf{j}$ and the second is at $(6 - t)\mathbf{i} + (8 - 3t)\mathbf{j}$. When $t = 2$, both particles are at $(4\mathbf{i} + 2\mathbf{j})$ m

10 $\mathbf{v} = 7\mathbf{i} + 6\mathbf{j}$

Review exercise 6

1 **a** scalar **b** vector **c** vector
  d vector **e** scalar

2 **a** **i** $3\mathbf{p}$ **ii** $2\mathbf{q} - 3\mathbf{p}$ **iii** $3\mathbf{p} - 3\mathbf{q}$
  b $\overrightarrow{FE} = \mathbf{p} - \mathbf{q}$, so $\overrightarrow{DF} = 3\overrightarrow{FE}$
    They are parallel vectors passing through F, so the points are collinear.

3 **a** $\mathbf{p} + \mathbf{q}$ **b** $2\mathbf{q}$ **c** $2\mathbf{p} + \mathbf{q}$ **d** $\mathbf{q} - \mathbf{p}$

4 **a** $\mathbf{q} - \mathbf{r}$ **b** $\dfrac{1}{2}(\mathbf{q} - \mathbf{r})$ **c** $\dfrac{1}{2}(\mathbf{q} + \mathbf{r})$
  d $\mathbf{r} - \mathbf{p}$ **e** $\dfrac{3}{4}(\mathbf{r} - \mathbf{p})$ **f** $\dfrac{1}{4}(\mathbf{p} + 3\mathbf{r})$

5 **a** $\theta = 74.9°$
  b Magnitude = 11.2 m s^{-1}

6 **a** $\mathbf{P} = 5.14\mathbf{i} + 6.13\mathbf{j}$
    $\mathbf{Q} = -1.74\mathbf{i} + 9.85\mathbf{j}$
    $\mathbf{R} = 5.64\mathbf{i} - 2.05\mathbf{j}$
  b Magnitude = 16.6 N
    Direction = 57.0°

7 **a** $OA = 5.39$ **b** $\overrightarrow{AB} = 4\mathbf{i} - 7\mathbf{j}$ **c** $AB = 8.06$

Assessment 6

1 **a** $\begin{pmatrix} 3 \\ 17 \end{pmatrix}$ **b** $\begin{pmatrix} -0.6 \\ 0.8 \end{pmatrix}$

2 **a** $\mathbf{d} = -10\mathbf{i} + 24\mathbf{j}$ **b** $|\mathbf{d}| = 26$

3 **a** 1346 m **b** 330°

4 **a** $x = 7, y = -4$ **b** $x = \dfrac{4}{3}$

5 **a** **i** 40 seconds **ii** 0.583 m s^{-1}
  b **i** Aim upstream at $\theta = 53.1°$ to bank
    ii 0.4 m s^{-1}
    iii 50 seconds

6 **a** **i** $\dfrac{2}{3}\mathbf{a}$ **ii** $\dfrac{3}{4}\mathbf{b}$ **iii** $\dfrac{1}{3}\mathbf{a} + \dfrac{1}{2}\mathbf{b}$
    iv $-\dfrac{2}{3}\mathbf{a} + \dfrac{1}{2}\mathbf{b}$ **v** $-\dfrac{1}{3}\mathbf{a} + \dfrac{1}{4}\mathbf{b}$
  b 2:1

7 $3\sqrt{5}, 116.6°$

8 **a** **i** $3\mathbf{i} - 2\mathbf{j}$ **ii** $9\mathbf{i} - 6\mathbf{j}$
  b $\overrightarrow{BC} = 3(3\mathbf{i} - 2\mathbf{j})$
    Since $\overrightarrow{BC}$ is a multiple of $\overrightarrow{AB}$, $\overrightarrow{AB}$ and $\overrightarrow{BC}$ are parallel, and since they have a point in common, A, B and C are collinear.

9 $\lambda = -4.8$ or 4

10 a **i** $2\mathbf{a} + \mathbf{c}$ **ii** $\mathbf{a} + \mathbf{c}$ **iii** $\mathbf{a} + \mathbf{c}$
    iv $\dfrac{3}{2}\mathbf{a} + \dfrac{1}{2}\mathbf{c}$ **v** $-\dfrac{1}{2}\mathbf{a} + \dfrac{1}{2}\mathbf{c}$
  b $\dfrac{4}{3}\mathbf{a} + \dfrac{2}{3}\mathbf{c}$

11 $\overrightarrow{PQ} = \mathbf{q} - \mathbf{p} = \begin{pmatrix} -3 \\ -5 \end{pmatrix} - \begin{pmatrix} 6 \\ 3 \end{pmatrix} = \begin{pmatrix} -9 \\ -8 \end{pmatrix}$
  $\overrightarrow{SR} = \begin{pmatrix} -9 \\ -8 \end{pmatrix}$
  $\overrightarrow{QR} = \mathbf{r} - \mathbf{q} = \begin{pmatrix} 1 \\ -3 \end{pmatrix} - \begin{pmatrix} -3 \\ -5 \end{pmatrix} = \begin{pmatrix} 4 \\ 2 \end{pmatrix}$
  $\overrightarrow{PS} = \begin{pmatrix} 4 \\ 2 \end{pmatrix}$
  Since $\overrightarrow{PQ} = \overrightarrow{SR}$ and $\overrightarrow{QR} = \overrightarrow{PS}$, $PQRS$ is a parallelogram

12 a $\mathbf{x} + 3\mathbf{y} = \begin{pmatrix} 1 \\ -1 \end{pmatrix} + \begin{pmatrix} 15 \\ 9 \end{pmatrix} = \begin{pmatrix} 16 \\ 8 \end{pmatrix}$
    $= 8 \times \begin{pmatrix} 2 \\ 1 \end{pmatrix}$
    Since $\mathbf{x} + 3\mathbf{y}$ is a multiple of $\mathbf{z}$, they are parallel
  b -8

13 a 343.4° **b** 5.1

Assessment chapters 1–6: Pure

1 **a** **i** 5^{mn} **ii** 3^{m+n-1} **iii** $7^{\frac{m}{2}}$
  b $x = 5$

2 $y = \dfrac{2}{3}$ or $\dfrac{1}{4}$, $x = -3$ or 2

3 **a** $b^2 - 4ac = -15 < 0$ so no real solutions
  b **i** $x = -1, -2$ **ii** $x = \dfrac{-3 \pm \sqrt{5}}{2}$
  c $k = \dfrac{25}{4}$

4 **a** $(x - 2)^2 + 8$
  b $x - 2 = \sqrt{-8}$ so no real solutions
    Or use discriminant.

c

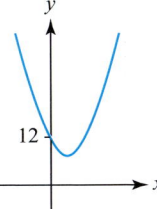

Minimum point is (2, 8)

5 a i $\mathbf{q} - \mathbf{p}$ **ii** $\frac{1}{2}(\mathbf{q} - \mathbf{p})$ **iii** $-\frac{1}{2}(\mathbf{q} + \mathbf{p})$

  b i $\sqrt{13}$ or 3.61   **ii** 2.55

  c Gradient of $\mathbf{p}$ is $\frac{3}{2}$

   Gradient of $\mathbf{q}$ is $-\frac{2}{3}$

   $\frac{3}{2} \times \left(-\frac{2}{3}\right) = -1$

   Therefore $\mathbf{p}$ and $\mathbf{q}$ are perpendicular.

6 a $-\frac{5}{2} - \frac{3}{2}\sqrt{3}$, $p = -\frac{5}{2}$ and $q = -\frac{3}{2}$ **b** $x = \frac{1}{2}$

7 a Centre is (3, −7)

   Radius is 5

  b

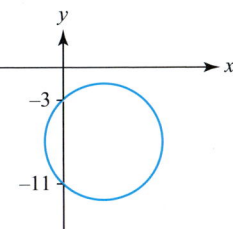

8 a $-\frac{1}{4}$ **b** $4x - y - 6 = 0$

  c (1, −2)   **d** $\sqrt{5}$

9 a $\mathbf{c} = -22\boldsymbol{i} - 6\boldsymbol{j}$

  b $|\mathbf{c}| = 22.8$, 195.3° from positive $\boldsymbol{i}$ direction

  c Parallel

   Since $\mathbf{c} = -2(11\boldsymbol{i} + 3\boldsymbol{j})$

10 e.g. $0 \times 1 = 0$ or $-3 \times -2 = 6$

  Which is even so this disproves the statement at the product of any two consecutive integers is odd.

11 a 126

  b i $1 - 6x + 15x^2 - 20x^3 + 15x^4 - 6x^5 + x^6$

    ii $1 + 8x + 24x^2 + 32x^3 + 16x^4$

12 a 4.1 m s^{-1} **b** 166°

13 a $p(3) = 3^3 - 2(3^2) - 5(3) + 6$

        $= 0$ so $x - 3$ is a factor of $p(x)$

  b $x = (3), -2, 1$

  c

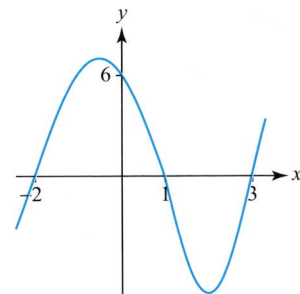

14 a

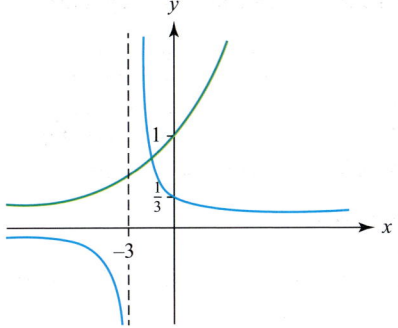

  b One as the curves intersect once only.

15 a 15.5 cm   **b** 83.4°   **c** 69.3 cm^2

16 a i 30°, 150°   **ii** 20.7°, 159.3°

  b

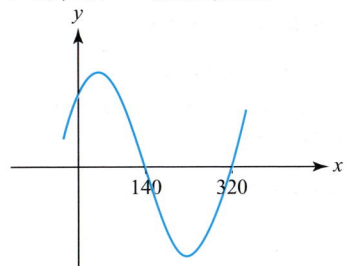

17 a 38.5°, 141.5°   **b** 80.4 cm^2

18 a $3x^2 - 10x^4$ **b** $\frac{1}{2}x^{-\frac{1}{2}} - x^{-2}$

19 a 10   **b** $x + 10y = 42$

20 a $\frac{4}{3}x^3 + 6x^2 + 9x + c$ **b** $2x^{\frac{5}{2}} - \frac{2}{3}x^{\frac{3}{2}} \; (+c)$

  c $2x^{\frac{1}{2}} + \frac{1}{3}x^{\frac{3}{2}} \; (+c)$

21 a (4, 0)   **b** $\frac{10}{3}\sqrt{5} - \frac{22}{3}$

22 a $\frac{\mathrm{d}y}{\mathrm{d}x} = 3x^2 - 4x - \frac{1}{x^2}$ **b** $y + 4 = 6(x + 1)$ **c** $\frac{1}{3}$

23 $y = 0.374x^{3.42}$

24 a i $\log_2 5$ **ii** $\log_2 54$ **iii** $\log_2 10$

  b $\log_3 17 = 2.58$

25 a

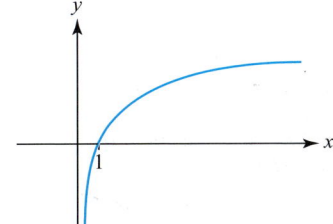

  Asymptote at $x = 0$

  b i $x = \frac{1}{64}$ **ii** $x = \frac{1}{4}$

26 a i $x \geq -2$ **ii** $-3 < x < 2$

  b $-2 \leq x < 2$

27 a $(x - 4)^2 + (y - 2)^2 = 26$ **b** $y - 7 = \frac{1}{5}(x - 3)$

28 a (4, −3) and (1, 3)   **b** $2x + y - 5 = 0$

29 Consider n^2:

  n odd $\Rightarrow n^2$ odd

  So m will be even $\Rightarrow m^2$ even

  $\therefore n^2 - m^2$ is odd

n even $\Rightarrow n^2$ even
So m will be odd $\Rightarrow m^2$ odd
$\therefore n^2 - m^2$ is odd

Alternatively, Let $m = n - 1$
$$n^2 - m^2 = n^2 - (n - 1)^2$$
$$= n^2 - n^2 + 2n - 1$$
$$= 2n - 1 \text{ which is odd}$$
So $n^2 - m^2$ is odd for any consecutive integers m and n

30 $k \le -8$, $k \ge 0$

31 $a = 4$, $b = -1$

32 a $(x + 4)(2x + 1)(x - 3)$

b

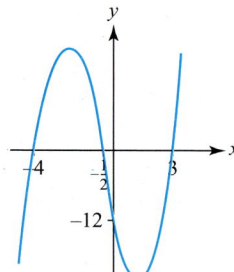

33 6.5 square units

34 405 m, 215°

35 a $176 + 80\sqrt{5}$ **b** $265 - 153\sqrt{3}$

36 a $64 + 192x + 240x^2 + 160x^3 + 60x^4 + 48x^5 + x^6$

b 64.192

37 a $\sin 3x + 2\cos 3x = 0$
$\sin 3x = -2\cos 3x$
$\dfrac{\sin 3x}{\cos 3x} = -2$
$\tan 3x = -2$

b $x = 39°, 99°, 159°$

38 a i

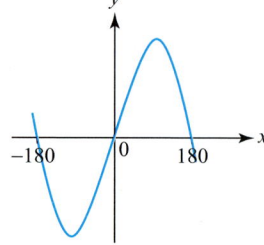

Maximum at $(90, 2)$
Minimum at $(-90, -2)$

ii

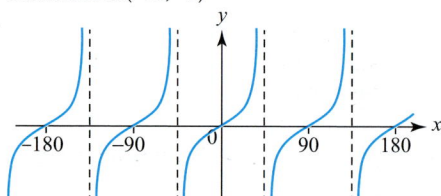

Asymptotes at $x = \pm45°$, $x = \pm135°$

b $x = 30°, 90°$

39 a $\dfrac{2}{3}\sqrt{2}$ **b** $2\sqrt{2}$

40 a 37.8° **b** 227.9 mm²

41 a 37.9 cm **b** 51 cm² **c** 32.6 cm

42 a $y = \dfrac{x^6 + 4x^3 + 4}{x^5}$
$$= x + \dfrac{4}{x^2} + \dfrac{4}{x^5}$$

$$= x + 4x^{-2} + 4x^{-5}$$

b i $\dfrac{dy}{dx} = 1 - 8x^{-3} - 20x^{-6}$ **ii** $\dfrac{d^2y}{dx^2} = 24x^{-4} + 120x^{-7}$

c $-\dfrac{22}{3}$

43 a $5\,\text{m s}^{-1}$ **b** 53.1°

44 a $\dfrac{dy}{dx} = 12x^2$ **b** 48

45 $x < -\dfrac{3}{5}$

46 $(0, 20)$ is a maximum
$(1, 15)$ is a minimum
$(-2, -12)$ is a minimum

47 a $x^2l = 3000 \Rightarrow l = \dfrac{3000}{x^2}$
$A = 4xl + x^2$
$$= 4x \times \dfrac{3000}{x^2} + x^2$$
$$= \dfrac{12000}{x} + x^2 \text{ as required}$$

b $x = 18.2$ cm

c $A_{\min} = 991$ cm²

d $\dfrac{d^2A}{dx^2} = \dfrac{24000}{x^3} + 2$
$\dfrac{d^2A}{dx^2}\bigg|_{x=18.2} > 0$ so a minimum

48 a $2x^5 - 4x^3 + x + c$

b $y = 2x^5 - 4x^3 + x - 25$

c $y = 2(-1)^5 - 4(-1)^3 + (-1) - 25$
$= -24$ so passes through $(-1, -24)$

49 405 m, 215°

50 $y = 2.2x^{1.5}$, $a = 2.2$, $n = 1.5$

51 a 5 years and 6 months

b Not suitable as it implies the car will ultimately have a negative value.

52 a

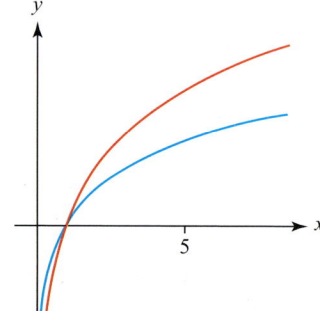

b $x = 9e^5$

53 a i If $n = 3$
Then $2^n + 1 = 8 + 1 = 9$
9 is divisible by 3 so not a prime number
Hence $2^n + 1$ is not a prime for all positive integers n

ii If $n = 1$
Then $3^n + 1 = 3 + 1 = 4$
$= 2 \times 2$ so even

If $n = 2$
Then $3^2 + 1 = 9 = 10$
$= 2 \times 5$ so even

If $n = 3$

Then $3^3 + 1 = 27 + 1 = 28$
$= 2 \times 14$ so even

If $n = 4$
Then $3^4 + 1 = 81 + 1 = 82$
$= 2 \times 41$ so even

So we have proved that $3^n + 1$ is even for all integers n in the range $1 \le n \le 4$

   iii $5^n + 10 = 5(5^{n-1} + 2)$ Which is a multiple of 5 for all positive integers n

b **i** Proof by counterexample
   ii Proof by exhaustion
   iii Direct proof

54 $A = -3, B = 16, C = 8$

55 a $n = 16$ **b** $a = -\dfrac{1}{2}$

56 a -2048 **b** 2944

57 a $50.8°, 129°, -50.8°, -129°$ **b** $180°$

58 a LHS $\equiv 3\sin^2 x - 4(1 - \sin^2 x)$
       $\equiv 3\sin^2 x - 4 + 4\sin^2 x$
       $\equiv 7\sin^2 x - 4$ as required
 b $57.7°, 122.3°, -57.7°, -122.3°$

59 $39.9°, 320.1°, 115.7°, 244.3°$

60 a $y = 10x - 16$
 b $P\left(\dfrac{16}{11}, -\dfrac{16}{11}\right)$

61 a $x = -1, 7, 3$ **b** 128

62 a $\dfrac{45}{8}$ cm
 b $V_{max} = \dfrac{2025}{32}\pi$ (198.8) cm^3
    $\dfrac{d^2 V}{dx^2} = -4\pi < 0$ so a maximum

63 a **i** $A = 11$ **ii** $k = 0.094$
 b

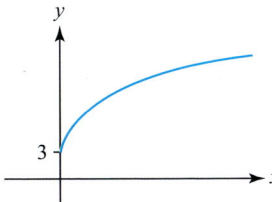

 c 4966
 d 567 organisms per hour
 e $11\,000$

64 $x = \ln\left(\dfrac{3}{2}\right), \ln 5$
    Or $0.405, 1.61$

65 a $x = 2$ **b** $y = 3$ **c** $x = 0.739, 27.2$

66 $x = \dfrac{1}{9}, y = 216$

67 $-3\mathbf{i} + 0\mathbf{j}$

Chapter 7

Exercise 7.1A Fluency and skills

1 a Force   **b** Mass
  c Speed or velocity   **d** Acceleration

2 a 8500 m   **b** 2300 mm
  c 4.82 m   **d** 1.65 km
  e 20 m s^{-1} **f** 50.4 km h^{-1}
  g 0.9 km h^{-1} **h** 24000 cm^2
  i 1400 g   **j** 1600 kg

3 a 120 km h^{-1} **b** 33.3 m s^{-1} (to 3 sf)

4 18 km

5 0.0154 m s^{-2}

6 45 kg

Exercise 7.1B Reasoning and problem-solving

1 29 min (to 2 sf)

2 31.5 km (to 3 sf)

3 36 s

4 236 kg (to 3 sf)

Exercise 7.2A Fluency and skills

1 a Between 5 and 15 s
 b Between 0 and 5 s
 c Between 15 and 20 s; the gradient is steepest

2 a It travels forward 12 m in 5 s, then backwards 18 m in the next 15 s
 b 2.4 m s^{-1} in first stage, then -1.2 m s^{-1} in second stage
 c **i** 1.5 m s^{-1} **ii** -0.3 m s^{-1}

3 a 37.5 m
 b -1 m s^{-2}

4 a Between 8 and 11 s   **b** -0.73 m s^{-2} (to 2 sf)
 c -2 m s^{-2}

5 a

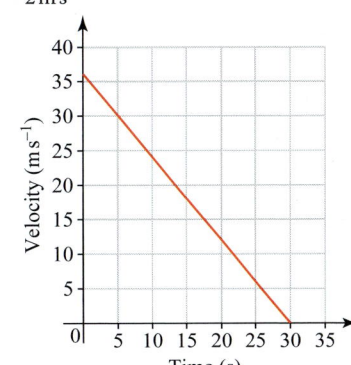

 b 540 m

6 a Displacement (m)

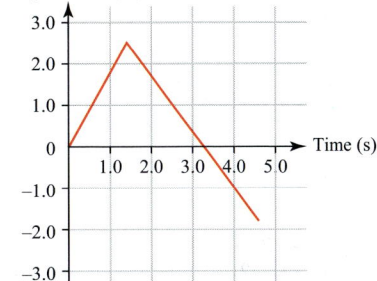

 b **i** 1.8 m s^{-1} (to 2 sf)
    ii -1.3 m s^{-1} (to 2 sf)
    iii 1.5 m s^{-1} (to 2 sf)

7 a 2.5 m s^{-2} **b** 360 m

Exercise 7.2B Reasoning and problem-solving

1 a

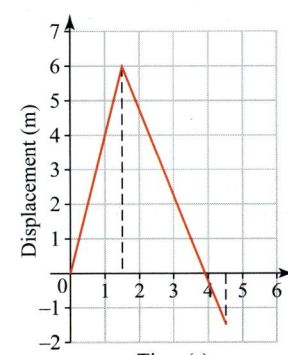

b **i** $3\,\mathrm{m\,s^{-1}}$ **ii** $-\dfrac{1}{3}\,\mathrm{m\,s^{-1}}$

2 **a** The cat goes forward 9 m in 6 s at constant speed, is still for 4 s, then (turns around and) goes backwards 15 m in 10 s at constant speed.

   b From 0 to 6 s, $v = 1.5\,\mathrm{m\,s^{-1}}$
      From 6 to 10 s, $v = 0\,\mathrm{m\,s^{-1}}$
      From 10 to 20 s, $v = -1.5\,\mathrm{m\,s^{-1}}$

   c **i** $1.2\,\mathrm{m\,s^{-1}}$ **ii** $-0.3\,\mathrm{m\,s^{-1}}$

   d The cat's velocity would not change instantaneously, so there would be curved sections of the graph where it's accelerating or decelerating.

3 **a**

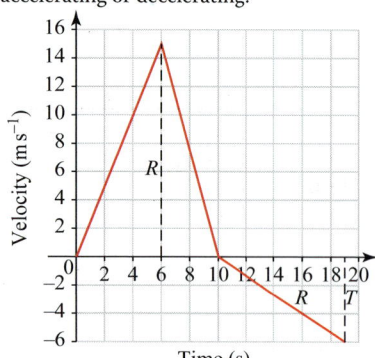

   b 75 m   **c** 35 s   **d** $-0.24\,\mathrm{m\,s^{-2}}$

4 **a**

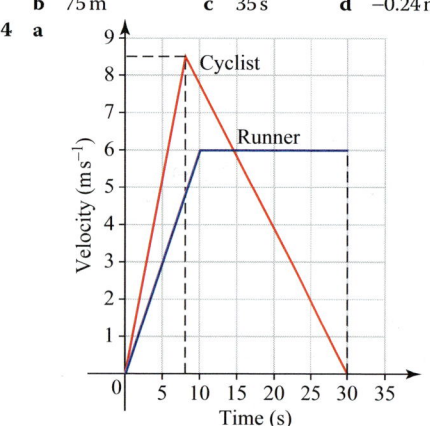

   b **i** 150 m   **ii** $10\,\mathrm{m\,s^{-1}}$

5 **a**

   b 75 m   **c** 7.5 s

   d E.g. It would be impossible for the cars to move with uniform acceleration due to factors like air resistance and the fact that human drivers aren't perfect.

6 Max is 7 min 33.6 s when the roadworks start more than 230 m from the start of the journey and end more than 230 m from the end of the journey.
Min is 7 min 28.8 s when the roadworks start at the beginning of the journey or end at the end of the journey.

Exercise 7.3A Fluency and skills

1 $u = v - at$ (or $v = u + at$)

2 **a** $18\,\mathrm{m\,s^{-1}}$ **b** $24\,\mathrm{m\,s^{-1}}$ **c** 9 s
3 **a** 36 m   **b** $3\,\mathrm{m\,s^{-2}}$ **c** 3 s
4 **a** 2 m   **b** 5 s   **c** 15 m
5 **a** $18.0\,\mathrm{m\,s^{-1}}$ (to 3 sf)   **b** 4.89 s (to 3 sf)
6 **a** $3\,\mathrm{m\,s^{-2}}$ **b** $30\,\mathrm{m\,s^{-1}}$
7 **a** $0.01\,\mathrm{m\,s^{-2}}$ **b** $0.058\,\mathrm{m\,s^{-1}}$
8 **a** $-4\,\mathrm{m\,s^{-2}}$ **b** 10 s
9 **a** 70 m   **b** $4\,\mathrm{m\,s^{-1}}$
10 **a** $102\,\mathrm{km\,h^{-1}}$ (to 3 sf)   **b** 1.16 s (to 3 sf)

Exercise 7.3B Reasoning and problem-solving

1 750 m
2 1440 m
   In second stage it travels $24 \times 240 = 5760$ m
   In third stage $0^2 = 24^2 - 2 \times 1.5s \Rightarrow s = 192$ m
   Distance $AB = 1440 + 5760 + 192 = 7392$ m

3 **a** $2.4\,\mathrm{m\,s^{-1}}$
   b 14.4 m
   c $(-)\,0.51\,\mathrm{m\,s^{-2}}$ (to 2 sf)
   d The ferry is of negligible size. It would need to decelerate quicker than this to prevent the front of the ferry crashing into the opposite bank (the values calculated are for the back of the ferry).

4 $-\dfrac{5}{72}\,\mathrm{m\,s^{-2}}$

5 **a** -25 m
   b 65 m

6 $5\,\mathrm{m\,s^{-2}}$

7 **a** B comes to a safe stop 19.375 m behind A (they don't collide).
   b The lorries have negligible size. If the lorries are longer than 19.375 m then they will actually collide.

8 See Ch7.3 Fluency and skills for full derivations.

9 **a** $23\dfrac{5}{6}$ m

   b $3\dfrac{2}{3}$ s

   c The actual distance between humps would be greater than the distance calculated by an amount equal to the length of the car. The time would be the same, as the distance between humps is irrelevant in its calculation.

Exercise 7.4A Fluency and skills

1 **a** $v = 8t - 3t^2$ **b** $4\,\mathrm{m\,s^{-1}}$
2 **a** $v = 2t - 4$ **b** $16\,\mathrm{m\,s^{-1}}$
   c $196\,\mathrm{m\,s^{-1}}$
3 **a** $a = 12t - 8$ **b** $28\,\mathrm{m\,s^{-2}}$
4 **a** $a = 3t + 7$ **b** $7\,\mathrm{m\,s^{-2}}$
5 **a** $14\,\mathrm{m\,s^{-1}}$ **b** $13\,\mathrm{m\,s^{-2}}$
6 **a** $v = 6t - 12$ **b** 2 s
7 **a** $a = 14 - 2t$ **b** 4 s
8 **a** $s = 24t - 3t^2$ **b** 36 m
9 **a** $v = 2t^3 + 3t + 4$ **b** $26\,\mathrm{m\,s^{-1}}$
10 **a** $12\dfrac{2}{3}$ m   **b** 22 m
11 **a** $-7\,\mathrm{m\,s^{-1}}$ **b** $-10\,\mathrm{m\,s^{-1}}$
12 **a** $3\,\mathrm{m\,s^{-1}}$ **b** 5 m

Exercise 7.4B Reasoning and problem-solving

1 **a** $v = 30t - 3t^2$ and $s = 15t^2 - t^3$ **b** 15 s **c** 500 m
2 **a** 6 s   **b** 72 m
3 45 m
4 **a** 4 m   **b** 2 s   **c** $12\,\mathrm{m\,s^{-1}}$ **d** 7.2 m
5 **a** **i** -48 m   **ii** $48\,\mathrm{m\,s^{-2}}$
   b $96\,\mathrm{m\,s^{-1}}$
   c 81 m
6 4 s

7 $s = 7\frac{1}{3}$ m and $s = 7\frac{1}{6}$ m

8 a $-9\,\text{m s}^{-1}$ **b** 6 m

9 a $v = \dfrac{dx}{dt} = \dfrac{1}{2700}t^2(t-30)^2$ **b** 30 s

   c 300 m   **d** $18.75\,\text{m s}^{-1}$

10 a Integrate: $[at]_0^t = [v]_u^v \Rightarrow v = u + at$

   b $\displaystyle\int_0^s ds = \int_0^t (u + at)\,dt$

     Integrate: $[s]_0^s = \left[ut + \dfrac{1}{2}at^2\right]_0^t \Rightarrow s = ut + \dfrac{1}{2}at^2$

Review exercise 7

1 0.473 km   **2** $12\,\text{m s}^{-1}$

3 a

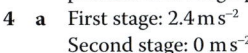

   b In the graph, the changes in speed are instantaneous (i.e. there is no acceleration). In reality, this would not be possible and the graph would curve slightly in places.

4 a First stage: $2.4\,\text{m s}^{-2}$
     Second stage: $0\,\text{m s}^{-2}$
     Third stage: $-2\,\text{m s}^{-2}$

   b i 27.5 m   **ii** 35.5 m
   c $2.75\,\text{m s}^{-1}$

5 a $0.4\,\text{m s}^{-2}$ **b** 75 m

6 a $v = 3t^2 - 16$ **b** $30\,\text{m s}^{-2}$

7 a $v = 2t^2 - 3t + 5$ **b** $38\frac{2}{3}$ m

Assessment 7

1 a

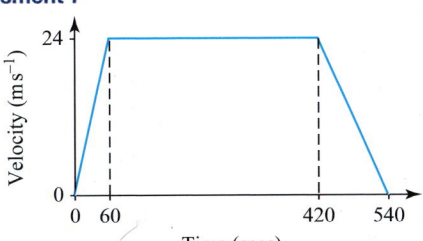

   b 10.8 km

2 a 15 s   **b** 10 s or 20 s

3 a $6\,\text{m s}^{-1}$ **b** 3 s

4 a $v = 3t^2 - 18t + 24$ **b** 2 s or 4 s   **c** 28 m

5 a 8 s   **b** $39\,\text{m s}^{-1}$

6 a

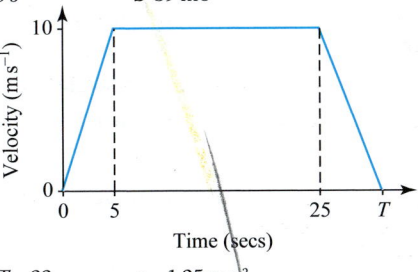

   b $T = 33$ **c** $-1.25\,\text{m s}^{-2}$

7 $29.2\,\text{m s}^{-1}$

8 a $20\,\text{m s}^{-1}$ **b** 15 s

9 a $2\,\text{m s}^{-1}$ **b** $\dfrac{5}{3}\,\text{m s}^{-2}$

   c constant velocity of $7\,\text{m s}^{-1}$ **d** $T = 12\frac{6}{7}$

10 a $v = t^2 - 6t + 10$ **b** 2 s or 4 s

   c $s = \dfrac{1}{3}t^3 - 3t^2 + 10t$ **d** 24 m

11 a

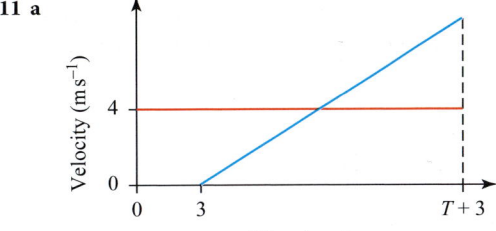

   b $T^2 - 4T - 12 = 0$
   c $T = 6$
   d $12\,\text{m s}^{-1}$

Chapter 8

Exercise 8.1A Fluency and skills

1 a   **b**   **c**   **d**

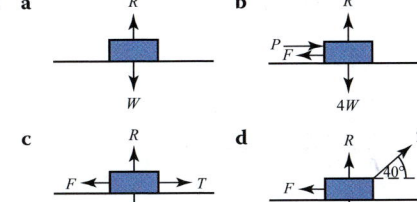

2 a Resolve vertically $R - W = 0$

   b Resolve horizontally $P - F = 0$
      Resolve vertically $R - 4W = 0$

   c Resolve horizontally $T - F = 0$
      Resolve vertically $R - W = 0$

3 a 40 N   **b** 40 N

4 a Resolve horizontally $90 - T = 0$
      Resolve vertically $Y - 50 = 0$

   b Resolve horizontally $Y + 20 - 30 = 0$
      Resolve vertically $40 - T = 0$

   c Resolve horizontally $50 - Y = 0$
      Resolve vertically $20 - X = 0$

   d Resolve horizontally $P + Z - X = 0$
      Resolve vertically $X - Q = 0$

5 a 900 N   **b** The speed decreases.

Exercise 8.1B Reasoning and problem-solving

1 a $(48\mathbf{i} + 18\mathbf{j})$
   b $R = 51.3$ N (to 3 sf), $\alpha = 20.6°$ (to 1 dp) above $\mathbf{i}$

2 a $(125\mathbf{i} + 125\mathbf{j})$ N
   b $R = 177$ N (to 3 sf)
     $\alpha = 45°$ above $\mathbf{i}$

3 a $a = -65, b = 85$
   b $R = 130\,\text{N}, \alpha = 67.4°$ above $-\mathbf{i}$
4 a $A = 20\,\text{N}, B = 30\,\text{N}$ **b** $C = 18\,\text{N}, D = 6\,\text{N}$
   c $E = 20\,\text{N}, F = 20\,\text{N}$ **d** $G = 10\,\text{N}, H = 20\,\text{N}$
5 $R = 24.2\,\text{N}$ on a bearing of $119.7°$
6 a $R = 5\,\text{N}, \alpha = 36.9°$ above the 12 N force.
   b $R = 26\,\text{N}, \alpha = 22.6°$ (to 1 dp) below the 94 N force.
7 a If $Y = 50$ then $Y \neq 20$
   b $4X = 3Y \Rightarrow Y = \dfrac{4}{3}X$

       $5X = 4Y \Rightarrow 5X = 4\left(\dfrac{4}{3}\right)X \Rightarrow X = 0$

       This contradicts the condition $X > 0$
8 a 300 N
   b The resistance to motion might decrease due to a change in wind strength or direction.
9 a 50 N   **b** 5 N
10 a Tension. The bar is pulled forwards by the engine and is pulled backwards by the resistance force on the carriage.
   b Yes, the bar is now in thrust. The bar is pushed forwards by the engine and pushed backwards by the resistance force on the carriage.
11 There is an overall upward force on the block. Therefore the rod must exert a downward force and so the rod is in tension.
12 $Y = 33.2\,\text{N}$ (to 3 sf), $X = 56.2\,\text{N}$ (to 3 sf)

Exercise 8.2A Fluency and skills

1 15 N   **3** $1.2\,\text{m s}^{-2}$
2 $1.5\,\text{m s}^{-2}$ **4** 712.5 N

5 a $\mathbf{a} = (1.8\mathbf{i} + 3.6\mathbf{j})\,\text{m s}^{-2}$ **b** $\mathbf{a} = \begin{pmatrix} 1.5 \\ 2.5 \end{pmatrix}\,\text{m s}^{-2}$

   c $\mathbf{a} = 0.28\,\text{m s}^{-2}$ **d** $\mathbf{a} = (2\mathbf{i} + 1.25\mathbf{j})\,\text{m s}^{-2}$
   e $\mathbf{a} = (60\mathbf{i} + 84\mathbf{j})\,\text{m s}^{-2}$
6 3900 N
7 a $R = 98\,\text{N}, a = 9\text{ms}^{-2}$ **b** $R = 196\,\text{N}, X = 30\,\text{N}$
8 $0.7\,\text{m s}^{-2}$
9 $m = 14\,\text{kg}$
10 a $3.75\,\text{m s}^{-2}$ **b** 8 s   **c** 120 m

Exercise 8.2B Reasoning and problem-solving

1 $m = 2800\,\text{kg}$ **2** 202.5 N
3 30 m   **4** 53 000 N
5 a 25 N   **b** 550 N (to 2 sf)
   c The resistance to motion (wind etc.) will decrease as she slows down, so the actual braking force will be higher than the answer in **b**.
6 $m = \dfrac{32}{7}$ and $P = \dfrac{42}{25}$
7 a 2400 N
   b The acceleration will not be constant over this time because the resistance to motion will decrease as the parachutist slows down.
8 2000 kg
9 a 49 N   **b** 0.95 s (to 2 sf)   **c** $1.9\,\text{m s}^{-1}$ (to 2 sf)
10 $a = 6$ and $b = 8$
    $3.75\,\text{m s}^{-2}$
11 $x = -13$ or $x = 4$ **12** $m = 20\,\text{kg}, R = 200\,\text{N}$

Exercise 8.3A Fluency and skills

1 a 0.098 N   **b** 98 000 N   **c** 10 kg (to 2 sf)
2 $a = 6.2\,\text{m s}^{-2}$
3 $14.7\,\text{m s}^{-1}$ (to 3 sf)
4 $20\,\text{m s}^{-1}$ (to 1 sf)
5 $m = 25\,\text{kg}, a = 6\,\text{m s}^{-2}$
6 $m = 4\,\text{kg}, a = 5\,\text{m s}^{-2}$
7 $3\,\text{m s}^{-2}$

8 8 kg
9 a 3900 N (to 2 sf)   **b** 4300 N (to 2 sf)   **c** 3100 N (to 2 sf)
10 120 N (to 2 sf)   **11** $R = 15.6\,\text{N}$ **12** 20 kg
13 2060 N (to 3 sf)
14 583 N (to 3 sf)

Exercise 8.3B Reasoning and problem-solving

1 $a = 1\,\text{m s}^{-2}, T = 13\,000\,\text{N}$ (to 2 sf)
2 $R = 50\,\text{N}, a = 5\,\text{m s}^{-2}$
3 $1 : 6$
4 2.5 m
5 4.0 s (to 2 sf)
6 a i $8.5g\,\text{N}$ **ii** $(8.5g + 17)\,\text{N}$
   b The rope is light and inextensible.
7 $a = 4\,\text{m s}^{-2}, s = 50\,\text{m}$
8 1.2 s (to 2 sf)
9 a 27 N   **b** $17\,\text{m s}^{-2}$ (to 2 sf)
10 a $3\,\text{m s}^{-2}$ **b** $5\,\text{m s}^{-2}$
11 $T = 3g$

Exercise 8.4A Fluency and skills

1 a 2000 N (to 2 sf)   **b** 400 N (to 2 sf)
2 $a = 2.9\text{m s}^{-2}$ (to 2 sf), $R = 250\,\text{N}$ (to 2 sf)
3 a $1.875\,\text{m s}^{-2}$ **b** 1600 N
4 $a = 2\,\text{m s}^{-2}, F = 9\,\text{N}$
5 a 3790 N   **b** 840 N   **c** 130 m (to 2 sf)
6 a $T = 5400\,\text{N}$ (to 2 sf), $R = 753\,\text{N}$ (to 3 sf)
   b 18.4 N (to 3 sf)
7 T_1 (both crates) = 3800 N (to 2 sf)
   T_2 (bottom crate) = 1600 N (to 2 sf)

Exercise 8.4B Reasoning and problem-solving

1 $T = 23.5\,\text{N}$ (to 3 sf), $a = 1.96\,\text{m s}^{-2}$ (to 3 sf)
2 a $a = 4.6\,\text{m s}^{-2}$ (to 2 sf), $T = 26\,\text{N}$ (to 2 sf)
   b $v = 1.4\,\text{m s}^{-1}$ (to 2 sf)
   c 0.43 m (to 2 sf)
   d 37 N (to 2 sf)
3 a $7\,\text{m s}^{-2}$ **b** 33 N   **c** $20\,\text{m s}^{-1}$ **d** 30 m
4 a $1.40\,\text{m s}^{-2}$ **b** 33.6 N   **c** $7.01\,\text{m s}^{-1}$ **d** 20 m
5 1.9 s (to 2 sf)
6 $T = \dfrac{2xyg}{x + y}, a = \left(\dfrac{y - x}{x + y}\right)g$

Review exercise 8

1 $(\mathbf{i} - \mathbf{j})\,\text{N}$
2 20.2 N (to 3 sf), $\alpha = 8.5°$ (to 2 sf) above $\mathbf{i}$
3 $(-20\mathbf{i} + 29.4\mathbf{j})\,\text{N}$
4 $B = 10\,000\,\text{N}$
5 a 7.2 N, 5.2 N (to 2 sf)   **b** 1.3 N, 0.95 N (to 2 sf)
6 $T = 4400\,\text{N}$ (to 2 sf), $R = 360\,\text{N}$ (to 2 sf)
7 a $T = 46\,\text{N}, a = 5.2\,\text{m s}^{-2}$ **b** 1.5 s (to 2 sf)
   c 0.36 m (to 2 sf)   **d** 16 N (to 2 sf)

Assessment 8

1 a 3000 N   **b** $4\,\text{m s}^{-2}$ **c** 62.5 m
2 a $2.0\,\text{m s}^{-2}$ (2 sf)   **b** 235.2 N
3 a $4\,\text{m s}^{-2}$ **b** $32\,\text{m s}^{-1}$ **c** 130 m (2 sf)
4 a $0.35\,\text{m s}^{-2}$ **b** 580 N
5 a

   b $R_1 = 1200\,\text{N}$ (to 2 sf), $R_2 = 980\,\text{N}, R_3 = 680\,\text{N}$

6 a 470 N (to 2 sf) **b** 390 N (to 2 sf) **c** 330 N (to 2 sf)

7 a $2900 - 700 - T = 1900\,a$, $T - 400 = 800\,a$

　　b i $\dfrac{2}{3}$ m s^{-2} **ii** $933\dfrac{1}{3}$ N

　　c 144 m

8 a $a = \dfrac{3g}{10}$ m s^{-2} **b** $v = \sqrt{\dfrac{9g}{5}}$ $(= 4.2$ m s$^{-1})$ **c** $\dfrac{3g}{4}$ m s^{-2}

　　d $v^2 = \left(\dfrac{9g}{5}\right) - 2 \times \left(\dfrac{3g}{4}\right) \times 1 \Rightarrow v = \sqrt{\dfrac{3g}{10}}$ m s^{-1}

9 a 1.7 m s^{-1} (to 2 sf) **b** 0.86 s (to 2 sf) **c** 160 cm (to 2 sf)

10 a 2.9 m s^{-2} (to 2 sf) **b** 25 N (to 2 sf) **c** 34 N (to 2 sf)

11 a 12 420 N **b** 1104 N
　　c -0.2 m s^{-2} **d** 768 N

12 a 1800 N **b** 4600 N **c** 5300 N

13 a 1400 N **b** 200 N

14 a 2 m s^{-2} **b** 240 N (to 2 sf)

15 a 3500 N **b** 9500 N
　　c 333 N (to 3 sf) **d** $t = 60$ s

Assessment Chapters 7–8: Mechanics

1 a
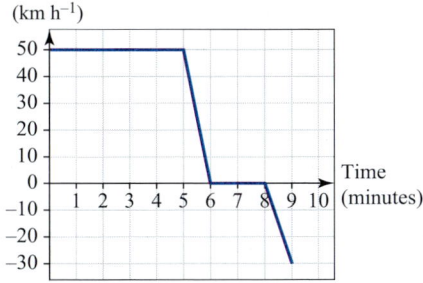

　　b 290 km **c** 260 km

2 a i 4 m s^{-2} **ii** 18 m
　　b 1.25 kg

3 a i 1.77 m **ii** 5.88 m s^{-1}
　　b Book is modelled as a particle, no air resistance, no spin, gravity constant etc.

4 a 140.3 m s^{-1} **b** 34 m s^{-2}

5 a i 5 s **ii** 5 cm s^{-1} **iii** 76 cm
　　b 9.33 cm s^{-1}

6 a 112 m **b** $v = 3t^2 - 3$ **c** 30 m s^{-2}

　　d
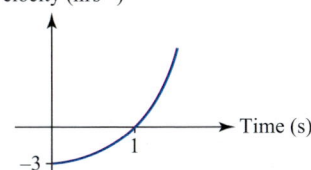

7 a 240 m **b** 12 m s^{-1}

8 a 2197.8 N **b** 268.62 N
　　c It will be larger the heavier the person.

9 a 6 m **b** 4 s

10 a i 1.5 m s^{-2} **ii** 5100 N
　　b 16.9 m

11 a
Velocity
(m s^{-1})

　　b 43.5 m
　　c 4.65 m s^{-1}

12 51.3 m

13 a 1.2 s **b** 8.4 m s^{-1}

14 a $t = 1$ or 1.5 **b** 1.33 m **c** 1.42 m

15 a 4.74 s **b** 20.2 m

16 1.72 m above the ground

17 a i 0.892 m s^{-2} **ii** 26.8 N **b** 4.18 m

Chapter 9

Exercise 9.1A Fluency and skills

1 Population is the adults in their extended family. Parameter is the average height. Sample is the adults in their extended family who live in their city. Statistics are the heights of the sampled adults.

2 Population is the trees of different species in the country's forest. The parameter for each species is the maximum height. The samples are the trees measured in the chosen forest. The statistics are the maximum heights of each species within that forest.

3 a To take a census, you need to collect data from every member of the population. This would mean finding out the favourite band of all 1000 students.
　　b Take the list of students in the school and order them from 1 upwards. Generate 40 different random numbers and choose those students.
　　c Taking a census guarantees an accurate view of the population. Taking a sample is quicker and cheaper than a census.

4 291 people

5 a Simple random sampling **b** Systematic sampling
　　c Stratified sampling

6 a 13 red bulbs, 17 blue bulbs, 6 green bulbs.
　　b Testing the lightbulbs involves damaging them. A census would test and damage every lightbulb in the population.

Exercise 9.1B Reasoning and problem-solving

1 a The average over three months will be biased depending on the season it was taken in. You would be best to conclude that the average temperature is 21 °C.
　　b Systematic sampling will give you only data values about a single month's average temperature and not the yearly average.

2 a True for Idea 1, but not for Idea 2.
　　b Idea 1 is fairest as each student has an equal probability of being chosen. Idea 2 is bad as each school is unlikely to have students whose views on location are similar to those of each other school.
　　c The teacher could use stratified sampling to make sure that the number of students chosen from each school is proportional to the number of students at that school. The teacher should then use a list of students at each school and use random numbers to randomly select the appropriate number of students from each school.

3 a To account for regional variations. This is cluster sampling.
　　b The same locations means they are more likely to experience similar weather conditions. The same set of months means you aren't comparing e.g. winter with summer, and so weather conditions should be more comparable.

4 a Lots of stations will mean that general patterns can be detected and distinguished from outliers. They shouldn't be too close together because spreading them evenly ensures a clearer picture of more of Britain.
　　b Accepting any submissions means they will have access to a lot more data without spending lots of money. They should be cautious because they haven't set this up themselves and there could be people who aren't as meticulous in their methods as the Met Office are.

Exercise 9.2A Fluency and skills

1 a i 8   **ii** 6.85   **iii** 7   **iv** $11 - 2 = 9$ **v** 2.41
  b i 68, 71   **ii** 68.8   **iii** 69   **iv** $75 - 62 = 13$ **v** 3.40
2 a $16 - 21$ **b** Mean 19.6
3 a 14.5, 30.5 and 16   **b** 2.1, 3.8 and 1.7
4 6
5 a 150.08 mg   **b** 4 mg
6 $\bar{x} = 7.41$
  $\sigma^2 = 0.425$ (to 3 dp)
7 a 3.76   **b** 3.70
8 $\sigma = 2.24$ OR $s = 2.26$
9 $\bar{x} = 15.96$
  $\sigma_x = 1.0648$

Exercise 9.2B Reasoning and problem-solving

1 a $M = 16.5$
  Range $= 4$
  IQR $= 1$
 b The range is affected by the outlier 19.5 so does not represent the spread of most of the data.
 c 14, 19
 d $14 < 14.5 < 19$, do *not* test the batch.
2 a $\bar{x} = 16.3$, $\sigma = 5.43$ **b** 33
3 a mean $= 5.22$, standard deviation $= 3.59$ **b** 3.47
4 a mean $= 19.8$, standard deviation $= 4.75$ (2 dp)
 b 2015 has a higher average maximum gust compared to 1987 but there is also greater variation in this variable.
5 a Median $= 12.5$, IQR $= 10.5$ **b** 36
 c Median will be reduced.
6 a No mode
  Median $= 107$
  Mean $= 109$
 b The median. There is no mode and the mean is distorted by the outlier 145. The median is not distorted by outliers.
 c $a = 102$
  The first team has a lower median score of 107 compared with 110 so are, in general, better golfers. The first team has a larger interquartile range of 9 compared with 7 so exhibit more diversity in achievement. The median and interquartile range are not affected by the outlier unlike the mean, standard deviation and range.

Exercise 9.3A Fluency and skills

1 Class 2 did better on average, the median is $82 > 73$ for Class 1. The Class 2 scores include an outlier, 50, but even with this included are generally more consistent with an interquartile range of $86 - 74 = 12 < 89 - 65 = 24$ for Class 1.
2 a 220 houses   **b** 208 houses
 c

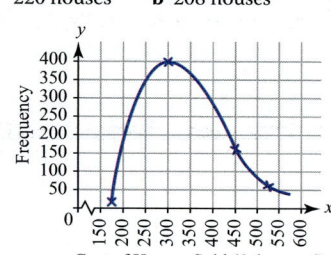

3 $Q_1 \approx 90$, $Q_3 \approx 110$
4 a

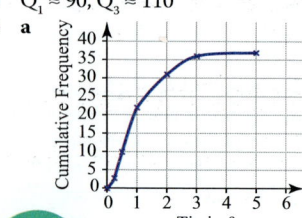

b £0.85 ± 5p
5 a $0.5 - 0.7$ frequency $= 6$
  $0.8 - 1.0$ frequency $= 2$
 b Width $= 0.4$ ($= 4$ blocks)
  Height $= 3.75$ blocks $= 0.375$ units
 c i 39%   **ii** 31%

Exercise 9.3B Reasoning and problem-solving

1 a A histogram is the better choice.
 b Box-and-whisker plot
2 $3 \leq x < 7$: 7; $7 \leq x < 10$: 10; $16 \leq x < 24$: 6; $24 \leq x < 31$: 2
3 a i 0.75   **ii** 0.625
 b The end of the upper whisker will move to 165 kg
4 a i $17 \leq T < 20$ has frequency 48; $20 \leq T < 21$ has frequency 9
  ii $15 \leq T < 17$: width $= 2$, height $= 13.5$; $21 \leq T < 25$: width $= 4$, height $= 0.25$
 b i 88   **ii** 49

Exercise 9.4A Fluency and skills

1 a Moderate negative correlation.
 b Moderate positive correlation.
 c Strong positive correlation.
 d Zero correlation.
2 Moderate negative correlation.
3 a

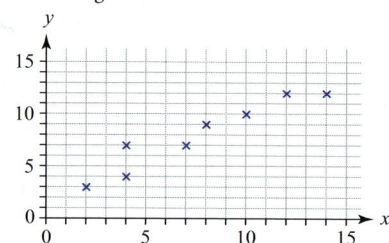

Strong positive correlation

 b

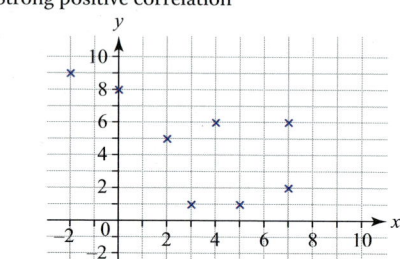

No correlation

4

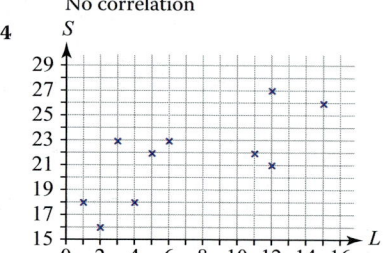

Moderate positive correlation

5

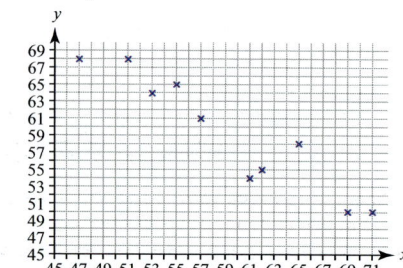

Moderate/strong negative correlation.

  Answers For full solutions go to http://www.oxfordsecondary.co.uk/edexcelalevelmaths-answers

Exercise 9.4B Reasoning and problem-solving

1 a Positive correlation, causal.
  b Positive correlation, causal.
  c Zero correlation.
  d Negative correlation, not causal.

2 a, b

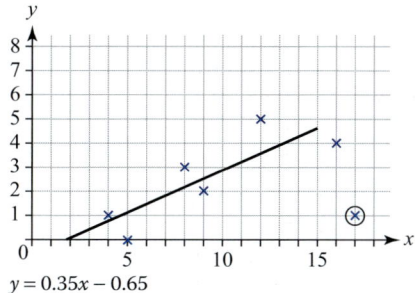

$y = 0.35x - 0.65$

  c 5
  d This x-value is outside the range of observed data so this is extrapolation. There is no evidence that this pattern will continue beyond the observed range.

3 a

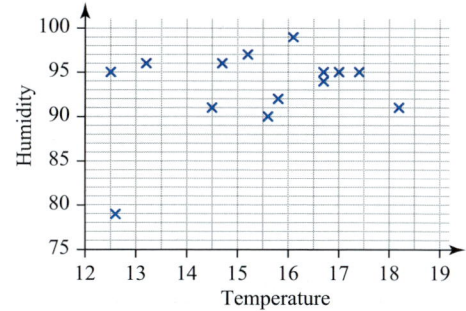

  b median = 95, IQR = 5
  c 79
  d Weak positive correlation. Reduce correlation coefficient to zero or weak negative.
  e $h = 92.06$

4 a i Positive **ii** Negative **iii** Negative
  b Causal. It seems likely that changes in humidity cause changes in rainfall, not simply that they are both changed by some third factor.

5 a

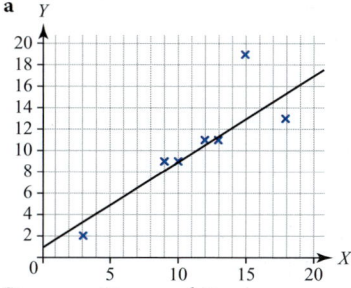

Strong positive correlation.
  b (15, 19) **c** 6.5

Review exercise 9

1 a Population: all LED light bulbs made in the factory.
    Sample: the 23 bulbs measured.
    Parameter: the temperature of a lit LED light bulb.
    Statistics: the temperatures of the 23 light bulbs lit.
  b Population: all casks of that type of whiskey made in the brewery.
    Sample: the nine casks measured.
    Parameter: the alcohol level of a cask of whiskey.
    Statistics: the alcohol content of the nine casks measured.

2 a Simple random sampling.
  b Opportunity sampling.
  c Opportunity sampling.

3

Class	Frequency (any multiple of this column)	Frequency Density
0–15	21	1.4
15–50	7	0.2
50–70	10	0.5
70–90	16	0.8
90–105	33	2.2
105–130	20	0.8
130–160	6	0.2
160+	0	0

4 a, b

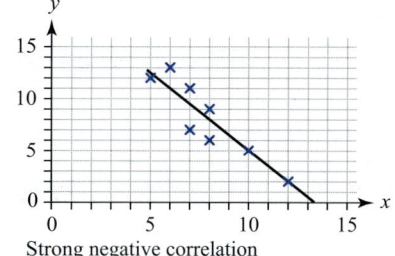

Strong negative correlation

  c 6 or 7

5 a

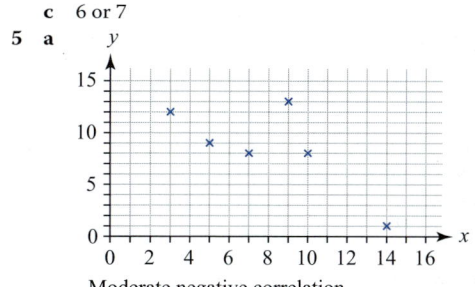

Moderate negative correlation

  b Incorrect point: (9, 13)
    More plausible point: (9, 6)
    Accept $y = 5$ to 7 also

6 Median value = 10.75, mean = 10.8, modal interval: $8 \le x < 12$, variance $\sigma^2 = 18.56$

7 a IQR = 6 **b** Outlier: 26

Assessment 9

1 a e.g. pick a random number between 1 and 10 and select that household and then select every 10th household thereafter.
  b e.g. non-response or not wishing to participate, incorrect or no phone number.
  c e.g. may change behaviour due to being monitored, households that agree to participate may be more likely to recycle.

2 a All the batteries produced.
  b You may be testing to destruction.
  c The mean/median length of life of the batteries in the sample.

3 a 22 is incorrect - impossible to have this many hours of sunshine per day in the UK.
    10.6
  b 16.4 **c** 11.45
    Median is not affected by outliers.

4 a Hurn: 2.2, Leuchars: 1.65
  b Hurn: 20.3°C, Leuchars: 19°C
  c The mean is higher at Hurn.
    The interquartile range is larger at Hurn.
    The warmest day in Hurn was nearly 3°C warmer than the warmest in Leuchars.
    OR The coolest day in Leuchars was 1.5°C cooler than the coolest day in Hurn.

d Leuchars: 15.1°C is an outlier, 22.1°C is not an outlier
Hurn: 25°C is an outlier.

5 Strong positive correlation, the higher the mean windspeed, the faster the gusts.

6 a 3
b 2.3, 1.49
c Median: 2, Q_1: 1, Q_3: 3
d Mode/median
e.g. any one of
Mode is the most likely number of assignments to be set.
Mode doesn't take into account fact that most students are set 3 or fewer assignments.
Median and mode are unaffected by outliers.
Mean is affected by the outlier who claims to have been set 8 homework assignments.

7 a i 91.1% **ii** 5.81
b i 87.8% **ii** 6.36
c Mean was higher in 1987 so May this year was more humid, however May 2015 has a more variation.
d

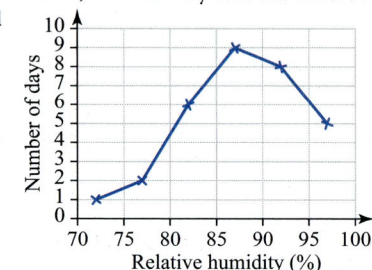

8 a e.g. sample will not be representative of the population as there are more girls than boys in each year/more year 12s than 13s.
b 20 year 12 girls
17 year 13 girls
13 year 12 boys
10 year 13 boys
c e.g. students in the common room may use the library less often.

9 A, C, D

10 a Temperature
b Moderate positive correlation.
A higher temperature implies a higher germination rate.
c Not necessarily – we do not have data for temperatures this high/high temperature may kill the seeds.

11 a Width is 10 mm, height is 28 mm
b i 0.22 mm **ii** 1.52 mm

12 a Method A – quota sampling
Method B – stratified random sampling
b Method B is preferable as each member is equally likely to be chosen, as opposed to Method A where members of larger schemes are less likely to be chosen.

13 a 13 **b** 8 **c** 24.4°C

14 a $a = 9$, $b = 4$ **b** $s = 4.05$ OR $\sigma = 3.99$
c Symmetric as mean and median are approximately the same.

15 a 89.5 **b** 7.5 **c** 16%

Chapter 10

Exercise 10.1A Fluency and skills

1 (1, 2), (1, 3), (1, 4), (2, 1), (2, 3), (2, 4), (3, 1), (3, 2), (3, 4), (4, 1), (4, 2), (4, 3)
2 a $a = 12$, $b = 8$, $c = 7$, $d = 25$, $e = 45$
b i $\dfrac{4}{9}$ **ii** $\dfrac{7}{45}$ **iii** $\dfrac{4}{9}$

3 a $\dfrac{3}{31}$ **b** $\dfrac{8}{31}$ **c** $\dfrac{20}{31}$
4 a 0.1 **b** 0.9
5 a 0.5 **b** 0.5 **c** 0.7 **d** 0
6 P(square number or prime) = P(1, 2, 3, 4 or 5) = $\dfrac{10}{12} = \dfrac{5}{6}$
7 a $\dfrac{1}{3}$
b

x	0	1	2	3	4	5
P($X = x$)	$\dfrac{1}{6}$	$\dfrac{5}{18}$	$\dfrac{2}{9}$	$\dfrac{1}{6}$	$\dfrac{1}{9}$	$\dfrac{1}{18}$

Exercise 10.1B Reasoning and problem-solving

1 a

Score	1	2	3
Probability	0.3	0.6	0.1

b i 0.9 **ii** 0.4
2 a 0 **b** 1 **c** $a = \dfrac{1}{2}$; $b = \dfrac{3}{2}$
d P($0.25 \leq X \leq 0.75$) = 0.5(f(0.25) + f(0.75)) × 0.5 = 0.25(f(0.25) + f(0.75))
P($0.5 \leq X \leq 1$) = 0.5(f(0.5) + f(1)) × 0.5 = 0.25(f(0.5) + f(1))
f(x) is an increasing function so f(0.25) < f(0.5) and f(0.75) < f(1)
$\Rightarrow$ P($0.25 \leq X \leq 0.75$) = 0.25(f(0.25) + f(0.75))
< 0.25(f(0.5) + f(1)) = P($0.5 \leq X \leq 1$)
3 a $a = 5$; $b = 30$; $c = 5$; $d = 30$; $e = 14$; $f = 61$
b i $\dfrac{14}{61}$ **ii** $\dfrac{45}{61}$ **iii** $\dfrac{7}{61}$
c $\dfrac{8}{17}$
4

	1	2	3	4	5	6
1	2	3	4	5	6	7
2	3	4	5	6	7	8
3	4	5	6	7	8	9
4	5	6	7	8	9	10
5	6	7	8	9	10	11
6	7	8	9	10	11	12

No. For fair dice P(sum of scores > 10) = $\dfrac{1}{12}$ so at least one dice is not fair.
5 Yes. $\dfrac{3}{4} \times \dfrac{1}{2} = \dfrac{3}{8}$
6 12
7 a i P(E) = 0.5, P(F) = 0.4, P(E or F) = 0.8 **ii** 0.2
b P(E) + P(F) − P(E and F) = 0.5 + 0.4 − 0.1 = 0.8 = P(E or F)

Exercise 10.2A Fluency and skills

1 a 0.132 **b** 0.837 **c** 0.832
2 a 0.00066 **b** 0.17 **c** 0.41 **d** 0.95
3 a 0.087 **b** 0.00038 **c** 0.29
4 a P($X = x$) = $^5C_x \times 0.4^x \times 0.6^{5-x}$
b

x	0	1	2	3	4	5
P($X = x$)	0.078	0.259	0.346	0.230	0.077	0.010

5 a $X \sim B\left(4, \dfrac{1}{6}\right)$
b i 0.001 **ii** 0.016 **iii** 0.502
6 0.016 (to 3 dp)

Exercise 10.2B Reasoning and problem-solving

1 a There should be a fixed number of independent and identical trials.
b i 0.690 (to 3 dp)
ii Independence of weather from one day to next is unlikely.
2 a Yes. $n = 4$, $p = \dfrac{3}{7}$
b Yes. If the population is sufficiently large when compared to the sample, as the patients are chosen at random the probability of getting a patient who will be prescribed antibiotics remains constant. $n = 8$, $p = 0.12$

c Yes. $n = 5$, $p = 0.6$

d No. Number of trials not fixed.

3 No. If it was, the probability of two zeros would be 0.0486

4 **a** 0.363 (to 3 dp), assuming independence between days.

b **i** 0.637 **ii** 0.363

5 You are more likely to be right. Let X be a random variable for the number of 6s in four throws.

$X \sim B\left(4, \dfrac{1}{6}\right)$

$P(X \geq 1) = 1 - P(X = 0) = 1 - \left(\dfrac{5}{6}\right)^4 = 0.518$ (to 3 sf)

6 0.939 (to 3 sf)

7 0.188 (to 3 dp)

8 0.5

9 Let X be a random variable for the number of girls born in five children.

$P(\text{girl}) = 0.5$ so $X \sim B(5, 0.5)$

$P(A) = P(X \text{ is not 0 or 5}) = 0.9375$

$P(B) = P(X \geq 3) = 0.5$

$P(A \text{ and } B) = P(X = 3 \text{ or } 4) = 0.46875$

$P(A) \times P(B) = 0.46875 = P(A \text{ and } B)$ so events are independent.

Review exercise 10

1

x	1	2	3
$P(X = x)$	p	$(1-p)p$	$(1-p)^2 p + (1-p)^3$

$p + (1-p)p + (1-p)^2 p + (1-p)^3$
$= p + p - p^2 + p - 2p^2 + p^3 + 1 - 3p + 3p^2 - p^3 = 1$

2 **a** 0.004 (to 3 dp) **b** 0.473 (to 3 dp)

3 **a** $X \sim B(5, 0.5)$

b **i** 0.31 (to 2 dp) **ii** 0.97 (to 2 dp) **iii** 0.47 (to 2 dp)

4 0.788 (to 3 sf)

5 0.993 (to 3 sf)

6 0.100 (to 3 sf)

7 Your friend. Let Y be the number of double 6s in 24 throws.

$P(Y \geq 1) = 1 - P(Y = 0) = 1 - \left(\dfrac{35}{36}\right)^{24} = 0.491$ (to 3 sf)

8 0.613 (to 3 sf)

9 **a** $X \sim B(12, 0.68)$ **b** 0.432 (to 3 sf)

Assessment 10

1 **a** 56212

b **i** $\dfrac{53}{94}$ **ii** $\dfrac{17}{46}$ **iii** $\dfrac{901}{4324}$

c Yes

2 **a** The probability that one event occurs is not affected by the occurrence or non-occurrence of any other event.
(Or $P(A \cap B) = P(A) \times P(B)$)

b $\dfrac{1}{84}$

c $\dfrac{1}{140}$

3 **a** $\dfrac{1}{20}$ **b** 0.35 **c** The second bag

4 **a** 0.0773 (to 3 sf) **b** 0.997 (to 3 sf) **c** 0.992 (to 3 sf)

5 **a** Each discarded sweet has an equal probability of being misshapen and each sweet's probability of being misshapen is not affected by any other sweet being misshapen.

b 0.0806 (to 3 sf) **c** 0.403 (to 3 sf) **d** 0.668 (to 3 sf)

6 0.991 (to 3 sf)

7 **a** 184

b **i** $\dfrac{43}{92}$ **ii** $\dfrac{65}{184}$ **iii** $\dfrac{25}{184}$

c No, since $\dfrac{43}{92} \times \dfrac{65}{184} \neq \dfrac{25}{184}$

8 **a** 0.483 (to 3 sf) **b** 0.875 (to 3 sf)

c 0.941 (to 3 sf) **d** 25.1 (to 3 sf)

9 0.335 (to 3 sf)

Chapter 11

Exercise 11.1A Fluency and skills

1 **a** Critical region: $X \leq 5$, acceptance region: $X \geq 6$

b Critical region: $X \geq 5$, acceptance region: $X \leq 4$

c Critical region: $X \leq 5$ or $X \geq 12$, acceptance region: $6 \leq X \leq 11$

2 **a** Accept H_0 **b** Reject H_0 **c** Accept H_0

3 **a** Reject H_0 **b** Accept H_0

4 **a** p is the probability that a game is lost. H_0: $p = 0.15$ and H_1: $p \neq 0.15$

b Critical region: $X \leq 2$ or $X \geq 14$, acceptance region: $3 \leq X \leq 13$

c **i** Accept H_0 **ii** Reject H_0 **iii** Reject H_0

5 **a** Critical region: $X \leq 33$, acceptance region: $X \geq 34$

b **i** Reject H_0 **ii** Accept H_0

6 **a** Critical region: $X \geq 9$, acceptance region: $X \leq 8$

b No.

7 **a** Critical region: $X \geq 5$, acceptance region: $X \leq 4$

b $p < 10$

Exercise 11.1B Reasoning and problem-solving

1 **a** H_0: $p = 0.6$ **b** H_0: $p = 0.7$ **c** H_0: $p = 0.4$
H_1: $p > 0.6$ $\quad$ H_1: $p < 0.7$ $\quad$ H_1: $p \neq 0.4$

2 **a** p is the probability that the daily mean windspeed is over 7 knots.
H_0: $p = 0.3$, H_1: $p \neq 0.3$

b $2 \leq n \leq 10$

3 **a** p is the probability that a voter supports the politician.
H_0: $p = 0.35$, H_1: $p > 0.35$

b Accept H_0. You conclude that there is not sufficient evidence, at the 5% significance level, to suggest the politician is underestimating her support.

c 41

4 **a** p is the probability that the number of hours of sunshine in a day in Leeming is fewer than 4
H_0: $p = 0.25$
H_1: $p < 0.25$

b The critical value at the 10% significance level is 6 so the critical region is $X \leq 6$. 5 is in the critical region so if James observes that Leeming gets fewer than 4 hours of sunshine on 5 days, you reject the null hypothesis. You conclude that there is evidence, at the 10% significance level, to say that the percentage of days on which Leeming gets fewer than 4 hours of sunshine is less than 25%.

c $n \geq 7$

d $x < 10$

5 **a** $a \geq 8$ and $b \leq 24$ **b** $x < 5$

Exercise 11.2A Fluency and skills

1 **a** $X \leq 1$ **b** Reject H_0

2 $X \leq 3$ or $X \geq 13$

3 **a** $P(X \leq 1) = 0.0480$ (to 4 dp). Reject H_0

b $P(X \leq 2) = 0.1514$ (to 4 dp). Accept H_0

4 **a** $P(X \leq 1) = 0.0274$ (to 4 dp), accept the null hypothesis.

b $P(X \geq 9) = 0.0468$ (to 4 dp), accept the null hypothesis.

5 **a** $P(X \geq 34) = 0.0462$. Reject H_0

b $P(X \leq 21) = 0.0758$. Accept H_0

6 $X \geq 8$

7 **a** Reject the null hypothesis.

b Accept the null hypothesis.

8 **a** **i** $X \geq 26$ **ii** $X \geq 23$ **iii** $X \geq 19$

b **i** Accept the null hypothesis.

ii Accept the null hypothesis.

iii Reject the null hypothesis.

c $2 < p < 20$

9 a i $X \geq 66$ **ii** $X \geq 64$ **iii** $X \geq 62$

   b i Accept H_0 **ii** Reject H_0 **iii** Reject H_0

   c 4

10 $y = 1$

11 a $X = 0$ **b** 28

12 $a = 5$ or 6

Exercise 11.2B Reasoning and problem-solving

1 a $H_0: p = 0.8$, $H_1: p < 0.8$, $X \leq 27$

   b Accept H_0: there was insufficient evidence to suggest the meteorologist's claim was an overestimation.

2 a $H_0: p = 0.6$, $H_1: p \neq 0.6$

    $P(X \geq 56) = 0.0056$ (to 4 dp)

   b Reject H_0: there was sufficient evidence to suggest that the company's claim was inaccurate.

3 a $H_0: p = 0.15$, $H_1: p \neq 0.15$ **b** Yes: accept H_0

    Reject H_0

4 a Under H_0, $X \sim B(n, 0.5)$

    i Accept H_0 **ii** Reject H_0

   b 18

5 a 0.1   **b** 10%

6 a $H_0: p = 0.1$, $H_1: p < 0.1$ **b** 1

7 a $n = 40$ **b** $X \leq 11$ or $X \geq 29$

Review exercise 11

1 $X \sim B(n, p)$

2 a $H_0: p = 0.2$, $H_1: p \neq 0.2$ **b** $X = 0$ or $X \geq 10$

   c 0.0211 (to 4 dp)   **d** The distribution is discrete.

3 a $H_0: p = 0.5$, $H_1: p > 0.5$ **b** 13

4 a $H_0: p = 0.2$, $H_1: p > 0.2$

   b Conclude that there is not enough evidence to suggest that the student was not guessing the answers.

5 a Fixed n and p; independent and identical trials; each trial results in exactly one of two possible outcomes, "success", "failure"; interested in the number of successes (or failures) only.

$$P(X = x) = \binom{n}{x} p^x (1-p)^{n-x}; x = 0, 1, 2 \ldots, n$$

   b

x	0	1	2	3	4	5	6
$P(X = x)$	0.0827	0.2555	0.3290	**0.2260**	**0.0873**	**0.0180**	**0.0015**

   c x-values 5 and 6. Significance level 1.95%

6 a $H_0: p = 0.1$, $H_1: p \neq 0.1$ **b** No.   **c** No.

Assessment 11

1 a $H_0: p = \dfrac{1}{2}$, $H_1: p \neq \dfrac{1}{2}$

   b X-values in the critical region are 0, 6

    Significance level $= P(X \leq 0$ or $X \geq 6) = 0.0312 = 3.12\%$

2 a $H_0: p = 0.4$, $H_1: p > 0.4$ **b** 0–7   **c** Yes

3 a $H_0: p = 0.5$ and $H_1: p > 0.5$; no reason to reject the null hypothesis   **b** Yes

4 a Cars chosen at random.

   b $H_0: p = 0.4$, $H_1: p < 0.4$. No reason to reject H_0

5 Reject H_0

   Size of critical region is 0.0834

6 a Random sample from large population, therefore identical trials. One reading is unlikely to affect another so independent trials. Trials each result in exactly one of two possible outcomes, "over $17.5\,^\circ$C" or "not over $17.5\,^\circ$C".

   b Missing values are 0.1910 and 0.0621

   c Accept H_0

7 a $H_0: p = 0.6$, $H_1: p < 0.6$ **b** 4

8 a Random sample. So that a binomial distribution is a valid model.

   b $H_0: p = 0.35$, $H_1: p > 0.35$, $X \sim B(30, 0.35)$

   c Accept H_0

9 a $k^n + nk^{n-1}(1-k)$ **b** $0 < k \leq 0.1$

Assessment chapters 9–11: Statistics

1 a Simple random sampling

   b Systematic sampling

   c Opportunity sampling

2 a 1.5

   b 6.5 mmol l^{-1}

   c e.g. The range/IQR is wider 2 hours after the meal than before the meal.

    e.g. The average blood glucose level is higher 2 hours after the meal than before the meal.

   d 75%

3 a 28   **b** 20   **c** 12 people   **d** 15

4 a (left to right, top to bottom) 57, 28, 40, 68, 87

   b $\dfrac{59}{184}$ or 0.321

   c Not independent as $P(\text{warm} \cap \text{dry}) \neq P(\text{warm}) \times P(\text{dry})$

5 a 0.0146   **b** 0.1256   **c** 0.1275

6 a i $\dfrac{1}{28}$ **ii** $\dfrac{291}{700}$ or 0.416

   b i $\dfrac{243}{3125}$ or 0.0778   **ii** $\dfrac{144}{625}$ or 0.230   **iii** 0.683

7 a $16\text{–}18\,^\circ$C

   b i $17.6\,^\circ$C   **ii** 3.61 or 3.62

   c Outliers are less than 10.4 or larger than 28.8; there are at least 4 and up to 12

8 a 22 is the error. 77 is the correct rate.

   b 103

   c Median = 68. Median unlikely to be affected by the outlier.

9 a Medium

   b Mean = 59.3 g, standard deviation = 9.7

   c Median = 58, IQR = 14.1

   d Mean will increase, standard deviation will decrease.

10 a i 0.9   **ii** 0.6

   b 0.1

11 a i 0.0282   **ii** 0.8497   **iii** 0.0016

   b Probably not suitable as trials unlikely to be independent.

12 a There is sufficient evidence to suggest that the probability of thermometers being faulty has increased.

   b 0.036

   c 0.036

13 a i Opportunity sampling; Will not be representative of the whole of the UK as all in one geographical area.

    ii Systematic random sampling; May not be representative of the whole of the UK as names may be similar within an area.

   b 18 from England; 4 from Wales; 6 from Scotland; 2 from Northern Ireland

14 a Width 2.5 cm; Height 2.4 cm

   b 15

   c 3.5 million to 26.75 million

15 a i 0.23   **ii** 0.95

   b 0.9861

   c Wind speed each day is independent of other days; Probability is constant each day.

16 9

17 a There is not sufficient evidence to reject her claim.

   b $9 < X < 22$

   c e.g. Probability of rain is independent from day to day; Binomial distribution is suitable; There is a clear definition of what constitutes rain; She has accurately recorded the exisitence or not of rain.

Chapter 12

Exercise 12.1A Fluency and skills

1   $(5n + 1)^4 - (5n - 1)^4 \equiv [(5n + 1)^2 - (5n - 1)^2][(5n + 1)^2 + (5n - 1)^2]$
$\equiv [(25n^2 + 10n + 1) - (25n^2 - 10n + 1)]$
$[(25n^2 + 10n + 1) + (25n^2 - 10n + 1)]$
$\equiv [20n][50n^2 + 2]$
$\equiv [40n][25n^2 + 1]$
This has a factor of 40, so $(5n + 1)^4 - (5n - 1)^4$ is divisible by 40

2   Let the three-digit number be $W =$ "xyz" with 100s digit x, 10s digit y, and units digit z
$W = 100x + 10y + z = 99x + 9y + (x + y + z)$
If the sum of the digits is divisible by 9, then $(x + y + z) = 9P$ for some integer P
So $W = 99x + 9y + 9P = 9(11x + y + P) = 9Q$ for some integer Q
Hence, W itself is divisible by 9

3   $p! \le 2^p$
$0! = 0 \le 1 = 2^0 \Rightarrow$ true for $p = 0$
$1! = 1 \le 2 = 2^1 \Rightarrow$ true for $p = 1$
$2! = 2 \le 4 = 2^2 \Rightarrow$ true for $p = 2$
$3! = 6 \le 8 = 2^3 \Rightarrow$ true for $p = 3$
So the statement is proved by exhaustion.

4   Let the numbers be n and $n + 1$
$(n + 1)^3 - n^3 \equiv n^3 + 3n^2 + 3n + 1 - n^3$
$\equiv 3n^2 + 3n + 1$
Case 1: n is even.
Then n^2 is also even and hence $3n^2$ is even
$3n$ is also even
So $3n^2 + 3n + 1$ would be even + even + odd, which is odd.
Case 2: n is odd.
Then n^2 is odd and hence $3n^2$ is odd
$3n$ is also odd
So $3n^2 + 3n + 1$ would be odd + odd + odd, which is still odd.
So the difference between the cubes of two consecutive integers is odd.

5   If $a = 5$, $b = -2$ and $c = -4$
Then $ab = 5 \times -2 = -10$ and $bc = -2 \times -4 = 8$
So, in this case $ab < bc$, which disproves the statement.

6   Suppose $p = \sqrt{5}$
$p^2 = (\sqrt{5})^2 = 5$, which is rational
$p = \sqrt{5}$, which is irrational
This counterexample disproves the statement.

7   $9^3 = 729$, which disproves the statement.

8   Contradiction statement: There is an even integer, n such that n^2 is odd.
If n is even, it can be written as $n = 2m$
Hence, $n^2 = (2m)^2 = 4m^2$
But 4 times any integer is even and so the statement is contradicted.
So if n^2 is odd, then n is odd.

9   Contradiction statement: There are integers a and b for which $a^2 - 4b = 2$
From this equation, $a^2 = 4b + 2 = 2(2b + 1)$
Any integer multiplied by 2 is even, so $2(2b + 1)$ is even and so a^2 must be even.
Since a^2 is even, then a must also be even, so we can write $a = 2m$ for some integer m
Substituting $a = 2m$ back into the original equation $a^2 - 4b = 2$ gives
$(2m)^2 - 4b = 2$
$\Rightarrow 4m^2 - 4b = 2$
$\Rightarrow 2m^2 - 2b = 1$
$\Rightarrow 2(m^2 - b) = 1$
Since $2(m^2 - b)$ is even, 1 must also be even

However, 1 is not even, and so the statement is contradicted.
So there are no integer values of a and b such that $a^2 - 4b = 2$

10 For $0° < x < 90°$, $\sin x > 0$ and $\cos x > 0$, so $\sin x + \cos x > 0$
Alternative statement: Let us also assume that $\sin x + \cos x < 1$
for $0° < x < 90°$
$\Rightarrow (\sin x + \cos x)^2 < 1$
$\Rightarrow \sin^2 x + 2\sin x \cos x + \cos^2 x < 1$
But $\sin^2 x + \cos^2 x = 1$, so $1 + 2\sin x \cos x < 1$
However, for $0° < x < 90°$, $\sin x > 0$ and $\cos x > 0$
$\Rightarrow 2\sin x \cos x > 0$
Hence, $1 + 2\sin x \cos x > 1$
This contradicts the statement.
Furthermore, if $x = 0°$ or $90°$ then $\sin x + \cos x = 1$
So for every real number x between $0°$ and $90°$,
$\sin x + \cos x \ge 1$

11 Alternative statement: m and n exist such that $\dfrac{m^2}{n^2} = 2$, and $\dfrac{m^2}{n^2}$ is a fully simplified fraction.
If $\dfrac{m^2}{n^2} = 2$, then $m^2 = 2n^2$
$2n^2$ must be even and so m^2 must also be even.
m^2 is even and so m must also be even.
so $m = 2k, k \in \mathbb{Z}$
$\Rightarrow (2k)^2 = 2n^2$
$\Rightarrow 4k^2 = 2n^2$
So n^2 must be even and hence n must also be even.
Since both m and n are even, they both have a factor of 2
Hence both m^2 and n^2 have a factor of 4
This is a contradiction and disproves the original statement.

12 Alternative statement: There is at least one integer greater than 1 which has no prime factors.
Let the *smallest* such integer be n
Case 1: n is prime
Then n has a prime factor: itself, since $n = n \times 1$
This contradicts the assumption, so n is not prime.
Case 2: n is not prime
Then n has a factor, f, where $f \ne n$ and $f \ne 1$
Because f is a factor of n, $f < n$
Since n is the *least* integer with no prime factor, f *does* have a prime factor, p
Since p is a prime factor of f, and f is a factor of n, so p must also be a prime factor of n
This contradicts the assumption that n does not have a prime factor.
So every integer greater than 1 does have at least one prime factor.

Exercise 12.1B Reasoning and problem-solving

1   Alternative statement: Suppose that m and n are integers and mn is odd but m and n are not both odd, that is, at least one of them is even.
Without loss of generality, suppose that m is even.
m is even $\Rightarrow m = 2p$ for some integer p
Hence $mn = 2pn$
Now pn must be another integer, say q
Hence $mn = 2q$
But $2q$ must be an even number, and so mn is even
This contradicts the original statement.
So if m and n are both integers and mn is odd, then both m and n must be odd.

2   Alternative statement: Suppose that m exists such that $m^2 < 2m$, but m does not lie in the range $0 < m < 2$
If $m \ge 2$ and $m^2 < 2m$, then $m < 2$ (dividing by m), which is a contradiction.

So $m^2 < 2m$ cannot be true, for $m \geq 2$

If m is zero, then $m^2 < 2m$ is not true.

If m is negative, then m^2 is positive and $2m$ is negative.

So $m^2 < 2m$ cannot be true for $m \leq 0$

The alternative statement is therefore disproved.

Hence, if $m^2 < 2m$ has any solutions, then $0 < m < 2$

3 Let $n = 2m + 1$ for integer m

Then $(-1)^n = (-1)^{2m+1}$

$\qquad\qquad = (-1)^{2m}(-1)^1$ [By laws of indices]

$\qquad\qquad = 1 \times (-1)^1$ [Since $(-1)^{2m} = 1$ as $2m$ is even]

$\qquad\qquad = -1$

4 Let $A = 60°$ and $B = 30°$

$\sin(A - B) = \sin(60 - 30) = \sin 30 = \dfrac{1}{2}$

$\sin A - \sin B = \sin 60 - \sin 30 = \dfrac{\sqrt{3}}{2} - \dfrac{1}{2} = \dfrac{\sqrt{3} - 1}{2} \neq \dfrac{1}{2}$

Hence $\sin(A - B) \neq \sin A - \sin B$ for all A and B

5 Contradiction statement: Suppose that, for integers m and n, there exists an integer k such that $(5m + 3)(5n + 3) = 5k$

Then $5k = (5m + 3)(5n + 3)$

$\Rightarrow 5k = 25mn + 15m + 15n + 9$

$\Rightarrow 5k = 5(5mn + 3m + 3n + 1) + 4$

$\Rightarrow 5k = 5x + 4$ where $x = (5mn + 3m + 3n + 1)$, an integer

Hence, $4 = 5(k - x) = 5p$ for some integer p

5 is a factor of the RHS, but not a factor of the LHS, so this is a contradiction.

So the statement is disproved, and there is no integer k such that $(5m + 3)(5n + 3) = 5k$ for integers m and n

6 Let the numbers be $2m$ and $2m + 1$

$(2m)^3 + (2m + 1)^3 = 8m^3 + 8m^3 + 12m^2 + 6m + 1$

$\qquad\qquad\qquad\quad = 16m^3 + 12m^2 + 6m + 1$

$\qquad\qquad\qquad\quad = 2(8m^3 + 6m^2 + 3m) + 1$

2 is a factor of $2(8m^3 + 6m^2 + 3m)$

Hence, dividing $(2m)^3 + (2m + 1)^3$ by 2 leaves a remainder of 1

7 $m < -\dfrac{1}{2}$

$1 > -\dfrac{1}{2m}$ ($\div m$ on each side. m is negative, so reverses the inequality sign)

$2 > -\dfrac{1}{m}$ ($\times 2$ on each side)

$3 > 1 - \dfrac{1}{m}$ ($+1$ on each side)

$1 - \dfrac{1}{m} < 3$

8 Contradiction statement: There is a smallest positive number. Let n be the smallest positive number.

If n is positive, then $\dfrac{n}{2}$ exists, and $0 < \dfrac{n}{2} < n$

So $\dfrac{n}{2}$ is a smaller positive number than n

This contradicts the statement that n is the smallest positive number.

So there is no smallest positive number.

9 Let the number, N, be 'abc'. Hence $N = 100a + 10b + c$

$11 \times (100a + 10b + c) = 1100a + 110b + 11c$

$\qquad\qquad\qquad\qquad = 1000a + 100a + 100b + 10b + 10c + c$

$\qquad\qquad\qquad\qquad = 1000a + (a + b)100 + (b + c)10 + c$

Written as a four-digit number, this is '$a\langle a + b\rangle\langle b + c\rangle c$', proving the rule.

The catch is when $a + b$ or $b + c$ comes to 10 or more, in which case a 'carry' has to be inserted.

10 Case 1: a is even $\Rightarrow a = 2m$ for some m

then $a^2 + 2 = (2m)^2 + 2$

$\qquad\qquad = 4m^2 + 2$

$\qquad\qquad = 2(2m^2 + 1)$

This has a factor of 2, but $2m^2$ is even and so $2m^2 + 1$ is odd

So $a^2 + 2$ cannot have the other required factor of 2

Hence, if a is even, it cannot be divided equally by 4 to give an integer.

Case 2: a is odd $\Rightarrow a = (2m + 1)$ for some m

then $a^2 + 2 = (2m + 1)^2 + 2$

$\qquad\qquad = 4m^2 + 4m + 3$

Both $4m^2$ and $4m$ are even and so $4m^2 + 4m + 3$ is odd

So $a^2 + 2$ cannot be divided equally by 4 to give an integer.

11 Alternative statement: Suppose there is a greatest odd integer, n

Then $n + 2 > n$ and $(n + 2)$ is also odd.

This is a contradiction, since n is the greatest odd integer.

Hence, there is no greatest odd integer.

12 If b is a factor of a then $a = rb$ for some integer, r

If c is a factor of b then $b = sc$ for some integer, s

So $a = rb = rsc = tc$ where t is an integer, equal to rs

Hence, c is a factor of a and Stephen is right.

13 Any positive rational number can be written as $\dfrac{a}{b}$ where a and b are either both positive or both negative

We can write $\dfrac{a}{b} = c\sqrt{2}$ for some positive number c

Since $\sqrt{2}$ is an irrational number, we must prove that c is also an irrational number.

Now $c = \dfrac{a}{b\sqrt{2}} = \dfrac{a\sqrt{2}}{2b}$ which is an irrational number.

Hence any positive rational number can be expressed as the product of two irrational numbers.

14 Every cube number is the cube of an integer.

Every integer, m, is either:

i a multiple of 3

ii or one less than a multiple of 3

iii or one more than a multiple of 3.

Hence these three cases are exhaustive. We will prove each in turn:

Case 1: Let m be a multiple of $3 \Rightarrow m = 3p$ for some integer p

$m^3 = 27p^3 = 9(3p^3) = 9k$ for some integer k

Case 2: Let m be one more than a multiple of $3 \Rightarrow m = 3p + 1$ for some integer p

$m^3 = (3p + 1)^3$

$\qquad = 27p^3 + 27p^2 + 9p + 1$

$\qquad = 9(3p^3 + 3p^2 + p) + 1$

$\qquad = 9k + 1$ for some integer k

Case 3: Let m be one less than a multiple of $3 \Rightarrow m = 3p - 1$ for some integer p

$m^3 = (3p - 1)^3$

$\qquad = 27p^3 - 27p^2 + 9p - 1$

$\qquad = 9(3p^3 - 3p^2 + p) - 1$

$\qquad = 9k - 1$ for some integer k

So every cube number can be expressed in the form $9k$ or $9k 1$ where k is an integer.

15 Alternative statement: There is a solution and $a^2 - b^2 = 1$ for some positive integers a and b

If this is true, then $a^2 - b^2 = 1$ so $(a - b)(a + b) = 1$

Now, since a and b are integers, then:

either $a - b = 1$ and $a + b = 1$

or $a - b = -1$ and $a + b = -1$

Solving $a - b = 1$ and $a + b = 1$ leads to the solution $a = 1$ and $b = 0$ which contradicts the original statement that a and b are both positive.

Answers   For full solutions go to http://www.oxfordsecondary.co.uk/edexcelalevelmaths-answers

Solving $a - b = -1$ and $a + b = -1$ leads to the solution $a = -1$ and $b = 0$ which again contradicts the original statement that a and b are both positive.

So there are no positive integer solutions to the equation $a^2 - b^2 = 1$

16 Assume that a is a non-square integer and that $\sqrt{a} = \dfrac{m}{n}$, where $m, n \in \mathbb{Z}$ are co-prime, $n \neq 0$

Then $\left(\dfrac{m}{n}\right)^2 = a$

Thus $m^2 = an^2$ and a divides m^2

So a must also divide m

So $m = ak$ for some integer k

Hence, $a^2k^2 = an^2$

$\Rightarrow ak^2 = n$ and a divides n^2

So a must also divide n

So $n = aj$ for some integer j

Hence, $\dfrac{m}{n} = \dfrac{ak}{aj}$ which implies m and n share a common factor, a

This contradicts our original assumption that m and n share no common factors.

Hence $\sqrt{a}$ is irrational.

17 The error occurs in the 5th line: "So $(a - b)(a + b) = b(a - b)$"

Dividing by $(a - b)$ is the same as dividing by 0, since $a = b$

Division by zero is undefined and so the argument becomes invalid.

Exercise 12.2A Fluency and skills

1 **a** $f(x) \in \mathbb{R}$
   b $f(x) \in \mathbb{R} : -32 < y < 28$
   c $f(x) \in \mathbb{R} : y > 0$
   d $f(x) \in \mathbb{R} : 0 \leq y \leq 225$

2 **a** $x \in \mathbb{R} : -2 < x \leq 3$
   b $x \in \mathbb{R} : x > 0$

3 **a** Domain $\{x \in \mathbb{R}\}$; range $\{f(x) \in \mathbb{R} : f(x) > 0\}$
   b Domain $\{x \in \mathbb{R} : x \neq -3\}$; range $\{f(x) \in \mathbb{R} : f(x) > 0\}$

4 **a** $f(x) = 2x^2$ is many-to-one
       E.g. $f(1) = 2$ and $f(-1) = 2$
   b $f(x) = 3^{-x}$ is one-to-one
       No two values of x give the same value for $f(x)$
   c $f(x) = x^4$ is many-to-one
       E.g. $f(1) = 1$ and $f(-1) = 1$
   d $f(x) = \sin^2 x$ is many-to-one in this interval
       E.g. $f(90) = 1$ and $f(270) = 1$
   e $f(x) = \dfrac{1}{x^2}$ is many-to-one
       E.g. $f(1) = 1$ and $f(-1) = 1$
   f $f(x) = -3x^3$ is one-to-one
       No two values of x give the same value for $f(x)$
   g $f(x) = \dfrac{1}{x - 3}$ is one-to-one
       No two values of x give the same value for $f(x)$
   h $f(x) = \cos x, 0° \leq x \leq 360°$ is many-to-one
       E.g. $f(0)$ and $f(360) = 1$
   i $f(x) = \cos x, 0° \leq x \leq 180°$ is one-to-one
       No two values of x give the same value for $f(x)$
   j $f(x) = \cos 2x, 0° \leq x \leq 180°$ is many-to-one
       E.g. $f(0)$ and $f(180) = 1$

5 **a** **i**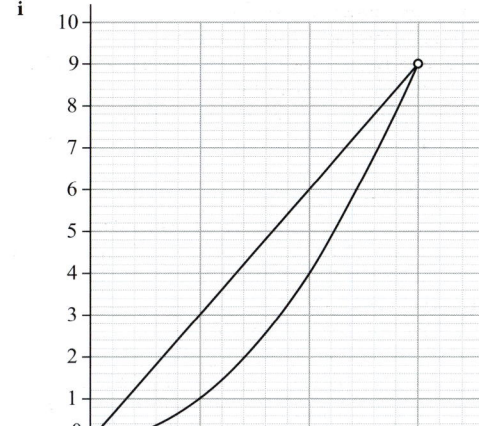
       ii Point of intersection $(0,0)$
   b **i**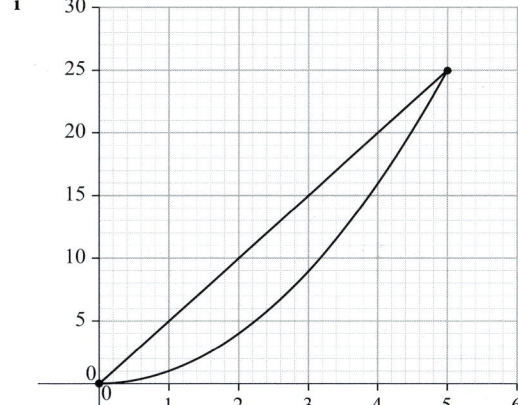
       ii Points of intersection $(0,0)$ and $(3,9)$
   c **i**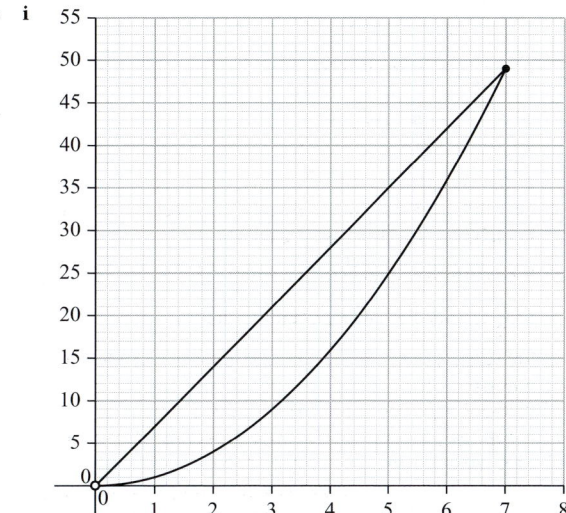
       ii Point of intersection $(3,9)$

6 **a** **i** $fg(1) = -11$
           $gf(-2) = -6$
           $ff\left(\dfrac{-2}{3}\right) = 22$

    ii $fg(1) = -1$

        $gf(-2) = \dfrac{-8}{5}$

        $ff\left(\dfrac{-2}{3}\right) = \dfrac{-2}{3}$

  b $fg(2) = 0$

      $gf(2) = -8$

      $fg(-4) = 36$

      $gf(-4) = -8$

7 **a** **i** The domain is $\{x \in \mathbb{R}\}$

        The range is $\{f(x) \in \mathbb{R}\}$

      ii $f(0) = 0$

         $f(-4) = -64$

         $f(4) = 64$

  b **i** The domain is $\{x \in \mathbb{R}\}$

        The range is $\{f(x) \in \mathbb{R}: y \geq -10\}$

      ii $f(0) = 6$

         $f(-4) = 54$

         $f(4) = -10$

8 13 or 17

9 **a** $y = 2x + 6$

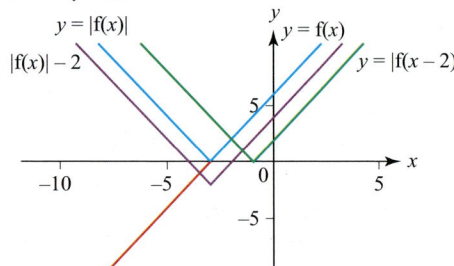

  b $y = -x - 5$

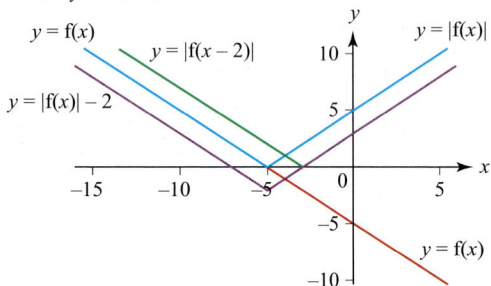

10 a **i** $0 \leq f(x) < 64$

      ii $f^{-1}(x) = \sqrt{x}, \, 0 \leq x < 64$

      iii

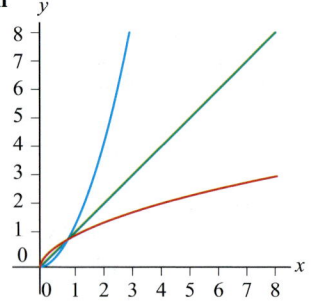

      iv $0 \leq f^{-1}(x) < 8$

b **i** $0 < f(x) < 512$

    ii $f^{-1}(x) = 2 + \sqrt[3]{x}, \, 0 < x < 512$

    iii

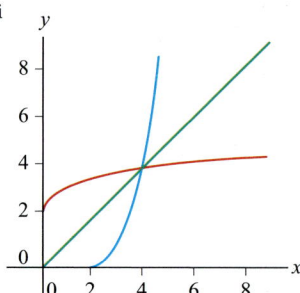

    iv $f^{-1}(x) \in (2, 10)$

c **i** $0 < f(x) < 1$

    ii $f^{-1}(x) = \log_2 x, \, 0 < x < 1$

    iii

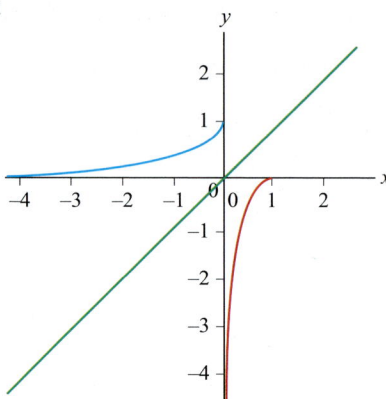

    iv $f^{-1}(x) < 0$

11 a $f(x) = x^2, \, g(x) = 2x$

    i $fg(x) = (2x)^2 = 4x^2$

    ii This transformation is a vertical stretch with sf 4

        (or horizontal stretch with sf ½).

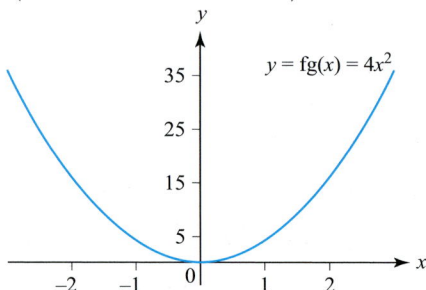

  b $f(x) = 4x^2, \, h(x) = x + 3$

    i $fh(x) = 4(h(x))^2$

            $= 4(x + 3)^2$

            $= 4(x^2 + 6x + 9)$

            $= 4x^2 + 24x + 36$

    ii This transformation is a translation through vector $\begin{pmatrix} -3 \\ 0 \end{pmatrix}$

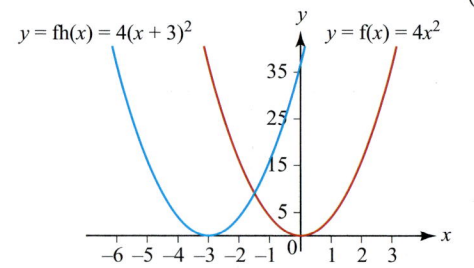

12 a $f(x) = (x + 3)^2, x \in \mathbb{R}$
b $g(x) = 4x, x \in \mathbb{R}$
c $gf(x) = 4(x + 3)^2, x \in \mathbb{R}$

$gf(x) = 4(x + 3)^2$

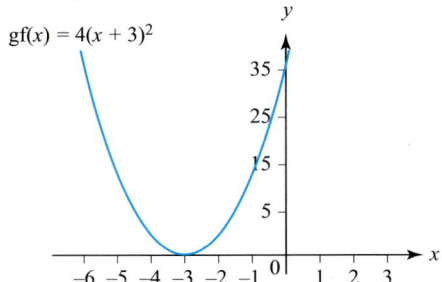

13 a i $x^2 - 6x + 13 \equiv (x - 3)^2 + 4$
Translate $y = x^2$ by the vector $\begin{pmatrix} 3 \\ 4 \end{pmatrix}$

ii $4x^2 + 12x + 8 \equiv (2x + 3)^2 - 1$
First, transform $y = x^2$ into $y = (2x)^2$ by a stretch sf $\dfrac{1}{2}$ parallel to the x-axis.
Then transform to $y = (2(x + 1.5))^2 = (2x + 3)^2$ by a translation by the vector $\begin{pmatrix} -1.5 \\ 0 \end{pmatrix}$
Then transform to $y = (2x + 3)^2 - 1$ by a translation by the vector $\begin{pmatrix} 0 \\ -1 \end{pmatrix}$

b i $(x + 2)^3 - 7$
Start with $y = x^3$
Use translation $\begin{pmatrix} -2 \\ 0 \end{pmatrix}$ to transform to $(x + 2)^3$
Use translation $\begin{pmatrix} 0 \\ -7 \end{pmatrix}$ to transform to $(x + 2)^3 - 7$

ii $(3x - 5)^3 + 6$
Start with $y = x^3$
Transform to $(3x)^3$. This is a stretch sf $\dfrac{1}{3}$ parallel to the x-axis.
Then transform to $(3x - 5)^3 = \left(3\left(x - \dfrac{5}{3}\right)\right)^3$. This is a translation $\begin{pmatrix} \frac{5}{3} \\ 0 \end{pmatrix}$
Then transform to $(3x - 5)^3 + 6$. This is a translation $\begin{pmatrix} 0 \\ 6 \end{pmatrix}$
OR
Transform to $27x^3$. This is a stretch sf 27 parallel to the y-axis.
Then transform to $27\left(x - \dfrac{5}{3}\right)^3$. This is a translation $\begin{pmatrix} \frac{5}{3} \\ 0 \end{pmatrix}$
Then transform to $27\left(x - \dfrac{5}{3}\right)^3 + 6 = (3x - 5)^3 + 6$. This is a translation $\begin{pmatrix} 0 \\ 6 \end{pmatrix}$

14 a $f(x) = e^x$ **b** $f(x) = \log(x - 1)^3$
15 a $gf(x) = (-x)^2, x \in \mathbb{R}$

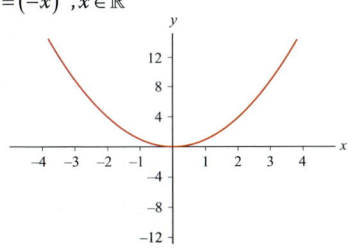

b $gf(x) = x^3, x \in \mathbb{R}$

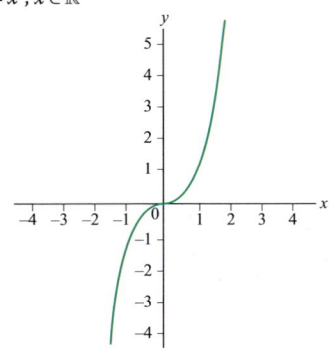

Exercise 12.2B Reasoning and problem-solving

1 a

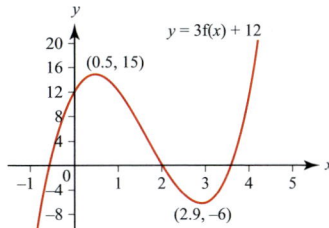

$y = 3f(x) + 12$
(0.5, 15)
(2.9, −6)

b Simple ways in which the domain could be restricted include $0.5 < x < 2.9$ or $x < 0.5$ or $x > 2.9$

2 a Transform w to $(w - 15)$ by the translation $\begin{pmatrix} 15 \\ 0 \end{pmatrix}$
Transform $(w - 15)$ to $|(w - 15)|$ by a reflection in the w-axis of the portion of the graph below the w-axis.
Transform $|(w - 15)|$ to $2|(w - 15)|$ by a stretch parallel to the S-axis, sf 2
Transform $2|(w - 15)|$ to $-2|(w - 15)|$ by a reflection in the w-axis
Transform $-2|(w - 15)|$ to $-2|(w - 15)| + 30$ by the translation $\begin{pmatrix} 0 \\ 30 \end{pmatrix}$

b

w	0	2	4	6	8	10	12	14	16	18	20	22	24	26	28	30
S	0	4	8	12	16	20	24	28	28	24	20	16	12	8	4	0

c

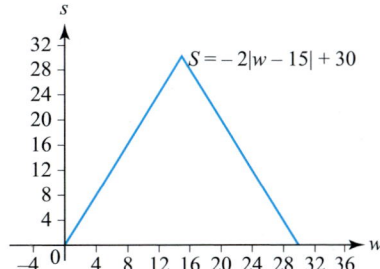

$S = -2|w - 15| + 30$

d £30 000

3 $y = |5x - 36|$

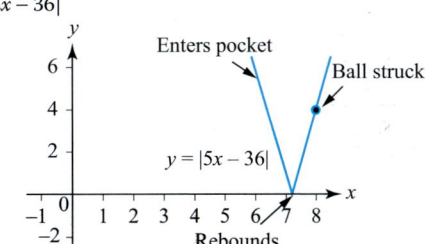

Enters pocket
Ball struck
$y = |5x - 36|$
Rebounds

4 $f(x) = \dfrac{-4}{3}|x - 16.5| + 22$

a

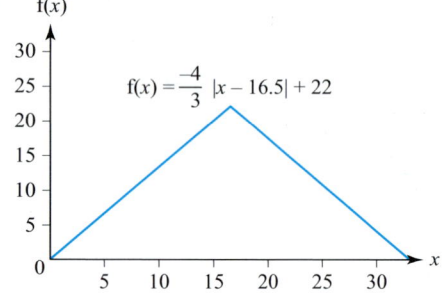

b Domain is $\{x \in \mathbb{R} : 0 \le x \le 33\}$, Range is $\{y \in \mathbb{R} : 0 \le y \le 22\}$

c 22 m

d 33 m

5 a There is no inverse function for f(x) because f(x) is MANY-ONE.

b There is no inverse function for g(x) because g(x) is MANY-ONE.

c The inverse function is $h^{-1}(x) = \sqrt{x} - 2, x \ge 0$

6 a Domain is $\{x \in \mathbb{R}; x \ne 2\}$
Range is $\{y \in \mathbb{R}; y \ne 0\}$

b $f^{-1}(x) = \dfrac{1}{x} + 2$

Domain of $f^{-1}(x)$ is $\{x \in \mathbb{R}; x \ne 0\}$; Range of $f^{-1}(x)$ is $\{y \in \mathbb{R}; y \ne 2\}$

c The domain of f(x) is the same as the range of $f^{-1}(x)$
The range of f(x) is the same as the domain of $f^{-1}(x)$

d

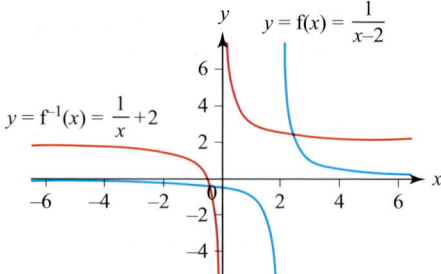

7 a, b, c and **d**

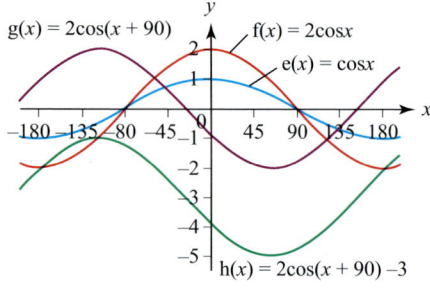

e f(x) is a vertical stretch of e(x) by sf 2

g(x) is a translation of f(x) by $\begin{pmatrix} -90 \\ 0 \end{pmatrix}$

h(x) is a translation of g(x) by $\begin{pmatrix} 0 \\ -3 \end{pmatrix}$

8 Reflection in the x-axis;

Translation by $\begin{pmatrix} 30 \\ 0 \end{pmatrix}$

9 a

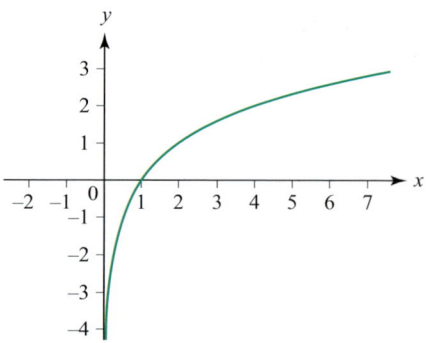

b

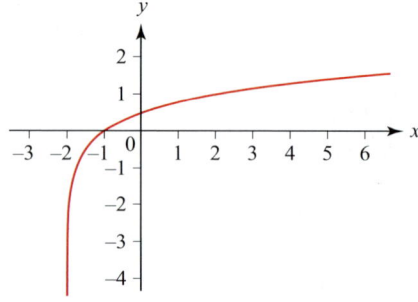

c $y = \dfrac{1}{2}\log_2(x + 2)$

10 a i Stretch, parallel to y-axis, sf 3; translation $\begin{pmatrix} 0 \\ -12 \end{pmatrix}$

ii $fg(e^x) = 3e^x - 12$

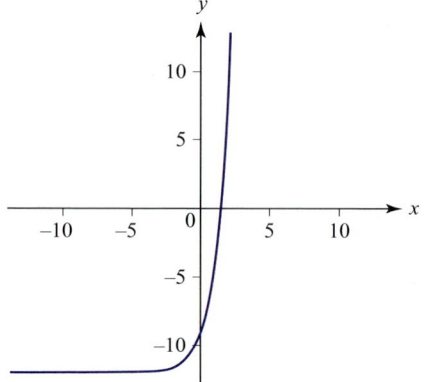

b i Stretch, parallel to y-axis, sf 3; translation $\begin{pmatrix} 0 \\ -4 \end{pmatrix}$

ii $gf(e^x) = 3e^x - 4$

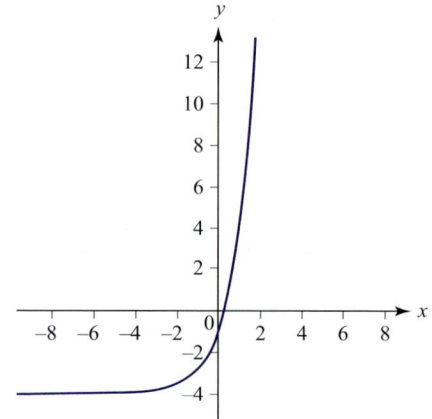

11 Her mistake is that $|4-x| \neq 4+x$

$$|f(x)| = \frac{1}{2}x \Rightarrow$$

$4-x = \frac{1}{2}x$ or $4-x = -\frac{1}{2}x$

$4 = \frac{3}{2}x$ or $4 = \frac{1}{2}x$

$x = \frac{8}{3}$ or $x = 8$

12 a Domain $\{x \in \mathbb{R} : x > 0\}$
Range $\{y \in \mathbb{R}\}$

b Saqib is not correct.
$fg(x) = (\ln x)^3$ is a function with domain $\{x \in \mathbb{R} : x > 0\}$
$gf(x) = \ln x^3$ is also a function if you restrict the domain to $\{x \in \mathbb{R} : x^3 > 0\}$, that is, $\{x \in \mathbb{R} : x > 0\}$

Exercise 12.3A Fluency and skills

1 a $t = 5 \rightarrow \left(5, \frac{4}{5}\right)$

$t = 2 \rightarrow (2, 2)$

$t = -3 \rightarrow \left(-3, \frac{-4}{3}\right)$

b $t = 5 \rightarrow \left(\frac{3}{25}, -10\right)$

$t = 2 \rightarrow \left(\frac{3}{4}, -4\right)$

$t = -3 \rightarrow \left(\frac{1}{3}, 6\right)$

c $t = 5 \rightarrow \left(-\frac{3}{2}, -\frac{3}{7}\right)$

$t = 2 \rightarrow (-3, 0)$

$t = -3 \rightarrow \left(\frac{-1}{2}, -5\right)$

d $t = 5 \rightarrow \left(\frac{15}{6}, 43\right)$

$t = 2 \rightarrow (3, 4)$

$t = -3 \rightarrow \left(\frac{1}{2}, \frac{-23}{3}\right)$

2 a $(y-1)^2 = 4x - 8$ **b** $y = -\frac{x^3}{2}$

c $y = \frac{3}{2x}$ **d** $y = \frac{48}{x^4}$

e $y = \frac{2(x-1)}{1+x}$ **f** $x^2 + y^2 = 4$

3 a
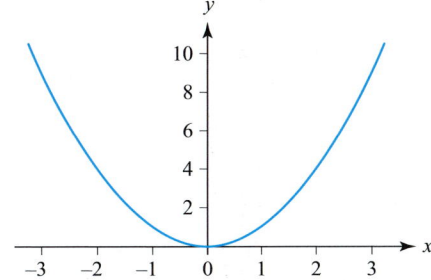

b
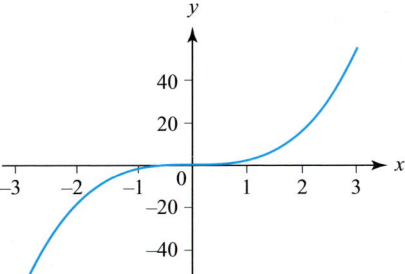

c
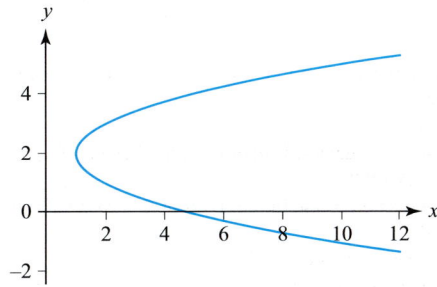

d
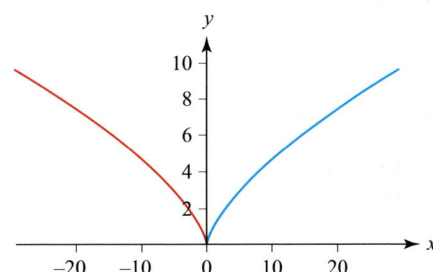

e
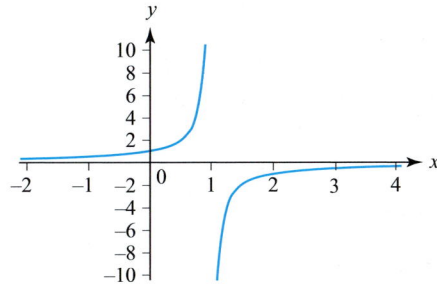

f
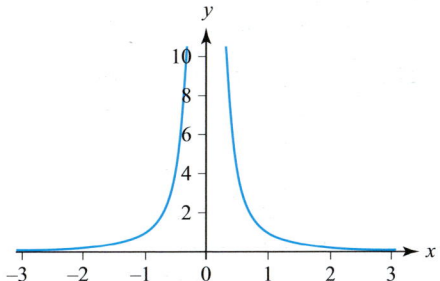

4 a Akeem is correct since, if $t = 2$ then the expression for x would be undefined.

b $y^2 = \frac{4}{x}$

$t > 2 \Rightarrow x > 0$ so domain is $\{x \in \mathbb{R} : x > 0\}$
$\Rightarrow y > 0$ so range is $\{y \in \mathbb{R} : y > 0\}$

5 $y^2 = (2at)^2 = 4a^2t^2$
$4ax = 4a(at^2) = 4a^2t^2$
So $y^2 = 4ax$

6 $25 = xy^2$ domain $x > 0$; range $y > 0$

7 $\sin^2\theta + \cos^2\theta = 1$

$x = 5\sin\theta; y = 5\cos\theta \Rightarrow \dfrac{x^2}{25} + \dfrac{y^2}{25} = 1$

So $x^2 + y^2 = 25$, which is a circle, centre O, radius 5

8 a $(x-8)^2 + (y-6)^2 = 49$ **b** $(x-3)^2 + (y+1)^2 = 25$

c $(x+4)^2 + (y-1)^2 = \dfrac{1}{4}$ **d** $(x+4)^2 + (y+3)^2 = 2$

9 a $\left(\dfrac{1}{2}, \dfrac{-3}{2}\right)$ **b** $\left(\dfrac{-19}{7}, \dfrac{-6}{25}\right)$

10 a $x = t^3$; $y = t^4$

b $x = t + 3; y = t^2 + 3t$

c $x = \dfrac{5}{1-t^2}; y = \dfrac{5t}{1-t^2}$

d $x = \dfrac{1 \pm \sqrt{1+4t^4}}{2t^4}; y = \dfrac{1 \pm \sqrt{1+4t^4}}{2t^3}$

e $x = \dfrac{-3t \pm \sqrt{9t^2+4t}}{2}; y = \dfrac{-3t^2 \pm t\sqrt{9t^2+4t}}{2}$

Exercise 12.3B Reasoning and problem-solving

1 a When $t = 3.16\,\text{s}$

b $2371.71\,\text{m}$

c $750\sqrt{5}\,\text{m}$

2 $x = -10\sin\theta; y = -10\cos\theta$

When $\theta = 90°$, child is at $(-10, 0)$

When $\theta = 135°$, child is at $(-5\sqrt{2}, 5\sqrt{2})$

When $\theta = 180°$, child is at $(0, 10)$

When $\theta = 270°$, child is at $(10, 0)$

3

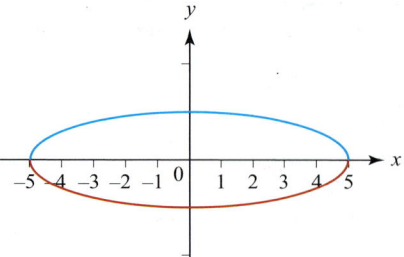

4 Yes he does succeed

5 a $125\,\text{m}$ **b** $t = 7\,\text{s}$ **c** $35\,\text{m}$

6 $y^2 = 4x^2(1 - x^2)$

7 a $x = \sin\theta + 6; y = \dfrac{3}{2}\cos\theta$

b $(x-6)^2 + \dfrac{4}{9}y^2 = 1$

8 a Curve A:

$y = 3 \times 3^{-2t}$

$\Rightarrow y = 3 \times \dfrac{5}{x}$

$\Rightarrow y = \dfrac{15}{x}$

Curve B:

$t = \dfrac{3}{x}$

$\Rightarrow y = 5 \times \dfrac{3}{x}$

$\Rightarrow y = \dfrac{15}{x}$

Hence curves A and B are the same.

b $y = \dfrac{15}{x}; x > 0$

9 $(2\sqrt{5}+7,\ 2\sqrt{5}+5)$ and $(7-2\sqrt{5},\ 5-2\sqrt{5})$

10 $(9, 6)$ and $(9, -6)$

11 $(5, -4)$ and $(29, -20)$

12 a $x = 1 + 5\cos\theta$ **b** $x = 1 - 5\sin\alpha$

$y = 3 + 5\sin\theta$ $\qquad y = 3 + 5\cos\alpha$

13 Circle is $x^2 + (y-4)^2 = 9$

For the parabola, $t = -\dfrac{x}{5}$

$\Rightarrow y = \dfrac{2x^2}{25}$

$\Rightarrow \dfrac{25y}{2} = x^2$

Parabola and circle intersect when

$\dfrac{25y}{2} + (y-4)^2 = 9$

$\Rightarrow y^2 + \dfrac{9}{2}y + 7 = 0$

Discriminant $= -\dfrac{31}{4} < 0 \Rightarrow$ no solutions

Hence the circle and parabola do not intersect.

Exercise 12.4A Fluency and skills

1 a $\dfrac{x(3x-11)}{(x-3)(x-5)}$ **b** $\dfrac{-2y(y+18)}{(y+8)(y-2)}$

c $\dfrac{x(7y-23)}{(y-3)(2y-7)}$ **d** $\dfrac{-z(z+57)}{(2z+9)(3z-4)}$

2 a i $\dfrac{2x-1}{x+2}$ **ii** $\dfrac{x+2}{2x-1}$

b 1

3 a $\dfrac{25(z+9)}{96(z+3)}$ **b** $\dfrac{3}{4w^2}$

c $\dfrac{3n-7}{2n+3}$ **d** $\dfrac{2(m-2)}{3m+1}$

4 a $8x + 14$ **b** $3x + 21$

5 a $x^2 + 5x - 1$

b $x^2 + 2x + \dfrac{5}{2}$ remainder $= -\dfrac{111}{2}$

c $x^2 + 5x - 7$ remainder $= -19$

d $2x^2 - \dfrac{1}{3}x - \dfrac{56}{9}$ remainder $= -\dfrac{52}{9}$

e $x^2 - 3x + 9$

6 a $\dfrac{107}{4}$ **b** 13

c $\dfrac{89}{9}$ **d** $\dfrac{27609}{8}$

7 $(x-1)(x+1)(x^2-3)$

8 a Remainder $= 6\left(\dfrac{4}{3}\right)^3 - 5\left(\dfrac{4}{3}\right)^2 - 16\left(\dfrac{4}{3}\right) + 16 = 0$

Hence, $(3x-4)$ is a factor

b Remainder $= 2\left(-\dfrac{1}{2}\right)^3 - \left(-\dfrac{1}{2}\right)^2 + 7\left(-\dfrac{1}{2}\right) + 4 = 0$

Hence, $(2x+1)$ is a factor

c Remainder $= 8(4)^3 - 36(4)^2 + 18(4) - 8 = 0$

Hence, $(2x-8)$ is a factor

9 a Remainder $= 6\left(-\dfrac{1}{3}\right)^3 + 5\left(-\dfrac{1}{3}\right)^2 + 13\left(-\dfrac{1}{3}\right) + 4 = 0$

Hence, $(3x+1)$ is a factor

b $(3x+1)(2x^2 + x + 4)$

10 a Remainder $= 3\left(\dfrac{2}{3}\right)^3 + 10\left(\dfrac{2}{3}\right)^2 - 23\left(\dfrac{2}{3}\right) + 10 = 0$

Hence, $(3x-2)$ is a factor

b $(3x-2)(x+5)(x-1)$

c $\dfrac{(x+5)(x-1)}{(x+2)}$

11 a $(x+2)(2x-1)(3x-2)$

b $(4x+3)(x-1)(x-2)$

c $x(6x-5)(3x^2 - 2x - 2)$

Exercise 12.4B Reasoning and problem-solving

1 $12x^3 + 2x^2 - 54x + 40$

2 $q = -4, p = -20$

3 $\dfrac{x}{x-1} + \dfrac{1}{(x-1)(x-2)} = \dfrac{x(x-1)(x-2) + (x-1)}{(x-1)^2(x-2)}$

$= \dfrac{x(x-2) + 1}{(x-1)(x-2)}$

$= \dfrac{x^2 - 2x + 1}{(x-1)(x-2)}$

$= \dfrac{(x-1)^2}{(x-1)(x-2)}$

$= \dfrac{x-1}{x-2}$

So the only vertical asymptote is $x = 2$

4 $x = a$ or $x = 2 - a$

5 $(x^2 - 9) = (x+3)(x-3)$ so we need to show that both $\mathrm{f}(3) = 0$ and $\mathrm{f}(-3) = 0$

$\mathrm{f}(3) = (3)^4 + 6(3)^3 - 4(3)^2 - 54(3) - 45 = 0$ so $(x-3)$ is a factor

$\mathrm{f}(-3) = (-3)^4 + 6(-3)^3 - 4(-3)^2 - 54(-3) - 45 = 0$ so $(x+3)$ is a factor

Hence, $(x^2 - 9)$ is a factor of $x^4 + 6x^3 - 4x^2 - 54x - 45$

6 $(x-2)(x+3) = x^2 + x - 6$

7 $(x-1)(x+4)$

8 $\mathrm{f}(2) = (2)^4 - 10(2)^3 + 37(2)^2 - 60(2) + 36$

$= 16 - 80 + 148 - 120 + 36 = 0$

$x^4 - 10x^3 + 37x^2 - 60x + 36 \equiv (x-2)(x^3 - 8x^2 + 21x - 18)$

$(2)^3 - 8(2)^2 + 21(2) - 18 = 0$

So $(x-2)(x-2)$ is a factor.

$x^4 - 10x^3 + 37x^2 - 60x + 36 \equiv (x-2)^2(x-3)^2$

9 a $30\left(-\dfrac{2}{5}\right)^3 + 7\left(-\dfrac{2}{5}\right)^2 - 12\left(-\dfrac{2}{5}\right) - 4 = 0$

$\Rightarrow (5x+2)$ is a factor

b $\dfrac{(5x+2)(3x-2)}{x}$

c i $x \in \mathbb{R}, x \neq 0; \mathrm{f}(x) \in \mathbb{R}$

ii $\mathrm{f}'(x) = \dfrac{15x^2 + 4}{x^2}$

$15x^2 + 4 > 4 > 0$ for all $x \neq 0$

$x^2 > 0$ for all $x \neq 0$

Positive ÷ positive = positive

$\therefore \mathrm{f}'(x) > 0$

10 a $(2x - 5)$ is a factor of $\mathrm{f}(x) \Rightarrow \mathrm{f}\left(\dfrac{5}{2}\right) = 0$

$a\left(\dfrac{5}{2}\right)^4 + b\left(\dfrac{5}{2}\right)^2 - 75 = 0$

$\Rightarrow \dfrac{625}{16}a + \dfrac{25}{4}b - 75 = 0$

Let $x = -\dfrac{5}{2}$

$\mathrm{f}\left(-\dfrac{5}{2}\right) = a\left(-\dfrac{5}{2}\right)^4 + b\left(-\dfrac{5}{2}\right)^2 - 75$

$= \dfrac{625}{16}a + \dfrac{25}{4}b - 75$

$= 0$

$\Rightarrow (2x + 5)$ is a factor of $\mathrm{f}(x)$

b $a = 4, b = -13, c = 3$

11 The remainder is not a constant term that can be added on at the end of the quotient.

$(4x^4 + 6x^3 + x - 8) \div (2x + 1) = 2x^3 + 2x^2 - 0.5 - \dfrac{7.5}{2x+1}$

12 $b = \dfrac{1}{2}$ or -4

13 a If $(x-a)$ is a factor of $\mathrm{f}(x)$ then $\mathrm{f}(x) = (x-a)\mathrm{p}(x)$

If $(x-a)$ is a factor of $\mathrm{g}(x)$ then $\mathrm{g}(x) = (x-a)\mathrm{q}(x)$

So $[\mathrm{f}(x) - \mathrm{g}(x)] \equiv (x-a)\mathrm{p}(x) - (x-a)\mathrm{q}(x)$

$\equiv (x-a)[\mathrm{p}(x) - \mathrm{q}(x)]$

$\equiv (x-a)\mathrm{r}(x)$

So $(x-a)$ is a common factor of $\mathrm{f}(x) - \mathrm{g}(x)$

b $(x-a)$ is a common factor of

$kx^3 + 3x^2 + x + 4 - (kx^3 + 2x^2 + 9x - 8) = 0$

or $x^2 - 8x + 12 = 0$ or $(x-2)(x-6) = 0$

So for these polynomials to have a common factor then $x = 2$ or 6

When $x = 2$, $k(2)^3 + 3(2)^2 + 2 + 4 = 0$ and

$k(2)^3 + 2(2)^2 + 9(2) - 8 = 0 \to k = \dfrac{-9}{4}$

When $x = 6$, $k(6)^3 + 3(6)^2 + 6 + 4 = 0$ and

$k(6)^3 + 2(6)^2 + 9(6) - 8 = 0 \to k = \dfrac{-59}{108}$

Exercise 12.5A Fluency and skills

1 $\dfrac{7}{(x-4)} - \dfrac{3}{(3x+2)}$

2 a $B = 4$ and $A = -4$

b $C = 2, D = 4, E = -5$

c $F = 3, H = 5$ and $G = 4$

3 a $B = 4, A = 3$

b $E = 17, C = -9$ and $D = 4$

c $G = 3, F = 2, H = -5$

4 $A = 3, C = -1, B = 2$

5 $\dfrac{1}{(1-x)^2} = \dfrac{1}{(1-x)^2}$ so there are no partial fractions.

$\dfrac{x}{(1-x)^2} = \dfrac{-1}{(1-x)} + \dfrac{1}{(1-x)^2}$ so there are partial fractions.

6 a $\dfrac{4}{x+3} - \dfrac{3}{x-2}$

b $\dfrac{2}{x-1} - \dfrac{2}{x+1} + \dfrac{6}{(x+1)^2}$

c $\dfrac{1}{2x-5} - \dfrac{2}{x+6} - \dfrac{5}{(2x-1)}$

7 a $\dfrac{2}{x} - \dfrac{2}{x+4}$ **b** $\dfrac{-1}{x+1} + \dfrac{1}{x-3}$

c $\dfrac{23}{18(x+5)} - \dfrac{23}{18(x-7)} - \dfrac{62}{3(x-7)^2}$

d $\dfrac{1}{x} + \dfrac{20}{11(2x-3)} - \dfrac{21}{11(x+4)}$

8 a $3 - \dfrac{2}{x-1} + \dfrac{4}{x-3}$ **b** $5 + \dfrac{1}{(x+1)} - \dfrac{4}{x+5}$

9 a $\dfrac{3}{x-\sqrt{3}} + \dfrac{1}{x+\sqrt{3}}$ **b** $\dfrac{5}{\sqrt{6}x-\sqrt{5}} - \dfrac{4}{\sqrt{6}x+\sqrt{5}}$

10 $\dfrac{a}{(a-b)(x-a)} + \dfrac{b}{(b-a)(x-b)}$

11 $Q = b - ac$; $P = a$

Exercise 12.5B Reasoning and Problem Solving

1 $\dfrac{5t-27}{(t-3)(t-7)}$ minutes

2 $\dfrac{1}{2x-1}$

3 A = 3, B = 5

4 a Let $a = b = c = d = 1$

Then $\dfrac{a}{x+c} + \dfrac{b}{x+d} \equiv \dfrac{1}{x+1} + \dfrac{1}{x+1} \equiv \dfrac{2}{x+1}$

Now $\dfrac{a+b}{(x+c)(x+d)} \equiv \dfrac{2}{(x+1)^2}$

Since $\dfrac{2}{(x+1)} \neq \dfrac{2}{(x+1)^2}$ for all x, then

$\dfrac{a+b}{(x+c)(x+d)} \neq \dfrac{a}{x+c} + \dfrac{b}{x+d}$ for all real a, b, c, d

b $a = \dfrac{a+b}{d-c}$ and $b = \dfrac{a+b}{c-d}$

5 Need three partial fractions in order to cope with the repeated $x+1$ term in the denominator

$\dfrac{14}{(x-5)(x+1)^2} \equiv \dfrac{A}{(x-5)} + \dfrac{B}{(x+1)} + \dfrac{C}{(x+1)^2}$

$14 = A(x+1)^2 + B(x-5)(x+1) + C(x-5)$

$x = 5 \Rightarrow A = \dfrac{7}{18}$

$x = -1 \Rightarrow C = -\dfrac{7}{3}$

$x^2: \quad 0 = \dfrac{7}{18} + B \Rightarrow B = -\dfrac{7}{18}$

$\dfrac{14}{(x-5)(x+1)^2} \equiv \dfrac{7}{18(x-5)} - \dfrac{7}{18(x+1)} - \dfrac{7}{3(x+1)^2}$

6 a $\dfrac{1}{r+1} - \dfrac{1}{r+2}$

b $\dfrac{1}{1(2)} + \dfrac{1}{2(3)} + \dfrac{1}{3(4)} + \dfrac{1}{4(5)} + \dots + \dfrac{1}{(n+1)(n+2)}$

$= \left(1 - \dfrac{1}{2}\right) + \left(\dfrac{1}{2} - \dfrac{1}{3}\right) + \left(\dfrac{1}{3} - \dfrac{1}{4}\right) + \left(\dfrac{1}{4} - \dfrac{1}{5}\right) +$

$\dots + \left(\dfrac{1}{n+1} - \dfrac{1}{n+2}\right)$

$= 1 + \left(\dfrac{1}{2} - \dfrac{1}{2}\right) + \left(\dfrac{1}{3} - \dfrac{1}{3}\right) + \left(\dfrac{1}{4} - \dfrac{1}{4}\right) +$

$\dots + \left(\dfrac{1}{n+1} - \dfrac{1}{n+1}\right) - \dfrac{1}{n+2}$

$= 1 - \dfrac{1}{n+2}$

c As $n \to \infty$, $\dfrac{1}{n+2} \to 0$ so the sum tends to 1

Exercise 12.6A Fluency and skills

1 a i $-3\mathbf{i} + 2\mathbf{j} + 4\mathbf{k}$ **ii** $11\mathbf{i} + 3\mathbf{j} - 11\mathbf{k}$
iii 7   **iv** $\sqrt{33}$
v $\dfrac{1}{7}(6\mathbf{i} - 3\mathbf{j} - 2\mathbf{k})$ **vi** $31.0°, 115.4°, 106.6°$

b $24\mathbf{i} - 12\mathbf{j} - 8\mathbf{k}$

c $p = 2, q = -3, r = 13$

2 $\mathbf{r} = 7.13\mathbf{i} + 0.697\mathbf{j} + 3.57\mathbf{k}$

3 a $27.1°$ or $152.9°$

b $\pm 10.7\mathbf{i} + 4.50\mathbf{j} + 3.11\mathbf{k}$

4 a $AB = 3\sqrt{2}$, $BC = 3\sqrt{2}$, $AC = \sqrt{36} = 6$

b $AB^2 + BC^2 = 18 + 18$
$= 36$
$= AC^2$
Triangle ABC satisfies Pythagoras' theorem
Hence, is right-angled at B

c Triangle is isosceles since $AB = BC$

5 a $6\mathbf{i} - 2\mathbf{j} + 3\mathbf{k}$

b 7 units

c $\dfrac{6}{7}\mathbf{i} - \dfrac{2}{7}\mathbf{j} + \dfrac{3}{7}\mathbf{k}$

6 $4.8\mathbf{i} + 4.1\mathbf{j} - 0.5\mathbf{k}$

7 a $\begin{pmatrix} 7 \\ 0 \\ -8 \end{pmatrix}$ **b** 3 units

c $\begin{pmatrix} -10 \\ 10 \\ 5 \end{pmatrix}$ **d** $132°$

e $\lambda = 2, \mu = -3$

Exercise 12.6B Reasoning and problem-solving

1 $AB = |\mathbf{b} - \mathbf{a}|$
$= |-\mathbf{i} - 7\mathbf{j} + 5\mathbf{k}|$
$= \sqrt{75}$
$BC = \sqrt{6}$
$AC = \sqrt{81}$
$AB^2 + BC^2 = AC^2$, hence ABC satisfies Pythagoras' theorem, and is right-angled at B

2 $\mathbf{d}$ could be at: $4\mathbf{i} - 2\mathbf{j} - \mathbf{k}$ or $2\mathbf{i} + 3\mathbf{k}$ or $4\mathbf{j} - 3\mathbf{k}$

3 $\mathbf{V} = \pm 2\sqrt{3}(\mathbf{i} + \mathbf{j} + \mathbf{k})$

4 a $26.6°$ **b** $50\,\mathrm{m}$

5 a $\mathbf{p} = \dfrac{1}{2}(\mathbf{a} + \mathbf{b})$, $\mathbf{q} = \dfrac{1}{2}(\mathbf{a} + \mathbf{d})$, $\mathbf{r} = \dfrac{1}{2}(\mathbf{b} + \mathbf{c})$

$\mathbf{s} = \dfrac{2\mathbf{p} + \mathbf{c}}{3} = \dfrac{1}{3}(\mathbf{a} + \mathbf{b} + \mathbf{c})$

$\mathbf{t} = \dfrac{1}{2}(\mathbf{q} + \mathbf{r}) = \dfrac{1}{4}(\mathbf{a} + \mathbf{b} + \mathbf{c} + \mathbf{d})$

Now $\overrightarrow{DT} = \mathbf{t} - \mathbf{d} = \dfrac{1}{4}(\mathbf{a} + \mathbf{b} + \mathbf{c} - 3\mathbf{d})$

and $\overrightarrow{DS} = \mathbf{s} - \mathbf{d} = \dfrac{1}{3}(\mathbf{a} + \mathbf{b} + \mathbf{c} - 3\mathbf{d})$

$\overrightarrow{DT}$ and $\overrightarrow{DS}$ are parallel and both pass through D, so D, T and S are collinear

b $3:1$

6 **a** 19.3

b $c = \dfrac{\mu\mathbf{a} + \lambda\mathbf{b}}{\lambda + \mu}$, $\mathbf{d} = \dfrac{-\mu\mathbf{a} + \lambda\mathbf{b}}{\lambda - \mu}$

$CD = |\mathbf{d} - \mathbf{c}|$

$\quad = \left| \dfrac{(\lambda + \mu)(-\mu\mathbf{a} + \lambda\mathbf{b}) - (\lambda - \mu)(\mu\mathbf{a} + \lambda\mathbf{b})}{\lambda^2 - \mu^2} \right|$

$\quad = \dfrac{2\lambda\mu}{\lambda^2 - \mu^2} |\mathbf{b} - \mathbf{a}|$

So $CD : AB = 2\lambda\mu : (\lambda^2 - \mu^2)$

Review exercise 12

1 Let n be any positive integer such that $n \geq 2$
Then $n! = n \times (n-1) \times (n-2) \times \dots \times 2 \times 1$
which has a factor of 2, so is even.

2 $385 = 5 \times 77 = 5 \times 7 \times 11$
$385 = 7 \times 55 = 7 \times 5 \times 11$
$385 = 11 \times 35 = 11 \times 5 \times 7$
So, whichever way you factorise 385 you end up with the same factors.

3 Assume x is rational, y is irrational, but $x - y$ is rational.
Then $x = \dfrac{p}{q}$, where $p, q \in \mathbb{Z}$, and $q \neq 0$.
Since $x - y$ is rational, $x - y = \dfrac{r}{s}$, for some $r, s \in \mathbb{Z}, s \neq 0$.
$y = x - \dfrac{r}{s}, = \dfrac{p}{q} - \dfrac{r}{s} = \dfrac{ps - qr}{qs}$
Since $ps - qr$ and qs are integers, and $qs \neq 0$, since $q \neq 0$ and $s \neq 0$, y is rational.
This contradicts our original statement.
Therefore $x - y$ is irrational.

4 **a** 14 **b** -125 **c** 26

5 **a** Transform $y = x$ into $y = 5x$ by a vertical stretch sf 5
Transform $y = 5x$ into $y = 5x - 4$ by a translation $\begin{pmatrix} 0 \\ -4 \end{pmatrix}$

b

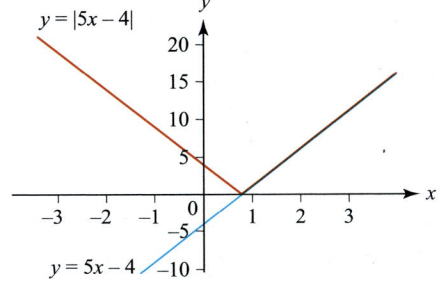

c Either 11 or 19
The sketch of $y = |5x - 4|$ shows these values when $x = 3$ or -3

6 **a** $f^{-1}(x) = \sqrt{\dfrac{5}{x}}, x > 0$ **b** $f^{-1}(x) = \dfrac{4x - 7}{3}, x \in \mathbb{R}$

c $f^{-1}(x) = \dfrac{4x - 2}{x - 1}, x \neq 1$

7 $xy + 20 = 0$

8 $(5, -1)$ and $(-7, -13)$

9 $x^3 - x^2 - x - 1$

10 $\dfrac{n^2 - 4n - 5}{8n^2 + 4n + 3}$

11 $(x - 3)(2x - 1)(2x + 1)$

12 $\dfrac{2}{x + 1} - \dfrac{2}{x - 5} + \dfrac{12}{(x - 5)^2}$

13 **a** $3\mathbf{i} - 3\mathbf{j} + 6\mathbf{k}$ **b** 7.35
c 35.3° **d** $4\mathbf{i} - \mathbf{j} + \mathbf{k}$

Assessment 12

1 $A = 3, B = -5, C = 2, D = -1, E = 3$

2 **a** $f(x) \geq -2$
b $f^{-1}(x) = \sqrt{x + 2}\ x \geq -2$
c

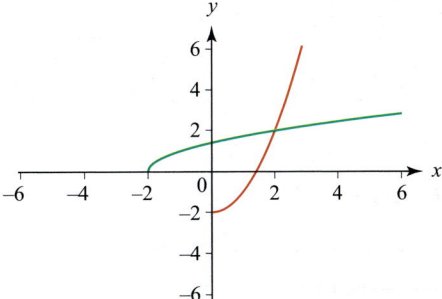

d $x = 2$

3 **a** $f^{-1}(x) = \dfrac{3x}{x - 1}, x \neq 1$

b $gf(x) = \dfrac{3x + 6}{x}, x \neq 0$

c $\dfrac{3x}{x - 1} = \dfrac{3x + 6}{x}$
$3x^2 = 3x^2 + 3x - 6$
$x = 2$

4 **a**

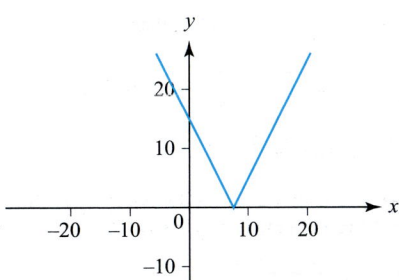

b $x = 9$ or $x = 6$
c $6 \leq x \leq 9$

5 **A** is one-to-one
B is many-to-one
C is one-to-one
D is many-to-one

6 Assume, for a contradiction, that n is even.
Then $n = 2N$ where N is an integer.
So $n^n = (2N)^{2N} = 2^{2N} \times N^{2N}$
So n^n is even, which contradicts our original assumption.
So our original assumption that n is even, must be wrong, and n must be odd.

7 **a** $\dfrac{3x - 7}{(x - 4)(x + 1)} - \dfrac{1}{(x - 4)} \times \dfrac{(x + 1)}{(x + 1)}$

$f(x) = \dfrac{3x - 7 - x - 1}{(x - 4)(x + 1)}$

$\quad = \dfrac{2(x - 4)}{(x - 4)(x + 1)}$

$\quad = \dfrac{2}{x + 1}$

b $\{x \in \mathbb{R} : x \neq -1\}$

c $f^{-1}(x) = \dfrac{2 - x}{x}, x \in \mathbb{R}, x \neq 0$

d $x = 1$ or -2

8 A True

Assume for a contradiction, there exists $x \neq -1$ such that

$$\frac{4x}{(x+1)^2} > 1$$

$$4x > (x+1)^2$$

$$x^2 - 2x + 1 < 0$$

$$(x-1)^2 < 0$$

But a square number cannot be negative, so this gives a contradiction, therefore $\frac{4x}{(x+1)^2} \leq 1$ for all $x \neq -1$

B False

$4! + 1 = 25$, which is not prime

C False

$7 \times 9 \times 11 = 693$, which is not a multiple of 15

D True

$n^3 - n = (n-1)n(n+1)$

In three consecutive numbers, at least one is bound to be even.

In three consecutive numbers, at least one is bound to be a multiple of 3

$2 \times 3 = 6$, hence $n^3 - n$ is divisible by 6

9 $3(x+1) - 4(x-2) = 2(x+1)(x-2)$

$0 = 2x^2 - x - 15$

$0 = (2x+5)(x-3)$

$x = -\frac{5}{2}$ or 3

10 $y = \dfrac{4x}{7x-1}$

11 a

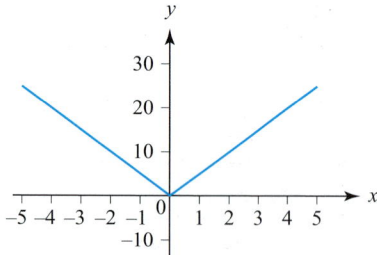

b

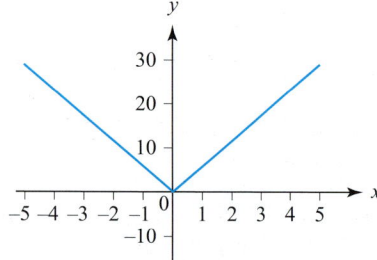

c Stretch, sf 2 in y-direction followed by translation $\begin{pmatrix} 0 \\ -1 \end{pmatrix}$

12 $A = 3, B = -2, C = 1, D = -3, E = 10$

13 a $\mathrm{fg}(x) = (3x-2)^2$

b $x = 1$

c $\mathrm{f}^{-1}(x) = \dfrac{1}{2}\ln x, x > 0$

d $x = \dfrac{1}{2}\ln 5$

14 a $1 = A$

$-2 = -B$

b For $x > 2$, both $\mathrm{f}(x) = \dfrac{1}{x-2}$ and $\mathrm{F}(x) = \dfrac{1}{x-1}$ are decreasing functions

Hence $\mathrm{f}(x) + 2\mathrm{F}(x)$ is also a decreasing function.

15 a Let a be an integer

By the principle of prime factorisation, there exist primes $p_1, p_2, ..., p_n$ such that $a = p_1 \times p_2 \times ... \times p_n$

Then $a^2 = (p_1 \times p_1) \times (p_2 \times p_2) \times ... \times (p_n \times p_n)$

Suppose 3 divides a^2

Since 3 is prime, at least one of $p_1, p_2, ..., p_n$ must be equal to 3

Since $a = p_1 \times p_2 \times ... \times p_n$ and one of $p_1, p_2, ..., p_n$ must be equal to 3, then it follows that 3 also divides a

b Suppose $\sqrt{3}$ is rational.

Let $\left(\dfrac{m}{n}\right)^2 = 3$ where m and n are integers which share no common factors.

Thus $m^2 = 3n^2$ and 3 divides m^2

Since 3 is prime, it must also divide m

So $m = 3k$ for some integer k

Hence, $9k^2 = 3n^2$

$\Rightarrow 3k^2 = n^2$ and 3 divides n^2

Since 3 is prime, it must also divide n

So $n = 3j$ for some integer j

Hence, $\dfrac{m}{n} = \dfrac{3k}{3j}$ which implies m and n share a common factor, 3

This contradicts our original assumption that m and n share no common factors.

Hence $\sqrt{3}$ is irrational.

16 a $x^4 = \sin^2 t$

$y^2 = 9\sin^2 t \cos^2 t$

$y^2 = 9\sin^2 t(1 - \sin^2 t)$

$y^2 = 9x^4(1 - x^4)$

$y = 3x^2\sqrt{(1 - x^4)}$

b Let $y = 2$

$3\sin t \cos t = 2$

$\sin t \cos t = \dfrac{2}{3}$

$\sin^2 t \cos^2 t = \dfrac{4}{9}$

$\sin^2 t(1 - \sin^2 t) = \dfrac{4}{9}$

$\sin^4 t - \sin^2 t + \dfrac{4}{9} = 0$

$(\sin^2 t)^2 - (\sin^2 t) + \dfrac{4}{9} = 0$

$b^2 - 4ac = (-1)^2 - 4 \times 1 \times \dfrac{4}{9}$

$= -\dfrac{7}{9}$

$b^2 - 4ac < 0 \Rightarrow$ no solutions for t

17 a i $\dfrac{4x+1}{(x+1)(x-2)} \equiv \dfrac{1}{(x+1)} + \dfrac{3}{(x-2)}$

ii $\dfrac{15-9x}{(x-1)(x-2)} \equiv \dfrac{6}{(x-1)} - \dfrac{3}{(x-2)}$

b $x = -2$ or $x = -3$

18 a $\begin{pmatrix} -5 \\ -12 \\ 15 \end{pmatrix}$

b $\dfrac{3}{\sqrt{394}}\begin{pmatrix} -5 \\ -12 \\ 15 \end{pmatrix}$

Answers For full solutions go to http://www.oxfordsecondary.co.uk/edexcelalevelmaths-answers

19 a i $\overrightarrow{AB} = \begin{pmatrix} 6 \\ -3 \\ 9 \end{pmatrix}$

ii $\overrightarrow{BC} = \begin{pmatrix} 2 \\ -1 \\ 3 \end{pmatrix}$

b $\overrightarrow{AB} = \begin{pmatrix} 6 \\ -3 \\ 9 \end{pmatrix}$

$= 3\begin{pmatrix} 2 \\ -1 \\ 3 \end{pmatrix}$

$= 3\overrightarrow{BC}$

So A, B and C are collinear.

20 a $x = 1$

b $y = 5$ or $y = -2$

Chapter 13
Exercise 13.1A Fluency and skills

1 a $1 - 3x + 6x^2 - 10x^3 + ...$ **b** $1 + \dfrac{1}{2}x - \dfrac{1}{8}x^2 + \dfrac{1}{16}x^3 + ...$

c $1 + \dfrac{2}{3}x - \dfrac{1}{9}x^2 + \dfrac{4}{81}x^3 + ...$ **d** $1 - 4x + 16x^2 - 64x^3 + ...$

e $1 + 6x + 27x^2 + 108x^3 + ...$ **f** $1 + \dfrac{1}{6}x - \dfrac{1}{36}x^2 + \dfrac{5}{648}x^3 + ...$

2 a $\dfrac{1}{16} - \dfrac{1}{8}x + \dfrac{5}{32}x^2 + ...$ **b** $\dfrac{1}{3} - \dfrac{2}{9}x + \dfrac{4}{27}x^2 + ...$

c $8 - 9x + \dfrac{27}{16}x^2 + ...$ **d** $\dfrac{1}{3} - \dfrac{2}{9}x + \dfrac{1}{9}x^2 + ...$

e $3 + \dfrac{1}{6}x - \dfrac{1}{216}x^2 + ...$ **f** $2 + \dfrac{8}{3}x + \dfrac{32}{9}x^2 + ...$

g $4 - \dfrac{9}{2}x + \dfrac{243}{32}x^2 + ...$ **h** $\dfrac{3}{2} + 2x + 2x^2 + ...$

3 a $1 - 3kx + 6k^2x^2 + ...$ **b** $2 + \dfrac{k}{4}x - \dfrac{k^2}{64}x^2 + ...$

c $1 - \dfrac{2}{k}x + \dfrac{3}{k^2}x^2 + ...$ **d** $\dfrac{1}{k} - \dfrac{2}{k^{\frac{3}{2}}}x + \dfrac{3}{k^2}x^2 + ...$

4 a $-\dfrac{1}{4} < x < \dfrac{1}{4}$, or $|x| < \dfrac{1}{4}$ **b** $-3 < x < 3$, or $|x| < 3$

c $-\dfrac{4}{5} < x < \dfrac{4}{5}$, or $|x| < \dfrac{4}{5}$ **d** $-\dfrac{3}{2} < x < \dfrac{3}{2}$, or $|x| < \dfrac{3}{2}$

5 a $x - 2x^2 + ...$ **b** $2 - 3x + 4x^2 + ...$

c $3 - 8x + 13x^2 + ...$ **d** $1 - 2x + ...$

6 a $\dfrac{1}{4} - \dfrac{1}{4}x + \dfrac{3}{16}x^2 + ...$ **b** $1 - \dfrac{3}{2}x + \dfrac{9}{4}x^2 + ...$

c $3 - 6x + 12x^2 + ...$ **d** $3 + \dfrac{3}{4}x + \dfrac{3}{16}x^2 + ...$

7 a 2 **b** -10

Exercise 13.1B Reasoning and problem-solving

1 a $1 + \dfrac{5}{2}x - \dfrac{25}{8}x^2 + ...$

b Let $x = 0.05$

$\sqrt{1 + 5(0.05)} \approx 1 + \dfrac{5}{2}(0.05) - \dfrac{25}{8}(0.05)^2$

$\text{LHS} = \sqrt{\dfrac{5}{4}} = \dfrac{\sqrt{5}}{2}$ $\text{RHS} = \dfrac{143}{128}$

$\therefore \dfrac{\sqrt{5}}{2} \approx \dfrac{143}{128}$

so $\sqrt{5} \approx \dfrac{143}{64}$

c Although when $x = 0.8$, $\sqrt{1 + 5x} = \sqrt{1 + 5(0.8)}$
$= \sqrt{5}$

the expansion of $\sqrt{1 + 5x}$ is only valid for $|x| < \dfrac{1}{5} = 0.2$

2 a $2 + \dfrac{1}{4}x - \dfrac{1}{64}x^2 + ...$

The full expansion is valid for $-4 < x < 4$

b i Let $x = 0.25$

$\sqrt{4 + 0.25} \approx 2 + \dfrac{1}{4}(0.25) - \dfrac{1}{64}(0.25)^2$

$\text{LHS} = \sqrt{\dfrac{17}{4}} = \dfrac{\sqrt{17}}{2}$ $\text{RHS} = \dfrac{2111}{1024}$

$\sqrt{17} \approx \dfrac{2111}{512}$

ii $\dfrac{1983}{512}$

c $\sqrt{17} - \sqrt{15} \approx \dfrac{2111}{512} - \dfrac{1983}{512}$

$= \dfrac{128}{512}$

$= \dfrac{1}{4}$

3 a $n = \dfrac{2}{3}, k = 2$

b 1.19

4 $(1 + px)^n = 1 + npx + \dfrac{n(n-1)}{2!}p^2x^2 + ...$

$1 + npx + \dfrac{n(n-1)}{2!}p^2x^2 + ... \equiv 1 + 12x + 24x^2 + ...$

Equate x coefficients: $np = 12$ (1)

Equate x^2 coefficients: $\dfrac{n(n-1)}{2}p^2 = 24$ (2)

Solve (1) and (2) simultaneously to give $p = 8$, $n = \dfrac{3}{2}$

$(1 + 8x)^{\frac{3}{2}} \approx 1 + 12x + 24x^2$

Let $x = 0.01$

$(1 + 8(0.01))^{\frac{3}{2}} \approx 1 + 12(0.01) + 24(0.01)^2$

$\text{LHS} = \left(\dfrac{27}{25}\right)^{\frac{3}{2}}$ $\text{RHS} = \dfrac{1403}{1250}$

$= \dfrac{81}{125}\sqrt{3}$

$\therefore \dfrac{81}{125}\sqrt{3} = \dfrac{1403}{1250}$

so $\sqrt{3} \approx \dfrac{1403}{810}$

5 a $3 - 5x + 11x^2 - 29x^3 + ...$

b $|x| < \dfrac{1}{3}$

6 a $1 + x^2 + ...$

b

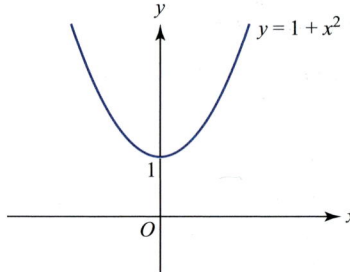

$y = 1 + x^2$

7 a 1.2

  b $1 - \dfrac{7}{4}x + \dfrac{9}{4}x^2 + \ldots$

  c $f(x) \approx 1 - \dfrac{7}{4}x + \dfrac{9}{4}x^2 + \ldots$

  $\therefore f'(x) \approx -\dfrac{7}{4} + \dfrac{9}{2}x + \ldots$

  Gradient of tangent at $P = f'(-0.1)$

  $$\approx -\dfrac{7}{4} + \dfrac{9}{2}(-0.1)$$

  $$= -2.2$$

  Tangent at P is approximated by the line with equation

  $$y - 1.2 = -2.2(x + 0.1)$$

  leading to $y = 0.98 - 2.2x$

8 a $\dfrac{1}{\cos^2 x} = \dfrac{1}{1 - \sin^2 x}$

  $$= (1 - \sin^2 x)^{-1}$$

  $$= 1 + (-1)(-\sin^2 x) + \dfrac{(-1)(-2)}{2!}(-\sin^2 x)^2$$

  $$+ \dfrac{(-1)(-2)(-3)}{3!}(-\sin^2 x)^3 + \ldots$$

  $$= 1 + \sin^2 x + \sin^4 x + \sin^6 x + \ldots$$

  The expansion is valid provided $-1 < \sin x < 1$ which holds provided $-90° < x < 90°$

  b The approximation $\dfrac{1}{\cos^2 x} \approx 1 + \sin^2 x$ is good only if $\sin x$ is close to 0

  The approximation $\dfrac{1}{\sin^2 x} \approx 1 + \cos^2 x$ is good only if $\cos x$ is close to 0

  Since $\sin^2 x + \cos^2 x \equiv 1$, it is not possible for both $\sin x$ and $\cos x$ to be close to 0

  for a given value of x (or, considering their graphs for $0° < x < 90°$, $\sin x \approx 0$ when $x \approx 0$ and $\cos x \approx 0$ when $x \approx 90°$)

Exercise 13.2A Fluency and skills

1 a $u_1 = 8, u_2 = 13, u_3 = 18, u_4 = 23$

  b $u_1 = -2, u_2 = 1, u_3 = 6, u_4 = 13$

  c $u_1 = 0, u_2 = 8, u_3 = 36, u_4 = 96$

  d $u_1 = -\dfrac{1}{5}, u_2 = -1, u_3 = 1, u_4 = \dfrac{2}{5}$

2 a $u_2 = 7, u_3 = 19, u_4 = 43$

  b $u_2 = 3, u_3 = 11, u_4 = 123$

  c $u_2 = \dfrac{1}{4}, u_3 = \dfrac{4}{5}, u_4 = \dfrac{5}{9}$

  d $u_2 = 4, u_3 = 1, u_4 = 4$

3 a Decreasing   **b** Increasing

  c Neither   **d** Neither

  e Decreasing   **f** Neither

4 a 3   **b** 3   **c** 4   **d** 4

5 a 77   **b** $n > 334$

6 a 13   **b** $u_4 = 2$

7 a $L = 5$ $[u_9 = 5.00\ (3\,\text{sf})]$

  b $L = -7.5$ $[u_{16} = -7.50\ (3\,\text{sf})]$

  c $L = 2.4$ $[u_6 = 2.40\ (3\,\text{sf})]$

  d $L = -\dfrac{5}{4}$ $[u_{13} = -1.25\ (3\,\text{sf})]$

  e $L = \dfrac{2}{3}$ $[u_{18} = 0.667\ (3\,\text{sf})]$

  f $L = 3 + \sqrt{5}$ $[u_{30} = 5.24\ (3\,\text{sf})]$

  g $L = 1 + \sqrt{3}$ $[u_6 = 2.73\ (3\,\text{sf})]$

  h $L = 2$ $[u_{792} = 2.00\ (3\,\text{sf})]$

Exercise 13.2B Reasoning and problem-solving

1 a 127   **b** 166

2 a $u_2 = au_1 + 7$

  $$= a(-1) + 7$$

  $$= 7 - a$$

  $u_3 = au_2 + 7$

  $$= a(7 - a) + 7$$

  $$= -a^2 + 7a + 7$$

  $-a^2 + 7a + 7 = 19$

  $a^2 - 7a + 12 = 0$

  b 64 or 83

3 a $a = 2, b = -6$ **b** 582

4 a i $u_4 = a - \dfrac{2a}{u_3}$

  $$= a - \dfrac{2a}{(-2)}$$

  $$= 2a$$

  So $u_1 = u_4$ (as the sequence has order 3)

  $$= 2a$$

  $u_2 = a - \dfrac{2a}{u_1}$

  $$= a - \dfrac{2a}{2a}$$

  $$= a - 1$$

  ii $u_3 = a - \dfrac{2a}{u_2}$

  $$= a - \dfrac{2a}{(a-1)}$$

  and $u_3 = -2$

  So $a - \dfrac{2a}{(a-1)} = -2$

  $$a(a-1) - 2a = -2(a-1)$$

  leading to $a^2 - a - 2 = 0$

  b 300

5 a $p_1 = 56$

  $p_2 = 0.75(p_1 + 8)$

  $$= 0.75(56 + 8)$$

  $$= 48$$

  $p_3 = 0.75(p_2 + 8)$

  $$= 0.75(48 + 8)$$

  $$= 42$$

  Decrease from 2nd to 3rd observations $= 48 - 42 = 6$ otters

  % decrease $= \dfrac{6}{48} \times 100 = 12.5\%$

b $p_n \geq 24$ for some n

$$p_{n+1} = 0.75(p_n + 8)$$
$$\geq 0.75(24 + 8)$$
$$= 24$$

So at least 24 otters were seen in the $(n+1)$th observation

c Using the same argument as in part **b**,

$p_{n+1} \geq 24$ implies $p_{n+2} \geq 24$,

which, in turn, implies $p_{n+3} \geq 24$ and so on

The number of otters seen from the nth observation onwards is always at least 24 and hence the population will never fall to zero.

6 a $m_2 = 4$

$$m_3 = a(m_2 - 1)$$
$$= a(3)$$
$$= 3a$$
$$m_4 = a(m_3 - 1)$$
$$= a(3a - 1)$$
$$m_4 = 10$$

So $a(3a - 1) = 10$

leading to $3a^2 - a - 10 = 0$

b 3 minutes

c 7 rounds

7 a 6 miles

b $e_2 = \dfrac{2}{3}e_1 + 2$

$$= \dfrac{2}{3}(1.5) + 2$$
$$= 3$$

$s_2 = \dfrac{1}{2}s_1 + k$

$$= \dfrac{1}{2}(2) + k$$
$$= 1 + k$$

Hypotenuse $= \sqrt{3^2 + (1+k)^2}$

$$= \sqrt{10 + 2k + k^2}$$

Total distance run $= 3 + (1 + k) + \sqrt{10 + 2k + k^2}$

So $4 + k + \sqrt{10 + 2k + k^2} = 12$

$$\sqrt{10 + 2k + k^2} = 8 - k, \text{ as required}$$
$$k = 3$$

c 20 miles

8 a $u_2 = 5p + 6$

$$\Rightarrow u_3 = p(5p + 6) + 6$$
$$= 5p^2 + 6p + 6$$
$$u_3 = 9.2 \therefore 5p^2 + 6p + 6 = 9.2$$
$$\Rightarrow 5p^2 + 6p - 3.2 = 0$$

b $L = 10$

9 a $p = 0.8, q = 0.4$ **b** $L = \dfrac{4}{3}$

Exercise 13.3A Fluency and skills

1 a 5, 7, 9, 11   **b** 9, 6, 3, 0

  c 3, 10, 17, 24   **d** $4, \dfrac{13}{2}, 9, \dfrac{23}{2}$

2 a $u_n = 4n + 3$

  b $u_n = 5 - 3n$

  c $u_n = \dfrac{3}{2}n + \dfrac{5}{2}$ or $u_n = \dfrac{1}{2}(3n + 5)$

d $u_n = \dfrac{2}{3}n + \dfrac{4}{3}$ or $u_n = \dfrac{2}{3}(n + 2)$

e $u_n = \dfrac{2}{5}n + \dfrac{1}{5}$ or $u_n = \dfrac{1}{5}(2n + 1)$

f $u_n = \sqrt{2}n$

3 a 150   **b** -139

  c $\dfrac{125}{2}$ **d** -102.4

4 a 585   **b** -144

  c 2623.5   **d** $100\sqrt{3}$

5 a 50   **b** 10

  c 100   **d** 10

6 a 480   **b** 306

  c 119   **d** 1080

Exercise 13.3B Reasoning and problem-solving

1 a $d = 5, a = 6$ **b** 61

2 a Find the first term a and the common difference d

$$u_n = a + (n - 1)d$$

$u_5 = 13$ so $a + 4d = 13$ ...$\boxed{1}$

$u_{10} = 7u_2$ so $a + 9d = 7(a + d)$

leading to $d = 3a$ ...$\boxed{2}$

Solve $\boxed{1}$ and $\boxed{2}$ to give $a = 1, d = 3$

$$u_n = 1 + (n - 1) \times 3$$
$$= 3n - 2$$
$$u_{18} = 3 \times 18 - 2$$
$$= 52$$

b $4: u_1(=1), u_2(=4), u_5(=13)$ and $u_{18}(=52)$

3 475

4 a $u_{16} = 2$ **b** 392

5 a $S_n = \dfrac{1}{2}n\big[2 \times 5 + (n - 1) \times 4\big]$

$$= \dfrac{1}{2}n(6 + 4n)$$
$$= n(2n + 3), \text{ as required}$$

b 77

6 a 17   **b** 790

7 Let c_n = number of complaints received in the nth month.

Let the first term be a and the common difference d

$$a = 152(= c_1)$$
$$d = c_2 - c_1$$
$$= 140 - 152$$
$$= -12$$

a $S_{12} = \dfrac{1}{2}(12)\big[2 \times 152 + 11 \times (-12)\big]$

$$= 1032$$

The total number of complaints received in the first year was 1032, as required.

b $S_6 = \dfrac{1}{2}(6)\big[2 \times 152 + 5 \times (-12)\big]$

$$= 732$$

Number of complaints in 2nd half of the year $= S_{12} - S_6$

$$= 1032 - 732$$
$$= 300$$

Reduction in the number of complaints from 1st to 2nd half of the year $= 732 - 300$

$$= 432$$

% reduction $= \dfrac{432}{732} \times 100\%$

$$= 59.01\%$$

The claim is accurate.

c In the following year,

$n = 13$ and $c_{13} = 152 + 12 \times (-12)$
$= 8$ (very low number of complaints, but still possible)

$n = 14$, $c_{14} = -4$ which is meaningless

So the model is not suitable for use in the following year.

8 a 3300   **b** £50 000

9 a i £142 500   **ii** £450 000

    b 13

    c The amount Jim earns each year is not likely to increase by the same amount each year forever. Sales made may increase or decrease over time.

10 $1^3 + 2^3 + 3^3 + \dots + n^3 \equiv (1 + 2 + 3 + \dots + n)^2$; $n = 24$

Exercise 13.4A Fluency and skills

1 a 4, 12, 36, 108   **b** 3, 12, 48, 192

    c $4, 2, 1, \dfrac{1}{2}$ **d** $2, \dfrac{2}{3}, \dfrac{2}{9}, \dfrac{2}{27}$

    e $\dfrac{5}{4}, \dfrac{5}{8}, \dfrac{5}{16}, \dfrac{5}{32}$ **f** $\dfrac{1}{2}, 1, 2, 4$

2 a $u_n = 5 \times 2^{n-1}$ **b** $u_n = 36 \times \left(\dfrac{2}{3}\right)^{n-1}$

    c $u_n = 2 \times (-3)^{n-1}$ **d** $u_n = (-8) \times \left(-\dfrac{3}{4}\right)^{n-1}$

    e $u_n = 7 \times \left(-\dfrac{7}{2}\right)^{n-1}$ **f** $u_n = (1) \times \left(\dfrac{1}{6}\right)^{n-1}$

3 a 6138   **b** 2059

    c 241   **d** -5320

4 a 1594 320   **b** 189

    c 10 200   **d** 32 784

Exercise 13.4B Reasoning and problem-solving

1 a $r = 4$, $a = \dfrac{1}{8}$

    b 2048

    c $u_n = 2^{2n-5}$ so $p = 2, q = -5$

2 a 1820   **b** 1 328 600

3 a 2   **b** 49 146

4 a 9310   **b** $\dfrac{3}{16}$ or 0.1875

5 a 125

    b i 10   **ii** 1.260

6 a 1.02

    b £2653 (nearest £)

    c £56 (nearest £)

7 a 0.75

    b v_n = value of car (in £) n years after its purchase

      First term $a = v_1$
$= 24\,000 \times 0.75$
$= 18\,000$

      $v_n = 18\,000 \times 0.75^{n-1}$

   so $v_3 = 18\,000 \times 0.75^2$
$= 10\,125$

      3 years after purchase the value of the car was £10 125, as required.

    c 8 whole years

8 a i 1600 miles

    ii 11 529 miles

    iii The total distance driven in Tom's lifetime must be less

than the sum to infinity of the series $3125 + 2500 + \dots$

$S_\infty = \dfrac{3125}{1 - 0.8}$
$= 15\,625$

Tom will drive less than 15 625 miles in his lifetime.

    b The assumption that the distance driven decreases by 20% every year is unrealistic.

9 a 6.4 mm

    b 160 cm³

    c 439 800 km, which is over 55 000 km more than the distance from the Earth to the Moon

10 b_n = amount due (in £) at end of nth month

A geometric model is appropriate since the amount due increased by 5% each month

So the common ratio $r = 1.05$

First term $a = 450$

$b_n = 450 \times 1.05^{n-1}$

and $S_n = \dfrac{450(1.05^n - 1)}{1.05 - 1}$

$S_{12} = \dfrac{450(1.05^{12} - 1)}{1.05 - 1}$
$= 7162.706\dots$

$\dfrac{S_{12}}{12} = \dfrac{7162.706\dots}{12}$
$= 596.892\dots$

Mean monthly bill in 1st year was £600 (nearest £10)

The bills are increasing in value so median bill = mean of the 6th and 7th terms

$b_6 = 450 \times 1.05^5$
$= 574.326\dots$

$b_7 = 450 \times 1.05^6$
$= 603.043\dots$

Mean of 6th and 7th terms $= \dfrac{574.326\dots + 603.043\dots}{2}$
$= 588.6848\dots$

Median bill for 1st year = £590 (nearest £10)

So the statement used only the mean.

11 21 hours

Review exercise 13

1 a $1 + \dfrac{4}{3}x + \dfrac{2}{9}x^2 - \dfrac{4}{81}x^3 + \dots$, $-1 < x < 1$

    b $1 - 6x + 24x^2 - 80x^3 + \dots$, $-\dfrac{1}{2} < x < \dfrac{1}{2}$

    c $27 - 18x + 2x^2 + \dfrac{4}{27}x^3 + \dots$, $-\dfrac{9}{4} < x < \dfrac{9}{4}$

    d $1 - \dfrac{3}{8}x + \dfrac{15}{128}x^2 - \dfrac{35}{1024}x^3 + \dots$, $-4 < x < 4$

2 a $\sqrt{3} \approx \dfrac{433}{250} (= 1.732)$

    b $\sqrt{3} \approx \dfrac{111}{64} (= 1.734375)$

3 a Increasing   **b** Decreasing

    c Increasing   **d** Decreasing

    e Neither   **f** Neither

4 a 2   **b** 4

    c 3   **d** 4

5 a nth term $= 6n + 1$; 480

    b nth term $= 19 - 4n$; -825

    c nth term $= 5n + 4$; 2445

    d nth term $= \dfrac{1}{2}n$; 637.5

e nth term $= 5.5n + 13$; 432.5

f nth term $= \dfrac{2}{3} - \dfrac{1}{6}n$; -19

6 a nth term $= 3 \times 6^{n-1}$; 1007769

b nth term $= 100 \times \left(\dfrac{1}{10}\right)^{n-1}$; 111

c nth term $= 200 \times (-0.95)^{n-1}$; 150

d nth term $= \log_2(3) \times 2^{n-1}$; 810

7 a 2415 **b** 6140

   c 40 **d** 375

Assessment 13

1 a i $\dfrac{1}{2}$ **ii** -1 **iii** 2

   b $\dfrac{1}{2}$

2 a Periodic   **b** Decreasing   **c** Increasing

3 $1 + 4x + 12x^2 + 32x^3 + \ldots$, valid for $|x| < \dfrac{1}{2}$

4 a £24 000   **b** £12 288   **c** year 9

5 a $u_2 = 4 - 3k$ **b** $u_3 = 4 - 4k + 3k^2$

   c $k = 2$ or $\dfrac{1}{3}$

6 a 0.8 **b** 2.56

7 a $800 + 2 \times 100 = 1000$

   b £35 000

   c 120

8 a $n = -\dfrac{2}{3}$, $a = 6$ **b** $-\dfrac{320}{3}$

9 a 0.5 **b** 240 **c** 480

10 a 2500 **b** 80

11 a $1 - x - \dfrac{1}{2}x^2 + \ldots$

   b $1 - \dfrac{1}{2}x + \dfrac{3}{8}x^2 + \ldots$

   c $\left(1 - x - \dfrac{1}{2}x^2\right)\left(1 - \dfrac{1}{2}x + \dfrac{3}{8}x^2\right)$

     $1 - \dfrac{3}{2}x + \dfrac{3}{8}x^2$

12 a -7 **b** 6 **c** 15

13 a $p = 1$ or $p = -1.8$

   b -25.6

   c 2

14 a $1 - \dfrac{1}{2}x - \dfrac{1}{8}x^2 - \dfrac{1}{16}x^3 + \ldots$

   b $\dfrac{887}{512}$

15 a i £2.40   **ii** £12   **iii** £36

   b £1188

16 a $\dfrac{a}{1-r} = 48$

    $a + ar = 45$

    $\dfrac{45}{1+r} = 48(1-r)$

    $1 - 16r^2 = 0$

   b 47.8125

17 a 15 km   **b** 120 km

   c 50th day   **d** 31st day

18 a $2 - \dfrac{x}{4} - \dfrac{x^2}{64} - \ldots$, valid for $|x| < 4$

   b 19.975

19 a £1000 + 1.03 × £1000

    $= £2030$

   b £3090.90

   c Year 31

20 a 8 **b** -3 **c** 18

21 a $A = -2$, $B = 1$

   b $\dfrac{5}{2}x + \dfrac{15}{4}x^2 + \dfrac{65}{8}x^3 + \ldots$

   c $|x| < \dfrac{1}{2}$

22 a $-\dfrac{33}{5} + -\dfrac{527}{25}x + -\dfrac{7873}{125}x^2 + \dfrac{118\,127}{625}x^3 \ldots$

   b $|x| < \dfrac{1}{3}$

23 a 1.5 m   **b** 5 m   **c** 14 m

24 $x = -6$ or $x = 5$

    $y = 36$ or $y = 25$

25 a 5 **b** 7.5 **c** -6

   d $4 + 2\sqrt{2}$ **e** $2 + 2\sqrt{2}$ **f** 1

Chapter 14

Exercise 14.1A Fluency and skills

1 a $114.6°$ **b** $171.9°$ **c** $28.6°$

2 a $\dfrac{\pi}{2} = 1.57$ radians   **b** $\dfrac{\pi}{3} = 1.05$ radians

   c $\dfrac{3\pi}{2} = 4.71$ radians

3 8 cm, 16 cm²

4

π		$\sin \pi$	$\cos \pi$	$\tan \pi$
deg	rad			
45	$\dfrac{\pi}{4}$	$\dfrac{1}{\sqrt{2}}$	$\dfrac{1}{\sqrt{2}}$	1
120	$\dfrac{2\pi}{3}$	$\dfrac{\sqrt{3}}{2}$	$-\dfrac{1}{2}$	$-\sqrt{3}$
135	$\dfrac{3\pi}{4}$	$\dfrac{1}{\sqrt{2}}$	$-\dfrac{1}{\sqrt{2}}$	-1
270	$\dfrac{3\pi}{2}$	-1	0	∞
360	2π	0	1	0

5 $DE = 51.8$ cm (3 sf)

   Angle $D = 1.73$ radians

   Angle $E = 0.630$ radians

6 $\dfrac{9}{2}\pi$ cm, $\dfrac{27}{2}\pi$ cm²

7 a

π	π (radians)	$\sin \pi$	$\tan \pi$
10°	0.1745	0.1736	0.1763
5°	0.0873	0.0872	0.0875
2°	0.0349	0.0349	0.0349

   b $2°$

8 1.5%

9 43.6 m

10 a 1 **b** $\dfrac{1}{4}$ **c** $\dfrac{1}{2}$

11 a $\theta = \dfrac{\pi}{6}$ or $\theta = \pi - \dfrac{\pi}{6} = \dfrac{5\pi}{6}$

   b $\theta = -\dfrac{5\pi}{6}$ or $\theta = \dfrac{5\pi}{6}$

   c $\theta = \dfrac{\pi}{4}$ or $\theta = -\pi + \dfrac{\pi}{4} = -\dfrac{3\pi}{4}$

Exercise 14.1B Reasoning and problem-solving

1 $\dfrac{2}{3}\pi$

2

r cm	π radians	s cm	A cm²
4	3	12	24
5	0.4	2	5
4	1.25	5	10
5	2.4	12	30

3 175 m

4 30.6 km

5 6980 km

6 π rad

7 0.25 rad

8 $4 + 2\pi$ cm, 2π cm²

9 a $\theta = \dfrac{\pi}{6}$ or $\dfrac{\pi}{3}$ **b** $\theta = \dfrac{\pi}{4} - \dfrac{\pi}{6} = \dfrac{\pi}{12}$

   c $\theta = \pm\dfrac{\pi}{2}$ or $\dfrac{\pi}{4}$ **d** $\theta = \dfrac{\pi}{4}$ or -0.464 rad

10 1.1 cm²

11 27.0 cm²

12 a $\left(\dfrac{7}{3}\pi + 4\right)$ cm   **b** $\dfrac{5}{3}\pi$ cm²

13 $\left(\sqrt{3} - \dfrac{\pi}{2}\right)r^2$

Exercise 14.2A Fluency and skills

1 The graphs have asymptotes.
The values are also ±1
The graphs are also positive (or negative).

2 a −1.06   **b** −0.839   **c** −2.92
  d −3.24   **e** −1.73   **f** 1.56

3

x	x rad	cosec x	sec x	cot x
30°	$\dfrac{\pi}{6}$	2	$\dfrac{2}{\sqrt{3}}$	$\sqrt{3}$
60°	$\dfrac{\pi}{3}$	$\dfrac{2}{\sqrt{3}}$	2	$\dfrac{1}{\sqrt{3}}$
90°	$\dfrac{\pi}{2}$	1	∞	0

4 a −1   **b** −2   **c** −2
  d $\dfrac{1}{\sqrt{3}}$ **e** $\sqrt{2}$ **f** −1

5 a $x = 30°$, $x = 150°$ **b** $x = 75.5°$, $x = -75.5°$
  c $x = 135°$, $x = -45°$ **d** $x = 80.5°$, $-60.5°$
  e $x = -24.5°$, $-175.5°$ **f** $x = -56.6°$, $123.4°$
  g $\theta = \pm 45°$, $\pm 105°$, $\pm 135°$ **h** $\theta = \dfrac{\pi}{12}$ or $\dfrac{5\pi}{12}$

6 a 1   **b** $\sin\theta$
  c $\cot\theta$ **d** $\tan\theta$

7 a 71.6°   **b** 75.5°   **c** 48.6°
  d 113.6°   **e** −63.4°   **f** −14.5°

8 a $\dfrac{\pi}{4}$ **b** $\dfrac{\pi}{3}$ **c** 0
  d $-\dfrac{\pi}{4}$ **e** $\dfrac{\pi}{6}$ **f** 0
  g $\dfrac{3\pi}{4}$ **h** $-\dfrac{\pi}{6}$

9 a $x = \dfrac{\pi}{6}$, $y = 1 + \dfrac{2}{\sqrt{3}}$, $x = \dfrac{5\pi}{6}$, $y = 1 - \dfrac{2}{\sqrt{3}}$
  b $\cos x = 0$, $x = 0$

10 a 30°, 150°   **b** 90°, 150°   **c** 66.4°

---

   d 35.6°   **e** 45°   **f** 5°
   g 0°, 90°   **h** 105°

11 a 0°   **b** 90°

12 a i Translation of $\begin{pmatrix} 45° \\ 0 \end{pmatrix}$

     ii Stretch (s.f. 2) parallel to y-axis and stretch (s.f. 3) parallel to x-axis

   b i Translation of $\begin{pmatrix} -45° \\ 0 \end{pmatrix}$ and stretch (s.f. 2) parallel to y-axis

     ii Stretch (s.f. 2) parallel to x-axis

   c i Reflection in x-axis and stretch $\left(\text{s.f. }\dfrac{1}{4}\right)$ parallel to the y-axis

     ii Translation of $\begin{pmatrix} 0 \\ 2 \end{pmatrix}$

13 a

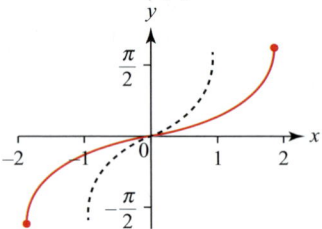

End points (± 2), $\left(\pm\dfrac{\pi}{2}\right)$

   b

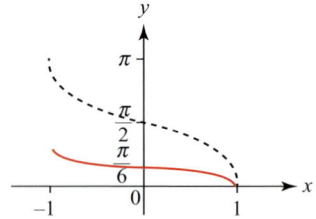

End points $\left(-1, \dfrac{\pi}{3}\right)$, $(1, 0)$

   c

End points $(\pm\infty)$, $(\pm\pi)$

   d

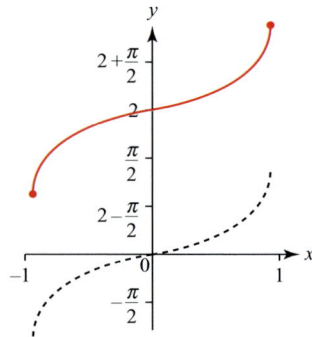

End points $\left(-1, 2 - \dfrac{\pi}{2}\right)$, $\left(1, 2 + \dfrac{\pi}{2}\right)$

Answers   For full solutions go to http://www.oxfordsecondary.co.uk/edexcelalevelmaths-answers

1 **a** $\theta = \pm\dfrac{\pi}{4}, \pm\dfrac{3\pi}{4}$ **b** $\theta = \pm\dfrac{\pi}{6}, \pm\dfrac{5\pi}{6}$

c $\theta = \pm\dfrac{\pi}{3}, \pm\dfrac{2\pi}{3}$ **d** $\theta = 0, \pm\dfrac{\pi}{3}, \pm\dfrac{2\pi}{3}$

e $x = \pm 0.955^c, \pm 2.19^c$ **f** $x = \pm 0.857^c, \pm 2.28^c$

g $\theta = \pm\dfrac{\pi}{8}, \pm\dfrac{3\pi}{8}, \pm\dfrac{5\pi}{8}, \pm\dfrac{7\pi}{8}$

h $\theta = \dfrac{\pi}{12}, \dfrac{5\pi}{12}, -\dfrac{\pi}{4}, -\dfrac{7\pi}{12}, -\dfrac{11\pi}{12}$

2 **a** $\theta = 45°, 71.6°, 225°, 251.6°$

b $\theta = 45°, 116.6°, 296.6°, 225°$

c $\theta = 90°, 221.8°, 318.2°$

d $\theta = 60°, 300°$

3 **a** $4y^2 + 16 = x^2$ **b** $9y^2 + 36 = 4x^2$

c $x^2(y^2 + 9) = 144$ **d** $1 + (1-x)^2 = (y-1)^2$

4 **a** **i** $\dfrac{5}{4}$ **ii** $\dfrac{3}{4}$

b **i** $-\dfrac{17}{15}$ **ii** $-\dfrac{15}{8}$

c **i** $-\dfrac{41}{40}$ **ii** $-\dfrac{9}{40}$

5 **a** $\cot^2\theta$ **b** $\sec^3\theta$ **c** $\operatorname{cosec}^2\theta$

6 **a** $\dfrac{2}{3}$ **b** 4

7 **a** $\text{LHS} = \dfrac{\sin^2 x}{\cos x} + \cos x$

$= \dfrac{\sin^2 x + \cos^2 x}{\cos x}$

$= \sec x$

$= \text{RHS}$

b $\text{LHS} = \cos\theta \times \dfrac{\sin\theta}{\cos\theta} \times \sin^2\theta$

$= \sin^3\theta$

$\text{RHS} = \cos^3\theta \times \dfrac{\sin^3\theta}{\cos^3\theta}$

$= \sin^3\theta$

$\therefore \text{LHS} = \text{RHS}$

c $\tan\theta + \cot\theta$

$\equiv \dfrac{\sin\theta}{\cos\theta} + \dfrac{\cos\theta}{\sin\theta}$

$\equiv \dfrac{\sin^2\theta + \cos^2\theta}{\cos\theta\sin\theta}$

$\equiv \sec\theta \times \operatorname{cosec}\theta$

d $(\sec\theta - \tan\theta)(\sec\theta + \tan\theta)$

$\equiv \sec^2\theta - \tan^2\theta$

$\equiv (1 + \tan^2\theta) - \tan^2\theta$

$\equiv 1$

Dividing by $\sec\theta + \tan\theta$ gives

$\sec\theta - \tan\theta \equiv \dfrac{1}{\sec\theta + \tan\theta}$

e $(1 + \sec\theta)(1 - \cos\theta)$

$\equiv 1 - \cos\theta + \sec\theta - 1$

$\equiv -\cos\theta + \dfrac{1}{\cos\theta}$

$\equiv \dfrac{1 - \cos^2\theta}{\cos\theta}$

$\equiv \dfrac{\sin^2\theta}{\cos\theta}$

$\equiv \tan\theta\sin\theta$

8 0.78 s

9 $\tan\theta = \dfrac{x}{\sqrt{1-x^2}}$

10 $\operatorname{arccot} x = 90° - \theta$

11 $f(\theta) = \dfrac{1 - \sin\theta + 1 + \sin\theta}{(1 + \sin\theta)(1 - \sin\theta)}$

$= \dfrac{2}{1 - \sin^2\theta}$

$= 2\sec^2\theta$

$x = \dfrac{\pi}{2}, \dfrac{5\pi}{6}$

12 **a**

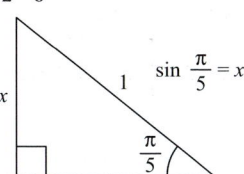

$\sin\dfrac{\pi}{5} = x$

Let $\arccos x = \alpha$

$x = \cos\alpha$

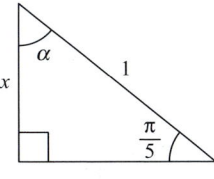

$\alpha + \dfrac{\pi}{5} = \dfrac{\pi}{2}$

$\alpha = \dfrac{\pi}{2} - \dfrac{\pi}{5} = \dfrac{3\pi}{10}$

$\arccos x = \dfrac{3\pi}{10}$

b $x = \dfrac{1}{2}$

13 **a** $\text{LHS} = \dfrac{\sin^2\theta + (1 + \cos\theta)^2}{\sin\theta(1 + \cos\theta)}$

$= \dfrac{\sin^2\theta + 1 + 2\cos\theta + \cos^2\theta}{\sin\theta(1 + \cos\theta)}$

$= \dfrac{2(1 + \cos\theta)}{\sin\theta(1 + \cos\theta)}$

$= 2\operatorname{cosec}\theta$

$= \text{RHS}$

b $\text{LHS} = \dfrac{1}{\dfrac{\cos\theta}{\sin\theta} + \dfrac{1}{\sin\theta}}$

$= \dfrac{\sin\theta}{1 + \cos\theta} \times \dfrac{1 - \cos\theta}{1 - \cos\theta}$

$= \dfrac{\sin\theta(1 - \cos\theta)}{1 - \cos^2\theta}$

$= \dfrac{1 - \cos\theta}{\sin\theta}$

$= \text{RHS}$

c Consider $(\sec\theta - \sin\theta)(\sec\theta + \sin\theta)$

$\sec^2\theta + \sec\theta\sin\theta - \sec\theta\sin\theta - \sin^2\theta$

$1 + \tan^2\theta - (1 - \cos^2\theta)$

$$\tan^2\theta - \cos^2\theta$$

Divide by $\sec\theta + \sin\theta$ $(\neq 0)$

$$\sec\theta - \sin\theta \equiv \frac{\tan^2\theta - \cos^2\theta}{\sec\theta + \sin\theta}$$

14 $\dfrac{4}{\cos^2\theta} - \dfrac{\sin^2\theta}{\cos^2\theta} = k$

$4 - \sin^2\theta = k(1 - \sin^2\theta)$

$(k-1)\sin^2\theta = k - 4$

$\mathrm{cosec}^2\theta = \dfrac{k-1}{k-4}$

$x = \dfrac{\pi}{2}$

Exercise 14.3A Fluency and skills

1 a $\dfrac{1}{4}\sqrt{2}(\sqrt{3}-1)$

  b i $\dfrac{1}{4}\sqrt{2}(\sqrt{3}-1)$

    ii $\dfrac{\sqrt{3}+1}{\sqrt{3}-1}$

    iii $-\dfrac{\sqrt{3}+1}{\sqrt{3}-1}$

2 a $\sin(35+10) = \sin 45°$

$$= \frac{1}{\sqrt{2}}$$

  b $\sin(70-10) = \sin 60°$

$$= \frac{\sqrt{3}}{2}$$

  c $\cos(40-10) = \cos 30°$

$$= \frac{\sqrt{3}}{2}$$

  d $\tan(74-45) = \tan 30°$

$$= \frac{1}{\sqrt{3}}$$

3 a $\sin(2\theta+\theta) = \sin 3\theta$ **b** $\cos(3\theta+\theta) = \cos 4\theta$

  c $\tan(2x+x) = \tan 3x$ **d** $\dfrac{1}{\tan(3x-x)} = \dfrac{1}{\tan 2x} = \cot 2x$

4 a $\sin(60°+x)$ or $\cos(30°-x)$ **b** $\tan(60°+x)$

  c $\sin(45°-x)$ or $\cos(45°+x)$ **d** $\tan(45°+x)$

  e $\cot(x+60°)$

5 a $\sin(90-A)$

$= \sin 90 \cos A - \cos 90 \sin A$

$= 1 \times \cos A - 0 \times \sin A$

$= \cos A$

  b $\sin(180-A)$

$= \sin 180 \cos A - \cos 180 \sin A$

$= 0 \times \cos A - (-1) \times \sin A$

$= \sin A$

  c $\cos(90-A)$

$= \cos 90 \cos A + \sin 90 \sin A$

$= 0 \times \cos A + 1 \times \sin A$

$= \sin A$

  d $\cos(180-A)$

$= \cos 180 \cos A + \sin 180 \sin A$

$= -1 \times \cos A + 0 \times \sin A$

$= -\cos A$

6 a

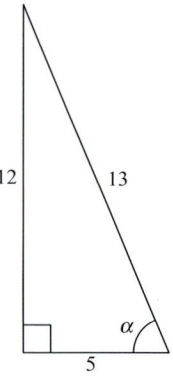

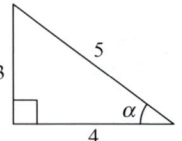

$\sin\alpha\cos\beta + \cos\alpha\sin\beta$

$$= \frac{12}{13} \cdot \frac{4}{5} + \frac{5}{13} \cdot \frac{3}{5}$$

$$= \frac{63}{65}$$

  b $\dfrac{\tan\alpha - \tan\beta}{1 + \tan\alpha\tan\beta}$

$$= \frac{\dfrac{12}{5} - \dfrac{3}{4}}{1 + \dfrac{12}{5}\cdot\dfrac{3}{4}}$$

$$= \frac{33}{20} \times \left(\frac{5}{14}\right)$$

$$= \frac{33}{56}$$

7 a $\sin 46°$ **b** $\cos 84°$ **c** $\tan 140°$

  d $\cos 100°$ **e** $\sin 6\theta$ **f** $\dfrac{1}{2}\sin 2\theta$

  g $\cos^2 20°$ **h** $\cos 8\theta$ **i** $2\cos^2\theta$

  j $\sin^2 25°$ **k** $\cos\dfrac{2\pi}{5}$ **l** $2\mathrm{cosec}\,2\theta$

  m $2\mathrm{cosec}\,2\theta$ **n** $\cot 8\theta$

8 a $\sin\left(2 \times \dfrac{\pi}{12}\right) = \sin\left(\dfrac{\pi}{6}\right)$

$$= \frac{1}{2}$$

  b $\cos\left(2 \times \dfrac{\pi}{8}\right) = \cos\left(\dfrac{\pi}{4}\right)$

$$= \frac{1}{\sqrt{2}}$$

  c $\cos\left(2 \times \dfrac{\pi}{4}\right) = \cos\left(\dfrac{\pi}{2}\right)$

$$= 0$$

  d $\dfrac{1}{\tan\left(2 \times 22\dfrac{1}{2}\right)} = \dfrac{1}{\tan 45°}$

$$= 1$$

9 a $\dfrac{24}{25}, \dfrac{1}{\sqrt{5}}$ **b** $\dfrac{120}{169}, \dfrac{1}{\sqrt{26}}$

Answers For full solutions go to http://www.oxfordsecondary.co.uk/edexcelalevelmaths-answers

10 a $\sin\dfrac{x}{2}=\dfrac{2}{3}$

$\cos\dfrac{x}{2}=\dfrac{\sqrt{5}}{3}$

$\tan\dfrac{x}{2}=\dfrac{2}{\sqrt{5}}$

b $\sin\dfrac{x}{2}=\dfrac{1}{\sqrt{10}}$

$\cos\dfrac{x}{2}=\dfrac{3}{\sqrt{10}}$

$\tan\dfrac{x}{2}=\dfrac{1}{3}$

11 a $\dfrac{\sqrt{5}}{7}$ **b** $\dfrac{3}{\sqrt{10}}$

12 a $\dfrac{1}{13}$ **b** $\dfrac{1}{2}$ **c** $\dfrac{5+\tan\beta}{1-5\tan\beta}$

Exercise 14.3B Reasoning and problem-solving

1 a $\theta=17.1°$ **b** $\theta=113.8°$ **c** $\theta=160.9°$

2 a $12\,\text{cm},\,13\,\text{cm}$

b $\sin A\hat{O}C=0.862$ to 3 sf

$\cos A\hat{O}C=0.508$ to 3 sf

$\tan A\hat{O}C=1.70$ to 3 sf

3 a $\text{LHS}=\dfrac{\sin A}{\cos A}+\dfrac{\cos A}{\sin A}$

$=\dfrac{\sin^2 A+\cos^2 A}{\cos A\sin A}$

$=\dfrac{2}{2\cos A\sin A}$

$=\dfrac{2}{\sin 2A}$

$=2\operatorname{cosec}2A$

$=\text{RHS}$

b $\text{LHS}=\dfrac{1}{\sin 2A}+\dfrac{\cos 2A}{\sin 2A}$

$=\dfrac{2\cos^2 A}{2\cos A\sin A}$

$=\cot A$

$=\text{RHS}$

$\text{LHS}=\dfrac{1}{\sin 2A}+\dfrac{\cos 2A}{\sin 2A}$

$=\dfrac{2\cos^2 A}{2\cos A\sin A}$

$=\cot A$

$=\text{RHS}$

$=\dfrac{2}{\sin 2A}$

$=2\operatorname{cosec}2A$

$=\text{RHS}$

c $\text{RHS}=\dfrac{1}{\cos A\sin A}$

$=\dfrac{2}{2\cos A\sin A}$

$=2\operatorname{cosec}2A$

$=\text{LHS}$

d $\text{LHS}=\dfrac{1}{\sin A}-\dfrac{\cos A}{\sin A}$

$=\dfrac{1-\cos A}{\sin A}$

$=\dfrac{2\sin^2\dfrac{A}{2}}{2\sin\dfrac{A}{2}\cos\dfrac{A}{2}}$

$=\tan\dfrac{A}{2}$

$=\text{RHS}$

4 $\sin\theta\cos\phi+\cos\theta\sin\phi=\cos\phi$

$\tan\theta\cos\phi+\sin\phi=\dfrac{\cos\phi}{\cos\theta}$

$\tan\theta+\tan\phi=\dfrac{1}{\cos\theta}$

$=\sec\theta$

5 Let $\arcsin\dfrac{1}{2}=\alpha$

So $\sin\alpha=\dfrac{1}{2}$

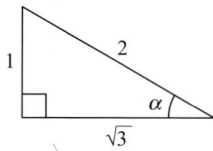

Let $\arcsin\dfrac{1}{3}=\beta$

So $\sin\beta=\dfrac{1}{3}$

$\sin(\alpha+\beta)=\dfrac{1}{2}\cdot\dfrac{\sqrt{8}}{3}+\dfrac{\sqrt{3}}{2}\cdot\dfrac{1}{3}$

$=\dfrac{\sqrt{8}+\sqrt{3}}{6}$

$\alpha+\beta=\arcsin\left(\dfrac{2\sqrt{2}+\sqrt{3}}{6}\right)$

6 a i $12+5\sqrt{5}$ **ii** $\dfrac{2-\sqrt{5}}{\sqrt{30}}$

b $\dfrac{\tan\theta+\sqrt{3}}{1-\sqrt{3}\tan\theta}=\dfrac{1}{3}$

$3\tan\theta+3\sqrt{3}=1-\sqrt{3}\tan\theta$

$(3+\sqrt{3})\tan\theta=1-3\sqrt{3}$

$\tan\theta=\dfrac{1-3\sqrt{3}}{3+\sqrt{3}}\times\dfrac{3-\sqrt{3}}{3-\sqrt{3}}$

$=\dfrac{12-10\sqrt{3}}{6}$

$=2-\dfrac{5}{3}\sqrt{3}$

7 a $\theta=0,\pi,2\pi,\dfrac{\pi}{6},\dfrac{5\pi}{6},\dfrac{7\pi}{6},\dfrac{11\pi}{6}$

b $\theta=\dfrac{7\pi}{6}$ or $\dfrac{11\pi}{6}$

c $\theta=\dfrac{\pi}{2},\dfrac{3\pi}{2},\dfrac{\pi}{4},\dfrac{5\pi}{4}$

d $\theta=0,\pi,2\pi,\dfrac{\pi}{3},\dfrac{5\pi}{3}$

e $\theta=0.332,\,1.239,\,3.473,\,4.381$ radians

f $\theta=0,\pi,2\pi,...$ or $\theta=\dfrac{\pi}{3},\dfrac{2\pi}{3},\dfrac{4\pi}{3},\dfrac{5\pi}{3}$

8 a $\theta = 0°, 360°, 120°$

   b $\theta = 0°, 360°, 78.4°, 281.6°$

   c $\theta = 120°$ or $0°$

   d $\theta = 180°, 90°, 270°$

9 $4y = x^2 - 8$

10 $36 + y^2 = 4x^2$ or $y^2 = 4(x^2 - 9)$

11 a $\tan(A+B) = \dfrac{\sin A \cos B + \cos A \sin B}{\cos A \cos B - \sin A \sin B}$

    Divide by $\cos A \cos B$ top and bottom

$$= \frac{\tan A - \tan B}{1 - \tan A \tan B}$$

   b $\tan(A-B) = \dfrac{\tan A - \tan B}{1 + \tan A \tan B}$

12 a $\dfrac{\sin\theta\cos\phi}{\cos\theta\cos\phi} + \dfrac{\cos\theta\sin\phi}{\cos\theta\cos\phi} \equiv \tan\theta + \tan\phi$

   b $\dfrac{2\sin\theta\cos\theta + \cos\theta}{1 + \sin\theta - (1 - 2\sin^2\theta)}$

$$\frac{\cos\theta(2\sin\theta + 1)}{\sin\theta(1 + 2\sin\theta)} \equiv \cot\theta$$

   c RHS $= \dfrac{1}{\cos 2\theta} + \dfrac{\sin 2\theta}{\cos 2\theta}$

$$= \frac{1 + 2\sin\theta\cos\theta}{\cos^2\theta - \sin^2\theta}$$

$$= \frac{\cos^2\theta + \sin^2\theta + 2\sin\theta\cos\theta}{\cos^2\theta - \sin^2\theta}$$

$$= \frac{(\cos\theta + \sin\theta)^2}{(\cos\theta + \sin\theta)(\cos\theta - \sin\theta)}$$

$$= \frac{\cos\theta + \sin\theta}{\cos\theta - \sin\theta}$$

$$= \text{LHS}$$

   d RHS $= \left(\dfrac{\tan\theta + \tan 45°}{1 - \tan\theta\tan 45°}\right)^2$

$$= \left(\frac{1 + \tan\theta}{1 - \tan\theta}\right)^2$$

    LHS $= \dfrac{1 + 2\sin\theta\cos\theta}{1 - 2\sin\theta\cos\theta}$

    Divide by $\cos^2\theta$

$$= \frac{\sec^2\theta + 2\tan\theta}{\sec^2\theta - 2\tan\theta}$$

$$= \frac{1 + \tan^2\theta + 2\tan\theta}{1 + \tan^2\theta - 2\tan\theta}$$

$$= \left(\frac{1 + \tan\theta}{1 - \tan\theta}\right)^2$$

$$\equiv \text{RHS}$$

   e $\dfrac{\cos^3\theta + \sin^3\theta}{\cos\theta + \sin\theta} = \dfrac{(\cos\theta + \sin\theta)(\cos^2\theta - \sin\theta\cos\theta + \sin^2\theta)}{\cos\theta + \sin\theta}$

$$= \cos^2\theta - \sin\theta\cos\theta + \sin^2\theta$$

$$= 1 - \sin\theta\cos\theta$$

$$= 1 - \frac{1}{2}\sin 2\theta$$

   f $\dfrac{2\sin\theta\cos\theta}{2\sin^2\theta} \equiv \dfrac{\cos\theta}{\sin\theta}$

$$\equiv \cot\theta$$

g $\dfrac{2\tan\theta}{1 + \tan^2\theta} \equiv \dfrac{2\sin\theta}{\cos\theta} \cdot \dfrac{1}{\sec^2\theta}$

$$\equiv 2\sin\theta \cdot \cos\theta$$

$$\equiv \sin 2\theta$$

h $\dfrac{2\sin\theta\cos\theta + \sin\theta}{2\cos^2\theta + \cos\theta}$

$$\equiv \frac{\sin\theta(2\cos\theta + 1)}{\cos\theta(2\cos\theta + 1)}$$

$$\equiv \tan\theta$$

13 $\cos^4\theta \equiv (1 - \sin^2\theta)^2$

$$\equiv 1 - 2\sin^2\theta + \sin^4\theta$$

   So $\cos^4\theta + 2\sin^2\theta - \sin^4\theta \equiv 1$

14 $\tan 45° = 1$

   So $\dfrac{\tan 45 - \tan 15}{1 + \tan 45 \times \tan 15}$

$$= \tan(45 - 15)$$

$$= \tan 30°$$

$$= \frac{1}{\sqrt{3}}$$

15 $\sin\theta \cdot \dfrac{\sin\theta}{\cos\theta} = 2 - \cos\theta$

$$1 - \cos^2\theta = 2\cos\theta - \cos^2\theta$$

$$2\cos\theta = 1$$

$$\cos\theta = \frac{1}{2}$$

$$\varphi = \frac{\pi}{3}$$

16 $3\dfrac{\cos\theta}{\sin\theta} = \dfrac{8}{\cos\theta}$

$$3\cos^2\theta = 8\sin\theta$$

$$3(1 - \sin^2\theta) = 8\sin\theta$$

$$3\sin^2\theta + 8\sin\theta - 3 = 0$$

$$\theta = 19.5° \text{ or } 160.5°$$

17 a $\cos^2 x + 2\cos x\cos y + \cos^2 y + \sin^2 x + 2\sin x\sin y + \sin^2 y$

$$\equiv 2 + 2(\cos x\cos y + \sin x\sin y)$$

$$\equiv 2 + 2\cos(x - y)$$

$$\equiv 2(1 + \cos(x - y))$$

$$\equiv 2 \times 2\cos^2\left(\frac{x-y}{2}\right)$$

$$\equiv 4\cos^2\left(\frac{x-y}{2}\right)$$

   b Let $x = 120°$, $y = 90°$

    So $\dfrac{x-y}{2} = 15°$

    $4\cos^2 15° = (\cos 120 + \cos 90)^2 + (\sin 120 + \sin 90)^2$

$$= \left(-\frac{1}{2} + 0\right)^2 + \left(\frac{\sqrt{3}}{2} + 1\right)^2$$

$$= \frac{1}{4} + \frac{3}{4} + \sqrt{3} + 1$$

$$= 2 + \sqrt{3}$$

    But $4\left(\dfrac{\sqrt{2} + \sqrt{6}}{4}\right)^2$

$$= \frac{1}{4}(2 + 2\sqrt{12} + 6)$$

$$= 2 + \sqrt{3}$$

So $\cos 15° = \dfrac{\sqrt{2} + \sqrt{6}}{4}$

18 a $\sin A + \sin B = 2\sin\left(\dfrac{A+B}{2}\right)\cos\left(\dfrac{A-B}{2}\right)$

$\dfrac{1}{2}[\sin(A+B) + \sin(A-B)] = \sin A \cos B$

$\sin(A+B) + \sin(A-B) = 2\sin A \cos B$

$A + B = x$

$A - B = y$

LHS $= \sin x + \sin y$

$2A = x + y \Rightarrow A = \dfrac{1}{2}(x+y)$

$2B = x - y \Rightarrow B = \dfrac{1}{2}(x-y)$

RHS $= 2\sin\dfrac{1}{2}(x+y)\cos\dfrac{1}{2}(x-y)$

as required.

b $x = 0, y = 0, \dfrac{\pi}{2}$

19 $\theta = 0.535\,\text{rad}, 3.68\,\text{rad}$

20 a $\dfrac{1 + \sin x + 1 - \sin x}{(1 - \sin x)(1 + \sin x)} = 8$

$\dfrac{2}{1 - \sin^2 x} = 8$

$2\sec^2 x = 8$

$\sec^2 x = 4$

b $\theta = 25°, 55°, 115°, 145°$

21 $k = 2$

22 $y = 4\cot t$

23 a $3(1 + \cot^2 x) - \cot^2 x = n$

$2\cot^2 x = n - 3$

$\tan^2 x = \dfrac{2}{n-3}$

$\sec^2 x = 1 + \dfrac{2}{n-3}$

$= \dfrac{n - 3 + 2}{n - 3}$

$= \dfrac{n - 1}{n - 3}$

($n \neq 3$ for $\sec^2 x$ to exist)

b $\theta = 15°, 120°, 210°, 300°, \dots$

24 The tides do not clash at 05:13 and 11:13

25 a i $\sin 3A = \sin(2A + A)$

$= \sin 2A \cos A + \cos 2A \sin A$

$= 2\sin A(1 - \sin^2 A) + (1 - 2\sin^2 A)\sin A$

$= 3\sin A - 4\sin^3 A$

ii $\cos 3A = \cos(2A + A)$

$= \cos 2A \cos A - \sin 2A \sin A$

$= (2\cos^2 A - 1)\cos A - 2(1 - \cos^2 A)\cos A$

$= 4\cos^3 A - 3\cos A$

b $\tan 3A = \dfrac{3\tan A - \tan^3 A}{1 - 3\tan^2 A}$

c $\sin 4A = 4\cos^3 A \sin A - 4\sin^3 A \cos A$

$\cos 4A = \cos^4 A - 6\cos^2 A \sin^2 A + \sin^4 A$

26 a Let $\arctan 3 = \alpha$ so $\tan\alpha = 3$

and $\text{arccot}\, x = \beta$ so $\cot\beta = x$

So $2\alpha = \beta$

$\tan 2\alpha = \tan\beta$

$\dfrac{2\tan\alpha}{1 - \tan^2\alpha} = \tan\beta$

$\dfrac{2 \times 3}{1 - 3^2} = \tan\beta$

$-\dfrac{3}{4} = \tan\beta$

$-\dfrac{3}{4} = \cot\beta$

But $\cot\beta = x$ so $x = -\dfrac{4}{3}$

b

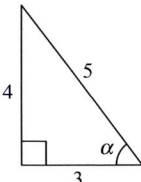

Let $\arcsin\dfrac{4}{5} = \alpha$ so $\sin\alpha = \dfrac{4}{5}$

and $\arccos\dfrac{12}{13} = \beta$ so $\cos\beta = \dfrac{12}{13}$

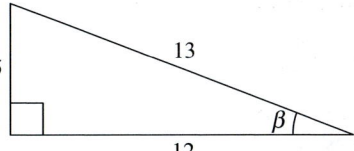

So

$\sin(\alpha - \beta) = \dfrac{4}{5}\cdot\dfrac{12}{13} - \dfrac{3}{5}\cdot\dfrac{5}{13}$

$= \dfrac{33}{65}$

$\alpha - \beta = \arcsin\left(\dfrac{33}{65}\right)$

So $x = \dfrac{33}{65}$

c $x = 90°\left(\text{or } \dfrac{\pi}{2}\right)$

Exercise 14.4A Fluency and skills

1 a $r = 13, \alpha = 67.4°$ **b** $r = 5, \alpha = 36.9°$

c $r = \sqrt{5}, \alpha = 63.4°$ **d** $r = 2\sqrt{5}, \alpha = 26.6°$

e $r = 5, \alpha = -53.1°$ **f** $r = 17, \alpha = -61.9°$

g $r = \sqrt{29}, \alpha = 21.8°$

2 a $r = \sqrt{2}$ **b** $r = \sqrt{2}$

3 a $\alpha = \dfrac{\pi}{3}$ (60°)   **b** $\alpha = \dfrac{\pi}{3}$ (60°)

4 a

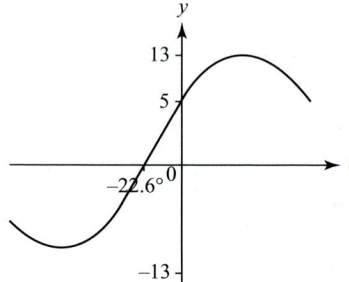

$r = 13, \alpha = 22.6°$

b $r = 6.4, \alpha = -51.3°$

5 The transformations are: a stretch (s.f. = 13) parallel to the

y-axis, followed by a translation of $\begin{pmatrix} 22.6° \\ 0 \end{pmatrix}$.

Exercise 14.4B Reasoning and problem-solving

1 a $\theta = 4.9°, 129.9°$ **b** $\theta = 17.6°, 229.8°$
  c $\theta = 82.0°, 334.1°$ **d** $\theta = 102.3°, 195.7°$
  e $\theta = 60°, 240°$ **f** $\theta = 122.0°$ or $302.0°$

2 a $\theta = \dfrac{\pi}{3}$ **b** $\theta = \dfrac{7\pi}{12}$ or $-\dfrac{\pi}{12}$

  c $\theta = \dfrac{\pi}{2}$ or $-\dfrac{\pi}{6}$ **d** $\theta = \dfrac{\pi}{12}$ or $\dfrac{-5\pi}{12}$

3 a $\theta = 0, 63.4°, 180°$ **b** $\theta = 83.6°, 108.7°$
  c $\theta = 54.6°, 157.9°$ **d** $\theta = 180°$

4 a Points of intersection are $\left(-15°, \dfrac{1}{\sqrt{2}} \right)$ and $\left(165°, -\dfrac{1}{\sqrt{2}} \right)$

  b $\sqrt{2} \sin(\theta + 45°) = \cos(\theta - 30°)$

$\sin\theta + \cos\theta = \cos\theta \cdot \dfrac{\sqrt{3}}{2} + \sin\theta \cdot \dfrac{1}{2}$

$\dfrac{1}{2}\sin\theta = \left(\dfrac{\sqrt{3} - 2}{2} \right)\cos\theta$

$\tan\theta = \sqrt{3} - 2$

$\theta = -15°$ or $165°$

$y = \dfrac{1}{\sqrt{2}}$ or $-\dfrac{1}{\sqrt{2}}$

5

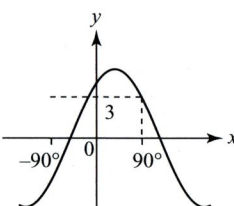

Let $3\sin x + 4\cos x = r\sin(x + \alpha)$
$= r\sin x \cos\alpha + r\cos x \sin\alpha$
Equate coefficients
$r\cos\alpha = 3$
$r\sin\alpha = 4$
$r = 5$
$\tan\alpha = \dfrac{4}{3}$
$\alpha = 53.1°$
So $y = 5\sin(x + 53.1°)$
$= 3$ when $\sin(x + 53.1°) = \dfrac{3}{5}$

$\sin(x + 53.1°) = 0.6$
$x + 53.1° = 36.9°$
or
$x + 53.1° = (180 - 36.9)°$
$= 36.9°$
or
$143.1°$
$x = -16.2°$
or
$90°$
For $-90° \le x \le 90°$
From graph, $y \ge 3$ for $-16.2° \le x \le 90°$

6

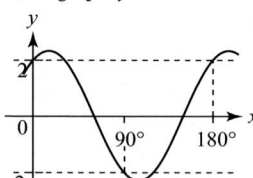

a For $y = 2$
$\sqrt{5} \sin(2x + 63.4°) = 2$
$2x + 63.4° = 63.4°$
$= 180 - 63.4°$
$= 360 + 63.4°$
$2x = 0, 53.2°, 360°, ...$
$x = 0, 26.6°, 180°$
for $0 \le x \le 180°$
From graph, $y > 2$ for $0 \le x \le 26.6°$

  b For $y > -2$
$\sin(2x + 63.4°) = -\dfrac{2}{\sqrt{5}}$
$2x + 63.4° = 180 + 63.4°$
$= 360 - 63.4°$
$= 540 + 63.4°, ...$
$2x = 180°, 233.2°, ...$
$x = 90°, 116.6°$
for $0 \le x \le 180°$
From graph, $y > -2$ for $90° \le x \le 116.6°$

7 a 12 amps
  b 0.26 seconds

8 a $\sqrt{2}\cos\theta + \sqrt{3}\sin\theta = r\cos\theta\cos\alpha + r\sin\theta\sin\alpha$
$r\cos\alpha = \sqrt{2}$
$r\sin\alpha = \sqrt{3}$
$r^2 = 5$
$r = \sqrt{5}$
$\tan\alpha = \sqrt{\dfrac{3}{2}}$

$\alpha = 50.8°$
$r\cos(\theta - \alpha) = \sqrt{5}\cos(\theta - 50.8°)$ has a max of $\sqrt{5}$ when
$\theta = 50.8°$

  b Min value $= \dfrac{1}{\sqrt{5}}$ when $\theta = 50.8°$

9 a Max of 2 when $\theta = 300°$
Min of -2 when $\theta = 120°$
  b Max of 25 when $\theta = 106.3°$
Min of -25 when $\theta = 286.3°$
  c Max of $\sqrt{13}$ when $\theta = 123.7°$
Min of $-\sqrt{13}$ when $\theta = 303.7°$
  d Max of 10 when $\theta = 161.6°$
Min of -10 when $\theta = 71.6°$

10 a $r = 25, \alpha = -73.7°$

$7\cos\theta - 24\sin\theta + 3 = 25\cos(\theta + 73.7°) + 3$

has a max value of $(25 \times 1) + 3 = 28$

and a min value of $(25 \times (-1)) + 3 = -22$

So $-22 \le 7\cos\theta - 24\sin\theta + 3 \le 28$

b A stretch of scale factor $\dfrac{1}{25}$ parallel to the y-axis, followed by a translation of $-73.7°$ parallel to the θ-axis

11 a i Max $= 33$ when $\theta = 157.4°$

ii Max $= 33$ when $\theta = 337.4°$

b Min $= 5$ when $\theta = 157.4°$

12 a Max $I = 2\sqrt{5}$ when $t = 74.1$ sec

b 30 seconds

Review exercise 14

1 a i $57.3°$ **ii** $143.2°$ **iii** $200.5°$

b i $\dfrac{\pi}{12}$ rad **ii** $\dfrac{2\pi}{5}$ rad **iii** $\dfrac{7\pi}{6}$ rad

2 a 1 **b** $\dfrac{1}{2}$ **c** 2θ

3 a 2 **b** $\sqrt{3}$ **c** $\sqrt{2}$

d -1 **e** $\dfrac{\pi}{4}$ **f** $\dfrac{\pi}{6}$

g $\dfrac{\pi}{2}$ **h** $-\dfrac{\pi}{4}$

4 a $\cos(40 + 10) = \cos 50°$

b $\dfrac{1}{\tan(100 + 35)} = \cot 135°$

5 a $x = 31.7°, 121.7°$

b $x = 10°, 130°$

c $x = 90°$ or $x = 19.5°, 160.5°$

d $x = 60°$

e $x = 15°$ or $75°$

f $x = 26.6°$

g $x = 45°$

6 a $2\pi \, \text{cm}^2$ **b** $2\pi - 4\sqrt{2} \, \text{cm}^2$

7 a 1 **b** 1

8 $\tan 15° = 2 - \sqrt{3}$

$\sec 75° = \sqrt{2}\left(\sqrt{3} + 1\right)$

9

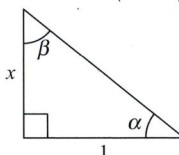

In the triangle

$\tan\alpha = x$

$\cot\beta = x$

and

$\alpha + \beta = \dfrac{\pi}{2}$

So

$\arctan x + \operatorname{arccot} x = \dfrac{\pi}{2}$

$\arctan x = \dfrac{\pi}{2} - \operatorname{arccot} x$

10 a $\theta = 106.1°$ **b** $\theta = 0°, 180°, 82.8°$

c $\theta = 0, 180°, 60°$ **d** $\theta = 30°, 150°$

11 a $7\sin\theta - 24\cos\theta = 25\sin(\theta - 73.7°)$

b $\theta = 110.6°, 216.8°$

12 a $7\cos\theta + 6\sin\theta = \sqrt{85}\sin(\theta + 49.4°)$

b Max is $\sqrt{85}$ when $\theta = 40.6°$

13 a $\text{LHS} = 2\sin^2 A$

$\text{RHS} = \dfrac{\sin A}{\cos A} \bullet 2\sin A\cos A = 2\sin^2 A$

$\text{LHS} = \text{RHS}$

b $\text{LHS} = 2\sin A\cos A \bullet \sec^2 A$

$= 2\sin A\cos A \dfrac{1}{\cos^2 A}$

$= 2\tan A$

$= \text{RHS}$

c $\text{RHS} = 2 \times \dfrac{1 - \tan^2 A}{2\tan A}$

$= \dfrac{1 - \tan^2 A}{\tan A}$

$= \dfrac{1}{\tan A} - \tan A$

$= \cot A - \tan A$

$= \text{LHS}$

14

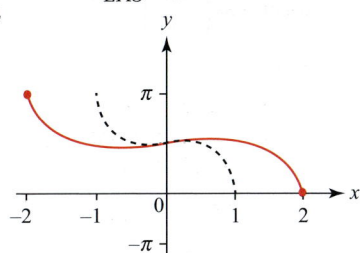

End points $(-2, \pi), (2, 0)$

15 $\dfrac{(1 - \sin x)^2 + \cos^2 x}{\cos x(1 - \sin x)} = \dfrac{1 - 2\sin x + 1}{\cos x(1 - \sin x)}$

$= \dfrac{2(1 - \sin x)}{\cos x(1 - \sin x)}$

$= 2\sec x$

So $\text{f}(x) = 2\sec x$

$x = 180°, 70.5°, 289.5°$

Assessment 14

1 a $x = 60°, -120°$

b

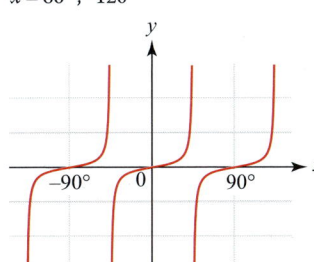

$x = \pm 45$

2 a $\dfrac{25\pi}{3}$ **b** $\dfrac{10\pi}{3}$

3 a 1.2 rad **b** 9.6 cm

4 a i $\sqrt{2}$ **ii** $-\sqrt{3}$

b

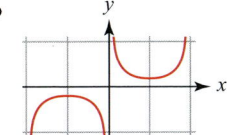

c $x = \pi, x = -\pi, x = 0$

$y \in \mathbb{R}, \ y > 1, \ y < -1$

5 a $x = 30°, 150°$
b $x = 135°, 225°$

6 a

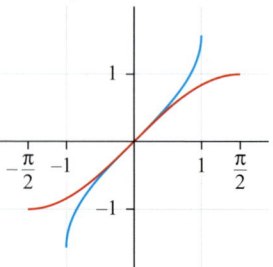

b Reflection in line $y = x$
c i $-\dfrac{\pi}{3}$ **ii** $\dfrac{\pi}{6}$

7 a $5\,\text{m}\,\text{s}^{-1}$ **b** $0, \dfrac{2\pi}{3}(2.09)\,\text{s}, \dfrac{14\pi}{3}(14.66)\,\text{s}, \dfrac{26\pi}{3}(27.2)\,\text{s}$

8 $x = \dfrac{\sqrt{2}}{2} - \dfrac{1}{2}$

9 a $\dfrac{\sin^2\theta}{\cos^2\theta} + \dfrac{\cos^2\theta}{\cos^2\theta} \equiv \dfrac{1}{\cos^2\theta}$
$\tan^2\theta + 1 \equiv \sec^2\theta$
$\sec^2\theta - \tan^2\theta \equiv 1$
b i $\sec\theta = \pm\sqrt{6}$ **ii** $\cos\theta = \pm\dfrac{1}{\sqrt{6}}$

10 a $\dfrac{\sqrt{6}}{4} - \dfrac{\sqrt{2}}{4}$ **b** $\dfrac{\sqrt{6}}{4} + \dfrac{\sqrt{2}}{4}$

11 Let $A = B = x$,
Then $\sin 2x = \sin x \cos x + \sin x \cos x$
$= 2\sin x \cos x$

12 $10\sin(\theta + 53.1)$

13 a 0.05 **b** $\dfrac{7}{8}$ **c** 0.5

14 a i

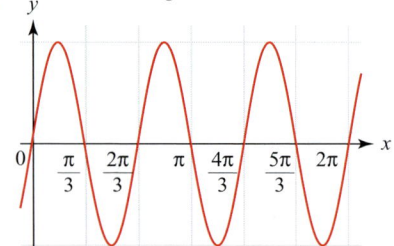

ii

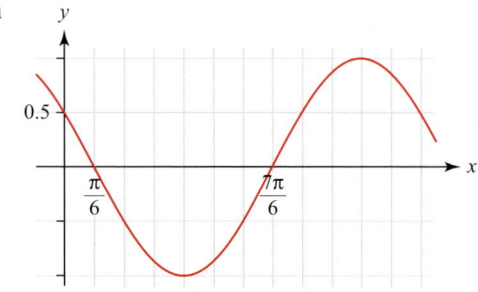

b $x = \dfrac{\pi}{12}, \dfrac{\pi}{4}, \dfrac{3\pi}{4}, \dfrac{11\pi}{12}$

15 a $13.2\,\text{cm}, 5.25\,\text{cm}$ **b** $60.3\,\text{cm}^2$

16 $50.4\,\text{cm}$

17 a 6500
b $10\,600$
c 3 weeks 3 days or 24 days

18 a i $4.9\,\text{cm}^2$ **ii** $9.8\,\text{cm}$
b i $0.51\,\text{cm}^2$ **ii** $5.53\,\text{cm}$

19 a i

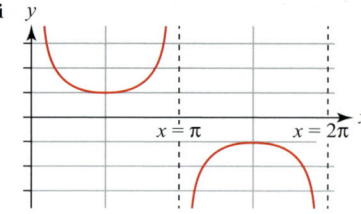

ii

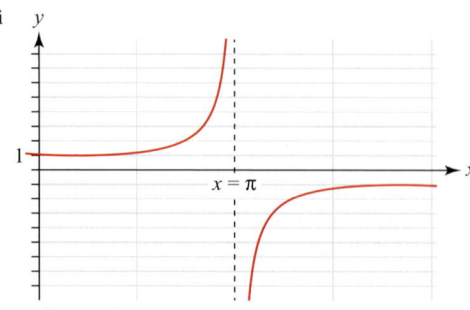

b i $y \in \mathbb{R}, y \le -2, y \ge 2$
ii $y \in \mathbb{R}, y \le -1, y \ge 1$

20 a

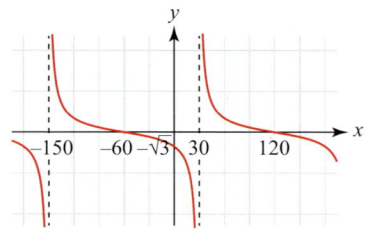

b $x = 30, x = -150$
$x = 108.7°, -71.3°$

21 i $x = \dfrac{2\pi}{3}, \dfrac{4\pi}{3}$
ii $x = \dfrac{3\pi}{8}, \dfrac{7\pi}{8}$

22 a

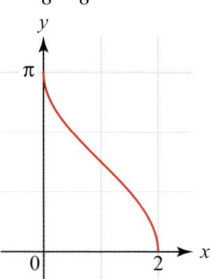

b $y \in \mathbb{R}, 0 \le y \le \pi$
c $\text{f}^{-1}(x) = 1 + \cos x$
Domain is $0 \le x \le \pi$
Range is $0 \le y \le 2$

23 a $x = \dfrac{1+\sqrt{3}}{2}$ **b** $x = 5 - \dfrac{\sqrt{3}}{3}$

24 a $2(\sec^2 x - 1) = \sec x - 1$
$2\sec^2 x - 2 = \sec x - 1$
$2\sec^2 x - \sec x - 1 = 0$
b $x = 0, 2\pi$

25 a $\dfrac{\sin^2\theta}{\sin^2\theta} + \dfrac{\cos^2\theta}{\sin^2\theta} \equiv \dfrac{1}{\sin^2\theta}$
$1 + \cot^2\theta \equiv \text{cosec}^2\theta$
$\text{cosec}^2\theta - \cot^2\theta \equiv 1$
b $\theta = 18.4°, 198.4°$

26 a $\sec^4 x - \tan^4 x$
$= (\sec^2 x - \tan^2 x)(\sec^2 x + \tan^2 x)$
$= \sec^2 x + \tan^2 x$
b $x = \pm 1.11\,\text{rad}, \pm 2.03\,\text{rad}$

27 a $\cos 3x = \cos(2x + x)$
$= \cos 2x \cos x - \sin x \sin 2x$
$= (2\cos^2 x - 1)\cos x - 2\sin x(2\sin x \cos x)$
$= 2\cos^3 x - \cos x - 2\sin^2 x \cos x$
$= 2\cos^3 x - \cos x - 2(1 - \cos^2 x)\cos x$
$= 2\cos^3 x - \cos x - 2\cos x + 2\cos^3 x$
$= 4\cos^3 x - 3\cos x$
b $x = \dfrac{\pi}{18}, \dfrac{11\pi}{18}, \dfrac{13\pi}{18}, \dfrac{23\pi}{18}, \dfrac{25\pi}{18}, \dfrac{35\pi}{18}$

28 a $\cos 2x = \cos x \cos x - \sin x \sin x$
$= \cos^2 x - \sin^2 x$
$= \cos^2 x - (1 - \cos^2 x)$
$= 2\cos^2 x - 1$
b $x = 120°, 180°, 240°$

29 e.g. $A = 30°, B = 60°$
$\cos(30 + 60) = \cos 90 = 0$
$\cos 30 + \cos 60 = \dfrac{\sqrt{3}}{2} + \dfrac{1}{2} \neq 0$
Therefore, in general,
$\cos(A + B) \neq \cos A + \cos B$

30 a $f(x) = 4\sqrt{5}\cos(x - 0.464)$
b $x = 1.78\,\text{rad}, 5.43\,\text{rad}$

31 a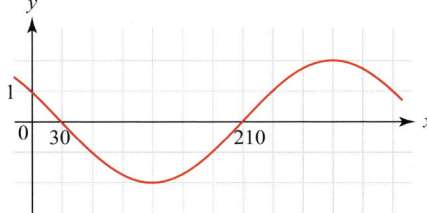

b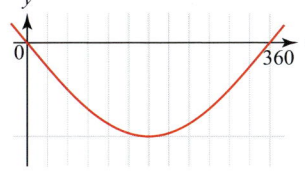

32 a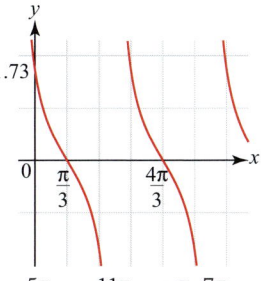

b $x = \dfrac{5\pi}{6}, x = \dfrac{11\pi}{6}, x = \dfrac{\pi}{6}, \dfrac{7\pi}{6}$

33 $16.8\,\text{cm}^2$

34 a $10.7\,\text{cm}$
b i $59.2\,\text{cm}^2$ **ii** $33.7\,\text{cm}$

35 a i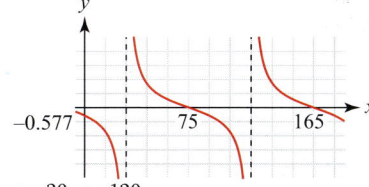
$x = 30, x = 120$

ii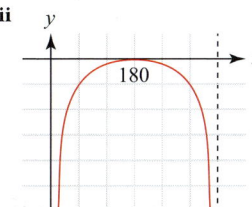

b $x = 0, x = 360$
$x = 8.3°, 98.3°, 128.3°$

36 a $\theta = 0.421, 3.56, 2.72, 5.86\,\text{rad}$
b $\theta = \dfrac{\pi}{12}, \dfrac{19\pi}{12}, \dfrac{7\pi}{12}, \dfrac{13\pi}{12}$

37 $x = 90°, 270°, y = 1, -1$

38 a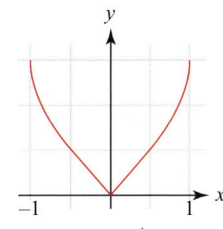

b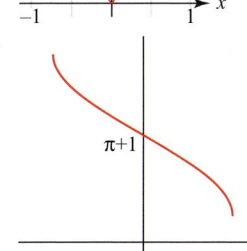

39 $\dfrac{5^2\sec^2 t}{25} - \dfrac{3^2\tan^2 t}{9} = \sec^2 t - \tan^2 t$
$= 1$

40 $x = 36.9°, 216.9°, 135°, 315°$

41 $\theta = 0.421, 2.721, \dfrac{\pi}{3}, \dfrac{2\pi}{3}\,\text{rad}$

42 $\text{LHS} = \dfrac{\cos^2 \theta}{\sin \theta} - \sin \theta$
$= \dfrac{\cos^2 \theta - \sin^2 \theta}{\sin \theta}$
$= \dfrac{1 - 2\sin^2 \theta}{\sin \theta}$
$= \operatorname{cosec} \theta - 2\sin \theta$

43 $x < \dfrac{\pi}{6}, x > \dfrac{5\pi}{6}$

44 a i $\text{LHS} = \dfrac{\cos x(1 - \cos x) - \sin x \sin x}{\sin x(1 - \cos x)}$
$= \dfrac{\cos x - \cos^2 x - \sin^2 x}{\sin x(1 - \cos x)}$

$$= \frac{\cos x - 1}{\sin x(1 - \cos x)}$$

$$= \frac{-1}{\sin x}$$

$$= -\cosec x \text{ as required}$$

    ii $x \neq 2n\pi$

  b $x = 3.48, 5.94 \text{ rad}$

45 a $\cos x = \cos\left(\dfrac{x}{2}\right)\cos\left(\dfrac{x}{2}\right) - \sin\left(\dfrac{x}{2}\right)\sin\left(\dfrac{x}{2}\right)$

$$= \cos^2\left(\frac{x}{2}\right) - \sin^2\left(\frac{x}{2}\right)$$

$$= 1 - \sin^2\left(\frac{x}{2}\right) - \sin^2\left(\frac{x}{2}\right)$$

$$= 1 - 2\sin^2\left(\frac{x}{2}\right) \text{ as required}$$

  b $x = 250.5°, 109.5°$

46 Earliest time 15:04

    Latest time 20:32

47 $x = 120°, 180°, 240°$

48 a $r = \sqrt{5}$ or 2.24

    $\alpha = 63.4$

  b Minimum is $\dfrac{1}{5}$

    $\theta = 26.6°$

  c Maximum is $\dfrac{1}{3 - \sqrt{5}}$

    $\theta = 206.6°$

Chapter 15
Exercise 15.1A Fluency and skills

1 a Concave for $x < 0$

    Convex for $x > 0$

  b Convex for $x < -2$

    Concave for $-2 < x < -1$

    Convex for $x > -1$

2 i a 1

    b Horizontal point of inflection ($x = 0$) as curve goes from convex to concave.

  ii a 1

    b Point of inflection on a decreasing section of the curve ($x \approx -0.5$), as curve goes from concave to convex.

  iii a 2

    b Point of inflection on a decreasing section of the curve ($x \approx -2$), as curve goes from concave to convex; horizontal point of inflection ($x = 0$), as curve goes from convex to concave.

  iv a 3

    b Point of inflection on a decreasing section of the curve ($x \approx -1.5$) as curve goes from concave to convex; point of inflection on an increasing section of the curve ($x \approx 1$) as curve goes from convex to concave; horizontal point of inflection ($x \approx 2$) as curve goes from concave to convex.

3 a Convex at $(1, 2)$

  b Concave at $(-1, -1)$

  c Convex at $(1, 2)$

  d Concave at $(1, 0)$

4 a -2

  b When $x = -2$, $\dfrac{dy}{dx} = (-2)^2 + 4 \times (-2) + 4 = 0$

    $\Rightarrow$ gradient is zero.

  c i Concave

    ii Convex

5 $(1, -1)$, on a decreasing section of the curve.

    $(0, 0)$, horizontal point of inflection.

6 $(-2, 0)$, on decreasing section of the curve.

7 $(0, 0)$, horizontal point of inflection. $(1, -7)$ on a decreasing section of the curve.

    $(-1, 7)$ on a decreasing section of the curve.

8 a $\dfrac{dy}{dx} = 2 - 2x$

    $\Rightarrow \dfrac{d^2y}{dx^2} = -2$

    Since it is a constant and negative, the curve is always concave.

  b $\dfrac{dy}{dx} = -\dfrac{1}{2\sqrt{x}}$

    $\Rightarrow \dfrac{d^2y}{dx^2} = \dfrac{1}{4(\sqrt{x})^3}$

    Since x is positive, the second derivative exists and is positive.

Exercise 15.1B Reasoning and problem-solving

1 a

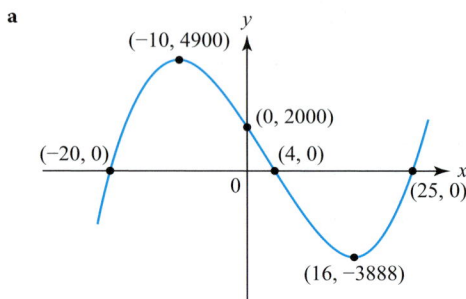

  b

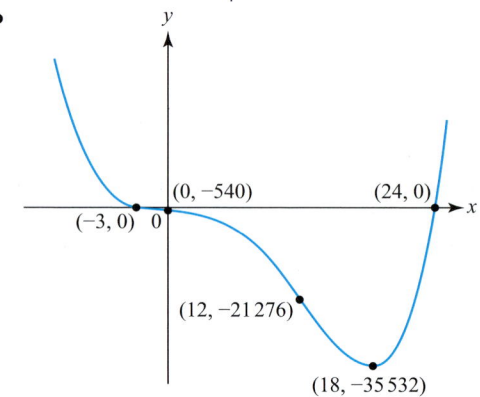

  c

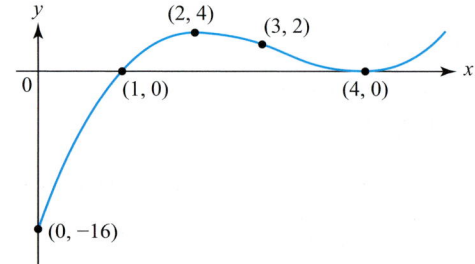

2 $y = ax^2 + bx + c$

    $\Rightarrow \dfrac{dy}{dx} = 2ax + b$

    $\Rightarrow \dfrac{d^2y}{dx^2} = 2a$

  a If $a > 0$ then $\dfrac{d^2y}{dx^2} > 0$ so curve is convex.

  b If $a < 0$ then $\dfrac{d^2y}{dx^2} < 0$ so curve is concave.

3 **a** $y = x^4 + 8x^3 + 25x^2 - 5x + 10$

$$\Rightarrow \frac{dy}{dx} = 4x^3 + 24x^2 + 50x - 5$$

$$\Rightarrow \frac{d^2y}{dx^2} = 12x^2 + 48x + 50$$

$$= 12(x^2 + 4x) + 50$$

$$= 12[(x+2)^2 - 2^2] + 50$$

$$\Rightarrow \frac{d^2y}{dx^2} = 12(x+2)^2 + 2 \text{ which is greater than zero for all } x$$

So y is convex for all x

b $y = 1 - 2x - 10x^2 + 4x^3 - x^4$

$$\Rightarrow \frac{dy}{dx} = -2 - 20x + 12x^2 - 4x^3$$

$$\Rightarrow \frac{d^2y}{dx^2} = -20 + 24x - 12x^2$$

$$= -12[x^2 - 2x] - 20$$

$$= -12[(x-1)^2 - 1^2] - 20$$

$$\Rightarrow \frac{d^2y}{dx^2} = -12(x-1)^2 - 8 \text{ which is less than zero for all } x$$

So y is concave for all x

c $y = 1 - 2x - 6x^2 + 4x^3 - x^4$

$$\Rightarrow \frac{dy}{dx} = -2 - 12x + 12x^2 - 4x^3$$

$$\Rightarrow \frac{d^2y}{dx^2} = -12 + 24x - 12x^2$$

$$= -12[x^2 - 2x] - 12$$

$$= -12[(x-1)^2 - 1^2] - 12$$

$$\Rightarrow \frac{d^2y}{dx^2} = -12(x-1)^2 \text{ which is less than } \textbf{or equal to} \text{ zero for all } x$$

So it is never greater than zero. So y is never convex for any value of x

4 $y = k(x-a)(x-b)(x-c)$

$$= k(x^3 - (a+b+c)x^2 + (ab+ac+bc)x - abc)$$

$$\Rightarrow \frac{dy}{dx} = k(3x^2 - 2(a+b+c)x + (ab+ac+bc))$$

$$\Rightarrow \frac{d^2y}{dx^2} = k(6x - 2(a+b+c))$$

At point of inflection, $\dfrac{d^2y}{dx^2} = 0$

$$\Rightarrow k(6x - 2(a+b+c)) = 0$$

$$\Rightarrow x = \frac{a+b+c}{3}$$

You know from the shape of cubic graphs that there is always a point of inflection, so the point where the second derivative is 0 must be that point.

5 Roots: $x = 0$ or $\pm\sqrt{\dfrac{5}{3}}$

Stationary points:
Horizontal point of inflection at (0, 0)
Maximum turning point at (−1, 2)
Minimum turning point at (1, −2)
Non-horizontal points of inflection, both on decreasing

sections of curve at $\left(\dfrac{1}{\sqrt{2}}, -\dfrac{7\sqrt{2}}{8}\right)$, and $\left(\dfrac{-1}{\sqrt{2}}, \dfrac{7\sqrt{2}}{8}\right)$

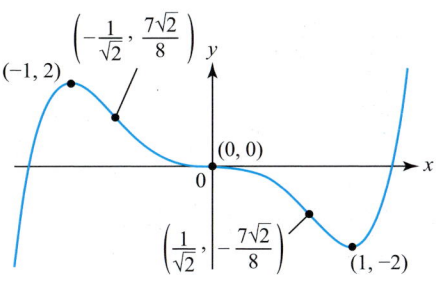

6 $y = -13x - 22$

7 When $x > \dfrac{a}{3}$

8 **a** **i** When $b^2 - 3ac < 0$

ii When $b^2 - 3ac = 0$

iii When $b^2 - 3ac > 0$

b **i** When $3ax + b < 0$

or $x < -\dfrac{b}{3a}$

ii When $3ax + b > 0$

or $x > -\dfrac{b}{3a}$

iii When $3ax + b = 0$

or $x = -\dfrac{b}{3a}$

9 $12ax^2 + 6bx + 2c = 0$ must have two roots.
The discriminant must be greater than zero.
So, $(6b)^2 - 4.12a.2c > 0$
giving $3b^2 - 8ac > 0$

Exercise 15.2A Fluency and skills

1 **a** $3\cos x$ **b** $-\dfrac{\sin x}{3}$ **c** $\sin x$ **d** $1 + \cos x$

e $2x - \sin x$ **f** $\cos x - \sin x$ **g** $3\cos x + 4\sin x$

2 **a** $\dfrac{dy}{dx} = 2\cos x$ **b** $\dfrac{dy}{dx} = 4x + \sin x$

$\dfrac{d^2y}{dx^2} = -2\sin x$ $\dfrac{d^2y}{dx^2} = 4 + \cos x$

c $\dfrac{dy}{dx} = 2\cos x + 3\sin x;$ **d** $\dfrac{dy}{dx} = \cos x - \sin x + 3x^2$

$\dfrac{d^2y}{dx^2} = -2\sin x + 3\cos x$ $\dfrac{d^2y}{dx^2} = -\sin x - \cos x + 6x$

e $\dfrac{dy}{dx} = \dfrac{-\sin x - \cos x}{2}$

$\dfrac{d^2y}{dx^2} = \dfrac{-\cos x + \sin x}{2}$

3 **a** **i** $\dfrac{3}{2}$ **ii** $-\dfrac{3\sqrt{3}}{2}$

b **i** -1 **ii** $-\sqrt{3}$

c **i** 0 **ii** $-\sqrt{2}$

d **i** 1 **ii** 0

e **i** 3 **ii** -4

4 **a** $\left(\dfrac{\pi}{2}, 2\right)$ and $\left(\dfrac{3\pi}{2}, 0\right)$

b $\left(\dfrac{\pi}{2}, 1\right)$ and $\left(\dfrac{3\pi}{2}, 1\right)$

5 4

6 Concave

7 **a** $2\cos x$ **b** $\sin x$ **c** $1 - \cos x$

Exercise 15.2B Reasoning and problem-solving

1 **a** $\cos x$ **b** $-\sin x$

   c $-\sin x$ **d** $\cos x$

2 **a** $3x - 6y = \pi - 3\sqrt{3}$

   b $\dfrac{d^2 y}{dx^2} = -\sin x$

     When $x = \dfrac{\pi}{3}, \dfrac{d^2 y}{dx^2} = -\sin \dfrac{\pi}{3}$

                           $= -\dfrac{\sqrt{3}}{2}$

     Since the result is negative, the curve is concave.

3 **a** $\dfrac{1}{2} - \dfrac{1}{2}\cos x$

   b $-12x^{-4} + \sin x$

   c $\cos x - \sin x$

   d $-\sin x + \cos x$

4 **a**   **i** -2 **ii** $\dfrac{1}{2}$ **iii** 3

      iv $\dfrac{3}{2}$ **v** 0

   b $-1 \le \cos x \le 1$

     $\Rightarrow -2 \le -2\cos x \le 2$

     $\Rightarrow \dfrac{x}{\pi} - 2 \le \dfrac{x}{\pi} - 2\cos x \le \dfrac{x}{\pi} + 2$

     $\Rightarrow \dfrac{x}{\pi} - 2 \le \dfrac{dy}{dx} \le \dfrac{x}{\pi} + 2$

     When $x > 2\pi, 0 < \dfrac{dy}{dx}$

     Since the derivative is always positive when $x > 2\pi$ then the function is always increasing.

   c $a > 2\pi$

5 **a** $11\,\text{m}$ **b** $10\,\text{m}\,\text{s}^{-1}$ **c** $0\,\text{m}\,\text{s}^{-2}$

6 **a** 2.40 hours per unit (3 sf)

   b When $x = 0, \pi, 2\pi$ units

   c $\dfrac{\pi}{2}, \dfrac{3\pi}{2}$

7 **a** $2\cos\left(x + \dfrac{\pi}{3}\right)$

   b   **i** 0.657 (3 sf)   **ii** -0.459 (3 sf)

   c Rising

8 Remember the limits: $\displaystyle\lim_{\theta \to 0}\left(\dfrac{\sin\theta}{\theta}\right) = 1$ and $\displaystyle\lim_{\theta \to 0}\left(\dfrac{\cos\theta - 1}{\theta}\right) = 0$

   a By definition $\dfrac{d(\sin 2x)}{dx} = \displaystyle\lim_{h \to 0}\left(\dfrac{\sin(2(x+h)) - \sin 2x}{h}\right)$

     $= \displaystyle\lim_{h \to 0}\left(\dfrac{\sin 2x \cos 2h + \cos 2x \sin 2h - \sin 2x}{h}\right)$

     $= \displaystyle\lim_{h \to 0}\left(\dfrac{\sin 2x(\cos 2h - 1) + \cos 2x \sin 2h}{h}\right)$

     $= \displaystyle\lim_{h \to 0}\left(\dfrac{2\sin 2x(\cos 2h - 1)}{2h}\right) + \displaystyle\lim_{h \to 0}\left(\dfrac{2\cos 2x \sin 2h}{2h}\right)$

     Note that as h tends to zero then so does $2h$

     $= \displaystyle\lim_{h \to 0}\left(2\sin 2x \dfrac{(\cos 2h - 1)}{2h}\right) + \displaystyle\lim_{h \to 0}\left(2\cos 2x \dfrac{\sin 2h}{2h}\right)$

     Treating $2h$ as θ

     $= 2\sin 2x \times 0 + 2\cos 2x \times 1$

     $= 2\cos 2x$

   b By definition $\dfrac{d(\sin ax)}{dx} = \displaystyle\lim_{h \to 0}\left(\dfrac{\sin(a(x+h)) - \sin ax}{h}\right)$

     $= \displaystyle\lim_{h \to 0}\left(\dfrac{\sin ax \cos ah + \cos ax \sin ah - \sin ax}{h}\right)$

     $= \displaystyle\lim_{h \to 0}\left(\dfrac{\sin ax(\cos ah - 1) + \cos ax \sin ah}{h}\right)$

     $= \displaystyle\lim_{h \to 0}\left(\dfrac{a\sin ax(\cos ah - 1)}{ah}\right) + \displaystyle\lim_{h \to 0}\left(\dfrac{a\cos ax \sin ah}{ah}\right)$

     Note that as h tends to zero then so does ah

     $= \displaystyle\lim_{h \to 0}\left(a\sin ax \dfrac{(\cos ah - 1)}{ah}\right) + \displaystyle\lim_{h \to 0}\left(a\cos ax \dfrac{\sin ah}{ah}\right)$

     Treating ah as θ

     $= a\sin ax \times 0 + a\cos ax \times 1$

     $= a\cos ax$

Exercise 15.3A Fluency and skills

1 **a** $4e^x$ **b** $1 - 2e^x$ **c** $5e^x - 6x$ **d** $-x^{-2} + e^x$

   e $-\dfrac{1}{x^2} + e^x$ **f** $e^x + \dfrac{2}{x^3}$ **g** $3x^2 + e^x + \sin x$ **h** $-\dfrac{e^x}{4}$

2 **a** $\dfrac{dy}{dx} = 3e^{3x}$ **b** $\dfrac{dy}{dx} = -8e^{-4x}$

      $\dfrac{d^2 y}{dx^2} = 9e^{3x}$ $\dfrac{d^2 y}{dx^2} = 32e^{-4x}$

   c $\dfrac{dy}{dx} = e^x - 2e^{2x}$ **d** $\dfrac{dy}{dx} = 3e^{0.5x}$

      $\dfrac{d^2 y}{dx^2} = e^x - 4e^{2x}$ $\dfrac{d^2 y}{dx^2} = 1.5e^{0.5x}$

   e $\dfrac{dy}{dx} = e^x - e^{-x}$ **f** $\dfrac{dy}{dx} = e^x + e^{-x}$

      $\dfrac{d^2 y}{dx^2} = e^x + e^{-x}$ $\dfrac{d^2 y}{dx^2} = e^x - e^{-x}$

   g $\dfrac{dy}{dx} = 1.25e^{1.25x} + 0.5e^{0.5x}$ **h** $\dfrac{dy}{dx} = e^x - 2e^{-2x}$

      $\dfrac{d^2 y}{dx^2} = 1.5625e^{1.25x} + 0.25e^{0.5x}$ $\dfrac{d^2 y}{dx^2} = e^x + 4e^{-2x}$

3 **a** $f'(x) = \dfrac{3}{x}$ **b** $f'(x) = -\dfrac{2}{x}$

      $f''(x) = -\dfrac{3}{x^2}$ $f''(x) = \dfrac{2}{x^2}$

   c $f'(x) = \dfrac{1}{x}$ **d** $f'(x) = \dfrac{3}{x} + \dfrac{1}{x} = \dfrac{4}{x}$

      $f''(x) = -\dfrac{1}{x^2}$ $f''(x) = -\dfrac{4}{x^2}$

   e $f'(x) = \dfrac{5}{x}$ **f** $f'(x) = -\dfrac{1}{x}$

      $f''(x) = -\dfrac{5}{x^2}$ $f''(x) = \dfrac{1}{x^2}$

   g $f'(x) = \dfrac{2}{x}$ **h** $f'(x) = -\dfrac{1}{x}$

      $f''(x) = -\dfrac{2}{x^2}$ $f''(x) = \dfrac{1}{x^2}$

   i $f'(x) = \dfrac{1}{2x}$

      $f''(x) = -\dfrac{1}{2x^2}$

4 **a** $5^x \ln 5$ **b** $2^x \ln 2$ **c** $6^x \ln 6$

   d $\dfrac{3^x \ln 3}{5}$ **e** $8^x \ln 8 - 7^x \ln 7$ **f** $5 - 5^x \ln 5$

   g $3 \times 4^x \ln 4$

5 a 3

b $1-2e$

c $1+6e$

d 2

e 8

f $\dfrac{1}{2}$

6 a 3

b $5e$

c $\ln 4$

d $5\ln 5$

e $\dfrac{1}{2}\ln 2$

f $1+36\ln 6$

g 3

h $1-\dfrac{2}{e}$

Exercise 15.3B Reasoning and problem-solving

1 a 4.27 years (3 sf)

b 14.25 pheasant pairs per year

c $\dfrac{d^2P}{dt^2}=0.9025e^{0.095t}>0$ for all t
Curve is always convex.

2 a $y=230e^{-0.134\times 0}-60=230-60$
$$=170$$

b Machine is losing £15.77 per year.

c Year 7

d **i** Year 10

ii Around £8 per year

e $\dfrac{dy}{dt}=-30.82e^{-0.134t}$

$\Rightarrow \dfrac{d^2y}{dt^2}=-0.134\times -30.82e^{-0.134t}$

$$=4.13e^{-0.134t}$$

which is positive for $0\le t\le 10$. So the curve is always convex during its lifetime.

3 a $A=82$
$k=0.093$

b **i** 7.6°C per minute

ii 3.0°C per minute

c 21.82 minutes

d 0.4°C per minute per minute

4 a 1085 m

b 11.7 metres per unit

c 12.1 metres per unit

5 a $c=250,\ k=-77.1$

b Decreasing by 19.3 tics per day

c **i** 25.3

ii Dropping by 3.05 tics per day

6 a $c=80,\ k=285.2$

b $w=285.2\ln t+80\Rightarrow \dfrac{dw}{dt}=\dfrac{285.2}{t}$

i 40.7 g/day

ii 20.4 g/day

c 11.3 g/day

d $\dfrac{dw}{dt}=\dfrac{285.2}{t}>0$ for $1\le t\le 30$, so the function is increasing in this domain.
$\dfrac{d^2w}{dt^2}=-\dfrac{285.2}{t^2}<0$ for $1\le t\le 30$, so the rate of growth is slowing down.

Exercise 15.4A Fluency and skills

1 a $\sin x+x\cos x$

b $e^x(1+x)$

c $e^x(\cos x-\sin x)$

d $3\ln x+3$

e $\cos x\cdot \ln x+\dfrac{\sin x}{x}$

f $e^{3x}(3x^2+2x+1)$

g $\dfrac{\ln 3x+2}{2\sqrt{x}}$

h $\dfrac{-\cos x+x\sin x}{x^2}$

i $2\left(x+\dfrac{1}{x^3}\right)$

j $e^{3x}\left(3\ln x+\dfrac{1}{x}\right)$

k $2x(1-3x+4x^2-5x^3)$

l $2x(1+2\ln x)$

m $-3\sqrt{x}\left(2+\dfrac{1}{2x^2}\right)$

n $\dfrac{\ln 3x+2}{2\sqrt{x}}$

o $-2\sin x\cos x$

p $4e^{2x+1}(x+1)$

2 a $\dfrac{d(\sin^2 x)}{dx}=\dfrac{d(\sin x\sin x)}{dx}$
Let $u=\sin x$ and $v=\sin x$
So $\dfrac{du}{dx}=\cos x$ and $\dfrac{dv}{dx}=\cos x$
Therefore, $\dfrac{dy}{dx}=\sin x\cos x+\sin x\cos x=2\sin x\cos x$
$=\sin 2x$

b $\dfrac{d(\cos^2 x)}{dx}=\dfrac{d(\cos x\cos x)}{dx}$
Let $u=\cos x$ and $v=\cos x$
So $\dfrac{du}{dx}=-\sin x$ and $\dfrac{dv}{dx}=-\sin x$
Therefore, $\dfrac{dy}{dx}=(\cos x)\cdot(-\sin x)+(\cos x)(-\sin x)$
$=-2\sin x\cos x\ =-\sin 2x$

c $\dfrac{d(\sin 2x)}{dx}=\dfrac{d(2\sin x\cos x)}{dx}$
Let $u=2\sin x$ and $v=\cos x$
So $\dfrac{du}{dx}=2\cos x$ and $\dfrac{dv}{dx}=-\sin x$
Therefore, $\dfrac{dy}{dx}=(\cos x).(2\cos x)+(2\sin x)(-\sin x)$
$=2(\cos^2 x-\sin^2 x)=2\cos 2x$

d $\dfrac{d(\cos 2x)}{dx}=\dfrac{d(\cos^2 x-\sin^2 x)}{dx}$
$=-2\sin x\cos x-2\sin x\cos x$
$=-4\sin x\cos x=-2\sin 2x$

3 a **i** 1

ii $\dfrac{6+\pi\sqrt{3}}{12}$

b **i** 0

ii $4e$

c **i** -1

ii 0

4 a $-\dfrac{2}{(x-1)^2}$

b $\dfrac{x\cos x-\sin x}{x^2}$

c $\dfrac{\cos x-\sin x}{e^x}$

d $\dfrac{x^2+2x-1}{(x+1)^2}$

e $-\dfrac{1}{\sin^2 x}$

f $\dfrac{1-\ln x}{x^2}$

g $-\dfrac{3\cos x}{\sin^2 x}$

h $\dfrac{\cos x+x\sin x}{\cos^2 x}$

i $\dfrac{x(4-3\sqrt{x})}{2(1-\sqrt{x})^2}$ **j** $\dfrac{3x^2-3x-1}{2x\sqrt{x}}$

k $\dfrac{e^{2x+3}(2x-3)}{x^4}$ **l** $\dfrac{1+\sin x}{\cos^2 x}$

5 a i -1 **ii** 1

  b i $\dfrac{1-e}{e^{e+1}}$ **ii** e^{-1}

  c i 0 **ii** $\dfrac{-4}{9}$ **iii** 0

Exercise 15.4B Reasoning and problem-solving

1 -2

2 $y=x+1$

3 $1-\dfrac{\pi}{2}$

4 a At $x=1$

  b Let $u=\ln x$ and $v=e^x$

  So $\dfrac{du}{dx}=\dfrac{1}{x}$ and $\dfrac{dv}{dx}=e^x$

  Therefore, $f'(x)=\dfrac{e^x\dfrac{1}{x}-e^x\ln x}{e^{2x}}$

  $=\dfrac{1-x\ln x}{xe^x}$

  At stationary point $f'(x)=0$

  $\Rightarrow \dfrac{1-x\ln x}{xe^x}=0$

  $\Rightarrow x\ln x=1$

  $\Rightarrow \ln x^x=1$

  $\Rightarrow x^x=e$

  Substituting $x=1.763$ gives $1.763^{1.763}$
  $=2.717\ldots$ which is e correct to 3 sf

  c $f'(x)=\dfrac{1-x\ln x}{xe^x}$

  $\Rightarrow f''(x)=\dfrac{xe^x\dfrac{d(1-x\ln x)}{dx}-(1-x\ln x)\dfrac{d(xe^x)}{dx}}{x^2e^{2x}}$

  $=\dfrac{xe^x(-1-\ln x)-(1-x\ln x)(xe^x+e^x)}{x^2e^{2x}}$

  Points of inflection occur when

  $\dfrac{xe^x(-1-\ln x)-(1-x\ln x)(xe^x+e^x)}{x^2e^{2x}}=0$

  $\Rightarrow xe^x(-1-\ln x)-(1-x\ln x)(xe^x+e^x)=0$

  $\Rightarrow x^2\ln x-2x-1=0$

  $\Rightarrow \ln x=\dfrac{2x+1}{x^2}$

  $\Rightarrow x=e^{\frac{2x+1}{x^2}}$

  $\Rightarrow e=x^{\frac{x^2}{2x+1}}$

  Substituting 2.55245 for x:

  $2.55245^{\frac{2.55245^2}{2\times2.55245+1}}=2.718284357\ldots$ which is e to 6 sf

5 a $\dfrac{d\sec x}{dx}=\dfrac{d\left(\dfrac{1}{\cos x}\right)}{dx}$

  Let $u=1$ and $v=\cos x$

  So $\dfrac{du}{dx}=0$ and $\dfrac{dv}{dx}=-\sin x$

Therefore, $\dfrac{dy}{dx}=\dfrac{\cos x\cdot 0+1\cdot\sin x}{\cos^2 x}$

  $=\dfrac{\sin x}{\cos x\cos x}$

  $=\sec x\tan x$

b $\dfrac{d\csc x}{dx}=\dfrac{d\left(\dfrac{1}{\sin x}\right)}{dx}$

  Let $u=1$ and $v=\sin x$

  So $\dfrac{du}{dx}=0$ and $\dfrac{dv}{dx}=\cos x$

  Therefore, $\dfrac{dy}{dx}=\dfrac{\sin x\cdot 0-1\cdot\cos x}{\sin^2 x}$

  $=\dfrac{-\cos x}{\sin x\sin x}$

  $=-\csc x\cot x$

c $\dfrac{d\cot x}{dx}=\dfrac{d\left(\dfrac{1}{\tan x}\right)}{dx}$

  Let $u=1$ and $v=\tan x$

  So $\dfrac{du}{dx}=0$ and $\dfrac{dv}{dx}=\sec^2 x$

  Therefore, $\dfrac{dy}{dx}=\dfrac{\tan x\cdot 0+1\cdot\sec^2 x}{\tan^2 x}$

  $=\dfrac{-\cos x\cos x}{\cos^2 x\sin^2 x}$

  $=-\csc^2 x$

6 a $3\sin^2 x\cos x$

  b $xe^x\cos x+(x+1)e^x\sin x$

7 a $T(x)=\dfrac{20}{100-x^2}$

  You can see that $100-x^2$ is decreasing for $0\le x<10$

  So $\dfrac{20}{100-x^2}$ must be increasing in this range.

  b 128 secs per mph

  c i 0.2 **ii** 0

  d As x approaches 10, the time taken to make the journey approaches infinity. This means that the closer the speed of the river is to 10 mph, the longer it takes, and if the speed of the river were to be 10mph, you would never complete the journey.

Exercise 15.5A Fluency and skills

1 a $y=u^2,\,u=\sin x$ **b** $y=\tan u,\,u=2x$

  c $y=u^{\frac{1}{2}},\,u=7x$ **d** $y=\ln u,\,u=\cos x$

  e $y=e^u,\,u=7x$ **f** $y=u^4,\,u=2x+1$

  g $y=u^3,\,u=3x-2$ **h** $y=u^{-\frac{1}{2}},\,u=\sin x$

  i $y=2u^{-\frac{1}{2}},\,u=x^3$ **j** $y=u^{\frac{1}{2}},\,u=x+\ln x$

  k $y=u^5,\,u=\sin x+\ln x$ **l** $y=u^{\frac{1}{3}},\,u=\cos x-\ln x$

2 a $15(3x+4)^4$ **b** $14(2x-1)^6$

  c $12x(x^2+1)^5$ **d** $-6(1-2x-3x^2)^2(1+3x)$

  e $\dfrac{1}{\sqrt{2x+1}}$ **f** $-\dfrac{5}{2\sqrt{3-5x}}$

  g $\dfrac{3x}{\sqrt{3x^2+4}}$ **h** $-\dfrac{2}{3\sqrt[3]{(1-2x)^2}}$

  i $\dfrac{2x+3}{2\sqrt{x^2+3x+4}}$ **j** $-\dfrac{4}{(2x+3)^3}$

For full solutions go to http://www.oxfordsecondary.co.uk/edexcelalevelmaths-answers

k $\dfrac{3}{2}(1-3x)^{-\frac{3}{2}}$

l $\dfrac{6(1-x)}{(x^2-2x+5)^2}$

m $-2\cos x\sin x$

n $-\dfrac{\sin x}{2\sqrt{\cos x}}$

o $3\cos(3x+2)$

p $5\sec^2(5x-1)$

q $-\dfrac{\sin\sqrt{x+1}}{2\sqrt{x+1}}$

r $-\cos(\cos x)\sin x$

s $e^{\sin x}\cos x$

t $\dfrac{e^{\sqrt{2x-1}}}{\sqrt{2x-1}}$

u $e^{(e^x)}\cdot e^x$

v $\cot x$

w $\dfrac{1}{2x+3}$

x $\dfrac{1}{x\ln x}$

y $\dfrac{\cos(\ln x)}{x}$

z $-\dfrac{1}{x(\ln x)^2}$

3 a 6   **b** $\dfrac{9}{8}$ **c** $\dfrac{3}{8}$

d -4 **e** $\dfrac{-3e^2}{4}$ **f** -1

g -12 **h** 3

4 a $2^{\sin x}\cos x\ln 2$

b $\dfrac{2}{(2x+1)\ln 10}$

c $2(x+1)3^{x^2+2x-1}\ln 3$

5 a $y=-5x+4$ **b** $y=4ex-3e$

c $8y-x=15$ **d** $y=2(x-1)$

e $y=4x-3$ **f** $y=2x-2+e$

6 a $-\dfrac{3}{4}$ **b** $4y-3x=13$

c $\dfrac{4}{3}$ **d** $\dfrac{-25}{27}$

Exercise 15.5B Reasoning and problem-solving

1 a $-2\sin\cos(\cos 2x)$

b $3\cos 3x\,e^{\sin 3x}$

c $5(\sin x+\cos 2x)^4(\cos x-2\sin 2x)$

d $\dfrac{2\cos(4x+1)}{\sqrt{\sin(4x+1)}}$

e $\dfrac{2x+1}{2\sqrt{(x-1)(x+2)}}$

f $e^x(x+1)\cos(xe^x)$

g $2(3x+2)e^{x(3x+4)}$

h $-x(x\cos x+2\sin x)\sin(x^2\sin x)$

i $-\dfrac{3}{(x-1)^2}\left(\dfrac{x+1}{x-1}\right)^{\frac{1}{2}}$

j $\dfrac{1}{(x+1)^2}\cos\left(\dfrac{x}{x+1}\right)$

k $\dfrac{1}{x(1-x)}$

l $\dfrac{x+1}{x}$

2 a $-\dfrac{\pi}{72}\sin\left(\dfrac{\pi D}{180}\right)$

b Minimum T of 16:30 pm occurs 180 days after the longest day.

c $\dfrac{d^2T}{dD^2}=\dfrac{\pi^2}{12\,960}>0$

$\Rightarrow$ a minimum turning point

3 $40\,\text{cm}^2\,\text{s}^{-1}$

4 a $\dfrac{b}{\sqrt{b^2-1}}-1$

b $b^2>b^2-1>0$

$\Rightarrow b>\sqrt{b^2-1}$

$\Rightarrow \dfrac{b}{\sqrt{b^2-1}}>1$

$\Rightarrow \dfrac{b}{\sqrt{b^2-1}}-1>0$

$\Rightarrow \dfrac{dx}{db}>0$

So x increases as b increases.

c $-\dfrac{1}{(b^2-1)\sqrt{b^2-1}}<0$

So graph of function is concave.

5 a $e^{0.001y}$

b 0.5% per annum

c $0.001\,e^{0.001y}$

6 a $3.3e^{-0.22t}$

b $h=15(1-e^{-0.22t})$

$\Rightarrow e^{-0.22t}=1-\dfrac{h}{15}$

So $\dfrac{dh}{dt}=0.22\times15e^{-0.22t}$

$\qquad =0.22\times15\left(1-\dfrac{h}{15}\right)$

$\qquad =0.22(15-h)$

7 a $\dfrac{2660e^{-0.7t}}{(19e^{-0.7t}+1)^2}$

b $\dfrac{200-N}{N}$

c $\dfrac{dN}{dt}=\dfrac{2660e^{-0.7t}}{\left(19e^{-0.7t}+1\right)^2}$

$\qquad =140\dfrac{19e^{-0.7t}}{\left(19e^{-0.7t}+1\right)^2}$

$\qquad =140\dfrac{\left(\dfrac{200}{N}-1\right)}{\left(\dfrac{200}{N}\right)^2}$

$\qquad =140\dfrac{200N-N^2}{40000}$

$\qquad =\dfrac{7}{2000}N(200-N)$

8 a $3\tan^2 x\sec^2 x$

b $\dfrac{\cos^2 x-\sin^2 x}{2\sqrt{\sin x\cos x}}=\dfrac{1}{\sqrt{2}}\cot 2x$

c $\dfrac{2\sin x\cos x\cos 7x-7\sin^2 x\sin 7x}{2\sqrt{\sin^2 x\cos 7x}}$

9 When $x=8$, $\dfrac{dV}{dt}=288\,\text{m}^3\,\text{minute}^{-1}$

Exercise 15.6A Fluency and skills

1 a i $f'(x)=6x^5$

$\qquad f^{-1}(x)=\sqrt[6]{x}$

ii $\dfrac{1}{6\left(\sqrt[6]{x}\right)^5}$

b **i** $f'(x) = \dfrac{2}{3\sqrt[3]{x}}$

$f^{-1}(x) = x^{\frac{3}{2}}$

ii $\dfrac{3\sqrt{x}}{2}$

c **i** $f'(x) = 2x$

$f^{-1}(x) = \sqrt{x-2}$

ii $\dfrac{1}{2\sqrt{x-2}}$

d **i** $f'(x) = 2(x+4)$

$f^{-1}(x) = \sqrt{x-1}-4$

ii $\dfrac{1}{2\sqrt{x-1}}$

e **i** $f'(x) = \dfrac{3}{2\sqrt{3x+2}}$

$f^{-1}(x) = \dfrac{x^2-2}{3}$

ii $\dfrac{2x}{3}$

f **i** $f'(x) = \dfrac{2}{3(2x+1)^{\frac{2}{3}}}$

$f^{-1}(x) = \dfrac{x^3-1}{2}$

ii $\dfrac{3x^2}{2}$

2 **a** $y = \cos^{-1} x \Rightarrow x = \cos y$

$\Rightarrow \dfrac{dx}{dy} = -\sin y$

$0 \le \cos^{-1} x \le \pi$

$\Rightarrow 0 \le y \le \pi$

$\Rightarrow \sin y \ge 0$

$\Rightarrow \sin y = \sqrt{1 - \cos^2 y}$

$\quad = \sqrt{1 - x^2}$

$\Rightarrow \dfrac{dy}{dx} = -\dfrac{1}{\sin y}$

$\quad = -\dfrac{1}{\sqrt{1-x^2}}$

b $y = \tan^{-1} x \Rightarrow x = \tan y$

$\Rightarrow \dfrac{dx}{dy} = 1 + \tan^2 y$

$\quad = 1 + x^2$

$\Rightarrow \dfrac{dy}{dx} = \dfrac{1}{1+x^2}$

c $y = \sec^{-1} x \Rightarrow x = \sec y = \dfrac{1}{\cos y}$

$\Rightarrow \dfrac{dx}{dy} = \dfrac{\sin y}{\cos^2 y}$

$\quad = \sin y \sec^2 y$

$0 \le \sec^{-1} x \le \pi$

$\Rightarrow 0 \le y \le \pi$

$\Rightarrow \sin y \ge 0$

$\Rightarrow \sin y = \sqrt{1 - \cos^2 y} = \sqrt{1 - \dfrac{1}{x^2}}$

$\dfrac{dx}{dy} = \sin y \sec^2 y$

$\quad = x^2 \sqrt{1 - \dfrac{1}{x^2}}$

$\quad = x\sqrt{x^2 - 1}$

$\Rightarrow \dfrac{dy}{dx} = \dfrac{1}{x\sqrt{x^2-1}}$

d $y = \cot^{-1} x \Rightarrow x = \cot y$

$\Rightarrow \dfrac{dx}{dy} = -\csc^2 y = -(1 + \cot^2 y)$

$\quad = -(1 + x^2)$

$\Rightarrow \dfrac{dy}{dx} = -\dfrac{1}{1+x^2}$

3 **a** **i** 3 **ii** $\dfrac{x-1}{3}$ **iii** $\dfrac{1}{3}$

b **i** -5 **ii** $\dfrac{1-x}{5}$ **iii** $-\dfrac{1}{5}$

c **i** $6x$ **ii** $\sqrt{\dfrac{x}{3}}$ **iii** $\dfrac{1}{2\sqrt{3x}}$

d **i** $\dfrac{1}{2}$ **ii** $2x-1$ **iii** 2

e **i** $-\dfrac{1}{(x+1)^2}$ **ii** $\dfrac{1}{x}-1$ **iii** $-\dfrac{1}{x^2}$

f **i** $\dfrac{1}{(x+1)^2}$ **ii** $\dfrac{x}{1-x}$ **iii** $\dfrac{1}{(1-x)^2}$

g **i** $-\dfrac{2}{x^3}$ **ii** $\dfrac{1}{\sqrt{x}}$ **iii** $-\dfrac{1}{2x\sqrt{x}}$

h **i** $\dfrac{1}{2\sqrt{x}}$ **ii** x^2 **iii** $2x$

i **i** $\dfrac{1}{2\sqrt{x}}$ **ii** $(x+1)^2$ **iii** $2(x+1)$

j **i** $\dfrac{1}{3}x^{-\frac{2}{3}}$ **ii** x^3 **iii** $3x^2$

k **i** $2e^{2x+1}$ **ii** $\dfrac{\ln x - 1}{2}$ **iii** $\dfrac{1}{2x}$

l **i** $\dfrac{1}{x}$ **ii** $\dfrac{e^x}{4}$ **iii** $\dfrac{e^x}{4}$

Exercise 15.6B Reasoning and problem-solving

1 $y = 3x - 5$

2 $6\sqrt{3}\,y - 6x = \pi\sqrt{3} - 3$

3 **a** $6\sqrt{3}\,y - 12x = \pi\sqrt{3} + 6$

b The gradient is undefined, so the tangent is parallel to the y-axis. Equation: $x = -2$

4 **a** $A = \dfrac{1}{2}r^2\theta$

$\quad = \dfrac{1.5^2}{2}\theta$

$\quad = \dfrac{9}{8}\theta$

$h = 1.5\sin\theta$

$\Rightarrow \theta = \sin^{-1}\dfrac{2h}{3}$

$A = \dfrac{9}{8}\sin^{-1}\dfrac{2h}{3}$

b $\dfrac{9}{4\sqrt{9 - 4h^2}}$

Answers For full solutions go to http://www.oxfordsecondary.co.uk/edexcelalevelmaths-answers

c $\dfrac{5}{4}$ m² per m

5 a $\dfrac{1.2}{\pi\sqrt{1-(2-0.2D)^2}}$

  b 0.382

Exercise 15.7A Fluency and skills

1 a $\dfrac{2x}{3y^2}$ **b** $-\dfrac{x+1}{y-3}$

  c $-\dfrac{2x}{y}$ **d** $-\dfrac{y^2}{x^2}$

  e $-\dfrac{2(x+y)}{2x+3}$ **f** $\dfrac{2xy^2}{y^2-1}$

  g -1 **h** $6e^x\sqrt{y}$

  i $-\dfrac{\cos x}{2\sin 2y}$ **j** $2x(y+1)$

  k $\dfrac{\cos y-\sec^2 x}{x\sin y}$ **l** $-\dfrac{y^{\frac{1}{4}}}{x^{\frac{1}{4}}}$

  m $\dfrac{\sqrt{1-y^2}\,(\sqrt{1-x^2}+1)}{\sqrt{1-x^2}\,(\sqrt{1-y^2}-1)}$ **n** $\dfrac{(x+y)(1+y)-1}{(1-x(x+y))}$

  o $\dfrac{2(x+y)}{e^{y+1}-2x}$ **p** $\dfrac{y(1+y^2)}{1+x(1+y^2)}$

  q $-\dfrac{y^3}{x^3}$

2 $xy+2x^2y^2=1$

  $\Rightarrow x\dfrac{dy}{dx}+y+4x^2y\dfrac{dy}{dx}+4xy^2=0$

  $\Rightarrow \dfrac{dy}{dx}=-\dfrac{4xy^2+y}{x+4x^2y}$

  When $x=1$ and $y=0.5$

  $\Rightarrow \dfrac{dy}{dx}=-\dfrac{4\times1\times0.25+0.5}{1+4\times1\times0.5}$

  $\qquad =-\dfrac{1}{2}$

3 a $-\dfrac{x}{y}$

  b When $x=3$
    $\Rightarrow y=4$ or -4
    At point (3, 4) the gradient $\dfrac{dy}{dx}=-\dfrac{3}{4}$
    and at point (3, -4) the gradient is $\dfrac{dy}{dx}=\dfrac{3}{4}$

  c When $x=0$, $y=5$
    $\Rightarrow \dfrac{dy}{dx}=-\dfrac{0}{5}$
    $\qquad =0$

4 a $(10,\pm2\sqrt{5})$

  b $\left(2,-\dfrac{5}{2}\right)$ and $(2,-1)$

  c $(0, 3)$

5 a $e^x+e^{2y}+3x=2-4y$
    When $x=0$
    $1+e^{2y}=2-4y$
    $\Rightarrow e^{2y}=1-4y$
    $\Rightarrow y=0$
    Therefore (0, 0) lies on the curve.

  b $-\dfrac{3+e^x}{2(e^{2y}+2)}$

c $-\dfrac{2}{3}$

6 a $(1+\ln x)(y-1)$

  b $\ln(y-1)=0$
    $\Rightarrow y-1=1$
    $\Rightarrow y=2$

  c 1

7 a $\dfrac{1-x}{y-2};\quad y\neq 2$

  b $\dfrac{4}{3}$

8 a $\dfrac{2x[3(x^2+y^2-1)^2-y^3]}{3y[x^2y-2(x^2+y^2-1)^2]}$

  b $-\dfrac{4}{3}$

Exercise 15.7B Reasoning and problem-solving

1 $y=-2x+10$

2 At points (1, 6) and (7, −2)

3 a Ellipse cuts x-axis when $y=0$
    $\dfrac{x^2}{a^2}+\dfrac{0^2}{b^2}=1$
    $\Rightarrow x=\pm a$
    Hence result.

  b $-\dfrac{b^2x}{a^2y}$

  c $(0, b)$ and $(0, -b)$

  d y is directly proportional to x

4 a $\dfrac{4x+7}{2(y+1)}$

  b $2y-3x=2$
    or $2y+3x=-6$

  c $\left(-\dfrac{4}{3},-1\right)$

5 a (3, −3) and (−3, 3)

  b Maximum: $2\sqrt{3}$; minimum: $-2\sqrt{3}$

  c $x=\mp2\sqrt{3}$

6 a i $3y-4x=11$

    ii $4y-3x=17$

  b (1, 5); the centre of the circle

7 a $11y-x=96$

  b $\left(-\dfrac{12\sqrt{7}}{7},-\dfrac{24\sqrt{7}}{7}\right)$
    $\left(\dfrac{12\sqrt{7}}{7},\dfrac{24\sqrt{7}}{7}\right)$

Exercise 15.8A Fluency and skills

1 a $\dfrac{1}{2}$ **b** 2

  c $\dfrac{1}{2t}$ **d** $-\tan\theta$

  e $-\dfrac{\sin(\theta+2)}{\cos(\theta+3)}$ **f** $-\tan\theta$

  g $-\tan(2\theta+1)$ **h** $\dfrac{e^t+e^{-t}}{e^t-e^{-t}}$

  i $-te^{-t}$ **j** $\dfrac{1}{t}$

  k $\dfrac{2t}{\cos t}$

2 **a** $\dfrac{\cos 3\theta - \cos\theta}{\sin 3\theta + \sin\theta}$ **b** $-\dfrac{4}{3}\tan 2\theta$

  c -1 **d** $\dfrac{\sqrt{1-\theta^2}}{1+\theta^2}$

  e -1 **f** $-\dfrac{1}{t\sqrt{3}}$

  g $-\tan t \sqrt{\tan t}$ **h** $-\dfrac{1}{\sin\theta}$

  i $-\dfrac{e^{\cos t}}{e^{\sin t}}\tan t$ **j** $\sec t$

  k $2t - 4$

3 **a** $(-\pi, 0)$ **b** π

4 **a** $x = 8, y = 2$ **b** $x = \dfrac{3}{4}, y = \dfrac{9}{4}$
    gradient $= \dfrac{1}{12}$ gradient $= 15$

  c $x = 1, y = 2$ **d** $x = 1, y = 1.5$
    gradient $= -2$ gradient $= -\sqrt{3}$

  e $x = 6, y = 4\sqrt{3}$ **f** $x = 7, y = 16$
    gradient $= \dfrac{8}{3\sqrt{3}}$ gradient $= \dfrac{3}{2}$

Exercise 15.8B Reasoning and problem-solving

1 **a** 768

  b $16x^3 - 4x^2 + 1$

2 **a** $x^2 + \dfrac{y^2}{25} = 1$

  b $y = \pm 5$

3 **a** $\dfrac{1 - 9t^2}{8t}$ **b** $\pm\dfrac{1}{3}$

  c $\left(\dfrac{14}{9}, \dfrac{7}{9}\right)$; minimum

    $\left(\dfrac{14}{9}, \dfrac{11}{9}\right)$; maximum

  d **i** $t = 0, \pm\dfrac{1}{\sqrt{3}}$

    ii $t = 0$ corresponds to the point $(2, 1)$ where the loop
    occurs (tangent vertical); $t = \pm\dfrac{1}{\sqrt{3}}$ corresponds to
    $\left(\dfrac{2}{3}, 1\right)$, the point where the curve intersects itself.
    $y = 1$ is an axis of reflective symmetry.

4 **a** $-\cot\theta$
  b $\sqrt{3}$
  c $x = 1 + 5\cos\theta$

    $\Rightarrow \cos\theta = \dfrac{x-1}{5}$

    $y = 2 + 5\sin\theta$

    $\Rightarrow \sin\theta = \dfrac{y-2}{5}$

    Squaring

    $\cos^2\theta = \dfrac{(x-1)^2}{25}$

    $\sin^2\theta = \dfrac{(y-2)^2}{25}$

    Adding gives

    $\dfrac{(x-1)^2}{25} + \dfrac{(y-2)^2}{25} = \cos^2\theta + \sin^2\theta$

      $= 1$

    Hence $(x-1)^2 + (y-2)^2 = 25$

5 **a** $1 - t$

  b **i** $t = 0, 2$
    ii $x = 0$ and $x = 20$

  c $x = 10t$

    $\Rightarrow t = \dfrac{x}{10}$

    $\Rightarrow y = 10\left(\dfrac{x}{10}\right) - 5\left(\dfrac{x}{10}\right)^2$

      $= x - \dfrac{1}{20}x^2$

    which is of the required form with $a = -\dfrac{1}{20}, b = 1, c = 0$

6 $(8\sqrt{2}, 0)$

Review exercise 15

1 Curve is convex at $\left(1, -\dfrac{2}{3}\right)$ a minimum turning point.

  Curve is concave at $(-3, 10)$ a maximum turning point.
  $\left(-1, 4\dfrac{2}{3}\right)$ is a point of inflection on a decreasing section of
  the curve.

2 **a** $(0, 1)$ and $(-3, 28)$

  b $(0, 1)$ is a minimum turning point
    $(-3, 28)$ is a point of inflection on a decreasing section of
    the curve.

3 **a** $f(x) = \sin 2x$

    $\Rightarrow f'(x) = \lim_{h \to 0} \dfrac{\sin 2(x+h) - \sin 2x}{h}$

    $\Rightarrow f'(x) = \lim_{h \to 0} \dfrac{\sin 2x \cos 2h + \cos 2x \sin 2h - \sin 2x}{h}$

    $= 2\lim_{h \to 0} \dfrac{\sin 2x(\cos 2h - 1) + \cos 2x \sin 2h}{2h}$

    $= 2\lim_{h \to 0} \sin 2x \dfrac{(\cos 2h - 1)}{2h} + 2\lim_{h \to 0} \cos 2x \dfrac{\sin 2h}{2h}$

    $= 2\sin 2x \lim_{h \to 0} \dfrac{(\cos 2h - 1)}{2h} + 2\cos 2x \lim_{h \to 0} \dfrac{\sin 2h}{2h}$

    As h tends to zero so does $2h$
    $\Rightarrow f'(x) = 2\sin 2x \times 0 + 2\cos 2x \times 1$
    $\Rightarrow f'(x) = 2\cos 2x$

  b **i** $-\dfrac{1}{\sqrt{2}}$

    ii $-\dfrac{5}{\sqrt{2}}$

4 **a** 9.69 cm s^{-1}

  b Falling

  c After $\dfrac{\pi}{4}$ seconds

5 **a** $8e^{2x+1}$ **b** $\dfrac{3}{3x+1}$

  c $2 \times 3^x \ln 3$ **d** $-2^{-x}\ln 2$

6 **a** $e^x \sin x(\sin x + 2\cos x)$

  b $\dfrac{3(1 - \ln x)}{2x^2}$

  c $\sqrt{x}\sec^2 x + \dfrac{\tan x}{2\sqrt{x}}$

  d $\dfrac{(3x^2 + 6)\cos x + (x^3 + 6x + 11)\sin x}{\cos^2 x}$

Answers For full solutions go to http://www.oxfordsecondary.co.uk/edexcelalevelmaths-answers

7 a $\dfrac{e^x[(1+x)\ln x - 1]}{(\ln x)^2}$

  b $\dfrac{-x - \cos x \sin x}{x^2 \sin^2 x}$

  c $\dfrac{-2x^2 \sin x - x \sin x + \cos x}{(2x+1)^2}$

8 a $20x(2x^2 - 3)^4$ **b** $x > 0$

9 a $\dfrac{1}{4}$ **b** 0

  c -6 **d** $\dfrac{e^3}{3}$

10 a $-3^{\cos x}\ln 3 \sin x$

  b $\dfrac{4^{\ln x}\ln 4}{x}$

11 a $-3x^{-3}y^3$ **b** $\tan x \tan y$

  c $\dfrac{1 - e^y - ye^x}{xe^y + e^x}$ **d** $\dfrac{x - e^y}{x(e^y \ln x - 2)}$

12 a $\dfrac{3}{6 - 4y}$ **b** $\dfrac{1}{6}$

13 $-\dfrac{1}{3}$

14 a $\dfrac{1}{e^y + 2}$ **b** $\dfrac{1}{3}$

15 a $\dfrac{-2x}{\sqrt{1 - x^4}}$ **b** $-\dfrac{1}{10\sqrt{2e^{\frac{x}{10}} - 1}}$

  c $\dfrac{1}{x\sqrt{x^2 - 1}}$

16 a 2 **b** $2t^2$

  c $-\dfrac{4}{3}\tan\theta$ **d** 2

17 a $-\dfrac{3}{t^3}$ **b** $-\dfrac{3}{64}$

18 a At $x = -0.3779645\ldots\left(= -\dfrac{1}{\sqrt{7}}\right)$ and $x = 0.3779645\left(= \dfrac{1}{\sqrt{7}}\right)$

  b $-\dfrac{7}{3\sqrt{3}}$

Assessment 15

1 a $6x + x^{-\frac{3}{2}}$

  b $\dfrac{193}{8}$ or 24.125

  c $y - 1 = -\dfrac{1}{7}(x - 1)$

2 $2\,\text{s}$

3 a i $-5\sin x$ **ii** $\dfrac{1}{x}$ **iii** $\dfrac{1}{4}e^x$

  b $x + 2y - 1 = 0$

4 a i $e^x + xe^x$ **ii** $2e^x + xe^x$

  b $ke^x + xe^x$

5 a $3\sin x + 3x\cos x$

  b $y = -3\pi(x - \pi)$

6 a i $\dfrac{2}{(x + 2)^2}$

  ii $\dfrac{6x\cos x + 3x^2 \sin x}{\cos^2 x}$

  iii $9x^2 e^x + (3x^3 + 5)e^x$

  b $\dfrac{x^2 - 10x - 15}{(x - 5)^2}$

7 a i $-3\sin 3x$ **ii** $6e^{3x}$ **iii** $2\cos(2x - 5)$

  b $\dfrac{\sqrt{3}}{2}$

8 a i $4y - \dfrac{4}{\sqrt{y}}$ **ii** $\dfrac{1}{4y - \dfrac{4}{\sqrt{y}}}$

b $y - 4 = -14(x - 16)$

9 $\dfrac{5y}{3y^2 - 5x}$

10 a 1

  b $y - 12 = \dfrac{3}{2}(x - 4)$

11 a $x = 0,\, y = 0$ or $x = -2,\, y = 8e^{-2}$

  b $(0, 0)$ is a minimum and $(-2, 8e^{-2})$ is a maximum.

12 a $x = 0$, there is a point of inflection on a decreasing part of the curve. At the point of inflection, the gradient of the curve is -4

  b

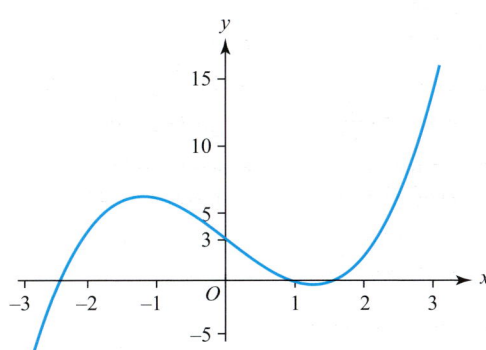

  c $x < 0$

13 a Stationary point at $(-1, 2e^{-1})$
  Either side of the stationary point $\dfrac{dy}{dx} > 0$
  So a point of inflection.

  b $\dfrac{d^2 y}{dx^2} = e^x(x^2 + 4x + 3)$

   $= e^x(x + 3)(x + 1)$

   $\dfrac{d^2 y}{dx^2} = 0$ when $x = -3$ and when $x = -1$

   When $x = -3$, $\dfrac{dy}{dx} = 4e^{-3} > 0$

   At $(-3, 10e^{-3})$ there is a point of inflection on an increasing part of the curve.

14 a $V = 2x^2 h$ where h is the height of the cuboid.

   $2x^2 h = 192$ gives $h = \dfrac{96}{x^2}$

   $S = 2(2x^2 + xh + 2xh)$

   $S = 4x^2 + 6x\left(\dfrac{96}{x^2}\right)$

   $S = 4x^2 + \dfrac{576}{x}$ $(k = 576)$

  b $208\ \text{cm}^2$

  c $\dfrac{d^2 S}{dx^2} = 8 + \dfrac{1152}{x^3}$

   which is always positive as $x > 0$
   Therefore a minimum.

15 $(-1, -6)$ is a point of inflection.

16 a $\dfrac{dy}{dx} = \lim_{h \to 0}\left(\dfrac{\sin(x + h) - \sin x}{x + h - x}\right)$

   $= \lim_{h \to 0}\left(\dfrac{\sin x \cos h + \sin h \cos x - \sin x}{h}\right)$

   $= \sin x \lim_{h \to 0}\left(\dfrac{\cos h - 1}{h}\right) + \cos x \lim_{h \to 0}\left(\dfrac{\sin h}{h}\right)$

   $= \cos x$

   Since $\lim_{h \to 0}\left(\dfrac{\cos h - 1}{h}\right) = 0$ and $\lim_{h \to 0}\left(\dfrac{\sin h}{h}\right) = 1$

b $\dfrac{dy}{dx} = \lim\limits_{h \to 0} \left(\dfrac{\cos(2x+h) - \cos 2x}{2x + h - 2x} \right)$

$\qquad = \lim\limits_{h \to 0} \left(\dfrac{\cos 2x \cos h - \sin h \sin 2x - \cos 2x}{h} \right)$

$\qquad = \cos 2x \lim\limits_{h \to 0} \left(\dfrac{\cos h - 1}{h} \right) - \sin 2x \lim\limits_{h \to 0} \left(\dfrac{\sin h}{h} \right)$

$\qquad = -\sin 2x$

$\qquad$ Since $\lim\limits_{h \to 0} \left(\dfrac{\cos h - 1}{h} \right) = 0$ and $\lim\limits_{h \to 0} \left(\dfrac{\sin h}{h} \right) = 1$

17 a $\sec^2 x$

$\quad$ **b** $\sec^2 x = (\cos x)^{-2}$

$\qquad$ $f''(x) = -2(\cos x)^{-3}(-\sin x)$

$\qquad\qquad = 2\dfrac{\sin x}{\cos x} \cdot \dfrac{1}{\cos^2 x}$

$\qquad\qquad = 2\tan x \sec^2 x$ as required

18 a $4^x \ln 4$

$\quad$ **b** $x = \dfrac{-1 - \ln 4}{\ln 4}$

19 a **i** $e^x \sin 5x + 5e^x \cos 5x$

$\qquad$ **ii** $10 e^x \cos 5x - 24 e^x \sin 5x$

$\quad$ **b** $x = 0.354$

$\qquad$ $e^{0.354}[10\cos(5 \times 0.354) - 24\sin(5 \times 0.354)]$

$\qquad = -36.3 < 0$ so a maximum.

20 a **i** $\dfrac{3e^{3t}(t^2 + 1) - 2te^{3t}}{(t^2 + 1)^2}$

$\qquad$ **ii** $3\ln t + 3$

$\qquad$ **iii** $-e^{-t} \sin 4t + 4e^{-t} \cos 4t$

$\quad$ **b** $2^x \ln 2 \tan x + 2^x \sec^2 x$

21 a $y = (\cos x)^{-1}$

$\qquad$ $\dfrac{dy}{dx} = -(\cos x)^{-2}(-\sin x)$

$\qquad$ $\dfrac{dy}{dx} = \dfrac{\sin x}{\cos^2 x} = \dfrac{1}{\cos x} \cdot \dfrac{\sin x}{\cos x}$

$\qquad$ $\dfrac{dy}{dx} = \sec x \tan x$ as required

$\quad$ **b** $2\sec^3 x - \sec x$

22 a £6749

$\quad$ **b** £174 per year

23 a $\dfrac{1}{1 + \ln 2}$

$\quad$ **b** $y = 2 + 2\ln 2 + 2(\ln 2)^2$

24 $\dfrac{dx}{dy} = -\sin y$

$\qquad$ $\dfrac{dy}{dx} = -\dfrac{1}{\sin y}$

$\qquad$ $\sin y = \sqrt{1 - \cos^2 y}$

$\qquad\qquad = \sqrt{1 - x^2}$

$\qquad$ (because $\sin y > 0$)

$\qquad$ Therefore, $\dfrac{dy}{dx} = -\dfrac{1}{\sqrt{1 - x^2}}$ as required

25 a $\dfrac{1 - xe^y}{x^2 e^y}$

$\quad$ **b** $\dfrac{4xy - \sin y}{x\cos y - 2x^2}$

26 $-\dfrac{1}{4}$

27 a -1

$\quad$ **b** $y = \dfrac{2x - 5}{x - 7}$

28 a $-\dfrac{3\sin t}{2\cos t}$

$\quad$ **b** $3\sqrt{3}\ x - 2y - 12 = 0$

29 $(0, -3)$ is a minimum.

30 a Convex

$\quad$ **b** Convex

31 $x < -1$ or $x > 1$

32 a Arc length is $r\theta = 6$

$\qquad$ So $r = \dfrac{6}{\theta}$

$\qquad$ Area of triangle $= \dfrac{1}{2}r^2 \sin \theta$

$\qquad$ $A = \dfrac{1}{2}\left(\dfrac{6}{\theta}\right)^2 \sin \theta$

$\qquad$ $A = \dfrac{18}{\theta^2} \sin \theta$

$\quad$ **b** $\dfrac{dA}{d\theta} = -\dfrac{36}{\theta^3} \sin \theta + \dfrac{18}{\theta^2} \cos \theta$

$\qquad$ $-36\sin\theta + 18\theta\cos\theta = 0$

$\qquad$ $36\sin\theta = 18\theta\cos\theta$

$\qquad$ $\dfrac{\sin\theta}{\cos\theta} = \dfrac{18\theta}{36}$

$\qquad$ $\tan\theta = \dfrac{\theta}{2}$ as required

33 $-6 + \ln 27$

34 a $y = (\sin 3x)^{-1}$

$\qquad$ $\dfrac{dy}{dx} = -(\sin 3x)^{-2}(3\cos 3x)$

$\qquad$ $\dfrac{dy}{dx} = -\dfrac{3\cos 3x}{\sin^2 3x} = -3\dfrac{\cos 3x}{\sin 3x} \cdot \dfrac{1}{\sin 3x}$

$\qquad$ $\dfrac{dy}{dx} = -3\cot 3x \csc 3x$ as required

$\quad$ **b** $9\cot^2 3x \csc 3x + 9\csc^3 3x$

35 a $\dfrac{1}{1 + \tan^2 y} = \dfrac{1}{1 + x^2}$

$\quad$ **b** $y - \dfrac{\pi}{4} = \dfrac{1}{2}(x - 1)$

$\quad$ **c** At B, $x = 0$

$\qquad$ $y - \dfrac{\pi}{4} = \dfrac{1}{2}(-1)$

$\qquad$ $y = \dfrac{\pi}{4} - \dfrac{1}{2}$

$\qquad$ At A, $y = 0$

$\qquad$ $-\dfrac{\pi}{4} = \dfrac{1}{2}(x - 1)$

$\qquad$ $x = 1 - \dfrac{\pi}{2}$ (negative)

$\qquad$ Area $= \dfrac{1}{2}\left(\dfrac{\pi}{2} - 1\right)\left(\dfrac{\pi}{4} - \dfrac{1}{2}\right)$

$\qquad\qquad = \dfrac{\pi^2}{16} - \dfrac{\pi}{8} - \dfrac{\pi}{8} + \dfrac{1}{4}$

$\qquad\qquad = \dfrac{1}{16}(\pi^2 - 4\pi + 4)$ as required

36 $\dfrac{1}{9}$

37 $\dfrac{1}{6\pi}$

38 Let $y = \sin^{-1} 2x$

Then $\sin y = 2x$

$\cos y \dfrac{dy}{dx} = 2$

$\dfrac{dy}{dx} = \dfrac{2}{\cos y}$

$= \dfrac{2}{\sqrt{1 - \sin^2 y}}$

(because $\cos y \geq 0$)

$= \dfrac{2}{\sqrt{1 - (2x)^2}}$ or $= \dfrac{2}{\sqrt{1 - 4x^2}}$

39 a $\dfrac{6x - y^2}{2xy + 2}$

b $\dfrac{5}{4}$ or $-\dfrac{15}{2}$

40 Let $y = a^x$

Then $\ln y = \ln a^x$

$\qquad = x \ln a$

Differentiate both sides with respect to x:

$\dfrac{1}{y} \dfrac{dy}{dx} = \ln a$

$\dfrac{dy}{dx} = y \ln a$

$\qquad = a^x \ln a$ as required

41 a $\dfrac{dx}{dt} = \dfrac{2}{(3 - t)^2}$

$\dfrac{dy}{dt} = \dfrac{2t}{3 - t} + \dfrac{t^2}{(3 - t)^2}$

or $\dfrac{2t(3 - t) + t^2}{(3 - t)^2} \left(= \dfrac{6t - t^2}{(3 - t)^2} \right)$

$\dfrac{dy}{dx} = \dfrac{6t - t^2}{(3 - t)^2} \cdot \dfrac{(3 - t)^2}{2}$

$\qquad = \dfrac{6t - t^2}{2}$

b $y = \dfrac{9}{2}x + \dfrac{2}{x} - 6$

42 a $\dfrac{dy}{d\theta} = \sec^2(\theta - 4)$

$\dfrac{dx}{d\theta} = \sec(\theta - 4)\tan(\theta - 4)$

$\dfrac{dy}{dx} = \dfrac{\sec^2(\theta - 4)}{\sec(\theta - 4)\tan(\theta - 4)}$

$\dfrac{dy}{dx} = \dfrac{\sec(\theta - 4)}{\tan(\theta - 4)}$

$\dfrac{dy}{dx} = \dfrac{1}{\cos(\theta - 4)} \cdot \dfrac{\cos(\theta - 4)}{\sin(\theta - 4)}$

$\dfrac{dy}{dx} = \text{cosec}(\theta - 4)$ as required

b $x^2 - y^2 = 1$

c When $x = 3$,

$9 - y^2 = 1$ gives $y = \sqrt{8} = 2\sqrt{2}$

Either differentiate the Cartesian equation to get

$2x - 2y\dfrac{dy}{dx} = 0$

$\dfrac{dy}{dx} = \dfrac{x}{y}$

So when $x = 3$ and $y = 2\sqrt{2}$,

$\dfrac{dy}{dx} = \dfrac{3}{2\sqrt{2}}$

Or use $3 = \sec(\theta - 4)$

Which gives $\dfrac{1}{3} = \cos(\theta - 4)$

So $\sin(\theta - 4) = \sqrt{1 - \left(\dfrac{1}{3}\right)^2} = \dfrac{2\sqrt{2}}{3}$

Therefore $\dfrac{dy}{dx} = \dfrac{1}{\dfrac{2\sqrt{2}}{3}} = \dfrac{3}{2\sqrt{2}}$

$y - 2\sqrt{2} = \dfrac{3}{2\sqrt{2}}(x - 3)$

$3x - 2y\sqrt{2} = 1$

Chapter 16

Exercise 16.1A Fluency and skills

1 a $4x^6 + c$ **b** $-7\cos x + c$ **c** $4\sin x + c$

d $\dfrac{2}{3}\tan x + c$ **e** $5e^x + c$ **f** $x^2 + \dfrac{2}{3}x^{\frac{3}{2}} + c$

g $-\dfrac{3}{x} + c$ **h** $3\ln|x| + c$ **i** $-\cos x + \tan x + c$

j $4\sin x + 3\cos x + c$

k $3\tan x + 2\sin x + c$

l $5e^x + c$

2 a $3x - 4e^x + c$ **b** $\dfrac{1}{2}\ln|x| + c$ **c** $\dfrac{11}{15}\ln|x| + c$

d $x - 3e^x + c$ **e** $7x + 6\cos x + c$ **f** $x^4 - 5\ln|x| + c$

g $x - 2\sin x + c$ **h** $\dfrac{x^2}{2} + \dfrac{1}{2x^2} + c$ **i** $\dfrac{x}{e} + \dfrac{e^x}{3} + c$

j $\dfrac{3x}{4} - \dfrac{1}{x^2} + \dfrac{\cos x}{3} + c$

k $-\dfrac{4}{x} - \dfrac{10}{3}x^{\frac{3}{2}} + c$

l $x + \dfrac{3}{5}x^{\frac{5}{3}} + c$

3 a 2 **b** 1 **c** $\dfrac{\sqrt{3} - 1}{2}$

d 2 **e** 1.718 **f** 2

g 8 **h** -12

4 a $71\dfrac{2}{3}$ **b** $\dfrac{\pi}{2} + 1$ **c** $\sqrt{3} - 1 + \dfrac{\pi}{2}$

d 2 **e** $e - \dfrac{1}{2}$ **f** $\dfrac{4}{3}$

5 a i $\dfrac{13}{4}$ **ii** $\displaystyle\int_1^2 6 - 2x\,dx$ **iii** 3

b i $\dfrac{165}{32}$ **ii** $\displaystyle\int_{-1}^2 x^2 + 1\,dx$ **iii** 6

6 a i 4.41 (to 3 sf)   **ii** $\displaystyle\int_1^4 -\sqrt{x}\,dx$ **iii** $\dfrac{14}{3}$

b i $\dfrac{197}{27}$ **ii** $\displaystyle\int_1^3 4x - x^2\,dx$ **iii** $\dfrac{22}{3}$

7 $\dfrac{2\pi}{3} + 3$

8 $\dfrac{1}{2}$

Exercise 16.1B Reasoning and problem-solving

1 a $\dfrac{2}{5}x^{\frac{5}{2}}+c$

   b $\dfrac{1}{3}x+\dfrac{1}{3}\ln x+c$

   c $\dfrac{1}{2}x^4+x^3+c$

   d $\dfrac{2}{5}x^{\frac{5}{2}}+x^2+c$

   e $\dfrac{1}{2}x^2+2x+\ln x+c$

   f $\dfrac{2}{3}x^{\frac{3}{2}}+x+c$

   g $\tan x+\sin x+c$

   h $\tan x-\cos x+c$

   i $\dfrac{1}{2}e^{2x}+e^x+c$

   j $\dfrac{1}{2}e^{2x}+3e^x-4x+c$

   k $\dfrac{1}{2}e^{2x}+\dfrac{1}{2}e^{-2x}+c$

   l $-\cos x+c$

   m $x+c$

2 a $4\dfrac{1}{2}$

   b $204\dfrac{17}{24}$

   c 30.375

   d $\dfrac{1}{3}$

3 a $\left(\dfrac{\pi}{4},\dfrac{1}{\sqrt{2}}\right)$ and $\left(\dfrac{5\pi}{4},-\dfrac{1}{\sqrt{2}}\right)$

   b $2\sqrt{2}$

   c $\sqrt{2}-1$

4 a $3\dfrac{1}{12}$

   b $3\dfrac{1}{12}$

   c $19\dfrac{1}{5}$

   d 0.274 (to 3 sf)

5 $512\ \text{m}^3$

6 a When $t=4$,

   $v_A=4\sqrt{4}=8\,\text{m s}^{-1}$

   $v_B=4^2-\left(\dfrac{4}{2}\right)^3$

   $=16-8$

   $=8\,\text{m s}^{-1}\ (=v_A)$

   b Distance $=\left|\displaystyle\int_0^4 4\sqrt{t}-\left(t^2-\left(\dfrac{t}{2}\right)^3\right)dx\right|$

   $=\left|\displaystyle\int_0^4 4t^{\frac{1}{2}}-t^2+\dfrac{t^3}{8}\,dx\right|$

   $=\left|\left[\dfrac{8}{3}t^{\frac{3}{2}}-\dfrac{t^3}{3}+\dfrac{t^4}{32}\right]_0^4\right|$

   $=\left|(21.3\ldots-21.3\ldots+8)-(0-0+0)\right|$

   $=8\,\text{m}$

7 a i x^2+2x+1

      ii x^3+3x^2+3x+1

      iii $x^4+4x^3+6x^2+4x+1$

      iv $x^5+5x^4+10x^3+10x^2+5x+1$

   b i $\dfrac{x^3}{3}+x^2+x+c$

      ii $\dfrac{x^3}{3}+x^2+x+\dfrac{1}{3}+c_1$

      iii $\dfrac{1}{3}(x^3+3x^2+3x+1)+c_1$

      iv $\dfrac{1}{3}(x+1)^3+c_1$

   c i $\dfrac{x^4}{4}+x^3+\dfrac{3}{2}x^2+x+c$

      ii $\dfrac{x^4}{4}+x^3+\dfrac{3}{2}x^2+x+\dfrac{1}{4}+c_1$

      iii $\dfrac{1}{4}(x^4+4x^3+6x^2+4x+1)+c_1$

      iv $\dfrac{1}{4}(x+1)^4+c_1$

   d $\displaystyle\int(x+1)^n\,dx=\dfrac{1}{n+1}(x+1)^{n+1}+c$

   e $\displaystyle\int(ax+1)^n\,dx=\dfrac{1}{a(n+1)}(ax+1)^{n+1}+c$

Exercise 16.2A Fluency and skills

1 a i $\sqrt{2x}+c$

      ii $\sqrt{4x-1}+c$

   b i $-\dfrac{1}{4}\cos 4x+c$

      ii $-\dfrac{1}{5}\cos(5x-2)+c$

   c i $\dfrac{1}{3}\sin 3x+c$

      ii $-\sin(2-3x)+c$

   d i $\dfrac{1}{5}\tan 5x+c$

      ii $\dfrac{1}{3}\tan(4+3x)+c$

   e i $\dfrac{1}{7}e^{7x}+c$

      ii $-2e^{-2x}+c$

   f i $\dfrac{1}{2}\ln|x|+c$

      ii $\dfrac{1}{2}\ln|4x+5|+c$

2 a $\dfrac{1}{33}(3x+2)^{11}+c$

   b $\dfrac{1}{45}(5x-1)^9+c$

   c $\dfrac{1}{707}(7x-3)^{101}+c$

   d $-\dfrac{1}{21}(3x-8)^{-7}+c$

   e $-\dfrac{1}{30}(1-3x)^{10}+c$

   f $-\dfrac{1}{8}(6-x)^8+c$

   g $-\dfrac{3}{8}(2x-1)^{-4}+c$

   h $\dfrac{1}{4}(10-x)^{-4}+c$

   i $\dfrac{1}{5}\cos(3-5x)+c$

   j $\dfrac{1}{2}\sin(4x-1)+c$

   k $\dfrac{1}{2}\sin(2x)+c$

   l $\dfrac{1}{4}\tan(4x+3)+c$

   m $\dfrac{3}{2}\tan(2x+1)+c$

   n $\dfrac{1}{5}e^{5x+2}+c$

   o $\dfrac{7}{3}\ln|3x+9|+c$

   p $-4\ln|8-x|+c$

3 a $\dfrac{1}{5}(x^2+3)^5+c$

   b $\dfrac{1}{4}(x^2+x-1)^4+c$

   c $\dfrac{2}{3}(2x^3-1)^{\frac{3}{2}}+c$

   d $\dfrac{1}{16}(2x^2-5)^4+c$

   e $-\dfrac{1}{6(x^2-7)^3}+c$

   f $-\dfrac{1}{x^2+3x-1}+c$

   g $-3\sqrt{1-x^2}+c$

   h $\dfrac{3}{2}(x+4)^{\frac{2}{3}}+c$

   i $-\dfrac{1}{6}\cos(3x^2+1)+c$

   j $\dfrac{1}{4}\tan^4 x+c$

   k $\ln|\sin x-1|+c$

   l $\dfrac{1}{2}\ln|3-2\cos x|+c$

   m $\dfrac{1}{4}e^{2x^2+1}+c$

   n $2e^{\sqrt{x}}+c$

   o $-e^{\cos x}+c$

   p $\dfrac{1}{2}(\ln x)^2+c$

4 a $\dfrac{1}{16}(2x^2-1)^4+c$

   b $\dfrac{2}{3}\left(\dfrac{x^3}{3}+\dfrac{x^2}{2}\right)^{\frac{3}{2}}+c$

   c $\dfrac{2}{3}\sqrt{x^3-3x^2}+c$

   d $-\dfrac{1}{2}\cos(x^2-2x+3)+c$

   e $\dfrac{1}{3}\ln\left|x^3+3x^2+9x+1\right|+c$

   f $-2\cos\sqrt{x}+c$

Answers For full solutions go to http://www.oxfordsecondary.co.uk/edexcelalevelmaths-answers

g $2\left(\dfrac{\left(\sqrt{x+1}\right)^3}{3}-\sqrt{x+1}+\sin\sqrt{x+1}\right)+c$

h $\dfrac{3}{4}(2x-1)^{\frac{2}{3}}+c$ **i** $-\cos x+\dfrac{\cos^4 x}{4}+c$

j $\tan x+\dfrac{\tan^3 x}{3}+c$ **k** $2\sqrt{1+\sin x}+c$

l $\ln\left|1+e^x\right|+c$ **m** $-\dfrac{1}{2}e^{\cos 2x}+c$

n $\dfrac{1}{3}\left(\ln(2x+3)\right)^{\frac{3}{2}}+c$

5 **a** $\dfrac{896}{3}$ **b** $\dfrac{1}{3}$ **c** $\dfrac{15}{8}$

  d $\ln\dfrac{5}{2}$ **e** 8 **f** $\dfrac{1}{2}\ln\dfrac{3}{2}$

  g $\dfrac{2}{\sqrt{3}}-1$ **h** $2-(12)^{\frac{1}{4}}$ **i** $\dfrac{38}{3}$

Exercise 16.2B Reasoning and problem-solving

1 **a** **i** $\dfrac{1}{2}x-\dfrac{1}{4}\sin 2x+c$ **ii** $\dfrac{1}{4}\sin 2x+\dfrac{1}{2}x+c$

  b **i** $\dfrac{1}{2}x-\dfrac{1}{8}\sin 4x+c$ **ii** $\dfrac{1}{8}\sin 4x+\dfrac{1}{2}x+c$

  iii $\dfrac{1}{2}x-\dfrac{1}{8}\sin(4x-2)+c$ **iv** $\dfrac{1}{8}\sin(4x+6)+\dfrac{1}{2}x+c$

  v $\dfrac{1}{2}x+\dfrac{1}{4}\sin(2-2x)+c$ **vi** $-\dfrac{1}{8}\sin(2-4x)+\dfrac{1}{2}x+c$

2 **a** $-\dfrac{1}{2}\cos 2x+c$ **b** $-\dfrac{1}{2}\cos 2x-2\cos x+c$

  c $\dfrac{1}{2}\sin 2x+c$

3 **a** $\dfrac{1}{2}\displaystyle\int\dfrac{\sec^2 u}{1+\tan^2 u}\,du$ **b** $\dfrac{1}{2}\displaystyle\int du$

  c $\dfrac{1}{2}u+c$ **d** $\dfrac{1}{2}\tan^{-1}\left(\dfrac{x}{2}\right)+c$

  e **i** $\dfrac{1}{3}\tan^{-1}\dfrac{x}{3}+c$ **ii** $\dfrac{1}{2}\tan^{-1}2x+c$

4 **a** $\displaystyle\int u\,du$ **b** $\dfrac{1}{2}u^2+c$ **c** $\dfrac{1}{2}[f(x)]^2+c$

  d **i** $\dfrac{1}{2}[x^2+x+5]^2+c$ **ii** $\dfrac{1}{2}[\sin x]^2+c$

  iii $\dfrac{1}{2}[\ln x]^2+c$ **iv** $\dfrac{1}{2}[e^{2x}+1]^2+c$

  e $u=f(x)\Rightarrow du=f'(x)\,dx$

  $\displaystyle\int\dfrac{f'(x)}{f(x)}\,dx=\int\dfrac{1}{u}\,du$

  $=\ln|u|+c$

  $=\ln|f(x)|+c$

  f **i** $\ln\left|\sin x\right|+c$ **ii** $\ln\left|\tan x\right|+c$

  iii $\ln\left|e^x+3\right|+c$ **v** $\ln\left|\ln x\right|+c$

5 **a** $\displaystyle\int e^{x\ln(a)}\,dx$

  b $dx=\dfrac{du}{\ln a}$

  c $\dfrac{1}{\ln a}a^x+c$

  d **i** $\dfrac{1}{\ln 4}4^x+c$

  ii $\dfrac{1}{2\ln 5}5^{2x}+c$

  iii $\dfrac{1}{\ln 3}3^{x+1}+c$

Exercise 16.3A Fluency and skills

1 **a** $3x\sin x+3\cos x+c$

  b $-2x\cos x+2\sin x+c$

  c $\dfrac{1}{2}x\sin x+\dfrac{1}{2}\cos x+c$

  d $-\dfrac{3}{2}x\cos 2x+\dfrac{3}{4}\sin 2x+c$

  e $\dfrac{1}{3}x\sin(3x+1)+\dfrac{1}{9}\cos(3x+1)+c$

  f $-\dfrac{x}{20}\sin(1-4x)+\dfrac{1}{80}\cos(1-4x)+c$

  g $-(2x+3)\cos x+2\sin x+c$

  h $(2x+1)\sin(x+1)+2\cos(x+1)+c$

  i $\dfrac{1}{4}(1-3x)\sin(4x)-\dfrac{3}{16}\cos(4x)+c$

2 **a** $3xe^x-3e^x+c$

  b $xe^{2x}-\dfrac{1}{2}e^{2x}+c$

  c $\dfrac{1}{4}xe^{2x+1}-\dfrac{1}{8}e^{2x+1}+c$

  d $\dfrac{1}{2}(x+2)e^{2x}-\dfrac{1}{4}e^{2x}+c$

  e $\dfrac{1}{2}(3-5x)e^{1+2x}+\dfrac{5}{4}e^{1+2x}+c$

  f $-\dfrac{1}{3}(3x-2)e^{1-3x}-\dfrac{1}{3}e^{1-3x}+c$

3 **a** $x^2\ln x-\dfrac{x^2}{2}+c$

  b $\dfrac{5}{2}x^2\ln x-\dfrac{5}{4}x^2+c$

  c $\left(\dfrac{x^2}{2}+x\right)\ln x-\left(\dfrac{x^2}{4}+x\right)+c$

  d $(x^2-x)\ln 2x-\left(\dfrac{x^2}{2}-x\right)+c$

  e $-\dfrac{\ln(3x)}{x}+\displaystyle\int\dfrac{1}{x^2}\,dx=-\dfrac{\ln(3x)}{x}-\dfrac{1}{x}+c$

  f $\dfrac{1}{3}(x+1)^3\ln x-\dfrac{1}{3}\left(\dfrac{x^3}{3}+\dfrac{3x^2}{2}+3x+\ln x\right)+c$

  g $2\sqrt{x}\ln 2x-4\sqrt{x}+c$

  h $\dfrac{2}{5}x^{\frac{5}{2}}+c$

  i $\dfrac{4}{3}x(x+1)^{\frac{3}{2}}-\dfrac{8}{15}(x+1)^{\frac{5}{2}}+c$

  j $x\tan x+\ln\left|\cos x\right|+c$

4 **a** 1 **b** $\dfrac{10}{9}$

  c $\dfrac{1}{\sqrt{2}}\left(\dfrac{\pi}{4}+2\right)-1$ **d** $\dfrac{4}{5}-\dfrac{\ln 5}{5}$

  e $5-10e^{-1}$ **f** 0

  g $\dfrac{1076}{15}$ **h** $\dfrac{\pi}{2}-1$

5 **a** $0\sin 0=\pi\sin\pi=2\pi\sin 2\pi=0$

  b **i** π **ii** 3π

  c 4π

Exercise 16.3B Reasoning and problem-solving

1 **a** $x\ln 4x-x+c$

  b $x\ln(3x+1)-x+\dfrac{1}{3}\ln\left|3x+1\right|+c$

  c $\dfrac{1}{8}x^8\ln(x^3)-\dfrac{3}{64}x^8+c$

d $-\cos x \ln(\cos x) + \cos x + c$

e $x \ln(1 - 5x) - x - \dfrac{1}{5} \ln|1 - 5x| + c$

f $-x \ln x + x + c$

g $x(\ln x)^2 - 2(x \ln x - x) + c$

h $x(\ln x^2) - 2x + c$

i $\dfrac{1}{4} x^2 - \dfrac{1}{4} x \sin 2x - \dfrac{1}{8} \cos 2x + c$

2 a $x^2 e^x - 2x e^x + 2 e^x + c$

b $-x^2 \cos x + 2x \sin x + 2 \cos x + c$

c $x^2 \sin x + 2x \cos x - 2 \sin x + c$

d $-(x+1)^2 \cos x + 2(x+1) \sin x + 2 \cos x + c$

e $(x^2 + 2x) \sin x + (2x+2) \cos x - 2 \sin x + c$

f $(1 - 3x)^2 e^x + 6(1 - 3x) e^x + 18 e^x + c$

g $-(x^2 + x + 1) e^{-x} - (2x+1) e^{-x} - 2 e^{-x} + c$

h $\dfrac{1}{8} x^2 (x+1)^8 - \dfrac{1}{36} x(x+1)^9 + \dfrac{1}{360}(x+1)^{10} + c$

i $\dfrac{-x^2}{x+1} + 2x - 2\ln(x+1) + c$

j $\dfrac{1}{6}(x+1)^2(x+3)^6 - \dfrac{1}{21}(x+1)(x+3)^7 + \dfrac{1}{168}(x+3)^8 + c$

k $-(x^2 + x)\cos x + (2x+1)\sin x + 2\cos x + c$

l $-\dfrac{1}{2} x^2 \cos 2x + \dfrac{1}{2} x \sin 2x + \dfrac{1}{4} \cos 2x + c$

m $\dfrac{1}{3} x^2 \sin 3x + \dfrac{2}{9} x \cos 3x - \dfrac{2}{27} \sin 3x + c$

3 a $54 e^{\frac{1}{3}} - 69$ **b** $\dfrac{\pi^2}{32} - \dfrac{1}{4}$ **c** 2π

4 a $x^2 e^x - 2x e^x + 2 e^x + c$

b $y = x^2 e^x - 2x e^x + 2 e^x = e^x(x^2 - 2x + 2)$

c The discriminant of $x^2 - 2x + 2$ is $(-2)^2 - 4.1.2 = -12$ so $x^2 - 2x + 2$ has no real roots and so the curve is always above the x-axis and $x^2 - 2x + 2 > 0$ for all x. Thus $y > 0$ for all x.

5 a $\dfrac{x^2}{2} \ln x - \dfrac{x^2}{4} + c$

b $2x^2 (\ln x)^2 - 2x^2 \ln x + x^2 + c$

6 a $(\pi - 0)^2 \sin 0 = (\pi - \pi)^2 \sin \pi = (\pi - 2\pi)^2 \sin 2\pi = 0$

b $2(\pi^2 - 4)$

7 $\dfrac{178}{60}$

8 a $P = -e^x \cos x + \displaystyle\int e^x \cos x \, dx$

b $P = -e^x \cos x + \left(e^x \sin x - \displaystyle\int e^x \sin x \, dx \right)$

c $P = -e^x \cos x + e^x \sin x - P$

d $\dfrac{e^x}{2}(\sin x - \cos x) + c$

Exercise 16.4A Fluency and skills

In each answer the form ln A has been used as the constant of integration.

1 a $\ln \left| A \dfrac{x}{x+1} \right|$ **b** $\ln \left| A \dfrac{x^2}{(x+3)^2} \right|$

c $\ln \left| A \dfrac{x}{(3x+1)} \right|$ **d** $\ln \left| A \sqrt{x(x+2)} \right|$

e $\ln \left| A \dfrac{(x-2)^2}{(x+2)^2} \right|$ **f** $\ln \left| A \dfrac{(x-4)}{(x+1)} \right|$

g $\ln \left| A \dfrac{(x+1)^2}{(x+2)^2} \right|$ **h** $\ln \left| A \dfrac{(x-4)^3}{(x+3)^3} \right|$

2 a $\ln A \left| \dfrac{(2x-3)}{(2x+1)} \right|^{\frac{1}{2}}$ **b** $\ln \left| A \dfrac{(3x+1)}{(x+2)} \right|$

c $\ln A \left| \dfrac{(3x+1)}{(x+5)} \right|^{\frac{1}{2}}$ **d** $\ln \left| A \sqrt{(x-1)(x+3)} \right|$

e $\ln \left| A(x+4)^3 (x-2) \right|$ **f** $\ln \left| A \dfrac{(3x-1)}{(2x+1)} \right|$

g $\ln \left| A(x-2)^4 (x+1)^2 \right|$ **h** $\ln \left(A \left| (2x+3) \right|^{\frac{2}{5}} \left| (x-1) \right|^{\frac{3}{5}} \right)$

i $\ln \left| A(x-1)(2x+1) \right|$ **j** $\ln \left(A \left| (2x+1)(2x-1) \right|^{\frac{3}{8}} \right)$

3 a $\ln \left(A \dfrac{\sqrt{|x-1||x+1|}}{|x|} \right)$ **b** $\ln \left(A \dfrac{|3x+2|^3 |x-1|^2}{|x|^5} \right)$

c $\ln \left(A \dfrac{|x||x-1|^4}{|4x+3|^5} \right)$ **d** $\ln \left(A \dfrac{|x-2|^2 |x+1|}{|x-1|^3} \right)$

e $\ln \left(A \dfrac{|2x-1||x-2|^2}{|x-1|^3} \right)$ **f** $\ln \left(A \dfrac{|x||x-2|^3}{|x-1|^3} \right)$

g $\ln \left(A \dfrac{|x+4|^{64} |x-1|}{|x+3|^{45}} \right)$ **h** $\ln \left(A \dfrac{|x-3|^{\frac{1}{2}} |2x-1|^{\frac{1}{6}}}{|x+1|^{\frac{1}{6}}} \right)$

i $\ln \left(A \dfrac{|x-1||x+2|^{\frac{1}{2}}}{|x|^{\frac{1}{2}}} \right)$ **j** $\ln \left(A \dfrac{|x|^{\frac{1}{2}}}{|1-x||x+2|^{\frac{1}{2}}} \right)$

4 a $-\dfrac{1}{5x} + \dfrac{9}{5(5-x)}$

   i $\dfrac{-\ln x}{5} - \dfrac{9\ln(x-5)}{5} + \ln A$

   ii $\dfrac{1}{5x^2} + \dfrac{9}{5(5-x)^2}$

b $\dfrac{1}{1-x} + \dfrac{2}{3x-1}$

   i $-\ln(1-x) + \dfrac{2\ln(3x-1)}{3} + \ln A$

   ii $\dfrac{1}{(1-x)^2} - \dfrac{6}{(3x-1)^2}$

c $-\dfrac{4}{1-2x} - \dfrac{3}{x-1}$

   i $2\ln(1-2x) - 3\ln(x-1) + \ln A$

   ii $\dfrac{-8}{(1-2x)^2} + \dfrac{3}{(x-1)^2}$

d $\dfrac{1}{2-3x} + \dfrac{1}{5x+1}$

   i $\dfrac{-\ln(2-3x)}{3} + \dfrac{\ln(5x+1)}{5} + \ln A$

   ii $\dfrac{3}{(2-3x)^2} - \dfrac{5}{(5x+1)^2}$

5 a $-\ln 6$ **b** $\ln 4$ **c** $\ln \left(\dfrac{36}{25} \right)$

   d $\ln \sqrt[4]{3}$ **e** $\ln \dfrac{3}{2}$ **f** $-\ln 4$

Exercise 16.4B Reasoning and problem-solving

1 a $\ln\left(A\left|\dfrac{x}{x+2}\right|^{\frac{1}{2}}\right)$ **b** $\ln\left(A\left|\dfrac{x}{x+5}\right|^{\frac{2}{5}}\right)$

c $\ln\left(A\left|\dfrac{x}{2x+1}\right|\right)$ **d** $\ln\left(A\dfrac{x^2}{|x+1|}\right)$

e $\ln\left(A\left|\dfrac{x-1}{x+1}\right|^{\frac{1}{2}}\right)$ **f** $\ln\left(A\left|\dfrac{x-3}{x+3}\right|^{2}\right)$

2 a $\ln\left(A\left|\dfrac{2x-1}{2x+1}\right|^{\frac{1}{2}}\right)$ **b** $\ln\left(A|(3x-2)(3x+2)|^{\frac{1}{6}}\right)$

c $\ln\left(A\left|\dfrac{x-3}{x-1}\right|^{\frac{1}{2}}\right)$ **d** $\ln\left(A\left|\dfrac{x-4}{x-1}\right|\right)$

e $\ln\left(A|x+2|^{\frac{2}{5}}|2x-1|^{\frac{1}{10}}\right)$ **f** $\ln\left(A\dfrac{|x-5|^{\frac{10}{3}}}{|x-2|^{\frac{4}{3}}}\right)$

g $\ln\left(A|x-2|^{\frac{3}{7}}|x+5|^{\frac{4}{7}}\right)$ **h** $\ln\left(A|2x-1|^{\frac{1}{18}}|x+4|^{\frac{13}{9}}\right)$

3 a $x+\ln\left(A\left|\dfrac{x-1}{x+2}\right|^{\frac{1}{3}}\right)$ **b** $x+\ln\left(A\left|\dfrac{x+1}{x+2}\right|^{4}\right)$

c $x+\ln\left(A\left|\dfrac{x-1}{x+3}\right|^{\frac{3}{4}}\right)$ **d** $2x+\ln\left(A\left|\dfrac{x-2}{x+2}\right|^{\frac{3}{4}}\right)$

e $2x+\ln\left(A\left|\dfrac{x+1}{x+3}\right|^{\frac{1}{2}}\right)$ **f** $3x+\ln\left(A\dfrac{|x-2|^{2}}{|x-1|}\right)$

4 a $\ln 4$ **b** $\ln 2$ **c** $\ln 16$

   d $\ln 9$ **e** $\ln 27$ **f** $\ln 81$

5 a $\ln\dfrac{8}{3}$ **b** $\ln\dfrac{8}{3}$

c The interval contains $x=1$ for which the function is undefined. The area is unbounded.

6 a $y=\dfrac{-8}{x(x-4)}$: lower curve

   $y=\dfrac{-9}{x(x-3)}$: upper curve

b i $\ln 9$ **ii** $\ln 64$

c $6\ln 2-2\ln 3$

7 a $4\ln 2-\ln 3$

b Any interval containing the x-values 1, 2 or 5

Exercise 16.5A Fluency and skills

1 $y=\dfrac{1}{2}x^2-2x+c$

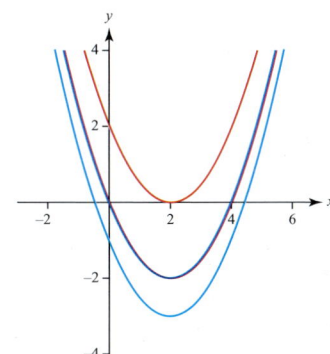

2 a $y=x^2-x+c$ **b** $y=x-\dfrac{x^3}{3}+c$

c $y=\dfrac{(\sqrt{2x+1})^3}{3}+c$ **d** $y=-\dfrac{3}{x}+c$

e $y=\ln Ax^2$ **f** $y=Ae^{3x}$

g $y=Ae^{\frac{5}{2}x^2}$ **h** $y=\pm\sqrt{x^2+2c}$

i $y=\pm\sqrt{(x+1)^2+2c}+3$ **j** $y=Ae^{\frac{x^2}{2}}-1$

k $y=5-Ae^{-x}$ **l** $y=\dfrac{Ae^{x^2+4x}-1}{2}$

m $y^2=x^3+c$ **n** $y=\sin^{-1}(-\cos x+c)$

o $y=\tan^{-1}\left(\dfrac{1}{2}e^{2x}+c\right)$ **p** $y=\ln(e^x+c)$

q $y=\dfrac{1}{3}\ln\left(\dfrac{3}{2}e^{2x+1}+c\right)$

3 a $y=2x^2-x+3$ **b** $y=2x-3x^3+10$

c $y=\ln|x|+3$ **d** $y=e^{6x-2}$

e $y=e^{\frac{1}{2}x^2-2}-3$ **f** $y=e^{\frac{1}{2}x^2+5x+1}$

g $y^2-y=\dfrac{1}{2}x^2+2x-6$ **h** $y=\cos^{-1}(1-\sin x)$

i $y=\ln\left(\dfrac{1}{2}e^{2x}+1-\dfrac{1}{2}e\right)$ **j** $y=2\sqrt{x}$

4 a i $-\dfrac{1}{8(y^2+1)^4}=x+c$

    ii $y=\sin^{-1}\left(e^{\frac{1}{2}x^2+c}\right)$

    iii $y=(\ln|x+c|)^2$

b i $-y\cos y+\sin y=-\cos x+c$

    ii $y=\sqrt{2(xe^x-e^x+c)}$

    iii $\dfrac{1}{3}y^3\ln y-\dfrac{1}{9}y^3=9x+c$

c i $y=A\left(\dfrac{2x-3}{x}\right)^2$

    ii $y=\dfrac{1}{\ln\left(A^{-2}\left|\dfrac{x+1}{x-1}\right|^2\right)}$

    iii $y=\left[\dfrac{1}{2}\ln\left(A\left|\dfrac{x-3}{x-2}\right|\right)\right]^2$

5 $y=\ln(\sin x+1)$

Exercise 16.5B Reasoning and problem-solving

1 a $\dfrac{dC}{dw}=k$ **b** $\dfrac{dS}{dt}=kt$

c $\dfrac{dA}{dT}=kA$ **d** $\dfrac{dT}{dt}=k(T-20)$

e $\dfrac{dD}{dP}=\dfrac{k}{P}$ **f** $\dfrac{dI}{dR}=k10^R$

g $\dfrac{dy}{dx}=\dfrac{k}{\sqrt{y}}$ **h** $\dfrac{dh}{dt}=\dfrac{k}{\sec^2 h}$

2 a $V=2t+0.3t^2+c$

b $V=2t+0.3t^2$

c $t=10$

3 a $P=50\sin\left(\dfrac{\pi t}{14}\right)+c$ **b** $P=50\sin\left(\dfrac{\pi t}{14}\right)+50$

c 89%   **d** Day 7

4 a $y^2 = kx^2 + 2c$

  b 12.5

  c $y^2 = -x^2 + 25$ or $x^2 + y^2 = 25$

    This is the equation of a circle centre the origin and radius 5. Here the origin is the meeting of the wall and the ground.

5 a $\dfrac{dN}{dt} = kN$

  b $N = e^{kt+c}$

  c $c = \ln 500$

  d $k = \ln 2$

    $N = 500 \times 2^t$

  e On the 6th day

  f Space and nutrients etc will prevent infinite growth.

6 a $\dfrac{dw}{dt} = k(2000 - w)$

  b $-7.58\,(3\,\text{sf})$

  c $0.0879\,(3\,\text{sf})$

  d $w = 2000 - e^{-0.0879t+7.58}$

  e On the 15th day

  f Will reach a point when weight doesn't increase further.

7 a $P = \dfrac{500e^{500(kt+c)}}{(1+e^{500(kt+c)})}$ **b** $P = \dfrac{500e^{1.69t-3.89}}{(1+e^{1.69t-3.89})}$

8 Students' own investigations.

Review exercise 16

1 a $\dfrac{4}{7}x^7 + c$ **b** $-\dfrac{2}{3}\cos(3x+1) + c$

  c $\dfrac{1}{2}\sin 2x + c$ **d** $\dfrac{1}{2}x^2 + \tan x + c$

  e $e^x - e^{-x} + c$

2 a 4   **b** $\dfrac{\sqrt{3}}{2}$ **c** 1

  d $\dfrac{\pi}{4} - 1$ **e** 3   **f** 6

3 a $\dfrac{9}{2}$ **b** $14\dfrac{17}{24}$

4 a $\dfrac{1}{12}(x^2+4)^6 + c$ **b** $\dfrac{1}{2}(x^4-2)^4 + c$

  c $\dfrac{1}{4}(x^2-4x+1)^2 + c$ **d** $\dfrac{2}{3}(\sin x)^{\frac{3}{2}} + c$

5 a 1   **b** $\dfrac{1}{2}$

6 a $-2x\cos x + 2\sin x + c$

  b $\dfrac{1}{2}(3x+1)e^{2x} - \dfrac{3}{4}e^{2x} + c$

  c $\dfrac{1}{3}x^2\sin 3x + \dfrac{2}{9}x\cos 3x - \dfrac{2}{27}\sin 3x + c$

7 a $4e^4 - e^4 - 3e^3 + e^3$

  b $2e^3 + 1$

8 a $(x^2-1)e^x + c$

  b $x^2(\ln x)^2 - x^2\ln x + \dfrac{1}{2}x^2 + c$

9 a $\ln\left(\dfrac{x-2}{x}\right)^4 + c$ **b** $\ln\left(\dfrac{x+2}{x+3}\right)^2 + c$

10 $\ln 18$

11 a $y = e^{\frac{x^2}{2}+x-\frac{3}{2}}$ **b** $y = \sin^{-1}\left(x^2 - \dfrac{1}{2}\right)$

12 a $\dfrac{dV}{dT} = kV$

---

b $c \approx 9.90$

    $k \approx -0.105$

  c $V = 20\,000e^{-0.105T}$

Assessment 16

1 a $\ln|x|\,(+c)$ **b** $-\cos(x-3)(+c)$

  c $\dfrac{1}{2}e^{2x}\,(+c)$ **d** $\dfrac{1}{2}\sin 2x\,(+c)$

2 $y = 2x - \dfrac{1}{2}\cos 6x + \dfrac{3}{2}$

3 a **i** $\ln|x+3|\,(+c)$

    ii $-\dfrac{1}{3}\cos^3 x\,(+c)$

    iii $\dfrac{1}{4}(x^2+4)^4\,(+c)$

  b $\dfrac{1}{2}\ln 2$

4 a **i** $xe^x - e^x\,(+c)$

    ii $-x\cos x + \sin x\,(+c)$

  b $-\dfrac{1}{2}x\cos 2x + \dfrac{1}{2}\displaystyle\int \cos 2x \; dx$

    $\left[-\dfrac{1}{2}x\cos 2x + \dfrac{1}{4}\sin 2x\right]_0^{\frac{\pi}{6}}$

    $\left(-\dfrac{1}{2}\dfrac{\pi}{6}\cos\dfrac{\pi}{3} + \dfrac{1}{4}\sin\dfrac{\pi}{3}\right) - (0+0)$

    $= -\dfrac{\pi}{12}\cdot\dfrac{1}{2} + \dfrac{1}{4}\dfrac{\sqrt{3}}{2}$

    $= \dfrac{1}{8}\sqrt{3} - \dfrac{1}{24}\pi$

5 a $y = A(x+1)$

  b $y = 4(x+1)$

6 a $A = 7$

    $B = 1$

  b $\dfrac{15}{2}\ln 3 - 7\ln 2$

7 a $\dfrac{(1+2x)}{2} - \dfrac{1}{2}\ln|1+2x| + c$

  b $1 - \dfrac{1}{2}\ln 3$

8 $\ln 4$

9 a $A = 2$

    $B = \dfrac{7}{5}$

    $C = -\dfrac{6}{5}$

  b $y = x^2 + \dfrac{7}{5}\ln|x+2| - \dfrac{6}{15}\ln|3x+1| - \dfrac{7}{5}\ln 2$

10 $y = \dfrac{A(x+2)}{1-2x}$

11 a $\dfrac{dN}{dt} = -kN$

  b $N = \dfrac{N_0}{2^{\frac{t}{T}}}$

---

12 a

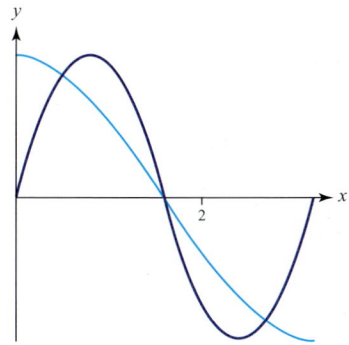

b $x = \dfrac{\pi}{2}$

or $x = \dfrac{\pi}{6}$ or $\dfrac{5\pi}{6}$

c $\dfrac{1}{2}$

13 a $y - 3e = 2e(x - 3)$

b $3xe^{\frac{x}{3}} - 9e^{\frac{x}{3}} + c$

c $9 - \dfrac{9e}{4}$

14 $2\ln\left|\dfrac{A\left(\sqrt{x} - 2\right)}{\sqrt{x} + 2}\right|$

or $\ln A\left(\dfrac{A\left(\sqrt{x} - 2\right)}{\sqrt{x} + 2}\right)^{2}$

15 $\dfrac{1}{24}(2x - 5)^{6} + \dfrac{1}{4}(2x - 5)^{5} + c$

16 a $-x^{2}\cos x + 2x\sin x + 2\cos x \ (+c)$

b $6\pi^{2} - 8$

17 a You are given $\dfrac{dV}{dt} = -c\sqrt{V}$

$V = \pi r^{2}h$ so $\dfrac{dV}{dh} = \pi r^{2}$

$\dfrac{dh}{dt} = \dfrac{dh}{dV} \cdot \dfrac{dV}{dt}$

$\dfrac{dh}{dt} = \dfrac{1}{\pi r^{2}} \cdot -c\sqrt{V}$

$= \dfrac{-c\sqrt{\pi r^{2}h}}{\pi r^{2}}$

$= -k\sqrt{h}$ where $k = \dfrac{c}{\sqrt{\pi r^{2}}}$

b $h = \left(\dfrac{A - kt}{2}\right)^{2}$

c When $t = 0$, $h = 2$ so $2 = \left(\dfrac{A}{2}\right)^{2}$

$\Rightarrow A = 2\sqrt{2}$

When $t = 2$, $h = 0$ so $0 = \left(\dfrac{2\sqrt{2} - 2k}{2}\right)^{2}$

$\Rightarrow k = \sqrt{2}$

$h = \left(\dfrac{2\sqrt{2} - \sqrt{2}t}{2}\right)^{2}$

$h = 2\left(1 - \dfrac{1}{2}t\right)^{2}$ as required

Chapter 17
Exercise 17.1A Fluency and skills

1 a $f(x) = 7 - 3x - x^{3}$

$f(1) = 3 > 0$, $f(2) = -7 < 0$

The continuous function $f(x)$ changes sign between $x = 1$ and $x = 2$

By the change of sign method, the equation $f(x) = 0$ has a root between 1 and 2

b $f(x) = x^{2} - \dfrac{1}{x} - 4$

$f(2.1) = -0.066.. < 0$, $f(2.2) = 0.385.. > 0$

The continuous function $f(x)$ changes sign between $x = 2.1$ and $x = 2.2$

By the change of sign method, the equation $f(x) = 0$ has a root between 2.1 and 2.2

c $f(x) = \sin(2x^{c}) - x^{2} + 3$

$f\left(\dfrac{1}{2}\pi\right) = 0.532.. > 0$, $f\left(\dfrac{2}{3}\pi\right) = -2.252... < 0$

The continuous function $f(x)$ changes sign between $x = \dfrac{1}{2}\pi$ and $x = \dfrac{2}{3}\pi$

By the change of sign method, the equation $f(x) = 0$ has a root between $\dfrac{1}{2}\pi$ and $\dfrac{2}{3}\pi$

d $f(x) = e^{x}\ln x - x^{2}$

$f(1.69) = -0.012... < 0$, $f(1.71) = 0.042... > 0$

The continuous function $f(x)$ changes sign between $x = 1.69$ and $x = 1.71$

By the change of sign method, the equation $f(x) = 0$ has a root between 1.69 and 1.71

2 a $f(x) = e^{x} - x^{3}$

i $f(1.85) = 0.028... > 0$, $f(1.95) = -0.386... < 0$

The continuous function $f(x)$ changes sign between $x = 1.85$ and $x = 1.95$

By the change of sign method, the equation $f(x) = 0$ has a root, α, between 1.85 and 1.95

ii $f(4.535) = -0.044... < 0$, $f(4.545) = 0.274... > 0$

The continuous function $f(x)$ changes sign between $x = 4.535$ and $x = 4.545$

By the change of sign method, the equation $f(x) = 0$ has a root between 4.535 and 4.545

b $\alpha = 1.9$ (1 dp), $\beta = 4.54$ (2 dp)

3 a i $f(x) = \dfrac{2}{x^{3}} - \dfrac{1}{x} - 2$

$f(0.85) = 0.08... > 0$, $f(0.95) = -0.71... < 0$

so by the change of sign method, the equation $f(x) = 0$ has a root between $x = 0.85$ and $x = 0.95$

ii 0.9 (1 dp)

b i $f(x) = e^{-x} + 2x - 1$

$f(-1.35) = 0.15... > 0$, $f(-1.25) = -0.009... < 0$

so by the change of sign method, the equation $f(x) = 0$ has a root between $x = -1.35$ and $x = -1.25$

ii -1.3 (1 dp)

c i $f(x) = x^{2}\sin x - 0.5$

$f(3.05) = 0.35... > 0$, $f(\pi) = -0.5 < 0$

so by the change of sign method, the equation $f(x) = 0$ has a root between $x = 3.05$ and $x = \pi$

ii 3.1 (1 dp)

4 a i $f(x) = x^{4} - 3x^{3} + 1$

$f(2.955) = -0.16... < 0$, $f(2.965) = 0.08... > 0$

so by the change of sign method, the equation $f(x) = 0$ has a root between $x = 2.955$ and $x = 2.965$

ii 2.96 (2 dp)

b **i** $f(x) = e^{\frac{1}{x}} - x^2$

$f(1.414) = 0.028... > 0, f(1.424) = -0.009... < 0$

so by the change of sign method, the equation $f(x) = 0$ has a root between $x = 1.414$ and $x = 1.424$

ii 1.4 (1 dp)

c **i** $f(x) = x^2 - \sqrt{x} - 2$

$f(1.8305) = -0.002... < 0, f(1.8315) = 0.001... > 0$

so by the change of sign method, the equation $f(x) = 0$ has a root between $x = 1.8305$ and $x = 1.8315$

ii 1.831 (3 dp)

d **i** $f(x) = 2 \ln x - \sec x$

$f\left(\frac{8}{5}\pi\right) = -0.007 < 0, f(5.02725) \approx 6 \times 10^{-4} > 0$

so by the change of sign method, the equation $f(x) = 0$ has a root between $x = \frac{8}{5}\pi$ and $x = 5.02725$

ii Root is between 5.02654... and 5.02725, so the root is 5.027 (3 dp)

e **i** $f(x) = e^{\cos x} - \cos(e^x)$

$f\left(-\frac{3}{5}e\right) = -0.03... < 0, f\left(-\frac{1}{2}\pi\right) = 0.021... > 0$

so by the change of sign method, the equation $f(x) = 0$ has a root between $x = -\frac{3}{5}e$ and $x = -\frac{1}{2}\pi$

ii −1.6 (1 dp)

5 **i** Equations **b** and **c**

ii Equation **c**

6 **a** Equation i ↔ C, Equation ii ↔ B, Equation iii ↔ A

b **i** 1.6 (1 dp)   **ii** 2.2 (1 dp)   **iii** 2.21 (2 dp)

7 **a** $f(x) = x^2 + 2x - 2a$

$f(0) = -2a < 0, f(a) = a^2 + 2a - 2a \Rightarrow f(a) = a^2 > 0$

Change of sign ⇒ equation $f(x) = 0$ has a root between 0 and a

b $f(x) = ax^2 + x - a^3$

$f(0) = -a^3 < 0, f(a) = a(a)^2 + a - a^3 \Rightarrow f(a) = a > 0$

Change of sign ⇒ equation $f(x) = 0$ has a root between 0 and a

c $f(x) = \cos\left(\frac{\pi}{a}x\right) - \frac{a}{\pi}x$

$f(0) = 1 > 0, f(a) = \cos \pi - \frac{a^2}{\pi} \Rightarrow f(a) = -1 - \frac{a^2}{\pi} < 0$

Change of sign ⇒ equation $f(x) = 0$ has a root between 0 and a

d $f(x) = x^3 + (a+1)x^2 - 2a^3$

$f(0) = -2a^3 < 0, f(a) = a^3 + (a+1)a^2 - 2a^3 \Rightarrow f(a) = a^2 > 0$

Change of sign ⇒ equation $f(x) = 0$ has a root between 0 and a

Exercise 17.1B Reasoning and problem-solving

1 **a** $f(x) = x^3 - x^2 - 1$

$f(1.4) = -0.216 < 0, f(1.5) = 0.125 > 0$

The continuous function $f(x)$ changes sign between $x = 1.4$ and $x = 1.5$ so the equation $f(x) = 0$ has a solution, α, in the interval (1.4, 1.5)

b The curve $y = x^3 - x^2$ and line $y = 1$ intersect exactly once, so there is exactly one real solution, α, to the equation $x^3 - x^2 = 1$

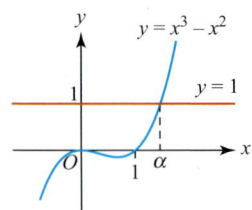

c $\beta = 0.7$ (1 dp)

2 **a** (1, 1)

b

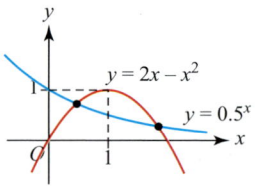

The curves $y = 2x - x^2$ and $y = 0.5^x$ intersect at exactly two points, so the equation $2x - x^2 = 0.5^x$ has exactly two solutions.

c $f(x) = 2x - x^2 - 0.5^x$

i $f(0.44) = -0.0507... < 0, f(0.48) = 0.0126... > 0$

The continuous function $f(x)$ changes sign between $x = 0.44$ and $x = 0.48$ so the equation $f(x) = 0$ has a root in the interval (0.44, 0.48)

Since $\beta > 0.5$, it follows that α lies in the interval (0.44, 0.48)

ii $f(1.84) = 0.015... > 0, f(1.88) = -0.046... < 0$

The continuous function $f(x)$ changes sign between $x = 1.84$ and $x = 1.88$ so the equation $f(x) = 0$ has a root in the interval (1.84, 1.88)

Since $\beta > 0.5$ and $\alpha < 0.48$, it follows that β lies in the interval (1.84, 1.88)

d $\beta - \alpha = 1.4$ (1 dp)

3 **a** $x = 1.8$ radians (1 dp)

b $f(2) = 0.557.. > 0, f(3) = -3.185... < 0$

So $f(x)$ changes sign across the interval (2, 3)

c If $f(x)$ was continuous on the interval (2,3) then, by the change of sign method, the equation $f(x) = 0$ would have a root β in this interval.

But then $f(x) = 0$ would have two roots, α and β, in the interval (0, π)

This would contradict the result of part **a**, which stated that $f(x) = 0$ had only one root, α, in the interval (0, π)

Hence $f(x)$ cannot be continuous on the interval (2, 3)

d $x = 2.6$ radians (1 dp)

4 **a** Line to be drawn is $y = 3 - \frac{1}{2}x$

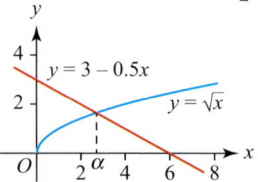

The line $y = 3 - \frac{1}{2}x$ and curve $y = \sqrt{x}$ intersect at exactly one point, so the equation $2\sqrt{x} = 6 - x$ has exactly one real solution, α

b $f(x) = 2\sqrt{x} - 6 + x$

$f(2.65) = -0.094... < 0, f(2.75) = 0.066... > 0$

The continuous function $f(x)$ changes sign between $x = 2.65$ and $x = 2.75$ so the equation $f(x) = 0$ has a root in the interval (2.65, 2.75)

Since α is the only root of this equation, $2.65 < \alpha < 2.75$ and therefore $\alpha = 2.7$ (1 dp)

c **i** $x = 8 - 2\sqrt{7}$, $a = 8$ and $b = -2$

ii Combining the results of parts **b** and **c i** gives

$8 - 2\sqrt{7} \approx 2.7$

$\Rightarrow \sqrt{7} \approx \frac{8 - 2.7}{2} = \frac{53}{20}$

$\Rightarrow \sqrt{7} \approx \frac{53}{20}$

5 **a** $f(0) = -1 < 0$ and $f(1) = -7 < 0$

Since $f(0)$ and $f(1)$ have the same sign, $f(x)$ does not change sign across the interval $(0, 1)$

b $f\left(\dfrac{1}{6}\right) = 0$

So $x = \dfrac{1}{6}$ is a root of the equation $f(x) = 0$

$f(x)$ does not change sign across the interval $(0, 1)$, but $\dfrac{1}{6}$ lies in $(0, 1)$ and $f\left(\dfrac{1}{6}\right) = 0$

Hence the equation $f(x) = 0$ must have an even number of roots between 0 and 1

Therefore there must be at least two roots of $f(x) = 0$ in the interval $(0, 1)$

c $y = 4\sin(\pi x)$ and $y = 6x + 1$ intersect twice only, so $x = \dfrac{1}{6}$ and $x = \dfrac{1}{2}$ are the only roots of $f(x) = 0$

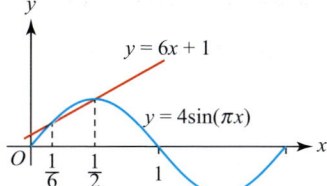

6 **a** True.

$f(a)$ has opposite sign to $f(b)$, and $f(c)$ has opposite sign to $f(b)$

Therefore $f(a)$ and $f(c)$ have the same sign.

b False.

$f(x)$ is continuous and changes sign across the interval (a, b), so $f(x) = 0$ has a root in the interval (a, b)

Since (a, b) is contained in (a, c), therefore $f(x) = 0$ has a root in the interval (a, c)

c True.

The equation $f(x) = 0$ has a root α in the interval (a, b) and a root β in the interval (b, c)

At some point between $x = \alpha$ and $x = \beta$ the graph of $y = f(x)$ must have a stationary point.

The x-coordinate of this point is a root of the equation $f'(x) = 0$

7 **a** $f(1) = -2$, $f(2) = 3$

The curve $y = f(x)$ passes through $A(1, -2)$ and $B(2, 3)$ so the gradient of the line AB is $\dfrac{3 - (-2)}{2 - 1} = 5$

Hence, the equation of the line through A and B is $y - 3 = 5(x - 2)$

The x-intercept of this line is an estimate for α.

When $y = 0$, $0 - 3 = 5(x - 2)$

$\Rightarrow x = -\dfrac{3}{5} + 2$

$\Rightarrow x = 1.4$, as required.

b $x = 1.6$ (1 dp) is another approximation for α

Use a change of sign across a suitable interval to show $\alpha = 1.6$ to 1 decimal place:

$f(1.55) = -0.37... < 0$, $f(1.65) = 0.19... > 0$

The continuous function $f(x)$ changes sign between $x = 1.55$ and $x = 1.65$ so $x = 1.6$ (1 dp) is a root of the equation $f(x) = 0$

Since α is the only root of this equation in the interval $(1, 2)$, therefore $\alpha = 1.6$ (1 dp), which agrees with the second estimate found.

c $\alpha = \dfrac{1 + \sqrt{5}}{2}$

1 **a** **i** $x_2 = 0.9$, $x_3 = 0.927$, $x_4 = 0.920$ (all to 3 dp where appropriate)

  ii $x_5 = 0.922$ (3 dp)

  b **i** $x_2 = 3$, $x_3 = 3.442$, $x_4 = 3.510$ (all to 3 dp where appropriate)

  ii $x_6 = 3.521$ (3 dp)

  c **i** $x_2 = -1$, $x_3 = -0.5$, $x_4 = -0.64$

  ii $x_7 = -0.618$ (3 dp)

  d **i** $x_2 = 1.6$, $x_3 = 1.425$, $x_4 = 1.511$ (all to 3 dp where appropriate)

  ii $x_{10} = 1.485$ (3 dp)

2 **a** **i** $\alpha = 2.12$ (2 dp)   **ii** $(2.115, 2.125)$

  b **i** $\alpha = 3.69$ (2 dp)   **ii** $(3.685, 3.695)$

  c **i** $\alpha = 2.36$ (2 dp)   **ii** $(2.355, 2.365)$

  d **i** $\alpha = 3.94$ (2 dp)   **ii** $(3.935, 3.945)$

  e **i** $\alpha = 2.20$ (2 dp)   **ii** $(2.195, 2.205)$

  f **i** $\alpha = 1.24$ (2 dp)   **ii** $(1.235, 1.245)$

3 **a** $N = 4$ iterations are required.

  b $x_1 = 100$

To 3 dp: $x_2 = 0.885$, $x_3 = 0.877$, $x_4 = 0.878$ (stop)

Only 3 iterations are required.

4 **a**

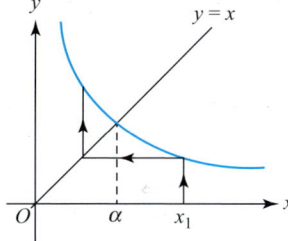

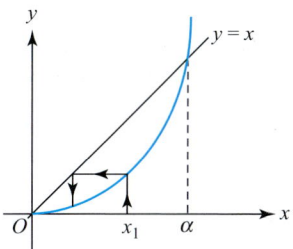

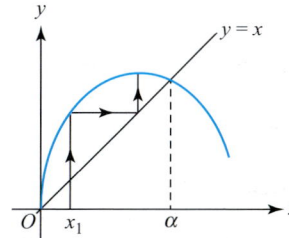

  b Fig 2, if continued, would definitely not illustrate convergence to α

5 **a**

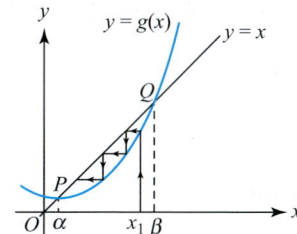

The staircase diagram shows the iterates converging to α even though the starting value is close to β

b

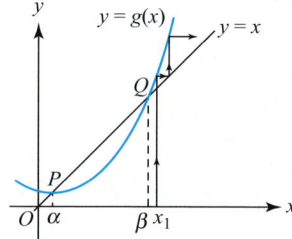

The staircase diagram shows the iterates diverging from β even though the starting value is close to this root.

Exercise 17.2B Reasoning and problem-solving

1 a $x^4 - 2x - 1 = 0 \Rightarrow x^4 = 2x + 1$

$$\Rightarrow x^2 = \frac{2x + 1}{x^2}$$

$$\Rightarrow x = \sqrt{\frac{2x + 1}{x^2}} = \frac{\sqrt{2x + 1}}{x}, \text{ as required}$$

b The iteration $x_{n+1} = \frac{\sqrt{2x_n + 1}}{x_n}$ converges to

1.3958... = 1.4 (1 dp)
Now f(1.35) = −0.37... < 0 and f(1.45) = 0.52... > 0
Since f(x) is continuous on (1.35, 1.45), change of sign $\Rightarrow$ 1.4 (1 dp) is a root of the equation f(x) = 0
∴ $\alpha = 1.4$ (1 dp)

c $x_1 = -0.5$, $x_2 = 0$, $x_3 =$ cannot be found (division by zero has been attempted)

2 a f(3.3545) = −0.01... < 0, f(3.3555) = 0.002... > 0
Since f(x) is continuous on (3.3545, 3.3555), change of sign $\Rightarrow$ 3.355 (3 dp) is a root of the equation f(x) = 0
∴ $\alpha = 3.355$ (3 dp)

b 12 iterations are needed to produce an estimate for α (correct to 3 dp) using (I)
7 iterations are needed to produce an estimate for α (correct to 3 dp) using (II)
Hence formula (II) converges to α roughly twice as quickly as formula (II)

c i $x^3 - 3x^2 - 4 = 0 \Rightarrow x^3 - 3x^2 = 4$

$$\Rightarrow x^2(x - 3) = 4$$

$$\Rightarrow x^2 = \frac{4}{x - 3}$$

$$\Rightarrow x = \sqrt{\frac{4}{x - 3}}$$

$$\Rightarrow x = \frac{2}{\sqrt{x - 3}}, \text{ as required}$$

ii $g'(x) = -\dfrac{1}{(x - 3)^{\frac{3}{2}}}$

$g'(x) < -1$ for values near the root α
Any starting value used in this iterative formula will lead to failure e.g. $x_1 = 4 \Rightarrow x_2 = 2$, $x_3 =$ cannot be found (square root of a negative number)
Hence this is not a suitable rearrangement for estimating α

3 a $T_A = T_B \Rightarrow 10e^{0.1t} = 16e^{-0.2t} + 25$

$$\Rightarrow e^{0.1t} = 1.6e^{-0.2t} + 2.5$$

$$\Rightarrow 0.1t = \ln(1.6e^{-0.2t} + 2.5)$$

$$\Rightarrow t = 10\ln(1.6e^{-0.2t} + 2.5), \text{ as required}$$

b $t = 10$ minutes (nearest minute)

c Use any suitable arrangement
e.g. $x_{n+1} = \sqrt[3]{2.5x_n^2 + 1.6}$, $x_1 = 0$
which gives $t \approx 10$ minutes

4 a

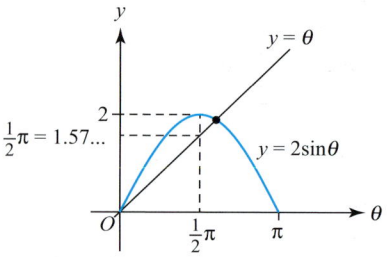

The line $y = \theta$ intersects the curve with equation
$y = 2\sin\theta$ exactly once between $\theta = \frac{1}{2}\pi$ and $\theta = \pi$
So there is exactly one solution to $\theta = 2\sin\theta$ between
$\theta = \frac{1}{2}\pi$ and $\theta = \pi$

b Area of $\triangle OAB$ = area of shaded segment

$$\Rightarrow \text{Area of } \triangle OAB = \frac{1}{2} \times \text{area of sector } OAB$$

$$\Rightarrow \frac{1}{2} \times 4 \times 4 \times \sin\alpha = \frac{1}{2} \times \frac{1}{2} \times 4^2\alpha$$

$$\Rightarrow \sin\alpha = \frac{1}{2}\alpha$$

$$\Rightarrow \alpha = 2\sin\alpha, \text{ as required}$$

c The iterates converge to 1.8954... = 1.90 (2 dp)
Now f(θ) = $\theta - 2\sin\theta$
f(1.895) $\approx -8 \times 10^{-4} < 0$, f(1.905) = 0.015... > 0
Change of sign $\Rightarrow$ 1.90 (2 dp) is a root of the equation f(θ) = 0
Since there is exactly one solution to $\theta = 2\sin\theta$ between
$\theta = \frac{1}{2}\pi$ and $\theta = \pi$
$\Rightarrow \alpha = 1.90$ to 2 dp

5 a Arc length $AB = r\alpha$
Cosine rule: Chord length $AB^2 = r^2 + r^2 - 2(r)(r)\cos\alpha$

$$\Rightarrow \text{Chord length } AB = \sqrt{2r^2 - 2r^2\cos\alpha}$$

(Arc length AB) = $2 \times$ (chord length AB)

$$\Rightarrow r\alpha = 2\sqrt{2r^2 - 2r^2\cos\alpha}$$

$$\Rightarrow r\alpha = r\sqrt{4}\sqrt{2 - 2\cos\alpha}$$

$$\Rightarrow \alpha = \sqrt{8 - 8\cos\alpha}, \text{ as required}$$

b The iterates converge to 3.7909... = 3.8 (1 dp)
Let f(θ) = $\theta - \sqrt{8 - 8\cos\theta}$
f(3.75) = −0.06... < 0, f(3.85) = 0.09... > 0
Change of sign $\Rightarrow$ 3.8 (1 dp) is a root of the equation f(θ) = 0
Since there is only one positive solution to f(θ) = 0, so
$\alpha = 3.8$ (1 dp)

c 1.2 cm² (1 dp)

d i Arc length PQ = chord length PQ

ii The student who claims the sequence converges to 0 is correct.

6 a The sequence is constant for any given starting value.

b The result of part **a** suggests that *any* positive number x is a root of the equation $x^{\ln 2} - 2^{\ln x} = 0$
This in turn suggests that $x^{\ln 2} - 2^{\ln x} \equiv 0$
To justify this identity, notice that

$\ln(x^{\ln 2}) \equiv \ln 2 \times \ln x$ (by the power rule for logs)
$\equiv \ln x \times \ln 2$
$\equiv \ln(2^{\ln x})$ (again, by the power rule for logs)
$\therefore x^{\ln 2} \equiv 2^{\ln x}$
$\Rightarrow x^{\ln 2} - 2^{\ln x} \equiv 0$

Also, note that $2 = e^{\ln 2}$ and so
$(2^{\ln x})^{\frac{1}{\ln 2}} = \left((e^{\ln 2})^{\ln x}\right)^{\frac{1}{\ln 2}}$

$= (e^{\ln x \times \ln 2})^{\frac{1}{\ln 2}}$

$= e^{\ln x}$

$= x$

which means that the iteration scheme reduces to $x_{n+1} = x_n$

Exercise 17.3A Fluency and skills

1   a $x_2 = 2.2$, $x_3 = 2.18$ (2 dp)
    b $x_2 = 2.1$, $x_3 = 2.09$ (2 dp)
    c $x_2 = -0.8$, $x_3 = -0.73$ (2 dp)
    d $x_2 = 3.45$, $x_3 = 3.42$ (2 dp)
    e $x_2 = 1.64$, $x_3 = 1.51$ (2 dp)
    f $x_2 = 4.54$, $x_3 = 4.56$ (2 dp)

2   a   The continuous function $f(x) = e^x + 3x - 4$ changes sign across the interval $(0.6765, 0.6775)$, and hence 0.677 (3 dp) is a root of the equation $f(x) = 0$
        Since α is the only root of this equation, it follows that $\alpha = 0.677$ (3 dp)
    b   The continuous function $f(x) = x^2 - 3e^{2x}$ changes sign across the interval $(-0.7885, -0.7875)$, and hence -0.788 (3 dp) is a root of the equation $f(x) = 0$
        Since α is the only root of this equation, it follows that $\alpha = -0.788$ (3 dp)
    c   The continuous function $f(x) = x^2 + 3\ln x$ changes sign across the interval $(0.8055, 0.8065)$, and hence 0.806 (3 dp) is a root of the equation $f(x) = 0$
        Since α is the only root of this equation, it follows that $\alpha = 0.806$ (3 dp)
    d   The continuous function $f(x) = \sin x + x - 3$ changes sign across the interval $(2.1795, 2.1805)$, and hence 2.180 (3 dp) is a root of the equation $f(x) = 0$
        Since α is the only root of this equation, it follows that $\alpha = 2.180$ (3 dp)
    e   The continuous function $f(x) = x - \cos^2 x - 3$ changes sign across the interval $(3.7095, 3.7105)$, and hence 3.710 (3 dp) is a root of the equation $f(x) = 0$
        Since α is the only root of this equation, it follows that $\alpha = 3.710$ (3 dp)
    f   The continuous function $f(x) = x^2 \ln x - 2$ changes sign across the interval $(1.8235, 1.8245)$, and hence 1.824 (3 dp) is a root of the equation $f(x) = 0$
        Since α is the only root of this equation, it follows that $\alpha = 1.824$ (3 dp)

3   a $f(x) = x^3 - 2x^2 - 7$
        $f'(x) = 3x^2 - 4x$
        $x_1 = 3 \Rightarrow x_2 = 2.8666...$
        $x_3 = 2.8574... = 2.857$ (3 dp)
        $f(2.8565) = -0.011... < 0$, $f(2.8575) = 0.001... > 0$
        The continuous function $f(x)$ changes sign across the interval $(2.8565, 2.8575)$, and hence 2.857 (3 dp) is a root of the equation $f(x) = 0$
        Since α is the only real root of this equation, it follows that $\alpha = 2.857$ (3 dp)
        Hence two iterations are sufficient to locate α to 3 decimal places.
    b   Yes: the iterates converge to $\alpha = 2.857$ (3 dp)

{11 iterations required}

4   a   i $f(x) = (x + \sin x)^2 - 1$
            $f'(x) = 2(x + \sin x)(1 + \cos x)$
            $x_1 = 1 \Rightarrow x_2 = 0.5785...$
            $\Rightarrow x_3 = 0.5141... = 0.51$ (2 dp)
            $f(0.505) = -0.02... < 0$, $f(0.515) = 0.01... > 0$
            The continuous function $f(x)$ changes sign across the interval $(0.505, 0.515)$, and hence 0.51 (2 dp) is a root of the equation $f(x) = 0$
            Since α is the only positive root of this equation, it follows that $\alpha = 0.51$ (2 dp)
            Hence two iterations are sufficient to locate α to 2 dp.
        ii $x_3 = 0.514$ (3 dp)
            $f(0.5135) = 0.009... > 0$, $f(0.5145) = 0.01... > 0$
            The continuous function $f(x)$ does *not* change sign across the interval $(0.5135, 0.5145)$ so α does not lie in this interval.
            $\therefore$ as an approximation to α, x_3 is not accurate to 3 decimal places.
    b   No: Newton-Raphson method with starting value $x_1 = 2$ does not produce a reliable estimate for α (since this iteration converges to a different root which is, in fact, $-\alpha$).

5   a $f(x) = \sin x + e^{-x}$
        $f(3.1825) \approx 5.9 \times 10^{-4} > 0$, $f(3.1835) \approx -4.5 \times 10^{-4} < 0$
        This change of sign means 3.183 (3 dp) is a root of the equation $f(x) = 0$
    b $h'(x) = -\sin x + e^{-x}$
        $h(1.2925) \approx 1.3 \times 10^{-4} > 0$, $h(1.2935) \approx -5.5 \times 10^{-4} < 0$
        This change of sign means 1.293 (3 dp) is a root of the equation $f'(x) = 0$

6   a   i $f(x) = 2\sqrt{x} - \dfrac{1}{x} + 1$

            $f'(x) = x^{-\frac{1}{2}} + x^{-2}$

            $x_1 = 1 \Rightarrow x_2 = 0$, $x_3 = 0 - \dfrac{f(0)}{f'(0)}$

            Neither $f(0)$ nor $f'(0)$ can be evaluated because this would involve division by zero.
            Hence x_3 cannot be calculated.
        ii $f(0.4315) = -0.003... < 0$, $f(0.4325) = 0.003... > 0$
            This change of sign means 0.432 (to 3 dp) is a root of the equation $f(x) = 0$
    b   i $f(x) = x^3 - 8\sqrt{x} + 2$

            $f'(x) = 3x^2 - 4x^{-\frac{1}{2}}$

            $x_1 = 1 \Rightarrow x_2 = -4$, $x_3 = -4 - \dfrac{f(-4)}{f'(-4)}$

            Neither $f(-4)$ nor $f'(-4)$ can be evaluated because they include $\sqrt{-4}$, which is not a real number.
            Hence x_3 cannot be calculated.
        ii $f(2.1305) = -0.006... < 0$, $f(2.1315) = 0.004... > 0$
            Also $f(0.0625) \approx 2 \times 10^{-4} > 0$, $f(0.0635) = -0.01... < 0$
            These changes of sign means 2.131 and 0.063 (to 3 dp) are roots of the equation $f(x) = 0$

Exercise 17.3B Reasoning and problem-solving

1   a $f(x) = 4x^3 - 12x^2 + 9x - 1$
        $f(0.125) = -0.054... < 0$ $f(0.135) = 0.006... > 0$
        This change of sign means 0.13 (2 decimal places) is a root of the equation $f(x) = 0$
        Since α is the only root of this equation, $\alpha = 0.13$ (2 dp)
    b   i $x_1 = 0.5$
            $f'(0.5) = 12(0.5)^2 - 24(0.5) + 9$
            $= 0$

Hence $x_2 = 0.5 - \dfrac{f(0.5)}{0}$ which cannot be evaluated.

So the Newton-Raphson method fails when $x_1 = 0.5$

ii $x_1 = 1 \Rightarrow x_2 = 1, x_3 = 1, \ldots$

The sequence is constant because we are already at a root and therefore it does not converge to α

c $\alpha = \dfrac{2 - \sqrt{3}}{2} = 0.1339745962\ldots$

Only four iterations are required to find α to 8 dp

2 a $x = 0$

b i $x = 1.1$ (1 dp)

ii $(0.6, -2.0)$ (both to 1 dp)

c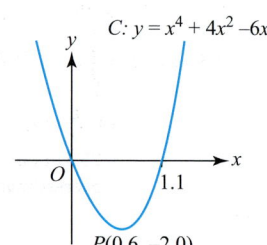
$C: y = x^4 + 4x^2 - 6x$
$P(0.6, -2.0)$

3 a $y = x \sin x + 2 \cos x \Rightarrow \dfrac{dy}{dx} = x \cos x + \sin x - 2 \sin x$

$= x \cos x - \sin x$

At a stationary point $\dfrac{dy}{dx} = 0 \Rightarrow x \cos x - \sin x = 0$

$\Rightarrow x = \dfrac{\sin x}{\cos x}$

$\Rightarrow \tan x - x = 0$

$\left\{ \text{as } \dfrac{\sin x}{\cos x} \equiv \tan x \right\}$

$\Rightarrow \tan \beta - \beta = 0$ {as P is the stationary point of C}

$\therefore \beta$ is a root of the equation $g(x) = 0$, where $g(x) = \tan x - x$

b $g(x) = \tan x - x$

$g'(x) = \sec^2 x - 1$

$= \tan^2 x$

$\therefore x_{n+1} = x_n - \dfrac{g(x_n)}{g'(x_n)}$ can be written as

$x_{n+1} = x_n - \left(\dfrac{\tan x_n - x_n}{\tan^2 x_n} \right)$

c $(4.49, -4.82)$ (each to 2 dp)

d Using 2nd derivative: $\dfrac{dy}{dx} = x \cos x - \sin x$

$\Rightarrow \dfrac{d^2 y}{dx^2} = x(-\sin x) + \cos x - \cos x$

$= -x \sin x$

Across the interval $(\pi, 2\pi)$, $x > 0$ and $\sin x < 0$

$\therefore \dfrac{d^2 y}{dx^2} > 0$ for all values $\pi < x < 2\pi$ and, in particular, when $x = \beta$

Hence P is a minimum point.

OR Using a logical approach:

P is the only stationary point in the interval $(\pi, 2\pi)$

$f(\pi) = -2$ and $f(2\pi) = 2$

Since $f(\beta) \approx -4.8$, P must be a minimum point.

4 a i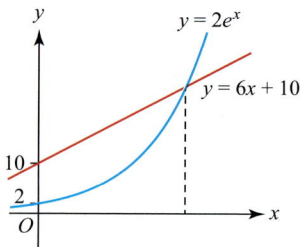
$y = 2e^x$
$y = 6x + 10$

For $x \geq 0$, the line with equation $y = 6x + 40$ and curve with equation $y = 2e^x$ intersect exactly once.

Hence the equation $2e^x = 6x + 40$ has exactly one positive solution.

ii $f(3.35) = -3.09\ldots < 0$; $f(3.45) = 2.30\ldots > 0$

The continuous function $f(x) = 2e^x - 6x - 40$ changes sign between $x = 3.35$ and $x = 3.45$, and hence $x = 3.4$ (1 dp) is the only positive solution of the equation $f(x) = 0$

b 2017

c £127 million

5 a Converges to $\sqrt{k}$

b Use the iteration $x_{n+1} = \dfrac{1}{2}\left(x_n + \dfrac{54321}{x_n} \right)$, $x_1 = 1$

c Let $f(x) = x^2 - k$

$f'(x) = 2x$

Newton-Raphson formula: $x_{n+1} = x_n - \left(\dfrac{x_n^2 - k}{2x_n} \right)$

$= \dfrac{1}{2}\left(2x_n - \left(\dfrac{x_n^2}{x_n} - \dfrac{k}{x_n} \right) \right)$

$= \dfrac{1}{2}\left(x_n + \dfrac{k}{x_n} \right)$

This sequence converges to the root of the equation $x^2 - k = 0$ that is, to $\sqrt{k}$

Exercise 17.4A Fluency and skills

1 a 23.19 (2 dp)   b 7.06 (2 dp)   c 4.14 (2 dp)

d 0.77 (2 dp)   e 0.52 (2 dp)   f 2.07 (2 dp)

2 a i 0.697 (3 dp)   ii underestimate

b i 3.268 (3 dp)   ii overestimate

c i 7.402 (3 dp)   ii overestimate

d i 3.715 (3 dp)   ii underestimate

3 a

x_i	1	1.1	1.2	1.3	1.4	1.5	1.6	1.7	1.8	1.9	2
y_i	1	1.26	1.62	2.13	2.87	3.95	5.55	7.98	11.72	17.59	27

b 6.9 (1 dp)

4 a i 12.211 (3 dp)   ii 11.214 (3 dp)

b Graph 1 pairs with the equation $y = 6 - e^{0.1x^2}$ and Graph II pairs with the equation $y = \dfrac{e^x}{x} - 2$

Graph 1 is concave. Each trapezium lies entirely under the curve, so the answer 11.213 is an underestimate, that is $\displaystyle\int_1^4 6 - e^{0.1x^2}\, dx > 11.213$

Graph 2 is convex. Each trapezium lies partly above the curve, so the answer 12.211 is an overestimate, that is $\displaystyle\int_1^4 \dfrac{e^x}{x} - 2\, dx < 12.211$

Answers For full solutions go to http://www.oxfordsecondary.co.uk/edexcelalevelmaths-answers

5 a i

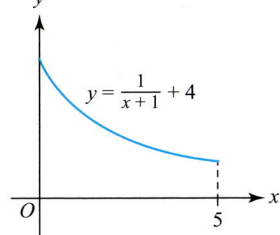

$y = \dfrac{1}{x+1} + 4$

ii 21.91 (2 dp)

iii The answer is an overestimate, due to the concave upwards shape of the curve.

When drawn, each trapezium lies partly above the curve.

iv $\displaystyle\int_0^5 \dfrac{1}{x+1} + 4 \, dx = \ln 6 + 20 \ \ (< 21.91, \text{ as required})$

b i

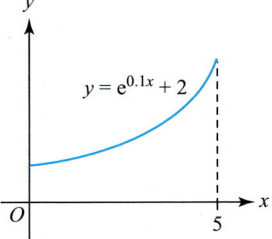

$y = e^{0.1x} + 2$

ii 16.49 (2 dp)

iii The answer is an overestimate, due to the concave upwards shape of the curve.

When drawn, each trapezium lies partly above the curve.

iv $\displaystyle\int_0^5 e^{0.1x} + 2 \, dx = 10\sqrt{e} \ \ (< 16.49, \text{ as required})$

Exercise 17.4B Reasoning and problem-solving

1 a 0.52 (2 sf)

b i 4.44 km² (3 sf)

ii Underestimate for the actual area of the field because 0.52110420 is an overestimate for

$\displaystyle\int_0^2 1 - \sqrt{\cos(0.25\pi x)} \, dx$ (due to convex curve).

2 a 4.04 (2 dp)

b 3.56 (2 dp)

c 3.5

d Using 5 intervals: $I \approx 4.04 \Rightarrow$ error $= 4.04 - 3.5 = 0.54$

Using 20 intervals: $I \approx 3.56 \Rightarrow$ error $= 3.56 - 3.5 = 0.06$

$\dfrac{0.54}{0.06} = 9$ so the answer (to 2 dp) obtained using 20 intervals is nine times more accurate than that obtained using five intervals.

3 a 6 metres

b 11.8 m²

c 6000 litres per second (1 significant figure)

4 a i 17.4 (1 dp)

ii

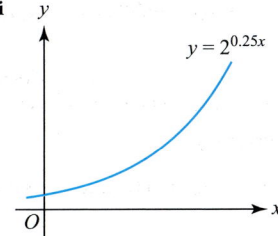

$y = 2^{0.25x}$

Due to the shape of the curve (convex), the answer found is an overestimate for the integral.

b $k = 1$

c A's distance from start after 8 seconds $= \displaystyle\int_0^8 1 + 2^{0.25t} \, dt$

$\approx \left[t\right]_0^8 + 17.4$

$= 25.4 \text{ m (an overestimate)}$

B's distance from start after

8 seconds $= 10\,\text{m} + \displaystyle\int_0^8 0.25t + 1 \, dt$

$= 10\,\text{m} + [0.125t^2 + t]_0^8$

$= 26\,\text{m}$

Hence B wins the race

5 a $S_n = \dfrac{a(r^n - 1)}{r - 1}$

b $h = \dfrac{1-0}{n} = \dfrac{1}{n}$

x_i	0	$\dfrac{1}{n}$	$\dfrac{2}{n}$	...	$\dfrac{n-1}{n}$	1
y_i	1	2	2^2	....	2^{n-1}	2^n

$\displaystyle\int_0^1 2^{nx} \, dx \approx \dfrac{1}{2}\left(\dfrac{1}{n}\right)[1 + 2^n + 2(2 + 2^2 + ... + 2^{n-1})]$

$2 + 2^2 + ... + 2^{n-1}$ is the sum of the first $(n-1)$ terms of a geometric series, first term and common ratio 2

Hence $2 + 2^2 + ... + 2^{n-1} = \dfrac{2(2^{n-1} - 1)}{2 - 1}$

$= 2^n - 2$

$\therefore \displaystyle\int_0^1 2^{nx} \, dx \approx \dfrac{1}{2}\left(\dfrac{1}{n}\right)[1 + 2^n + 2(2^n - 2)]$

$= \dfrac{1}{2n}[1 + 2^n + 2^{n+1} - 4]$

$= \dfrac{1}{2n}[2^n(1 + 2) - 3]$

$= \dfrac{3}{2n}(2^n - 1)$, as required

c i $n = 16$

To justify, Let $f(x) = 2^x - 1 - 4096x$

$f(15.5) = -17148.04.. < 0$, $f(16.5) = 25096.9... > 0$

Change of sign $\Rightarrow f(x) = 0$ has a root between 15.5 and 16.5

By sketching the graphs of $y = 2^x - 1$ and $y = 4096x$, it is clear that they intersect only once for positive x

Hence, 16 is a root of the equation $2^x - 1 = 4096x$ (to the nearest integer)

ii 47.8 (1 dp)

6 a 0.299 (3 dp)

b Letting $u = -\dfrac{3}{x} \Rightarrow \dfrac{dx}{du} = \dfrac{3}{u^2}$

Change limits: $x = 3 \Rightarrow u = -1$
$x = 2 \Rightarrow u = -1.5$

$$\int_2^3 e^{-\frac{3}{x}} \, dx = \int_{-1.5}^{-1} \dfrac{3}{u^2} e^u \, du$$

c 0.301 (3 dp)

d $I = 0.30$ (2 dp)

7 a **i** $I \approx 3.1312$ **ii** $I \approx 3.1349$ **iii** $I \approx 3.1390$

b As n increases, the estimates for I seem to approach π

c By using the substitution, $I = \pi$

Review exercise 17

1 a **i** $f(x) = x^3 + 3 - 5x^2$
$f(0) = 3 > 0, f(1) = -1 < 0$
The continuous function $f(x)$ changes sign across the interval $(0, 1)$
∴ the equation $f(x) = 0$ has a root between $x = 0$ and $x = 1$
∴ the equation $x^3 + 3 = 5x^2$ has a solution between $x = 0$ and $x = 1$

ii $f(x) = x^3 + 3 - 5x^2$
$f(-0.8) = -0.712 < 0, f(-0.7) = 0.207 > 0$
The change of sign shows that the equation $x^3 + 3 = 5x^2$ has a solution between $x = -0.8$ and $x = -0.7$

iii $f(x) = 2^x - e^{\sqrt{x-1}} - 3$
$f(2.15) = -0.15... < 0, f(2.25) = 0.108... > 0$
The change of sign shows that the equation $2^x - e^{\sqrt{x-1}} = 3$ has a solution between $x = 2.15$ and $x = 2.25$

iv $f(x) = x \sin x - \cos(\pi x) - 1$
$f(3) = 0.42... > 0, f(\pi) = -0.09... < 0$
The change of sign shows that the equation $x \sin x = \cos(\pi x) + 1$ has a solution between $x = 3$ and $x = \pi$

b The solution to the equation $2^x - e^{\sqrt{x-1}} = 3$ (part **iii**)

2 a $x = 1 \pm \sqrt{2}$

b $f(1.5) = -3.5 < 0, f(2.4) = 0.1 > 0$

c The equation $f(x) = 0$ does not have a root between $x = 1.5$ and $x = 2.4$ since
$1 - \sqrt{2} = -0.41... < 1.5$ and $1 + \sqrt{2} = 2.414... > 2.4$
However, $f(x)$ changes sign between $x = 1.5$ and $x = 2.4$
Therefore, $f(x)$ cannot be continuous on the interval $(1.5, 2.4)$

3 a

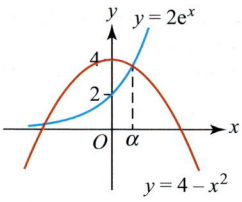

The two curves $y = 2e^x$ and $y = 4 - x^2$ intersect exactly once for positive x
∴ the equation $2e^x = 4 - x^2$ has exactly one positive solution

b $2e^x = 4 - x^2 \Rightarrow e^x = 2 - 0.5x^2$
$\Rightarrow x = \ln(2 - 0.5x^2)$ as required

c Let $f(x) = 2e^x - 4 + x^2$
$f(0.5985) = -0.003.... < 0, f(0.5995) = 0.001... > 0$
This change of sign $\Rightarrow 0.599$ (3 dp) is a root of the equation $f(x) = 0$

Since the equation $f(x) = 0$ has exactly one positive solution, $\alpha = 0.599$ (3 dp) (2 dp)

4 a $x = 0.481$ (3 dp) **b** $x = 5.236$ (3 dp)
c $x = 1.073$ (3 dp) **d** $x = 2.308$ (3 dp)

5 a $f(2.7655) = -0.003... < 0, f(2.7665) \approx 4 \times 10^{-4} > 0$
This change of sign $\Rightarrow 2.766$ (3 dp) is a root of the equation $f(x) = 0$

b $f'(1) = 2(1) - 2(1)^{-\frac{1}{2}}$
$= 0$
The starting value is the x-coordinate of a stationary point of the curve with equation $y = f(x)$
This means division by zero occurs when using the Newton-Raphson formula to calculate x_2 (or, graphically, the tangent to the graph at the first approximation never intersects the x-axis, which is required for the Newton-Raphson method to work).

6 a $n = 5, h = \dfrac{2 - 0}{5} = 0.4$

x_i	0	0.4	0.8	1.2	1.6	2
y_i (3 dp where appropriate)	1	1.741	3.031	5.278	9.190	16

$$\int_0^2 4^x \, dx \approx \dfrac{1}{2}(0.4)\{1 + 16 + 2(1.741 + ... + 9.190)\}$$
$$= \dfrac{1}{2}(0.4)\{55.48\}$$
$$= 11.096$$
$$= 11.10 \text{ (2 dp)}$$

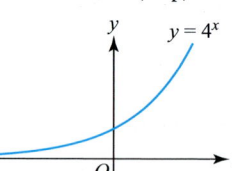

This answer is an overestimate for $\displaystyle\int_0^2 4^x \, dx$ because of the concave upwards shape of the graph with equation $y = 4^x$

b $2^{2x+3} = 2^{2x} \times 2^3$
$= (2^2)^x \times 8$
$= 4^x \times 8$
$$\therefore \int_0^2 2^{2x+3} \, dx = 8 \times \int_0^2 4^x \, dx$$
$$\approx 8 \times 11.096$$
$$= 88.768$$
$$= 88.8 \text{ (3 sf)}$$

Assessment 17

1 $1^3 + 2(1)^2 - 3(1) - 2 = -2, 2^3 + 2(2)^2 - 3(2) - 2 = 8$
Change of sign, hence root in interval.

2 $f(3) = 0.42, f(3.5) = -1.23$
Change of sign, hence root in interval.

3 a $2^3 - 4(2) - 1 = -1, 2.5^3 - 4(2.5) - 1 = 4.6$
Change of sign, hence root in interval.

b $x_2 = \sqrt{4 + \dfrac{1}{2}} = 2.12, x_3 = \sqrt{4 + \dfrac{1}{2.12}} = 2.11$

4 a 1.31

b $e^{1.305} + 1.305 - 5 = -0.007, e^{1.315} + 1.315 - 5 = 0.04$

Change of sign, hence must be correct to 2 decimal places.

5 a $1^3 - 3(1) + 1 = -1$, $2^3 - 3(2) + 1 = 3$

Change of sign, hence root in interval.

b 1.55

6 a −0.91

b f(−0.915) = 0.04, f(−0.905) = −0.055

Change of sign, hence must be correct to 2 decimal places.

7 1.13

8 −6.66

9 a $f(0) = 5(0) - e^0 = -1$, $f(0.5) = 5(0.5) - e^{0.5} = 0.85$

Change of sign, hence root in (0, 0.5)

b Because the curve intersects the x-axis between 2 and 3.

c $f(2.45) = 5(2.45) - e^{2.45} = 0.66$,
$f(2.55) = 5(2.55) - e^{2.55} = -0.06$

Change of sign, hence must be correct to 1 decimal place.

10 a

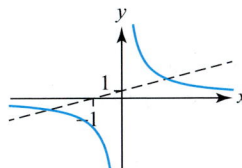

b Two as they intersect twice.

c Let $f(x) = x + 1 - \dfrac{4}{x}$, $f(1.5) = 1.5 + 1 - \dfrac{4}{1.5} = -0.17$, $f(1.6) = 0.1$

Change of sign so root in interval.

d $x = 1.56, -2.56$

11 a Three as the curves intersect 3 times.

b Let $f(x) = e^x \sin x - x - 2$, $f(1.2) = e^{1.2}\sin 1.2 - 1.2 - 2 = -0.10$,
$f(1.3) = 0.236$

Change of sign so root in interval.

c Root in interval (2.8, 2.9)

12 a $x^3 = 5 + 3x^2$, $x^2 = \dfrac{5}{x} + 3x$, $x = \sqrt{\dfrac{5}{x} + 3x}$

b $x_5 = 3.42$

13 a $x_2 = 0.85$, $x_3 = 0.77$, $x_4 = 0.89$, $x_5 = 0.71$

b The sequence is diverging from the root.

c $x_2 = 0.83$, $x_3 = 0.81$, $x_4 = 0.83$, $x_5 = 0.82$

14 a

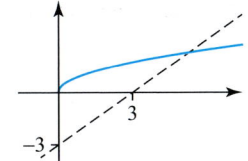

b 5.30

c i

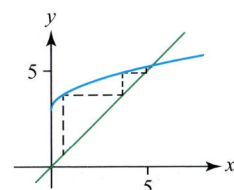

ii Staircase diagram.

15 a

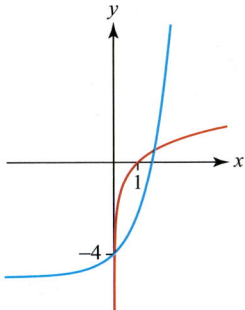

b Two as curves intersect twice.

c Let $f(x) = \ln x - e^x + 5$, $f(1.6) = \ln 1.6 - e^{1.6} + 5 = 0.517$,
$f(1.8) = -0.462$

Change of sign so root in interval.

d 1.71

16 a $x^3 + 4x - 3 = 0$
$$4x = 3 - x^3$$
$$x = \frac{(3x - x^3)}{4}$$
$$a = \frac{1}{4}, b = 3$$

b $x_5 = 0.67$

c i

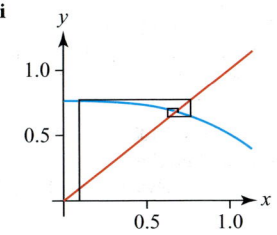

ii Cobweb diagram.

17 2.33

18 a 2.168

b Overestimate as curve is convex.

c i 2.097   **ii** 3.4%

19 a Three as the curve intersects the x-axis three times.

b $x^3 = 5x + 3$, $x^2 = 5 + \dfrac{3}{x}$, $x = \pm\sqrt{5 + \dfrac{3}{x}}$

c 2.49

d −0.66

e $f(-1.835) = (-1.835)^3 - 5(-1.835) - 3 = -0.004$,
$f(-1.825) = 0.047$

Change of sign, hence must be correct to 3 significant figures.

20 a There is an asymptote $x = 3$ so although there is a change of sign in the interval, there is no root.

b f(0) and f(1) are both negative because there are two roots $\left(\dfrac{1}{2} \text{ and } \dfrac{2}{3}\right)$ between them.

21 a 0.76   **b** 2.91

22 a 0.568

b Underestimate as curve is concave. Hence, actual area will be significantly higher than 0.5

23 a f(1) = −1, f(2) = 0.39, f(1.5) = −0.39, f(1.7) = −0.10,
f(1.9) = 0.22

So solution is in interval (1.7, 1.9)

b 1.763

Assessment chapters 12–17: Pure

1 $6 - 2\sqrt{5}$

2 $x = 1, 0.585$

3 a $k > -1 + \sqrt{8}$, $k < -1 - \sqrt{8}$
  b $x = \pm 1, \pm \sqrt{2}$

4 $(5, -1)$ and $(-2, 6)$

5 $-\dfrac{1}{3} - \dfrac{\sqrt{7}}{3} \le x \le -\dfrac{1}{3} + \dfrac{\sqrt{7}}{3}$

6 $\theta = 30°, 150°$ or $\theta = 199.5°, 340.5°$

7 $y = \ln 27$, $x = \ln\left(\dfrac{1}{9}\right)$

8 a $\left(-\dfrac{3}{2}, 0\right), \left(\dfrac{5}{3}, 0\right)$

  b
  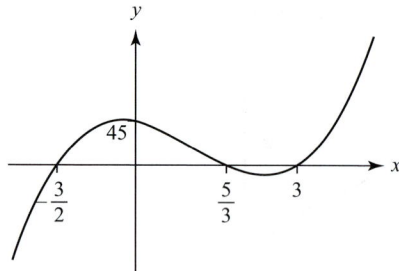

  c 101.7 square units

9 $\left(-1, -\dfrac{17}{2}\right)$, a minimum

  $\left(3, \dfrac{47}{2}\right)$, a minimum

  $\left(2.5, \dfrac{757}{32}\right)$, a maximum

10 a $A = 2$, $k = \dfrac{1}{3}\ln 2$

  b 23.9 days

  c 10.7 days

  d
  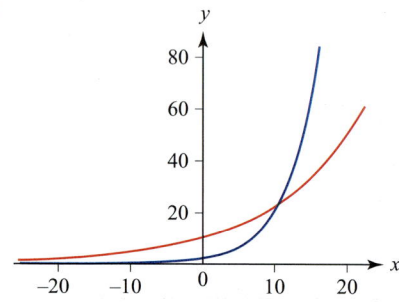

  e Only realistic for small values of t.
  Justification, e.g. Students who catch one infection may be off school so not catch other infection/may have weakened immune system so more likely to catch other infection.
  Numbers of cases cannot increase indefinitely.
  There is only a finite population/some students may have a natural immunity and not catch an infection.

11 a $f\left(-\dfrac{1}{2}\right)$

  $= 6\left(-\dfrac{1}{2}\right)^3 - 19\left(-\dfrac{1}{2}\right)^2 - 51\left(-\dfrac{1}{2}\right) - 20$

  $= 0$
  Therefore $2x + 1$ is a factor of $f(x)$

  b $x = \left(-\dfrac{1}{2}\right), 5, -\dfrac{4}{3}$

12 a $\dfrac{3x - 1}{(x + 3)(2x + 1)} = \dfrac{2}{x + 3} - \dfrac{1}{2x + 1}$

  b $\dfrac{3x - 5}{x^2 - 25} = \dfrac{2}{x + 5} + \dfrac{1}{x - 5}$

13 a $f(x) = \dfrac{2x - 14}{(x - 3)(x + 1)} + \dfrac{2(x + 1)}{(x - 3)(x + 1)}$

  $= \dfrac{4x - 12}{(x - 3)(x + 1)}$

  $= \dfrac{4(x - 3)}{(x - 3)(x + 1)}$

  $= \dfrac{4}{x + 1}$

  b i $x > 3$ **ii** $0 < f(x) < 1$

  c $f^{-1}(x) = \dfrac{4}{x} - 1$

    Domain is $0 < x < 1$

14 a $\dfrac{3}{5}$ or 0.6   **b** $x = 3, 1$

  c Range of $gh(x)$ is $gh(x) \ne 3$
    Range of $hg(x)$ is $hg(x) \ne \pm 1$
    So not the same (can use graphs to illustrate).

15 $1 + \dfrac{1}{2}x - \dfrac{3}{8}x^2 + \dfrac{7}{16}x^3 + \dots$

16 a -3 **b** -67

17 a $5.18\,\text{cm}^2$ **b** $13.4\,\text{cm}$

18 a i
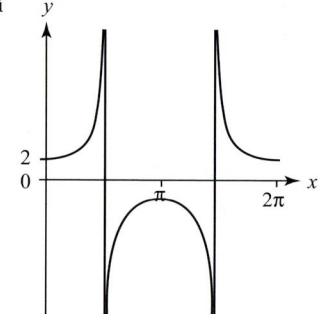

  ii
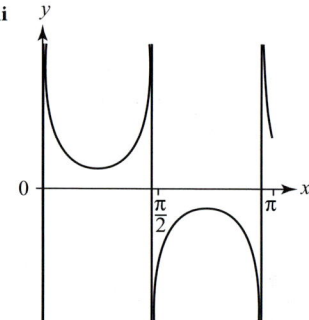

  b $x = 56.3°, 236.3°$

19 a $x = \dfrac{\sqrt{2}}{2}$ **b** $x = \dfrac{\sqrt{3}}{4}$

20 a i $3\cos x$ **ii** $\ln x + 1$

  b $\sqrt{3} - \dfrac{\pi}{6} - \dfrac{1}{2}$

21 a $x = -\dfrac{1}{2}$, $y = -\dfrac{1}{2}e^{-1}$

  b $\dfrac{d^2 y}{dx^2} = 2e^{2x} + 2e^{2x} + 4xe^{2x}$

    When $x = -\dfrac{1}{2}$,

$$\frac{d^2y}{dx^2} = 4e^{-1} - 2e^{-1}$$

$$= 2e^{-1}$$

> 0 so a minimum

22 a   i $-\cos x\,(+c)$ **ii** $3\ln|x|\,(+c)$

   b $2\ln 5$ or $\ln 25$

23 2

24 a $3^3 - 6(3) - 12 = -3$

       $3.5^3 - 6(3.5) - 12 = 9.875$

       Change of sign so root in interval $(3, 3.5)$

   b $x_2 = 3.16$, $x_3 = 3.13$

   c $3.125^3 - 6(3.125) - 12 = -0.23$

       $3.135^3 - 6(3.135) - 12 = 0.0015$

       Change of sign so 3.13 is correct to 2 decimal places

25 0.527

26 Assume there exists an even integer n such that n^2 is odd.

Then $n = 2k$ for some integer k.

So $n^2 = (2k)^2$

$\qquad = 4k^2$

$\qquad = 2(2k^2)$ hence it is even

Which is a contradiction so we have proved the statement.

27 $\dfrac{2x^2 + 4x + 3}{2x^2 - x - 1}$

$= \dfrac{A(x-1)(2x+1) + B(2x+1) + C(x-1)}{(2x+1)(x-1)}$

$2x^2 + 4x + 3 = A(x-1)(2x+1) + B(2x+1) + C(x-1)$

$A = 1$, $B = 3$, $C = -1$

28 a

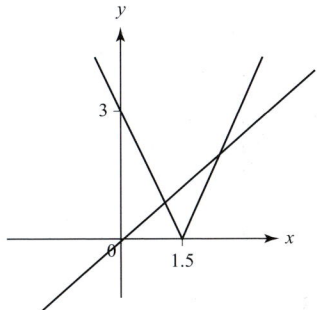

   b 2 as they will intersect twice – either sketch to demonstrate or explain that gradient of $y = x$ is less than gradient of $y = -(3 - 2x)$.

   c $x \le 1$, $x \ge 3$

29 a $f^{-1}(x) = \dfrac{1}{3}(e^x - 1)$

   b

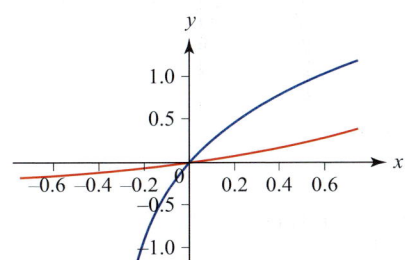

   c Domain is $x \in \mathbb{R}$

       Range is $y > -\dfrac{1}{3}$

30 a $\dfrac{1}{x-1} + \dfrac{-1}{x+3} + \dfrac{2}{(x+3)^2}$

   b $\ln|x-1| - \ln|x+3| - \dfrac{2}{x+3}\,(+c)$

31 13.3

32 a $x_2 = -1$, $x_3 = 3$ **b** $3n - 6$

33 a 8   **b** $\dfrac{2}{9}$

34 a $S_n = a + (a+d) + (a+2d) + \ldots + (a + (n-2)d) + (a + (n-1)d)$

Also,

$S_n = (a + (n-1)d) + (a + (n-2)d) + \ldots$

$\qquad + (a + 2d) + (a + d) + a$

So $2S_n = (2a + (n-1)d) + (2a + (n-1)d) + \ldots$

$\qquad\qquad + (2a + (n-1)d) + (2a + (n-1)d)$

$\Rightarrow 2S_n = n(2a + (n-1)d)$

$\Rightarrow S_n = \dfrac{n}{2}(2a + (n-1)d)$ as required

   b 12

35 a   i

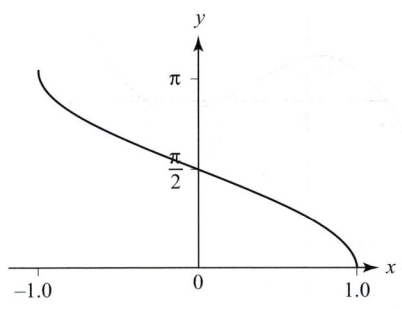

      ii

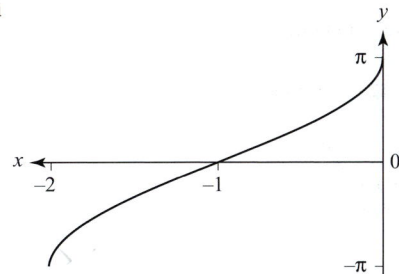

   b   i $0 \le y \le \pi$ **ii** $-\pi \le y \le \pi$

   c $f^{-1}(x) = \sin\left(\dfrac{x}{2}\right) - 1$

       Domain is $-\pi \le x \le \pi$

36 a

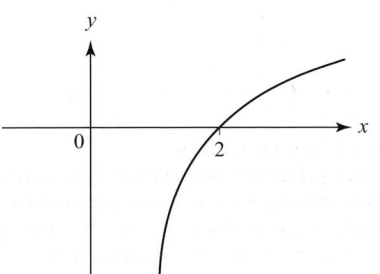

       Asymptote at $x = 1$

   b $\dfrac{3}{2}$

37 a $\sec 2x = \dfrac{1}{\cos 2x}$

$\approx \dfrac{1}{1 - \dfrac{(2x)^2}{2}}$

$= \dfrac{2}{2 - (2x)^2}$

$= \dfrac{2}{2 - 4x^2}$

$$= \frac{1}{1-2x^2}$$

b $1+2x^2+4x^4$

c 1.0204

38 a $x = 0.421, 2.72, 3.56, 5.86$

  b $x = 1.85$

39 a LHS $\equiv 3(1+\tan^2\theta)-7\tan^2\theta$
   $\equiv 3+3\tan^2\theta-7\tan^2\theta$
   $\equiv 3-4\tan^2\theta$
   $\equiv$ RHS

  b $\theta = 0.669, 3.81, 2.47$

40 a $\cos 2x \equiv \cos x\cos x - \sin x\sin x$
   $\equiv \cos^2 x - \sin^2 x$

   $\equiv 1-2\sin^2 x$ as required

  b $x = 90°, -30°, -150°$

41 a $\sqrt{17}\sin(\theta+0.245)$

  b Max is $\sqrt{17}$
   Occurs when $\theta = 1.33°$

42 $-3 < x < 1$

43 $y = 2x+1$

44 a $\dfrac{(x+2)\cos x - \sin x}{(x+2)^2}$

  b $\dfrac{(x+2)\cos x - \sin x}{(x+2)^2} = 0$
   $\Rightarrow (x+2)\cos x - \sin x = 0$
   $\Rightarrow (x+2) = \tan x$
   $\Rightarrow \arctan(x+2) = x$ as required

  c $x = 1.27$

45 a $\dfrac{1}{t}$

  b $y-5 = -2(x-1)$

  c $x = \left(\dfrac{y-1}{2}\right)^2 - 3$

46 When $y = 0$, $\dfrac{dy}{dx} = \dfrac{2}{3}$
   When $y = \dfrac{3}{2}$, $\dfrac{dy}{dx} = \dfrac{5}{6}$

47 $\cos(x+h) - \cos x$
   $= \cos x\cos h - \sin x\sin h - \cos x$
   when h small
   $\approx \cos x\left(1-\dfrac{h^2}{2}\right) - h\sin x - \cos x$
   $= -\dfrac{h^2}{2}\cos x - h\sin x$
   $\dfrac{dy}{dx} = \lim_{h\to 0}\left(\dfrac{\cos(x+h)-\cos x}{(x+h)-x}\right)$
   $= \lim_{h\to 0}\left(\dfrac{-\dfrac{h^2}{2}\cos x - h\sin x}{h}\right)$
   $= \lim_{h\to 0}\left(-\dfrac{h}{2}\cos x - \sin x\right)$
   $= -\sin x$

48 a **i** $-x\cos x + \sin x \; (+c)$
   ii $x\ln x - x \; (+c)$

  b $-\dfrac{1}{4}(x^2+1)^{-2}(+c)$

49 $\ln\left(\dfrac{27}{4}\right)$

50 $y = e^{\sin x - \frac{1}{2}}$

51 a

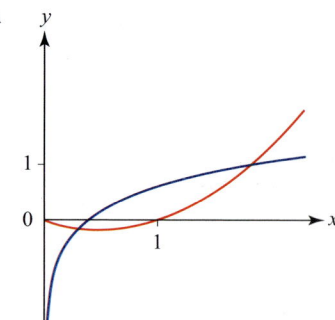

  b $x = 0.298$ (or $x = 1.87$)

  c e.g.
   $0.2975^2 - 0.2975 - \ln 0.2975 - 1 = 0.003$
   $0.2985^2 - 0.2985 - \ln 0.2985 - 1 = -0.0004$
   Change of sign so root in interval, hence solution is
   correct to 3 significant figures.

52 a $x = -1.343$

  b 1.9 is near a stationary point so process could be
   unstable/next approximation is 19.6 which is nowhere
   near a root.

53 a One solution as the graphs intersect once

  b Let $f(x) = \sqrt{x} - e^{-x}$
   Then $f(0.4) = \sqrt{0.4} - e^{-0.4} \; (= -0.04)$
   $f(0.5) = \sqrt{0.5} - e^{-0.5} (= 0.1)$
   Change of sign so solution is between 0.4 and 0.5

  c $x = 0.43$
   $f(0.425) = \sqrt{0.425} - e^{-0.425} \; (= -0.002)$
   $f(0.435) = \sqrt{0.435} - e^{-0.435} \; (= 0.012)$

   Change of sign so 0.43 is correct to 2 dp

54 a

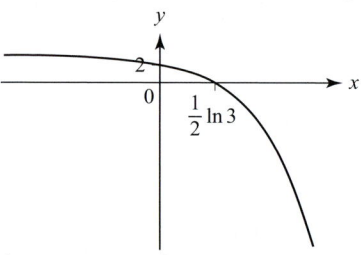

   Asymptote at $y = 3$

  b $f^{-1}(x) = \dfrac{1}{2}\ln(3-x)$

   Domain is $x < 3$
   Range is $f^{-1}(x) \in \mathbb{R}$

55 $\sqrt{27}$

56 $d = 2.5, a = 1$

57 $\dfrac{12}{5}$ or 2.4

58 a **i** Geometric, $a = \dfrac{2}{3}$, $r = \dfrac{1}{3}$
   ii 1

  b Arithmetic, $\dfrac{n}{2}(n+1)(\ln 3)$

59 a $2 - \dfrac{5}{12}x - \dfrac{25}{288}x^2 + \ldots$ **b** 19.9582

60 Let $\dfrac{A+B}{2} = P$ and $\dfrac{A-B}{2} = Q$

Then $P + Q = A$ and $P - Q = B$

LHS $= \cos(P+Q) - \cos(P-Q)$

$\equiv (\cos P \cos Q - \sin P \sin Q) - (\cos P \cos Q + \sin P \sin Q)$

$\equiv -2\sin P \sin Q$

So $K = -2$

61 $0.554 \le x \le 2.12$

62 $x = 214.7°, 325.3°$

63 $\left(53.1°, \dfrac{1}{5}\right)$

64 a LHS

$\equiv (\sin^2 x + \cos^2 x)^2 - 2\sin^2 x \cos^2 x$

$\equiv 1 - 2(\sin x \cos x)^2$

$\equiv 1 - 2\left(\dfrac{1}{2}\sin 2x\right)^2$

$\equiv 1 - 2\left(\dfrac{1}{4}\sin^2 2x\right)$

$\equiv 1 - \dfrac{1}{2}\sin^2 2x$

$\equiv 1 - \dfrac{1}{2}\left(\dfrac{1-\cos 4x}{2}\right)$

$\equiv 1 - \dfrac{1}{4}(1 - \cos 4x)$

$\equiv \dfrac{1}{4}(4 - 1 + \cos 4x)$

$\equiv \dfrac{1}{4}(3 + \cos 4x) \equiv$ RHS

b $x = \pm 0.48, \pm 1.09, \pm 2.05, \pm 2.66$

65 a i

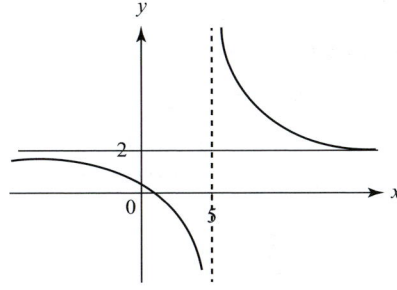

$x = 5, y = 2$

ii

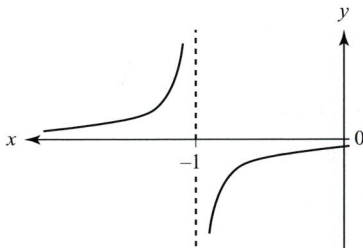

$x = -1, y = 0$

b $b = -4, a = 8$

c i $y = \dfrac{2x-2}{x-5}$ **ii** $y = -\dfrac{8}{1+x}$

66 a $\dfrac{2}{27}$ **b** $y = \dfrac{1}{2}\sqrt{x} + 10 + \dfrac{50}{\sqrt{x}}$

67 $\dfrac{5}{2}, -\dfrac{20}{23}$

68 149 cm², $r = 9.73$ cm

69 $-\ln 2$

70 Let $y = \arccos 3x$

Then $\cos y = 3x$

$-\sin y \dfrac{dy}{dx} = 3$

$\Rightarrow \dfrac{dy}{dx} = -\dfrac{3}{\sin y}$

$= -\dfrac{3}{\sqrt{1-\cos^2 y}}$

$= -\dfrac{3}{\sqrt{1-(3x)^2}}$

$= -\dfrac{3}{\sqrt{1-9x^2}}$ as required

71 $f(\theta) = \displaystyle\int \dfrac{\sin \theta}{\cos \theta}\, d\theta$

$= -\ln|\cos\theta| + c$

$= \ln|\cos \theta|^{-1} + c$

$= \ln|\sec \theta| + c$

$f\left(\dfrac{\pi}{3}\right) = \ln 6$

$\Rightarrow \ln 2 + c = \ln 6$

$\Rightarrow c = \ln 3$

$\therefore f(\theta) = \ln|\sec \theta| + \ln 3$

72 a $y e^y - e^y = \dfrac{1}{2}e^{2x} + c$

b $x = \dfrac{1}{2}\ln(2y e^y - 2e^y + 1)$

73 $\dfrac{1}{216}(3x-1)^8 + \dfrac{2}{189}(3x-1)^7 + \dfrac{1}{162}(3x-1)^6 (+c)$

74 4

75 a 3.955

b Not true, estimation is an over-estimate since curve is convex.

76 13.5 square units

77 $\dfrac{\pi}{4} + \dfrac{\sqrt{3}}{4}$

Chapter 18
Exercise 18.1A Fluency and skills

1 a $(13\mathbf{i} - 3\mathbf{j})\,\text{m s}^{-1}$ **b** $(18\mathbf{i} + 1.5\mathbf{j})\,\text{m}$

2 a $(13\mathbf{i} + 6\mathbf{j})\,\text{m s}^{-1}$ **b** $(-\mathbf{i} + 2\mathbf{j})\,\text{m s}^{-2}$ **c** $16.5\mathbf{i}\,\text{m}$

3 $(24\mathbf{i} + 4\mathbf{j})\,\text{m}$

4 $(6\mathbf{i} + 3\mathbf{j})\,\text{m s}^{-1}$

5 a $\mathbf{s} = (36\mathbf{i} - 6\mathbf{j})$ **b** $\mathbf{s} = (80\mathbf{i} - 104\mathbf{j})$

c $\mathbf{a} = (-2\mathbf{i} + 3\mathbf{j})$ **d** $\mathbf{v} = (3\mathbf{i} + 11\mathbf{j})$

6 $(13.5\mathbf{i} + 4\mathbf{j})\,\text{m}$

7 Speed $= 11.4\,\text{m s}^{-1}$

Direction $= 37.9°$ to the x-direction

8 $(25\mathbf{i} + 13\mathbf{j})\,\text{m}$

Exercise 18.1B Reasoning and problem-solving

1 a Speed $= 38.5\,\text{m s}^{-1}$

Direction $= 27.9°$ to the x-axis

b $152\,\text{m}$

2 $(41\mathbf{i} + 8\mathbf{j})\,\text{m}$

3 $12.2\,\text{m}$

4 $\mathbf{u} = (6\mathbf{i} + 2\mathbf{j})$

$\mathbf{v} = (6\mathbf{i} + 2\mathbf{j}) + (-\mathbf{i} - 7\mathbf{j}) \times 2$

$= (4\mathbf{i} - 12\mathbf{j})$

Initial speed $u = \sqrt{6^2 + 2^2}$

$= 2\sqrt{10}$

$$= \sqrt{4^2 + (-12)^2}$$

Final speed $= 4\sqrt{10}$

$$= 2u$$

Initial direction along a line with gradient $m_1 = \dfrac{2}{6}$

$$= \dfrac{1}{3}$$

Final direction along a line with gradient $m_2 = \dfrac{-12}{4}$

$$= -3$$

$m_1 m_2 = -1$, so directions are perpendicular.

5 $\mathbf{v} = (10\mathbf{i} - 3\mathbf{j})\,\mathrm{m\,s^{-1}}$

 $\mathbf{s} = (28\mathbf{i} + 4\mathbf{j})\,\mathrm{m}$

6 Speed $= 60.2\,\mathrm{m\,s^{-1}}$

 $\mathbf{s} = (3\mathbf{i} + \mathbf{j}) \times 2 + \dfrac{1}{2} \times (16\mathbf{i} + 24\mathbf{j}) \times 4$

 $= (38\mathbf{i} + 50\mathbf{j})$

 Final position $\mathbf{r} = (2\mathbf{i} + 6\mathbf{j}) + (38\mathbf{i} + 50\mathbf{j})$

 $= (40\mathbf{i} + 56\mathbf{j})$

 $\mathbf{r} = \dfrac{8}{7}\mathbf{v}$ so $\mathbf{v}$ has same direction as $\mathbf{r}$.

 Particle is moving directly away from O.

7 $\mathbf{u} = (-3\mathbf{i} + 2\mathbf{j})$, $\mathbf{a} = (2\mathbf{i} - 4\mathbf{j})$

8 **a** $(8\mathbf{i} + 12\mathbf{j})\,\mathrm{m}$

 b $(8\mathbf{i} - 15\mathbf{j})\,\mathrm{m\,s^{-1}}$

 c $(-2\mathbf{i} + 3.75\mathbf{j})\,\mathrm{m\,s^{-2}}$

 d $(54\mathbf{i} - 54\mathbf{j})\,\mathrm{m}$

9 **a** $\overrightarrow{OP} = (120t - 2t^2)\mathbf{i} + (400 - 4t^2)\mathbf{j}\,\mathrm{m}$

 b $10\,\mathrm{s}$

 c $1000\,\mathrm{m}$

Exercise 18.2A Fluency and skills

1 **a** $\mathbf{v} = 4\mathbf{i} + 4t\mathbf{j}$

 b $\mathbf{v} = (2t - 4)\mathbf{i} + (3t^2 - 4t)\mathbf{j}$

 c $\mathbf{v} = -2\sin t\mathbf{i} + 4\cos t\mathbf{j}$

 d $\mathbf{v} = e^t\mathbf{i} + \dfrac{1}{t+1}\mathbf{j}$

2 **a** $\mathbf{v} = (3t - t^2 + 3)\mathbf{i} + (t^2 - 1.5t^4)\mathbf{j}$

 b $\mathbf{v} = (2 + 2\sin 2t)\mathbf{i} + (5 - 4\cos 2t)\mathbf{j}$

3 **a** $\mathbf{a} = -10\mathbf{j}$

 $\mathbf{s} = (15t + 4)\mathbf{i} + (20t - 5t^2 - 3)\mathbf{j}$

 b $\mathbf{a} = -8\cos t\,\mathbf{i} + 8e^{2t}\,\mathbf{j}$

 $\mathbf{r} = (8\cos t - 7)\mathbf{i} + (2e^{2t} + 1)\mathbf{j}$

 c $\mathbf{a} = (2 - 12t)\mathbf{i} + (6t - 12t^2)\mathbf{j}$

 $\mathbf{r} = (t^2 - 2t^3 + 2)\mathbf{i} + (t^3 - t^4)\mathbf{j}$

4 $(5\mathbf{i} + 9\mathbf{j})\,\mathrm{m\,s^{-1}}$

5 $-\pi\mathbf{i}\,\mathrm{m\,s^{-1}}$

6 $(6\mathbf{i} + 11\mathbf{j})\,\mathrm{m}$

7 **a** $(3(t^2 + 1)\mathbf{i} + 2(t + 2)\mathbf{j})\,\mathrm{m\,s^{-1}}$

 b $17\,\mathrm{m\,s^{-1}}$

 c $28.1°$ to the x-direction

8 $\mathbf{v} = \left(12\mathbf{i} + \dfrac{8}{3}\mathbf{j}\right)\mathrm{m\,s^{-1}}$, $\mathbf{a} = \left(12\mathbf{i} - \dfrac{8}{9}\mathbf{j}\right)\mathrm{m\,s^{-2}}$

9 **a** $-18\sin 3t\mathbf{i} + 16\cos 2t\mathbf{j}$

 b $(2\sin 3t + 2)\mathbf{i} - (4\cos 2t + 1)\mathbf{j}$

Exercise 18.2B Reasoning and problem-solving

1 **a** $3.16\,\mathrm{m\,s^{-1}}$ **b** $8\mathbf{i} + \dfrac{34}{3}\mathbf{j}$

2 $\sqrt{\dfrac{656}{5}}\,\mathrm{m}$

3 **a** $\mathbf{v} = (3\mathbf{i} + (4 - 4t)\mathbf{j})\,\mathrm{m\,s^{-1}}$

 $\mathbf{a} = -4\mathbf{j}\,\mathrm{m\,s^{-2}}$

 b $5\,\mathrm{m\,s^{-1}}$

c $1\,\mathrm{s}$

d No; to be stationary $\mathbf{v}$ must be $0\mathbf{i} + 0\mathbf{j}$, but x-component is always 3.

4 **a** $\dfrac{\mathrm{d}\mathbf{v}}{\mathrm{d}t} = \mathbf{a} \implies \mathbf{v} = \displaystyle\int \mathbf{a}\ \mathrm{d}t = \mathbf{a}t + \mathbf{c}$

 When $t = 0$, $\mathbf{v} = \mathbf{u}$, so $\mathbf{c} = \mathbf{u}$

 giving $\mathbf{v} = \mathbf{u} + \mathbf{a}t$

 $\mathbf{s} = \displaystyle\int \mathbf{v}\ \mathrm{d}t$

 b $= \displaystyle\int \mathbf{u} + \mathbf{a}t\ \mathrm{d}t$

 $= \mathbf{u}t + \dfrac{1}{2}\mathbf{a}t^2 + \mathbf{c}$

 When $t = 0$, $\mathbf{s} = \mathbf{0}$, so $\mathbf{c} = \mathbf{0}$

 giving $\mathbf{s} = \mathbf{u}t + \dfrac{1}{2}\mathbf{a}t^2$

5 $\mathbf{v} = (70\mathbf{i} - 6\mathbf{j})\,\mathrm{m\,s^{-1}}$

 $\mathbf{r} = (134\mathbf{i} + 21\mathbf{j})\,\mathrm{m}$

6 **a** $t = \dfrac{\pi}{2}$

 b $\mathbf{v} = (\mathbf{i} - 2\mathbf{j})$, $\mathbf{a} = \mathbf{0}$

7 **a** $24.1\,\mathrm{m\,s^{-1}}$

 b $30.0\,\mathrm{m}$

8 **a** $\mathbf{v} = \displaystyle\int \mathbf{a}\ \mathrm{d}t$

 $= -2\sin 2t\mathbf{i} + 2\cos 2t\mathbf{j} + \mathbf{c}$

 When $t = 0$, $\mathbf{v} = 2\mathbf{j}$, so $\mathbf{c} = \mathbf{0}$

 $\mathbf{v} = -2\sin 2t\mathbf{i} + 2\cos 2t\mathbf{j}$

 giving speed $= \sqrt{4(\sin^2 2t + \cos^2 2t)}$

 $= 2\,\mathrm{m\,s^{-1}}$

 Therefore constant speed.

 b Distance $= 1\,\mathrm{m}$

 Path is a circle, radius $1\,\mathrm{m}$, about the origin.

9 **a** $t = 4.20\,\mathrm{s}$

 b $t = 2.11\,\mathrm{s}$

10 **a** $t = 2$

 b $\mathbf{r} = \dfrac{8}{7}\mathbf{v}$

 $\mathbf{v}$ and $\mathbf{r}$ point the same way, so particle is moving away from O

Exercise 18.3A Fluency and skills

1 **a** $1.76\,\mathrm{s}$ **b** $21.6\,\mathrm{m}$ **c** $0.878\,\mathrm{s}$ **d** $3.78\,\mathrm{m}$

2 **a** $1.77\,\mathrm{s}$ **b** $8.83\,\mathrm{m}$ **c** $0.883\,\mathrm{s}$ **d** $3.82\,\mathrm{m}$

3 **a** $17.2\,\mathrm{m}$

 b Speed $= 15.4\,\mathrm{m\,s^{-1}}$

 Direction $= -2.57°$, that is, $2.57°$ below the horizontal

4 **a** **i** $(30\mathbf{i} + 35.1\mathbf{j})\,\mathrm{m}$ **ii** $(60\mathbf{i} + 60.4\mathbf{j})\,\mathrm{m}$

 b $8.16\,\mathrm{s}$

 c $245\,\mathrm{m}$

 d $y = \dfrac{4x}{3} - \dfrac{49x^2}{9000}$

5 $\alpha = 30.0°$

6 $31.6\,\mathrm{m\,s^{-1}}$

Exercise 18.3B Reasoning and problem-solving

1 **a** $64.8\,\mathrm{m}$

 b Assumes stone is a particle, no air resistance, beach is horizontal.

2 **a** $22.6°$

b The available headroom would be reduced by the radius of the ball, so the maximum angle would be less.

3 a $\dfrac{50V^2 \sin^2 40°}{981}$ **b** $21.8\,\text{m s}^{-1}$ **c** $47.7\,\text{m}$

4 $2.67\,\text{s}$

5 $28.5\,\text{m}$

6 a Speed $= 54.3\,\text{m s}^{-1}$
　　Direction $= 46.3°$
b Speed $= 42.3\,\text{m s}^{-1}$
　　Direction $= 27.6°$
c $5.61\,\text{s}$

7 $v_y = u \sin \alpha - gt = 0$ for max height $\Rightarrow t = \dfrac{u \sin \alpha}{g}$

$y = ut \sin \alpha - \dfrac{1}{2} gt^2 \Rightarrow h = u\left(\dfrac{u \sin \alpha}{g}\right) - \dfrac{1}{2} g\left(\dfrac{u \sin \alpha}{g}\right)^2$

$\qquad\qquad\qquad\qquad\qquad = \dfrac{u^2 \sin^2 \alpha}{2g}$

8 a $u = 60\,\text{m s}^{-1}$, $v = 9.20\,\text{m s}^{-1}$
b $0.33\,\text{m}$

9 a $x = ut \cos \alpha \Rightarrow t = \dfrac{x}{u \cos \alpha}$

$y = ut \sin \alpha - \dfrac{1}{2} gt^2$

$\quad = u\left(\dfrac{x}{u \cos \alpha}\right) \sin \alpha - \dfrac{1}{2} g\left(\dfrac{x}{u \cos \alpha}\right)$

$\quad = x \tan \alpha - \left(\dfrac{g \sec^2 \alpha}{2u^2}\right) x^2$

b Let range be R. Projectile passes through (a, b), (b, a) and $(0, R)$.

$b = a \tan \alpha - \left(\dfrac{g \sec^2 \alpha}{2u^2}\right) a^2$ ...[1]

$a = b \tan \alpha - \left(\dfrac{g \sec^2 \alpha}{2u^2}\right) b^2$ ...[2]

$0 = R \tan \alpha - \left(\dfrac{g \sec^2 \alpha}{2u^2}\right) R^2$ ...[3]

Substitute $\tan \alpha$ from [3] into [1] and [2]:

$b = aR\left(\dfrac{g \sec^2 \alpha}{2u^2}\right) - \left(\dfrac{g \sec^2 \alpha}{2u^2}\right) a^2$ ...[4]

$a = bR\left(\dfrac{g \sec^2 \alpha}{2u^2}\right) - \left(\dfrac{g \sec^2 \alpha}{2u^2}\right) b^2$ ...[5]

Divide [5] by [4]:

$\dfrac{a}{b} = \dfrac{bR - b^2}{aR - a^2} \Rightarrow R = \dfrac{a^3 - b^3}{a^2 - b^2} = \dfrac{(a-b)(a^2 + ab + b^2)}{(a-b)(a+b)}$

$\qquad\qquad\qquad\qquad\qquad = \dfrac{a^2 + ab + b^2}{a+b}$

Exercise 18.4A Fluency and skills

1 a $7.21\,\text{m s}^{-2}$
b $33.7°$
2 a $4.47\,\text{m s}^{-2}$
b $26.6°$
3 $P = 32.5\,\text{N}$, $Q = 19.7\,\text{N}$
4 a $17.2\,\text{N}$
b $11.1\,\text{m}$
5 a $68.4\,\text{N}$
b $1.40\,\text{m s}^{-2}$
6 a $38.0°$
b $0.959\,\text{m s}^{-2}$

c $4.39\,\text{m s}^{-1}$
7 a $2.80\,\text{m s}^{-2}$
b $21.3\,\text{N}$
8 Distance $= 77.5\,\text{m}$
　Bearing $= 025°$ (to nearest degree)
9 Magnitude $= 9.97\,\text{m s}^{-2}$
　Bearing $= 108°$ (to nearest degree)
10 a $37.6\,\text{N}$
b $4.46\,\text{m s}^{-2}$

Exercise 18.4B Reasoning and problem-solving

1 a $62.5\,\text{N}$
b $2.02\,\text{m s}^{-2}$
2 $7.70\,\text{m s}^{-2}$
3 $T_1 = 4098\,\text{N}$, $T_2 = 2698\,\text{N}$
4 a Resolve vertically:
$R - mg + mg \sin \theta = 0$
$R = mg(1 - \sin \theta)$ ...[1]
Resolve horizontally:
$mg \cos \theta - 0.75R = ma$ ...[2]
Substitute from [1] into [2]:
$ma = mg \cos \theta - 0.75mg(1 - \sin \theta)$
$a = \dfrac{g(3 \sin \theta + 4 \cos \theta - 3)}{4}$

b $\dfrac{\mathrm{d}a}{\mathrm{d}\theta} = \dfrac{g}{4}(3 \cos \theta - 4 \sin \theta)$
$\quad = 0$ for turning point

$3 \cos \theta = 4 \sin \theta \Rightarrow \tan \theta = \dfrac{3}{4} \Rightarrow \theta = 36.9°$

To prove a maximum, $\dfrac{\mathrm{d}^2 a}{\mathrm{d}\theta^2} = \dfrac{g}{4}(-3 \sin \theta - 4 \cos \theta)$ which is negative for $\theta = 36.9°$. Hence, a maximum.

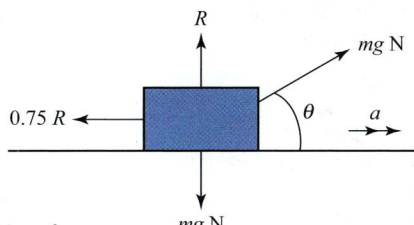

5 $2.84\,\text{m s}^{-2}$
6 $18.2\,\text{m}$
7 $24.7\,\text{m}$
8 $T_1 = 508\,\text{N}$, $T_2 = 584\,\text{N}$

Review exercise 18

1 a $\mathbf{s} = 25\mathbf{i} + 10\mathbf{j}$
b $\mathbf{s} = 30\mathbf{i} - 54\mathbf{j}$
c $\mathbf{a} = -2\mathbf{i} + 3\mathbf{j}$
d $\mathbf{v} = 4.2\mathbf{i} - 5.2\mathbf{j}$
2 $(21\mathbf{i} + 43\mathbf{j})\,\text{m}$
3 $\mathbf{v} = 10\mathbf{i} + 20\mathbf{j}$
　$\mathbf{a} = 6\mathbf{i} + 22\mathbf{j}$
4 $-13\mathbf{i} + 10\mathbf{j}$
5 $\mathbf{a} = 4 \sin 4t\,\mathbf{i} + 4 \cos 2t\,\mathbf{j}$
When $t = \dfrac{\pi}{4}$, $\mathbf{v} = \mathbf{i} + 2\mathbf{j}$ and $\mathbf{r} = 3\mathbf{i} + \mathbf{j}$

$\mathbf{v} = \int \mathbf{a}\,\mathrm{d}t$

$\quad = -\cos 4t\,\mathbf{i} + 2 \sin 2t\,\mathbf{j} + \mathbf{c}$

When $t = \dfrac{\pi}{4}$, $\mathbf{v} = \mathbf{i} + 2\mathbf{j}$

$$\mathbf{i}+2\mathbf{j}=-\cos\pi\mathbf{i}+2\sin\frac{\pi}{2}\mathbf{j}+\mathbf{c}$$

$$\mathbf{i}+2\mathbf{j}=\mathbf{i}+2\mathbf{j}+\mathbf{c}$$

$$\mathbf{c}=0$$

$$\mathbf{v}=-\cos4t\mathbf{i}+2\sin2t\mathbf{j}$$

$$\mathbf{r}=\int\mathbf{v}\,\mathrm{d}t$$

$$=-\frac{1}{4}\sin4t\mathbf{i}-\cos2t\mathbf{j}+\mathbf{c}$$

When $t=\dfrac{\pi}{4}$, $\mathbf{r}=3\mathbf{i}+\mathbf{j}$

$$3\mathbf{i}+\mathbf{j}=-\frac{1}{4}\sin\pi\mathbf{i}-\cos\frac{\pi}{2}\mathbf{j}+\mathbf{c}$$

$$3\mathbf{i}+\mathbf{j}=\mathbf{c}$$

$$\mathbf{r}=\left(-\frac{1}{4}\sin4t+3\right)\mathbf{i}-\left(\cos2t+1\right)\mathbf{j}$$

When $t=\dfrac{\pi}{2}$

$$\mathbf{r}=\left(-\frac{1}{4}\sin2\pi+3\right)\mathbf{i}-\left(\cos\pi-1\right)\mathbf{j}$$

$$\mathbf{r}=3\mathbf{i}+2\mathbf{j}$$

Distance from origin is $|\mathbf{r}|=\sqrt{3^2+2^2}$
$$=\sqrt{13}$$

6 **a** $1.03\,\mathrm{s}$
   b $11.3\,\mathrm{m}$
   c $0.517\,\mathrm{s}$
   d $1.31\,\mathrm{m}$

7 **a** $91.7\,\mathrm{m}$
   b $y=x-\dfrac{9.8x^2}{900}$

8 **a** $P=29.3\,\mathrm{N}$
   b $54.4\,\mathrm{m}$

Assessment 18

1 **a** $10\mathbf{i}+24\mathbf{j}\,\mathrm{m\,s^{-1}}$
   b $26\,\mathrm{m\,s^{-1}}$

2 $2\sqrt{3}\,g=33.9\,\mathrm{N}$ (to 3 sf)

3 **a** $\mathbf{i}-2\mathbf{j}$
   b $6\sqrt{2}\,\mathrm{m}=8.49\,\mathrm{m}$ (to 3 sf)

4 **a** $44.1\,\mathrm{m}$
   b $305\,\mathrm{m}$

5 **a** $122.5\,\mathrm{m}$
   b $100\,\mathrm{m}$
   c $67.8°$ below horizontal

6 **a** $\mathbf{i}-2\sqrt{3}\,\mathbf{j}\,\mathrm{m\,s^{-1}}$
   b $-2\sqrt{3}\,\mathbf{i}+8\mathbf{j}\,\mathrm{m\,s^{-2}}$

7 **a** $-4\mathbf{i}+9\mathbf{j}\,\mathrm{m\,s^{-1}}$
   b $-\mathbf{i}+2\mathbf{j}\,\mathrm{m\,s^{-2}}$
   c $1.5\,\mathrm{s}$

8 **a** $3g\,\mathrm{N}$
   b $3\sqrt{3}\,g\,\mathrm{N}$

9 **a** $\begin{pmatrix}4.8\\5.6\end{pmatrix}\mathrm{N}$
   b $30\,\mathrm{m}$

10 **a** $12.1\,\mathrm{m}$
   b $7.00\,\mathrm{m\,s^{-1}}$
   c $31.0°$ below horizontal

11 **a** $\begin{pmatrix}16\\1\end{pmatrix}\mathrm{m\,s^{-2}}$
   b $\begin{pmatrix}0\\-\dfrac{1}{\sqrt{2}}\end{pmatrix}\mathrm{m\,s^{-1}}$
   c $\begin{pmatrix}2\\0\end{pmatrix}\mathrm{m}$

12 **a** $\mathbf{S}_\mathrm{P}=(4+3t)\,\mathbf{i}+(11-8t)\,\mathbf{j}$
   b $\mathbf{S}_\mathrm{Q}=(9-7t)\,\mathbf{i}+(3.5+12t)\,\mathbf{j}$
   c $\mathbf{S}_\mathrm{P}-\mathbf{S}_\mathrm{Q}=(-5+10t)\,\mathbf{i}+(7.5-20t)\,\mathbf{j}$
      $\left|\mathbf{S}_\mathrm{P}-\mathbf{S}_\mathrm{Q}\right|^2=(-5+10t)^2+(7.5-20t)^2$
   d $12.24\,\mathrm{pm}$
   e $\dfrac{\sqrt{5}}{2}\,\mathrm{km}$

13 **a** $x=u\cos\alpha\times t$
      $t=\dfrac{x}{u\cos\alpha}$
      $y=u\sin\alpha\times t-\dfrac{1}{2}gt^2$
      $y=u\sin\alpha\times\left(\dfrac{x}{u\cos\alpha}\right)-\dfrac{1}{2}g\left(\dfrac{x}{u\cos\alpha}\right)^2$
      $y=x\tan\alpha-\dfrac{gx^2}{2u^2}\sec^2\alpha$
   b $76.0°$ or $63.4°$

Chapter 19
Exercise 19.1A Fluency and skills

1 **a** **i** $X=-1.83\,\mathrm{N}$ (to 3 sf)
         $Y=3.17\,\mathrm{N}$ (to 3 sf)
      ii $R=3.7\,\mathrm{N}$ (to 2 sf)
         $\theta=30°$ to the nearest degree
   b **i** $X=42.8\,\mathrm{N}$ (to 3 sf)
         $Y=-1.67\,\mathrm{N}$ (to 3 sf)
      ii $R=43\,\mathrm{N}$ (to 2 sf)
         $\theta=92°$ to the nearest degree

2 $R=625\,\mathrm{N}$
   $\mu=0.485$ (to 3 sf)

3 $R=570\,\mathrm{N}$ (to 2 sf)
   $W=990\,\mathrm{N}$ (to 2 sf)

4

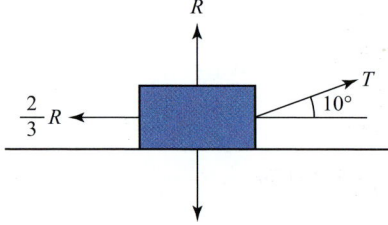

750 N

$(\rightarrow)$ $T\cos10°-\dfrac{2}{3}R=0$ (1)

$(\uparrow)$ $R+T\sin10°-750=0$ (2)

From (2) $R=750-T\sin10°$
and so (1) gives $T\cos10°-\dfrac{2}{3}\left(750-T\sin10°\right)=0$

So
$$T=\frac{750\times\frac{2}{3}}{\left(\cos10°+\frac{2}{3}\sin10°\right)}$$
$$=450\,\mathrm{N}\text{ (to 2 sf)}$$

$R=670\,\mathrm{N}$ (to 2 sf)

5 Magnitude $=113\,\mathrm{N}$ (to 3 sf)
   Bearing $=246°$ (to 3 sf)

6 $T = 156\,\text{N}$ (to 3 sf)

$\mu = 0.397$ (to 3 sf)

7 $W = 36.2\,\text{N}$ (to 3 sf)

$\mu = 0.567$ (to 3 sf)

Exercise 19.1B Reasoning and problem-solving

1 $S = 58.7\,\text{N}$ (to 3 sf)

$R = 40.2\,\text{N}$ (to 3 sf)

2 $W = 116\,\text{N}$ (to 3 sf)

3 a $102\,\text{N}$ (to 3 sf)

b $184\,\text{N}$ (to 3 sf)

4 $\mu = \dfrac{2}{3}$

5 $k = \sqrt{3} + 3$

6 $S = 40.0\,\text{N}$, $T = 56.6\,\text{N}$ (to 3 sf)

7 $\mu = \dfrac{1}{k}$

8 $\mu = 0.331$

9 a $m = 2$ **b** $m = 8$

10 $1320\,\text{N}$

Exercise 19.2A Fluency and skills

1 $X = 11.5\,\text{N}$ (to 3 sf)

$a = 11.5\,\text{m s}^{-2}$ (to 3 sf)

2 $146\,\text{m s}^{-2}$ (to 3 sf), $4°$

3 $2.95\,\text{m s}^{-2}$ (to 3 sf)

4 0.743 (to 3 sf)

5 $2.84\,\text{m s}^{-2}$ (to 3 sf) at $13.0°$ above the 55 N force

6 $0.89\,\text{m s}^{-2}$ (to 2 sf)

7 0.28 (to 2 sf)

Exercise 19.2B Reasoning and problem-solving

1 $0.72\,\text{m s}^{-2}$ (to 2 sf)

2 0.581 (to 3 sf)

3 $950\,\text{N}$ (to 2 sf)

4

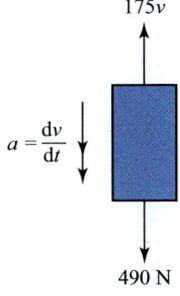

$(\downarrow) \quad 50\dfrac{\mathrm{d}v}{\mathrm{d}t} = 490 - 175v$

$\dfrac{\mathrm{d}v}{\mathrm{d}t} = 9.8 - 3.5v$

$= -3.5(v - 2.8)$

$v = 10\mathrm{e}^{-3.5t} + 2.8$

5

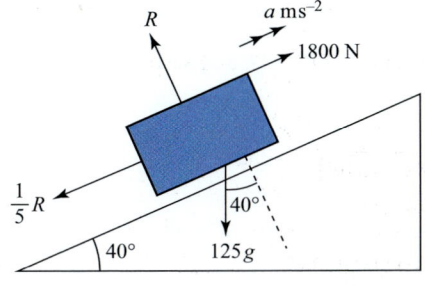

$\left(\nwarrow\right) R - 125g\cos 40° = 0$

So $R = 125g\cos 40°$

$\left(\nearrow\right) 1800 - \dfrac{1}{5}R - 125g\sin 40° = 125a$

$a = \dfrac{825.904\ldots}{125}$

$= 6.6\,\text{m s}^{-2}$ (to 2 sf)

$t = 1.7\,\text{s}$ (to 2 sf)

6 $X = 411\,\text{N}$ (to 3 sf)

$a = 8.71\,\text{m s}^{-2}$ (to 3 sf)

7

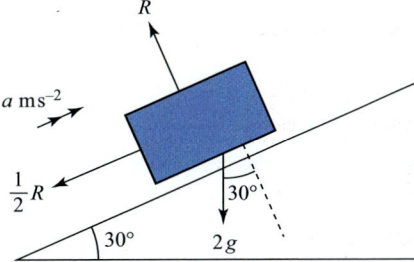

a $\left(\nwarrow\right) R - 2g\cos 30° = 0$

So $R = 2g\cos 30°$

$\left(\swarrow\right) -\dfrac{1}{2}R - 2g\sin 30° = 2a$

So

$a = -\dfrac{18.287\ldots}{2}$

$= -9.1\,\text{m s}^{-2}$ (to 2 sf)

So deceleration is $9.1\,\text{m s}^{-2}$

b $0.49\,\text{m}$ (to 2 sf)

c $0.66\,\text{m s}^{-2}$ (to 2 sf)

d $s = 0.492\ldots$ $u = 0$ $a = 0.656\ldots$

Using $v^2 = u^2 + 2as$ gives

$v = \sqrt{0^2 + 2 \times 0.656\ldots \times 0.492\ldots}$

$= 0.8\,\text{m s}^{-2}$ (to 1 sf)

8 $v = 30(1 - \mathrm{e}^{-0.15t})$

As $t \to \infty$, $\mathrm{e}^{-0.15t} \to 0$ and so $v \to 30\,\text{m s}^{-1}$

9 a

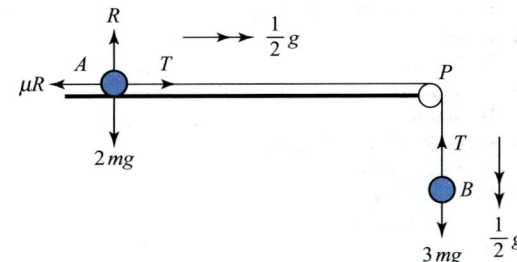

$3m \qquad (\downarrow) \qquad 3mg - T = 3m \times \dfrac{1}{2}g \qquad (1)$

$2m \qquad (\uparrow) \qquad R - 2mg = 0 \qquad (2)$

$\qquad\qquad (\to) \qquad T - \mu R = 2m \times \dfrac{1}{2}g \qquad (3)$

(2) in (3) gives $(\to) \qquad T - 2mg\mu = mg$

(1) $\qquad\qquad\qquad 3mg - T = \dfrac{3mg}{2}$

Adding these up gives

$3mg - 2mg\mu = \dfrac{3mg}{2} + mg$

$= \dfrac{5mg}{2}$

and so $\mu = \dfrac{1}{4}$

$$T = \dfrac{3mg}{2}$$

  b $3h$ m

Exercise 19.3A Fluency and skills

1 **a** -16 N m   **b** -30 N m

    c -1 N m   **d** -39 N m

2 2 N m

3 **a** The object will not turn.

    b The object will turn clockwise.

4 80 N

5 10 N m anticlockwise

6 350 N

7 $p = -3$ or $p = 2$

8 $x = -6$

Exercise 19.3B Reasoning and problem-solving

1 **a** At the edge of the table

    b 50 N

    c 100 N

2 **a** 1 m   **b** 10 kg

3 **a** 1220 N   **b** 50 N

4 **a** $80g$ at C, $70g$ at D

    b **i** 70 kg   **ii** 85.7 kg (to 3 sf)

5 **a** 56.3 (to 3 sf)   **b** 153 N (to 3 sf)

6 **a** 0.260 (to 3 sf)   **b** 206 N (to 3 sf)

7 **a** $(\mathbf{i} - 3\mathbf{j})$ **b** $(\mathbf{i} + 3\mathbf{j})$

Review exercise 19

1 0.08

2 **a** $20g$ at C, $15g$ at D   **b** 3.18 kg (to 3 sf)

3 **a** 1.34 m s^{-1} (to 3 sf)   **b** 1.25 m s^{-1}

4 **a** $\mu < \dfrac{2}{5}$ **b** 2 s

5 **a** $48.2°$ (to 3 sf)   **b** 52.6 N (to 3 sf)

Assessment 19

1 **a** $\mathbf{S} = -8\mathbf{i} + 15\mathbf{j}$ **b** $|\mathbf{S}| = 17$

2 **a** $F = 10g$ **b** $10\sqrt{3}\,g$

3 **a** 122 **b** 0.723

4 Magnitude $= 4\sqrt{5} = 8.94$

   Direction is $63.4°$ below $\mathbf{i}$

5 **a** 45 N   **b** 15 N

6 **a** $\dfrac{5g\sqrt{3}}{2}$ **b** $g \le P \le 4g$

7 **a**

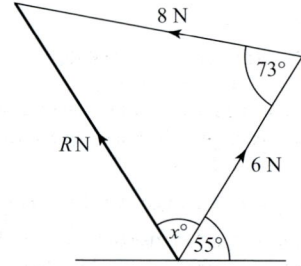

    8.48

    b $330.6°$

8 **a** $\dfrac{5\sqrt{2}}{(\sqrt{3} + 1)}\,g$ or equivalent exact form

    b $\dfrac{10}{(\sqrt{3} + 1)}\,g$ or equivalent exact form

9 **a** 1.8 m s^{-2} **b** 5.7 m s^{-1}

10 **a** 80 **b** 4.56 m

11 **a** $27°$ **b** 110 N

12 11.4

13 **a** $25g$

    b $F = T\cos x$

       $F = \dfrac{4}{5} \times T$

       $F = 20g$

       $2R = 30g + 20g + 20g$

       $R = 35g$

       $20g = \mu \times 35g$

       $\mu = \dfrac{4}{7}$

14 **a** $10\sqrt{3}\,g$ **b** $8\sqrt{3}\,g$ **c** $12g$

15 **a** -3.48 m s^{-2}

    b 2.30 m

    c $mg\sin 15° > \dfrac{mg\cos 15°}{10}$

    Component of weight down the slope exceeds $F_{\max}$ so the particle does start to move back down the slope.

16 **a** 40 N   **b** 60 N   **c** $4\dfrac{1}{3}$ m

17 **a** $\dfrac{g}{10}$ **b** $\dfrac{11}{5}g$ **c** $\dfrac{36}{5}g$

18 **a** **i** $\dfrac{7}{6}Mg$ **ii** $\dfrac{Mg}{3}$

    b **i** 1.5 **ii** $\dfrac{5Mg}{2}$

19 **a** $\dfrac{2g}{5}$ **b** $\sqrt{\dfrac{4g}{5}}$ **c** $\dfrac{1}{2}$ m

Assessment: Mechanics

1 **a**

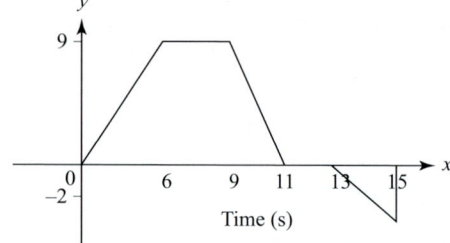

    b 61 m

2 **a** $\dfrac{9}{8}g$ **b** $\dfrac{3}{2}g$ **c** $\dfrac{9}{8}g$

3 $X = \dfrac{3\sqrt{3}}{2}$ N, $Y = \dfrac{3}{2}$ N

4 1225 m

5 **a** **i** $42g$ or 411.6 N   **ii** $0.8g$

    b 0.4 m

6 **a** **i** $0.4g$ or 3.92 m s^{-2} **ii** $\dfrac{12}{25}g$ or 4.71 N

    b **i** String has no mass and does not stretch.

       ii There is no frictional force.

7 **a** $\begin{pmatrix} -1 \\ 2 \end{pmatrix}$ **b** 5.59 m

8 **a** 20.6 m s^{-1} **b** 22.0 s

9 **a** -1.73 **b** $\dfrac{t^2}{2} + \dfrac{1}{2}\sin 2t - \dfrac{\pi^2}{2}$

10 **a** $6\mathbf{i} + 2\mathbf{j}$ **b** $\left(\dfrac{t^3}{3} + 3\right)\mathbf{i} + (t^2 - 5)\mathbf{j}$

11 **a** **i** 6.18 m s^{-1} **ii** 5.10 m s^{-2}

    b 54 m

12 a $\theta = 52.6$, $P = 17.1$
b $\theta = 72.1$, $P = 14.3$
13 a **i** $F = 30\,\text{N}$ **ii** $F = 48.1\,\text{N}$
b $1.7\,\text{m}\,\text{s}^{-2}$
14 a $R_D = 2.4\,\text{N}$, $R_C = 0.6\,\text{N}$
b Beam is light so has no mass.
15 a At A: $R = 173\,\text{N}$
At B: $R = 238\,\text{N}$
b Weight acts at the centre of the rod.
16 a $\begin{pmatrix} -1 \\ 0.375 \end{pmatrix}\text{m}\,\text{s}^{-2}$ **b** $4.27\,\text{m}\,\text{s}^{-1}$ **c** $291°$
17 a $3.46\,\text{m}\,\text{s}^{-1}$ **b** $\begin{pmatrix} -4 \\ 12\sqrt{3} \end{pmatrix}$
18 a $1.06\,\text{s}$ **b** $1.38\,\text{m}$
c $3.18\,\text{m}$
d Ball is a particle/no spin. No air resistance/wind.
19 0.618
20 a $7.99\,\text{N}$ **b** $1.31\,\text{m}\,\text{s}^{-2}$
21 a **i** $R_B = 7g$ or $68.6\,\text{N}$
$R_A = 14g$ or $137.2\,\text{N}$
ii $55.7\,\text{cm}$
b The value of x will decrease as centre of mass will be $2.5\,\text{cm}$ from end of the shelf.
22 $T_2 = 1.50\,\text{N}$, $T_1 = 6.19\,\text{N}$
23 $3.85\,\text{m}$
24 a **i** $\dfrac{25}{1.5\cos\theta}$ **ii** $\theta = 27.7°$
b $17.7\,\text{m}\,\text{s}^{-1}$ at $19.7°$ above the horizontal
25 a **i** $37880\,\text{N}$ **ii** $10823\,\text{N}$
b $2.85\,\text{s}$
c $22130\,\text{N}$
26 a **i** $8.1°$, $81.9°$ **ii** $0.61\,\text{s}$, $4.24\,\text{s}$
b Would travel less distance due to air resistance.
27 a $T_B = 26g$, $T_A = 4g$ **b** $3.47\,\text{m}$
28 $72.9\,\text{m}$
29 a $14.9° < \theta < 86.1°$
b $14.9° < \theta < 18.6°$ or $69.8° < \theta < 86.1°$
30 $90\,\text{m}$

Chapter 20
Exercise 20.1A Fluency and skills

1 a $\dfrac{1}{2}$ **b** $\dfrac{1}{8}$ **c** $\dfrac{4}{11}$

2

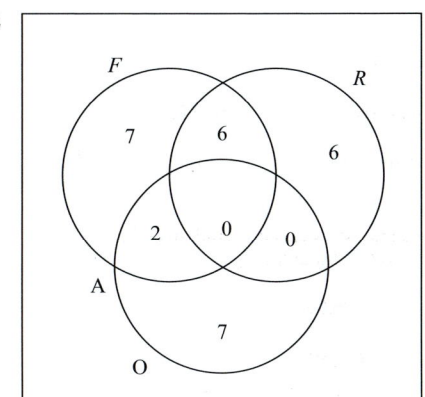

a $\dfrac{3}{7}$ **b** $\dfrac{19}{28}$ **c** $\dfrac{2}{9}$

3 a $\dfrac{1}{6}$ **b** $\dfrac{1}{2}$ **c** $\dfrac{1}{3}$

4 $\dfrac{48}{91}$

5 a $\dfrac{3}{20}$ **b** $\dfrac{5}{8}$

6 a $\dfrac{4}{7}$ **b** $\dfrac{1}{3}$ **c** $\dfrac{3}{7}$

d $\dfrac{2}{3}$ **e** 0.8

7 a $\dfrac{1}{2}$ **b** $\dfrac{1}{6}$ **c** $\dfrac{1}{3}$ **d** $\dfrac{1}{4}$

e $\dfrac{7}{12}$ **f** $\dfrac{2}{3}$

8 a $\dfrac{7}{20}$ **b** 0.45

Exercise 20.1B Reasoning and problem-solving

1 a

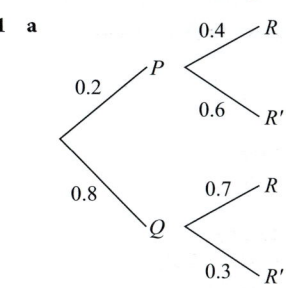

$P(R) = 0.64$

b $\dfrac{1}{8}$

2 a 0.3194 **b** 0.844 (3 dp)
3 a 0.215 **b** 0.377 (3 dp)

4 a $\dfrac{7}{31}$ **b** $\dfrac{7}{12}$

5 a $\dfrac{19}{61}$ **b** $\dfrac{32}{61}$ **c** $\dfrac{19}{32}$

6 a $\dfrac{2}{3}$ **b** $\dfrac{1}{2}$

7 a $\dfrac{35}{72}$ **b** $\dfrac{2}{7}$

8 a $\dfrac{5}{39}$ **b** $\dfrac{5}{13}$

9 $\dfrac{101}{201}$

Exercise 20.2A Fluency and skills

1 a $\mu = 3.025$, $\sigma^2 = 1.074$ (3dp) OR $s^2 = 1.102$ (3dp)
b Using $X \sim B(n, p)$, mean = 3, variance = 1.2
Using sample, mean = $3.025 \approx 3$, $\sigma^2 = 1.074 \approx 1.2$ OR $s^2 = 1.102 \approx 1.2$. The values given by the sample suggest this model is a good fit for the data.

2 $X \sim B\left(300, \dfrac{1}{3}\right)$. Proportion of scores greater than 2 is $\dfrac{1}{3}$.
A binomial model is appropriate because there is a fixed number of independent trials (300), each trial has two possible outcomes ('greater than 2' or 'less than or equal to 2'), and the same dice is used each time so the probabilities should remain constant.

3 Using $B(3, 0.2)$, $P(X = x) = 0.512, 0.384, 0.096, 0.008$ for $x = 0, 1, 2, 3$
Good fit to data $\Rightarrow$ good model

Exercise 20.2B Reasoning and problem-solving

1 a $X \sim B(30, 0.5)$. A binomial model is appropriate because there is a fixed number of independent trials (30), each trial has two possible outcomes ('head' or 'not head') and the same coin is used each time so the probability should remain constant. Heads and not heads are assumed to be equally likely which gives $p = 0.5$.

b The coin may not be fair, in which case $p = 0.5$ would be inaccurate.

2 a i 0.5

ii 0.75

Leuchars is on the east coast of Scotland so Town A in England will be to the south of it, and may be a great distance away. This means the weather in Town A might not be similar to the weather in Leuchars.

b From the Town A data $Q_1 = 14.6°C\,(1dp)$, $Q_2 = 16.2°C\,(1dp)$, $Q_3 = 17.9°C\,(1dp)$. The median, lower quartile and upper quartile are all about 2.4°C higher in Town A than Leuchars but the interquartile range is the same at 3.3°C. This suggests Town A was warmer than Leuchars in general that summer but the variation in temperature was about the same. The higher temperature might be due to the more southern location.

3 a Sian assumes that the later pages in her notes will take the same amount of time to review as the earlier pages, but these are likely to cover more challenging material so may take longer. Sian assumes she will work at the same rate over 8 hours that she did over 90 minutes, but it's likely that over a longer period she will need more breaks or get tired and work more slowly. Sian assumes a very small standard deviation but over a period of three weeks a number of unexpected events might happen which could change her working pattern, such as illness or family responsibilities.

b Sian's mean will reduce to account for a slower working pace. Sian's standard deviation will increase if she considers the high likelihood of unexpected events.

Exercise 20.3A Fluency and skills

1 a 0.8413   **b** 0.0918   **c** 0.8186
2 a 0.7734   **b** 0.2902   **c** 0.8186
3 a 0.65   **b** −2   **c** 0.65
4 a $\dfrac{5}{4}$ **b** 2   **c** 0

   d $\dfrac{25}{17}$ **e** $-1\dfrac{7}{23}$ **f** −10

5 10.46 (2dp)
6 16.95 (2dp)
7 0.7499 (2dp)

Exercise 20.3B Reasoning and problem-solving

1 a $\mu_R = 13.43\,(2dp)$
b $\sigma_S = 7.90\,(2dp)$
c $\mu_T = 12.92$
   $\sigma_T = 3.96\,(2dp)$

2 a Let T be a random variable for the time taken, in seconds, for a randomly chosen runner. $T \sim N(730, 80^2)$.

b 10 minutes 27 seconds

3 a $X \sim N(\mu, \sigma^2) \Rightarrow P(\mu - 2\sigma < X < \mu + 2\sigma) = P(-2 < Z < 2)$
$$= 2P(0 < Z < 2)$$
$$= 2(0.9772 - 0.5)$$
$$= 0.95$$

b For a normal population, mean ≈ mode ≈ median and

~95% of data lies within 2 standard deviations of the mean.
For this data $275 \approx 270 \approx 265$ and $93\% \approx 95\%$
Reaction data gives very good fit to these statistics. Therefore, evidence suggests that a normal distribution would provide a good model.

4 a $\mu = 8.25$, $\sigma = 2.91$

b 69.8%

c This distribution could be modelled by a Normal distribution as close to 68% of values lie within one standard deviation of the mean, the distribution is roughly symmetrical and the mean 8.25 is close to the modal class $6 \le m < 8$.

5 9.31 (2dp)

6 a i 0.1587

ii 0.1587

iii $P(W > \mu + \sigma) = P\left(Z > \dfrac{(\mu + \sigma) - \mu}{\sigma}\right)$
$$= P(Z > 1)$$
$$\approx P(1 < Z < \mu + 5\sigma)$$
$$= P(1 < Z < 5)$$
$$= 0.1587$$

iv $P(W < \mu - n\sigma) = P\left(Z < \dfrac{(\mu - n\sigma) - \mu}{\sigma}\right)$
$$= P(Z < -n)$$

b 1.1

7 a Median is approximately equal to mean.
mean ± standard deviation = (1008, 1028), the proportion within this range is $\dfrac{45}{61} = 74\%$ which is close to the Normal distribution proportion of 68%

b 0.375

8 T – waiting time in minutes. $T \sim N(\mu, \sigma^2)$

a $P(T > 5) = P\left(Z > \dfrac{5 - \mu}{\sigma}\right)$
$$= 0.14$$
$$\dfrac{5 - \mu}{\sigma} = 1.0801$$
$P(T < 1) = P\left(Z < \dfrac{1 - \mu}{\sigma}\right)$
$$= 0.08$$
$$\dfrac{1 - \mu}{\sigma} = -1.4065$$
$$\mu = 3.26, \ \sigma = 1.61$$
$$P(T > 6) = 4.439\% > 2.5\%$$
Requirement not met.

b Maximum mean waiting time is 2.84 minutes (2dp)

Exercise 20.4A Fluency and skills

1 a 0.0804   **b** 0.7866   **c** 0.7463
2 a i $P(22.5 < Y < 23.5)$
ii $P(X = 23) = 0.0611$
   $P(22.5 < Y < 23.5) = 0.0604$
   $0.0611 \approx 0.0604$
b i $P(Y < 50.5)$
ii $P(X \le 50) = 0.999...$
   Lower bound $= 27 - 5 \times \sqrt{14.85} = 7.73$
   $P(Y < 50.5) \approx P(7.73 < Y < 50.5) = 0.999...$
   $P(X \le 50) \approx P(7.73 < Y < 50.5)$
c i $P(Y > 11.5)$
ii $P(X \ge 12) = 1 - P(X \le 11)$

$= 0.999986...$

Upper bound $= 27 + 5 \times \sqrt{14.85} = 46.27$

$P(11.5 < Y < 46.27) = 0.999970...$

$0.999986... \approx 0.999970...$

d **i** $P(31.5 < Y < 51.5)$

  ii $P(32 \leq X \leq 51) = P(X \leq 51) - P(X \leq 31)$

$\qquad\qquad\qquad\qquad\quad = 1 - 0.8783...$

$\qquad\qquad\qquad\qquad\quad = 0.12165...$

$\qquad P(31.5 < Y < 51.5) = 0.12145...$

$\qquad 0.12165... \approx 0.12145...$

e **i** $P(X > 17) = P(X \geq 18)$

$\qquad P(Y > 17.5)$

  ii $P(X > 17) = 1 - P(X \leq 17)$

$\qquad\qquad\qquad = 1 - 0.00608...$

$\qquad\qquad\qquad = 0.99391...$

$\qquad$ Upper bound $= 27 + 5 \times \sqrt{14.85} = 46.27$

$\qquad P(17.5 < Y < 46.27) = 0.99315...$

$\qquad 0.99391... \approx 0.99315...$

f **i** $P(X < 40) = P(X \leq 39)$

$\qquad P(Y < 39.5)$

  ii $P(X < 40) = P(X \leq 39)$

$\qquad\qquad\qquad = 0.99941...$

$\qquad$ Lower bound $= 27 - 5 \times \sqrt{14.85} = 7.73$

$\qquad P(Y < 39.5) \approx P(7.73 < Y < 39.5) = 0.99940...$

$\qquad 0.99941... \approx 0.99940...$

3 0.608

4 0.029 (3dp)

5 0.428

Exercise 20.4B Reasoning and problem-solving

1 Median value (20) far from mean.

Large standard deviation suggesting negative lifetimes.

2 $X \sim B(30, 0.4)$

  a $Y \sim N(12, 7.2)$

$\qquad P(12 - \sqrt{7.2} < Y < 12 + \sqrt{7.2})$

$\qquad = 0.682$

  b Very good approximation as about 68% expected if exactly normal.

3 **a** If coin is fair, $P(7) = 0.1172$

$\qquad$ No reason to suggest coin is biased.

  b X – number of heads

$\qquad$ If fair coin, $X \sim B(10000, 0.5)$

$\qquad Y \sim N(5000, 2500)$

$\qquad P(Y \geq 5100) = 0.023$

$\qquad$ Coin is probably biased.

4 **a** 0, minimum possible value equals mode.

  b **i** Pascal (Pa) 1 hPa = 100 Pa

$\qquad$ **ii** 0.28 (2dp)

$\qquad$ **iii** 0.48 (2dp)

5 Y – vegetable content in g

Assume $Y \sim N(98, 100)$ so $P(Y > 98) = 0.5$

X – number of pies out of 10 with vegetable content over 98 g

$X \sim B(10, 0.5)$

Mean of X is $10 \times 0.5 = 5$, variance $10 \times 0.5 \times 0.5 = 2.5$

Data has mean 3.1 and variance $2.1^2 = 4.41$

Discrepancy suggests normal model is not a good one.

6 X – number of passengers who turn up

$X \sim B(N, 0.94)$ where N is the number of tickets sold

$Y \sim N(0.94N, 0.0564N)$

$P(\text{accommodate all passengers}) = P(X \leq 300) = P(Y < 300.5)$

$$= P\left(Z < \frac{300.5 - 0.94N}{\sqrt{0.0564N}} \right) > 0.99$$

$$\frac{300.5 - 0.94 \times N}{\sqrt{0.0564 \times N}} > 2.326$$

Test $N = 309$ and $N = 310$

$$\frac{300.5 - 0.94 \times 309}{\sqrt{0.0564 \times 309}} = 2.405$$

$$\frac{300.5 - 0.94 \times 310}{\sqrt{0.0564 \times 310}} = 2.176$$

Therefore $N = 309$

Review exercise 20

1 **a** $\dfrac{1}{2}$ **b** $\dfrac{23}{45}$

2 **a**

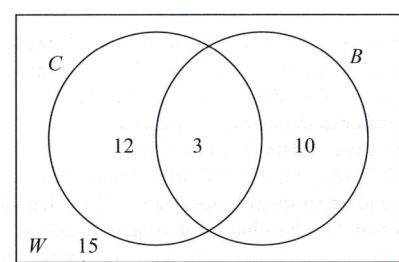

  b **i** $\dfrac{3}{40}$ **ii** $\dfrac{13}{25}$

3 $\dfrac{5}{8}$

4 **a**

x	0	1	2	3	4
P(X = x)	0.3164	0.4219	0.2109	0.0469	0.0039

  b

x	0	1	2	3	4
expected f	25.312	33.752	16.872	3.752	0.312

Good comparison with data suggests spinner is fair.

5 **a** **i** 0.1241   **ii** 0.8342

  b $a = 16.675$

6 0.74988

7 Median $= 279\,s$

Mode, median and mean approximately equal and about $\dfrac{2}{3}$ of

all observations within one standard deviation of the mean so

Normal is a good model.

8 0.2973 (4dp)

Assessment 20

1 **a**

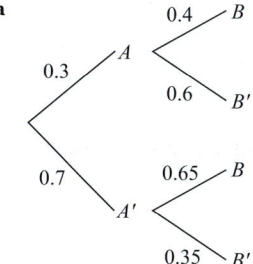

  b 0.365

c

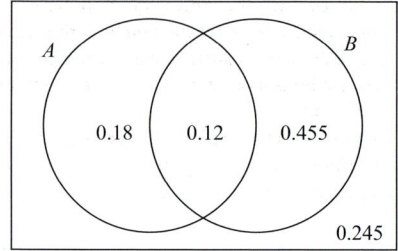

2 a i $\dfrac{11}{92}$ **ii** $\dfrac{10}{11}$ **iii** $\dfrac{141}{184}$ **iv** $\dfrac{20}{141}$

b The probability that the day had a heatwave given that it was dry is very low. You wouldn't assume there was a heatwave, even though the probability it was dry given that there was a heatwave is very high.

3 a 6 minutes

b 1

c It's not reasonable as delays are likely inevitable. The assumption that the buses never run late is almost definitely a false one.

4 a 0.163   **b** 0.223   **c** 0

5 a 0.775

b i 0.841   **ii** 0.159   **iii** 0.225

6 a 0.236   **b** 0.432

c No as there are weather conditions and psychological factors to consider

OR

Yes as a competition archer is likely to be reasonably consistent between each arrow being fired

d 2.16

7 a $\mu = 0.99$, $\sigma^2 = 0.07$

b $\mu - \sigma = 0.99 - 0.26 = 0.73$

$\mu + \sigma = 0.99 + 0.26 = 1.25$

$\dfrac{0.8 - 0.73}{0.8 - 0.7} = 0.7$

$0.7 \times 5 = 3.5$

masses in the $0.7 \leq m < 0.8$ category lie within one standard deviation of the mean.

$\dfrac{1.25 - 1.2}{1.3 - 1.2} = 0.5$

$0.5 \times 6 = 3$

masses in the $1.2 \leq m < 1.3$ category lie within one standard deviation of the mean.

$3.5 + 13 + 15 + 3 = 34.5$ masses lie within one standard deviation of the mean.

$\dfrac{34.5}{50} = 69\%$

This distribution could be modelled by a Normal distribution as close to 68% of values lie within one standard deviation of the mean, the distribution is roughly symmetrical and the mean 0.99 is close to the modal class $1.0 \leq m < 1.2$

c i 0.5153   **ii** 0.8677   **iii** 0.9226

d (0.21, 1.77)

8 a i $\dfrac{5}{3}$ **ii** $\dfrac{25}{18}$ **iii** 3.25, 0.0855   **iv** 0.930

b $S \sim N\left(\dfrac{n}{6}, \dfrac{5n}{36}\right)$

c i 0.8344   **ii** 0.8246   **iii** 0.8207

Chapter 21
Exercise 21.1A Fluency and skills

1 a 0.6137

b Reject H_0: the p-value is less than the significance level, the actual result is more surprising than we need.

2 a The jewellery auction house is investigating to see if there is any correlation present. This is a two-tailed test.

b Accept H_0: the p-value is more than the significance level, so the actual result is not surprising enough.

3 Reject H_0: the PMCC for the sample is more negative than the critical value.

4 Reject H_0: the critical value for this test is -0.2605. Since the PMCC for the sample is larger in size than the critical value, the result is significant.

Exercise 21.1B Reasoning and problem-solving

1 The PMCC is not further from zero than the critical value so the result is not significant. Accept H_0, no correlation between levels of taxation and the amount individuals give to charity.

2 The p-value is lower than the significance level. Reject H_0, positive correlation between the Maths and English scores of primary school students.

3 a 5% chance of rejecting the null hypothesis if it is actually true.

b One flavour.

c Since the result exceeds the critical value for positive correlation, the scientists may conclude that there is positive correlation between that one flavour and being carcinogenic.

d As you expect one of the 20 flavours to show positive correlation it is not surprising when one of them actually does. It would not be reasonable to reach the conclusion that any of the flavours are carcinogenic without conducting further tests.

4 The test statistic is more positive than the critical value. There is sufficient evidence to reject the null hypothesis.

5 a The test statistic is less negative than the critical value. There is insufficient evidence to reject the null hypothesis so it is accepted. You conclude that there is no correlation between daily total sunshine and daily mean windspeed.

b Camborne has about the same amount of sunshine but is much windier. The PMCC (-0.1512) is actually significant in Camborne.

6 The p-value for the test statistic (38.2%) is much larger than the significance level of the test so the result is not significant. There isn't sufficient evidence to reject the null hypothesis. You conclude that there is no correlation between the geographical location of a school and the likelihood that a student there will get at least one A at A level.

7 a There is evidence to suggest that A and B are positively correlated.

b There is evidence to suggest that B and C are positively correlated.

c There is no evidence to suggest that C and A are positively correlated.

d All three can be true simultaneously unless there is perfect correlation between the variables. It does not appear that there is perfect correlation between these variables. It doesn't follow that A and B being correlated and B and C being correlated makes C and A correlated.

Exercise 21.2A Fluency and skills

1 $H_0: \mu = 19$
$H_1: \mu \neq 19$

2 $H_0: \mu = 43$
$H_1: \mu \neq 43$

3 a $H_0: \mu = 32$
$H_1: \mu \neq 32$

b -2

c ± 1.645

d Reject H_0: the test statistic is larger in size than the critical value.

4 **a** $H_0: \mu = 25$

$H_1: \mu \neq 25$

b 1.697

c ± 1.645

d Reject H_0: the test statistic is greater than the critical value.

5 **a** 1.714 **b** 0.0433

c Reject H_0: since the p-value is smaller than the significance level, the result is significant.

6 **a** -1.304 **b** 0.0961

c Reject H_0: the p-value is smaller than the significance level.

Exercise 21.2B Reasoning and problem-solving

1 **a** $H_0: \mu = 4.38, H_1: \mu > 4.38$

b $t = \dfrac{6.76 - 4.38}{\sqrt{14.0}} = 0.6361$ (to 4 sf).

Critical value is $\Phi^{-1}\left(1 - \dfrac{5}{100}\right) = 1.645$

c The critical value is larger than the test statistic so you do not reject the null hypothesis. There is insufficient evidence to suggest that the average UK temperature in June 2015 was particularly high.

2 **a** $H_0: \mu = 125, H_1: \mu > 125$

b $z = 3.175$ (4 sf)

The p-value is 0.1%

c You reject the null hypothesis as the p-value is less than the significance level.

d You conclude that there is sufficient evidence to suggest that the mean mass of a banana that summer is greater than 125 g.

e It is not reasonable as we've only found evidence that it's greater than 125 g. 131 g would likely be a reasonable guess but it's very unlikely to be exactly that amount.

3 **a** $H_0: \mu = 93.59, H_1: \mu \neq 93.59$

b $\dfrac{93.13 - 93.59}{\sqrt{\dfrac{35.32}{30}}} = -0.4239$

c ± 1.645

d The test statistic is nearer to zero than the critical value so there is insufficient evidence to reject the null hypothesis.

4 **a** Statistician one claims $H_0: \mu = 16$ and $H_1: \mu \neq 16$

Statistician two claims $H_0: \mu = 17$ and $H_1: \mu \neq 17$

b The result is not significant for either statistician, and both conclude that they were correct. Increasing the sample size would give a better chance of yielding a result that favours one view over the other.

Review exercise 21

1 $H_0: \rho = 0$

$H_1: \rho \neq 0$

2 **a** $H_0: \rho = 0$

$H_1: \rho > 0$

b Reject H_0: the PMCC is greater than the critical value.

3 **a** Accept H_0: the p-value is greater than the significance level.

b There is no negative correlation between the amount of the chemical and the number of bacteria present in a petri dish.

4 **a** $H_0: \rho = 0$

$H_1: \rho > 0$

b The PMCC is less than the critical value so there is insufficient evidence to reject the null hypothesis. You conclude that there is no positive correlation between levels of annual household income and the amount spent per year on books.

5 Reject H_0: the p-value is lower than the significance level. Conclusion: the long jump and high jump scores achieved by heptathletes are correlated.

6 $H_0: \mu = -5$

$H_1: \mu > -5$

7 **a** $H_0: \mu = 30$

$H_1: \mu \neq 30$

b Accept H_0: the p-value is larger than the significance level.

c It is reasonable to believe that the mean length of a stick of rock in the population the sample is drawn from is 30 cm.

Assessment 21

1 $H_0: \mu = 20$

$H_0: \mu \neq 20$

$z = \dfrac{22 - 20}{\sqrt{\dfrac{16}{15}}} = 1.93649$, p-value = 2.64%

Accept H_0, there is no evidence to suggest the mean is not 20.

2 $H_0: \mu = 40$; $H_1: \mu < 40$

Reject H_0; evidence supports mean is less than 40

3 No reason to reject H_0; evidence supports mean equals 22

4 **a** 2020 hours

b No reason to reject H_0; evidence supports mean equals 1930

5 Reject H_0; evidence suggests that the mean is less than 12

6 **a** $H_0: \mu = 17.7$

$H_0: \mu > 17.7$

b test statistic = 1.3387 (4 dp), critical value = 1.2815 (4 dp)

c Reject H_0. There is evidence that the temperature is higher than the rest of the country.

7 **a** $H_0: \mu = 0$

$H_0: \mu > 0$

b Reject H_0. There is evidence of positive correlation between rainfall and humidity.

8 **a** The test statistic is further from zero than the critical value, reject the null hypothesis.

b The test statistic is further from zero than the critical value, reject the null hypothesis.

c The mean is not -13.6 but one is larger and one is smaller. The scientists should do a further test with a larger sample.

9 Reject H_0; evidence supports mean volume greater than 568 ml.

10 No reason to reject H_0

11 **a** $H_0: \rho = 0$; $H_1: \rho > 0$

b Reject H_0

12 No reason to reject H_0

13 Reject H_0 and the manufacturer's claim.

14 **a** $\bar{x} > \mu_0 + 2.054 \dfrac{\sigma}{\sqrt{n}}$

b Reject H_0: evidence supports mean breaking point greater than 80 kN.

Assessment chapters 20–21: Statistics

1 **a** Expect a negative correlation since Perth is in the southern hemisphere and Beijing in the northern hemisphere.

b Positive since they are geographically close.

2 a (left to right, top to bottom): 12, 187, 126, 313, 5, 140

  b i $\dfrac{32}{350}$ or $\dfrac{16}{175}$

   ii $\dfrac{140}{350}$ or $\dfrac{2}{5}$

   iii $\dfrac{187}{350}$

  c $\text{P(mixed handed)} \times \text{P(man)} = \dfrac{5}{350} \times \dfrac{3}{5}$

   $= \dfrac{3}{350}$

   $= \text{P(mixed handed and man)}$

   So yes

3 a Parameter is the underlying mean for the whole population of the town. The statistic is the mean of the sample that we are calculating.

  b Could be biased due to location being similar/people being out at work/non-response etc

  c 4

4 a i 2.01 mm

   ii 3.67 mm (or 3.69)

   iii 0.789 mm

   iv 1.83 mm

  b Median, as not effected by outliers

5 a 0.4

  b

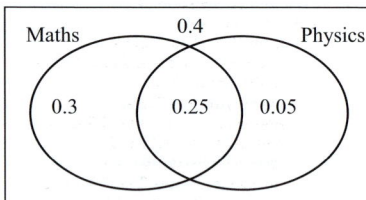

  c $\dfrac{5}{6}$

  d No; $\text{P(Maths | Physics)} \neq \text{P(Maths)}$

6 a i $\dfrac{16}{30}$ or $\dfrac{8}{15}$ **ii** $\dfrac{3}{16}$

  b

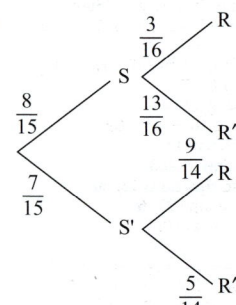

7 a i $\dfrac{1}{35}$ **ii** $\dfrac{6}{35}$ **iii** $\dfrac{4}{19}$

  b i $\dfrac{19}{105}$ **ii** $\dfrac{7}{15}$

8 a i 0.24   **ii** 0.54

b

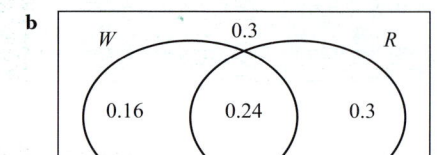

9 a i 0.9772

   ii 0.1587

   iii 0.2286

  b 2.893

10 $H_0: \rho = 0$, $H_1: \rho > 0$

  Reject H_0

11 a $H_0: \rho = 0$, $H_1: \rho < 0$

   Reject H_0

  b $r < -0.4226$

12 H_0 will be rejected when $r < -0.4438$ or $r > 0.4438$

13 a $\dfrac{11}{30}$ (or 0.36...)   **b** $\dfrac{1}{12}$ (or 0.083...)

14 a 14 meerkats

  b i 27.1 cm   **ii** 3.4 cm

  c Any one of: Continuous data; Approximately symmetrical; All data lies within 3 sd of the mean; Approximately 67% lies within 1 sd of the mean

  d i 0.8749   **ii** 0.2697

  e $n \geq 15$

15 a 11.6   **b** 10.4, 12.8

16 a 0.133   **b** 0.0781   **c** 0.0505

17 a Yes, data is continuous and the distribution is almost symmetric.

  b i 0.7791

   ii According to the box and whisker diagram, 75% of the days had a temperature under 16 °C, so this is quite close to our result of 77.91

  c i 0.282   **ii** 0.3503   **iii** 0.4041

18 a Could be modelled by binomial, $X \sim B(8, 0.5)$

  b No; not a fixed number of trials.

  c No; probability is not constant/trials are not independent.

19 $H_0: \rho = 0$, $H_1: \rho < 0$

  Accept H_0

20 a $H_0: \mu = 22.1$, $H_1: \mu > 22.1$

   Reject H_0

  b $z > 1.2816$

21 a $\sigma = 5.41$; $\mu = 132$ **b** 3.30

22 a 2.27

  b 0.0392

  c i 0.819   **ii** 0.986

23 a i 0.8708   **ii** 0.459

  b Normal distribution is appropriate since n is large and np and $nq > 5$

24 a $\sigma = 14.7$; $\mu = 90.9\%$

  b 0.3897

  c 0.281

25 a Yes as fixed number of trials (10 people), two outcomes (vegetarian/not), trials are independent, probability is constant.

  b i 0.0746   **ii** 0.0115

  c $H_0: \mu = 10$, $H_1: \mu \neq 10$

   Accept H_0; The probability of being vegetarian is not different in town B.

26 $H_0: \mu = 30$, $H_1: \mu < 30$

  Reject H_0 / there is enough evidence to support the customer's claim.

27 a $H_0: \mu = 6.68$, $H_1: \mu < 6.68$, Critical value is -1.96

  b 51

Index